C.H. NESTLER

COLLOID SCIENCE

COLLOID SCIENCE

BY

A. E. ALEXANDER

AND

P. JOHNSON

OXFORD
AT THE CLARENDON PRESS

Oxford University Press, Amen House, London E.C. 4
GLASGOW NEW YORK TORONTO MELBOURNE WELLINGTON
BOMBAY CALCUTTA MADRAS CAPE TOWN
Geoffrey Cumberlege, Publisher to the University

FIRST PUBLISHED IN TWO VOLUMES 1949
Reprinted lithographically as one volume in Great Britain
at the University Press, Oxford, 1950
from corrected sheets of the first edition

FOREWORD

I HAVE been asked to write an introduction to this book. A word of apology is needed as personal matters intrude. Some twenty years ago it was my duty to attempt to build up a laboratory for teaching and research which would serve as a bridge between the biological sciences and physics and chemistry. It was clearly understood that the foundation should be on the side of the 'exact' sciences, but the bridge should be one which would carry passengers in either direction with equal facility. It was an unfortunate fact that whilst the biological sciences were largely descriptive in character, those portions of physics and physical chemistry which appeared to have the closest connexion with the world of organized matter, the colloids, were usually treated in a similar phenomenological manner. Evidently it was desirable to put first things first, using the simplest of systems capable of rigorous and exact treatment before constructing and examining the more complex models. During the intervening years many experiments on a pragmatic basis in methods of presentation and in choice of material have been made.

This volume by two of my co-workers, Dr. Alexander and Dr. Johnson, reveals how the subject-matter has developed. The reader as he turns the pages will note how indebted we are to those masters of their trade, Donnan, Freundlich, Gibbs, Hardy, Langmuir, Perrin, and Svedberg, none of whom found any satisfaction in the mere description of either form or movement.

Recent years have witnessed the development of polymers. The application of the laws of physics and chemistry to macro-molecular systems involves new considerations, fresh experimental methods, novel colligative relationships and new forms of kinetic expressions; these are all properly the domain of colloid science.

The study both of the interface and of the protein molecule, especially under non-equilibrium conditions, will assist in illuminating the path along which that more mysterious and more complex entity, Biophysics, will shortly travel.

The book has been given the title *Colloid Science*. I do not agree with all that the authors have written (this is only fit and proper in a rapidly expanding subject), but I heartily commend it.

ERIC K. RIDEAL

22 *July* 1947

PREFACE

THIS book was written to meet the need for a modern treatment of colloidal systems, intermediate in level between the elementary, purely descriptive text-books and the specialized accounts of certain aspects available in monographs and the 'recent advances' type of article. We hope it will prove useful not only for teaching purposes, but also for academic and industrial research workers and for biologists. The general approach is from the fundamental angle, although certain industrial and biological applications receive mention.

Its twenty-eight chapters are sub-divided into three main parts. Part I, the historical and general survey, also includes chapters on stability and the application of thermodynamics. Part II deals with the principal experimental methods and their theoretical basis, covering those applicable to the system as a whole (osmotic pressure, diffusion, ultracentrifuge, electrophoresis, viscosity, etc.), as well as those used for studying the interface (surface chemistry). Part III, entitled 'The principal colloidal systems', has separate chapters devoted to sols, gels and pastes, foams, emulsions, colloidal electrolytes, clays and zeolites, proteins, polymers, and membranes.

In connexion with Part II we have concentrated largely on the newer methods which have not been adequately presented in the form we wanted. For example, the space devoted to surface chemistry is certainly not commensurate with its importance in colloid science, but several excellent accounts were already available.

In such a wide survey, covering all the principal branches of colloid science, some unification of the existing terminology, notation, and symbols was clearly essential. Concerning symbols, we have in general followed the recommendations of the Joint Committee of the Faraday and Chemical Societies, except in cases where certain ones have come into rather accepted usage. For convenience a list of the *principal* symbols employed has been appended.

The title of 'colloid science', rather than the more common one of 'colloid chemistry', was chosen for two reasons. Firstly, modern colloids cover a very much wider field than the latter term would suggest; secondly, 'Colloid Science' is the name of the department, founded and so richly endowed by Professor Rideal, with which it has been our good fortune to be associated. Our debt to Professor Rideal, teacher, counsellor, and friend, cannot adequately be expressed.

PREFACE

To our many research associates, at home and abroad, we are greatly indebted for reading sections of the manuscript, for discussions and enlightenment on difficult points, and for additional help in the form of photographs and diagrams incorporated in the text. We also wish to thank the following for permission to reproduce diagrams:

Dr. H. A. Abramson, Drs. G. S. and M. E. Adair, Professor N. K. Adam, Dr. T. F. Anderson, Professor W. T. Astbury, Dr. R. M. Barrer, Dr. S. Brunauer, Dr. N. A. de Bruyne, Dr. H. B. Bull, Dr. C. W. Bunn, Professor G. L. Clark, Dr. G. Cockbain, Dr. D. J. Crisp, Dr. R. B. Dean, Professor P. Debye, Professor P. Doty, Dr. W. Dickinson, Professor J. T. Edsall, Professor P. J. Flory, Dr. M. M. Frocht, Professor R. M. Fuoss, Dr. G. Gee, Professor W. D. Harkins, Dr. G. S. Hartley, Professor E. Hauser, Dr. J. Hillier and the R.C.A., Dr. M. L. Huggins and the Eastman Kodak Co., Dr. E. W. Hughes, Dr. E. Hutchinson, Dr. D. O. Jordan, Professor R. A. Kekwick, Dr. A. King, Dr. I. Langmuir, Professor O. Lamm, Professor W. C. Lewis, Professor L. G. Longsworth, Dr. T. L. McMeekin, Professor H. Mark, Dr. L. Marton, Professor K. H. Meyer, Dr. R. C. Murray, Professor H. Neurath, Professor J. L. Oncley, Dr. K. Pankhurst, Dr. K. O. Pedersen, Professor P. Putzeys, Professor J. M. Robertson, Dr. J. R. Robinson, Professor F. J. W. Roughton, Dr. J. H. Schulman, Dr. B. T. Shaw, Professor R. Signer, Dr. E. Stenhagen, Professor T. Svedberg, Dr. L. R. G. Treloar, Dr. E. J. Verwey and Messrs. Philips Ltd., Dr. R. C. Warner, Dr. A. F. Wells, Professor R. Whytlaw-Gray, Dr. A. Wissler, Dr. H. J. Woods, the British Standards Institution, Interscience Publishers, Messrs. Methuen, Messrs. Reinhold, Messrs. Sharples Centrifuges, Messrs. John Wiley, *Acta Physicochimica, U.R.S.S., Annals of the New York Academy of Sciences, Biochemical Journal, Chemistry and Industry, Comptes rendus Laboratoire Carlsberg, Industrial and Engineering Chemistry, Helvetica Chimica Acta, Journal of the American Chemical Society, Journal of Biological Chemistry, Journal of Chemical Physics, Journal de Chimie Physique, Journal of Colloid Science, Journal of Physical Chemistry, Nature, Polymer Bulletin, Proceedings of the Royal Society, Proceedings of the Physical Society, Reports on Progress in Physics, Science, Transactions of the Faraday Society,* and *Zeitschrift für physikalische Chemie.*

We owe much to members of the staff of our laboratories for secretarial assistance, and a special debt to our wives for help with the manuscript and diagrams, and in other less tangible ways. The Clarendon Press, despite present-day difficulties, has lived up to its best tradition, and our cordial thanks go to all concerned in the production of this book.

DEPARTMENT OF COLLOID SCIENCE,   A. E. A.
    CAMBRIDGE.
THE ROYAL INSTITUTION,   P. J.
    LONDON.

CONTENTS

PRINCIPAL SYMBOLS xxi

Part I. HISTORICAL AND GENERAL SURVEY

Chapter I. EARLY HISTORY (UP TO 1910) 1
Colloids in early civilizations 1
Principal investigations in the early nineteenth century . . . 2
  Reuss (1809)—electrokinetic phenomena. Brown (1827)—the motion of suspended particles. Ascherson (1838)—adsorption of proteins onto oil droplets.
Early studies of inorganic colloids 3
  Selmi (1845-50)—preparation and certain properties. Graham (1861-4)—fundamental work on colloids. Introduction of nomenclature. Faraday (1857)—optical properties of gold sols.
Major developments towards the end of the nineteenth century . . 6
  Optical properties of colloids. Preparation and purification of colloids. Theoretical contributions. Electrokinetic phenomena.
Major developments during the early years of the twentieth century . 12
  Classifications of colloidal systems. The formation and growth of nuclei. Interaction of colloids with electrolytes and other colloids. The ultramicroscope. Brownian movement and particle size determination.

Chapter II. THE PRINCIPAL ADVANCES SINCE 1910 . . . 21
Introduction 21
Surface chemistry 21
  Adsorbed and insoluble films.
Membrane equilibria 25
Size, shape, and charge of colloidal particles 26
  Direct methods: microscope, ultramicroscope, electron-microscope, sedimentation velocity, sedimentation equilibrium, diffusion.
  Indirect methods: viscosity, streaming birefringence, X-ray and electron diffraction. Electrophoresis.
Rheology 30
Colloidal electrolytes 30
Classical colloids 31
Future developments 32

Chapter III. MODERN VIEWS OF THE COLLOIDAL STATE . . 34
Introduction 34
Effects of subdivision of matter upon: 34
  (a) Surface area and surface energy. (b) Solubility, vapour pressure, and chemical reactivity. (c) Colour.
The question of 'molecular weight' or 'particle mass' in colloidal systems . 40
Molecular weight and molecular weight averages . . . 41
Origin of the charge in colloidal systems 43
  (a) Ionization. (b) Adsorption. (c) Contact.
Magnitude and distribution of charge 48

Chapter IV. THERMODYNAMICS AND ITS APPLICATION TO COLLOIDAL SYSTEMS 50
Introduction 50

CONTENTS

THE PRINCIPAL THERMODYNAMIC FUNCTIONS	50
First and second laws	50
Energy and entropy	52
Single-valued and non single-valued functions	52
Heat content	53
Gibbs (G) and Helmholtz (F) free energies.	
General criteria of equilibrium	56
Clapeyron-Clausius equation	
Partial differential coefficients of free energy	58
Gibbs-Helmholtz equation	
Free energy changes	60
Single perfect gas. Imperfect gas. Mixture of perfect gases.	
MANY-COMPONENT SYSTEMS	62
Partial quantities	62
Partial molar free energy and chemical potential. Partial molar volume. Duhem-Margules equation. Differential coefficients of partial quantities.	
Equilibrium conditions in heterogeneous system.	65
Chemical potentials in many-component systems	67
Ideal and non-ideal systems. Activities and activity coefficients—Osmotic coefficients. Use of the Duhem-Margules equation. Solubility. Ions and strong electrolytes.	
APPLICATION TO CERTAIN PROBLEMS	73
Osmotic pressure and membrane equilibria	73
Osmotic pressure and thermodynamic functions. Approximate osmotic pressure equation. Donnan membrane equilibria. Ion pressure differences and particle charge from membrane potentials.	
Sedimentation equilibrium	83
Surface phases and Gibbs's adsorption isotherm	84
Vapour pressure and solubility of colloidal particles	86
Electrification.	
Polymer-solvent systems	91
Solubility. Swelling. Fractionation.	
Mechanical properties of polymers	97

CHAPTER V. THE APPLICATION OF DEBYE-HÜCKEL THEORY TO COLLOIDAL PARTICLES. STABILITY AND ITS ORIGIN . . . 100

APPLICATION OF DEBYE-HÜCKEL THEORY TO COLLOIDAL PARTICLES	100
General theory (outline).	100
Limitations and extensions of the elementary theory	104
The breakdown of the ionic strength principle	108
The inconsistencies of the Boltzmann equation	109
Consideration of colloidal electrolytes	110
(a) Mobility. (b) Magnitude of atmospheric effects.	
Consideration of large colloidal particles (ca. $0.01\mu - 1\mu$)	114
STABILITY AND ITS ORIGIN	114
(a) Lyophobic colloids	115
Potential and charge distribution for two interacting double layers. Free energy considerations. Potential curves.	
(b) Lyophilic colloids	124
General considerations. Solvation.	

CONTENTS

PART II
EXPERIMENTAL METHODS AND THEIR THEORETICAL BASIS
Section A. SYSTEMS AS A WHOLE

CHAPTER VI. INTRODUCTION 128
Important data for colloidal systems 128
Application of analytical methods to colloidal systems 130
   (a) Trace element and group analysis. (b) Combining weights. (c) End-group analysis.
Application of colligative methods to colloidal systems 136
   Osmotic pressure.
The partial specific volume factor 137
   Determination, meaning, and use.
Solvation 140
   General occurrence and forces involved. Hydration of protein crystals. Other estimates of protein hydration.

CHAPTER VII. OSMOTIC PRESSURE AND MEMBRANE PHENOMENA . 150
OSMOTIC PRESSURE 150
Applicability of van't Hoff relation 150
   Simple molecules and high polymers.
Linear polymers. 153
   Abnormal entropy of mixing. Calculated osmotic pressure-concentration equation.
Donnan equilibria and osmotic pressure 157
   Use of nearly iso-electric solutions and high salt concentrations.
Aqueous systems 160
   Osmometers—membranes, accuracy. Criteria of osmotic equilibrium. Results.
Non-aqueous systems 169
   Osmometers and their operation, membranes. Results—comparison with theory.
Limits of the osmotic method 176
New techniques 177
   Osmotic balance. Light-scattering method.
MEMBRANE PHENOMENA. 180
Measurement of membrane potentials 180
   Direct and indirect methods.
Results 183
   Membrane potential and protein concentration. Membrane potentials and ion pressure differences. Membrane potentials and protein charge.

CHAPTER VIII. SEDIMENTATION EQUILIBRIUM 188
Introduction 188
SEDIMENTATION EQUILIBRIUM IN GENERAL 188
In the atmosphere 189
In colloidal solutions under gravity (Perrin) 190
In colloidal solutions in the ultracentrifuge 190
SEDIMENTATION EQUILIBRIUM IN THE ULTRACENTRIFUGE . . 191
Experimental 191
Theoretical 194
   Derivation of ideal sedimentation equilibrium equation. Average molecular weights.

Monodisperse systems 197
Polydisperse systems 198
Factors complicating sedimentation equilibrium 199
   (a) Deviations from ideal behaviour. (b) Charge effects. (c) Duration of sedimentation equilibrium experiment.
Results 204
Advantages and disadvantages 206

CHAPTER IX. THE DYNAMIC METHODS 208
THEORETICAL BASIS 208
Treatment of colloidal particles in terms of geometrically simple models . . 209
Frictional Constants 210
   Spherical particles: Translational and rotational. Asymmetric particles: Translational and rotational; Frictional ratios; Perrin's equations.
Solvation—Effect on frictional ratios 214
Charge effects and their elimination 215
THE EXAMINATION OF CONCENTRATION GRADIENTS IN LIQUID COLUMNS . 216
Light absorption methods 217
   Use of visible and ultra-violet light; calibration.
Refractive index methods 219
   Schlieren methods: Simple scanning, diagonal, and other modifications; Experimental requirements; Different types of boundary behaviour; Simultaneous use of absorption and refractive index methods.
   Scale method.

CHAPTER X. TRANSLATIONAL DIFFUSION 233
Introduction 233
Theoretical 233
   (a) Fick's First Law. (b) Fick's Second Law. (c) A particular solution of the differential equation of diffusion.
Experimental 239
   (a) Porous disc method. (b) Free boundary method: (i) Diffusion cells; (ii) Calculation of results—(1) Absorption method, (2) Refractive index methods.
Diffusion in non-interacting systems 248
   (a) Monodisperse systems. (b) Polydisperse systems.
Diffusion in strongly interacting systems 250
   Skew curves.
Factors complicating diffusion 254
   (a) Charge effects. (b) Boundary anomalies. (c) Diffusion in equilibrium systems.
Diffusion and molecular properties 255
   (a) Diffusion and osmotic pressure—diffusion coefficient and frictional constant $(D = RT/F)$. (b) Frictional constants and molecular properties: (i) Spherical molecules; (ii) Non-spherical molecules—Frictional ratios and their interpretation, solvation, molecular asymmetry. (c) Influence of temperature and solvent viscosity on diffusion coefficients.

CHAPTER XI. SEDIMENTATION VELOCITY. 266
Introduction—Principles of the method 266
Experimental—Construction and operation of ultracentrifuge . . 267
Sedimentation of simple non-interacting systems 271
   (a) Monodisperse systems—Effect of diffusion. (b) Paucidisperse systems. (c) Polydisperse systems. (d) Sedimentation constant and concentration.

CONTENTS

Factors complicating sedimentation and special cases 276
  (a) Charge effects. (b) Sedimentation in organic solvents. (c) Boundary anomalies. (d) Sedimentation in equilibrium systems: (i) Rapid; (ii) Slow; (iii) Intermediate.
Sedimentation of strongly interacting systems 283
  Cellulose nitrate in acetone: (i) Experimental results and their interpretation; (ii) Meaning of sedimentation constant—Concentration dependence.
Sedimentation and molecular properties 287
  (a) Molecular weight calculations: (i) Extrapolation of sedimentation constants and diffusion coefficients; (ii) Correction of sedimentation constants; (iii) Nature of calculated molecular weight; (iv) Results. (b) The use of frictional ratios in estimating molecular shape.

CHAPTER XII. ELECTROPHORESIS AND ALLIED PHENOMENA. . 295
ELECTROPHORESIS 295
Introduction 295
Electrochemical potentials 296
  (a) Origin of the charges at interfaces. (b) Electrokinetic potentials and the electrical double layer. (c) Electrokinetic and thermodynamic potentials.
Theoretical 300
  (a) Small non-conducting particles—thick double layer; Hückel's equation. (b) Large non-conducting particles—thin double layer; Smoluchowski's equation. (c) Spherical particles of intermediate size; Henry's treatment. (d) Calculation of charge for spherical particles; Calculation of density of charge for large particles. (e) Asymmetric particles of intermediate size
Experimental 308
  (A) *The microscopic method*: (a) Electro-osmosis in closed tubes: Cylindrical cells—Rectangular cell—Stationary levels—Electro-osmotic mobility—Double-tube cell. (b) Experimental. (c) Chief Results: Verification of Smoluchowski's equation; Stability of colloids; Application to biological systems; pH–mobility curves for proteins.
  (B) *The moving boundary method*: (a) The development of the method. (b) The apparatus and its operation. (c) Important devices. (d) Boundary migration in an electrical field: Monodisperse systems—Determination of mobility—Effects of diffusion; Paucidisperse systems—Determination of mobility and composition—Resolution; Polydisperse systems; Mobility and solute concentration. (e) Refinements of the method: Electrophoretic preparations; Double centre section cell. (f) Boundary anomalies: δ and ϵ boundaries. (g) Advantages of the moving boundary method. (h) Chief results: pH–mobility curves—Comparison with results from micro method.
Net charge per molecule from electrophoresis and other methods 332
  (a) Net charge from electrophoresis. (b) Net charge from titration curves. (c) Net charge from membrane potentials.
ELECTRO-OSMOSIS, STREAMING AND SEDIMENTATION POTENTIALS . . 338
  (a) Electro-osmosis: Experimental investigation and comparison with electrophoresis. (b) Streaming potentials: Experimental investigation—Calculation of ζ potential and dipole moment per unit surface area. (c) Sedimentation potentials.

CHAPTER XIII. VISCOSITY 345
Introduction 345
  Definitions: Dynamic and kinematic viscosity; Fluidity; Newtonian behaviour; Non-Newtonian behaviour—Structural viscosity, Particle orientation.

Experimental methods 349
  (a) Capillary flow method: Variation in velocity gradient across a tube; Poiseuille law and viscosity determination—corrections; Verification of Newtonian law.
  (b) Concentric cylinder method: Total variation in velocity gradient; Viscosity determination.
Different functions of viscosity coefficient 357
  Relative, specific, and intrinsic viscosity.
THE VISCOSITY OF COLLOIDAL SOLUTIONS 358
Spherical particles 358
  (a) Very dilute solutions: Einstein's equation—assumptions involved; Experimental verification. (b) Moderate concentrations: Guth and Simha equation; Interaction term. (c) Solvation.
Asymmetric particles 361
  (a) Theoretical approach: Equations of Jeffery, Kuhn, Huggins, Simha, etc.; Viscosity and orientation; Solvation and asymmetry. (b) Viscosity and molecular weight: Staudinger law—evidence for and against modified forms of Staudinger law; Types of average molecular weight from viscosity. (c) Viscosity-concentration relations.
Effects of solvent and temperature on viscosity 374
Electro-viscous effects.

CHAPTER XIV. ROTATIONAL DIFFUSION 380
Introduction 380
Theoretical 380
  (a) Equations of rotational diffusion: Rotational diffusion coefficient and Brownian motion; Rotational frictional constants and relaxation times.
  (b) Rotational diffusion and molecular dimensions: Spherical particles; Ellipsoids of rotation.
Orientation in a velocity gradient 387
  (a) Mechanism of orientation. (b) Theoretical treatment of orientation: Boeder's theory for two dimensions; Extension to three dimensions. (c) The magnitude of streaming birefringence: 'Wiener' type of double refraction. (d) Orientation in a concentric cylinder apparatus: Experimental—Apparatus and determination of extinction angle and double refraction. (e) Results: Large asymmetric particles; Corpuscular proteins; Linear polymers. (f) Effect of polydispersity. (g) Importance of deformation double refraction. (h) Limitations of streaming birefringence method.
Orientation in an electrical field 412
  (a) Dispersion of the dielectric constant and of the conductance: Qualitative explanation. (b) Experimental methods. (c) Dielectric increments and their evaluation: Low frequency, High frequency, and Total increments. (d) Determination of molecular weight and shape, and dipole angle. (e) Dielectric increments and dipole moments—Electrical symmetry of protein molecules. (f) Limitations of the dielectric dispersion method.

CHAPTER XV. OPTICAL, X-RAY, ELECTRON DIFFRACTION, AND OTHER METHODS 426
THE OPTICAL PROPERTIES OF COLLOIDAL SOLUTIONS . . . 426
Non- or weakly absorbing, insulating, particles 427
Strongly absorbing, conducting, particles 428
Polarization properties—Effects of particle size, shape, and anisotropy . 430
  Depolarization ratios and their use in investigating colloidal particles.

CONTENTS

Molecular weight and shape 434
   Use of Rayleigh formula to calculate molecular weight. Determination of molecular dimensions from disymmetry of scattering.
Infra-red absorption spectra 439
DIFFRACTION OF X-RAYS 440
Introductory—Experimental and theoretical 440
Procedure in single crystal examination 443
The use of intensity measurements in structure determinations . . 445
   Structure factor—Determination by experiment and calculation. Electron density in terms of structure factor.
Crystallite and particle size 453
Results 454
   Crystalline proteins. Fibrous proteins and other fibrous materials. Structure of soaps in solution.
Electron microscope 458
Electron diffraction 461

CHAPTER XVI. PLASTIC FLOW AND ELASTICITY . . . 464
Introduction and general survey 464
Forces and deformations 464
Elasticity 466
Relaxation time 467
Experimental methods 468
   I. Plastometers: (*a*) Compression and recovery plastometers. (*b*) Rotation plastometers—Couette, Ungar, Stormer, Goodeve. (*c*) Capillary plastometers. (*d*) Penetrometers. (*e*) Farinograph.
   Flow-rate *v.* applied stress curves.
   II. Elastometers: (*a*) Extension or compression under known loads. (*b*) Couette method. (*c*) Photo-elastic method. (*d*) Other methods.
Classification of rheological properties of real materials . . . 476
   Plasticity. Thixotropy. Dilatancy.
Interpretation of rheological data 480
   (*a*) The analytical approach: Qualitative classification; Quantitative classification. (*b*) The integralist approach.

SECTION B. STUDY OF THE INTERFACE

CHAPTER XVII. INSOLUBLE MONOLAYERS AT THE AIR/WATER AND OIL/WATER INTERFACES 489
AIR/WATER MONOLAYERS 489
Surface pressure and surface area measurements 489
Surface potential 492
Mechanical properties 494
Multilayers (built-up films) 496
Principal results and applications of air/water monolayers . . . 497
   (I) Phases and phase changes: (*a*) Gaseous and vapour films. (*b*) Liquid-expanded films. (*c*) Condensed films. (*d*) Thermodynamics of surface phases. (*e*) Kinetics of spreading.
   (II) Applications to other fields. (*a*) Structure of complex organic molecules. (*b*) Diffusion through interfaces. (*c*) Reactions at interfaces. (*d*) Biological problems.

Oil/water monolayers 508
Surface pressure and surface area measurements. 508
Interfacial potentials 509
Mechanical properties 510
Principal results and applications of oil/water monolayers . . . 510

CHAPTER XVIII. ADSORBED FILMS AT THE GAS/LIQUID AND LIQUID/
LIQUID INTERFACES 513
Surface and interfacial tension methods 513
   (a) Static methods: Capillary rise; Sessile drop; Pendent drop; Maximum bubble pressure; Ring and other detachment methods; Drop-weight and drop-volume; Wilhelmy plate. (b) Dynamic methods: Jet methods; Air electrode method.
Surface area determination from Gibbs's adsorption isotherm . . . 519
Phase boundary potentials 521
Mechanical properties of adsorbed films 522
Duplex films 523
Thermodynamics of adsorption 524
Principal results and applications of adsorbed films 524
   (a) Experimental tests of Gibbs's adsorption isotherm. (b) The molecular structure of the surfaces of solutions. (c) Equation of state and phases of adsorbed monolayers. (d) The phenomenon of 'surface ageing'. (e) Biological applications.
Electrocapillary phenomena 531

CHAPTER XIX. THE SOLID/LIQUID INTERFACE 536
Determination of surface area of solids 536
   (a) Direct measurement. (b) Sedimentation methods. (c) X-ray methods. (d) Rates of solution or evaporation. (e) Permeability method. (f) From heats of wetting. (g) Heat conductivity method. (h) Adsorption methods: (i) Dye-stuffs; (ii) Organic compounds readily estimated; (iii) Radio-active indicators; (iv) Electrolytic method; (v) 'Retardation volume' method; (vi) Methods based on adsorption of gases or vapours.
Adsorption isotherms 548
Thermodynamics of adsorption 549
Heats of adsorption and wetting by calorimetric methods . . . 550
Phase boundary potential changes 551
Work of adhesion between solids and liquids 551
Measurement of contact angles 552
Effect of adsorbed films upon macroscopic properties of solid/liquid systems . 553

Part III

THE PRINCIPAL COLLOIDAL SYSTEMS

Chapter XX. DILUTE SUSPENSIONS (SOLS)	556
Introduction	556
Suspensions in liquid media (chiefly aqueous)	556

(a) Preparation: (1) Dispersion processes—Mechanical, Electrical, Peptization, other methods; (2) Aggregation processes—Formation of supersaturated solutions, Formation of nuclei, Growth of nuclei. (b) Properties: (1) Electrical mobility; (2) Viscosity; (3) Optical properties. (c) Effect of addition of salts, non-electrolytes, and other colloids. (d) Kinetics of coagulation.

Suspensions in organic media	576
Suspensions in gaseous media (aerosols)	576

(a) Preparation: (1) Dispersion methods; (2) Aggregation methods—Formation of nuclei, Condensation upon nuclei. (b) Measurement of size, charge, and concentration of aerosols. (c) Coagulation of aerosols.

Chapter XXI. GELS AND PASTES	585
General introduction and definitions	585
Gels	585
Terminology and classification	586
Formation of gels	587

(a) By variation of temperature. (b) By addition of precipitating liquids. (c) By addition of salts. (d) Chemical reactions leading to gel formation.

Changes occurring upon gel formation	590

(a) Thermal. (b) Optical. (c) Electrical conductivity. (d) Dielectric constant. (e) Volume. (f) Mechanical properties (viscosity and elasticity).

The properties of gels	596

(a) Elasticity and rigidity. (b) Optical. (c) Swelling. (d) Diffusion and reactions in gels. (e) Stability; Syneresis. (f) 'Free' and 'bound' water in gels. (g) Vapour pressures over rigid gels.

The structure of gels	608

(a) Rigid gels (silica gel). (b) Elastic gels. (c) Thixotropic gels.

Pastes	611
General introduction	611
Measurement of adhesion between solid and liquid, and between solid and solid	612

(a) From the sedimentation volume. (b) From the 'adhesion number'. (c) Direct measurement of friction. (d) From heats of wetting.

Rheological properties of pastes	618

(a) Qualitative measurements. (b) Quantitative measurements.

Chapter XXII. FOAMS	624
Definitions and units	624
Foaming agents	624
Formation and destruction of foams	626
Measurement of foaming power and foam stability	626

(a) Static methods. (b) Dynamic methods. (c) Observations on single bubbles.

Factors in the formation and stability of foams	628

CONTENTS

Discussion of the principal types of foaming agents 630
    (a) Soaps: Soap films in single bubbles; Foams from soap solutions. (b) Proteins: Insoluble monolayers of proteins; Adsorbed films of proteins; Foams from protein solutions. (c) Solid powders. (d) Other stabilizing materials.
Flotation 640
Foams in non-aqueous media 641
Fractionation by adsorption on foam 642

Chapter XXIII. EMULSIONS 643
General introduction 643
Emulsifying agents 644
    (a) Agents for oil-in-water emulsions. (b) Agents for water-in-oil emulsions.
Preparation of emulsions 646
Determination of emulsion type 648
Measurement of emulsion stability 649
Factors in the formation and stability of emulsions . . . 651
Theories of emulsion type (O/W or W/O) 651
Discussion of the principal types of emulsifying agents . . . 652
    (a) Soaps: Emulsions of the O/W type; Emulsions of the W/O type; Information from monolayer studies. (b) Proteins: Insoluble monolayers; Adsorbed films; Emulsions. (c) Solid powders. (d) Electrolytes. (e) Other stabilizers.
The application of monolayer studies to emulsion systems . . 656
Physical properties of emulsions 659
Certain technical and biological aspects of emulsions . . . 660
'Creaming' in emulsions 664
Breaking of emulsions 665

Chapter XXIV. COLLOIDAL ELECTROLYTES . . . 667
Definition and principal classes 667

Soaps 669
The structure of the micelles 670
    (a) The spherical micelle; Molecular considerations; Application of law of mass action to micelle formation. (b) The laminar micelle: Molecular considerations.
The physical properties of dilute soap solutions 676
    (a) Colligative properties (particularly vapour pressure and freezing-point). (b) Conductivity. (c) Surface behaviour. (d) Solubility (the Krafft phenomenon). (e) Solubilization in dilute soap solutions. (f) Detergent action. (g) Diffusion. (h) Density. (i) Hydrolysis. (j) Other properties.
Concentrated systems 689
The physical properties of concentrated soap solutions . . 690
Three-component systems 694
Four-component systems 695
Some biological effects of soaps 695
Soaps in non-aqueous media 696

Dyestuffs 697
Methods for measuring the degree of aggregation . . . 698
The aggregation of dyes 699
Colloid aspects of the dyeing process 699
Some biological uses of dyestuffs 702

CONTENTS

CHAPTER XXV. CLAYS AND ZEOLITES 704
General considerations 704
The principal types of aggregation in silicates 704
CLAYS 706
Principal types 706
Structure and shape of clay minerals 706
The properties of clay dispersions 712
    (a) Ion-exchange. (b) Dispersion, swelling, and flocculation. (c) Mechanical properties (viscosity, plasticity, thixotropy, dilatancy). (d) Streaming birefringence.
Applied colloidal chemistry of clays 720
ZEOLITES 721
Structure of principal types 722
Sorptive properties 722
Ion-exchange 726

CHAPTER XXVI. PROTEINS 730
INTRODUCTION 730
Definition and general features 730
Classification 732
Isolation of Proteins 732
THE PHYSICO-CHEMICAL PROPERTIES OF PROTEINS IN SOLUTION . . 735
Molecular weights—homogeneity of proteins 735
The Svedberg multiple law and Bergmann–Niemann hypothesis . . 736
Molecular shape 738
Applications of the electrophoretic method 741
    Serum proteins. Control of fractionation. Determination of molecular charge and shape.
THE STRUCTURE OF CORPUSCULAR PROTEIN MOLECULES . . . 745
Sub-structure and dissociation reactions 745
Information from X-ray measurements 746
    Astbury's layer structure.
Viruses 751
IMPORTANT GENERAL PROPERTIES OF PROTEINS 752
Specificity 752
Denaturation 753
Side-chain reactivity 754

CHAPTER XXVII. POLYMERS 756
Introduction 756
PRINCIPAL TYPES OF POLYMERS 757
(a) Saturated hydrocarbons 757
    Polythene. Poly-isobutene. Vinyl polymers. Silicones.
(b) Unsaturated Hydrocarbons 762
    Rubber. Gutta-percha. Butadiene polymers. Polychloroprene.
(c) Carbohydrates 766
    Cellulose and its derivatives. Starch and glycogen.
(d) Linear poly-esters, and poly-amides 774
(e) Branched or three-dimensional polymers 774
    Phenol-formaldehyde. Urea-formaldehyde.

CONTENTS

Physical Properties 775
Amorphous, crystalline, rubber-like, and plastic states . . . 775
The rubber-like state 779
    Elastic and thermodynamic properties. The statistical length of flexible chain-like molecules. Treatment of a network of chain-like molecules. Effect of cross-linkages—Vulcanization.

The Interactions of High Polymers with Low Molecular Weight Liquids 786
Calculations of free energy of dilution and comparison with experiment . 786
Solubility and swelling of high polymers. 793
Fractionation of high polymers 796
    Experimental and theoretical.

Polymers in Solution 798
Molecular weights and shapes in solution 798
The flexibility of chain-like molecules 800
Difficulties in the examination of polymers in solution . . . 801

Chapter XXVIII. MEMBRANES 804
Artificial membranes 804
(a) Preparation of principal types 804
    (i) Membranes from natural and synthetic polymers. (ii) Thin oil films.
(b) Structure 807
(c) Permeability 808
    (i) Permeability of rigid membranes: Gases and vapours; Non-electrolytes (in solution); Electrolytes; Colloidal molecules and organisms. (ii) Permeability of thin oil films.

Natural membranes 820
(a) Structure 820
(b) Permeability 822
    Gases. Non-electrolytes (in solution). Electrolytes.
(c) Bio-electric potentials 826
(d) Theories of cell permeability and specificity 827

APPENDIXES 830
    I. The relation between displacement and diffusion coefficient . 830
    II. The preparation of thimble-type cellulose nitrate membranes. . 831
    III. A method for the partial denitration of cellulose nitrate membranes. 832
    IV. Dimensions of standard U-tube viscometers. . . . 833

INDEXES 834

LIST OF PLATES

Figs. 11.1 and 11.2, 12.13 a and b, and 25.9 face pages 267, 320, and 720 respectively; Figs. 15.11, 15.13 a and b, 15.15–22 are between pages 456 and 457. Table 11.1 faces page 288.

PRINCIPAL SYMBOLS

a activity,
radius (special),
semi-axis of revolution of rotational ellipsoid.
a   ampere; (ma. milliampere).
A amplitude,
area.
Å   Ångström unit.
b equatorial semi-axis of ellipsoid.
c concentration,
with subscript: specific heat,
velocity of light.
C capacity (condenser),
couple,
with subscript: molecular heat capacity.
d distance (e.g. thickness of double layer),
differential.
∂ partial differential.
D dielectric constant,
diffusion coefficient.
e base of natural logarithms,
electronic charge.
E elastic (Young's) modulus,
energy (internal, activation),
potential (electrode, membrane).
f activity coefficient,
distribution function,
frictional coefficient per molecule.
F frictional coefficient per mole,
Helmholtz free energy,
Faraday's constant,
structure factor.
g acceleration due to gravity,
osmotic coefficient.
G Gibbs free energy,
optical magnification factor,
velocity gradient (rate of shear).
h height,
number of equivalents of bound acid or base per mole.
H heat content.
i electric current.
I ionic strength,
moment of inertia,
intensity of light.
J axial ratio a/b,

J twice the ionic strength.
k velocity constant of reaction,
Boltzmann's constant.
K equilibrium constant,
bulk modulus of elasticity.
l length.
l.   litre.
L inductance,
latent heat (per mole),
total length.
m mass.
m_0 molecular weight of repeat unit in a polymer.
m.   metre.
M molecular weight,
with subscript: molecular weight average.
n number (e.g. of moles),
refractive index,
shear modulus.
N mole fraction,
Avogadro's number.
p polarization (per c.c.),
pressure (usually vapour pressure).
P polarization (per mole),
pressure (usually hydrostatic),
probability.
q axial ratio b/a,
heat absorbed,
optical disymmetry coefficient.
Q electrical charge.
r radius (general).
R electrical resistance,
gas constant per mole.
s sedimentation constant.
S entropy,
solubility,
stress.
t time.
T absolute temperature.
u mobility.
u_0 mobility of liquid.
$U_K(U_A)$ mobility of cation (anion) in a membrane.
v velocity,
volume (including specific volume).
v_0 velocity of liquid.
$\bar{v}$ partial specific volume.

PRINCIPAL SYMBOLS

V Volta potential, volume.
$\bar{V}$ partial molar volume.
ΔV surface potential.
V volt (mV millivolt).
w weight, work.
W weight fraction, work (*with subscript*: of adhesion, cohesion).
x amount of adsorption, linear dimension.
X electrical field strength.
z valency of ion.
Z coordination number, optical displacement, tension.

GREEK

α polarizability, refractive index increment.
γ ratio of specific heats, surface tension.
Γ surface concentration excess.
δ resolving power, very small increment.
Δ double refraction.
ζ electrokinetic potential, rotational frictional constant.
η viscosity (η_0 solvent).
η_r relative viscosity.
η_{sp} specific viscosity.
$[\eta]$ intrinsic viscosity.
θ angle (e.g. contact angle),

θ fraction of surface covered by adsorbate.
Θ rotational diffusion coefficient.
κ Debye-Hückel function, specific conductance.
λ wavelength, logarithmic decrement.
Λ equivalent conductance.
μ chemical potential, coefficient of friction, dipole moment, micron, Poisson's ratio.
ν frequency, kinematic viscosity.
Π osmotic pressure, surface pressure.
π ratio of circumference to diameter.
ρ density, volume charge density, depolarization ratio.
Σ sum.
σ surface density of charge, standard deviation, strain.
τ relaxation time, thickness of adsorbed film, turbidity.
ϕ centrifugal force or potential, fluidity, volume fraction.
χ extinction angle.
ψ angle of isocline, electrical potential (general).
ω angular velocity.

PART I
HISTORICAL AND GENERAL SURVEY

I
EARLY HISTORY (UP TO 1910)

Colloids in early civilizations

ALTHOUGH the term *colloid* has not yet attained its centenary—it was first used by Graham in 1861—and although the scientific study of colloidal systems can only be reckoned to have begun about half a century earlier, yet the *art* of preparing and processing colloidal materials goes back many thousands of years, in fact as far back as civilization itself. Mention of a few of the everyday materials now classed as colloids will serve to illustrate this point: clays, soils, milk, butter and cheese, dough and pastry, inks and paints, as well as fibres such as silk, cotton, and wool. Early civilizations also knew of the swelling pressures of colloids, as they employed the swelling which dry wooden wedges undergo when moistened to split rocks and stone. In addition to these domestic uses a number of terrestrial phenomena of a colloidal nature, such as fogs, mists, clouds, and smokes, must have forced themselves on the attention of early man, although he would scarcely have recognized a common basis—other than divine—in this diversity.

What is the common property of such a variety of systems and phenomena? Anticipating the scientific work of Graham and others we can define a colloidal substance as one in a peculiarly fine state of subdivision, as a consequence of which the properties of the surfaces or inter-phases (interfaces) play a more or less dominating role. The presence of an interface clearly entails the presence of at least two phases, the continuous one being termed the *dispersion medium* and the other the *disperse phase*. For example, in milk particles of fat are dispersed throughout an aqueous phase, in butter the converse obtains; in smokes small solid particles and in mists small water droplets are dispersed with air as the continuous phase, and so on. In each case the disperse phase occurs in units large compared with atoms or simple molecules, but not so large that gravitational sedimentation occurs rapidly. As we shall see more clearly later, the term *colloidal* is no longer regarded as applicable only to certain substances, but refers to a state of matter which can be taken up quite generally. We therefore speak of the *colloidal state*, just as we do of the gaseous, liquid, and solid states.

Whilst the development of the art of handling colloidal materials must have played a great part in early civilizations, yet no experiments which can properly be termed scientific are on record much before the beginning of the nineteenth century. The alchemists of the Middle Ages were familiar with *potable gold* (a gold suspension still used medicinally), and the preparation of gold ruby glass goes back to the sixteenth century, but their colloidal nature was not appreciated. As would be expected most of the early work dealt with the preparation and properties of suspensions (*sols* as Graham subsequently termed them) of gold and silver, and of sulphur (on account of its occurrence in natural waters). The celebrated *Purple of Cassius*, for example, discovered in 1663, was a gold sol prepared by reducing a soluble gold salt with stannous chloride.

Principal investigations in the early nineteenth century

Before suspensions were systematically studied, however, very notable contributions upon three very different aspects of colloidal systems had been made by Reuss (1809), Brown (1827), and Ascherson (1838).

Reuss packed a clay plug into a U tube, filled the arms with water, and applied a potential across the plug. He observed that the small suspended particles of clay moved towards the positive electrode and that, after some time, the water-level stood higher in the arm connected to the negative electrode. The movement, under an applied electrical field, of the colloidal particle is termed *electrophoresis* (or cataphoresis), and that of the liquid *electro-osmosis*.

The unceasing movement of very small (colloidal) particles which can be seen under the microscope was pointed out in 1827 by the English botanist Brown, using an aqueous pollen suspension, and in his honour the phenomenon is termed the *Brownian movement*. Many criticized his observations on the grounds that the movement was an artefact arising from external vibrational or thermal influences; alternatively it was ascribed to the organic nature of the substances he used. (It would not seem unreasonable in those days to ascribe one of the usual properties associated with life, namely movement, to particles of biological origin.) However, later workers observed the phenomenon with colloids of a purely inorganic nature, and it was also found to be present in liquid inclusions in minerals, where external influences would be at a minimum. The final vindication of Brown's early observations came some eighty years later with the theoretical interpretation of Brownian motion by Einstein and the experimental work of Perrin and Svedberg, which is discussed later.

In 1838 an outstanding paper, which even now makes interesting reading, was given to the Paris Academy of Sciences by Ascherson, then lecturer in medicine at the University of Berlin. Pondering upon the physiological role of fats, which he showed were often present as microscopic droplets surrounded by a visible skin, he was led to study the behaviour of droplets of oils (such as olive oil) in an aqueous medium containing dissolved proteins (e.g. egg-white or serum). To his great pleasure, as he records, he found 'that coagulation in form of a membrane occurs inevitably and instantaneously when albumin comes into contact with a liquid fat'. These membranes, which would effectively prevent the drops from coalescing, were often thick enough to be visible (i.e. many molecules thick), and the protein constituting them appeared to have lost completely its former solubility in water. This *denaturation* of proteins on adsorption was later taken up by Melsens and more particularly by Ramsden at the end of the century, but even now, despite its biological and industrial importance, our understanding of the phenomenon is by no means complete.

Early studies of inorganic colloids

The first systematic study of inorganic colloids (in this case silver chloride, prussian blue, and sulphur) is due to Francesco Selmi, an Italian professor (published 1845–50). Of the various methods tried for the preparation of colloidal silver chloride he found double decomposition in very dilute solution to be the best, and in this connexion his quotation from Berzelius's *Treatise of Chemistry* shows that the spontaneous dispersion of silver chloride precipitates upon thorough washing was known earlier. (Graham later termed this latter phenomenon *peptization*.) Selmi showed that salts would coagulate these colloidal bodies and that they differed in their precipitating action, although it was over forty years before the effect of salts was put on a quantitative basis by Schulze and Hardy. Nevertheless he realized that the dispersed particles were not in a state of molecular subdivision, a conclusion supported by the absence of 'the phenomenon common to all true solutions, viz. the absorption of caloric or the production of cold by the dissolving substance, a phenomenon caused beyond doubt by its passing from the solid state of aggregation to that of extreme molecular division'. This argument was based upon the absence of any detectable temperature change upon peptizing the precipitate by washing or upon coagulating the suspension with salts.

Mention has already been made of Graham, particularly his contributions to the terminology of colloids. Thomas Graham (1805–69) is usually regarded as the founder of modern colloid science, although his contributions to other fields were also of great importance. He was appointed professor in his native city of Glasgow at the early age of twenty-five, later moved to University College, London, and in 1855 became Master of the Mint. A prolific investigator he published in all forty-six papers, of which two are of particular interest here: 'Liquid Diffusion applied to Analysis' (1861), and 'On the Properties of Silicic Acid and Other Analogous Colloidal Substances' (1864).

Graham's conception of the properties characterizing the colloidal state, such as low diffusivity, absence of crystallinity, and of ordinary chemical relations, as well as its biological significance, are conveyed by a quotation from the former paper:

The property of volatility, possessed in various degrees by so many substances, affords invaluable means of separation, as is seen in ever-recurring processes of evaporation and distillation. So similar in character to volatility is the diffusive power possessed by all liquid substances, that we may fairly reckon upon a class of analogous analytical resources to arise from it. The range also in the degree of diffusive mobility exhibited by different substances appears to be as wide as the scale of vapour tensions. Thus hydrate of potash may be said to possess double the velocity of diffusion of sulphate of potash, and sulphate of potash again double that of sugar, alcohol and sulphate of magnesia. But the substances named belong all, as regards diffusion, to the more 'volatile' class. The comparatively 'fixed' class, as regards diffusion, is represented by a different order of chemical substances, marked out by the absence of the power to crystallize, which are slow in the extreme. Among the latter are hydrated silicic acid, hydrated alumina and other metallic peroxides of the aluminous class, when they exist in the soluble form; with starch, dextrin, and the gums, caramel, tannin, albumen, gelatin, vegetable and animal extractive matters. Low diffusibility is not the only property which the bodies last enumerated possess in common. They are distinguished by the gelatinous character of their hydrates. Although often largely soluble in water, they are held in solution by a most feeble force. They appear singularly inert in the capacity of acids and bases, and in all the ordinary chemical relations. But, on the other hand, their peculiar physical aggregation with the chemical indifference referred to appears to be required in substances that can intervene in the organic processes of life. The plastic elements of the animal body are found in this class. As gelatin appears to be its type, it is proposed to designate substances of this class as colloids, and to speak of their peculiar form of aggregation as the colloidal condition of matter. Opposed to the colloidal is the crystalline condition. Substances affecting the latter form will be classed as crystalloids. The distinction is no doubt one of intimate molecular constitution.

That he realized the colloidal state could be obtained by the aggregation of molecular units is shown very well by the following quotation (1864),

The inquiry suggests itself whether the colloid molecule may not be constituted by the grouping together of a number of smaller crystalloid molecules and whether the basis of colloidality may not really be this composite character of the molecule.

This is the idea generally accepted to-day, although it is also realized now that individual molecules may in some instances reach colloidal dimensions.

The term colloid (from the Greek word for glue) was but one of Graham's numerous contributions to the nomenclature of colloids.

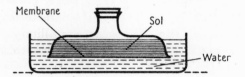

FIG. 1.1. Simple dialyser.

Others still in use to-day are the terms *sol*, *gel*, *peptization*, and *syneresis*. A sol, as already mentioned, is a dispersion of a solid in discrete units, the type of dispersion medium being shown by a prefix as in hydrosol, alcosol, aerosol, etc. (for water, alcohol, and air respectively). If, however, the solid particles formed a bridged structure with some mechanical stability then the system was termed a gel (hydrogel, alcogel, aerogel, etc.). Peptization—from analogy with peptic digestion—referred to the spontaneous dispersion of a precipitate to form a colloid, as with the gelatinous precipitates of silicic acid and aluminium hydroxide on the respective addition of traces of alkali and acid. (The peptization of silver chloride and prussian blue on thorough washing was mentioned above.) Syneresis was the term suggested by the study of silicic acid to cover the phenomenon of the spontaneous shrinkage of a gel to form a more concentrated gel and free exuded dispersion medium.

Graham emphasized the value of dialysis through membranes such as parchment for the separation of crystalloids from his colloidal solutions, although this technique had been known earlier. (A simple dialyser still used to-day is shown in Fig. 1.1.) He was also familiar with adsorption phenomena, such as the removal of iodine from aqueous solutions by charcoal, and realized that this was a surface phenomenon due to the large specific surface (the surface area per unit volume) which arises from the fine state of subdivision in colloidal materials.

As previously mentioned, metal sols, particularly of gold and silver, were a favourite subject for the early investigators; their study during the latter half of the nineteenth century being associated particularly with the names of Faraday, Muthmann, Carey Lea, Pauli, and Bredig.

Faraday published a detailed account of the preparation and optical properties of gold sols in 1857. Reduction of soluble gold salts, such as by phosphorus, leaves the metal either as a very thin film or as a sol which is frequently ruby-red in colour. The ruby sols, which had all the appearances of solutions, nevertheless were shown to contain particles since a cone of light became visible when passed through them. (This is the famous *Tyndall cone* or *beam*, since it was later studied in more detail by Tyndall.) The similarity in optical properties between the films and the sols (e.g. the light reflected from both was of a golden character), as well as the behaviour with chemical reagents, led him to the conclusion that the particles consisted essentially of metallic gold. He also showed that the colour observed depended upon the position of the observer relative to the light source, i.e. upon the degree to which the light was reflected or transmitted. From a study of reducing agents other than phosphorus Faraday suggested that *Purple of Cassius* was probably a mixture of finely divided gold with stannic oxide, and that the colour of gold ruby glass also arose from the fine particles of gold.

Addition of small amounts of salts to the ruby sols produces a colour change to blue or violet. Faraday showed, from the increased rate of settling, that this change was due to aggregation and that it could be prevented by the presence of gelatine. This last phenomenon is termed *protection*, and will be encountered again later.

Major developments towards the end of the nineteenth century

The detailed practical study of light scattering by colloidal suspensions was carried out by Tyndall at the Royal Institution in London, and the first theoretical analysis was given by Rayleigh at about the same time (1871). One rather interesting use to which Tyndall applied the technique was in the study of air-borne bacteria, and his experiments set out in his 'Essays on the Floating Matter of the Air in Relation to Putrefaction and Infection' (first published in 1881) are of historical interest. It may also be mentioned that Leonardo da Vinci, as long ago as the end of the fifteenth century, had pointed out that with smoke the blueness increased with the fineness of the particles, and that the smoke from the burning of wet green wood was white since the particles were large enough to reflect light like a solid body.

In considering the colours of colloidal particles there are two extreme cases, non-conductors (or dielectrics, e.g. sulphur) and conductors (e.g. metals). For the former case Rayleigh showed that for particles small compared with the wavelength of light the intensity of scattered light should be proportional to v^2/λ^4, where v is the volume of a particle and

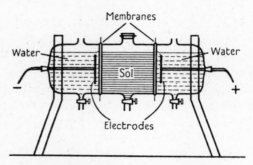

FIG. 1.2. Simple form of electro-dialyser.

λ the wavelength of light. Hence the blue end of the spectrum should preponderate in the scattered light and the red end in the transmitted, as is found experimentally. (The blue colour of the sky is explained on the same basis, except that the scattering is due to the air molecules and not to colloidal particles.) With metallic dispersions the conditions are clearly very different—witness Faraday's gold sols mentioned above—and it appears that the colours are largely determined by absorption. For example, the colour of a ruby gold sol arises from the strong absorption in the green part of the spectrum. Further extensions to the theory came later, with the work of Mie (1908) and of Debye (1909).

Returning to the preparation, purification, and constitution of metal sols, Muthmann showed in 1887 that those of silver consisted of dispersed metal and not of sub-oxides (as many had claimed), and that they could be rendered more stable by dialysing away the salts present. Subsequently the purification of sols was much facilitated by a combination of electrolysis and dialysis, a simple form of 'electro-dialyser' being shown in Fig. 1.2. In this the applied potential attracts the ions of any salts present to the electrodes and this helps the ordinary diffusion process. A modification of electrodialysis, developed chiefly by Pauli and termed *electro-decantation*, is shown diagrammatically in Fig. 1.3. The sol is concentrated at one side and at the bottom of the middle compartment by the combination of electrophoretic and gravitational forces, when the supernatant liquid can be changed by the decantation principle.

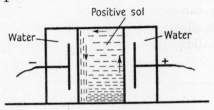

Fig. 1.3. Purification and concentration of sols by electro-decantation.

Another method, developed principally by Bechhold at about this time for the concentration and purification of colloids, was *ultrafiltration*, in which the dispersion medium is forced by pressure through an

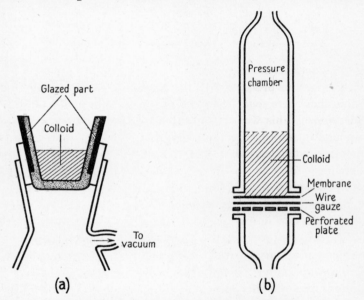

Fig. 1.4. Purification and concentration by ultrafiltration. (a) Simple laboratory apparatus. (b) Type for high-pressure work.

unglazed porcelain filter, usually impregnated by materials such as collodion to give it a suitable pore size. Fig. 1.4 (a) shows the principle of a simple laboratory type; Fig. 1.4 (b) a later development for use at high pressures.

At about the same time as Muthmann, Carey Lea in America was investigating the reduction products of silver salts and their connexion with the photographic process, photography being at that time in a state of rapid, if empirical, development. The colour changes of silver halides occurring under the influence of light are of course familiar to all chemists; they can also be obtained by chemical means, such as reduction of soluble silver salts with ferrous citrate, or by boiling colloidal silver with silver halide. The colour arises in all cases from the presence

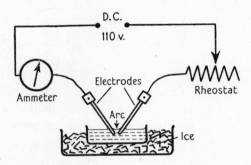

FIG. 1.5. Bredig arc method for preparation of metal sols.

of colloidal particles of silver embedded in the silver halide. (Carey Lea thought the various products which he obtained were allotropic forms of silver, but they were really colloidal silver particles in different degrees of subdivision.) This work eventually led Lorenz to suggest that the latent image in photography was none other than a photohalide containing silver particles too small to produce a recognizable colour but large enough to serve as nuclei during the subsequent chemical reduction (development) process.

The preparation of metal sols by means of an electric arc struck between electrodes immersed in water was due originally to Bredig (1898). A simple apparatus for this purpose is shown in Fig. 1.5. In the presence of a trace of electrolyte which is frequently caustic soda, fine stable suspensions can be obtained consisting chiefly of metal in the case of the noble elements such as gold, silver, and platinum, but with electropositive elements (e.g. zinc and aluminium) most probably of oxide. Svedberg subsequently improved the technique by introducing the use of alternating current and high current densities; by this means it is possible to obtain dispersions in non-aqueous media, such as sodium in benzene.

The part played by water in aqueous colloidal systems, particularly the hydroxides of the elements silicon, tin, aluminium, chromium, and iron, was investigated in great detail by van Bemmelen, then professor of inorganic chemistry at Leyden, and described in a paper published in 1888. He came to the conclusion that these colloidal 'hydrous oxides' (as they were subsequently termed), even when dry to the touch, were not definite chemical compounds, but held their contained water by adsorption. Thus on desiccation the vapour pressure and the velocity of dehydration decreased progressively with the colloidal hydroxides,

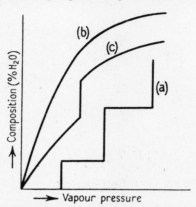

FIG. 1.6. Vapour pressure–composition curves for (a) A system forming true hydrates. (b) A system showing only adsorption. (c) A system showing a combination of (a) and (b).

whereas true hydrates gave a series of uniform steps, each of which corresponds to the coexistence of two of the hydrated phases (see Fig. 1.6). The composition of these colloidal materials was found to depend upon the molecular aggregation of the colloid, the vapour pressure, and the temperature.

Great advances in the study of colloids were made towards the close of the nineteenth century, due not only to their increasing interest but to the concurrent growth of other sciences, physical chemistry in particular. These other sciences not only provided new experimental techniques, but also a better theoretical background, of which the thermodynamic studies of Willard Gibbs (1876) are of course pre-eminent. Of Gibbs's contributions to colloids, perhaps the most noted is his adsorption isotherm, which enables the amount of adsorption at a liquid surface to be calculated from measurements of surface tension. Other contributions, such as the theory of membrane phenomena and the effect of particle size on solubility, will be discussed below. Owing largely to his abstract method of presentation the importance of Gibbs's work was not appreciated at the time, and it was over thirty years before his adsorption isotherm was tested experimentally. Shortly after

Gibbs a number of similar conclusions were reached independently by J. J. Thomson (1880), by the methods of generalized dynamics.

Also at about the same time as Gibbs and Thomson, Helmholtz (1879), famous for his researches on electricity, gave the mathematical theory for electrokinetic phenomena, the general name applied to those phenomena in which an electrical potential difference is associated with

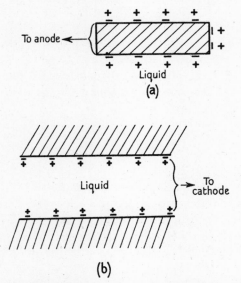

FIG. 1.7. Electrokinetic phenomena: (a) electrophoresis, (b) electro-osmosis.

tangential motion at an interface between a fluid (usually water) and some other phase. Mention has already been made of the work of Reuss on electrophoresis and electro-osmosis, two of the four possible types of electrokinetic phenomena which are as follows:

1. Electrophoresis (cataphoresis). Movement of particles with respect to a liquid by an applied field. (Fig. 1.7 (a).)
2. Electro-osmosis. Movement of liquid with respect to a fixed solid (e.g. a porous diaphragm or capillary tube) by an applied field. (Fig. 1.7 (b).)
3. Streaming potential. The potential difference set up by the movement of a liquid relative to the solid by the application of a mechanical force (e.g. when a liquid is forced through a porous membrane).
4. Sedimentation potential (or Dorn effect). The potential difference

set up in a liquid by the movement of suspended particles through it in response to an applied force (e.g. between the top and bottom of a tube when a suspension settles out under gravity).

The third of these phenomena was recorded by Quincke in 1859, the last by Dorn in 1880. It will be seen that in two of the four cases tangential motion of the phases is produced by an applied potential whereas in the other two the potential is produced by the tangential motion.

Helmholtz based his theory on the assumption that at an interface a separation of charge occurred, giving rise to an electrical double layer as shown in Fig. 1.7. Application of electrostatic theory then gave the simple formula for the velocity of electrophoresis (v), or the average velocity of flow of the liquid (v_0),

$$v \text{ (or } v_0) = \frac{DX\zeta}{4\pi\eta},$$

where X is the applied electrical field, D and η the dielectric constant and viscosity of the medium, and ζ the potential difference between interface and liquid (i.e. across the double layer). Nernst and later investigators identified the electrical double layer with the space-charge arising from the non-uniform distribution of ions at interfaces. Gouy (1910) pointed out that the mobile part of the double layer must be diffuse on account of thermal agitation, and so extend into the liquid for a distance much greater than the atomic order expected from the above picture. Further developments came later with the application of the Debye-Hückel theory of strong electrolytes, as will be discussed in more detail in Chapters V and XII.

The name of Helmholtz is also associated with the phenomenon, investigated particularly by Lippmann (1873), termed *electrocapillarity*, in which the interfacial tension between mercury and an aqueous solution is altered by means of a potential applied across the interface. This phenomenon he also interpreted on the basis of a condensed double layer of charge, one being on the mercury, the other in the aqueous phase.

Major developments during the early years of the twentieth century

The first decade of the present century witnessed notable advances in the elucidation of the conditions leading to the formation and destruction of the colloidal state, as well as in the study of the Brownian move-

EARLY HISTORY (UP TO 1910)

ment and the size of colloidal particles. Introduction of new nomenclature and systems of classification also assisted in a more general way.

Perrin (1905) introduced the terms *hydrophilic* and *hydrophobic* to differentiate aqueous suspensions of markedly differing properties, as typified by gelatine and the colloidal metals respectively. The general terms *lyophilic* (i.e. systems where the disperse phase has a high affinity for the dispersion medium) and *lyophobic* (low affinity between disperse phase and medium) were given shortly after by Freundlich. Although a complete range of intermediate types is known to exist, these terms are still in common use and the differences between the extreme types may be briefly enumerated.

Lyophilic
1. High concentrations of disperse phase frequently stable.
2. Unaffected by small amounts of electrolytes. 'Salted out' by large amounts.
3. Stable to prolonged dialysis.

4. Residue after desiccation will take up dispersion medium spontaneously.
5. Coagulation gives a gel, or jelly.
6. Usually give a weak Tyndall beam.
7. Surface tension generally lower than dispersion medium.
8. Viscosity frequently much higher than that of medium.

Lyophobic
1. Only low concentrations of disperse phase are stable.
2. Very easily precipitated by electrolytes.

3. Unstable on prolonged dialysis (due to removal of the small amount of electrolyte necessary for stabilization).
4. Irreversibly coagulated on desiccation.

5. Coagulation gives definite granules.
6. Very marked light scattering and Tyndall beam.
7. Surface tension not affected.

8. Viscosity only slightly increased.

(The differences in viscosity often arise in part from the different concentrations possible in the two cases.)

A different basis of classification was given in 1907 by Wolfgang Ostwald (son of Wilhelm, one of the founders of physical chemistry), who pointed out the theoretically possible types of colloidal systems with respect to the three states of aggregation—gaseous, liquid, and solid. This system, which is set out on p. 14, although useful, did not distinguish between the extreme types of suspension just discussed above.

Disperse phase \ Dispersion medium	Gas	Liquid	Solid
Gas	..	G/L (foam)	G/S (solid foam)
Liquid	L/G (mist)	L/L (emulsion)	L/S (solid emulsion)
Solid	S/G (smoke)	S/L (suspension)	S/S (solid suspension)

In the abbreviated notation used above the discontinuous or disperse phase precedes the continuous phase (the dispersion medium), thus, for example, S/G means a solid dispersed in a gas, and G/L a gas dispersed in a liquid.

The basic conditions for the separation of a phase in the colloidal state were clarified by the early work of Tammann (1903) and von Weimarn

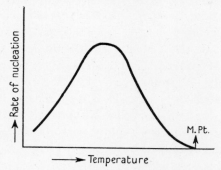

FIG. 1.8. Rate of formation of nuclei as a function of the temperature.

(1911). Tammann measured how the degree of supercooling of a pure substance (e.g. glycerol) affected the rate of formation of nuclei and the linear velocity of crystallization in a capillary tube. In both cases a maximum was found (see Fig. 1.8), which can best be interpreted on molecular kinetic considerations as follows. In supercooled melts only molecules with sufficiently low kinetic energy will be able to form the nucleus upon which the solid phase can separate out. As the temperature is lowered the chance encounters suitable for nucleus formation (nucleation) become more probable, and if this were the only factor involved the rates of nucleation and crystallization should increase indefinitely with lowered temperature. However, at some stage the viscosity of the medium, which increases exponentially with decreasing temperature, so slows down molecular movement that the nucleation rate actually diminishes. (Glass provides a good example of a supercooled liquid with negligible rate of crystallization.) Closely related to the above is the work of Noyes and Whitney (1897), Nernst (1904), and others, on the rates of dissolution from crystals and of crystallization from solutions. It was found that the latter process, like crystallization from pure melts, was frequently much affected by traces of added substances (e.g. lyophilic colloids, dyestuffs) owing to their adsorption.

von Weimarn investigated the conditions required to give colloidal solutions by condensation methods, and found the principal one to be a high initial supersaturation, so that a large number of small nuclei were formed simultaneously. By this means almost any substance could be obtained in the colloidal state. The various methods used to give the requisite degree of supersaturation will be discussed when the preparation of suspensions is considered (Chapter XX).

Turning now to the question of the interaction of colloids with electrolytes, non-electrolytes, and other colloids, mention was made above of the observations of Selmi and Faraday on the power of salts to coagulate hydrophobic sols. The quantitative work of Schulze (1882) on arsenious sulphide sols, later put on a more general basis by Hardy, showed that coagulation is brought about by the ion carrying the opposite charge to that on the colloidal particle, and the coagulating power of an ion increased very rapidly with its charge. For example the concentrations of Na^+, Ca^{++}, and Al^{+++} required to precipitate a negatively charged sol were approximately in the ratio $1:1/30:1/900$. The kinetics of coagulation of hydrophobic sols were later measured by von Smoluchowski (1916), who distinguished between the two cases of rapid coagulation (when $\zeta = 0$) and slow coagulation (when ζ has been lowered but not to zero). The quantitative relationship between the zeta potential and stability is considered in more detail in Chapter V.

The effect of mixing two colloids was found to depend upon the type of colloid (hydrophobic or hydrophilic) and upon the charge carried. Mixing two hydrophobic sols of the same charge in general produces no visible change, but if they carry opposite charges, such as arsenious sulphide (negative) and ferric hydroxide (positive), mutual precipitation occurs unless one is in considerable excess. Addition of an excess of a hydrophilic colloid such as gelatine to a hydrophobic sol stabilizes the latter and renders it less sensitive to salts. The comparative protecting action of various hydrophilic colloids upon ruby gold sols can readily be measured by utilizing the change to blue which indicates the first stage of coagulation. This forms the basis of Zsigmondy's 'gold number'

(Zsigmondy, 1901), which is defined as the minimum amount (mg.) of protective colloid required to prevent the colour change of 10 c.c. red gold sol on addition of 1 c.c. of a 10 per cent. sodium chloride solution. A few typical values obtained by Gortner (1920) are given below.

Protective colloid	Gelatine	Gum arabic	Sodium oleate	Dextrin
Gold number	0·005–0·0125	0·10–0·125	2–4	125–150

As already mentioned, to precipitate hydrophilic colloids, such as the proteins, requires relatively high salt concentrations. It was pointed out by Hofmeister (1888) that with a series of salts containing a common

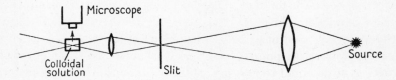

FIG. 1.9. Schematic diagram of the ultramicroscope.

ion (e.g. the sodium halides), the efficacy ran parallel to the hydration of the solid salts and to their solubility in water, indicating that precipitation was due to dehydration of the colloid, i.e. a true 'salting-out' effect. A series of cations or anions in order of coagulating power (e.g. $Li^+ > Na^+ > K^+$; $Cl' > Br' > I'$) is therefore termed the Hofmeister or lyotropic series.

A notable advance in the technique for examining colloidal systems became possible with the ultramicroscope of Siedentopf and Zsigmondy (1903). The fundamental difference from previous microscopic usage was the mounting of the microscope at right angles to the incident beam (see Fig. 1.9), so that the colloidal particles, due to the light they scatter laterally, stand out as points of light on a dark background. This at once extended the limit of visibility from about 2,000 Å ($0.2\ \mu$) with the ordinary microscope down to about 100 Å in some cases, but it must be realized that although such minute particles are visible by this means yet their size and shape are not resolved. The scattering depends upon the difference in refractive indices between particles and medium and so is more pronounced with the lyophobic type of sols, such as the metals, for which this difference is greatest. The ultramicroscope showed immediately that many systems previously thought to be molecularly dispersed were in fact optically resolvable, and contained particles with dimensions within the range of ca. 100–2,000 Å. Even such fine colloidal particles must, however, be constituted of a large number of atoms (or molecules) since atomic dimensions are only a few Ångström units. (Thus the finest particle of gold so detectable would contain at least ten thousand atoms.) This statement does not, of course, exclude the existence of colloidal solutions in which the individual particle is a molecular entity; indeed such cases are now commonplace (e.g. many proteins and polymers).

Zsigmondy, using a fine gold suspension in his ultramicroscope, confirmed the phenomenon of Brownian movement. Wiener had suggested over forty years previously that it arose from the irregular bombardment of the particles by the molecules of the dispersion medium; this idea formed the basis for the mathematical treatment of Einstein (1905) and von Smoluchowski (1906), and was confirmed experimentally by Perrin, Svedberg, and others.

von Smoluchowski pointed out that the mean kinetic energy of a particle must be equal to the mean kinetic energy of the solvent molecules which bombard it, i.e. from an energetic standpoint *the particle must behave as if it were a molecule of a dissolved substance*. This has been confirmed, and the study of Brownian movement has provided one of the most striking pieces of evidence in favour of the kinetic theory and of the real existence of molecules.

Einstein considered rotational as well as translational Brownian movement, since by the law of equipartition of energy the mean energy of rotation should be equal to that for translation. For translation he deduced (see Appendix I) the formula

$$\bar{x}^2 = 2Dt, \qquad (1.1)$$

where $\bar{x}$ is the mean displacement in the x direction in time t, and D the coefficient of diffusion (i.e. the number of particles crossing unit area in unit time, under unit concentration gradient). Einstein also showed (see Chapter X) that

$$D = \frac{RT}{Nf} = \frac{RT}{F}, \qquad (1.2)$$

where R is the molar gas constant, N Avogadro's number, and f and F the frictional coefficients per molecule and mole respectively. For spherical particles under certain conditions $f = 6\pi\eta r$ (Stokes's law) so that

$$\frac{\bar{x}^2}{2t} = \frac{RT}{N} \cdot \frac{1}{6\pi\eta r}, \qquad (1.3)$$

where r is the radius of the particle and η the viscosity of the medium.

For the case of rotational Brownian movement of spherical particles Einstein deduced an analogous relation

$$\frac{\bar{\theta}^2}{2t} = \frac{RT}{N} \cdot \frac{1}{8\pi\eta r^3}, \qquad (1.4)$$

where $\bar{\theta}$ is the mean angular displacement in time t (see Chapter XIV).

Eqns. (1.3) and (1.4) enabled Perrin and Svedberg, by measuring $\bar{x}$ or $\bar{\theta}$ as a function of time for particles of known size, to determine Avogadro's number N. The values so obtained were in fair accord with one another and with those from other methods. Conversely, knowing N, which is now known more accurately by other means (e.g. Millikan's oil-drop method), we can find r (if the particles are assumed spherical). This is important since although colloidal particles down to ca. 100 A can be discerned in the ultramicroscope, yet as we have pointed out, their size and shape are not resolvable much below $1\,\mu$.

Perrin pointed out that the distribution under gravity of particles in a suspension should resemble the falling-off in density of the atmosphere with height, and so should obey an equation similar to the well-known hypsometric formula
$$p = p_0 e^{-hgM/RT}.$$
In a colloidal suspension when 'sedimentation equilibrium', as it is termed, is set up between the opposing processes of diffusion and sedimentation under gravity, the above equation becomes (see Chapter VIII)

$$C_1 = C_2 e^{-hgM/RT}, \qquad (1.5)$$

where C_1 and C_2 are the concentrations of particles at two heights h_1 and h_2, such that $h = h_1 - h_2$, and M is now the 'apparent mass' of the colloidal particle, since for a liquid dispersion medium allowance must be made for the buoyancy correction. This formula was shown by Perrin (1908) to hold over small distances (ca. 0·1 mm.) in dilute suspensions. Having found M in this way it is clear that the radius of the particles can be found if they are assumed to be spherical.

In addition to the above methods, particle sizes can in some cases be found from the 'sedimentation velocity', in which the limiting velocity of fall (v) through aqueous or other media is measured by observations upon either a single particle or the edge of the sedimenting suspension. Equating the force due to gravity with that due to the viscous drag we have, for spherical particles,

$$\tfrac{4}{3}\pi r^3(\rho_P - \rho_M)g = 6\pi\eta rv, \qquad (1.6)$$

where ρ_P and ρ_M are the densities of particle and medium respectively. Hence r can be found from measurements of v. For particles down to about 10 μ the settling under gravity is measurable, but for particles much smaller than this the rates become too slow for accurate work and diffusion causes the boundary to become excessively indistinct. Under such circumstances a powerful centrifugal field has to be used instead of the gravitational one. This will be mentioned in the next chapter in connexion with the work of Svedberg, and discussed in detail in Chapter XI.

REFERENCES

ASCHERSON,† 1840, *Archiv Anat. Physiol. Lpz.*, p. 44.
BEMMELEN, VAN,† 1888, *Rec. trav. chim. Pays-Bas*, **7**, 37.
BREDIG, 1898, *Z. angew. Chem.* **11**, 951.
BROWN, 1828, *Phil. Mag.* **4**, 161.
—— 1829, ibid. **6**, 161.
DEBYE, 1909, *Ann. Phys. Lpz.* **30**, 57.
DORN, 1880, *Wied. Ann.* **10**, 70.
EINSTEIN, 1905, *Ann. Phys. Lpz.* **17**, 549.
—— 1906, ibid. **19**, 371.
FARADAY,† 1857, *Philos. Trans.*, p. 145.
GIBBS, 1876, *Collected Works*, **1**, Longmans, Green & Co., New York, 1928.
GORTNER, 1920, *J. Amer. Chem. Soc.* **42**, 595.
GOUY, 1910, *J. Phys.* **9**, 457.
GRAHAM, 1861, *Philos. Trans.* **151**, 183.
—— 1864,† *J. Chem. Soc.*, p. 618.
HARDY, 1900, *Proc. Roy. Soc.* **66**, 110.
—— 1900, *J. phys. Chem.* **4**, 235; *Z. phys. Chem.* **33**, 385.
HELMHOLTZ, 1879, *Wied. Ann.* **7**, 337.
HOFMEISTER, 1888, *Arch. exp. Path. Pharmak.* **24**, 247.
LEA, CAREY,† 1889, *Amer. J. Sci.* **37**, 476.
LIPPMANN, 1873, *Ann. Phys. Lpz.* **149**, 546.
MIE, 1908, *Ann. Phys. Lpz.* **25**, 377.
MUTHMANN,† 1887, *Ber. dtsch. chem. Ges.* **20**, 983.
NERNST, 1904, *Z. phys. Chem.* **47**, 52.
NOYES and WHITNEY, 1897, *Z. phys. Chem.* **23**, 689.
OSTWALD, WO., 1907, *Kolloidzschr.* **1**, 291.
PERRIN, 1905, *J. Chim. phys.* **3**, 50.
—— 1908, *C.R. Acad. Sci. Paris*, **146**, 967; **147**, 530, 594.
QUINCKE, 1859, *Pogg. Ann.* **107**, 1.
RAMSDEN, 1903, *Proc. Roy. Soc.* **72**, 156.
—— 1904, *Z. phys. Chem.* **47**, 336.
RAYLEIGH (then STRUTT), 1871, *Phil. Mag.* **41**, 107, 274, 447.
REUSS, 1809, *Mémoires de la Société Imperial des Naturalistes de Moskou*, **2**, 327.
    (Quoted by Svedberg, 1928, *Colloid Chemistry*, 2nd ed., Chemical Catalog Co.)
SCHULZE, 1882, *J. prakt. Chem.* **25**, 431.
—— 1883, ibid. **27**, 320.
SELMI,† 1845, *Nuovi Ann. d. Scienze Naturali di Bologna*, Serie II, t. IV, 146.

† Reprinted in Hatschek, 1925, *The Foundations of Colloid Chemistry*, Ernest Benn, Ltd.

SELMI, 1847, ibid., Serie II, t. VIII, 401.
—— (with SOBRERO), 1850, *Ann. Chim. et Phys.* **28**, 210.
SIEDENTOPF and ZSIGMONDY, 1903, *Ann. Phys. Lpz.* **10**, 1.
SMOLUCHOWSKI, VON, 1906, *Ann. Phys. Lpz.* **21**, 756.
—— 1916, *Phys. Z.* **17**, 557, 585.
TAMMANN, 1903, *Kristallisieren und Schmelzen*, Leipzig.
THOMSON, 1880, *Application of Dynamics to Physics and Chemistry*, Macmillan.
TYNDALL, 1881, *Floating Matter of the Air*, Longmans, Green & Co.
WEIMARN, VON, 1911, *Grundzüge der Dispersoidchemie*, T. Steinkopf, Leipzig.
WIENER, 1863, *Pogg. Ann.* **118**, 79.
ZSIGMONDY, 1901, *Z. anal. Chem.* **40**, 697.

Further references to the early work will be found in:
FREUNDLICH, 1926, *Colloid and Capillary Chemistry*, 1st English ed., Methuen.
THOMAS, 1934, *Colloid Chemistry*, McGraw-Hill.

II
THE PRINCIPAL ADVANCES SINCE 1910

Introduction

FOR the purposes of subdivision the year 1910 is a convenient choice, for although there is naturally no sudden change just at that point, yet it does mark, in a general way, the end of 'classical colloids' and the beginning of the more modern approaches. This change was undoubtedly considerably influenced by concurrent advances in physics and chemistry, such as the electronic structure of the atom and its immediate bearing upon theories of valency and force fields around molecules, the ideas of molecular size and shape as revealed by X-rays, and the nature of solutions, particularly the interionic attraction theory of Debye and Hückel. An increasing stimulus has also come from the industrial and biological interest of colloidal systems, such as the soaps and polymers on the one hand, and the proteins, membranes, and interfacial reactions on the other.

The period since 1910 has seen the introduction of new and powerful techniques for studying the colloidal system as a whole, such as the ultracentrifuge, electrophoresis, X-ray diffraction, etc., and in addition an entirely new development, the study of the interface. (This last is generally termed *surface chemistry* although *surface physical chemistry* would be a more appropriate title.) It has also been a period of consolidation during which the exploratory tracks of the earlier workers have broadened into a well-beaten highway. This makes it impossible to do more than indicate the main trends of recent advances and a few of the more noted exponents.

Surface chemistry

The field of surface chemistry has, of course, its origins in the study of surface tension, the effects of which are of everyday occurrence. The measurement and theory of surface tension were topics of great interest for the physicists of the nineteenth century, and this resulted in the collection of a large amount of data upon solutions and pure liquids. It was found that metals and molten salts had surface tensions much higher than that of water ($\gamma_{H_2O} \simeq 70$ dynes/cm. at ordinary temperature), whereas most organic compounds gave considerably lower values. In aqueous systems most inorganic salts were found to increase the surface tension, whereas it was depressed by most organic compounds, particularly by the soaps which were of much interest then in connexion with the study of soap bubbles and foams. With aqueous solutions of an homologous series, such as the lower fatty acids, it was shown by Traube (1891) that the concentration required to produce an equal lowering is reduced to about one-third by every additional carbon atom in the chain (Traube's rule). The spreading of one liquid over another, particularly over water, was also measured and it was found, for example, that the hydrocarbons showed little or no spreading tendency whereas many of their simple derivatives, containing groups such as —OH, —NH$_2$, —COOH, etc., spread rapidly over a water surface. The interpretation of such results in terms of molecular orientation and the fields of force around molecules was developed later, chiefly by Hardy, Langmuir, and Harkins, as discussed in more detail below.

In addition to the gas/liquid interface some attention was also directed in these early days to the liquid/liquid and solid/liquid interfaces. As regards the first of these the oil/water interface was interesting in relation to the formation and stability of emulsions, and the mercury/water interface in connexion with the *electro-capillary phenomenon,* as previously mentioned. The solid/liquid interface was chiefly studied by measuring the uptake of organic compounds (e.g. fatty acids, dyestuffs) by powerful adsorbents such as charcoal and silica-gel. In dilute solutions an adsorption isotherm of the type

$$x/m = ac^{1/n} \tag{2.1}$$

was usually found, where x was the uptake by a mass m of adsorbent, c the concentration of adsorbate in solution, and a and n constants. The above is an empirical equation usually ascribed to Freundlich (1906); Langmuir later showed how the simple kinetic theory leads to the equation

$$\frac{x}{m} = \frac{abc}{1+bc}, \tag{2.2}$$

which frequently fits the results just as well. Here a and b are constants, and the other terms have the same significance as above. In addition to the above work on the adsorptive powers of solids a great deal of empirical data on the solid/liquid interface had been obtained in connexion with the separation of ores by the *flotation* process, which by 1910 was already of industrial importance.

Mention has already been made of the adsorption isotherm of Gibbs. Beginning with the work of Donnan and his school (e.g. Donnan and Barker, 1911) this has been the subject of much experiment and discussion. The early tests of the isotherm employed air or oil bubbles which, after ascending through a solution containing an easily estimated solute (e.g. fatty acids, dyestuffs) were allowed to break, and the adsorption thus found was compared with that calculated from boundary tension measurements and the Gibbs equation. The agreement between theory and experiment varied widely, but we now know that the apparently anomalous cases arose either from experimental error or complications due to the solutes used (see Chapter XVIII). The recent investigations of McBain using an ingenious microtome method which skims off the surface layer have further supported the validity of the Gibbs equation.

In his treatment of the thermodynamics of interfaces Gibbs (1876) had included the case of a film insoluble in both phases which, however, he regarded as 'of secondary interest' (compared with those adsorbed from solutions). Some years later (1891) Miss Pockels showed how insoluble films on water surfaces, such as were given by castor-oil or olive-oil, could be handled by means of waxed paper or glass barriers. Using a large surface area and a small amount of oil no lowering of the surface tension was observed until the area was reduced to a certain value, and Rayleigh (1899) suggested that at this point the molecules were just in contact in a layer 1 molecule thick. Further studies of these oil films were made by Devaux, Marcelin, Labrouste, and others, but it was Langmuir (1917) who first gave the fundamental theory and methods upon which the present ideas of molecular orientation and structure of insoluble monolayers are based.

A few years before Langmuir's work Hardy (1913) had discussed the concept of the varying field of force around a long-chain molecule such as a fatty acid, and the effect of this varying field upon orientation at an interface. Hardy was a physiologist whose biological interests led him more and more into the study of interfaces, of which his work on lubrication is perhaps the best known, although he made other notable contributions, such as the investigation of monolayers by their influence on the stability of bubbles blown beneath them. His two last papers, entitled respectively 'Note on the Central Nervous System of the Crayfish' and 'Problems of the Boundary State', provide a very fitting indication of his range of interests.

Langmuir, on the other hand, approached surface chemistry from the viewpoint of the physicist. He measured the adsorption of gases such as oxygen, nitrogen, and hydrogen by tungsten filaments (in connexion with the behaviour of metal filament lamps), and showed that the initial amounts of gas were held very strongly (and often irreversibly) until a layer monomolecular in thickness was formed. After that point further adsorption was relatively small and was readily reversible, showing that the initial monolayer had almost completely satisfied the residual field or valencies at the surface to which adsorption is ascribed.

This work indicated the short range of molecular forces and Langmuir applied this concept to the question of adsorption and orientation of long-chain compounds at an air/water interface. He developed a *film balance* by means of which the surface pressure or force, due to the lowering of surface tension by an insoluble monolayer present on one side of a movable barrier, could be directly measured, a technique which, with some modifications, is in use to-day (see Chapter XVII). This method showed that paraffin-chain derivatives with a water-attracting group (as shown by their effect on the solubility of the shorter chain members), such as —COOH, —OH, —NH$_2$, etc., formed a monomolecular layer on a water surface with molecular orientation as depicted in Fig. 2.1. In this manner the affinity of the *polar* group for water is satisfied as well as the *escaping tendency*, or insolubility, of the paraffin-chains. (Groups such as —COOH, —CO, —OH, etc., are usually referred to as polar, since their centres of positive and negative charge do not coincide, giving rise to small electric doublets or dipoles, e.g. $\overset{+}{>}\text{C}=\overset{-}{\text{O}}$, $\overset{-}{\text{O}}—\overset{+}{\text{H}}$.) The aqueous solubility of compounds containing such groups is ascribed largely to the interaction of such dipoles with the water dipoles.

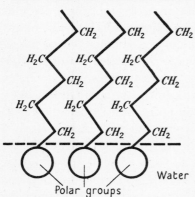

FIG. 2.1. Orientation of a long-chain polar compound at the air/water interface.

The paraffin-chains when close-packed on the surface were shown, from the amount of material occupying a given area, to have a cross-sectional area of about 20 Å^2/molecule, a value confirmed some ten years later by X-ray measurements on crystals.

The systematic study of the relationship between the lowering of surface tension (termed the force) and the area (per molecule) was started by Adam and his co-workers in 1921, and this showed that gaseous, vapour, liquid, and solid states existed in these two-dimensional monolayers just as in three-dimensional matter. Additional information was subsequently obtained by other workers from the study of mechanical properties, chiefly the surface viscosity, and from the *surface potential*—the change in potential at the air/water boundary produced by the monolayer. The use of monolayer techniques for studying reactions at interfaces is due largely to Rideal's school at Cambridge; another large school of 'surface chemists' under Harkins at Chicago chiefly developed accurate methods for the study of both adsorbed and insoluble films. Several developments are due to Russian scientists, Frumkin's work on adsorbed films and electro-capillarity being particularly well known; other notable contributions were made by French and Dutch workers. In fact the subject of surface physical chemistry is now very extensive, and has shed a good deal of light not only upon colloidal systems as such, e.g. emulsions, foams, soaps, and proteins, but also upon reactions at interfaces and many biological problems. Examples will be found at various points throughout this book.

Membrane equilibria

Amongst his other contributions to the thermodynamics of interfaces Gibbs had discussed the system of electrolytes in which one ion was restrained in some way from diffusing to all parts of that system, and he showed that this would lead to an unequal distribution of all the other ions present. This particular state of unequal ionic distribution is the characteristic feature of the 'Donnan equilibrium' as it is called, since it was first confirmed experimentally by Donnan (1911). The usual system contains a mixture of ions, one of which is of colloidal size and prevented from diffusion by a membrane permeable to all the other ions (see Fig. 2.2). (Hence the term 'membrane equilibrium' which is commonly applied to such systems.) Proteins in the presence of simple salts, enclosed in a cellophane or collodion sack, are frequently used. In agreement with theory it is found that when the ratio of salt to colloid is small the salt is held largely in the colloid-free compartment, whereas when the ratio is high the salt is equally distributed throughout. An unequal distribution of salt contributes to the osmotic pressure of colloidal systems, and in addition gives rise to a potential difference, termed a 'membrane potential', between the two solutions separated by the membrane. The above concepts are of great importance

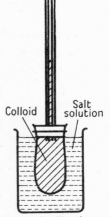

Fig. 2.2. Simple apparatus for the study of membrane equilibria.

in many colloidal phenomena, such as the osmotic pressure of protein solutions, the swelling of gels, the adsorption of ions by colloidal particles, the behaviour of membranes containing ionized groups, the charge and valence of dissolved proteins, and so on, as will be discussed in later chapters.

Size, shape, and charge of colloidal particles

Throughout the early study of colloids the questions of size and shape frequently recurred. Thus it was found that the properties of such systems as clay pastes, paints, polymers, and gels depended to a great extent upon the size, shape, and size distribution of the colloidal constituents. The biologist was also interested in these questions, not only in connexion with the soluble proteins but also with gels and membranes. Direct measurement with the microscope was not feasible below $ca.$ $0 \cdot 2\,\mu$ (or $ca.$ $0 \cdot 1\,\mu$ using ultra-violet light), and although the ultramicroscope can demonstrate the presence of much smaller particles yet their size cannot readily be measured. (The ultramicroscope can, however, give some indication of the *shape* of particles, since with rods or flat discs a very noticeable twinkling is observed.) Determination of the colligative properties was of no use in general owing to the minute changes to be measured, although the osmotic pressure method can be used with smaller colloidal particles, such as some proteins and polymers. Other methods for particle size determination, based upon the velocity of sedimentation under gravity, sedimentation equilibrium, or free diffusion, have been mentioned above in connexion with the Brownian movement. These techniques have been developed considerably during the last thirty years, the first two particularly by Svedberg at Uppsala, who was awarded a Nobel Prize in 1926 for his achievements in the realm of colloids.

Sedimentation under gravity has a practical limit, usually about $1\,\mu$ (in exceptional cases down to about $0 \cdot 1\,\mu$), owing to the extremely slow rate of fall for smaller particles, but for many industrial materials such as paints and clays this limitation is frequently not important and the method is still widely used. Svedberg replaced the gravitational field by a very powerful centrifugal one and thus extended the method to particles of 50 A or less in diameter. This method alone will only give the true particle mass (or molecular weight), if the particles are spherical. In other cases, particularly with linear polymers such as cellulose derivatives, it must be combined with diffusion or sedimentation equilibrium measurements.

Svedberg also applied centrifugal fields to the sedimentation equilibrium method, using the centrifugal force (instead of gravity) to balance the diffusion (or osmotic) tendency. This method has the advantage of giving the molecular weight regardless of the shape of the particle, but the times required for attainment of equilibrium are relatively long (frequently several days). If, in addition, diffusion or sedimentation velocity measurements are made, some idea of the shape of the molecule can be found as well.

Methods for the quantitative study of diffusion have been based either upon the use of a sintered disc, in the pores of which a convectionless concentration gradient is set up between the solvent and the solution, or by the study of *free diffusion*, in which a boundary is formed between solvent and solution and the diffusion followed by suitable optical or analytical means. Of these the second has been considerably improved by the development of new types of cell and more accurate optical techniques. As pointed out above, determination of the diffusion coefficient gives the molar frictional constant F, from which the particle size can be calculated only for spherical particles. If, however, this is combined with sedimentation velocity or sedimentation equilibrium measurements, then, as we have mentioned, the size of the molecule as well as information about its shape can be obtained.

As regards size distribution some idea of this can be obtained from a detailed study of sedimentation and diffusion behaviour, but much more information may be obtained by fractionating the original sample and examining each fraction separately. Such fractionation may prove to be very difficult in practice, particularly with high molecular weight linear polymers.

By the means outlined above considerable information upon size and shape has been obtained for a great variety of materials, proteins of all kinds (animal and vegetable origin, enzymes, antibodies, viruses, etc.), natural and synthetic polymers, etc., and this forms the basis for much of the discussion of colloidal systems.

A number of other methods giving rather less direct information about size and shape have been developed during the last thirty years or so, based upon viscosity, streaming birefringence, dielectric phenomena, X-ray and electron diffraction, etc.

According to the theory developed by Einstein (1906) the viscosity (η_0) of a medium should be increased by the addition of suspended spherical particles to a new value (η) given by

$$\eta = \eta_0(1+2{\cdot}5\phi), \tag{2.3}$$

where ϕ is the volume fraction occupied by the particles. It is clear that, for a given value of ϕ, η would be independent of the size of the particles, and so could give no information as to their size and shape. However, the derivation of this expression assumed spherical particles, an assumption which is often not the case in colloidal systems, particularly with solutions of long-chain polymers. For the linear polymers Staudinger (1930) suggested an empirical relation between the specific viscosity η_{sp} $\{\eta_{sp} = (\eta - \eta_0)/\eta_0\}$ and the chain-length or molecular weight (M), viz.

$$\eta_{sp} = kcM, \tag{2.4}$$

where c is the concentration of polymer and k a constant for a given polymer type. This equation has been the subject of much discussion from both the theoretical and practical angles, but in slightly modified forms is in wide use to-day, both for industrial control and in research, owing to the ease and accuracy with which viscosities can be determined. Opinions still differ, however, as to whether or not the method is capable of giving a true molecular weight.

Mention has been made above of how asymmetrical particles affect the Tyndall scattering as seen in the ultramicroscope. By causing the particles to orientate in some direction and by the use of polarized light, the effect can be magnified. Orientation can be brought about by an external electric or magnetic field, but more usually by velocity gradients in liquids. Thus when a vanadium pentoxide sol in a capillary tube is viewed stationary through crossed nicols the field appears dark, but it lights up on flow owing to orientation of the rod-shaped particles along the stream-lines (streaming birefringence). Freundlich and his co-workers developed this method for many inorganic colloids, and more recently it has been extended, using a Couette type of viscometer, to a number of linear polymers, particularly asymmetric protein molecules and cellulosic derivatives.

The application of X-ray and electron diffraction methods to colloidal systems has so far been limited to certain fields owing to inherent difficulties both of a practical and theoretical nature. The former method has been used to great advantage in the study of fibrous materials, such as silk, wool, cellulose (e.g. Astbury, 1933), cellulosic derivatives and some synthetic polymers, and more recently viruses and protein crystals. The broadening of the X-ray diffraction lines when the particles approach colloidal dimensions has been used to estimate the particle size in certain crystalline colloids, and to follow their growth, and more recently the study of low-angle scattering has given promising results with amorphous colloidal materials. Nevertheless the principal contribution of X-ray work to colloids has been indirect, from its information upon crystal structures, bond lengths and angles, structure of liquids, etc. The use of electron diffraction in colloids has been almost entirely concerned with the investigation of adsorbed films upon solid surfaces.

We may conclude this review of methods for size and shape determination with the most recent and most publicized—the electron microscope. The resolving power of any microscope is limited by the wavelength of the light used, and since the de Broglie wavelength of electrons is only a few Ångströms as compared with 4,000–7,000 A for visible light or 2,000–4,000 A for ultra-violet, it is clear that the use of an electron beam could theoretically increase the resolving power by a factor of one hundred or more. At present the limit is about 30–50 A in favourable cases (inorganic colloids). One of the drawbacks as far as many colloidal systems are concerned is the necessity of exposing the specimen to a very high vacuum, but for inorganic colloids such as paints, clays, carbon black, etc., the method is extremely valuable.

It was mentioned in Chapter I that the hydrophobic colloids, such as the metal sols, owe their stability primarily to their ζ potential, since when this is reduced by addition of salts coagulation is found to occur. The theory of complete ionization of strong electrolytes advanced by Debye and Hückel in 1924 gave a great stimulus to this aspect, and was subsequently applied by them and others to particles of colloidal dimensions. There are really two problems involved here—the calculation of the ζ potential from measurements of electrokinetic phenomena, and the dependence of the stability upon the magnitude of the ζ potential.

As regards the experimental study of electrokinetic phenomena the chief developments have been in connexion with electrophoresis. For particles visible in the ultramicroscope the mobility can be measured directly in various types of cell, such as the micro-capillary apparatus of Mattson (1922) or the flat rectangular cell of Abramson (1932). For smaller particles, particularly the globular proteins, Tiselius has developed the 'moving-boundary' method, resuscitating an old optical technique, the 'schlieren' (shadow) method of Töpler, for observing the boundary of the migrating protein. This for the first time provided accurate mobility measurements on dissolved proteins and showed that some proteins homogeneous in the ultracentrifuge were not so electrophoretically, and vice versa. Subsequently the apparatus was increased in size and is now used to prepare pure electrophoretically homogeneous proteins.

Rheology

The scientific study of the mechanical properties, such as flow, ductility, and plasticity, of concentrated colloidal systems has been termed *rheology*. Some of the principal systems included are the soil colloids (e.g. sand, clays), paints, doughs, cheese, bitumen, plasticized polymers, and gel systems; this range of materials gives some idea of its industrial interest. As mentioned above the questions of the size, shape, and size distribution of the colloidal constituents are often of decisive importance since the conditions make interaction between the particles almost inevitable, although many other factors, such as the interaction with the dispersion medium, frequently play a part. In addition their mechanical properties frequently depend upon the past treatment and age of the specimen, and this naturally makes their characterization and elucidation a more difficult matter. Of the above systems the polymers have probably been most systematically studied, and striking advances have been made in the theory of high elasticity (rubber and rubber-like materials), swelling, solubility and fractionation, and in the mode of action of plasticizers.

The investigation of the structure of gels, and the nature of the sol–gel transformation, led to the discovery of a new type of gel, termed *thixotropic*, since it can be liquefied by shaking. (With the usual type as given by gelatine or agar the sol–gel change is of course dependent upon the temperature.) Thixotropic gels are given by the oxides of aluminium and iron, by bentonite, and—most noted of all—by quicksand. (This consists of sand made thixotropic by the presence of clay.) Sand itself is normally *dilatant*, i.e. it hardens on shear, the opposite behaviour to a thixotropic gel. Another common example of a dilatant system is a thick paste of starch, as can readily be demonstrated by rapidly withdrawing some rigid object, such as a fork, from the mixture.

Colloidal electrolytes

The soaps, together with the dyestuffs, constitute the principal members of the class of colloids termed *colloidal electrolytes*, that is ionized substances of relatively simple structure whose solutions show colloidal properties over a certain concentration range. The alkali metal salts of the higher fatty acids are the oldest and still the most widely used soaps, now often termed *detergents* or *wetting agents*, but beginning with the preparation of hexadecyl sulphonic acid ($C_{16}H_{33}SO_3H$) and its salts by Reychler (1913), a great number of synthetic compounds are now known and are prepared commercially. Various terms have been suggested to cover all the various types, but none appear very satisfactory so we now propose to use the term 'soap' generically to include all these compounds, ordinary soaps being referred to as 'fatty acid soaps'.

The study of soap films and bubbles was of great interest in the last century, and Krafft in 1895 had pointed out the abnormal solubility increase which soaps show over a small temperature range (the 'Krafft point'), but the study of the colloidal properties of soap solutions only begins about 1910 with the work of McBain. McBain showed that soap solutions combined a low concentration of osmotically active particles, often much less than that calculated for no ionic dissociation, with a relatively high electrical conductivity, and he pointed out that this apparent anomaly could be explained by the formation of aggregates, termed *micelles*, of soap ions. These micelles would clearly exert little osmotic activity but could still have considerable conductivity. The equilibrium between micelle and the simple long-chain ions appears to obey the law of mass action, and this explains the disappearance of colloidal properties on sufficient dilution (Jones and Bury, 1927; Hartley, 1936). Although some discussion still centres upon the structure of the micelle and whether one or two types exist, it is generally agreed that the physical properties of soap solutions, such as surface activity, conductivity, osmotic coefficients, solubilization of organic compounds, the Krafft phenomenon, etc., are due, directly or indirectly, to the occurrence of micelles.

Classical colloids

This survey of advances since about 1910 may be concluded with a brief mention of recent work in the field of *classical* colloids, e.g. metal sols, emulsions, foams, smokes, etc.

The elucidation of the structure of really pure metal sols in water has been considerably advanced by the work of Pauli (1923) on gold sols, and of Pennycuik (1927) on those of platinum. The origin of the charge in these hydrophobic sols is still a subject of discussion; some favour adsorption of ions, others the dissociation of ionogenic groups present on the surface. In the case of platinum sols for example, Pennycuik has presented strong evidence for the presence of hexahydroxy platinic acid on the surface.

As regards foams and emulsions a vast literature now appertains to them, but despite this many points need clarification. Their fundamental study has, however, been simplified by the availability of pure stabilizing agents, particularly the synthetic soaps, which eliminate many of the complicating factors present in earlier work. As might be expected the application of the techniques and results of surface chemistry has helped notably in these fields.

The early work of C. T. R. Wilson around 1900 in connexion with the detection of ions by the *cloud-chamber* technique laid the basis for the study of mists. Subsequently the general study of aerosols (i.e. smokes and mists) has been largely stimulated by their practical uses, as in agricultural and insecticidal sprays, and in connexion with camouflage and poison-gas dispersal in war-time.

Future developments

Having briefly surveyed the more outstanding developments of the past we may attempt a few speculations upon probable future trends.

There are two broad aspects which, for purposes of classification, may be separated, namely industrial development and fundamental research. Recent times, particularly the war years, have witnessed an increasing industrial interest in the fundamentals of colloid science, not unnaturally, since these may frequently be applied directly to a wide range of industrial problems. Linked to this has been the expansion in the number and activities of the Research Associations, jointly sponsored by the Government and by industry, several of which deal very largely with colloidal materials (e.g. pottery, paints, printing inks, textiles and fibres, etc.). It may reasonably be expected that this aspect will gather increasing momentum, as it cannot yet be said that the fundamentals of colloid behaviour—even as far as they are known—have been widely appreciated or applied by those who direct our industrial effort.

Turning now to the more fundamental side, three main avenues can be envisaged, although once again there are no hard and fast distinctions between them. The first will be the study of colloidal systems by the more precise physical methods now available, and of great prominence here will be the application of modern ideas of solutions, solids, and surface forces to fields which have as yet not proved particularly amenable, such as gels, thixotropy, emulsions, soaps, etc. The second will be the expansion of surface chemistry, which despite its achievements still presents many points requiring clarification, more particularly with regard to the liquid/liquid and solid/liquid interfaces. The third avenue will include those aspects of colloids more directly concerned with biology, which may be termed 'bio-colloids' or 'bio-physical chemistry', as the terms biophysics and biochemistry have already acquired their own rather definite meanings in recent years. This would involve the application of the above techniques, preferably as a co-operative effort between biologist and colloid scientist, to definite biological problems, a field which has already indicated its potentialities.

REFERENCES

In addition to those given below a more detailed survey of the literature will be found under the relevant chapters.

ABRAMSON, 1932. See *Electrophoresis of Proteins*, by Abramson, Moyer, and Gorin, 1942, Reinhold.
ADAM, 1921, *Proc. Roy. Soc.* A **99**, 336, and many later publications.
ASTBURY, 1933, *The Fundamentals of Fibre Structure*, Oxford.
DONNAN, 1911, *Z. Elektrochem.* **17**, 572.
—— and BARKER, 1911, *Proc. Roy. Soc.* A **85**, 557.
EINSTEIN, 1906, *Ann. Phys. Lpz.* **19**, 289.
FREUNDLICH, 1906, *Z. phys. Chem.* **57**, 385.
GIBBS, 1876, *Collected Works*, vol. i, Longmans, Green & Co., New York, 1928.
HARDY, 1913. See *Collected Works*, 1936, Cambridge.
HARTLEY, 1936, *Aqueous Solutions of Paraffin-chain Salts*, Hermann et Cie, Paris.
JONES and BURY, 1927, *Phil. Mag.* **4**, 841.
KRAFFT and WIGLOW, 1895, *Ber. dtsch. chem. Ges.* **28**, 2566.
LANGMUIR, 1917, *J. Amer. Chem. Soc.* **39**, 1848.
McBAIN. See article in the *Third Colloid Report of the British Association*, 1920.
MATTSON, 1922, *Kolloidchem. Beihefte*, **14**, 227, and later papers.
PAULI et al., 1923, ibid. **17**, 294, and later papers.
PENNYCUIK, 1927, *J. Chem. Soc.*, p. 2600, and later papers.
POCKELS, 1891, *Nature* (London), **43**, 437.
RAYLEIGH, 1899, *Phil. Mag.* **48**, 321.
REYCHLER, 1913, *Bull. Soc. chim. Belg.* **27**, 110, 113, 217.
STAUDINGER et al., 1930, *Ber. dtsch. chem. Ges.* **63**, 222, and later papers.
TRAUBE, 1891, *Liebigs Ann.* **265**, 27.

III
MODERN VIEWS OF THE COLLOIDAL STATE

Introduction

IT is hoped that the historical review of the previous two chapters has given some indication of the wide variety of systems classed as colloids. Thus the term now covers not only the 'classical colloids' such as suspensions, foams, emulsions, etc., but also fibres, polymers, membranes, soil and plant colloids, and many biological systems, particularly those concerned with cells and bacteria. Colloids is a boundary science in a more profound sense than the literal one of being largely concerned with interfaces, for it fringes upon, and has application to, an unusually large number of scientific disciplines, chiefly of course, physics and chemistry, but also agriculture, geography, geology, mineralogy, metallurgy, etc., as well as very many aspects of biology.

This diversity makes the subject one of great interest and general appeal, although it has the drawback of making it difficult to keep in touch with developments in all its borderline repercussions. It also adds to the difficulty of giving a clear-cut definition of the terms 'colloid' and 'colloidal state'. As pointed out earlier the original idea that a 'colloid' was a particular type of substance was soon abandoned in favour of the view that it was a similarity of state which colloidal systems possessed in common. The colloidal state was defined as matter in a peculiarly fine state of subdivision, much larger than atomic or simple molecular dimensions, but much smaller than particles visible to the unaided eye. The ratio of surface area to volume increases with decreasing size of the particles and by virtue of the large interface presented by matter in the colloidal state the properties of the surfaces, such as the surface energy, are always important and frequently predominate. We will therefore inquire in more detail how the properties of matter are affected by subdivision into particles of colloidal size (i.e. $ca.\ 1\mu$–$ca.\ 50$ A).

Effects of subdivision of matter

The most obvious effect is, of course, upon the surface area and hence the surface energy, other properties affected being the solubility, vapour pressure, chemical reactivity, colour, etc. These will be discussed in turn.

(a) Surface area and surface energy

A cube of 1 cm. edge has a surface area of 6 cm.^{2} and a surface energy of 6γ ergs, where γ is the surface tension or surface energy (in dynes/cm.

or ergs/cm.^{2}) of the material composing the cube against the surrounding medium. When this cube is subdivided into cubes of 100 A side, a size typical of many colloidal particles, we now have 10^{18} cubes and therefore a total surface energy of $6 \times 10^6 \gamma$ (ergs), i.e. the surface energy has been increased by a factor of a million. The value of γ depends, of course, upon the particular system; against air or water organic liquids usually give values up to about 50 ergs/cm.^{2}, whereas salts and metals usually give values higher by about a factor of ten. Taking γ as 10 ergs/cm.^{2}, this gives a total surface energy for the 10^{18} cubes of 6×10^7 ergs, so that each individual cube has a surface energy of 6×10^{-11} ergs. (Cf. $kT \simeq 4 \times 10^{-14}$ ergs at room temperature.)

The large specific surface (i.e. surface area per unit volume) which thus obtains in the colloidal state accounts for the great activity of certain metal catalysts and for the great adsorptive powers of charcoal, silica gel, etc., possibly also for certain enzymatic activities, and it is undoubtedly an important factor in gel systems, where the concentration of the colloidal gelling agent may be as little as 0·5–1 per cent. Taking, for example, the simple case of a gelling agent consisting of fine rods 10 A in diameter and forming a cubic grid structure, then at a concentration of 1 per cent. each small cube has a side of *ca.* 180 A, i.e. an appreciable fraction of the solvent molecules must come within the range of influence of the gelling agent. (Gelled systems consist as a rule of a loose network structure built up of anisometric particles, gelatine being an obvious example.)

(b) *Solubility, vapour pressure, and chemical reactivity*

These are classed together, since they are but different manifestations arising from a common origin, namely the increase in free energy or chemical potential brought about by the surface energy of the particles. This effect will be presented in a qualitative manner here and deduced rigidly by thermodynamics in Chapter IV.

When a cm. cube is broken down into colloidal particles of 100 A size the surface energy increases by *ca.* $6 \times 10^6 \gamma$ ergs, since the original surface energy is negligible in comparison with that after subdivision. It is clear, therefore, that this system possesses extra energy over and above that of the original macroscopic solid, and it is to this that the increased chemical reactivity, vapour pressure, and solubility are due. The colloidal phase is thus seen to bear the same relation to its original bulk phase as a metastable phase to its more stable form (e.g. yellow and red phosphorus), where the differences in chemical reactivity,

36 MODERN VIEWS OF THE COLLOIDAL STATE

vapour pressure, solubility, etc., are of course well known. As a general rule, however, the differences observed in colloidal systems are very much less than for the example quoted since the energy differences are less pronounced.

As will be shown (p. 86), the solubility (S_r) of colloidal spherical particles of radius r is related to the bulk solubility (S_∞) (i.e. for particles where $r \to \infty$) by the relation

$$\frac{RT}{M}\ln\frac{S_r}{S_\infty} = \frac{2\gamma}{\rho r}, \qquad (3.1)$$

where ρ and γ are respectively the density and surface energy of the solid and M its molecular weight.

Some idea of the changes in solubility to be expected from eqn. (3.1) can be found by inserting reasonable values. Thus taking $M = 250$, $\rho = 1$, $R = 8\cdot4\times10^7$ ergs, $T = 300°$ K., we find, for $r = 0\cdot1\,\mu$,

$S_r/S_\infty = 1\cdot02$, $1\cdot22$, and $7\cdot41$ for $\gamma = 10$, 100, and 1,000 ergs/cm.^{2} respectively.

If $r = 1\,\mu$, then $S_r/S_\infty = 1\cdot02$ and $1\cdot22$ for $\gamma = 100$ and 1,000 ergs/cm.^{2} respectively. The combination $r = 0\cdot1\,\mu$, $\rho = 10$ gives the same result as that with $r = 1\,\mu$, $\rho = 1$.

Experimental investigations

Eqn. (3.1) was first examined experimentally by Hulett (1901, 1904), using calcium and barium sulphates and mercuric oxide, and he observed an increase in solubility with decreasing particle size as predicted by that equation.

A more extensive test was later made by Dundon and Mack (1923, 1924), and they also utilized their results to determine the surface energies of the particular solid/liquid interfaces employed, an application which has considerable interest in the field of surface chemistry. Some of their results are given in Table 3.1, and it will be seen that the values of γ so obtained are of the order which might be expected from measurements upon mobile interfaces and upon related solid/liquid interfaces.

TABLE 3·1

Substance:	PbI_2	$CaSO_4\cdot2H_2O$	$BaSO_4$	CaF_2
Particle size, μ	0·4	0·2–0·5	0·1	0·3
Solubility increase, %	2	4·4–12	80	18
Calculated surface energy, γ	130	370	1,250	2,500

Three recent publications have, however, cast much doubt upon the experimental validity of the above investigations. Cohen and Blekkingh (1940) could detect no change in the solubility of very pure barium sulphate with varying particle size, provided that very great care was taken in making and filtering the saturated solutions. (The solubility was determined by conductance and polarographic methods.) A similar conclusion was reached by Balarev (1941) using a number of compounds including barium sulphate, of particle size $0 \cdot 1\mu$ up to 300μ. With particle sizes below about 2μ a measurable increase in solubility was expected from eqn. (3.1) but again no such effect was detectable.

These results at first sight appear to invalidate eqn. (3.1), and as this equation can be deduced rigorously from thermodynamics, also to cast doubt upon the applicability of thermodynamics. As with the Gibbs adsorption equation such an unpalatable conclusion indicates the necessity of critically examining both the theoretical and experimental aspects of the problem, and in this case also a reasonable explanation can be suggested.

Of the likely sources of error two seem to warrant much more detailed consideration than they have yet been accorded:

(a) on the experimental side—the problem of preparing and maintaining clean surfaces;

(b) on the theoretical side—the question of the effect arising from the charge carried by the colloidal particles.

Experimental difficulties

As regards (a), studies of both fluid and solid surfaces have shown how difficult it is to maintain a clean surface for any length of time. In the case of fine powders the surface cannot readily be cleaned and the amount of impurity required for an adsorbed monolayer is exceedingly small. (For example, 1 mg. of protein would suffice to reduce the surface energy of 10^5 cm.^{2} to a very small fraction of its original value. 10^5 cm.^{2} is the surface area of $\frac{1}{6}$ c.c. when subdivided into $0 \cdot 1 \mu$ cubes.) If by such means the surface energy is considerably reduced, then the changes in solubility may become indetectable.

Effect of charge on solubility

In aqueous media at any rate, all colloids are normally charged to some extent, and this must, from thermodynamical considerations, affect the chemical potential of the colloidal material and hence its solubility. It is shown in Chapter IV (p. 89) that eqn. (3.1) then becomes

$$\frac{RT}{M}\ln\frac{S_r}{S_\infty} = \frac{1}{\rho}\left(\frac{2\gamma}{r}+p'\right), \tag{3.2}$$

where p' is the increase in pressure on the drop due to its charge. In the calculation of p' two cases were considered, constant charge and constant surface density of charge.

(a) *Constant charge.* If the charge (Q) carried by the particle remains constant when its size is varied, then by eqn. (4.180)

$$p' = -\frac{Q^2}{8\pi Dr^4}\frac{(2rd+d^2)}{(r+d)^2},$$

where d is the thickness of the double layer. In this case the charge acts in opposition to the surface energy term, and the particle can remain in equilibrium with a plane surface (i.e. $S_r = S_\infty$) if $p'+2\gamma/r = 0$.

As far as colloidal systems are concerned the only condition where constancy of charge seems at all likely is for condensation of water and other vapours on gaseous ions, which will be discussed in some detail in Chapter XX.

(b) *Constant surface density of charge.* For colloidal suspensions in aqueous media it seems much more likely that the surface density of charge (σ) should remain constant as the size varies, rather than the total charge. In this case, by eqn. (4.184), $p' = \dfrac{2\pi\sigma^2 d}{D}\dfrac{(2r+3d)}{(r+d)^2}$, so that the charge *increases* the chemical potential and hence the solubility.

The above considerations show that effects arising from the charges carried by aqueous suspensions are unlikely to resolve the discrepancy between experiment and the predictions of eqn. (3.1), since only in the improbable event of the total charge (Q) remaining constant would these effects tend to reduce the solubility. Even if Q did remain constant, however, the simple calculations below show that the term p' is unlikely to be of sufficient magnitude to influence appreciably the surface energy term.

Thus, taking the reasonable value of 100 mV (i.e. 0·1/300 e.s.u.) for the ζ potential, and d to be 300 Å (i.e. assuming $N/10{,}000$ uni-univalent salt, or equivalent ionic strength, to be present), then the charge can be calculated from eqn. (12.16), which becomes $\zeta = Qd/Dr^2$ when $d \ll r$. Since $p' = -4\pi\sigma^2 d/Dr = -Q^2 d/4\pi Dr^5$ when $d \ll r$ (eqn. (4.182)), we find, taking $D = 80$ (water), that $p' = -2\cdot 4\times 10^3$ dynes/cm.² for $r = 1\,\mu$, and *ca.* $-2\cdot 4\times 10^4$ for $r = 0\cdot 1\,\mu$.

For the case $d \gg r$, i.e. in the complete absence of electrolytes,

$\zeta = Q/Dr$ and $p' = -Q^2/8\pi Dr^4$ (eqn. (4.181)). Taking $\zeta = 0.1$ V as before, we find $p' = -35$ dynes/cm.^{2} for $r = 1\,\mu$, and -3.5×10^3 for $r = 0.1\,\mu$.

When $r = 1\,\mu$, $2\gamma/r = 2\times 10^5$ and 2×10^6 dynes/cm.^{2} for $\gamma = 10$ and 100 dynes/cm. respectively.

When $r = 0.1\,\mu$, $2\gamma/r = 2\times 10^6$ and 2×10^7 dynes/cm.^{2} for $\gamma = 10$ and 100 dynes/cm. respectively.

Comparing the p' and $2\gamma/r$ terms it is clear that charge effects are unlikely, in aqueous suspensions, to compensate appreciably for the increase in solubility due to surface energy, even if the assumption of constant charge were correct.

The alternative assumption that the surface density of charge (σ) is constant can only tend to increase the solubility of small particles, as pointed out above. In this case, just as with an assumed constant charge, the effects are likely to be negligible in comparison with the surface energy term, for aqueous suspensions at any rate. That this is so can readily be seen by comparing the relevant equations for p' in the two cases.

The conclusion that the charges carried by colloidal particles are unlikely to influence the solubility in either direction seems to point inescapably to surface contamination as being the most likely reason for eqn. (3.1) not being obeyed experimentally.

Other phenomena

Other phenomena which have been explained by surface energy considerations are the effect of high subdivision upon the melting-point, the heat of solution, and the vapour pressure. Thus Meissner (1920) found that the melting-points of several organic compounds (e.g. stearic acid, tristearin, and azobenzene) were lower by several tenths of a degree when the particle size was very small, and Lipsett, Johnson, and Maas (1928) observed the heat of solution of sodium chloride of diameter $1.3\,\mu$ to be -900 cal./mole as compared with -928.6 cal./mole for coarse crystals. The effect of particle size on vapour pressure, as mentioned earlier, is associated particularly with C. T. R. Wilson's cloud chamber work in which ions are detected by their ability to act as condensation nuclei, an aspect which will be treated in more detail in Chapter XX. As would be expected, the change of vapour pressure only becomes appreciable with drops of *ca.* $1\,\mu$ and less.

As regards the possibility of surface energy affecting chemical reactivity it would appear, as with physical properties, a rather difficult matter to detect any significant effect, owing to difficulties in preparing really clean colloidal particles and measuring their surface area. Unless the latter were known very accurately it would clearly be impossible to say whether the reaction rate per unit area was actually greater for particles of colloidal size.

(c). *Colour*

Some aspects of the early investigations of the colour of colloidal systems were mentioned in Chapter I. It was pointed out that with non- or weakly-absorbing insulating particles (e.g. sulphur, gum mastic, many smokes) the colour was primarily determined by their size, so that with particles very small relative to the wavelength of the incident light the scattered light tends to be blue and the transmitted yellowish-red. On the other hand, with conductors, of which sols of the noble metals have been the most widely studied, a much more varied behaviour is to be observed and the colour is usually determined more by absorption than by the particle size. These optical properties of colloids, including polarization effects, will be considered in more detail in Chapter XV.

The question of 'molecular weight' or 'particle mass' in colloidal systems

So far, in discussing colloidal particles, which in the early investigations were invariably of non-uniform size (polydisperse), we have talked of the size and shape of a 'particle' rather than of a 'molecule'. It is important to consider this question a little more carefully, and to see to what extent the latter term is applicable to colloidal systems.

In ordinary physico-chemical terminology the term 'molecular weight' refers to a definite entity—the smallest particle possessing the chemical properties of the substance and from which the substance can be regarded as being built up. For many solutions, such as sucrose or acetic acid in water, no difficulty as to the meaning of molecular weight arises, as the value obtained, by colligative properties, for example, is the simple formula weight and remains sensibly constant over quite a wide range of concentration. With fatty acids in organic solvents, however, the dimerization which occurs there results in the calculated molecular weight varying with the concentration and having values of up to double that in aqueous solution. Further, in the solid state many substances have been shown by X-ray analysis to exist as giant molecules in one, two, or three dimensions (e.g. high molecular weight linear polymers, graphite and diamond or silica respectively).

Colloidal systems comprise examples ranging from the metal sols on the one hand, in which each particle consists of a crystallite containing a great many identical atoms, to proteins and polymers on the other, where the colloidal particle is believed to consist of a single molecule and to which therefore the term 'molecular weight' could rigorously apply. The proteins have molecular weights from several millions down to a few thousand, below which the system could scarcely be termed colloidal, and thus they include particles similar as regards size to those existing in fine gold sols. It is also well established that certain proteins, upon suitable change of pH, salt concentration, etc., dissociate into two, four, or even eight equal units, and that the change is reversible, so that even with the proteins the molecular weight may not have such a unique value either. Thus when quoting a value for the 'molecular weight' it is clearly essential to specify the conditions under which it was obtained.

These general remarks have been made in order to draw attention to some of the difficulties which may arise if we attempt to give a too rigorous definition of the term 'molecular weight' for colloidal systems. However, no such difficulty arises with the concept of 'particle mass', and logically it would seem better to use the term 'particle' rather than 'molecule' when discussing size and shape of colloidal materials. This is usually followed when referring to inorganic colloids (e.g. sols of metals, hydroxides, clays) and to some of the polydisperse organic type (e.g. sols of gums, oils, waxes), but with proteins and polymers generally it is more customary to talk of a 'molecule' even though the system may be polydisperse.

The question of 'shape' does not normally enter into many physico-chemical considerations, especially those involving equilibria, since with simple molecules it is unlikely to change very much on passing from one phase to another. With a colloidal particle, however, the shape may be altered radically by changes in its environment, and this may influence certain aspects of the behaviour of the system as a whole. For example, linear polymers such as rubber are believed to be curled up in solution to varying degrees depending upon the nature of the solvent, being more extended in 'good' solvents (i.e. those with a strong affinity for rubber) than in 'poor' ones (see Chapter XXVII). This will influence the thermodynamic properties to some extent but will be shown up more strongly in kinetic properties such as diffusion and viscosity.

Molecular weight and molecular weight averages

As pointed out above, we can often talk of a true molecular weight, particularly with the corpuscular or globular type proteins such as serum albumin, serum globulin, haemoglobin, etc., since a uniform (monodisperse) material can be prepared with a constant molecular weight over a range of conditions. The majority of polymers, however, particularly the synthetic linear type, whatever the method of preparation, contain a wide range of molecular weights. These polydisperse materials can always be separated into more uniform fractions by processes such as fractional precipitation or solution, but although the range of molecular weights in any one fraction can be reduced, a strictly homogeneous material can never be obtained, so that the molecular weight measured is invariably an average one.

Now the mean value obtained will depend upon the method of averaging, which in turn is dependent upon the relative weights given to unit quantity of the constituent fractions. As we shall show below, several different averages are involved in the principal experimental methods such as osmotic pressure, viscosity (using Staudinger's relation or one of its modifications), sedimentation velocity, sedimentation equilibrium, diffusion, etc.

Thus any method based upon the colligative properties of solutions, of which osmotic pressure is the only one used practically for colloidal materials, merely gives the *number* of particles present, since a long-chain or high molecular weight molecule has no more effect on colligative properties than a smaller one. The molecular weight so found (M_N) is thus termed a *number average* and is shown below to be given by

$$M_N = \frac{\sum_i n_i M_i}{\sum_i n_i}, \tag{3.3}$$

where n_i is the number of gram moles of molecular weight M_i and the summation is taken over all values of i. For an ideal system the osmotic pressure is given by $\Pi_i = (n_i/V)RT$, where n_i is the number of moles of component i in volume V; hence

$$\Pi_{total} = \sum_i \Pi_i = \frac{RT}{V} \sum n_i.$$

Total weight of solute $w = \sum_i w_i = \sum_i n_i M_i$ as $n_i = w_i/M_i$; hence

$$\Pi_{total} = \frac{w}{M_N} \frac{RT}{V} = \frac{\sum_i n_i M_i}{M_N} \frac{RT}{V}.$$

$$\therefore M_N = \frac{\sum_i n_i M_i}{\sum_i n_i} \quad \text{as required.}$$

In addition to osmotic pressure, chemical methods, such as end group analysis, also give this type of average, arising from the equal weighting of small and large molecules.

With viscosity, on the other hand, a long-chain molecule will, according to Staudinger's treatment (p. 368), make a bigger contribution than a shorter one, and the molecular weight (M_W) thus calculated is usually regarded as a *weight average*, in which each molecular size is weighted according to its weight fraction in the mixture:

$$M_W = \frac{\sum\limits_i w_i M_i}{\sum\limits_i w_i} = \frac{\sum\limits_i (n_i M_i) M_i}{\sum\limits_i n_i M_i} = \frac{\sum\limits_i n_i M_i^2}{\sum\limits_i n_i M_i}, \qquad (3.4)$$

where w_i is the weight fraction of component i.

It is obvious that for any polydisperse system $M_N < M_W$, and the ratio M_W/M_N is commonly used to give an indication of the degree of polydispersity.

The sedimentation equilibrium method, being based upon thermodynamics like the colligative properties, might be expected to yield the same molecular weight (i.e. M_N). This is not so because of the particular experimental method of observation and evaluation (optically the variation of refractive index with concentration (dn/dc) is the same for a large range of molecular weights). One method of calculation gives the weight average M_W, but still another type, the Z average, can be obtained by a different treatment of the results. The Z average is defined by

$$M_Z = \frac{\sum\limits_i n_i M_i^3}{\sum\limits_i n_i M_i^2}. \qquad (3.5)$$

(For further discussion see p. 197.)

It will be shown later (Chapter XI) that a molecular weight value can be obtained by combining the results from sedimentation velocity and diffusion. For a polydisperse material the value thus found generally lies between M_N and M_W, the exact value depending upon the frequency distribution with respect to molecular weight, sedimentation constant or diffusion coefficient, as independent variable.

Only in the case of homogeneous substances do all the different averages coincide. For polydisperse systems, therefore, it is usual to separate into as narrow fractions as possible, so that the differences between M_N, M_W, and M_Z are reduced to a minimum.

Origin of the charge in colloidal systems

Before considering the attempts to correlate stability of colloidal suspensions with the charge carried by the particles (see Chapter V) it is necessary to consider the origin of the charge of colloidal systems in general. Unless otherwise mentioned, only systems in aqueous media are in question.

It may be emphasized at the outset that since the charges in colloidal systems arise spontaneously, the process involved (which gives rise to a separation of charge and hence to the double layer, as pointed out in Chapter I) must correspond to an overall reduction in free energy of the system (see Chapter IV). Now a certain amount of electrical energy has to be expended in charging up the double-layer 'condenser', and hence there must be another and larger compensating term in the expression for the free energy. This second part is, of course, the chemical part of the free energy expression (see p. 53). If, for example, $c_i^{\text{surface}} > c_i^{\text{soln}}$ at equilibrium, then the *initially* adsorbed molecules must have a lower free energy (chemical potential) than those free in solution. The process of attainment of equilibrium, leading to $\mu_i^{\text{surface}} = \mu_i^{\text{soln}}$, is then accompanied by a decrease in the total free energy of the system. The same argument should apply to the *chemical* part of the free energy change in the case of an ion, and if this exceeds the increase in the electrical part of the free energy (i.e. $Q\psi$, where Q is the charge involved and ψ the potential difference set up by the charge separation) then

$$c_{\text{ion}}^{\text{surface}} > c_{\text{ion}}^{\text{soln}}.$$

Thus, to take a simple example, an AgI sol is respectively negatively and positively charged in the presence of KI and $AgNO_3$ because the I' ions in the first case, and the Ag^+ ions in the second, have a chemical preference with respect to the surface (probably a lattice packing effect).

(a) Ionization

It is a commonplace that the solution of salts in water is accompanied by a separation of charged ions (often preformed in the crystal), made possible by the combination of interaction with solvent (hydration), entropy changes, and the high dielectric constant of water. With many hydrophilic colloids, such as soaps and proteins, the charging process arises from precisely the same fundamental causes, the only difference being the much larger size and charge (or valency) of the colloidal ions as compared with simple electrolytes. The small ions set free into the solution when the colloidal particle ionizes are given the special name 'gegenions' or 'counter-ions'.

The ionized (or ionizable) groups are —COONa in the case of the fatty acid soaps, and amongst others —COOH and —NH_2 in the case of the proteins. The colloidal ion thus carries in the case of fatty acid soaps a negative charge (the —COO' being attached to the paraffin chains which aggregate to give the colloidal micelle); in the case of proteins a negative charge at high pH (owing to ionization of —COOH → —COO'), and a positive one at low pH values (arising from —NH$_2$ → —N$\overset{+}{\text{H}}_3$). The electrical state of the protein molecule can thus be depicted as e.g. $R\!\!<\!\!\begin{smallmatrix}\text{COO}'\\ \text{NH}_2\end{smallmatrix}$ Na$^+$, in the presence of NaOH at high pH, and as $R\!\!<\!\!\begin{smallmatrix}\text{COOH}\\ \overset{+}{\text{NH}}_3\end{smallmatrix}$ Cl', in the presence of HCl at low pH. (R denotes the residue of the protein molecule.) At some pH the contributions to the charge arising from the acidic groups will be equal and opposite to those from the basic ones, and the net charge will be zero provided no other ions are adsorbed (cf., however, p. 336). This pH is termed the *iso-electric point* of the protein. (The protein is probably ionized at the iso-electric point, being in the zwitterionic form $R\!\!<\!\!\begin{smallmatrix}\text{COO}^-\\ \text{NH}_3^+\end{smallmatrix}$, but the *net* charge is zero.)

(b) Adsorption

Whilst the above charging process presents no difficulties in the case of compounds with ionizable groups it is clear that a different one must be operative in most hydrophobic suspensions and with hydrophilic colloids held to be free of ionogenic groups (e.g. starches and pure carbohydrates). Even systems such as pure paraffin droplets or air bubbles in the purest conductivity water are found to migrate in an electric field, showing that a separation of charges has occurred, the particle being negatively charged in both the above cases. Since ionizable groups must be absent the charge is usually ascribed to preferential adsorption of one ion (in this case OH'). The majority of non-ionogenic substances studied (e.g. long-chain esters and alcohols, paraffin wax, nitro-cellulose, carborundum, cellulose, etc.) show this negative charge both in pure water and in dilute solutions of simple salts (e.g. NaCl), and although the presence of salts complicates the picture, yet even here preferential adsorption of OH' ion seems to be chiefly responsible for the charge (Ham, 1941). In the presence of ions of higher valence (e.g. Al^{3+}, [Fe(CN)$_6$]$^{4-}$) or with a tendency to be capillary active (e.g. I', CNS', organic ions, dyes) the adsorption of these ions, rather than of OH', appears to determine the charge.

Factors affecting adsorption. Why should ions tend to become adsorbed at an interface and why should this tendency for anions appear as a rule to be greater than for cations? These effects are usually attributed to polarization (and may be described as an ion-induced dipole interaction). Thus an ion near a material surface of any kind (polar or non-polar) will induce a separation of charge (dipole) in the surface atoms, and the induced dipole in turn exerts an attractive force on the ion. If the binding energy (E) is large compared with the thermal energy of the ion (i.e. $E \gg kT$) then the ion will be held at the surface and in the case of a particle will move kinetically with it. Various factors determine the magnitude of the binding energy, such as the charge, size, hydration, and polarizability of the ion, and the polarizability of the material comprising the interface. (For its calculation see, for example, Slater, 1939.)

Hydration not only increases the size of the ion but reduces the force between an ion and its induced dipole owing to the high dielectric constant of water. Anions are usually less hydrated, and owing to their larger number of electrons per nuclear charge more polarizable than cations, and these reasons are thought to be responsible for the greater adsorption of anions at interfaces. (The special position occupied by OH′ may arise from its unsymmetrical structure as compared with H_3O^+ or simple monatomic anions such as Cl′.)

Thus the radii of Na^+ and Cl′ in the crystal are about 1·0 A and 1·8 A respectively, in water about 2·5 A and 1·95 A, clearly showing the difference due to hydration. The degree of hydration has been calculated from measurements of partial molar volumes or from conductivity, and it would appear that all monatomic cations, except for the large Rb^+ and Cs^+, are completely hydrated (i.e. surrounded by a tightly packed layer of water molecules). With Rb^+ and Cs^+ it seems that their size is so large that their surface field is not sufficiently strong to bind the water dipoles. Monatomic anions are in general large, and except for F′ appear to be only slightly, if at all, hydrated. Polyatomic anions are usually unhydrated, while all multivalent monatomic ions appear to be hydrated, as would be expected.

With organic salts the tendency of one ion to be preferentially adsorbed will also be affected by the non-ionogenic part, and in extreme cases, such as the soaps and dyestuffs, it is this latter part which, owing to its tendency to be squeezed out by the water dipoles, determines which ion is adsorbed and hence the charge on an interface. Thus an oil droplet in pure water, originally negative, becomes more negative on the addition of a long-chain anionic soap (e.g. $C_{16}H_{33}COO'Na^+$), but the charge is reversed by extremely low concentrations of a cationic detergent (e.g. $C_{16}H_{33}\overset{+}{N}(CH_3)_3Br'$). Should the original interface contain groups capable of forming hydrogen bonds with the adsorbing ion, then the strength of attachment is increased: this is shown very strikingly with droplets of long-chain amines or alcohols in the presence of soaps, the hydrogen bond being between the —NH_2 or —OH and the —COO' group (Alexander and McMullen, unpublished).

Finally, as the most extreme type of preferential adsorption of one ion, is the case where this ion forms a definite chemical compound with some constituent of the interface. An example of this is supposed to be the adsorption of the picrate ion on silk (Sookne and Harris, 1939).

Hydrophobic suspensions. With certain hydrophobic suspensions, particularly some of the metal sols, the question whether the charge arises from preferential ionic adsorption, or from the dissociation of complex ionogenic groups present on the surface, has been hotly debated. As an example of the latter theory, we may cite the case of Bredig platinum sols, for which Pennycuik (1927–39) has presented evidence for the surface containing platinum oxides and an acid, $H_2Pt(OH)_6$, which is believed to give rise to the charge, the sol being formulated

$$[x\text{Pt}.y\text{PtO}_2.\text{Pt}(OH)_6]'' 2H^+.$$

Other examples are the Bredig gold sols which according to Pauli (1932, 1939) should be formulated $[x\text{Au}.\text{AuCl}_2]' H^+$, and the sulphur sols where it seems that polythionates are always present and thus can be formulated as complexes such as $[xS.S_6O_6H]' H^+$, etc.

It may be pointed out, however, that even in such cases as the above the two theories are not as mutually exclusive as they might at first sight appear to be, since the final state may be the same whichever mechanism is assumed. This is clearly shown by the two simple examples given below.

(a) $\quad [xS.S_6O_6H_2] \longrightarrow [xS.S_6O_6H]' H^+.$

$\quad [xS] H^+ S_6O_6H' \longrightarrow [xS.S_6O_6H]' H^+.$

(b) $\quad [xAgI.KI] \longrightarrow [xAgI.I]' K^+.$

$\quad [xAgI] K^+ I' \longrightarrow [xAgI.I]' K^+.$

(c) *Contact*

Another charging mechanism, which has been postulated especially for non-aqueous media, is that the charge is due to the frictional contacts between particles and medium, arising from the Brownian movement of the former, just as an amber rod is electrified by rubbing with a silk cloth. As yet, however, there is little direct experimental evidence in favour of this theory, and an alternative explanation, favoured by many workers, is that the charges arise from adsorption of traces of ions which are never completely removed even from non-polar media.

Magnitude and distribution of charge

So far we have only discussed the factors which lead to the interface acquiring a charge, which may be positive or negative depending upon the conditions, but it is clear that the magnitude of this charge will also be of great importance. If, for example, 50 long-chain anions aggregated to form an ionic micelle, this should behave like a large ion of valence 50. In point of fact, as will be discussed in Chapter XXIV, these large colloidal ions bind a fraction (which may be as much as 80–90 per cent. depending on the conditions) of their gegenions in order to reduce their net charge. These gegenions are bound by electrostatic forces solely (i.e. as ion-pairs on the Bjerrum view), and move kinetically with the particle, which thus behaves as one of lower valence.

In the case of very large particles, the extreme case being a plane interface, the total charge is clearly less convenient to consider than the charge density (i.e. the charge per unit area of the surface). This quantity can be obtained from the measurement of ζ potentials (see Chapter XII).

One interesting question arises in connexion with the proteins, which in general will have a certain number of both positively and negatively charged groups ionized simultaneously. Are these charges randomly arranged or are they concentrated in groups at various points? Some indication about this can be obtained from the study of rotational diffusion, as will be shown in Chapter XIV.

REFERENCES

BALAREV, 1941, *Kolloidzchr.* **96**, 19; **97**, 300.
COHEN and BLEKKINGH, 1940, *Proc. Acad. Sci. Amst.* **43**, 32, 189, 334.
—— 1940, *Z. phys. Chem.* A **186**, 257.
DUNDON, 1923, *J. Amer. Chem. Soc.* **45**, 2658.
—— and MACK, 1923, ibid. **45**, 2479.
HAM, 1941, *Trans. Faraday Soc.* **37**, 194.
HULETT, 1901, *Z. phys. Chem.* **37**, 385.
—— 1904, ibid. **47**, 357.
LIPSETT, JOHNSON, and MAAS, 1928, *J. Amer. chem. Soc.* **50**, 2701.
MEISSNER, 1920, *Z. anorg. Chem.* **110**, 169.

PAULI et al., 1932, *Naturwissenschaften*, **20**, 551.
—— —— 1939, *Trans. Faraday Soc.* **35**, 1178.
PENNYCUIK, 1927, *J. Chem. Soc.*, p. 2600.
—— 1930, *J. Amer. chem. Soc.* **52**, 4621.
—— 1939, ibid. **61**, 2234.
SLATER, 1939, *Introduction to Chemical Physics*, McGraw-Hill.
SOOKNE and HARRIS, 1939, *J. Research Nat. Bur. Standards*, **23**, 299.

IV
THERMODYNAMICS AND ITS APPLICATION TO COLLOIDAL SYSTEMS

Introduction

IT is not possible in this chapter to present a rigorous and complete treatment of thermodynamics, but before considering how thermodynamics can be applied to colloidal systems it is essential to outline briefly the development and use of the more important functions. We shall assume that the definitions of the more elementary terms are known, e.g. a thermodynamic system, reversible and irreversible processes, etc., as well as some knowledge of differentiation and partial differentials. For more detailed accounts of the basic thermodynamics the reader is referred to specialized treatises, of which those of Guggenheim (1933), Butler (1934), and Partington (1940) are to be recommended.

THE PRINCIPAL THERMODYNAMIC FUNCTIONS

The First and Second Laws of Thermodynamics form the basis of the greater part of this science and must therefore be considered briefly.

The First Law, known in a more general form as the Law of the Conservation of Energy, is usually stated in the form

$$dE = \delta q + \delta w \tag{4.1}$$

for an infinitesimal change, or as

$$\Delta E = q + w \tag{4.1a}$$

for a finite one.

Here E refers to the total (internal) energy of the system, q to the heat *absorbed*, and w to the work done *on* the system. In several well-known works the symbol w refers to work done *by* the system, and is therefore given a negative sign in the equation analogous to (4.1) above.

It follows from eqn. (4.1) that two special cases arise:

(1) where no energy change is involved (e.g. as in cyclic processes) then
$$\delta q + \delta w = 0, \tag{4.2}$$

and (2) where a change occurs at constant volume, other work (than $p_i dV$) being excluded, then $w = 0$, and

$$(dE)_V = \delta q. \tag{4.3}$$

Thus the increase in total energy is equal to the amount of heat absorbed at constant volume. Such heat absorption must cause an increase in

temperature dT, and if E refers to 1 gm. molecule of working substance, then the molecular heat capacity at constant volume (C_V) may be defined as
$$\left(\frac{dE}{dT}\right)_V = C_V. \tag{4.4}$$

The Second Law of Thermodynamics may conveniently be treated in terms of the entropy (S). Thus it states that for an infinitesimal change in a process which occurs naturally or spontaneously (i.e. an irreversible process)
$$dS > \frac{\delta q}{T}, \tag{4.5}$$
whilst for an infinitesimal change in a system at equilibrium (i.e. a reversible process), at temperature T,
$$dS = \frac{\delta q}{T}. \tag{4.5a}$$
For a finite reversible change, the change in entropy is given by a summation
$$\Delta S = \int \frac{\delta q}{T}. \tag{4.6}$$

For a reversible process at constant temperature T we may substitute eqn. (4.5a) into (4.1) giving
$$dE = T\,dS + \delta w, \tag{4.7}$$
which summarizes the first and second laws. For a finite reversible process, at constant temperature
$$\Delta E = T\,\Delta S + w. \tag{4.7a}$$
Where the *only work done is the expansion of a gas against a constant pressure*, p,
$$\delta w = -p\,dV,$$
and substituting in (4.7) we have
$$dE = T\,dS - p\,dV, \tag{4.8}$$
a very commonly occurring equation, with which we shall shortly deal further. In general form eqn. (4.8) may be written
$$E = \phi(S, V), \tag{4.9}$$
$\phi(S, V)$ being some function of the entropy and volume of the system.

Summarizing, the First Law of Thermodynamics gives information on the energy change occurring during a process; the Second Law, on the other hand, enables us to see in which direction a process will go, because a process is only naturally possible if $dS > \delta q/T$. This will become clearer when we formulate equilibrium conditions at a later stage (p. 56).

The two functions energy (E) and entropy (S), arising from the First and Second Laws of Thermodynamics respectively, are essentially different in nature from the quantities q and w which we have also used. E and S are single-valued functions of the state of the system, which may be defined by certain variables, e.g. the pressure p, the temperature T, and the composition of the system. Thus for any change in the system, ΔE and ΔS depend only on the initial and final values of the variables, being quite independent of the precise path by which the change was accomplished.

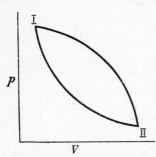

FIG. 4.1. Diagram illustrating the difference between single-valued and non single-valued thermodynamic functions.

On the other hand q, the heat absorbed, and w, the work done on the system, depend on the precise path by which a process occurs, e.g. reversibly or irreversibly. They are not therefore single-valued functions of the variables defining the system (hence the inequality in eqn. (4.5) for changes other than reversible ones), as may be illustrated by considering volume changes in a gas.

Thus the work done by a gas expanding reversibly from state I to state II is obtained by integrating $p\,dV$ over the desired limits, the complete integration being represented by the area under the pV curve (Fig. 4.1). Such an area obviously depends, not only on the positions of states I and II, but also upon the particular path chosen to reach II from I. The quantity w is *not* therefore a definite function of the variables defining the system. Further from eqn. (4.1 a), since ΔE is a single-valued function, it also follows that q as well as w is *not* a single-valued function.

If E is a single-valued function of the variables, (x, y, z) say, of the system, then any small change in E may be expressed as

$$dE = \left(\frac{\partial E}{\partial x}\right)_{y,z} dx + \left(\frac{\partial E}{\partial y}\right)_{x,z} dy + \left(\frac{\partial E}{\partial z}\right)_{x,y} dz. \qquad (4.10)$$

This is known as a perfect differential, where $(\partial E/\partial x)_{y,z}$, $(\partial E/\partial y)_{x,z}$ and $(\partial E/\partial z)_{x,y}$ are partial differential coefficients.

The actual choice of the variables defining a phase is somewhat arbitrary (Guggenheim, 1933), but it is usual to take entropy, S, volume,

APPLICATION TO COLLOIDAL SYSTEMS

V, and the chemical composition, given by the number of gm. moles of each component, n_i, respectively as x, y, and z. Then eqn. (4.10) becomes

$$dE = \left(\frac{\partial E}{\partial S}\right)_{V,n_i} dS + \left(\frac{\partial E}{\partial V}\right)_{S,n_i} dV + \sum_i \left(\frac{\partial E}{\partial n_i}\right)_{S,V,n_j} dn_i, \quad (4.11)$$

where n_j refers to all components other than i.

Excluding for the moment any variation in the composition of the system (i.e. considering only the first two terms on the R.H.S. of eqn. (4.11)), we may compare eqns. (4.8) and (4.11), from which we see that

(a) $\left(\frac{\partial E}{\partial S}\right)_{V,n_i} = T$

and

(b) $\left(\frac{\partial E}{\partial V}\right)_{S,n_i} = -p$

$\quad (4.12)$

Making use of Gibbs's definition of the *chemical potential*, μ_i, of a species i we put

$$\mu_i = \left(\frac{\partial E}{\partial n_i}\right)_{S,V,n_j} \quad (4.13)$$

We may then write in general for eqn. (4.11), for changes in which the only work done is in the expansion of a system against pressure p,

$$dE = T\,dS - p\,dV + \sum_i \mu_i\,dn_i. \quad (4.14)$$

This equation and modifications, designed to include work other than that involved in volume changes, are basic to the thermodynamic treatment of all systems. We shall refer back to it frequently, and shall now mention the modifications necessary in certain cases of importance in the study of colloids.

Where we have electrically charged species and the possibility of doing electrical work, the chemical potential (μ_i) must be replaced by the *electrochemical potential* written usually as $\bar{\mu}_i$, which contains an additional energy term of the type $ze\psi$, where z is the valency of the ion, e is the electronic charge, and ψ the electrical potential; i.e.

$$\bar{\mu}_i = \mu_i + ze\psi. \quad (4.15)$$

This modification we shall have occasion to use in connexion with membrane equilibrium and in considering the distribution of ions around colloidal particles.

Another important modification arises in systems in which gravitational energy cannot be neglected. Normally gravitational effects are not important, but if we make use of the high centrifugal fields of the ultracentrifuge, it is essential to consider centrifugal effects in the thermodynamic treatment. A work or energy term is again involved, of the type

$$\text{mass} \times \text{(difference in gravitational (or centrifugal) potential)},$$

e.g. $M\,\delta\phi$

where M is the molecular weight and $\delta\phi$ the difference in gravitational potential.

The importance of surface energy in colloidal systems, resulting from their state of high subdivision, is well known, and provision must be made for it in a complete thermodynamic treatment of colloidal systems.

Energy, entropy, volume, pressure, chemical potential, and composition, the quantities of eqn. (4.14), referring to three-dimensional systems, have their counterparts in a consideration of surface phases, and the analogue of eqn. (4.14) for surfaces may be written

$$dE_{\text{surface}} = T^\sigma dS^\sigma + \gamma\, dA + \sum \mu_i^\sigma \, dn_i^\sigma, \qquad (4.16)$$

where γ is the surface tension of the surface (analogous to $-p$), A is the area of surface (analogous to V), and the other quantities of the equation refer only to the surface.

We shall not, for the moment, develop these special cases further, but shall return to them at a later stage. It is necessary now to consider some other thermodynamic functions which are frequently required.

We have seen for a change at constant volume, if no other work is involved (eqn. (4.3)), that the increase in total energy is identical with the heat absorbed by the system. Consider now a change occurring at constant pressure between states I and II of the system, work other than that occurring as a result of volume change being excluded. Then from eqn. (4.1a)

$$\Delta E = E_{\text{II}} - E_{\text{I}} = q - p(V_{\text{II}} - V_{\text{I}}). \qquad (4.17)$$

Rearranging,

$$q = (E_{\text{II}} + pV_{\text{II}}) - (E_{\text{I}} + pV_{\text{I}}). \qquad (4.18)$$

Defining a new quantity, *heat content*, H, as

$$H = E + pV \qquad (4.19)$$

and substituting in eqn. (4.18) above, we have

$$(q)_p = \Delta H, \qquad (4.20)$$

i.e. the increase in heat content is equal to the amount of heat absorbed at constant pressure (cf. eqn. (4.3)). Such heat absorption must cause an increase in temperature, dT, and if H refers to 1 gm. molecule of working substance, the molecular heat capacity at constant pressure (C_p) may be defined by the relation

$$\left(\frac{dH}{dT}\right)_p = C_p. \tag{4.21}$$

Further, if we restrict eqn. (4.20) to an infinitesimal change we may substitute $\delta q = T\,dS$ from eqn. (4.5a), thus obtaining

$$T\,dS = dH$$

or
$$dS = \frac{dH}{T} = \frac{C_p}{T}dT. \tag{4.22}$$

Thus the functions total energy (E), for changes at constant volume, and heat content (H), for those at constant pressure, are closely analogous. Further, though absolute values of E and H cannot yet be assigned, we are usually interested only in their changes, which, as we see, may be given by heat absorptions under the appropriate conditions. For these purposes it is sufficient, usually, to define arbitrarily E and H as zero in a certain specified state, e.g. solid, liquid, or gas, at a given temperature and pressure.

Two other functions are required, the Helmholtz Free Energy (F) and the Gibbs Free Energy (G), defined by the equations

$$F = E - TS, \tag{4.23}$$

$$G = H - TS = E + pV - TS. \tag{4.24}$$

From these definitions, since E, T, S, p, and V are all single-valued functions, then F and G are themselves single-valued functions of the variables defining the system.

Thus for a change from state I to state II at constant temperature T

$$\Delta F = F_{II} - F_{I} = \Delta E - T\,\Delta S. \tag{4.25}$$

But from eqn. (4.7a),

$$\Delta E - T\,\Delta S = w = \left(w' - \int p\,dV\right), \tag{4.26}$$

where w' is the work done on the system, other than that involved during volume changes ($-\int p\,dV$). For constant volume therefore ($\Delta V = 0$),

$$\Delta F = w = w' \tag{4.27}$$
$$= -(\text{maximum work done } by \text{ system})$$

or

$$+(\text{maximum work done } on \text{ system}).$$

Thus the increase in F for a change at constant temperature and volume is equal to the reversible work done on the system.

This property of the Helmholtz free energy for changes occurring at constant volume is the basis of its importance and explains why it was at first called the 'work function'.

Turning now to the Gibbs free energy G, for a change from state I to state II at constant temperature and pressure we have

$$\Delta G = G_{II} - G_I = \Delta E + p\Delta V - T\Delta S. \qquad (4.28)$$

For conditions of constant temperature and pressure, eqn. (4.28) reduces to

$$\Delta E - T\Delta S + p\Delta V = w'$$

or
$$\Delta G = w' \qquad (4.29)$$
$$= -(\text{net work done } by \text{ system at constant pressure}).$$

Thus the increase in G for a process at constant temperature and pressure is equal to the net work (i.e. work excluding that involved in volume changes) performed reversibly *on* the system. This is the fundamental property of the Gibbs free energy function.

General criteria of equilibrium

The properties of F and G which we have just discussed allow us to formulate equilibrium conditions in terms of changes in them.

Thus a system at constant temperature and volume can only be in equilibrium if there is no way in which it might do work and thus decrease its Helmholtz free energy, i.e. for equilibrium at constant temperature and volume

$$dF \geqslant 0. \qquad (4.30)$$

Similarly a system at constant temperature and pressure can only be in equilibrium if it cannot by any means do net work and so decrease its Gibbs free energy. Thus for equilibrium at constant temperature and pressure

$$dG \geqslant 0. \qquad (4.31)$$

Eqns. (4.30) and (4.31), the general criteria of equilibria in systems at constant volume and constant pressure respectively, are extensively applied. The formulation of these same conditions, in terms of the actual properties (e.g. S, T, n_i) of a system, gives the equilibrium relation for these properties which is often of great value. We shall see several examples of this procedure later in the chapter.

APPLICATION TO COLLOIDAL SYSTEMS

The application of the equilibrium conditions to a system of more than one phase of a single component

Let us now consider an isolated system, at constant total volume, containing more than one homogeneous phase of a single pure substance, e.g. liquid and its vapour. It can be shown (Butler (1934), p. 19) that for thermal and mechanical equilibrium in such a system, the temperature and pressure must be constant throughout. Under such conditions the criterion of equilibrium is

$$dG \geqslant 0$$

for all possible variations of the system.

Let $\bar{G}'$ and $\bar{G}''$ be the Gibbs free energies per gm. mole of two phases, and let us consider the transfer of dn moles from one phase to the other. Then the total increase in Gibbs free energy is

$$dG = \bar{G}'' dn - \bar{G}' dn$$
$$= (\bar{G}'' - \bar{G}') dn.$$

Since dn may be positive or negative, the condition of equilibrium is only obeyed if $\bar{G}'' = \bar{G}'$, i.e. the Gibbs free energies per mole of the two phases are equal. In general it is also true that several pure phases can only coexist if their molar Gibbs free energies are equal. (As we shall see later, $\bar{G}$ is given the special name of *chemical potential* (μ) in view of its importance in determining chemical equilibria in general.)

Let us consider now a small shift in the equilibrium position of a two-phase system of a single pure component. Then anticipating eqn. (4.38) for the change at constant composition $\left(\text{i.e. } \sum_i \mu_i dn_i = 0\right)$ in each phase we have

$$dG' = -S' dT + V' dp$$
and
$$dG'' = -S'' dT + V'' dp.$$

Since the system after the change is still in equilibrium, it follows that $dG' = dG''$, and we have

$$-S' dT + V' dp = -S'' dT + V'' dp,$$

which on rearranging gives

$$\frac{dp}{dT} = \frac{S'' - S'}{V'' - V'}. \tag{4.32}$$

$S'' - S'$ may be replaced by q/T, where q is the heat absorbed on changing 1 gm. molecule of substance from state I to state II, giving

$$\frac{dp}{dT} = \frac{q}{T(V'' - V')}. \tag{4.33}$$

This is the well-known Clapeyron-Clausius equation which is applicable to any two phases in equilibrium. Thus, for a liquid and its saturated vapour, it gives the rate of change of vapour pressure with temperature in terms of q, in this case the latent heat of evaporation per gm. molecule at temperature T, and V'' and V' the molar volumes of the gaseous and liquid states. Similarly the equation may be used to calculate the rate of change of freezing-point with the pressure on a solid-liquid system in equilibrium.

A useful approximation and simplification of eqn. (4.33) is often made for the case of a liquid-vapour system, in which the volume of 1 gm. molecule of liquid is assumed negligible compared with that of the vapour which is treated as a perfect gas. Thus we put

$$V'' = \frac{RT}{p}, \quad \text{and} \quad V' = 0,$$

when eqn. (4.33) becomes

$$\frac{dp}{dT} = \frac{qp}{RT^2} \quad \text{or} \quad \frac{d\ln p}{dT} = \frac{q}{RT^2}. \tag{4.34}$$

Assuming q to be a constant, this equation may be integrated between limits, giving

$$\ln(p_2/p_1) = -\frac{q}{R}\left[\frac{1}{T_2} - \frac{1}{T_1}\right], \tag{4.35}$$

from which the vapour pressure (p_2) at any temperature T_2 may be calculated from that (p_1) at T_1. The constancy of q may only be assumed, however, for small temperature ranges.

The variation of solubility with temperature is treated on essentially the same lines. Thus if the solution is ideal (i.e. obeys Raoult's law, see p. 68), from eqn. (4.34) we may write

$$\frac{d\ln N_s}{dT} = \frac{q}{RT^2}, \tag{4.36}$$

where N_s is the mole fraction of solute in the saturated solution and q is the molar heat of solution of the solute (positive when heat absorbed on dissolving).

Since multi-phase equilibria occur often at constant pressure, the heat absorbed per mole on changing state is to be identified as ΔH, the increase in heat content. Eqns. (4.33)–(4.36) are often therefore written in terms of ΔH instead of q.

Partial differential coefficients of free energy

Consider any infinitesimal change in the Gibbs free energy. Differentiating (4.24) we obtain

$$dG = dE + p\,dV + V\,dp - T\,dS - S\,dT. \tag{4.37}$$

Substituting for dE from eqn. (4.14), we have
$$dG = V\,dp - S\,dT + \sum_i \mu_i\,dn_i, \tag{4.38}$$
or
$$G = \phi(T, p, n_i), \tag{4.39}$$
with

(a) $\quad \left(\dfrac{\partial G}{\partial p}\right)_{T,n_i} = V,$

(b) $\quad \left(\dfrac{\partial G}{\partial T}\right)_{p,n_i} = -S,$ $\qquad (4.40)$

(c) $\quad \left(\dfrac{\partial G}{\partial n_i}\right)_{T,p,n_j} = \mu_i.$

In eqn. (4.40 c) we have an alternative definition (cf. eqn. (4.13)) of chemical potential which is often used and which gives rise to the alternative name of partial molar Gibbs free energy and the alternative symbol $\bar{G}_i$. We now see more clearly the reason for the significance of chemical potential, since it acts as a pointer to the direction in which a chemical reaction tends to go; pressure and temperature likewise determine the direction of mechanical movement and heat flow.

In a similar way, from the Helmholtz free energy, we may derive the following relation:
$$dF = -S\,dT - p\,dV + \sum_i \mu_i\,dn_i, \tag{4.41}$$
or
$$F = \phi(T, V, n_i),$$
where

(a) $\quad \left(\dfrac{\partial F}{\partial V}\right)_{T,n_i} = -p,$

(b) $\quad \left(\dfrac{\partial F}{\partial T}\right)_{V,n_i} = -S,$ $\qquad (4.42)$

(c) $\quad \left(\dfrac{\partial F}{\partial n_i}\right)_{T,V,n_j} = \mu_i,$

providing yet another definition of chemical potential.

We have already stated, however, that we are usually interested in changes in thermodynamic functions rather than in their absolute values. Thus
$$\Delta G = G_2 - G_1.$$
$$\therefore \left(\dfrac{\partial(\Delta G)}{\partial T}\right)_p = -(S_2 - S_1) \tag{4.43}$$
and
$$\left(\dfrac{\partial(\Delta G)}{\partial p}\right)_T = (V_2 - V_1). \tag{4.44}$$

But from the definition of G (see eqn. (4.24))
$$\Delta G = \Delta H - T\Delta S.$$
On substituting for ΔS, we have
$$\Delta G - \Delta H = T\left(\frac{\partial(\Delta G)}{\partial T}\right)_p. \tag{4.45}$$
Dividing throughout by T^2 and rearranging, we have
$$\frac{1}{T}\left(\frac{\partial(\Delta G)}{\partial T}\right)_p - \frac{\Delta G}{T^2} = \frac{\partial(\Delta G/T)_p}{\partial T} = -\frac{\Delta H}{T^2}. \tag{4.46}$$

Eqns. (4.45) and (4.46) are two forms of the Gibbs-Helmholtz equation, from which a knowledge of the variation of ΔG with temperature may be utilized to calculate the corresponding value of ΔH. Further, from the definition of ΔG (eqn. (4.24)), ΔS for the same process may also then be calculated. Examples of this type of procedure will be provided in Chapters VII and XXVII.

Similar relations concerning the Helmholtz free energy are also of use. Thus
$$\left(\frac{\partial(\Delta F)}{\partial T}\right)_V = -(S_2 - S_1) \tag{4.47}$$
and
$$\left(\frac{\partial(\Delta F)}{\partial V}\right)_T = -(p_2 - p_1). \tag{4.48}$$
But, since
$$\Delta F = \Delta E - T\Delta S,$$
we have, on substituting for ΔS,
$$\Delta F - \Delta E = T\left(\frac{\partial(\Delta F)}{\partial T}\right)_V. \tag{4.49}$$
Dividing throughout by T^2 and rearranging we have
$$\frac{1}{T}\left(\frac{\partial(\Delta F)}{\partial T}\right)_V - \frac{\Delta F}{T^2} = \frac{\partial(\Delta F/T)_V}{\partial T} = -\frac{\Delta E}{T^2}. \tag{4.50}$$

For systems at constant volume, eqns. (4.49) and (4.50) may be utilized in a manner analogous to that indicated above for the corresponding equations in Gibbs free energy.

The free energy of a perfect gas
Isothermal changes

It is necessary first to consider the entropy change when a perfect gas expands reversibly at constant temperature from a pressure p_1 to p_2. For such a finite change, we have by eqn. (4.6)
$$\Delta S = \frac{q}{T}.$$

APPLICATION TO COLLOIDAL SYSTEMS

But for an isothermal change, since $\Delta E = 0$, from (4.2) we have for 1 gm. mole

$$q = -w = \int_{p_1}^{p_2} p\, dV = \int_{V_1}^{V_2} RT \frac{dV}{V} = RT \ln \frac{V_2}{V_1} = RT \ln \frac{p_1}{p_2}. \quad (4.51)$$

Substituting in (4.6), we obtain

$$\Delta S = R \ln(p_1/p_2). \quad (4.52)$$

If, now, we allow p_1 to have the value unity, and let S^0 be the corresponding entropy value, then

$$S - S^0 = \Delta S = R \log 1/p_2 = -R \ln p_2, \quad (4.53)$$

or
$$S = S^0 - R \ln p, \quad (4.53\,a)$$

where S refers to the entropy/gm. molecule at pressure p.

Non-isothermal changes

If absorption of heat at constant pressure occurs reversibly, the increase in entropy may be calculated by integrating eqn. (4.22) over the required limits of temperature:

$$\Delta S = \int_{T_1}^{T_2} \frac{C_p}{T}\, dT = \int_{T_1}^{T_2} C_p\, d\ln T, \quad (4.54)$$

and by analogy with eqn. (4.53 a),

$$S = S^0_{T_1} + \int_{T_1}^{T_2} C_p\, d\ln T. \quad (4.55)$$

A combination of eqn. (4.53 a) and (4.55) may now be used to calculate the entropy change involved between any two states, (p_1, T_1) and (p_2, T_2), of the system, since the change may be imagined to occur in two parts: (i) an isothermal expansion at temperature T_1 from pressure p_1 to p_2 followed by (ii) a reversible absorption of heat from temperature T_1 to T_2 at constant pressure. Thus

$$\Delta S^{p_2\,T_2}_{p_1\,T_1} = R \ln (p_1/p_2) + \int_{T_1}^{T_2} C_p\, d\ln T, \quad (4.56)$$

or
$$S_{p_2\,T_2} = S_{p_1\,T_1} + R \ln (p_1/p_2) + \int_{T_1}^{T_2} C_p\, d\ln T. \quad (4.56\,a)$$

Introducing eqn. (4.53 a) into eqn. (4.24) we have

$$G = E + pV - TS^0 + RT \ln p. \quad (4.57)$$

The first three terms of this equation, referring to a perfect gas, depend

on temperature only (independently of pressure) and may be considered together as the Gibbs free energy at temperature T and unit pressure (G^0). Thus
$$G = G^0 + RT \ln p \qquad (4.58)$$
and, in general, for a change in 1 gm. molecule of a perfect gas at constant temperature from p_1 to p_2, the Gibbs free energy change is given by
$$\Delta G_{p_1}^{p_2} = RT \ln(p_2/p_1), \qquad (4.59)$$
the arbitrary constant G^0 disappearing in the latter equation.

Imperfect gases

The above treatment assumed perfect gas behaviour and is therefore applicable only where a real gas is approximating to such behaviour.

For real gases, G. N. Lewis has suggested the use of the fugacity, usually denoted by p^*, and defined by
$$G = G^0 + RT \ln p^*. \qquad (4.60)$$
G^0, being a function of temperature only,
$$\frac{p^*}{p} \to 1 \quad \text{as} \quad p \to 0.$$
p^* is obviously the pressure which a gas would exert if it were behaving ideally, and the ratio p^*/p is a measure of the imperfectness of the gas.

Mixture of perfect gases

For a mixture of perfect gases we may apply Dalton's Law of Partial Pressures, i.e. the total pressure of the mixture is merely the sum of the partial pressures of the components, each partial pressure being that which would be exerted if the component occupied the same volume alone under the same conditions. Thus the partial pressure of any one component is unaffected by the presence of the other components, and it therefore follows that the free energies of the components in the mixture may be written down as for pure components. Thus
$$\left. \begin{array}{l} G_1 = G_1^0 + RT \ln p_1, \\ G_2 = G_2^0 + RT \ln p_2, \quad \text{etc.} \end{array} \right\} \qquad (4.61)$$

Many Component Systems
Partial molar quantities

In the last section we saw that it is possible in the case of a mixture of perfect gases to calculate the Gibbs free energy of each component without considering the effect of other components. This possibility is not, however, generally available, and other methods have been developed. We are now concerned with such methods.

If we examine eqns. (4.38) and (4.41) it will be observed that each term has the dimensions of energy and is a product of two quantities (e.g. p and dV, and T and dS) which are of different character. One set of quantities may, with good reason, be called *intensity factors*, being independent of the amount of substance considered (p, T, μ_i) and the other set, *capacity factors*, depending directly as they do upon the amount of material considered (V, S). The products of intensity and capacity factors, as well as the sum of such products, are also capacity factors, so that, as well as the simpler quantities volume and heat capacity, we may include total energy, Gibbs and Helmholtz free energies.

Such capacity factors are also single-valued functions, and a small change in them can be written as in the following equation for a change in the Gibbs free energy at constant temperature and pressure:

$$dG = \left(\frac{\partial G}{\partial n_1}\right)dn_1 + \left(\frac{\partial G}{\partial n_2}\right)dn_2 + \left(\frac{\partial G}{\partial n_3}\right)dn_3 + \ldots$$
$$= \bar{G}_1 dn_1 + \bar{G}_2 dn_2 + \bar{G}_3 dn_3 + \ldots, \quad (4.62)$$

where $\partial G/\partial n_1$, written often as $\bar{G}_1$, refers to the increase in G caused by the addition to the system of dn_1 moles of component 1, the other components being kept constant. Similarly, for $\partial G/\partial n_2 = \bar{G}_2$ and $\partial G/\partial n_3 = \bar{G}_3$, and other such differential coefficients, which are known as the partial molar Gibbs free energies of the different components.

It should be noted that for a change at constant temperature and pressure eqn. (4.38) reduces to

$$dG = \sum_i \mu_i dn_i \quad (4.63)$$

which, by recalling that $\mu_i = (\partial G/\partial n_i)_{T,p,n_j}$ from (4.40c), reduces to eqn. (4.62). This identity of partial molar Gibbs free energies and chemical potentials should be remembered, since it will be further utilized.

Partial molar quantities of the other capacity factors are defined similarly and are often used in many component systems. Later we shall have occasion to use the partial molar volume $\partial V/\partial n_1 = \bar{V}_1$, which may be considered as the increase in the volume of the system which occurs when one mole of component 1 is added to such a large amount of the system that the relative amounts of the components remain unaltered. For a binary system containing n_1 gm. moles of component 1 and n_2 gm. moles of component 2,

$$dV = \bar{V}_1 dn_1 + \bar{V}_2 dn_2. \quad (4.64)$$

It is now possible to add small quantities, dn_1 and dn_2, of the two components such that
$$\frac{dn_1}{dn_2} = \frac{n_1}{n_2},$$
thus keeping the composition of the system constant, and with it the partial molar volumes or any other partial molar quantities. If sufficient amounts of the two components are so added, then for the additional volume we may write
$$V = \bar{V}_1 n_1 + \bar{V}_2 n_2, \qquad (4.65)$$
an equation which must apply equally well also to the original solution. The same result is achieved mathematically by the integration of eqn. (4.64), but it is of use to understand the physical implications of such a procedure.

Eqn. (4.65) is generally true for other partial molar quantities in systems containing any number of components, thus
$$G = E - TS + pV = \bar{G}_1 n_1 + \bar{G}_2 n_2 + \bar{G}_3 n_3 + \ldots \qquad (4.66)$$
Differentiating eqn. (4.66) we have
$$dG = \bar{G}_1 dn_1 + n_1 d\bar{G}_1 + \bar{G}_2 dn_2 + n_2 d\bar{G}_2 + \ldots, \qquad (4.67)$$
but from eqn. (4.38), writing $\mu_i = \bar{G}_i$, we have
$$dG = V dp - S dT + \bar{G}_1 dn_1 + \bar{G}_2 dn_2, \quad \text{etc.} \qquad (4.68)$$
Subtracting from (4.67) we have
$$S dT - V dp + n_1 d\bar{G}_1 + n_2 d\bar{G}_2 + n_3 d\bar{G}_3 + \ldots = 0, \qquad (4.69)$$
which is known as the Duhem-Margules equation. At constant temperature and pressure this equation reduces to
$$\sum_i n_i d\bar{G}_i = \sum_i n_i d\mu_i = 0. \qquad (4.70)$$
Other partial molar quantities may occur in similar equations.

The use of such equations may be readily seen for a binary system, when by equation (4.70)
$$n_1 d\bar{G}_1 + n_2 d\bar{G}_2 = 0. \qquad (4.71)$$
$$\therefore \quad d\bar{G}_2 = -\frac{n_1}{n_2} d\bar{G}_1.$$
On integrating we have
$$\int_I^{II} d\bar{G}_2 = -\int_I^{II} \frac{n_1}{n_2} d\bar{G}_1. \qquad (4.72)$$

Thus the variation in the Gibbs free energy of one component may be used in the Duhem-Margules equation to calculate the corresponding variation in the Gibbs free energy of the other component (see p. 70).

APPLICATION TO COLLOIDAL SYSTEMS

Differential coefficients of partial molar quantities

Differential relations between partial molar quantities are often used and are simply obtained by differentiation of equations of the type of (4.40) and (4.42) with respect to n_i at constant temperature and pressure. Thus eqns. (4.40 a) and (4.40 b) become respectively

$$\left(\frac{\partial \bar{G}_i}{\partial p}\right)_{T,n_i,n_j} = \left(\frac{\partial \mu_i}{\partial p}\right)_{T,n_i,n_j} = \bar{V}_i \tag{4.73}$$

and
$$\left(\frac{\partial \bar{G}_i}{\partial T}\right)_{p,n_i,n_j} = \left(\frac{\partial \mu_i}{\partial T}\right)_{p,n_i,n_j} = -\bar{S}_i, \tag{4.74}$$

and a small variation in $\bar{G}_i$ from eqn. (4.38) becomes

$$d\bar{G}_i = d\mu_i = \bar{V}_i dp - \bar{S}_i dT + \sum_s \frac{\partial \mu_s}{\partial n_i} dn_s. \tag{4.75}$$

Equilibrium conditions in a heterogeneous system of many components

We have already mentioned (p. 57) that the condition for thermal and mechanical equilibrium of a heterogeneous system is equality of temperature and pressure. We now wish to find out the condition of chemical equilibrium in a system for which the temperature and pressure are constant throughout the system and the change. In order to do this we apply the general criterion of equilibrium for a system at constant pressure, viz., eqn. (4.31),

$$dG \geqslant 0.$$

Let us consider two phases containing respectively:

n_1' and n_1'' gm. moles of component 1,

n_2' and n_2'' gm. moles of component 2,

n_3' and n_3'' gm. moles of component 3, etc.

Using eqn. (4.62) we have, for the two phases,

$$dG' = \mu_1' dn_1' + \mu_2' dn_2' + \ldots + \mu_n' dn_n', \tag{4.76}$$
and
$$dG'' = \mu_1'' dn_1'' + \mu_2'' dn_2'' + \ldots + \mu_n'' dn_n''. \tag{4.77}$$

For equilibrium $(dG' + dG'') \geqslant 0.$

Restricting ourselves to consideration of a change in which small quantities of the components leave one phase and enter the other, then we have
$$\left.\begin{aligned} dn_1' &= -dn_1'', \\ dn_2' &= -dn_2'', \\ dn_n' &= -dn_n''. \end{aligned}\right\} \tag{4.78}$$

Substituting eqns. (4.76), (4.77), and (4.78) in the equilibrium condition above we have

$$(\mu'_1-\mu''_1)dn'_1+(\mu'_2-\mu''_2)dn'_2+\ldots+(\mu'_n-\mu''_n)dn'_n \geqslant 0. \qquad (4.79)$$

Since $dn'_1, dn'_2, \ldots, dn'_n$ may be positive or negative, then eqn. (4.79) can only be true in general if

$$\mu'_1 = \mu''_1, \quad \mu'_2 = \mu''_2, \quad \ldots, \quad \mu'_n = \mu''_n. \qquad (4.80)$$

Thus for chemical equilibrium the chemical potentials of the different components must be the same in both phases. In the general case of more than two phases it is also true that any component must have the same chemical potential in any phase in which it occurs.

It is quite easy to show that where

$$\mu'_n > \mu''_n$$

then component n will pass from phase ' to phase ". It follows, therefore, that if any component does not occur in a certain phase then its chemical potential cannot be less than in other phases in which it does occur. Otherwise there would be a flow of material into the first phase which would result in a finite concentration of the component.

One special case ought to be considered here. We have assumed the system to have a constant temperature and pressure throughout. If, however, we have parts of the system separated by a semi-permeable membrane, then for equilibrium, though the temperature will be constant throughout, we may not have the same pressure on the two sides of the membrane. However, for those species to which the membrane is permeable there must be equality of the chemical potentials on the two sides of the membrane, and in this connexion we shall see, when we consider osmotic pressure in detail, that the inequality in the pressure on the two sides of the membrane has considerable importance.

For those species which are 'mechanically' constrained from passing through the membrane it is clear that there is no relation possible between the chemical potentials on opposite sides of the membrane.

It must be noted that the simple equalities of eqn. (4.80) are dependent upon the validity of those of eqn. (4.78), which is not invariably true. In general, however, dn'_1 and dn''_1 are related in some simple way and we may write

$$\left. \begin{array}{l} \dfrac{dn''_1}{dn'_1} = -x_1, \\[1em] \dfrac{dn''_2}{dn'_2} = -x_2, \quad \text{etc.,} \end{array} \right\} \qquad (4.81)$$

APPLICATION TO COLLOIDAL SYSTEMS

where x_1 and x_2 are simple fractions. Eqn. (4.79) then becomes

$$(\mu_1' - x_1\mu_1'')\,dn_1' + (\mu_2' - x_2\mu_2'')\,dn_2' + \ldots + (\mu_n' - x_n\mu_n'')\,dn_n' \geqslant 0, \quad (4.82)$$

which is generally true only when

$$\mu_1' = x_1\mu_1'', \quad \mu_2' = x_2\mu_2'', \quad \ldots, \quad \mu_n' = x_n\mu_n'', \quad \text{etc.} \quad (4.83)$$

It is clear that this case is very similar to the general one of a chemical equilibrium

$$aA + bB \rightleftharpoons cC + dD, \quad (4.84)$$

in which it is easy to show, by considering a small variation in the system, that the equilibrium condition is

$$a\mu_A + b\mu_B = c\mu_c + d\mu_d. \quad (4.85)$$

Chemical potentials in many component systems

We have seen that for a mixture of perfect gases the Gibbs free energy of any component is given by the expression (eqn. (4.61))

$$G_1 = G_1^0 + RT \ln p_1,$$

where G_1^0 is the Gibbs free energy at temperature T and unit pressure.

We now see that it is more correct for many component systems in general to use either partial molar Gibbs free energies ($\bar{G}_1$) or (as we are doing) chemical potentials μ_1 which are the same, giving

$$\mu_1 = \mu_1^0 + RT \ln p_1, \quad (4.86)$$

though for the special case of perfect gases, in which the components do not interact, eqns. (4.61) and (4.86) are equally valid.

It should be noted that whilst a perfect gas is usually defined as one for which the relation

$$pV = RT \quad (4.87)$$

holds for 1 gm. molecule, it can equally well be defined as one in which E and H are functions of T only, or for which the chemical potential may be written

$$\mu = \mu^0(T) + RT \ln p, \quad (4.88)$$

where $\mu^0(T)$ is a function of temperature only. The definitions are clearly equivalent since any one involves the others.

If we now consider a two component system containing both liquid and gaseous phases, we know from our previous consideration of equilibrium conditions that for each component the chemical potentials in the two phases must be identical, i.e.

$$\left.\begin{array}{l} \mu_1(\text{liquid}) = \mu_1(\text{vapour}), \\ \mu_2(\text{liquid}) = \mu_2(\text{vapour}). \end{array}\right\} \quad (4.89)$$

If we now assume that the vapours behave as perfect gases we may write
$$\mu_1(\text{vapour}) = \mu_1^0(\text{vapour}) + RT \ln p_1, \qquad (4.90)$$
where p_1 is the partial pressure of component 1 in the vapour.

In conformity with the expression for the chemical potential of a gas, let us now write for the chemical potential in the liquid
$$\mu_1(\text{liquid}) = \mu_1^0(\text{liquid}) + RT \ln a_1, \qquad (4.91)$$
where a_1 is known as the activity of component 1 in the liquid, and $\mu_1^0(\text{liquid})$ is the chemical potential in the liquid in a standard state. Then by equating (4.90) and (4.91) and rearranging, we obtain
$$\mu_1^0(\text{liquid}) - \mu_1^0(\text{vapour}) = RT \ln(p_1/a_1). \qquad (4.92)$$
Since the L.H.S. of this equation is a constant dependent only on our definition of standard states in the liquid and vapour phases, it therefore follows that $p_1/a_1 = $ constant.

If we now define $a_1 = 1$ for pure component 1 (fixing also the value of $\mu_1^0(\text{liquid})$, the standard chemical potential, as the value of $\mu_1(\text{liquid})$ for pure liquid component 1), we have from eqn. (4.92) for a pure liquid and its vapour
$$\mu_1^0(\text{liquid}) - \mu_1^0(\text{vapour}) = RT \ln p_1^0, \qquad (4.93)$$
from which it is apparent that
$$\frac{p_1}{p_1^0} = a_1. \qquad (4.94)$$
It therefore follows that the chemical potential of a liquid may be written
$$\mu_1(\text{liquid}) = \mu_1^0(\text{pure liquid}) + RT \ln(p_1/p_1^0). \qquad (4.95)$$
Thus it is apparent that vapour pressure measurements may be successfully used to measure the difference in chemical potential between a liquid in its pure state and in a solution containing other components. Similarly such differences may be estimated for two solutions of different concentrations. If the additional components are involatile then measurement of total vapour pressure suffices for such determinations but, otherwise, partial vapour pressure data are required.

For systems obeying Raoult's law (ideal systems)
$$\frac{p_1}{p_1^0} = \frac{n_1}{n_1 + n_2} = N_1, \qquad (4.96)$$
where N_1 is the mole fraction of component 1. Substituting in eqn. (4.95) we obtain
$$\mu_1(\text{liquid}) = \mu_1^0(L) + RT \ln N_1. \qquad (4.97)$$

For dilute solutions of 1, N_1 becomes proportional to its molar concentration, m_1, and we have

$$\mu_1(\text{liquid}) = \mu_1'^0(L) + RT \ln m_1, \qquad (4.97\text{a})$$

where, however, we must now use a new standard state $\mu_1'^0(L)$.

In general, however, Raoult's law does not hold, but we may put

$$f_1 = \frac{p_1}{p_1^0} \cdot \frac{1}{N_1} = \frac{a_1}{N_1}, \qquad (4.98)$$

where f_1 is known as an activity coefficient, which is unity for pure component 1. Eqn. (4.95) then becomes

$$\mu_1(L) = \mu_1^0(L) + RT \ln(N_1 f_1). \qquad (4.99)$$

Where Raoult's law holds, it is apparent that knowledge of the composition of the system enables a calculation of changes in chemical potential, but in general, deviations from the law cannot be neglected and a knowledge of activity coefficients from vapour pressure measurements or other data (e.g. see Glasstone, 1937, p. 116) is essential also.

It is important to realize that the definition of $a_1 = 1$ for pure liquid 1 is contained in eqn. (4.95), and that alternative definitions, leading to a different equation could have been made. Thus in many cases we define $f_1 = 1$ (i.e. $a_1 = N_1$) for vanishingly small concentrations of 1. The standard chemical potential, $\mu_1^0(L)$, is then no longer the chemical potential of pure liquid 1. Such a change in the definition of activity corresponds only to a change of our zero level of chemical potential, which is usually unimportant since we are normally concerned only with changes in chemical potential between specified states.

Osmotic coefficient

Deviations from ideal behaviour can clearly be described in terms of the activity coefficient f by means of eqn. (4.99). Alternatively, however, it is possible to make use of the osmotic coefficient g defined by the equation

$$\mu_1(L) = \mu_1^0(L) + gRT \ln N_1, \qquad (4.100)$$

in which $g \to 1$ as $N_1 \to 1$.

Comparing (4.99) with (4.100), the osmotic and activity coefficients are seen to be related by the expression

$$\ln f_1 = (1-g) \ln \frac{1}{N_1}. \qquad (4.101)$$

The use of the Duhem-Margules equation

We have seen that for a binary system the Duhem-Margules equation (4.70) may be used to calculate a change in chemical potential of one component from a knowledge of the corresponding variation in that of the other component. Thus measurement of partial vapour pressures for one component may be used to give information on the change in chemical potential and of vapour pressure of the second component without assuming ideal behaviour or Raoult's law. Thus differentiating eqn. (4.95) and substituting in eqn. (4.71) we have

$$n_1 d\ln(p_1/p_1^0) + n_2 d\ln(p_2/p_2^0) = 0$$

or, since p_1^0 and p_2^0 are constant,

$$n_1 d\ln p_1 + n_2 d\ln p_2 = 0. \qquad (4.102)$$

Thus an increase in the partial vapour pressure of one component (say, 1) leads to a decrease in that of the other component (2) which is inversely proportional to its molecular concentration (n_2).

Solubility

The criteria of equilibrium already outlined are clearly fundamental to the treatment of solubility. A solution becomes saturated with a given component when the chemical potential in solution (μ_s) is identical with that of the pure substance (μ_s'), i.e.

$$\mu_s' = \mu_s,$$

and for any infinitesimal change in the equilibrium position we have

$$d\mu_s' = d\mu_s. \qquad (4.103)$$

Applying eqn. (4.75) we have

$$-S_s' dT + V_s dp' = -S_s dT + V_s dp + \frac{\partial \mu_s}{\partial N_s} dN_s, \qquad (4.104)$$

where N_s is the mole fraction of the solute in solution. At constant pressure, this reduces to

$$\left.\begin{array}{r}(S_s - S_s') dT = \dfrac{\partial \mu_s}{\partial N_s} dN_s \\[6pt] \dfrac{L_s}{T} dT = \dfrac{\partial \mu_s}{\partial N_s} dN_s,\end{array}\right\} \qquad (4.105)$$

or

where L_s is the molar heat of solution (i.e. heat absorbed when one mole of solute is dissolved).

Applying eqn. (4.97) (i.e. assuming an ideal solution) we obtain

$$\frac{L_s}{T} dT = RT d\ln N_s,$$

i.e.

$$\frac{d\ln N_s}{dT} = \frac{L_s}{RT^2}, \qquad (4.106)$$

an equation which expresses the change in solubility with temperature in terms of the molar heat of solution. The similarity of this equation with the approximate form of the Clapeyron-Clausius equation should be noted.

Activities of ions and strong electrolytes

As yet we have only considered the chemical potentials of neutral molecules, and we must now consider strong electrolytes. Regarding a strong binary electrolyte as completely dissociated we can write down for the chemical potentials (μ_+ and μ_-) of the component ions

$$\left.\begin{array}{l}\mu_+ = \mu_+^0 + RT \ln a_+,\\ \mu_- = \mu_-^0 + RT \ln a_-,\end{array}\right\} \quad (4.107)$$

in which the activities of the ions are so defined that a_+ and a_- approach the molar concentrations m_+ and m_- respectively as m_+ and m_- approach zero.

For the whole molecule (eqn. (4.91))

$$\mu = \mu^0 + RT \ln a,$$

where a is the activity of the salt ($a \neq m$ in dilute solution). But it must be true that

$$\mu = \mu_+ + \mu_-.$$

$$\therefore \quad \mu^0 + RT \ln a = \mu_+^0 + \mu_-^0 + RT \ln a_+ a_-.$$

We define the activity of the salt such that

$$\mu^0 = \mu_+^0 + \mu_-^0 \quad \text{and} \quad a = a_+ a_-.$$

Making use of
$$a_\pm = \sqrt{(a_+ a_-)} = \sqrt{a}, \quad (4.108)$$

eqn. (4.91) may be written

$$\mu = \mu^0 + 2RT \ln a_\pm. \quad (4.109)$$

In the general case of a strong electrolyte of formula $A_{\nu_A} B_{\nu_B}$, which dissociates completely, this becomes

$$\mu = \mu^0 + (\nu_A + \nu_B) RT \ln a_\pm, \quad (4.110)$$

where
$$a_\pm = (a_+^{\nu_A} a_-^{\nu_B})^{1/(\nu_A + \nu_B)}. \quad (4.111)$$

The activity coefficients of ions, e.g. $f_+ = a_+/m_+$ and $f_- = a_-/m_-$, where m_+ and m_- are the molar concentrations of the respective ions, are often used, as is also the mean activity coefficient, defined by $f_\pm = a_\pm/m_\pm$. From the definition of ionic activities it follows that f_+, f_-, and $f_\pm$ become unity at infinite dilution of the strong electrolyte.

Variation of the activities of ions with salt concentration has been the subject of considerable important work. Lewis and Randall stated the important rule, for *dilute solutions*, that the activity coefficient of a salt is the same in all solutions of the same ionic strength (I), the latter being defined as

$$I = \tfrac{1}{2} \sum_i c_i z_i^2, \qquad (4.112)$$

where c_i is the molar concentration of the ion i per litre, and z_i its valency.

For solutions for which $I > 0\cdot 01$, ions have specific effects on activity coefficients and the rule no longer holds.

Debye and Hückel gave a theoretical foundation to this rule by working out the effect on the thermodynamic properties of electrical forces in dilute solutions of strong electrolytes. Thus for the activity coefficient of an ion they obtained the equation

$$-\ln f_i = \frac{z_i^2 e^2 \kappa}{2DkT}, \qquad (4.113)$$

where κ is the well-known Debye-Hückel function, considered elsewhere (p. 102), which is proportional to $\sqrt{I}$. It follows, therefore, that for very dilute solutions at constant temperature and constant dielectric constant the mean activity coefficient is given by

$$-\ln f_{\pm} = \text{constant}.\sqrt{I}, \qquad (4.114)$$

thus providing a quantitative expression of the Lewis and Randall rule.

The extension of the Debye-Hückel treatment to cover more concentrated solutions and certain colloidal systems is discussed in Chapter V.

Free energy and heat changes in heterogeneous equilibria

We shall later have occasion to consider adsorption from solution onto solid or liquid surfaces. The free energy associated with the process will indicate whether adsorption (i.e. an excess of solute on the surface) does or does not tend to occur. Regarding the solution as one phase (denoted by superscript α), and the surface as another (denoted by β), the problem is clearly identical with the general one for the distribution of a component between two phases (partition coefficients).

At equilibrium $\mu_s^\alpha = \mu_s^\beta$,

where s denotes the solute. If the activity coefficients of s in both phases are unity, then

$$\mu_s^\alpha = \mu_s^0(T, P)^\alpha + RT \ln(C_s)^\alpha,$$

$$\mu_s^\beta = \mu_s^0(T, P)^\beta + RT \ln(C_s)^\beta,$$

where C_s denotes the concentration of s.

$$\therefore \quad \mu_s^0(T, P)^\beta - \mu_s^0(T, P)^\alpha = -RT \ln\{(C_s)^\beta/(C_s)^\alpha\} \qquad (4.115)$$

$$= -RT \ln K, \qquad (4.115\text{a})$$

where K is the equilibrium constant for the reaction

$$s \text{ in } \alpha\text{-phase} \rightleftharpoons s \text{ in } \beta\text{-phase}.$$

If the reaction goes 100 per cent. from left to right then the *increase* in free energy (ΔG) is given by

$$\Delta G = \mu_s^\beta - \mu_s^\alpha$$
$$= \mu_s^0(T, P)^\beta - \mu_s^0(T, P)^\alpha + RT \ln\{(C_s)^\beta/(C_s)^\alpha\}.$$
$$\therefore \quad \Delta G = -RT \ln K + RT \ln\{(C_s)^\beta/(C_s)^\alpha\}. \tag{4.116}$$

If the reaction begins with $(C_s)^\alpha = 1$ and ends with $(C_s)^\beta = 1$ then

$$\Delta G^0 = -RT \ln K, \tag{4.117}$$

ΔG^0 being termed the *standard free energy*, or *standard affinity* for the reaction considered. (An ideal solution at unit concentration is said to be in its *standard state*.)

Since
$$K = \frac{(C_s)^\beta_{\text{equil.}}}{(C_s)^\alpha_{\text{equil.}}}$$

it follows that, if the surface concentration of the solute exceeds its bulk concentration, then ΔG^0 is negative, i.e. the reaction, in which solute at unit concentration goes from solution to the adsorbed phase, is accompanied by a decrease in free energy of the system.

In addition to the free energy of adsorption we are usually also interested in the associated heat change (ΔH^0) and the entropy change (ΔS^0), which are related as usual by the eqn. (4.24), viz.

$$\Delta G^0 = \Delta H^0 - T \Delta S^0.$$

ΔH^0 (and hence ΔS^0) is found from the temperature coefficient of the equilibrium constant K, since from eqn. (4.46),

$$\frac{d \ln K}{dT} = \frac{d}{dT}\left(-\frac{\Delta G^0}{RT}\right) = \frac{\Delta H^0}{RT^2}, \tag{4.118}$$

where ΔH^0 is the heat absorbed when 1 mole of s is transferred from phase α (solution) to phase β (the surface).

So far we have discussed only ideal systems; for non-ideal ones (i.e. $f_s \neq 1$) see Guggenheim, 1933, p. 122.

APPLICATION OF THERMODYNAMIC METHODS TO CERTAIN PROBLEMS

Osmotic pressure and membrane equilibrium

We have seen (p. 66) that where two phases are separated by a semi-permeable membrane, the conditions of equilibrium are

1. Constancy of temperature throughout the system.
2. Identity of the chemical potential (or electrochemical potentials for charged species) for all diffusible species.

The presence of the membrane makes it mechanically possible to have a pressure difference between the two phases at equilibrium.

Consider on one side of the membrane a binary solution containing solvent S_1 and solute S_2, and on the other side the pure solvent. The presence of solute on one side, by itself, causes a lowering of the partial vapour pressure of the solvent there and therefore of its chemical potential which for equilibrium must be compensated. From eqns. (4.73) and (4.95) we have

$$\left(\frac{\partial \mu_1}{\partial P}\right)_{T, n_1, n_2} = \bar{V}_1 = RT \frac{\partial \ln p_1}{\partial P} \tag{4.119}$$

and we see that the increase of chemical potential with pressure is a positive quantity, the partial molar volume of the solvent. Thus the chemical potential and partial vapour pressure of the solvent, lowered by the presence of the second component according to eqn. (4.95), may be restored to their original values by the application of a certain hydrostatic pressure. This is the basis of the application of thermodynamics to osmotic pressure.

Thus at equilibrium the total variation of the chemical potential of the solvent is given by

$$RT \frac{\partial \ln p_1}{\partial P} dP + RT \frac{\partial \ln p_1}{\partial n_1} dn_1 = 0,$$

i.e.
$$\frac{\partial \ln p_1}{\partial P} dP + \frac{\partial \ln p_1}{\partial n_1} dn_1 = 0$$

and
$$\frac{\bar{V}_1}{RT} dP + \frac{\partial \ln p_1}{\partial n_1} dn_1 = 0. \tag{4.120}$$

If the pressure on the pure solvent is P_0 then the pressure on the solution in excess is the osmotic pressure (Π).

Substituting in (4.120) and integrating we have

$$\frac{1}{RT} \int_{P_0}^{P_0+\Pi} \bar{V}_1 \, d\Pi = -\int_{p_1^0}^{p_1} d\ln p_1 = \ln \frac{p_1^0}{p_1}. \tag{4.121}$$

If $\bar{V}_1$ is constant then we have

$$\Pi \bar{V}_1 = RT \ln \frac{p_1^0}{p_1} = -RT \ln a_1. \tag{4.122}$$

More accurately the variation in partial molar volume of the solvent with pressure can be expressed in terms of the compressibility κ. Thus

$$\bar{V}_1 = \bar{V}_1^*(1 - \kappa_1 \Pi), \tag{4.123}$$

where $\bar{V}_1^*$ is the value of $\bar{V}_1$ at zero pressure. Introducing into eqn. (4.121) and integrating we have

$$\Pi \bar{V}_1^*(1-\tfrac{1}{2}\kappa_1 \Pi) = RT \ln(p_1^0/p_1). \tag{4.124}$$

This equation is thermodynamically exact, the only assumptions being concerned with the variation of the partial molar volume of the solvent with pressure and with the behaviour of the solvent vapour as a perfect gas. It has been shown to hold quite generally over very large ranges of osmotic pressure (see Table 4.1).

TABLE 4.1. *Cane Sugar at 0° C. in Water* (*Berkley, Hartley, and Burton,* 1919)

Conc. gm./100 gm. H_2O	Π (atms.) calc.	Π (atms.) obs.
56·50	43·91	43·84
81·20	67·43	67·68
112·00	100·53	100·43
141·00	134·86	134·71

Relations between osmotic pressure and other thermodynamic functions

From eqn. (4.91) we have

$$\begin{aligned}-RT \ln a_1 &= \mu_1^0 - \mu_1 \\ &= G_1 - \bar{G}_1 \\ &= -\Delta \bar{G}_1,\end{aligned} \tag{4.125}$$

where $\Delta \bar{G}_1$ is the increase in partial Gibbs free energy of the solvent in the solution over that in the pure solvent. Hence, substituting eqn. (4.125) in eqn. (4.122), we obtain

$$\Pi \bar{V}_1 = -\Delta \bar{G}_1 = T \Delta \bar{S}_1 - \Delta \bar{H}_1. \tag{4.126}$$

It is apparent therefore that a knowledge of $\Delta \bar{S}_1$ and $\Delta \bar{H}_1$, the increases in partial molar entropy and heat content of the solvent in the solution over the values in pure solvent at a given temperature, allows a direct calculation of osmotic pressure.

We shall see, in connexion with linear polymers, that this approach is valuable. In some of these systems it is found that $\Delta \bar{H}_1 \approx 0$, and that $\Delta \bar{S}_1$, which may be calculated statistically from certain model structures, is much the more important term, having a value much higher than ideal. A very important difference between the osmotic pressure-concentration dependence for linear polymers and the more symmetrical macromolecules thus arises, and is fully confirmed experimentally (p. 156). It is to be noted that the occurrence of $\Delta \bar{S}_1$ values greater than ideal must by eqns. (4.125) and (4.126) cause the activity to have much smaller values than would be expected by Raoult's law (see Fig. 4.5).

It is also to be observed from (4.126) that osmotic pressure measurements may be used for the determination of changes in partial molar Gibbs free energy or chemical potential. This is of very considerable importance in very dilute solutions where the direct measurement of vapour pressure is not suitable. Further, the temperature coefficient of osmotic pressure may be used (eqn. (4.46)) to calculate $\Delta \bar{H}_1$, and therefore $\Delta \bar{S}_1$ values also, so that all the thermodynamic functions of eqn. (4.126) may be evaluated from osmotic pressure measurements alone.

Simplified osmotic pressure equation

For those systems which obey Raoult's law, we may substitute eqn. (4.96) into eqn. (4.122), giving

$$\Pi \bar{V}_1 = RT \ln \frac{1}{N_1} = RT \ln \frac{n_1 + n_2}{n_1}$$

$$= RT \ln\left(1 + \frac{n_2}{n_1}\right). \tag{4.127}$$

For dilute solutions, $n_1 \gg n_2$. Therefore

$$\ln(1 + n_2/n_1) \approx n_2/n_1, \qquad n_1 \bar{V}_1 \approx V_{\text{total}}$$

and
$$\Pi(\bar{V}_1 n_1) = RT n_2.$$

It therefore follows that

$$\Pi = \frac{n_2 RT}{V} = \frac{c}{M} RT, \tag{4.128}$$

where V is the total volume and $c/M =$ conc. in gm. molecules of component 2/litre.

It is clear that eqn. (4.128) has been obtained from the thermodynamically accurate eqn. (4.122) only by postulating

(1) the validity of Raoult's law;
(2) very dilute solutions of component 2 in component 1, allowing a simplification of the logarithmic term.

These somewhat drastic assumptions, which will be considered later (p. 151), are not generally true, and eqn. (4.128) is therefore of restricted application.

Donnan membrane equilibrium for charged species

It was pointed out (p. 53) that, where electrical work is involved, chemical potentials are replaced by electrochemical potentials containing an electrical work term of the type $ze\psi$. Such a term arises in the following way.

APPLICATION TO COLLOIDAL SYSTEMS

If dn gm. moles of an ion are transferred from phase I to phase II, then the change in Gibbs free energy arising from differences in chemical potential is
$$dn(\mu'' - \mu').$$
The electrical work done on the system in such a transfer is
$$zF\,dn(\psi'' - \psi'),$$
where ψ' and ψ'' are the electrical potentials of phases I and II. Since for equilibrium $\delta G(\text{overall}) \geqslant 0$,
$$dn(\mu'' - \mu') + zF\,dn(\psi'' - \psi') \geqslant 0.$$
Since dn may be positive or negative, the equality sign must hold and we have
$$\mu'' + zF\psi'' = \mu' + zF\psi', \tag{4.129}$$
where the quantity on each side of the equation is the electrochemical potential of the ion for the appropriate phase.

In the case of the transfer of equal quantities of positive and negative ions of the same valency, the two electrical terms ($zF\psi$ and $-zF\psi$) cancel and the terms involving chemical potential only remain. Thus it is correct in such a case to consider the transfer of the undissociated molecule, for
$$\mu_{\text{NaCl}} = \mu_{\text{Na}} + \mu_{\text{Cl}}$$
$$= (\mu_{\text{Na}} + F\psi) + (\mu_{\text{Cl}} - F\psi).$$

When on one side of a membrane we have a charged ion to which the membrane is impermeable, it is apparent, from the condition of electrical neutrality, that there must be an unequal distribution of the diffusible ions on the two sides of the membrane. A similar situation arises whenever a charged ion is prevented in any way from diffusing to all parts of the system, e.g. in a gel or at an interface, and the general phenomenon is known as the Donnan Membrane Equilibrium.

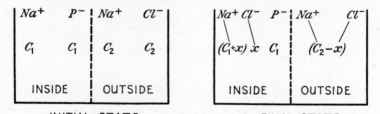

FIG. 4.2. The Donnan membrane equilibrium.

Consider the system shown diagrammatically in Fig. 4.2, containing, inside the membrane, a solution of the sodium salt of a polyvalent ion (e.g. protein at a pH above the iso-electric point), and, outside, a solution of NaCl. For convenience, let the volumes on the two sides of the membrane be unity.

Chlorine ions, accompanied by sodium ions to preserve electrical neutrality, must pass from the outside to the inside of the membrane until at equilibrium there must be identity of:

(1) chemical potentials for uncharged species

and (2) electrochemical potentials for charged species.

Let the amount of [Cl⁻] passing across be x. Thus applying condition 1 to the undissociated [NaCl], we have from eqn. (4.109)

$$RT \ln a'_{\pm} = RT \ln a''_{\pm},$$

i.e.

$$\left. \begin{array}{c} a'_{Na} a'_{Cl} = a''_{Na} a''_{Cl}, \\ \dfrac{a'_{Na}}{a''_{Na}} = \dfrac{a''_{Cl}}{a'_{Cl}}. \end{array} \right\} \quad (4.130)$$

or

At very low concentrations, activities may be replaced by the appropriate concentrations.

Substituting concentrations in eqn. (4.130) we have

$$(c_1+x)x = (c_2-x)^2.$$

$$\therefore \quad x = \frac{c_2^2}{c_1+2c_2}.$$

Thus
$$\frac{[\text{NaCl}]_{\text{outside}}}{[\text{NaCl}]_{\text{inside}}} = \frac{c_2-x}{x} = 1+\frac{c_1}{c_2}. \quad (4.131)$$

This result is of considerable importance in connexion with the use of osmotic pressure in aqueous solutions. Clearly, unless

$$[\text{NaCl}]_{\text{outside}} = [\text{NaCl}]_{\text{inside}},$$

there will be an osmotic pressure (Π_i), due to the unequal distribution of diffusible ions, which is additional to that of the colloidal material, and which must either be eliminated or suitably allowed for (p. 79).

If we now apply condition 2 to the Na ions then we have

$$\mu'' + F\psi'' = \mu' + F\psi'.$$

$$\therefore \quad \psi'' - \psi' = \frac{1}{F}(\mu' - \mu'')$$

$$= \frac{RT}{F} \ln \frac{a'_{Na}}{a''_{Na}} = \frac{RT}{F} \ln \frac{a''_{Cl}}{a'_{Cl}}. \quad (4.132)$$

In the general case eqn. (4.130) becomes

$$\left(\frac{[\text{Anion}]_{\text{inside}}}{[\text{Anion}]_{\text{outside}}} \right)^{1/z_A} = \left(\frac{[\text{Cation}]_{\text{outside}}}{[\text{Cation}]_{\text{inside}}} \right)^{1/z_C} \quad (4.133)$$

and eqn. (4.132) becomes

$$\psi'' - \psi' = \frac{RT}{z_A F} \ln \frac{a'_-}{a''_-} = \frac{RT}{z_C F} \ln \frac{a'_+}{a''_+}, \qquad (4.134)$$

where z_A and z_C are the respective valencies of the diffusible anions and cations.

It should be noted that the membrane potential (see Fig. 4.3) is equal in magnitude but opposite in sign to the electrode potential which

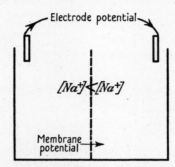

FIG. 4.3. Membrane and electrode potentials.

would be obtained if reversible electrodes were placed in the two compartments which then acted as a concentration cell without any membrane effect (see also p. 181).

The calculation of ion pressure differences and of particle charge from membrane potential measurements

Where unequal distribution of diffusible ions cannot be avoided, it is necessary to make an appropriate allowance before information on the molecular weight of the colloidal molecule (e.g. protein) may be obtained.

The total osmotic pressure (Π_T) may be written

$$\Pi_T = \Pi_p + \Pi_i, \qquad (4.135)$$

where Π_p is the partial osmotic pressure of the protein ions and Π_i is the pressure due to the unequal distribution of diffusible ions across the membrane.

A more detailed expression for the total osmotic pressure may be obtained thermodynamically, as shown by Adair (1929). Thus the general form of the Duhem-Margules equation (4.69) may be written

$$-S\,dT + V\,dP - \sum n_i\,d\mu_i = 0, \qquad (4.69)$$

which for constant temperature becomes

$$dP = \frac{n_p}{V}d\mu_p + \frac{n_a}{V}d\mu_a + \ldots + \frac{n_i}{V}d\mu_i \tag{4.69a}$$

$$= c_p RT\, d\ln a_p + c_a RT\, d\ln a_a + \ldots + c_i RT\, d\ln a_i, \tag{4.69b}$$

dP being the excess hydrostatic pressure on the colloidal solution inside the membrane, and n_p, μ_p, c_p, and a_p respectively the number of gm. moles, the chemical potential, the concentration in moles per litre, and the activity of the protein.

$$n_a,\ldots,n_i, \quad \mu_a,\ldots,\mu_i, \quad c_a,\ldots,c_i, \quad a_a,\ldots,a_i$$

are the corresponding quantities for the diffusible ions occurring in the colloidal solution.

Now the activities of the diffusible ions inside the membrane are related to those outside by eqn. (4.134), which may be written

$$RT\ln a_i' = RT\ln a_i'' - z_i FE, \tag{4.134a}$$

where a_i'' is the activity of the ion outside which is kept constant, and $E = (\psi' - \psi'')$ is the membrane potential. It follows, therefore, that $RT\, d\ln a_i = -z_i F\, dE$, which may be substituted in eqn. (4.69b) giving

$$dP = c_p RT\, d\ln a_p - F\, dE(z_a c_a + \ldots + z_i c_i). \tag{4.136}$$

Since, however, the protein or colloidal solution is electrically neutral

$$c_p z_p = -(c_a z_a + \ldots + c_i z_i) \tag{4.137}$$

and eqn. (4.136) becomes

$$dP = c_p RT\, d\ln a_p + c_p z_p F\, dE \tag{4.138}$$

$$= c_p RT\, d\ln a_p + c_p z_p RT\, du, \tag{4.138a}$$

where $\quad u = E \times F/RT.$

(At $0°$ C., $u = E/23\cdot 53$ when E is in millivolts.) Integrating, assuming the mean valency of the protein ions to be constant, we obtain

$$\Pi_{\text{total}} = RT\int_0^{c_p} c_p\, d\ln a_p + RT\int_0^u c_p z_p\, du, \tag{4.139}$$

in which the first term on the right is to be identified with Π_p, and the second with Π_i, i.e.

$$\Pi_p = RT\int_0^{c_p} c_p\, d\ln a_p, \tag{4.140}$$

$$\Pi_i = RT\int_0^u c_p z_p\, du. \tag{4.141}$$

APPLICATION TO COLLOIDAL SYSTEMS

At infinite dilution Π_i may also be written

$$\Pi_i = RT(\sum c'_i - \sum c''_i), \tag{4.142}$$

where $\sum c'_i$ and $\sum c''_i$ represent the sums of the concentrations of the diffusible ions inside and outside the membrane respectively.

At higher concentrations equation (4.142) becomes

$$\Pi_i = RT(g' \sum c' - g'' \sum c''), \tag{4.143}$$

where g' and g'' are the mean osmotic coefficients in the protein solution and dialysate respectively (see p. 69).

The product $c_p z_p$ is the equivalent concentration of the colloidal material, and may be obtained independently of molecular weight by analytical measurements, e.g. of the quantity of acid or base bound, provided no other ions are also bound.

According to Adair, eqn. (4.141) should, more correctly, be written with c_p replaced by m_p, i.e.

$$\Pi_i = RT \int_0^u m_p z_p \, du, \tag{4.144}$$

where m_p is the corrected concentration of protein in gm. moles per litre of *solvent*. The product $m_p z_p$ may then be written in terms of the composition of the liquid outside the membrane as we show below.

If the activity coefficients of ions inside and outside the membrane are equal (i.e. for small membrane potentials, e.g. $E < 0.002$ volts) then eqn. (4.134a) may be written in the form

$$\frac{a'_i}{a''_i} = \frac{N'_i}{N''_i} = r_i = e^{-z_i FE/RT} \tag{4.134b}$$
$$= e^{-z_i u}.$$

The ratio, r_i, of the mole fraction of a given ion species i inside to that outside the membrane may, in dilute solutions, be identified with the ratio of the corresponding corrected concentrations, i.e.

$$r_i = \frac{N'_i}{N''_i} = \frac{m'_i}{m''_i}. \tag{4.145}$$

Substituting in (4.134b) and expanding the exponential we obtain

$$m'_i = m''_i(1 - z_i u + \tfrac{1}{2} z_i^2 u^2 - \tfrac{1}{6} z_i^3 u^3 + \ldots) \tag{4.146}$$

But from the condition of electrical neutrality for each side of the membrane, we have

$$m_p z_p = - \sum m'_i z_i. \tag{4.147}$$

Making use of (4.146) this becomes

$$m_p z_p = - \sum m''_i z_i (1 - z_i u + \tfrac{1}{2} z_i^2 u^2 - \ldots),$$

and neglecting powers of u higher than the first, we obtain

$$m_p z_p = -\sum m_i'' z_i + u \sum m_i'' z_i^2$$
$$= uJ, \qquad (4.148)$$

where $J = \sum m_i'' z_i^2$ is twice the ionic strength. It should, however, be noted that in view of the neglect of higher powers in the exponential series, eqn. (4.148) holds only at low values of u and, therefore, of the protein concentration.

Introducing (4.148) into equation (4.144), the expression for the ion pressure difference, we obtain

$$\Pi_i = RT \int_0^u uJ\, du = \tfrac{1}{2} RT J u^2$$
$$= \tfrac{1}{2} RT J \left(\frac{EF}{RT}\right)^2$$
$$= \frac{JF^2}{2RT} E^2. \qquad (4.149)$$

Thus Π_i may be evaluated in terms of the ionic composition of the liquid outside the membrane (dialysate) and the measured membrane potential. A more approximate estimate (Π_i^*) may also be made using eqn. (4.143) which Adair (1929) shows may be written

$$\Pi_i^* = RT g_i'' (\sum c_i'' e^{-z_i u} - \sum c_i''). \qquad (4.150)$$

g_i'', the mean osmotic coefficient of the diffusible ions in the dialysate, is usually obtained from freezing-point measurements.

Further, from eqn. (4.148) we have

$$z_p = \left(\frac{uJ}{m_p}\right)_{m_p \to 0} = \frac{M}{10} J \left(\frac{u}{c_v}\right)_{c_v \to 0}, \qquad (4.151)$$

where c_v = grams protein per 100 c.c. of solvent and
M = the molecular weight of the protein.

When E is expressed in millivolts, eqn. (4.151) becomes at 0° C.

$$z_p = 0{\cdot}00425 MJ \left(\frac{E}{c_v}\right)_{c_v=0}, \qquad (4.152)$$

$(E/c_v)_{c_v=0}$ being obtained by extrapolating to zero protein concentration at a constant dialysate composition.

Clearly, therefore, a knowledge of membrane potentials is of considerable importance, allowing as it does a calculation of ion pressure differences and the mean valence of colloidal particles. The measurement of such potentials is considered in Chapter VII.

Sedimentation equilibrium

Although gravitational and centrifugal effects may often be neglected, there are some important cases in which they must be considered.† ·In such cases, in a manner analogous to the consideration of electrical effects, we replace chemical potentials for a given species by a new function $(\mu_1 + M_1\phi)$, where ϕ is the gravitational potential. At equilibrium

$$\mu_1 + M_1\phi = \text{constant}$$

or
$$d\mu_1 + M_1\, d\phi = 0. \tag{4.153}$$

A small change in $d\mu_1$ at constant temperature may be written (eqn. (4.75)) in the form

$$d\mu_1 = \overline{V}_1\, dP + \sum_i \frac{\partial \mu_i}{\partial n_1} dn_i,$$

which for an ideal solution reduces to

$$d\mu_1 = \overline{V}_1\, dP + RT\, d\ln N_1 \tag{4.154}$$

and for non-ideal solutions to

$$d\mu_1 = \overline{V}_1\, dP + RT\, d\ln(N_1 f_1). \tag{4.155}$$

Introducing (4.154) into (4.153) we have

$$\overline{V}_1\, dP + RT\, d\ln N_1 + M_1\, d\phi = 0. \tag{4.156}$$

If we now consider the effect of a centrifugal field we may write $d\phi = -\omega^2 x\, dx$ and $dP = \rho\omega^2 x\, dx$, where x is the distance from the axis of rotation and ρ is the density of the solution (assumed constant).

Introducing into eqn. (4.156) we have, on rearranging,

$$M_1 \omega^2 x\, dx \left(1 - \frac{\overline{V}_1}{M_1}\rho\right) = RT\, d\ln N_1. \tag{4.157}$$

On integrating between limits, we have

$$M_1\left(1 - \frac{\overline{V}_1}{M_1}\rho\right)\omega^2 \int_{x_1}^{x_2} x\, dx = RT \int_{N_1}^{N_2} d\ln N_1,$$

giving
$$M_1 = \frac{2RT \ln(N_2/N_1)}{\{1-(\overline{V}_1/M_1)\rho\}\omega^2(x_2^2 - x_1^2)}. \tag{4.158}$$

For non-ideal solutions, mole fraction N is replaced by the product Nf and for dilute solutions of M_1, mole fractions may be replaced by concentrations, giving

$$M_1 = \frac{2RT \ln(c_2/c_1)}{\{1-(\overline{V}_1/M_1)\rho\}\omega^2(x_2^2 - x_1^2)}, \tag{4.159}$$

† For a very rigorous treatment see Guggenheim (1933), p. 156.

which is the equation commonly employed in the experimental determination of molecular weight by the Sedimentation Equilibrium Method. It will be shown later that this equation is very similar to that used in describing the fall off in the density of the atmosphere with height (p. 189).

Surface phases and the Gibbs adsorption isotherm

We shall first consider only systems with effectively planar interfaces for which the conditions of mechanical and thermal equilibrium, like those for three-dimensional systems, are the constancy of pressure and temperature throughout. Thus let us consider a planar interface (SS) at which two homogeneous phases are initially brought together (Fig. 4.4). Such an interface does not remain completely sharp, but extends over one or more molecular layers in which the properties vary more or less continuously from one phase to the other.

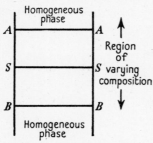

FIG. 4.4. Planar interface.

Consider now two additional planar surfaces AA and BB, on either side of SS such that each is in a region of constant properties of the two bulk phases. The composition of each homogeneous phase is, as normally, stated in terms of the number of moles n'_i, and n''_i of each component. That of the surface (n_i^σ) is given in terms of the excess of each component over the value calculated assuming that the two phases are completely homogeneous up to SS, i.e.

$$n_i^\sigma = n_i - (n'_i + n''_i), \quad \text{etc.} \tag{4.160}$$

By dividing by the area (A) of interface, we derive the *surface excess* or *surface concentration*,

$$\Gamma_i^\sigma = \frac{n_i^\sigma}{A}, \tag{4.161}$$

with which we shall be concerned. Although the quantity is termed surface concentration, it is to be noted that it is dimensionally different from the usual concentration.

In a similar way surface energy and entropy are defined by

$$E^\sigma = E - (E' + E'') \tag{4.162}$$

and
$$S^\sigma = S - (S' + S''). \tag{4.163}$$

Since in general n' and n'' are not equal, it follows from (4.160) that the value of n_i^σ, and therefore of Γ_i^σ, depends upon the exact position of

SS. This is usually chosen arbitrarily as that position which makes Γ^σ for one component (usually the component termed the solvent) zero. The other surface excesses as well as surface energy and entropy are then defined completely.

In a manner similar to that already employed in the three-dimensional case, it may be shown that a small change (dE^σ) in the surface energy may be expressed as

$$dE^\sigma = T^\sigma dS^\sigma + \gamma dA + \sum_i \mu_i^\sigma dn_i^\sigma, \qquad (4.16)$$

which is a complete differential. Integration of eqn. (4.16) (which is equivalent, physically, to an increase in the area under constant conditions) gives

$$E^\sigma = T^\sigma S^\sigma + \gamma A + \sum_i \mu_i^\sigma n_i^\sigma, \qquad (4.164)$$

which on being differentiated becomes

$$dE^\sigma = T^\sigma dS^\sigma + S^\sigma dT^\sigma + \gamma dA + A\, d\gamma + \sum_i \mu_i^\sigma dn_i^\sigma + \sum_i d\mu_i^\sigma n_i^\sigma. \qquad (4.165)$$

Comparison of (4.16) and (4.165) gives

$$S^\sigma dT^\sigma + A\, d\gamma + \sum_i n_i^\sigma d\mu_i^\sigma = 0, \qquad (4.166)$$

an equation which is analogous to the Duhem-Margules equation (4.69).

Dividing by A, we obtain at constant temperature, on rearranging,

$$-d\gamma = \sum_i \Gamma_i\, d\mu_i^\sigma, \qquad (4.167)$$

which is the usual form of the Gibbs adsorption isotherm.

For a binary solution, eqn. (4.167) becomes

$$d\gamma = -\Gamma_1 d\mu_1^\sigma - \Gamma_2 d\mu_2^\sigma \qquad (4.168)$$

and by fixing the interface to give $\Gamma_1 = 0$, we have

$$d\gamma = -\Gamma_2 d\mu_2^\sigma. \qquad (4.169)$$

Now, for complete equilibrium, in addition to constancy of pressure and temperature throughout the system, it is also necessary that the chemical potentials of a given species occurring throughout the system without constraint must be identical, i.e.

$$\mu_i' = \mu_i'' = \mu_i^\sigma. \qquad (4.170)$$

Making use of (4.170) and (4.99) in (4.169) we obtain

$$d\gamma = -\Gamma_2 RT\, d\ln(f_2 N_2), \qquad (4.171)$$

where f_2 and N_2 refer to one of the bulk phases. Eqn. (4.171) is usually written

$$\Gamma_2 = -\frac{1}{RT} \frac{d\gamma}{d\ln(f_2 N_2)}. \qquad (4.171\text{a})$$

For ideal solutions $f_2 = 1$ and we have

$$\Gamma_2 = -\frac{1}{RT}\frac{d\gamma}{d\ln N_2} \qquad (4.172)$$

and, in dilute solutions for which the concentration c_2 is proportional to mole fraction N_2,

$$\Gamma_2 = -\frac{1}{RT}\frac{d\gamma}{d\ln c_2}. \qquad (4.172\,\mathrm{a})$$

From this equation it is apparent that where surface tension increases with solute concentration, the surface excess is a negative quantity. The experimental confirmation and the use of the equation will be treated fully in Chapter XVIII.

Vapour pressure and solubility of colloidal particles

Mention was made in Chapter III of the fact that the vapour pressure and solubility can be modified when the particle size approaches colloidal dimensions, owing to the large specific surface energy (i.e. surface energy/unit volume) which obtains in the colloidal state. In addition to the surface energy the effect of charge also needs consideration, since dispersions, particularly in aqueous and gaseous media, are usually charged to some extent. As already pointed out, the changes in vapour pressure and solubility are merely the visible manifestations of changes in the chemical potential, so it is only necessary to consider the effect of surface energy and charge upon this latter quantity. We shall assume that the surface energy and charge effects are additive so that they may be treated separately.

Effect of surface energy upon the chemical potential

Owing to the contractile tendency arising from the free surface energy (or its mathematical equivalent—the surface tension) there must, for equilibrium, be a compensating excess pressure (p_e) inside a drop. For simplicity we shall consider spherical drops. At equilibrium

$$p_e\,dV = \left(\frac{\partial E_s}{\partial V}\right)dV,$$

where dV is a small increase in the volume and E_s is the free surface energy of the drop. Since

$$E_s = 4\pi r^2\gamma \quad\text{and}\quad V = \tfrac{4}{3}\pi r^3,$$

where γ is the surface tension and r is the radius of the drop, we obtain

$$\frac{\partial E_s}{\partial r} = 8\pi r\gamma, \qquad \frac{\partial V}{\partial r} = 4\pi r^2.$$

$$\therefore \quad \frac{\partial E_s}{\partial V} = \left(\frac{\partial E_s}{\partial r}\right)\left(\frac{\partial r}{\partial V}\right) = \frac{8\pi r \gamma}{4\pi r^2} = \frac{2\gamma}{r},$$

and
$$p_e = \frac{2\gamma}{r}. \tag{4.173}$$

This equation is of frequent occurrence in the discussion of surface tension (e.g. Chapter XVIII).

With one-component condensed systems (solids or liquids), assumed incompressible, we have

$$\frac{\partial \mu_i}{\partial P} = \overline{V}_i, \tag{4.73}$$

where the partial molar volume $\overline{V}_i$ is, by the above assumption, independent of pressure and a function only of temperature. Hence, integrating,

$$\mu_i = \mu_i^*(T) + P\overline{V}_i, \tag{4.174}$$

where $\mu_i^*(T)$ is a function of T only.

In the case of the colloidally dispersed material the pressure term will contain the extra quantity p_e, so denoting the bulk and colloidal material by the superscripts b and c respectively, we have

$$\mu_i^b = \mu_i^*(T) + P\overline{V}_i,$$
$$\mu_i^c = \mu_i^*(T) + (P+p_e)\overline{V}_i.$$

$$\therefore \quad \mu_i^c - \mu_i^b = p_e \overline{V}_i = \frac{2\gamma}{r}\overline{V}_i \tag{4.175}$$

$$= \frac{2\gamma}{r}\frac{M}{\rho}, \tag{4.175a}$$

since $\overline{V}_i = M/\rho$, where M is the molecular weight and ρ the density of the dispersed material.

Eqns. (4.175) and (4.175a) show how the surface energy affects the chemical potential, which is what we require.

Effect of charge upon the chemical potential

Consider a drop of radius r and charge Q (positive), with a compensating double layer at a distance d from the surface of the drop. Let the excess pressure inside the drop at equilibrium be p'.

$$\text{Potential of drop} = \psi = \frac{Q}{Dr} - \frac{Q}{D(r+d)} = \frac{Qd}{Dr(r+d)}, \tag{4.176}$$

where D is the dielectric constant of the medium.

$$\text{Electrostatic energy of drop} = E' = \int_0^Q \psi \, dQ$$

$$= \frac{1}{2}\frac{Q^2}{Dr}\frac{d}{r+d}. \tag{4.177}$$

At equilibrium,
$$p'\, dV = \left(\frac{\partial E'}{\partial V}\right) dV$$
$$= \left(\frac{\partial E'}{\partial r}\right)\left(\frac{\partial r}{\partial V}\right) dV.$$
$$\therefore\ p' = \left(\frac{\partial E'}{\partial r}\right)\frac{1}{4\pi r^2}. \tag{4.178}$$

Two principal cases have to be considered (d is assumed constant throughout):

(a) total charge (Q) constant as r is varied;
(b) surface density of charge (σ) constant as r is varied ($4\pi r^2 \sigma = Q$).

(a) *Q constant*

Differentiating eqn. (4.177) we obtain
$$\frac{\partial E'}{\partial r} = -\frac{1}{2}\frac{Q^2}{D}\left[\frac{2rd+d^2}{r^2(r+d)^2}\right]. \tag{4.179}$$
$$\therefore\ p' = -\frac{1}{2}\frac{Q^2}{D}\left[\frac{2rd+d^2}{r^2(r+d)^2}\right]\frac{1}{4\pi r^2}$$
$$= -\frac{Q^2}{8\pi D r^4}\frac{(2rd+d^2)}{(r+d)^2}. \tag{4.180}$$

Two limiting cases arise:

(1) When $d \gg r$, $\quad p' = -\dfrac{Q^2}{8\pi D r^4}.$ \hfill (4.181)

(2) When $d \ll r$, $\quad p' = -\dfrac{Q^2 d}{4\pi D r^5}.$ \hfill (4.182)

(b) *σ constant*

Substituting $Q = 4\pi r^2 \sigma$ in (4.177) and differentiating with respect to r, we obtain
$$\left(\frac{\partial E'}{\partial r}\right) = \frac{8\pi^2 \sigma^2 d}{D}\left[\frac{2r^3 + 3dr^2}{(r+d)^2}\right]. \tag{4.183}$$
$$\therefore\ p' = \frac{2\pi\sigma^2 d}{D}\frac{(2r+3d)}{(r+d)^2}. \tag{4.184}$$

As above we have the two limiting cases:

(1) When $d \gg r$, $\quad p' = \dfrac{6\pi\sigma^2}{D} = \dfrac{3Q^2}{8\pi D r^4}.$ \hfill (4.185)

(2) When $d \ll r$, $\quad p' = \dfrac{4\pi\sigma^2 d}{Dr} = \dfrac{Q^2 d}{4\pi D r^5}.$ \hfill (4.186)

APPLICATION TO COLLOIDAL SYSTEMS

The effect of the charge upon μ_i can be incorporated just as for the surface energy effect, giving from (4.175) for the change in μ_i due to the charge only

$$\mu_i^c - \mu_i^b = p'\overline{V_i} = p'\frac{M}{\rho}, \qquad (4.187)$$

where p' is given by eqns. (4.180) or (4.186) depending upon the system which is appropriate.

Considering now the case involving *both* surface energy and charge effects we have therefore, assuming additivity,

$$\mu_i^c - \mu_i^b = \left(\frac{2\gamma}{r} + p'\right)\overline{V_i} = \left(\frac{2\gamma}{r} + p'\right)\frac{M}{\rho}. \qquad (4.188)$$

From the above treatment we see that if
(a) the charge on the drop (i.e. Q) is constant, then p' is negative and the charge *opposes* the surface energy effect;
(b) the charge density (i.e. σ) is constant, then p' is positive, and the charge *enhances* the surface energy effect.

So far only the first of these cases (Q constant) appears to have been considered in connexion with colloidal systems, and then only for the special case, $d \gg r$, for which $p' = -Q^2/8\pi Dr^4$. The principal application of this equation is in connexion with condensation of water vapour on gaseous ions (p. 578). However, for aqueous suspensions the second case (σ constant) would appear to be much the more likely, and its bearing upon solubility, vapour pressure, etc., is dealt with in Chapter III.

Effect of surface energy and charge upon solubility and vapour pressure

Consider now the equilibria between colloidal particles and the bulk disperse phase with their saturated solutions (assumed ideal). For the bulk material (in equilibrium with its saturated solution)

$$\mu_i^b = \mu_i^{\text{soln.}} = \mu_i^0(T, P) + RT \ln N_i^b,$$

and for the colloid particles

$$\mu_i^c = \mu_i^{\text{soln.}} = \mu_i^0(T, P) + RT \ln N_i^c,$$

where N_i^b and N_i^c denote respectively the mole fraction of i in solution for equilibrium with bulk and colloidal materials. Subtracting, we obtain

$$\mu_i^c - \mu_i^b = RT \ln (N_i^c/N_i^b), \qquad (4.189)$$

since $\mu_i^0(T, P)$ is the same for both solutions, being the same function of the temperature and pressure only. Hence, from (4.188), we have

$$RT \ln (N_i^c/N_i^b) = \left(\frac{2\gamma}{r} + p'\right)\frac{M}{\rho}, \qquad (4.190)$$

which is the general expression required.

For uncharged particles (4.190) reduces to

$$\frac{RT}{M}\ln\frac{N_i^c}{N_i^b} = \frac{2\gamma}{\rho r}$$

or, as usually given,
$$\frac{RT}{M}\ln\frac{S_r}{S_\infty} = \frac{2\gamma}{\rho r}, \qquad (4.191)$$

where S_r and S_∞ are the respective solubilities for the colloidal particles of radius r and bulk material ($r = \infty$). For dilute solutions, ordinary concentration units (e.g. gm./litre) can be used instead of mole fractions.

For equilibria of colloidal particles and the bulk disperse phase with the vapour phase, instead of $\mu_i^{\text{soln.}}$, we have μ_i^{vapour} which is given by

$$\mu_i^{\text{vapour}} = \mu_i^*(T) + RT\ln p_i,$$

p_i being the saturation vapour pressure. By the same argument as before we therefore find

$$RT\ln\frac{p_i^c}{p_i^b} = \frac{M}{\rho}\left(\frac{2\gamma}{r}+p'\right) \qquad (4.192)$$

for the general case of charged drops of radius r, and

$$\frac{RT}{M}\ln\frac{p_i^c}{p_i^b} = \frac{2\gamma}{\rho r} \qquad (4.193)$$

for the same system when uncharged.

The application of eqns. (4.190) and (4.192) to the solubility and vapour pressure of colloidal particles is given in Chapters III and XX respectively.

The effect of an applied potential upon surface tension (Lippmann's equation)

If an external potential is applied across an interface a change in surface tension is usually observed. For deducing the relationship between the applied potential (**E**) and the change in surface tension the interface is regarded as a parallel plate condenser with the potential difference **E** between the plates. The increase in free energy when the area is increased by an amount dA will be given by

$$dG = \gamma\, dA + \mathbf{E}\, dQ + \sum_i \mu_i\, dn_i, \qquad (4.194)$$

where dQ is the charge associated with area of surface dA, being related by the expression $dQ = \sigma\, dA$, where σ is the surface density of charge.

The first term ($\gamma\, dA$) is the work done against the surface tension (γ), the second ($\mathbf{E}\, dQ$) the increase in electrical energy, the third the chemical contribution.

Integrating (cf. p. 85),
$$G = \gamma A + \mathbf{E}Q + \sum \mu_i n_i. \tag{4.195}$$

Differentiating generally,
$$dG = \gamma\, dA + A\, d\gamma + \mathbf{E}\, dQ + Q\, d\mathbf{E} + \sum_i \mu_i\, dn_i + \sum_i n_i\, d\mu_i, \tag{4.196}$$

and subtracting (4.194) from (4.196), we obtain (cf. eqn. (4.69))
$$A\, d\gamma + Q\, d\mathbf{E} + \sum_i n_i\, d\mu_i = 0. \tag{4.197}$$

At constant composition this reduces to
$$A\, d\gamma + Q\, d\mathbf{E} = 0$$
or $\qquad (\partial\gamma/\partial\mathbf{E})_{n_{i,j\ldots}} = -Q/A = -\sigma. \tag{4.198}$

This is Lippmann's equation.

Differentiating (4.198) with respect to $\mathbf{E}$,
$$(\partial^2\gamma/\partial\mathbf{E}^2)_{n_{i,j\ldots}} = -C, \tag{4.199}$$

since $(\partial\sigma/\partial\mathbf{E}) = C$, where C is the capacity per unit area of the double layer 'condenser'.

An outline of the more practical aspects of electro-capillary phenomena is given later in Chapter XVIII (p. 531).

The application of thermodynamics to problems involving a low molecular weight solvent and a polymer

As yet thermodynamic results have been mainly applied to single- and two-component systems or to two liquids separated by a semi-permeable membrane. However, the criteria of equilibria which we have developed are applicable to many phase systems generally and, of such systems, we now wish to consider those involving polymers and simple liquids. Amongst such equilibria we shall consider briefly the question of the solubility and swelling of polymeric materials and their separation into fractions homogeneous in molecular weight or other properties. A more detailed treatment will be given at a later stage (Chapter XXVII).

First of all, however, we ought to add a word of warning. As has been indicated, thermodynamics is concerned with equilibrium states of a system, e.g. a liquid and its saturated vapour, but not a liquid and the vapour above it which has not yet reached its equilibrium pressure. Thus where the time factor is important, it is essential to make use thermodynamically only of those properties which have reached equilibrium values.

With polymeric materials and colloidal systems generally, their peculiar molecular structure and coarse aggregation, as compared with simple solutions, often results in very slow rates of reaching equilibrium both in chemical and physical processes. Thus, if due care is not taken, it is quite easy to mistake for an equilibrium state one from which the rate of change is so small as to be observable only with difficulty. In such cases, where possible, it is wise to arrange experiments so that equilibrium is approached from opposite directions, e.g. from low and high concentrations. The equilibrial state must be independent of the method of attaining it.

In this section we do not intend to go into the details of the treatment of polymer-liquid systems, but merely wish to outline the thermodynamic treatment.

If a small quantity of polymer (2) be added to a liquid (1) in which it dissolves, it follows that the chemical potential of the polymer in solution (μ_2) is smaller than in the pure component (μ_2^0), i.e.

$$\mu_2 < \mu_2^0,$$

and making use of eqn. (4.91) and of the identity of chemical potentials and partial molar Gibbs free energies (eqn. (4.40 c)), we have

$$\mu_2 - \mu_2^0 = RT \ln a_2 = \Delta \bar{G}_2 = \Delta \bar{H}_2 - T \Delta \bar{S}_2 < 0, \qquad (4.200)$$

i.e. $$\ln a_2 < 0 \quad \text{or} \quad T \Delta \bar{S}_2 > \Delta \bar{H}_2. \qquad (4.200\,\text{a})$$

Similarly, if on adding a small quantity of liquid (1) to the pure polymer, the liquid dissolves in or is taken up by the polymer, it follows that

$$\mu_1 < \mu_1^0$$

and $$\mu_1 - \mu_1^0 = RT \ln a_1 = \Delta \bar{G}_1 = \Delta \bar{H}_1 - T \Delta \bar{S}_1 < 0, \qquad (4.201)$$

i.e. $$\ln a_1 < 0 \quad \text{or} \quad T \Delta \bar{S}_1 > \Delta \bar{H}_1. \qquad (4.201\,\text{a})$$

If $\ln a_1$ decreases continuously with increasing mole or volume fraction (N_2 and ϕ_2) of the polymer (or $\ln a_2$ with increasing N_1 or ϕ_1) (Fig. 4.5 (a) and 4.5 (b)), as for the system rubber–benzene, then the coexistence of two phases of different composition is impossible, since there could not be identity of the chemical potentials of the components in the two phases. Such a plot of $\ln a_1$ against mole fraction, we have seen, can be obtained by direct vapour pressure measurements of the solvent (component 1) over the mixture or, more indirectly, by the evaluation of the increases in partial heat content ($\Delta \bar{H}_1$) and entropy ($\Delta \bar{S}_1$). Further details will, however, be provided in Chapters VII and XXVII.

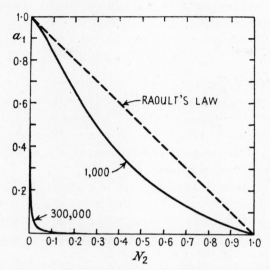

FIG. 4.5 (a). Activity of solvent (1) as a function of molecular fraction of solute (2) for rubber–benzene system.
Left-hand curve: molecular weight of rubber = 300,000.
Middle curve: molecular weight of rubber = 1,000.

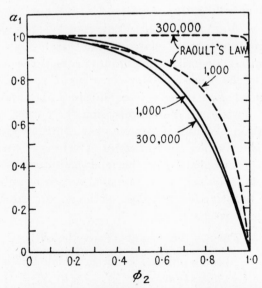

FIG. 4.5 (b). Activity of solvent (1) as a function of volume fraction of solute (2) for rubber–benzene system.
Full lines—experimental curves.
Broken lines—calculated from Raoult's law.

Two phases containing different concentrations of the two components can only coexist in *equilibrium* if the chemical potentials of both components are alike in the two phases, i.e.

$$\mu_1' = \mu_1'' \quad \text{and} \quad \mu_2' = \mu_2''.$$

This is the basis of the thermodynamic treatment of the equilibrium swelling of polymers by organic liquids, of their limited solubility, and of fractionation procedures, in which a solid or gel phase is caused to separate out from a previously homogeneous solution of the polymer by some alteration of the properties of the solvent, e.g. addition of precipitant, or lowering of temperature.

With the rubber–benzene system the identity of chemical potentials in phases of different composition is impossible since, as we have seen, $\ln a_1$ decreases continuously with increasing mole fraction of polymer. With other systems this is only possible if either a maximum or a minimum exists in the curve of $\ln a_1$ or $\Delta \bar{G}_1$ (and $\ln a_2$ and $\Delta \bar{G}_2$) against concentration.

A good example has been shown by Gee (1942) to occur with rubber and mixtures of benzene and methyl alcohol. Here Δg_1 and Δg_2 are the increase in total Gibbs free energy caused by the addition of 1 gm. of solvent (1) and of rubber (2) respectively to a large amount of the mixture ($\Delta g_1 = \mu_1/M_1$, $\Delta g_2 = \mu_2/M_2$). Values of Δg_1 and Δg_2 corresponding to the same composition are related by the Duhem-Margules equation (4.69) and it may be shown that a maximum in one curve is accompanied by a minimum in the other and vice versa (see Fig. 4.6).

The compositions of two phases in equilibrium together are given by the extremes of horizontal lines intersecting the curves in the central region (cf. equation of state of vapour near central region). Since at these positions the chemical potentials or activities of each component are the same in the two phases, our thermodynamic criteria of equilibria are obeyed. It is to be noted that in such a case as rubber–benzene, Δg_1 and Δg_2 are negative for all values of the volume fraction ϕ_2 of the polymer, and hence we have

$$T\Delta \bar{S}_1 > \Delta \bar{H}_1 \quad \text{and} \quad T\Delta \bar{S}_2 > \Delta \bar{H}_2. \tag{4.202}$$

However, cases of phase equilibria occur in which this is not true for the whole of the concentration range ($N_2 = 0$ to $N_2 = 1$). Rather, for certain portions of the plot of $\Delta \bar{G}_1$ against composition, we have

$$\Delta \bar{G}_1 = \Delta \bar{H}_1 - T\Delta \bar{S}_1 > 0,$$
and
$$\Delta \bar{G}_1 = RT \ln a_1 > 0 \quad (\text{i.e. } a_1 > 1). \tag{4.203}$$

This is shown in the upper curve of the family shown in Fig. 4.7, obtained by Huggins (1943) from the theoretical evaluation of $\Delta \bar{G}_1$ from $\Delta \bar{H}_1$ and $\Delta \bar{S}_1$. The quantity μ, used by Huggins in this diagram may be taken to indicate the magnitude of $\Delta \bar{H}_1$, high positive values (e.g. $\mu = 1$) thus giving $\Delta \bar{H}_1 > T\Delta \bar{S}_1$, in accordance with eqn. (4.203). It

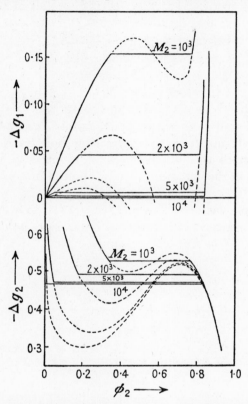

Fig. 4.6. Gibbs free energies of dilution (Δg_1) and solution (Δg_2) as functions of volume fraction of polymer (2) for rubber in benzene-methyl alcohol mixture.

must, however, be noted that the portion of the curve for which $\Delta \bar{G}_1$ and $a_1 > 1$ cannot be experimentally realized, and for compositions which fall within this region, two phases coexist, one being almost pure solvent and the other having a composition given by the intersection of the line $a_1 = 1$ with the activity curve. For the latter phase, since $a_1 = 1$, it follows that

$$T\Delta \bar{S}_1 = \Delta \bar{H}_1.$$

For compositions outside the range in which $a_1 \geqslant 1$, one swollen polymer phase only will exist as Fig. 4.7 demonstrates.

These last considerations, characterized by $\Delta \bar{H}_1 > T\Delta \bar{S}_1$ at certain concentrations, are applicable to those polymer-liquid systems in which swelling of the polymer occurs without solution and in the relations of

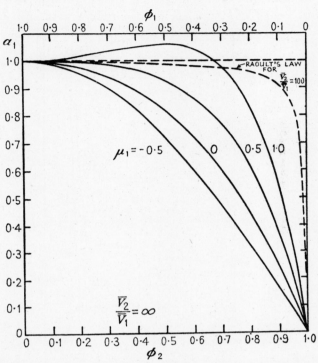

FIG. 4.7. Activity of solvent (1) as a function of the volume fraction of polymer (2) in a polymer-solvent mixture. Polymer assumed to be of infinite molecular weight.

cross-linked polymers (e.g. vulcanized rubber) generally with organic liquids.

Similarly in the case of fractionation, another example of phases in equilibrium, it is possible, by the use of the same criteria of equilibrium, to work out the composition of a phase which separates out under certain conditions from a previously homogeneous solution of a mixed polymer. For this purpose it is necessary to possess information on the variation with concentration and temperature of the free energy and chemical potential of the different species within the polymer. Considerable difficulties are, however, encountered in treating this problem theoretically (see p. 796) and a satisfactory treatment is not yet available.

APPLICATION TO COLLOIDAL SYSTEMS

In all these polymer-liquid systems, the theoretical advances occurring during the last five or six years have greatly helped our understanding. By the use of statistical methods it is possible to calculate the entropy of mixing of a polymer with a liquid, making use of model polymer structures to suit particular types of system. A combination of these results with semi-empirical expressions for heat content allows a calculation of Gibbs free energies and chemical potentials from which we can make predictions regarding the solubility, swelling, and fractionation of high polymers. Since the work is so important we shall return to it when we discuss high polymers in more detail (Chapter XXVII).

Thermodynamics and the mechanical properties of polymers

We have seen that in general the Helmholtz free energy may be written

$$F = \phi(T, V, n_i) = E - TS \tag{4.23}$$

and a small change in it

$$dF = -p\, dV - S\, dT + \sum_i \mu_i\, dn_i. \tag{4.41}$$

These expressions may readily be modified to cover the case of the extension of a rubber-like solid which occurs without volume change. Thus, since no change in composition occurs, the term involving chemical potentials may be omitted, and further, since no change in volume or pressure occurs during extension, we may replace p and V by tension Z (with opposite sign) and length L. Thus we may write

$$F = E(L, T) - TS(L, T). \tag{4.204}$$

Differentiating with respect to length L at constant temperature,

$$\left(\frac{\partial F}{\partial L}\right)_T = Z(L, T) = \left(\frac{\partial E}{\partial L}\right)_T - T\left(\frac{\partial S}{\partial L}\right)_T. \tag{4.205}$$

But, from eqn. (4.42 b),

$$-\left(\frac{\partial S}{\partial L}\right)_T = \frac{\partial}{\partial L}\left(\frac{\partial F}{\partial T}\right)_L = \left(\frac{\partial Z}{\partial T}\right)_L, \tag{4.206}$$

and substituting above we have

$$Z(L, T) = \left(\frac{\partial E}{\partial L}\right)_T + T\left(\frac{\partial Z}{\partial T}\right)_L. \tag{4.207}$$

This equation has been used as the basis of experiments on rubber, from which the tension and its variation with temperature for a constant length of specimen have been measured (e.g. see Fig. 4.8). Thus it has been shown empirically (e.g. Meyer and Ferri (1935), Hauk and Neumann (1938)) that there is a linear variation of tension at constant length with

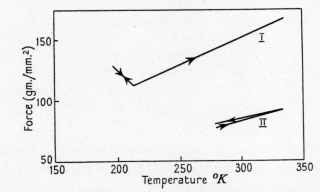

Fig. 4.8. Force-temperature curves of moderately vulcanized rubber (8 per cent. S) at 350 per cent. (curve I) and 170 per cent. (curve II) elongation. (*From* the figure of Meyer and Ferri (1935).)

temperature, which held up to extensions in excess of 400 per cent. over a 50° range of temperature, i.e.

$$Z = b + aT, \qquad (4.208)$$

where a and b were empirically determined. Comparison with eqn. (4.207) gives

$$\left(\frac{\partial Z}{\partial T}\right)_L = -\left(\frac{\partial S}{\partial L}\right)_T = a \quad \text{and} \quad \left(\frac{\partial E}{\partial L}\right)_T = b. \qquad (4.209)$$

It was thus shown that the constant b (i.e. the increase of internal energy with extension) was small at low values of the extension, confirming the kinetic view (see p. 782) that the stretching of rubber causes an uncoiling of coiled chains rather than a distortion of normal bonds. Similarly the value of a was considerably larger, i.e. $\partial S/\partial L$ had a large negative value, indicating that, at low extensions, resistance to stretching is due to the decrease in entropy on extension, rather than to energy effects. At very high extensions, however, entropy changes become less and energy changes more important.

Theoretically the above conclusions may also be checked by the observation of the heat changes during extension. From the definition of entropy (eqn. (4.5)) for isothermal extensions we have

$$\left(\frac{\partial S}{\partial L}\right)_T = \frac{1}{T}\left(\frac{\partial q}{\partial L}\right)_T \qquad (4.210)$$

and for the temperature change during an adiabatic extension (see Guggenheim (1933), p. 39, eqn. (182))

$$\left(\frac{\partial T}{\partial L}\right)_S = \frac{T}{c_L}\left(\frac{\partial Z}{\partial T}\right)_L, \qquad (4.211)$$

where c_L is the heat capacity at constant length. But, substituting

$$-\left(\frac{\partial S}{\partial L}\right)_T = \left(\frac{\partial Z}{\partial T}\right)_L$$

from (4.206), we obtain

$$\left(\frac{\partial T}{\partial L}\right)_S = -\frac{T}{c_L}\left(\frac{\partial S}{\partial L}\right)_T. \qquad (4.212)$$

Thus the temperature changes on adiabatic extension may be used to calculate the changes in entropy during isothermal extension.

However, the temperature changes are small and not easily measurable and as yet good agreement between the two methods has not been obtained. It is, however, of importance to see the link between the two types of experiment.

REFERENCES

ADAIR, 1929, *Proc. Roy. Soc.* A **126**, 16.
BERKLEY, HARTLEY, and BURTON, 1919, *Philos. Trans.* **218**, 295.
BUTLER, 1934, *The Fundamentals of Chemical Thermodynamics*, Part II, Macmillan & Co.
GEE, 1942, *Trans. Faraday Soc.* **38**, 276.
GLASSTONE, 1937, *The Electrochemistry of Solutions*, 2nd edn.
GUGGENHEIM, 1933, *Modern Thermodynamics*, Methuen.
HAUK and NEUMANN, 1938, *Z. phys. Chem.* A **182**, 285.
HUGGINS, 1943, *Industr. Engng. Chem.* **35**, 216.
MEYER and FERRI, 1935, *Helv. chim. Acta*, **18**, 570.
PARTINGTON, 1940, *Chemical Thermodynamics*, Constable.

V

THE APPLICATION OF DEBYE-HÜCKEL THEORY TO COLLOIDAL PARTICLES. STABILITY AND ITS ORIGIN

The Application of Debye-Hückel Theory to Colloidal Particles

The general nature of the influence of the charge carried by a suspended colloidal particle upon its stability has been indicated earlier, and in the next section some mention will be made of the quantitative attempts to correlate stability with charge (or ζ potential). Closely related, in the sense that electrical effects are of major importance, are the surface activity and the aggregation of soaps and other colloidal electrolytes (see Chapter XXIV).

For the consideration of Debye-Hückel effects, i.e. those effects of electrical origin due to the charge carried by the colloidal particle and its equivalent gegenions, colloidal systems can for simplicity of treatment be divided into two size ranges:

(a) particles up to *ca.* 100 A diameter;

(b) particles from *ca.* $0.01\,\mu$ (100 A) to *ca.* $1\,\mu$ diameter.

The first range would include the colloidal electrolytes (soaps and dyestuffs) and the majority of proteins, the second most of the hydrophobic suspensions (for which electrical effects are so important), such as the metal sols and oil suspensions. The reason for the above subdivision is that only with the first class is it possible to observe osmotic effects sufficiently accurately for comparison with the predictions of the Debye-Hückel theory or its subsequent extensions. With the second class the only experimental quantity available for comparison with theory is the mobility (or other related electrokinetic quantities). The calculation of activity coefficients will be restricted to colloidal electrolytes since these are likely to carry much higher charges than the other members of the group and thus to represent the extreme case of type (a). As regards mobilities both size ranges will be considered.

Compared with simple electrolytes all colloidal particles are of very much greater radius and usually of very much higher valence (e.g. a soap micelle of say 40 A diameter may contain 50 long-chain univalent ions and so should have a valence of 50), and before considering them it is essential to outline briefly the major points of the Debye-Hückel theory as applied to low-valence electrolytes.

General theory (outline)

The 'anomalies of strong electrolytes', and in particular the complete breakdown of the Ostwald Dilution Law which is based upon the simple Law of Mass Action (i.e. without activity coefficients), led to numerous attempts to calculate the deviations from ideal behaviour from considerations of the electrical (Coulomb) forces involved. As compared with those between non-electrolytes the forces between ions fall off with distance much less rapidly, so that a purely random arrangement which is taken as the standard for 'ideality' in the thermodynamic treatment of solutions (see p. 71) is only obtained at very much lower concentrations in solutions of electrolytes. (E.g. in 0·01 M. solution the activity coefficient of acetic acid is still sensibly unity, whereas that for KCl is 0·90.) In electrolyte solutions, since the amounts of positive and negative charges must necessarily be equal, the separation of any one ion from the rest divides the solution into two parts with charges equal in magnitude but opposite in sign.

The concept of the 'ion atmosphere' introduced by Debye and Hückel (1923) enabled the effect of the Coulomb forces upon the activity coefficient of an ion to be calculated in a relatively simple manner, although as far as colloidal particles are concerned the idea of a diffuse 'atmosphere' or double layer of gegenions surrounding the particle or charged surface had been introduced much earlier (Gouy, 1910). The simplification introduced by Debye and Hückel was to select one ion (the central ion c), and to consider the distribution around it of the equivalent opposite charge or 'ion atmosphere', under the opposing influences of the Coulomb forces tending to form an ordered arrangement (as in an ionic lattice), and the thermal forces tending to give a purely random (ideal) distribution.

The Boltzmann equation is first used to relate the concentration (c_i) of ions of type i with the electrical potential (ψ) at a distance r from the central ion, i.e.
$$c_i = \bar{c}_i e^{-z_i e \psi / kT}, \tag{5.1}$$
where $\bar{c}_i$ is the concentration of ions of type i when $\psi \to 0$ (i.e. when $r \to \infty$). For the present purposes it is convenient to measure concentrations in gm. equivalents/c.c., reckoned negative for negative ions. (Using gm. equivalents the condition for electrical neutrality is $\sum_i \bar{c}_i = 0$, whereas using gm. ions it is $\sum_i z_i \bar{c}_i = 0$.)

The potential is related to the charge density (ρ) by the Poisson differential equation
$$\nabla^2 \psi = -\frac{4\pi \rho}{D}, \tag{5.2}$$

where the L.H.S. can be replaced by the alternative forms

$$\frac{1}{r^2}\frac{d}{dr}\left(r^2\frac{d\psi}{dr}\right) \quad \text{or} \quad \frac{d^2\psi}{dr^2}+\frac{2}{r}\frac{d\psi}{dr}.$$

The charge density $\qquad \rho = N\sum_i c_i e,$

hence, substituting for c_i from (5.1), we obtain

$$\frac{1}{r^2}\frac{d}{dr}\left(r^2\frac{d\psi}{dr}\right) = -\frac{4\pi Ne}{D}\sum \bar{c}_i e^{-z_i e\psi/kT}. \tag{5.3}$$

This is the fundamental differential equation of Debye and Hückel.

If now $z_i e\psi \ll kT$, i.e. the electrical forces are small compared with the thermal ones and so the deviations from ideality are also small, then the exponential can be replaced by $(1-z_i e\psi/kT)$ and eqn. (5.3) becomes

$$\frac{1}{r^2}\frac{d}{dr}\left(r^2\frac{d\psi}{dr}\right) = \frac{4\pi Ne^2}{DkT}\sum \bar{c}_i z_i \psi, \tag{5.4}$$

since by the condition of electrical neutrality the first term is zero.

This is readily integrated, yielding

$$\psi = \frac{Ae^{-\kappa r}}{r}+\frac{Be^{\kappa r}}{r}, \tag{5.5}$$

where $\kappa^2 = 8\pi Ne^2 I/DkT$ and the ionic strength $I = \frac{1}{2}\sum z_i c_i$. (If, as is more usual, concentrations are in gm. ions/litre, then the ionic strength $I = \frac{1}{2}\sum z_i^2 c_i$ and $\kappa^2 = 8\pi Ne^2 I/1000DkT$.) [For water at 25° C., $\kappa = 0.327 \times 10^8\sqrt{I}$ cm.$^{-1}$ Thus $1/\kappa$ which has the dimensions of a length $\simeq 3/\sqrt{I}$ A.] B must clearly be zero since $\psi \to 0$ as $r \to \infty$. The constant A is obtained as follows. Eqn. (5.6) below, which is based on Coulomb's law, gives the gradient of potential, i.e. the force on a unit charge placed at r, due to the charge within the sphere of radius r, which we will assume to include part of the atmosphere as well as the central ion, the net valence being $\bar{z}$. ($\bar{z} < z_c$ as the central ion and atmosphere have opposite charges.)

$$\frac{d\psi}{dr} = -\frac{\bar{z}e}{Dr^2}. \tag{5.6}$$

When $r = a$, the distance of closest approach of two ions, then $\bar{z} = z_c$, the valence of the central ion, and $(d\psi/dr)_{r=a} = -z_c e/Da^2$. Differentiating eqn. (5.5) and substituting $r = a$, we find

$$A = \frac{z_c e}{D}\frac{e^{\kappa a}}{1+\kappa a},$$

COLLOIDAL PARTICLES. STABILITY AND ITS ORIGIN

and eqn. (5.5) becomes

$$\psi_{c,r} = \frac{z_c e}{D} \frac{e^{\kappa a}}{1+\kappa a} \frac{e^{-\kappa r}}{r} \tag{5.7a}$$

$$= \frac{z_c e}{D(1+\kappa a)} \frac{e^{\kappa(a-r)}}{r}, \tag{5.7b}$$

where $\psi_{c,r}$ is the potential at distance r due to the ion c and its atmosphere. Eqn. (5.7) is the well-known solution of Debye and Hückel.

When $r = a$, $\quad \psi_{c,a} = \dfrac{z_c e}{D}\left(\dfrac{1}{a} - \dfrac{1}{a+1/\kappa}\right),$ \hfill (5.8)

i.e. the residual charge or 'atmosphere' acts as if distributed over the surface of a sphere centred on the ion c and of radius $1/\kappa$ (strictly speaking of radius $a+1/\kappa$, where a is the distance of closest approach of two ions). The above equation also shows that only when $1/\kappa \to \infty$, i.e. when the concentration tends to zero, does $\psi_{c,a}$ have its ideal value of $z_c e/Da$. Otherwise $\psi_{c,a}$ is lowered, showing that the removal of the central ion from its equilibrium position requires work.

The relationship between the potential and the measurable properties of the solution can be obtained as follows. In the absence of its atmosphere the central ion would behave ideally, and hence (see p. 71)

$$\mu_c^{\text{actual}} - \mu_c^{\text{ideal}} = kT \ln f_c,$$

where μ_c is the chemical potential and f_c the activity coefficient of ion c. This difference in chemical potential (or free energy) will be equivalent to the electrical work done in building up the atmosphere on addition of a *single ion* to the assembly of ions. The discharged ion may be imagined added to the assembly and its charge then gradually increased from zero to $z_c e$.

$$\therefore \quad kT \ln f_c = \int_0^{z_c e} \psi \, dQ \quad \text{(where } \psi \text{ is the potential due to the atmo-}$$

sphere, and the element of charge added $dQ = e\,dz$)

$$= -\int_0^{z_c e} \frac{z_c e}{D} \frac{1}{a+1/\kappa} e \, dz \quad \text{(on substituting for } \psi \text{ and } dQ\text{)}$$

$$= -\frac{1}{2} \frac{z_c^2 e^2}{D} \frac{\kappa}{1+\kappa a}.$$

$$\therefore \quad -\ln f_c = \frac{z_c^2 e^2}{2DkT} \frac{\kappa}{1+\kappa a}. \tag{5.9}$$

Considering both ions of valencies z_+ and z_- eqn. (5.9) becomes

$$-\ln f_{\pm} = \frac{z_+ z_- e^2}{2DkT} \frac{\kappa}{1+\kappa a}, \tag{5.10}$$

where $f_{\pm}$ is the mean activity coefficient.

When $\kappa a \ll 1$, $\qquad -\ln f_{\pm} \propto z_+ z_- \sqrt{I}$, $\qquad (5.11)$

a relation which has been verified by many investigators with electrolytes of low valence type.

Limitations and extensions of the elementary theory

The treatment given below is based upon the simplified accounts of Scatchard (1933), Onsager (1933), and in particular Hartley (1935), of whom the last named has considered in detail the question of ionic atmosphere effects in colloidal electrolytes.

Of the assumptions made in deducing eqn. (5.10) the most obvious one was that all the $z_i e\psi/kT$'s were $\ll 1$, and since this has been frequently assumed when applying Debye-Hückel theory to colloidal particles its validity needs consideration. From eqn. (5.7) $\psi_{c,r}$ is approximately equal to $z_c e/Dr$ for small values of r, so that

$$\frac{z_i e\psi}{kT} \simeq \frac{z_i z_c e^2}{DkTr} \tag{5.12 a}$$

$$\simeq \frac{z_i z_c}{r} \cdot 7 \cdot 10^{-8} \quad \text{for water at } 25^\circ \text{ C.} \tag{5.12 b}$$

Hence even with a 1.1 (uni-univalent) electrolyte r must exceed 7×10^{-8} cm. (i.e. 7 A) before $z_i e\psi/kT$ is < 1, and 70 A before $z_i e\psi/kT$ is < 0.1, which we will take as the standard for $z_i e\psi/kT \ll 1$. At first sight, therefore, it appears surprising that the simple theory is so successful.

In order to examine this question Hartley has developed Scatchard's method of analysing the distribution around the central ion. In Fig. 5.1 the abscissa represents the distance from the central ion as a function of κr (i.e. in units of $1/\kappa$), and the ordinates are respectively:

(1) $\bar{z}/z_c$, i.e. the ratio of valence of sphere of radius r to valence of central ion. When $a = 0$, $\bar{z}/z_c = (1+\kappa r)/e^{\kappa r}$, by differentiating (5.7) and comparing with (5.6).

(2) $\dfrac{1}{z_c} \dfrac{d\bar{z}}{d(\kappa r)}$, i.e. the fraction of the ion atmosphere contained in a shell of radius r divided by the thickness of the shell. When $a = 0$,

$$\frac{1}{z_c} \frac{d\bar{z}}{d(\kappa r)} = \frac{\kappa r}{e^{\kappa r}}.$$

The maximum of this curve occurs at $\kappa r = 1$, i.e. when $r = 1/\kappa$.

(3) $\dfrac{1}{z_c}\dfrac{1}{\kappa r}\dfrac{d\bar{z}}{d(\kappa r)}$, i.e. the fractional effect on the potential at the central ion produced by that part of the atmosphere further away than r.

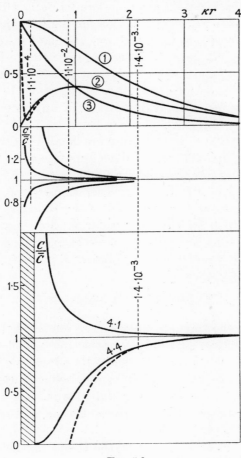

Fig. 5.1.

When $a = 0$, $\dfrac{1}{z_c}\dfrac{1}{\kappa r}\dfrac{d\bar{z}}{d(\kappa r)} = \dfrac{1}{e^{\kappa r}}$. The total area under curves (2) and (3) is therefore unity. For finite values of a, the portions of the curve to the left of κa must be eliminated; i.e. the ordinate must be displaced from $\kappa r = 0$ to $\kappa r = \kappa a$.

Consider first the case of a 1.1 electrolyte in $N/10{,}000$ aqueous solution. $1/\kappa$ is about 300 A (see p. 102), and from eqn. (5.12 b) $ze\psi/kT = 0.1$ at about 70 A. The Debye-Hückel approximation will therefore hold for values of κr above ca. 0.2. This is indicated in Fig. 5.1

as the dotted ordinate $1\cdot1.10^{-4}$ and it will be seen that over 80 per cent. of the contribution of the atmosphere to the potential is made by that part of it to which the Debye-Hückel eqn. (5.7) would apply. Errors in this equation for $\kappa r < 0\cdot2$ cannot therefore seriously affect the overall picture. At higher concentrations and for higher valence types the portion of the atmosphere for which our standard ($z_i e\psi/kT < 0\cdot1$) is valid will decrease, as shown by the dotted ordinates in Fig. 5.1 for 1.1 electrolyte at $N/100$ and 1.4 electrolyte at $N/1,000$. Under such conditions increasing deviations from eqn. (5.10) are to be expected and these will become increasingly important in any practical range of concentration.

It is clearly essential to consider in more detail the distribution close to the central ion, and this can be done approximately by using eqns. (5.1) and (5.12) since the atmosphere terms in eqn. (5.7) are small for low values of r. The calculated concentrations (c) of the anion and cation around a central ion (expressed as the ratio $c/\bar{c}$, where $\bar{c}$ is the average concentration for the whole solution) are shown in the lower part of Fig. 5.1. Using these results to reconstruct curve (2) of Fig. 5.1 a radical change is observed at low values of r, namely the appearance of a sharp minimum when

$$r_{\min} = \frac{z_+ z_- e^2}{2DkT} \simeq 3\cdot5 z_+ z_- \text{ A.} \tag{5.13}$$

The dotted curve shown is an approximate one for 1.1 electrolyte in $N/100$ aqueous solution. It will be seen that more of the atmosphere is to be found near the central ion, where its effect on the potential is greater than would be calculated from the Debye-Hückel approximation. The error due to this cause is negligible in very dilute solutions but clearly will increase with the concentration and with increasing valence.

Bjerrum (1926), in his extension of the Debye-Hückel theory, treats this inner part of the atmosphere (i.e. where $z_i e\psi/kT$ is $not \ll 1$) separately. He regards an oppositely charged ion found within the radius at the minimum as forming an ion-pair (i.e. a temporary association held together solely by electrostatic forces), and calculates the number of such pairs from the law of mass action. This picture has a reasonable physical basis since the energy required to separate ions so close together exceeds $4kT$. The effect of the electrostatic field of an associated pair on the remaining ions is ignored, and Debye's theory is applied to the free ions which are assumed to have an effective diameter of $r_{\min}$, since if they come closer than $r_{\min}$ they cease to count as free.

COLLOIDAL PARTICLES. STABILITY AND ITS ORIGIN

The problem of the complete solution of the fundamental differential equation (5.3) has been attacked in two ways. Gronwall, La Mer, and Sandved (1928) obtained a complete solution in the form of a converging infinite series, of which the first term was identical with the solution of Debye and Hückel. They completed the numerical calculations of sufficient terms of this series to make it applicable to symmetrical electrolytes (e.g. KCl, $MgSO_4$), and later (La Mer, Gronwall, and Grieff, 1931) extended the treatment to unsymmetrical valence types.

Müller (1927) attacked the problem by graphical approximation of the unexpanded exponential (eqn. (5.3)). For the present discussion this method is to be preferred on account of its simplicity. All of the above three methods give answers which are practically identical, and show a better agreement with experiment than the original Debye-Hückel approximation.

One consequence of the Debye-Hückel approximation is that the *total* number of ions at any point of the atmosphere is constant, and hence the charge contribution from each ion is directly proportional to its valence. This can readily be shown by considering the distribution around one ion, say the cation.

Then $c_+ = \bar{c}_+\left(1 - \dfrac{z_+ e\psi}{kT}\right)$, from the Debye-Hückel approximation for eqn. (5.1).

Charge contribution from cation $= c_+ - \bar{c}_+ = -\bar{c}_+ \dfrac{z_+ e\psi}{kT}$ (negative as z_+ is positive).

Similarly, charge contribution from anion $= c_- - \bar{c}_- = -\bar{c}_- \dfrac{z_- e\psi}{kT}$ (positive as z_- is negative).

$$\therefore \frac{\text{charge due to cation}}{\text{charge due to anion}} = \frac{z_+}{z_-},$$

since, as the concentrations are in equivalents, $\bar{c}_+ = \bar{c}_-$.

Number of cations leaving $= \dfrac{\text{charge contribution}}{z_+ e} = \bar{c}_+ \dfrac{\psi}{kT}$.

Number of anions entering $= \dfrac{\text{charge contribution}}{z_- e} = \bar{c}_- \dfrac{\psi}{kT}$

$=$ number of cations leaving since $\bar{c}_+ = \bar{c}_-$.

This proves the statements given above.

Thus considering any point in the atmosphere around the tetravalent cation in the case of a 4.1 electrolyte, $\frac{4}{5}$ of the charge should come from the *emigration* of cations and $\tfrac{1}{5}$ from the *immigration* of the univalent anions. While this is true for the outer parts of the atmosphere it is certainly not so near the central ion where the ionic concentrations may deviate considerably from their bulk values, because the ratio $c/\bar{c}$ for the tetravalent ion cannot be reduced below zero whereas that for the univalent can, and does, exceed $5/4$ (the ratio calculated for zero concentration of tetravalent ion). The dotted curve in the lowest graph of Fig. 5.1 shows how the concentration of tetravalent ion *should* vary to keep its contribution four times that of the univalent ion; the thick lines give the distribution of the univalent and tetravalent ions about the tetravalent ion. The former reaches a limiting value of about 7 (when $\kappa r = \kappa a$), as contrasted with the value of $5/4$ from Debye-Hückel theory. In the intermediate region between the shaded area and the dotted ordinate the tetravalent ion is unable to make its requisite contribution to the atmospheric charge, and this provides some compensation for those errors previously discussed which tend to make f lower than calculated from the Debye-Hückel approximation. As we shall see below, this effect is very marked with high-valent ions of colloidal size.

The breakdown of the ionic strength principle

According to eqn. (5.11) $-\ln f$ is proportional to κ, which by definition is proportional to the square root of the ionic strength. When considering deviations from the simple theory, however, i.e. when taking into account higher powers in the exponential in eqn. (5.3), further functions of the valencies and concentrations must clearly occur. This may be clarified by reference to Fig. 5.2, which shows the distribution around a central tetravalent *cation* of two solutions, one of $4^+ . 1^-$ electrolyte, the other $1^+ . 4^-$ electrolyte, of the same ionic strength, and which should therefore give similar atmospheres and activity coefficients. Now the *total* charge of the atmosphere must be the same in both cases, but with the latter solution, owing to the high concentration of 4^- ion near the central 4^+ ion, the atmosphere is closer, and therefore more effective, than for the former. Hence a 4.1 electrolyte present in large excess of a 1.4 electrolyte will have a lower activity coefficient than when present alone at the same total concentration and *the same ionic strength*. This phenomenon will be more marked with the more unsymmetrical electrolytes, so that very high-valent micelles would become almost indifferent to the valence of ions of like sign, but extremely sensitive to that of ions of opposite sign. A similar conclusion on a closely related system was

COLLOIDAL PARTICLES. STABILITY AND ITS ORIGIN

reached by Cassie and Palmer (1941), namely that the effect of salts on monolayers of long-chain anions will be independent of the nature of the anions in solution (see also p. 682). Also the effect of salts upon the coagulation of colloids, as expressed by the Schulze-Hardy rule (p. 15), can clearly be interpreted on the above considerations.

Although the ionic strength principle would thus appear to break down completely with multivalent colloidal ions, yet there are conditions of great practical interest when it is still applicable, namely when the colloidal particle or ion is present in an *excess of simple electrolytes*. Under such conditions practically all the ions can approach to within the same distance of the particle and κ (or I) *relates only to the simple salts*. $1/\kappa$, as calculated from the usual expression, then gives the *thickness* of the atmosphere, since the atmospheric charge behaves as if concentrated at this distance from the *surface* of the colloidal ion (provided, of course, that all the $z_i e\psi/kT$'s are small).

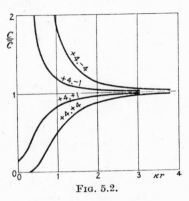

Fig. 5.2.

The inconsistencies of the Boltzmann equation

Onsager (1933) pointed out that the Boltzmann equation in the form used above (eqn. (5.1)) is incorrect, particularly when the deviations from the simple Debye-Hückel theory are considerable, i.e. in the regions close to ions and for the more concentrated solutions.

Considering a $j.k$ valent electrolyte, then according to eqn. (5.1) the distribution of j ions around a k ion, and that of k ions around a j ion, should be identical, since they are given respectively by

$$c_j = \bar{c}_j e^{-z_j e\psi_{k,r}/kT}, \qquad (5.14\text{a})$$

$$c_k = \bar{c}_k e^{-z_k e\psi_{j,r}/kT}, \qquad (5.14\text{b})$$

and the ψ's are related by the equation

$$z_j \psi_{k,r} = z_k \psi_{j,r}. \qquad (5.15)$$

(Each of the quantities $z_j \psi_{k,r}$ and $z_k \psi_{j,r}$ represents the work required to separate two such ions, distance r apart, to infinity.) This similarity in distributions holds for large values of r but, as shown below, breaks down as soon as the concentrations diverge at all considerably from their bulk values.

The reason for the failure of the Boltzmann equation is that eqn. (5.1) is strictly valid only if the solutions are ideal. If the solutions are sufficiently concentrated to require second-order terms in eqn. (5.4) then the ions find themselves in different local concentrations in different parts of the atmosphere, which necessitates the introduction of an activity coefficient into eqn. (5.1). This can be seen by considering the electrochemical potential ($\bar{\mu}_i$) of an ion i at two points of potential ψ^{I} and ψ^{II} (see also p. 77):

$$\bar{\mu}_i = \mu_i^{\mathrm{chem.}} + \mu_i^{\mathrm{electr.}} = \mu_i^{\mathrm{chem.}} + z_i e \psi.$$

For equilibrium $\bar{\mu}_i^{\mathrm{I}} = \bar{\mu}_i^{\mathrm{II}}$, which on substituting from the above becomes

$$(c_i f_i)^{\mathrm{I}} = (c_i f_i)^{\mathrm{II}} e^{-z_i e(\psi^{\mathrm{I}} - \psi^{\mathrm{II}})/kT}.$$

If state II refers to bulk solution, $\psi^{\mathrm{II}} = 0$ and $(c_i f_i)^{\mathrm{II}} = \bar{c}_i \bar{f}_i$, i.e.

$$c_i f_i = \bar{c}_i \bar{f}_i e^{-z_i e \psi / kT}. \tag{5.16}$$

This becomes identical with eqn. (5.1) only if $f_i = \bar{f}_i$, which will clearly not be the case if the local concentrations of i differ very considerably from their bulk values, as with concentrated solutions and in particular with high-valent electrolytes.

Consideration of colloidal electrolytes

With colloidal micellar ions of the size and charge found in colloidal electrolytes the above considerations indicate that the charge residing on the inner portions of the atmosphere arises mainly from a great increase in concentration of the gegenions, the high-valence ions (owing to their similar charges) being almost completely absent, so that the inner portions of the atmosphere are reasonably symmetrical. The atmosphere thus has a more obvious physical reality than with low-valence ions.

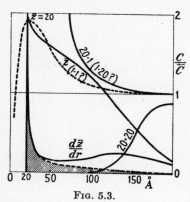

Fig. 5.3.

Fig. 5.3 shows the concentrations of the ions of a $N/1{,}000$ solution of a 1.20 electrolyte around the 20-valent ion of radius 20 A, as calculated by Müller's method (Hartley, 1935). It is clear that within a distance of about 80 A the probability of finding another 20-valent ion is negligible, so that only in the outer regions, where the volumes are large and the potential effects in any case small, will temporary asymmetry due to the movements of micellar ions be frequent.

COLLOIDAL PARTICLES. STABILITY AND ITS ORIGIN

The variation of $\bar{z}$ (i.e. the net valence of central ion and atmosphere) and $d\bar{z}/dr$ with distance r are also given in Fig. 5.3, the shaded area under the latter giving the contribution to the atmospheric charge caused by the *immigration* of the 1-valent ions; this is seen to be about one-third of the total charge (in marked disagreement with the ionic strength principle as discussed above). The tendency of the 1-valent ions to congregate around the multivalent ion results in their concentration in this region becoming greater than the bulk concentration, as indicated by the tentative distribution curve (1.1?) in Fig. 5.3. The atmospheres of these congregated gegenions will in turn affect the potential of the central ion, but it is shown by Hartley that their contributions to $\ln f$ will be less than $\frac{1}{20}$ of the 20-valent ion, and hence can be neglected to a first approximation.

(a) *Mobility of colloidal ions*

Two retarding influences of the atmosphere upon mobility are generally recognized—the 'time of relaxation' effect, owing to the central ion being slightly ahead of its atmosphere, and the 'electrophoretic' effect, which arises from the movement of the atmosphere in the opposite direction to the central ion (see also Chapter XXIV). The latter term increases directly with the valence of the central ion, whereas the former increases much less rapidly and for high-valent ions is assumed negligible for the present treatment.

Considering an elementary shell of the atmosphere of radius r, thickness dr, and valence $d\bar{z}$ $\left(d\bar{z} = 4\pi r^2 \, dr \, N \sum_i c_i\right)$, this will move according to Stokes's law with a velocity given by

$$dv = \frac{Xe \, d\bar{z}}{6\pi\eta r}, \qquad (5.17)$$

where X is the field strength and η the viscosity. The total electrophoretic contribution (Δv) of the atmosphere is obtained by integrating (5.17) from $r = \infty$ to $r = a$. Hence

$$\Delta v = \frac{Xe}{6\pi\eta} \int_\infty^a \frac{d\bar{z}}{r} = \frac{DX}{6\pi\eta} \int_\infty^a \frac{e \, d\bar{z}}{Dr}. \qquad (5.18)$$

Now $\int_\infty^a e \, d\bar{z}/Dr$ is the contribution of the atmosphere to the potential of the central ion, and hence the *potential and mobility must be influenced in an exactly parallel degree however the ions in the atmosphere are distributed*.

In the absence of atmosphere effects the velocity (v_c) of the central ion would be given by

$$v_c = \frac{Xez_c}{6\pi\eta a}. \tag{5.19}$$

Hence
$$\frac{v_c - \Delta v}{v_c} = \frac{\psi_{c,a}}{\psi_{c,a}^0}, \tag{5.20}$$

where $\psi_{c,a}^0$ is the potential of the central ion due to its own charge (i.e. $z_c e/Da$), and $\psi_{c,a}$ as before is its potential in the presence of the atmosphere.

If the conditions are such that the ionic strength principle holds (i.e. $z_i e\psi/kT$'s are all $\ll 1$), then $\psi_{c,a}$ is given by eqn. (5.8) and eqn. (5.20) becomes

$$\frac{v_c - \Delta v}{v_c} = \frac{1}{1+\kappa a}, \quad \text{or} \quad \frac{\Delta v}{v_c} = \frac{\kappa a}{1+\kappa a} \tag{5.21}$$

corresponding to the Debye-Hückel equation (5.9) for $\ln f$. Also the observed velocity (v) is given by

$$v = v_c - \Delta v = \frac{v_c}{1+\kappa a} = \frac{Xez_c}{6\pi\eta a}\frac{1}{1+\kappa a},$$

i.e.
$$v = \frac{DX\zeta}{6\pi\eta}, \tag{5.22}$$

where
$$\zeta = \frac{z_c e}{Da}\frac{1}{1+\kappa a}. \tag{5.22a}$$

Eqn. (5.22) is the well-known equation for electrophoretic velocity for *small* spherical particles (see Chapter XII), and, as the above discussion shows, eqn. (5.22a) is strictly valid only when the Debye-Hückel approximation is permissible.

The more general case of eqn. (5.20) can be dealt with by calculating $\psi_{c,a}$ by methods such as the graphical one of Müller (1927).

(b) *The magnitude of atmospheric effects in colloidal electrolytes*

Using Müller's graphical method, Hartley (1935) has computed the magnitude of atmospheric effects on the assumptions of spherical ions and spherically symmetrical atmospheres as well as those stated above with regard to the atmosphere effects on the gegenions. For a 1.10 electrolyte of $a = 10$ A, and a 1.20 electrolyte of $a = 20$ A, the atmosphere effects and hence the deviations from ideality are, at concentrations of $N/100$ and below, always *less than would be predicted from the simple Debye-Hückel theory* (taking a as the distance of closest approach of large ion and a gegenion). Fig. 5.4 compares, for the 1.20 case, the

COLLOIDAL PARTICLES. STABILITY AND ITS ORIGIN 113

values of the activity coefficient (f), the osmotic coefficient (g), and the multivalent ion mobility (expressed as $(v_c - \Delta v)/v_c$), the Debye-Hückel values being shown as the dotted lines. If the a values are increased the deviations from the theory might at first become greater and later diminish, but would certainly not be reversed. The effect of decreasing the valence whilst keeping a constant will be similar. For the 20-valent

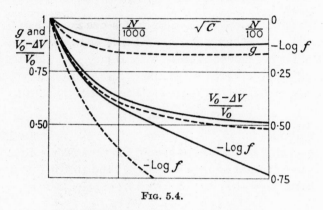

Fig. 5.4.

micellar ion in $N/1{,}000$ solution a has to be reduced to about 15 A before the mobility is equal to that calculated from eqn. (5.21), and to about 13 A before $-\ln f$ equals that from eqn. (5.10). As micelles with more concentrated charges than these are improbable Hartley concludes that *atmosphere effects in* pure *solutions will be less than calculated by simple Debye-Hückel theory (eqns. (5.10) and (5.21)), taking the a value as the distance of closest approach of a micelle and simple ion of opposite charge.*

When the colloidal ion is present in a relative *excess of simple electrolyte* the atmospheric effects are, in contrast with the case of pure solutions given above, probably always greater than predicted by eqns. (5.10) or (5.21). Thus with the 1.20 electrolyte of radius 20 A present in small amount in $N/1{,}000$ 1.1 electrolyte the graphical method gives for $-\log f$ a value of 0.335, eqn. (5.10) one of 0.265. Comparison of the former value with that of 0.421 for the pure 1.20 electrolyte in $N/1{,}000$ solution, of ionic strength ten times greater, shows how completely the ionic strength principle fails here.

Some mention of the application of the above concepts to the interpretation of certain properties of colloidal electrolyte solutions, such as osmotic pressure, diffusion and conductivity, will be given in Chapter XXIV.

Consideration of larger colloidal particles ($0.01\,\mu$–$1\,\mu$)

As previously pointed out, it is only the question of atmosphere effects upon the *mobility* that is of any practical significance with colloidal particles in this size range. For example a 1 per cent. (by volume) gold sol of diameter $0.1\,\mu$ would only give an osmotic pressure of about 10^{-2} mm. water, and any comparison with Debye-Hückel theory is out of the question.

The mobility of a colloidal particle in this size range may be modified by its distorting effect upon the applied field as well as by atmospheric effects. The former has been treated by Henry (see Chapter XII) and for non-conducting spheres of radius a his solution for the velocity reduces to $v = DX\zeta/4\pi\eta$ for large particles ($\kappa a \gg 1$), and to

$$v = DX\zeta/6\pi\eta$$

for very small ones ($\kappa a \ll 1$). (κ refers only to the simple salts present, no allowance being made for the colloid for the reasons already given.)

If the Debye-Hückel approximation is valid (i.e. if $\zeta < ca.$ 25 mV), then the charge $Q(=z_c e)$ can be found from eqn. (5.22a)

$$\zeta = \frac{Q}{Da}\frac{1}{1+\kappa a}.$$

For very many colloidal particles, however, the ζ potentials are very much greater than 25 mV (e.g. oil droplets with adsorbed soap may be as high as 250 mV), and hence the charge calculated from the above expression must be more or less in error.

STABILITY AND ITS ORIGIN

As pointed out in Chapter I, it is convenient to subdivide colloidal systems into the hydrophobic (or more generally the lyophobic) and the hydrophilic (lyophilic) types, although all types of intermediate behaviour are actually encountered. The former, as typified by an aqueous gold sol, are readily precipitated by low concentrations of salts, particularly in the presence of water-soluble organic compounds such as alcohol, whereas the latter type, such as a dilute gelatine solution, are much less sensitive and require much higher concentrations of precipitating agents. For this reason it is desirable to consider the extreme types separately, but before doing so brief mention must be made of the ways of measuring stability.

The ultimate measure of stability in colloidal systems is the rate of decrease of the interfacial (free) energy with time, and this can be measured by the changes in interfacial area if the surface energy per unit area (i.e. γ) is independent of particle size. (This is generally assumed to be the case.) The reduction in surface area and surface energy with time is seen most clearly in the breaking of an emulsion. Measurement of stability thus requires a size-frequency analysis of the colloidal particles at suitable intervals, from which the rate of change of interfacial area (energy) can be calculated. The experimental methods used will be given in Chapter XX (for suspensions) and in Chapter XXIII (for emulsions).

(a) Lyophobic colloids

We will consider here only suspensions in aqueous media, since these have been by far the most studied and also offer the best hopes of theoretical analysis.

The tendency of salts to coagulate hydrophobic sols has been mentioned earlier in connexion with the work of Selmi, Faraday, and in particular of Schulze and Hardy. The Schulze-Hardy rule states that the coagulation is brought about by the ion of opposite charge to that on the colloid, and that the efficacy of this ion increases (markedly) with its valence.

These early investigations naturally led to attempts to correlate stability with the charge on the colloid particle, or to the ζ potential which is closely related to the charge. Powis (1916) examined the effect of salts on the ζ potential and stability of suspensions, and he concluded that for stability a fairly definite ζ potential (*ca.* 20–30 mV, depending on the system) is required; below this value rapid coagulation occurs. However, later work on this and other hydrophobic colloids indicates that the transition is a much more gradual one. Nevertheless there is no doubt about the general parallelism between the ζ potential and stability, and since the advent of the Debye-Hückel theory of strong electrolytes many attempts to formulate quantitative theories have been made. It is usual to assume that the behaviour is controlled by the van der Waals (London) attractive forces and the electrical repulsive ones arising from the charges on the colloidal particles and the surrounding ionic atmospheres. In addition to the question of stability of hydrophobic suspensions these general ideas have formed the basis of certain treatments of gelation and thixotropy (see Chapter XXI).

Fig. 5.5, based upon the work of Hamaker (1938), indicates how the potential energy of two particles is believed to vary with their separation, the dotted curve indicating the net result due to the repulsive and the attractive contributions. The particles will be stable when the maximum in the resultant energy curve is large compared with their translational kinetic energy (i.e. $E_{max} \gg kT$), but unstable when E_{max} is comparable or less than this quantity. The problem then reduces to the calculation of the electrical (repulsive) forces arising from the charges and the double layers, and of the van der Waals (attractive) forces.

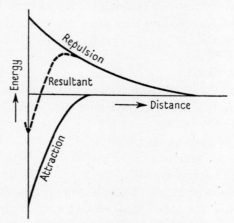

Fig. 5.5. Variation of potential energy of two colloidal particles with separation.

The problem of the electrical forces arising from the charges and double layers of two particles (e.g. two spheres or two parallel plates) has been attempted by numerous authors, with some very diverging conclusions. Recently Verwey (1945) has made a critical survey of the previous work, and has given what is probably the most accurate solution yet available, differing radically from all the earlier workers with the exception of Derjaguin (for earlier references see Verwey, 1945). An outline of his method and conclusions is given below. (See also Overbeek and Verwey, 1946; Levine, 1946.)

Potential and charge distribution for two interacting double layers

The problem of the interaction of two particles with their adherent double layers can be examined in the simple cases of two parallel plates or two spherical particles. Only the first of these is discussed by Verwey as it is the easier to handle mathematically, and is a good approximation for particles large compared with the thickness of the double layer (i.e. $\kappa a \gg 1$), such as obtains for most hydrophobic suspensions containing some electrolyte, particularly when near the state of agglomeration. For very fine suspensions practically free of electrolytes as in dialysed sols (e.g. $a = 100$ A, $\kappa = 10^5$ cm.$^{-1}$, i.e. $\kappa a = 0 \cdot 1$), then the model of two spherical particles becomes the better one. In discussing the problem of stability in systems with so much electrolyte that they are on the point of flocculating, the parallel plate model is clearly the more relevant.

As regards which theory of the double layer should be adopted Verwey chooses the Gouy-Chapman on account of its simplicity, but points out that extension to Stern's theory is comparatively easy. It is assumed that the ions are point charges and that their concentrations at any point in the double layer are determined by the average electrical potential at that point, i.e. a Boltzmann distribution is assumed (cf. p. 109). In the case of a single double layer the potential (ψ) falls from a value of ψ_0 at the surface ($x = 0$), to a value of zero at large distances from the surface (i.e. $x = \infty$). (NOTE: the potential ψ_0 is not to be confused with the ζ potential, which is the potential at the surface of shear.) The value ψ_0 is determined by the equilibrium between surface and solution of the ions responsible for the surface charge (e.g. I' ions in the case of a negative AgI sol, and very probably OH' ions for non-ionogenic substances such as hydrocarbons and their derivatives), and it is assumed that for a constant concentration of these ions ψ_0 is constant and is given by a Nernst equation (i.e. in the above example by $\psi_0 = \psi_0' + (RT/F)\ln\{I'\}$, where $\{I'\}$ is the activity of the I' ions in the bulk of the solution and ψ_0' a constant). Thus with dilute suspensions ψ_0 is assumed not to change on addition of indifferent electrolyte, i.e. the ions of the indifferent electrolyte are adsorbed equally (if at all), and do not change the bulk activity of the potential-determining ions.

Considering now the case of two interpenetrating double layers, the electric potential function and charge distribution as a function of their separation are calculated, using the above assumptions and an additional one that the added electrolyte shall be of symmetrical valence type. The Poisson equation for a flat plate becomes

$$\frac{d^2\psi}{dx^2} = -\frac{4\pi\rho}{D} = -\frac{4\pi}{D}(c_+ z_+ e + c_- z_- e). \tag{5.23}$$

Applying the Boltzmann equation this becomes

$$\frac{d^2\psi}{dx^2} = \frac{4\pi cze}{D}(e^{ze\psi/kT} - e^{-ze\psi/kT}) = \frac{8\pi cze}{D}\sinh\frac{ze\psi}{kT}, \tag{5.24}$$

where c = concentration of cations (anions) per c.c., and $z = z_+ = -z_-$.

For one double layer the boundary conditions are $\psi = 0$ and

$$d\psi/dx = 0 \quad \text{for} \quad x = \infty.$$

For two double layers of distance $2d$ between the plates, these conditions are replaced by $\psi = \psi_d$, and $d\psi/dx = 0$ (by symmetry) for $x = d$ (see Fig. 5.6 (a)). Substituting $y = ze\psi/kT$, $y_0 = ze\psi_0/kT$, $y_d = ze\psi_d/kT$ and integrating (5.24) once, we obtain

$$\frac{d\psi}{dx} = -\sqrt{\left(\frac{16\pi ckT}{D}(\cosh y - \cosh y_d)\right)}. \qquad (5.25)$$

Now $d\psi/dx$ and σ (the surface charge/cm.^{2}) are related by the equation

$$\left(\frac{d\psi}{dx}\right)_{x=0} = -\frac{4\pi\sigma}{D},$$

and hence

$$\sigma = -\frac{D}{4\pi}\left(\frac{d\psi}{dx}\right)_{x=0} = \sqrt{\left(\frac{DckT}{\pi}(\cosh y_0 - \cosh y_d)\right)}. \qquad (5.26)$$

The convention in eqn. (5.24) is that approach of ions of the same sign increases the energy of the system, so that σ and ψ have the same sign in eqn. (5.26). Thus if the colloid is negatively charged (e.g. AgI in the presence I' ions), then ψ will be negative and $-z_i e\psi$ will be positive for cations and negative for anions. Eqn. (5.25) still contains the unknown potential ψ_d at the point midway between the plates. By making the substitutions

$$y = \frac{ze\psi}{kT}, \; dy = \frac{ze}{kT}d\psi, \; \kappa^2 = \frac{8\pi cz^2e^2}{DkT} \quad \text{(the Debye-Hückel function)},$$

it becomes

$$\frac{dy}{dx} = -\kappa\sqrt{\{2(\cosh y - \cosh y_d)\}}.$$

Integration gives the required relation between y_d and d:

$$\int_{y_0}^{y_d} -\frac{dy}{\sqrt{\{2(\cosh y - \cosh y_d)\}}} = \int_0^d \kappa \, dx = \kappa d. \qquad (5.27)$$

Eqn. (5.27) yields an elliptic integral of the first kind which may be solved numerically with the aid of tables, or in some cases by approximations. If ψ_0 is assumed constant (see below) and independent of d, then decreasing d increases y_d or ψ_d (i.e. the potential midway between the plates). Fig. 5.6 (a) shows schematically the electrical potential function for two different values of the plate separation ($2d$).

Eqn. (5.25) shows that the effect of decreasing d (i.e. the plate separation) is to decrease $d\psi/dx$ and hence the surface charge density (σ), as shown in Fig. 5.6 (b). For $d = 0$ the charge becomes zero.

The physical requirement for the assumption that ψ_0 remains constant on approach of two particles is that perfect equilibrium between

COLLOIDAL PARTICLES. STABILITY AND ITS ORIGIN 119

surfaces and solution is always maintained. If this is so the electrical potential will remain constant over the whole surface of the particle, being determined by the concentration of the potential-determining ions far away in the bulk solution. In the case of the approach of two negatively charged AgI particles, for example, this requirement means that I′ ions present on the approaching crystal planes must be released

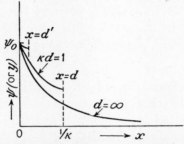

FIG. 5.6 (a). Electrical potential function (schematic) for the case of two interpenetrating double layers, according to the model of Gouy-Chapman (neglecting the dimensions of the ions). The figure gives the electrical potential function for two different values of the plate distance ($2d$). Because of the symmetry with respect to the central plane $x = d$, only the left-hand part of the curve has been given. For $d = \infty$ the curve passes over into that for one double layer using the theory of Gouy.

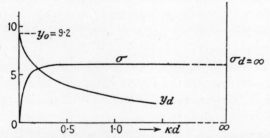

FIG. 5.6 (b). Value of the electrical potential at a point midway between the two plates (y_d or ψ_d) and of the surface charge (σ) as a function of the plate distance, for the case where $y_0 = 9.2$, (hence $\psi_0 = 230$ mV for uni-univalent electrolyte, etc.).

and diffuse away rapidly compared with the translational movement of the particles, the converse happening when the particles move apart. In view of the difference in size and diffusion constants between simple ions and colloidal particles this seems very probable and hence the assumption, $\psi_0 = $ constant, reasonable. If, however, the regulation of the charge to maintain equilibrium cannot occur rapidly enough then it might be more reasonable to assume the charge density σ to be constant rather than the potential ψ_0. In this case, according to eqn. (5.26), ψ_0 increases to infinity when the plates approach one another.

Free energy considerations

To see whether the ionic interaction induces an attraction or repulsion we must determine the potential energy of the two plates as a function of their separation, taking into account only the effects due to changes in the double layers, and compare this with the free energy of two single isolated double layers ($d = \infty$). This may be done in two ways:

(1) by calculating the force acting between the plates and hence the potential curve by integrating this with respect to d;

(2) by determining the free energy of the interpenetrating double layers directly.

Method (1) would appear to present difficulties (see Verwey, 1945, p. 39) although it does show that the force between the plates is always repulsive.

In the second method, as in the Debye-Hückel theory of electrolytes, the problem is to calculate the free energy of the arrangement of the ions under the influence of their mutual electrostatic forces. In the case of the double layer system we have to bring about its removal everywhere, i.e. to discharge the surface as well as to remove the charge at each point throughout the solution (by some reversible and isothermal process), and to calculate the energy associated with the entire process. Hence integration is necessary not only over the discharging process but also over the whole space occupied by the double layer system.

Another difference from electrolyte theory arises from the assumption that the double layer potential (ψ_0) is constant and the surface charge adjusts itself according to the separation of the particles by transference of potential-determining ions from solution to surface or vice versa. The movement of an ion from solution, where it is hydrated, to the surface lattice where it is not, must be accompanied by a change in its chemical potential (see Chapter IV and also p. 44). It is this difference in chemical potential ($\Delta\mu$) which enables ionic adsorption to occur (thus giving rise to the surface charge) and which, in the equilibrium state, balances the electrical potential difference thus set up, i.e. $e\psi_0 = \Delta\mu$ for univalent potential-determining ions. Thus when discharging the double layer the associated chemical energy must also be considered. In the discharging process the excess ions on the surface, as well as the ions in solution, are assumed to be gradually discharged in small steps of $e\,d\lambda$, where $d\lambda$ is a small fraction. This process is much simplified if ψ_0 is assumed constant (cf. Levine, 1946). For that purpose the difference in chemical potential ($\Delta\mu$) is gradually decreased in steps of $\Delta\mu\,d\lambda$ (by letting the μ on the surface approach its value in solution). Hence for each degree of discharging (measured by the quantity λ, decreasing from 1 to 0), equation $\lambda e\psi_0 = \lambda \Delta\mu$ will hold, so that ψ_0 is constant throughout the whole process. (The condition $\lambda e\psi_0 = \lambda \Delta\mu$ needs further examination if hydration effects come in (Levine, 1946).) Thus when the charge on a surface ion is decreased by $e\,d\lambda$ the amount of electrical energy liberated ($\psi_0 e\,d\lambda$) is exactly counterbalanced by the amount of chemical energy required ($\Delta\mu\,d\lambda$). Accordingly the amount of chemical energy is automatically included if we calculate the amount of *electrical* energy necessary for discharging the ions in the *solution*.

To determine the free energy/cm.² double layer for a system of two double layers we consider the space charge between $x = 0$ and $x = d$, using $\rho_\lambda(x)$ and $\psi_\lambda(x)$ to denote respectively the space charge and potential corresponding to the charge $\lambda z e$, ρ and ψ referring to the initial state (i.e. when $\lambda = 1$). The excess number of ions in a layer dx is then $(\rho_\lambda/\lambda z e)\,dx$, and the amount of energy required to decrease the charge on these ions by a fraction $ze\,d\lambda$ is therefore $\rho_\lambda \psi_\lambda (d\lambda/\lambda)\,dx$. Hence the amount of energy involved in the whole process is

$$G = \int_0^1 \frac{d\lambda}{\lambda} \int_0^d \psi_\lambda\, \rho_\lambda\, dx. \qquad (5.28)$$

This equation is soluble on the basis of eqns. (5.24) and (5.25), yielding

$$G = -\frac{2ckT}{\kappa}\Big\{\tfrac{1}{2}\kappa d(3e^{y_d} - 2 - e^{-y_d}) + 2\sqrt{(2\cosh y_0 - 2\cosh y_d)} +$$

$$+ \int_{y_d}^{y_0} \frac{e^{-y} - e^{y_d}}{\sqrt{(2\cosh y - 2\cosh y_d)}}\, dy\Big\}, \qquad (5.29)$$

where the integral leads to an elliptic integral of the second kind, for which tables are available (see Verwey, 1945). Eqns. (5.29) and (5.27) enable G to be calculated for any value of d. For $d = \infty$ eqn. (5.29) becomes

$$G_\infty = -\frac{2ckT}{\kappa}\{4\cosh \tfrac{1}{2}y_0 - 4\}. \qquad (5.30)$$

The potential energy (E_R) due to the interaction of the two double layers is then given by
$$E_R = 2(G - G_\infty). \qquad (5.31)$$

Fig. 5.7 shows its value in the presence of 1.1 electrolyte as a function of d, for different values of y_0 (i.e. of ψ_0). When $\kappa d > 1$, the $\log E_R \sim d$ relationship is practically linear, so that E_R increases approximately exponentially with decreasing plate distance. It will be seen that $(G-G_\infty)$ is always positive so that the double layer interaction always gives rise to *repulsion* between the two plates.

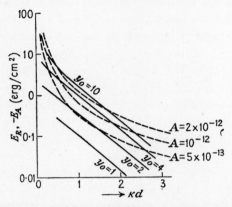

FIG. 5.7. Repulsive potential due to the interaction of the double layers (E_R, full lines), and attractive potential due to van der Waals (London) forces ($-E_A$, dotted lines) both on a logarithmic scale as a function of the distance between the plates, for various values of the attraction constant A ($z = 1$). The figure applies to $\kappa = 10^7$.

Potential curves

The most important attractive force counteracting the double layer repulsion is believed by Hamaker, Verwey, and others to be of the van der Waals (London) type. Hamaker (1938) has made quantitative estimates of the van der Waals attraction potential (E_A) for various cases, such as two spherical particles, two parallel plates etc., and for the latter case he gives

$$E_A = -\frac{A}{48\pi}\frac{1}{d^2}, \qquad (5.32)$$

where A is a constant of not very precisely known value, and which depends to some extent on the materials of the system. According to quantum theory A is expected to lie between 10^{-11} and 10^{-14}, and in general between 10^{-12} and 10^{-13}.

Potential curves showing the variation of total energy (E_R+E_A) as a function of plate separation can then be constructed from the values of E_R and E_A given by eqns. (5.31) and (5.32) for various values of ψ_0 (or y_0), κ, and A. Two types are found (see Fig. 5.8):

(*a*) Those with a maximum above the horizontal axis. Provided this maximum is not too low (i.e. $>$ a few kT) this type corresponds to stable suspensions since the potential at the maximum is sufficient to

COLLOIDAL PARTICLES. STABILITY AND ITS ORIGIN 123

prevent agglomeration. These curves also show a shallow minimum at greater distance, but this may arise from uncertainties in eqn. (5.32) and so may have no physical significance.

(b) Those which always lie below the horizontal axis. These correspond to unstable systems which coagulate spontaneously since the particles always attract one another.

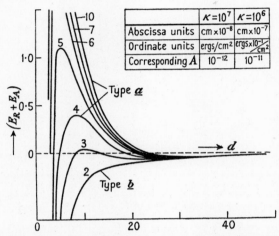

FIG. 5.8. The total energy (E_R+E_A) of the two plates with respect to each other (per cm.² plate surface) as a function of the plate distance, for various values of y_0 from 2 to 10. Two cases—type a: curves with a maximum of positive energy (energy barrier), and type b: the attraction potential prevails for all plate distances. The transition occurs here between $y_0 = 2$ and $y_0 = 3$. The figure applies to 1.1 valent electrolyte.

The type into which any given system will fall, and hence its stability for a given combination of ψ_0, κ, and A, can be found by use of Fig. 5.7, where E_R and $-E_A$ are plotted against d for a number of cases. The potential curve will show a maximum (i.e. type a) if the E_A and E_R curves intersect twice. By this means the effect of varying conditions (e.g. of concentration and valence of electrolytes, ψ_0, A, etc.) upon the change over from 'stability' (i.e. E_R+E_A positive at some point) to 'instability' (i.e. E_R+E_A always negative) can be found, as shown in Fig. 5.9. If A is taken as 2×10^{-12}, then a suspension with $\psi_0 = 150$ mV should have a coagulation limit for $z = 1$ (i.e. 1.1 valent electrolyte) of about 100 millimole/l., for $z = 2$ of about 2·3 millimole/l., and for $z = 3$ of about 0·2 millimole/l. The experiment alvalues depend somewhat upon the particular system but are of this order (e.g. 180, 3, and 0·1 millimole/l. for K^+, Ba^{++}, and La^{+++} with a negative AgI sol), so that the above theory provides a quantitative explanation for the Schulze-

124 THE APPLICATION OF DEBYE-HÜCKEL THEORY TO

Hardy rule. In addition it can explain other phenomena, such as the greater stability of large particles compared with smaller ones of the same material (Verwey, 1945).

Verwey's theory thus appears to provide a sound theoretical basis for the stability of hydrophobic suspensions, although whether it is generally applicable to all systems seems less certain. According to Levine (1946)

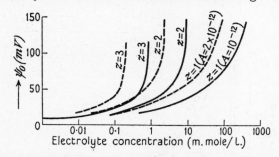

FIG. 5.9. Corresponding values of the surface potential (ψ_0) and the electrolyte concentration for which the total potential curve passes over from type a to type b, for different valency of the ions (symmetrical electrolytes) and for two values of the attraction constant A. (Dotted curve for $A = 2 \times 10^{-12}$, full curve for $A = 10^{-12}$.) For sufficiently large particles the curves separate the 'stable area' (upper left-hand part of figure) from the 'unstable area'.

it is limited to systems in which the potential determining ion after adsorption is indistinguishable from the rest of the colloidal particle (e.g. AgI sol and I' ion), and not to those where this condition is not satisfied (e.g. paraffin or metal sols stabilized by OH' or soap ions).

(b) Lyophilic colloids

According to the classification given in Chapter I the lyophilic type of colloids is less sensitive to changes in the dispersion medium, and is frequently formed spontaneously merely by bringing the constituents together. Those in aqueous media (hydrophilic type), e.g. proteins, starches, agar, etc., are usually unaffected by the concentration of simple salts (e.g. M.NaCl) which suffice to reduce the ζ potential to very small values. It is clear then that the ζ potential cannot play the dominating role in stability which it does in hydrophobic suspensions. Nevertheless it does make a definite contribution, since after the net charge has been reduced to zero the colloid is now more sensitive to precipitating agents. (E.g. to alcohol or acetone in the case of agar or gelatine.) Conversely after addition of alcohol, acetone, etc., such hydrophilic colloids become more sensitive to salts. The effect of salts on the charge (or ζ potential) can be treated by the methods given above for hydrophobic suspensions.

COLLOIDAL PARTICLES. STABILITY AND ITS ORIGIN

In organic sols, such as rubber, polystyrene, etc., in benzene, charge effects are clearly improbable, and thus with all types of lyophilic systems other factors determining stability are operative—the most important is usually believed to be *solvation* (i.e. hydration in aqueous systems).

Before going on to consider the evidence supporting the idea of solvation as the cause of stability, and what is meant by the term *solvation*, it is important to emphasize one difference between the lyophobic and lyophilic types (in particular between the hydrophobic and hydrophilic). The fact that the latter type are formed (or redisperse) spontaneously shows that the free energy of formation (ΔG) is negative (see Chapter IV), whereas the converse is true for the lyophobic type. Thus lyophobic suspensions appear to be thermodynamically unstable and are only prevented from coagulating by some sort of energy barrier, which appears to be electrical in origin in the case of aqueous systems. The free energy term (ΔG) is made up of two parts (p. 55), an energy term (ΔH) which is a measure of the difference between particle/particle and solvent/particle interactions, and an entropy term (ΔS), related to the difference in possible configurational modes in the two states (i.e. coagulated and dispersed systems). If the attraction between solvent and particle is large compared with that between the particles, then ΔH (as measured by heats of wetting and so on—see Chapter IV and p. 550) is large and negative, and the dispersion will be stable (ΔG negative) unless the entropy changes counterbalance (i.e. are large and negative). The question of entropy changes arises particularly with solutions of long-chain polymers, and in such cases ΔS usually provides the requisite free energy decrease for stability even if ΔH is zero or slightly unfavourable (see Chapter XXVII).

Solvation

The notion that solvation was responsible for the stability of lyophilic colloids arose chiefly from the observation that the precipitating action of salts, upon gelatine and other proteins, ran parallel to their solubility in water and to their tendency to become hydrated. A series of cations or anions arranged in order of efficacy is termed the Hofmeister or lyotropic series, as mentioned earlier, in Chapter I. (Cf. the effect of salts on hydrophobic colloids, where the coagulating power of a salt is determined almost entirely by the valence of the ion of opposite sign to the colloidal particle.) The ions of the salt were believed to act by the competitive process of dehydrating the surface of the colloid and thus reducing its stability. The sensitization or precipitation which may be brought about by adding precipitants miscible with the dispersion medium (e.g. alcohol with aqueous agar sols, petrol ether with rubber in benzene) has also been ascribed to desolvation, which is, however, merely a convenient way of expressing the reduced affinity between colloid and medium.

The original ideas of the solvation layer were that it was relatively thick (from a molecular viewpoint) and prevented coalescence by its mechanical 'buffering' effect. With our increasing knowledge of the forces at surfaces and of solvation, in particular hydration, the existence of thick solvation layers seems most improbable, and the fact that, with simple liquids, the viscosity of very thin films is not abnormal (e.g. Bastow and Bowden, 1935) confirms this belief. It seems unlikely, as pointed out earlier (p. 46), that the hydration sphere of simple ions ever exceeds a single molecular layer, and in the case of organic media there seems no reason to expect any marked difference. Hydration has been most studied in the case of the proteins, and in the case of haemoglobin, for example, the hydration (gm. water/gm. dry protein) is found to be about 30 per cent. (Adair and Adair, 1936). This value is very close to the amount calculated on the basis of a spherical molecule of diameter 50 A (see Chapter XXVI) covered with a monolayer of water (*ca.* 3 A thick), although this picture may be modified if water molecules can penetrate into the internal spaces of the protein.

It might appear at first sight that ions of colloidal size would, owing to their high charge, be more hydrated than those of low valence, but theoretical considerations lead to the opposite conclusion. Thus, despite the particle's high charge, the potential at its surface is unlikely to be greater than at that of a univalent ion, although the potential does fall off less rapidly with distance from the surface, a factor which, together with that due to the large surface area, accounts for the large number of gegenions held by the colloidal particle (p. 110). At the surface the potential will be proportional to z/r (where z = valence and r = radius of particle), the radial field of force to z/r^2, and the divergence of the field to z/r^3. Since z/r is of the same order for the colloidal particle and a single ion, it is clear that the two latter quantities will be very much less with colloidal than with simple ions. The orientation of the water dipoles is determined by the field of force, and the divergence of the field is responsible for attracting the oriented dipoles, hence both the orientation and attraction will be less near the surface of the colloidal ion. Thus any marked increased hydration of the colloid due to its high valence seems improbable.

Some further mention of solvation will be made in connexion with its effect upon the apparent disymmetry constant of proteins and other polymers (see p. 260).

REFERENCES

More detailed references to the development of Debye-Hückel theory will be found in the papers of Scatchard, Onsager, and Hartley, referred to below.

For a discussion of the earlier work on the stability of hydrophobic systems see the symposium *Hydrophobic Colloids*, 1938, D. B. Centen, Amsterdam.

ADAIR and ADAIR, 1936, *Proc. Roy. Soc.* B **120**, 422.
BASTOW and BOWDEN, 1935, ibid. A **151**, 220.
BJERRUM, 1926, *Ergebn. exakt. Naturw.* **6**, 125.
DEBYE and HÜCKEL, 1923, *Phys. Z.* **24**, 185, 305.
GOUY, 1910, *J. Phys.* (4), **9**, 457.
GRONWALL, LA MER, and SANDVED, 1928, *Phys. Z.* **29**, 358.
HAMAKER, 1938, *Hydrophobic Colloids*, loc. cit., p. 16.
HARTLEY, 1935, *Trans. Faraday Soc.* **31**, 31.
LA MER, GRONWALL, and GRIEFF, 1931, *J. Phys. Chem.* **35**, 2245.
LEVINE, 1946, *Trans. Faraday Soc.* **42** B, 102, 128.
MÜLLER, 1927, *Phys. Z.* **28**, 324.
ONSAGER, 1933, *Chem. Rev.* **13**, 73.
OVERBEEK and VERWEY, 1946, *Trans. Faraday Soc.* **42** B, 117, 126.
SCATCHARD, 1933, *Chem. Rev.* **13**, 7.
VERWEY, 1945, *Philips Research Reports*, **1**, 33.
—— and OVERBEEK, 1948, *Theory of the Stability of Lyophobic Colloids*, Elsevier.

PART II
EXPERIMENTAL METHODS AND THEIR THEORETICAL BASIS

Section A
SYSTEMS AS A WHOLE

VI
INTRODUCTION

Important data for colloidal systems

To understand the nature of colloidal systems it is clearly necessary to possess information on the properties and structure of the individual particles (or molecules), to which colloidal behaviour is to be attributed, as well as on their interactions with one another and with the dispersion medium. Largely as the result of developments since 1920 it is now possible to make quantitative estimates of the size, shape, and electrical properties of individual colloidal particles, in systems for which, previously, our knowledge was very scanty. Section A is devoted to the relevant methods of investigation which fall naturally into three different groups; the equilibrium methods (Chapters VII and VIII), the dynamic methods (Chapters IX to XIV), and miscellaneous methods largely concerned with the use of X-rays and electrons, visible and infra-red light (Chapter XV). Finally (Chapter XVI) a brief account of plastic flow and elasticity is included.

We list below the type of information which these methods are designed to provide:

1. Molecular (or particle) Weight, which with density data gives molecular (or particle) volume.
2. Molecular Shape.
3. Molecular Charge and Dipolar Properties.
4. Interactions between the molecules (or particles) and with the solvent.
5. Internal Structure of the molecules.

By the term 'colloidal particle' we include all particles or molecules for which at least one linear dimension lies between the limits 10^1–10^5 A

INTRODUCTION

(Fig. 6.1), the latter being roughly the limit of visibility with the ordinary compound microscope.

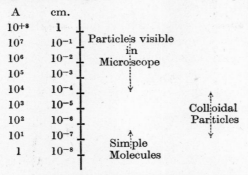

FIG. 6.1. The colloidal state.

Whilst Section A is concerned with any particles within this size range, it is often found convenient for the purposes of illustration to consider systems for which the following properties are either completely or closely realized:

(1) Each particle is identical with every other, in every property, for a given sample of the material (monodisperse systems).
(2) The same properties are easily reproduced.

Whether the first condition is ever *completely* true in any system is not easily established, but for many systems containing corpuscular proteins in aqueous solution, the deviations from uniformity, if they do occur, must occur over an extremely small range. Such proteins, whose properties are usually strictly reproducible, are normally isolated by simple processes so that for the purpose of comparing different experimental methods of investigating colloidal particles, they are almost ideal reference substances. The frequent mention of proteins in this section of the book is not to be taken therefore as inferring any restriction of the experimental methods to these substances, but merely as a sign of their special suitability for illustrative purposes. Their obvious biological importance has, however, very understandably made them a favourite object of study and the large body of information which has accumulated will be reviewed in a later chapter (XXVI). For many other colloidal systems, particles with physical and chemical properties varying over wide ranges may occur (polydisperse systems); nor are such properties readily reproduced in different samples of the material. Accordingly in the investigations of such systems, the information which may be gained on a particular sample of material, may not be of general significance. Important investigations have, however, been made, chiefly upon fractionated materials, as will be seen in later chapters.

As discussed in Chapter III, the use of the term 'molecular weight' in describing monodisperse protein systems is fairly free from objection: however, it will also be used in the following chapters in connexion with the proteins in general and with other polymeric substances. It should be realized that in some cases the use of the terms 'micellar' or 'particle' weight may be more correct. In certain cases, also, the choice between 'colloidal molecule' and 'colloidal particle' may present some difficulty and must be decided (not always unambiguously) for each individual system.

Before considering the newer experimental methods, it is of interest to consider the information on colloidal particles which may be yielded by the classical methods, normal to low molecular weight organic chemistry, and their modifications. Certain general questions, relevant to most of the newer methods, will then be discussed.

Analytical techniques

The use of analytical methods alone in low molecular weight chemistry can give only an equivalent weight, which being that weight of substance which contains one atom or group of the type estimated, is a simple fraction of the real molecular weight and is often termed a minimal molecular weight. Similarly, in applying analytical methods to polymeric molecules, only a minimal molecular weight (M_e) is obtained, which is related to the real molecular weight (M) by the expression

$$M = n \times M_e, \tag{6.1}$$

n being a small integer. Providing n is small, its rough evaluation by means other than analytical is sufficient to decide on its particular integral value, which, combined with the accurate analytical value of M_e in eqn. (6.1), gives an accurate molecular weight.

For polymeric materials in which M, and therefore M_e also, is large (n being small), the element or grouping estimated must occur only in small quantity. (Small values of M_e are otherwise obtained.) In estimating such small quantities the analytical technique must therefore be capable of great precision. It follows also that the colloidal substance must be capable of careful purification, especially from material also containing the atom or group to be estimated: precise analysis would otherwise be useless. The stringent nature of these conditions, especially the latter, rules out the use of the method for many colloidal systems, and in practice it can only be applied to stable polymeric molecules, of which examples are quoted below. Clearly for hydrophobic sols, e.g. those of the metals, the careful purification prior to analysis would by itself make the method impossible, and its limited applicability, in general, is readily understood.

(a) Trace element analysis

One of the most useful analytical methods has been that in which an element, occurring in small quantity, is estimated accurately in deriving a minimal molecular weight. The estimation of iron in the haemoglobins is one of the best examples of this. Containing approximately 0·3 per cent. of iron, the haemoglobins, which are of considerable stability, may be carefully purified from extraneous iron, and estimated for combined iron by precision methods. Table 6.1, compiled by Cohn, Hendry, and

TABLE 6.1. *Minimal Molecular Weights of the Haemoglobins*

Type	Method	Amount of constituent present in per cent.	Weight combining with or containing 1 atom or molecule in gm.	Assumed number of atoms or molecules	Minimal molecular weight (M_e)
Horse	Iron content	0·335	16,669	1	16,669
	Sulphide sulphur content	0·190	16,878	1	16,878
	Sulphur content	0·390	8,223	2	16,446
Pig	Iron content	0·40	13,960	1	13,960
	Sulphur content	0·48	6,681	2	13,362
Cat	Iron content	0·35	15,954	1	15,954
	Sulphur content	0·62	5,172	3	15,516
Ox	CO-combining capacity	..	16,721	2	33,442
	Iron content	0·336	16,619	2	33,238
	Sulphur content	0·45	7,127	5	35,635
	Sulphur content	0·48	6,681	5	33,405
	Arginine content	4·24	4,107	8	32,856
Fowl	Iron content	0·335	16,669	2	33,338
	Sulphur content	0·86	3,729	9	33,561
Dog	Iron content	0·336	16,619	3	49,857
	Sulphide sulphur content	0·335	9,573	5	47,865
	Sulphur content	0·568	5,646	9	50,814

Prentiss (1925), contains such analytical figures and the derived minimal molecular weight values; for comparison, similar figures for sulphide and total sulphur are given. The good agreement, for any given haemoglobin, of the different minimal values shows the possibilities of the method. It is probable that a repetition of some of the analyses, making use of recent improvements in analytical technique, would lead to still better agreement.

In subsequent chapters (especially Chapter VII) it is shown how the value of n, the integer by which the commonest minimal molecular weight ($ca.$ 16,700) must be multiplied to give the true molecular weight, was determined to be 4 for several haemoglobins, and the actual molecular weight approximately 67,000.

The determination of copper in the haemocyanins, the respiratory pigments of certain lower animals, has been used similarly. However, different species show greater variation in copper content and minimal molecular weight than in the case of the haemoglobins. Further, the minimal molecular weight is a much smaller fraction of the real molecular weight and is therefore much less valuable in molecular weight determination.

In certain proteins, analysis for *trace* amino acids may similarly be employed to give minimal molecular weight. Thus, as shown in Table 6.1, the arginine content of ox haemoglobin gives a value agreeing well with that from the iron and sulphur contents, and in the case of the wheat protein, glutenin (Table 6.2), other trace amino acids have been utilized. Where, as in this latter case, reliable data from other methods are not easily available, such analytical results assume greater importance.

TABLE 6.2. *Minimal Molecular Weight of Glutenin (from Wheat)*
(Cohn, Hendry, and Prentiss, 1925)

Method	Amount of constituent present in per cent.	Weight containing 1 gm. molecule	Assumed number of molecules	Minimal molecular weight
Tryptophane content	1·68	12,149	3	36,447
β-Hydroxyglutamic acid content	1·8	9,061	4	36,244
Tyrosine content	4·5	4,024	9	36,216

To be of use, the particular amino acid being estimated must clearly occur in small quantity and be capable of accurate determination; for these reasons tyrosine, cystine, and tryptophane are generally the most useful. In individual cases other acids may also be useful, e.g. arginine in ox haemoglobin.

(b) *Combining weights (including acid and base titrations)*

Equivalent or minimal molecular weights may also be determined as the weight of material which combines or reacts with 1 gm. molecule of monovalent acid or base or other chemical reagent. In the case of the proteins, the free acidic and basic groups contained by each molecule are capable of reacting normally with added substances so that such minimal molecular weights, determined from maximum acid and base binding capacities (see p. 334), are readily obtained. The disadvantage is, in this case, the large number of reactive acidic and basic groups per protein molecule; the minimal molecular weight is thus too small a fraction of the true molecular weight to be of much value. For instance, the egg albumin molecule, of molecular weight approximately 44,000, possesses *ca.* 45–9 free acid groups: the minimal molecular weight of the protein anion will therefore be roughly one-fiftieth of the true molecular weight.

Reactions with acidic or basic dyes have similar disadvantages in molecular weight determinations. In a later chapter (XII) it will, however, be shown that acid and base titration curves of proteins have considerable importance in providing estimates of the electrical charge of the molecule under different conditions of hydrogen ion and salt concentration.

A very useful application of combining weights has been made in the case of the respiratory proteins. Oxygen and carbon monoxide combine reversibly with haemoglobin and the maximum absorption of each gas is strictly proportional to the amount of protein present. Quantitatively it was shown (see Greenberg, 1938) that 1 gm. of haemoglobin combines with a maximum of 1·34 c.c. of oxygen or carbon monoxide at N.T.P. and that the weight of protein which combines with 1 gm. mole of either gas is therefore

$$M_e = \frac{22400}{1\cdot 34} = 16717.$$

This value is in good agreement with minimal molecular weight values calculated from the iron and sulphur content of haemoglobin (Table 6.1), so that 1 mole of gas is absorbed for every atom of iron in the protein. A similar comparison in the case of the haemocyanins shows that 1 atom of oxygen is bound for every atom of combined copper contained by the protein.

(c) *End-group analysis*

In applying the method of end-group analysis it is necessary either that the structure of the polymeric units be known, or that some structure be postulated, upon which the interpretation of results is closely dependent. The classical example is the study, by Haworth and co-workers, of the chain-length or degree of polymerization of cellulose (see e.g. Haworth and Machemer, 1932). Assuming that cellulose is a linear array of glucose residues joined by 1:4 links, then the structure of the completely methylated derivative is given by:

[Structure diagram of methylated cellulose showing 2,3,4,6 tetramethyl end unit and 2,3,6 trimethyl repeating units]

At *one* end of each chain there is a 2,3,4,6 tetramethyl unit, whilst all other glucose residues occur as the 2,3,6 trimethyl derivative. (It should be noted that, at the other end of the chain, the glucose residue exists as a trisubstituted methyl glucoside.) Clearly a knowledge of the number of 2,3,4,6 tetra substituted derivatives in a methylated cellulose gives the number of separate cellulose chains, and with the total amount of material, the mean number average (see p. 42) molecular weight or chain-length.

To estimate the mean chain-length of cellulose the method is applied as follows. The cellulose is completely methylated by the mildest method available, to minimize hydrolysis of the glucose–glucose links, and the methylated cellulose is then hydrolysed under conditions such that glucose–glucose links are broken but substituting methyl groups are not affected. Substituted glucose derivatives (some glucosides) are thus obtained, converted (if not already) into their corresponding methyl glucosides, and fractionally distilled to give the amount of tetra substituted methyl glucoside.

The presence of tetramethyl glucoside confirmed that the cellulose chains were not endless, and the quantity occurring was used to calculate degrees of polymerization. Haworth and Machemer thus obtained the value 100 to 200 β-glucose units, corresponding to a molecular weight of 20,000 to 40,000. Haworth regarded such estimates only as lower limits, for degradation during methylation could not be entirely neglected, and other experimental difficulties probably introduced further inaccuracy.

The use of reducing power and other methods to estimate terminal glucose residues is very similar to the method just described and will not be further considered.

Carothers and co-workers (1933) made a precise end-group determination for poly ω-hydroxydecanoic acid,

$$HO[-(CH_2)_9CO-O-]_n(CH_2)_9CO-OH,$$

by titrating the terminal carboxyl groups with alcoholic potash in chloroform solution. The number-average molecular weights so obtained were compared in some cases with sedimentation equilibrium values from the ultracentrifuge. For one sample of material, which formed the approximate upper limit in molecular weight for the titration method, values of 25,700 and 26,700 were obtained by end-group and ultracentrifuge methods respectively.

Many other procedures for end-group determinations have been used, e.g. the so-called Zerewitinoff method (e.g. Bolland, 1941) by which OH or SH groups in a polymer molecule are estimated by reaction with Grignard reagent and measurement of the volume of the gas (usually methane) evolved. By such methods information concerning reaction mechanism, as well as molecular weights, has been provided.

All end-group determinations depend on a knowledge of the structure of the polymeric molecule, and in many cases this constitutes a serious disadvantage of the method. Thus in a case where the structure is assumed to be of the linear chain type, any occurrence of branching, of closed chains, or other different structures (unless quantitatively allowed for) must introduce error. It should also be noted that, as for all number-average determinations (see p. 42), the presence of small amounts of highly degraded materials has an effect out of all proportion to its weight fraction upon the number-average molecular weight; thus for highly polydisperse materials, such values may not be a very useful criterion of the high polymer portion contained. In such cases the advantages of fractionation (p. 43) into nearly monodisperse fractions are clear. The use of end-group determinations is seen to be not without difficulties even in favourable cases, and the method has now been largely superseded by others, to be described in later chapters.

With other analytical techniques, the end-group method is not applicable to colloidal particles generally, but only to certain systems (e.g. colloidal *molecules*) in which the necessary stability of the colloidal particle during isolation and purification occurs. Clearly it would be advantageous for many systems if the particle could be examined as it occurs in the colloidal system (i.e. without isolation and purification), and it is with experimental methods which make this possible that subsequent chapters largely deal. In some cases we shall find that such an examination can be made by means of the colligative methods, whose extensions for use in colloidal problems we shall now discuss. In other cases entirely new procedures have been evolved.

The colligative methods

The *colligative* phenomena, comprising the lowering of vapour pressure and of freezing-point, the elevation of the boiling-point, and osmotic pressure, all possess the same sound thermodynamical basis, and the possibility of applying them to the study of colloids is therefore of paramount importance. Consider a simple colloidal solution, e.g. an aqueous solution of haemoglobin of molecular weight 68,000, containing 1 gm. of dry protein per 100 gm. of water (i.e. 10 gm./1,000 gm. of water). In order to calculate the order of magnitude of the colligative phenomena, let us assume ideal behaviour, and consider simply the lowering of the freezing-point. The ideal lowering of the freezing-point for 1 gm. molecule of any solute in 1,000 gm. of water (i.e. the molecular depression of the freezing-point) is 1·86°. The ideal freezing-point depression for the haemoglobin solution is therefore

$$\frac{10}{68000} \times 1\cdot 86 = 0\cdot 0027° \text{ C.},$$

a quantity so small that it cannot be measured with any accuracy by normal methods. Much more serious, however, is the fact that traces of low molecular weight impurity (e.g. salts) in the solid haemoglobin would give depressions quite as large (or even larger) than that of the protein (e.g. 0·1 per cent. of impurity of molecular weight 70 in the solid protein would give an equal depression).

Similar difficulties are associated with the elevation of the boiling-point and more so with the lowering of the vapour pressure, and, unless some revolutionary improvements in technique arise, it is clear that their application to the determination of particle size in colloidal systems cannot be of much help.

However, the use of osmotic measurements offers greater possibilities in view of greater sensitivity, and because the effects of low molecular weight impurity may be eliminated. Thus, for the haemoglobin solution already considered, the osmotic pressure due to the protein is given by

$$\frac{10}{68000} \times 22\cdot 4 \times 760 \times 13\cdot 6 \text{ cm. of water,}$$

i.e. *ca.* 3·4 cm. of water.

Pressures of such a magnitude are clearly measurable with some accuracy. The elimination of the osmotic effects of low molecular weight impurity is discussed in detail in Chapter VII; it will suffice to say here that this is accomplished by the use of either iso-electric colloids or suitably large diffusible salt concentrations. Of the colligative methods, therefore, only that of osmotic pressure has so far been extensively employed to obtain colloidal molecular weights, but its success warrants the allotment of a whole chapter to its description (Chapter VII). It will be seen that its application is chiefly to low and medium molecular weight polymers. However, one of the newer methods, that of sedimentation equilibrium, which also possesses a thermodynamic basis, is of considerable use for the higher molecular weight systems where the sensitivity of the osmotic method is inadequate. In addition, the so-called dynamic methods give much valuable information on very high molecular weight systems, and, with the equilibrium methods, are described in Section A.

The partial specific volume factor

Common to most of the methods to be described is the question of partial specific volume which, following the definition of partial *molar* volume already given (p. 63), is defined for a component i as the increase in volume when 1 gm. of the component i is added to a very large volume of the system, i.e.

$$\bar{v}_i = \frac{\partial V}{\partial w_i}, \qquad (6.2)$$

where V is the total volume of the system and w_i is the weight of component i in the mixture.

The specific volume (v) of the system is given by

$$v\left(=\frac{1}{\rho}\right) = \frac{V}{w_1+w_2+\ldots}. \qquad (6.3)$$

For a system containing two components 1 and 2 (or which may be treated thus) the partial specific volume is often expressed in terms of weight fractions W_1 and W_2 which, in terms of component weights w_1 and w_2 may be written

$$W_1 = \frac{w_1}{w_1+w_2} \quad \text{and} \quad W_2 = \frac{w_2}{w_1+w_2}. \qquad (6.4)$$

Making use of eqns. (6.3) and (6.4) in (6.2), we obtain (see Butler, 1934, p. 65)

$$\bar{v}_1 = v - W_2\left(\frac{dv}{dW_2}\right), \qquad (6.5)$$

a relation in which the partial specific volume of one component, under given conditions, is expressed in terms of experimental quantities obtained by plotting the specific volume against weight fraction of the other component. Details of the determination of partial specific volumes have already been described (e.g. Kraemer, 1940, p. 57; Lewis and Randall, 1923) and will not be discussed here.

For a component dissolving in a solvent without interaction, and in the absence of electrostriction effects (such as the decreased volume arising from high local electrical fields, usually occurring in ionic solutions), the partial specific volume is clearly related reciprocally to the density of the pure component, i.e.

$$\bar{v}_i = \frac{1}{\rho_i}. \tag{6.6}$$

For this reason the partial specific volume is prominent in the quantitative treatment of sedimentation velocity, the effective density of a solute in solution being

$$(\rho_{solute} - \rho_{solvent}) \quad \text{or} \quad \rho_{solute}(1 - \bar{v}_{solute}\rho_{solvent}).$$

More will be said in this connexion in Chapter XI. Partial specific volume is also used considerably in converting experimental concentrations, which are usually expressed according to weight, into volume concentrations and volume fractions. The latter modes of expressing solution concentrations are often encountered in the next few chapters, especially in the theoretical treatment of the properties of solutions (e.g. viscosity-concentration relations, p. 371).

If the solute interacts with the solvent, i.e. is 'solvated', the increase in volume on introducing a given amount into the system may not represent the actual volume of the solute as it occurs in the system. In such a case the kinetic unit whose properties largely govern the behaviour of the solution must include solvating solvent, and the real volume of the disperse phase is therefore greater than that of the anhydrous solute, which alone is given by the increase in total volume (in absence of electrostriction). If electrostriction effects are of importance, then the total volume change cannot be related to either the solvated or unsolvated solute volume. However, it is probable that electrostriction effects are small in all polymer systems except those containing molecules of extremely high charge (see Cohn and Edsall, 1941, p. 373). In the remainder of this book it is therefore assumed that partial specific volumes, as determined by means of eqn. (6.5) and similar methods, refer to the unsolvated solute. Arising from this, it will be seen in Chapter XI, that although the sedimenting species is undoubtedly in many cases solvated, the molecular weight calculated from sedimentation velocity measurements is approximately that of the unsolvated molecule.

INTRODUCTION

In general, partial specific volumes are obtained experimentally, as already indicated, from density measurements; typical values are included in Table 6.3. It should, however, also be realized that if the composition of a given polymer is known accurately, then, making use of the conclusions of Kopp, Traube, and others regarding the approximate additivity of molecular volume, it is possible to calculate the partial specific (and partial molar) volume of a polymer from the volumes of the constituent groups and atoms. In the case of certain proteins, from the incomplete data on amino acid composition available, Cohn and Edsall (1941, p. 375) have made such calculations which are seen from Table (6.3) to agree reasonably with the observed values.

TABLE 6.3. *Calculation of Partial Specific Volumes of Proteins from Amino Acid Composition*

(Cohn and Edsall, 1941, p. 375)

Amino acid	Specific volume of amino acid residue v	Volume per cent. of amino acid residue $vW \times 100$				
		Insulin	Egg albumin	Edestin	Gelatin	Zein
Glycine	0·64	..	..	1·85	12·40	..
Alanine	0·74	..	1·31	2·12	5·13	5·78
Serine	0·63	1·86	..	0·17	1·72	0·53
Threonine	0·70	1·58	..	..	0·83	..
Valine	0·86	..	1·82	4·08	..	1·37
Leucine and isoleucine	0·90	23·29	8·32	16·23	5·51	19·40
Proline	0·76	..	2·66	2·63	12·63	5·80
Oxyproline	0·68	..	..	1·18	8·44	0·47
Phenylalanine	0·77	..	3·48	2·12	0·96	5·21
Methionine	0·75	..	3·45	1·57	..	1·55
Cystine	0·61	7·05	1·01	0·77	0·10	0·51
Tryptophane	0·74	..	0·90	0·98	..	..
Tyrosine	0·71	7·98	2·54	2·90	..	3·77
Histidine	0·67	6·34	0·88	1·43	1·74	0·48
Arginine	0·70	1·91	3·55	10·52	5·45	1·00
Lysine	0·82	0·90	3·57	1·71	4·25	..
Aspartic acid	0·60	..	4·20	6·23	1·73	0·93
Glutamic acid	0·66	9·10	3·16	1·21	1·12	..
Hydroxyglutamic acid	0·60	..	0·73	..	..	1·33
Glutamine	0·67	8·33	6·21	10·94	2·30	18·55
Volume per cent. of amino acid residues $- \sum v_i W_i \times 100$		68·34	47·79	68·64	64·31	66·68
Weight per cent. protein of amino acid residues $- \sum W_i \times 100$		93·43	65·20	93·30	90·47	88·74
Specific volume Calculated $\sum v_i W_i / \sum W_i$		0·73	0·73	0·74	0·71	0·75
Observed		0·749	0·749	0·744	..	..

Solvation

We have seen that where solvation of any solute occurs, then its volume concentration as deduced from anhydrous partial specific volumes does not represent the total volume fraction of the solvated component. Clearly, therefore, a knowledge of the difference between solvated and unsolvated volumes, and of solvation generally, is basic to a more complete understanding of the behaviour of solutions.

Little is known in detail of the extent of solvation in individual systems, but the general features are explicable in ordinary terms. Perhaps the best-known cases of hydration are those of the alkali metals, in which the hydration decreases so rapidly with increase in atomic number that ionic conductivities increase with increasing atomic number, the reverse of what would be expected from the sizes of the unhydrated ions (Glasstone, 1937). Thus lithium, with the lowest atomic number and smallest unsolvated radius, would be expected to possess the highest conductivity; in practice it has the lowest owing to its being the most strongly hydrated. Potassium, rubidium, and caesium, of much higher atomic number and unsolvated radii, have considerably increased conductivities owing to their smaller degrees of hydration. Little is definitely known, even in this case, about the precise mechanism by which solvating liquid is held, but it is likely that, in addition to the formation of specific compounds (e.g. ion hydrates, which also occur in crystal hydrates), solvent is held by the interaction of the ions with solvent dipoles both originally present and induced. The high solvation of small ions is readily explained on this basis.

The solvation of colloidal materials is probably to be explained on similar principles. Thus in addition to solute-solvent compounds, which may form at suitably active points on the heterogeneous surface of the colloidal particle, weaker interactions, such as ion-dipole, dipole-dipole, and van der Waals forces, will also occur, probably with all possible gradations between them.

On general grounds, the strongly bound layer is unlikely to exceed a monolayer, which probably exists independently of Brownian or any other motion undergone by the particle, and must therefore be included as part of the kinetic unit.

Experimental data on the solvation of colloidal materials is very meagre, but in certain systems enough is known to establish its order of magnitude. In the next few pages we shall outline briefly work done on the hydration of proteins, which, as we have said, have an important advantage over many other colloidal systems in that their properties,

INTRODUCTION

and especially their state of subdivision, are readily reproducible. For a more detailed review see Adair (1937).

Probably the most precise estimates of protein hydration have been carried out on protein crystals, so that strictly the results so obtained are not applicable to the molecules in solution. It has, however, been argued by Adair and Adair (1936) that such estimates probably represent minimum values for proteins dissolved in dilute solution, and are not much different from the latter. Since we have no direct information on the hydration of dilute solutions of proteins, we have no alternative to accepting this view, which does not appear unreasonable.

Early work on the hydration of crystalline egg albumin was carried out by Sorensen and Hoyrup (1917) on material which was recrystallized several times and then precipitated from a strong ammonium sulphate solution. A weighed amount of the crystals, plus adhering mother liquor, was then analysed for ammonia nitrogen (a_b per 100 gm.) and protein nitrogen (p_b per 100 gm.). Similar quantities (a_f and p_f) were also obtained for the filtrate. Sorensen then assumes that, of the water occurring within the crystals, part is associated with ammonium sulphate as a solution of the same concentration as the filtrate, the remainder being held by solvation forces in the protein crystal; in other words, the ammonium sulphate occurring with the crystals occurs purely in association with the water present and is not specifically held by the protein. It will be seen that it is in this assumption that Sorensen differs from some recent workers.

The total crystal protein nitrogen per 100 gm. of precipitate is obtained by subtracting from p_b the protein nitrogen of the adhering mother liquor, i.e.

$$\text{crystal protein nitrogen}/100 \text{ gm. of precipitate} = p_b - \frac{a_b}{a_f} p_f. \quad (6.7)$$

Further,

$$\text{protein hydrate}/100 \text{ gm. of precipitate} = 100 - \frac{a_b}{a_f} 100. \quad (6.8)$$

If, then, r is a factor by which the crystal protein nitrogen is to be multiplied to give the weight of crystal hydrate, we have

$$r = 100 \frac{(1 - a_b/a_f)}{p_b - (a_b/a_f)p_f} = 100 \frac{(a_f - a_b)}{a_f p_b - a_b p_f}. \quad (6.9)$$

By purely analytical means, Sorensen worked out values of r for several different precipitations at slightly differing ammonium sulphate concentrations, and considers the constancy of r to be evidence for his basic assumption that ammonium sulphate is not specifically held by the protein.

Another factor r_a, similar to r but referring to anhydrous protein and given by

$$r_a \times \text{protein nitrogen} = \text{weight of anhydrous protein},$$

was then determined for salt-free egg albumin, which was dried very thoroughly. The weight (w) of water/gm. of dry protein in the protein hydrate is then given by

$$w = \frac{r - r_a}{r_a}. \tag{6.10}$$

For crystalline egg albumin, Sorensen and Hoyrup obtained the value $w = 0.228$.

Adair and Adair (1936) have extended these calculations using data from Sorensen and others, and compiled Table (6.4), in which values for the hydration of several proteins are contained. It will be noted that such values are of the same order as that of Sorensen and Hoyrup.

TABLE 6.4. *Hydration of Crystalline, Precipitated, and Coagulated Proteins by Sorensen's Method of Proportionality*

(From Adair and Adair, 1936)

Protein	Medium	pH	r	r_a	w
Egg albumin crystals	Ammonium phosphate	4.8	7.89	6.4	0.233
,,	,,	5.5	8.12	6.4	0.269
,,	Ammonium sulphate	..	7.86	6.4	0.228
Egg albumin coagulated	,,	..	7.65	6.4	0.195
,,	Sucrose	..	7.49	6.4	0.171
Serum albumin crystals	Ammonium sulphate	..	8.35	6.41	0.303
Pseudo-globulin precipitated	,,	..	8.88	6.61	0.343
Horse haemoglobin crystals	,,	..	7.82	5.99	0.306

Recently, however, McMeekin and Warner (1942) have criticized Sorensen's method and results on the basis of their own investigation of the hydration of crystalline β-lactoglobulin. In a first series of experiments the complications of salt absorption were avoided by crystallizing from water. The large crystals so obtained were quickly wiped free of water on cotton flannel and the decreasing weight obtained as a function of time. A plot of the weight of a crystal against time (Fig. 6.2) was a smooth curve from which the weight extrapolated to zero time could be readily obtained. Control experiments with inorganic crystal hydrates of known composition demonstrated the accuracy of the extrapolation procedure. The weight of the dried protein crystal was obtained after prolonged oven drying.

INTRODUCTION

It was thus shown that the crystals at zero time contained, on the average, 0·84 gm. of water per gm. of dried protein, a value considerably in excess of those previously obtained.

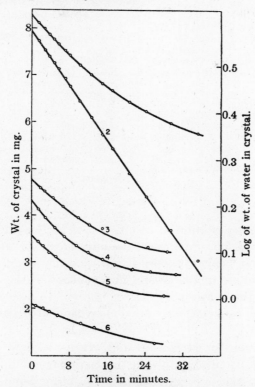

FIG. 6.2. Rate of loss of water by β-lactoglobulin crystals at room temperature. Curves 1, 4, and 5, crystals from salt-free solution; Curve 3, crystal from ammonium sulphate solution; Curve 6, crystal of $Na_2SO_4.10H_2O$; Curve 2, plot of logarithm of weight of water in the crystal shown in Curve 1.

In further experiments crystallization was carried out from concentrated ammonium sulphate solutions, and analysis for ammonium sulphate, water, and protein was carried out (Table 6.5). It was thus shown that the total amount of water in the crystals was not appreciably different from that in the salt-free preparations.

McMeekin and Warner therefore assume that the ammonium sulphate does not occur merely in solution in the crystals as unchanged filtrate. They calculate hydrations in terms of total water held and obtain the values of Table 6.5 in the seventh column which are much in excess of those calculated previously by the Sorensen method. However, it should be noted that the difference does not reside in the experimental data,

TABLE 6.5. *Composition of β-Lactoglobulin Crystals in the presence of Ammonium Sulphate Solutions at pH 5·2 and 25°*

Ammonium sulphate in filtrate wt. %	Hydrated crystal (extr. to zero time), mg.	Dry crystal mg.	Ammonium sulphate in crystal mg.	Ammonium sulphate in crystal water, wt. %	Concn. ammonium sulphate in crystal expressed as % of concn. in filtrate	Water per gm. dry protein, gm.	
						McMeekin and Warner	Sorensen and Hoyrup
28·9	(a) 3·30	1·97	0·415	23·8	82·3	0·86	0·20
	(b) 5·85	3·55	0·719	23·8	82·3	0·82	0·19
30·2	(a) 7·80	4·82	0·915	(23·5)†	(77·8)†	(0·76)†	0·22
	(b) 2·64	1·62	0·328	24·3	80·4	0·79	0·20
31·6	(a) 4·78	2·88	0·683	26·4	83·4	0·86	0·19
	(b) 6·55	4·00	0·910	26·3	83·2	0·82	0·19
				Average	82·3	0·83	0·20

† Not included in the average.

but is one of interpretation of results, for if McMeekin and Warner's data are calculated on Sorensen's assumption (i.e. that the water of hydration is that held in excess of the quantity associated with the ammonium sulphate in the crystal in a solution of the same strength as the filtrate), the figures of the last column of Table 6.5 are obtained, which are of the same order as previously calculated values (e.g. as in Table 6.4).

The justification for McMeekin and Warner's view that the total water is water of hydration lies in the fact that a similar quantity is associated with salt-free crystals. On the other hand, it seems not unreasonable to consider that some of the water is associated with ammonium sulphate rather than with protein.

McMeekin and Warner also carried out density determinations in bromobenzene-xylene mixtures, of wet and anhydrous β-lactoglobulin crystals, both salt-free and salt-containing, and compared these with density values calculated on the basis of known partial specific volumes in dilute solutions and the analytical figures for the crystals, already discussed. The good agreement obtained is especially significant in the case of the salt-free crystal, which, in the wet state, yielded experimentally the very low density 1·146, as against 1·155 required for the high hydration value of 0·84 gm. of water per gm. of dry protein. Crystals equilibrated against 30·2 per cent. ammonium sulphate solution were found, on the other hand, to have a density 1·214, comparing reasonably with the value 1·225 calculated from analytical figures.

INTRODUCTION

Adair and Adair (1936) estimated the hydration of several crystalline, amorphous, and denatured protein preparations in concentrated aqueous media, from the lowering of density (caused by bound water) from the value for anhydrous protein, obtained as the reciprocal of the partial specific volume in dilute aqueous solution. The solid materials were suspended in concentrated salt solutions, whose concentrations were adjusted until on centrifuging they neither rose nor sank. The density of the suspension media were then determined by normal pyknometry, and an accuracy of $\pm 0\cdot 001$ in the final crystal densities was obtained. It was shown that, whilst the density values obtained under given conditions were readily reproducible, definite variations occurred with changes in the composition and pH of the suspension media. Such variations are readily explained on the basis of salt adsorption resulting from the permeability of the crystal and the attainment of a dynamic equilibrium as suggested by Adair and Adair.

The density values obtained, a selection of which is contained in Table 6.6, were all lower than the apparent densities of anhydrous protein in dilute aqueous solutions (i.e. the reciprocal of the partial specific volumes), which are usually taken as $1\cdot 34$.

TABLE 6.6. *The Density and Hydration of Certain Protein Crystals*
(From Adair and Adair, 1936)

Protein	Medium	Temp., °C.	Density $\rho = (1/\bar{v})$	Hydration w
Edestin	Satd. ammonium sulphate and sucrose	21·1	1·317	0·063
,,	Phosphate, pH 5·0	22·0	1·288	0·143
,,	Citrate, pH 8·3	24·0	1·288	0·143
,,	Citrate, pH 6·6	20·9	1·290	0·137
,,	Citrate, pH 4·3	24·4	1·308	0·086
Serum albumin	Half-satd. ammonium sulphate and sucrose, pH 4·8	18·0	1·276	0·192
,,	Two-thirds satd. ammonium sulphate and sucrose, pH 4·8	18·0	1·279	0·182
,,	Citrate, pH 6·6	20·8	1·246	0·304
,,	Phosphate, pH 5·0	20·3	1·237	0·344
,,	Phosphate, pH 3·2	21·5	1·278	0·185
Egg albumin	Phosphate, pH 5·0	19·0	1·239	0·324
,,	Phosphate, pH 3·2	20·0	1·268	0·208
Serum globulin	Phosphate, pH 5·0	20·0	1·236	0·331

On calculating the hydration from the formula

$$w = \left(\frac{\rho_p - \rho}{\rho - \rho_1}\right) \frac{\rho_1}{\rho_p}, \qquad (6.11)$$

where ρ, $\rho_p (= 1/\bar{v}_p)$, and ρ_1 are the densities of the crystal, of anhydrous protein, and water respectively, it therefore follows that positive values of w will be obtained. Table 6.6 contains the hydration values so obtained, which are of the same order of magnitude as those of Table 6.4 except in the case of edestin, the protein from hemp-seed.

The use of formula (6.11) involves the assumptions:
(1) That the specific volumes of water and anhydrous protein are the same in the crystal as those normally obtained.
(2) That no salt adsorption occurs.

The errors involved in deviations from assumption (1) are likely to be small, but in the light of McMeekin and Warner's work, it appears that salt adsorption may increase crystal densities far above the salt-free values and thus cause low calculated hydration values. The particularly high density of edestin in ammonium sulphate–sucrose solution is probably to be attributed to sucrose adsorption, leading also to a very small calculated hydration.

In summarizing the crystal hydration determinations, it is probably true to say that the salt-free experiments of McMeekin and Warner are more free from objection than any others, but it must also be said that the results of hydration determinations by other methods, though possibly of a less precise character, do agree reasonably amongst themselves in giving much smaller hydration values. It is clearly necessary that the method of McMeekin and Warner should be applied to other protein systems showing apparent low hydration. Should high results be obtained also in these cases, then it would appear inescapable that salt adsorption largely invalidates the Sorensen analytical and the simple density methods.

Valuable information concerning crystal hydration may also be obtained from a combination of X-ray and density measurements on wet and dry crystals. Thus from a knowledge of the dimensions of their unit cells, of the number of molecules per unit cell, and of their densities, the molecular weights of the anhydrous and hydrated protein may be calculated and the degree of hydration obtained (see Chapter XV). Crowfoot and Riley (1939), for lactoglobulin, give as anhydrous and hydrated molecular weights the values 37,600 and 52,400, from which the crystal hydration may be calculated as 0·39 gm. of water per gram of dry protein, a value somewhat higher than most other estimates but lower than that of McMeekin and Warner. It is probable that much more information of this type will be available in the immediate future.

Neurath and Bull (1936) made estimates of the hydration of native and denatured egg albumin by observing the contraction in volume when water was added to the dry protein preparations. Volume contractions were plotted against the composition of the mixtures, and from the curves the hydration of native egg albumin was estimated as 0·36 gm. per gram of dry protein. From similar experiments with alcohol–water mixtures they conclude that part of the water is loosely bound and another part more strongly.

Moran (1935, and other papers) also estimated the hydration of egg albumin and other protein preparations by freezing protein–water mixtures in collodion sacs surrounded by ice. Water passed through the sacs as the temperature fell, until equilibrium was reached, at which the composition and temperature were recorded; this was repeated for a series of temperatures down to $-20°$ C. or lower. The equilibrium composition was also recorded on subjecting the mixtures to high pressure, and it was shown that when the composition was plotted against the activity of water, the two sets of observations fell approximately on the same curve, which showed a distinct flat portion in which composition varied only slightly with water activity. At this stage the water content of the systems, containing either native or denatured egg albumin and water, was 0·26 gm. per gram of dry protein; a figure in reasonable agreement with that of Sorensen. However, Moran also showed that in the presence of sodium chloride, there was a considerable lowering of the hydration from this value.

An interesting estimate of protein hydration has been made by Bull and Cooper (1943), who compared diffusion and viscosity data for several proteins. In explaining their method it is necessary to anticipate the contents of subsequent chapters. Values of $(D_0/D)_A$ (Chapter X) have been plotted against $(\eta_{sp}/\phi)_A$ (Chapter XIII), where η_{sp} is the specific viscosity of the protein solution and ϕ is the protein volume fraction (see p. 358), the subscript A denoting the assumption of unhydrated molecules. Anticipating further results (Chapter XIII), (η_{sp}/ϕ) should, by Einstein's equation, have the value 2·5 for spherical molecules in solution, whether hydrated or unhydrated, provided the correct volume concentration is utilized. Further, for $\eta_{sp}/\phi = 2·5$, D_0/D should be unity, provided D_0 is calculated correctly, allowing for increased molecular volume due to hydration. However, the curve of $(D_0/D)_A$ against $(\eta_{sp}/\phi)_A$ does not, in practice, pass through the point $(\eta_{sp}/\phi)_A = 2·5$, $(D_0/D)_A = 1$ for the reason that hydration was neglected in assessing ϕ and D_0. Bull and Cooper therefore determined the mean increase in volume concentration of the solute which would displace the curve to pass through this point, and obtained the value 0·283 c.c. per c.c.

of dry protein or 0·21 gm. water per gm. of dry protein. This calculation, assuming the same hydration for all the protein data covered, is necessarily approximate, but is interesting in providing another approximate confirmation of Sorensen's estimate of hydration.

It will also be seen later, in discussing results from sedimentation velocity and diffusion measurements, that for many proteins a maximum hydration value of 0·30 or 0·40 gm. of water per gram of dry protein appears likely; again close to Sorensen's estimate.

Before leaving the subject of hydration we emphasize again that the hydration of a crystal is not necessarily the same as that of a molecule in solution, and that the latter may vary considerably with the composition and temperature of the solution. Adair and Adair (1936) argue that, in view of the apparent penetration into protein crystals of salt solutions, the environment of a molecule in the crystal immersed in a solvent is very similar to that of a molecule dissolved in the same solution. Further, since protein solutions usually involve the use of salt solutions more dilute than would be possible for the suspension of crystals (i.e. higher water activities), then crystal hydration would appear to set a lower limit to solution hydration. However, it must also be remembered that the water of hydration of a crystal is not necessarily the same as that which is so strongly bound to the protein molecule as to act as part of the same kinetic unit. The authors feel that it is impossible *a priori* to decide which should be the greater, and for this reason probably the most satisfactory procedure in determining solution hydration would undoubtedly involve examination of and measurements on the protein in solution. McBain's (1936) suggestion that hydration should be calculated from the density of the solution in which sedimentation just does not occur (e.g. by eqn. (6.11) used by the Adairs) has therefore much to commend it, though, again, the hydration calculated thus does not necessarily hold for solutions of different salt concentration. McBain's method has not yet been applied, but if the calculated hydration were found to be independent of, or insensitive to, the composition of the solvent, then it would appear to have considerable possibilities.

Sufficient has been written here to show the present uncertainty of estimates of hydration. The weight of available evidence appears to favour an average value of about 0·3 gm. of water per gram of dry protein, with an upper limit of about 0·4, but future revision of these values is quite possible.

It is of interest that the possession of a single complete solvating layer of water molecules by a spherical protein molecule of diameter 50 A (approximately haemoglobin) corresponds roughly to 0·30 gm. of water per gram of dry protein (see p. 126). The hydration of larger molecules would, on this view, be expected to be smaller, a conclusion which has experimental support in the low crystal hydration of edestin (molecular weight *ca.* 300,000), as compared with that of substances of lower molecular weight also included in Table 6.6.

REFERENCES

ADAIR, 1937, *Ann. Rev. Biochem.* **6**, 163.
ADAIR and ADAIR, 1936, *Proc. Roy. Soc.* B **120**, 422.
BOLLAND, 1941, Publication No. 8 (III), British Rubber Producers' Research Association.
BULL and COOPER, 1943, in *Surface Chemistry*, American Association for the Advancement of Science, p. 150.
BUTLER, 1934, *The Fundamentals of Chemical Thermodynamics*, Part II, Macmillan.
CAROTHERS and VAN NATTA, 1933, *J. Amer. chem. Soc.* **55**, 4714.
COHN and EDSALL, 1941, *Proteins, Amino Acids and Peptides*, Reinhold.
COHN, HENDRY, and PRENTISS, 1925, *J. biol. Chem.* **63**, 721.
CROWFOOT and RILEY, 1939, *Nature*, London, **144**, 1011.
GLASSTONE, 1937, *The Electrochemistry of Solutions*, 2nd edn., Methuen, p. 45.
GREENBERG, 1938, in *The Chemistry of the Amino Acids and Proteins*, edited by C. L. A. Schmidt, published by Thomas, p. 334.
HAWORTH and MACHEMER, 1932, *J. Chem. Soc.* **134**, 2370.
KRAEMER, 1940, in *The Ultracentrifuge* by Svedberg, Pedersen, and others, Oxford.
LEWIS and RANDALL, 1923, *Thermodynamics*, McGraw-Hill.
McBAIN, 1936, *J. Amer. chem. Soc.* **58**, 315.
McMEEKIN and WARNER, 1942, ibid. **64**, 2393.
MORAN, 1935, *Proc. Roy. Soc.* B **118**, 548.
NEURATH and BULL, 1936, *J. biol. Chem.* **115**, 519.
SORENSEN and HOYRUP, 1917, *C.R. Lab. Carlsberg*, **12**, 169.

VII
OSMOTIC PRESSURE AND MEMBRANE PHENOMENA
Osmotic Pressure

In the last chapter we saw that, of the colligative methods for determining the molecular weight of low molecular weight substances in dilute solution, only the osmotic pressure method may readily be extended experimentally to the domain of colloidal and high molecular weight substances. The lowering of the vapour pressure and of the freezing-point, and the elevation of the boiling-point, for such cases are normally much too small to be observed experimentally, and, even if measurable, are unreliable owing to the comparatively large effect of traces of low molecular impurity (e.g. salts, solvent, etc.). Not only does the osmotic method provide measurable effects, but owing to the permeability of the membrane, the absence of trace impurities from the high molecular weight material is not essential. The last point will be explained further in this chapter.

Applicability of the van't Hoff relation

The calculation of molecular weights from the observed osmotic pressures of dilute solutions of low molecular weight materials is usually made from the van 't Hoff equation

$$\Pi = \frac{c}{M}RT, \qquad (7.1)$$

where Π is the osmotic pressure, c is the concentration of solute in gm. per litre, and M is the molecular weight of the solute in the solution. Although used in the preceding chapter to calculate the order of the magnitude of osmotic pressures in colloidal systems, the assumptions involved in the derivation of the equation require careful study before deciding whether it is really applicable to such systems, and, if so, under what conditions.

It has been shown in Chapter IV that osmotic pressure is described accurately by the thermodynamically derived equation

$$\Pi = \frac{RT}{\overline{V}_1} \ln \frac{p_1^0}{p_1}, \qquad (7.2)$$

where $\overline{V}_1$ is the partial molar volume of the solvent, and p_1^0 and p_1 are the solvent vapour pressures over the pure solvent and solution respectively. The chief assumptions made in deducing this equation are that the solvent is incompressible, and that its vapour may be treated as a perfect gas. The good agreement between experimental results and those calculated by the equation provide strong justification for the validity of these assumptions (see Table 4.1, p. 75).

The van 't Hoff equation (7.1) is derived from eqn. (7.2), as shown in Chapter IV, by assuming

(1) The validity of Raoult's law.
(2) Dilute solutions, i.e. nearly pure solvent.

Assumption (1) enables p_1/p_1^0 to be equated to the mole fraction of the solvent N_1 $(=(1-N_2))$, and assumption (2) makes it possible to write $\ln(1-N_2) = -N_2$, and the total volume of the system as the volume of the pure solvent present. Since osmotic pressure measurements are usually made in solutions containing only a few per cent. of solute (at the most), assumption (2) is reasonably correct and it remains to consider the validity of Raoult's law for such dilute solutions.

Restricting ourselves for the moment to a consideration of systems in which solute and solvent molecules are of approximately equal size, it is well known that mixtures obeying Raoult's law over appreciable ranges of concentration are exceptional. It can be shown (Hildebrand, 1936, p. 57) that such mixtures occur only when the separate liquid components are very similar physically and chemically and mix without heat or volume change. Clearly in such cases (e.g. n-hexane and n-heptane or ethyl bromide and ethyl iodide) a molecule of one component may be replaced by one of the other without greatly affecting intermolecular forces, and it is not therefore difficult to see that Raoult's law will be nearly obeyed over large concentration ranges.

In general, however, the substitution of one type of molecule for another in a mixture causes definite changes in the intermolecular force fields, and heat and volume changes on mixing liquids are therefore observed. Positive or negative deviations from Raoult's law, as illustrated by the systems methylal ($CH_2(OCH_3)_2$)–carbon disulphide and acetone–chloroform respectively, in Figs. 7.1 (a) and 7.1 (b), are usually observed in real systems. The dotted lines in each figure represent the partial vapour pressures calculated from Raoult's law. In each case it should be observed that Raoult's law is obeyed very closely for the nearly pure liquids, and it is generally true that 'Raoult's law is a limiting law for the solvent regardless of the nature of the solute' (Hildebrand, 1936, p. 45). For such systems, therefore, at concentrations for which assumption (2) is valid, Raoult's law can usually also be considered valid, and the van 't Hoff relationship can be safely applied. It is clear therefore that, for dilute solutions of low molecular weight solutes, the van 't Hoff law is expected to be valid on theoretical grounds.

Now consider solutions of molecules which are large compared with those of the solvent. The factors which are additional to those discussed

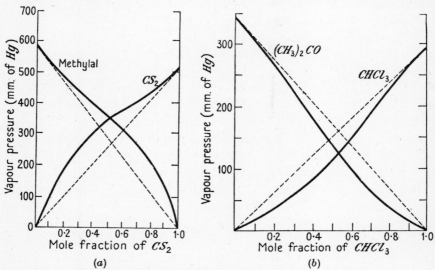

Fig. 7.1. Deviations from Raoult's law (from Zawidzki (1900)). Dotted curves indicate ideal behaviour. (a) Positive. Methylal–carbon disulphide. (b) Negative. Acetone–chloroform.

for solutions of low molecular weight solutes are readily understood in terms of the entropy of dilution by the equation

$$\Pi \bar{V}_1 = T\Delta \bar{S}_1 - \Delta \bar{H}_1. \tag{4.126}$$

Where the heat of dilution is very small, we may write simply

$$\Pi \bar{V}_1 = T\Delta \bar{S}_1. \tag{7.3}$$

For systems in which the molecules of solvent and solute may be treated as approximately equal spheres, the entropy of dilution, $\Delta \bar{S}_1$, may be calculated statistically (Fowler and Guggenheim, 1939, p. 163) to be

$$\Delta \bar{S}_1 = -R \ln N_1, \tag{7.4}$$

which when substituted into (7.3) gives

$$\Pi \bar{V}_1 = -RT \ln N_1, \tag{7.5}$$

an expression which is identical with that obtained from eqn. (7.2) by assuming Raoult's law. Thus the assumption that the entropy of dilution is given by eqn. (7.4) is identical with the assumption of Raoult's law, and deviations from the law may be considered alternatively as deviations from the ideal entropy of dilution. This approach is very useful, since it is possible by statistical mechanical methods to calculate the entropy of dilution assuming various types of molecular model for the solute.

Linear polymers

Such calculations have been performed by Flory (1941, 1942, 1945), Huggins (1941 and 1942), and others, for long flexible chains of spherical sub-molecules, each sub-molecule being similar to a solvent molecule, and a substantial measure of agreement has been obtained. The entropy of mixing is obtained statistically by comparing the total number of possible configurations of the chain molecules in the solute-solvent mixture with the number available initially, followed by the use of the Boltzmann equation connecting the entropy and the probability of a given state. An outline of this derivation is given in Chapter XXVII. We shall merely state the results at this stage. For a chain molecule containing x units or segments, each of which is replaceable by β solvent molecules, Flory calculates the entropy of dilution to be

$$\Delta \bar{S}_1 = -\frac{R}{\beta}\left[\ln \phi_1 + \left(1 - \frac{1}{x}\right)\phi_2\right], \tag{7.6}$$

where ϕ_1 and ϕ_2 are the volume fractions of solvent and polymer respectively. For $x = 1$, and $\beta = 1$, this expression correctly reduces to the ideal form of eqn. (7.4), but where x is large, then $\Delta \bar{S}_1$ values much larger than ideal are obtained. Thus for a solution containing 10 per cent. by volume of a long-chain polymer of molecular weight 100,000, for which $x = 1000$, $\beta = 1$, and the solvent is of molecular weight 100, we obtain from (7.6)

$$\Delta \bar{S}_1 \approx -R[\ln 0{\cdot}90 + 0{\cdot}10]$$
$$\approx 0{\cdot}005R.$$

From the ideal expression (7.4), however, we obtain

$$\Delta \bar{S}_1 = -R \ln N_1$$
$$= -R \ln\left[\frac{0{\cdot}90/100}{0{\cdot}90/100 + 0{\cdot}10/100000}\right] = -R \ln \frac{0{\cdot}9000}{0{\cdot}9001}$$
$$\approx 0{\cdot}00011R.$$

Thus the ideal entropy of dilution is considerably smaller than that calculated on the basis of Flory's equation (7.6), leading to solvent activities which are considerably smaller than ideal. Such differences are very clearly shown in Fig. 4.5(a) (due to Huggins (1943)) in which the solvent activity $a_1 (= p_1/p_1^0)$ is plotted against mole fraction of polymer (N_2) for the rubber–benzene system. The two full curves, calculated on the basis of polymer molecular weights of 1,000 and 300,000, deviate markedly from Raoult's law, depicted as a dotted line. (For details of the calculation see Chapter XXVII.) The deviations from Raoult's law may also be shown by plotting the solvent activity against volume fraction (ϕ_2) (rather than mole fraction) of polymer, a method which is more useful since osmotic measurements are almost invariably determined as a function of concentration expressed on a volume (or weight) fraction, rather than a mole fraction basis. Fig. 4.5(b) shows clearly that high polymers obey Raoult's law only at infinite dilution, and that at any finite volume fraction, the deviation from Raoult's law is usually smaller, the smaller the molecular weight of the solute.

If the heat changes on mixing polymer and solvent are negligible, then by introducing the expression for $\Delta \bar{S}_1$ (eqn. (7.6)) into eqn. (7.3) we obtain a relation connecting osmotic pressure and solute volume fraction, i.e.

$$\Pi \bar{V}_1 = -\frac{RT}{\beta}\left[\ln \phi_1 + \left(1 - \frac{1}{x}\right)\phi_2\right]. \tag{7.7}$$

Putting

$$\ln \phi_1 = \ln(1-\phi_2) \approx -\phi_2 - \frac{\phi_2^2}{2} \quad \text{(valid at low solute concentrations)},$$

$$\phi_2 = \frac{c}{\rho_2}, \quad \text{where } c = \text{grams of polymer per c.c.},$$

and

$$x\beta \bar{V}_1 \rho_2 = M_2, \quad \text{the molecular weight of the polymer,}$$

we obtain

$$\frac{\Pi}{c} = \frac{RT}{M_2} + kc, \tag{7.8}$$

where $k = RT/2\beta \rho_2^2 \bar{V}_1$. It should be noted that for a given material k is a constant independent of molecular weight.

According to this equation, a plot of Π/c (sometimes called the reduced osmotic pressure) against c should give, at low concentration, a straight line with a slope, k, and an intercept on the Π/c axis of RT/M_2. Further, each of a series of fractions of differing molecular weight (M_2', M_2'', M_2''', etc.) should also give straight line plots, parallel to one another (i.e. same k value) but possessing different intercepts (RT/M_2', RT/M_2'', RT/M_2''', etc., respectively), which could be used to determine molecular weights.

Before considering eqn. (7.8) further it should, however, be realized that allowance must be made if heat effects (i.e. $\Delta \bar{H}_1$) are not negligible, as so far assumed. It is usually assumed that $\Delta \bar{H}_1$ may be written as a power series in ϕ_2 beginning with the term in ϕ_2^2, i.e.

$$\Delta \bar{H}_1 = k_2' \phi_2^2 + k_3' \phi_2^3 + \ldots, \tag{7.9}$$

k_2' and k_3' being constants. At low solute concentrations, however, the volume fraction ϕ_2 is proportional to the concentration in gm./c.c. and we may write

$$\Delta \bar{H}_1 = k_2 c^2 + k_3 c^3 + \ldots. \tag{7.10}$$

Incorporating both (7.6) and (7.10) in (4.126) we obtain

$$\frac{\Pi}{c} = \frac{RT}{M_2} + (k - k_2)c - k_3 c^2 + \ldots, \tag{7.11}$$

which at low concentrations reduces to (dropping subscripts on M_2)

$$\frac{\Pi}{c} = \frac{RT}{M} + (k - k_2)c. \tag{7.12}$$

It should be noted from this equation that allowance for heat of dilution does not affect the value of the intercept on the Π/c axis in a plot of Π/c v. c, but merely the slope of the lines. Thus with varying solvents, i.e. varying heats of dilution and values of k_2, the slopes of the Π/c v. c plots for a given material vary, but according to eqn. (7.12) the extrapolation limit RT/M should be independent of such variations. Fig. 7.2 demonstrates these points for three fractions of a given substance dissolved in two different solvents.

The absence of solvent effects on the value of the intercept RT/M is of fundamental importance for the determination of molecular weights by osmotic pressure, and it may be emphasized that theoretically this arises from the absence of a term in the first power of the concentration (i.e. $k_1 = 0$) in eqn. (7.10) for the heat of dilution. Support for this expression comes experimentally from direct measurements of the heat of mixing of simple liquids, and also from the fact that the value of the intercept on the Π/c axis in plotting Π/c v. c is found to be substantially independent of the nature of the solvent.

The determination of $\lim_{c \to 0} \Pi/c$ by extrapolation of the Π/c v. c curve to zero concentration is clearly of great importance in molecular weight determinations and will be further discussed later. It should, however, be observed now that the uncertainties attending extrapolation may be minimized if a solvent or solvent mixture is chosen to make $(k - k_2)$, the coefficient of the term in c in eqn. (7.12), either zero or small. Gee (1942) thus states that rubber solutions in a mixed benzene–methyl alcohol solvent containing 15 per cent. methyl alcohol give Π/c values which are nearly independent of concentration at low concentrations, and from which an accurate extrapolation is therefore possible (see Fig. 27.9).

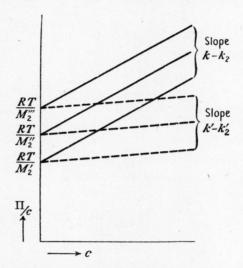

Fig. 7.2. Calculated form of Π/c v. c curves for three different polymer fractions (M_2', M_2'', M_2''') in two different solvents.

Since, as we have seen, ideal behaviour in polymer solutions occurs only at infinite dilution, the van 't Hoff relation must in these systems be replaced by
$$\lim_{c \to 0}(\Pi/c) = RT/M, \qquad (7.13)$$
or
$$M = RT/\lim_{c \to 0}(\Pi/c). \qquad (7.14)$$

Eqn. (7.13) and its derivation apply strictly only to solute molecules approximating to the model assumed, i.e. a linear flexible chain molecule. Whilst entropy and osmotic pressure calculations of the type indicated cover the general features of the behaviour of many linear polymers in dilute solution, it is likely that refinements for individual systems will be developed. Thus in the case of a linear polymer which is folded into a permanent, compact, and nearly spherical configuration (e.g. corpuscular proteins), the entropy of dilution ($\Delta \bar{S}_1$) is much smaller than that given by eqn. (7.6) and much nearer the ideal van 't Hoff value. This is clearly due to the much smaller effective volume of the single polymer molecule and to the consequent reduction in the number

OSMOTIC PRESSURE AND MEMBRANE PHENOMENA 157

of available configurations in solution. Deviations from the ideal van 't Hoff osmotic pressure relation are therefore much less severe (i.e. $(k-k_2)$ of eqn. (7.12) small), and the dependence of Π/c upon c is much less important than for solutions of extended linear polymer molecules. Fig. 7.3, due to Meyer (1942, p. 594), demonstrates the point very clearly in a comparison of osmotic data for cellulose nitrate in acetone and haemoglobin in aqueous solution, the two polymers being approximately of the same molecular weight. The osmotic behaviour of haemoglobin is typical of all corpuscular proteins except those of very asymmetric shape, and indeed the presence or absence of a pronounced Π/c v. c dependence is a useful criterion of molecular shape. Whilst for the corpuscular proteins extrapolation to zero concentration is not so important as for solutions of linear polymers, yet it is essential for accurate molecular weight measurement.

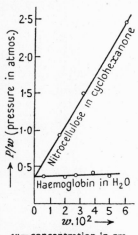

FIG. 7.3. Reduced osmotic pressure of nitrocellulose and haemoglobin as a function of the concentration by weight.

An equation has been suggested empirically (Burk and Greenberg, 1930) to cover the observed osmotic pressures of protein solutions up to high protein concentrations:

$$\Pi = K\frac{100c_v}{100-hc_v}, \qquad (7.15)$$

where c_v is the solute concentration per 100 c.c. of solvent, and K and h are constants. This equation differs from the van 't Hoff relation only as a result of the term hc_v, and it has been suggested that the value of h is to be regarded as a measure of solute hydration. Such a statement presupposes the absolute correctness of the van 't Hoff equation, even at high solute concentration, if hydration is allowed for. We have already seen, however, that for reasons other than hydration, the van 't Hoff relation is not expected to hold except at low concentration, so that it is sounder to consider h as merely an empirical constant and eqn. (7.15) an empirical equation.

Donnan equilibria and osmotic pressure determinations

The thermodynamic treatment of Donnan membrane equilibria has already been considered (p. 76) and will now be used to discuss the methods for eliminating Donnan disturbances. The method of allowing for unequal distribution of diffusible ions is considered later (p. 184).

Whenever the high molecular weight solute, to which the osmometer membrane is impermeable, possesses an electrical charge, equilibrium requires a certain unequal distribution of the small permeable or diffusible ions. Arising from the equilibrium condition that the electrochemical potentials of diffusible ions must be the same on the two sides of the membrane, we obtain for a system containing the sodium salt of a protein (sodium proteinate), sodium chloride, and water (Fig. 4.2) the relation

$$\frac{[\text{NaCl}]_{\text{outside}}}{[\text{NaCl}]_{\text{inside}}} = 1 + \frac{c_1}{c_2}, \qquad (7.16)$$

where c_1 and c_2 are respectively the concentrations (in equivalents per litre) of the protein solution inside the membrane and the sodium chloride solution initially outside, the volumes of the two solutions being equal. The difference in the concentration of the diffusible ions on the two sides of the membrane gives rise to an ion pressure difference (Π_i) which has the effect of increasing the observed osmotic pressure, as is shown in Fig. 7.4, containing osmotic data for haemoglobin solutions. From eqn. (7.16) the magnitude of this pressure difference can be decreased either by decreasing c_1, the equivalent protein concentration, or by increasing c_2, the initial external salt concentration. For protein concentrations not greater than 2–3 gm. per 100 c.c., the use of 0·10 molar buffered solutions is usually sufficient to reduce the ion pressure difference to below experimental error. The use of lower protein concentrations makes possible the use of correspondingly small salt concentration. Alternatively, the equivalent protein concentration may be effectively reduced as demonstrated in Fig. 7.4 by working near the iso-electric point. In absence of a net charge on the non-diffusible ion (as at the iso-electric point), there should be no unequal distribution of diffusible ions and therefore no ion pressure difference. However, aggregation and precipitation, which are liable to occur to a great extent at pH values near the iso-electric point, may cause complications. For this reason it is usually safer to work at a pH at least one unit from the iso-electric point where charge effects are small and the danger of aggregation reduced.

Summarizing, therefore, we see that in general, for osmotic work in aqueous media, it is necessary to minimize Donnan effects by the use of relatively high concentrations of diffusible ions, low concentrations of non-diffusible ions, and of pH's not far removed from the iso-electric point. Under such conditions, observed osmotic pressures are independent of salt concentration and of the presence of diffusible impurities (occurring initially with the colloidal material before dissolving) owing to the permeability of the membrane. It will be recalled that such impurities are very troublesome in the other colligative methods, giving observed effects much larger than that of the high molecular weight colloid.

In the case of colloidal solutions in organic solvents, charged ions, both of diffusible and non-diffusible types, are usually absent, and

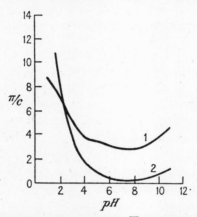

FIG. 7.4. Reduced total osmotic pressure (Π/c) and ion pressure (Π_i/c) as functions of pH at constant ionic strength (0·1) for dilute solutions of haemoglobin and its derivatives. (Pressure in mm. of Hg: c in gm. dry protein per 100 c.c. of solution.)

Curve 1. Π/c (observed). Curve 2. $1\Pi_i/c$ (calculated from membrane potential measurements by equation (4.149)). (From the figure of Adair and Adair (1938).)

Donnan effects do not therefore occur. However, for other reasons already discussed (especially the existence of marked deviations from ideal behaviour), it is necessary to work at low colloid concentration only in making molecular weight determinations. This requirement holds generally for all osmotic pressure determinations in colloidal systems, and reliable results can only be obtained if it is strictly obeyed.

Having now discussed theoretically the use of osmotic pressure determinations in colloidal systems it is necessary to consider the more practical aspects. A 1 per cent. solution of a colloidal molecule of molecular weight 68,000 gives at ordinary temperatures an osmotic pressure of about 3 cm. of water (p. 136). From eqn. (7.1) the osmotic pressure, Π, for a given weight concentration c is inversely proportional to the molecular weight of the solute; solutes of molecular weight 350,000 and 1,000,000 in 1 per cent. solutions would therefore give, ideally, osmotic pressures of ca. 0·6 cm. and 0·2 cm. respectively. Since high solute concentrations are barred theoretically, it is clear that the accurate use of the osmotic method for high molecular weight colloids depends very much on the precise measurement of very small heads of pressure. To understand the procedures applied to this problem it is necessary now to consider the design of osmometers, of which two main types, suitable for aqueous and organic solvent work respectively, have been devised. In view of these different types, and also as a result of the greater importance of Donnan effects and less serious nature of deviations from Raoult's law in aqueous systems, it is convenient to discuss aqueous and non-aqueous systems separately.

Aqueous systems

The measurement of osmotic pressure in aqueous colloidal systems was first carefully investigated by Sorensen and co-workers (1917), using solutions of egg albumin. The experimental procedure was carefully examined and the effects of varying the salt and hydrogen-ion concentration investigated, a molecular weight of 34,000 being obtained.

Adair (1925a and later papers) subsequently developed both the theory and practice of the method, which is now capable of considerable precision even in its simplest forms. He has described several types of osmometer, giving details of their construction and use and has stated the criteria which must be satisfied before readings of osmotic pressure could be accepted as true equilibrium values. A simple form of osmometer suitable for pressures up to ca. 30 cm. of water, developed by Adair[†] from an earlier model (Adair, 1925), is shown diagrammatically in Fig. 7.5 (a).

The thimble type of collodion membrane, M (of diameter about 1 cm.), is tied securely by rubber thread upon one end of a short length of closely fitting, good quality, rubber pressure tubing (R) supported by a short length of thin-walled glass tubing. Into the other end of the pressure tubing is forced a suitable length of capillary tubing (C) of uniform bore[‡] (1 to 3 mm. internal diameter), in which the osmotic head is to be observed. By careful choice of the size of the membrane, rubber pressure tubing, and capillary tubing, firm and leak-proof joints, which are an obvious but not always easily achieved requirement, can be obtained.

The membrane and support are filled with protein or colloidal solution by pouring from a capillary pipette through the pressure tubing support

[†] The authors are grateful to Dr. G. S. Adair, upon whose information a considerable portion of this chapter is based.

[‡] Precision bore capillary tubing ('Veridia') is available from Messrs. Chance Bros., Ltd., Glass Works, Smethwick, 40, Nr. Birmingham.

OSMOTIC PRESSURE AND MEMBRANE PHENOMENA

taking care to remove all air bubbles. The capillary tube is then inserted, without introducing air bubbles, so that the solution rises inside the tube as an unbroken column, the position of its meniscus being controlled by adjusting the capillary tube in its rubber sleeve. After rinsing externally with the solvent medium, the membrane and support are first immersed in a large tube of this medium and the whole

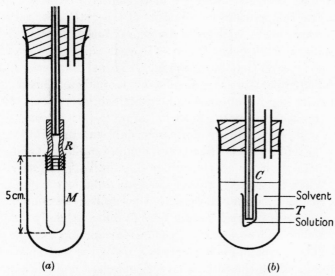

FIG. 7.5. (a) The Adair osmometer. (b) The Adair capillary correction apparatus.

apparatus is then placed in a constant temperature bath ($\pm 0\cdot 2°$ C.) and allowed to come to equilibrium, when the difference in liquid levels (h) inside and outside the capillary tube is measured. This value is greater than the osmotic pressure by a small column of liquid of height (h_c) rising into the tube as a result of capillary effects. This correction must be carefully evaluated by a special measurement after dismantling the osmometer, and is further discussed later. It is advisable also to measure the colloid concentration and density (ρ) of the solution *after* the measurement of the equilibrium head, rather than to accept the initial values, which may have changed as a result of entry or loss of solvent and adsorption of colloid by the membrane. Analytical, refractive index, colorimetric, or any available method may be applied in estimating the concentration of colloidal solute. The osmotic pressure is given in absolute units by

$$\Pi = (h - h_c)\rho g. \qquad (7.17)$$

The use of capillary rather than wide-bore tubing to observe the equilibrium osmotic head of liquid makes possible the establishment of equilibrium in very short times, since the volume of solvent which must be transferred to attain equilibrium is thus much reduced. However, whilst the use of capillary tube reduces the equilibration time, it also introduces difficulties. Thus the smaller the capillary bore the greater is the capillary correction and the more sensitive is the solution head of pressure to temperature fluctuations of the bath. For these reasons it is not practicable in aqueous systems to reduce the internal diameter of the capillary much below 1 mm. For such tubing the capillary correction is not greater than 2 cm. of water, thermal fluctuations are unimportant for a temperature control of $\pm 0.2°$ C., and equilibrium is rapidly achieved (with the membranes to be described later, in *ca.* 24 hours).

When the total osmotic head is large (e.g. > 12 cm. of water) compared with the capillary correction, only an approximate value of the latter is required, and fine-bore (1 mm.) capillary tubing, with a large and possibly inaccurately measurable capillary rise may therefore be used. When the correction is of the same order as the observed head then it is clearly necessary to measure the capillary correction with the greatest possible precision. Larger bore (3 mm.) capillary tubing, with a much smaller and accurately measurable capillary rise, is therefore essential. The correction is measured by using the same piece of capillary tubing in which the total head was observed and the same colloid solution, as far as possible under conditions identical with those occurring during observation of the osmotic head. Failure to comply with these requirements involves considerable error in determining the correction, which in the case of proteins is liable to considerable variation, probably owing to denaturation of the material adsorbed at the meniscus followed by contamination of the walls of the capillary. Adair (private communication) recommends a procedure illustrated by Fig. 7.5 (*b*), in which the colloid solution is contained in a small glass tube, T, fitting closely over the capillary tube, C, which with tube T is then fitted into the external osmometer tube containing solvent medium. The colloid solution diffuses from T only very slowly and the head of liquid in the capillary, which may be observed for several weeks, represents the capillary rise correction under conditions approximating closely to those holding in the osmometer. Adair observes, with this arrangement, a reproducibility of ± 0.1 cm. in a capillary rise of *ca.* 2 cm. in tubing of 1 mm. internal diameter, and ± 0.02 cm. in tubing of 3 mm. internal diameter.

It is worth noting here that whilst the order of magnitude of the capillary correction may be calculated by the formula (see p. 514)

$$h_c = \frac{2\gamma \cos \theta}{r \rho g}, \qquad (7.18)$$

where γ is the surface tension of the liquid at the meniscus in the capillary, θ is the contact angle (assumed zero for aqueous liquid), and r is the radius of the capillary, accurate values of h_c cannot usually be computed for aqueous systems owing largely to uncertainty in our knowledge of γ.

Clearly the accuracy with which low osmotic pressures (*ca.* 1 cm. of water) may be measured in osmometers of the type described is strictly limited, and it follows that the apparatus is not very suitable for the measurement of very high molecular weight. Thus the error involved by a capillary rise uncertainty of 0·1 cm. represents 15 per cent. of the actual osmotic pressure of a 1 per cent. solution of a high polymer of molecular weight 350,000 (*ca.* 0·6 cm. of water).

Greater accuracy than this is clearly very desirable, and we now proceed to discuss some improvements in osmometer design and technique directed towards this end.

Recent improvements in osmometry

The chief defects of the simple Adair type osmometer arise from uncertainty in the capillary rise correction. Accordingly osmometers have been constructed in which the osmotic head is due to a column of an organic liquid (often toluene containing an oil-soluble dye) for which the capillary correction is smaller and much more reproducible than for aqueous solutions.

The simple osmometer may be adapted to give the osmotic head in terms of a column of toluene as in Fig. 7.6 (*a*). The two lengths of capillary C_1 and C_2 are joined by a bulb B of relatively large internal diameter (*ca.* 1 cm.) containing the interface between the aqueous colloidal solution which fills the membrane and tube C_1 and the lighter organic liquid whose head in tube C_2 is to be observed. At equilibrium the height of the toluene meniscus in C_2 above the external liquid is observed as in the simple osmometer. The capillary correction is then obtained by removing the membrane on its rubber mount from the capillary C_1, keeping the same level of external liquid. Variations in the level of the aqueous liquid in bulb B are negligible, owing to the large internal diameter of bulb B relative to the bore of the capillaries, so that the real osmotic head of toluene is given by subtracting the height of toluene above the external liquid in the capillary rise experiment from that observed at osmotic equilibrium.

This simple improvement is capable of an accuracy greater than the normal Adair osmometer, but the instrument is not easy to fill or handle, and Adair has therefore developed another, illustrated in Fig. 7.6 (b), which though not so simply made, is simpler in use. Glass taps have

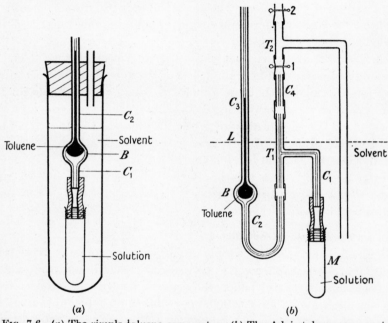

FIG. 7.6. (a) The simple toluene osmometer. (b) The Adair toluene osmometer for aqueous solutions.

been deliberately omitted in favour of rubber clips (at 1 and 2) as less likely to introduce leaks. Further, the volume of liquid contained, which is susceptible to temperature change from the thermostat, is kept low so that temperature fluctuations should not cause large variation in the observed head of liquid. By working near 0° C. temperature changes cause opposite variations in the volume of the aqueous and organic liquids, so that by suitable adjustment of the liquid volumes the instrument can be made almost completely temperature independent. This is a great advantage over several other types of 'toluene' osmometer which have been suggested and for which a very accurate thermostat is essential.

The colloidal solution fills the membrane, support, capillaries C_1, C_2, and C_4, T-piece T_1, and the lower part of bulb B, whilst the rest of the bulb and part of C_3 contains dyed toluene. T-piece T_2 and the wide-bore tubing leading to the external solvent contain this solvent. The diffusion of protein or colloid solution from the capillaries into the wide-bore tubing is so slow as to be neglected.

With clip 1 closed, the osmometer with the observation tube C_3 and connecting parts forms a closed system, so that the height of the liquid in C_3 (h) above the external liquid level (L) is a measure of the osmotic pressure plus capillary rise. With clip 2 closed and 1 open, the osmometer is now connected directly to the external liquid and the new height of liquid in C_3 above L (h_c) is a measure only of capillary rise. The true osmotic pressure is therefore given in absolute units by $(h-h_c)\rho g$. As in the simpler toluene osmometer, the capillary correction is allowed for without being completely evaluated.

Alternatively the capillary rise correction may be performed by removing the membrane from tube C_1, and with the same external liquid level the new height of the toluene meniscus is subtracted from the previous equilibrium height. Osmotic pressures accurate to ± 0.02 cm. of toluene are obtained with the osmometer, and the final equilibrium position is usually obtained within 24 hours.

A useful method of using an osmometer to measure very small osmotic pressures has been recently used by Adair (personal communication). An osmometer of any type, containing inside the membrane any stable colloidal solution which exerts an easily measurable osmotic pressure, is equilibrated against the buffer solvent outside until the observed head is quite steady. If the external buffer is now replaced by the unknown colloidal solution in the same buffer, which possesses only a small osmotic pressure, then when equilibrium is again reached the osmotic head observed will be lowered from its initial value by the *true* osmotic head of the external colloidal solution. The capillary rise correction is thus avoided, and for the small osmotic pressures of very dilute colloidal solutions or of solutions of very high molecular weight materials this modification is a very useful addition to the ordinary technique.

Other types of osmometer for aqueous solutions, usually of more complex construction than those described above, have been used, amongst which we may note those of Oakley (1935), Bourdillon (1939), and Bull (1943): details are available in the references provided. Useful reviews of osmometers are also given by Spurlin (1943) and Weissberger (1946).

Preparation of Membranes. Since the osmotic method is completely dependent upon membranes being truly semi-permeable, membrane preparation is of considerable importance. Suitable membranes, whilst preventing the escape of colloidal solute, should allow rapid diffusion of the solvent (including simple salts), and should not cause loss of colloid through adsorption. For aqueous work the membranes are usually of cellulose nitrate (collodion) and are prepared by suitably depositing films of the material from solution, usually in an alcohol-ether mixture. A very suitable cellulose nitrate, recommended by Adair, is HL 120/170, obtainable from Messrs. I.C.I. Ltd.,† and the following solution is recommended:

$3\frac{1}{2}$ gm. of HL 120/170,
50 c.c. absolute alcohol,
50 c.c. anhydrous ether,
5 or 6 c.c. ethylene glycol.

Sheet membranes may be made by pouring this solution upon a mercury surface inside a stainless steel ring, and when dry may be removed on the ring from the mercury. The more usual thimble membranes are prepared on a mould consisting of a cylindrical tube of suitable size whose 'closed' rounded end is perforated by a fine pin-hole. The tube is rotated slowly about a horizontal axis and whilst rotating the solution is poured over, giving a thin even film of cellulose nitrate. Several coats are applied and each is dried for a few minutes in front of an electric fire before application of the next.

The membrane may be removed from its mould, after immersing in water, by application of pressure to the open end of the mould, and is stored in any aqueous solution which will inhibit the growth of bacteria (usually a strong solution of ammonium sulphate). After mounting the membranes, they are tested by filling with water and applying pressure. Freedom from pin-holes is essential, and the membrane surface should be uniform in its permeability to water. Membranes of a generally suitable permeability allow the passage through of water at a rate of *ca.* 0·05 c.c. per minute under an applied pressure of 45·0 cm. of Hg.

Further details of the preparation of membranes are given in Appendix II or may be obtained from Adair (1925).

Attainment of Equilibrium. The rate at which osmotic equilibrium is established depends on the permeability of the membrane and its surface area. For membranes with dimensions approximating to those of Fig. 7.5 (*a*), and prepared as above, the osmotic head is within 2 or 3 per cent. of its equilibrium value after 24 hours. Membranes of greater permeability may be used, especially for very high molecular weight

† *Address*: I.C.I. (Explosives) Ltd., Nobel House, Stevenston, Ayrshire.

solutes, but in these cases it is usually advisable to test carefully for the leakage of the solute through the membrane.

Attention may also be drawn to the possibility of using the so-called 'dynamic' method of obtaining rapidly the equilibrium osmotic head. Thus, if the rate of movement of the meniscus is plotted against its height, a smooth curve is obtained from which it is possible to extrapolate to the level corresponding to zero rate of movement, i.e. the equilibrium height. This method is used very generally, as shown later, for osmotic pressure work in organic solvents, but for aqueous systems it is not so satisfactory, probably owing to the more frequent occurrence of 'false' equilibria and to the greater importance of capillary effects. However, it does allow approximate values of the osmotic pressure to be obtained in only a few hours.

Criteria of osmotic equilibrium

In making measurements of osmotic pressure it is necessary to ensure that true equilibrium has been attained rather than a 'false equilibrium', which may arise from a variety of causes of which the sticking of the meniscus in the capillary and slow salt diffusion are the most frequent. In order to avoid mistakes, Adair (1925) suggests three criteria of osmotic equilibrium:

1. The pressure must be steady over long periods. Thus the variation in the observed pressure over a time long compared with that required for establishment of equilibrium must be small. If equilibrium is nearly attained in 24 hours, then for at least a week afterwards the variations in the pressure observed should be small and not all in one direction.
2. The pressures should be reversible, i.e. the same equilibrium position of the meniscus should be obtained independently of whether it is approached from above or below.
3. The pressures should be reproducible, e.g. same pressures should be obtained in different osmometers and using different membranes.

In the most careful osmotic work in aqueous solutions these (or equivalent) criteria are fulfilled and except where the colloidal material is of very high molecular weight, the molecular weight values so obtained are of the highest accuracy and reliability.

Results of osmotic pressure determinations

Amongst aqueous systems the greatest interest centres around solutions of lyophilic colloids (e.g. starches, carbohydrates, nucleic acids,

168 OSMOTIC PRESSURE AND MEMBRANE PHENOMENA

proteins, etc.) for which the molecular weight may be an important physical property. In other cases the molecular or particle weight may not have the same significance, being often extremely sensitive to trace impurities.

The most detailed work has been carried out on proteins, for which the earliest really accurate investigation was carried out on haemoglobin

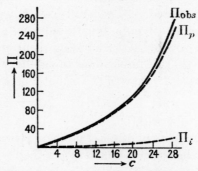

Fig. 7.7. Π_{obs} = observed osmotic pressure of haemoglobin in mm. of Hg at 0° C. Π_p = partial pressure of the protein ions. Π_i = diffusible ion pressure difference, calculated by equation (4.149). c = concentration of haemoglobin in gm. of dry protein per 100 c.c. of solution, in equilibrium with phosphate buffer mixture at pH 7·8. (*From* diagram of Adair, (1928).)

solutions by Adair (1925 *a* and *b*, 1928), some of whose results are shown in Fig. 7.7. At low protein concentration the observed osmotic pressure was found to be independent, within wide limits, of both the concentration and type of salt present, and the molecular weight of haemoglobin was calculated to be 67,000, a value four times the minimal value obtained from the iron content of haemoglobin (see p. 131). Further, it was shown, in salt-free solutions, in which Donnan effects were eliminated by suitable pH adjustment, that the molecular weight was substantially unaltered: it therefore followed that the state of aggregation of haemoglobin was not affected by increasing salt concentration. Nor was there any reason to consider that the state of aggregation varied with the protein concentration, the observed osmotic pressures at high concentrations being explicable in terms of deviations from the van 't Hoff law common also to low molecular weight solutes. It has since been shown that the deviations from ideal behaviour may vary somewhat with the nature and concentration of the diffusible salts present with the protein.

Since Adair's work many osmotic investigations of the molecular weight of proteins have been made and fair agreement amongst different workers has been recorded. Table (7.1) contains a selection of the results available; for more complete data see Cohn and Edsall (1943, p. 391).

TABLE 7.1. *Molecular Weights of Proteins from Osmotic Data*

Protein	Molecular weight	Reference
Egg albumin	43,000	Marrack and Hewitt (1929)
	46,000	Taylor, Adair, and Adair (1932)
	45,000	Bull (1941)
Haemoglobin	67,000	Adair (1928)
Serum albumin	72,000	Adair and Robinson (1930)
Serum globulin	175,000	Adair and Robinson (1930)
Amandin	206,000	Burk (1937)
Excelsin	214,000	Burk (1937)
Haemocyanin, crab	550,000	Roche, Roche, Adair, and Adair (1935)
octopus	710,000	Roche, Roche, Adair, and Adair (1935)
snail (Helix)	1,800,000	Roche, Roche, Adair, and Adair (1935)

The osmotic investigations of the haemocyanins probably represent an upper limit to the molecular weight which can be investigated by direct osmotic pressure measurements. Thus for a solution of Helix haemocyanin containing 1 gm. per 100 c.c., the osmotic pressure was *ca.* 0·13 cm. of water. Since accurate measurements of such a pressure were impossible, molecular weight values were estimated provisionally from observed osmotic pressures at concentrations of 3 gm. of protein per 100 c.c. of solution; as Roche *et alia* (1935) point out, these are not of great accuracy.

Osmotic measurements on proteins have also been made in special solvents, e.g. concentrated aqueous solutions of urea and glycerol, and, whilst in certain cases interesting changes in the state of aggregation have been noted, in others no changes could be detected. These investigations are of considerable importance in considerations of protein structure and will be discussed in Chapter XXVI.

Non-aqueous systems

The principles of osmotic pressure measurement in organic solvent systems are not different from those involved in aqueous systems, but the techniques have diverged appreciably, owing to differences in the importance of various factors.

Thus, generally, in solutions of macromolecules in organic solvents with which we shall be chiefly concerned, the solute molecule does not possess a resultant electrical charge. Donnan effects are therefore absent, and the osmotic pressure arises entirely from the colloidal solute. True osmotic pressures are therefore obtained in solutions containing only the *pure* organic solvent and the macromolecule (cf. p. 158).

In many cases also, especially where the solute is a linear or chain-like polymer, the deviations from ideal behaviour are, as we have already seen, very serious, and it is necessary to make osmotic pressure measurements down to the smallest workable concentrations ($<$ 1 per cent., if possible) so that extrapolation of Π/c to zero concentration becomes as accurate as possible (see p. 155).

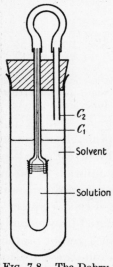

Fig. 7.8. The Dobry type of osmometer.

A further important feature is the reduced capillary effects in organic solvents, which makes the dynamic method of obtaining the equilibrium osmotic pressure much more reliable. This type of measurement has been utilized considerably in organic solvents and especially suitable types of osmometer have been devised, which effectively reduce the troubles arising from the solvent volatility and the occurrence of ageing effects in macromolecular systems. Thus in the osmometer described by Fuoss and Mead (1943), an osmotic reading may be obtained in less than 20 minutes after filling. Such rapid estimations are made possible by the use of very permeable membranes of very large area. So high is the permeability to solvent in some cases that the membranes are not truly impermeable to the macromolecules, but during the course of such rapid determinations the loss of solute may be negligible.

The dynamic type of osmometer is undoubtedly the most popular for work in organic solvents at the present time, but the best forms require very careful machining and are not readily available in many laboratories. Accordingly, before describing the dynamic instrument, a brief description is given of a simple modification of the Adair aqueous osmometer for work in organic solvents, used originally by Dobry (1935). Collodion thimble membranes, prepared as already described, are mounted directly upon a closely fitting glass tube sealed on to the capillary observation tube, by binding with surgical silk upon a layer of cellophane. The mounted membrane is then partially denitrated (to about 6 per cent. nitrogen) by any suitable procedure (e.g. see Appendix III) to prevent solubility in, but to retain permeability by, organic solvents. The osmometer is then set up as in Fig. 7.8 with a connexion between the observation capillary C_1, and the short tube C_2 to prevent loss of solvent by evaporation. Such osmometers were used by Campbell and Johnson (1945), also Campbell (1946) for determining the osmotic behaviour of solutions of cellulose nitrate (per cent. $N_2 \approx 12.2$) in acetone. Equilibrium was nearly obtained overnight and final values after three days.

The dynamic osmometer

One of the most useful osmometers for organic solvents is that of Fuoss and Mead (1943), developed from the original version of Montonna and Jilk (1941), one form of which is shown diagrammatically in Fig. 7.9 (a). Only an outline of its construction and use can be included here. The osmometer was designed to make possible the rapid establishment of equilibrium, important factors being the large membrane area and the fine-bore capillary observation tubes; the volume of liquid transferred in attaining equilibrium is thus very small. Special fast membranes have also been evolved (see p. 174).

The flat membrane of diameter *ca.* 5 inches is clamped between the two half-cells shown in Fig. 7.9 (a), each of which is constructed from a stainless steel (or nickel-plated brass) plate. The inside surfaces of the half-cells shown in Fig. 7.9 (b) each possess an outside flat ring between which the membrane acting as a gasket effectively seals the cell; within the ring the concentric cuttings (2 mm. wide and 2 mm. deep) are connected by a vertical cutting joining the inlet and outlet of each half-cell, in which the two liquid components of the system are contained. Both half-cells are connected by suitable joints to identical pieces of precision glass capillary, of internal diameter *ca.* 0.4 mm., in which liquid levels are observed, and to the valve block and needle valve system shown in Fig. 7.9 (a). Unless the surface tensions of solution and solvent are very different, no correction for capillary rise is necessary, so that the osmotic pressure is given directly as an observed head of liquid in the observation tubes. The greatest of care is required in the manufacture of the apparatus if no leakage is to occur at the valve blocks and the capillary joints. In view of the narrow bore of the observation tubes the smallest leak cannot be tolerated. It has been found convenient in the authors' laboratory to pack the instrument well with cotton wool in a small box which is then contained inside a glass-fronted air thermostat whose temperature is accurate to $\pm 0.05°$ C. The temperature is recorded by a thermometer in a boring in one of the half-cells and liquid levels are observed by a cathetometer from outside the thermostat.

Solvent (in the right half) and solution are poured into the cell through the wide-bore filling tubes, and air bubbles between the half-cells are thus swept out. Further bubbles may then be removed from the valve

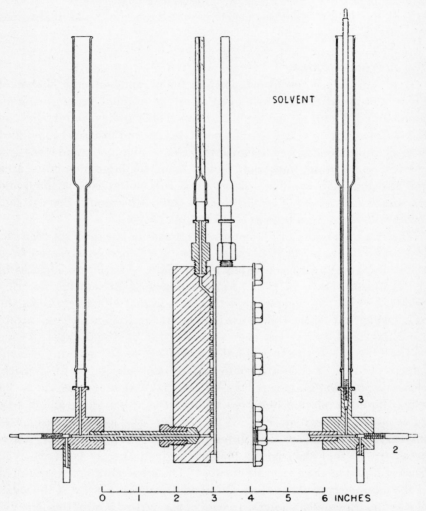

Fig. 7.9(a). A dynamic osmometer. Vertical section of assembled osmometer. (Fuoss and Mead (1943).)

blocks by opening valves 1 and 2. In view of the small liquid volume on both sides of the membrane, osmotic pressures determinations can be carried out with economy of materials, 20 c.c. of solution being required for one filling. The solution level in its capillary may be raised

OSMOTIC PRESSURE AND MEMBRANE PHENOMENA   173

by addition of liquid in the left-hand filling tube and lowered by opening valve 1 and that of the solvent by opening valves 3 and 2 respectively.

Normally the osmometer is worked by adjusting the solution level to a convenient zero point, and the solvent level to a position 1 or 2 cm. above (or below) the expected equilibrium position. Solvent enters (or

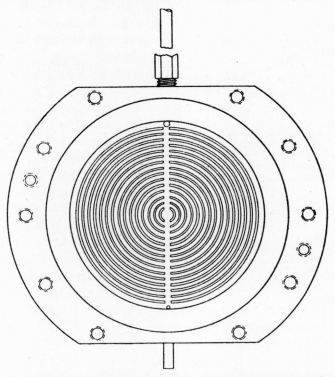

FIG. 7.9(b). A dynamic osmometer. The inner surface of each half-cell. (Fuoss and Mead (1943).)

leaves) the solution side, the solvent level varying correspondingly, whilst the solution level, owing to its connexion to its wide filling tube, does not vary noticeably. The solvent level is therefore read every half-minute until it is changing very slowly. It is then set 1 or 2 cm. below (or above) the expected final equilibrium position; similar readings of its position against time are taken and with the previous set of observations plotted as in Fig. 7.10. If the final equilibrium position of the solvent level has been estimated accurately the two curves will be almost symmetrical about a horizontal line and a plot of the half-sum of the observations against time will be a straight line almost independent of time as in Fig. 7.10. If the estimate has been very inaccurate the two curves will be displaced sideways relatively to one another, and only at the longer times will the half-sum of the observations at a given time become independent of time. The half-sum value thus obtained will make possible an accurate estimate of the final osmotic pressure which may be used in a repeat determination to give an even more precise estimate. An accuracy of ± 0.05 cm. or better is usually claimed for the final osmotic pressure head.

With suitably fast membranes the whole osmotic pressure determination may be completed in well under 1 hour, which for unstable materials and volatile solvents is a very important consideration. Further, in such short times, the slight permeability of the membrane to the macromolecule or colloid does not introduce serious errors.

It should be noted that the instrument may, with suitably slower membranes, be used for the direct observation of the final equilibrium position. In view of the permeability to solute, the faster membranes cannot be used in such determinations.

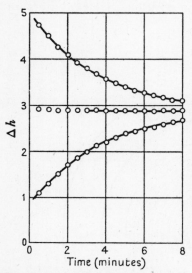

Fig. 7.10. Curves of observed head (Δh) against time. (Fuoss and Mead (1943).)

Membranes for dynamic osmometers. Many workers have described the preparation of special types of membranes (e.g. Carter and Record, 1939; Flory, 1943; and Melville *et alia*, 1946). Fuoss and Mead prepare flat cellulose nitrate membranes by pouring a solution of the material upon a mercury surface and allowing slow evaporation. The membrane, after washing with water, is then partially denitrated by treatment with ammonium sulphide, washed with water and organic solvents in a definite sequence, and stored under methyl amyl ketone.

Melville *et alia* (1946) have utilized as membrane material the cellulose secretion of a micro-organism, *Acetobacter xylinum*, growing at 30° C. on a special medium. The permeability is controlled by suitable treatment with organic solvents, and membranes with solvent permeabilities similar to those of Fuoss and Mead have been prepared. It is claimed that such membranes are suitable for osmotic determinations on vinyl polymers of molecular weight greater than 10,000, and that for such materials the rate of penetration of the membrane by polymer is small. When fresh they have no asymmetry pressure (i.e. the pressure difference developed when solvent is placed on both sides), a feature which is sometimes disturbing with partially denitrated cellulose nitrate membranes.

Results

The osmotic pressure-concentration dependence of solutions of linear polymers has already been discussed theoretically, making use of the calculated results on the entropy of mixing of such polymers with low molecular weight solvents. Theory predicts (Fig. 7.2) that for a given polymer the plot of Π/c v. c should at low concentrations be linear, with slope independent of the precise molecular weight of the polymer but dependent on the polymer-solvent system, and with intercept on the Π/c axis $\{$i.e. $\lim_{c \to 0} (\Pi/c)\}$ independent of the solvent but inversely proportional to the polymer molecular weight and having the value RT/M.

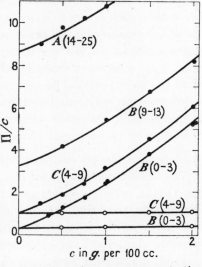

FIG. 7.11. Osmotic pressure-concentration ratio (Π/c) v. concentration (c) for polyisobutylene fractions in cyclohexane (●) and benzene (○) solution.

These predictions are qualitatively correct as shown, for example, by the results of Flory (1943) on polyisobutylene fractions in cyclohexane and benzene (Fig. 7.11). Clearly the different fractions in cyclohexane give parallel curves of Π/c v. c and the two higher molecular weight (i.e. low osmotic pressure) fractions give identical values of $\lim_{c \to 0} \Pi/c$ in the two solvents used, cyclohexane and benzene, though the slopes in these solvents are markedly different. Other workers confirm these conclusions (e.g. Meyer and Wertheim, 1941; Alfrey, Bartovics, and Mark, 1943).

One important uncertainty, however, remains. Thus a close examination of Fig. 7.11 shows that the curves with cyclohexane as solvent are not strictly linear, but possess definite curvations, an observation which has been confirmed by other workers. Thus Dobry (1935 *a* and *b*) showed for cellulose nitrate in several solvents, and Gee and Treloar (1942) for rubbers in benzene-methyl alcohol mixtures that the Π/c v. c plot was not accurately linear. Recent investigations of rubber in chloroform (Bywater and Johnson, 1946) have given non-linear plots also. On the other hand, some other workers (e.g. Doty, Wagner, and Singer, 1947; Spurlin, Martin, and Tennent, 1946) consider the Π/c v. c curves to be linear within experimental error, and further work is required to settle the point. If, as now seems possible, this departure from linearity is of general occurrence the values of $\lim_{c\to 0}(\Pi/c)$ based on linear extrapolation must be, to an extent depending upon the system concerned, inaccurate. To reduce such errors it is necessary, despite severe experimental difficulties, to make osmotic pressure measurements at the lowest possible concentrations (ca. 0·1 per cent.). It is worth noting in this connexion, that if for a series of fractions the Π/c v. c curves are strictly parallel, then the extrapolation procedure for those high molecular weight fractions showing only small and inaccurately measured osmotic pressures is made easier and more accurate than if no relation existed between the slope of the curves. Thus the accuracy of $\lim_{c\to 0}(\Pi/c)$ may be better than that of the poorest individual pressure readings.

The deviations from linear Π/c v. c plots have been discussed by Spurlin (1943), who considers one possible cause to be interaction and association occurring between chain molecules. An examination and refinement of the statistical theory of solutions of long-chain molecules possibly along the lines suggested by Spurlin would clearly be of value.

From a practical viewpoint there is need for a detailed comparison of molecular weight values determined by osmotic and other methods, data of a very fragmentary nature only being available up to now.

Practical limits of the osmotic method

In both aqueous and organic solvent osmometers the final osmotic pressure head is usually subject to an error of at least ± 0.02 cm. For a 1 per cent. solution of a polymer of molecular weight 350,000, the ideal osmotic pressure is calculated to be ca. 0·7 cm. of water, so that at the best the error in the final calculated molecular weight is not less than ± 3 per cent. For higher molecular weights the error is correspondingly larger. In many cases also the accuracy in determining the osmotic head is not better than ± 0.05 cm., which clearly excludes the accurate measurement of the higher molecular weights. A somewhat optimistic limit to the molecular weights derivable with accuracy might be placed at 500,000, for which an error of ± 0.02 cm. in reading the osmotic head of a 1 per cent. solution alone involves an error of *ca.* 5 per cent. in the calculated molecular weight. Nor would it appear that the osmometers already considered are capable of substantial refinement. Alternative experimental approaches are therefore of importance, and two recent developments are outlined below.

The osmotic balance

As we have seen, accuracy in the usual osmometers, which is dependent on measurement of the height of small liquid columns, has the

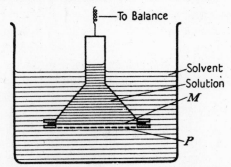

FIG. 7.12. The osmotic balance (diagrammatic).

approximate limit of ± 0.01 cm. Svedberg suggested that by weighing small columns of liquid (instead of measuring their height) on the usual type of analytical balance, the osmotic method would be improved. Thus, assuming a cross-sectional area of 1 cm.^{2}, unit liquid density, and a weighing accuracy of ± 0.0001 gm., height differences to ± 0.0001 cm. should be measurable, i.e. a hundredfold increase in accuracy over the usual osmometer. Clearly the above calculation is somewhat optimistic, for the use of an observation tube of area of cross-section 1 sq. cm. would involve large transfers of solvent, and therefore very long equilibrium times. Smaller-bore tubes would clearly be advantageous, but the differences in height which could just be weighed would be correspondingly larger.

Jullander (1945), working in Svedberg's laboratory, has examined the possibilities of Svedberg's suggestion and has devised the osmotic balance illustrated diagrammatically in Fig. 7.12. The cell, constructed of aluminium and glass and containing the solution, may be described as an inverted funnel, the membrane M being supported over the base of the funnel by a perforated plate P. Suspended from an analytical balance by the short stem of the funnel the cell is partially immersed in solvent contained in a wide vessel.

After assembling and filling the cell, weights are placed upon the other scale-pan until the meniscus inside the cell coincides with that outside. As osmosis occurs solvent enters the cell and more weights must be added to keep the cell in its original position. In view of the long times needed for the establishment of equilibrium, the dynamic method has been applied in determining the equilibrium position.

It would appear that although great sensitivity has been obtained, this has been possible only by the most careful experimentation. Thus in order to avoid temperature corrections it is necessary that the temperature in the balance case should not vary by more than 0·005° C. during 2–3 days. Since, however, the method does make possible the osmotic pressure measurement of very dilute solutions (e.g. 0·02 gm. cellulose nitrate/100 c.c.), it is an important addition to osmotic technique and will probably be extensively tried out in the next few years.

Light-scattering measurements

Theoretically it can be shown that a completely pure, homogeneous medium should not scatter light at all. In practice, however, scattering always occurs, and this is attributable in the case of a pure solvent to the thermal fluctuations in density which occur. In the case of a solution, local fluctuations in concentration also occur and give rise to increased light scattering. Scattering may be quantitatively described in terms of the turbidity, τ, defined as the fractional decrease in the intensity of the incident light due to scattering, on passing through one centimetre of solution. It has been shown (e.g. Debye, 1944; Zimm, Stein, and Doty, 1945) by methods due originally to Einstein, that the total turbidity (τ_T) due to fluctuations in density and concentration may be written

$$\tau_T = \tau_d + \tau_c \tag{7.19}$$

or

$$\tau_T = \frac{32\pi^3 n^2}{3\lambda^4 N}\left[RT\kappa\left(\rho\frac{\partial n}{\partial \rho}\right)^2 + \frac{M_1 c}{\rho(-\partial \ln a_1/\partial c)}\left(\frac{\partial n}{\partial c}\right)^2\right], \tag{7.20}$$

where n is the refractive index of the solution for the wavelength λ (cm.$^{-1}$ *in vacuo*), N is Avogadro's number, κ is the isothermal compressibility of the solvent of density ρ and molecular weight M_1, and c is the concentration of solute in grams per cm.^{3} of solution. The first term within the bracket gives the turbidity (τ_d) arising from density fluctuations in the solvent, and clearly must be subtracted from the total turbidity to give that due only to concentration fluctuations (τ_c), with which we are now concerned.

A connexion between τ_c and the activity (a_1) or partial Gibbs free

OSMOTIC PRESSURE AND MEMBRANE PHENOMENA

energy of the solvent $\bar{G}_1$ is to be expected qualitatively, for, clearly, the extent to which local fluctuations in concentration occur is dependent upon the free energy changes involved. It follows also from eqn. (4.122) that

$$-\frac{\partial \ln a_1}{\partial c} = \frac{\bar{V}_1}{RT}\frac{\partial \Pi}{\partial c}; \qquad (7.21)$$

so that we may write

$$\tau_c = \frac{32\pi^3 n^2 c (\partial n/\partial c)^2}{3\lambda^4 N (1/RT)(\partial \Pi/\partial c)}. \qquad (7.22)$$

The connexion between turbidity and osmotic pressure indicated in eqn. (7.22) is demonstrated clearly by Fig. 7.13 (a), due to Debye (1947),

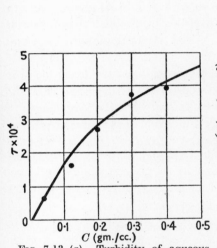 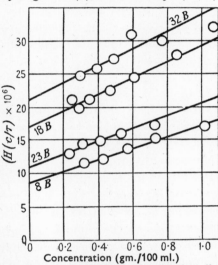

Fig. 7.13 (a). Turbidity of aqueous sucrose solutions. Curve calculated from osmotic pressure data. Points determined by 90° scattering.

Fig. 7.13 (b). Turbidity data for the cellulose acetate fractions.

in which experimental turbidity measurements (·) are compared with the curve calculated by the equation from experimental osmotic pressure measurements on aqueous sucrose solutions.

Assuming the variation in osmotic pressure with concentration to be given by eqn. (7.12), we obtain on substituting into (7.22) and rearranging

$$\frac{c}{\tau_c} = \frac{3\lambda^4 N}{32\pi^3 n^2 (\partial n/\partial c)^2}\left[\frac{1}{M_2} + \frac{2(k-k_2)}{RT}c + \ldots\right] \qquad (7.23)$$

$$= \frac{1}{H}\left[\frac{1}{M_2} + \frac{2(k-k_2)}{RT}c + \ldots\right], \qquad (7.24)$$

where

$$H = \frac{32\pi^3 n^2 (\partial n/\partial c)^2}{3\lambda^4 N}.$$

A plot of Hc/τ_c against concentration should therefore give a straight line of slope, $2(k-k_2)/RT$, and intercept on the Hc/τ_c axis of $1/M_2$, from which molecular weight may be obtained. As in the usual osmotic pressure methods, the slope should be independent of molecular weight for a given solvent-polymer system.

Measurements of turbidity over a range of concentrations, and the more routine determination of the quantities involved in the term H, thus lead to a determination of M_2, which may be shown to be a weight average value.

Experimentally the determination of small quantities of scattered light has not proved easy, but considerable progress has been made by Stein and Doty (1946) and other workers. Fig. 7.13 (b) gives turbidity data for cellulose acetate fractions in acetone, from which the molecular weights 43,000, 55,000, 87,700, and 114,000 were calculated (Stein and Doty).

One further aspect only of light-scattering measurements may be mentioned here. Eqn. (7.24) was derived on the assumption that the colloidal solute molecules were small compared with the wavelength of light. If, however, this is not so for any dimension of the solute particles, then, owing to interference of light scattered from different parts of the particles, there is a reduction in the scattered light which must be corrected for in calculating M_2. Such particles also give unsymmetrical scattering about a direction perpendicular to the incident beam, and observations on the angular dependence of the scattering may be used to derive molecular dimensions. Further details and suggested uses of the method are given in Chapter XV.

MEMBRANE PHENOMENA
Membrane potentials

We have already seen (p. 82) that membrane potentials may be used to calculate the partial osmotic pressures arising from the unequal distribution of diffusible ions (Π_i) and the mean particle charge in a colloidal solution. In this section we consider firstly the measurement of membrane potentials and, secondly, their application in these two directions.

Measurement

Consider the system shown diagrammatically in Fig. 7.14, in which a protein chloride solution (e.g. 1 per cent. solution of gelatin chloride) placed inside a membrane is allowed to come to equilibrium with a solution of hydrochloric acid, placed outside.

OSMOTIC PRESSURE AND MEMBRANE PHENOMENA 181

From the condition of equilibrium, it follows that if electrodes reversible with respect to H+ or Cl- (e.g. hydrogen or silver-silver chloride) are placed on the two sides of the membrane, then no difference in potential between them will be recorded.

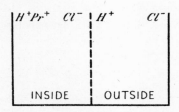

FIG. 7.14. Membrane equilibrium between protein chloride solution and dilute hydrochloric acid.

If, however, the hydrogen or chloride ion concentrations of the two solutions are measured separately, definite differences are observed. Further, if the two solutions are used in a concentration cell in which liquid junction effects are eliminated (e.g. by a saturated KCl bridge) the E.M.F. (E_c) observed between reversible electrodes is just that which is calculated for such a cell, viz.

$$E_c = \frac{RT}{F} \ln \frac{[\text{H}^+]^{\text{I}}}{[\text{H}^+]^{\text{II}}}, \qquad (7.25)$$

or

$$E_c = \frac{RT}{F} \ln \frac{[\text{Cl}^-]^{\text{II}}}{[\text{Cl}^-]^{\text{I}}}, \qquad (7.25\,\text{a})$$

where the superscripts I and II refer to inside and outside the membrane.

The absence of such a potential in the membrane system is to be explained by the presence of the additional membrane potential (E_m) which is equal and opposite (see Fig. 4.3) to that (E_c) of the system acting as a concentration cell without a membrane, i.e.

$$E_m = -E_c$$

and

$$E_m = -E_c = -\frac{RT}{F} \ln \frac{[\text{H}^+]^{\text{I}}}{[\text{H}^+]^{\text{II}}}, \qquad (7.26)$$

in agreement with the expression for membrane potential determined by an alternative method (eqn. (4.134)). Clearly, therefore, *it is impossible to measure a membrane potential by the use of electrodes which are reversible to any ionic component of the system.*

One method of measuring the membrane potential has already been indicated. Thus if in the cell depicted in Fig. 7.14 $[\text{H}^+]^{\text{I}}$ and $[\text{H}^+]^{\text{II}}$, or

[Cl$^-$]^I and [Cl$^-$]^{II}, are estimated by electrometric or other means, then the value of the membrane potential may be calculated by eqn. (7.26).

A more direct method was utilized by Loeb (1924), who made an extensive investigation of membrane potentials and demonstrated that an explanation was possible on the basis of the Donnan membrane equilibrium. His direct method of measurement is illustrated in Fig. 7.15.

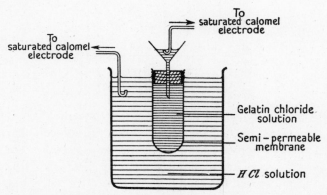

FIG. 7.15. Loeb's direct method of measuring membrane potentials. (From figure by Loeb (1924).)

The protein solution (e.g. 1 per cent. gelatin), contained in a collodion bag tied upon a rubber stopper through which passed a glass manometer tube, was allowed 20 hours or more to come to equilibrium with a suitable acid solution (e.g. aqueous hydrochloric acid). The glass manometer tube was then replaced by a funnel and the potential between identical calomel electrodes placed in the solutions inside and outside the membrane was observed (Fig. 7.15). Such a potential was opposite in direction to that which would arise from the system acting as a concentration cell (with reversible electrodes) in the absence of the membrane, and of a magnitude in agreement with eqn. (7.26). Loeb therefore considered it to be the membrane potential. Table 7.2, due to Loeb, shows pH values inside and outside the membrane, and the observed and calculated potentials for the system gelatin chloride–hydrochloric acid; the agreement between the two potentials is within experimental error.

An explanation of the success of Loeb's direct method is a matter of some difficulty. Loeb describes calomel electrodes as 'indifferent' but does not explain this term. It is, of course, true that the calomel electrode (containing the usual standard KCl solution) is not reversible

TABLE 7.2. *The Influence of* pH *on Calculated and Observed Membrane Potentials for Gelatin Chloride Solutions at Equilibrium*

C.c. of 0·1 N HCl in 100 c.c. of 1 per cent. iso-electric gelatin

	1	2	4	6	8	10	12·5	15	20	30	40	50
pH inside	4·56	4·31	4·03	3·85	3·33	3·25	2·85	2·52	2·13	1·99	1·79	1·57
pH outside	4·14	3·78	3·44	3·26	2·87	2·81	2·53	2·28	2·00	1·89	1·72	1·53
pH inside — pH outside	0·42	0·53	0·59	0·59	0·46	0·44	0·32	0·24	0·13	0·10	0·07	0·04
Calculated membrane potential	24·7	31·0	34·5	34·5	27·0	25·8	18·8	14·0	7·6	5·9	4·1	2·3
Observed membrane potential	24·0	32·0	33·0	32·5	26·0	24·5	16·5	11·2	6·4	4·8	3·7	2·1

(From Loeb, 1924, p. 183.)

to hydrogen or chloride ions in the sense that the potential difference set up when it dips into another solution does not depend upon [H+] or [Cl−] of that solution. It may therefore be imagined that the use of the two calomel electrodes damps out the potential due to concentration differences, leaving only the membrane potential. Any electrode, whose potential is a constant and which is provided with a suitable salt bridge to eliminate diffusion potentials, could be used in place of a calomel electrode (e.g. Ag | AgCl | saturated KCl). The calomel electrode, owing to its convenience and reliability, is, however, almost exclusively used.

Adair and Adair (1934) used calomel electrodes to make careful measurements of membrane potentials in systems in which small quantities of purified protein solutions (e.g. 1·5–5 c.c.) were equilibrated against relatively large volumes of a dialysate of well-defined pH and salt concentration. Constancy of the observed osmotic head or of the measured membrane potential was used as the criterion of equilibrium. Some of the results obtained are discussed briefly in the following pages.

Results

Membrane potential and protein concentration

For a constant pH and dialysate composition the observed membrane potential should, at low values, be proportional to the corrected protein concentration as eqn. (4.152) shows. Adair and Adair (1934) verified this conclusion for many systems including that for which results are shown in Fig. 7.16.

In addition it is clear from this figure that with varying pH the slope of the lines (E/c_v) and therefore the mean valence of the protein ions changes; the reversal in the sign of the membrane potential on passing through the iso-electric point (pH = 6·8–7·0) for carboxyhaemoglobin is also apparent.

At higher values of the membrane potential (> 2 mV) distinct deviations from the linear E v. c_v relation are encountered, as is to be anticipated when the activity coefficients of the diffusible ions on the two sides of the membrane are no longer equal (see Chapter IV, p. 81).

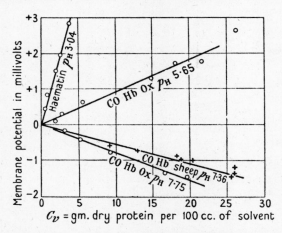

Fig. 7.16. Membrane potentials of CO-haemoglobin and acid haematin at different concentrations. Dialysates of ionic strength 0·01, compositions given in mols. per litre:

pH 3·04 CH$_3$COONa 0·0025, CH$_3$COOH 0·0975, NaCl 0·0975; pH 5·65 CH$_3$COONa 0·100, CH$_3$COOH 0·0105; pH 7·36 (NH$_4$)$_2$HPO$_4$ 0·03, NH$_4$H$_2$PO$_4$ 0·01; pH 7·75 Na$_2$HPO$_4$ 0·0321, KH$_2$PO$_4$ 0·00357.

Membrane potentials and ion pressure differences

Formulae for the accurate and approximate calculation of ion pressure differences have already been obtained (eqns. (4.149) and (4.150)). Table 7.3, due to Adair (1929), shows the application of these formulae to the calculation of the partial osmotic pressures in sheep haemoglobin solutions of different concentration equilibrated against a constant composition of dialysate. The accurate (Π_i) and approximate (Π_i^*) values of the ion pressure difference agree well at the lower protein concentrations (< 10 gm./100 c.c.), but appreciable differences are observed at the higher concentrations.

Since for calculating protein molecular weights measurements on dilute protein solutions only are utilized, it is clear, in this connexion, that the somewhat inaccurate method of estimating ion pressure differences (formula (4.150)) is quite good enough in correcting for the unequal distribution of diffusible ions.

TABLE 7.3. *The Partial Osmotic Pressures of Haemoglobin Ions at* $0°C.$, *in solutions equilibrated with a mixture composed of* $0·1$ *mols.* KCl, $0·0613$ *mols.* NaH_2PO_4, *and* $0·00533$ *mols.* KH_2PO_4. $pH = 7·8$

(From Adair, 1929.)

Grams protein per 100 c.c. solution	Grams protein per 100 c.c. solvent	Mols. protein per 1,000 c.c. solvent m_p	Membrane potential, millivolts E_m	Observed osmotic pressure Π_{obs}	Ion pressure difference Π_i	Partial pressure of protein Π_p	Provisional value Π_p^*	Provisional value Π_p°
0·68	0·685	0·000103	−0·02	1·9	0·005	1·9	0·003	1·9
2·21	2·26	0·000338	−0·07	6·2	0·06	6·1	0·04	6·2
2·90	2·98	0·000447	−0·10	8·6	0·13	8·5	0·07	8·5
3·58	3·71	0·000555	−0·10	11·2	0·13	11·1	0·07	11·1
5·00	5·25	0·000787	−0·20	14·9	0·51	14·4	0·31	14·6
8·00	8·67	0·001300	−0·24	28·2	0·74	27·5	0·50	27·7
8·12	8·81	0·001320	−0·28	28·8	1·0	27·8	0·61	28·2
10·00	11·09	0·00166	−0·50	40·2	3·2	37·0	2·00	38·2
12·00	13·57	0·00203	−0·40	48·7	2·1	46·7	1·20	47·5
15·50	18·22	0·00272	−0·60	67·4	4·6	62·8	2·80	64·6
19·40	23·87	0·00357	−0·75	103·6	7·2	96·4	4·40	99·2
19·80	24·48	0·00367	−0·70	104·2	6·3	97·9	3·80	100·4
20·00	24·78	0·00371	−0·87	110·9	9·7	101·2	5·70	105·2
24·00	31·23	0·00468	−1·05	155·0	14·1	140·9	8·50	146·5
25·00	32·95	0·00493	−1·10	179·0	15·5	163·5	9·50	169·5
28·00	38·37	0·00574	−1·35	242·0	23·3	218·7	14·20	227·8
29·00	40·27	0·00603	−1·45	264·0	26·9	237·7	16·50	248·1
34·41	51·51	0·00771	−1·98	382·8	50·2	332·6	30·60	352·2

Membrane potentials and protein charge

The mean charge on a protein or colloidal ion may be calculated from the limiting value at low concentration of the ratio (E/c_v) by means of eqn. (4.152). For sheep carbon monoxide-haemoglobin, $M = 67,000$, and at $0·5°$ C. eqn. (4.152) becomes

$$z_p = 284J\left(\frac{E}{c_v}\right), \tag{7.27}$$

from which the values of the valence of the protein ion given in Table 7.4 have been calculated by Adair and Adair (1940).

In view of the possibility that positively charged haemoglobin ions may combine with chloride or phosphate ions, values of the protein charge calculated, for example, from eqn. (4.137), as that charge required to neutralize the excess negative charge of diffusible ions occurring within the membrane (estimated by chemical analysis) are liable to be too highly positive. As Adair points out, it is likely that more reliable values of the mean charge are obtained by direct membrane potential measurements.

From the total net charge on a protein under given conditions of pH and salt concentration, it is possible (as is shown in Chapter XII),

TABLE 7.4. *Valence of Carbon Monoxide-Haemoglobin*
(from Membrane Potentials)
(From Adair and Adair, 1940.)

Buffers	Ionic strength 0·02		Ionic strength 0·10	
	pH	Apparent valence	pH	Apparent valence
Na_2HPO_4, NaH_2PO_4	6·28	+2·2	6·03	+0·3
,, ,,	6·77	+1·0	6·60	−1·9
,, ,,	7·46	−1·6	7·41	−4·0
,, ,,	7·79	−2·9	8·04	−6·4
Na_2CO_3, $NaHCO_3$	10·09	−11·3	9·11	−12·7
,, ,,	..	..	9·95	−16·4

assuming a definite molecular model, to calculate the electrophoretic mobility, which may be compared with experimentally observed values (p. 308). The comparison given on p. 338 for a range of pH values shows a good measure of agreement between the calculated and observed mobilities, giving support to the values of the net charge obtained from membrane potentials.

REFERENCES

ADAIR, 1925 a, *Proc. Roy. Soc.* A **108**, 627.
—— 1925 b, ibid. A **109**, 292.
—— 1928, ibid. A **120**, 573.
—— 1929, ibid. A **126**, 16.
ADAIR and ADAIR, 1934, *Bio-chem. J.* **28**, 199.
—— —— 1938, *C.R. Lab. Carlsberg, Sér. chim.* **22**, 8.
—— —— 1940, *Trans. Faraday Soc.* **36**, 23.
—— and ROBINSON, 1930, *Bio-chem. J.* **24**, 1864.
ALFREY, BARTOVICS, and MARK, 1943, *J. Amer. chem. Soc.* **65**, 2319.
BOURDILLON, 1939, *J. biol. Chem.* **127**, 617.
BULL, 1941, ibid. **137**, 143.
—— 1943, *Physical Biochemistry*, Wiley.
BURK, 1937, *J. biol. Chem.* **120**, 63.
—— and GREENBERG, 1930, ibid. **87**, 197.
CAMPBELL, 1946, Thesis, Cambridge University.
CAMPBELL and JOHNSON (unpublished).
COHN and EDSALL, 1943, *Proteins, Amino Acids and Peptides*, Reinhold Publ. Corp.
DEBYE, 1944, *J. applied Physics*, **15**, 338.
—— 1947, *J. phys. coll. Chem.* **51**, 18.
DOBRY, 1935 a, *J. Chim. Phys.* **32**, 46.
—— 1935 b, *Bull. Soc. chim. Fr.* [5] **2**, 1882.
DOTY, WAGNER, and SINGER, 1947, *J. phys. coll. Chem.* **51**, 32.
FLORY, 1941, *J. chem. Phys.* **9**, 660.
—— 1942, ibid. **10**, 51.
—— 1943, *J. Amer. chem. Soc.* **65**, 372.
—— 1945, *J. chem. Phys.* **13**, 453.

FOWLER and GUGGENHEIM, 1939, *Statistical Thermodynamics*, Cambridge.
FUOSS and MEAD, 1943, *J. phys. Chem.* **47**, 59.
GEE, 1942, *Rep. Progr. Chem.* **39**, 7.
—— and TRELOAR, 1942, *Trans. Faraday Soc.* **38**, 147.
HILDEBRAND, 1936, *Solubility*, 2nd edn., Reinhold Publ. Corp.
HUGGINS, 1941, *J. chem. Phys.* **9**, 440.
—— 1942, *J. phys. Chem.* **46**, 151.
—— 1943, *Industr. Engng. Chem.* **35**, 216.
JULLANDER, 1945, *Ark. Kemi Min. Geol.*, No. 8, **21** A.
LOEB, 1924, *Proteins and the Theory of Colloidal Behaviour*, McGraw-Hill, 2nd edn., p. 178.
MARRACK and HEWITT, 1929, *Bio-chem. J.* **23**, 1079.
MELVILLE, MASSON, CRUICKSHANK, and MENZIES, 1946, *Nature*, London, **157**, 74.
MEYER, 1942, *Natural and Synthetic High Polymers*, Interscience.
—— and WERTHEIM, 1941, *Helv. chim. Acta*, **24**, 217.
MONTONNA and JILK, 1941, *J. phys. Chem.* **45**, 1374.
OAKLEY, 1935, *Trans. Faraday Soc.* **31**, 136.
ROCHE, ROCHE, ADAIR, and ADAIR, 1935, *Biochem. J.* **29**, 2576.
SORENSEN, 1917, *C.R. Lab. Carlsberg*, **12**, 262.
SPURLIN, 1943, in *Cellulose and Cellulose Derivatives*, edited by Ott, Interscience, p. 910.
—— MARTIN, and TENNENT, 1946, *J. Polymer Sci.* **1**, 63.
STEIN and DOTY, 1946, *J. Amer. chem. Soc.* **68**, 159.
TAYLOR, ADAIR, and ADAIR, 1932, *J. Hyg.*, Cambridge, **32**, 340.
WEISSBERGER, 1946, *Physical Methods*, vol. i, Interscience, p. 253.
ZAWIDZKI, 1900, *Z. phys. Chem.* **35**, 129.
ZIMM, STEIN, and DOTY, 1945, *Polymer Bulletin* 1, p. 90.

VIII
SEDIMENTATION EQUILIBRIUM

Introduction

In the last chapter we saw, especially in aqueous systems, that the osmotic method does not possess the sensitivity required for the determination of very high molecular weights. For this reason there is need of additional methods possessing, if possible, the same thermodynamic basis and of sufficient sensitivity for use with those systems of very small osmotic pressure. Sedimentation equilibrium provides such a method, by which molecular weights of several millions can be determined with great accuracy. Indeed, certain experimental requirements are less severe the higher the molecular weight.

Since, as we have said, the method of sedimentation equilibrium possesses a thermodynamic basis, it is important to realize that it is indeed concerned, as its name suggests, with true equilibrium states. It should not be confused with the method of sedimentation velocity (described in Chapter XI), which, being concerned with the *rate* of sedimentation, is fundamentally different. The sedimentation equilibrium method is therefore liable to complications and deviations from ideal behaviour of a type very similar to those observed in the osmotic and other thermodynamically based methods. These will be considered in some detail later.

Sedimentation Equilibrium in General

In the main, we shall be considering equilibria in which sedimentation occurs under the influence of a centrifugal field, but the principles involved in its treatment are essentially similar to those describing sedimentation equilibria under gravity. These latter types will be considered first as they are in many ways simpler.

The falling off in the pressure or density of the atmosphere with increasing height above the earth's surface is well known and arises as a result of the presence of the earth's gravitational forces. Such forces, imparting a downward component of velocity to all the molecules in the atmosphere, would by themselves cause the concentration of the atmosphere in a condensed layer on the earth's surface. However, the movements of thermal agitation counteract such a process, and an equilibrium state is reached in which the settling tendency of the atmosphere due to gravity is just balanced by the opposing process of

SEDIMENTATION EQUILIBRIUM

thermal agitation or translational diffusion, tending to equality of concentration throughout the system. Such a balance is only possible when the distribution of density is of the type occurring in the atmosphere (i.e. density falling off with increasing height), when diffusion processes opposing the density or concentration gradient act in opposition to sedimentation. At any point in the atmosphere at equilibrium, therefore, the settling tendency due to gravity is balanced by diffusion.

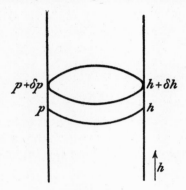

Fig. 8.1.

Consider in the atmosphere a vertical cylinder of unit cross-section and two horizontal planes at heights h and $h+\delta h$ cutting it (Fig. 8.1). Let the atmospheric pressure at these heights be respectively p and $p+\delta p$. The increase in pressure δp corresponding to a decrease in height δh is clearly equal to the weight of the molecules in the cylinder between the two planes, or
$$\delta p = -\rho g\, \delta h, \tag{8.1}$$
where ρ is the density of the atmosphere, the negative sign of the equation arising as a result of the decrease in pressure with increasing height.

Assuming the atmosphere to obey the perfect gas law, ρ may be expressed as
$$\rho = \frac{M}{V} = \frac{Mp}{RT}, \tag{8.2}$$
where M is the effective molecular weight of the gases of the atmosphere, which being inserted into (8.1) gives
$$\delta p/p = -\frac{Mg}{RT}\delta h. \tag{8.3}$$

Integrating between limits we obtain
$$\ln(p_1/p_2) = \frac{Mg}{RT}(h_2-h_1). \tag{8.4}$$

This equation quantitatively describes the fall off with height of atmospheric pressure, density, or concentration

(since $p_1/p_2 = \rho_1/\rho_2 = c_1/c_2$).

It is also basic to Perrin's method of determining the Avogadro number from observations of the distribution with height under gravity of colloidal particles in a colloidal solution, and to the sedimentation equilibrium method. Perrin counted, under the microscope, the number of particles of gamboge (or mastic) at different depths in an aqueous suspension; for one investigation the radii and density were respectively ca. 0.3×10^{-4} cm. and 1·2. The term Mg, which has the value ca. $30g$ for the atmosphere, is therefore to be replaced in this work of Perrin by

$$\tfrac{4}{3}\pi r^3(\rho_{\text{solute}}-\rho_{\text{solvent}}) \times Ng \approx 10^{10}g \text{ to } 10^{11}g.$$

In sedimentation equilibrium in the ultracentrifuge, the gravitational field is replaced by a centrifugal field of the order of $10,000g$, and molecular weights of the order of 10^5 are investigated. In this case Mg is to be replaced by

$$\frac{M}{\rho_{\text{solute}}}(\rho_{\text{solute}}-\rho_{\text{solvent}}) \times 10,000g \approx 10^8 g$$

where ρ_{solute} and ρ_{solvent} are taken to be 1·33 and 1·0 respectively.

Giving p_1/p_2 a definite value, e.g. $p_1/p_2 = \tfrac{1}{2}$, the corresponding values of (h_2-h_1) for the different systems are calculated by eqn. (8.4) and shown in Table 8.1.

TABLE 8.1. *Values of* (h_2-h_1) *for which* $p_1/p_2 = \tfrac{1}{2}$

System	Field	Molecular weight M	Sedimenting force per mole of particles	(h_2-h_1), cm.
Atmosphere	Gravity (g)	ca. 30	$30g$ dynes	6×10^5
Perrin determinations	Gravity (g)	10^{10}–10^{11}	$10^{10}g$,,	1.8×10^{-4}
Sedimentation equilibrium	Centrifugal field	10^5	10^8g ,,	1.8×10^{-2}

Thus, in Perrin's method, the use of colloidal particles of relatively large size, and in sedimentation equilibrium in the ultracentrifuge the combined use of materials of high molecular weight and of fields much greater than that of gravity, cause appreciable falling off of concentration over distances which are minute compared with those for which corresponding changes occur in the atmosphere.

In order to obtain molecular weights from observations of sedimentation equilibria in the ultracentrifuge it is necessary to determine the concentration distribution of colloidal solute over quite short distances in the sedimenting column; direct counting, as in Perrin's method, is clearly out of the question in such systems.

SEDIMENTATION EQUILIBRIUM IN THE ULTRACENTRIFUGE
Experimental

The design of low-speed ultracentrifuges, suitable for the study of sedimentation equilibria generally (and of the sedimentation velocity for *very* high molecular weight materials), has been almost entirely due

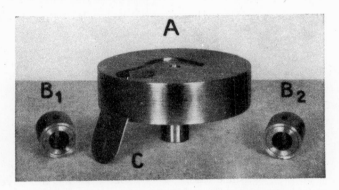

FIG. 8.2. Rotor and cells for low-speed ultracentrifuge with direct motor drive.
A. Rotor; B_1, B_2. Cells; C. Upper sector diaphragm.

to Svedberg and co-workers. Operating at speeds considerably smaller than those of the sedimentation velocity ultracentrifuge (see p. 270), it was possible to make use of conventional high-speed electric motors to obtain the required centrifugal fields.

The test liquid is contained in a (usually) sector-shaped cavity in a cylindrical steel cell provided with transparent quartz windows, which with a second similar cell fits into the rotor, a solid cylindrical mass of steel. Fig. 8.2, due to Svedberg (1940), contains photographs of the latest forms of cells and rotor for the equilibrium ultracentrifuge. Since the cells are very similar to those of the sedimentation velocity ultracentrifuge (p. 267) they will not be further discussed here. When filled with test liquid the cell is inserted into its cavity in the rotor with the sector piece alined radially in the rotor so that the direction of sedimentation near the walls is parallel to them. The second cell, which usually contains a solution of a non-sedimenting substance of suitable light absorption by which calibration of the light absorption system can be performed, is fitted similarly into the rotor in a diametrically opposite position.

The rotor, mounted directly upon the shaft of a high, variable-speed motor, is surrounded by a water-cooled brass casing through which hydrogen at atmospheric pressure flows. The temperature of the motor is carefully controlled so that no appreciable temperature gradients will occur in the rotor. Rotational speeds up to 15,000 r.p.m., corresponding

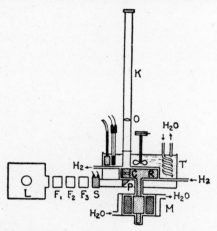

FIG. 8.3. Scheme of low-speed ultracentrifuge with direct motor drive.

M. Motor
R. Rotor
C. Cell
P. Prism
O. Objective
K. Camera
T. Thermostat
L. Lamp
F_1, F_2, F_3. Filters
S. Shutter

to a centrifugal field (for $x = 5$ cm.) greater than $10,000g$, are used, and determined accurately either by the frequency of the driving current (in the case of some three-phase A.C. motors) or by stroboscopic methods. The rotation must occur without accompanying vibration or precessional motions, and must be capable of being maintained for several days (in some cases several weeks, see p. 203) in order to ensure the complete establishment of equilibrium. The speed of rotation chosen in a given experiment is such that the ratio of the concentrations at the bottom of the cell and at the meniscus is between 2 and 5.

The optical system (Fig. 8.3), built around the rotor, consists of a light source (usually a quartz mercury lamp) L, filters F_1, F_2, and F_3, a shutter S, a prism P by which light is totally reflected through the cell, a camera objective O (usually of quartz), and camera K. By this means an image of the cell is formed on the photographic plate. If the colloidal solute absorbs light, either in the visible or ultra-violet part of the spectrum, then by the use of a suitable light source and exposure time in taking photographs, the blackening on the latter at different distances (x) from the axis of rotation may be used photometrically to obtain the relative distribution of concentration in the cell. By making use of the absorption by the second calibrating solution, the absolute concentration distribution in the cell may be worked out. Although only relative concentrations are required for molecular weight calculation by eqn. (8.6), absolute concentrations are often useful, e.g. in con-

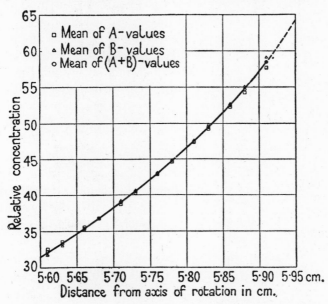

FIG. 8.4. The (c, x) curve calculated from the photometer curves.

firming the total amount of the solute by integration over the whole column. Further discussion of optical systems is contained in Chapter IX, and full experimental details are given by Svedberg and Pedersen (1940).

In the case of the proteins, which are usually colourless and upon which the greater part of sedimentation equilibrium measurements has been made, there is a suitable light absorption in the ultra-violet which is usually utilized in the light absorption method. On the other hand, where absorption in the visible occurs (e.g. solutions of dyestuffs or haemoglobin) this is usually utilized. Fig. 8.4 shows the relative concentration distribution in a sedimentation equilibrium experiment on a dilute solution (0·1 gm./litre) of congo-red in 0·1 M.NaCl (see Svedberg and Pedersen, 1940, p. 309). The A and B values refer merely to different light absorption exposures. The results of Fig. 8.4 are typical for sedimentation equilibrium experiments and the calculations made from them are discussed later.

Another optical method, the refractive index scale method, based upon differences in refractive index occurring in the cell, is also available. In this method, which is outlined in Chapter IX and which is thoroughly described by Svedberg and Pedersen (1940), a graduated scale placed behind the cell with the scale lines perpendicular to the central radius passing through the cell sector is photographed through the cell. The occurrence of refractive index gradients in the test liquid causes a displacement (Z) of the scale line images on the photographic plate, which is proportional to the refraction gradient (dn/dx) through which the light has passed. To each scale line may be assigned definite positions $x_1, x_2,...$ in the cell, and as shown later, the (Z, x) curves obtained from the scale method may be employed directly to calculate the solute molecular weight (p. 196).

Alternatively, however, from the (Z, x) curves it is possible to calculate the distribution of refractive index (and therefore of concentration) by a suitable integration process (Svedberg and Pedersen, 1940, p. 312), giving the results in a form identical with those of the light absorption method.

Exposures of the light absorption 'scale' type are taken when equilibrium is being approached, the final exposure being accepted only when no further changes in them can be observed. For a protein molecule of molecular weight 40,000 this occurs (e.g. Pedersen, 1936) only after about 100 hours, and even longer times are required for molecules of larger molecular weight and more asymmetric shape (see also p. 203).

Theoretical

The distribution of concentration in the ultracentrifuge cell at equilibrium is precisely analogous to the distribution of density in the earth's atmosphere. Thus putting for Mg the term

$$(M/\rho_{\text{solute}})(\rho_{\text{solute}} - \rho_{\text{solution}})\omega^2 x,$$

where ω is the angular velocity and $\delta h = -\delta x$ (the negative sign in the latter relation arising since the centrifugal field acts in the direction of increasing x values, whilst gravity acts towards lower h values) in (8.3), we obtain

$$\frac{\delta p}{p} = \frac{\delta c}{c} = \frac{M(1-\bar{v}\rho)\omega^2 x}{RT}\delta x, \qquad (8.5)$$

which on integration and rearrangement gives

$$M = \frac{2RT\ln(c_2/c_1)}{(1-\bar{v}\rho)\omega^2(x_2^2-x_1^2)}, \qquad (8.6)$$

where $\bar{v}$ is the partial specific volume of the solute, ρ is the density of the solution, and c_1 and c_2 are the concentrations at distances x_1 and x_2 respectively in the cell. Eqn. (8.6) is identical with (4.159), derived thermodynamically for an ideal system in a state of sedimentation equilibrium. It should be recalled that in deriving eqn. (4.159) the perfect gas law ($pv = RT$) was assumed also.

It is clear that sedimentation equilibrium in the ultracentrifuge is not at all different in principle from other types of sedimentation equilibrium. Normally, however, eqn. (8.6) is derived more directly by considering in detail the opposing processes of sedimentation and diffusion. Thus consider any plane in the cell, perpendicular to the centrifugal field at a distance x from the axis of rotation; for simplicity the cell is assumed rectangular (Fig. 8.5). Let us assume further that the solution in the cell is of a homogeneous polymer at such a concentration (c_0) that individual macromolecules are free from interactions with like molecules in all parts of the cell; thus they can migrate independently.

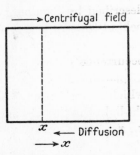

FIG. 8.5. Opposing sedimentation and diffusion in the ultracentrifuge cell.

Anticipating the results of Chapter XI (p. 287) the component of velocity of each solute molecule due to the centrifugal field at plane x is

$$M\frac{(1-\bar{v}\rho)\omega^2 x}{F}, \qquad (8.7)$$

where F is the frictional force per mole (p. 210).

The quantity of solute crossing unit area of the plane at x due to sedimentation (δm_S) in a short interval of time, δt, is then

$$\delta m_S = M\frac{(1-\bar{v}\rho)\omega^2 x}{F} c\, \delta t, \qquad (8.8)$$

where c is the concentration at x.

From Fick's first law (p. 233) we obtain an expression for the quantity of solute (δm_D), crossing unit area of the plane due to diffusion in the same time interval:

$$\delta m_D = -D\frac{dc}{dx}\delta t. \qquad (8.9)$$

By eqn. (10.27), the diffusion coefficient $D = RT/F$, which being substituted into (8.9) gives

$$\delta m_D = -\frac{RT}{F}\frac{dc}{dx}\delta t. \qquad (8.10)$$

At equilibrium the total amount of solute crossing any plane in the cell is zero, i.e.
$$\delta m_S + \delta m_D = 0. \tag{8.11}$$

$$\therefore \quad \frac{M(1-\bar{v}\rho)\omega^2 x}{F} c\, \delta t = +\frac{RT}{F}\frac{dc}{dx}\delta t,$$

and, rearranging,
$$M = \frac{RT(dc/dx)}{(1-\bar{v}\rho)\omega^2 xc}, \tag{8.12}$$

which on integrating gives
$$M = \frac{2RT\ln(c_2/c_1)}{(1-\bar{v}\rho)\omega^2(x_2^2-x_1^2)}, \tag{8.6}$$

an expression identical with that derived by other methods. It should be realized that the assumption of Fick's law (eqn. (10.1)) and eqn. (10.27) is equivalent to the assumption of ideal behaviour, as will become clear from Chapter X (p. 258). Eqn. (8.6) is to be regarded only as a limiting law for ideally behaving systems which in real non-ideal systems is to be replaced by

$$M = \frac{2RT\ln(c_2 f_2/c_1 f_1)}{(1-\bar{v}\rho)\omega^2(x_2^2-x_1^2)}, \tag{8.13}$$

f_1 and f_2 denoting activity coefficients at the concentrations c_1 and c_2 respectively. The latter equation is required under any conditions for which osmotic or other colligative properties indicate deviations from ideal behaviour.

In using the ideal equation to calculate the molecular weight M, in addition to the partial specific volume of the solute and the solution density ρ, the solute concentration at different x values is required. Light absorption diagrams give such information directly and, as already stated, it is possible to integrate a (Z, x) diagram to give it also. However, it has been shown (Svedberg and Pedersen, 1940, p. 314) that, alternatively, the solute molecular weight may be calculated by the expression
$$M = \frac{2RT\ln(Z_2 x_1/Z_1 x_2)}{(1-\bar{v}\rho)\omega^2(x_2^2-x_1^2)}, \tag{8.14}$$

where Z_1 and Z_2 are the scale line displacements corresponding to levels x_1 and x_2, respectively, in the cell.

For a monodisperse solute eqns. (8.6) and (8.14) give (within experimental error) the same value of M. For polydisperse solutes the two equations are not, however, identical. If M, as obtained by eqn. (8.6), for a small range of x values is integrated over the whole column of liquid, a weight average value M_W ($\sum n_i M_i^2 / \sum n_i M_i$) is obtained. Integration of M values obtained from eqn. (8.14) over the whole liquid column gives, on the other hand, the 'Z average' ($\sum n_i M_i^3 / \sum n_i M_i^2$), which is only identical with the weight average for a monodisperse solute (see p. 43). It should be realized that the molecular weight value calculated by either equation (8.6) or (8.14) for a small interval in x (M_x) is neither a weight nor a Z average for the solute. Such values can only be obtained by integration of M_x values over the cell. This will be more clearly understood when the nature of the results from sedimentation equilibria is discussed.

Sedimentation equilibrium in monodisperse systems

For simplicity, let us again consider the sedimentation equilibrium occurring in a solution of a monodisperse polymer, which is sufficiently dilute that interactions between like solute molecules do not occur. A dilute solution of the dyestuff, congo-red (0·1 gm. per litre), in a salt solution of well-defined composition and pH is probably such a system. The ideal equation (8.6) should therefore be applicable and the results of Table 8.2 below have been thus obtained (Svedberg and Pedersen, 1940, p. 310) from experimental data of the kind shown diagrammatically in Fig. 8.4.

TABLE 8.2. *Sedimentation Equilibrium in a Solution of Congo-red*

Initial concentration of congo-red = 0·1 gm. per litre; solvent = 0·1 M. NaCl; speed of rotation = 299·6 r.p.s.; $\bar{v} = 0·60$; $\rho = 1·0023$; $(1-\bar{v}\rho) = 0·399$.

x_2	x_1	$x_2^2 - x_1^2$	c_2	c_1	c_2/c_1	$\log_{10} \dfrac{c_2}{c_1}$	M
5·87	5·84	0·3513	53·60	50·46	1·062	0·0261	5,900
5·84	5·81	0·3495	50·46	47·57	1·061	0·0257	5,830
5·81	5·78	0·3477	47·57	44·79	1·062	0·0261	5,960
5·78	5·75	0·3459	44·79	42·18	1·062	0·0261	5,990
5·75	5·72	0·3441	42·18	39·76	1·061	0·0257	5,930
5·72	5·69	0·3423	39·76	37·46	1·061	0·0257	5,960
5·69	5·66	0·3405	37·46	35·36	1·059	0·0249	5,800
5·66	5·63	0·3387	35·36	33·36	1·060	0·0253	5,930
						Mean:	5,910

It should be noted there is no regular drift in M with changing x values: observed variations are random and within experimental error. This behaviour is observed only if the solute is monodisperse and thermodynamically ideal. If the solute is not ideal under the conditions of the experiment, i.e. activity coefficients depart from the value unity, eqn. (8.13) is required; as in osmotic pressure determinations, the use of the ideal equation under such conditions gives molecular weight values which decrease with increasing concentration (or with increasing values of x for $\rho_{solute} > \rho_{solvent}$ in sedimentation equilibrium).

Departure from the condition of monodispersity also gives rise to a drift of M values with varying x, which is now discussed below.

Sedimentation equilibrium in polydisperse systems

If a solution contains colloidal molecules (or particles) of differing mass (but identical chemical composition) under conditions such that

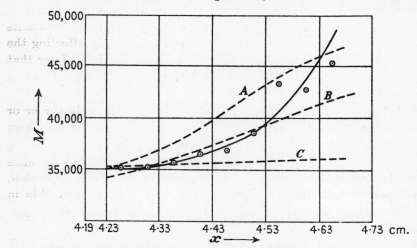

FIG. 8.6. Relation between molecular weight and distance.

Full line: calculated curve for a mixture of molecules, 94 per cent. of 34,000 and 6 per cent. of 170,000. Circles: experimental data. Dotted lines: hypothetical mixtures of molecules.
A. 40 per cent. of 17,000, 10 per cent. of 34,000, 10 per cent. of 51,000, and 40 per cent. of 68,000.
B. 30 per cent. of 17,000, 30 per cent. of 34,000, 20 per cent. of 51,000, and 20 per cent. of 28,000.
C. 95 per cent. of 34,000 and 5 per cent. of 68,000. (*From* Svedberg and Nichols (1926).)

any solute molecule is free from interaction with other like molecules, then each type of unit will set up its own equilibrium distribution of concentration independently of the other types. As is clear from eqn. (8.6), for solutes of density greater than the solvent, the larger the solute molecular weight, the higher its concentrations at the larger x values, and the lower at low x values. For a polydisperse solute, therefore, relatively larger amounts of the high and smaller amounts of the low molecular weight portions will occur at larger x values and the converse at low x values. Thus at low x values the concentration distribution will be due mainly to the lower molecular weight material and at high x values to the higher portions. Molecular weights calculated from eqn. (8.6) will, therefore, show an increase with increasing x values. Fig. 8.6 gives diagrammatically the results of molecular weight calculations for a sedimentation equilibrium investigation of unelectrodialysed egg albumin. A prominent increase in M with increasing x occurs which is fitted closely by the full curve calculated on the basis 94 per cent. of solute with $M = 34,000$ and 6 per cent. with $M = 170,000$. Although this particular molecular weight distribution agrees more closely with experimental results than the others for which the dotted curves A, B, and C are computed, it is clear that an accurate determination of the molecular weight of the different molecular species occurring in a polydisperse system is not possible, for both the percentages and the molecular weights of the different constituents are capable of considerable variation without noticeably affecting the correspondence with experimental results. It will be shown later that in general a much more clear-cut result is obtained by the sedimentation velocity method for such systems.

If the solute is of lower density than the solvent (i.e. polystyrene or rubber in chloroform) then the drift in M values with distance from axis of rotation is in the opposite direction, i.e. M values increasing with decreasing x will be observed. Whatever the relative densities of solute and solvent, however, it is generally true that, by itself, polydispersity introduces a drift in calculated molecular weights in the direction of increasing molecular weight with increasing concentration.

Whilst some such increase or decrease in calculated M values with increasing x is to be expected generally, it should be noted that other factors may mask it. Thus the deviations from ideal behaviour at the higher solute concentrations introduce, as we have seen, a lowering in the molecular weight, which is larger the higher the solute concentration. This effect is considered in further detail below, but it may now be noted that in view of the opposing nature of the two effects it is theoretically possible for M values independent of x to occur in a polydisperse non-ideal system. It is, therefore, of considerable importance that several experiments covering a range of initial concentrations should be performed.

Factors complicating sedimentation equilibrium

(a) Deviations from ideal behaviour

The ideal equation (8.6) of sedimentation equilibrium applies only under certain conditions which may vary from system to system. Under such conditions any of the various equations expressing ideal behaviour (e.g. $pV = RT$, van't Hoff equation, Raoult's law, etc.) must be simultaneously obeyed. Departures from eqn. (8.6) may therefore be considered in terms of these different ways of expressing ideal behaviour, and also in terms of the effect of solute concentration on the processes of sedimentation and diffusion (as is shown later). It should be realized, in view of the range of concentrations occurring in the solution at equilibrium, that behaviour varying from almost ideal in one portion of the cell to strongly non-ideal in another may occur simultaneously. Since, in general, we deal with solutes which are more dense than the solvent, the higher concentrations will occur at higher distances from the axis of rotation and the greatest imperfections are to be expected there. Since deviations from the van 't Hoff equation have been considered in the previous chapter, it is convenient now to make use of the conclusions reached there.

For dilute solutions of low molecular weight solutes it is to be anticipated that the ideal sedimentation equilibrium relation will apply as does the van 't Hoff osmotic pressure equation. Although not usually applied to low molecular weight solutes, useful studies of sedimentation equilibria in solutions of inorganic salts (NaCl, LiCl) and other low molecular weight materials have been made (Pedersen, 1935) in the high velocity ultracentrifuge (Ch. XI) and results in agreement with anticipated behaviour have been obtained.

For polymeric substances of compact and nearly spherical molecular form, in solution at concentrations lower than 2–3 gm. per 100 c.c. of solution, the deviations from van 't Hoff's law are not serious, and at these concentrations in the sedimentation equilibrium cell the ideal relation is usually applicable. Departures from perfect behaviour may be detected, especially in the case of a monodisperse solute, by a decrease in the calculated molecular weight with increasing values of the concentration (i.e. usually with increasing x values). Such a decrease is closely analogous to the decreased molecular weight values obtained from osmotic pressure measurements at higher concentrations. In the case of polydisperse solutes, however, the increase in the calculated molecular weight with increasing x values, which normally occurs, may prevent the observation of the opposite trend arising from imperfect behaviour. In such circumstances it is wise to carry out sedimentation equilibrium experiments over a range of low solute concentrations and to extrapolate to zero concentration.

For solutions of polymeric substances of very asymmetric form (e.g. cellulose and its derivatives, rubber, synthetic linear polymers), the deviations from ideal behaviour are thought to be small only at very great dilution in sedimentation equilibrium, as in osmotic pressure.

Such deviations may be considered thermodynamically in terms of activity coefficients (eqn. (8.13)) or, as has been done by Gralen (1944, p. 75), in terms of sedimentation constants, s (p. 272), and diffusion coefficients (D) which are dependent on concentration. Thus putting

$$s = \frac{s_0}{1+kc} \tag{8.15}$$

and
$$D = D_0(1+k_1 c), \tag{8.16}$$

where s_0 and D_0 denote the sedimentation constant and diffusion coefficient at zero solute concentration, and introducing in place of s and D in the kinetic derivation of the sedimentation equilibrium expression (p. 195), we obtain

$$M = \frac{2RT[\log(c_2/c_1)+(k+k_1)(c_2-c_1)+(kk_1/2)(c_2^2-c_1^2)]}{(1-\bar{v}\rho)\omega^2(x_2^2-x_1^2)}. \tag{8.17}$$

Clearly this equation is alternative to (8.13), containing activity coefficients. Table 8.3, due to Gralen (1944), gives experimental details and results for the examination, by the sedimentation equilibrium method, of a solution of a sample of cellulose nitrate in amyl acetate; the last two columns contain also the molecular weight values calculated by eqns. (8.6) and (8.17).

TABLE 8.3. *Sedimentation Equilibrium of Cellulose Nitrate (from unbleached American linters)*

Concentration = 0·102 gm./100 c.c. in amyl acetate; dn/dc = 0·00085; Speed = 2,400 r.p.m.; k = 4·2, k_1 = 4·2; M (sedimentation v. diffusion) = 780,000.

x	Z	$\frac{dn}{dx} \cdot 10^5$	c	M (8.6)	M (8.17)
4·86	28	16·2	0·0638	426,000	708,000
4·91	28	16·2	0·0734	362,000	642,000
4·96	28	16·2	0·0829	330,000	615,000
5·01	28	16·2	0·0925	291,000	578,000
5·06	28·5	16·5	0·1021	273,000	575,000
5·11	30	17·4	0·1121	234,000	560,000
5·16	32·5	18·9	0·1228	289,000	648,000
5·21	36	20·9	0·1345	270,000	653,000
			0·1451		
M_w:				300,000	618,000

These results demonstrate clearly the difficulties attending the use of the sedimentation equilibrium method for asymmetric polymer molecules. Thus the two columns of molecular weight values diverge widely and, accepting the new equation as an improvement, it is apparent that the ideal equation is not applicable quantitatively. The decreasing molecular weights (calculated by eqn. (8.6)) with increasing x values is further evidence of the importance of the deviations from ideal behaviour. Nor is the expected increase in M with increasing x shown clearly by calculations from eqn. (8.17). For cellulose nitrate, a well-known example of a polydisperse polymer, such an increase should be prominent and it may be found that eqn. (8.17) needs further refinement. The better agreement with the molecular weight as determined by sedimentation and diffusion measurements (p. 287) is, however, strongly in favour of the new equation.

In Chapter XI the variation of sedimentation constant with polymer concentration is discussed in detail, and it is shown there that the sedimentation constant for very high polymers may be strongly concentration dependent even at the lowest usable concentrations. Whereever such behaviour with respect to sedimentation constant or diffusion coefficient is observed, it is clear from eqn. (8.17) that similar deviations from ideal sedimentation equilibrium must be anticipated. Further, since the equilibrium method is not suitable for use with extremely low polymer concentrations (< 0.100 gm./100 c.c.), it would appear that, unless an equation of the type of (8.17) or some other equation is shown reliable, the method cannot be used for the accurate determination of the molecular weights of high polymeric molecules of asymmetric form. Gralen (1944) very reasonably considers values of M greater than 100,000 to be unreliable if calculated from the ideal equilibrium equation. Extension of the method therefore awaits the verification and possibly the refinement of eqn. (8.17) or the determination of appropriate activity coefficients for use in eqn. (8.13).

(b) Charge effects

Consider a solution of a colloidal electrolyte, consisting of a large multivalent colloidal ion and its many neutralizing simple ions, in water. If a centrifugal field be applied to the solution, the colloidal ion will, by virtue of its great mass, tend to separate from its ionic atmosphere or gegenions, but such a separation will be resisted by the potential difference arising from the separation of opposite charges. It is clear, therefore, that the equilibrium distributions of the colloidal and simple ions will not be independent of one another, and in practice a single sedimentation equilibrium distribution, in which no observable separation of the colloidal and simple ions occurs, is observed. This distribution is characteristic of neither ion alone, and it may be shown (Svedberg and Pedersen, 1940, p. 54) that for an $(n, 1)$ electrolyte the molecular weight calculated from it by eqn. (8.6) is $1/(1+n)$ times the actual molecular weight of the colloidal ion.

It will be seen in later chapters that the separate processes of sedimentation and diffusion are, under similar conditions, complicated by the presence of the ionic atmosphere of a colloidal ion. If charge effects are absent in one type of process, then they are absent under the same conditions from the others.

Just as in osmotic pressure determinations Donnan effects are eliminated by the addition of neutral salt, so in sedimentation equilibria the addition of sufficient neutral salt (with or without an ion common to the colloidal electrolyte), to give constant activity coefficients for its two ions throughout the cell, removes the difficulties arising from charge effects. Eqn. (8.6) then becomes valid and correct molecular weights for the colloidal ion are obtained from it.

Further, just as Donnan effects do not occur for iso-electric, or uncharged colloids, and especially for solutions of macromolecules in organic solvents, so also under such conditions charge effects do not complicate sedimentation equilibria.

(c) *Duration of a sedimentation equilibrium experiment*

We have already seen that, for high polymers of asymmetric form, sedimentation equilibrium is complicated by the magnitude of the deviations from ideal behaviour even in very dilute solutions. A further complication of a more practical nature occurs also in such systems in that the equilibrium state is attained only very slowly. Signer (see Svedberg and Pedersen, 1940, p. 440) has thus stated that for a polystyrene of $M = 270,000$, three to four weeks is required, and clearly this is a serious limitation of the method.

The slowness in such cases is to be attributed to the small diffusion coefficients of the high polymer asymmetric molecules; any molecule, however, asymmetric or spherical, of low diffusion coefficient gives rise to the same difficulty.

Weaver (1926), in dealing with the attainment of equilibrium in gravitational fields, considered that the time required should be not greater than that ($t_{\max}$) in which sedimentation through twice the height of the liquid column occurs. Applying to sedimentation equilibrium in the ultracentrifuge, we have

$$t_{\max} = \frac{2(x_n - x_0)}{\omega^2 s_{20} \cdot \frac{1}{2}(x_n + x_0) 3600} \text{ hours,} \qquad (8.18)$$

where x_0 and x_n are the distances from the axis of rotation of the meniscus and bottom of the cell respectively, and s is the sedimentation constant of the solute. The value of ω is chosen to give a suitable concentration distribution in the cell and from eqn. (8.6) is given by

$$\omega^2 = \frac{2RT\ln(c_n/c_0)}{M(1-\bar{v}\rho)(x_n^2-x_0^2)},$$

giving $$t_{\max} = \frac{(x_n-x_0)^2 M(1-\bar{v}\rho)}{s_{20} 1800 . RT\ln(c_n/c_0)}. \quad (8.19)$$

Making use of eqn. (11.10) eqn. (8.19) becomes

$$t_{\max} = \frac{(x_n-x_0)^2}{D_{20} 1800 \ln(c_n/c_0)} \text{ hours.} \quad (8.20)$$

Putting $\quad x_n-x_0 = 0.5$ cm.,

$c_n/c_0 = 4,$

$D_{20} = 7 \times 10^{-7}$ c.g.s. units (see Chapter X)

we obtain $\quad t_{\max} \approx 142$ hours.

For molecules with diffusion coefficients as high as 7×10^{-7} c.g.s. units (corresponding approximately to a spherical protein molecule of $M = 70{,}000$), equilibrium is, in practice, usually attained in shorter times, and it is likely that $t_{\max}$ is indeed an upper estimate. The very long equilibrium times for high molecular weight linear polymers, for which D_{20} may be ten times smaller, are readily understandable on the basis of eqn. (8.20).

Results

The first sedimentation equilibrium investigation was made by Svedberg and Fahraeus (1926), and the method quickly showed its possibilities by confirming Adair's (1924) conclusion, derived from osmotic pressure determinations, that the molecular weight of carbon monoxide-haemoglobin is four times the minimal value based on the iron content of the molecule.

Investigations on many other well-known proteins and colloidal systems (e.g. Faraday gold sols, colloidal ferric oxide) quickly followed, and simultaneously the difficulties of the method were discovered, eliminated, and important refinements introduced. It would appear that for those substances of not too asymmetric molecular shape and for which molecular weights are too high for accurate estimation by osmotic methods (especially for $M > 200{,}000$), the sedimentation equilibrium method provides one of the most convenient and accurate methods for molecular weight determination. Table (8.4) contains a selection of molecular weight values for well-known proteins, and, for comparison, the osmotically determined values are given alongside. Whilst longer equilibration times are required for very high molecular weight molecules ($M > 10^6$) *of nearly spherical shape*, the method has been applied successfully to Bushy Stunt Virus protein (McFarlane and Kekwick, 1938) of molecular weight 7,600,000, and reasonable agreement with results from sedimentation velocity measurements has been obtained (see Table 11.1, facing p. 288).

TABLE 8.4. *Protein Molecular Weights by Sedimentation Equilibrium and Osmotic Methods*

Protein	Molecular weight	
	Sedimentation equilibrium	Osmotic pressure
Egg albumin	40,500	45,000
Haemoglobin (horse)	68,000	67,000
Serum albumin (horse)	68,000	72,000
Serum globulin (horse)	150,000	175,000
Amandin	330,000	206,000

The difficulties associated with molecular weight determination for high polymeric *chain-like* molecules have, as yet, to be overcome, and it is doubtful if accuracy can be claimed for molecular weight values greater than 100,000. A considerable amount of work has, however, been done in this field amongst which we may mention that by Stamm (1930) on cellulose, Kraemer and Lansing (1933) on poly-ω-hydroxydecanoic acid, Signer and Gross (1934) on polystyrene, Mosimann (1943) and Gralen (1944) on cellulose derivatives. It has been shown by several workers that for the lower molecular weight linear polymers ($M < 100,000$) molecular weight values in reasonable agreement with those from other methods were obtained. Thus Kraemer and Lansing (1933) obtained for poly-ω-hydroxydecanoic acid the values 27,000, 25,200, and 26,000 from sedimentation equilibrium, end-group titration, and the sedimentation velocity-diffusion method (at zero solute concentration) respectively. Similar agreements were also reported by Stamm (1930) for cellulose of molecular weight 55,000, and Signer and Gross (1934) claimed that their results were in agreement with those determined by the Staudinger viscosity law (p. 368). Much more, however, remains to be done in the check of one method against others of independent origin; the latter will themselves be considered in subsequent chapters.

Advantages and disadvantages of the sedimentation equilibrium method

The chief advantage of the method lies in the soundness of the thermodynamic basis it possesses. Under suitable conditions results of unquestionable validity are therefore obtained. A further important property is that molecular weights are obtained from sedimentation equilibrium and density data alone. It will be seen later in connexion with the dynamic methods that the results of two quite independent methods must generally be combined to provide similar data.

Against the equilibrium method the difficulties associated with the frequent and sometimes very large deviations from ideal behaviour must be quoted. As we have seen, these difficulties are so serious for asymmetric molecules that very high molecular weights so obtained are of doubtful quantitative significance. A further disadvantage in the case of paucidisperse (i.e. containing a few different M values) and polydisperse systems is the lack of definite information which is provided from sedimentation equilibrium measurements. A much clearer cut result is obtainable usually from sedimentation velocity measurements. Again, even in those systems for which ideal behaviour occurs, molecular weight information only is obtained, no indications of molecular shape being derivable.

Such information is commonly obtained from frictional constants, as is explained in detail later (p. 292), and it will be recalled that in our derivation of the ideal sedimentation equilibrium expression by consideration of the opposing processes of sedimentation and diffusion, the frictional constant cancelled out. It is evident, therefore, that quantitative information concerning molecular shape might be obtained from the separate consideration of sedimentation velocity or diffusion.

A disadvantage of a more practical nature is the length of time which an equilibrium investigation requires, especially in the case of substances of high molecular weight or asymmetric molecular form. For materials of doubtful stability this may be very serious.

Summarizing, it is evident that valuable, but by no means complete, information on macromolecules and colloidal particles is given by the sedimentation equilibrium method. The obvious gaps in our picture of the nature of the disperse phase can be filled only by the simultaneous application of different experimental methods; it is with such methods that we shall be concerned in the next few chapters.

REFERENCES

ADAIR, 1924, *Proc. Camb. phil. Soc.* (Biol.), **1**, 75.
GRALEN, 1944, *Sedimentation and Diffusion Measurements on Cellulose and Cellulose Derivatives*, Dissertation, Uppsala.
KRAEMER and LANSING, 1933, *J. Amer. chem. Soc.* **55**, 4319.
MCFARLANE and KEKWICK, 1938, *Bio-chem. J.* **32**, 1607.
MOSIMANN, 1943, *Helv. chim. Acta*, **26**, 369.
PEDERSEN, 1935, *Nature*, London, **135**, 304.
—— 1936, *Bio-chem. J.* **30**, 961.
SIGNER and GROSS, 1934, *Helv. chim. Acta*, **17**, 335.
STAMM, 1930, *J. Amer. chem. Soc.* **52**, 3047.
SVEDBERG and FAHRAEUS, 1926, ibid. **48**, 430.
—— and NICHOLS, 1926, ibid. **48**, 3081.
—— and PEDERSEN, 1940, *The Ultracentrifuge*, Oxford.
WEAVER, 1926, *Phys. Rev.* **27**, 499.

IX
THE DYNAMIC METHODS
Theoretical Basis

It is proposed, in this chapter, to discuss theoretical and practical questions common to several of the dynamic methods of investigating colloidal molecules and particles in solution, with which subsequent chapters deal. Only general principles are considered here, more detailed treatments being deferred until each method is considered separately.

In the last two chapters, devoted to osmotic pressure and sedimentation equilibrium, it will be recalled that a state of equilibrium is observed from which calculations based upon thermodynamic theory are used to obtain molecular weights. Theoretically the chief uncertainty in these methods, especially in the case of very asymmetrical particles, is the importance of solute-solute interaction effects (at the concentrations which must be used experimentally) in the thermodynamical treatment of the equilibrium states. A further disadvantage is that no information on molecular shape can be obtained.

The dynamic methods, on the other hand, are not concerned with equilibrium states and thus possess no thermodynamic basis, but, if correctly applied, can provide information concerning not only molecular weight but also molecular shape. In these methods we are concerned (except in the method of viscosity) with the rate at which the colloidal particles or molecules undergo certain types of motion under the influence of appropriate applied forces. Such forces may arise from a centrifugal field or a static electrical field as in the methods of sedimentation velocity and electrophoresis respectively, or from differences of chemical potential in different parts of a system as in diffusion, in which cases translational motion of the colloidal particles occurs and is observed. Alternatively the rotation of asymmetric molecules and particles may be investigated by means of a study of their orientation in a velocity gradient (for example, as in a sheared liquid) or in a rapidly alternating electrical field. Since thermodynamics is not concerned with rate processes, we require other theoretical approaches which fundamentally are less sound. Satisfactory tests are, however, provided by comparing the results obtained for the same system by differently derived methods.

Treatment of colloidal particles in terms of geometrically simple models

In treating theoretically the motion of colloidal molecules it is necessary to idealize the system in certain ways. Thus it is usual to consider only those systems in which each colloidal molecule is free to carry out the appropriate motion without interference from like molecules. Absence of this condition not only introduces restricted motion, but also a modified effective molecular weight, both of which effects will vary with the particular system and cannot therefore be incorporated into a simple treatment. Theory applies rigidly, therefore, only to sufficiently dilute solutions.

Further, it is necessary to treat each colloidal molecule in terms of some geometrically simple model which is susceptible to hydrodynamic treatment (e.g. a sphere, an ellipsoid, or a rod). This requirement may at first sight appear improbable in real systems, for a molecule of approximately spherical or ellipsoidal shape is unlikely to possess a regular or homogeneous surface; rather will it contain many surface irregularities (depressions and protuberances) as well as a mixture of chemically non-polar and polar portions. This view is undoubtedly true, but the particle will usually be surrounded by a solvating layer, which will probably partly offset the effect of surface irregularities. Further, it is possible to visualize a hydrodynamically equivalent surface, which will include the effect of any surface irregularities.

For those cases in which the real molecule approximates closely to the simple geometrical model assumed, the dimensions determined experimentally are to be identified with those of the actual molecule. In this category are placed many corpuscular proteins which are thought to approximate closely to spherical or ellipsoidal shape. In other cases, e.g. where a molecule consists of a long flexible chain of continually changing shape (as for many linear polymers), it can be treated in terms of a mean hydrodynamically equivalent ellipsoid, but the calculated dimensions of the latter may then have little relation to the actual dimensions of the molecule.

A further possibility also arises in the case of flexible linear molecules. It would seem likely on *a priori* grounds, and there are strong indications experimentally (for example, see p. 407), that the mean statistically determined shape of the molecule may be affected by the experimental conditions during observation. Thus, under high velocity gradients, stretching of curled-up molecules occurs, and clearly the hydrodynamically equivalent model must change correspondingly. This is an inherent complexity when dealing with linear polymers. The simple theory involving a single hydrodynamically equivalent model applies therefore only under carefully specified conditions.

Frictional constants

Accepting the view that the colloidal entity may be treated as a smooth body of geometrically simple shape, and of large dimensions compared with those of the surrounding solvent molecules, let us consider the further treatment of the motion of such bodies. Clearly, having postulated the independent motion of each particle, the frictional resistance to motion through the solvent medium will be conditioned only by the properties of the solvent, the type of motion, and by the external properties of size and shape of the individual particle itself.

Since frictional forces are in general greater, the higher the particle velocity, it is clear that for a given applied force, a particle initially at rest or moving slowly will be accelerated until it reaches a steady or 'terminal' velocity, at which the applied force exactly balances the frictional forces. Quantitatively the frictional resistance is usually assumed to be proportional to the velocity of the particle, i.e.

$$\text{force (or couple)} = fv, \qquad (9.1)$$

where v is the velocity of the appropriate motion, and f, the frictional coefficient or constant of the particle, represents the force required to attain unit velocity. Such a relation is true only for small values of v, additional terms involving v^2 and possibly higher powers being required at higher velocities, as was realized by Stokes (1850) for spherical particles. It is considered likely, however, that in the dynamic methods to be discussed here the velocities are low enough for eqn. (9.1) to hold.

As was mentioned above, the frictional resistance depends upon the properties of both the solvent and particle, as well as upon the particle velocity, and the frictional constant per particle, f, has been quantitatively expressed in these terms for various types of particle by making use of hydrodynamical calculations. Since in such calculations the solvent is considered not as a particulate medium but as a continuum through which the colloidal particle moves, it is essential for the validity of the treatment that the dimensions of the particles should be large compared with those of the solvent molecules.

Spherical particles

For the case of spheres, the expressions for translational and rotational motion are due to Stokes (1850), who derived them hydrodynamically assuming the layer of solvent adjacent to the solid sphere to have the same velocity as the sphere.

For translational motion, his result may be given as

$$f = 6\pi\eta r, \tag{9.2}$$

where r is the radius of the spherical particle and η the viscosity coefficient of the solvent, providing $vr \ll \eta$, a condition additional to those requiring low particle velocities and relatively large colloidal particles. The radius, r, may also be expressed in terms of the molecular (or particle) weight and partial specific volume, giving

$$f = 6\pi\eta \left(\frac{3M\bar{v}}{4\pi N}\right)^{\frac{1}{3}}, \tag{9.3}$$

an expression which is used considerably to describe the translational motion of spherical molecules.

In deriving these expressions, the sphere was assumed to be moving in a large body of liquid, away from the effects of container walls. If this condition is not fulfilled, then correction factors must be inserted into the simple expression (for example, see Newman and Searle, 1933). Such corrections are not as a rule necessary in dealing with the motion of colloidal particles, for which the radius of the particles is usually small compared with the dimensions of the containing vessel. An important case in which the corrections must be included is in the determination of fairly high liquid viscosities by the rate of fall of small spheres (e.g. steel ball-bearings) through the liquid contained in a long tube. By equating the gravitational forces acting on the sphere with the frictional resistance given by the uncorrected form of Stokes's law we have

$$\tfrac{4}{3}\pi r^3 g(\rho_{sp}-\rho) = 6\pi\eta rv, \tag{9.4}$$

where ρ_{sp} is the density of the sphere and ρ that of the solvent. On rearranging we have

$$\eta = \frac{2}{9}(\rho_{sp}-\rho)\frac{r^2 g}{v}. \tag{9.5}$$

To correct for the wall and end effects of the container, the modified expression

$$\eta = \frac{2}{9}\frac{(\rho_{sp}-\rho)r^2 g}{v\{1+2\cdot 4(r/R)\}\{1+3\cdot 3(r/h)\}}, \tag{9.6}$$

where R is the radius of the container and h is the total height of liquid, is required (Newman and Searle, 1933). It is evident that by timing the fall of a sphere over a given distance (usually about 15 cm. over the middle portion of a tube not less than 26 cm. long and 32 mm. in internal diameter, as suggested by B.S. 188: 1937) and use of eqn. (9.6), the viscosity may readily be calculated. From eqn. (9.6) it is also clear that for colloidal particles where r/R and r/h are very small the correction terms are not necessary.

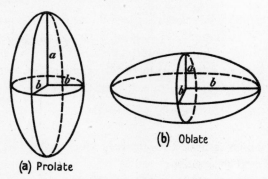

FIG. 9.1. Ellipsoids of revolution. (a) Prolate. (b) Oblate.

In the rotational case Stokes's formula for spheres may be written

$$f_{\text{rot}} = \zeta = 8\pi\eta r^3 \tag{9.7}$$

or
$$\zeta = 6\pi\eta V, \tag{9.8}$$

where the rotational frictional constant is usually denoted by ζ (see p. 382) and V is the volume of the spherical particle. It will be seen that these expressions are of considerable use in dealing with the phenomena of rotational diffusion as investigated by streaming birefringence and dielectric dispersion (Chapter XIV).

Asymmetric particles

Those colloidal molecules and particles which deviate appreciably from spherical shape are usually considered as elongated (prolate) or flattened (oblate) ellipsoids of rotation, with semi-axis of revolution of length a, and equatorial semi-axis of length b (Fig. 9.1). Theoretically it would be preferable to treat them as generalized ellipsoids, possessing three axes, a, b, and c, of different lengths, but such a procedure introduces too much complexity to be generally applied. These ellipsoids of rotation are commonly described by their axial ratio a/b (or its reciprocal), and their frictional constants are usually expressed quantitatively in terms of the axial ratio and the frictional constant (f_0) of a sphere of the same molecular weight and volume, i.e.

$$f = f(f_0, a/b). \tag{9.9}$$

It should be realized that, unlike spheres, ellipsoids have by reason of their asymmetry the possibility of more than one frictional constant according to the particular motion involved. Thus in the translational case the frictional constant depends upon the orientation of the ellipsoid, the maximum difference being obtained between motions parallel to the long and short axes. In the case of the ellipsoid of rotation two frictional coefficients are required for motion parallel and perpendicular to the axis of rotation, whilst in the general case three frictional constants are required for motion parallel to each of the three different axes. If, however, orientation is completely random, which is normally true in the translational phenomena of sedimentation velocity, diffusion, and electrophoresis, then an intermediate frictional constant is applicable (see p. 260).

Similarly, in rotational motion different frictional constants are required for rotation about different axes—three for generalized ellipsoids and two for an ellipsoid of revolution, as is usually assumed. In certain cases, e.g. orientation in a velocity gradient (p. 385), the problem simplifies still further, since we are concerned only with one type of rotation, namely, the slow rotation of a long axis about a shorter one, which alone is observed.

As shown later, the frictional constants in translational and rotational motion (f and ζ respectively) are related directly to the appropriate diffusion coefficients by the general equation

$$\text{diffusion coefficient} = \frac{kT}{\text{frictional constant per molecule}}. \quad (9.10)$$

If the molecular weight of the particle is known, then the frictional constants for translation or rotation of the hypothetical sphere of identical molecular weight and volume (f_0 and ζ_0 respectively) may be assessed by Stokes's equations (eqns. (9.3) and (9.8) respectively), and the ratio of the observed to the hypothetical frictional constant, or frictional ratio (f/f_0 or ζ/ζ_0), is thus obtained. From eqn. (9.10) we therefore have

$$\frac{\text{hypothetical diffusion coefficient}}{\text{real diffusion coefficient}} = \frac{\text{real frictional constant}}{\text{hypothetical frictional constant}}. \quad (9.11)$$

Largely owing to Perrin (1934, 1936), frictional ratios have been expressed in terms of axial ratios for prolate and oblate ellipsoids of rotation in translational and rotational motion. From Perrin's equations (see p. 261) it is possible to construct curves of the frictional ratio against numerical values of the axial ratio. If, therefore, the deviation of the frictional ratio from the value unity is attributable only to molecular asymmetry, then by the use of such curves the axial ratio is readily obtained from it. Knowing the molecular volume (from molecular weight and partial specific volume) the corresponding dimensions of the ellipsoid of rotation readily follow. The warnings given earlier (p. 209), that in the case of complicated molecules such dimensions for the hydrodynamically equivalent ellipsoid may bear no simple relation to those of the actual molecule, must, however, be emphasized.

Summarizing, the non-orientated motion of asymmetric colloidal molecules or particles is thus seen to require for its description two properties of the individual molecule; usually the molecular weight and frictional constant (or frictional ratio). In other words, the motion of a colloidal molecule under a given force depends not only upon the molecular volume (or weight), but also upon the molecular shape, and possibly, as we shall show, upon the degree of solvation. Having (in the simplest case) two unknown quantities, it is clear that two independent experimental measurements are required to evaluate them. Various combinations of methods have been used, particularly that of sedimentation and diffusion. It has also been found convenient to employ molecular weight data from equilibrium methods such as osmotic pressure in interpreting the results of dynamic measurements.

In addition to the above general principles, special procedures have been evolved for certain systems, which will be treated individually in the appropriate chapters.

It should be noted that non-spherical particles may also be treated as hydrodynamically equivalent rods and discs (rather than ellipsoids). For some purposes this may be more convenient, and some preliminary work on the electrophoresis of charged asymmetric particles has been carried out on this basis.

Solvation

It has been shown that values of the frictional ratio deviating from unity (usually > 1) arise as a result of particle asymmetry. However, before determining axial ratios from experimentally measured frictional ratios it is necessary to inquire whether other factors are also involved.

Solvation has already been mentioned as probably smoothing out surface irregularities on colloidal particles. Retention of solvent by a particle will also in general increase the frictional constant (if only by reason of increased volume) beyond that of the unsolvated particle (f_0) which is calculated from the unsolvated molecular weight values usually obtained. In the case of proteins, for example, some 0·30 to 0·40 gm. of water is probably strongly bound by 1 gm. of dry protein. Such values lead for spherical molecules to frictional ratios of 1·12 to 1·15 respectively (see p. 293). Clearly therefore deviations of the frictional ratio from unity may be attributed to solvation as well as to asymmetry, and reliable estimates of molecular shape from experimentally measured ratios can only be attempted if proper allowance is made for solvation. Such estimates, however, must necessarily remain approximate until the extent of hydration is less uncertain. It is also possible that solvation may affect the shape of a particle, especially where it occurs unequally over the particle surface, or more especially where the polymer is of the linear flexible-chain type discussed later (p. 374).

Charge effects

If the colloidal particles possess a net charge, or strong dipolar properties, then their motion, in ionizing media, must necessarily affect simple ions in their neighbourhood. At equilibrium, simple ions of opposite charge are on the average distributed symmetrically around the central ion, upon which there is therefore no resultant force. However, when motion of the central ion with respect to its atmosphere occurs the latter becomes asymmetric about the central ion, in view of the fact that its redistribution requires a finite time. The resultant force upon the central ion is no longer zero therefore, but it increases in magnitude with the velocity of the particle relative to its atmosphere. Such forces are sometimes described in terms of 'diffusion potentials', the potentials set up in this way by the separation of opposite charges.

These considerations apply to both translational and rotational motion. In the latter case, which is important, for example, for zwitterions rotating in an alternating electrical field, any charged group of the colloidal ion must give rise to its own ionic atmosphere, whose symmetry is affected by any motion of the central particle.

The details of the effects observed differ with the circumstances and are therefore discussed separately later. Common, however, to all the methods is the means by which the effects are minimized. Thus, in common with Donnan effects, diffusion potentials and accompanying effects are usually eliminated by the addition of sufficient neutral salt to the system. For 1 per cent. protein solutions, for example, 0·2 M. NaCl or KCl is usually sufficient to reduce the so-called charge effects

216 THE DYNAMIC METHODS

to below experimental error; more concentrated colloidal solutions require correspondingly higher salt concentrations.

The Examination of Concentration Gradients in Liquid Columns

Of the dynamic methods, three, namely, sedimentation velocity, translational diffusion, and electrophoresis by the moving boundary method, are experimentally similar in that they are concerned with

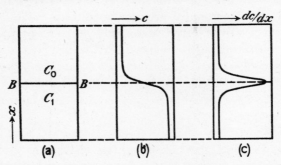

Fig. 9.2. (a) Horizontal boundary in a liquid column. (b) Concentration distribution over boundary. (c) Distribution of concentration gradient over boundary.

concentration changes occurring over effectively horizontal boundaries in vertical liquid columns.† The observation of such changes in concentration or concentration gradients is therefore of some importance, and we now proceed to discuss the principles upon which the methods of examination are based. Although differences in detail occur, the three methods employ essentially the same optical principles, which have been utilized also in other ways (e.g. in adsorption analysis, see p. 546).

Consider a horizontal boundary BB (Fig. 9.2 (a)) set up in a vertical tube between solutions of different concentration c_0 and c_1 such that the denser (and usually the more highly refracting) solution is placed below for stability. A typical boundary is that between an aqueous buffer medium above, and a 1 per cent. protein solution in the same buffer placed below. Such a boundary is never infinitely sharp on formation but, rather, occupies a finite volume over which gravity keeps the density and therefore the concentration and refractive index constant in any horizontal plane. This constancy of properties over horizontal planes of the boundary is most important, since it leads to a great simplification in the problem of their examination. Diffusion processes

† In the sedimentation velocity method, the plane of the boundary is not necessarily horizontal but is perpendicular to the direction of the centrifugal force which takes the place of gravity in this method.

cause the spreading of even a stationary and mechanically isolated boundary so that the concentration distribution in the column becomes of the type shown in Fig. 9.2 (b). The gradient of concentration, dc/dx, derived from the c v. x curve, where x represents distance along the column, is found to vary with x, as shown in Fig. 9.2 (c), the position of the 'peak' occurring at the same x value as the point of inflexion in the c v. x curve; ideally the latter occurs at the position of initial formation of the boundary, for which the concentration is constant at the value $c = (c_0+c_1)/2$. These facts arise directly from the occurrence of ideal diffusion over the boundary, as shown in Chapter X. In real systems they may not be quantitatively true, but the properties of most boundaries are, at any rate, qualitatively represented by Fig. 9.2 (a), (b), and (c).

It is possible to derive some information on concentration gradients by careful removal and analysis of liquid from different layers of the boundary region, but this operation is difficult experimentally and the data obtained are incomplete. Optical methods, however, have the great advantage that, by their use, the boundary may be observed continuously without disturbing it.

Two types of optical method are available, depending, respectively, on the differences in light absorption and in refractive index occurring over the boundary. Although refractive index methods have largely superseded those making use of light absorption, we shall consider the latter briefly, since for certain systems they are particularly suitable. The optical systems as described will be somewhat idealized, since only the chief optical principles involved are dealt with here. Features specific to single applications of a method (e.g. as required by the mechanical construction of the ultracentrifuge, or the thermostatting in electrophoresis and diffusion) are not discussed.

Light absorption methods

The use of light absorption in observing a boundary is not different in principle from its use in determining concentration in general. Light of a suitable wavelength is passed through a known thickness of the unknown solution, and for a known exposure the resultant blackening on a photographic plate is compared with that occurring under similar conditions for solutions of known and comparable concentration. The unknown concentration may then be estimated and, if required, greater accuracy may be obtained by repeating the procedure described, using for calibration purposes known solutions for which the concentrations are chosen to be much closer to that of the unknown on the basis of the first comparison.

In a similar way suitable parallel light is passed horizontally through

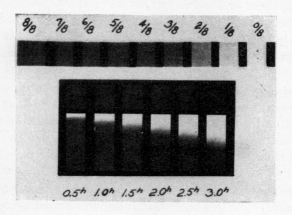

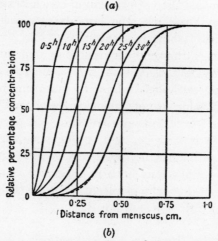

FIG. 9.3. (a) gives the sedimentation picture for CO–haemoglobin at pH 7·56 in the first high-speed ultracentrifuge. Centrifugal force, 115,000 times gravity. Time between exposures, 30 minutes. (b) gives the concentration distribution in the cell at the corresponding times. The dotted curve for 2 and 3 hours represents the theoretical diffusion curve of a substance with $D_{20} = 6 \cdot 13 \times 10^{-7}$. (Svedberg and Nichols (1927)).

the liquid boundary and causes blackening of a photographic plate which is focused upon the boundary. Since the concentration over a given horizontal plane in the boundary is constant, the blackening of the photographic plate at a conjugate height on the image of the cell gives a measure of this concentration. The blackening over the boundary region is therefore compared with that obtained under identical conditions for known concentrations of the same solution, and on this basis the concentration distribution in the horizontal layers over the boundary region is evaluated.

Visible light can be used for those solutions which possess absorption in the visible part of the spectrum (e.g. haemoglobin solutions), but many systems, especially amongst the proteins, possess no such absorption, and ultra-violet absorption may then be employed. In the latter case the components of the optical system (including the windows of the observation cell) must be of quartz, since ordinary glass does not transmit ultra-violet light. In the case of proteins there is the added disadvantage that exposure to ultra-violet light may introduce irreversible changes which become troublesome if long exposures are required.

Fig. 9.3 (a) shows typical light absorption diagrams (in the visible) for carboxy-haemoglobin sedimenting in the ultracentrifuge, with calibration exposures alongside at known protein concentrations. Comparison of the boundary region blackening (usually photometrically) with the calibration exposures then yields the curves of Fig. 9.3 (b) corresponding to the different times of sedimentation.

The successful use of the light absorption method requires the most careful standardization of conditions of exposure, of development, and of comparison with calibration exposures. (For details see Svedberg and Pedersen, 1940, or works concerned with optical methods of examination, such as Gibb, 1942.) In recent years, however, this method has been largely displaced by the refractive index methods discussed below which are capable of attaining greater accuracy with considerably less labour.

Refractive index methods

Two related methods are available:
 (a) The Schlieren Method (and its modifications).
 (b) The Scale Method.

The schlieren method, in one of its various forms, is now the most extensively used of all the optical methods mentioned, chiefly as a result of its freedom from tedious calculations in providing very detailed information. It is not, however, as accurate as the scale method, but the latter involves considerable calculation and is not therefore used unless information of the greatest precision is required.

Basic to the use of both refractive index methods is the relation between concentration and refractive index. For many systems (for example, see Adair and Robinson, 1930) it is found that

$$\frac{n-n_0}{c} = \text{constant } (\alpha), \qquad (9.12)$$

where n and n_0 are the refractive indices of solution and pure solvent, c is the concentration of solute in the solution, expressed usually in grams per 100 c.c. of solution, and the constant α is known as the

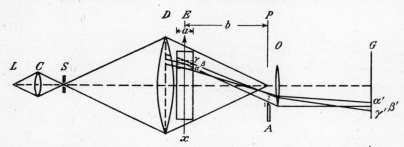

FIG. 9.4. The elements of the schlieren optical system. (Longsworth (1939).)

specific refraction increment. The need for expressing c in terms of amount of solute in a given *volume* of solution (not in a given weight or volume of pure solvent) should be noted. For other methods of expressing c, α is no longer constant.

From eqn. (9.12), Δn, the refractive index change between solution and solvent (or between two solutions), is proportional to the corresponding change of concentration Δc, so that the curves of n and dn/dx against x are geometrically similar to those of c and dc/dx against x shown, for example, in Fig. 9.2 (b) and (c).

The principles of the schlieren method will now be described and its various possible modifications mentioned. The scale method, which to some extent is similar in principle to the schlieren method, will then briefly be discussed.

Schlieren methods

The basic schlieren method is shown diagrammatically in Fig. 9.4. An image of the narrow slit S, illuminated by means of the light source L and the condensing lens C, is brought to a focus by the schlieren lens D on the axis of the system and in the plane of the movable opaque diaphragm A (the schlieren diaphragm), possessing a sharp upper horizontal edge. The cell E, with plane parallel walls, placed as near D as possible so that it may be uniformly illuminated, does not affect the position of the image of S, provided the liquid contained is homogeneous throughout. The photographic objective O, placed immediately behind the schlieren diaphragm, forms an image of the cell E upon the ground-glass screen or photographic plate, G. Thus there are two linked lens systems, one forming an image of the slit S in the plane P of the schlieren diaphragm and the other forming an image of the cell on the plane of G.

If the cell contents are homogeneous throughout, then all light passing through them is brought to a focus on the axis of the optical system, in plane P. If, however, the refractive index varies in the cell, then light will be deflected so that it no longer intersects the axis in plane P. Thus consider a liquid boundary, of the type already discussed, occurring in the cell over the region $\alpha\beta\gamma$, with the fluid of higher refractive index below, the peak of the concentration gradient curve (i.e. $(dc/dx)_{max}$) occurring at position β and lower values of dc/dx at α and γ. As already pointed out, the position at which $(dc/dx)_{max}$ occurs is also that of $(dn/dx)_{max}$, in view of the proportionality between refractive index and concentration changes.

Consider now the passage of a nearly horizontal light ray through this boundary region in which the refractive index gradient is confined to the vertical dimension. Such a ray will pass through only a small range of values of dn/dx, and it can be shown (Svedberg and Pedersen, 1940) that, for moderate values of dn/dx, the ray is deflected downwards through an angle given by

$$\delta = a(dn/dx), \tag{9.13}$$

where dn/dx is the mean value of the refractive index gradient through which the ray passes, and a is the thickness of the liquid column in the direction of the optical axis.

The deflected ray must now strike the plane P below the axis of the system, the linear displacement in the plane being, for small δ values, given by

$$\Delta = b\delta = ab(dn/dx), \tag{9.14}$$

where b is the optical distance $\left(\text{i.e. } \sum \dfrac{\text{distance}}{n}\right)$ in the direction of the optical axis between plane P and a central vertical plane through the cell. Maximum values of the displacement in plane P will clearly be caused by passage through the highest values of dn/dx, i.e. through region β of the boundary, whilst in positions for which dn/dx is smaller, smaller displacements will be suffered.

If the diaphragm A is well below the path of deflected and undeflected rays, all light rays reaching the objective O will still be brought to a correct focus on G whether they were deflected in passing through the cell or not. This results merely from the property of the objective, that each point on the object (in this case the cell E) gives rise to a conjugate point in the image, irrespective of the inclination of light rays leaving the object. If, however, the diaphragm A is raised gradually towards the axis, those rays which have suffered the maximum displacement will first be cut off, and prevented from reaching the image plane. Such rays must have passed through the planes of maximum refractive index gradient in the region β of the cell and therefore a horizontal black line in the image of the cell, conjugate to these planes, must occur (Fig. 9.5). As A is further raised, rays passing through smaller values of dn/dx, i.e. in planes on either side of β, will also be cut off and the dark horizontal line in the cell image will broaden on both sides till its edges are conjugate to the planes in the cell at which (dn/dx) has the value related to the particular displacement Δ of the diaphragm edge below the optical axis, by eqn. (9.14). Thus, since Δ is proportional to dn/dx in those planes which are conjugate to the edges of the schlieren band (a and b in eqn. (9.14) both being constant), by taking photographs of the cell at different known positions of the schlieren diaphragm, the refractive index gradient at different levels of the cell may be explored, as shown diagrammatically in Fig. 9.5, in which each photograph corresponds to a given value of Δ, and therefore of dn/dx, which occurs at those planes conjugate to both edges of the schlieren band. If the separate photographs are taken at regular intervals of Δ, then a continuous curve of refractive index gradient against height x in the cell may be obtained by drawing through the central points of the edges of the schlieren bands, as indicated in Fig. 9.5 by the dotted curve.

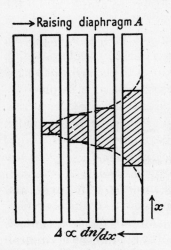

Fig. 9.5. The formation of horizontal schlieren bands.

Alternative and more direct ways of constructing continuous curves of dn/dx v. x are, however, available, and we shall consider them under two headings:

1. The Scanning Method.
2. The Diagonal Schlieren Method, or Cylindrical Lens Method.

The scanning method. This method is due chiefly to Longsworth (e.g. 1939, 1946) A narrow vertical slit (0·2 mm. wide) masks the image of the cell at G, and in obtaining the continuous dn/dx v. x curve, the photographic plate is moved continuously horizontally (i.e. in a direction perpendicular to the plane of Fig. 9.4), whilst the diaphragm edge to which it is geared is moved vertically upwards. Movements are

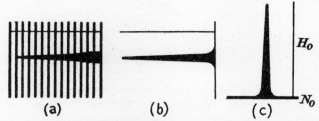

FIG. 9.6. Horizontal schlieren bands (a) and scanning photographs (b and c) of a single boundary. (Longsworth (1946).)

carried out by a constant-speed motor, and in order that horizontal distances along the plate shall really be proportional to Δ and therefore to dn/dx (as in Fig. 9.5), the gearing between plate and diaphragm must be constant throughout any one photograph. The ratio of the horizontal to vertical dimensions on the photographic plate may, however, be altered where required merely by altering the gearing connexion between plate and diaphragm. Fig. 9.6, due to Longsworth (1946), shows in (a) a series of separate exposures taken at different heights of the schlieren diaphragm, and in (b) the same boundary examined by the scanning procedure. Clearly (b) may be imagined to be derived from (a) by making each vertical photograph in (a) very narrow, and by taking a much larger number of them. Fig. 9.6 (c) merely gives (b) in the conventional manner with the peak vertical. H_0 is merely an index mark on the cell and N_0 is the base-line position, corresponding to homogeneous solution in the cell.

The diagonal schlieren or cylindrical lens method. This method was first developed by Philpot (1938), and involves the use of an adjustable diagonal edge instead of the horizontal one, A, and a cylindrical lens with its axis vertical placed between the objective O and the ground-glass screen at G. The cylindrical lens is so placed that in the horizontal dimension an image of the plane P of the diagonal edge is now formed at the ground-glass screen, whilst in the vertical dimension the image plane is still conjugate to the cell. Thus fine vertical lines in plane P (but not horizontal ones) and horizontal ones in the cell (but not vertical) are focused in the camera.

The insertion of a diagonal edge in the plane P instead of the horizontal one introduces the fact that the amount of light cut off at one side of the edge is different from that cut off at the other. Thus in Fig. 9.7, referring to plane P, II is the undeviated slit image, $I_M I_M$ is the image of maximum deviation arising from a pencil of maximum deflected rays, and AA the inclined edge making an angle θ with the

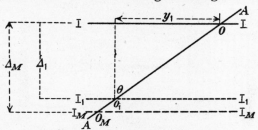

Fig. 9.7. The diagonal schlieren method.

vertical. It is clear that to the right of the point of intersection (O) of AA and II, no light will reach the objective, whilst to the left of the point of intersection (O_M) of $I_M I_M$ with AA, no light will be cut off. Between these two points Δ varies from 0 to $\Delta_{\max}$ ($= ab(dn/dx)_{\max}$) respectively, and by focusing the photographic plate in the horizontal dimension upon the plane of the diaphragm, the schlieren band varies in width from infinity at points on the photographic plate conjugate to O (i.e. base-line) to zero at points conjugate to O_M. In other words, a peak of the type of Fig. 9.2(c) is obtained, whose maximum is at a horizontal position on the plate conjugate to point O_M and whose baseline (corresponding to the cut-off of undeviated light) is conjugate in its horizontal dimension to point O. It remains, however, to prove that the horizontal dimension of the peak is indeed proportional to the refractive index gradient at conjugate levels in the cell.

Consider light passing through the horizontal layer of the boundary in the cell at level x_1. All such light suffers a linear deviation in the plane of the diaphragm of

$$\Delta_1 = ab(dn/dx)_1, \qquad (9.14\,\text{a})$$

giving rise to the deviated image $I_1 I_1$. In the absence of the diaphragm the light would all reach G, and would be brought to a focus at a horizontal level conjugate to x_1 (say X_1). In the presence of a diaphragm, however, only light to the left of O_1, the point of intersection of AA with $I_1 I_1$ with $y = y_1$, reaches the plane of G. Therefore on the screen at G at level X_1 there is a boundary between lightness and darkness at a point conjugate to O_1 (say Y_1).

Quantitatively, let the vertical dimension (X) at G be related to the vertical dimension (x) of the cell by the magnification factor G_x in the equation
$$X = G_x x \qquad (9.15)$$
and the horizontal dimension (Y) at G to the horizontal dimension (y) of the schlieren diaphragm by the magnification factor G_y, in
$$Y = G_y y. \qquad (9.16)$$
From Fig. 9.7
$$y_1 = \Delta_1 \tan \theta, \qquad (9.17)$$
and substituting from (9.14 a) we have
$$y_1 = ab(dn/dx)_1 \tan \theta. \qquad (9.17\,a)$$
Substituting (9.17 a) in (9.16), we have
$$Y_1 = G_y ab \tan \theta (dn/dx)_1. \qquad (9.18)$$
In general, therefore,
$$Y = G_y ab \tan \theta (dn/dx), \qquad (9.19)$$
or, for a given optical setting of the diaphragm,
$$Y \propto dn/dx; \qquad (9.19\,a)$$
in other words, the horizontal dimension of the peak on screen G is proportional to the refractive index gradient at a conjugate height in the cell. Thus, since in the vertical dimension the image is conjugate to the cell, the curve obtained on the ground-glass screen or photographic plate gives a plot of the refractive index gradient throughout the cell.

From eqn. (9.19) it is clear that the scale upon which dn/dx is recorded at G, or the sensitivity of the method, is dependent upon the value of the angle θ between the diagonal edge and the vertical. If $\theta = 90°$, $\tan \theta = \infty$, and infinite Y values are given by eqn. (9.19). This corresponds clearly to the use of the horizontal diaphragm in which parallel schlieren bands are obtained. For values of θ near $90°$ very tall peaks and large areas are obtained, but at the expense of definition. On the other hand the use of small values of θ (e.g. $< 45°$) gives shorter but well-defined peaks whose contours can be accurately measured.

Variation of the angle θ corresponds, as will be recognized, with change of the gearing between the photographic plate and the horizontal diaphragm in the scanning method.

Peak areas and concentrations. Rearranging eqn. (9.19) we have

$$dn = \frac{1}{G_y \, ab \tan \theta} Y \, dx \qquad (9.20)$$

$$= \frac{1}{G_x G_y \, ab \tan \theta} Y \, dX. \qquad (9.20\text{a})$$

Integrating between the limits X_2 and X_1 we have

$$[\Delta n]_{X_1}^{X_2} = \frac{1}{G_x G_y \, ab \tan \theta} \int_{X_1}^{X_2} Y \, dX, \qquad (9.21)$$

where $\int_{X_1}^{X_2} Y \, dX$ clearly represents the area under the peak between X_1 and X_2. Thus the difference in refractive index between any two levels of the cell is proportional to the area under the curve between conjugate levels. If X_1 and X_2 be chosen to lie in the homogeneous fluids on opposite sides of the boundary, then Δn corresponds to the total refractive index change over the boundary, and may be evaluated from this area by (9.21) if the constants G_x, G_y, a, b, and θ are known. Alternatively $1/(G_x G_y \, ab \tan \theta)$ may be evaluated (for given θ) by forming a boundary, across which Δn is known, followed by measurement of the curve area, by planimeter or other means.

Peak areas may, therefore, be used in evaluating refractive index changes over a boundary, and clearly, if the specific refraction increment (α) is known (eqn. (9.12)), then the concentration change over the boundary may also be obtained from the expression

$$\Delta c = \frac{1}{\alpha K} \int Y \, dX, \qquad (9.22)$$

where $K = G_x G_y \, ab \tan \theta$. The use of variations in θ in achieving different sensitivity is clear from eqn. (9.22).

This diagonal schlieren apparatus is of great importance, especially where paucidisperse systems, i.e. systems containing a few different components, are involved. It has thus frequently been possible to estimate the composition of a polymer mixture where no other method was convenient. Whilst we have discussed peak areas in terms of the diagonal schlieren method, very similar considerations also apply to the scanning method, since, again, the horizontal dimension is proportional to dn/dx (the proportionality factor here involving the gear ratio as well as a and b) and the vertical dimension to height x in the cell. Peak areas may therefore be evaluated and used to determine refractive index and concentration changes as described above. The scanning and diagonal schlieren methods provide identical information in theory, but there are points in favour of each. The absence of moving parts and the actual appearance of schlieren diagrams on a ground-glass screen are advantages in favour of the diagonal method; on the other hand the possibility of imperfections in the cylindrical lens, leading to the distortion of the diagram, has to be admitted. It seems, however, that in general, with the good-quality cylindrical lenses now available, most workers prefer the diagonal schlieren method.

Certain modifications of the diagonal schlieren method. Various workers have used diagonal schlieren systems similar to the one just described except that the diagonal edge in these systems is replaced by other optical components.

In the apparatus already described, deviated light is to a large extent cut off, whilst undeviated light mostly reaches the ground-glass screen. The schlieren peak occurs, therefore, as a dark area on a light background. If, on the other hand, the diagonal edge is inverted, then the undeviated light is largely cut off whilst deviated light reaches the screen, so that the schlieren peak now appears as a light region on a dark background.

In some cases, however, it is an advantage to have the schlieren peak in outline only. Svensson (1939) therefore introduced the use of a diagonal slit in place of a diaphragm, and it is easy to show that the schlieren peak appears now only in outline as a bright line on a dark background. The appearance may be reversed by using a thin diagonal wire instead of the slit, as the only obstacle in the path of the light. The peak then appears, again in outline, as a dark line on a light background.

Each of these systems possesses advantages under particular circumstances and, since the other optical components are not affected, they may be made readily interchangeable.

Important requirements of schlieren methods. For the use of the schlieren methods as described, it is essential that the refractive index should be constant over any horizontal plane, i.e. that the refractive index gradient should occur only in a vertical direction. Where the lower component has a much higher density than the upper, gravity tends to preserve the constancy of density, refractive index, and other properties in a horizontal plane, but if the boundary becomes disturbed by convection, mechanical vibration, or any other cause, refractive index variations in a horizontal plane may well occur and will introduce complications. It is essential, therefore, that boundary disturbances be minimized in order to obtain uncomplicated schlieren diagrams.

Another important requirement concerns the inclination of the light through the cell. In order to achieve high resolving power, a given light ray should pass through only a very small range of dn/dx values and therefore through only a very thin horizontal layer of liquid. A horizontal (or nearly horizontal) parallel beam of light meets this requirement and under such conditions the resolving power of the method can be very high. In order, also, to avoid complications due to incorrect focusing, long focal-length lenses are used (especially the schlieren lens, D, and the objective, O), and if these lenses are uncorrected, strictly monochromatic light is required.

The accurate adjustment of the schlieren optical system requires the most careful alinement and focusing of the components. The whole procedure has been fully described by Longsworth (1946) and by Svensson (1946).

In order to understand the possibilities of the schlieren methods the different types of boundary behaviour may briefly be summarized. Each is more fully described in subsequent chapters.

1. Stationary, Diffusing, Single Boundary. This is the simplest case, in which the only effects to be observed are those of diffusion, leading to a broadening and shortening of the peak with time, the area remaining constant.

2. Moving and Diffusing, Single Boundary. In addition to the broadening of diffusion, the rate of movement of the boundary under the applied forces may be observed, and some type of mobility determined. The diffusion broadening processes may, however, in such a case be complicated by additional effects.

3. Moving and Diffusing, Multiple Boundary. In addition to the possibilities mentioned in (2), the movement of the boundary may result in a splitting of the initially single boundary into several components for which mobilities may be determined and concentrations measured.

4. Polydisperse Boundaries. The most complex case is that of a polydisperse boundary, from whose rate of movement an average mobility may be determined, but whose spreading may be due to the combined effects of diffusion and polydispersity.

Simultaneous use of absorption and refractive index methods. In certain cases, e.g. 1 per cent. solution of haemoglobin, a boundary may be observed by the light absorption and refractive index methods simultaneously. Fig. 9.8 contains sedimentation diagrams for a 0·5 per cent. solution of carboxy-haemoglobin in phosphate buffer at pH 8·6. Since the protein has a very strong absorption in the visible part of the spectrum, transmitting only red light, the position of the boundary is marked in the diagrams by the transition from the unexposed to the

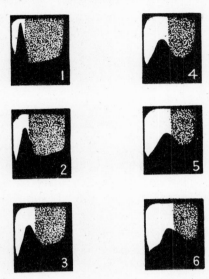

Fig. 9.8. Absorption and diagonal schlieren patterns superimposed in sedimentation diagrams of carboxy-haemoglobin.

0·50 gm. of protein per 100 c.c. of solution in phosphate buffer at pH 8·6. Speed—*ca.* 1,266 r.p.s. Centrifugal field—*ca.* 182,000*g*. Interval between exposures approximately 15 minutes.

exposed portions. However, over the same boundary there is a corresponding change in the refractive index, so that the schlieren methods may be used to detect the boundary position. Fig. 9.8 therefore shows the schlieren peaks superimposed upon the absorption patterns. Although in this case the care required in exposing, printing, and developing for the accurate use of the absorption method has not been observed, the diagrams clearly show the great advantage of the schlieren over the absorption method in deciding accurately upon a boundary position.

The simultaneous use of the two types of diagram may prove very helpful in mixtures of proteins with widely differing absorptions, e.g. a coloured and a colourless protein. The refractive index method alone merely gives a plot of the total refractive index gradient and concentration through the cell, with no means of subdivision into those portions of the total effect for which each individual component is responsible.

If, however, by means of light absorption, information on the concentration throughout the cell of one component is also obtained, that for the second readily follows.

Scale method

The scale method is more accurate than the absorption or schlieren methods, but, as will become clear, its use involves a considerable amount of tedious measuring and computing.

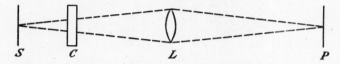

FIG. 9.9. The scale method.

A uniformly illuminated transparent scale, S, with horizontal lines (distant approx. 0·02 cm.) is placed behind the observation cell C, and an image of S is formed by the camera objective L upon the ground-glass or photographic plate P (Fig. 9.9). If the cell contains a homogeneous fluid, then an undistorted image of S is formed at P. If, on the other hand, the cell contains a liquid boundary over which a change of refractive index occurs, then some of the scale lines in the image are displaced. The actual displacements can be measured by means of a micro-comparator on the two photographs obtained in the presence and absence of the boundary.

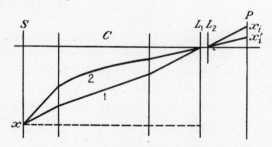

FIG. 9.10. The displacement of scale lines.

The displacement of the scale lines on the image plane by the boundary may be understood from Fig. 9.10, a modification of that due to Lamm (1937). L_1 and L_2 are here the principal planes of lens L, and ray (1) is typical of those passing from a given scale line at x through a homogeneous medium in the cell to give an image at x_i. Ray (2) shows the deviation occurring on passing through a refractive index gradient (with higher refractive index below), the line image becoming displaced to a point below at x_i'. It can be shown (see Lamm, 1937, or Svedberg and Pedersen, 1940, p. 254) that the displacement $Z = x_i - x_i'$ may be written

$$Z = Gb\delta, \qquad (9.23)$$

where G is the photographic enlargement factor, b is the optical distance from the scale to the centre of the cell, and δ is the deviation of the ray

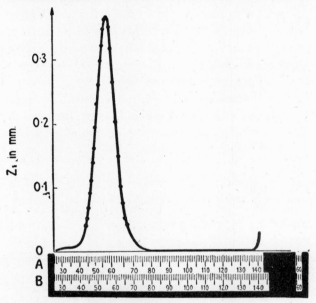

FIG. 9.11. Illustration of the relationship between the scale image and the sedimentation diagram.

Upper part: The sedimentation diagram. *A*. The normal exposure. *B*. The reference exposure. Observe the distortion of scale A between the lines 40–70. (*From* Svedberg and Pedersen (1940).)

in passing through the refractive index gradient in the cell. Substituting for δ from eqn. (9.13) we have

$$Z = Gab(dn/dx), \qquad (9.24)$$

a result similar to, but not identical with, eqn. (9.14) in the schlieren method. Any given displacement results from passage through a thin layer of liquid in the boundary region, whose height is proportional to the actual position of the displaced line x_i'. Thus distances z on the photographic plate are multiplied by the factor

$$F = \frac{1}{G}\frac{l-b}{l} \qquad (9.25)$$

where l is the optical distance between the scale and the first principal plane of L. It is essential, as in the schlieren system, that rays deviated by the boundary region should pass through a small vertical thickness of the column only, so that a given line displacement is representative of dn/dx over a well-defined position in the column. Long focal-length objectives and nearly parallel light through the cell must therefore be used.

From experimental photographs it is therefore possible to plot a curve of Z ($\propto dn/dx$) against height in the cell which is very similar to schlieren curves. However, a single scale line curve of the type shown in Fig. 9.11 requires a considerable amount of measuring and computing, whereas the schlieren method gives such curves directly. It is clear, therefore, that the greater detail and freedom from errors of the scale method—the latter a result of the simpler optical arrangements—are obtained only at the expense of much labour. In general it is only where dn/dx v. x curves of the most precise character are required, as, for example, in diffusion experiments, that the scale method is now used.

In the next few chapters many examples of the use of the different optical systems in investigating boundaries will be encountered.

REFERENCES

ADAIR and ROBINSON, 1930, *Bio-chem. J.* **24**, 993.
B.S. 188: 1937, i.e. British Standards 188: 1937, obtainable from British Standards Institution, 28 Victoria St., London, S.W.1.
GIBB, 1942, *Optical Methods of Chemical Analysis*, McGraw-Hill Book Company.
LAMM, 1937, *Nova Acta Soc. Sci. upsal.* **10**, No. 6, Ser. IV, 15.
LONGSWORTH, 1939, *Ann. N.Y. Acad. Sci.* **39**, 188.
—— and MACINNES, 1939, *Chem. Rev.* **24**, 271.
—— 1946, *Industr. Engng. Chem.* (Anal.), **18**, 219.
NEWMAN and SEARLE, 1933, *The General Properties of Matter*, Benn, p. 207.
PERRIN, 1934, *J. Phys. Radium* (7), **5**, 497.
—— 1936, ibid. **7**, 1.
PHILPOT, 1938, *Nature*, London, **141**, 283.
STOKES, 1850, see *Mathematical and Physical Papers*, III, Cambridge, 1901.
SVEDBERG and PEDERSEN, 1940, *The Ultracentrifuge*, Oxford.
SVENSSON, 1939, *Kolloidzschr.* **87**, 181.
—— 1946, *Ark. Kemi Min. Geol.*, No. 10, **22**.

X
TRANSLATIONAL DIFFUSION

Introductory

It has been shown in Chapter IV that the chemical potential of a substance in solution is dependent upon its activity, and, for dilute solutions, upon its concentration. Further, from thermodynamics, we know that a substance will tend to move from a region of high to one of low chemical potential and that this movement will continue as long as the difference in potential exists. Thus, if two dilute solutions containing the same solute at different concentrations are brought together, there will occur a transfer of solute from the more to the less concentrated solution, until there is uniform concentration throughout.

Thermodynamics cannot, however, make any pronouncement on the mechanism or rate by which concentration equalization occurs. It is the purpose of this chapter to discuss the latter question in some detail, in systems which are independent of external force fields. In such systems concentration equalization can only occur as a result of a process involving the thermal motions of both solute and solvent molecules, and termed diffusion.

Theoretical

(a) Fick's First Law

Diffusion was first put upon a quantitative basis by Fick (1855), who was led to state the law bearing his name by analogy with the processes involved in heat conduction. Thus the amount of substance dm diffusing in the x direction in time dt across an area A (Fig. 10.1) is proportional to the concentration gradient dc/dx in the plane of the area and may be written

$$dm = -DA \frac{dc}{dx} dt. \qquad (10.1\,\mathrm{a})$$

In terms of the number of diffusing molecules (n) we have

$$dn = -DA \frac{dc'}{dx} dt, \qquad (10.1\,\mathrm{b})$$

where c' is now a molecular concentration.

D, the diffusion coefficient (or less correctly the diffusion constant), is thus the constant of proportionality in eqn. (10.1), which measures the amount of material diffusing in unit time across unit area when the concentration gradient is unity. It will later be shown that D is not strictly a constant, but that at constant temperature it may vary with concentration; the measured value of D is often, therefore, to be regarded as a mean for the concentration range covered, unless special methods of calculation are adopted (p. 238). The negative sign in eqn. (10.1) is required, since diffusion occurs in a direction opposite to that of increasing concentration. Inspection of eqn. (10.1 a) shows that

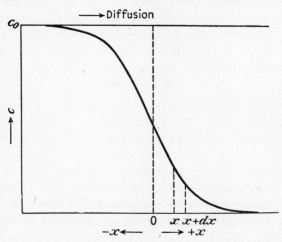

Fig. 10.1. Diffusion and the concentration gradient.

if dm, the amount of material diffusing, and c, the concentration, are both expressed in terms of the same unit, e.g. grams or gram molecules, then D, with dimensions length^2 time^{-1} (e.g. cm.^{2} sec.$^{-1}$) will be independent of the unit used. Failure to do this will, however, seriously affect the value of D. Although empirically derived and confirmed in the first place, the law may also be deduced osmotically for dilute solutions, as we shall show later (p. 256).

(b) *Fick's Second Law*

There is another fundamental equation in diffusion, sometimes regarded as Fick's Second Law, which is derived from the First Law by eliminating the dependent variable, m.

Consider the infinitesimal volume between planes at x and $x+dx$ (Fig. 10.1). The increase in the amount of diffusing substance in this volume in any given time, dt, is given by the difference between the amounts entering and leaving by the planes at x and $(x+dx)$. This may be written

$$dm - \left(dm + \frac{\partial(dm)}{\partial x} dx\right) = -\frac{\partial(dm)}{\partial x} dx.$$

TRANSLATIONAL DIFFUSION

Such an increase in diffusing substance corresponds to a concentration increase of
$$-\frac{\partial(dm)/\partial x}{A\,dx}\,dx,$$
which may also be written as $(\partial c/\partial t)\,dt$.

Thus
$$\frac{\partial c}{\partial t}\,dt = -\frac{1}{A}\frac{\partial(dm)}{\partial x}$$

and differentiating eqn. (10.1 a) with respect to x, assuming D to be independent of concentration, and substituting for $\partial(dm)/\partial x$, we have

$$\frac{\partial c}{\partial t} = D\frac{\partial^2 c}{\partial x^2}. \tag{10.2}$$

This equation, Fick's Second Law, is the general differential equation of diffusion, the solution of which is dependent upon the particular boundary conditions which are imposed.

(c) *Particular solutions of the diffusion equation*

In this chapter we are concerned only with free diffusion across a horizontal boundary between a solution of a colloidal material and the pure solvent, the denser component being placed below to give stability to the boundary. We should point out that although we shall usually speak of the diffusion from a solution to the pure solvent, this should be taken (with slight modifications only) to cover also the important case of diffusion between two solutions of different concentration. Mathematically we may state the boundary conditions as follows:

At $t = 0$,
$$\text{for}\quad x > 0,\quad c = 0\quad(\text{or } c_0'),$$
$$x < 0,\quad c = c_0,$$

where c_0 is the initial concentration.

If we assume that:

(1) the diffusion coefficient is independent of concentration (as above),
(2) the experimental cell is so long that at the end of the measurements the initial concentrations for large positive and negative values of x are unaltered,

the general diffusion equation may be integrated, and the solution is given by

$$c = \frac{c_0}{2}\left(1 - \frac{2}{\sqrt{\pi}}\int_0^y e^{-v^2}\,dy\right), \tag{10.3}$$

where
$$y = \frac{x}{2\sqrt{(Dt)}} \tag{10.4}$$

and c is to be regarded as the concentration at time t and distance x from the original boundary.

It is of importance to examine the properties of this solution. At $x = 0$, for all values of time t (except $t = 0$) $c = c_0/2$, i.e. at the position in which the boundary was originally formed the concentration is constant at half the initial concentration. For x values other than zero, the concentration differs from $c_0/2$ by the value of the integral term

$$\phi(y) = \frac{2}{\sqrt{\pi}} \int_0^y e^{-y^2} dy = 1 - \frac{2c}{c_0}, \tag{10.5}$$

which may be identified as the well-known probability integral and whose value varies from 0 to 1 as y increases from 0 to infinity. Values of $\phi(y)$ for different y may be obtained from statistical tables of the type of Table 10.1 (Svedberg and Pedersen, 1940, p. 19).

TABLE 10.1. *Numerical Values for the Function $\phi(y)$ for Different Values of c/c_0*

($\phi(y)$ is positive for $c/c_0 < 0.5$ and negative for $c/c_0 > 0.5$.)

c/c_0		$\phi(y)$	y	c/c_0		$\phi(y)$	y
0·0005	or 0·9995	0·999	2·3268	0·030	or 0·970	0·940	1·3301
0·001	0·999	0·998	2·1852	0·040	0·960	0·920	1·2379
0·002	0·998	0·996	2·0352	0·050	0·950	0·900	1·1632
0·004	0·996	0·992	1·8753	0·060	0·940	0·880	1·0994
0·005	0·995	0·990	1·8214	0·080	0·920	0·840	0·9936
0·006	0·994	0·988	1·7764	0·100	0·900	0·800	0·9061
0·008	0·992	0·984	1·7034	0·150	0·850	0·700	0·7329
0·010	0·990	0·980	1·6450	0·200	0·800	0·600	0·5951
0·012	0·988	0·976	1·5961	0·250	0·750	0·500	0·4769
0·014	0·986	0·972	1·5538	0·300	0·700	0·400	0·3708
0·016	0·984	0·968	1·5164	0·400	0·600	0·200	0·1791
0·020	0·980	0·960	1·4522	0·500	0·500	0	0

These values are interpolated from a table given in E. Czuber, *Wahrscheinlichkeitsrechnung*, I (B. G. Teubner, Leipzig, 1908), p. 385.

For positive values of y (or x), $\phi(y)$ is positive and from eqn. (10.3) it follows that $c < c_0/2$, whilst for negative y, $\phi(y)$ is negative and $c > c_0/2$. Typical solutions of the general equation are given in Fig. 10.2(*b*) for different t values. In accordance with eqn. (10.3), the concentration gradient over the boundary region is observed to become less steep with increasing values of t, and at infinite values the concentration is uniform throughout the cell (at $c_0/2$ if the original volumes of the solution and solvent are identical). The curve of c against x may thus be imagined to 'rotate' around the point $x = 0$, $c = c_0/2$, also becoming increasingly flattened with increasing time.

This picture of diffusion is also of use in some cases of forced diffusion

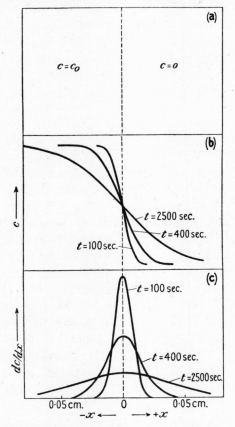

FIG. 10.2. (a) The original boundary. (b) Distribution of concentration at different times (t) after formation of the boundary. Calculated from equation (10.3) for $D = 2 \cdot 5 \times 10^{-7}$ c.g.s. units. (c) Concentration gradient over the boundary region for different times. Calculated from equation (10.6) for $D = 2 \cdot 5 \times 10^{-7}$ c.g.s. units.

where during diffusion the whole of the solute molecules are subject to some steady force, leading to a definite component of velocity in the direction of the force. Thus in sedimentation it will be shown (p. 273) that ideally the effects of diffusion and sedimentation are additive, the presence of sedimentation merely causing the transport of the normally behaving diffusion curve along the cell in the direction of the centrifugal field.

Eqns. (10.3) and (10.4) form the basis of one type of experiment devised for the determination of diffusion coefficients, which will be described further shortly (p. 244). Another important method makes use of concentration gradient curves some of which, derived from c v. x curves of Fig. 10.2 (b) for different values of t are contained in Fig. 10.2 (c). The position of the peak of these concentration gradient curves clearly corresponds in this ideal case to the initial position of the boundary at $x = 0$, for which a point of inflexion (giving critical values of dc/dx) occurs in the concentration curves. The mathematical form of the curves is obtained by differentiating eqn. (10.3) with respect to x, yielding the equation

$$\frac{dc}{dx} = \frac{c_0}{2\sqrt{(\pi Dt)}} e^{-x^2/4Dt}. \tag{10.6}$$

This type of function is well known in statistical mathematics as the Normal Probability or Gaussian Distribution Function, and the properties of ideal Gaussian distribution curves are utilized, as we shall show later, in calculating diffusion coefficients.

It will be recalled that the diagonal schlieren and scanning methods give directly, and the scale method indirectly, a record of the variation of the concentration gradient across a boundary (providing $\Delta n \propto \Delta c$) and by making use of eqn. (10.6) expressing this variation, the diffusion coefficient may be calculated. Details will be given later.

Concentration-dependent systems. It should be realized that, in deriving eqns. (10.3) and (10.6) from the general differential equation of diffusion (10.2), it was necessary to assume that the diffusion coefficient was independent of concentration. If, however, this assumption is invalid, the above treatment is not applicable. In such circumstances it can readily be shown that eqn. (10.2) is to be replaced by

$$\frac{\partial c}{\partial t} = \frac{\partial}{\partial x}\left(D \frac{\partial c}{\partial x}\right). \tag{10.7}$$

Following Boltzmann (1894), if c is a function of $z = x/\sqrt{t}$, this equation may be written in the form

$$-\frac{z}{2}\frac{dc}{dz} = \frac{d}{dz}\left(D\frac{dc}{dz}\right), \tag{10.8}$$

which, on integration at constant t, becomes

$$D(c) = -\frac{1}{2t}\frac{dx}{dc}\int_0^c x\,dc = -\frac{1}{2t}\frac{dx}{dc}\int_{-\infty}^x x\,\frac{dc}{dx}\,dx, \tag{10.9}$$

$D(c)$ being a differential diffusion coefficient corresponding to a definite concentration c, rather than an integral value, appropriate to a range of concentrations, which would be obtained by the application of procedures based upon eqns. (10.3) and (10.6) to concentration-dependent systems.

From eqn. (10.9) the evaluation of $D(c)$ involves an integration between definite limits, which may readily be accomplished graphically from experimental curves of the refractive index or concentration gradient over the boundary region. An example of the use of this method is provided later.

Experimental methods

Two main types of experiment have been devised to investigate free diffusion occurring across a horizontal boundary in liquid systems. These are:

(a) The Porous Disc Method, in which a concentration gradient occurs in the pores of a sintered glass disc, diffusion being followed by removal of samples and estimation of diffusing substance by any analytical means.

(b) The Free Boundary Method, in which the changes occurring over the free boundary region are investigated optically.

Since the free boundary method is much the more free from theoretical objections, especially in the case of colloidal systems, we shall largely be concerned with it here. In view, however, of the much greater experimental simplicity of the porous disc method and the fact that for some substances accurate diffusion coefficients have been obtained by its use, it will be briefly considered.

(a) The porous disc method

In one simple form of this method, developed originally by Northrop and Anson (1929), a sintered glass disc, with pore size of 5–15 μ, mounted at the end of a short glass tube (Fig. 10.3 (a)) closed by a stopcock, separates the two homogeneous liquid components of the diffusion system (the heavier above). Diffusion occurs through the pores of the disc in response to the concentration gradient existing there, which, by virtue of the protection afforded by pores of the size specified above, may possess very large values and still remain unaffected by convectional or mechanical disturbances. It is not therefore necessary to eliminate these disturbances (as in the free boundary method) and this constitutes one of the important advantages of the method. Homogeneity of the two liquids is effected by the slight convection occurring as a result of the density differences set up during diffusion or by some form of stirring the two liquids, e.g. the axial rotation of the cell (Fig. 10.3 (b)) used by Hartley and Runnicles (1938), which contains a glass bead in each compartment.

If the concentration difference across the disc is c and the corresponding mean pore length h, then the mean concentration gradient in the

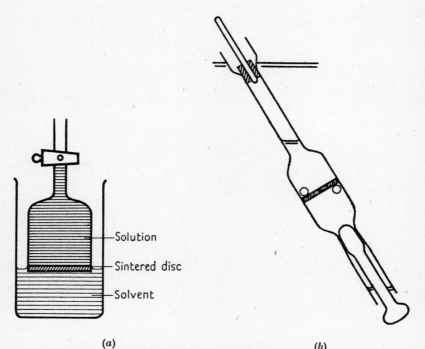

Fig. 10.3. (a) The simple porous disc apparatus. (b) The porous disc apparatus of Hartley and Runnicles (1938).

pores is c/h. For an effective pore area, A, of the disc, by the use of Fick's First Law (eqn. (10.1)) we obtain (omitting the negative sign)

$$D = \frac{h\,dm}{Ac\,dt} \qquad (10.10)$$

or for a finite change

$$D = \frac{h\,\Delta m}{Ac\,\Delta t}. \qquad (10.10\,\text{a})$$

The quantities h and A are, however, not directly measurable, and it is the chief assumption of the method that the factor h/A, which is obtained by calibration with low molecular weight substances of known diffusion coefficient, is the same for all solutes, even of colloidal type.

The method is applied by placing the disc between known volumes of liquids of known composition, and by analysis of aliquot samples removed from time to time from one or both liquids. For small concentration changes (i.e. a few per cent. of total) the simple eqn. (10.10a) may be used, in which for c is substituted the mean concentration difference for the interval. For larger concentration changes the integrated form of this equation must be used, which may be written (assuming D constant)

$$D = \frac{v_u v_l}{(v_u+v_l)} \frac{h}{At} \ln \frac{c_u^0}{c_u^0 - c_l(1+v_l/v_u)}, \qquad (10.11)$$

where v_u and v_l are the volumes of solution and solvent, c_u^0 is the initial concentration of the solution, and c_l is the concentration of solute at time t in the originally pure solvent.

Concentrations may be expressed in terms of any convenient units since we are concerned with ratios of concentrations in eqn. (10.11) or with $\Delta m/c$ in eqn. (10.10a). The method is particularly useful in connexion with dilute solutions of biologically active materials, for which relative concentrations can sometimes be assessed by biological testing where no other method is sufficiently sensitive (e.g. enzymes and viruses). If the assumption that D is independent of concentration (involved in deriving eqn. (10.11)) is invalid, then diffusion coefficients so calculated are mean values for a range of concentrations.

Calibration of the sintered disc has proved much more troublesome than might have been expected. Low molecular weight substances in aqueous solution, hydrochloric acid, and sodium and potassium chlorides have been used, making use of diffusion coefficient values for these substances obtained by the free boundary method described below. However, for none of the substances mentioned is the diffusion coefficient independent of concentration, and a completely satisfactory calibration method has not yet been established. The suggestions of Hartley and Runnicles (1938) and of Gordon *et alia* (1939) do, however, represent a substantial advance on previous methods.

The chief advantages of the method are:

(a) The simplicity of the technique involved, arising from the stability of the concentration gradient in the pores to convectional and vibrational disturbances.

(b) Owing to the unlimited choice of methods of analysing samples of solution, the method may be used under such conditions that the optical methods involved in Free Boundary Diffusion are much too insensitive.

Against these advantages must be set fundamental disadvantages which have, as yet, not been sufficiently investigated.

(a) The calibration of the cell by very low molecular weight substances is not necessarily valid for high molecular weight substances, especially if molecules of the latter are markedly asymmetric. The use of a known substance of high molecular weight would be more satisfactory, but even so, molecular shape effects might cause difficulty (e.g. diffusion of a long rod-like molecule through a narrow channel might be quite different from that of a spherical molecule of the same diffusion coefficient as determined by the free boundary method).

(b) The appearance of air bubbles or any tendency to adsorption in the pores would produce marked changes in the calibration factor. Thus definite indications of irregularities due to adsorption have been observed by Hartley and Runnicles (1938) in studying the diffusion of certain paraffin-chain salts.

(b) *Free boundary method*

In the free boundary method the solution and solvent (the denser below), initially separated in a vertical tube by some type of barrier, are allowed to come to temperature equilibrium in an accurately controlled thermostat ($\pm 0\cdot 001$–$0\cdot 003°$ C.) provided with good optical observation windows. The barrier is then carefully removed, so forming a sharp boundary between the solvent and solution, which must be kept as free as possible from all forms of disturbance. Vibrational and convectional disturbances particularly must be eliminated, or spuriously high diffusion coefficients are obtained. Further, any possibility of complication by the diffusion of electrolytes or low molecular weight constituents of the solvent must be removed by a prolonged dialysis of the solution against the solvent before boundary formation.

(i) *Diffusion cells*

Several successful diffusion cells, differing largely in the type of barrier employed for the initial separation of the liquids, have been used. Perhaps the simplest is the cylindrical U-tube (Fig. 10.4 (a)) employed by Svedberg (1925) and others. A three-way tap near the base of one limb allows the liquids in the two limbs to be levelled as well as separated before boundary formation. The U-tube arrangement shown in Fig. 10.4 (b), which is a modification by Campbell and Johnson (1944) of an apparatus designed by Lamm (1937), is suitable for use with organic solvents since the use of lubricants (which would be soluble) near the site of boundary formation is avoided. A glass sintered disc, D, impermeable to mercury, is sealed into the end of the observation tube (O) which is filled with the denser component of the diffusion system.

The mercury, just covering the sintered disc and the rubber bung (B) upon which the upper vessel (V) is fitted securely, is in turn covered by the lighter liquid. On the attainment of temperature equilibrium, the mercury covering the disc is run out by the tap (T), and the boundary formed in the pores of the disc is hydrostatically forced down the observation tube until in a position suitable for observation.

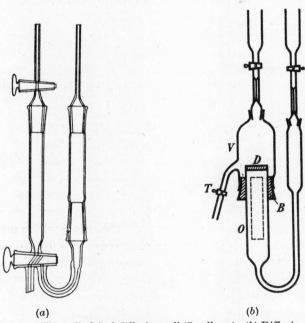

(a) (b)

Fig. 10.4. (a) The cylindrical diffusion cell (Svedberg). (b) Diffusion cell for organic solvents.

In recent years, diffusion cells have been designed as much with the object of satisfying the requirements of the scale and schlieren optical systems as of producing sharper and more stable boundaries. Cells of rectangular cross-section have therefore been especially favoured.

The Lamm cell (Fig. 10.5(a)) consists of a single rectangular cavity in a stainless steel or plastic spacer, pressed between two plates of glass. The cavity is divided into upper and lower compartments, containing solvent and solution, by a movable slide which, in position, completely seals off the liquids from one another. On removing the slide, a sharp boundary is formed. A modification of this design has recently been reported by Claesson (1946).

Fig. 10.5(b) is a diagrammatic version of the Neurath (1941) cell, which, in its method of boundary formation, is reminiscent of the

Tiselius (1937) electrophoresis cell (see also p. 319). Two U-shaped blocks of stainless steel (A and A_1), pressed between plate-glass windows (B and B_1), can be moved horizontally across one another by means of a screw (G) to give, when in alinement, one long rectangular cavity, or, out of alinement, two short cavities which are completely isolated.

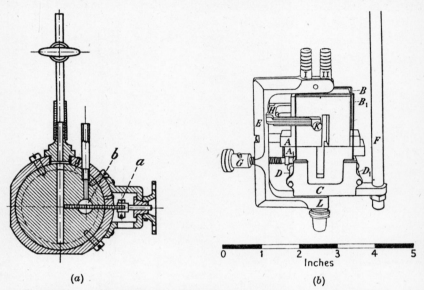

FIG. 10.5. (a) Metal diffusion cell with slide. Two circular glass plates are screwed against middle pieces of stainless steel, forming a plane parallel cell for the diffusion column. The slide is tightened by wool fat at a or by mercury at b. (b) The Neurath diffusion cell.

The liquids are contained in the two cavities, initially out of alinement, and the boundary is formed by sliding the two blocks into coincidence.

By careful manipulation, sharp boundaries may be formed in all the cells described, and the subsequent changes occurring as a result of diffusion may be observed by both absorption and refractive index methods.

(ii) *Calculation of diffusion coefficients*

Absorption method. As will be clear from Chapter IX, the absorption method yields, for different times after boundary formation, a quantitative plot (of the type of Fig. 10.2 (b)) of the concentration of diffusing substance over the boundary region. Such a plot is represented mathematically by eqn. (10.3), but, in calculating diffusion coefficients, it is convenient to make use of the rearranged form, eqn. (10.5). For a given value of c (e.g. $3c_0/4$ or $c_0/4$) the value of $\phi(y)$ is defined, and by the use (backwards) of statistical tables the appropriate value of y is obtained. Substitution of x and t, appropriate to the value of c chosen, in (10.4) then gives the diffusion coefficient directly.

This calculation, however, assumes the formation at $t = 0$ of a perfect boundary. Any imperfection in the boundary formation or the slight disturbance caused in moving a boundary to its observation position will give a 'zero error' in the observed times. This may be overcome by the use only of time differences. Thus, making use of the c v. x curves for times t_1 and t_2 and choosing the same concentration c (i.e. $y_1 = y_2$), we have

$$y^2 = \frac{x_1^2}{4Dt_1}, \qquad y^2 = \frac{x_2^2}{4Dt_2}.$$

Rearranging and subtracting we have

$$D = \frac{1}{4y^2} \frac{x_2^2 - x_1^2}{t_2 - t_1}. \qquad (10.12)$$

Such a calculation may be repeated for different concentrations of solute in the boundary region, and the presence of any systematic trends in diffusion coefficient with changing concentration may be detected.

Some of the best examples of the precise use of the absorption method in diffusion are to be found in the investigation by Tiselius and Gross (1934) of the diffusion coefficients of certain proteins.

Refractive index methods. The use of the scale or schlieren optical methods gives ultimately, as we have seen (p. 225), a plot of the refractive index gradient dn/dx over the boundary region, from which we have to calculate diffusion coefficients. Fig. 10.6 contains a typical diffusion diagram for a monodisperse system. We have seen that eqn. (10.6) represents the variation in concentration gradient with distance from the boundary and it follows therefore, if $\Delta n \propto \Delta c$, that the modified expression

$$\frac{dn}{dx} = \frac{n - n_0}{2\sqrt{(\pi Dt)}} e^{-x^2/4Dt} \qquad (10.13)$$

is the mathematical function representing the experimental curves. We shall first consider only the simplest types of calculation of diffusion coefficients making use of the properties of this function. The more difficult case involving concentration-dependent diffusion coefficients is mentioned later.

1. *Area method.* At $x = 0$, the exponential term becomes unity, and dn/dx ($= H$) has its maximum value, H_m, given by

$$(dn/dx)_{x=0} = H_m = \frac{n - n_0}{2\sqrt{(\pi Dt)}}. \qquad (10.14)$$

Rearranging, and making use of eqn. (9.21), we have

$$D = \frac{(n-n_0)^2}{4\pi t(H_m)^2} = \frac{A^2}{4\pi t(H_m)^2},\qquad(10.15)$$

where A is the area under the experimental curve (Fig. 10.7).

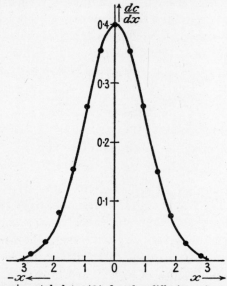

FIG. 10.6. Experimental data (●) for the diffusion of sucrose in aqueous solution plotted in normal coordinates and compared with the ideal concentration gradient curve. (Equation (10.6).)

Time after boundary formation: 90 minutes. Concentration: 1·03 gm. of sucrose per 100 c.c. of solution.

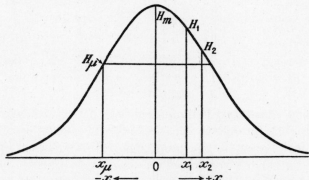

FIG. 10.7. The ideal curve of concentration gradient against distance.

It is usual in applying this method to plot H_m against $1/\sqrt{t}$, when a straight line passing through (or near) the origin is obtained with a slope $A/2\sqrt{(\pi D)}$. Evaluation of A by a planimeter or other methods then determines D directly.

It is to be noted here that although in the experimental curves it is essential to plot the x dimension in absolute units or to correct suitably, dn/dx may be plotted in any unit since the quotient A/H_m is involved in the final result, in which any proportionality constant disappears.

2. *Inflexion method.* The points of inflexion (denoted by subscript μ) of the gradient curve are given by equating the second differential of the refractive index gradient ($d^3n/dx^3 = d^2H/dx^2$) to zero, when it is readily shown that
$$x_\mu^2 = 2Dt. \tag{10.16}$$
Substituting in (10.13) we have
$$H_\mu = \frac{(n-n_0)}{2\sqrt{(\pi Dt)}} e^{-\frac{1}{2}} = \frac{H_m}{\sqrt{e}}. \tag{10.17}$$

In practice, therefore, the points of inflexion are located by eqn. (10.17) making use of the easily measurable H_m, and the values of the x coordinate are then read from the curve, giving D directly.

3. *Successive analysis method.* This method is somewhat analogous to that used in the calculation of diffusion coefficients from the results of the light absorption method. Writing eqn. (10.13) in the logarithmic form, we have
$$\ln H = \ln H_m - \frac{x^2}{4Dt}.$$

If H_1 and H_2 are the values of dn/dx corresponding to x_1 and x_2, by subtraction and rearranging, we obtain
$$D = \frac{x_2^2 - x_1^2}{4t \ln (H_1/H_2)}. \tag{10.18}$$

The method is usually applied by drawing an equal number of equally spaced ordinates on each side of the median, and by calculating the diffusion coefficient for each interval. Since the solute concentration varies continuously over the boundary, this procedure is equivalent to the calculation of mean diffusion coefficients for different concentration intervals.

4. *Statistical method.* It is proposed merely to give an outline of this method, and it will be necessary to assume some acquaintance with the statistical treatment of the normal probability curves (e.g. Aitken, 1942).

Eqn. (10.13) may be written, in conformity with statistical practice,
$$\frac{dn}{dx} = \frac{n-n_0}{\sigma\sqrt{(2\pi)}} e^{-x^2/2\sigma^2}, \tag{10.19}$$
where
$$\sigma^2 = 2Dt, \quad \text{or} \quad D = \frac{\sigma^2}{2t}. \tag{10.20}$$

σ, the standard deviation, is defined as the square root of the second moment of the curve (in absolute units) about the ordinate for which the first moment μ_1 is zero.

The first and second moments (μ_1 and μ_2 respectively in the general case) may be defined as follows. Let the experimental curve be divided into equally spaced vertical strips of height $S_1, S_2, S_3,..., S_i,...$ numbered $s_1, s_2, s_3,..., s_i,...$ respectively from an origin near the centre of the base-line. The first and second moments about this origin may then be written

$$\mu_1' = \frac{\sum (s_i S_i)}{\sum (S_i)} \tag{10.21}$$

and

$$\mu_2' = \frac{\sum (s_i^2 S_i)}{\sum (S_i)}. \tag{10.22}$$

The coordinate about which the first moment is zero is given by

$$x_0 = s_0 - \mu_1' \tag{10.23}$$

and the second moment about this coordinate may quite simply be shown to be

$$\mu_2 = \mu_2' - (\mu_1')^2 \tag{10.24}$$

However, this value is dependent upon the width (w) of the strips used and to convert into absolute units (μ_2^0) we make use of the relation

$$\mu_2^0 = \mu_2 w^2. \tag{10.25}$$

μ_2 is obtained graphically, making use of the above relations, from the experimental refractive index gradient curves, and with eqn. (10.25) gives the diffusion coefficient D.

Details of this and the other methods described may be obtained from Lamm (1937) and Neurath (1942).

Diffusion in non-interacting systems

(a) *Monodisperse systems*

In the case of monodisperse systems in which the diffusing molecules exert no mutual effect (non-interacting systems), the ideal laws of diffusion, given above, are obeyed, and the experimental curves may be represented accurately by eqns. (10.3) and (10.6). In an important experimental test of the monodisperse character of a system, the experimental curves are compared in detail with the ideal Gaussian probability curves, and the degree of coincidence gives a sensitive measure both of the monodispersity and lack of mutual interaction of the diffusing molecules (see Fig. 10.6).

In spite of the diversity of the methods which can be employed in calculating diffusion coefficients, it should be noted that for these monodisperse, non-interacting systems all final results should be identical, with little or no concentration dependence. Good examples of such behaviour are to be found amongst the more highly symmetrical of the protein molecules, e.g. egg albumin. Table 10.2, taken from the work of Polson (1939), contains the results of diffusion measurements on egg

TABLE 10.2. *Diffusion Coefficients of Egg Albumin*
(Polson, 1939)

Protein concentration (gm. per 100 c.c.)	Time (sec.)	$D_A \cdot 10^7$	$D_\mu \cdot 10^7$
		($cm.^2/sec.$)	
0·7	28,800	7·66	7·57
	86,400	7·28	7·42
	115,200	7·30	7·52
		Mean 7·50	
		Corrected mean 7·71	
0·83	79,200	7·75	7·61
	86,400	7·43	7·43
		Mean 7·52	
		Corrected mean 7·73	
0·88	28,800	7·06	7·72
	43,200	7·58	7·35
	86,400	8·17	7·66
	115,200	7·50	7·46
	172,800	7·46	7·55
		Mean 7·55	
		Corrected mean 7·76	
0·91	21,600	8·30	7·58
	29,700	7·49	7·41
	43,200	7·89	7·50
		Mean 7·50	
		Corrected mean 7·71	
1·4	86,400	7·67	7·50
	115,200	7·50	7·50
	129,600	7·55	7·34
		Mean 7·44	
		Corrected mean 7·64	

albumin in acetate buffer (0·02 m. CH_3COONa; 0·02 m. CH_3COOH; 0·2 m. NaCl; pH = 4·6; $\eta/\eta_0 = 1·028$) at 20° C. D_A and D_μ are diffusion coefficients calculated respectively by the area and inflexion point methods. Mean values corrected to the viscosity of water at 20° C. are also included.

(b) *Polydisperse systems*

If the diffusing substance is not monodisperse, but contains molecules with different diffusion coefficients (though identical chemical compositions), then, providing the different components diffuse independently of one another, the refractive index and refractive index gradient curves will be composite, i.e. made up by the superposition of the different ideal curves of the individual components. In such a case the gradient curves must deviate from the ideal Gaussian curves, though they may be quite symmetrical about the median line (in the absence of interaction effects). Qualitatively it may be seen that at greater distances from the position of boundary formation the experimental curves will be largely due to the more rapidly diffusing molecules, and it is to be expected, therefore, that the application of the Method of Successive Analysis and similar methods (p. 247) will show an increase in the calculated diffusion coefficient with increasing values of $|x|$ but no continuous trend with changing concentration.

Further, in the case of polydisperse systems the observed diffusion coefficient is, of necessity, some type of average value, and it has been shown (Gralen, 1941) that different treatments of the experimental results may yield different averages. Thus the application of the statistical method outlined above yields a weight average diffusion constant $\left(D_W = \sum_i c_i D_i \big/ \sum_i c_i\right)$, whereas the area method gives a result (D_A) between the weight and number averages ($D_W < D_A < D_N$).

In view of the well-defined character of the diffusion coefficient average obtained from the statistical method, this has been widely used in the investigation of polydisperse systems. Gralen (1944) also attempted to use the ratio D_W/D_A to characterize polydispersity in the case of cellulose and its derivatives, but the difficulties involved were such as to render it unreliable. It appears, in general, at the present time that much more information can be obtained from a polydisperse system by careful resolution into fractions of maximum homogeneity and examination of the fractions by diffusion and other methods, than by attempting to make a detailed analysis of experiments carried out on unfractionated material.

Diffusion in interacting systems

If the effective volume of the diffusing molecules is large as a result of molecular asymmetry, then intermolecular interference will be of importance at much lower concentrations than for spherical molecules, and the frictional resistance to diffusion will be no longer independent of solute concentration (as may occur for highly symmetrical molecules). Since the diffusion coefficient is inversely proportional to the frictional constant (p. 258) it also must become concentration-dependent.

Further, as will be shown later, diffusion may be considered to occur in response to the osmotic gradient in the system, and since deviations from ideal osmotic behaviour are pronounced even at low concentrations for molecules of asymmetric form (see p. 175), similar deviations in the driving force of diffusion are to be anticipated. The concentration-dependence of the diffusion coefficient may thus be visualized in terms of at least a concentration-variable frictional constant and a non-ideal osmotic pressure.

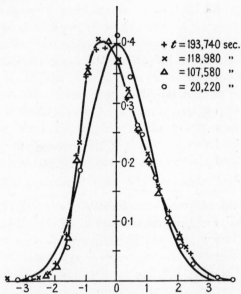

FIG. 10.8 (a). Experimental data, for different times t, for the diffusion of rubber into chloroform, reduced to normal coordinates and compared with ideal Gaussian curve.
Initial concentration c_0 of rubber in $CHCl_3$ = 0·957 gm. per 100 c.c. of solution.

Since solute concentration varies continuously along a direction perpendicular to the plane of the boundary, the diffusion coefficient in these interacting systems must also change in a corresponding manner. Thus the values calculated from one side of the concentration gradient 'peak' will in general be different from those on the other side, or, in other words, the peak must be asymmetrical or skew. Skewness of the experimental diffusion curves of the concentration gradient is to be taken therefore as a general indication of the dependence of the diffusion coefficient on concentration. (Cf. skewness of sedimenting peaks when sedimentation constant shows dependence upon concentration.)

Such a dependence usually arises, as we have suggested above, from the highly asymmetric nature of the diffusing molecules, as in the diffusion of rubber fractions (Fig. 10.8 (a)) in chloroform solution

(Bywater and Johnson, 1946). Another possible cause was suggested by Lamm (1937) and is discussed later.

The range of diffusion coefficient values represented in a single diffusion curve depends for a given substance only on the concentration difference between the two liquid phases. In order, therefore, to obtain a diffusion coefficient corresponding to a small concentration range, it is useful to carry out diffusion experiments between two solutions differing only slightly in concentration (instead of between a solution and pure solvent). In this way the variation of diffusion coefficient with concentration may be investigated accurately over large concentration ranges as demonstrated in the work of Polson (1939) on egg albumin and other proteins at high protein concentration. Table 10.3, below, contains Polson's figures for the diffusion coefficient of egg albumin in acetate buffer (0·01 m. CH_3COOH; 0·01 m. CH_3COONa; 0·1 m. NaCl; $\eta/\eta_0 = 1·013$; pH 4·57), determined by observation of diffusion between two solutions (a and b) whose concentrations are given in the first two columns of the table.

TABLE 10.3. *Diffusion Coefficients of Egg Albumin at Higher Concentrations*

(Polson, 1939)

Protein concentration (gm. per 100 c.c.)		Time in hours	$D_\mu \cdot 10^7$ $cm.^2/sec.$
Solution a	Solution b		
3·0	2·0	17·25	7·33
		39·00	7·35
		23·00	7·30
			Mean 7·33
			Corrected mean 7·43
6·0	5·0	8·0	6·86
		21·0	6·71
			Mean 6·78
			Corrected mean 6·86
9·0	8·0	24·0	6·03
			Corrected mean 6·11

Thus for egg albumin, and for other highly symmetrical proteins also, the diffusion coefficient undergoes considerable change at high concentrations of diffusing substance. Although this method of investigating concentration effects is more accurate the smaller the concentration difference between the two solutions, strictly an integration based upon eqn. (10.9), as already suggested, should be applied. In more highly concentration-dependent systems such a procedure is essential, as, for example, in the case of the diffusion of rubber and rubber-like materials in solution. Fig. 10.8 (b), obtained thus by graphical integration, gives the diffusion coefficient as a function of concentration for the rubber fraction in chloroform to which Fig. 10.8 (a) also refers. Within the limits of experimental error, the D v. c relation is a linear one, a conclusion which has been drawn also for other systems (e.g. see Beckmann and Rosenberg, 1945). If this result is shown to be true in general,

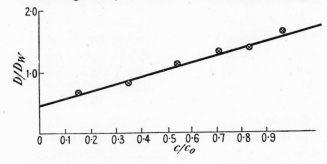

FIG. 10.8 (b). Plot of D/D_W against c/c_0 obtained graphically from the experimental data shown in Fig. 10.8 (a). D is the differential diffusion coefficient corresponding to the concentration c; D_W ($= 2\cdot 5 \times 10^{-7}$ c.g.s. units) being the weight average value determined by eqn. (10.20).

then simplified methods of determining differential diffusion coefficients may be employed (e.g. Gralen, 1944); further work is required, however, to establish this point. It is certainly probable that much diffusion work, interpreted as yet on the basis of concentration-independent diffusion coefficients, will be refined by the more reliable but somewhat tedious methods based upon eqn. (10.9).

Summarizing, the chief advantage of the free boundary method lies in the soundness and freedom from assumption of the theoretical principles underlying the method. The accuracy of the optical methods employed and their suitability in providing results in a form convenient for calculation must also be mentioned. The disadvantages of the method arise from the stringent experimental requirements, especially that of rigorous temperature control and freedom from vibrational disturbances.

The occurrence of diffusion at a sedimenting boundary will be considered later (p. 272), but it is of importance to realize at this stage that such diffusion is merely a special case of free boundary diffusion. Thus the 'peak broadening' occurring in the sedimentation diagrams of Fig. 9.8 (p. 229) is not fundamentally different from that occurring in the typical diffusion diagram of Fig. 10.6. It is therefore possible to make use of the methods just described, modified suitably, to calculate diffusion coefficients from sedimentation diagrams. Details of the methods may be obtained in Svedberg and Pedersen (1940). Difficulties in applying the method are mentioned on p. 273, and these usually prevent really accurate determinations.

Factors complicating diffusion

(a) *Diffusion of charged colloids*

We have already (p. 215) discussed generally the effect of the ionic atmosphere (gegenions) on the motion of a charged colloidal ion. In the special case of diffusion, separation of the colloidal ion from its atmosphere sets up a *diffusion potential* which, resisting further separation, may cause either a retardation or acceleration of the large ion according to the conditions of the experiment. Such effects may be suppressed by the use of sufficiently high salt concentrations (usually *ca.* 0·2 m.).

Further, in the study of diffusion in such systems, especially when followed by refractive index methods, it is of importance to ensure that the diffusion of the colloidal ion is not complicated by that of the salts present. It is therefore necessary, before boundary formation, to dialyse the two component liquids against one another at the temperature of the proposed diffusion experiment until equilibrium is reached. No gross transport of salt can therefore occur after boundary formation, and diffusion of the colloidal material only will be observed.

For diffusion in organic solvents, in which the solute is usually uncharged, charge effects are usually absent, and no preliminary dialysis is required. This simplification is usually at least partly offset, however, by the difficulty of finding suitable lubricating materials for the diffusion cell.

(b) *Boundary anomalies*

Diffusion curves deviating from the ideal Gaussian form have been shown to occur in the case of polydisperse systems, even of the non-interacting type, and where the diffusion coefficient depends upon the concentration. Since, in the former case, the concentration gradient curve is symmetrical and merely the resultant of the superposition of several ideal curves, this behaviour is strictly not anomalous. In the latter case, however, skew curves are obtained, and in general such skewness is an important indication of concentration effects. In addition to the skewness arising from asymmetry of the diffusing molecules another possible cause was pointed out by Lamm (1937). He found that, for solutions of starch in pure water, the diffusion curves were skew, but they became symmetrical if electrolyte was added. The skewness was therefore reasonably attributed to charge effects arising through failure to suppress diffusion potentials by the presence of sufficient electrolyte. Such charge effects are larger the higher the concentration of the diffusing molecule for a given electrolyte concentration, and hence a diffusion coefficient-concentration dependence arises, which gives rise to the skewness of the experimental curves. Such a mechanism is, of course, possible only in ionizing solvents.

It will be seen that similar boundary asymmetry also occurs in sedimenting boundaries.

(c) *Diffusion in equilibrium systems*

It remains to discuss briefly the case in which the differently diffusing entities of a polydisperse system form the components of an equilibrium system. Such systems are discussed more fully in Chapter XI in connexion with the sedimentation velocity method, from which information is more readily obtained.

On the one hand, in the case of very slowly established equilibria (half-lives for the forward and reverse processes of several days at least) the system is not sensibly different from normal polydisperse non-interacting ones. On the other hand, the case where equilibrium is very rapidly established (e.g. half-lives of the order of a few seconds) will not be different at first sight from ideal monodisperse systems, and the observed diffusion coefficient will be intermediate between those of the equilibrium forms, the precise value being dependent on the exact equilibrium position.

It would appear from the work of Tiselius and Gross (1934) that dilute aqueous solutions of carboxy-haemoglobin are representative of the latter type of rapidly established equilibria (see also p. 279). Here it was shown that in solutions below 0·7 per cent. the diffusion coefficient increased sharply with dilution, and this is usually attributed to the increasing dissociation of the haemoglobin molecule.

Analysis of the diffusion curves of intermediate systems is of considerable complexity. Since their investigation is simpler and more profitable by the sedimentation velocity method, further discussion will be deferred until Chapter XI.

Diffusion and molecular properties

Diffusion coefficients, in dilute aqueous solution, of substances ranging from very low to very high molecular weight are contained in Table 10.4, in order of increasing molecular weight.

It is apparent that with increasing molecular weight the diffusion coefficient grows steadily smaller, until for the colloidally disperse materials in which we are chiefly interested the value is of the order of 10^{-7} cm.^{2}/sec. and equalization of concentration after boundary formation occupies a time of at least several days.

TABLE 10.4. *Diffusion Coefficients in Aqueous Solution*

Substance	Molecular weight	Concentration (moles/litre)	Temperature °C.	$D.10^5 (cm.^2/sec.)$	Reference
H_2	2	Very dilute	21	5·2	
O_2	32	,,	18	2·02	
HNO_3	63	0·05	20	2·62	
HCl	36·5	0·1	19	2·5	
NaCl	58·5	0·05	20	1·39	
Oxalic acid	90	0·1	20	1·41	(a)
Ethyl alcohol	46	Very dilute	15	1·00	
Glycerol	92	0·125	20	0·83	
Mannitol	182	0·3	20	0·56	
Maltose	342	0·1	20	0·42	
Lactalbumin	17,400	Very dilute	20	0·106	
Lactoglobulin	40,000	,,	20	0·073	
Serum albumin (horse)	70,000	,,	20	0·061	
Serum globulin (horse)	167,000	,,	20	0·040	(b)
Urease	480,000	,,	20	0·035	
Haemocyanin (Helix pomatia)	6,600,000	,,	20	0·0138	

(a) From *International Critical Tables* (1929), 5, p. 63.
(b) From Svedberg and Pedersen (1940), *The Ultracentrifuge*, Oxford, p. 406.

In view of the relation between diffusion coefficient and the displacement in a given time under Brownian motion (p. 17),

$$D = \frac{\bar{x}^2}{2t}, \qquad (1.1)$$

such a result is not unexpected, for it is a well-known experimental fact that with increasing particle size Brownian displacements become smaller. In the next section we shall be concerned with the quantitative relation between diffusion coefficient and molecular or particle dimensions.

(a) *Diffusion and osmotic pressure*

In a homogeneous colloidal solution, in which the effects of gravitational and other external force fields may be neglected, Brownian displacements of a colloidal particle in opposite directions are equally probable; no macroscopic concentration change can, therefore, occur, although local concentration changes on a very small scale ('perturbations') undoubtedly are observable (see p. 178). When, however, a boundary is produced at which an abrupt concentration change occurs, it is only through the thermal or Brownian motions of the solvent and solute molecules that the ensuing transport of material and concentration equalization of diffusion can occur.

It follows, therefore, that superimposed upon the normal Brownian motion, there is another motion in the direction required for equalizing concentration, for which a definite force is responsible. Such a force is undoubtedly related to osmotic pressure, for diffusion in solution merely represents a process involved in the attainment of an equilibrium state, which is alternative to the development of an equilibrium osmotic pressure where diffusion is prevented by the presence of a semi-permeable membrane.

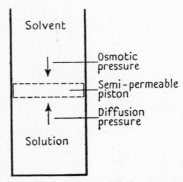

FIG. 10.9. Diffusion and osmosis.

However, it should be observed that a fundamental difference exists between diffusion and osmotic pressure, in that we are concerned in diffusion processes with the escaping tendency of the solute, and, in osmotic phenomena, with the escaping tendency of the solvent.

According to Einstein, however, the forces involved in osmosis and diffusion should be equal in magnitude but opposite in direction. This relationship may be illustrated by the idealized system depicted in Fig. 10.9, in which a solution, in a cylinder of cross-sectional area A, is separated from the solvent by a semi-permeable membrane acting as a piston.

Solvent molecules will tend to pass from the solvent through the piston into the solution, so increasing the solution volume and raising the piston, until a head of liquid with a pressure equal to the osmotic pressure (Π) prevents any such further movement. The equilibrium osmotic force acting downwards on the piston is then $A\Pi$, or, in the case of two solutions separated by a semi-permeable membrane, $A\,\delta\Pi$.

In the absence, however, of the piston, equilibrium is reached by diffusion occurring as a result of the diffusion 'pressure' of the solute, acting in a direction opposite to that of the previous osmotic pressure, and given, therefore, according to Einstein, by $-A\,\delta\Pi$.

Consider, therefore, the solute molecules inside a small element of volume, of thickness dx, of the cylinder. The number of such molecules is $NAc\,dx$, using c as a molar concentration.

Force acting on them is $-A\,d\Pi$.

Assuming the van 't Hoff equation (7.1),
$$d\Pi = RT\,dc,$$
and substituting for $d\Pi$, we have
$$\text{total force acting} = -ART\,dc.$$
$$\therefore\quad \text{force/molecule} = -\frac{ART\,dc}{NAc\,dx} = -\frac{RT}{Nc}\frac{dc}{dx}.$$

If f is the frictional constant per molecule, then
$$\text{velocity } (v) \text{ acquired by a molecule} = -\frac{RT}{Ncf}\frac{dc}{dx}.$$

The number of moles of solute crossing the small element of volume in time dt is, therefore,
$$dn = cvA\,dt$$
$$= -\frac{RT}{Nf}\frac{dc}{dx}A\,dt. \qquad (10.26)$$

This equation is to be identified with Fick's first law (eqn. (10.1b)), when, identifying coefficients,
$$D = \frac{RT}{Nf} \qquad (10.27)$$
or
$$D = \frac{RT}{F}, \qquad (10.27\,\text{a})$$
where F is the frictional constant per gram molecule of the solute.

It should be noted that eqn. (10.1b), or Fick's first law, is thus derivable from elementary osmotic theory, and that such a derivation involves the assumption of the van 't Hoff osmotic equation. Deviations from this equation must therefore involve departures from the constancy of the diffusion coefficient.

It will be recalled that eqn. (10.27) has been utilized in deriving sedimentation equilibrium relationships (p. 195) and it will be further used (p. 288) when combining the data from sedimentation velocity and diffusion experiments in molecular weight calculations. In both cases the equation is merely used to substitute for the frictional constant in terms of the measured diffusion coefficient. The frictional constants f and F, have, however, important meanings for their own sakes, since they may be related in certain idealized cases to molecular size and dimensions. It is now proposed to deal with this question.

(b) Frictional constants and molecular properties

The relation between frictional constant and molecular dimensions has only been worked out for certain idealized models (see p. 209), and of these we shall be concerned only with spheres and ellipsoids of rotation.

(i) *Spherical molecules.* It has already been shown (p. 211) that under certain conditions, which are approximately observed for spherical colloidal particles in solution, the frictional constant is given by Stokes's law as

$$f = 6\pi\eta r, \qquad (9.2)$$

where η is the viscosity of the solvent, and r is the radius of the particle. Substitution of this result in eqn. (10.27) yields the important equation

$$D = \frac{RT}{6\pi\eta r N} \qquad (10.28)$$

known as the Sutherland-Einstein equation.

Expressing r in terms of the molecular (or particle) weight and the partial specific volume, we obtain

$$D = \frac{RT}{6\pi\eta N(3M\bar{v}/4\pi N)^{\frac{1}{3}}}. \qquad (10.29)$$

The latter equation gives a quantitative explanation of the observation, from Table 10.4, that the diffusion coefficient of approximately spherical molecules falls with increasing molecular weight.

For spherical particles, therefore, the determination of the diffusion coefficient allows the calculation of the molecular or particle weight and radius, a method which was used by Svedberg (1928) in calculating the dimensions of colloidal gold particles. For one sol, he thus obtained the value $r = 12.9$ A as against 13.3 A obtained by another independent method involving the use of the particles as crystallizing centres.

(ii) *Non-spherical molecules.* We now proceed to consider non-spherical molecules, of which the simplest are those of ellipsoidal shape. In the first place it should be realized that such molecules require different frictional constants for different types of orientation. Thus for a generalized ellipsoid, motion parallel to any one of the three axes is possible and three different diffusion and frictional constants are required to describe such motions. Thus

$$D_1 = \frac{RT}{Nf_1}, \qquad D_2 = \frac{RT}{Nf_2}, \qquad D_3 = \frac{RT}{Nf_3}, \qquad (10.30)$$

where the suffixes 1, 2, 3 refer to motion parallel to axes 1, 2, and 3 respectively.

If, however, the molecule is quite randomly orientated, then it was shown by Perrin (1936) that the observed diffusion coefficient (D) is related to the component diffusion and frictional constants by the equation

$$D = \tfrac{1}{3}(D_1+D_2+D_3) = \frac{RT}{3N}\left(\frac{1}{f_1}+\frac{1}{f_2}+\frac{1}{f_3}\right). \qquad (10.31)$$

It is with such randomly orientated molecules that we shall be concerned. For ellipsoids of rotation, $f_2 = f_3$. In using frictional constants to derive information on molecular shape, molecules are treated as ellipsoids of rotation with axial ratio denoted by b/a, where $2b$ is the equatorial diameter and $2a$ the axis of rotation (see Fig. 9.1).

We have seen that determination of the diffusion coefficient and the use of eqn. (10.27) gives immediately the frictional constant, f, of the molecule. It will also be shown later (p. 289) that the molecular weight may be obtained by a combination of diffusion and sedimentation velocity data and that this value corresponds to the anhydrous or unsolvated molecule even though the sedimenting species carries appreciable quantities of strongly adhering solvent. Using these data the frictional constant (f_0) of a hypothetical *unsolvated spherical* molecule of the same molecular weight and density as that investigated may be calculated from

$$f_0 = 6\pi\eta\left(\frac{3M\bar{v}}{4\pi N}\right)^{\frac{1}{3}}. \qquad (9.3)$$

The ratio f/f_0, *the frictional ratio* (sometimes written D_0/D, where D_0 is the diffusion coefficient corresponding to f_0), has been extensively used in attempts to investigate molecular size and shape. If the actual molecule is accurately spherical and unsolvated, then the frictional ratio is unity, for f as well as f_0 is given by eqn. (9.3).

Two main causes, however, contribute to give f/f_0 values greater than unity:

(1) The occurrence of solvation (hydration in aqueous systems).
(2) The occurrence of molecules deviating from spherical shape.

The presence of a concentric spherical sheath of solvating molecules causes an increase in the volume of the kinetic unit and by eqn. (9.2) an increase in the frictional constant of a spherical molecule. Similarly, it can be shown quantitatively by hydrodynamic methods, and qualitatively it is a matter of experience, that the frictional force increases with increasing deviations from spherical shape. Oncley (1941) therefore suggested the separation of f/f_0 into the solvation and asymmetry factors by the following equation,

$$(f/f_0) = (f/f_e)(f_e/f_0), \qquad (10.32)$$

where (f/f_e) represents the increase in frictional constant arising by solvation, and f_e/f_0 represents the similar effect caused by the asymmetry of the molecule. From eqn. (9.3) the frictional constant of a spherical molecule is proportional to the cube root of its molecular volume ($M\bar{v}$),

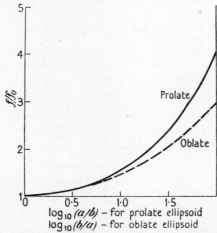

Fig. 10.10. The relation between the frictional ratio and the axial ratio for ellipsoids of rotation, according to Perrin's equations.

and following Kraemer (1940) we may write, assuming solvating liquid to have the same density as the bulk liquid,

$$f/f_e = \left(1+\frac{w}{\bar{v}\rho}\right)^{\frac{1}{3}}, \qquad (10.33)$$

where w is the weight of solvent of density ρ solvating 1 gm. of pure solute of partial volume, $\bar{v}$.

The relationship between (f_e/f_0) and the axial ratio of the ellipsoid of rotation has been deduced by Perrin (1936) and similar expressions were independently obtained by Herzog, Illig, and Kudar (1934). We give Perrin's relations below:

prolate ellipsoids
(i.e. $b/a < 1$)
$$f_e/f_0 = (D_0/D_e) = \frac{(1-b^2/a^2)^{\frac{1}{2}}}{\left(\frac{b}{a}\right)^{\frac{2}{3}} \ln\left[\frac{1+(1-b^2/a^2)^{\frac{1}{2}}}{b/a}\right]}, \qquad (10.34)$$

and for

oblate ellipsoids
(i.e. $b/a > 1$)
$$f_e/f_0 = D_0/D_e = \frac{(b^2/a^2-1)^{\frac{1}{2}}}{\left(\frac{b}{a}\right)^{\frac{2}{3}} \tan^{-1}\left(\frac{b^2}{a^2}-1\right)^{\frac{1}{2}}}. \qquad (10.35)$$

These relations are shown diagrammatically in Fig. 10.10, which shows clearly the slow increase in f_e/f_0 with increasing axial ratio a/b

for prolate ellipsoids, and with increasing values of b/a for oblate ellipsoids. Thus an axial ratio of 10, for a prolate ellipsoid, gives a frictional constant approximately 50 per cent. greater only than for a spherical molecule of the same molecular weight and density.

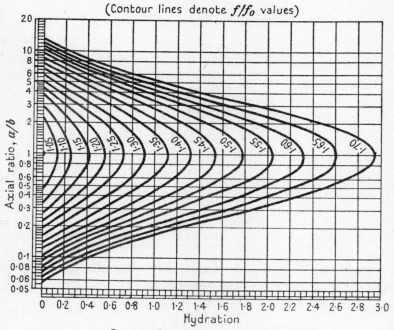

Fig. 10.11. Values of axial ratio and hydration in accord with various frictional ratios (contour lines denote f/f_0 values).

Oncley (1941) has shown, for proteins, the combined effects of hydration and asymmetry in the contours of Fig. 10.11, in which axial ratio a/b is plotted logarithmically as ordinate and hydration as abscissa. Contours are drawn through points of equal frictional constant.

If, therefore, we know the frictional ratio and either the amount of hydration or the axial ratio of a molecule, then we may use such contours to calculate the other unknown. However, the separate evaluation of solvation and asymmetry factors is not an easy problem, and it is not possible, yet, to cite any case which has been worked out in detail and completely confirmed. Many frictional ratios have been worked out, however, on the basis of zero solvation, and a selection of results is given in Table 11.2; these axial ratio values undoubtedly represent upper limits of asymmetry. However, from several lines of investigation (see p. 148) a hydration factor of approximately 30 per cent. (i.e. 0·3 gm. of water associated with 1 gm. of dry protein) appears probable, and the axial ratio values modified by the incorporation of this factor are included also in Table 11.2. It seems probable that more accurate individual estimates of hydration will be available in the future and more information on asymmetry will also be forthcoming from those methods involving molecular orientation (rotational diffusion). It is not, therefore, being unduly optimistic to say that, in the very near future, for those types of molecule for which treatment as an ellipsoid of rotation is valid, we shall shortly obtain more accurate information regarding their solvation and molecular dimensions.

The question of the validity of treating colloidal molecules and particles as ellipsoids of rotation is of fundamental importance and will have to be checked for individual systems. Too little is known of the structure of many colloidal molecules and particles to make definite predictions, but certain indications may be stated.

For linear and branched polymers, as well as colloidal aggregates of no uniquely defined molecular configuration, it is very unlikely that treatment as ellipsoids of rotation can be other than a very rough approximation which will probably yield at best only upper estimates of the molecular asymmetry. It is especially unlikely that for long flexible chains, enclosing and transporting solvent, or cross-linked to form extended networks in solution, that any quantitative information will be forthcoming from experimentally measured frictional ratios, as long as these are interpreted on the basis of ellipsoids of rotation. On the other hand, though much more remains to be done, it appears, from the agreement of different lines of investigation, that, for uniquely defined configurations of not too great asymmetry (especially corpuscular proteins), the treatment is at worst a reasonable approximation.

(c) *The influence of temperature and solvent viscosity on diffusion coefficients*

From eqn. (10.27) it is apparent that the diffusion coefficient is proportional to the absolute temperature. This connexion, however, presupposes that the frictional constant is independent of temperature changes whereas, in practice, alteration of the temperature of a colloidal solution causes a change in the frictional constant of the solute if only as a result of the change in *solvent* viscosity. From Stokes's law the frictional constant for spherical molecules is proportional to the viscosity coefficient of the solvent, and it is usual to assume the same proportionality in general. Providing, therefore, no changes in the solvation or shape of the solute molecules occur, then the diffusion coefficients D_1 and D_2 under two sets of conditions 1 and 2 may be related by the expression

$$D_2 = D_1 \frac{T_2}{T_1} \frac{\eta_1}{\eta_2}, \qquad (10.36)$$

where η_1 and η_2 refer to *solvent* viscosities at temperatures T_1 and T_2 respectively.

This equation holds satisfactorily for monodisperse systems up to concentrations of 2 or 3 per cent., and over the moderate temperature ranges which alone have been investigated. Thus, under these conditions, it appears that the frictional constant is mainly conditioned by *solvent* viscosity and not by the macroscopic viscosity of the system. At low solute concentration, however, and especially for nearly spherical solute molecules, solvent and solution viscosities are not appreciably different, and the distinction is not therefore important.

In this connexion the decrease in diffusion coefficient at higher protein concentrations (p. 252) ought to be noted, since in such cases it appears that factors other than solvent viscosity appear to affect the frictional constant. The use of total *solution* viscosity has been tried in correcting by eqn. (10.36) to standard conditions, but leading as it does to diffusion coefficients *increasing* rapidly with concentration, it has been abandoned. To give a constant diffusion coefficient some intermediate viscosity value (between that of solvent and solution) is required, so that it would appear that solute interactions to some extent cause retarded diffusion.

In the case of polydisperse systems it would also appear probable that high molecular weight solute might be operative in determining the frictional constant of low molecular weight material and possibly vice versa. A sound approach to such problems would be provided by the study of mixtures of two or three monodisperse systems of well-defined character, when the factors determining frictional constant might be established.

REFERENCES

AITKEN, 1942, *Statistical Mathematics*, Oliver & Boyd.
BECKMANN and ROSENBERG, 1945, *Ann. N.Y. Acad. Sci.* **46**, 329.
BOLTZMANN, 1894, *Ann. Phys. Lpz.* **53**, 959.
BYWATER and JOHNSON, 1946, unpublished experiments.
CAMPBELL and JOHNSON, 1944, unpublished experiments.
CLAESSON, 1946, *Nature*, London, **158**, 834.

FICK, 1855, *Ann. Phys. Lpz.* **170,** 59.
GORDON, JAMES, and HOLLINGSHEAD, 1939, *J. chem. Phys.* **7,** 89.
GRALEN, 1941, *Kolloidzschr.* **95,** 188.
—— 1944, *Sedimentation and Diffusion Measurements on Cellulose and Cellulose Derivatives*, Dissertation, Uppsala.
HARTLEY and RUNNICLES, 1938, *Proc. Roy. Soc.* A **168,** 420.
HERZOG, ILLIG, and KUDAR, 1934, *Z. phys. Chem.* A **167,** 329.
KRAEMER, 1940, in *The Ultracentrifuge*, by Svedberg and Pedersen, Oxford.
LAMM, 1937, *Nova Acta Soc. Sci. upsal.* (4), **10,** No. 6.
NEURATH, 1941, *Science*, **93,** 431.
—— 1942, *Chem. Rev.* **30,** 357.
NORTHROP and ANSON, 1929, *J. gen. Physiol.* **12,** 543.
ONCLEY, 1941, *Ann. N.Y. Acad. Sci.* **41,** 121.
PERRIN, 1936, *J. Phys. Rad.* (7), **7,** 1.
POLSON, 1939, *Kolloidzschr.* **87,** 149.
SVEDBERG, 1925, *Kolloidzschr.* (*Ergänzungsband*), **36,** 53.
—— 1928, *Colloid Chemistry*, 2nd edn., American Chemical Society Monograph, No. 16.
—— and PEDERSEN, 1940, *The Ultracentrifuge*, Oxford.
TISELIUS, 1937, *Trans. Faraday Soc.* **33,** 524.
TISELIUS and GROSS, 1934, *Kolloidzschr.* **66,** 12.

XI
SEDIMENTATION VELOCITY

Introductory

IN the method of sedimentation velocity, a powerful centrifugal field is applied to a solution of the macromolecule or colloid, and the velocity of sedimentation of the solute in the direction of the field is measured. Such information, we shall see later, combined with data from diffusion experiments, makes possible not only a calculation of the molecular weight of the solute but also (in certain cases) provides an estimate of the dimensions and shape of its individual molecules. By itself, information from sedimentation velocity measurements is often a useful criterion of molecular size, and a study of the influence of concentration upon sedimentation behaviour may be used to investigate the presence of solute-solute interactions.

Clearly this method is an extension of that commonly used industrially, of estimating particle size in suspensions by the rate of sedimentation under gravity. For a given density difference between the particles and the suspending medium, and provided the particles do not vary too greatly in shape, it is clear that the larger the particles the more rapid their fall under gravity. Such sedimentation can be usefully employed only for particles above a certain limiting size, which varies according to the density difference between the particle and its medium, but usually corresponds to a particle diameter of 1 μ. For smaller particles, the opposing transfer of material by diffusion balances that occurring by sedimentation under gravity, and the method can no longer be employed (see p. 195).

If, instead of gravity, we employ a much stronger centrifugal field, the rate of sedimentation is increased and it becomes possible to investigate the sedimentation of particles with diameters considerably less than 1 μ. As in the simple gravity method, it is required that the rate of sedimentation should be much larger than that of diffusion for the successful use of the method.

It will be recalled that in the sedimentation equilibrium method, a state of balance between the processes of sedimentation and diffusion is obtained, from which molecular weights are calculated. In the sedimentation velocity method, on the other hand, we are interested in the rate of sedimentation only, which is arranged deliberately to be much larger than the rate of diffusion, and, therefore, to a large extent unaffected by it.

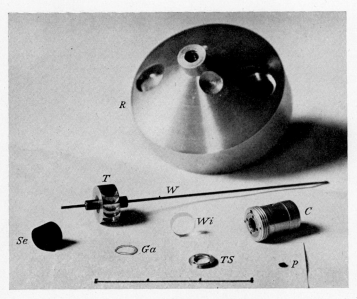

FIG. 11.1. Ultracentrifuge Rotor, Cell and Components.

  R. Rotor.   *Se*. Cell sector piece.
  T. Air turbine.   *Wi*. Cell window (quartz).
  W. Flexible wire drive. *Ga*. Cell gasket.
  C. Cell.   *TS*. Cell tightening screw.
     P. Rubber sealing plug.

FIG. 11.2. Ultracentrifuge Head (*H*) with rotor attached, and Rotor Chamber, *Rc*.
(Sharples Centrifuges, Ltd.)

Briefly, the sedimentation velocity ultracentrifuge provides a means by which an intense centrifugal field can be applied to a small quantity of test liquid (containing macromolecules), and in which convectionless sedimentation can be observed.

Before proceeding to consider sedimentation in more detail, the practical side of ultracentrifuge construction and design will be outlined.

Practical details†

It is proposed to consider only the ultracentrifuges in which optical examination of the sedimenting system can be made throughout the sedimentation.‡ In these instruments a small quantity of the test solution ($\frac{1}{4}$ to 1 c.c.) is contained in a sector-shaped cavity in a cell, provided with transparent windows, which is screwed firmly into the rotor, a large circular or oval solid block of metal, which can be spun at high speeds without vibration (Fig. 11.1). By having the cell sector accurately alined as a sector of the rotor, radial sedimentation can take place without any convectional disturbances which otherwise arise when the sector is not so alined, as a result of sedimenting material striking and being reflected from the sides of the cavity.

The name of Svedberg is outstanding in all aspects of ultracentrifuge work, and in his laboratory, oil-turbine ultracentrifuges of the highest performance have been designed, built, and operated with great success. Such instruments probably represent the peak of ultracentrifuge construction, giving precision and resolving power of the highest order under centrifugal fields of great intensity. However, these ultracentrifuges are very expensive and therefore unattainable for most laboratories.

Largely as a result of its considerably smaller cost the air-driven ultracentrifuge has been more widely installed in many laboratories and good results have been obtained. Developed from the spinning-top of Henriot and Huguenard (1927) by Beams (1937, 1938), Pickels (1937) and co-workers, such machines providing vibrationless rotation at very high speeds are now available commercially for preparative as well as analytical work. Figs. 11.2 and 11.3 contain photographs and a diagram of such an analytical instrument§ which has been in use in the Department of Colloid Science, Cambridge, since 1939. Although one of the

† For full descriptions of different types of ultracentrifuge see Svedberg and Pedersen (1940).

‡ McBain (1939) has devised an 'opaque' centrifuge in which sedimentation is followed by analytical means at the end of an experiment.

§ Made by Sharples Centrifuges, Ltd.

smaller air-turbine ultracentrifuges, this machine has a performance only slightly inferior to that of the more expensive instruments.

The rotor (R), of duralumin, weighing about $1\frac{1}{2}$ lb. and 8·85 cm. in diameter, hangs by a flexible steel wire (W) from the small duralumin

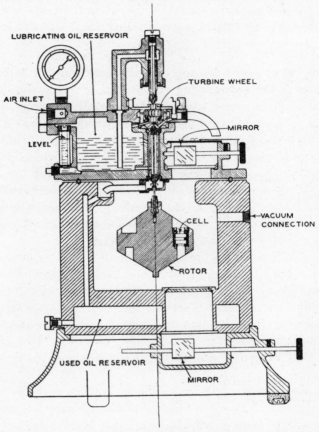

Fig. 11.3. Cross-section of ultracentrifuge assembled.

air turbine (Fig. 11.4) supported at rest by the stator. The oil glands (G) surrounding the steel wire serve not only to guide the wire drive but also isolate the thermostatted rotor chamber surrounding R from the air-turbine region above the glands. It is thus possible to evacuate the rotor chamber without air leakage from the compressed air drive.

Light from a tungsten ribbon lamp (L) focused upon a slit (S) passes through the schlieren lens (Sc) built into the base of the rotor chamber, through the cell (C) and is then reflected by a mirror (M) on to the optical bench containing the usual other elements of the diagonal schlieren system (see p. 220).

In operating the ultracentrifuge two air lines (depicted by arrows in Fig. 11.4) are utilized; one, of 5 lb./sq. in., is directed vertically upwards upon the under surface of the turbine, and is of just sufficient power to lift it with the rotor hanging below, whilst the other, of 10–25 lb./sq. in., is directed horizontally from several jets upon the flutings on the turbine, and provides the driving pressure. The use of the air support or cushion avoids the usual difficulties of high-velocity lubrication, and its incorporation represented an important advance in ultracentrifuge construction.

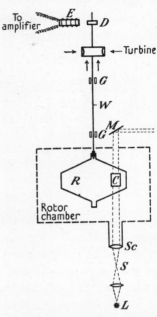

Fig. 11.4. The air-driven ultra-centrifuge (diagrammatic).

By evacuation of the rotor chamber, the air friction at the surface of the rotor is reduced, and hence, also, the driving pressures required to obtain a given speed. In the instrument described hydrogen gas is streamed slowly through the rotor chamber at a pressure of 10 mm. of mercury, in order to remove any heat generated in the rotor, hydrogen giving the maximum thermal conductivity with the minimum frictional resistance.

The speed of rotation of the rotor is determined electrically (Fig. 11.4). A small brass disc (D) containing a small steel segment, and mounted centrally upon the extension of the driving shaft (W) above the turbine, rotates past the poles of an electromagnetic generator (E). The alternating current thus produced is led to a two-valve amplifier and the amplified current is used to drive a phonic wheel connected to a timing disc through a reduction gear of ratio 10,000:1. Rotation of the disc is timed by a stop-watch and, after precautions have been taken to ensure that the phonic wheel is running on the fundamental frequency of the alternating current, this gives an accurate measure of the speed of rotation of the rotor.

The rotor, with cell and contents, must be carefully balanced, though more tolerance is allowed in the air-driven instrument as a result of the flexibility of its drive than with the oil-turbine ultracentrifuge.

Limitations on speed arise from the strength of the rotor material; in the case of the instrument described above there is a maximum operational speed of 85,000 r.p.m. (*ca.* 250,000g), which allows the study of proteins, for instance, down to a molecular weight of 17,000. Such a spherical molecule would have a diameter of roughly 20 A. A comparison with sedimentation under gravity thus shows that the ultracentrifuge extends

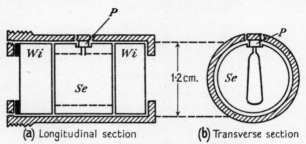

Fig. 11.5. The ultracentrifuge cell.

the usefulness of sedimentation velocity methods to particles smaller by a factor of about 1,000 (i.e. 1 μ or 10^4 A to 10 A).

The test liquid contained in the cell sector-piece, Se, of radial dimension 1·0 cm., angle 5°, and depth 1·0 cm., is totally enclosed (see Figs. 11.5 (*a*) and 11.5 (*b*)) by two quartz windows (Wi) at the ends of the sector-piece and above the sector filling hole by the small rubber plug P, which, during an experiment, is forced upon the seating surrounding the hole (sealing it) by the centrifugal field. Careful sealing of the cell by special methods is essential to avoid the leaks which otherwise develop at high speed under vacuum.

The chief disadvantages of the small instrument described are:
1. No direct method of temperature control or measurement for the rotor and cell.
2. Small length of sedimentation (< 1 cm.), caused by the smaller type of cell, resulting in poorer resolution than in the larger instruments (see p. 274).
3. Lack of precise knowledge of the axis of rotation, owing to the flexibility of the wire drive.

With regard to (1), the constancy of the temperature of the double-walled rotor chamber, especially at low gas pressures inside it, is probably not sufficient to ensure a highly constant rotor and cell temperature. However, calibration runs, using substances of well-known sedimentation characteristics, have indicated that the rotor temperature does not deviate appreciably from the temperature of its chamber.

Providing the exact position of the axis of rotation does not vary during a run, which is probable, (3) should merely cause a small error in the estimation of the centrifugal field, and therefore in the sedimentation velocity under unit field (sedimentation constant).

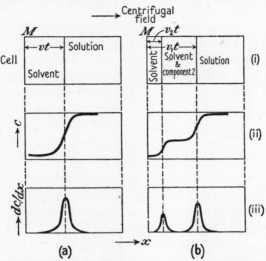

FIG. 11.6. Sedimentation in the ultracentrifuge cell. (*a*) Monodisperse solute. (*b*) Mixture of two monodisperse solutes.

We now proceed to consider in more detail the actual process of sedimentation inside the cell.

Sedimentation of non-interacting systems

In this section it is intended to study those systems in which the macromolecule is free from interaction with like molecules and so can migrate independently. Such a condition is probably approximately obeyed with the more symmetrical molecules except at *very high* concentrations (> 3–5 gm./100 c.c. of solution).

Although, as has been mentioned, a sector-shaped cell is used for convectionless sedimentation, it is convenient for illustrative purposes to consider a rectangular cell with the centrifugal field parallel to one side (Fig. 11.6).

(*a*) *Sedimentation in a solution of a homogeneous macromolecule.*

A solution of a macromolecular substance, whose molecules are all alike in size and shape and in which diffusion is very slow, represents the simplest case of sedimentation.

From the condition of homogeneity, all such molecules have the same terminal velocity (v) in the direction of the field for a given value of the latter. Thus, in time t, all macromolecules will have moved a distance vt (assuming v constant† over the cell) (Fig. 11.6 (a) (i)), and it is clear that at this instant t those molecules initially to the left of the line drawn at a distance vt from the meniscus (M) will have crossed or be situated on this line, termed the boundary. Pure solvent will therefore be contained to the left of the boundary and, for the idealized cell now considered, unchanged solution to the right. An abrupt change in concentration at the position of the boundary (Fig. 11.6 (a) (ii)) will therefore occur. Since diffusion processes are always present to some extent, such a boundary can never be infinitely sharp, but it is usually correct to say that the smaller the diffusion constant the sharper the boundary. Fig. 11.6 (a) (iii) shows the corresponding plot of dc/dx, the concentration gradient, along the cell dimension, x, in the direction of the centrifugal field.

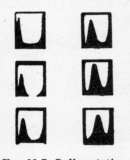

FIG. 11.7. Sedimentation diagrams for egg albumin in phosphate buffer at pH = 7·4. Protein concentration = 0·97 g./100 c.c. of solution.

Such a plot is given directly (if $\Delta n \propto \Delta c$) by the diagonal schlieren method, so that by taking photographs at increasing values of the time t it is possible to examine the progress of the 'peak' down the cell and, taking the position of the boundary to be given by the top of the peak (p. 217), to measure v for different values of x (see Figs. 11.7 and 9.8). Thus the sedimentation constant (s), defined by

$$s = \frac{v}{\omega^2 x} = \frac{dx/dt}{\omega^2 x}, \qquad (11.1)$$

i.e. the velocity of sedimentation under unit centrifugal field, may be calculated throughout the course of the sedimentation.

Effect of diffusion. In the absence of diffusion, a homogeneous substance must give rise to an infinitely thin sedimenting boundary. However, diffusion, present to some extent in all systems and tending to smooth out any concentration difference, gives rise to a boundary occupying a finite distance along the x dimension and in which dc/dx is not infinite at any point. Thus with increasing time an initially very sharp boundary becomes increasingly broad (Fig. 11.7).

It can readily be shown that the superposition of slow diffusion introduces no serious complications in the treatment of sedimentation.

† In practice, the centrifugal field (and therefore v) increases with increasing distance from the axis of rotation, but this does not affect the main conclusions of the treatment above.

Thus Fick's first law (p. 233) becomes, for a sedimenting system (see Cohn and Edsall, 1943, p. 419),

$$dm = A\left[c\frac{dx}{dt} - D\frac{dc}{dx}\right]dt, \qquad (11.2)$$

in which the separate effects of sedimentation and diffusion are added. Making use of the frictional constant as already defined (eqn. (9.1)), eqn. (11.2) becomes

$$dm = A\left[\frac{\phi c}{f} - D\frac{dc}{dx}\right]dt, \qquad (11.3)$$

where ϕ is the centrifugal force per molecule. Fick's second law then becomes

$$\frac{\partial c}{\partial t} = D\frac{\partial^2 c}{\partial x^2} - \frac{1}{f}\frac{\partial(\phi c)}{\partial x}. \qquad (11.4)$$

This equation is basic to both the sedimentation equilibrium and velocity methods. In the first method, at equilibrium, $\partial c/\partial t = 0$ at all points, and this condition may be utilized, on substituting for ϕ, to derive the normal expression connecting molecular weight and the concentration distribution (eqn. (8.6)). In the velocity method $\partial c/\partial t \neq 0$ over the boundary region, and it is apparent that the total observed effect is merely the sum of the separate processes of sedimentation and diffusion. Thus, ideally, the curves of c and dc/dx against x in sedimentation are simple diffusion curves of the normal type already discussed (p. 236), and the presence of rapid sedimentation merely transports the whole curve 'revolving' about the $c_0/2$ point along the centrifuge cell (p. 237). It therefore follows that the position of the boundary for sedimentation as well as for diffusion is given by $c = c_0/2$, which, for ideal curves, coincides with the top of the migrating peak.

For such ideal systems it is clear that diffusion coefficients may be calculated from the broadening of sedimentation diagrams as in usual diffusion experiments. However, for good reasons, such calculations are liable to be inaccurate. Thus the duration of a sedimentation velocity run is considerably shorter than that of normal diffusion measurements and more serious is the fact that any artificial sharpening or broadening of the peak arising from non-ideal sedimentation (see p. 279) would be attributed to diffusion and thus cause very erroneous values of the diffusion coefficient.

(b) *Sedimentation in paucidisperse systems*

It is proposed to consider now the sedimentation in a solution of two or more homogeneous macromolecular substances.

Let the (say) two macromolecules have, under the conditions of experiment, the velocities of sedimentation v_1 and v_2. After time t, their boundaries will thus occur at distances $v_1 t$ and $v_2 t$ from the meniscus, the separation of the two being

$$(v_1 - v_2)t \quad \text{(Fig. 11.6}(b)\text{).}$$

Therefore, at small values of t only a single boundary will be observed, but as t increases the separation will increase, the initially sharp single boundary will broaden, and, providing v_1 and v_2 are sufficiently different, it will eventually split into two boundaries moving independently. Fig. 11.8 is typical of such a system.

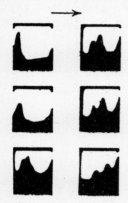

Fig. 11.8. Sedimentation diagrams for the ground-nut globulins in 0·05 M. phosphate buffer at pH 8. Protein concentration $\approx$ 0·7 gm./100 c.c.

It is clear that the larger t, and, therefore, the length through which sedimentation occurs, the greater the resolution of two different boundaries. In some cases where diffusion is rather rapid, the increased separation of boundaries at longer times may be more than counterbalanced by diffusion broadening. In general, however, better resolution is obtained for longer lengths of sedimentation. It will be recalled (p. 270) that the short sedimentation path was mentioned as one of the disadvantages of the air-driven ultracentrifuge.

In a similar way, more than two components may, under favourable circumstances, be resolved, providing the total solute concentration is low enough for independent migration of the different systems to occur. Further, if good resolution of the peaks occurs, then an accurate estimation of the peak areas and therefore of Δn, the refractive index change, for each component may be obtained. Knowledge of dn/dc, the rate of change of refractive index with concentration for each component, would then make possible the calculation of the concentrations of the differently sedimenting species. Thus we have the possibility of analysing a paucidisperse system on the basis of sedimentation velocity. Such a result should be contrasted with the rather ill-defined results obtainable for such a system by the sedimentation equilibrium method (p. 199). The presence of diffusion, however, usually limits such sedimentation analysis to two or three components.

(c) *Sedimentation in polydisperse systems*

If, instead of discrete values v_1, v_2, etc., we have a continuous range of v values, it is clear by extension of the case of paucidisperse systems, that instead of the single sharp peak of homogeneous systems or the discrete peaks of paucidisperse systems, we shall have one very broad peak in which the spreading of polydispersity is added to the usual diffusion spreading. This result again presupposes the condition of independent migration of the individual macromolecules, a condition which will be shown not to hold for linear polymer molecules (p. 286) at what are usually considered as low concentrations.

The extent of polydisperse boundary spreading depends upon the range of sedimentation constants involved, and, as mentioned above, upon the length of path through which sedimentation has occurred. It is of importance to realize that boundary spreading may be due to the two factors of diffusion and polydispersity, in both of which increasing time leads to increased spreading. Methods of determining to which factor a given boundary spreading is due are therefore of importance, and of these, two may be mentioned here:

1. Diffusion coefficients for the substance under the same conditions as in the sedimentation are determined by normal diffusion methods. Such a value for a polydisperse system will, of course, be an average. From this constant, the spreading at different times during sedimentation may be calculated and compared with that experimentally observed. If the latter is *considerably* in excess of the former, it may be taken as a strong indication of polydispersity.

The method must be used with care, and the presence of anomalous boundary effects may in some cases prevent its use (e.g. p. 278).

2. If sedimentation velocity experiments are carried out at a range of speeds, a given length of sedimentation is traversed in a time which is shorter, the higher the speed. Now the boundary broadening of polydispersity is dependent upon the length of the sedimentation path (i.e. upon $v \times t$ rather than upon t alone) and therefore, for a given path length, is independent of the time of sedimentation or of the speed of rotation. However, diffusion, which we have seen occurs independently of sedimentation in the ideal case, remains dependent only upon the time. At higher speeds, therefore, when the time taken to traverse a given sedimentation path is shorter, diffusion broadening will be smaller, but polydisperse spreading unaltered. Hence the occurrence of sharper boundaries at higher speeds is a sign that spreading is due to diffusion, whilst the lack of any such change at higher speeds suggests polydispersity.

Other types of information may enable us to decide on the cause of boundary spreading in a given case. Thus, for example, the occurrence, in normal diffusion experiments, of dc/dx against x curves deviating widely from ideal Gaussian curves (p. 249), or the possibility of fractionating a polymeric system into portions differing widely in physical properties, may provide additional evidence.

(d) *Sedimentation and concentration*

In this section we have considered the sedimentation of polymer molecules in systems such that each molecule can sediment uninfluenced by others sedimenting similarly. It is important to know how far such a condition may be fulfilled in general and how in any given case it is possible to investigate this question.

If molecules are, indeed, migrating independently, then it is clear that the resistance to their motion (= frictional constant × velocity) is conditioned only by the solvent, being independent of the presence in solution of other like molecules. Thus, where the condition of independent molecular migration holds, the sedimentation constant, which is inversely proportional (p. 288) to the frictional constant of the individual molecule, will be found to be independent of the concentration of the macromolecule. On the other hand, the existence of a marked dependence of sedimentation constant on concentration may be a strong indication of intermolecular interference between the sedimenting molecules.

Few systems, if indeed any, are without any concentration dependence of sedimentation constant, but amongst the corpuscular proteins (see p. 739) this dependence is often quite small (Fig. 11.9 (a)). The more asymmetric the macromolecule, the more marked this dependence becomes (for example, cf. serum albumin and serum globulin), and extreme cases are encountered amongst the so-called linear polymers such as cellulose nitrate (Fig. 11.9 (b)). It will be seen clearly later that the theoretically valuable sedimentation constants are those involving independent molecular migration; it therefore follows that it is necessary for any system to establish whether or not a marked concentration dependence exists, and if so, to measure sedimentation constants at concentrations which are sufficiently low to permit an accurate extrapolation to zero concentration.

Factors complicating sedimentation

(a) *Charge effects*[†]

Any charged colloidal ion in solution is surrounded by its 'ionic atmosphere' containing an excess of oppositely charged ions. Motion of

† For detailed treatment see Svedberg and Pedersen (1940), p. 23.

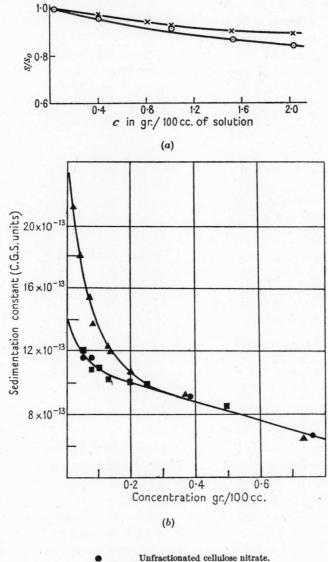

FIG. 11.9. (a) Plot of s/s_0 (s_0 being the value of s, the sedimentation constant, at infinite dilution obtained by extrapolation) against concentration for horse serum albumin (×) and globulin (⊙). (b) Sedimentation constant-concentration relation for cellulose nitrates.

the colloidal ion under the influence of a centrifugal field would, however, tend to separate it from its slowly sedimenting atmosphere, so setting up a potential gradient in solution which would retard the rapidly sedimenting colloid and accelerate the slowly moving ions of its atmosphere.

Such an effect, the primary charge effect, will obviously be the more important, the larger the charge on the colloidal ion, and the lower the salt concentration of the solution. Experimentally it is of importance to eliminate it so that undisturbed sedimentation may be observed. This is usually achieved either by working with an iso-electric (i.e. uncharged) colloid or, more usually, by the addition of neutral salts. In the latter case, for a 1 per cent. protein solution, a neutral salt concentration of 0·2 M. is usually sufficient; for higher protein concentrations a correspondingly higher salt concentration is required.

Of smaller magnitude and importance is the secondary charge effect, which is also dependent upon the charge of the sedimenting species. If, in addition to the colloidal particle, several ions of differing sedimentation constants and signs of charge occur, their separation during sedimentation will introduce a potential gradient in which the sedimentation of the colloidal particle may be either accelerated or retarded. Such an effect may occur even in concentrated salt solutions and is eliminated only for a charged colloid by the elimination of the potential gradient from which it arises. This can be done only if ions of approximately equal sedimentation constant are employed. As in the primary salt effect, by working with iso-electric colloidal particles, the secondary effect may also be eliminated.

(b) *Sedimentation in organic solvents*

Macromolecules dissolved in organic solvents are usually uncharged and sedimentation occurs, therefore, without any complication from charge effects. Some other differences of a more practical character exist, amongst which we may mention:
1. The greater effects of compressibility in organic solvents leading to an increase in solvent density with distance from the meniscus.
2. Special cell requirements. The solution being examined must be completely enclosed by materials not attacked or softened by organic solvents. For aqueous systems, prevented from evaporating by a layer of oil, the problem is considerably simpler.

(c) *Boundary anomalies*

In the previous section attention was drawn to the frequent dependence of sedimentation constant upon concentration. It is of importance to consider how this affects the nature of a sedimenting boundary. Consider, therefore, a monodisperse system (Fig. 11.6(a)). On the advancing edge of the peak, the concentration of macromolecule is considerably higher than on the 'trailing' edge. The existence of a sedimentation constant-concentration dependence therefore involves the motion of the two edges of the peak at different speeds.

Sharpening of the boundary must therefore occur if sedimentation constant decreases with increasing concentration (e.g. serum albumin and serum globulin), the magnitude of the effect depending upon the extent of variation of the sedimentation constant. For most of the corpuscular proteins, in which sedimentation constant-concentration dependences are not marked, the spreading of diffusion occurring simultaneously is sufficient to outweigh such sharpening.

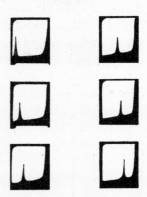

FIG. 11.10. Sedimentation diagram of cellulose nitrate fraction in acetone. Polymer concentration = 0·73 gm./100 c.c.

Where such dependences are much larger, e.g. in solutions of linear polymers, such pronounced sharpening of the sedimenting peak occurs that diffusion appears to be completely absent (Fig. 11.10).

If, on the other hand, for any reason the opposite change in sedimentation constant with concentration occurs, i.e. the sedimentation constant decreases with decreasing concentration, then it follows similarly that an increased spreading of the boundary will occur. Such cases, which will be discussed later (p. 281), may arise from the tendency of a macromolecule to dissociate at low concentration.

Both cases are of importance in demonstrating that effects other than of diffusion may be involved in the spreading of a sedimenting boundary (see p. 273). The accurate calculation of diffusion coefficients from boundary spreading in the ultracentrifuge can only be undertaken, therefore, in the absence of such complications or if adequate allowances can be made.

(d) Sedimentation of the components of an equilibrium system

Let us consider a simple equilibrium system of the type

$$A_2 \underset{k_2}{\overset{k_1}{\rightleftharpoons}} 2A,$$

in which a large molecule A_2 dissociates to give equal halves. The sedimentation constant of A_2 (s_{A_2}) will be different from (usually larger than) that of A (s_A), so that if A_2 and A can exist separately they will be readily distinguishable by their sedimentation behaviour. It is convenient, in dealing with the sedimentation of a mixture of A_2 and A molecules in equilibrium to consider the following three cases:

(i) Those in which the time required for the establishment of equilibrium is small compared with the duration of a sedimentation velocity experiment.

(ii) Those in which the time required for the establishment of equilibrium is long compared with the duration of a sedimentation velocity experiment.

(iii) Intermediate systems in which the time required for establishment of equilibrium is comparable with the duration of a sedimentation velocity experiment.

(i) *Rapid equilibria.* In the first category rapid equilibration rates arise as a result of rapid forward and reverse reactions. It follows, therefore, that during a short time of sedimentation any A molecule exists partly as a single A molecule and partly associated with another in a double molecule.

Further, all molecules will tend to spend the same fractions of their lives in the two forms, being subject to the same environment. One sedimentation constant only will therefore be observed (not two) whose value will lie between the hypothetical constants of the double and single molecules, the actual value depending upon the statistically determined fraction of the total time spent by a molecule in the two forms. Thus if, during a short time t, on an average a molecule spends times t_2 and t_1 in the forms A_2 and A_1 where $t = t_1 + t_2$, then the total distance travelled in time t under unit field is

$$st = s_2 t_2 + s_1 t_1, \qquad (11.5)$$

where s is the experimentally observed sedimentation constant, or

$$s = s_2 \frac{t_2}{t} + s_1 \frac{t_1}{t}. \qquad (11.5\,\text{a})$$

A similar extended expression holds for more than two components in rapid equilibrium, and in general

$$s = s_1 \frac{t_1}{t} + s_2 \frac{t_2}{t} + s_3 \frac{t_3}{t} + \cdots. \qquad (11.6)$$

For such rapidly equilibrating systems, therefore, a single discrete peak will be observed which (apart from other complications) moves in the direction of the field and spreads at rates between those expected for the different forms. From a single experiment, therefore, the system will behave in a manner not noticeably different from a single monodisperse one.

If, however, other experiments are performed under such conditions that the position of equilibrium has changed (i.e. t_1/t and t_2/t have values different from the original values), then measured sedimentation and diffusion constants will also change. Thus if the simple dissociation above obeys the law of mass action, dilution should push the equilibrium across to the single A molecules, and thus lower the observed sedimentation constant. Similarly, the change in the sedimentation constant (corrected to standard solvent viscosity and temperature—see p. 290) caused by other changes in the solvent medium or in the temperature may provide evidence of the existence of an equilibrium system.

It should be noted, however, that all such changes in the sedimentation and diffusion constants must not be interpreted on this basis. We have already seen that, in the case of the more asymmetrical molecules, sedimentation constants may vary considerably with concentration as a result of solute-solute interaction effects. Evidence for dissociative equilibria from variation in the sedimentation constant can only be accepted if other effects are shown not to be responsible.

It would seem that, under certain conditions (see Svedberg and Pedersen, 1940, p. 355) horse haemoglobin, whose sedimentation constant 'reaches a maximum value at a concentration between 0·5 and 1 per cent.', is representative of rapid equilibrium systems, a possibility which is supported by the observation of Tiselius and Gross that the diffusion coefficient of the protein increases with dilution at concentrations below 0·8 per cent.

(ii) *Slowly established equilibria.* If equilibrium is established only very slowly, we may treat the two equilibrium forms (A_2 and A) as stable molecules. The whole system therefore approximates to a paucidisperse system. The double and single molecules with different sedimentation constants, will cause, first, excessive peak spreading during sedimentation, and, later, if the constants are sufficiently different, peak splitting to give sub-peaks which move down the cell quite independently.

The globulins from the ground-nut (Johnson, 1946a) form such a system. A large molecule (of molecular weight *ca.* 400,000) may be shown to dissociate into apparently equal halves under certain conditions. Under other conditions, however, the large parent molecules, alone or mixed with the sub-molecules, will exist for several months without any observable change in their proportions. This must, it is thought, be due to the slowness in the establishment of equilibrium. The normal sedimentation and spreading of the peaks is in agreement with what is to be expected on the basis of such a conclusion.

(iii) *Intermediate equilibrium systems.* Here we consider systems in which equilibration occurs in a time comparable with the duration of a sedimentation velocity experiment. The sedimentation behaviour of such systems may be seen as a modification of either of the previously discussed types of equilibrium. Let us consider modifications of slowly established equilibrium systems.

Thus consider a boundary due to single A molecules, sedimenting monodispersely as a result of the slow rate of the association process (i.e. small k_2). An A_2 boundary would similarly be monodisperse because of the slow dissociation reaction (i.e. small k_1). We may now imagine that to the system is added a very small amount of a suitable catalyst, which increases both velocity constants without affecting their ratio, controlling the equilibrium position. The catalyst concentration may be so chosen that although both processes are still slow, a small finite amount of the second component appears amongst the otherwise monodisperse material responsible for the two boundaries. The small number of double (A_2) molecules appearing in the A boundary cannot immediately revert into single ones, in view of the smallness of k_1; they must therefore sediment characteristically of double molecules and must cause polydisperse boundary spreading. The occurrence of a finite rate of production of A molecules in the material comprising an A_2 boundary region must give a similar polydisperse effect.

As the catalyst concentration increases from very low values and with it the separate values of the velocity constants, the amount of the second component in each boundary grows larger, and polydisperse spreading more pronounced. Eventually, such spreading may be so large that the two separate boundaries are no longer visible and instead we have one very broad peak, in which all sedimentation constants varying between s_{A_2} and s_A are represented.

As the catalyst concentration increases yet further, a situation arises eventually in which the half-life of both processes (forward and reverse) is of the same order as the duration of a sedimentation velocity experiment (*ca.* 1–2 hrs.). More rapid equilibrium rates will therefore cause reduced spreading of the broad single boundary since the deviations of the sedimentation constants of individual molecules from the statistically determined mean constant of eqn. (11.6) (i.e. the spread of effective sedimentation constant) will decrease.

Finally, at very rapid equilibrium rates we have the *apparently* monodisperse boundary of the rapid equilibrium systems already discussed. Such systems, with those involving very slow equilibrium rates, form the two extremes of the range of equilibrium systems

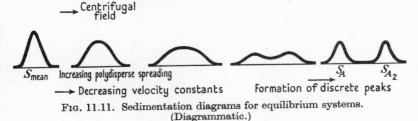

Fig. 11.11. Sedimentation diagrams for equilibrium systems. (Diagrammatic.)

depicted in Fig. 11.11, which differ from one another only in the separate values of the velocity constants k_1 and k_2, the equilibrium position being unaltered throughout.

Sedimentation of strongly interacting systems

In this section we wish to consider in some detail the sedimentation of systems of macromolecules which, through pronounced asymmetry, occupy a large effective volume in solution. Hence, even at extremely low concentrations, the molecules interfere with one another, so that the condition of independent migration of molecules, holding for the more symmetrical types, no longer holds here.

We have seen (p. 276) that for non-interacting systems the resistance to motion of a single molecule is a function of the properties of the solvent, being independent largely of the presence of other like particles. For interacting systems, however, it is evident that properties other than those of the solvent are important and we may expect that the rate of sedimentation under unit field will no longer be independent of the concentration of macromolecule.

Cellulose nitrate in acetone

(i) *Experimental results and their interpretation.* A good example of such a system is cellulose nitrate in acetone (Mosimann, 1943; Gralen, 1944; Campbell and Johnson, 1944). Commercial samples of cellulose nitrate are highly polydisperse with regard to chain-length or degree of polymerization, as may be shown by fractionation procedures (see p. 796). Thus Fig. 11.12 shows the different viscosity behaviour of solutions of fractions obtained by fractional precipitation with water from a dilute solution in acetone of a commercial cellulose nitrate (nitrogen = 12·2 per cent.). The considerable range of viscosity covered by the fractions, obtained from this single-stage fractionation, amongst which nitrogen content was sensibly constant, demonstrates clearly the polydisperse

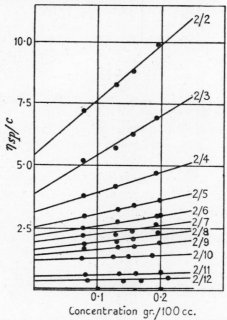

FIG. 11.12. η_{sp}/c v. c for fractions of a commercial cellulose nitrate in acetone (●).
N.B. For comparison the corresponding curve for human serum albumin (■) in aqueous solution is included.

nature of the original material. Further, it was not to be expected that the fractions from a single-stage fractionation process would be monodisperse, but rather that these would contain material with properties continuously distributed about a mean value representing the greater part of the fraction. From experience with non-interacting systems, it might have been expected, therefore, that considerable boundary spreading during sedimentation, due to the added effects of polydispersity and diffusion, would occur.

Figs. 11.13 (a) and 11.13 (b) contain the sedimentation diagrams for acetone solutions of the original unfractionated material and of a high viscosity fraction. Both sets of diagrams, obtained at concentrations (< 1 per cent.) considered low in dealing with non-interacting systems, clearly show the absence of any appreciable boundary spreading. In view, therefore, of the known composition of the materials, then under the conditions of experiment, either cellulose nitrate molecules of differing chain-length have identical sedimentation constants and very slow diffusion rates, or normal boundary spreading is prevented by large anomalous boundary effects (e.g. see p. 278). Both causes might be effective to some extent.

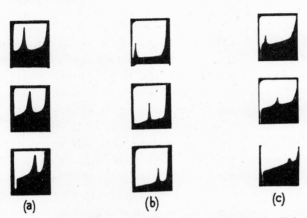

FIG. 11.13. (a) Sedimentation diagrams for an unfractionated cellulose nitrate in acetone at a concentration of 0·76 gm. per 100 c.c. of solution. (b) Sedimentation diagrams for a high viscosity fraction of cellulose nitrate in acetone at a concentration of 0·73 gm. per 100 c.c. of solution. (c) Sedimentation diagrams for the same high viscosity fraction of cellulose nitrate in acetone at a concentration of 0·122 gm. per 100 c.c. of solution.

Further information was obtained by studying the sedimentation behaviour of cellulose nitrate fractions over a range of concentrations. Fig. 11.9 (b), containing some of the results obtained, shows the existence of a sedimentation constant-concentration dependence, which is the greater the higher the intrinsic viscosity of the fraction. Further, it appears that differences in sedimentation constant which are apparent at very low concentration are levelled out at slightly higher concentrations ($\approx$ 0·5 gm./100 c.c.). At such concentrations therefore, in spite of the polydisperse nature of the sedimenting material, no polydisperse spreading can occur; added to this we have a pronounced sharpening of the moving boundary, caused by the existence of a concentration dependence of the type shown in Fig. 11.9 (b). The strikingly sharp sedimenting boundaries of Figs. 11.10 and 11.13 are thus understandable. It should be noted that, since the change in sedimentation constant with concentration is to be explained in terms of a change in the frictional constant of the sedimenting molecule, this must also contribute to the concentration-dependence of the diffusion coefficient, though, as we have seen (p. 250), other factors are also involved here.

At lower concentrations, for which differences in sedimentation constants are observable and anomalous boundary effects less important, appreciable boundary spreading during sedimentation is to be expected. Experiment has confirmed this prediction. A comparison of Figs. 11.13 (b) and 11.13 (c), showing the sedimentation of the same cellulose nitrate fraction at different concentrations in acetone, illustrates this point.

It is of importance to realize that the above sedimentation behaviour is expected to be of quite general occurrence. All types of linear polymers, synthetic as well as natural, in which the extended form of the molecular chain is preserved in solution, would be expected to behave similarly. Cellulose and its long-chain derivatives, which on stereochemical grounds are expected to exist in the form of nearly completely extended chains, would be expected to show interaction effects to a much larger degree than those substances, e.g. rubber, in which the long primary valency chain may tend to curl into a more compact configuration. An extreme example of the latter type of configuration is provided by the corpuscular proteins, in which the polypeptide chain is usually regarded as being folded and held in a unique, compact, corpuscular form, giving rise to little or no interaction effects. On the other hand, the fibrous proteins, or corpuscular proteins in which the polypeptide chain has by suitable treatment become freed from its native foldings and considerably extended, would be expected to give increased interaction effects. Tobacco mosaic virus protein (Lauffer, 1944), for example, whose kinetic units in solution are known from electron microscope and other evidence to be very asymmetric, shows strong sedimentation-concentration dependence.

(ii) *The meaning of the sedimentation constant-concentration dependence.* From the sedimentation constant-concentration curves of Fig. 11.9 (b) certain conclusions may be drawn regarding the solute-solute interactions at different concentrations. We have seen that different fractions show their individuality only at very low concentrations, whereas at higher concentrations differences are submerged. The latter fact can be due only to the existence of such strong interaction between the molecular chains that no independent migration of a molecule is possible. On the contrary, it is possible that an interlocking or entangled network of molecules exists which moves as a whole under the influence of the centrifugal field. The motion of such a network would not depend upon the precise length of the molecules composing it, providing they were long enough to take part in the structure. On such grounds, therefore, the observation of a sedimentation constant independent of chain-length at the higher concentrations (> 0·5 gm./100 c.c.) is explained. It is important to realize that such behaviour is not dependent upon, although it may be enhanced by, the existence of strong intermolecular forces, but can arise solely from the asymmetric nature and the consequent large effective volume of a dissolved macromolecule.

With decreasing concentration and fewer intermolecular entanglements, individual chain molecules will sediment influenced to a smaller degree by the contact or approach of neighbouring molecules. Gradually, therefore, with decreasing concentration, differences in sedimentation constant between molecules of different chain-length might be expected to appear, and at very low concentrations the condition of independent migration may even hold. In view, however, of the large concentration dependence at the lowest experimentally available concentrations of cellulose nitrate, it is doubtful in this case if independent migration can be observed.

Knowledge of molecular interaction has an important bearing also on the application of thermodynamic methods. We have seen that the simple osmotic relations and the use of sedimentation equilibria presuppose the independent existence of the macromolecules. The large deviations from this condition, indicated by sedimentation velocity measurements at the higher concentrations, very largely invalidate the thermodynamic methods in their simplest form over this range, and it is of great importance to know how far the lower concentration (0·1–0·2 gm./100 c.c.) interactions affect thermodynamic properties. Experimental data are not, however, available.

Sedimentation and molecular properties

As yet, we have been concerned, rather qualitatively, with sedimentation in different types of system, without relating sedimentation constants to the properties—molecular weight and dimensions—of the sedimenting entity. It is our purpose here to deal with this question.

Consider, therefore, the sedimentation of a monodisperse polymer of molecular weight M and partial specific volume $\bar{v}$ $(= 1/\rho_P)$ in a solvent of density ρ. We assume that the effective kinetic unit of polymer is indeed the single molecule of mass M/N, and that interaction with like molecules is therefore negligible. For an angular velocity ω of the rotor, the centrifugal field at a point distant x from the axis of rotation is, in absolute units, $\omega^2 x$. Therefore the centrifugal force acting on a single molecule is†

$$(\text{volume of molecule})(\rho_P - \rho)\omega^2 x.$$

On p. 210 we have seen that the acting force is to be equated to the product of the frictional constant (f_s) and the terminal velocity (in this case the velocity of sedimentation). Thus

$$f_s(dx/dt) = \frac{M}{N}(1-\rho\bar{v})\omega^2 x. \tag{11.7}$$

Rearranging, we have

$$M = \frac{Nf_s}{(1-\rho\bar{v})}\frac{dx/dt}{\omega^2 x}, \tag{11.8}$$

and making use of the definition of sedimentation constant, i.e. the sedimentation velocity under unit centrifugal field, we obtain

$$M = \frac{F_s}{(1-\rho\bar{v})}s. \tag{11.9}$$

We have also seen that the frictional constant is simply related to the diffusion coefficient by the equation

$$D = \frac{RT}{F_D}. \tag{10.27a}$$

It should be noted that we have here symbolized the frictional constants in sedimentation and diffusion differently because under certain conditions (e.g. if orientation occurred in either case) they are not necessarily identical. However, normally, if

(1) f_s and f_D refer to the same solvent medium, and

(2) no orientation of molecules occurs during sedimentation,

it is justifiable to regard the two constants as identical. With regard to (1) it is possible to make use of sedimentation and diffusion measurements done under slightly different conditions by suitably correcting one set of data to the same solvent temperature and viscosity as the other (see pp. 264, 290). It is, however, important to ensure that temperature changes do not introduce other effects, e.g. changes in molecular configuration, which cannot be included in the ordinary correction procedure. Molecular orientation can arise in the absence of electrical fields only if appreciable velocity gradients occur (see p. 387). Such gradients are absent in diffusion experiments and are negligibly small in sedimentation so that f_D and f_s both refer to random orientation of the colloidal molecules.

† There appears to be no universal agreement as to whether ρ should refer to the solvent or solution density. However, the distinction is, in practice, unimportant.

Assuming, therefore, that $f_D = f_s = f$, and substituting from eqn. (10.27 a) in (11.9) we have

$$M = \frac{RTs}{D(1-\bar{v}\rho)}. \tag{11.10}$$

This equation and eqn. (11.9), from which it was derived by substituting for the frictional constant in terms of the diffusion coefficient, are fundamental to the sedimentation velocity method, and it is of value to recall the assumptions implicit in them. We have assumed:
(1) the validity of the law of motion contained in eqn. (9.1), and
(2) that single molecules sediment unaffected by other like molecules.

It should be noted that no assumptions have been made as to the shape of the sedimenting molecules, and, therefore, that provided correct substitutions are made, correct molecular weight values will be obtained regardless of molecular shape. From the molecular weight, the frictional constant, f_0, for the hypothetical unsolvated spherical molecule of the same molecular weight, volume, and density can thus be calculated (eqn. (9.3)) and hence the frictional ratio f/f_0 obtained. We have already seen, however, that it is only for certain idealized types of molecule that such values may be utilized to give quantitative information on molecular shape.

(a) *Molecular weight calculations*

(i) *Extrapolation of sedimentation constants and diffusion coefficients.* The calculation of molecular weight by eqn. (11.10) resolves itself into the determination of:
1. The partial specific volume of the solute ($\bar{v}$), and solvent density.
2. The sedimentation constant and diffusion coefficient appropriate to independent motion of the solute molecules.

Partial specific volumes are determined by means of density measurements of a very high order of accuracy (for example, see Kraemer, 1940 a), but suitable methods are available. Considerable care is also required in the use of sedimentation constants and diffusion coefficients owing to variation with concentration.

Assumption (2) above (p. 276) is strictly true only at infinite dilution of the sedimenting material, and it is, therefore, necessary to extrapolate sedimentation constant and diffusion coefficient values to zero solute concentration in obtaining values for substitution in eqn. (11.10). This involves the use of sedimentation and diffusion measurements over a wide range of low concentrations, especially for those systems which show appreciable concentration effects.

For many corpuscular proteins and other highly symmetrical molecules, at concentrations of about 1 per cent., sedimentation constants and diffusion coefficients are dependent only to a small extent upon concentration, and extrapolated values at zero solute concentration can be obtained without much error. Recent work has, however, tended to show that such concentration effects have perhaps been underestimated (Johnson, 1946 b), and that some molecular weight values may, for this reason, need revision.

For linear polymers, however, and all asymmetrical molecules, concentration effects are extremely important (for example, see Fig. 11.9 (b)) and accurate extrapolation, though often difficult, is essential. In some cases it is doubtful if interaction effects are sufficiently small at any workable concentrations to justify extrapolation procedures, and for such cases there is need of data from quite independent experimental approaches. The system, cellulose nitrate in acetone, is of this type in which, without independent evidence, little confidence is to be placed in the accuracy of extrapolation methods used.

(ii) *Correction of sedimentation constants.* Sedimentation constants, measured usually at 20° C., are usually recorded for this temperature and corrected also to the viscosity and density of the pure solvent at 20° C. (s_{20}^0). In such a correction procedure it is assumed that temperature influences the sedimentation constant only by altering the properties (i.e. viscosity and density) of the solvent. Alteration in solvent viscosity thus causes a corresponding change in the frictional constant of the solute whilst density changes affect the term $(1-\bar{v}\rho)$ in the denominator of eqns. (11.9) and (11.10). The complete correction expression may thus be written

$$s_{20}^0 = s \frac{\eta}{\eta_{20}^0} \frac{(1-\bar{v}_{20}^0 \rho_{20}^0)}{(1-\bar{v}\rho)}, \qquad (11.11)$$

where s_{20}^0 is the sedimentation constant corresponding to η_{20}^0, and ρ_{20}^0 the viscosity and density respectively of pure water at 20° C., and $\bar{v}_{20}^0$ is the partial specific volume of the protein under these conditions. The symbols without sub- or superscript refer to the experimental conditions of temperature and solvent viscosity.

This correction procedure should be contrasted with that for diffusion coefficients (p. 263) in which it will be recalled that there exists a direct proportionality between the coefficient and absolute temperature, in addition to the factors correcting for change in solvent viscosity.

(iii) *Nature of calculated molecular weight.* Since the sedimenting entity, upon the properties of which the sedimentation constant depends, is solvated, it might be expected that the molecular weight calculated from eqn. (11.10) would refer also to the solvated molecule. Such, however, is not the case, and in fact molecular weights, as normally calculated, refer very nearly to the unsolvated molecule. This curious result arises as a result of the nature of the partial specific volume ($\bar{v}$) term (see Chapter VI) used in the equation.

Let w_s gm. of solvent (2) combine with 1 gm. of pure macromolecule (1), when we may write for the partial specific volume of the compound ($\bar{v}_{12}$)

$$\bar{v}_{12} = \frac{w_s \bar{v}_2 + \bar{v}_1}{w_s + 1}, \qquad (11.12)$$

and for its molecular weight

$$M_{12} = (w_s + 1) M_1, \qquad (11.13)$$

again assuming no change in the density of either material when solvation occurs.

If now $\bar{v}_{12}$ were substituted for $\bar{v}$ in eqn. (11.10) in calculating molecular weight, then M_{12} would be obtained, since the sedimenting particle is solvated. However, the methods of obtaining partial specific volumes yield $\bar{v}_1$ for the unsolvated polymer, and on substituting in eqn. (11.10) we get a molecular weight, M_1^a, related to M_{12} by the expression (derived from eqn. (11.10))

$$M_1^a(1 - \bar{v}_1 \rho) = M_{12}(1 - \bar{v}_{12} \rho) \quad (= F_s s). \qquad (11.14)$$

Introducing (11.12) and (11.13) in (11.14) we obtain

$$M_1 = M_1^a \frac{(1 - \bar{v}_1 \rho)}{(1 - \bar{v}_1 \rho) + w_s(1 - \bar{v}_2 \rho)}.$$

Introducing now the values of $\bar{v}_1$ and $\bar{v}_2$ (from eqn. (6.5)) in terms of weight fractions we have

$$M_1 = M_1^a \left(\frac{1 - W_1}{1 - W_1 - w_s W_1} \right) \qquad (11.15)$$

In this equation solvation enters only as the product of w_s and the weight fraction (W_1) of the solute, so that for small values of W_1 (e.g. several gm. protein/100 c.c. solution) considerable values of w_s (e.g. up to 1 gm. solvent/gm. macromolecule) give low values for the product $w_s W_1$, and hence values for the bracketed function differing only slightly from unity. The calculated molecular weights (M_1^a) are usually, therefore, not appreciably different from the unsolvated values, unlike the results usually obtained from investigations of colloidal molecules in solution. We ought to point out, however, that the theory developed above is a simplified one for binary systems, and that a more general treatment has been worked out by Kraemer (1940 b), by which it is possible to take account of the absorption of salts or other components of a mixed solvent. Such absorption, if present, may cause effects of a much more serious character than mere solvation, but the theory has, as yet, not been applied.

It should also be noted that in the case of polydisperse substances, the average molecular weight obtained does not correspond exactly to one of the simple types of average (see p. 42), but usually lies between a weight and a number average (see Jullander, 1945). The precise average depends upon the particular distribution curve with respect to one of the molecular properties, e.g. molecular weight, sedimentation constant, or diffusion coefficient.

(iv) *Results.* The main part of sedimentation velocity and diffusion measurements has been applied to a study of the proteins, for the good reason that, among such substances, monodisperse systems of highly reproducible character are readily prepared. Other polymeric materials are in the main polydisperse with widely varying properties.

In Table 11.1, due to Svedberg and Pedersen (1940, p. 406), a large amount of the physico-chemical data concerning well-known proteins has been collected. In addition to information from sedimentation velocity and diffusion measurements and relevant data, the table also contains iso-electric points and values of the slope of the pH v. electrophoretic mobility curve near the iso-electric point (see Chapter XII).

Values of M_s, calculated from the sedimentation velocity and diffusion data of the preceding columns, should be compared with M_e, from sedimentation equilibrium, and $M_{calc.}$, obtained according to Svedberg's simple multiple rule (see p. 736). The consistent agreement between the M_s and M_e values, determined as they are by methods which are fundamentally quite independent, is a valuable confirmation of the soundness of the methods.

A detailed discussion of these results will be postponed until Chapter XXVI, but before leaving the subject, some general conclusions regarding the shape of protein molecules should be stated to illustrate the methods of determining molecular shape.

(b) *The use of frictional ratios to determine molecular shape*

The great majority of the frictional ratio values lie between 1·0 and 1·5, and it is of great interest to know whether such values are to be attributed to hydration or to molecular asymmetry. Our knowledge of the hydration of proteins (see p. 148) is not in an advanced state, but undoubtedly water is immobilized on protein molecules, and the figure of 0·3 gm. water per gram of dry protein would seem a reasonable one (see p. 263). From Fig. 10.11 the higher figure of 0·40 gm./gm. of protein gives by itself the frictional ratio value 1·15, which is therefore probably an upper estimate of the possible contribution of hydration. It is unlikely, for this reason, that frictional ratios greater than 1·15 can be explained by hydration; molecules possessing such values must undoubtedly deviate from spherical shape. For ratios smaller than 1·15, although hydration may be solely responsible, deviations from spherical symmetry cannot be excluded until the exact contribution from hydration is known. It should, however, be realized that, even after correcting for the effect of solvation on the frictional ratio, the calculated axial ratio refers to the solvated entity as it occurs in solution, unlike molecular weight values derived from sedimentation velocity and diffusion.

In Table 11.2 is considered a selection of proteins. We include, in addition to the frictional ratios, values of the axial ratios worked out for prolate and oblate ellipsoids of rotation, assuming 30 per cent. hydration. For comparison the corresponding values, worked out assuming zero hydration, are also included.

TABLE 11.2

Protein	f/f_0 (a)	Axial Ratio, a/b			
		30% Hydration (b)		Zero Hydration (b)	
		Prolate	Oblate	Prolate	Oblate
Pepsin	1·08	..	..	2·8	0·37
Lactoglobulin	1·26	3·2	0·30	5·1	0·18
Egg albumin	1·16	1·9	0·52	3·8	0·26
Haemoglobin (man)	1·16	1·9	0·52	3·8	0·26
Serum albumin (horse)	1·27	3·4	0·30	5·3	0·18
Serum globulin (horse)	1·44	5·7	0·16	8·3	0·10
Excelsin	1·13	1·5	0·67	3·4	0·29
Amandin	1·28	3·5	0·27	5·5	0·17
Thyroglobulin (pig)	1·43	5·7	0·16	8·1	0·11

(a) From Svedberg and Pedersen (1940), p. 406.
(b) Calculated from the curves of Fig. 10.11.

On the basis of these considerations it is clear that relatively few proteins can be truly spherical, but that many are only moderately asymmetrical (e.g. $a/b < 5$). A very few proteins also, with frictional ratios of the order of two, are considerably elongated ($a/b \approx 15$–20). Amongst other polymers, however, much greater degrees of asymmetry are encountered, as, for instance, in the case of cellulose nitrate which may have axial ratios as great or greater than 100 (Campbell and Johnson, 1944). Other cellulose derivatives, rubber, and many synthetic polymers undoubtedly have asymmetric molecules, for which we may expect much valuable information from the application of sedimentation and diffusion methods within the next few years. (See Chapter XXVII for detailed treatment.)

REFERENCES

BEAMS, 1937, *J. Applied Phys.* **8**, 795, and other papers.
—— LINKE, and SOMMER, 1938, *Rev. sci. Instrum.* **9**, 248.
BUTLER, 1934, *The Fundamentals of Chemical Thermodynamics*, Macmillan.
CAMPBELL and JOHNSON, 1944, *Trans. Faraday Soc.* **40**, 221.
COHN and EDSALL, 1943, *Proteins, Amino Acids and Peptides*, Reinhold.
GRALEN, 1944, Dissertation, *Sedimentation and Diffusion Measurements on Cellulose and Cellulose Derivatives*, Uppsala.
HENRIOT and HUGUENARD, 1927, *J. Phys. Radium,* **8**, 433.
JOHNSON, 1946 a, *Trans. Faraday Soc.* **42**, 28.
—— 1946 b, unpublished experiments.
JULLANDER, 1945, *Ark. Kemi Min. Geol.*, No. 8, **21**A.
KRAEMER, 1940 a, in *The Ultracentrifuge*, by Svedberg and Pedersen, p. 57.
—— 1940 b, *J. Franklin Inst.* **229**, 531.
LAUFFER, 1944, *J. Amer. chem. Soc.* **66**, 1188.
MCBAIN, 1939, *Chem. Rev.* **24**, 289.
MOSIMANN, 1943, *Helv. chim. Acta*, **26**, 61.
PICKELS and BAUER, 1937, *J. exp. Med.* **65**, 565.
SVEDBERG and PEDERSEN, 1940, *The Ultracentrifuge*, Oxford.
TISELIUS and GROSS, 1934, *Kolloidzschr.* **66**, 12.

XII
ELECTROPHORESIS AND ALLIED PHENOMENA
Electrophoresis
Introductory

In general it appears that when two phases come into contact, there occurs, by some means, a flow of electricity and a resultant separation of charges leading to the establishment of a potential difference across the interface between the phases. We shall be concerned here with those cases of interfacial potential between liquids and solids or between liquids and particles dissolved or suspended in them. Such potentials in colloidal systems are not fundamentally different from those set up in the neighbourhood of simple ions, and it is often merely a matter of convenience that electrokinetic data in colloidal systems are interpreted in terms of interfacial potentials rather than electrical charge, which is more commonly employed in describing the properties and behaviour of simple ions. It should, however, be noted that in many colloidal systems the particle charge has lost the uniqueness of value attaching to it for simple ions.

Interfacial or electrokinetic potentials underlie the occurrence of all the electrokinetic phenomena—electrophoresis, electro-osmosis, streaming and sedimentation potentials, which have been defined and briefly discussed in Chapter I. Of these four different manifestations of the existence of electrokinetic potentials, the first three have been the object of sufficient quantitative investigations to permit a check of the basic theory. It is found that where complications have been removed, electrokinetic potentials, calculated by the different methods, have shown a good measure of agreement (see p. 299). So much so, that where good agreement is not obtained, possible sources of complication are immediately examined. We shall indicate methods for comparing electrophoretic and electro-osmotic measurements, and at the end of the chapter shall give a brief description of electro-osmosis and streaming potentials. In the main, however, this chapter is concerned with electrophoresis.

Before undertaking the theoretical treatment of electrokinetic phenomena, it is necessary to recall the origin of the charge residing on phases in contact (see Chapter III) and to consider electrokinetic potentials in detail, aspects which are of common importance in all electrokinetic phenomena.

Electrochemical potentials

(a) *Origin of the charges at interfaces*

It has been pointed out in earlier chapters that colloidal particles in aqueous media are usually charged, the separation of charge occurring spontaneously and thus arising from a diminution in the free energy of the system. A similar separation of charges occurs at all interfaces, and the various methods by which it might occur have been discussed in Chapter III; of these the most important are the dissociation of ionogenic groups and the preferential adsorption of one ion from the solution.

It will be noted that, whatever the theory of the origin of the charge, the whole system is electrically neutral, and that therefore, in addition to the possible motion of colloidal particles in an electrical field, the motion of the gegenions or ionic atmosphere of the particle must be considered, and will tend to occur in the opposite direction. It will be shown later how the effects of this ionic atmosphere are included in the quantitative theoretical treatment of electrophoresis.

(b) *Electrokinetic (or ζ) potentials and the electrical double layer*

It is necessary now to consider in detail the distribution of charge on and near a colloidal particle, and also the electrical potential to which such charges give rise, which is operative in electrokinetic phenomena (the electrokinetic or ζ potential).

For illustration, let us consider a large, spherical, negatively charged colloidal particle in an aqueous conducting liquid. Let the charge be symmetrically distributed (Fig. 12.1). As a result of the influence of the charge on the water dipoles, such a particle will be surrounded by a layer of strongly bound water molecules, approximately one molecule thick, which is carried about by the particle and must therefore be considered as part of the same kinetic unit. Amongst the strongly bound water molecules there will also occur a certain number of positively charged ions, which are very strongly attracted by the colloidal particle (see Chapter III), and which can therefore rarely escape into the surrounding liquid. They must also be included in the kinetic unit, and therefore effectively reduce the net charge on the particle. Thus the surface of shear of the particle is on the exterior of the solvated layer, the net or effective charge is the total charge within the surface of shear, and the potential effective in electrokinetic phenomena, the ζ potential, is the work required to bring unit charge from infinity to the surface of shear. Such, then, is the nature of the colloidal particle in solution, and it should be remembered that, for electrokinetic purposes, the colloidal particle plus adsorbed constituents is the effective entity, and it is not possible from electrokinetic considerations to subdivide the effective kinetic unit into its component parts.

In the atmosphere surrounding the negative colloidal particle there will be a preponderance of positive charges, in quantity just sufficient

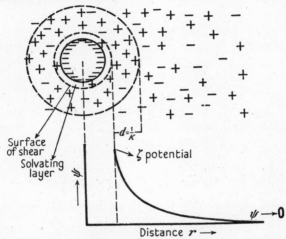

Fig. 12.1. The electrokinetic potential (ζ) and the electrical double layer.

on the average to balance out the net negative particle charge. These surrounding charges, being relatively distant from the particle, are not immobilized (like those in the solvating layer), but owing to thermal energy are in a constant state of motion into and from the main body of the liquid. Thus the behaviour of the ionic atmosphere is statistically determined, and, occupying as it does a finite volume surrounding the particle, it is known as the 'diffuse double layer'. However, for certain purposes it is useful to think of a shell of charges equivalent in action to a diffuse layer, and the distance between this shell and the surface of shear is known as the thickness of the double layer (d). It is to be expected that this distance should be related to the ionic concentration in the solution, being smaller for large concentrations of ions and approaching infinity as infinite dilution is approached. As mentioned in Chapter V, it can be shown (see also Abramson *et alia*, 1942) that with colloidal systems of low charge, in the presence of salts, the thickness of the double layer is to be identified with the reciprocal of the Debye-Hückel function κ:

$$\kappa = \frac{1}{d} = \sqrt{\frac{8\pi e^2 N^2 I}{1000 DRT}}, \qquad (12.1)$$

where e = electronic charge, N = Avogadro number, I = ionic strength, and D = dielectric constant. It is of interest to note that in addition to universal constants, d is dependent only upon the ionic strength at

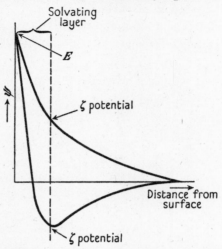

Fig. 12.2. Electrokinetic (ζ) and thermodynamic (E) potentials.

a given temperature (ζ potential and charge being absent). Thus for water at 25° C.

$$d = \frac{3 \cdot 05 \times 10^{-8}}{\sqrt{I}} \text{ cm.} \qquad (12.2)$$

Thus at $I = 0 \cdot 01$, $d = 30 \cdot 5$ A, and at $I = 1$, $d = 3 \cdot 05$ A. This calculation is approximate in that ions have been assumed point charges, an assumption which causes greater error at lower values of d.

(c) *Electrokinetic (ζ) and thermodynamic (Nernst) potentials*

As we have seen, in view of the stability of the solvating layer on a colloidal particle, the electrokinetic potential refers to the potential at the outside of the solvating layer and hence may be profoundly modified by the nature of the solvent or any other strongly adsorbed components. Thus in the case of the strong adsorption of substances of very different electrical character, it is possible to have curves of potential (ψ) against distance from the 'bare' surface of the colloidal particle (not surface of shear) as different as those of Fig. 12.2. It is clear that the ζ potentials are of opposite sign in the two cases shown, although the potential (E) between a point at infinity and the particle surface is unchanged. The latter potential is commonly known as the thermodynamic or Nernst potential, and is operative in concentration and galvanic cells. Thus the potential between a metal electrode and the solution in which it is immersed, a thermodynamic potential, is given by E of Fig. 12.2. The entire independence of the two types of potential is quite general and must be emphasized.

A good practical example of the distinction between thermodynamic and electrokinetic potentials, described by Freundlich (1926) and later by Abramson et alia (1942), makes use of a glass electrode in an aqueous medium which, it will be recalled, acts effectively as a hydrogen electrode. Thus the electrode potential (E) is given by

$$E = E_0 + \frac{RT}{F} \ln a_H, \qquad (12.3)$$

where E_0 is a constant equal to the potential when the activity of the hydrogen ions (a_H) equals unity, F being the Faraday.

Thus at constant external pH, the thermodynamic potential is constant. If the external pH is chosen to be between the iso-electric points of two proteins, say serum albumin and serum globulin (iso-electric points at pH 4·9 and 5·6 respectively), at pH 5·2, then at this pH the two proteins will possess opposite electrical charges (negative and positive respectively). On separate addition of the two proteins to the solution containing the glass electrode, strong adsorption occurs, giving potential-distance curves as different as those of Fig. 12.2, and opposite electrokinetic behaviour will be obtained. This arises from the adsorption of the two oppositely charged proteins. In spite, however, of such electrokinetic differences, it is quite possible for the pH, and therefore the thermodynamic potential, to remain unaltered by the protein (especially by the use of buffered solutions).

The aspects of colloidal charge and electrokinetic potential, with which we have been concerned over the last few pages, are of importance in all electrokinetic phenomena and one test of the correctness of this view has been provided by comparison of the ζ potential for a given system as obtained by measurements on electrophoresis, electro-osmosis, and streaming potentials (sedimentation potentials have not yet been utilized quantitatively). Thus Bull (1935) in showing, for electro-dialysed gelatin and recrystallized egg albumin adsorbed upon pyrex, that the ζ potentials from the three types of measurement were, within the limits of experimental error, identical, provided strong evidence for the basic electrokinetic theory. Other methods of testing this theory are indicated elsewhere in this chapter.

It is now our intention to consider the relation between net charge and ζ potential on the one hand, with electrokinetically determined quantities on the other. For this purpose, we shall have electrophoresis directly in mind, since, of electrokinetic phenomena, this has been most used in the past owing to the fact that it is much more readily applied to many colloidal systems of interest; further, it has been developed and utilized to a very large degree in the last decade. However, a considerable portion of the theoretical treatment of electrophoresis applies also to electro-osmosis and to a smaller extent to streaming potentials.

Electrophoresis—theoretical

We have seen that in an ionizing liquid, a colloidal particle is surrounded by an ionic atmosphere composed chiefly of oppositely charged single ions, and that in considering the motion of the large particle it is necessary also to consider the effect of the atmosphere. The complete solution of the problem is considerably too complex and lengthy to include here, but recent treatments may be seen in Abramson *et alia* (1942) or in the chapter by Mueller in Cohn and Edsall (1943). However, in certain extreme cases, the ionic atmosphere may be simply treated, and results of importance obtained. These cases (*a*) and (*b*) will be considered as well as an indication of the nature of the more general solution (*c*).

(a) Small non-conducting particles

If the thickness of the diffuse double layer is much larger than the radius of the particle (i.e. d/a large), or in the case of media containing few or no ions (insulating media), only the charged non-conducting particle need be considered in deriving an expression for its electrophoretic mobility (u) defined as the velocity of migration under unit potential gradient.

If the particle possesses a net charge Q, then if X is the field strength (or potential gradient) the force producing migration is QX. The particle will therefore attain a terminal velocity such that the frictional resistance of the medium in which it moves just balances the electrical force. If the particle is spherical and Stokes's law can be applied, then

$$QX = 6\pi\eta av, \qquad (12.4)$$

where η = viscosity of medium, a = radius of particle, and v = velocity of migration.

Further, introducing the mobility, u, and rearranging, we obtain

$$u = \frac{v}{X} = \frac{Q}{6\pi\eta a}. \qquad (12.5)$$

But for a charged sphere, the ζ potential may be related to the net charge Q by

$$\zeta = \frac{Q}{Da}, \qquad (12.6)$$

where D is the dielectric constant of the dispersion medium. Introducing (12.6) into equation (12.5) gives

$$u = \frac{D\zeta}{6\pi\eta}, \qquad (12.7)$$

known as Hückel's equation.

For non-spherical particles Stokes's law cannot be applied, nor can the simple relation (12.6) between ζ potential and net charge. It can, however, be shown that in general

$$u = \frac{\zeta D}{\eta} C, \qquad (12.8)$$

where C, a constant, depends upon the shape and orientation of the particles. In the special case of rod-shaped particles, for migration in directions parallel and perpendicular to the orientation direction,

$$u = \frac{\zeta D}{4\pi\eta} \quad \text{and} \quad u = \frac{\zeta D}{8\pi\eta} \qquad (12.9)$$

respectively.

In summarizing this case the dependence of mobility upon shape, size, and orientation of the migrating particles should be stressed.

(b) *Large non-conducting particles*

In the second extreme case we consider the electrical migration of non-conducting particles for which the thickness of the double layer is much smaller than the radius of curvature at any point on the surface (i.e. $d \ll a$).

Thus it is possible to consider the particle with its double layer as a parallel plate condenser whose plates are at a distance apart given by the thickness (d) of the double layer. Let the plates have a charge σ per unit area.

When a steady state is reached in which the particle is moving at a constant speed through the liquid, the frictional forces must be equal to the electrical forces. (Note here that the problem in which we have a fixed solid and moving liquid is very similar.) For unit area of condenser surface the electrical force is $X\sigma$.

From the definition of viscosity (force/unit area $= \eta \times$ velocity gradient), the frictional forces may be written $\eta \times dv/dl$, where l is a distance measured perpendicular to the plates (see Chapter XIII).

If, now, the velocity of the liquid within the double layer varies linearly with l and is zero at $l = 0$, then dv/dl may be replaced by v/d, where v is the velocity of migration. Therefore

$$X\sigma = \eta \frac{v}{d}. \qquad (12.10)$$

Making use of the electrostatic expression

$$\sigma = \frac{D\zeta}{4\pi d} \qquad (12.11)$$

and substituting in (12.10) we have, on rearranging,

$$u = \frac{v}{X} = \frac{D\zeta}{4\pi\eta}. \qquad (12.12)$$

This is generally referred to as Smoluchowski's equation. For water at 25° C., this expression reduced to

$$\zeta = 12\cdot 85u, \qquad (12.13)$$

from which it is apparent that mobility determinations provide a ready means for the determination of ζ potentials. It should be noted that application of a slightly modified method gives an expression for electro-osmotic mobility identical with that above (eqn. (12.12)) for electrophoretic mobility. A method of checking this result experimentally will be indicated later (p. 312).

It is clear that in the above simplified derivation of eqn. (12.12), Smoluchowski's equation, certain definite assumptions have been made, some of which are also involved in Smoluchowski's treatment, the first derivation of the equation. As stated by Henry (1931), Smoluchowski's assumptions are:

(1) that the usual hydrodynamical equations for the motion of a viscous fluid may be assumed to hold both in the bulk of the liquid and within the double layer;
(2) that the motion is 'stream-line motion' and slow enough for the 'inertia terms' in the hydrodynamic equation to be neglected;
(3) that the applied field may be taken as simply superimposed on the field due to the electrical double layer; and
(4) that the thickness of the double layer (i.e. the distance normal to the interface over which the potential differs appreciably from that in the bulk of the liquid) is small compared with the radius of curvature at any point on the surface.

Of these assumptions, Henry considers that the first is probably valid, though the correct values of the dielectric constant and viscosity required in the calculations may not be those of the bulk liquids, and the second also except in exceptional circumstances. The third he considers to be an approximation, which has proved very difficult to improve upon, and the last reasonable for large particles of not too irregular shape.

On the basis of the same assumptions Henry (1931) re-investigated the problem theoretically, and for large insulating spheres and cylinders (axial and transverse) confirmed the Smoluchowski equation, considering further that the equation holds also for any shape of insulating particle of microscopically visible size.

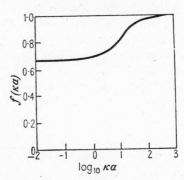

Fig. 12.3. Graph of $f(\kappa a)$ against $\log \kappa a$. (*From* Henry (1931).)

His treatment, which was rigid and the most comprehensive to date, covered, as well as very large particles, the cases of the small and intermediate sized spheres. On this basis, he was able to predict the accuracy and range of validity of the Smoluchowski equation.

To summarize, we may say that sound theoretical treatments lead to the important result that the mobility of a large insulating particle is independent of its size, shape, and orientation, and is simply given by the Smoluchowski equation.

(c) *Spherical particles of intermediate size*

Henry's complete theoretical treatment of electrophoretic mobility covered all sizes of spherical particles on the basis of the first three of the Smoluchowski assumptions, and he made a further assumption that the electrical distribution in the ionic atmosphere is of the Debye-Hückel exponential type (see Chapter V). He thus derived the general equation

$$u = \frac{D\zeta}{4\pi\eta} f(\kappa a), \qquad (12.14)$$

where $f(\kappa a)$ (often called the Henry factor) is a function of the product of the particle radius a and the Debye-Hückel function κ, which he evaluated for different values of κa. Fig. 12.3, due to Henry, gives a plot of $f(\kappa a)$ against $\log_{10} \kappa a$, from which it is apparent that at high values of κa (100–1200), $f(\kappa a) \approx 1$, so that Smoluchowski's equation is approached, whilst at small values of κa (≈ 1), $f(\kappa a) \approx \frac{2}{3}$, when

$$u \approx \frac{D\zeta}{6\pi\eta}, \qquad (12.7)$$

the Hückel equation, now holds. More precisely, for $\kappa a \geqslant 300$, Smoluchowski's equation holds to within 1 per cent., and for $\kappa a \leqslant 0.5$, Hückel's equation is similarly accurate. Table 12.1, due to Henry, in addition to showing the dependence of the thickness of the double layer ($d = 1/\kappa$) on the concentration of salts of differing type, includes the minimum (s) and maximum (H) limits of particle diameter for the validity of the Smoluchowski and Hückel equations respectively to an accuracy of 1 per cent.

TABLE 12.1

(Henry, 1931.)

Electrolyte	Conc. (gm. moles/litre)	κ at 20° A^{-1}	$1/\kappa = d$ A at 20° C.	Diameters (A)	
				s	H
Pure water		1.0×10^{-4}	10,000	600×10^4	1.0×10^4
Uni-univalent	10^{-5}	1.0×10^{-3}	1,000	60×10^4	1.0×10^3
	10^{-3}	1.0×10^{-2}	100	6×10^4	1.0×10^2
	10^{-1}	1.0×10^{-1}	10	6×10^3	1.0×10
Uni-bivalent	10^{-5}	1.8×10^{-3}	560	34×10^4	560
	10^{-1}	1.8×10^{-1}	5.6	34×10^2	5.6
Bi-bivalent	10^{-5}	2.1×10^{-3}	480	29×10^4	480
	10^{-1}	2.1×10^{-1}	4.8	29×10^2	4.8

For microscopically (not ultramicroscopically) visible particles, the Smoluchowski equation is taken to be accurately obeyed. In general, however, particles in *stable* colloidal solutions have diameters smaller than the minimum required for Smoluchowski's equation so that their mobilities vary with size, shape, and orientation. The latter statement covers not only the cases of macromolecules in solution, e.g. proteins, carbohydrates, and synthetic polymers, but also most of the inorganic colloids and colloidal electrolytes.

(d) *Calculation of charge for spherical particles*

For spherical particles in an insulating medium, or very small spheres generally (case a), the net charge (Q) can be obtained directly from mobility measurements by means of eqn. (12.5) providing we know the radius of the particle and the viscosity of the suspending medium. Considering now those particles (case c) for which the Hückel equation (eqn. (12.7)) does not apply, it is necessary first to express the ζ potential in terms of the net charge, which is then substituted in eqn. (12.14).

The Boltzmann equation (see Chapter V) is used to provide the density of charge in a given element of volume at potential ψ and by substituting this into the Poisson equation, integrating, and making use of the appropriate boundary conditions, we obtain (see Chapter V)

$$\psi = \frac{Q}{Dr}\frac{e^{-\kappa(r-a)}}{1+\kappa a}, \qquad (12.15)$$

where r is the distance from the centre of the particle of radius a. For $r = a$ and $\psi = \zeta$,

$$\zeta = \frac{Q}{Da}\frac{1}{1+\kappa a}. \qquad (12.16)$$

It should be noted that this expression assumes that ions other than the central one are point charges and that all the $ze\psi/kT$ terms are $\ll 1$ (see Chapter V). Substituting into eqn. (12.14), we have

$$u = \frac{Q}{4\pi\eta a}\frac{f(\kappa a)}{1+\kappa a}. \qquad (12.17)$$

Thus knowing the particle radius, and evaluating κ and hence $f(\kappa a)$ from the ionic concentrations of the system, which also determine the viscosity η of the suspending medium, the net charge Q may be obtained.

The presence of gegenions of finite radius introduces the fact that the nearest distance of approach of the centres of such ions to the centre of the central particle is now $(a+r_i)$ (instead of a for point charges), where r_i is the average radius of the gegenions. Arising from this, it can be shown (see Abramson *et alia*, 1942) that

$$\zeta = \frac{Q}{Da}\left(\frac{1+\kappa r_i}{1+\kappa a+\kappa r_i}\right), \qquad (12.18)$$

which being substituted in eqn. (12.14) gives

$$u = \frac{Q(1+\kappa r_i)f(\kappa a)}{4\pi\eta a(1+\kappa a+\kappa r_i)}. \qquad (12.19)$$

In the use of this equation for calculating net charge only one more constant, the average radius of the gegenions (r_i), is required than in the simpler form of eqn. (12.17). The difference between the two equations for a system in which $a = 25$ A (e.g. haemoglobin molecule) and $r_i = 2\cdot 5$ A has been calculated by Gorin, whose results are given in Table 12.2 below.

TABLE 12.2

(Gorin, in Abramson *et alia* (1942), p. 123.)

I	κa	κr_i	$\dfrac{1+\kappa r_i}{1+\kappa a+\kappa r_i}$ (I)	$\dfrac{1}{1+\kappa r}$ (II)	$\left(\dfrac{\mathrm{I}}{\mathrm{II}}\right)$
0·001	0·259	0·0259	0·7984	0·7942	1·005
0·005	0·588	0·0588	0·6830	0·6297	1·021
0·02	1·159	0·1159	0·4905	0·4631	1·059
0·1	2·592	0·2592	0·3269	0·2784	1·174
0·2	3·871	0·3871	0·2608	0·2053	1·270

The importance of introducing the finite size of the ions in the atmosphere at the higher salt concentrations is well illustrated by these data.

Calculation of charge density for large particles. For all large particles of a given substance immersed in the same medium, it is likely that the charge per unit surface σ rather than the total charge is a constant. In such cases, therefore, the charge per unit surface is of greater interest than the total charge, Q, and its calculation is of considerable importance. The calculation may be made simply using eqn. (12.11), but more accurately by making use of the expression

$$\sigma = \sqrt{\left(\frac{NDkT}{2000\pi}\right)} \sqrt{\{\sum c_c(e^{-z_c e \zeta/kT}-1) + \sum c_a(e^{+z_a e \zeta/kT}-1)\}}, \quad (12.20)$$

where c_c and c_a are the concentrations in moles per litre of cations and anions of valency z_c and z_a respectively, and σ has the same sign as ζ.

This equation is derived by substituting for the volume charge density near an infinite charged plane surface (from the Boltzmann equation, p. 101) in the Poisson equation (p. 101) and integrating, making use of the boundary conditions,

$$\frac{d\psi}{dx} = 0, \quad \psi = 0 \text{ at } x = \infty \text{ (Abramson, 1934)}.$$

Eqn. (12.20) simplifies considerably for ions of a common valency (z) becoming

$$\sigma = 2\sqrt{\left(\frac{NDkT}{2000\pi}\right)} \sqrt{c} \sinh\frac{ze\zeta}{2kT}, \quad (12.21)$$

which for $\sinh ze\zeta/2kT = ze\zeta/2kT$ (i.e. small values of $ze\zeta/2kT$) reduces to eqn. (12.11). For water at 25° C., eqn. (12.21) reduces to

$$\sigma = 3\cdot53 \times 10^4 \sqrt{c} \sinh\frac{ze\zeta}{2kT}, \quad (12.22)$$

giving σ in electrostatic charges per sq. cm. for ζ in millivolts. For a fuller discussion of the calculation of the density of surface charge, the reader is referred to either of Abramson's books.

(e) *Asymmetric particles of intermediate size*

We have seen that Smoluchowski's equation is valid for *large* particles of any shape and orientation, but where it is not valid we have as yet considered spherical particles only. In this section we shall consider briefly the treatment of non-spherical particles of the intermediate size range in which mobility is dependent upon size, shape, and orientation. In view of the fact that many colloidal particles (especially the so-called globular proteins and other macromolecules in solution) are not accurately spherical and are contained in this size range, the case is important.

The treatment, which has been recently developed by Gorin, has not yet been extensively applied. To make possible the mathematical formulation and solution of the problem, it was necessary, as in other treatments of asymmetrical molecules, to postulate certain idealized molecular models, and of these, one, the long non-conducting cylindrical molecule of length, l, and diameter, $2a$, was chiefly considered. Only an outline of the methods applied can be included here.

It is usual to make use of a general electrophoretic equation

$$u = \frac{\zeta D}{\eta} C, \qquad (12.23)$$

in which C is a constant dependent upon particle size, shape, and orientation. (The equation also covers, of course, the particles for which Smoluchowski's equation applies, in which case $C = 1/4\pi$.) C has been evaluated for randomly orientated cylinders in terms of its value for cylinders orientated parallel and perpendicular to the field, assuming that for every one cylinder orientated axially, two would be in the transverse position. On this basis, C has been tabulated for different values of κa. Thus if C is obtained for the appropriate value of κa, then ζ may readily be obtained from mobility measurements.

Further, it may be shown that the ζ potential of such a cylinder is related to the net charge Q by the equation

$$\zeta = \frac{2Q}{D(l+2a)} \left[\frac{K_0(\kappa a + \kappa r_i)}{(\kappa a + \kappa r_i) K_1(\kappa a + \kappa r_i)} + \ln\left(\frac{a + r_i}{a}\right) \right], \qquad (12.24)$$

where K_0 and K_1 are Bessel functions and the other symbols have their previous meanings. The values of K_0 and K_1 have been calculated and tabulated for different κa values.

If, therefore, the dimensions of a given rod-like molecule are known, or probable dimensions postulated, then knowledge of the ζ potential gained from eqn. (12.23) may be utilized to calculate the net charge. By comparison with values of the net charge determined independently, it is possible to assess the accuracy of the molecular dimensions used.

In practice, the calculations may be simplified by the use of Gorin's calculations of the ratio (R) of the mobilities of spheres and cylinders of the same molecular volume and charge, at different ionic strengths (Fig. 12.4). It should be noted that only at $I = 0$ is R equal to the ratio of the corresponding diffusion coefficients or sedimentation constants. At finite ionic strengths the retardation caused by the ionic atmosphere is greater, the greater the molecular asymmetry. An example of such calculations will be provided later. Having now considered the relation between observable mobilities and the fundamental properties of the colloidal particle (ζ potential, charge, size, shape, etc.) we proceed to describe the experimental methods used in mobility determinations before discussing some of the actual results obtained.

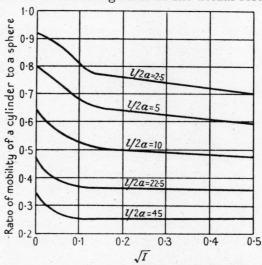

FIG. 12.4. The ratio of mobility of a cylinder to a sphere of the same molecular volume at various ionic strengths and length/breadth ratios.

Experimental study of electrophoresis

Two methods have, in the main, been utilized in the experimental study of electrophoresis:

(A) The Microscopic (or Ultramicroscopic) Method.

(B) The Moving Boundary Method.

Both methods are widely used, and although to some extent their results may be compared, they are to be regarded as complementary, both having problems of special suitability.

(A) *The microscopic method*

This method is especially suited to the study of the electrophoretic movement of microscopically or ultramicroscopically visible particles which do not settle too rapidly under gravity (i.e. particle diameters from *ca.* 10^2 to 10^5 A). A large body of information on the electrophoretic behaviour of hydrophobic suspensions (e.g. metal sols, oil sols, etc.) as well as of biological materials such as bacteria, fungi, living cells, etc., has been obtained. Further, since many larger colloidal particles of substances such as quartz, paraffin, or collodion, upon immersion in a solution of a hydrophilic colloid (e.g. proteins, soaps, etc.) become covered with a layer of this material which controls the electrophoretic behaviour of the particles, it is possible to make use of the microscopic method in obtaining information of some relevance to dissolved colloids which are themselves invisible in the ultramicroscope. However, the unqualified assumption that results from the microscopic investigation of film-covered particles are directly applicable to dissolved molecules of the film-forming material is to be deplored for many reasons, which will be discussed later.

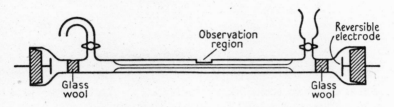

Fig. 12.5. A cylindrical micro-electrophoresis cell.

The micro-electrophoresis cell consists usually of a single horizontal tube, of either rectangular or cylindrical cross-section, provided at each end with a porous plug leading to a reversible electrode (Ag–AgCl, Cu–CuSO$_4$, etc.), and with a tap for filling and cleaning. For low salt concentrations (below *ca.* $M/100$) platinum electrodes may be used in place of the reversible type mentioned above, which facilitates the cleaning of the cell, since without removal of electrodes, chromic acid or other cleaning fluid may be poured into it. Fig. 12.5 shows a simple but convenient form of the cell, of cylindrical cross-section. Cells of rectangular cross-section, used considerably by Abramson and co-workers, require a large ratio of width to depth, a point which will be explained shortly. Mounted horizontally on the microscope stage, the particles suspended in the liquid contained in the cell are observed by microscopic or ultramicroscopic methods (Fig. 1.9). On applying the electrical field (*ca.* 10 volts/cm.) the velocity of the particles is measured by timing their movement across the plane of a calibrated graticule placed in the microscope eyepiece. The whole cell is contained in an enclosure at constant temperature ($\pm 0.2°$ C.), but in order to ensure that the effects of convection are eliminated, the direction of the current can be reversed and the same particle timed in both directions. The potential gradient is obtained by calculation using Ohm's law. Thus

$$X = \frac{IR_{sp}}{A}, \qquad (12.25)$$

where I is the measured current flowing through the cell, R_{sp} is the specific resistance of the liquid, and A is the cross-sectional area of the observation tube.

Abramson has stressed the importance of using this method of calculating the potential gradient rather than that from the total voltage applied and the length of the capillary. Constancy of the potential gradient is assured by careful control of the current.

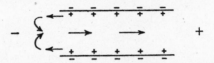

FIG. 12.6. Electro-osmosis in a closed tube.

(a) *Electro-osmosis in closed tubes*

In addition to the electrokinetic potential at the interface between colloidal particles and liquid, a similar potential usually exists between the walls of the tube and the liquid. Thus together with the electrophoretic movement of the particles, electro-osmotic movement of the liquid near the walls of the tube must occur. Further, since there is no resultant transport of liquid, the liquid moving electro-osmotically near the tube walls must return along its centre, as in Fig. 12.6. Thus, levels must exist at which the electro-osmotic and return flows just cancel, and only at such levels (the stationary levels) can electrophoretic movement, uncomplicated by electro-osmosis, be observed. The importance of the stationary levels is thus evident. At any other level, the observed movement of a particle is determined by the algebraic summation of the electrophoretic and electro-osmotic effects. Thus if the particles and the wall have the same sign of charge, electrophoresis near the walls of the tube will be opposed by the electro-osmotic flow, but near the centre of the tube the two effects will combine to give an increased particle velocity. Curve 1 of Fig. 12.7 shows the distribution of particle velocity for such a case. If, however, particles and walls have opposite charges, then the opposite effect occurs, viz. increased velocity at the walls and decreased velocity at the centre, as shown by curve 2 of Fig. 12.7.

Cylindrical cell. The stationary levels for a cylindrical cell, which is more amenable to hydrodynamic treatment than the rectangular form, may be readily calculated (Mattson, 1933). Thus the velocity distribution of a liquid in a cylindrical tube is of the type

$$v_l = c(r^2 - C), \qquad (12.26)$$

where v_l is the velocity of the liquid at a distance r from the axis of the tube, and c and C are constants. Since electro-osmosis in a closed tube of internal radius a causes no resultant transport of material, we have on integrating
$$\int_0^a v_l(2\pi r)\,dr = 0,$$
i.e.
$$\int_0^a v_l r\,dr = 0. \tag{12.27}$$

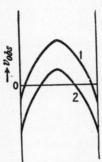

Fig. 12.7. The effect of electro-osmosis on observed velocities of electrophoresis in a cylindrical tube.
Curve 1. Particles and tube walls with same sign of charge.
Curve 2. Particles and tube walls with opposite sign of charge.

Substituting from (12.26) and integrating, we have
$$c\left(\frac{a^4}{4} - C\frac{a^2}{2}\right) = 0,$$
or
$$C = a^2/2. \tag{12.28}$$
Inserting in (12.26),
$$v_l = c\left(r^2 - \frac{a^2}{2}\right). \tag{12.29}$$

Thus $v_l = 0$, for $r = a/\sqrt{2} = 0{\cdot}707a$, and the stationary levels are thus in positions distant $0{\cdot}146$ of the cell's diameter from the floor and ceiling of the cell. The constant c is readily determined by using the fact that for $r = a$, $v_l =$ the electro-osmotic velocity, v_0.

Thus we have, substituting in (12.29),
$$v_0 = c\frac{a^2}{2} \quad \text{or} \quad c = v_0\frac{2}{a^2}, \tag{12.30}$$
which being introduced into (12.29) gives
$$v_l = v_0\left(\frac{2r^2}{a^2} - 1\right), \tag{12.31}$$
the usual expression for the liquid velocity in a cylindrical tube.

The observed velocity (v_{obs}) of a particle at any position in the cell is given by
$$v_{\text{obs}} = v + v_l$$
$$= v + v_0\left(\frac{2r^2}{a^2} - 1\right), \tag{12.32}$$

where v is the real electrophoretic velocity (with respect to the liquid). Applying this equation therefore, we have, at the centre of the tube, $r = 0$, and
$$v_{\text{obs}} = v - v_0; \tag{12.33}$$
at the wall, $r = a$, and
$$v_{\text{obs}} = v + v_0. \tag{12.34}$$

If, therefore, the electrophoretic mobility is known from observations at the stationary levels, it is possible also from further observations at the centre and walls of the tube, to calculate the electro-osmotic velocity v_0 by equations (12.33) and (12.34). We have seen (p. 302) that for similar surfaces and microscopic particles, electro-osmotic and electrophoretic mobilities should be identical in magnitude, though opposite in sign. This conclusion would therefore be confirmed by the observation that the observed electrophoretic velocities at the walls and axis of the tube are respectively zero and twice that at the stationary levels, but this is true only if the particle and wall surfaces are identical.

Rectangular cell. The theoretical treatment of the rectangular cell is complex unless the ratio of the width of the cell to the depth is great (> 20), owing to the complicating electro-osmotic effects of the side walls. We shall merely summarize here the results for the simplified type of rectangular cell. The velocity of the liquid at different depths, x, in such a cell of total depth, x_1, is given by
$$v_l = v_0\left[1 - 6\left\{\frac{x}{x_1} - \left(\frac{x}{x_1}\right)^2\right\}\right]. \tag{12.35}$$
Stationary levels ($v_l = 0$) are given at
$$x = 0.211 x_1 \quad \text{and} \quad 0.789 x_1. \tag{12.36}$$
At $x = 0$ and $x = x_1$ (i.e. at the walls)
$$v_{\text{obs}} = v_0 + v, \tag{12.37}$$
and at $x = x_1/2$,
$$v_{\text{obs}} = v - v_0/2. \tag{12.38}$$

As in the case of the cylindrical cell, observations at stationary levels are used to provide electrophoretic velocities, and for systems in which particles and walls have similar surfaces, further observations at the walls and axis of the tube may be used to check the electro-osmotic velocity. It should be noted that if the ratio width/depth is not great, then corrections must be employed to eqn. (12.36) to obtain the true stationary levels in a flat cell. However, a limit to the shallowness of the cell (as also to the diameter of the cylindrical cell) arises as a result of the large velocity gradients which occur for small depths. Thus for very small depths the rate of change of $v_{obs.}$ with depth becomes very large and focusing needs to be more critical than is usually possible.

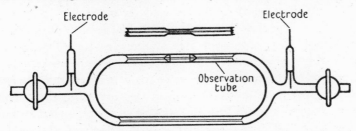

Fig. 12.8. All glass double-tube electrophoresis cell (modification by Bradbury and Jordan (1942)).

Before leaving the question of electro-osmotic flow in micro-electrophoresis cells, it is of interest to note the double-tube cell of Smith and Lisse (1936), in which electro-osmotic back flow does not occur in the observation tube, but in a second tube of larger radius (see modified form in Fig. 12.8). No movement of the liquid then occurs along the axis of the smaller observation tube. This condition is possible only when the lengths (l) and radii (a) of the tubes (1 and 2) obey the following relation

$$\frac{l_2}{l_1} = \left(\frac{a_2}{a_1}\right)^2 \left[\left(\frac{a_2}{a_1}\right)^2 - 2\right]. \qquad (12.39)$$

In this case the single stationary level (for which $v_{obs.} = v$) occurs in the centre of the observation tube, at which the velocity gradient in the liquid is zero, another advantage over single tube cells. In spite of its advantages, however, this cell has not been widely adopted.

(b) *Experimental points*

It is usual, in using the method, first to focus upon the ceiling of the cell (from above), and then to make particle velocity determinations at different levels by racking down the microscope by known amounts, thus focusing on these levels. Because of the finite depth of focus, convectional disturbances, and Brownian effects (for ultramicroscopic particles), a range of velocities is obtained by observing several particles in each direction (10 at least), and the average is taken.

Although simpler hydrodynamically, the cylindrical tube introduces optical complications. Thus not only does image distortion occur but

Henry (1938) has shown that an important correction must be applied in determining the stationary levels of the cylindrical cell, owing to the focusing action of the cell walls. This correction is considerably reduced if the centre of the tube is found from measurements of the apparent velocity at various depths, whereas if the stationary level is found by interpolation from the positions of the top and bottom of the capillary tube, the error may easily be 20 per cent. or more.

Before proceeding now to consider the chief results obtained by the micro-electrophoresis technique, we summarize its advantages. The method is inexpensive, simple to operate, and so rapid that neither the particle nor its environment can change perceptibly during observation, though if required, single particles may, by reversing the current, be kept under observation for long times. The size, shape, and orientation, as well as the mobility of a particle, may be observed, and particles of different mobilities viewed simultaneously, an important consideration in the examination of bacteria and living cells, etc. An important advantage over the moving boundary method, to be described shortly and for which low salt concentrations introduce troublesome complications, is in the large range of salt concentration which can be tolerated ($M/5$ down to distilled water). It is thus apparent that the method has a wide range of applicability, and in illustration, we now proceed to review those results obtained by it which are of special interest in colloid science.

(c) *Chief results of micro-electrophoresis technique*

One of the more important results emerging from the use of micro-electrophoresis has been the confirmation of the theoretical conclusions regarding the electrophoretic migration of microscopically visible particles. Thus it has been shown, chiefly by Abramson (1931) and co-workers, that glass, quartz, and clay particles of size between 3 and 15μ and of widely varying shape migrate with speeds independent of such variations. Similarly leucocytes, blood platelets, and red cells and their irregularly shaped aggregates migrate with the same speed. Even more convincing is the fact that microscopically visible particles of many substances (e.g. spherical mastic particles and oil globules of size $1-5\mu$, needles of asbestos and m-aminobenzoic acid from $12-150\mu$ long and very thin) when coated with the same protein all migrate, irrespective of orientation, with the same speed. On the basis of such data it is now generally accepted that the more qualitative result of Smoluchowski's treatment of large particles, viz. that electrophoretic mobility is independent of particle size, shape, and orientation, is proved. It remains now to consider the quantitative aspect. Controversy has centred round the factor 4π in the denominator of the expression (eqn. (12.12)) for electrophoretic mobility u; the alternative value 6π having been suggested by Hückel. The former factor is undoubtedly true for electro-osmotic mobility (u_0), and a quantitative test of Smoluchowski's equation is therefore obtained by the comparison of the two mobilities for a given type of surface (see p 312). Thus if the Smoluchowski equation for electrophoresis is correct, then $u = -u_0$. If, on the other hand, Hückel's equation (eqn. (12.7)) is correct, then $1 \cdot 5u = -u_0$. A considerable amount of evidence has now accumulated that $u = -u_0$. Thus Abramson (1931), from a large number of experiments using two different proteins to coat quartz, glass, and even benzyl alcohol, as well as different potential gradients and electrolytes, found as a mean $u_0/u = -1 \cdot 01$. Henry (1931) also concluded that $u = -u_0$ within 3 or 4 per cent. for wax particles of diameter 2 to 6μ. Further, the conclusion of Bull (1935) that the ζ potentials calculated for electrophoresis and electro-osmosis were equal (see p. 299) assumed the Smoluchowski equation, further verifying the validity of the 4π constant. Another check of the Smoluchowski equation, depending upon a comparison of electrophoretic and electro-osmotic results is described later (p. 339) after the electro-osmotic method has been considered further.

Summarizing, therefore, it is now generally accepted that the micro-electrophoresis technique has amply confirmed the validity of Smoluchowski's conclusions regarding the electrokinetic migration of microscopic charged particles.

Another field of application has been concerned with the stability of colloids, a subject which has already been considered (p. 114). It has long been known that lyophobic colloids are very susceptible to the influence of even traces of electrolyte, and it has been possible to gain considerable insight into the mechanism of salt precipitation. Thus from the electrophoretic mobilities of non-ionogenic microscopic particles at varying salt concentrations, ζ potentials may be calculated, and hence the surface charge density (σ) obtained from eqn. (12.20). Fig. 12.9 shows typical data of charge density for different concentrations of salts, which are very well represented by simple adsorption isotherm equations (see p. 22, eqn. (2.2)), from which the curves were drawn. The important conclusion emerges that such colloids are not usually discharged by salt addition, since increasing salt concentration does not cause at any stage a decrease in charge density, but, rather, a rapid increase at low salt concentrations, followed by an increasingly slow rise at higher salt concentration until finally a limiting charge density is reached. The instability of hydrophobic colloids, caused by

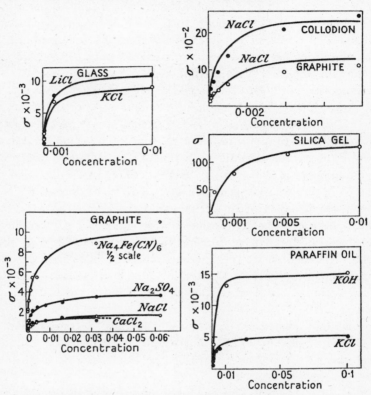

Fig. 12.9. Illustrations of the regularity of the shape of the charge density-concentration curve for a variety of surfaces. (Abramson *et alia* (1942).)

the presence of salts, is a result, not of lowered charge density, but of lowered ζ potential, a function not only of the charge density but also of the thickness of the electrical double layer. Thus, rearranging eqn. (12.11), we have

$$\zeta = \frac{4\pi\sigma}{D}d. \qquad (12.40)$$

It is clear, therefore, that where σ is nearly independent of salt concentration, further increase in salt concentration, causing a decrease in the thickness of the double layer, must lead to a lowering of ζ and to coagulation. This view is confirmed also by Bull and Gortner (1931) by means of streaming potential measurements at cellulose surfaces.

The more detailed treatment of colloidal stability is contained in Chapter V.

Similar investigations have been concerned with the mobility of blood cells, bacteria, etc., in an electrical field, and the effect of the addition of salts, surface active agents, and other materials on surface properties. Since such experiments may be carried out under conditions simulating the natural ones, their results are of considerable importance in attempts to explain the properties of natural membranes.

Finally, we ought to draw attention to the extensive investigation of electrophoretic mobilities of protein-coated particles in aqueous solutions. It will be shown later that in many cases mobilities so obtained are in good agreement with those for the same dissolved protein, and the micro-electrophoresis method is commonly used therefore to determine electrophoretic mobility–pH curves. Since the moving boundary method is also used to provide similar data, we shall defer consideration of the results and their comparison until the moving boundary method has been outlined. We should, however, add the warning that the identity of mobilities for adsorbed and dissolved protein cannot be regarded as holding generally, and its assumption without experimental proof is unjustified and dangerous.

(B) *The moving boundary method*

(a) *Development of the method*

Whilst the micro-electrophoresis technique is especially suited to the study of microscopically visible particles, the moving boundary method has as its main field of application the electrophoresis of *dissolved* materials, both colloidal and otherwise. Although requiring more expensive equipment, and a more specialized technique, the moving boundary method has tended in recent years to displace the micro method where both were possible, and its development in the last decade has been very rapid. In this section we can only outline the method, its advantages and disadvantages, and its application to certain systems. For fuller treatments the reader is referred to Abramson, Moyer, and Gorin (1942), to the excellent review accounts by Longsworth and MacInnes (1939) and Longsworth (1942), and, for a recent, very thorough account, to a thesis by Svensson (1946).

Originally used by Lodge (1886), for the study of the migration of simple inorganic ions in an electrical field, the method has since been considerably improved for this purpose, and forms an accurate method for the determination of transport numbers of simple electrolytes

(MacInnes and Longsworth, 1932). We are, however, concerned with the method as developed for use with colloidal systems from the original work of Picton and Linder (1897).

Although some progress was made in subsequent years (especially in the hands of Hardy and Svedberg), the modern form of the moving boundary apparatus for dissolved colloids may rightly be said to date from 1930, when Tiselius (1930), working in the Department of Physical Chemistry at Uppsala, laid the foundations for the accurate use of the method to determine colloidal mobilities. Continuing this work, Tiselius (1937) designed an ideally suitable apparatus which, with minor modifications, provides the method utilized in most electrophoretic investigations. It is true to say that Tiselius has played a part in the development of electrophoretic techniques similar to that of Svedberg in the development of the ultracentrifuge, and, for these contributions alone, the debt owed by colloid science to the laboratories at Uppsala is considerable.

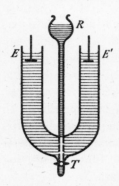

FIG. 12.10. Cylindrical electrophoresis cell. (Burton.)

(b) *The apparatus and its operation*

Fig. 12.10 illustrates the apparatus of Burton (e.g. 1938), typical of many others, in which the colloidal solution in the reservoir R is initially separated by tap T from the lighter supernatant fluid against which the boundaries are to be formed. By cautiously opening the tap, the colloidal solution entering the two limbs forms a boundary in each (between the supernatant and colloid solutions) whose movement under the electrical field applied by the platinum electrodes E and E' can be followed. An improved version of this apparatus shown in Fig. 12.11 has been used by W. C. M. Lewis and co-workers (for example, see Price and Lewis, 1933; Dickinson, 1941) in investigating the electrophoretic behaviour of dispersions and emulsions (e.g. of lecithin and hydrocarbons in aqueous media). In this apparatus contamination of the boundaries by electrode products was avoided by placing the electrodes X and Y at considerable distances from the main tube, the potential across which was measured by means of the small platinum electrodes L and M connected into the appropriate high resistance circuit shown. Adjustment of this potential to a constant value was performed by the potentiometer P. Another feature was the addition to the colloidal solution of 2 per cent. of sucrose, which was claimed not to affect electrophoretic mobilities, but to give the required density difference over the boundary for boundary stability. Movement of the boundary was observed visually. It would appear that the apparatus does provide

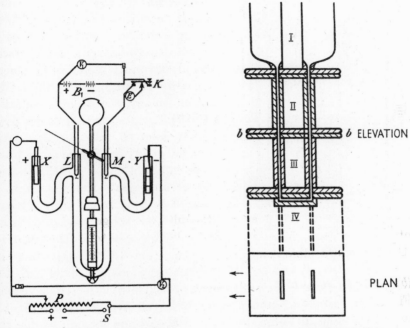

FIG. 12.11. Cylindrical electrophoresis cell. (Price and Lewis (1933).)

FIG. 12.12. The Tiselius electrophoresis cell (diagrammatic).

a simple inexpensive method of investigating the electrophoretic mobilities of lyophobic suspensions.

Tiselius retained the advantages of the U-tube apparatus in his cell, the basic form of which is illustrated in Fig. 12.12, and which is merely a U-tube of rectangular cross-section built in sections. Fig. 12.13 contains photographs of the separate cell sections and of the assembled cell in its supporting stand. Each section is built upon one or two groundglass plates, through which apertures, identical with the rectangular internal bore of the sections, have been cut. The plates are well greased before use so that when sections are assembled upon one another they can be moved sideways relative to one another. Thus in Fig. 12.12, showing the alined and communicating sections, section III (or section II) may be moved to the right or left out of alinement and communication with the rest of the tube, and may be replaced when required.

Other movements may be made similarly. Such movements are usually made, after fitting the alined cell into a stand of the type shown in Fig. 12.13 (b) by the use of the air pumps shown, or by a rack and pinion mechanism in the case of many American models (see facing p. 320)

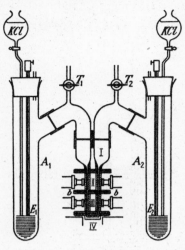

Fig. 12.14. Assembled electrophoresis cell and electrode vessels.

The colloidal solution is usually denser than the solvent, and for stable boundaries must therefore be placed beneath. Accordingly, in filling the assembled cell, the colloidal solution is first introduced until it covers completely joint bb, (Fig. 12.12). After removal of any entrapped air bubbles (especially in IV) section III is moved out of alinement, and section II is washed several times with solvent. Finally fresh solvent is added until nearly filling section I. The cell in its stand is then introduced into the carriage (Fig. 12.14) supporting the electrode vessels (A_1 and A_2), to which connexions are made with the two limbs of section I. More solvent is poured into the electrode vessels, until the side arms and tubes leading to the cell are full. The reversible silver-silver chloride electrodes, E_1 and E_2, are now introduced into the electrode vessels and saturated potassium chloride solution poured through long tubes dipping to the bottom of the electrode vessels until they are covered. The whole apparatus, on its carriage, is now transferred to a thermostat, working usually near 0° C., and provided with good quality observation windows. After connecting the two sides of the apparatus by means of taps T_1 and T_2 to ensure equal hydrostatic pressures in the plane bb, (Fig. 12.14), and allowing a considerable time ($\frac{1}{2}$–1 hour) to ensure complete temperature equilibrium within the cell, the section III, previously out of alinement, is now carefully pushed into complete alinement, when colloid solution and pure solvent are brought into contact in the plane bb, forming two boundaries there. Without previous hydrostatic equilibration, such boundaries may, on boundary formation, be forced (jerkily) away from this position and disturbed, but if the operations have been carefully done, the boundaries should be sharp and stable, as may be shown if, by slow injection of a little solvent into one electrode vessel, the boundaries are forced (one

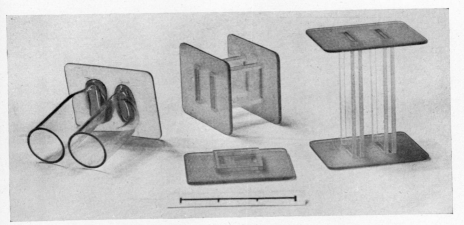

FIG. 12.13a. Sections of Tiselius Electrophoresis Cell (including double length centre section).

FIG. 12.13b. Electrophoresis Cell in Supporting Stand.
(*Scale in inches*)

ELECTROPHORESIS AND ALLIED PHENOMENA

up and the other down) from their initial positions of formation where the joint bb prevents their examination. Further movement of the boundaries in the same directions is then caused by application of an electrical field (in the appropriate direction), usually from a transformer-rectifier system fed by the mains, which by means of a potentiometer can supply 0–500 volts at currents of up to 30 m.amps.

The boundaries are observed usually by some form of schlieren apparatus, giving rise to peaked diagrams similar to those obtained in ultracentrifugal work (Chapter XI). Their movement under an electrical field, similar in many respects to that of ultracentrifuge boundaries under a centrifugal field, is similarly observed and will be further discussed. As will be realized, for the success of the moving boundary method it is essential that the boundary be stable, its movement uncomplicated, and representative only of the colloidal material. The cell described was designed with a view to satisfying these aims, and we shall now consider briefly the means by which this is accomplished.

(c) *Important devices in the moving boundary technique*

(i) Osmotic equality between the colloid solution and the supernatant liquid is achieved by prolonged dialysis (3 or 4 days) against one another. The effects of salt diffusion across each boundary and the consequent blurring involved are therefore minimized.

(ii) Since heat is generated in the U-tube owing to the passage of current, density differences are caused which introduce the possibility of boundary disturbances by convectional streaming. Such disturbances are minimized in the main by two devices:

(1) The rectangular section of the U-tube is such that the ratio of surface area to volume is much larger than in tubes of nearly square or circular cross-section. The rate of dissipation of heat is therefore very greatly increased over that occurring for similar current densities in circular tubes, and temperature rises in the liquids are thus reduced.

(2) Acting on a suggestion of Tiselius, electrophoresis is usually conducted at the temperature of maximum density of the solvent (near 4° C. for aqueous solutions), since at such temperatures the rate of change of liquid density with temperature, $d\rho/dT$, is zero or small. Thus the temperature changes in the cell which cannot be avoided give rise to minimum convectional effects.

Where advantage may be taken of (1) and (2), it has been found possible to make use of a total power of 6 watts (i.e. 20 m.amps. at

300 volts) in a Tiselius cell of normal size. In certain work, however, where it is not possible to work at low temperatures (e.g. solutions of certain rather insoluble proteins) the maximum power possible is lowered, but in an accurately controlled thermostat at 20° C. ($\pm 0.002°$ C.) in the authors' laboratory it has been possible to make use of a power of $1\frac{1}{2}$ watts without observing boundary disturbances.

(iii) Of importance also for undisturbed boundary migration are the following considerations:

(1) The use of reversible electrodes, since any gas evolution would cause pronounced boundary disturbances. It is to be noted, however, even in the case of reversible electrodes, that volume changes occur there as a result of the electrochemical reactions proceeding during the passage of the current, for which a small correction term is applied in the most accurate work. Further, in order to make possible the prolonged examination of a slowly moving boundary, the volume of buffer between the electrodes and observation cell is made deliberately large, so that the products of the electrode reactions can reach the boundary regions only after very long times.

(2) The essential condition for the quantitative use of the moving boundary method for determining mobilities, as shown by Tiselius (1930), is constancy of the conductivity and mobility throughout the tube. However, the conductivity of a colloidal solution is different from (usually less than) that of the supernatant liquid with which it is in osmotic equilibrium, the difference being larger the higher the colloid and the smaller the electrolyte concentrations. The condition of Tiselius can be approximately fulfilled, therefore, only for relatively high salt concentration (e.g. 0·05 M. buffers are usually suitable) and low colloid concentration (e.g. $< 2-3$ gm./100 c.c. in the case of proteins, and even lower for colloids of lower molecular weight). For colloidal materials in or near the iso-electric condition, conductivity differences are reduced (see Membrane Equilibrium, p. 158) and more extreme conditions can be tolerated. The 'boundary anomalies' arising when appreciable variations in conductivity occur over the boundary regions by departure from the condition of low colloid and high salt concentration, will be considered shortly.

(3) The use of the schlieren optical system for boundary observation makes possible undisturbed boundary migration. Further, the sensitivity of the method makes possible the use of the small colloid concentrations which are so desirable, as we have seen, for the avoidance of boundary anomalies.

(d) Boundary migration in an electrical field

In many ways the motion of a boundary under an electrical field is similar to that under centrifugal forces. There is the initial difference, however, in that special methods have had to be evolved for the preparation of the boundaries in electrophoresis, whilst a boundary forms automatically during favourable conditions of sedimentation. Owing to the restriction (as yet) of the electrophoretic method to aqueous systems, it has had a less varied application than the centrifugal method; in general it is also true to say that electrophoretic investigations have been directed more towards obtaining information on specific systems than to providing fundamental information on the electrophoretic processes involved. Thus, whilst a large volume of literature is concerned with the detection and isolation of the different electrophoretically behaving species in many naturally occurring systems (see Chapter XXVI), the exact determination of the effect of colloid concentration and of the increasing solute-solute interactions on electrophoretic mobility has been largely neglected. We shall be concerned largely with migration in systems with little or no solute-solute interaction, a condition which holds, as we have seen, for nearly all corpuscular proteins except at high concentration (> 3 gm./100 c.c.).

Monodisperse systems. In the ideal case (constant conductivity and mobility throughout the cell) the terminal velocity acquired by a single charged colloidal particle in the electrical field is identical with that of all other such particles in the cell. Thus both boundaries move at a constant rate towards the same electrode, one, the ascending boundary, moving upwards into the space lately occupied by solvent and the other, the descending, into solution. Further, in the absence of diffusion, a homogeneous colloid should migrate with boundaries becoming neither sharper nor broader than when originally formed, and with the two boundaries symmetrical and identical in shape. In practice, however, diffusion is never absent, and, where other complications do not occur, the *irreversible* spreading of diffusion can usually be observed. Fig. 12.15 (a) shows typical photographs, taken by the diagonal schlieren optical system, of the two electrophoretically migrating boundaries of a nearly monodisperse protein preparation.

Reversible boundary spreading has been observed also for haemocyanin solutions by Horsfall (1939), who showed that the similar boundary spreading occurring in both limbs could be reversed by reversing the current, distinguishing the effect from diffusion spreading. These observations, which have had some confirmation, were attributed to slight electrical inhomogeneity in the protein preparation and might alternatively, therefore, be considered under polydisperse systems.

The position of the boundary is taken for ideal symmetrical peaks at the peak maximum (for theoretical reasons, see p. 236), or for more unsymmetrical types at the ordinate dividing the peak into equal areas. From the rate of movement of the boundary image on the camera plate, and the appropriate magnification factor, the real rate of boundary migration (v) under the known electrical field is known. The electrophoretic mobility (u), or rate of migration under unit electrical field is therefore calculated by the expression

$$u = \frac{vA\kappa_P}{i}, \qquad (12.41)$$

where A is the area of cross-section of the cell, κ_P is the specific conductivity of the protein solution, and i is the current in amperes. The mobility u is then given in cm.^{2} volt^{-1} sec.$^{-1}$ (usually as μ/sec./V./cm.).

Paucidisperse systems. If a small number of monodisperse non-interacting systems are mixed (giving a paucidisperse system) then, under the electrical field, the different molecules will usually acquire different velocities, and as in the sedimentation of paucidisperse systems (see p. 273) the originally sharp single peak will spread and finally split, giving rise to subsidiary peaks moving with velocities characteristic of the different types of molecule. From the schlieren photographs of such boundaries at different times it is possible, for those peaks which are sufficiently resolved, to determine mobilities, and further, as was explained in the case of sedimentation, to work out the composition of the whole colloidal system in terms of the electrophoretically different components. It is clear, in those cases where good resolution cannot be obtained, that the overlapping of peaks causes difficulty in determining both the exact position and area of a peak. The attainment of high resolution is therefore of the greatest importance, and a notable improvement in this respect is obtained by the use of the double-length cell section, which is described later. Here the total length of migration is double that in the normal cell, and much better resolution is obtained. However, it should be noted that if migration is very slow, the occurrence of appreciable diffusion spreading may more than counterbalance the advantages arising from the increased length of migration. The serum diagrams of Fig. 12.15(b) are typical of a system which is resolved only with great difficulty, but approximate mobilities and figures for the total composition in terms of the differently

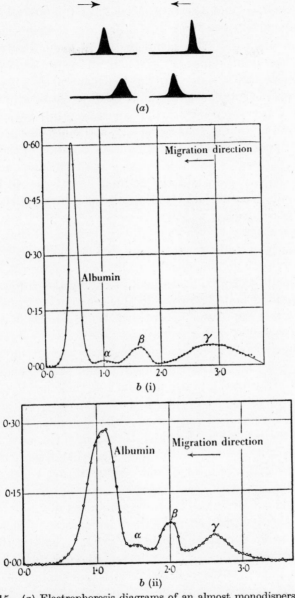

FIG. 12.15. (a) Electrophoresis diagrams of an almost monodisperse groundnut globulin in borate buffer at pH 9·0 and ionic strength, 0·15. Protein concentration = 1 per cent. Times of migration, 75 and 125 mins.
(b) Electrophoresis diagrams of normal human serum. Phosphate buffer: pH 8, $I = 0·1$. 4·9 volts/cm. Abscissa: distance in U-tube in cm. Ordinate: scale line displacement in mm.
(i) Anode limb, 152 mins. after starting current.
(ii) Cathode limb, 124 mins. after starting current. (*From* Kekwick (1939).)

migrating species can be obtained. It is of interest to note that by ultracentrifugal analysis, only two differently sedimenting species are observed in this case. This surprising result demonstrates the importance, in examining colloidal systems, of using several techniques simultaneously. Not only are the deficiencies of any one method detected, but the correlation of results by different lines of investigation provides the necessary information for testing our basic assumptions regarding the motion in solution of colloidal molecules or particles.

It is worth noting that if after complete resolution of the components of serum the current is reversed, then the resolved components will tend to come together again and the original boundary will be reformed, though usually more diffuse than initially as a result of the irreversible spreading of diffusion.

Polydisperse systems. We have already mentioned the case of reversible boundary spreading, supposedly due to slight electrical inhomogeneities. If such inhomogeneities are more important, i.e. a wider range of mobilities is involved, then, providing the different molecules can migrate independently, considerable boundary spreading, which is reversed by reversal of the current, will be shown. It is probable that in the denaturation of many proteins, the loss of specific molecular configurations causes such inhomogeneity as a result of both altered electrical charge and frictional constant, and such spreading is to be anticipated. However, it is also possible here that the increased molecular asymmetry introduced on denaturation may partly invalidate the condition of independent migration of the molecules when (as in sedimentation) diminished boundary spreading might occur.

Mobility and concentration. We have seen that, even for corpuscular proteins, diffusion coefficients and sedimentation constants are not entirely independent of concentration, and the decreased values at higher protein concentration are to be attributed to solute-solute effects causing an increase in the effective frictional constant. Such an increase is expected to be operative also in electrophoresis, and this view has been qualitatively confirmed (Johnson and Shooter, 1946) for the serum proteins. A quantitative comparison with sedimentation data for the same proteins is much to be desired.

(e) *Refinements of the method*

Electrophoretic preparations. If we consider an easily resolved two-component system (A and B), it is clear that the region between the two subsidiary ascending boundaries will contain only the faster component (say A), whilst between the two descending boundaries in the other limb only the slower component (B) will occur (Fig. 12.16). Thus if a pipetting device can be inserted into the cell at the correct points without causing mixing of the liquids, samples of pure A and pure B in solution could be obtained. The sectional construction of the cell, however, makes possible a better method.

After separation of the A and B boundaries, it is possible to inject liquid into the electrode vessel connected to the limb in which the boundaries are ascending at such a rate that the movement of the faster A boundary is retarded and the B boundary stopped completely. By such 'compensation', the distance separating the two boundaries may increase until it covers the whole length of a single central section of the cell. This separation could be obtained without compensation only by means of a very long cell.

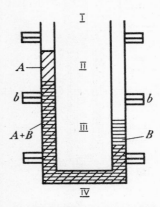

FIG. 12.16. The separation of electrophoretically different components.

At such a stage the B boundaries are carefully adjusted to the centre of the bb joint, and by means of the air pumps sections II and III are moved out of alinement with themselves and with the rest of the cell. The latter can now be dismantled and the pure A and B solutions removed without fear of mixing.

Various devices for injecting liquid into the electrode vessels or for providing alternative mechanisms have been designed, and details are provided in the references already quoted.

The double centre-section cell. The resolution of a paucidisperse system is, for equal diffusion effects, greater the larger the migration path. An effective increase in path length can be obtained by use of the compensation mechanism just described, when components may be held stationary or even reversed, and others then allowed to separate by distances typical of long migration paths. However, mobility data in such cases necessarily become less accurate. An alternative scheme makes use in the cell of a double centre-section of total height equal to the combined heights of two normal centre sections (Figs. 12.12 and 12.17). Boundaries are initially formed at joints aa' and bb' (instead of at the same level) and the electrical field must be applied in a direction necessary to bring both boundaries into the long cell. The increased path length thus introduced makes possible both increased resolution and increased accuracy in mobility determinations. However, this new cell is not suitable for preparative work, nor for use on unknown systems where components of opposite charge might be encountered, in which case some components would migrate into sections I and IV from the initial boundaries in joints aa' and bb' respectively, and be lost to observation.

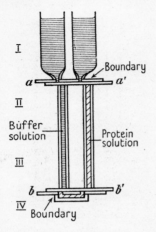

Fig. 12.17. Formation of boundaries in electrophoresis cell with double-length centre section.

(f) *Boundary anomalies*

In the ideal case for a monodisperse colloid (p. 323) the ascending and descending boundaries should be mirror images of one another and should move at identical rates. Such behaviour, we have seen, is conditional on the constancy of the conductivity of the liquids and of the mobility of the colloid throughout the cell.

It is clear that the assumption of complete constancy of conductivity throughout the cell can only be approximate, for the colloid solution and the aqueous solvent which have been dialysed into equilibrium against one another will, as a result of the Donnan equilibrium, possess different concentrations of diffusible electrolytes, the difference being larger the higher the colloid and the lower the electrolyte concentration. It is not therefore surprising that, even for monodisperse systems, anomalous boundary effects have been reported. Typical effects are shown in Fig. 12.18 (a), in which the ascending boundary is shown to be sharper and more rapidly moving than the descending one, and in which two 'false' nearly stationary boundaries, δ and ϵ, appear. It is also observed that the area of the main ascending peak (A) is smaller than the main descending one, but that the total areas on the ascending and descending sides are identical. This behaviour has been shown to be quite general and an explanation in terms of conductivities has been proposed (Longsworth and MacInnes, 1939). A brief outline only can be included here.

As the descending boundary (Fig. 12.18 (b)) moves from its original position O to the position D, the composition of the liquid in the space vacated by the protein is in some way altered to a value B' different from that of the original buffer (B), giving rise to a gradient (ϵ) of salt concentration near O, which moves slowly, usually in the direction of the field. In a similar way the entry of protein molecules at the ascending side into buffer previously free from colloid gives rise in this liquid to an 'adjustment' of both protein and salt concentration occurring there (to P') causing another gradient of both colloid and salt concentration (δ) which is also found to move very slowly.

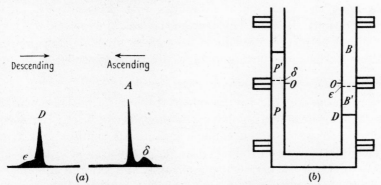

FIG. 12.18. (a) Boundary anomalies. (One per cent. gum arabic solution in phosphate/NaCl buffer at pH 7·8 and $I = 0·1$.) (b) The δ and ϵ boundary anomalies.

Since the δ boundary is partly due, at least, to a gradient in protein concentration it gives rise, therefore, to a larger peak than the ϵ boundary, which is thought to be due to a salt gradient alone. The observation that peak A is of smaller area than peak D is therefore understood since there must be identical total changes in protein concentration in the two limbs. It is also observed experimentally that

$$\kappa_{B'} > \kappa_P \text{ and } \kappa_{P'} < \kappa_B,$$

where κ is specific conductivity and the subscripts B and P refer to original buffer and protein solutions and B' and P' to those of adjusted buffer and protein solutions respectively.

For each boundary, therefore, the conductivity below the boundary is less than that above it. Thus, since the current is constant throughout the cell, the field strength is higher below than above each boundary, which in the case of the descending boundary causes the advancing edge to be faster than the trailing edge giving a diffuse peak, and for the ascending boundary an advancing edge which is slower than the trailing one, giving a sharp peak, both being definitely asymmetric.

Thus, in terms of the experimentally investigated concentration adjustments occurring in the boundary regions, a consistent explanation of the main boundary anomalies is obtained, but the theoretical side of the 'adjustment' processes is not well understood.

Whilst we cannot enter further into discussion of boundary anomalies, it ought to be noted that Svensson (1946) considers the δ and ϵ boundaries to be merely two of a whole family of possible false boundaries numbering $(n-x-1)$, where n is the total number of different ions and x the number of true boundaries (i.e. giving true mobilities).

It will be noted also that we have not considered anomalous effects arising from the variation of mobilities with concentration, or with other conditions. In view, however, of the lack of experimental results, this will not be further considered, though by analogy with the results of sedimentation some such anomalies are to be expected.

(g) Advantages of the moving boundary method

Before discussing the results of electrophoretic measurements, the advantages of the moving boundary method may be briefly summarized. Applicable to any charged colloids in aqueous solution, the method provides information on the composition of the system in terms of the electrophoretically separable components, and on the mobility and electrical homogeneity of these components over wide pH and moderate salt concentration ranges. As in the case of the sedimentation velocity method, the exploration of mobility-concentration relations may provide important data on solute-solute and solute-solvent interactions. Further, the use of the method for preparative purposes opens up new fields of investigation.

(h) Chief results

Certain results of moving boundary mobility determinations as applied to particular systems will be discussed in Chapter XXVI, dealing with proteins; we propose here merely to outline the chief features.

The net charge on a protein molecule in solution is to be attributed largely to the balance between the dissociation of side chain carboxyl groups (giving COO^-) and the proton uptake of the side-chain amino groups (giving NH_3^+). Proteins of different chemical constitution will therefore possess different net charges in solutions of the same pH, and, further, the rate of change of net charge with varying pH will also be dependent on the amino acid composition of the molecule. Since mobility is directly related to net charge, it is apparent that the pH-mobility curve is determined by the composition of the protein molecule and is therefore of great interest, providing a link between chemical composition and physical properties. pH-mobility curves are usually determined by the moving boundary method, but, as was mentioned earlier, the micro-electrophoresis method may be employed to study the electrophoresis of protein covered particles as a function of pH. Fig. 12.19 (a) shows a comparison of electrophoretic data for serum albumin, in 0·02 M. acetate buffer at 20° C., obtained by the

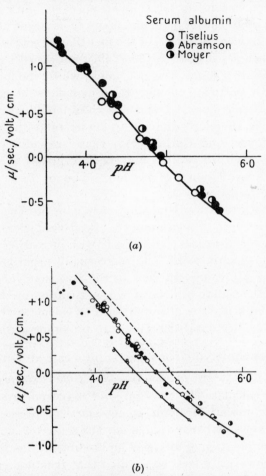

FIG. 12.19. (a) Comparison of moving boundary data of Tiselius for horse serum albumin with data for the adsorbed protein of Abramson and Moyer, at $I = 0.02$ and 20° C. (Abramson, Moyer, and Gorin (1942).) (b) A comparison of dissolved and adsorbed egg albumin. The small open circles are data for the electric mobility of dissolved egg albumin in M/50 acetate buffer at 20° C. The large circles are values for the electric mobility of particles of 'Pyrex' glass, mineral oil, carbon, collodion, and quartz coated with films of egg albumin in solutions of the same ionic strength. The curve for adsorbed egg albumin is essentially parallel to that of dissolved egg albumin over the range of pH investigated. However, it is definitely shifted to the right on the pH scale. The iso-electric point for the adsorbed protein is at pH 4·82 instead of pH 4·55. The dashed curve is for surface-denatured albumin.

moving boundary and micro-electrophoresis methods. In this case it is clear that within the limits of experimental error both methods give the same pH-mobility curve. Dissolved and adsorbed egg albumin (Fig. 12.19 (b)) do not, however, give identical curves. It is clear that before any general conclusions on the comparison of mobilities for dissolved and adsorbed proteins can be made, many more accurate comparisons are required. In view, however, of the fact that dissolved protein molecules are of such a size that their mobility varies with size and shape (i.e. Smoluchowski's equation is not valid), whilst the micro method is concerned with particles obeying the Smoluchowski equation, good agreement between dissolved and adsorbed mobilities would not be expected. It is therefore difficult to understand the statement by Abramson, Moyer, and Gorin (1942) that 'Any difference between dissolved and adsorbed protein may be interpreted as manifesting a change in the availability of certain acidic or basic groups upon adsorption ....'.

The general features of pH-mobility curves for proteins are, however, shown by both types of measurement. Positively charged at low pH values, the charge decreases continuously with increasing pH, becoming zero at the iso-electric point, and increasingly negative with further increase in the pH value. In view of the fact that values of the net charge per molecule may be determined by titration methods and from membrane equilibria, it is of importance to calculate absolute charge values from mobility data with a view to providing a comparison.

Net charge per molecule from electrophoresis and other methods
(a) Net charge of dissolved protein molecules from electrophoresis

In order to calculate the net charge of dissolved protein molecules from mobility measurements it is necessary to possess information regarding the molecular shape and size. The source of such information is usually a combination of diffusion and sedimentation measurements, from which, however, it will be recalled that it is not yet possible to calculate unambiguous molecular dimensions owing to the uncertainty of the solvation factor. However, two possible extremes could explain observed results:

(i) strongly solvated spherical, or

(ii) unsolvated cylindrical (or asymmetric) molecules.

Something between these extremes is most probable, of course, but the extreme models may be utilized in calculating the limits of the net charge per molecule.

Dissolved spherical protein molecules fall into the intermediate size category for which Henry's treatment is valid. The net charge, Q, of a spherical molecule is related to the experimentally measured mobility by eqn. (12.19), which, being rearranged, may be written

$$Q = \frac{4\pi\eta a(1+\kappa a+\kappa r_i)}{f(\kappa a)(1+\kappa r_i)}u = \nu e, \qquad (12.19\,\text{a})$$

where ν is the valency of the protein ion, and e is the electronic charge.

The radius a of a spherical molecule is determined from the diffusion coefficient by eqn. (10.28), and the Debye-Hückel function κ from the salt concentration by eqn. (12.1). Knowing κa, $f(\kappa a)$ may be obtained from the curves of Henry (e.g. Fig. 12.3), and r_i, the average radius of the ions in the atmosphere of the protein molecule, from the composition of the solution and the component ionic diameters. Inserting these values, we obtain for the serum albumin molecule, assumed spherical, under the conditions of Table 12.3, the relation

$$\nu = 9\cdot 65u, \qquad (12.42)$$

connecting the net charge (in terms of the electronic charge) and the experimental mobility. Thus the pH-mobility curve may be transformed directly into one connecting net charge and pH. Such data are contained in the second column of Table 12.3.

In calculating the net charge for an unsolvated cylinder, we make use of Gorin's method (p. 307). From the experimentally determined molecular weight, the molecular volume and radius of the hypothetical unsolvated spherical molecule (for which f_0 is the frictional constant) is readily determined. The net charge of such a molecule ($\nu_0 e$) can be calculated in terms of mobility as we have just shown. Thus for serum albumin under the conditions stated in Table 12.3,

$$\nu_0 = 6\cdot 70u. \qquad (12.43)$$

But from the frictional constant the axial ratio of the unsolvated molecule may be obtained, using Perrin's equation for a prolate ellipsoid of revolution. Identifying the ellipsoidal and cylindrical models, the axial ratio value may be used to derive from the calculations of Gorin, the ratio (R) of the mobility of the cylindrical molecule (u) to that of a sphere (u_0) of the same molecular volume. From this value and an expression of the type of (12.43) above, the relation between the net charge of the unsolvated cylinder, and its mobility is obtained. In the case of serum albumin, at $I = 0\cdot 02$, the ratio $u/u_0 = 0\cdot 65$, when we have

$$\nu = \frac{6\cdot 70}{0\cdot 65}u = 10\cdot 3u. \qquad (12.44)$$

Table 12.3, drawn from the data of Abramson, Moyer, and Gorin (1942), contains the detailed values used in the calculations for serum albumin as well as a comparison with values of the net charge calculated at two values of the ionic strength from titration experiments, which will be considered shortly. The good agreement between the valency values calculated on the assumption of a cylindrical molecule with

TABLE 12.3. *The Net Charge of Serum Albumin*

Buffer: sodium acetate-acetic acid, $T = 20°$ C., $I = 0.02$, $r_i = 2.5$ A.

$D = 6.10 \times 10^{-7}$. $\therefore$ a (hydrated sphere) $= 34.8$ A.
$f/f_0 = 1.27$; $l/2a$ (unhydrated rod) $= 5$; $R = 0.65$.
$M = 70,000$. $\therefore$ a (unhydrated sphere) $= 27.4$ A.
$\kappa = 4.67 \times 10^6$. $f(\kappa a) = 1.04$.

		Valency			
		Electrophoresis		Titration	
pH	Mobility u	Sphere	Cylinder	$I = 0$	$I = 0.02$
4·19	0·847	8·2	8·7	8·7	13·5
4·36	0·588	5·7	6·1	6·0	8·6
4·67	0·168	1·6	1·7	1·7	2·3
4·81	0·000	0	0	0	0
4·97	−0·225	−2·2	−2·3	−1·6	−2·5
5·28	−0·439	−4·2	−4·5	−4·6	−6·6
5·49	−0·595	−5·7	−6·1	−6·2	−8·8

those from titration data at $I = 0$† is strong evidence of the asymmetry of the molecule of serum albumin. A comparison similar to that shown in Table 12.3 for serum albumin is given diagrammatically in Fig. 12.20 for egg albumin. Again it would appear that the egg albumin molecule approximates much more to an unsolvated rod-like particle than to a heavily hydrated sphere. It should be noted that this use of electrophoretic data provides a method of distinguishing between the effects of solvation and asymmetry, which could not be accomplished using sedimentation and diffusion data alone; another reason for the use together on a single system of several types of investigation (see p. 326).

(b) *The net charge from titration curves*

At the iso-electric point of a protein, the net charge is zero, and if combination with ions other than hydrogen and hydroxyl can be neglected (as at very low salt concentration), then it can be said (neglecting charged groups other than COO^- and NH_3^+) that the number of negatively charged carboxyl groups is equal to the number of positively charged amino groups (also the iso-ionic point). If now acid (e.g. HCl) be added, hydrogen ions (H^+) are bound by the protein in an amount

† Titration data at finite ionic strengths are complicated by the uptake by protein of undissociated acid (see p. 336). More reliable values of the net charge are thus obtained from titration data extrapolated to zero ionic strength.

which is given by the difference between the quantity of acid added and that remaining free in solution. Similarly the quantity of hydroxyl ions (OH⁻) bound is given by the difference between the quantity of

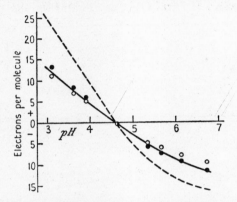

FIG. 12.20. Agreement between charge of egg albumin calculated from electric mobility data assuming unhydrated rods, ●, and hydrated spheres, ○, and that obtained from titration curves (smooth curve, titration at $I = 0$; dashed curve at $I = 0.1$). Data of Longsworth (1941), and Cannan et alia (1941).

alkali added and the free alkali. In both cases, free acid and alkali are related to pH by the equation

$$\mathrm{pH} = -\log[\mathrm{H}^+]f_{\mathrm{H}^+} = \log\left(\frac{[\mathrm{OH}^-]f_{\mathrm{OH}^-}}{K_W}\right), \qquad (12.45)$$

where f_{H^+} and f_{OH^-} are the activity coefficients of the H⁺ and OH⁻ ions, and K_W is the ionic product of water.

Where the binding of other ions can be neglected, the bound acid or alkali alone determines the net charge on a protein molecule. Thus if h equivalents of acid be bound/mole of protein at a given pH, then h is the net positive charge per molecule in terms of the electronic charge. Similarly, the number of equivalents of base bound per mole of protein gives the net negative charge. Hence the curves of acid or base bound as a function of pH (titration curves) are of considerable importance, and, like pH-mobility curves, are an expression of the amino-acid composition of the protein. The subject has been reviewed by Cannan (1942) from whose article details of the actual determination of titration curves may be found. Fig. 12.21 shows the titration curve at 0° C., and $I = 0.1$ for egg albumin, pH being plotted as abscissa and the number of bound equivalents of acid per gram of protein as left-hand ordinate (Longsworth, 1941). On the same diagram the circles represent mobility data plotted on the scale of the right ordinate. To understand the choice of scales on the ordinates and the detailed comparison of the titration and mobility data, it is necessary to consider the effects of the binding of ions other than H^+ and OH^-.

If ions other than H^+ and OH^- are bound by the protein, then the

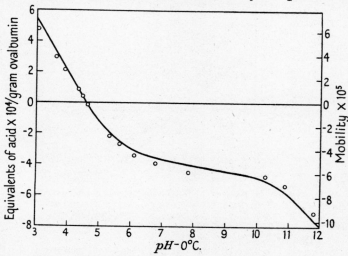

Fig. 12.21. Comparison of titration and mobility data at 0° C. for egg albumin. (Longsworth (1941).)

acid or base bound no longer gives the net charge per molecule, but a correction must now be applied to the titration data to represent the effect of these other ions on the net charge. Thus the binding of chloride ions (increasing the negative charge by one electronic charge) will shift the iso-electric point to lower pH values, and the pH of zero acid or base bound (the iso-ionic point) will no longer be identical with the iso-electric point but will now be that at which the net charge is -1 (electronic charge). If such ion binding is independent of pH, then over the whole titration curve, the net charge is given by $(h-1)$, which merely corresponds to a vertical displacement along the mobility axis of the titration curve. The extent of the required vertical displacement is conveniently decided by bringing the point $h = 0$ on the titration curve upon the electrophoretically determined iso-electric point. The ordinate on the titration curve is now proportional to net charge, and mobility and titration data may now be directly compared by a simple choice of the ordinate scales. It should, however, be remembered that this is conditional on 'other' ion binding being independent of pH, a conclusion which at best would be expected to hold over small pH ranges only.

In Fig. 12.21 the titration curve has been displaced vertically to put $h = 0$ at pH = 4·58 (the iso-electric point at $I = 0·1$), and its ordinate scale adjusted to bring the curve through the points representing mobility data. Apart from minor variations, the mobility and titration data agree well over the whole pH range, at the extremes of which irreversible reactions are beginning to occur. In view of the rather unlikely assumption (for large pH ranges) that 'other' ion binding is independent of pH, the agreement is surprising.

The binding by protein of undissociated acid represents another complication, since such acid contributes to the bound acid value (h) without affecting the net change. Gorin and Moyer (see Abramson, Moyer, and Gorin, 1942) have attempted to eliminate such discrepancies by extrapolating the experimental titration curve to zero ionic strength and zero protein concentration. The success of this procedure is evident from the titration figures of Table 12.3, in which good agreement with net charge values from electrophoresis is obtained only at $I = 0$. Fig. 12.20, containing a similar comparison for egg albumin, shows that titration data for $I = 0$ compare very well with the electrophoretic net charge values obtained on the assumption of unhydrated rods or cylinders. It will be recalled that a similar conclusion was obtained from Table 12.3 for serum albumin.

In connexion with titration curves, it is of interest that from the known amino acid composition of certain proteins, it has been possible to construct theoretical titration curves in good agreement with those determined experimentally (Cannan, Kibrick, and Palmer, 1941). Thus it follows also that for proteins of known composition it is possible to calculate the net charge under given conditions and to compare, therefore with electrophoretically determined values. The intimate nature of the connexion between chemical composition and physical properties is thus demonstrated once more.

(c) *Net charge from membrane potentials*

It has been shown (Chapter VII) that membrane potentials may be utilized to calculate the net charge of a colloidal ion, so that such measurements provide another check of electrophoretic calculations. Adair and Adair (1940) have thus calculated the valencies of egg albumin and haemoglobin at different pH values and ionic strengths (see Table 7.4). In the case of haemoglobin they have also calculated mobility as a function of pH from membrane potential data, assuming the molecules to be hydrated spheres of radius 30·7 A, and have compared with electrophoretically determined mobilities. Fig. 12.22, containing such comparisons at two values of the ionic strength, shows an encouraging measure of agreement. Abramson, Moyer, and Gorin (1942), by interpolating, obtain from the data of Adair and Adair (1940), the valence $-13\cdot 8$ for egg albumin in phosphate buffer at

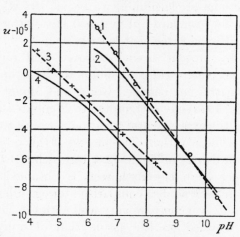

FIG. 12.22. Comparison of observed and theoretical mobilities of haemoglobin ions. Ordinates, mobilities $\times 10^5$; abscissae, pH value. Curve 1 and circles observed mobility, ionic strength 0·02, $u = 0$ at pH 7·3. Curve 2 theoretical mobility, ionic strength 0·02, $u_t = 0$ and E (membrane potential) $= 0$, pH 7·05. Curve 3 and crosses observed mobility, ionic strength 0·10, $u = 0$ at pH 6·8. Curve 4 theoretical mobility, ionic strength 0·10, $u_t = 0$ at pH 6·1. Curves 3 and 4 are shifted two pH units to the left.

$I = 0\cdot 1$ and pH 7·10, in comparison with the values $-11\cdot 9$ and $-14\cdot 1$, calculated from electrophoretic data, obtained under the same conditions on the assumption of a hydrated sphere and unhydrated rod respectively. From this calculation it would therefore appear that the membrane potentials, like titration data, support the nearly unhydrated rod model for egg albumin. Much more, however, remains to be done before reliable conclusions can be drawn.

Electro-osmosis, streaming, and sedimentation potentials

(a) *Electro-osmosis*

In determining theoretically an expression for the electrophoretic mobility of large particles ($d \ll a$), it was pointed out that by a very similar treatment, the electro-osmotic mobility (u_0), involving the motion of a liquid relative to a fixed solid, was given by an identical expression

$$u_0 = \frac{v_0}{X} = \frac{\zeta D}{4\pi\eta}, \qquad (12.46)$$

ELECTROPHORESIS AND ALLIED PHENOMENA

where v_0 is the electro-osmotic velocity. This expression may be used to describe electro-osmotic effects in a single tube (e.g. a single closed tube, p. 310) or in a porous plug, membrane, or diaphragm. The electro-osmotic velocity is related to the volume of liquid (V) flowing through the capillary or plug per second by

$$v_0 = \frac{V}{A}, \qquad (12.47)$$

where A is the effective area of cross-section through which liquid flows. Combining (12.46) and (12.47) we therefore obtain

$$V = \frac{\zeta DX.A}{4\pi\eta}, \qquad (12.48)$$

which may be rearranged to give

$$\zeta = \frac{4\pi\eta}{D} \frac{V}{XA}, \qquad (12.49)$$

and

$$\zeta = \frac{4\pi\eta}{D} \frac{V}{E} \frac{l}{A}, \qquad (12.49\,a)$$

putting $X = E/l$, where E is the potential applied across the capillary or plug which has an effective length l.

Eqn. (12.49) should be compared with a similar one involving electrophoretic mobility,

$$\zeta = \frac{4\pi\eta}{D} u. \qquad (12.12)$$

It thus becomes clear that, for a given system, involving constant values of ζ, η, and D, the factor $V/E \times l/A$ in electro-osmosis ought to be identical with the electrophoretic mobility. A comparison of these quantities is therefore equivalent to a comparison of ζ potentials in electro-osmosis and electrophoresis, and consequently provides a test of the Smoluchowski equation. In view of doubts regarding the correct values of D and η for the double layer, many workers prefer such a comparison, which is free from the objections involved in the calculation of actual ζ potential values. Ham and Douglas (1942) made such comparisons for octadecyl alcohol in aqueous media. Electrophoretic data on dispersions of the material in aqueous solutions, with particle diameter $ca.$ 5×10^{-5} cm. at pH values between 2 and 12, were obtained by the Lewis and Price modification of the moving boundary method already discussed (p. 318). Electro-osmotic data were obtained on an apparatus developed from that of Fairbrother and Balkin (1931) and shown diagrammatically in Fig. 12.23. A diaphragm (A) of octadecyl alcohol (0·5 cm. $\times$ 3 cm.^{2}) is formed between platinum gauze electrodes in a wide tube, which, with the calibrated capillary tube D (of 2 mm. bore) and the connecting side-arms, forms a closed circuit filled with the required aqueous medium except for a small air bubble in D to be used as an indicator of the flow of liquid. A potential (up to 220 volts D.C.)

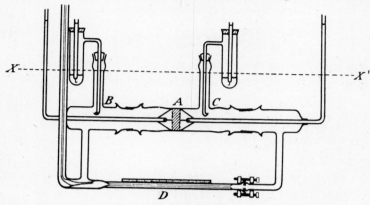

Fig. 12.23. Electro-osmosis cell. (Ham and Douglas (1942).)

was applied by means of the calomel electrodes at B and C, and the flow of liquid caused by electro-osmosis in the diaphragm was observed. The ratio of effective length to cross-section of the diaphragm, l/A, was obtained by filling with 0·1 M. KCl, of known specific conductivity, and measuring the total conductivity across the diaphragm. Neglecting any conductivity of the solid, which was shown to be small even for conducting solids, l/A was then readily obtained. Table 12.4 contains

TABLE 12.4

pH	$\dfrac{Vl}{EA} \times 10^5$ $cm.^2\ volt^{-1}\ sec.^{-1}$	$u \times 10^5$ $cm.^2\ volt^{-1}\ sec.^{-1}$	$\dfrac{Vl}{EA}\bigg/u$
2	+ 3·7	+ 4·0	0·93
3	− 3·1	− 2·9	1·07
4	− 5·1	− 4·9	1·04
5	− 6·6	− 6·4	1·03
6	−10·0	−10·2	0·98
7	−16·1	−16·4	0·98
8	−22·3	−22·0	1·01
9	−26·1	−26·1	1·00
10	−27·7	−27·9	0·99
11	−28·1	−27·9	1·01
12	−28·2	−27·9	1·01

results from electrophoresis and electro-osmosis and their comparison, from which it is clear that the ζ potentials calculated by eqns. (12.12) and (12.49) are identical within the limits of experimental error.

However, just as the Smoluchowski equation for electrophoresis breaks down for particles below a certain size (see Table 12.1), the simple equations describing electro-osmosis also fail when capillary or pore size becomes too small. This is probably due partly to the fact

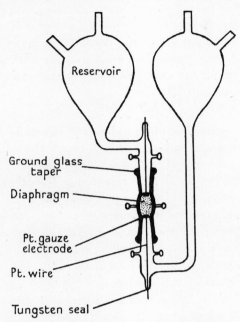

FIG. 12.24. Solid-liquid streaming potential cell.

that for small pore sizes it is unjustifiable to treat the double layer as a parallel plate condenser and partly also to the overlapping of electrical double layers. Such effects, which cause complication in streaming potential work as well as in electro-osmosis, are clearly discussed by Bull (1943) and will not be further considered here.

(b) *Streaming potentials*

We have seen that in electro-osmosis the application of an electrical field causes the motion of a liquid with respect to a solid. Conversely, the motion of a liquid with respect to a solid, e.g. through a porous plug or capillary tube, may produce a potential difference between the ends, known as the streaming potential.

A suitable apparatus for the study of streaming potentials devised by Gortner and his co-workers (see Gortner, 1938, p. 160) is shown diagrammatically in Fig. 12.24. The solid material to be investigated is packed in the form of a plug between the two platinum gauze electrodes and liquid is forced through from the reservoir by its own hydrostatic head of pressure. The potential set up across the electrodes is measured by a high resistance circuit, e.g. quadrant or valve electrometer. Suitable cells for investigating electrokinetic potentials across liquid-liquid interfaces by streaming potential measurements have also been devised and used (e.g. Gortner, 1938).

The magnitude of the streaming potential is dependent upon the ζ potential between the liquid and solid and may therefore be used to determine it. Thus it can be shown (see Gortner, 1938, p. 158) that

$$\zeta = \frac{4\pi\eta\kappa_s E_s}{PD}, \qquad (12.50)$$

where E_s is the streaming potential across the ends of a capillary or plug, P is the pressure forcing liquid through the capillary, and κ_s is the specific conductivity of the liquid in the capillary (or in the pores of a diaphragm across which E_s is measured), the other symbols having their usual significance.

It is to be noted that, arising from eqn. (12.50), the streaming potential is independent of the length and cross-section of the capillaries, so that the packing of the plug or diaphragm is not critical. In calculating ζ, P must be substituted in absolute units (dynes/cm.²), and κ_s and E_s in absolute electrostatic units, when ζ is obtained also in electrostatic units. (N.B. 1 e.s.u. = 299·86 volts, 1 reciprocal ohm = 9×10^{11} e.s.u.) Alternatively, for water at 20° C., eqn. (12.50) reduces to

$$\zeta = 1\cdot06 \times 10^5 \times \frac{E_s \kappa_s}{P}, \qquad (12.51)$$

where ζ and E_s are in millivolts, κ_s is in reciprocal ohms, and P is in cm. of Hg.

We have already referred (p. 299) to the work of Bull (1935), who determined ζ potentials by eqn. (12.50) for protein covered pyrex particles and compared with those determined by electrophoresis and electro-osmosis. As an alternative, however, to the calculation of ζ, the electric moment per sq. cm. of the double layer (i.e. $d\sigma$) may be obtained. Thus substituting in eqn. (12.11) for ζ from eqn. (12.50) we obtain

$$d\sigma = \frac{\eta\kappa_s E_s}{P}, \qquad (12.52)$$

the quantity $d\sigma$ being clearly analogous to the dipole moment of a molecule.

Gortner and his school made extensive investigations of the streaming potentials arising from different types of interface and their results are contained in many original papers. The method is not, however, free from criticism, and Bull (1943) has outlined the difficulties associated with it.

Whilst the method of streaming potentials does not possess the accuracy nor the possibilities of electrophoresis, it is of considerable use in investigating solid surfaces (e.g. fibrous or powdered materials) whose examination by other methods is not convenient.

(c) *Sedimentation potentials*

Attempts have been made to make the sedimentation potential (i.e. the potential set up in a liquid caused by the motion through it of charged particles), the subject of quantitative study (see Gortner, 1938, p. 161), but the results obtained have not, in the main, been conspicuously successful. The magnitude of the potentials arising from the fall of solids through non-conducting liquids (e.g. 80 volts by quartz in toluene) should be noted.

Gortner attributes failure to the irregular path taken by small particles in falling through liquids. Washburn and Quist (1940), who have developed a method in which they claim to have adequate control of the flow of the solid and protection against contamination, have obtained results conforming to the relevant electrokinetic equation, and from which reasonable values of the potential were calculated. An extension of this work is clearly desirable.

REFERENCES

ABRAMSON, 1931, *J. phys. Chem.* **35**, 289.
—— 1934, *Electrokinetic Phenomena*, Chemical Catalog Co.
—— MOYER, and GORIN, 1942, *Electrophoresis of Proteins*, Reinhold.
ADAIR and ADAIR, 1940, *Trans. Faraday Soc.* **36**, 23.
BRADBURY and JORDAN, 1942, *Bio-chem. J.* **36**, 287.
BULL, 1935, *J. phys. Chem.* **39**, 577.
—— 1943, *Physical Biochemistry*, Wiley, p. 160.
BULL and GORTNER (1931), *J. phys. Chem.* **35**, 309.
BURTON, 1938, *The Physical Properties of Colloidal Solutions*, 3rd ed., Longmans, p. 174.
CANNAN, KIBRICK, and PALMER, 1941, *Ann. N.Y. Acad. Sci.* **41**, 243.
—— 1942, *Chem. Rev.* **30**, 395.
COHN and EDSALL, 1943, *Proteins, Amino Acids, and Peptides*, Reinhold.
DICKINSON, 1941, *Trans. Faraday Soc.* **37**, 140.
FAIRBROTHER and BALKIN, 1931, *J. Chem. Soc.*, p. 389.
FREUNDLICH, 1926, *Capillary and Colloid Chemistry*.
GORTNER, 1938, *Outlines of Biochemistry*, 2nd edn., Wiley.
HAM and DOUGLAS, 1942, *Trans. Faraday Soc.* **38**, 404.
HENRY, 1931, *Proc. Roy. Soc.* A. **133**, 106.
—— 1938, *J. chem. Soc.*, p. 997.

HORSFALL. 1939, *Annals N.Y. Acad. Sci.* **39**, 203.
JOHNSON and SHOOTER, 1946, unpublished.
KEKWICK, 1939, *Bio-chem. J.* **33**, 1122.
LAUFFER and GORTNER, 1938, *J. phys. Chem.* **42**, 641.
LODGE, 1886, *Rep. Brit. Assn.*, p. 389.
LONGSWORTH, 1941, *Ann. N.Y. Acad. Sci.* **41**, 267.
—— 1942, *Chem. Rev.* **30**, 323.
—— and MACINNES, 1939, *Chem. Rev.* **24**, 271.
—— —— 1940, *J. Amer. chem. Soc.* **62**, 705.
MACINNES and LONGSWORTH, 1932, *Chem. Rev.* **11**, 171.
MATTSON, 1933, *J. phys. Chem.* **37**, 223.
PAULI, 1935, *Trans. Faraday Soc.* Symposium on *Colloidal Electrolytes*, p. 11.
PICTON and LINDER, 1897, *J. chem. Soc.* **71**, 568.
PRICE and LEWIS, 1933, *Trans. Faraday Soc.* **29**, 775.
SMITH and LISSE, 1936, *J. phys. Chem.* **40**, 399.
SVENSSON, 1946, *Ark. Kemi Min. Geol.* **22**, No. 10.
TISELIUS, 1930, *Nova Acta Soc. Sci. upsal.*, IV, 7, No. 4.
—— 1937, *Trans. Faraday Soc.* **33**, 524.
WASHBURN and QUIST, 1940, *J. Amer. chem. Soc.* **62**, 3169.

XIII
VISCOSITY

Introductory

In this chapter it is proposed to consider only the viscosity of dilute colloidal solutions, from which it is often possible to determine, indirectly, information about the nature of the disperse phase. Viscosity in concentrated solutions is complicated by many additional factors, and though its study is of the greatest fundamental and technological importance, it is as yet incapable of quantitative explanation in terms of the structure of, and interactions between, the colloidal particles. For this reason the viscous behaviour of concentrated solutions is considered separately (Chapter XVI).

The viscosity of a liquid is that property which is responsible for the internal resistance offered to the relative motion of different parts of the liquid. Such internal resistance or friction, involved in the motion of one layer of molecules with respect to the next, is closely connected with the interactions (van der Waals forces, dipole interactions, etc.) between separate molecules of the liquid and with the structure of the liquid phase. A quantitative description of the viscosity of a pure liquid in terms of the properties of a single molecule is therefore dependent upon our knowledge of the liquid state, which, for good reasons, has lagged considerably behind that of the solid and gaseous phases.

Newton assumed the shearing forces (τ) between two parallel planes in a liquid in relative motion to be proportional to the area of the planes and to the velocity gradient between them, writing his well-known law for infinitely near planes, in the form

$$\tau = \eta \frac{dv}{dl} A, \qquad (13.1)$$

where dv/dl is the velocity gradient perpendicular to the planes, A is the area of the planes, and η, the coefficient of viscosity, is the proportionality constant involved. The dimensions of the viscosity coefficient are $ML^{-1}T^{-1}$ and the absolute c.g.s. unit is the poise (gm. cm.$^{-1}$ sec.$^{-1}$). A liquid has a viscosity of 1 poise if a steady tangential force of 1 dyne produces a relative velocity of 1 cm. per sec. between two parallel planes of area 1 cm.^{2}, separated by 1 cm. and immersed in the liquid.

Since the work done in overcoming frictional resistance is transformed into heat, a connexion must exist between the viscosity coefficient and the kinetic energy transformed into heat. Thus we have the useful alternative definition of the viscosity coefficient:

$$q = \eta \left(\frac{dv}{dl}\right)^2, \qquad (13.2)$$

where q is the kinetic energy transformed into heat in unit time for unit volume (ergs per sec. per cm.^{3}). This view of the viscosity coefficient as a measure of the energy dissipated is sometimes more fruitful than the commoner one in forming a physical picture of the processes affecting viscosity, e.g. those responsible for increased viscosity on introducing colloidal molecules into a pure liquid, which we shall consider later. Another type of viscosity, the kinematic (ν), as distinct from the normal dynamic viscosity, is defined by

$$\nu = \eta/\rho, \qquad (13.3)$$

where ρ is the density of the liquid. This function has the advantage of being directly related to flow time in an Ostwald viscometer, but for our purposes such viscosities must be converted into the dynamic form.

The quantity, fluidity, ϕ, defined as the reciprocal of the dynamic viscosity coefficient

$$\phi = \frac{1}{\eta}, \qquad (13.4)$$

and measured in reciprocal poises (cm. sec. gm.$^{-1}$), is also sometimes utilized instead of the viscosity.

For most pure liquids and many solutions the Newtonian law (eqn. (13.1)), expressing the proportionality between shearing force and velocity gradient, has been completely confirmed for stream-line or laminar flow: such fluids are therefore said to be Newtonian in their viscous behaviour.

By stream-line flow we mean a steady state such that at any point of the system the velocity of the moving liquid does not vary with time in magnitude or direction. The occurrence of irregular or time variable eddy formation is therefore excluded. In flow through a tube this involves (except at the ends) a liquid velocity parallel to the walls of the tube and of magnitude depending only upon the distance from the walls (see p. 350). Cylindrical laminae of liquid move relatively to one another in a direction parallel to the length of the tube.

At a certain high critical liquid velocity (v_c) orderly stream-line fluid motion is replaced by disorderly turbulent flow involving dissipation of energy in eddy formation, etc., when the Newtonian law is no longer obeyed. Reynolds showed experimentally that the critical velocity could be expressed in the form

$$v_c = \frac{k\eta}{\rho r}, \qquad (13.5)$$

where k, a constant, is for narrow tubes approximately 1,000. It is necessary, therefore, in applying the Newtonian law, that only liquid velocities well below the critical should be utilized. For water ($\eta \approx 0\cdot01$ poise, $\rho \approx 1$) in a tube of radius $r = 0\cdot02$ cm., $v_c \approx 500$ cm./sec., a value considerably in excess of those reached in capillary viscometers.

However, for many of the more concentrated colloidal solutions and even dilute solutions of certain very elongated macromolecules (e.g. tobacco mosaic virus) the proportionality between shearing force and velocity gradient is definitely absent at low liquid velocities, the 'apparent' viscosity coefficient decreasing markedly with increasing velocity gradients. Such fluids are said to show anomalous viscosity or non-Newtonian flow and two main reasons have been advanced to explain their behaviour.

1. The occurrence in solution of some type of structure possessing properties similar to those of an elastic solid, and for which an 'elastic' constant or a 'rigidity' is required. Thus, as pointed out by Mark (1940, p. 268), 'So long as the shearing stresses are smaller than this flow limit, it will only be a matter of energy being stored in the liquid and there will be no movement. Only when the shearing stress exceeds the flow limit, will a velocity gradient be produced in the observed system.'

Such properties it is thought, may arise not only in those systems in which the solute units have strong affinities for one another, but also when, owing to extreme asymmetry, the units interfere with and restrict one another's movements.

2. The orientation of very asymmetric macromolecules or particles, caused by velocity gradients in solution (see p. 388 for the mechanism of such orientation), causes a decrease in energy dissipation and therefore in the 'apparent' viscosity. Also involved in the same effect may be the stretching and flattening of kinked molecules in the velocity gradient. A clear discussion of these types of system is given by Mark (1940).

Orientational phenomena form the subject of the next chapter in which it will be shown how the degree of orientation under given conditions can give information on the nature of the orientated units. The study of 'elastic' behaving solutions is covered by the science of rheology (see Chapter XVI). For our present purposes, however, we are concerned only with the elimination of these causes of anomaly, so that true viscosity values may be obtained. Structural viscosity may

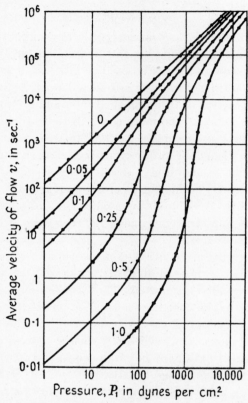

FIG. 13.1. Flow curves for solutions of cellulose tri-nitrate (13·7 per cent. N.) in butyl acetate.
Ordinate: mean velocity of flow, in sec.$^{-1}$
Pressure, in dynes per cm.^{2}

be eliminated by the use of dilute solutions (below *ca*. 1 per cent. usually) in which the solute molecules are so attenuated that structure formation is impossible. Fig. 13.1 demonstrates, for cellulose tri-nitrate in butyl acetate, the increasing occurrence of flow anomalies at higher concentrations. At zero concentration the velocity of flow is directly proportional to the pressure, a result which arises directly from the integration of eqn. (13.1) over a capillary tube. In certain systems, e.g. dilute gelatine gels, however, the tendency to structural viscosity may be so great that at the lowest usable concentrations it may still occur. In such cases the viscosity values obtained are of doubtful value.

Where orientation occurs it is usual to attempt to obtain either complete orientation or complete randomness, cases which can both be treated theoretically. Complete randomness or absence of orientation occurs at very low velocity gradients for which molecular distortion is usually negligible also. Complete orientation may be approached at high velocity gradients, but the possible introduction of distortional effects must also be considered, and it is often not clear how the effects of distortion and orientation, both appearing to give decreased apparent viscosities, may be separated. These questions will be discussed further in Chapter XIV.

Summarizing, in order to obtain the true viscosities of colloidal systems it is necessary to study dilute solutions under conditions for which the orientation by flow of the disperse units is either complete or totally absent.

The experimental methods

For the precise measurement of the viscosities of *dilute* solutions two chief methods are available:

(a) The Capillary Flow Method, in which the flow of the liquid through a suitable tube is timed accurately.

(b) The Concentric Cylinder or Couette Method, in which the liquid, contained between two concentric cylinders rotating relatively to one another, transmits a torque whose measurement may be used to obtain the viscosity of the liquid.

Of these methods, the former is both simpler in operation and more precise in its results, but it suffers from the fundamental disadvantage of having velocity gradients, varying over wide ranges, from zero on the axis of the capillary to a maximum at the walls. Thus, in the case of very large asymmetric particles, the velocity gradient range of an Ostwald viscometer may cover the states of zero, intermediate, and complete orientation, in which case the measured viscosity, though accurate, may be of little theoretical significance. In the concentric cylinder apparatus, on the other hand, the range of velocity gradients is much less, and the mean value readily controlled by the rate of rotation of the moving cylinder; such an instrument is usually used for the study of solutions in which orientation occurs at low velocity gradients.

A short account of each instrument now follows. A third method of measuring viscosity, suitable for the more viscous liquids and solutions, makes use of the rate of fall of small steel (or glass) spheres through the liquid in a tube. An outline of the method, depending upon the validity of Stokes's law, has already been given (p. 211) and since it is not very suitable for use with dilute solutions it will not be described further. For detailed accounts the reader should refer to the monograph by Barr (1931).

(a) *Capillary flow method*

Let us consider the flow of liquid of viscosity coefficient η, through a cylindrical tube, of radius a and length l, between whose ends there is a difference of pressure P. Then it can readily be shown (e.g. see Newman and Searle, 1933) that the distribution of velocity (v) in the liquid across the tube is given by

$$v = \frac{P}{4\eta l}(a^2 - r^2), \qquad (13.6)$$

where r is the distance from the tube axis. At $r = 0$, along the axis, the velocity has its maximum value $Pa^2/4\eta l$, and at the walls, $r = a$, the value zero. Differentiation of (13.6) with respect to r gives

$$\frac{dv}{dr} = -\frac{Pr}{2\eta l}, \qquad (13.7)$$

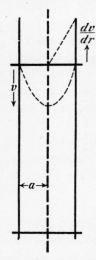

FIG. 13.2. The distribution of velocity and velocity gradient in liquid flowing through a cylindrical tube.

from which the velocity gradient is seen to vary linearly with the distance r from zero at the axis of the tube (Fig. 13.2). To obtain an approximate idea of the usual magnitude of this gradient, the following values, which are roughly typical of a standard Ostwald viscometer for dilute aqueous solutions, may be used:

$P = 10$ cm. of water $\approx 10^4$ dynes; $\quad a \approx 0 \cdot 02$ cm.

$\eta = 0 \cdot 01$ poise; $\hspace{3cm} l \approx 10$ cm.

$$\therefore \quad \frac{dv}{dr} \approx \frac{10^4 \cdot 10^{-2} \cdot 2}{2 \cdot 10^{-2} \cdot 10} \approx 10^3.$$

It will be useful to recall this value later when considering the question of particle orientation.

By integration of eqn. (13.6) over the whole tube, the total volume of liquid (V) flowing through the tube in unit time may be shown to be

$$V = \frac{\pi P a^4}{8\eta l}, \qquad (13.8)$$

an expression often known as Poiseuille's law.

For viscometers in which a liquid flows under its own head of pressure, eqn. (13.8) may be simplified into

$$\eta = C\rho t, \qquad (13.9)$$

where t is the time of flow of a given volume of liquid in the viscometer between two definite points, and C is a constant depending upon the construction of the viscometer. It therefore follows that for two liquids of viscosities η_1 and η_2, densities ρ_1 and ρ_2, and times of flow t_1 and t_2,

$$\eta_1 = \eta_2 \frac{\rho_1 t_1}{\rho_2 t_2}, \qquad (13.10)$$

an equation by which an unknown viscosity may be determined from flow times, densities, and the use of a liquid of known viscosity.

One of the most convenient and widely used of capillary viscometers is that due originally to Ostwald, of which a modern version recommended by the British Standard (188: 1937) is shown in Fig. 13.3. The dimensions of the table in Appendix IV are suggested in the construction of a range of such viscometers. Liquid is introduced into the right-hand limb up to the mark at G, and is then forced through the capillary de into the bulb BC until above the mark at B. Measurements of the time required for the meniscus to fall between marks B and C, placed at points of equal internal diameter, are then made by stop-watch to an accuracy of $\frac{1}{10}$ sec. Other types of capillary viscometer are also used, but since these differ only in minor experimental detail from the Ostwald they will not be described. Eqns. (13.8), (13.9), and (13.10) are, however, accurately obeyed only under certain conditions which we shall now consider.

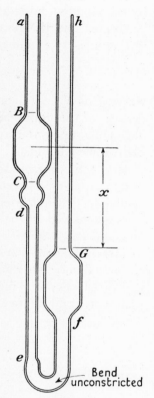

FIG. 13.3. Standard U-tube viscometer. (Suitable dimensions given in Appendix IV.)

Corrections and precautions

(1) *Kinetic energy correction.* In deriving eqn. (13.8) from the Newtonian law it was implicitly assumed that the acting pressure was effective only in overcoming viscous resistance, whereas in practice part of the pressure is used in imparting acceleration (at the entrance of the tube) and kinetic energy to the flowing liquid. The correction

for the pressure required for liquid acceleration is usually made by introducing for l a corrected length l^*; for the kinetic energy possessed by the liquid an integration of this energy over the whole tube may be performed and the total pressure responsible thus evaluated. Introducing into (13.8) and rearranging we therefore obtain

$$\eta = \frac{P\pi a^4}{8Vl^*} - \frac{V\rho k}{8\pi l^*}, \qquad (13.11)$$

where P is the total pressure between the ends of the tube, ρ is the density of the liquid, and k is a constant (near unity) which, though depending upon the apparatus, is usually evaluated by calibration with known liquids.

Although occurring to some extent in any capillary viscometer, it is usual to design a viscometer with a view to making the correction term negligible, when the simple equations (13.8) and (13.10) are valid. This is achieved by making V/l^* small, i.e. by employing slow rates of flow and long capillaries, conditions which also ensure the occurrence of stream-line motion only. The magnitude of the kinetic energy correction may be roughly calculated on the basis of the following figures which are typical of an Ostwald viscometer suitable for use with water and dilute aqueous solutions.

Time of flow $\approx$ 200 sec.; total volume of liquid flowing $\approx$ 5 c.c.

$l^* = 10$ cm.; $\rho = 1$; $k = 1$.

$\therefore$ kinetic energy correction $= \dfrac{\frac{5}{200} \cdot 1 \cdot 1}{8 \cdot \pi \cdot 10} \approx 0 \cdot 0001$;

viscosity of water $\approx 0\cdot 01$ poise.

In this case the correction amounts to 1 per cent., and to reduce it further longer times of flow and longer capillaries are required. The use of much finer bore capillaries (which would increase flow times) is not recommended, in view of the troublesome nature of dust particles which sometimes become fixed in fine capillaries, causing very irregular flow times. Since Ostwald viscometers are nearly always calibrated by means of a liquid with a known viscosity approximating to that of the unknown, kinetic energy corrections of the same order occur in both cases; the error involved in the unknown viscosity through neglect of the correction is usually, therefore, negligible unless the viscometer possesses an unusually large correction term. The extent of the error involved is usually checked by timing the flows of two liquids of accurately known viscosities, when $\rho_1 t_1/\rho_2 t_2$ should be accurately equal to the known η_1/η_2 value.

(2) *Surface tension effects.* In view of the reduction in effective pressure caused by capillary active substances filling a tube up to a given height, only liquids of approximately equal surface tension should be compared or special forms of correction devised.

(3) *Miscellaneous.* The viscometer must be supported inside a thermostat, accurate to $\pm 0.05°$ C. (or better), in a nearly vertical manner which can readily be reproduced after every filling so that the effective head of liquid is constant. Constancy of the thermostat temperature is necessary because of the almost universally high negative temperature coefficient of absolute viscosity (e.g. the viscosity of water decreases by *ca.* 20 per cent. between 20° and 30° C.).

A further essential, especially in the case of capillaries of very fine bore, is the removal of dust and fibrous material by filtration of liquids just prior to introduction into the viscometer. A useful, rapid method of doing this efficiently, is to suck up the liquid into a pipette through a small sintered glass filter fitted over the end of the pipette.

Verification of Newtonian law in capillary viscometers

For negligible kinetic energy corrections, Poiseuille's law (eqn. (13.8)) is merely the integrated form of the Newtonian law, so that a test of proportionality between volume of liquid flowing per second and the acting pressure provides a confirmation of the Newtonian law. In a capillary viscometer this is conveniently accomplished by applying a range of external pressures and measurement of the appropriate rates of flow. For simple liquids and many dilute solutions in stream-line motion the proportionality has been amply confirmed (see Fig. 13.4).

The device of varying the applied pressure in a capillary viscometer has also been considerably used to vary the mean velocity gradient in the tube, by which means the state of orientation in dilute colloidal solutions may be investigated. Proportionality between V and P (eqn. (13.8)) indicates no change, over the range investigated, of the orientational state, and except for very large particles (e.g. tobacco mosaic virus particles in solution, of length 2,800 A) no appreciable orientation. Fig. 13.4 demonstrates such proportionality for dilute solutions of cellulose nitrate in acetone, for which orientation is most probably absent.

However, the concentric cylinder apparatus with its narrower range and easier control of velocity gradients is more suitable for such work, especially for very large colloidal particles whose orientation may be complete at velocity gradients lower than the minimum available in capillary viscometers. The concentric cylinder apparatus will, therefore, be briefly described now.

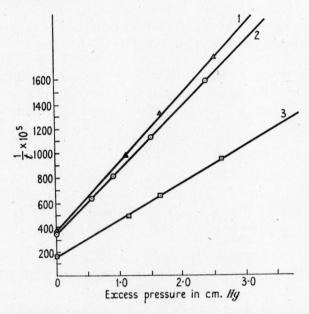

FIG. 13.4. Experimental data for three cellulose nitrate solutions in acetone at concentrations of *ca.* 0·1 gm. per 100 c.c., demonstrating the proportionality between rate of flow through an Ostwald viscometer and the applied pressure.

Abscissa: excess pressure in cm. of Hg; Ordinate: $1/t \times 10^5$, t being the time of flow in seconds.

(b) *Concentric cylinder or Couette viscometer*

In the usual form of apparatus, depicted in Fig. 13.5, the outer cylinder, containing the liquid, is rotated and the inner is suspended freely by a torsion wire. The torque transmitted by the liquid from the outer rotating to the inner cylinder is measured by a mirror and scale arrangement in the usual way. Temperature constancy of the liquid is ensured by pumping water from a thermostat through a liquid jacket (not shown in Fig. 13.5) surrounding the outer cylinder. The latter often has a glass base to permit optical examination (usually by plane polarized light) of the liquid in the annulus (see p. 397). For further details, the account of Edsall (1942) should be consulted.

Taylor (1936) showed that the above arrangement is considerably superior to that in which the inner cylinder rotates, in that stream-line flow occurs up to much higher speeds. For quantitative work, therefore the former arrangement is in general preferable. However, the difference between the two types diminishes as the gap width between the cylinders decreases, and for high speeds and small gaps a rotating inner cylinder is often used, since the mechanical construction is simpler.

For stream-line flow the velocity gradient in the liquid at a distance r from the axis of the system is (Hatschek, 1928, p. 33)

$$\frac{dv}{dr} = 2\omega \frac{1/r^2}{1/R_1^2 - 1/R_2^2}, \quad (13.12)$$

where ω is the angular velocity of the outer rotating cylinder and R_1 and R_2 are the radii of the inner and outer cylinders respectively. Stream-line flow in this apparatus involves the orderly motion of liquid in circular paths concentric with the cylinders, with velocities increasing from zero at the surface of the inner cylinder to a maximum at the outer one. It is clear from eqn. (13.12) that the velocity gradient varies across the gap between

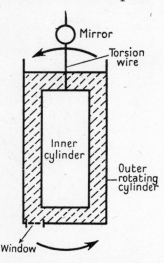

Fig. 13.5. The concentric cylinder or Couette viscometer.

the cylinders but, since R_1 and R_2 do not differ very much in general, the variation is small compared with that occurring in capillary viscometers.

With a decrease in the width of the gap the possible variation in velocity gradient decreases, and if

$$d = R_2 - R_1 \ll R_1$$

then it becomes effectively constant at the value

$$\frac{dv}{dr} \approx \frac{R_1 \omega}{d} \left(\approx \frac{R_2 \omega}{d} \right). \quad (13.13)$$

In much work on the orientation of asymmetric colloidal particles (Chapter XIV) instruments are used with very narrow gaps (≈ 0.02 cm.) and relatively large cylinder radii (≈ 2.5 cm.), giving very high velocity gradients, for which eqn. (13.13) is applicable. For normal viscosity work, however, gaps of several millimetres are more useful, so that the variation in velocity gradient over the gap is appreciable; much smaller absolute values are also used. In this respect we quote, for illustration, figures relating to the apparatus used by Robinson (1939) in an investigation of dilute solutions of tobacco mosaic virus.

$$2R_1 = 3.45 \text{ cm.}; \quad d = 1.135 \text{ cm.};$$
$$\text{length of inner cylinder} = 20.7 \text{ cm.}$$

The velocity gradient calculated by eqn. (13.12) for unit angular velocity varies from 1·14 at the surface of the inner cylinder to 3·14 c.g.s. units at the outer cylinder (cf. with Ostwald viscometer, p. 350).

For stream-line motion between the cylinders and negligible end

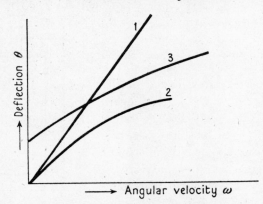

FIG. 13.6. Different types of behaviour in a Couette viscometer. (Deflexion θ v. angular velocity ω.)

Curve 1. Newtonian liquid; Curve 2. Non-Newtonian. Probably particle orientation; Curve 3. Non-Newtonian. Solution 'rigidity'.

effects, the torque on the inner stationary cylinder caused by the rotation of the outer is given by

$$\tau = c\theta = \frac{4\pi l \eta \omega}{(1/R_1^2 - 1/R_2^2)}, \qquad (13.14)$$

where θ is the angular deflexion of the cylinder, c is the torsional constant of the wire, and l is the length of the inner cylinder which is covered by liquid.

Eqn. (13.14) may be rearranged to give

$$\theta = \eta \frac{4\pi}{c(1/R_1^2 - 1/R_2^2)} l\omega = K\eta l\omega, \qquad (13.15)$$

where $K = 4\pi/c(1/R_1^2 - 1/R_2^2)$ depends only upon the torsion wire and the dimensions of the apparatus.

For a liquid whose viscosity coefficient is independent of rate of shear (or velocity gradient) the plot of θ against ω should therefore be a straight line passing through the origin, whose slope has the value $K\eta l$ (curve 1, Fig. 13.6).

The constant K is usually obtained by calibration with a liquid of known viscosity, when unknown viscosities may subsequently be determined merely from the slope of the θ v. ω plot. If the plot is not a straight line passing through the origin, then the liquid is non-

VISCOSITY

Newtonian. A line passing through the origin with a slope decreasing with increasing values of θ may indicate, for a dilute solution, the occurrence of appreciable solute orientation (curve 2), whilst the intersection of the line with the θ axis at finite positive values indicates rigidity at low velocity gradients (curve 3). Fig. 13.6 shows diagrammatically the different possible types of θ v. ω plot. Where there are definite indications of the occurrence of orientation at higher rates of shear, viscosities corresponding to no orientation are obtained by extrapolation of apparent viscosities to zero angular velocity.

End correction. In deriving eqn. (13.14) the torque (τ'), arising as a result of the motion of the end of the outer cylinder relative to that of the inner (i.e. end effect) was neglected. Since this is a constant for a given liquid and given separation of a pair of cylinders, and is independent of the length (l) of inner cylinder covered, we can write for two different lengths l_1 and l_2:

$$\tau_1 + \tau' = \frac{4\pi l_1 \eta}{(1/R_1^2 - 1/R_2^2)} \omega + \text{constant} \qquad (13.16)$$

and

$$\tau_2 + \tau' = \frac{4\pi l_2 \eta}{(1/R_1^2 - 1/R_2^2)} \omega + \text{constant}. \qquad (13.17)$$

Subtracting (13.17) from (13.16) gives

$$\tau_1 - \tau_2 = \frac{4\pi(l_1 - l_2)\eta}{(1/R_1^2 - 1/R_2^2)} \omega = c(\theta_1 - \theta_2). \qquad (13.18)$$

End effects, if important, are therefore eliminated by filling the annulus to different heights and by the use of eqn. (13.18). It should be noted that anomalous surface effects, e.g. those arising from the formation of a rigid surface skin, are eliminated by the same procedure.

Summarizing briefly, we have seen how, experimentally, Newtonian and non-Newtonian behaviour may be detected. In the former case, viscosity coefficients corresponding to a range of velocity gradients are readily and conveniently determined in an Ostwald type viscometer; in the latter case, the viscosity coefficients corresponding to zero velocity gradient are obtained by extrapolating apparent viscosities to zero gradient; in this case the experimental measurements are most suitably done in a concentric cylinder type of viscometer.

Different functions of viscosity

With solutions we are usually concerned less with absolute values than with the viscosity increments due to the added solute. For this reason, the different functions accepted for this purpose, which we shall use later in this chapter, are given below.

η_0 denotes the viscosity of the pure solvent and
η that of the solution.

$\eta/\eta_0 = \eta_r$, termed the *relative viscosity*, is conveniently determined from densities and timing measurements using a capillary viscometer.

$(\eta/\eta_0 - 1) = (\eta_r - 1) = \eta_{sp}$ is termed the *specific viscosity*.

$\lim_{c \to 0} \eta_{sp}/c = [\eta]$ is known as the *intrinsic viscosity*, where c is usually measured in grams per 100 c.c. of solution. It may be defined alternatively by the equivalent expression $\lim_{c \to 0} \ln(\eta_r/c)$.

The Viscosity of Colloidal Solutions

The introduction of colloidal material into a solvent causes an increase in viscosity which varies widely with the nature of the colloidal particles. Accordingly considerable attention has been focused on the study of viscosity-concentration relationships with a view to deriving information on such properties as the size and shape of the particles.

Spherical particles

(a) Very dilute solutions

Progress was first made in the study of dilute suspensions of large rigid spheres in which each sphere may be considered surrounded completely by a large volume of unaltered solvent. Increased viscosity arises as a result of distortion of the stream-lines of the flowing liquid around the spheres. Outside the influence of the suspended particles the stream-lines in the solvent are considered unaltered. On such a picture Einstein (1906) calculated the change in viscosity hydrodynamically, making the further assumptions:

1. The suspended particles are large compared with those of the solvent, but small compared with the dimensions of the apparatus.
2. The suspended particles are rigid and wet completely by the solvent which is incompressible.
3. The solution is so dilute that overlapping of the disturbed regions of flow around each suspended particle does not occur.
4. The effects of gravitation and inertia are negligible and turbulence is excluded.

His result, valid therefore only at low solute concentrations, is usually written
$$\eta/\eta_0 = \eta_r = 1 + 2 \cdot 5\phi, \tag{13.19}$$

VISCOSITY

where ϕ is the volume fraction of the disperse phase (i.e. volume of particles/total volume). A plot of η_r against ϕ should thus have a slope equal to 2·5 at low concentration.

Alternatively the equation may be written in the well-known form

$$\frac{\eta_r - 1}{\phi} = \frac{\eta_{sp}}{\phi} = 2\cdot 5, \qquad (13.19\,\text{a})$$

from which a plot of η_{sp}/ϕ against ϕ should give a line parallel to the ϕ axis cutting the η_{sp}/ϕ axis in the value 2·5.

From this equation it follows that the relative viscosity at low concentration is independent of the nature of the disperse phase and of the particle size, a result which is possible only because the increased viscosity is not directly caused by the suspended particles, but arises indirectly as a result of the disturbance of the solvent flow. The equation cannot therefore be used for determining particle size.

The most satisfactory experimental tests of Einstein's equation have been performed by Eirich (1936 and 1937) and co-workers, using suspensions of glass spheres, and of certain spherical spores and fungi. Table 13.1 contains typical results of their viscosity measurements (as summarized by Mark, 1940) made by different experimental methods. With the exception of those from the capillary flow method (attributed to inertia effects), these figures demonstrate the validity of the 2·5 coefficient, up to volume concentrations of 2–3 per cent.

TABLE 13.1. *Test of Einstein's Equation*

Viscometer	Radius of spheres, cm. $\times 10^{-4}$	η_{sp}/ϕ	$\dfrac{\eta_{sp} - 2\cdot 5\phi}{\phi^2}$
Concentric cylinder (Couette) .	3	2·5 ($\pm 0\cdot 2$)	9·0 ($\pm 2\cdot 0$)
	80	2·5 ($\pm 0\cdot 2$)	12·0 ($\pm 4\cdot 0$)
	160	2·5 ($\pm 0\cdot 2$)	13·0 ($\pm 2\cdot 0$)
Capillary tube . . .	3	2·7 ($\pm 0\cdot 3$)	6·0 ($\pm 3\cdot 0$)
	80	1·8 ($\pm 0\cdot 3$)	..
	160	2·0 ($\pm 0\cdot 3$)	..
Falling ball . . .	80	2·4 ($\pm 0\cdot 5$)	..

The success of Einstein's equations, demonstrated by other work also, implies the validity, under the appropriate experimental conditions, of the Einstein assumptions listed above. It should be noted, however, that not all colloidal particles are spherical or rigid, so that deviations from the 2·5 coefficient in some systems might be anticipated on these grounds.

(b) Moderate concentrations

It was also shown, however, as is evident from Table 13.1, that the simple equation holds only at small concentrations, the value 2·5 being the limiting slope of the plot of η_r against ϕ, and that further terms were required at higher concentrations. An extension of the Einstein theoretical treatment required modification of his assumptions, the most important being consideration of the mutual interactions of the disturbed flow region around each suspended sphere. In a new treatment Guth and Simha (1936) considered these effects and derived the equation

$$\eta_r = 1+2\cdot5\phi+14\cdot1\phi^2+..., \qquad (13.20)$$

in which a term involving the square of the volume concentration and the coefficient 14·1 appears. Again, by this equation, the viscosity is independent of the particle size, but this is an approximation, since minor factors involving the particle size were neglected in deriving the equation.

In an experimental check of eqn. (13.20), it is convenient to plot $\eta_{sp}/\phi \left(= \dfrac{\eta_r - 1}{\phi} \right)$ against ϕ, for, from the rearranged form,

$$\eta_{sp}/\phi = 2\cdot5+14\cdot1\phi, \qquad (13.20\text{a})$$

the intercept on the η_{sp}/ϕ axis should be 2·5 and the slope of the line 14.1.

The last column of Table 13.1 demonstrates the approximate correctness of the new coefficient, and Smith (1942) has shown, for rubber latex emulsions, the correctness of the equation up to concentrations of 30 per cent.†

In order to obtain a function of viscosity which is quite independent of solute-solute interactions, it is clearly necessary to make measurements over a range of low concentrations and to extrapolate to zero concentration. This statement is of even greater importance in the case of asymmetrical molecules or particles when interaction effects achieve greater importance.

(c) Solvation

Since many colloidal materials are strongly solvated in solution it is of importance to know how viscosity is thus affected. Solvation involves the strong binding of solvent molecules, usually in a layer not more than a molecule thick, the particle plus solvent layer thus acting as the kinetic unit in solution. An increase in volume concentration is thus clearly involved which if unaccounted for will cause deviations from eqns. (13.19) and (13.20). It is therefore possible to calculate the degree of solvation for spherical colloidal particles by comparing the 'dry' volume concentrations with those required to conform with the appro-

† Quoted by Bull (1943), the original paper being inaccessible to the authors.

VISCOSITY

priate equation. Such calculations were, at one time, made for colloidal materials of very diverse character, and the results of Table 13.2 from Kraemer (1931) show the divergent values obtained for the ratio of hydrodynamic to dry specific volume (ϕ/v).

TABLE 13.2. *Ratios of Hydrodynamic to Dry Specific Volumes*

Substance	ϕ/v	Substance	ϕ/v
Sucrose in H_2O	1·6	Clay in water	9
Egg albumin in H_2O	0·9	Starch in water	20
Sulphur in water	1·2	Iso-electric gelatin at 40°	6
		,, ,, ,, 20°	30
Gamboge in water	1·25	Cellulose nitrate in ethyl acetate	80
Diluted rubber latex	1·00	Rubber in benzene	300–500

For the first column of substances the values of ϕ/v are near enough to unity to be reasonable. In the second column, however, the very large values introduced by the high solution viscosities are clearly due to causes other than solvation, and it is now accepted that two main causes are responsible.

1. The occurrence of interlacing networks of solute molecules or liquid structure. The big difference between solutions of iso-electric gelatin at 20° and 40° C. is most reasonably explained by aggregation of the long-chain molecules at the lower temperature. It should be realized that this could be readily detected as a deviation from Newtonian behaviour.

2. The deviation of colloidal particles from spherical shape. In this respect it is known that many linear polymers exist in solution as asymmetric units, for which the assumptions involved in deriving (13.19) and (13.20) are necessarily invalid. Deviations from Newtonian behaviour are readily detected for very large asymmetric particles (e.g. old vanadium pentoxide sols, tobacco mosaic virus), but for many asymmetric colloidal molecules, such deviations are detected with more difficulty only at very high velocity gradients. However, such systems showing apparent Newtonian behaviour, give viscosities far in excess of those required by eqns. (13.19) and (13.20) for spherical particles.

In the dilute solutions considered here the high viscosities arising from the asymmetric nature of the colloidal particles are very common, and we shall now proceed to discuss them in more detail. Structural viscosity is assumed to be absent.

Asymmetric particles

The introduction of spherical colloidal particles into a pure solvent leads to increased viscosity mainly as a result of the extra dissipation of energy involved by flow distortion in the solvent around each particle. With asymmetric particles additional causes are also involved. Thus the rotation of the asymmetric particles, enforced by the velocity gradient, introduces increased extra energy dissipation, a mechanism which is clearly much less important the more nearly spherical the suspended particles. The deformation of the particles in the velocity gradient and the much greater interactions occurring in suspensions of asymmetric particles provide further possible mechanisms.

In the following section interaction effects are not considered so that the treatment must be considered applicable only to results at vanishingly low concentration, i.e. obtained by extrapolation to zero concentration. Such interactions introduce considerable complexity and have not as yet been satisfactorily treated theoretically.

(a) Theoretical approach

The first attempt to calculate the viscosity of dilute suspensions of rigid ellipsoidal particles was made by Jeffery (1923), using an extension of Einstein's methods. Since, however, such particles rotate continuously in suspension, and, further, since the energy dissipation is dependent upon the angle made by the particle with the stream-lines, indeterminancy in the final equations could only be avoided by assumptions regarding the orientational state of the particles.

Jeffery's equations have now been superseded by several others obtained by making various assumptions regarding the nature of the suspended particles and their behaviour in the flowing solvent.

Kuhn (1933 and 1934) derived expressions for the viscosity of dilute suspensions of rigid chains of spherical particles in the cases of overwhelming and negligible Brownian motion. In the former case (applicable to many solutions of high polymers) he obtained by modifying the Einstein equation,

$$\eta_{\mathrm{sp}}/\phi = 2 \cdot 5 + \frac{J^2}{16}, \tag{13.21}$$

where J is the ratio of the long (L) to the short (s) axis of the chains.

For high axial ratios this reduces to

$$\eta_{\mathrm{sp}}/\phi \simeq \frac{J^2}{16}. \tag{13.22}$$

The inclusion of the Einstein term has been criticized (e.g. Huggins, 1938), but in its support it should be said that the equation gives axial ratio values in good agreement with those from the most recent equations.

Huggins (1938, 1939), using the model of Kuhn and improving his method, obtained for the case of dilute suspensions of long rod-like molecules in strong Brownian motion (no orientation):

$$\eta_{sp}/\Phi = \left(\frac{\pi N}{24000}\right) l_M^2 a n^2, \qquad (13.23)$$

where Φ is the concentration in submoles. per litre (a submolecule being merely the repeating unit in the chain, e.g. $C_6H_5CH.CH_2$—in polystyrene), N is the Avogadro number, and l_M and a are the length and radius respectively of the spherical submolecules of which there are n per chain.

In the case of polyatomic chains of carbon atoms Kuhn (1934) had shown that the statistical mean length of the chain is proportional to $\sqrt{n}$, i.e.

$$L = \text{const.}\sqrt{n}, \qquad (13.24)$$

where n is the number of carbon atoms in the chain, the constant depending upon the type of molecule and the conditions. If we can assume the transverse dimension of such molecules to be constant, then the axial ratio becomes proportional to $\sqrt{n}$ (or $\sqrt{M}$) and inserting into Kuhn's equation (13.22) we obtain

$$\eta_{sp}/\phi \propto n \text{ (or } M). \qquad (13.25)$$

Huggins generalized this treatment and obtained for rigid, randomly kinked, molecules an analogous expression,

$$\eta_{sp}/\Phi = kBl_M^2 an, \qquad (13.26)$$

where B is a constant depending upon the angle between the two bonds joining each submolecule to its neighbours, and k, another constant, has the values $2 \cdot 94 \times 10^{20}$ and $2 \cdot 82 \times 10^{20}$ for the cases of overwhelming and weak Brownian motion respectively.

In deriving this expression each molecule is considered rigid in its statistically most probable form, which explains the very small effect of orientation. Thus a very much kinked chain would have almost no preferential direction of orientation in a velocity gradient. However, the possible stretching of such chains, which is neglected in this treatment, might cause the effect of orientation to be much greater (see p. 407).

The recent treatment by Simha (1940) of rigid ellipsoidal particles of axial ratio J ($= a/b$, see p. 212), in such overwhelming Brownian motion that preferential orientation is absent, avoids any indeterminacy and is probably the most satisfactory yet for this case. His final expression is

$$\eta_{sp}/\phi = \frac{J^2}{15(\ln 2J - \tfrac{3}{2})} + \frac{J^2}{5(\ln 2J - \tfrac{1}{2})} + \frac{14}{15}, \qquad (13.27)$$

an equation whose first term is identical with the right-hand side of the equations of Eisenschitz (1933) and of Burgers (1938) for little or no orientation. The inclusion of the further terms of Simha's equation gives a larger increase of viscosity with axial ratio than the other two equations, a result to be attributed to the greater emphasis on the complete homogeneity of distribution of particle axes assumed by Simha.

Other equations, appropriate to other conditions, have been proposed, for information on which the reader is referred to original papers or to the reviews by Mark (1940) and Eirich (1940). Common to all these equations is the fact that axial ratios so derived refer to the *solvated* unit as it occurs in the solution.

An attempt to investigate experimentally the effect of axial ratio on suspension viscosity was made by Eirich, Margaretha, and Bunzl (1936) using model rod-like suspensions of glass and silk. The particles were so large (lengths of a few tenths of a millimetre) that Brownian motion was ineffective in preventing orientation, so that a quantitative comparison with the theoretically devised equations assuming strong Brownian motion was not possible. However, the strong dependence of viscosity on axial ratio was demonstrated clearly, and there was some measure of agreement with the theoretical equations in which strong Brownian motion was not assumed.

All equations agree that specific viscosity rises with axial ratio, though quantitatively large differences exist as is shown by Fig. 13.7, in which η_{sp}/ϕ v. J is plotted for the different equations mentioned, all of which assume strong Brownian motion and inappreciable orientation. Table 13.3 also demonstrates the differences in the mean axial ratios, calculated by the different equations, from the viscosity data (also included) for certain cellulose nitrate fractions of approximately uniform nitrogen content (Campbell and Johnson, 1944).

Viscosity and degree of orientation. Much remains to be done concerning the dependence of viscosity on velocity gradient or degree of orientation. In this respect the hydrodynamic treatment by Burgers (1938), though possibly inaccurate in detail, is complete and useful. Only a brief outline can be included here. In all cases, for suspensions of elongated ellipsoidal or rod-like particles, the specific viscosity may be written

$$\eta_{sp}/\phi = K, \qquad (13.28)$$

where K depends upon the type of particle. For spheres, $K = 2 \cdot 5$, in accordance with Einstein's equation, whilst for very long rods,

$$K = \frac{J^2}{(\ln 2J - \tfrac{3}{2})} F, \qquad (13.29)$$

VISCOSITY

where F depends upon the orientational state of the particles. Where the effective rate of shear (G) is much less than the rotational diffusion

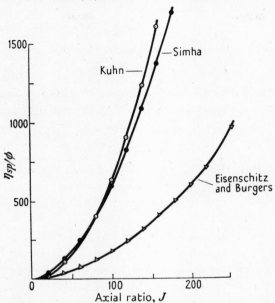

Fig. 13.7. η_{sp}/ϕ against the axial ratio, J, for well-known theoretically derived equations.

Table 13.3. *Intrinsic Viscosities and Calculated Axial Ratios of Cellulose Nitrate Fractions*

	Axial ratios				
			Huggins		Burgers and Eisenschitz
$(\eta_{sp}/c)_{c=0}$	Kuhn	Simha	Flexible	Rigid	
5·37	113	118	560	24·4	216
3·36	101	104	446	21·8	188
3·10	90	91	355	19·4	183
2·51	81	81	290	17·5	162
2·16	76	76	254	16·4	151
1·90	69	68	213	15·1	136
1·57	62	61	172	13·5	120
1·35	62	60	171	13·5	119
1·15	56	54	141	12·2	108
0·55	37	33	61	8·0	65
0·33	31	26	41	6·6	52

coefficient (Θ), i.e. $G/2\Theta \leqslant 1$, and orientation is slight, $F = \frac{1}{15}$ and we have

$$\eta_{sp}/\phi = \frac{J^2}{15(\ln 2J - \frac{3}{2})} \qquad (13.30)$$

(see Chapter XIV), in which K (R.H.S.) is identical with the first term of Simha's equation. However, where $G/2\Theta \approx 1$, i.e. weaker Brownian motion and appreciable orientation,

$$K \approx \text{const.} \times L, \tag{13.31}$$

where L is the length of the suspended particle, and on introducing into (13.28),

$$\eta_{sp}/\phi \approx \text{const.} \times L. \tag{13.32}$$

For homologous series of long, rod-like, particles of constant cross-section (i.e. $M \propto L$) eqns. (13.30) and (13.32) reduce respectively to

$$\eta_{sp}/\phi \propto M^2, \tag{13.30a}$$
$$\eta_{sp}/\phi \propto M. \tag{13.32a}$$

In this case an important difference arises from the orientational state of the suspension (cf. Huggins's results for randomly kinked chains, p. 363), as is to be expected when particles of the disperse phase are of pronounced asymmetry.

Asymmetry and solvation. Eqns. (13.21) to (13.32) are concerned with the increase in specific viscosity caused by particle asymmetry, and the calculations of Table 13.3 were made directly from these equations. Strictly, however, account should have been taken of solvation, which we have seen increases the effective volume concentration of the disperse phase and leads to increased viscosity. Thus, just as the factors of solvation and asymmetry are usually both involved in giving frictional ratios greater than unity (p. 260), so in a similar manner both factors are involved in giving values of $\lim_{\phi \to 0} \eta_{sp}/\phi$ greater than the Einstein value 2·5. Oncley (1941) has constructed, chiefly for proteins, contours of constant $(\eta_{sp}/\phi)_{\phi=0}$ values on a diagram with hydration as abscissa and axial ratio plotted logarithmically as ordinate (Fig. 13.8). From these contours, calculated from the now generally accepted Simha relation, on the basis of rigid ellipsoids of revolution, an estimate of either hydration or asymmetry can be used to obtain the other unknown for a given value of $(\eta_{sp}/\phi)_{\phi=0}$ (cf. contours of constant frictional ratio, f/f_0, p. 262). Assuming once more (see also p. 293) the value 0·40 as an upper limit to the weight of water immobilized by 1 gm. of any protein, the upper limit to the value of the intrinsic viscosity of a solution of spherical molecules (whose volume fraction is calculated on the basis of dry solute) is approximately 3·8. Table 13.4 contains axial ratio values for several proteins obtained from viscosity data collected by Mehl, Oncley, and Simha (1940) by means of Oncley's curves (Fig. 13.8) on the basis of zero and 30 per cent. hydration for elongated and flattened ellipsoids of revolution. These results should be compared with those of Table 11.2, which were obtained on similar assumptions from sedimentation and diffusion data.

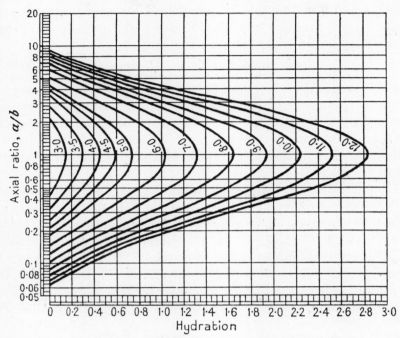

Fig. 13.8. Values of axial ratio and hydration in accord with various values of η_{sp}/ϕ for proteins. (Oncley (1941).)

TABLE 13.4 *Axial Ratios from Viscosity*

Protein	$(\eta_{sp}/\phi)_{\phi=0}$	30% hydration		Zero hydration	
		Prolate	Oblate	Prolate	Oblate
Pepsin	5·2	3·2	3·6	4·5	5·9
Haemoglobin	5·3	3·2	3·5	4·6	5·9
Egg albumin	5·7	3·5	4·0	5·0	6·2
Lactoglobulin	6·0	3·7	4·5	5·1	6·9
Serum albumin	6·5	4·0	5·0	5·5	7·5
Amandin	7·0	4·5	5·5	6·0	8·5
Serum globulin	9·0	5·7	8·3	7·5	11·6
Thyroglobulin	9·9	6·1	7·4	7·8	13·0

(b) *Viscosity and molecular weight*

The ease and accuracy with which viscosity measurements can be made has brought Staudinger's law relating viscosity and molecular weight into great prominence. Originally stated in the form

$$\eta_{sp}/c = K_m M, \qquad (13.33)$$

where K_m is a constant to be determined for each polymer-solvent system, it was soon apparent that, in view of the common variation of η_{sp}/c with c, the original form should be replaced by

$$\lim_{c \to 0} \eta_{sp}/c = [\eta] = K_m M. \qquad (13.34)$$

The concentration units used by Staudinger and co-workers are sub-molecules per litre, and their constants K_m, depending upon the concentration unit are different from, but proportional to, those of other workers who use mainly grams/100 c.c. of solution. Unless otherwise stated, it should be taken that c is expressed as grams/100 c.c. of solution in other parts of this chapter.

The original experimental evidence for the Staudinger law was based on molecular weights of medium and short chain-length polymers obtained by normal colligative methods. Later, additional confirmation was claimed on the basis of osmotic pressure measurements made on solutions of higher molecular weight materials. However, criticism has been directed at both lines of evidence: firstly, on the grounds that extrapolation from low to high molecular weights of a different order is not justifiable; secondly, that the osmotic pressure measurements were carried out at concentrations where large deviations from the ideal van 't Hoff equation occur so that inaccurate osmotic molecular weights were obtained. Further, since osmotic methods yield, for polydisperse materials, a number average molecular weight, whilst (if the Staudinger law be true) viscosity yields a weight average value, a check of the law by osmotic pressure is reliable only if monodisperse or highly fractionated materials are available. It is doubtful if the fractions used by Staudinger and co-workers were of the well-defined character required.

These criticisms were undoubtedly at least partly true, so that the quantitative side of the Staudinger law was open to doubt, though it has often been used to obtain a rough idea of molecular weights.

In view of the application of hydrodynamic methods to the viscosity of suspensions it is of importance to consider the Staudinger law from this point of view. Huggins (1943) summarizes his results for solute molecules which are large relative to those of the solvent and in strong Brownian motion, by stating that η_{sp}/c for a homologous series is:

(1) constant for spherical molecules;

(2) proportional to n^2 for rod-like molecules; and
(3) proportional to n for randomly kinked molecules.

The last relation only is to be identified with the Staudinger law. Not all workers would, however, accept these alternatives. Carter, Scott, and Magat (1946), for instance, suggest that two extreme types of polymer molecules exist: one, entrapping solvent within itself, would conform to $[\eta] \propto M^{\frac{1}{2}}$, and the other, in which at least part of the neighbouring solvent flows freely past the polymer, to $[\eta] \propto M$.

As pointed out by Burgers, for suspensions in strong Brownian motion, a linear relation between the molecular weight and $(\eta_{sp}/c)_{c=0}$ is dependent upon the existence of some non-linear relation between the length of the molecule and molecular weight. Such a relation may exist, in certain cases, e.g. for randomly kinked chain molecules (where $L \propto \sqrt{M}$), but in general it cannot be true. On the other hand, for rod-like particles in weak Brownian motion it appears, from Burgers's treatment, that the Staudinger law should hold.

Summarizing, therefore, hydrodynamic methods agree that though the Staudinger law may hold in certain individual cases and under certain conditions, the general validity of the law is improbable.

Experimentally it has been shown by various authors, that though eqn. (13.34) may hold approximately over small molecular weight ranges, other terms and modifications are required for larger ranges. Thus Meyer and van der Wyk (1935) and Fordyce and Hibbert (1939), working respectively on well-defined straight chain hydrocarbons and polyoxyethylene glycols [HO.$CH_2CH_2O(CH_2CH_2O)_nCH_2CH_2OH$] (the latter with molecular weights up to about 8000), suggested the introduction of a new constant term in the Staudinger equation, giving

$$[\eta] = K_i[M] + K_0, \qquad (13.35)$$

the constants K_i and K_0 varying with the solvent-polymer system and with temperature as in Table 13.5.

TABLE 13.5. *Values of K_i and K_0 (eqn. 13.35)*

System	Temp. °C.	K_i	K_0
Polyoxyethylene glycol in CCl_4	20	0.83×10^{-5}	0.034
n-Paraffins in CCl_4.	25	10.40×10^{-5}	0.011

More recently, a different equation for higher molecular weights ($> 10{,}000$) has been used by Flory (1943) and Alfrey, Bartovics, and Mark (1943) amongst others, and good experimental evidence has been provided. Thus Flory, working on carefully fractionated polyisobutylenes whose molecular weights were measured osmotically, showed that the relation between intrinsic viscosity and molecular weight, for molecular weights varying between 10,000 and 1,000,000, was well represented by

$$[\eta] = KM^\alpha, \tag{13.36}$$

where K and α are constants characteristic of a given polymer-solvent system and the temperature, α possessing values varying between 0·5 and 1·5. Table 13.6 contains a selection of values of K and α taken from those compiled by Mark (1945), where detailed references will be found. It is probable, however, that some of the values may be modified in the light of further work.

TABLE 13.6. *Values of K and α for Certain Systems*

System	T °C.	$K \times 10^4$	α
Cellulose nitrate in acetone	27	0·38	1·0
Cellulose acetate in acetone	25	9·1	0·67
Polyisobutylene in di-isobutylene	20	3·6	0·64
Polymethylmethacrylate in acetone	30	1·6	0·67
Polyvinyl acetate in acetone	30	2·8	0·67
Polyvinyl alcohol in water	30	5·9	0·67
Polyvinyl chloride in cyclohexanone	25	0·11	1·0
Cellulose in cuprammonium oxide	25	18	0·72

Although there is no doubt that eqns. (13.35) and (13.36) represent substantial improvements on the Staudinger law (e.g. Flory (1943) points out that the Staudinger law for high molecular weight polyisobutylenes gives values ten times too low), yet neither is sufficient by itself to cover the whole range of molecular weights, each being applicable only for its own special range. It seems unlikely that any simple equation could cover satisfactorily the whole molecular weight range.

Average molecular weights from viscosity. It was stated on p. 43 that the average molecular weight obtained from viscosity measurements was different from the number average value, and where Staudinger's law holds it can readily be shown to be the weight average value (e.g. Kraemer, 1941).

Thus if $(\eta_r - 1)_i$ be the viscosity increment for the molecules of the ith component, of molecular weight M_i, of a polydisperse system, then by Staudinger's law

$$(\eta_r - 1)_i = KM_i c_i, \tag{13.37}$$

where c_i is the concentration of the ith solute component in *dilute* solution and K is the appropriate Staudinger constant.

The total viscosity increment is given by

$$(\eta_r - 1) = \sum (\eta_r - 1)_i = K \sum M_i c_i \tag{13.38}$$

and the intrinsic viscosity (assuming infinitely dilute solution) by

$$[\eta] = \left[\frac{(\eta_r - 1)}{c}\right]_{c=0} = \frac{K \sum M_i c_i}{\sum c_i} = \frac{K \sum w_i M_i}{\sum w_i}, \quad (13.39)$$

where w_i, the weight fraction of the ith component, is proportional to the corresponding concentration c_i. Thus the molecular weight obtained by the Staudinger law is

$$M = \frac{\sum_i w_i M_i}{\sum_i w_i} = \frac{\sum n_i M_i^2}{\sum n_i M_i}, \quad (13.40)$$

which is the weight average molecular weight.

In a similar manner it is readily shown that where eqn. (13.35) rather than the Staudinger law holds, the same weight average value is obtained. However, if eqn. (13.36) involving a fractional power α of the molecular weight is valid, then applying the same method as above the viscosity average molecular weight, M_ν, obtained is given by

$$M_\nu = \left[\frac{\sum w_i M_i^\alpha}{\sum w_i}\right]^{1/\alpha} = \left[\frac{\sum_i n_i M_i^{\alpha+1}}{\sum n_i M_i}\right]^{1/\alpha}. \quad (13.41)$$

For $0 < \alpha < 1$ it can be seen that

$$M_N < M_\nu < M_W,$$

so that viscosity measurements would yield molecular weights between weight and number averages. On the other hand for $\alpha > 1$, molecular weights greater than weight averages and nearer Z averages would be obtained (p. 43).

Determination of $[\eta]$. In determining values of $[\eta]$ suitable for the calculation of molecular weights, it is usual to measure the viscosities of four or five solutions whose concentrations are not greater than 1 gm. per 100 c.c. of solution. η_{sp}/c or $\ln(\eta_{sp}/c)$ is then plotted against c, giving usually an approximately straight line, which may readily be extrapolated to zero concentration giving $[\eta]$.

(c) *Viscosity-concentration relations*

We have seen that in determining intrinsic viscosities it is usual to make measurements of viscosity at such dilutions that extrapolation to zero concentration is as reliable as possible. Some knowledge of the nature of viscosity concentration relations is clearly helpful in this respect. Experimentally it is found that the plot of η_{sp}/c against c is well represented by an equation of the type

$$\eta_{sp}/c = a + bc, \quad (13.42)$$

where a and b are constants for a given polymer and solvent. Fig. 11.12 shows a family of such plots for cellulose nitrate fractions, in which it is apparent that with increasing values of the constant a ($\equiv [\eta]$) the slope of the curves (b) grows larger (Campbell and Johnson, 1944). Quantitatively, it was shown that within the limits of experimental error

$$\sqrt{b} = Aa, \qquad (13.43)$$

where A was a constant, and inserting in (13.42) we obtain

$$\eta_{\text{sp}}/c = a + (Aa)^2 c, \qquad (13.44)$$

or in terms of intrinsic viscosities,

$$\eta_{\text{sp}}/c = [\eta] + A^2 [\eta]^2 c, \qquad (13.45)$$

an expression similar to that used by Mead and Fuoss (1942) for polyvinyl chloride solutions and for which theoretical support has been given by Huggins (e.g. 1943). For very high molecular weight materials or at slightly higher concentration, where curvature of the η_{sp}/c v. c lines occurs, it is necessary to introduce a further term in c^2 into eqn. (13.45). The complete equation

$$\eta_{\text{sp}}/c = [\eta] + k'[\eta]^2 c + \text{higher terms in } c, \qquad (13.46)$$

where we have written $A = k'$ to conform with modern nomenclature, is now widely applied to polymer systems. Recalling that for long particles in strong Brownian motion, $[\eta] \propto J^2$ (p. 362), eqn. (13.46) is seen to be very similar to the equation

$$\eta_{\text{sp}}/c = n_1 J^2 + n_2 J^4 c + \ldots, \qquad (13.47)$$

where n_1 and n_2 are numerical coefficients and J the axial ratio, which was derived theoretically by Guth (1936) also assuming predominant Brownian motion.

Many other empirical equations expressing viscosity-concentration relationships reduce to a similar form. Thus Schulz and Blaschke (1941) suggest

$$[\eta_{\text{sp}}/c]_{c=0} = [\eta] = \frac{\eta_{\text{sp}}/c}{1 + K\eta_{\text{sp}}}, \qquad (13.48)$$

which reduces to

$$\eta_{\text{sp}}/c = [\eta] + K[\eta]^2 c + K^2 [\eta]^3 c^2 + \ldots, \qquad (13.48\text{a})$$

an equation only superficially different from eqn. (13.46).

If the term involving c in (13.46) can be considered analogous to the similar term in eqn. (13.20 a) for interacting spheres, then it is naturally interpreted in terms of solute-solute interactions. Further, since such interactions are expected to increase in importance with increasing

VISCOSITY 373

values of $[\eta]$, which is at worst a rough measure of molecular size, the presence of $[\eta]$ in the interaction term is understandable. Though this view can be accepted to some extent, the complexities of the more concentrated suspensions of asymmetric particles are so great that probably other effects are also involved.

Whilst eqn. (13.46) without terms higher than that in c holds only for dilute solutions ($c \not> 1$ gm./100 c.c.), other equations have been suggested to cover larger concentration ranges (e.g. see Mark, 1940, or Pfeiffer and Osborn, 1943). Of these we shall mention two of the more outstanding. Martin (1942) empirically developed an equation holding up to concentrations of 5 per cent. for solutions of cellulose and its derivatives, and of rubber, polystyrene, and other polymers, which may be written

$$\eta_{sp}/c = [\eta]e^{K[\eta]c} \tag{13.49}$$

or

$$\log(\eta_{sp}/c) = \log[\eta] + K[\eta]c. \tag{13.50}$$

Clearly a plot of $\log(\eta_{sp}/c)$ v. c yields, for systems obeying this equation, a straight line whose intercept on the $\log(\eta_{sp}/c)$ axis is $\log[\eta]$ and whose slope is $K[\eta]$.

Expansion of the exponential term, and neglect of terms in c higher than the first power (which is valid for low values of c), yields an equation identical with (13.46), which we have seen is generally applicable for low polymer concentrations. The identity of the two equations at low concentration reflects very favourably on the use of Martin's equation.

Philipoff's equation,

$$\eta_r = \left(1 + \frac{[\eta].c}{8}\right)^8, \tag{13.51}$$

has been applied considerably, partly because it has only one characteristic constant ($[\eta]$) and also because it does seem to be applicable up to very high concentrations (e.g. up to 25 per cent. for ethyl cellulose). Industrially, therefore, the equation is valuable. However, solutions of higher concentrations are covered only at the expense of agreement at low concentration as might be anticipated, especially with a one-constant equation. Eqn. (13.51) may be expanded into the power series

$$\eta_r = 1 + [\eta]c + \text{higher terms}, \tag{13.52}$$

which at low concentration reduces to $\eta_{sp} = [\eta]c$.

Estimations of $[\eta]$ from the more concentrated solutions are made by plotting $(\eta_r)^{\frac{1}{8}}$ against concentration when the slope of the line obtained gives $[\eta]/8$.

However, although equations covering high concentration ranges are undoubtedly of use, especially industrially, the practice of obtaining intrinsic viscosities for fundamental purposes by the application of such equations is not encouraged. In view of the complexities of concentrated solutions, it is perhaps accidental that certain equations should fit results so well, and the only safe method of obtaining information characteristic of single molecules or particles is by the use of systems in which they exist in or approach this state.

Effect of temperature and solvent

In this section we are concerned with the effects of temperature and solvent, not on absolute, but upon relative viscosity and its derived functions.

From previous discussion it is clear that changes in the relative viscosity of very dilute solutions may be caused by changes in:
(1) Solvation, causing variation in the volume fraction of solute.
(2) Particle or molecular weight.
(3) Particle or molecular shape (without accompanying changes in particle or molecular weight).
(4) The extent of solute-solute interactions, which for very asymmetric particles may exist down to the lowest workable solute concentrations.
(5) The degree of orientation of asymmetric particles.

Within the bounds of colloidal systems, examples of all of these changes occurring singly or together are known. A complete account would be out of place here, where we must restrict ourselves to considering those dilute solutions of macromolecules in which changes in molecular weight do not occur (these having already been considered), and in which the effects of solute-solute interactions are not preponderant.

Solvent effects

A change in the solvent may involve changes in solvation or molecular shape. With regard to the former, increases or decreases may occur according to the particular system, and in view of the lack of data on solvation, reliable quantitative predictions of the effect cannot be made, though it is expected to be small.

With regard to variations in molecular shape certain predictions can be made. If a strong affinity exists between a flexible macromolecule and the solvent, then it is to be expected that the strong solvation which occurs will tend to open out a previously curled-up chain, making it more accessible to the solvent. On the other hand, if no such affinity exists, strong solvation will not occur, but rather the different parts of the molecule will tend to 'solvate' themselves, giving a more compact and curled-up configuration. 'Good' solvents may therefore be expected to lead to the extension of flexible macromolecules and to high relative viscosities and bad solvents to compact configurations and reduced viscosities. These points are well illustrated by the figures in Table 13.7 obtained by Alfrey, Bartovics, and Mark (1942) for polystyrene in toluene and toluene-alcohol mixtures, toluene being a good solvent and the alcohols precipitants or poor solvents in this case. Similar results were also obtained for rubber.

TABLE 13.7. η_{sp} for Polystyrene

Solvent	Temperature	
	25° C.	60° C.
Toluene	0·370	0·350
Toluene + 10% methyl alcohol	0·320	0·317
„ + 20% „ „	0·160	0·185
„ + 10% amyl alcohol	0·336	0·340
„ + 33% „ „	0·170	0·210

Such a reduction of the specific viscosity in poor solvents is, however, dependent upon the flexibility of the macromolecule. Thus it was shown with cellulose acetate, which, with cellulose and other derivatives, possesses relatively inflexible chains of modified glucose residues, that η_{sp} was not sensibly different in good and bad solvents.

It is of importance to draw attention here to the parallel investigation by light scattering of the molecular shape of polymers in good and bad solvents (see p. 439) and to the considerable quantitative discrepancy between the two types of data. Until this is explained, some doubt must be associated with both methods, and more so with the use of viscosity in determining molecular dimensions.

Temperature effects

Table 13.7 also contains η_{sp} values at different temperatures, which at first sight may seem puzzling, since the temperature increase from 25° to 60° C. in some cases causes a decrease and in others an increase in viscosity. However, a more careful consideration of the effects of temperature leads to a satisfactory explanation.

Where orientation is inappreciable (as in most solutions of macromolecules with molecular weights of 100,000–300,000 except under very high velocity gradients), temperature changes can affect viscosity through its effects on both degree of solvation and molecular shape. Increased temperature usually involves decreased solvation and hence *slight* decreases in relative viscosity. However, increased temperature also causes more vigorous internal Brownian motion, which may result for flexible macromolecules in increased or decreased asymmetry according to whether the molecules at the lower temperature were respectively in a compact or extended form. For very flexible molecules such effects will outweigh those arising from changes in solvation, though for inflexible chains only the effects of decreased solvation at the higher temperatures will be observed.

For polystyrene (a flexible molecule) in a good solvent η_{sp} decreases with increase in temperature, whilst in a poor solvent (e.g. toluene+33 per cent. amyl alcohol) η_{sp} shows a corresponding increase. In general, dilute solutions of flexible macromolecules in good solvents have negative temperature coefficients of viscosity, and, in poor solvents, positive temperature coefficients, whilst rigid macromolecules in good and bad solvents tend only to have negative temperature coefficients. On the basis of this general result, the effect of temperature on viscosity may be used to obtain information on the flexibility of a molecule in solution.

In discussing temperature effects we have so far assumed no preferred molecular orientation. However, where orientation is appreciable but not complete, temperature changes may introduce a new effect.

Thus the increasing effectiveness of Brownian motion at higher temperatures leads to decreased orientation and, therefore, to an increased relative viscosity, which may overwhelm the normal effects discussed above. Such a case is provided by dilute aqueous solutions of the rigid, rod-like, tobacco mosaic virus particles investigated by Robinson (1939), which were mentioned earlier as showing anomalous flow. Robinson showed that at low velocity gradients (where orientation was not complete), increasing temperature caused an increase in relative viscosity (Fig. 13.9) and simultaneously decreased double refraction (see p. 401). This case is of considerable importance in demonstrating conclusively that streaming birefringence and anomalous viscosity may be due, not to structure in solution (for which a negative temperature coefficient of viscosity would exist), but to orientation of asymmetrical particles.

Our explanation of solvent and temperature effects has been based upon the existence of individual particles or molecules in solution, and upon the changes occurring in these entities. Whilst this is of importance, it is also very probably true that, at experimentally suitable concentrations, interaction effects are never completely absent from solutions of asymmetric particles, but since no theoretical treatment is yet available, their importance is uncertain. It is likely that such effects are dependent on the solvent composition and upon temperature, but knowledge of these must await future investigation.

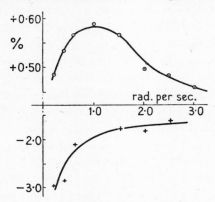

FIG. 13.9. Temperature coefficients of relative viscosity (○) and of double refraction flow (+); 0·02 per cent. tobacco mosaic virus.

Electro-viscous effects

We have been concerned with the effects of the size and shape of suspended particles upon the suspension viscosity. However, another common property of suspended particles may contribute under certain conditions. Thus it has been shown experimentally (e.g. Hardy, 1906) for many proteins in dilute solution, that the relative viscosity varies with pH, possessing a minimum near the iso-electric point and considerably higher values at higher or lower pH values, a clear indication of the effect of the electrical charge of the protein molecules.

Smoluchowski (1916) published, without derivation, the equation

$$\eta_{\rm sp}/\phi = 2 \cdot 5 \left[1 + \frac{1}{\kappa_{\rm sp}\,\eta_0\,a^2} \left(\frac{\zeta D}{2\pi}\right)^2 \right], \tag{13.53}$$

relating the viscosity of a suspension of charged spherical particles of radius a to their electrokinetic potential ζ. η_0 and D are, respectively, the viscosity and dielectric constant of the dispersion medium and $\kappa_{\rm sp}$ is the specific conductivity of the suspension.

A slightly modified form of this equation, which predicts an increase in the viscosity of the suspension as a result of the electrical charge on the suspended particles, was later obtained for non-conducting, uniformly charged spheres with a thin electrical double layer by Krasny-Ergen (1936), who in addition made the assumptions involved in obtaining the Einstein equation.

Attempts at testing the equation quantitatively have not been very successful (e.g. Bull, 1940), and further work is certainly required. It should be noted, in the case of proteins in relatively weak electrolyte solutions upon which tests have been attempted and in which electro-viscous effects are obtained, that it is doubtful if the thickness of the electrical double layer is small compared with the dimensions of a protein molecule; thus one of the important assumptions upon which eqn. (13.53) is based is probably invalid. Systems of larger particles are probably, therefore, more suitable for the test. It is apparent from the equation that when the specific conductivity of the suspension is high the *electro-viscous* term becomes small, which suggests the method of eliminating such effects. Thus, enough neutral salt is added to the suspension (usually 0·1–0·2 M.) until the viscosity is independent of pH, when the electro-viscous term is negligible, and only the Einstein term in eqn. (13.53) remains. For asymmetric particles the same procedure is applied, and the usual equation is then utilized as in uncharged suspensions. No analogue of eqn. (13.53) for asymmetric particles has yet been developed.

REFERENCES

ALFREY, BARTOVICS, and MARK, 1942, *J. Amer. chem. Soc.* **64**, 1557.
—— —— —— 1943, ibid. **65**, 2319.
BARR, 1931, *Monograph on Viscosity*, Oxford Univ. Press.
BRITISH STANDARD 188: 1937, *Determination of Viscosity of Liquids in Absolute Units*, British Standards Institution, 28, Victoria St., London, S.W. 1. (2s. post free.)
BULL, 1940, *Trans. Faraday Soc.* **36**, 80.
—— 1943, *Physical Biochemistry*, Wiley, p. 258.
BURGERS, 1938, *Second Report on Viscosity and Plasticity*, Acad. Sciences, Amsterdam.
CAMPBELL and JOHNSON, 1944, *Trans. Faraday Soc.* **40**, 221.
CARTER, SCOTT, and MAGAT, 1946, *J. Amer. chem. Soc.* **68**, 1480.
EDSALL, 1942, in *Advances in Colloid Science*, vol. i, Interscience, p. 269.
EINSTEIN, 1906, *Ann. Phys. Lpz.* **19**, 289; 1911, **34**, 591.
EIRICH, 1940, *Rep. Progr. Phys.* **7**, 329.
—— BUNZL, and MARGARETHA, 1936, *Kolloidzschr.* **74**, 276.
—— MARGARETHA, and BUNZL, 1936, ibid. **75**, 20.
—— and GOLDSCHMIDT, 1937, ibid. **81**, 7.
EISENSCHITZ, 1933, *Z. phys. Chem.* A. **163**, 133.
FLORY, 1943, *J. Amer. chem. Soc.* **65**, 372.
FORDYCE and HIBBERT, 1939, ibid. **61**, 1910.
GUTH, 1936, *Kolloidzschr.* **74**, 147.
—— and SIMHA, 1936, ibid. **74**, 266.
HARDY, 1906, *J. Physiol.* **33**, 251.
HATSCHEK, 1928, *The Viscosity of Liquids*, Bell.
HUGGINS, 1938, *J. phys. Chem.* **42**, 911; 1939, **43**, 439.

HUGGINS, 1943, in *Cellulose and Cellulose Derivatives*, edited by Ott, Interscience, p. 943.
JEFFERY, 1922-3, *Proc. Roy. Soc.* A. **102**, 161.
KRAEMER, 1931, in *Physical Chemistry*, by H. S. Taylor, vol. ii, 2nd edn., Macmillan, p. 1567.
—— 1941, *J. Franklin Inst.* **231**, 1.
KRASNY-ERGEN, 1936, *Kolloidzschr.* **74**, 172.
KUHN, 1933, ibid. **62**, 269.
—— 1934, ibid. **68**, 2.
MARK, 1940, *High Polymers*, vol. ii, Interscience.
—— 1945, in *Physical Methods*, vol. i, edited by Weissberger, p. 135.
MARTIN, 1942, *American Chemical Society Meeting*, Memphis, April 20-4, quoted by Pfeiffer and Osborn, 1943.
MEAD and FUOSS, 1942, *J. Amer. chem. Soc.* **64**, 277.
MEHL, ONCLEY, and SIMHA, 1940, *Science*, **92**, 132.
MEYER and V. DER WYK, 1935, *Helv. chim. Acta.* **18**, 1067.
NEWMAN and SEARLE, 1933, *The General Properties of Matter*, 2nd edn., Benn.
ONCLEY, 1941, *Ann. N.Y. Acad. Sci.* **41**, 121.
PFEIFFER and OSBORN, 1943, in *Cellulose and Cellulose Derivatives*, edited by E. Ott, Interscience, p. 956.
ROBINSON, 1939, *Proc. Roy. Soc.* A. **170**, 519.
SCHULZ and BLASCHKE, 1941, *Z. phys. Chem.* B. **50**, 305.
SIMHA, 1940, *J. phys. Chem.* **44**, 25.
SMITH, 1942, *Rubb. Chem. Technol.* **15**, 301.
SMOLUCHOWSKI, 1916, *Kolloidzschr.* **18**, 190.
TAYLOR, 1936, *Proc. Roy. Soc.* A, **157**, 546.

XIV
ROTATIONAL DIFFUSION

Introductory

In the last chapter we saw that the orientation of asymmetrical suspended particles, increasing with the velocity gradient, caused reduced dissipation of energy in flow, and the lowered viscosity coefficients associated with anomalous or non-Newtonian behaviour. Knowledge of the orientational state of a suspension is therefore a necessity before any fundamental interpretation of viscosity results can be attempted. In addition, however, such knowledge is of great importance for its own sake, for the degree of orientation, under given conditions, of any given type of asymmetric molecule or particle may be used to derive what is known as a 'rotational' or 'rotary diffusion coefficient' (or a related function) from which information on particle shape and size may be derived. The study of orientational phenomena thus provides another method, quite independent of those which we have discussed in previous chapters, of deriving information on the nature of colloidal molecules or particles. The comparison of data from different experimental approaches is essential, not only in ensuring the correctness of individual results, but, of more importance, in providing a check of our fundamental theories of the behaviour of molecules and particles in solution.

The thermal energy of the solvent molecules is known to be responsible for the well-known Brownian motion, in which microscopically visible particles may be observed to move in zig-zag fashion through a liquid. Less spectacular, but none the less real, is the rotational form of Brownian motion, by which colloidal particles, struck unequally on all sides, are caused to rotate. Such motion was studied experimentally by J. Perrin (1909). Just as translatory Brownian motion may be discussed quantitatively in terms of a translational diffusion coefficient, so rotary diffusion coefficients are used to describe the processes of rotary Brownian motion. The two types of diffusion are, for the most part, treated theoretically by similar methods and in such cases we shall in this chapter merely state results. In discussing the possibly more obscure rotational diffusion it is often helpful to recall the analogy, but it may be seen to break down at certain points.

Theoretical

Consider a solution of non-spherical colloidal molecules which is so dilute that the interactions of these molecules may be neglected.

ROTATIONAL DIFFUSION

Molecular orientation may be described by the inclination, with respect to some fixed direction in space, of a definite molecular axis, e.g. the long axis of a rod-like molecule. Restricting ourselves to the two-dimensional case, let θ be the angle made by the long axis of a molecule with a fixed X direction. If all molecules can be orientated so that, say, $\theta = 0$ and then suddenly all released together, by analogy with translational diffusion, the rotational diffusion coefficient gives a measure (inverse) of the time required for the orientation of the molecules to become completely random. In defining the rotational diffusion coefficient quantitatively, it is necessary to make use of a distribution function to describe the relative numbers of molecules orientated on the average in different directions (i.e. an 'angular concentration'). Thus if Δn be the number of molecules per unit volume whose long axes lie within the two directions θ and $\theta + \Delta\theta$, then the distribution function, $f(\theta)$, may be written

$$f(\theta) = \lim_{\Delta\theta \to 0} \frac{\Delta n}{\Delta\theta}. \tag{14.1}$$

For a random distribution of orientations $f(\theta)$ is a constant, independent of θ, or in other words $df/d\theta = 0$. If, however, there are certain preferred directions of orientation, then $f(\theta)$ will have a maximum at the corresponding values of θ. The number of molecules, dn, per unit volume whose long axes move in time, dt, through the position for which the orientation is θ to higher values of θ is, by analogy with the translational case,

$$dn = -\Theta \frac{df}{d\theta} dt \tag{14.2}$$

(cf. eqn. (10.1b) of Chapter X: $dn = -DA(dc'/dx)dt$), where Θ is the rotational diffusion coefficient. The negative sign occurs in eqn. (14.2) since rotational diffusion tends to cause movement of molecules from directions of high to that of low angular concentration or against the gradient (as in translational diffusion). Eqn. (14.2) is to be regarded as the analogue of Fick's first law for rotational diffusion, but the absence of a term corresponding to area A appearing in Fick's first law and the different dimensions of Θ, T^{-1} (units-sec.$^{-1}$) should be noted.

The number of molecules, n, may be eliminated by the following procedure. The number of molecules for which the angle of orientation moves in unit time through the value θ to higher values is by (14.2)

$$\left(\frac{dn}{dt}\right)_\theta = -\Theta \frac{df}{d\theta}.$$

The number for angle $\theta + d\theta$ is similarly

$$\left(\frac{dn}{dt}\right)_{\theta+d\theta} = -\Theta\left(\frac{df}{d\theta} + \frac{\partial^2 f}{\partial \theta^2}d\theta\right).$$

The increase per unit volume of particles with orientations between θ and $\theta + d\theta$ is thus

$$\left(\frac{dn}{dt}\right)_{\theta} - \left(\frac{dn}{dt}\right)_{\theta+d\theta} = +\Theta\frac{\partial^2 f}{\partial \theta^2}d\theta,$$

which may be written alternatively as $(\partial f/\partial t)\,d\theta$.

$$\therefore \quad \frac{\partial f}{\partial t} = \Theta\frac{\partial^2 f}{\partial \theta^2}. \tag{14.3}$$

This equation, clearly the analogue of Fick's second law, is the fundamental differential equation of rotational diffusion, with whose solutions we are concerned experimentally.

Before considering the equation further, however, it will be convenient to consider other relations in which the rotational diffusion coefficient occurs. We have already mentioned the connexion with rotational Brownian motion and this may be quantitatively expressed by the equation

$$\Theta = \frac{1}{2}\frac{\bar{\theta}^2}{t}, \tag{14.4}$$

where $\bar{\theta}^2$ is the mean square of the angular displacement in time t. This equation, for which the translational analogue exists (eqn. (1.1)), was used by J. Perrin in estimating Θ for colloidal particles by direct observation under a microscope.

From the definition of Θ (eqn. (14.2)) it is clear that the rotational diffusion coefficient is concerned with the rotational mobility about axes perpendicular to the chosen axis, and therefore must depend upon the shape and dimensions of the molecule. It is possible to speak of the frictional constant for a given type of rotation, and it can be shown (as in the translational case) that Θ is proportional directly to the thermal energy (kT) of a single molecule and inversely to its frictional constant, ζ, for the appropriate type of rotation: i.e.

$$\Theta = \frac{kT}{\zeta}, \tag{14.5}$$

where ζ is defined by $\quad C = \zeta\omega, \tag{14.6}$

C being the couple required to maintain rotation of the molecule with an angular velocity of ω (cf. translation frictional constant f, for which force $= f\,dx/dt$).

In the general case of an ellipsoid, the frictional constants for rotation about the a, b, and c axes are respectively denoted by ζ_a, ζ_b, and ζ_c, to each of which the rotational diffusion coefficients Θ_a, Θ_b, and Θ_c are related by the equations

$$\Theta_a = \frac{kT}{\zeta_a}, \qquad \Theta_b = \frac{kT}{\zeta_b}, \qquad \Theta_c = \frac{kT}{\zeta_c}. \qquad (14.7)$$

For ellipsoids of revolution, in which $\zeta_b = \zeta_c$, there are only two different rotational diffusion coefficients, of which we are mainly interested in that involving rotation around a short axis (see p. 385). In view of this simplification, it is usual to treat actual colloidal molecules as ellipsoids of revolution, an assumption which, from the agreement of the results of independent investigations making use of it, seems to be at least approximately valid.

As yet, the functions which we have introduced to describe rotational diffusion have had very obvious parallels in the translational case, and it is possible to treat the problem fully without introducing further quantities. However, it is customary and convenient in treating certain orientational phenomena to make use of another quantity, the relaxation time (τ), which gives a measure of the time required for the complete orientation of a system of molecules to disappear under Brownian agitation. Thus if X is the direction of orientation, and θ is the angle (in any plane) made by the axis of the molecule with this direction, then τ is defined (Debye, 1929) as the time required for the mean value of $\cos \theta$ to fall to $1/e$ (e = exponential constant = $2 \cdot 718...$). Clearly τ is greater, the greater the appropriate frictional constant and the smaller the diffusion coefficient. Further, the relaxation time τ_a corresponding to the rotation of the a axis (N.B. τ_a refers to the rotation *of* the a axis, whilst Θ_a and ζ_a refer to the rotation *about* the a axis) is related to the two diffusion coefficients Θ_b and Θ_c, since the a axis can rotate about the b and c axes. Similarly τ_b is related to Θ_a and Θ_c, and τ_c to Θ_a and Θ_b. Arising from the definition of relaxation time above, these relations may be expressed simply as:

$$\tau_a = \frac{1}{\Theta_b + \Theta_c}, \qquad \tau_b = \frac{1}{\Theta_a + \Theta_c}, \qquad \tau_c = \frac{1}{\Theta_a + \Theta_b}, \qquad (14.8)$$

which reduce for ellipsoids of revolution ($b = c$) into

$$\tau_a = \frac{1}{2\Theta_b}, \qquad \tau_b = \tau_c = \frac{1}{\Theta_a + \Theta_b}. \qquad (14.9)$$

Relations between relaxation times and frictional constants are readily obtained by a combination of eqns. (14.7) and (14.8).

It will be shown later in this chapter that relaxation times are used considerably in the treatment of dielectric dispersion, whilst the alternative functions are usually utilized in describing the phenomena associated with orientation in a velocity gradient.

Rotational diffusion and molecular dimensions

It is of importance now to consider in more detail the relations existing between the quantities describing rotational diffusion macroscopically, and the properties of the individual molecules or particles, chiefly size and shape, which are responsible.

Spherical particles. The rotational frictional constant for spheres (see p. 212) was given by Stokes as

$$\zeta = 8\pi\eta r^3, \tag{9.7}$$

which being inserted into eqn. (14.5) gives

$$\Theta = \frac{kT}{8\pi\eta r^3}, \tag{14.10}$$

an equation which was ingeniously verified by J. Perrin (1909) for the rotation of spherical mastic particles of suitably large radius ($r = 6 \cdot 5 \times 10^{-4}$ cm.), by observing microscopically the movement of impurities on their surfaces.

Asymmetric particles. For prolate and oblate ellipsoids of revolution the equations of F. Perrin (1934) are usually utilized. As in the translational case, a hypothetical spherical molecule of the same volume and density as the molecule investigated is visualized and the 'frictional ratio' of the real frictional constant to that of the hypothetical molecule is expressed in terms of molecular dimensions. Thus the volume of an ellipsoid (Fig. 9.1) with axis of revolution $2a$ and equatorial axis $2b$ is

$$V = \frac{4\pi}{3}ab^2. \tag{14.11}$$

The rotational diffusion coefficient Θ_0 of the hypothetical spherical molecule is thus given by

$$\Theta_0 = \frac{kT}{8\pi\eta\{V/(4\pi/3)\}} = \frac{kT}{8\pi\eta ab^2}. \tag{14.12}$$

Elongated ellipsoids of revolution. For an elongated or prolate ellipsoid of revolution ($b/a = q < 1$) the rotation of the a axis is characterized by the relaxation time τ_a, the diffusion coefficient Θ_b ($= \Theta_c$), and frictional constant ζ_b, which are related to τ_0, Θ_0, and ζ_0 by the equation

$$\frac{\Theta_0}{\Theta_b} = \frac{\zeta_b}{\zeta_0}\left(=\frac{\tau_a}{\tau_0}\right) = \frac{2(1-q^4)}{\frac{3q^2(2-q^2)}{\sqrt{(1-q^2)}}\ln\left\{\frac{1+\sqrt{(1-q^2)}}{q}\right\}-3q^2}, \quad (14.13)$$

an expression always greater than unity and increasing rapidly with increase in a/b. For large values of a/b it reduces to

$$\Theta_b = \Theta_0 \frac{3b^2}{2a^2}\left(2\ln\frac{2a}{b} - 1\right), \quad (14.14)$$

which on substituting for Θ_0 from eqn. (14.12) becomes

$$\Theta_b = \frac{3kT}{16\pi\eta a^3}\left\{2\ln\left(\frac{2a}{b}\right) - 1\right\}. \quad (14.15)$$

This equation contains both the axial ratio and the length of the ellipsoid, but since $\ln(2a/b)$ varies slowly with a/b, Θ_b is approximately inversely proportional to [length]³ and independent of the thickness of the ellipsoid. A very approximate value of a/b may first be substituted to obtain a rough measure of a, after which further refinement may be made (see p. 400) by choice of a more accurate a/b value.

For the motion of the b (or c) axis of an elongated ellipsoid of revolution, which may occur around the a and c (or a and b) axes and which is characterized by the relaxation time τ_b, and the sum of the diffusion coefficients Θ_a and Θ_b, Perrin gives

$$\frac{\tau_b}{\tau_0} = \frac{2\Theta_0}{\Theta_a+\Theta_b} = \frac{4(1-q^4)}{\frac{3q^2(2q^2-1)}{\sqrt{(1-q^2)}}\ln\left\{\frac{1+\sqrt{(1-q^2)}}{q}\right\}+3}, \quad (14.16)$$

which for large values of a/b ($= 1/q$) reduces to

$$\tau_b/\tau_0 \approx \frac{4}{3\{1-(b^2/a^2)\ln(2a/b)\}}. \quad (14.17)$$

On introducing the value of τ_0 from (14.12) this becomes

$$\tau_b \approx \frac{16\pi ab^2\eta}{3kT\{1-(b^2/a^2)\ln(2a/b)\}}. \quad (14.18)$$

It is evident that the relaxation time for the rotation of the equatorial axis is strongly dependent upon the length of both axes of the ellipsoid of revolution, and its use to obtain molecular dimensions is therefore more difficult than for the rotation of the axis of revolution.

τ_b, varying between $4\tau_0/3$ for infinite values of a/b and τ_0 when $a/b = 1$, is thus very much smaller than τ_a for large values of a/b and hence is not so readily observed experimentally.

Flattened ellipsoids of revolution. For oblate or flattened ellipsoids of rotation ($a/b < 1$, $q > 1$) Perrin derived the following equations:

$$\tau_a/\tau_0 = \frac{\Theta_0}{\Theta_b} = \frac{2(1-q^4)}{\dfrac{3q^2(2-q^2)}{\sqrt{(q^2-1)}} \tan^{-1}\sqrt{(q^2-1)} - 3q^2}, \quad (14.19)$$

and

$$\tau_b/\tau_0 = \frac{2\Theta_0}{\Theta_a+\Theta_b} = \frac{4(1-q^4)}{\dfrac{3q^2(2q^2-1)}{\sqrt{(q^2-1)}} \tan^{-1}\sqrt{(q^2-1)} + 3}. \quad (14.20)$$

For $b/a \gg 1$, or disc-like ellipsoids of revolution, both (14.19) and (14.20) reduce to

$$\frac{\tau_a}{\tau_0} \approx \frac{\tau_b}{\tau_0} \approx \frac{2b/a}{3\tan^{-1}(b/a)} \approx \frac{4b}{3\pi a}, \quad (14.21)$$

which gives, on inserting the value of τ_0 from eqn. (14.12)

$$\tau_a \approx \tau_b \approx \frac{16\eta b^3}{3kT}. \quad (14.22)$$

From the identity of τ_a and τ_b at high values of b/a it also follows that $\Theta_a = \Theta_b$ and

$$\frac{1}{2\Theta_a} \approx \frac{1}{2\Theta_b} \approx \frac{16\eta b^3}{3kT}. \quad (14.23)$$

Thus as b/a grows larger, the relaxation times for the rotations of the a and b axes, becoming large, approach one another, become proportional to the cube of the equatorial semi-axis b, and independent of the axis a; for the rotational diffusion coefficients inverse relationships apply.

Summarizing, we see that it is possible to relate rotational diffusion coefficients to the molecular dimensions of model structures, which may be postulated as a result of other experimental evidence. It is important to recall, however, that the relations involved assume the unrestricted rotation of individual suspended molecules or particles, and experimental proof that this is so is very desirable. Observations over a range of low solute concentrations are therefore most useful, and the constancy of the rotational diffusion coefficients over such a concentration range (as in the translational case) is usually taken as sufficient to prove the unimportance of solute-solute interaction. However, whilst in using rotational diffusion coefficients to calculate molecular dimensions it is essential that the coefficient refers to rotation unhampered by other like molecules, the inverse proportionality between the diffusion coefficient and the cube of the molecular dimension required means that in these cases the final calculated dimension is not sensitive to minor errors in the coefficient. Molecular dimensions so calculated refer to the solvated molecule as it occurs in the solution, like most results derived from the examination of colloidal molecules in solution.

Experimentally, rotational diffusion in solutions of colloidal molecules has been investigated chiefly in two ways:

(a) From a study of the orientation of asymmetric molecules in a velocity gradient.

(b) From a study of the variation in dielectric constant of the solution with the frequency of the alternating field.

We shall consider these methods in turn, indicating typical results for each.

Orientation in a velocity gradient

In Chapter XIII it was stated that a decrease in the apparent viscosity of a dilute suspension of asymmetric particles at the higher velocity gradients was usually due to appreciable particle orientation under the influence of the velocity gradient. In support of this view the existence of the unusual positive temperature coefficient of relative viscosity under conditions of appreciable but not overwhelming orientation was quoted. However, a more convincing proof of these views is available experimentally. The base of the outer rotating cylinder of a concentric cylinder viscometer (Fig. 13.5) is made of glass and a nearly parallel beam of plane polarized light, directed through the liquid in the annular space, is observed through a crossed Nicol. When the outer cylinder is stationary, the field appears completely dark for all dilute solutions and pure liquids. If, however, the liquid is a dilute (e.g. 0·02 per cent.) aqueous solution of tobacco mosaic virus or a vanadium pentoxide sol, on rotating the outer cylinder and viewing through the crossed Nicol, the whole annulus, except for a black cross, the cross of isocline (Fig. 14.1), becomes bright, in complete contrast to its appearance when the cylinder is stationary. With increasing rate of rotation, the arms of the cross move nearer to coincidence with the planes of vibration of the polarizer and analyser and the intensity of the light transmitted increases. These phenomena, associated with streaming (or flow) birefringence, are completely absent (except at tremendously high rates of shear) for pure liquids and accurately Newtonian solutions. Streaming birefringence may also be examined qualitatively by observing the suspensions under crossed Nicols when flowing through a capillary tube, or wherever the liquid is subject to a velocity gradient (e.g. by stirring).

In view of the low concentrations at which streaming birefringence can be observed, it seems most probable that the optical anisotropy brought about by the velocity gradient arises from the orientation of anisotropic particles (see also p. 376). The close correspondence between

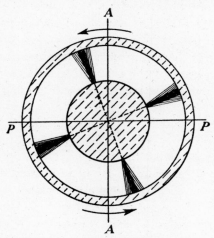

Fig. 14.1. The cross of isocline (diagrammatic).

experimental observation and the results of a theoretical treatment of particle orientation in a velocity gradient, which will be outlined later, confirm this explanation. Another factor to be considered, at high velocity gradients, is the possible distortion of asymmetric molecules, affecting both the measured viscosity and flow birefringence of the system. In the next few pages, devoted to orientational phenomena, the complications which may arise if molecular deformation occurs will be considered.

(a) *Mechanism of orientation*

The orientational state of solute molecules in a solution subject to a velocity gradient is dependent upon the balance between the orientating action of the gradient and the disorientating action of rotational diffusion (or thermal agitation). An increase in temperature causes increased thermal agitation and rotational diffusion, and at a constant velocity gradient leads to a state of decreased orientation. Similarly an increase in the velocity gradient at constant temperature results in increased orientation.

Consider an idealized system of long rods suspended in a liquid in which there is a velocity gradient (as in a concentric cylinder viscometer). Fig. 14.2 shows one rod which at a given instant occupies a position making an angle θ with the stream-lines. Since the suspending liquid adjacent to the two ends is moving with different velocities, the rod will be subject to a turning moment which will have a maximum value when the rod is perpendicular to the stream-lines ($\theta = 90°$) and a mini-

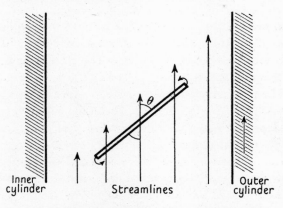

FIG. 14.2. Particle orientation in a velocity gradient.

mum value (near zero) when parallel to them ($\theta = 0°$). In the absence of other factors, therefore, the velocity gradient tends to cause orientation of the rods parallel to the stream-line, i.e. in the position at which the turning moment, and therefore the angular velocity of rotation, is smallest. Thus, as well as being transported bodily by the flowing liquid, the suspended rods are caused to swing into the direction of the stream-lines, in a manner similar to the movement lengthwise of logs in a rapidly moving stream. Such then is the action of the velocity gradient.

Thermal forces will, however, impart irregular rotations in all directions, and merely because of the greater number of the rods congregating in the direction of the stream-lines, more will diffuse away from such directions than from any others. The presence of rotational Brownian motion clearly, therefore, reduces the intensity of orientation, to a greater extent the more violent the thermal agitation. Further, in addition to a decrease in the *extent* of orientation, the *direction* of maximum orientation is also affected by the presence of rotational Brownian motion, as is shown below.

For negligible Brownian motion the orientation is complete and along the stream-lines. For appreciable Brownian motion the suspended rods are caused to rotate in clockwise and anti-clockwise directions from the position of maximum orientation. Referring to Fig. 14.2, anti-clockwise rotations will, in this case, be aided by the velocity gradient and increased rotational speeds obtained, whilst clockwise rotations will be hindered. Some direction, clockwise with respect to the stream-lines for which θ has a small positive value, say θ_{max}, at which the couple due to thermal forces is just balanced by that due to the velocity gradient will therefore be one of minimum angular velocity and, consequently of maximum orientation. As G/Θ, the ratio of the velocity gradient or rate of shear G to the rotational diffusion coefficient, decreases, this position will move clockwise away from the stream-lines, i.e. θ_{max} will increase, corresponding to higher values of the turning moment of the velocity gradient at which equality with the thermal couple is reached. As G/Θ increases, θ_{max} will decrease, becoming effectively zero at very high G/Θ values. It should be noted further that as θ_{max} increases from low values, the extent of orientation must decrease progressively owing to the greater effectiveness of the rotational disorientation process. The dynamic nature of the orientation process, even when a steady state is reached, should be noted carefully. The direction of orientation is not one in which the molecules remain static, but one through which rotation takes place most slowly. Further, since at any instant more molecules are alined in this direction than in any other, the number leaving the position is greater than for any other position, but equal to the number entering for a steady state.

It is necessary now to consider these questions in a quantitative form so that experimental measurements may be related to such fundamentals as rotational diffusion coefficient and degree of orientation.

(b) *Theoretical treatment of orientation*

As we have already seen (Chapter XIII, p. 362) Jeffery (1923), in considering the viscosity of suspensions of ellipsoidal particles, quantitatively treated the question of the rotation of such particles in a velocity gradient. For ellipsoids of revolution of axial ratio a/b, he gave for the angular velocity of rotation

$$\frac{d\theta}{dt} = -\frac{G(a^2\sin^2\theta + b^2\cos^2\theta)}{a^2+b^2}, \qquad (14.24)$$

where θ is the angle between the ellipsoid axis of rotation and the streamlines, and G is the velocity gradient. For prolate ellipsoids ($a > b$), $d\theta/dt$ is a maximum when $\theta = 90°$ and a minimum for $\theta = 0°$. In the case of very elongated ellipsoids, for which $a \gg b$, eqn. (14.24) reduces to

$$\frac{d\theta}{dt} = -G\sin^2\theta, \qquad (14.25)$$

which expresses quantitatively our earlier qualitative statements regarding the rotation of long rods in a velocity gradient.

The number of rods per unit volume which move in unit time through the position of orientation θ to higher values is $(f\, d\theta/dt)_\theta$. The same quantity at position $(\theta+d\theta)$ is $(f\, d\theta/dt)_{\theta+d\theta}$. The difference between these quantities gives for the effect of the velocity gradient the increase per unit volume, in the number of rods with orientations between the positions θ and $\theta+d\theta$ in unit time. Thus we have

$$\left(\frac{\partial f}{\partial t}\right)_s = (f\omega)_\theta - (f\omega)_{\theta+d\theta} = -\frac{\partial(f\omega)}{\partial \theta}, \qquad (14.26)$$

where $d\theta/dt = \omega$, and $(\partial f/\partial t)_s$ refers to the effects of the velocity gradient alone.

To obtain the total value of $(\partial f/\partial t)$ it is necessary to add the contribution of rotational diffusion from eqn. (14.3), giving

$$\left(\frac{\partial f}{\partial t}\right)_{\text{total}} = \Theta \frac{\partial^2 f}{\partial \theta^2} - \frac{\partial(f\omega)}{\partial \theta}. \qquad (14.27)$$

When the system is in a steady state $\partial f/\partial t = 0$, and on integrating (14.27) we obtain

$$\Theta \frac{\partial f}{\partial \theta} - f\omega = c_1, \qquad (14.28)$$

which is usually written, using eqn. (14.25),

$$\frac{\partial f}{\partial \theta} + \alpha f \sin^2\theta = C, \qquad (14.29)$$

where
$$\alpha = \frac{G}{\Theta}. \qquad (14.30)$$

Boeder (1932) obtained and solved this equation for two-dimensional orientation, giving his results in the graphical form of Fig. 14.3 (taken from Robinson, 1939). The curves corresponding to different values of α confirm quantitatively the qualitative statements of p. 390 regarding the direction and degree of orientation. Thus, as α increases, the degree of orientation increases and the direction of maximum orientation moves into coincidence with the stream-lines. The angle between these directions ($\theta_{\max}$) is later shown to be equal to an experimentally observable quantity, the extinction angle χ, whilst the degree of orientation is proportional to the magnitude of the observed double refraction. From the curves it is clear that as χ changes from nearly $0°$ to nearly $45°$ the degree of orientation changes from almost complete to almost random orientation. It may also be seen that, corresponding to the direction of maximum orientation and moving with it, there is, at an angle to it of approximately 90°, a trough corresponding to a direction of minimum orientation which at high velocity gradient moves into a position making an angle of approximately 90° with the stream-lines. Such a trough is obviously a result of the maximum value of the

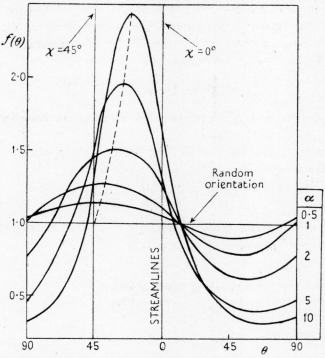

Fig. 14.3. Statistical orientation distribution curves, adapted from Boeder. (*From* Robinson (1939).)

turning moment of the velocity gradient in the position at 90° with the stream-lines.

For very large asymmetric particles (e.g. vanadium pentoxide and tobacco mosaic virus particles) possessing low values of the rotational diffusion coefficient, α is relatively large even for low values of G, so that appreciable orientation and values of χ deviating considerably from 45° are obtained at low velocity gradients. However, for small suspended particles (e.g. low molecular weight cellulose derivatives) and molecular solutions, the rotational diffusion coefficient is large (corresponding to violent thermal agitation), and even at high values of G, α is still small, and orientation is not appreciably different from random nor the extinction angle from 45°.

Clearly, from a measurement of the extinction angle, the value of α

may be determined from Boeder's curves, and from the value of the applied velocity gradient also Θ can be obtained. This provides one method of obtaining the rotational diffusion coefficient. A better method, however, makes use of a theoretical analysis of the problem of orientation in three dimensions, of which only the relevant results can be quoted here.

Boeder obtained for the orientation of rod-like particles in three dimensions the approximate result for low values of α:

$$\chi = \tfrac{1}{2}\tan^{-1}\frac{6}{\alpha} = \frac{\pi}{4} - \frac{\alpha}{12}\left(1 - \frac{\alpha^2}{108} + \ldots\right), \tag{14.31}$$

where χ is in radians.

More recently Peterlin and Stuart (1939) have investigated the problem of the orientation of ellipsoids of revolution and, again, for low values of α obtain

$$\chi = \frac{\pi}{4} - \frac{\alpha}{12}\left[1 - \frac{\alpha^2}{108}\left(1 + \frac{24}{35}\frac{a^2-b^2}{a^2+b^2} + \ldots\right)\right], \tag{14.32}$$

an expression differing from that of Boeder in introducing the lengths of axes of the ellipsoid. For very low values of α both (14.31) and (14.32) reduce to

$$\chi = \frac{\pi}{4} - \frac{\alpha}{12}, \tag{14.33}$$

from which a linear relation between χ and α is to be expected. Where (14.33) holds, it affords a very simple method of calculating α and hence Θ from observations of the extinction angle.

(c) *The magnitude of streaming birefringence*

We have seen that the application of a velocity gradient to a suspension of large asymmetric particles may cause orientation, which for anisotropic particles would be detected by birefringence of the suspension. However, it was shown by Wiener (1912) that a suspension of optically isotropic ellipsoids, small in size compared with the wave-length of light and orientated in an isotropic solvent of different refractive index, should also be birefringent to an extent depending upon the difference between the two refractive indices, the birefringence disappearing only when the refractive indices were equal. These predictions are supported by the results of Fig. 14.4, due to Weber (1934–5), who made birefringence measurements upon myosin and gelatin fibres immersed in different media. For gelatin, which is isotropic, the birefringence becomes zero only when its refractive index is equal to that of the surrounding medium; under other conditions birefringence is observed. For myosin, on the other hand, which is intrinsically optically anisotropic, the birefringence reaches a minimum, which is not zero, at a certain value of the outside refractive index. At this minimum the observed birefringence is to be attributed only to the

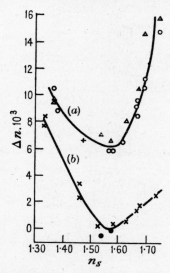

Fig. 14.4. Birefringence of myosin (a) and gelatin fibres (b), as a function of the refractive index (n_s) of the surrounding medium. (*From* diagram by Edsall (1942).)

inherent properties of the fibre, whilst under other conditions the 'Wiener' type of birefringence is also operative.

Peterlin and Stuart (1939) have generalized Wiener's treatment for suspensions of particles whose dimensions lie between 100 and 1,000 A, obtaining the equation

$$\frac{n_e - n_0}{n} = \phi \frac{2\pi}{n^2}(g_1 - g_2) f\left(\alpha, \frac{a}{b}\right), \qquad (14.34)$$

where $n_e - n_0$ is the magnitude of the double refraction, n is the refractive index of the stationary liquid, ϕ is the volume concentration of the disperse phase, $g_1 - g_2$ is termed the optical anisotropy factor, and $f(\alpha, a/b)$ is a complex function of α and the axial ratio. For small values of α, $f(\alpha, a/b)$ may be written

$$f(\alpha, a/b) = \frac{\alpha}{15}\left(\frac{a^2 - b^2}{a^2 + b^2}\right)\left[1 - \frac{\alpha^2}{72}\left(1 + \frac{6}{35}\frac{a^2 - b^2}{a^2 + b^2}\right) + \ldots\right]. \qquad (14.35)$$

It may therefore be assessed from the results of extinction angle observations and particle dimensions.

The optical anisotropy factor is given by Peterlin and Stuart as

$$g_1 - g_2 = \frac{1}{4\pi} \frac{(n_1^2 - n_2^2) + e(n_1^2 - n_s^2)(n_2^2 - n_s^2)/n_s^2}{\left[\dfrac{n_1^2 + 2n_s^2}{3n_s^2} - \dfrac{2e}{3}\dfrac{n_1^2 - n_s^2}{n_s^2}\right]\left[\dfrac{n_2^2 + 2n_s^2}{3n_s^2} + \dfrac{e}{3}\dfrac{n_2^2 - n_s^2}{n_s^2}\right]}, \quad (14.36)$$

where n_1 and n_2 are the principal refractive indices of the ellipsoids, n_s is the refractive index of the solvent, and e is a factor depending upon the axial ratio of the ellipsoids, having the following values:

$$a \gg b, \quad e = 0\cdot 5;$$
$$a = b, \quad e = 0;$$
$$a \ll b, \quad e = -1.$$

Of the two terms in the numerator of (14.36), the first, depending only upon the inherent birefringence of the suspended particles, is zero for optically isotropic particles. The second, however, depends also upon particle shape, becoming zero for spherical particles, and also for n_2 or $n_1 = n_s$. This expression accounts quantitatively for results like those of Fig. 14.4, in which a minimum double refraction is observed at a certain value of the solvent refractive index.

It will be clear that a determination, from the observed optical effects, of the inherent birefringence of suspended particles is by no means a simple problem. However, more qualitative indications, especially of the magnitude of the minimum birefringence observed over a range of solvent refractive indices, are sometimes useful. Thus Edsall and co-workers (1940) found that the double refraction of flow of myosin is drastically affected, not only by the usually accepted denaturing agents, but also by mild reagents like the chlorides of calcium, magnesium, and barium. It is probable that in such cases the observed birefringence may be a sensitive criterion of the state of aggregation of a protein preparation and for this reason affords a useful method of examination.

(d) Orientation in a concentric cylinder apparatus

It is now of importance to consider this treatment in its relation to the concentric cylinder apparatus. In Fig. 14.5 SL corresponds to the direction of the stream-lines in the liquid at some point H, situated on one arm (B) of the cross of isocline. Since point H is dark when viewed between crossed Nicols, the vibration directions of the suspended molecules at H must be parallel or perpendicular to those of the polarizer PP and analyser AA. If these vibration directions can be assumed (as in many materials) to coincide with the geometrical axes of the rod-like molecules, then clearly the rods must be orientated in the direction IJ, perpendicular to PP, and making an angle $\theta_{\max}$ with the stream-lines. Clearly, by geometry, $\angle BOP$ and the smaller of the angles made by the arms of the brushes with directions PP and AA are all equal to $\theta_{\max}$.

Experimentally it is readily possible to measure the angles between

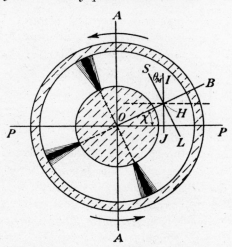

Fig. 14.5. Particle orientation and the cross of isocline.

the arms of the cross and the polarizer or analyser vibration directions. The smaller of these angles, known as the *extinction angle* χ, being equal to $\theta_{\max}$, is of considerable importance since, for asymmetric particles, in which the optical and geometric axes coincide, it gives a measure of the position of maximum orientation with respect to the stream-lines, which can readily be used as shown above to calculate the rotational diffusion coefficient. Whilst it is more common now to use the extinction angle to describe the position of the cross of isocline, it should be noted that its complement, the 'angle of isocline', denoted by ψ, has also been used and will be encountered in the literature. Since from Chapter XIII we saw that the velocity gradient varies over a wide annulus between concentric cylinders, the extinction angle must vary across it and some curvature of the arms of the cross of isocline may therefore be observed in such cases. It is, therefore, usual to limit the field for quantitative work to a small width of the annulus or to use only narrow gaps (between the cylinders) which are in any case necessary if high velocity gradients without turbulence are required.

Experimental. Considerable thought has been devoted to the design and construction of concentric cylinder instruments and their optical systems. The aims have been to secure very high velocity gradients smoothly and to ensure that the streaming double refraction observed is not complicated by polarization effects introduced by stray light reflections. Detailed descriptions of good instruments will be found in the papers by Edsall and co-workers (e.g. 1942, 1944, 1945), by Signer and Gross (1933) and Sadron (1936), and the difficulties of the method are discussed by Snellman (1945). Here we can merely outline the method briefly.

As previously mentioned, the simple concentric cylinder viscosity apparatus may be used for observation of streaming birefringence if a transparent window be placed in the base of the outer rotating cylinder and an optical system of the type shown in Fig. 14.6 is provided. Monochromatic light from the illuminated diaphragm, D, is made parallel by the lens L_1, and after passing through the polarizing Nicol P, the plane-polarized parallel beam passes through the streaming liquid between the cylinders, through the analysing Nicol A, and is focused, by lens L_2 into the observing telescope (T). The absence of lenses between the Nicols is intended to avoid possible complications owing to slight optical anisotropy of the lens material. The Nicol prisms may be rotated separately or together, and various other devices may be employed both in determining the

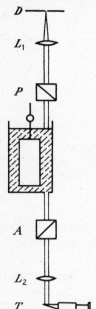

FIG. 14.6. Optical system for observing streaming birefringence in the concentric cylinder apparatus.

extinction angle and the magnitude of the double refraction. We mention here simple procedures which have been commonly employed for these purposes.

With the cylinders at rest the Nicols are crossed and clamped together so that they can now only rotate together. The outer cylinder is now rotated at suitable speed, and the Nicols both rotated until one arm of the cross of isocline appears in the field of vision. At this stage the optic axes of the suspended particles in the field of view are parallel or perpendicular to the polarizer and analyser vibration directions. It is now necessary to find the angle between these directions and the streamlines. Rotation is now stopped and a quartz wedge is inserted in a position between A and the outer cylinder so that its optical axis (parallel to the length of the wedge) is parallel to the stream-lines. The field is now light again and the crossed Nicols are again rotated by a measured amount until there is darkness. This amount, which measures the angle between the optic axes of the particles and the stream-lines (or its complement), is the extinction angle.

An alternative procedure after locating one arm of the cross of isocline in the field of view, is to rotate the outer cylinder at an equal but opposite speed. The Nicols are then turned together to give darkness through an angle which is twice the extinction angle (or twice its complement).

The magnitude of the double refraction is usually obtained by the use of a compensator, usually a quarter-wave (or $\lambda/4$) plate. After locating part of the cross of isocline in the field of view, the Nicols are turned through $45°$, when two beams of light of equal intensity but plane-polarized perpendicularly and suffering from a phase difference (Δp), given by

$$\Delta p = \frac{d}{\lambda_0}(n_e - n_0), \qquad (14.37)$$

emerge from the streaming liquid. d is the path length, n_e and n_0 are the refractive indices for vibrations parallel and perpendicular to the optic axis, and λ_0 is the wavelength *in vacuo* of the light employed. From a determination of Δp, $(n_e - n_0)$ is therefore easily calculated.

The quarter-wave plate is inserted between the analyser and cylinders with its vibration planes parallel to those of the analyser or polarizer, when the two plane-polarized and out of phase beams (or elliptically polarized beam) from the liquid are transformed into plane-polarized light, whose plane of polarization has been rotated by an angle Δ, given by

$$\Delta = \frac{\Delta p \times 360°}{2} = 180 \frac{d}{\lambda_0}(n_e - n_0). \qquad (14.38)$$

Δ is, therefore, simply determined as the angle through which the analyser alone must now be rotated to give darkness. Further details of these and of other methods may be found in Edsall (1942) and the optical principles involved are discussed in optics text-books (e.g. Hartshorne and Stuart, 1934, and Robertson, 1929).

In recent years several workers have constructed concentric cylinder instruments with very narrow gaps (e.g. 0·5 mm. and less) between the cylinders, in which the high rates of rotation possible give velocity gradients up to 30,000 sec.$^{-1}$ and higher. In such instruments it has usually been necessary to fix securely the non-rotating cylinder (which has commonly been the outer one) so that simultaneous viscosity measurements could not be obtained. It has thus been possible to cause the orientation of many of the more mildly asymmetric molecules (e.g.

serum globulin by Sadron and co-workers, 1939), whose molecular shape and dimensions are known also from other experimental approaches. Whilst a detailed description of the instruments and technique used would be out of place, we must mention the considerable experimental

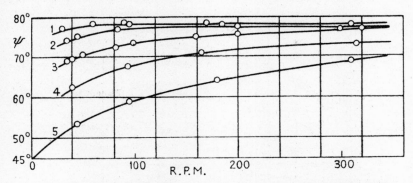

Fig. 14.7. Angle of isocline (ψ) v. speed of rotation for various concentrations, given in mg. nitrogen per litre. 1. 2240, 2. 1790, 3. 1345, 4. 450, 5. 225.

difficulties, especially those of an optical character, which have been met and overcome (see references above).

(e) Results

For very large asymmetric particles (e.g. vanadium pentoxide, native myosin, or tobacco mosaic virus) which are appreciably orientated at low velocity gradients, several important theoretical predictions have been confirmed. Thus it is to be expected that the extinction angle should steadily decrease with increasing velocity gradient from 45° until, at high gradients, values not very different from 0° should be obtained. Simultaneously the observed double refraction should increase with increasing velocity gradient, rapidly at first but at a steadily decreasing rate, and should tend to a saturation value when the extinction angle is small, due to the nearly complete orientation which then obtains. An important exception to such a saturation limit may arise in the case of solutions in which molecular distortion occurs at very high velocity gradients. We shall consider this question later.

In a now classic investigation of the muscle protein, myosin, von Muralt and Edsall (1930) laid the foundations for much future work. Figs. 14.7 and 14.8, taken from their paper, show the variation in the angle of isocline ψ ($\psi = 90° - \chi$) and double refraction with increasing rate of rotation. The variation in angle of isocline with concentration at a given rate of rotation is very prominent, the much higher values (i.e. smaller extinction angles) at higher concentration being probably a result of decreased effective rotational diffusion caused by the mutual interference of the asymmetric myosin molecules. For the most dilute solution from Fig. 14.7, $\psi = 53°$ ($\chi = 37°$) at 47 r.p.m., which corre-

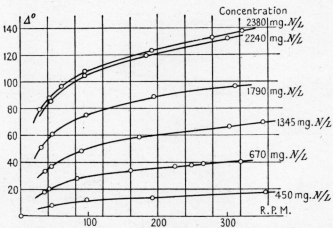

Fig. 14.8. Double refraction (Δ) of rabbit myosin as a function of velocity gradient at several protein concentrations (denoted on each curve in mg. of nitrogen per litre). (Muralt and Edsall (1930).)

sponds to a velocity gradient of $G = 10$ sec.$^{-1}$ Substituting $\chi = 37°$ in (14.31) we obtain $\alpha = 1·3$ and $\Theta = 7$ sec.$^{-1}$

Substituting for Θ in eqn. (14.15) and putting $a/b = 100$ and $\eta = 0·016$, the viscosity of water at 3° C., Edsall (see Cohn and Edsall, 1943, p. 536) obtains $a = 5,800$ A, which corresponds to a total length of molecule of $2a = 11,600$ A. It is easy to show that wide variations in the value of a/b used cause quite small changes in the calculated value of a.

Estimates of the molecular dimensions of the tobacco mosaic virus unit in solution, made similarly by Mehl (1938), are of more than usual interest, since the electron microscope provides a direct method of confirmation in this case (see p. 461). As yet, however, it is doubtful if a comparison of the same specimen under controlled conditions has been made, and ageing and aggregate effects have probably combined to give the very much higher lengths of the particles obtained from streaming birefringence measurements.

In a later investigation by Robinson (1939) on tobacco mosaic virus in aqueous solution, the solute concentration was between 0·02 per cent. and 0·005 per cent. by weight. Under these conditions the extinction angle was independent of the protein concentration and the double refraction observed at a given velocity gradient was proportional to it, as is shown in Table 14.1, due to Robinson, where the double refraction is expressed in scale divisions of the analyser, 200 being equivalent to a path difference of 1 wavelength of sodium light. Robinson took

TABLE 14.1

Concentration (c) %	Double refraction (D.R.)	D.R./c
0·02	2·78	139
0·0143	2·04	143
0·010	1·43	143
0·0083	1·20	145
0·0077	1·125	146
0·0067	0·975	145
0·00525	0·71	136

viscosity measurements simultaneously with optical readings, Fig. 14.9, from his results, demonstrating the decreasing viscosity observed with increasing orientation.

The increased relative viscosity and decreased double refraction of

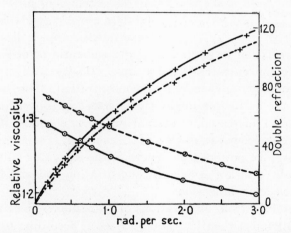

FIG. 14.9. 0·02 per cent. tobacco virus: relative viscosity (⊙) and double refraction of flow (+). ——— at 14·4° C. ----- at 19·6° C.

higher temperatures, which we have mentioned already (p. 376) is clearly visible. On the basis of his optical measurements of extinction angle and double refraction Robinson calculated the curve of relative viscosity against angular velocity, and from Fig. 14.10 this is seen to compare very favourably with the observed curve. Robinson's investigation seems to be one of the most complete yet accomplished, and is free from the objection, common to many others, that viscosity measurements were obtained in capillary viscometers at ill-defined velocity gradients.

However, mechanical difficulties prevent the determination of viscosities in the narrow-gap, concentric cylinder instruments which are required for the study of particles which are smaller and less asymmetric than myosin and tobacco mosaic virus. In such cases there is no alternative to the determination of viscosities separately by additional experiments, usually in capillary viscometers of high mean velocity gradient.

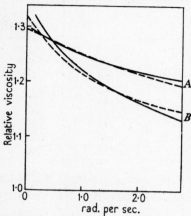

FIG. 14.10. Observed (———) and calculated (------) relative viscosities. A, for 0·02 per cent. tobacco mosaic virus at 14·4° C.; B, for 0·04 per cent. mercury sulphosalicylic acid at 18·7° C. (Robinson (1939).)

Foster and Edsall (1945) have made such an investigation, chiefly in propylene glycol solution, of the maize protein, zein, for which molecular data are also available from other experimental approaches. Fig. 14.11 shows the variation in extinction angle and ratio of observed double refraction to concentration (Δ/c) against velocity gradient. Although Δ/c is apparently independent of concentration, there is a definite decrease in the extinction angle with increase in concentration, corresponding to an increased rotational diffusion constant (Θ) at lower concentrations. However, in view of the fact that Θ is inversely proportional to the third power of the length of the elongated ellipsoid postulated, a small variation of Θ with concentration does not lead to very different molecular dimensions.

Θ was obtained from observations of χ at known velocity gradients (eqn. (14.33)) as between 10^{-1} and 10^{-3} sec.$^{-1}$ according to the solvent viscosity. Assuming the molecule to be a prolate ellipsoid of revolution, its length was calculated to be between 300 and 400 A for the different preparations, with an axial ratio of about 20. Williams and Watson (1938) obtained, for the length, the value 425 A, and an axial ratio 25 from sedimentation-diffusion data.

It is of interest that calculation of molecular dimensions assuming an oblate or flattened ellipsoid gives for the b dimension the value 240 A. Combined with the molecular weight, 50,000, of the protein and partial specific volume, this dimension involves the value 2 A for the total thickness of the ellipsoid $2a$. Not only is this an impossibly small value, but it leads to a quite erroneous value of the axial ratio. It is probable,

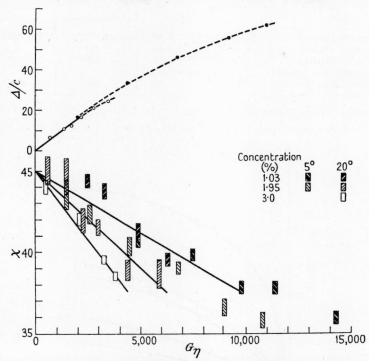

FIG. 14.11. Dependence of Δ/c (upper curve) and χ (lower curve) on $G\eta_0$ (η_0 = viscosity of solvent) for zein fraction (II-1) in propylene glycol. In the upper curve the open circles correspond to measurements at room temperature, the full circles to low temperature (5°), all for 1·95 per cent. protein solution. The probable errors in the extinction angle measurements are indicated by the vertical height of the rectangles.

therefore, that the zein molecule approximates more to the prolate ellipsoid of revolution in solution.

Table 14.2, compiled by Edsall (in Cohn and Edsall, 1943), shows further comparisons of the molecular lengths of ultracentrifugally monodisperse proteins obtained assuming prolate ellipsoids of revolution.

In view of the uncertainties attaching to both types of measurement the results are not unsatisfactory. The three different sets of values for Helix haemocyanin refer to the parent molecule and its dissociation products, and the nearly identical length of these molecules suggest the lengthwise splitting of the parent molecule.

TABLE 14.2. *The Lengths of Certain Protein Molecules*

Protein	Molecular Weight	f/f_o	Length 2a	
			Sedimentation	Double refraction
Haemocyanin (Helix)	8,900,000	1·45	1,130	890
Haemocyanin (Helix)	4,300,000	1·40	820	890
Haemocyanin (Helix)	1,030,000	1·79	820	960
Antipneumococcus serum globulin (horse)	910,000	1·86	960	1,280

Several investigations of streaming birefringence in solutions of linear polymers have now been made, of which we may quote those of Signer

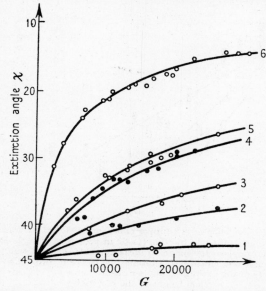

FIG. 14.12. Extinction angle of six cellulose nitrates (of different molecular weights) in solution in cyclohexanone, as a function of the velocity gradient, G.

	Conc. of solution (sub-moles/litre)	$[\eta_{sp}/c]_{c \to 0}$ for cellulose nitrate
Curve 1	0·035	17
,, 2	0·015	33
,, 3	0·010	55
,, 4	0·005	100
,, 5	0·005	115
,, 6	0·002	245

N.B. Concentration c, is given here in the Staudinger form, viz. sub-moles (i.e. the nitrated glucose unit) per litre. (*From* diagram by Signer and Gross (1933).)

and co-workers (1933, 1936, see also Wissler, 1940) on polystyrene and cellulose derivatives, and of de Rosset (1941) on polymethyl methacrylate. Fig. 14.12, from Signer and Gross (1933), showing the extinction angle as a function of velocity gradient for different cellulose nitrate fractions, is typical of the results obtained. For low molecular weight linear polymers, even at very high velocity gradients, extinction angles not differing much from 45° are obtained, whilst for the higher molecular weight fractions, much lower values were obtained at much lower gradients.

Measurements of double refraction were of considerable interest. Thus for low and medium molecular weight polystyrenes ($M < 300,000$), the observed double refraction ($n_e - n_0$) was proportional to the velocity gradient G (Fig. 14.13 (a)) over a large range of values, showing no signs of a saturation maximum, and being greater the higher the polymer concentration. For a given mean molecular weight Signer and Gross (1933) found that the function $(n_e - n_0)/G\eta c$ was approximately constant, especially in the case of the lower molecular weight fractions. Further, Signer and Gross state that this function must, according to Boeder's theory, be inversely proportional to the rotational diffusion coefficient; relative values of the coefficient were calculated, therefore, on this assumption. Such values surprisingly enough were found inversely proportional to the mean molecular weights as determined by Staudinger's law for the lower molecular weight fractions, and to corrected molecular weight values for those of quite high molecular weight. Table 14.3 contains experimental data for the cellulose nitrate fractions referred to in Fig. 14.12, which demonstrate the conclusions of Signer and Gross. m_0 refers to the molecular weight of the basic unit in the chain (i.e. nitrated glucose residue), so that the quantity $0.1 m_0[\eta]$ is $[\eta_{sp}/c]_{c \to 0}$, expressed in the Staudinger form (p. 368), with c in basic

TABLE 14.3. *Cellulose Nitrate Fractions in Cyclohexanone*

$0.1 m_0[\eta]$	Molecular weight from viscosity	$\frac{n_e - n_0}{G\eta c} \cdot 10^{10}$	Θ_{rel} from $(n_e - n_0)/G\eta c$	Corrected mol. wt.	$M_{corr} \Theta_{rel} 10^6$
11	8,500	148	50	8,500	0.42
17	13,000	217	34	12,000	0.41
33	25,000	599	12.2	35,000	0.43
55	41,000	964	7.6	56,000	0.43
100	75,000	2,140	3.4	125,000	0.43
115	88,000	2,520	2.9	150,000	0.43
245	220,000	7,330	1	400,000	0.40

moles per litre, and is directly comparable with values of $[\eta_{sp}/c]_{c \to 0}$ attached to Fig. 14.12. A further interesting point arose from the examination of the double refraction of the higher molecular weight polystyrene fractions in solution, and was later confirmed and extended by Signer and Sadron (1936). Thus at a certain point on the straight line representing ($n_e - n_0$) against G, a discontinuity occurs (Fig. 14.13 (b))

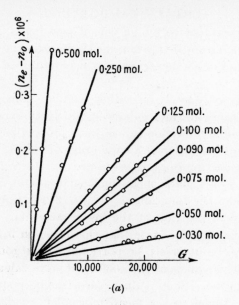

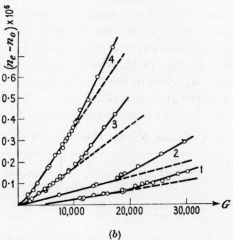

Fig. 14.13. (a) Double refraction of different concentrations of low molecular weight polystyrene in cyclohexanone solution, as a function of the velocity gradient.

$\{[\eta_{sp}/c]_{c \to 0}$, for polymer ($c$ in sub-moles/litre) $= 5 \cdot 6\}$.

(*From* Signer and Gross (1933).)

(b) Streaming double refraction of high molecular weight polystyrene in solution, as a function of velocity gradient.

Polymer Concentration.

Curve 1. 1/70 sub-moles per litre in cyclohexanone at 22° C.
  „   2. 1/40   „   „   „   „   22° C.
  „   3. 1/20   „   „   „   „   19° C.
  „   4. 1/20 sub-moles per litre in tetralin at 19° C.

(*From* Signer and Sadron (1936).)

after which the slope of the line is constant at a considerably increased value. The discontinuity, for solutions of different concentration, is characterized by a constant value of $G\eta$, where η is the viscosity of the solution, and it was also observed that a similar discontinuity in the extinction angle against velocity gradient curve occurred at the same point. These observations were reasonably attributed to deformation and stretching of the previously curled chain which sets in when the stress in solution reaches a given value at the points of discontinuity observed. In support of this view are the observations of Signer and Gross (1933) that no such points of discontinuity are observed even for the highest molecular weight cellulose nitrates which show proportionality between $(n_e - n_0)$ and G at low velocity gradients and signs of saturation double refraction at higher gradients. On stereochemical grounds this was to be expected, for further stretching in this way of the nearly completely stretched chain of glucose residues is unlikely. We shall refer later to the general importance of molecular deformation in streaming birefringence work.

Rotation diffusion coefficients were also calculated from extinction angle measurements by Signer and Gross (1933), of which Table 14.4 contains those for two cellulose nitrate samples in butyl acetate and cyclohexanone. The molecular dimensions at the base of the table were calculated by Edsall from the data above by eqn. (14.15).

TABLE 14.4. *Rotational Diffusion Coefficients of Cellulose Nitrates from the Extinction Angle at* $20°$ *C.*

	Cellulose nitrate $0 \cdot 1 m_0[\eta] = 55$		Cellulose nitrate $0 \cdot 1 m_0[\eta] = 270$ or 245	
Gradient sec.$^{-1}$	Butyl acetate	Cyclohexanone	Butyl acetate	Cyclohexanone
5,000	18,200 sec.$^{-1}$	13,300 sec.$^{-1}$	2,800 sec.$^{-1}$	1,400 sec.$^{-1}$
10,000	17,500	12,500	2,800	1,900
15,000	17,400	13,600	2,500	2,300
20,000	17,800	14,800	2,600	2,500
25,000	18,000	15,700	2,800	2,700
Length $(2a)$, A	900	700	2,000	1,700

The increase in rotational diffusion coefficient with increasing velocity gradient, which is especially noticeable in solutions in cyclohexanone, is most probably to be attributed to polydispersity, as is later explained.

Recently Wissler (1940), working in Signer's laboratory, has made a careful study by the streaming birefringence method of gelatin and gelatin fractions, fractionated methyl cellulose and cellulose nitrate, and of other macromolecules. He has thus made estimates of molecular dimensions which he compared with those from viscosity and sedimentation velocity-diffusion investigations. In Fig. 14.14, based upon Wissler's data, the total length of the cellulose nitrate molecule in solution is plotted against the degree of polymerization, as determined by the

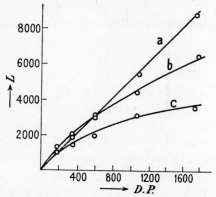

FIG. 14.14. Length (L) of the cellulose nitrate molecule as a function of the degree of polymerization. (D.P.)
(a) Maximum calculated length.
(b) Length from viscosity.
(c) Length from sedimentation-diffusion. (*From* diagram by Wissler (1940).)

different methods. A considerable discrepancy between the two sets of experimental values is thus indicated at the higher degrees of polymerization.

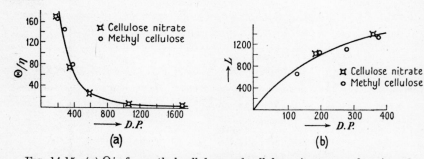

FIG. 14.15. (a) Θ/η for methyl cellulose and cellulose nitrate as a function of the degree of polymerization. (D.P.)
(b) Length (L) of the methyl cellulose and cellulose nitrate molecules as a function of the degree of polymerization. (*From* diagrams by Wissler (1940).)

In comparing the two cellulose derivatives (methyl cellulose and cellulose nitrate) a further interesting result was observed. Thus the curves expressing the variation of the corrected rotational diffusion coefficient (Θ/η) and length of the molecule as functions of the degree of polymerization were, within experimental error, identical for methyl cellulose and cellulose nitrate (Figs. 14.15 (a) and (b)). The work by

Wissler (1940) is especially important since it is one of the first attempts to compare the many different experimental methods of investigating the nature of dispersed molecules or particles. Further investigations of this type are greatly needed.

Before concluding this review of results, attention must be drawn to two special effects which may introduce complexity.

(f) *Effect of polydispersity*

If a solution contains high molecular weight asymmetric particles of very different chemical and (therefore) optical properties as well as of different molecular weight and shape, it is to be expected theoretically, and it has been confirmed experimentally by Sadron and co-workers (1939 and previous papers), that the curves of χ and double refraction against velocity gradient are very different from those of monodisperse systems in solution.

Qualitatively we expect that as the velocity gradient increases from very low values, orientation will first occur for the larger and more asymmetric molecules, and the extinction angle and double refraction will be characteristic of these. However, as higher gradients are reached, smaller and less asymmetric molecules will be affected and the resultant optical effects will include contributions from all species with appreciable orientation.

If the different species are optically opposite in character (i.e. positively and negatively birefringent), the extinction direction may cross the stream-lines from positive values at low velocity gradients to negative ones at higher rates. In such systems the usual analysis of experimental results to give rotational diffusion coefficients is clearly impossible, and in view of such possible complications great care must always be exercised. Sadron (1938) has developed equations for the behaviour of such systems in terms of the components present, which appear to give reasonable agreement with experiment, and further development may be expected.

For polydisperse systems of a homologous series and of constant optical properties, the observed behaviour will not be so complex. Thus at low velocity gradients, the extinction angle measurements will yield the smaller rotational diffusion coefficients of the more asymmetric fractions, and with increasing velocity gradients increasing coefficients will be obtained. Some of the figures of Table 14.4 demonstrate this point clearly. Further, since at the higher gradients hitherto unorientated molecules and particles will be appreciably orientated, the increase in double refraction at the higher gradients will be greater than to be expected from its readings at low gradients. The attainment of a saturation maximum will therefore tend to be absent or delayed.

It would appear that in this type of investigation, as in many others (e.g. p. 135), the advantages of careful fractionation are so considerable as to outweigh any difficulties in the fractionation process.

(g) Importance of deformation double refraction

As yet, the effects of streaming double refraction have been explained almost entirely on the basis of particle and molecule orientation. However, it was also mentioned that Signer and Gross (1933) obtained evidence that at higher velocity gradients deformation of polystyrene molecules in dilute solution occurred. Previously Kuhn (1932) had worked out a theory of double refraction of flow based on the distortion and subsequent orientation of the distorted molecules, but Signer and Gross did not consider it applicable to their results, and attributed the phenomena with which they were mainly concerned as due to particle orientation. Distortional effects were introduced only to explain the special behaviour of high molecular weight polystyrenes. However, Zvetkov and Frisman (1945), working on dilute solutions of polyisobutylene† in petroleum ether (120°–140°), have obtained further evidence of molecular deformation at high velocity gradients and even of chain-breaking leading to irreversible viscosity behaviour. Optical measurements (extinction angle and double refraction) were made in a concentric cylinder apparatus capable of producing velocity gradients up to 80,000 sec.$^{-1}$ and viscosities at high mean velocity gradients were measured in capillary viscometers. Typical results are contained in Figs. 14.16 and 14.17. The absence of any signs of a saturation limit to the double refraction and increasing slopes of the $(n_e - n_0)$ v. G curves at high gradients are strong evidence, it is felt, for the occurrence of molecular distortion. Further, treatment for 5 minutes of some of the solutions in the concentric cylinder apparatus at a velocity gradient of 20,000 sec.$^{-1}$ was found to have caused a permanently decreased viscosity, which is not easily explained except on the basis of irreversible chain rupture. Strong dependences of relative viscosities on velocity gradient (Fig. 14.16) were also observed except for the fractions of lowest molecular weight.

It seems clear, therefore, in some cases, and particularly for those linear, flexible, chain-like molecules of no uniquely defined configuration,

† It is believed that further similar work has been performed on rubber solutions, but the particular paper is not yet accessible to the authors.

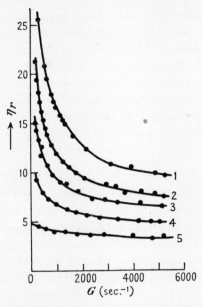

FIG. 14.16. Relative viscosity as a function of the velocity gradient for solutions of poly-isobutylene in petroleum ether.

Curve	1	2	3	4	5
[Volume Fraction $\times 10^2$]	0.84	0.67	0.56	0.42	0.28

(*From* Zvetkov and Frisman (1945).)

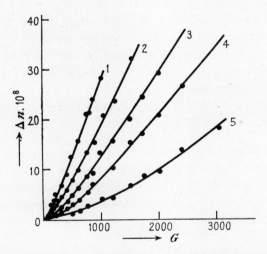

FIG. 14.17. Double refraction as a function of velocity gradient for solutions of poly-isobutylene in petroleum ether. Polymer concentrations as in Fig. 14.16. (*From* Zvetkov and Frisman (1945).)

that molecular deformation may contribute to streaming birefringence effects. However, for most corpuscular proteins, in which the polypeptide chain appears to be uniquely folded and rigidly held, in view of the reasonable correspondence of the results obtained by methods with and without high velocity gradients, it seems that molecular deformation does not occur appreciably and that orientational effects are almost entirely responsible for observed behaviour. In such cases the methods involving orientation in velocity gradients may be confidently applied to the determination of molecular size and shape.

(*h*) *Limitations of the streaming birefringence method*

In the preceding pages we have seen how the streaming birefringence method may be employed to study the orientation of particles with rotational diffusion coefficients as high as $20,000$ sec.$^{-1}$ (e.g. low molecular cellulose nitrate of Table 14.4) and as low as 1 sec.$^{-1}$ (e.g. certain myosins). Although it may be possible to extend the method to lower diffusion coefficients than 1 sec.$^{-1}$ (or longer relaxation times than 1 sec.), practical difficulties (especially the onset of turbulence) limit further extension to values of the diffusion coefficient much higher than $20,000$ sec.$^{-1}$

However, the method of dielectric dispersion which will now be described is most suitably applied to particles with larger rotational diffusion coefficients, and shorter relaxation times. Thus electrical fields with frequencies as high as 10^{11} cycles per sec. may be employed, by which relaxation times as low as that of water (*ca.* 10^{-11} sec.) may be investigated, as well as those corresponding to the more commonly encountered frequencies (e.g. 50 cycles per sec.). It is clear therefore that this method is extremely useful in covering a range of relaxation times complementary to that for which the streaming birefringence method is eminently suited.

We now propose, therefore, to consider the method of dielectric dispersion and some of its applications.

Orientation in an electrical field

The dielectric constant (D) of a medium is usually defined as the ratio of the capacity of a condenser (C) containing the medium between its plates to that of the same condenser (C_0) containing air (or more correctly *in vacuo*):

$$D = \frac{C}{C_0}. \tag{14.39}$$

The presence of the medium effectively causes a reduction in the intensity of the field between the condenser plates as a result either of the formation of induced dipoles (distortion polarization) in the medium, or of the orientation in it of permanent dipoles already existing there (orientation polarization).

These effects are quantitatively expressed for gases (and dilute solutions in non-polar liquids) in the well-known equation

$$P = \frac{4\pi}{3} N\left(\alpha_0 + \frac{\mu^2}{3kT}\right) = \frac{D-1}{D+2} V, \qquad (14.40)$$

where P is known as the molar polarization, α_0 is the polarizability of the medium, μ is the dipole moment, and V is the molar volume. The term, $\mu^2/3kT$, derived by Debye, represents the average moment per molecule due to the orientation of dipoles in a field of unit intensity, whilst α_0 is the value of the dipole per molecule induced by the same field. Although distortion or induced polarization always occurs to some extent, much larger effects arise from orientation polarization, with which we are now concerned. Approximately, therefore, we may say that the value of the dielectric constant of a polar liquid gives a measure of the number and type of dipoles orientated *under the conditions of measurement*.

(a) Dispersion of the dielectric constant

The dielectric constant, which is usually measured in an alternating current circuit, is found for certain systems to be strongly dependent upon the frequency, a phenomenon commonly known as *dielectric dispersion*. Thus, for a solution containing both large and small polar molecules (e.g. a protein and an amino acid), the type of dielectric constant-frequency curve shown in Fig. 14.18, due to Oncley (1942), is obtained. It will be noted that there are three flat portions (A, C, and E) for which the dielectric constant is independent of the frequency, and that these are separated by two portions B and D at 10^5–10^7 and 10^9–10^{11} cycles per sec. respectively, in which the constant decreases rapidly with increasing frequency. In the absence of large polar molecules there is no fall in dielectric constant between 10^5–10^7, whilst the fall between 10^9 and 10^{11} cycles per sec. occurs whenever low molecular weight polar solute or solvent is present. It is therefore reasonable to associate the lower frequency fall in dielectric constant with the presence of the large polar molecules, and that at higher frequency with the presence of small polar molecules.

The curve may be interpreted in detail in terms of relaxation times. As we have seen, a molecule of given molecular weight and shape has, under given conditions, a definite rotational diffusion coefficient (or relaxation time), determined by the rotational frictional constant of the molecule (as is expressed in eqn. (14.5)). Briefly, the longer the relaxation time (τ) the greater the frictional constant. Now the time required for molecules to orientate (or disorientate) themselves in a given field

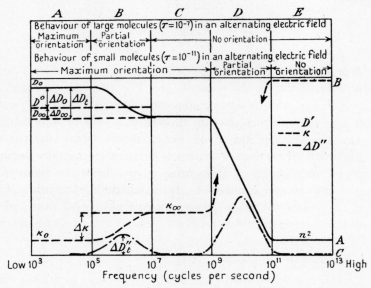

FIG. 14.18. Schematic diagram of anomalous dispersion of the dielectric constant (D'), the specific conductivity (κ), and the dielectric adsorption ($\Delta D'' = 1 \cdot 80(\kappa - \kappa_0)/\nu$) for two widely separated critical frequencies. (*From* Shack, J., Ph.D. dissertation, Harvard University, 15 (1939). *See also* Oncley, J. L., Ferry, J. D., and Shack, J., *Ann. New York Acad. Sci.*, **40**, 371 (1940).)

is determined by the relaxation time or frictional constant of the molecule. At very low applied frequencies (10^2–10^3) the time required for orientation is much shorter than the period of the applied field. All molecules can therefore become completely orientated in the electrical field and can follow its reversals. They can therefore make their maximum contribution to the measured dielectric constant. No change in dielectric constant is therefore to be expected, nor is it observed, as long as the period of the applied field is long compared with τ. However, as the frequency is increased, becoming comparable with $1/\tau$, only a fraction of the molecules manage to keep in step with the changing field. The maximum possible contribution to the measured dielectric constant is not therefore made, and a lower value is recorded. As the frequency is further increased fewer and fewer molecules contribute to the dielectric constant, which finally falls to a value (region C) arising from the orientation of small polar molecules only and from distortional effects.

With still further increasing frequency the dielectric constant remains constant until the time required for the rotation of the smaller molecules is comparable with the period of the applied field, when a decrease (region D) similar to that described above for the large polar molecules sets in. The lowest value of the dielectric constant, observed at the highest frequencies, represents only distortional polarization and, as Fig. 14.18 shows, is usually small compared with that involving orientation. For many proteins, relaxation times are between 10^{-6} and 10^{-8} sec., whilst for simple peptides and smaller polar molecules they are about 10^{-10} and 10^{-11} sec. respectively. For a solution of a protein in water, containing other low molecular weight substances, the regions B and D will be clearly separated (as in Fig. 14.18) when analysis of the curves is much facilitated. However, it should be realized that intermediate relaxation times can occur (e.g. those of partially degraded proteins) and overlapping of 'dispersion' regions is not uncommon.

This section is concerned only with the dielectric dispersion in solution of large polar molecules like the proteins, for which the relevant portions of the dielectric constant-frequency curves (A, B, and C) are well separated from those parts involving the orientation of other small polar molecules present. The requisite frequencies range approximately from 10^3 to 10^8 cycles per sec.

Conductivity measurements over a range of frequencies show a similar dispersion of conductivity, occurring at the same frequencies as that of the dielectric constant for a given solution (Fig. 14.18). It is not proposed to treat this subject in detail, but we mention it here since it can provide evidence confirming that obtained from dielectric investigations. The explanation of the effect is probably similar to that of the Debye frequency effect, whereby the conductivity of electrolyte solutions increases at high frequencies owing to the disappearance of the retardation due to the ionic atmosphere of any ion. Thus the rapid reversals of the field effectively prevent the formation of the asymmetric ionic atmosphere common to an ion in slow motion and to which a retardation at low frequencies is due. In a similar way the slow rotation of a dipole is accompanied by retardation effects which disappear at higher frequencies to give increased conductance.

(b) *Experimental methods*

Several methods have been devised for measuring dielectric constants of solutions over a range of frequencies, some merely developments of the usual methods of measuring capacity which are described in standard text-books on electricity, whilst others have been especially developed. The methods are listed and outlined by Oncley (1942, see also his article in Cohn and Edsall (1943, p. 543)) where detailed references are provided and sources of error (especially electrode polarization) discussed. Owing to their very technical nature we shall mention here only two of the most generally used methods.

1. The capacities of a cell filled first with air and then with test liquid may be measured by a bridge method in terms of standard resistances and capacities. The accuracy of the determination depends largely on the quality of the resistance standards.

2. In the resonance method (e.g. Wyman, 1930) the cell occupies part of an alternating current circuit, whose resonance frequency ν is given by $1/\{2\pi\sqrt{(LC)}\}$, where L is the inductance and C the capacity. By determining the resonance frequency of the circuit with the cell first in air (ν_0) and then immersed in test liquid, the dielectric constant is obtained from the equation

$$D = \frac{C}{C_0} = \frac{\nu_0^2}{\nu^2}. \tag{14.41}$$

The resonance frequencies are measured by comparison with standard frequencies.

(c) *Dielectric increments and their evaluation*

The contribution to the total dielectric constant (in region A) caused by the orientation of the large polar molecules is given by

$$\Delta D_t = D_0 - D_\infty, \tag{14.42}$$

where D_0 and D_∞ are the dielectric constants in regions A and C of Fig. 14.18. Since ΔD_t depends upon the concentration, it is usual to make use of the quantity known as the *total dielectric increment*, $\Delta D_t/c$ where c is the concentration in grams per litre.

ΔD_t can be evaluated by measuring the dielectric constant of the same solution in the A and C frequency ranges. Alternatively, however, it may be obtained by making use of the dielectric increments, $\Delta D_0/c$ and $\Delta D_\infty/c$ at low and high frequencies respectively, of the solution over the pure solvent. ΔD_0 and ΔD_∞ may be defined as

$$\Delta D_0 = D_0 - D^0, \tag{14.43}$$

and

$$\Delta D_\infty = D_\infty - D^0, \tag{14.44}$$

where D^0 is the dielectric constant of the pure solvent which has the same value in the low-frequency (region A) and high-frequency (region C) ranges.

Subtracting (14.44) from (14.43) and dividing by c we have

$$\frac{\Delta D_0}{c} - \frac{\Delta D_\infty}{c} = \frac{D_0 - D_\infty}{c} = \frac{\Delta D_t}{c}, \qquad (14.45)$$

from which it is clear that the total dielectric increment is given by the difference of the low-frequency and high-frequency increments.

High-frequency increment. It should be realized, as Fig. 14.18 shows, that although not very different, ΔD_t and ΔD_0 are not identical, the inequality arising, as may be seen in eqns. (14.42) and (14.43) from the fact that D_∞ and D^0 are not the same. In general D^0 is slightly larger than D_∞ owing to the smaller contribution in the latter of the reduced number of solvent molecules in the solution, some of which have been displaced by those of the high molecular weight material whose orientation contributes nothing to the dielectric constant in the high-frequency (C) region. It follows, therefore, from eqns. (14.42) and (14.43) that

$$\Delta D_0 < \Delta D_t,$$

but the difference, ΔD_∞, is not great. The latter may be estimated experimentally from eqn. (14.44) by measuring the dielectric constants of the solution and pure solvent in the high-frequency range, or it may be computed theoretically in terms of the solvent displaced from a given volume of solution. Thus if 1 gm. of the polymeric material occupies a volume v in solution, then, assuming the dielectric constant of the solute to be unity (i.e. neglecting distortional polarization) the reduction in dielectric constant from the pure solvent value is

$$-\frac{\Delta D_\infty}{c} = \frac{(D^0 - 1)v}{1000}. \qquad (14.46)$$

But v may be expressed as the sum of the volumes of anhydrous material and bound solvent, i.e.

$$v = \bar{v} + w/\rho_0, \qquad (14.47)$$

where w is the weight of solvent of density, ρ_0, associated with 1 gm. of anhydrous polymer. Putting $\bar{v} = 0.75$, $w = 0.30$, $\rho_0 = 0.997$, and $D^0 = 78$ Debye units for water at 25° C., we have

$$-\frac{\Delta D_\infty}{c} = 0.081, \qquad (14.48)$$

a value which is confirmed qualitatively by experimental measurement (see Table 14.5). Comparison with values of $\Delta D_0/c$ shows that, although small, $\Delta D_\infty/c$ is by no means negligible.

Low-frequency increment. The low-frequency increment, $\Delta D_0/c$, is obtained by measuring, at low frequencies (region A), the dielectric constant D^0 of the pure solvent, and D_0, of the solution over a range of concentrations. In dilute solution ΔD_0 is found to vary linearly with concentration c, as illustrated in Fig. 14.19, so that $\Delta D_0/c$ is given by the slope of the lines. At higher concentrations and with highly polar

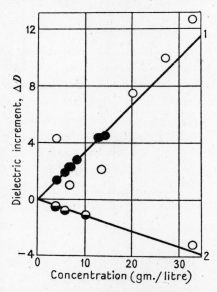

Fig. 14.19. Low (1) and high (2) frequency dielectric increments of carboxyhaemoglobin. Data collected by Oncley.

molecules, deviations from linearity occur, due probably to electrostatic interactions. The increment must therefore be obtained only at low concentrations where the ΔD_0 v. c curve is linear.

Further, since the dipole moment of the polymer molecule must vary with the net charge, variations in $\Delta D_0/c$ with pH are to be expected for many substances and have been verified experimentally (see Oncley in Cohn and Edsall, 1943). Thus in the case of proteins it is essential to make measurements of dielectric increments under well-defined conditions of pH and salt concentrations. Table 14.5, due to Oncley, contains experimental values of high-frequency, low-frequency, and total dielectric increments as well as other relevant data for several well-known proteins.

From experimental curves of the type shown in Fig. 14.18 two types of information may be obtained:

(1) The frequency range over which dielectric dispersion occurs, which is related to the molecular size and shape of the solute.
(2) Dielectric increments, which are clearly connected with the dipole moment of the dissolved molecules.

We are chiefly concerned with (1), although both aspects are of considerable interest.

TABLE 14.5. *Dielectric Increments and Dipole Moments for Certain Protein Molecules*

Protein	Solvent	Temperature °C.	$\Delta D_0/c$	$-\Delta D_\infty/c$	$\Delta D_t/c$	Mol. wt. (M)	Dipole moment (μ) Debye units
CO-Haemoglobin (horse)	Water	25	0·33	0·09	0·42	67,000	480
CO-Haemoglobin (pig)	,,	20	..	..	0·30	(67,000)	410
Myoglobin	,,	25	0·15	(0·06)	0·21	17,000	170
Insulin	80% aqueous propylene glycol	25	..	..	0·38	40,000	360
Insulin	90% aqueous propylene glycol	25	..	..	0·29	40,000	310
Insulin	100% propylene glycol	25	..	..	0·26	40,000	300
β-Lactoglobulin	M/2 and M/4 glycine	25	1·51	(0·07)	1·58	40,000	730
β-Lactoglobulin	M/2 and M/4 glycine	0	1·84	(0·08)	1·92	40,000	770
Egg albumin	Water	25	0·10	0·07	0·17	44,000	250
Horse serum albumin (carbohydrate-free)	,,	25	0·17	0·07	0·24	70,000	380
Horse serum γ-pseudo globulin	,,	25	1·08	(0·06)	1·14	142,000	1,100
Horse serum γ-pseudo globulin	,,	0	1·26	(0·06)	1·32	142,000	1,300
Edestin	2M glycine	25	0·7	(0·1)	0·8	310,000	1,400
Gliadin	56% aqueous ethanol	25	..	..	0·10	42,000	190
Secalin	54% aqueous ethanol	25	..	..	1·0	24,000	440
Zein	72% aqueous ethanol	25	..	..	0·45	40,000	380

(d) *Determination of molecular weight, molecular shape, and dipole angle*

A considerable part of the theory underlying the treatment of dielectric dispersion is due to Debye (e.g. 1929). For a single relaxation time, Debye showed that the dielectric constant measured at a frequency ν is related to the low- and high-frequency values, D_0 and D_∞ respectively, by an equation which for highly polar media may be written (Wyman, 1936)

$$D = D_\infty + \frac{D_0 - D_\infty}{1 + \nu^2/\nu_c^2}, \qquad (14.49)$$

where ν_c, the critical frequency, is simply related to the relaxation time τ, by $\nu_c = 1/2\pi\tau$ and represents that frequency at which the dielectric constant is half-way between D_0 and D_∞, i.e. $(D_0 + D_\infty)/2$.

From measured values of D_0 and D_∞, D may be calculated as a function of ν if the value of ν_c is assumed. The value of ν_c which gives good agreement with the experimental curve is accepted as the true critical frequency from which the relaxation time and rotational diffusion coefficient may be obtained. In general, however, a molecule cannot be described in terms of one relaxation time only, but rather at least two such times, τ_a and τ_b, with corresponding dielectric increments, ΔD_a and ΔD_b, appropriate to an ellipsoid of rotation are required. Eqn. (14.49) may be extended into the general form

$$D = D_\infty + \frac{\Delta D_a}{1+\nu^2/\nu_a^2} + \frac{\Delta D_b}{1+\nu^2/\nu_b^2} + \ldots, \quad (14.50)$$

where
$$\nu_a = \frac{1}{2\pi\tau_a} \quad \text{and} \quad \nu_b = \frac{1}{2\pi\tau_b}. \quad (14.51)$$

But, by means of the Perrin and derived equations (eqns. (14.13)–(14.23)), we have seen that τ_a and τ_b may be evaluated for an ellipsoid of rotation in terms of its axial ratio and the relaxation time τ_0 of a sphere of the same molecular weight and volume. Thus, postulating certain values of a/b and ν_0 ($=1/2\pi\tau_0$) and arbitrarily dividing the experimentally observed ΔD_t into components ΔD_a and ΔD_b along the axes of the ellipsoid, the curve of D against frequency may be calculated. Figs. 14.20 (a) and (b) show such curves, the effect of varying axial ratio being shown in (a) and varying ratio, $\Delta D_a/\Delta D_b$, in (b). Such curves may be compared with experimental curves, the closest fits identified, and further refinements carried out. This procedure, if carefully carried out, thus establishes a/b, τ_0, and $\Delta D_a/\Delta D_b$ for the unknown molecule.

From τ_0 ($=1/2\Theta_0$) the solvated molecular volume (Mv) may be evaluated from eqn. (14.12), which may be written

$$\Theta_0 = \frac{1}{2\tau_0} = \frac{RT}{6(Mv)\eta}, \quad (14.12\,\text{a})$$

or
$$Mv = \frac{RT\tau_0}{3\eta}. \quad (14.12\,\text{b})$$

Further, knowing v, from the anhydrous partial specific volume and degree of solvation (eqn. (14.47)), the molecular weight is readily obtained.

The magnitude and direction of the resultant dipole of a molecule which has components along two axes at right angles is given by vectorial summation of the components. Thus, if θ be the angle between the resultant and the a axis, termed the dipole angle, then

$$\tan\theta = \mu_b/\mu_a = \sqrt{\frac{\Delta D_b}{\Delta D_a}}, \quad (14.52)$$

where the proportionality between dielectric increment and the square of the dipole moment is assumed. The derivation of this result is discussed later (p. 423).

ROTATIONAL DIFFUSION

The ratio $\Delta D_a/\Delta D_b$, determined by comparison of experimental and theoretical dispersion curves, thus provides a measure of the direction of the resultant dipole with respect to the geometrical axes of the molecule.

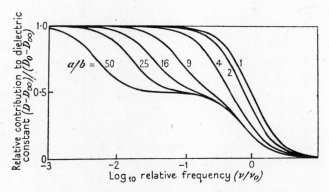

FIG. 14.20. (a) Dielectric dispersion curves for elongated ellipsoids of revolution (according to Perrin), with $\Delta D_a/\Delta D_b = 1$ and $\theta = 45°$

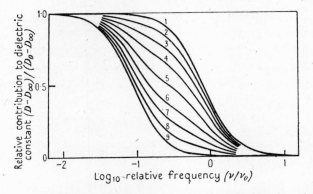

FIG. 14.20. (b) Dielectric dispersion curves for ellipsoids of revolution with $a/b = 9$. Curves 1–9 are for $\Delta D_a/\Delta D_b = 0, 0{\cdot}1, 0{\cdot}25, 0{\cdot}5, 1, 2, 4, 10$, and ∞ respectively.

Dispersion of conductivity. For the dispersion of conductivity we have the general equation, analogous to (14.50),

$$\kappa = \kappa_0 + \frac{\Delta \kappa_a (\nu/\nu_a)^2}{1+(\nu/\nu_a)^2} + \frac{\Delta \kappa_b (\nu/\nu_b)^2}{1+(\nu/\nu_b)^2} + ..., \qquad (14.53)$$

where $\Delta \kappa_a$, $\Delta \kappa_b$, etc., are the conductivity increments corresponding to the critical frequencies ν_a, ν_b, etc., respectively. Clearly, theoretical conductivity dispersion curves can be calculated in a manner similar to that already indicated for dielectric dispersion, and values of the axial ratio, a/b, molecular weight (from τ_0), and ratio $\Delta\kappa_a/\Delta\kappa_b$ obtained. Further, a given conductivity increment is related to the corresponding dielectric increment by equations of the type

$$\Delta\kappa_a = \frac{\Delta D_a \nu_a}{1\cdot 80},\qquad(14.54)$$

where conductivities are expressed in reciprocal ohms/cm. and frequencies in megacycles. The detailed comparison of dielectric and conductivity curves can, therefore, be readily carried out.

It is apparent, therefore, that the careful use of the dielectric (and conductivity) dispersion method provides information not only on the size and shape of polar molecules, which may be directly compared with similar information obtained by other methods, but also on the direction of the resultant electrical moment. Further information on the electrical state of a molecule may also be obtained by examination of the magnitude of dielectric increments, as will shortly be shown. Table 14.6, due to Oncley (1940), contains the calculated results of dielectric dispersion measurements for three well-known proteins for which information from sedimentation, diffusion, and viscosity is also tabulated. Although in certain respects approximate (e.g. no correction for hydration), the calculations show a fair measure of agreement which is very encouraging. The comparison of measurements made by the dielectric and other methods is also considered in Chapter XXVI, devoted to a consideration of our knowledge of proteins.

(e) Dielectric increments and dipole moments

For solutions in polar liquids eqn. (14.40) is not valid, and it appears probable that an equation

$$D = j + hp,\qquad(14.55)$$

where p is the polarization of the liquid per c.c. and j and h are constants, is valid (Wyman, 1936). From work involving the use of solutions of substances of known dipole moments, suitable values of j and h have been given by Wyman as -1 and $8\cdot 5$ respectively. Modification of these values is, however, quite probable.

By dividing p into the contributions of solute and solvent according to their volume concentrations, and introducing for the molar polarization (P_s) of the solute, the Debye value

$$P_s = \frac{4\pi}{9}\frac{N\mu^2}{kT},\qquad(14.56)$$

it can be shown (Cohn and Edsall, 1943, p. 151) that the dipole moment may be related to the total dielectric increment by the equation

$$\mu^2 = \frac{9000kTM}{4\pi Nh}[\Delta D_0/c - \Delta D_\infty/c]. \qquad (14.57)$$

TABLE 14.6. *Data from Dielectric and other Measurements for Certain Proteins*

Measurements	Serum albumin	Serum pseudo-globulin	Edestin
Dielectric:			
Low-frequency increment per gram, $\Delta D_0/c$	0·17	1·1	0·7
Total increment per gram, $\Delta D_t/c$.	0·24	(1·2)	(0·8)
Electric moment, μ, in Debye units	380	1,200	1,400
Relaxation time for equivalent sphere, $\tau_0 \times 10^8$ (in water at 25° C.)	6·0	22	21
Axial ratio, a/b	6	9	9
Dipole angle, θ	45°	45°	45°
Ultracentrifuge and Diffusion:			
Molecular weight, M . . .	70,000	142,000	310,000
Partial specific volume, $\bar{v}$ . .	0·748	0·730	0·744
Sedimentation constant, $s_{20}^0 \times 10^{13}$.	4·5	6·5	12·8
Diffusion coefficient, $D_{20}^0 \times 10^7$.	6·2	4·1	3·9
Frictional ratio, f/f_0 . . .	1·25	1·5	1·21
Relaxation time for equivalent sphere (no hydration), $\tau_0 \times 10^8$ (in water at 25° C.). . . .	5·7	11·3	25
Axial ratio, a/b (assuming no hydration and an elongated ellipsoid) .	5	9	4
Viscosity:			
Viscosity coefficient, η_{sp}/ϕ . .	6·5	10	..
Axial ratio, a/b (assuming no hydration and an elongated ellipsoid) .	6	8	..

Oncley (in Cohn and Edsall, 1943) has summarized the methods available for obtaining the value of h, for which, as stated above, the value 8·5 has been suggested. Knowing also the molecular weight M (from the dispersion curve or other methods), μ can clearly be obtained from values of the total dielectric increment, whose determination has already been discussed. Dipole moment values in Debye units (i.e. 10^{-18} e.s.u.) for several protein molecules are contained in Table 14.5 with the dielectric data from which they were derived. Such values are very large compared with those of simple molecules (usually a few Debye units, e.g. for ethyl alcohol $\mu = 1·7$ D.), but this is not necessarily to be interpreted as due to the possession of large charges at the extremes of the dipolar arm, for the diameters of the protein molecules are themselves large. A qualitative view of the electrical symmetry of the molecules may be obtained by assuming them spherical and unhydrated, and calculating their diameter. The charge (z) required at each end of such a diameter to give the observed dipole moment is then readily obtained. Such values for a number of protein molecules are found to be very small compared with the maximum positive or negative charge (z_{max}) which they can possess. Thus for serum albumin, haemoglobin, and edestin z/z_{max} has been calculated respectively to be 0·02, 0·03, and 0·01. The positive and negative groups on these molecules must therefore be distributed in such a way as to give a resultant moment much smaller than that obtained if the opposite charges were concentrated at the two extremes of a molecular diameter. The molecules are therefore spoken of as possessing a high degree of electrical symmetry.

(*f*) *Limitations of the dielectric dispersion method*

We have, as yet, considered dipole orientation in an electrical field only as it occurs through the rotation of a whole molecule. For small molecules, and polymers in which a specific configuration is rigidly held, this is undoubtedly sufficient, as the success of the methods demonstrates. However, for many solid polymeric materials with polar side groups, e.g. linear polymers of the type of polyvinyl chloride

$$-CH_2.\overset{|}{\underset{Cl}{CH}}.CH_2.\overset{|}{\underset{Cl}{CH}}.CH_2.\overset{|}{\underset{Cl}{CH}}.CH_2.\overset{|}{\underset{Cl}{CH}}-$$

or three-dimensional polymers like the bakelites which contain unreacted hydroxyls, orientation of dipoles by rotation of the complete polymer molecule is impossible. In such cases, dipole orientation occurs locally about bonds at or in the neighbourhood of each dipole. Thus it has been shown for plasticized and unplasticized polymers that large ranges of relaxation times are required to explain the dielectric constant-frequency curves, corresponding to the different modes of orientation possible (for a review see Fuoss, 1943). For polyvinyl chloride this range is at least from 10^{-1} to 10^{-6} sec.

Whilst in dilute systems of large molecules, molecular rotation is never restricted to the same extent as in solid materials, it cannot be concluded that local orientation of dipoles, involving rotation of parts of the whole molecule, is always absent. It is probable, rather, that further work on dilute solutions of linear polymers may show such effects, and their possibility should not therefore be excluded.

REFERENCES

BOEDER, 1932, *Z. Phys.* **75**, 259.
COHN and EDSALL, 1943, *Proteins, Amino Acids, and Peptides*, Reinhold.

DEBYE, 1929, *Polar Molecules*, Reinhold.
DE ROSSETT, 1941, *J. chem. Phys.* **9**, 766.
EDSALL and MEHL, 1940, *J. biol. Chem.* **133**, 409.
—— 1942, in *Advances in Colloid Science*, vol. i, edited by E. O. Kraemer, Interscience, p. 269.
—— *et alia*, 1944, *Rev. sci. Instrum.* **15**, 243.
—— and FOSTER, 1945, *J. Amer. chem. Soc.* **67**, 617.
FUOSS, 1943, in *The Chemistry of Large Molecules*, Interscience.
HARTSHORNE and STUART, 1934, *Crystals and the Polarizing Microscope*, Arnold, London.
JEFFERY, 1923, *Proc. Roy. Soc.* A. **102**, 161.
KUHN, 1932, *Z. phys. Chem.* A. **161**, 427.
MEHL, 1938, *Cold Spring Harbor Symposia Quant. Biol.* **6**, 218.
MURALT and EDSALL, 1930, *J. biol. Chem.* **89**, 315, 351.
ONCLEY, 1940, *J. phys. Chem.* **44**, 1103.
—— 1942, *Chem. Rev.* **30**, 433.
PERRIN, J., 1909, *C.R. Acad. Sci., Paris*, **149**, 549.
PERRIN, F., 1934, *J. Phys. Radium* (7), **5**, 497.
PETERLIN and STUART, 1939, *Z. Phys.* **112**, 1, 129.
ROBERTSON, 1929, *Introduction to Physical Optics*, 2nd edn., Chapman & Hall.
ROBINSON, 1939, *Proc. Roy. Soc.* A **170**, 519.
SADRON, 1936, *J. Phys. Radium* (7), **7**, 263.
—— 1938, ibid. **9**, 38.
—— *et alia*, 1939, *J. Chim. phys.* **36**, 78.
SIGNER and GROSS, 1933, *Z. phys. Chem.* A **165**, 161.
—— 1936, *Trans. Faraday Soc.* **32**, 296.
—— and SADRON, 1936, *Helv. chim. Acta*, **19**, 1324.
SNELLMAN, 1945, *Arkiv Kemi Min. Geol.* **19** A, No. 30.
WEBER, 1934–5, *Pflüg. Arch. ges. Physiol.* **235**, 205.
WIENER, 1912, *Abh. sächs. Ges. (Akad.) Wiss.* **32**, 507.
WILLIAMS and WATSON, 1938, *Cold Spring Harbor Symposia Quant. Biol.* **6**, 208.
WISSLER, 1940, Inaugural Dissertation, Bern.
WYMAN, J., 1930, *Phys. Rev.* **35**, 623.
—— 1936, *Chem. Rev.* **19**, 213.
ZVETKOV and FRISMAN, 1945, *Acta Physicochimica, U.R.S.S.*, No. 1, **20**, 61.

XV

OPTICAL, X-RAY, ELECTRON DIFFRACTION, AND OTHER METHODS

The Optical Properties of Colloidal Solutions

THE colours of many suspensions, particularly of the metals, are so striking that the interest that they aroused in the nineteenth-century physicists is not surprising. Some mention of the early work of Faraday, Tyndall, and Rayleigh was given in Chapter I. Now we shall consider the optical properties of colloidal systems in more detail, and shall inquire what information they can provide about the size, shape, and interactions of colloidal particles.

Colloidal suspensions (e.g. sulphur or gold sols) differ from ordinary molecular solutions in being readily visible by the light which they scatter laterally, giving rise to the well-known 'Tyndall beam', by which a beam of unpolarized light passing through the suspension is clearly visible from the side. Such scattering is the result of the presence of colloidal particles and can, in fact, be used to estimate the solution concentration. Two methods are used. In the first, the method of turbidimetry, an ordinary visual or photo-electric colorimeter is used to estimate the light transmitted in the incident direction by the colloidal solution, a calibration being made by means of a solution of approximately the same particle size and of known concentration.

With suspensions of slight turbidity, a second method is used, in which the intensity of scattered light, usually at 90° or 45° to the line of illumination, is measured directly in a nephelometer. Limitations similar to those in turbidity determinations apply here also. Both methods require careful standardization of conditions for their accurate use. For details, practical books should be consulted.

Turbidity and nephelometry now find considerable uses in analysis where the material to be estimated is present in small amount and can be made into a fine colloidal suspension. Silver and soluble halides (estimated as silver halide), phosphates, sulphates, and many organic compounds can thus be estimated, as well as bacterial suspensions.

When considering scattering phenomena in detail, it is usual to consider the two extreme types of particle, namely non-absorbing non-conductors (or dielectrics, e.g. sulphur, gum mastic, etc.) and strongly absorbing conductors (e.g. gold and other metals).

Non-absorbing insulating particles

The light scattered by very fine sols of sulphur or mastic ($< 1\,\mu$) is predominantly blue in colour, and hence that which is transmitted is yellowish-red. Further, if the incident beam is unpolarized, light scattered at 90° to the incident direction is polarized almost completely, the proportion of polarized light decreasing symmetrically about the 90° position.

Rayleigh (Strutt, 1871, and other papers) worked out the theory of scattering by spherical particles whose radius is small compared with the wavelength of the incident light (λ). Normal reflection and refraction (dependent upon surfaces with dimensions greater than λ) do not therefore occur. By the method of dimensions (and later more rigidly from the electromagnetic theory of light) he showed that the intensity of scattered light (I), at a distance x from a disturbing particle of volume v is given by

$$I = \text{constant}\,\frac{v^2}{\lambda^4 x^2}. \tag{15.1}$$

His complete expression for the scattered intensity, per unit volume, in a direction making an angle θ with the incident unpolarized light may be written

$$I = I_0 \left(\frac{D'-D}{D}\right)^2 \frac{\nu\pi^2 v^2}{\lambda^4 x^2}(1+\cos^2\theta), \tag{15.2}$$

where I_0 is the incident intensity, D' and D are the optical densities (from the old elastic solid theory of light) of the particles and the dispersion medium, and ν is the number of particles per unit volume.

For the scattered intensity perpendicular to the incident beam we thus have

$$I = I_0 \left(\frac{D'-D}{D}\right)^2 \frac{\nu\pi^2 v^2}{\lambda^4 x^2}. \tag{15.3}$$

If $(D'-D)/D$ be now replaced by its equivalent in the electromagnetic theory of light, $2(n-1)/\nu v$, we obtain from (15.2)

$$I = I_0 \frac{4\pi^2(n-1)^2}{\lambda^4 x^2 \nu}(1+\cos^2\theta), \tag{15.4}$$

where n is the refractive index of the medium as modified by the particles. This equation applies to the scattering by small particles in a gaseous phase, but Debye (1947) has shown that it may be extended to infinitely dilute colloidal solutions by introducing $n_1(n-n_1)$ for $(n-1)$ in eqn. (15.4), where n_1 is the refractive index of the solvent. Thus we have

$$I = I_0 \frac{4\pi^2 n_1^2(n-n_1)^2}{\lambda^4 x^2 \nu}(1+\cos^2\theta). \tag{15.4a}$$

The intensity of scattered light being inversely proportional to the fourth power of the wavelength, light of short wavelengths is therefore preferentially scattered, giving the blue colour characteristic of many colloidal solutions, and leaving the transmitted light yellowish-green or red owing to the loss of the blue rays. Eqn. (15.2) also demonstrates that for a constant volume fraction of the disperse phase (i.e. $vv =$ constant) the intensity of the scattered light is proportional to the volume of a single particle, and for a constant number of particles to the square of this volume. Clearly, therefore, the larger particles are much more effective in light scattering than the smaller ones, and the need for removal of coarse dust particles and impurities in making quantitative measurements on fine suspensions is obvious.

The results of scattering experiments with sols of non-conductors of small particle size agree well with theoretical predictions, as shown, for example, by Casperson (1932, 1933), using sulphur and gum mastic, and by the recent detailed study of sulphur sols by La Mer and Barnes (1946). The latter workers prepared and used highly monodisperse sols, and, in addition to verifying the main conclusions of Rayleigh's theory, observed 'spectra' of several orders arising from dispersion of the scattering, and tested theoretical predictions concerning sols of particle size comparable with the wavelength of light (see p. 431).

Strongly absorbing conducting particles

Metal sols are clearly not all bluish by scattered light and yellowish-red by transmitted light; gold sols, for example, may be ruby red, violet, blue, or green by transmitted light according to the method of preparation, age, etc. In its simple form, therefore, Rayleigh's theory for feebly absorbing insulators does not apply, but it was extended by Garnett (1904, 1906) and by Mie (1908) to metallic particles by using the complex indices of refraction (involving the absorption coefficient κ and the ordinary refractive index, and given by $n(1-i\kappa)$), as obtained from measurements on the ordinary metals.

Underlying these extensions is the fact that gold possesses an absorption maximum in the green, and a reflection maximum in the yellow. With small particles absorption predominates, but as the particle size increases, reflection becomes more important, leading to an increased scattering intensity and to colour changes. According to Mie, a fine gold sol containing spherical particles, of radius $ca.$ 200 A, should show a maximum in both the absorbed and the scattered light at $ca.$ 5300 A, the latter being only a small fraction of the former (Fig. 15.1 (*a*) (Freundlich

(1926), p. 383)). The ruby-red colour of such sols by transmitted light is thus readily understood. With increasing particle size, however, the

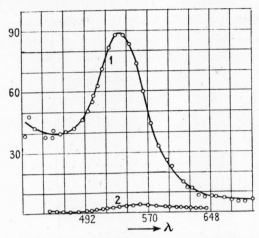

FIG. 15.1. (a) Absorption coefficient (curve 1) and amount of light radiated laterally (curve 2) by a red gold sol.

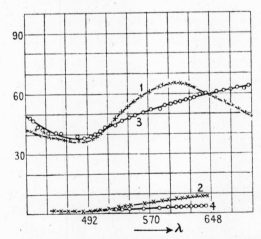

FIG. 15.1. (b) Curves 1 and 2: absorption coefficient and amount of light radiated laterally by normal blue gold sols (almost spherical particles). Curves 3 and 4: absorption coefficient and amount of light radiated laterally by anomalous blue gold sols (non-spherical particles). (Freundlich (1926).)

theory predicts an absorption flattened and shifted towards the red, with an increased intensity of scattered radiation (Fig. 15.1 (b) (Freundlich (1926), p. 383)). The disappearance of the ruby-red colour and onset of the blue, characteristic of the coarser gold sols, is thus explained.

Mie's theory was also able to treat successfully the change in the state of polarization of the scattered light with the size of the scattering particles, as will be shown below.

Under certain conditions (see Freundlich, 1926, p. 385) it is possible to prepare unusual gold sols of very small particle size (i.e. linear dimensions < 500 A) as shown, for example, by their polarization properties (see below), but which are nevertheless blue in colour. Such systems,

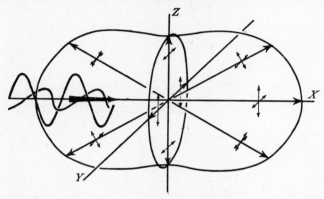

FIG. 15.2. Scattering from a small, spherical particle with unpolarized incident light. The incident light is analysed into two perpendicularly polarized beams out of phase. The scattered light, as indicated by the arrows, also generally contains two independent components, except for light in the X–Y plane. The scattering is symmetrical around the X-axis.

whose absorption and scattering curves (curves 3 and 4 of Fig. 15.1 (b)) are clearly anomalous, are believed to contain non-spherical particles, since, according to Gans (1912), who extended Mie's theory to discs and rod-like particles, the absorption in such cases should be shifted to longer wavelengths. Gold sols with non-spherical particles should therefore tend to be blue in colour.

Polarization properties

In their polarization properties, insulating and metallic particles do not show differences as fundamental as those already discussed. As Rayleigh pointed out, for very small spherical particles (linear dimensions < 500 A), light scattered at 90° to the incident direction is almost completely polarized, the completeness of the polarization as well as the scattered intensity falling off with inclination to the 90° direction. Fig. 15.2 (Zimm, Stein, and Doty (1945)) shows diagrammatically the main features of such light scattering. Clearly, the completeness of the polarization at 90° results from the absence of incident longitudinal vibration components, whilst in other scattering directions the incident light has vibration components in planes other than the vertical one.

The presence of such a polarization maximum at 90° to the incident direction was thus taken in the case of the 'anomalous' blue gold sols to indicate that the particle dimensions were less than 500 A. As we have seen, the colour of such sols is usually attributed to the asymmetry of the particles.

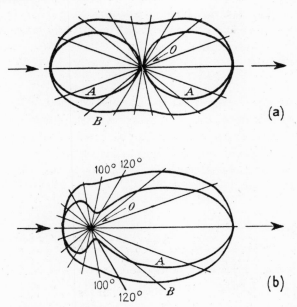

Fig. 15.3. Scattering from small spherical (a) and larger spherical particles (b). OA and OB give the relative proportions of unpolarized and total scattered light.

Mie (1908) showed, for spherical particles, that as the diameter of spherical scattering particles became comparable with the wavelength of light the completeness of polarization in the '90° position' fell off, and further, that the position of maximum polarization and scattered intensity moved towards the exit direction of the light. Fig. 15.3 (a) and (b), due to Mie, summarizes the main features of scattering for very small and larger spherical particles respectively, OA and OB being proportional to the unpolarized and total scattered light. The main features of the theory of Mie for the larger particles have now been experimentally verified, especially by La Mer and Barnes (1946).

Incomplete polarization at the 90° position can also arise, even in the case of diatomic gases and small molecules or particles, from the anisotropy of the scattering entities, as well as from the asymmetry of external shape. In view of this effect of particle size, shape, and anisotropy on the completeness of the polarization at 90° to the incident direction, the use of polarization measurements in making estimates of particle size and shape suggests itself, and such measurements have in fact been utilized. For this purpose it is usual to define the depolarization ratio (ρ), which, for light scattered horizontally from a horizontal incident beam, is the ratio of the intensities of the horizontally (H) and vertically (V) vibrating components. For initially unpolarized light, indicated by the subscript u,

$$\rho_u = \frac{H_u}{V_u}. \tag{15.5}$$

For small isotropic spheres, as we have mentioned, $\rho_u = 0$. In other cases finite values of ρ_u arise as a result of particle size and anisotropy, and in order to separate these effects two other depolarization ratios ρ_v and ρ_h, for horizontally incident light vibrating in vertical and horizontal planes respectively, have been defined (e.g. see Gans, 1912, and Krishnan, 1937) as

$$\rho_v = \frac{H_v}{V_v} \tag{15.6}$$

and

$$\rho_h = \frac{V_h}{H_h}. \tag{15.7}$$

Since unpolarized light may be treated as the sum of the horizontally and vertically vibrating components, eqn. (15.5) becomes

$$\rho_u = \frac{H_u}{V_u} = \frac{H_v + H_h}{V_v + V_h}. \tag{15.8}$$

For most polymer solutions the three polarization ratios are not completely independent, for, providing the solution contains a large number of randomly distributed scattering particles and no optically active molecules, we have, by Rayleigh's Law of Reciprocity (see Zimm, Stein, and Doty, 1945),

$$H_v = V_h, \tag{15.9}$$

and substituting in (15.8) we obtain

$$\rho_u = \frac{1 + 1/\rho_h}{1 + 1/\rho_v}, \tag{15.10}$$

a relation by which the third depolarization ratio may be calculated if the other two are known. Since in many cases the horizontal scattering from horizontally incident and vibrating light is very feeble, the relation is often used in calculating ρ_h from the more easily measured values of ρ_u and ρ_v.

Experimentally, the depolarization ratios are obtained by measuring the intensities of the vertical and horizontal vibration components in light scattered at 90° to the incident direction for the cases of unpolarized, vertically polarized, and horizontally polarized light. A recent apparatus (Doty and Kaufman, 1945) is shown diagrammatically in Fig. 15.4. Light from the mercury lamp, S, is made parallel by the

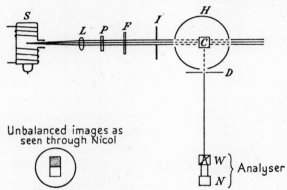

FIG. 15.4. Apparatus for the measurement of depolarization.

lens L, polarized by the polaroid P (when required), and made monochromatic by the filter F. The diaphragm I defines the beam entering the cell C, surrounded by the housing H, and light scattered at 90° is defined by a second diaphragm D, before reaching the Wollaston prism W separating the vertically and horizontally vibrating components to give two images. The tangent of the angle (θ) made by the vibration plane of the Nicol prism N with respect to the vertical, after rotating to give identical intensity of the images, gives directly the ratio of the amplitudes of the vertically and horizontally vibrating components. It therefore follows that

$$\tan^2\theta = \frac{V}{H}. \qquad (15.11)$$

Where the scattering is very weak, the uncertainty in the corresponding depolarization value may be great, but an improved value may be obtained from eqn. (15.10) if the other two depolarizations are more accurately known.

As we have seen, ρ_u is zero only for small isotropic spherical particles, finite values arising from both the appreciable size of the scattering particles with respect to the wavelength of light, and from their anisotropy in shape or structure. However, the different effects of size and anisotropy on the other two depolarization ratios make possible the use of depolarization measurements in investigating the size and form of colloidal particles. Table 15.1, due to Zimm, Stein, and Doty (1945), summarizes the effect of size and anisotropy on the different scattered intensities and depolarization ratios. The plus sign denotes a finite value which, in the case of the depolarization ratios, is between 0 and unity.

TABLE 15.1. *Scattered Intensities and Depolarization Ratios*

Type of particle	H_v	V_v	H_h	V_h	ρ_u	ρ_v	ρ_h
Small isotropic	0	+	0	0	0	0	0
Large ,,	0	+	+	0	+	0	0
Small anisotropic	+	+	+	+	+	+	1
Large ,,	+	+	+	+	+	+	+

Clearly a complete depolarization investigation and comparison with Table 15.1 allows of a qualitative decision as to the nature of a system of scattering particles. Thus Doty and Kaufman (1945) have examined the form of the cellulose acetate, polyvinyl chloride, and polystyrene molecules in solution, as functions of the molecular weight, concentration, and nature of the solvent (see also Chapter XXVII). Krishnan (1937, 1938) has measured the depolarization for different metal sols, emulsions, and other colloidal systems as functions of the wavelength of the incident light. By comparing these results with the predictions of the theory of Gans (1912), he was then able to make semi-quantitative estimates of the shape and size of the scattering particles.

Molecular weights and shapes from optical examination

The preceding section indicated that some information on particle size and shape, usually of a qualitative nature, may be obtained from depolarization measurements. The ultramicroscope, in certain cases, affords similar information. Thus, for asymmetrical particles in rotary Brownian motion, the intensity of the light scattered towards the observer varies with the orientation of the particles, being a maximum when their direction of greatest refractive index is perpendicular to the direction of scattering. The rotation of such particles thus gives rise to a 'twinkling' effect which is a useful indication of particle asymmetry. If a colloidal suspension, flowing through a tube in which appreciable velocity gradients exist, is thus examined ultramicroscopically it may be possible to decide whether the asymmetry is of the rod or disc type (Freundlich, 1926, p. 405).

Recent developments have, however, made possible the accurate determination of molecular weights and dimensions from optical measurements. Thus Putzeys and Brosteaux (1935) measured the scattering at 90° to the incident direction caused by solutions of several globular proteins, quoting the following equation as an extension of the Rayleigh formula (eqn. (15.3)) to correct for the depolarization:

$$\frac{I_{90°}}{I_0} = \frac{9\pi^2}{2\lambda^4 N} \left[\frac{n_2^2 - n_1^2}{n_2^2 + 2n_1^2}\right]^2 \frac{6(1+\rho_u)}{6-7\rho_u} \bar{v}^2 cM, \qquad (15.12)$$

where I_0 is the intensity of the incident unpolarized light and $I_{90°}$ that of the light scattered at 90° to the incident beam at unit distance, n_1 and n_2 are defined as the refractive indices of the solvent and protein respectively (refractive index being assumed an additive property), $\frac{6(1+\rho_u)}{6-7\rho_u}$ is the depolarization correction for which ρ_u is determined as already indicated, and $\bar{v}$, M, and c are respectively the partial specific volume of the protein, the molecular weight, and the concentration (in grams per c.c.) of the protein.† Rearranging, eqn. (15.12) becomes

$$\frac{I_{90°}/I_0 \times 1/c}{\left[\dfrac{n_2^2-n_1^2}{n_2^2+2n_1^2}\right]^2 \dfrac{6(1+\rho_u)}{6-7\rho_u} \bar{v}^2} = KM, \qquad (15.13)$$

an equation similar to that used by Putzeys and Brosteaux except for the omission of a correction factor for absorption and for proportionality factors. These workers made careful scattering measurements on solutions of egg albumin, amandin, excelsin, and haemocyanin at a series of concentrations, and plotted a function proportional to the left-hand side of eqn. (15.13) divided by the known molecular weight, against concentration. As is shown by Fig. 15.5 (Putzeys and Brosteaux), containing results of measurements on total, green ($\lambda = 5{,}461$ A), and yellow ($\lambda = 5{,}769\text{–}5{,}790$ A) light, the behaviour of the different proteins is well represented by a single line for which the extrapolated value at zero concentration varies only slightly with the wavelength used. Ideally, K should be independent of concentration; its variations require further investigation.

The identical effect of concentration on the plotted values in the four cases is worth noting, and this concentration dependence is probably to be attributed to the inadequacy of eqn. (15.12) at the concentrations used.

In a further paper the same authors (1941) have used amandin as a

† This equation differs from that derivable from (15.4a) not only in the different definition of $I_{90°}$, leading to a changed numerical factor, but also in the changed refractive index term. In view of recent work (e.g. Debye, 1947), it would appear that the term $\left[\dfrac{3}{n_2^2+2n_1^2}\right]^2$ should be omitted.

reference substance to calculate K, assuming a molecular weight of 330,000. Table 26.2 compares the molecular weights of several other proteins obtained in this manner with the values from other techniques.

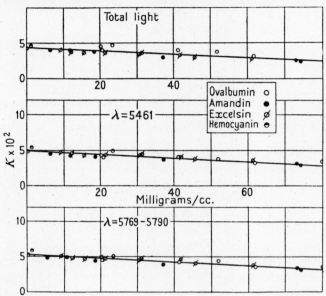

Fig. 15.5. Plot of K (15.13) against concentration (mg. per c.c.) for total, green ($\lambda = 5{,}461$ A), and yellow ($\lambda = 5{,}769\text{--}5{,}790$ A) light.

This comparison demonstrates unquestionably the value of the light-scattering method.

An alternative method of utilizing light-scattering measurements for molecular weight calculations has already been indicated in Chapter VII, p. 178, and a careful comparison of the two methods would seem to be very desirable.

Experimentally the different workers have used similar arrangements; the recent arrangement of Stein and Doty (1946), shown diagrammatically in Fig. 15.6, suffices to illustrate the general principles. Light from the mercury lamp A passes through the filter B (transmitting $\lambda = 5{,}461$ A), is made parallel by the lens C, and then passes through the diaphragm D before reaching the colloidal solution in the square cell G itself immersed in a water chamber H. A small fraction of the light leaving the filter is reflected by the glass plate E, placed at 45° to the incident beam, upon the opalescent glass plate F. Light scattered at 90° from the solution is focused by lens J into one side of a Zeiss Pulfrich photometer K, and is compared with that entering the other side from the opalescent plate, acting as a reference standard. As an absolute standard the use of a magnesium carbonate prism, mounted in place of the cell G at 45° to the incident beam, is recommended.

Quantitative estimates of molecular dimensions have also been attempted from light-scattering measurements (Doty, Affens, and Zimm, 1946). For particles small compared with the wavelength of light, the intensity of the scattered light is found to be symmetrically distributed about the 90° position (Fig. 15.2). If, however, a linear dimension of the scattering particle is greater than $ca.\ \lambda/20$, it is no longer possible to consider the particle as a point source of radiation, and, in fact, light scattered from different parts of the particles interferes, resulting in an increased forward and decreased backward scattering as was shown by Mie (Fig. 15.3 (b)). In addition to this intramolecular effect intermolecular interference may occur also, especially at the higher solute concentrations, and precautions must be taken to eliminate it in utilizing experimental measurements.

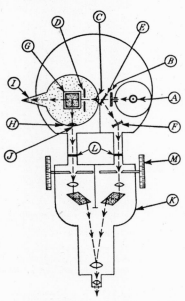

FIG. 15.6. Apparatus for the investigation of light scattering. (In addition to parts of apparatus mentioned in text, I is a cone for adsorption of light by multiple reflection, L is a neutral filter, and M is a slit area adjusting dial.)

If various model structures are postulated, the variation in scattered intensity with the angle of scattering may be calculated and compared with experimental observations. Thus, for randomly coiling chains (probably a suitable model for the rubber molecule), or long thin rod-like particles (e.g. tobacco mosaic virus), the ratio of the intensities (I_1 and I_2) in two given directions, making angles θ_1 and θ_2 respectively with the incident direction, may be expressed in terms of these angles and the dimensions of the chain or rod respectively. In this connexion use is made of the dissymmetry coefficient, q, defined by

$$q = \frac{I_1}{I_2} - 1, \qquad (15.14)$$

where θ_1 and θ_2 (to which I_1 and I_2 respectively refer) are related by

$$\theta_2 - 90° = 90° - \theta_1. \qquad (15.15)$$

To eliminate intermolecular interference the value of q at zero concentration is obtained by extrapolation from low concentration measurements, and this value is then compared with calculated curves for the appropriate model in which the dissymmetry coefficient is plotted against

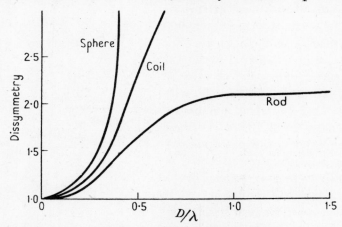

FIG. 15.7. Dissymmetry ($= q+1$), for $\theta = 50°$ and $\theta = 130°$, as a function of D/λ, where D is the diameter, in the case of the sphere, the root mean square separation of the ends for a randomly kinked coil, and the length for a rod.

some function of the molecular dimensions (Fig. 15.7 from Doty, 1947). Estimates of the dimensions of certain macromolecules or particles have thus been obtained.

If the molecular dimensions so calculated are to be reliable, it is essential

(1) that the difference in refractive index between particle and solvent is not too great;
(2) that intermolecular interference has been eliminated;
(3) that the model structure on which the estimate is based accurately represents the scattering units.

Condition (1) is not violated for many polymer solutions, according to Doty, Affens, and Zimm (1946), whilst the method of satisfying (2) has already been indicated. On stereochemical and other grounds a suitable model can usually be chosen with confidence. The method appears, therefore, to offer great possibilities, but it is clearly advisable to test it against established ones wherever possible. A useful test against the electron microscope has thus been accomplished by Doty, Affens, and Zimm (1946) using tobacco mosaic virus, for which the values 2,750 A and 2,800 A respectively were obtained for the length of the particles by the light-scattering and electron microscope methods. The same workers also obtained information on the shape of the polystyrene molecule in solution which was compared with that from viscosity. From Table 15.2, containing such light-scattering and viscosity results, it is clear that the differences in the dimensions of the polymer molecules in good and bad solvents as determined by the dissymmetry of scattering are not sufficient to explain observed viscosity differences on current viscosity theory (cf. p. 375).

TABLE 15.2. *Molecular Data on Polystyrene Fractions*

Fraction	Solvent	Weight average molecular weight	Extended length (A)	Root mean square length (A)	Intrinsic viscosity
RT—H	Toluene	4×10^6	96,000	2,370	7·6
,,	Ethylene dichloride	,,	,,	2,390	5·3
,,	Butanone	,,	,,	2,100	2·3
,,	Butanone-isopropanol	,,	,,	1,900	1·6
BZO–4	Ethylene dichloride	$1·75 \times 10^6$	42,000	1,100	2·16
,,	Butanone	,,	,,	1,230	1·45
,,	Butanone-isopropanol	,,	,,	990	0·84

Infra-red absorption spectra

The study of absorption in the infra-red part of the spectrum has been extremely successful in determining the molecular dimensions of many low molecular weight materials, and in establishing the presence or absence of various chemical bonds and structural units; its application to problems of high polymers or colloid science has, however, presented many difficulties both of a theoretical and a practical nature. Despite this, distinct contributions have been made in the field of high polymers (e.g. see Faraday Discussion, 1945).

In the case of colloidal aggregates and high polymer molecules the great complexity of the rotational fine structure in the infra-red rules out its use in determining molecular or aggregate dimensions. Further, since in many colloidal problems, as we have already seen (p. 13), removal of the dispersion medium may change or destroy the system, examination of the system as a whole is desirable, but in view of the usually troublesome absorption of the solvent (especially if of the aqueous type), this is frequently not practicable. High refractive index differences between dispersion and disperse phase frequently add further difficulty by causing considerable scattering of light.

Understandably, therefore, progress has chiefly occurred in the study of pure high polymeric materials, usually prepared in the form of thin films. Efforts have, as yet, been mainly directed towards the identification of bonds or groupings (e.g. O—H, C—H, N—H, —C—C—, —C=C—) by their absorption at frequencies which are relatively unaltered from the values observed in simple molecules. In this way it has been possible to confirm certain known structural features as well as to identify bonds or groupings hitherto unsuspected. For instance, although the predominantly straight-chain paraffinic nature of the polythene molecule is confirmed by infra-red measurements, it is also indicated that methyl and carbonyl groups appear, the former to the extent of one per fifty methylene groups (Thompson and Torkington, 1945). Many of the important synthetic and natural high polymers have been similarly examined (see p. 764), and information on polymerization and other reaction mechanisms has been obtained. Much remains to be done in this field, and other approaches by the infra-red method to colloidal and high-polymer problems have been mentioned (e.g. Sutherland, 1945) which may eventually provide a valuable addition to colloidal techniques. A useful description of many aspects of the infra-red method is contained in the Faraday Discussion (1945) already mentioned.

Diffraction of X-rays

The application of X-ray diffraction methods to structural problems has been discussed recently by several authors (e.g. Bunn, 1945; Clark, 1940). Here only the chief stages in the general method will be outlined as an introduction to those features which are of special interest in colloidal and allied problems. It is not surprising, in view of the difficulties associated with simpler structures, that the extension of X-ray methods to colloids has so far been very limited.

The successful use of the X-ray method is dependent upon the occurrence of some type of ordered structure, which may vary from the nearly perfect three-dimensional order of many low molecular weight single crystals to the almost unordered state of amorphous substances. Amongst high molecular weight materials only the proteins are so far known to form single crystals. In other cases order may occur in either two dimensions or in one, with the possibility of regions of almost complete disorder. It is clear, therefore, that in applying the method to such a range of materials, uniformly complete results are not obtained. We shall first consider the application of X-ray methods to single crystals, and shall then indicate its use in less perfectly ordered structures.

X-rays are produced by directing a stream of electrons, accelerated by a potential of the order 10–30 kV, upon a metallic target or anti-cathode, when X-rays are emitted in all directions. A well-defined beam is picked out by suitable stops and utilized. Fig. 15.8 illustrates a convenient form of X-ray tube recommended for research purposes. X-rays as thus produced contain several sharp frequencies superimposed upon a background of continuous 'white' radiation. For some purposes they are utilized unaltered, but for most work almost monochromatic radiation is required. This is often obtained by passage through suitable filters consisting of thin films of certain metals. Thus, the often used copper $K\alpha$ radiation, consisting of two lines at $\lambda = 1\cdot5405$ and $1\cdot5443$ A, with only small amounts of other wavelengths, is produced by passing the mixed radiation through a thin nickel film. More strictly monochromatic X-rays are, however, obtained (though with reduced intensity) by reflection of the 'white' radiation from suitably large crystals (e.g. pentaerythritol, erythritol, the sugars, urea nitrate, or other substances) at the appropriate Bragg angle corresponding to the desired wavelength.

There are several methods of taking X-ray photographs, the choice largely depending upon the nature and the form of the substance to be examined. For single crystals the rotation method is commonly used, in which the crystal is rotated, or oscillated through a definite arc, about each crystallographic axis in turn, the incident beam of X-rays being directed at right angles to this axis. The diffracted X-rays are received on a cylindrical photographic film whose axis coincides with the axis of rotation. If the crystallographic axes have been correctly and accurately chosen, then the diffracted 'spots' fall upon well-defined horizontal lines, the layer lines, numbered outwards from the equatorial line (0, 1, 2, etc.). As will be shown, the spacings of these layer lines provide a ready means of determining the unit cell dimension along the appropriate rotational axis.

A modification of the usual rotation photograph occurs in the case of fibrous materials (e.g. hair, wool, cotton, etc.) in which order may exist only in the direction of the fibres. The X-ray beam is in this case directed perpendicularly to the length of the specimen, and in view of the lack of order in the plane at right angles to this length (i.e. all orientations about the main fibre axis occur), diagrams similar to those of the rotation method are obtained with the specimen stationary. The repeat distances along the fibre length are given in the usual way from the layer line distances (p. 444). In many colloidal problems, where long lattice spacings (several hundred A) are of especial interest,

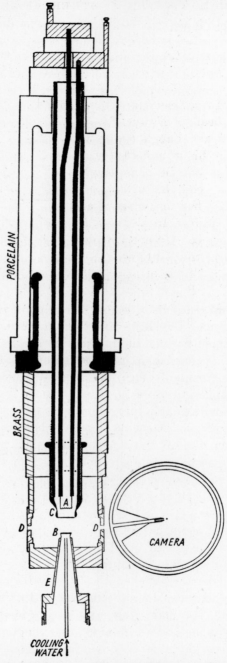

FIG. 15.8. A demountable X-ray tube. A, Filament; B, Target; C, Focusing shield; D, Windows; E, Target holder.

(*From* Bunn (1945).)

The Bragg law

Consider a set of regularly spaced parallel lattice planes P, P', P'', etc., upon which a beam of X-rays of wavelength λ is incident at an angle θ with the planes (Fig. 15.9). Light 'reflected' from successive planes is out of phase by an amount depending upon the difference in distance traversed. Thus for planes P and P' this difference is given by $NP'+P'N'$, N and N' being the feet of perpendiculars dropped from P upon the X-ray beam incident on and reflected from plane P'. By simple geometry, $NP'+P'N' = 2d\sin\theta$, where d is the distance between successive planes.

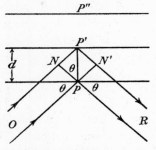

Fig. 15.9. The diffraction of X-rays by a set of parallel planes.

Clearly, reflections from different planes will interfere constructively and a strong diffracted beam will be produced if

$$n\lambda = 2d\sin\theta, \qquad (15.16)$$

n being an integer. This is the well-known Bragg law. Using this relation it is possible, knowing λ and n (the order of the reflection), and by determining θ experimentally, to calculate the distance d between any set of lattice planes. In the rotation method the 'spots' appear at those stages of the rotation for which the incident angle θ is related to the different lattice distances, d, by the Bragg relation. In fibres the lack of order about the fibre axis gives similar photographs without the need for rotation. In the case of very finely divided crystalline materials (for which 'powder' photographs are usually obtained), all possible crystallographic orientations occur and diffracted rings (rather than spots) arise, any one of which corresponds to a constant angle of diffraction (2θ) and a constant distance d.

Procedure in the examination of single crystals

Below we indicate, in outline, the sequence of operations normally undertaken in the structure determination of a single crystal.

1. An essential preliminary to X-ray measurements on crystals is the examination of their crystallographic, optical, physical (e.g. pyro, piezo-electric, and magnetic properties, dielectric constant, thermal and

electrical conductivity), and mechanical properties (e.g. elastic constants) (see Bunn, 1945, p. 278). By this means it should then be possible:

(a) to assign the crystal to one of the thirty-two crystal classes or point groups;

(b) to identify the crystallographic axes and to measure their relative lengths.

2. Rotation photographs are taken about the crystallographic axes determined in 1 (b). From the layer-line distances (y_1, y_2, etc.) from the equatorial line the dimension of the unit cell in the direction of the rotational axis (say c) is readily obtained from

$$c = \frac{n\lambda}{\cos \phi}, \qquad (15.17)$$

n being the number of the layer line and ϕ being the angle whose tangent is r/y_n, r being the radius of the cylindrical camera. It should be noted that ϕ is not in general related to the angle (2θ) made by a single diffracted beam with the incident direction. Unit cell dimensions may be checked, for the simpler crystal classes, from the distance d_{hkl} between planes (hkl) by the expression

$$d_{hkl} = \frac{1}{\sqrt{(h^2/a^2 + k^2/b^2 + l^2/c^2)}}, \qquad (15.18)$$

a, b, and c being the dimensions of the unit cell along the x, y, and z directions.

3. From the measured crystal density, ρ, and the volume of the unit cell (V) (derived directly from its dimensions), the number of molecules per unit cell (n) may be determined if the molecular weight (M) is known, for

$$\rho = \frac{nM}{NV}, \qquad (15.19)$$

N being Avogadro's number.

Alternatively, the molecular weight may be calculated if n be known. This procedure has been usefully employed amongst the proteins (p. 454).

4. From the crystal class already deduced and an examination of X-ray photographs for missing reflections (e.g. Bunn, 1945, p. 245) it should now be possible to assign the crystal to one of the 230 space groups, and, knowing n, to decide whether atoms or molecules occur in special (e.g. at a centre of symmetry) or general positions. A good example occurs in the case of the metallic phthalocyanines (e.g. $NiC_{32}H_{16}N_8$), whose crystals contain only two molecules per unit cell. The unit cell contains, therefore, only two metal atoms and, in order

to conform with the symmetry of the relevant space group, it is necessary that each should occur at a centre of symmetry (Robertson, 1935).

5. The experimentally measured intensities of the different reflections are utilized. From such intensities, and other experimental data (see p. 447), it is possible to calculate the magnitude (not the sign) of the corresponding structure factors $F(h,k,l)$, which give a measure of the scattering power of the material in the (hkl) planes. The structure factors may then be compared with values calculated from a unit cell structure postulated on the basis of all the other evidence available. When a good fit is obtained, the signs of the calculated structure factors are accepted and used with the experimentally determined magnitudes to calculate the electron density, $\rho(x, y, z)$, throughout the unit cell by the Fourier method. From this electron density map the positions of the different atoms can usually be precisely located.

Alternatively, from the experimental $F(h,k,l)$ magnitudes, a Patterson summation (which does not require the sign of $F(h,k,l)$) may be performed which may give enough data about the structure to make possible a direct Fourier summation. In either case, the presence of symmetry elements introduces great simplification.

6. Finally, a check of the derived structure is carried out by examination of its bond lengths and of the physical and chemical properties expected from it.

Of the steps outlined above, most have been well described elsewhere. Step 5 is, however, of considerable difficulty, and since no simplified account is (to our knowledge) available, and, in view of its relevance to the X-ray investigation of single crystals of polymeric substances, it seems desirable to discuss it further here. Readers who find the following account difficult may omit it without losing the sequence of the chapter.

The use of intensities in X-ray structure determinations

The following section draws considerably upon the comprehensive account by Robertson (1938). In Fig. 15.9 consider some distribution of electrons representing scattering matter about point P, and let a beam of X-rays be incident at angle θ upon the plane through P. We have already seen that light scattered from the different parallel planes, at an angle θ to these planes, interferes constructively when the Bragg relation (p. 443) is obeyed. If, now, an electron is placed at point P and if the X-ray beam, OP, of amplitude A, has its electric vector perpendicular to plane OPR, then, by classical electromagnetic theory, the amplitude of the scattered wave $(\bar{A})$ at a distance r is given by

$$\bar{A} = A \frac{e^2}{mrc^2}, \qquad (15.20)$$

m being the mass of the electron, e its charge, and c the velocity of light.

For an electron at a distance x above the plane through P, the amplitude of the scattered wave will be given by a similar equation, but by reference to equation (15.16) the wave will be out of phase with the radiation from P by the phase angle $2\pi\left(\dfrac{2x \sin \theta}{\lambda}\right)$, and its contribution to the scattered intensity may therefore be written $\bar{A} \cos 2\pi\left(\dfrac{2x \sin \theta}{\lambda}\right)$. However, for any X-ray reflection we have, by Bragg's law,

$$\frac{n}{d} = \frac{2 \sin \theta}{\lambda}$$

which, being substituted into the previous expression, gives for the contribution to the scattered intensity by an electron at a distance x above point P,

$$\bar{A} \cos 2\pi \frac{nx}{d}.$$

Putting x as a main crystallographic direction and $d = a$, this becomes

$$\bar{A} \cos 2\pi \frac{nx}{a}.$$

The whole scattered intensity due to scattering matter between the limits $+a/2$ and $-a/2$ is therefore written

$$\bar{A} \int_{-a/2}^{+a/2} \rho(x) \cos 2\pi \frac{nx}{a} dx = F(h,k,l)\bar{A}, \qquad (15.21)$$

where $\rho(x)$ is the probability of finding an electron at any position x, $F(h,k,l)$ being termed the structure factor for the planes under discussion, being merely the ratio of the observed amplitude to that which would be observed if the scattering matter consisted of only one electron placed at $x = 0$.

Extending (15.21) to cover the scattering matter in the complete unit cell of axial lengths a, b, and c, we have by analogy

$$F(h,k,l)\bar{A} = \bar{A} \int_{-a/2}^{+a/2} \int_{-b/2}^{+b/2} \int_{-c/2}^{+c/2} \rho(x,y,z) \cos 2\pi \left(\frac{hx}{a} + \frac{ky}{b} + \frac{lz}{c}\right) \frac{V}{abc} dx\,dy\,dz,$$

$$(15.22)$$

where V is the volume of the unit cell and the factor $\dfrac{V}{abc}$ is required to cover the case where the three axes are not mutually perpendicular. In common with other phenomena for which it is necessary to consider the phase as well as the amplitude (e.g. alternating current, see Owen, 1937, p. 12) eqn. (15.22) is usually written in terms of complex quantities, thus:

$$\bar{A}\int_0^a\int_0^b\int_0^c \rho(x,y,z)\exp 2\pi i\left(\frac{hx}{a}+\frac{ky}{b}+\frac{lz}{c}\right)\frac{V}{abc}\,dxdydz = \bar{A}F(h,k,l). \quad (15.23)$$

The structure factor for the unit cell may therefore be written

$$F(h,k,l) = \frac{V}{abc}\int_0^a\int_0^b\int_0^c \rho(x,y,z)\exp 2\pi i\left(\frac{hx}{a}+\frac{ky}{b}+\frac{lz}{c}\right)dxdydz, \quad (15.24)$$

a relation demonstrating the connexion between the distribution of electrons in the unit cell and the structure factor for any given set of reflecting planes.

Experimental determination of structure factors

From its definition, it is clear that the structure factor of a given set of planes must be intimately connected with the experimental intensity of the reflection from the planes. Amongst other quantities it is to be expected that electronic constants, the glancing angle θ, and the nature of the radiation will also be involved.

It can be shown that, for a small crystal of volume δV and negligible X-ray absorption, $F(h,k,l)$ is related to the energy, E, received on the photographic plate when the crystal is rotated with an angular velocity ω, by the expression

$$\frac{E\omega}{I_0} = \left[N\frac{e^2}{mc^2}F(h,k,l)\right]^2\lambda^3\frac{(1+\cos^2 2\theta)}{2\sin\theta}, \quad (15.25)$$

I_0 being the intensity of the incident unpolarized beam (energy/sec./cm.^{2}).

This equation holds under certain conditions which are commonly obeyed, and may therefore be employed directly to obtain $F^2(h,k,l)$. It should be realized that only the magnitude is thus obtained. The absence of the sign, as will shortly be seen, introduces considerable difficulty in utilizing intensity measurements, and though no universal method of determining it exists, procedures of great usefulness in individual cases have been evolved.

Calculation of structure factors

If the electronic distribution in the unit cell is known (e.g. from the positions of all the atoms) then, theoretically, it should be possible to calculate $F(h,k,l)$ utilizing eqn. (15.24) directly.

In practice the various atomic contributions are calculated separately and summed. Thus we may write

$$F(h,k,l) = S_1 f_1 + S_2 f_2 + ..., \qquad (15.26)$$

where f_1 is the atomic structure factor of atom 1, f_2 is the atomic structure factor of atom 2, etc., and S_1 and S_2 are the corresponding geometric structure factors. The atomic structure factor is a function of the type of atom considered and of $\sin\theta/\lambda$ (see Bunn, 1945, p. 202), but does not contain any phase contribution, the latter being contained entirely in the geometric factor. Values of the atomic factors as functions of $\sin\theta/\lambda$ may be obtained experimentally from X-ray measurements on substances of known structure, or, alternatively, concordant values may be obtained by calculation from the electronic structure of atoms. If several atoms of the same type occur in the unit cell, then it is usual to sum them together in the geometric structure factor. Thus

$$S_1(h,k,l) = \sum \exp 2\pi i \left(\frac{hx}{a} + \frac{ky}{b} + \frac{lz}{c}\right), \qquad (15.27)$$

where the summation is carried out over all atoms of type 1 in the unit cell.

The same procedure is repeated for different types of atom in the cell, and, using (15.26), $F(h,k,l)$ is obtained.

Summarizing, if some unit cell structure is postulated, then it is possible to calculate the structure factors for the different reflections and to compare them in magnitude with those obtained directly from experimental intensity measurements, the degree of correspondence giving a qualitative indication of the correctness of the structure assumed.

For many simple crystals (e.g. simple inorganic salts and organic compounds) this is a convenient procedure, but where a structure is more complicated (e.g. the phthalocyanines) there are too many variables to allow of reliable 'trial and error' structures. In such cases the use of Fourier series, a brief outline of which follows, may be of considerable assistance.

Expression of electron density in terms of a Fourier series

Any crystal, whatever its symmetry, possesses a periodic structure, and it therefore follows that the electron density $\rho(x,y,z)$ must be expressible as a Fourier series

$$\rho(x,y,z) = \sum_{-\infty}^{+\infty}\sum_{-\infty}^{+\infty}\sum_{-\infty}^{+\infty} A_{pqr} \exp 2\pi i \left(\frac{px}{a} + \frac{qy}{b} + \frac{rz}{c}\right), \qquad (15.28)$$

p, q, and r being integers and A_{pqr} the coefficient of the general term. Introducing into (15.24) and integrating (Robertson, 1938), all terms of the series become zero except that for which $p = -h$, $q = -k$, and $r = -l$, which gives

$$F(h, k, l) = V A_{-h-k-l},$$

or
$$A_{-h-k-l} = \frac{F(h, k, l)}{V}. \tag{15.29}$$

Eqn. (15.28) therefore becomes

$$\rho(x, y, z) = \sum_{-\infty}^{+\infty} \sum_{-\infty}^{+\infty} \sum_{-\infty}^{+\infty} \frac{F(h, k, l)}{V} \exp - 2\pi i \left(\frac{hx}{a} + \frac{ky}{b} + \frac{lz}{c} \right). \tag{15.30}$$

Thus the electron density may be expressed as a Fourier series in which the coefficients are the appropriate structure factors divided by the volume of the unit cell, the summation being carried out over all values of (h, k, l).

If the phase change on scattering is the same for all parts of the unit cell, and if the latter possesses a centre of symmetry, then eqn. (15.30) reduces to

$$\rho(x, y, z) = \sum_{-\infty}^{+\infty} \sum_{-\infty}^{+\infty} \sum_{-\infty}^{+\infty} \pm \frac{F(h, k, l)}{V} \cos 2\pi \left(\frac{hx}{a} + \frac{ky}{b} + \frac{lz}{c} \right). \tag{15.31}$$

Such a triple summation involves a prohibitive amount of computation and the two-dimensional form,

$$\rho(x, y) = \sum_{-\infty}^{+\infty} \sum_{-\infty}^{+\infty} \pm \frac{F(h, k, 0)}{A} \cos 2\pi \left(\frac{hx}{a} + \frac{ky}{b} \right), \tag{15.32}$$

in which A is the area of the cell face upon which the projection is made, is commonly used. In this case $\rho(x, y)$ gives the electron density per unit area in a projection along the Z axis, and it may be evaluated from intensity measurements belonging only to one zone of the crystal and obtained from a single rotation photograph (i.e. $F(h, k, 0)$ values from the equatorial line of a rotation photograph about the c axis). A further projection of the same type along another axis is then usually sufficient to give the whole structure completely. Alternatively, one-dimensional summations, e.g.

$$\rho(x) = \sum_{-\infty}^{+\infty} \pm \frac{F(h, 0, 0)}{d} \cos 2\pi \frac{hx}{a}, \tag{15.33}$$

d being the spacing of the fundamental plane (100), in which $\rho(x)$ gives the electron density per unit length projected upon the x direction, may provide the additional information.

As indicated, three-dimensional summations should cover all (h,k,l) values to infinity, but clearly only those values corresponding to experimentally observable reflections may be included. Omission of the higher terms for which the corresponding reflections are either too weak, or unobservable for any other reason (see Robertson, 1938, p. 346) constitutes one of the difficulties of the method. It would appear that each structure examination must be considered individually as to the error incurred through the impossibility of carrying summations to infinity. In Fig. 15.10, due to Robertson, each projection was obtained from a series of 281 terms, which were evaluated at 1,800 points.

Determination of sign of $F(h,k,l)$. It has been seen that only the magnitude but not the sign of the structure factor is obtained from experimental intensity measurements. There is no general method of determining the sign of the structure factor and it follows, therefore, that the complete structure determination by Fourier methods is not always possible. However, in certain special cases complete solutions have been obtained.

In the case of comparatively simple structures with only a few parameters (e.g. simple salts or organic compounds of low molecular weight) the sign of the structure factor may be obtained by calculation from a structure postulated on general grounds, which gives numerical agreement with observed structure factors. The signs of the calculated structure factors may then be accepted.

For more complicated structures the introduction of a heavy metal into the molecule and into a well-defined lattice position has proved very useful. Such heavy metals are of much greater scattering power than the elements normally occurring in organic compounds (C, H, O, N, etc.) and if introduced without altering the remainder of the structure, they cause large changes in all the structure factors. From the positions of the heavy metal atoms, the sign of their contributions is readily calculated, and, therefore, from the actual observed change in structure factor on introduction of the metal, the sign of the structure factors of the metal-free compounds may be deduced. A good example of this technique occurs in the structure determination of the phthalocyanines (Robertson, 1936), where as already mentioned (p. 444) the two nickel atoms in the unit cell of nickel phthalocyanine occur at centres of symmetry $((0,0,0), (\frac{1}{2},\frac{1}{2},0))$. For any set of planes (h,k,l) with $(h+k)$ even, we may write

$$F(\text{Ni derivative}) - F(\text{metal-free compound}) = 2F(\text{Ni}). \quad (15.34)$$

(N.B. For $(h+k)$ odd, the metal atoms make no contribution for the particular space group involved.) Since $F(\text{Ni})$ is large and always

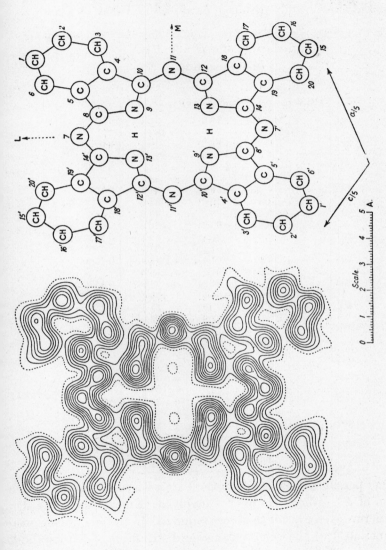

FIG. 15.10. Projection along the b axis, showing one complete phthalocyanine molecule. The plane of the molecule is steeply inclined to the plane of the projection, the M direction making an angle of 46° with the b axis, and the L direction 2·3°. Each contour line represents a density increment of one electron per $Å^2$, the one-electron line being dotted.

positive, according as $F(h, k, l)$ for the metal-free compound is positive or negative, an increase or decrease respectively in the structure factor will occur on introducing the metal. The sign of F (metal-free compound) is therefore readily deduced, as shown in Table 15.3 below.

TABLE 15.3. *Determination of the Sign of Structure Factors by the Introduction of a Heavy Metal*

(Robertson, 1936)

hkl	F (Ni derivative) $-$ F (metal-free)	$= 2F$ (Ni)	$\sin \theta$
001	$(+112 \cdot 0) - (+77 \cdot 6)$	$= 34 \cdot 4 \leftarrow$	0·062
	$(+112 \cdot 0) - (-77 \cdot 6)$	$= 189 \cdot 6$	
$20\bar{1}$	$(+136 \cdot 5) - (+96 \cdot 3)$	$= 40 \cdot 2 \leftarrow$	0·079
	$(+136 \cdot 5) - (-96 \cdot 3)$	$= 232 \cdot 8$	
200	$(+ 55 \cdot 0) - (+ 3 \cdot 9)$	$= 51 \cdot 1$	0·092
	$(+ 55 \cdot 0) - (- 3 \cdot 9)$	$= 58 \cdot 9$	
$20\bar{2}$	$(+ 71 \cdot 4) - (+16 \cdot 6)$	$= 54 \cdot 8 \leftarrow$	0·107
	$(+ 71 \cdot 4) - (-16 \cdot 6)$	$= 88 \cdot 0$	
002	$(- 36 \cdot 5) - (-85 \cdot 0)$	$= 48 \cdot 5 \leftarrow$	0·123
	$(+ 36 \cdot 5) - (-85 \cdot 0)$	$= 121 \cdot 5$	
201	$(- 43 \cdot 0) - (-92 \cdot 1)$	$= 49 \cdot 1 \leftarrow$	0·135
	$(+ 43 \cdot 0) - (-92 \cdot 1)$	$= 135 \cdot 1$	
$20\bar{3}$	$(+112 \cdot 6) - (+67 \cdot 4)$	$= 45 \cdot 2 \leftarrow$	0·156
	$(+112 \cdot 6) - (-67 \cdot 4)$	$= 180 \cdot 0$	
$40\bar{2}$	$(- 46 \cdot 0) - (-84 \cdot 8)$	$= 38 \cdot 8 \leftarrow$	0·157
	$(+ 46 \cdot 0) - (-84 \cdot 8)$	$= 130 \cdot 8$	
$40\bar{1}$	$(+129 \cdot 0) - (+77 \cdot 9)$	$= 51 \cdot 1 \leftarrow$	0·159
	$(+129 \cdot 0) - (-77 \cdot 9)$	$= 206 \cdot 9$	
310	$(+ 20 \cdot 4) - (-19 \cdot 2)$	$= 39 \cdot 6 \leftarrow$	0·213
	$(+ 20 \cdot 4) - (+19 \cdot 2)$	$= 1 \cdot 2$	
004	$(+ 55 \cdot 6) - (+17 \cdot 0)$	$= 38 \cdot 6 \leftarrow$	0·246
	$(+ 55 \cdot 6) - (-17 \cdot 0)$	$= 72 \cdot 6$	
$20\bar{5}$	$(+ 73 \cdot 5) - (+36 \cdot 0)$	$= 37 \cdot 5 \leftarrow$	0·270
	$(+ 73 \cdot 5) - (-36 \cdot 0)$	$= 109 \cdot 5$	
$60\bar{6}$	$(+ 65 \cdot 7) - (+30 \cdot 7)$	$= 35 \cdot 0 \leftarrow$	0·322
	$(+ 65 \cdot 7) - (-30 \cdot 7)$	$= 96 \cdot 4$	
$40\bar{7}$	$(+ 73 \cdot 1) - (+42 \cdot 4)$	$= 30 \cdot 7 \leftarrow$	0·367
	$(+ 73 \cdot 1) - (-42 \cdot 4)$	$= 115 \cdot 5$	
024	$(+ 73 \cdot 5) - (+47 \cdot 8)$	$= 25 \cdot 7 \leftarrow$	0·408
	$(+ 73 \cdot 5) - (-47 \cdot 8)$	$= 121 \cdot 3$	

A further possibility makes use of the Patterson summation in which a function of F^2, rather than of F is involved. Knowledge of the sign of the structure factor is not therefore required. In the one-dimensional case, a function $A(u)$, the weighted average distribution of density about the point x, is defined by

$$A(u) = \frac{1}{a} \int_0^a \rho(x)\rho(x+u)\, dx, \qquad (15.35)$$

where $\rho(x+u)$ is the electron density about x as a function of the parameter u, and is weighted by the scattering matter between x and $x+dx$. Expanding $\rho(x)$ and $\rho(x+u)$ as Fourier series, eqn. (15.35) becomes

$$A(u) = \frac{1}{d^2} \sum_{-\infty}^{+\infty} F^2(h, 0, 0) \exp 2\pi i \frac{hu}{a}, \qquad (15.36)$$

the Patterson analogue of eqn. (15.33). Large values of $A(u)$ arise when both $\rho(x)$ and $\rho(x+u)$ are large, so that for atoms distant u, a maximum in $A(u)$ occurs for this value.

The use of intensity data alone in this summation may, especially in some of the simpler structures, give enough information to make possible a complete Fourier analysis. Where the structure is more complex, it is not always possible to decide to which pair of atoms a given maximum refers.

For further discussion of Patterson summation methods see Robertson (1938) and the references given there.

Crystallite and particle size

Before considering the results of X-ray examinations of colloidal materials, mention must be made of two X-ray methods of investigating submicroscopic structure by determination of crystallite or particle size. Both methods are, however, fraught with many difficulties.

It is found experimentally that the X-ray diffracted spots become broadened and less sharp when the crystallite size is reduced below a certain level, usually about the upper limit of the colloidal range (particle diameter $ca.$ 10^{-6} cm.). The effect arises from the decrease in the number of reflecting planes and has its optical equivalent in the relationship between the number of lines in a grating and the sharpness of the spectral lines produced. For spherical crystallites free from strain and lattice imperfections, and of uniform size, the increased breadth β (usually at half-intensity) of the diffracted spot due to crystallite size is related to the crystallite diameter, t, by

$$\beta \text{ (radians)} = \frac{C\lambda}{t} \sec \theta, \qquad (15.37)$$

where C is a constant and θ is the Bragg angle. Experimentally it is difficult to distinguish between this broadening and the similar effects of lattice strains and imperfections, as well as the finite breadth of spot arising from the experimental technique. Further, the assumption of spherical crystallites is of very doubtful validity. Results obtained are thus of an approximate nature only, and are usually regarded as providing a lower limit to the crystallite size, which in any case cannot be identified with particle size (a particle may contain several crystallites). The method has been used to follow the growth of crystals in amorphous precipitates as shown in Fig. 15.11 for the case of stannic oxide. The increasing sharpness of the lines with increasing particle size is clearly shown.

Another method utilizes the low-angle scattering of X-rays, treating the problem in a manner analogous to the scattering of light by colloidal solutions which we have already considered (see Clark, 1940, p. 504). For closely similar particles randomly distributed within a much larger volume, the calculations yield the radius of gyration, R, defined by

$$R = \sqrt{\frac{\sum mr^2}{\sum m}}, \qquad (15.38)$$

m being the number of atoms and r the distance of each atom from the centre of gravity of the particles. Knowledge of the particle volume also allows a qualitative estimate of particle shape. Clearly, heterodispersity introduces complications and the method, like the previous one, is only semi-quantitative.

Of great importance, also, in certain colloidal systems, is the study of X-ray diffraction at very small angles, since this yields information about any very long spacings which may be present (e.g. those of several hundred A). The use of vacuum cameras, to eliminate the X-ray scattering by gas molecules has constituted an important advance in this technique (see p. 443). Such studies have been especially successful in connexion with fibrous materials, virus solutions, and soap solutions.

Results

Of all polymeric materials, single crystals have so far been found only amongst proteins, several well-known members of which have now been carefully examined by the X-ray methods. Table 15.4, due to Fankuchen (1945), summarizes the results for certain proteins.

As was to be expected for such complex substances, diffraction patterns of great complexity though sharp detail are obtained. Reflections corresponding to very long (> 100 A) as well as to very short spacings (e.g. ≈ 2 A) have been observed, the latter demonstrating the ordered nature of the protein structure down to interatomic distances, especially in the case of wet crystals. From such diagrams unit cell dimensions are readily obtained, and from density measurements and the number of molecules per unit cell, the molecular weights also included in Table 15.4 were derived. As already noted (p. 146) valuable estimates of crystal hydration may be obtained by comparison of such molecular

TABLE 15.4. *X-ray Data on Crystalline Proteins* (Fankuchen, 1945)

	Best non-X-ray molecular weights	Condition wet or dry	'a' in Å (10^{-8} cm.)	'b'	'c'	β in degrees	$c \times \sin\beta$	Volume in Å^3	n	Volume per molecule (V)	Density (ρ)	Molecular weight = $V\rho/1\cdot 65n$	Molecular weight corrected for residual water	Space group	Smallest observed spacing
Ribonuclease (orthorhombic)	13,000 to 15,000	wet dry	36·6	40·5	52·3	90	52·3	77,300	4	19,300	1·341	15,700	13,700	$P2_12_12_1$	2
Ribonuclease (monoclinic)	13,000 to 15,000	wet dry	30·8 28·7	38·5 29·3	53·5 45·2	107 100	51 44·5	60,000 37,400	2 2	30,000 18,700	.. ..	.. ..	.. ..	$P2_1$ $P2_1$	2 3·7
Insulin†	35,100 to 40,900	wet dry	144 130	83 74·8	34 30·9	90 90	34 30·9	404,000 298,000	6 6	67,000 50,000	1·28 1·315	52,400 39,500	.. 37,400	R3 R3	2·4 7
Lactoglobulin (tabular form)	37,900 to 41,800	wet dry	67·5 60	67·5 63	154 110	90 90	154 110	702,000 416,000	8 8	88,000 52,000	1·257 1·27	67,000 40,000	.. ..	$P2_12_12_1$ $P2_12_12_1$	2·4 20
Lactoglobulin (needle form)	37,900 to 41,800	wet dry	67·5 56	67·5 56	133·5 130	90 90	133·5 130	608,000 408,000	8 8	76,000 51,000	.. 1·30	.. 40,100	.. ..	$P4_12_1$ ..	.. 20
γ-Chymotrypsin	27,000	wet dry	69·5 63·0	69·5 63·0	97·5 74·5	90 90	97·5 74·5	471,000 298,000	8 8	58,900 37,200	1·33 1·277	30,100	.. ..	$P4_12_1$ $P4_32_1$	2·5 10
Pepsin	41,000	wet dry	49·6 45	67·8 62·5	66·5 57·5	102 112	65 53·5	219,000 151,000	2 2	109,000 75,500	1·277 1·31	84,500 60,000	.. 54,000	$P2_1$ $P2_1$	2 5
Chymotrypsin	35,500 to 39,200	wet dry	116	67	461	90	461	3,580,000	54	66,500	1·32	53,000	..		
Horse methaemoglobin	66,700	wet dry	110 102	63·8 51	54·2 47	112 130	50·2 36	352,000 188,000	2 2	176,000 94,000	1·242 1·27	132,000 72,000	.. 66,700	C2 C2	2 13
Horse serum albumin	70,000 to 73,000	wet dry	96·7 74·5	.. ..	145 130	120 120	.. ..	1,170,000 610,000	6 6	195,000 102,000	1·27 1·34	150,000 82,800	.. ..	H H	4 20
Tobacco seed globulin	(300,000)	wet dry	123	123	123	90	123	1,860,000	4	465,000	1·287	362,000	322,000	F	
Excelsin	294,000	wet dry	149	86	208	90	208	2,670,000	6	445,000	1·31 ?	350,000	305,800	R3	
Bushy stunt virus	7,600,000 to 10,600,000	wet dry	394 318	394 318	394 318	90 90	394 318	61,000,000 32,000,000	2 2	30,500,000 16,000,000	1·286 1·35	24,000,000 13,000,000	.. ..	I I	

† The values given here for insulin are referred to a hexagonal unit cell. The unit cell described on the basis of a rhombohedral lattice gives $a = 44\cdot 4$, $\alpha = 114°\ 28'$, $n = 1$, for dry insulin; and $a = 49\cdot 4$, $\alpha = 114°\ 16'$, and $n = 1$, for wet insulin.

weights for the wet and dry crystals. Further, some estimate of the shape of the individual molecules can be made from the fact that a known number must fit into the unit cell; in the case of lactoglobulin and haemoglobin, for example, it has been deduced that the molecules, though not perfectly spherical, do not deviate far from spherical shape (Crowfoot, 1941).

In several cases, by the use of experimental intensity measurements, two-dimensional Patterson summations of considerable complexity have been accomplished, from which further evidence concerning the existence in the crystal of the 'Svedberg' molecule has been provided. In the case of haemoglobin Perutz (1943) has even attempted *direct* Fourier summations, from which he deduces 'that the haemoglobin molecule is a platelet having the approximate dimensions $36 \times 64 \times 48$ A ... and that it is probably made up of four parallel layers of scattering matter which are a little less than 9 A apart'. It is likely that haemoglobin and other crystalline proteins will be more exhaustively examined by such methods in the near future.

Important X-ray studies of the virus proteins, especially by Bernal and co-workers (1941), have been reported, but space forbids their detailed discussion here. As in the case of the globular proteins, however, spacings of both an inter- and intra-molecular character appear to have been identified (see p. 751).

Amongst fibrous materials, the fibrous proteins have been a favourite object of study especially for Astbury and his collaborators (see e.g. Astbury, 1942). Such proteins fall into two main groups: (1) the keratin-myosin group and (2) the collagen group, of which most of the published work has concerned the first. Amongst members of the keratin-myosin group the elastic behaviour, and especially the range of elasticity, varies widely, but the X-ray examination of the fully stretched forms (β-keratin) demonstrates the common existence of a prominent spacing of approximately 7 A along the fibre axis (giving, therefore, reflections along the meridian), corresponding to two amino-acid residues. In unstretched keratin fibres (α-keratin) this spacing is replaced by one of 5·1 A along the fibre axis, and various types of folding of the polypeptide chain have been suggested to explain it (see p. 749). Supercontracted forms of the keratin fibres also exist which are probably similar in structure to that form of the muscle protein, myosin, which occurs in muscular contraction.

In the case of β-keratin, prominent equatorial reflections corresponding to spacings of 4·65 and 10 A, in the directions perpendicular to the

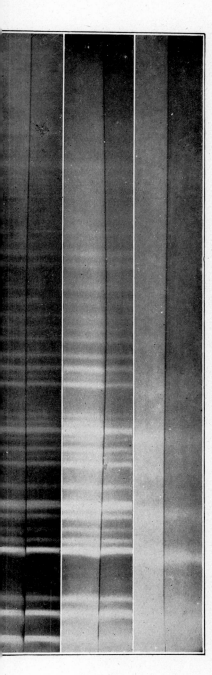

Fig. 15.11. Diffraction patterns of three samples of SnO$_2$, with decreasing particle size from top to bottom. (Clark.)

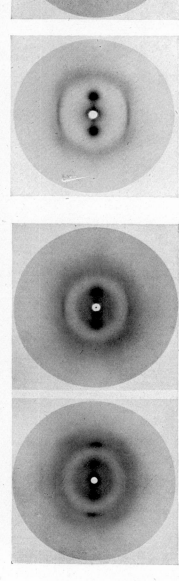

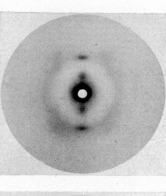

Fig. 15.13a (i) and (ii)
(a) Unstretched (i) and stretched (ii) horn.

Fig. 15.13b (i) and (ii)
(b) Unstretched (i) and stretched (ii) (80% extension) Cotswold wool.

Typical X-ray diagrams of $\alpha-$ and $\beta-$keratin.

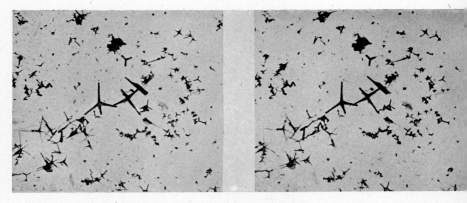

FIG. 15.15. Stereoscopic picture of zinc oxide crystals, collected on a collodion membrane. Magnification 4,700×. (R.C.A. Laboratories.)

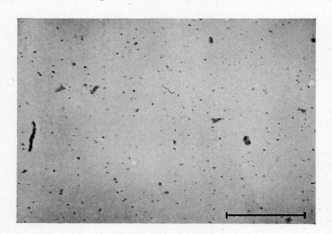

FIG. 15.16. Electron-micrograph of a red gold sol. (Astbury and Reed.)

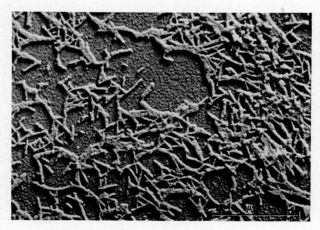

FIG. 15.17. Electron-micrograph of tobacco mosaic virus (gold shadowed). (Astbury and Reed.)

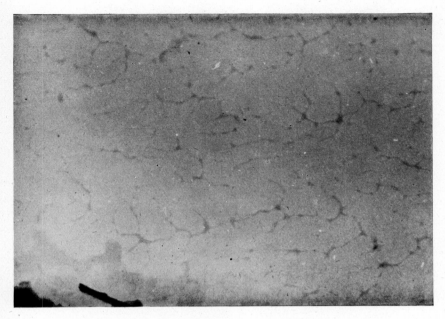

Fig. 15.18. Electron-micrograph of dextran molecules (65,000×). (Ingelman and Siegbahn (1944).)

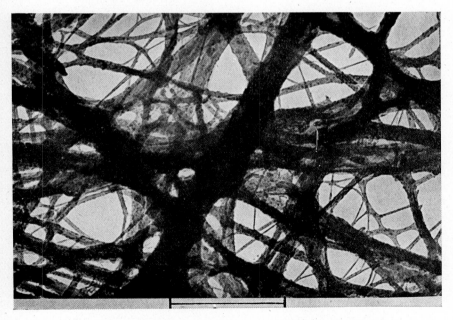

Fig. 15.19. Electron-micrograph of sodium laurate curd. (Marton.)

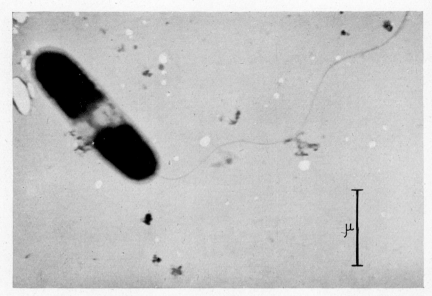

FIG. 15.21. Electron-micrograph of soil bacillus with monotrichate flagellum (most probably in act of division). (Astbury and Reed.)

FIG. 15.20. Electron-micrograph of Kaolinite. (Shaw.)

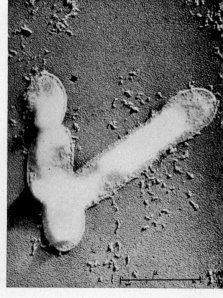

FIG. 15.22. Electron-micrograph of bacterial cells with fragments of a muscle protein (gold shadowed). (Astbury and Reed.)

fibre axis and to each other, have also been observed. Of these dimensions the former is considered to correspond to the distance between neighbouring chains in a direction perpendicular to the plane containing the main chain and the side chains ('backbone' spacing) and the latter

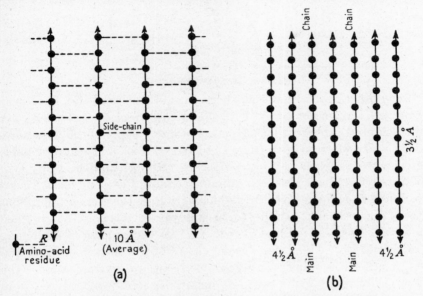

FIG. 15.12. (a) A polypeptide 'grid' in the β-configuration. (b) A regular aggregate (crystallite) of such grids.

('side-chain' spacing) to the inter-chain distance in this plane (Fig. 15.12). In the unstretched α-keratin the equatorial spacing, corresponding to about 4·3–4·6 A, is not so well defined. Fig. 15.13, due to Astbury, shows the chief features of the α- and β-keratin diagrams, from which our information concerning their structure has been largely drawn.

The fibrous proteins of the collagen group, which includes the tissue proteins of tendon, cartilage, etc., have been found to give the quite different spacings (Astbury, 1940):

	Spacing in A
Along fibre axis	2·86
Backbone	4·4
Side-chain	10·4–11·5 (according to moisture content)

According to Astbury these new features arise from the quite different amino-acid content of the collagens, notably the high content of proline and hydroxyproline, giving rise to a modified polypeptide chain (cis-configuration) in which the repeat distance per amino-acid residue is *ca.* 2·86 A.

$$\cdots -\underset{\underset{O}{\|}}{C}-\underset{\underset{O}{\|}}{\overset{\overset{N-CH}{\diagup \diagdown}}{C}}-NH-\underset{\underset{O}{\|}}{C}-\underset{R}{\overset{CH_2\ NH}{\diagup\diagdown}}CH-\underset{\underset{O}{\|}}{C}-\underset{\underset{O}{\|}}{\overset{\overset{N-CH}{\diagup \diagdown}}{C}}-\cdots$$

$\longleftarrow \cdots 3 \times 2\cdot 86 \text{ A} \cdots \longrightarrow$

$\longleftarrow$—————Fibre axis—————$\longrightarrow$

Suggested structure for collagens.

Many other fibrous materials, both natural and synthetic, have been examined, though not always with the success which has attended work on the fibrous proteins. The structure of many types of cellulose fibres is now well understood (e.g. Marsh and Wood, 1942) and is in reasonable agreement with their physical properties. Rubber and the many synthetic polymers have also been examined (Bunn, 1946) and in such cases the diagrams indicate structures which vary from the almost amorphous ones of the unstretched rubber to the highly ordered type of the stretched material. As indicated in Chapter XXVII, the process of stretching straightens out the kinked molecular chains of the unstretched material and, if other conditions are favourable, leads to crystallization. The structure of many high polymeric materials, mainly as deduced from X-ray and other measurements, will be considered in more detail in Chapter XXVII.

Another valuable application of the X-ray method has been to the structure of soap solutions and the related phospholipoids, whose anomalous physical properties are attributed to the formation of aggregates or micelles, containing large numbers of single molecules. At the higher soap concentrations ($\approx$ 15 per cent.) the micelles are believed to be in the form of lamellae which are sufficiently large to give X-ray reflections, from which micellar dimensions and structure can be deduced. Details are given in Chapter XXIV.

The electron microscope

Of the techniques involving the diffraction of X-rays and electrons, the electron microscope is the most direct in providing information on the size and shape of colloidal particles. Although the complete theory of the instrument is too complex to consider here, the chief uses and limitations may be understood if it is treated by analogy with the simple optical microscope.

The resolving power (δ), or the smallest distance by which two small particles may be separated and still remain separately visible is, from diffraction theory,

$$\delta = \frac{0 \cdot 61 \lambda}{\text{N.A.}}, \qquad (15.39)$$

where λ is the wavelength of the light used, and N.A. is the numerical aperture (= the refractive index × sine of half the angle subtended by the objective in the object plane).

In the simple optical microscope the numerical aperture is not much greater than unity, so that for $\lambda = 5{,}000$ A we have $\delta \approx 3{,}000$ A. Clearly the resolving power may be increased (i.e. δ decreased) by using light of shorter wavelength, and, indeed, the use of ultra-violet light for this purpose has been successful. However, such improvements were until recently limited, owing to the difficulty of obtaining suitable lens materials.

From the de Broglie relation,

$$\lambda = \frac{h}{mv} \approx \frac{12 \cdot 2}{\sqrt{V}}, \qquad (15.40)$$

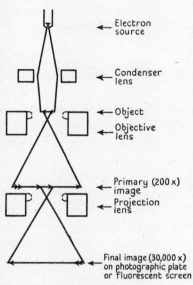

FIG. 15.14. The electron microscope (diagrammatic).

h being Planck's constant, m and v the mass and velocity respectively of an electron, and V the accelerating voltage, it may be calculated that the wavelength corresponding to an electron with an energy of 60,000 electron volts is 0·05 A. The resolution for such a wavelength would be extremely great even for very low numerical apertures; for example, for N.A. = 0·001, $\delta = 30$ A. The discovery of suitable 'lenses' for beams of electrons thus made possible the electron microscope with its unrivalled possibilities of high resolving power.

Cylindrically symmetrical magnetic or electrical fields are used as the 'lenses' of the microscope, the general construction of which is very similar to the optical analogue (see Fig. 15.14). A beam of electrons from a heated filament is focused by the condenser upon the object placed above the objective which forms a magnified image of the object (200×), itself giving rise to a further magnified (up to 30,000×) image, by means of the projection lens, upon either a photographic plate or fluorescent screen. The whole lens system is enclosed in a vacuum chamber to prevent the scattering of electrons by gas molecules.

The focal lengths (f) of magnetic and electrostatic lenses may be calculated in terms of the accelerating potential and the magnetic or electrostatic field (see e.g. Anderson, 1942; Gabor, 1944). Each type of lens possesses characteristic advantages and disadvantages, and corresponding microscopes have therefore been evolved. Object and image distances from the lenses r_0 and r_i respectively are calculated by the familiar optical formula

$$\frac{1}{f} = \frac{1}{r_0} + \frac{1}{r_i}. \tag{15.41}$$

Aberration effects are minimized by the use of small numerical apertures. The depth of focus is surprisingly large ($\approx 10^4$ A) so that the whole object appears in focus as a projection on the image plane.

Particle dimensions other than those in the image plane may be estimated by taking stereoscopic pictures, i.e. sets of two pictures inclined at slightly different angles, which are then viewed together. Fig. 15.15 is a pair of such pictures of zinc oxide crystals which, when mounted and viewed through a stereoscope, appear to stand out in three dimensions. As in the optical instrument, both bright and dark field illumination may be used.

To minimize aggregation, specimens for examination are commonly prepared by depositing from dilute solution (e.g. 10^{-8} gm. per c.c. for some protein solutions) upon thin collodion films (25 A) which show no observable structure. In other cases, e.g. smokes, the particles may be examined after being deposited on a wire gauze screen. Many other methods are possible according to the nature of the specimen; for details the original papers should be consulted.

Recently an important increase in resolving power has been achieved by means of the atom-shadowing technique (Williams and Wyckoff, 1944), in which the mounted specimen is coated with a thin layer of strongly scattering heavy metal, e.g. gold, before being examined. In the case of gold shadowing, a beam of atomic gold is directed at glancing incidence upon the plane of the collodion mount, when only the raised parts of the specimen are coated, long uncoated shadows remaining. Not only does this make possible the determination of dimensions other than those in the image plane, but the greatly improved contrast thus achieved leads to increased resolution.

Since the comparatively recent introduction of commercial electron microscopes, a large number of systems has been examined of which only a few representative examples can be mentioned here. Gold sols (Fig. 15.16), with a great range of particle size, have been carefully studied (von Borries and Kausche, 1940), and found to be crystalline over the whole observable range, thus confirming previous indications that the colloidal particle and bulk phase have fundamentally the same structure. The size distribution in any given suspension was very closely Gaussian, and for a mixture of two different suspensions the observed distribution curve proved to be the sum of the two separate ones, showing the slow rate of change of particle size.

The investigations of particle size and shape in the case of tobacco mosaic virus are important, since estimates from the indirect dynamic methods are available also. A direct check of the latter methods thus becomes possible. In the electron microscope it was confirmed that the particles are rod shaped (Fig. 15.17) and fairly uniform in both shape and size (Stanley and Anderson, 1941). Thus 70 per cent. of the observed particles had lengths and widths respectively within 7 per cent. of 2,800 A and 150°. It will be recalled that light scattering measurements (p. 438) gave a length of 2,750 A. Further, using for the density of the particles the value 1·33, the molecular weight is given by

$$(2800 \times \pi \times 75^2 \times 10^{-24}) \times 1 \cdot 33 \times 6 \cdot 02 \times 10^{24} = 39 \cdot 8 \times 10^6,$$

a value in good agreement with that, $42 \cdot 6 \times 10^6$, obtained by the dynamic methods and quoted by Stanley and Lauffer (1939) for the virus. Other carefully controlled comparisons of the different methods of obtaining particle size and shape are one of the most urgent requirements of present-day colloid science.

An important study by Ingelman and Siegbahn (1944) of dextran molecules deposited from 0·002 per cent. solution has made use of a resolving power of 30 A. Long branched chains of thickness 30–100 A were observed, in reasonable agreement with the predictions of physicochemical measurements. Certain observations are yet, however, to be explained, for example, the appearance of nodes or pronounced swellings at distances apart of about 800 A (Fig. 15.18).

Further examples of the use of the electron microscope are shown in Figs. 15.19–15.22.

Electron diffraction

In this technique a stream of electrons, accelerated by a high potential (e.g. 60 kV, giving $\lambda = 5 \times 10^{-2}$ A) is directed upon the specimen, and the diffraction pattern is registered on a photographic plate, placed normal to the beam, as a series of concentric rings. As in the electron microscope, high vacuum is required to prevent the scattering of electrons by gas molecules.

Since electrons have much less penetrating power than X-rays the method is chiefly suitable for examining thin films of material (ca. 1,000 A) or surface layers. The nature of metal and catalyst surfaces is therefore ideally studied by the method. The breadth of the diffracted rings may be used to calculate particle size as in the X-ray method.

An interesting application has been to the examination of multi-layers of several long-chain molecules (vinyl acetate, methyl and ethyl acrylate, and methacrylate) formed on the surface of an aqueous liquid by the Langmuir-Blodgett technique (Coumoulos, 1943). It was concluded that the main hydrocarbon chain tends to lie flat in the water surface with the side chains orientated perpendicularly to this surface. A prominent spacing (of 7 A for polyvinyl acetate) is said to arise from the length of the side chains (S), which controls the spacing between adjacent monolayers.

$$\begin{array}{c}-CH_2-CR-CH_2-CR-CH_2-CR-\\ |||\\ C{=}OC{=}OC{=}O\\ |||\\ OOO\\ |||\\ R'R'R'\end{array} \updownarrow S$$

Other smaller spacings at 1·2 and 2·2 A arise, according to Coumoulos, from the close packing of the side chains with the consequent distortion of the main chain.

REFERENCES

ANDERSON, 1942, in *Advances in Colloid Science*, vol. i, Interscience, p. 353.
ASTBURY, 1940, *Nature*, London, **145**, 421.
—— 1942, *J. chem. Soc.*, p. 337.
BERNAL and FANKUCHEN, 1941, *J. gen. Physiol.* **25**, 111, 147.
VON BORRIES and KAUSCHE, 1940, *Kolloidzschr.* **90**, 132.
BUNN, 1945, *Chemical Crystallography*, Oxford.
—— 1946, in *Advances in Colloid Science*, vol. ii, p. 95, Interscience.
CASPERSON, 1932, ibid. **60**, 151.
—— 1933, ibid. **65**, 162.
CLARK, 1940, *Applied X-rays*, 3rd edn., McGraw-Hill Book Co.
COUMOULOS, 1943, *Proc. Roy. Soc.* A **182**, 166.
CROWFOOT, 1941, *Chem. Rev.* **28**, 215.
DEBYE, 1947, *J. phys. coll. Chem.* **51**, 18.
DOTY, 1947, *J. Chim. phys.* (in press).
DOTY and KAUFMAN, 1945, *J. phys. Chem.* **49**, 583.
—— AFFENS, and ZIMM, 1946, *Trans. Faraday Soc.* **42** B, 66.
FANKUCHEN, 1945, *Advances in Protein Chemistry*, vol. ii, p. 387, Academic Press.
FARADAY DISCUSSION, 1945, *Trans. Faraday Soc.* **41**, 171.
FREUNDLICH, 1926, *Colloid and Capillary Chemistry*, Methuen.

GABOR, 1944, *The Electron Microscope*, Hulton Press, Ltd.
GANS, 1912, *Ann. Phys. Lpz.* **37**, 881.
GARNETT, 1904, *Philos. Trans.* A **203**, 385.
—— 1906, ibid. A **205**, 237.
INGELMAN and SIEGBAHN, 1944, *Ark. Kemi Min. Geol.*, Bd. 18 B, No. 1.
KRISHNAN, 1937, *Proc. Ind. Acad. Sci.* **5** A, 94, 305, 551.
—— 1938, ibid. **7** A, 98.
LA MER and BARNES, 1946, *J. coll. Sci.* **1**, 71.
MARSH and WOOD, 1942, *An Introduction to the Chemistry of Cellulose*, 2nd edn.. Chapman & Hall, p. 100.
MCBAIN, 1942, in *Advances in Colloid Science*, vol. i, Interscience, p. 125.
MIE, 1908, *Ann. Phys. Lpz.* **25**, 377.
OWEN, 1937, *Alternating Current Measurements*, Methuen Monograph.
PERUTZ and BOYES-WATSON, 1943, *Nature*, London, **151**, 714.
PUTZEYS and BROSTEAUX, 1935, *Trans. Faraday Soc.* **31**, 1314.
—— —— 1941, *Meded. Koninklijke Vlaamsche Acad. Belgie*, Jaargang III, No. 1.
ROBERTSON, 1936, *J. chem. Soc.*, p. 1195.
—— 1938, *Rep. Progr. Phys.* **4**, 332.
STANLEY and ANDERSON, 1941, *J. biol. Chem.* **139**, 325.
—— and LAUFFER, 1939, *Chem. Rev.* **24**, 303.
STEIN and DOTY, 1946, *J. Amer. chem. Soc.* **68**, 159.
STRUTT, 1871, *Phil. Mag.* **41**, 107, 447; 1899, **47**, 375.
SUTHERLAND, 1945, in *Colloid Science Course*, Cambridge University Summer School, p. 86.
THOMPSON and TORKINGTON, 1945, *Trans. Faraday Soc.* **41**, 246.
WILLIAMS and WYCKOFF, 1944, *J. App. Phys.* **15**, 712.
ZIMM, STEIN, and DOTY, 1945, *Polymer Bulletin*, **1**, 90.

XVI
PLASTIC FLOW AND ELASTICITY

The analysis of the flow behaviour of concentrated colloidal dispersions has not attained anything like the precision possible with the dilute systems considered in Chapter XIII. Nevertheless the last two decades have seen marked progress and some outline of the general methods of approach seems very desirable, particularly when we come to consider gels, pastes, and polymers.

Other than the fact that they exhibit properties intermediate between those characteristic of the liquid and solid states, the materials with which we shall be concerned here have little in common. They are frequently, but not invariably, colloids in the wider sense of that term, and include many examples of pastes, such as potter's clay, dough, putty, and paints; many natural and synthetic polymers, particularly when plasticized; and certain miscellaneous substances, such as honey, cheese, cream, bitumen, and so on.

The study of the flow and deformation properties of these materials, more particularly their stress–strain–time relations, and the influence of other variables, such as temperature, chemical composition, etc., forms the science of 'rheology', which, as mentioned in Chapter II, has such enormous industrial implications and applications.

These diverse but industrially important materials have been in the past, and still are to a great extent now, assessed subjectively by the senses of sight and touch, and by their general handling properties. Attempts to analyse their behaviour in terms of the usual objective conceptions of classical physics have so far only had limited success, and this has led certain rheologists into the realm of psycho-physics. For example it has been suggested that properties denoted by such everyday terms as 'firmness' or 'tack' may prove to be a 'Gestalt', i.e. may display properties other than can be derived from the parts in summation. Such questions are beyond the scope of the present book, but a very stimulating discussion of the application of Gestalt psychology to rheology has been given recently by Scott Blair (1944).

Forces and deformations

The behaviour of rheological systems is generally considered in terms of the relevant stresses and strains. 'Stress', denoted by S, has been defined by Ewing as 'the mutual action between two bodies or between two parts of a body whereby each of the two exerts a force upon the other'.

PLASTIC FLOW AND ELASTICITY 465

The term 'strain' (σ) is used by most rheologists to cover all deformations, whether recoverable instantaneously or slowly, or non-recoverable as in viscous flow where all the strain energy is dissipated. In all cases strain is defined relative to a hypothetical condition under which the

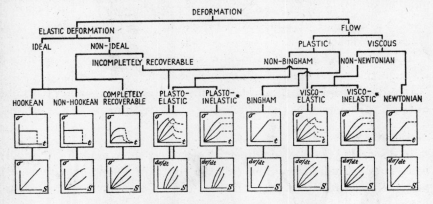

Fig. 16.1. (1) σ represents shear strain; S represents shear stress; t represents time. (2) Continuous curves represent deformation under stress. (3) Dotted curves represent behaviour after removal of stress. (4) When two or more curves appear these indicate alternative types of behaviour. (5) In cases marked * either the σ v. t curve or the $d\sigma/dt$ v. S curve may be a straight line, but not both, since in this case the systems would be Bingham or Newtonian, which are given separate categories. (6) The upper series of diagrams of σ v. t represents deformation under constant S. The lower series, in which $d\sigma/dt$ or σ is plotted against S, is for an arbitrary time.

body was supposed 'free from the action of external forces'. For example, in the extension of a rod of initial length l_0,

$$\sigma = \frac{\Delta l}{l_0} \quad \text{for small strains,}$$

and

$$\int_{l_0}^{l} \frac{dl}{l} = \log_e \frac{l}{l_0} \quad \text{for large strains.}$$

Fig. 16.1 shows the preliminary classification of strains drawn up by the British Rheologists' Club in 1942. All deformations are, for convenience, divided into elastic (recoverable) deformations and flow. Further subdivisions are made as follows:

Elastic deformations into 'ideal' (no time effect) and 'non-ideal' (showing time effect).

Ideal elastic deformations into 'Hookean' ($S \propto \sigma$) and 'Non-Hookean'.

Non-ideal elastic deformations into 'completely recoverable' and 'incompletely recoverable'.

'Flow' deformations divide into 'viscous' and 'plastic', the former again into 'Newtonian' and 'Non-Newtonian', and the latter into 'Bingham' and 'Non-Bingham'. Of these Newtonian and Bingham flow are formally similar in showing linear relationships for σ·v. t and $d\sigma/dt$ v. S. 'Non-Newtonian' is further subdivided into 'Visco-elastic' and 'Visco-inelastic', and 'Non-Bingham' into 'Plasto-elastic' and 'Plasto-inelastic'.

The elastic deformation and flow sides of Fig. 16.1 are thus linked through the subsections of 'plasto-elastic' and 'visco-elastic'.

In order to study the S-σ-t relationships it is clearly necessary to fix one of these parameters, or the differential of one with respect to another.

Elasticity

In discussing the elastic properties of matter three moduli are usually considered—the bulk modulus (K), the elastic or Young's modulus (E), and the shear modulus (n). All three can be expressed in the general form:

$$\text{modulus} = \frac{\text{stress}}{\text{strain}} = \frac{S}{\sigma}. \tag{16.1}$$

The special forms, for small strains, are as follows:

$$K = S \bigg/ \frac{\Delta v}{v_0}, \tag{16.2}$$

where S is the compressive force per unit area and Δv the decrease in volume in a block of initial volume v_0;

$$E = S \bigg/ \frac{\Delta l}{l_0}, \tag{16.3}$$

where S is the applied load per unit cross-sectional area and Δl the extension of a rod of initial length l_0;

$$n = S/\theta, \tag{16.4}$$

where S is the tangential stress per unit area, and θ the angular deformation.

These are not independent quantities but are related, in the case of isotropic systems, by the relation

$$\frac{1}{E} = \frac{1}{9K} + \frac{1}{3n}. \tag{16.5}$$

One other important relation is

$$E = 2n(1+\mu), \tag{16.6}$$

where μ is Poisson's ratio, defined as the ratio of the lateral contracting

PLASTIC FLOW AND ELASTICITY

strain to the elongation strain when a rod is stretched by a force applied only at its ends (see, for example, Newman and Searle, 1932).

For most colloidal systems, such as gels, pastes, and other rheological bodies, μ is equal to $\frac{1}{2}$, since the volume change is negligible, i.e.

$$E = 3n. \tag{16.7}$$

This relation also follows from eqn. (16.5), since for these materials $K \gg E$ or n. (In the case of a 10 per cent. gelatine gel, for example, n is ca. 3×10^5 dynes/cm.^{2}, whereas for most liquids K is ca. 10^8–10^{10} dynes/cm.^{2})

Relaxation time

For an elastic body obeying Hooke's law the deformation (strain) is independent of time, i.e.

$$\frac{dS}{dt} = n\frac{d\sigma}{dt}, \tag{16.8}$$

where S denotes shearing stress, σ the strain, t the time, and n the shear modulus.

For a purely viscous body the rate of shear is proportional to the applied stress, i.e.

$$\frac{d\sigma}{dt} \propto S, \quad \text{or} \quad S = \eta\frac{d\sigma}{dt}, \tag{16.9}$$

where η is the coefficient of viscosity (p. 345).

The problem of relaxation in materials intermediate between solids and liquids was first attempted by Maxwell (1867), in the manner clearly shown by the following well-known quotation. (The symbols have been altered to correspond to those used in this book. C in the final equation denotes a constant.)

A distortion or strain of some kind which we may call σ is produced in the body by displacement. A state of stress or elastic force which we may call S is thus excited. The relation between the stress and the strain may be written $S = n\sigma$, where n is the coefficient of elasticity for that particular kind of strain. In a solid body free from viscosity S will remain equal to $n\sigma$ and

$$\frac{dS}{dt} = n\frac{d\sigma}{dt}.$$

If, however, the body is viscous, S will not remain constant, but will tend to disappear at a rate depending on the value of S and on the nature of the body. If we suppose this rate proportional to S, the equation may be written

$$\frac{dS}{dt} = n\frac{d\sigma}{dt} - \frac{S}{\tau}$$

which will indicate the phenomena in an empirical manner. For if σ be constant,

$$S = n\sigma e^{-t/\tau},$$

showing that S gradually disappears, so that if the body is left to itself it gradually loses any internal stress, and the pressures are finally distributed as in a fluid at rest.

If $d\sigma/dt$ is constant, that is, if there is a steady motion of the body which continually increases the displacement,

$$S = n\tau \frac{d\sigma}{dt} + Ce^{-t/\tau}$$

showing that S tends to a constant value depending on the rate of displacement. The quantity $n\tau$ by which the rate of displacement must be multiplied to get the force may be called the *coefficient of viscosity*. It is the product of a coefficient of elasticity, n, and a time τ, which may be called the *time of relaxation* of the elastic force. In mobile fluids τ is a very small fraction of a second, and n is not easily determined experimentally. In viscous solids τ may be several hours or days, and then n is easily measured. It is possible in some bodies that τ may be a function of S, and this would account for the gradual untwisting of wires after being twisted beyond the limit of perfect elasticity. For if τ diminishes as S increases, the parts of the wire furthest from the axis will yield more rapidly than the parts near the axis during the twisting process, and when the twisting force is removed the wire will at first untwist till there is equilibrium between the stresses in the inner and outer portions. These stresses will then undergo a gradual relaxation; but since the actual value of the stress is greater in the outer layers, it will have a more rapid rate of relaxation, so that the wire will go on gradually untwisting for some hours or days, owing to the stress on the interior portions maintaining itself longer than that of the outer parts. This phenomenon was observed by Weber in silk fibres, by Kohlrausch in glass fibres, and by myself in steel wires.

Maxwell's assumption that the internal stress is relieved at a rate proportional to its value at that moment thus gives the simple relation

$$\tau = \eta/n, \qquad (16.10)$$

and from the relation $\qquad S = n\sigma e^{-t/\tau} \qquad (16.11)$

we see that τ is the time for the internal stress to fall to $1/e$th of its original value. On the above theory τ should be independent of S and in the c.g.s. system it will be measured in seconds.

Using the Couette torsion apparatus described on p. 474 and keeping the cylinder at constant deflexion (i.e. θ and therefore σ constant), the plot of the stress, given by the applied torque (i.e. $\delta - \theta$), against the time should be logarithmic if Maxwell's assumption holds. Experimentally this is frequently not the case and a number, or spectrum, of relaxation times is required in order to satisfy the observed behaviour.

Experimental methods
I. *Plastometers*

Instruments used to study flow in rheological materials are generally termed 'plastometers' or 'rheometers', rather than 'viscometers'. The reason is that the nature of the flow differs fundamentally from that

PLASTIC FLOW AND ELASTICITY

in dilute solutions and suspensions (see Chapter XIII), as it arises almost entirely from particle/particle contacts in the case of pastes and from polymer/polymer associations in the case of plasticized polymers or concentrated polymer solutions. The 'coefficient of internal friction', formally analogous to the viscosity in normal fluids, is consequently very high in these systems—of the order 10^3–10^5 poises or higher.

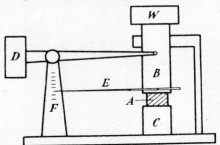

Fig. 16.2. Principle of a balanced plastometer.

A very useful survey of plastometers has been given by Nieuwenburg (1938), from which much of the following summary has been drawn.

(a) Compression and recovery plastometers

The material to be investigated, usually in the form of a cylinder, is compressed by suitable means, and the decrease in height measured as a function of the applied force and the time. With suitably balanced instruments it is also possible to follow the 'recovery', if any, on removal of the load.

Fig. 16.2 illustrates the principle of a balanced plastometer, in which the sample disc A is compressed between two pistons B and C, the weight of the upper one being compensated by the weight D. The load W is applied to B and the deformation measured either by means of a travelling microscope or the lever and scale E, F. The whole apparatus is placed in an air thermostat for temperature control. Various methods have been used to allow for the change in area on compression, since the situation is clearly much simpler if the deformation is measured at a constant rather than a varying stress. (For a survey see Scott Blair, 1944.)

(b) Rotation plastometers

These have the great advantage over the type previously considered in that the displacement may be increased indefinitely without appreciably changing the form of the sample under investigation. Of the forms in common use the Couette, Stormer, Ungar, and Goodeve modifications require special mention.

470 PLASTIC FLOW AND ELASTICITY

(i) *Couette and Stormer types* (Fig. 16.3). These are similar in principle to that described in Chapter XIII (p. 355) except for their more robust construction, clearly essential in view of the higher stresses involved. This instrument is suitable for the determination of the shear

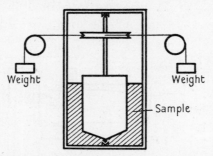

Fig. 16.3. Principle of Couette and Stormer plastometers.

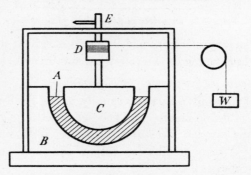

Fig. 16.4. Ungar plastometer (diagrammatic).

elastic modulus and the yield value, as well as for the study of the flow behaviour. (See below, p. 474.)

Trouble is sometimes experienced owing to slip at the walls, but this can generally be overcome by fine vertical ribbing of the inner and outer cylinders. If end corrections are serious they can be eliminated as in the normal Couette viscometer by making measurements with the inner cylinder immersed to different depths.

(ii) *Ungar type* (Fig. 16.4). Ungar shears the material (A) between hemispheres instead of two cylinders, the outer one (B) being thermostatted and the inner one (C) rotated by means of a weight (W) acting on a drum (D). (Some modifications employ cones instead of hemispheres.) The separation of the hemispheres, and hence the shear-rate, is readily varied by means of a locking nut at the top of the rotating spindle.

PLASTIC FLOW AND ELASTICITY

When filled with a Newtonian liquid the following formula applies:

$$\omega = \frac{Wga}{4\pi\eta} \frac{r_2^3 - r_1^3}{r_1^3 r_2^3}, \tag{16.12}$$

where ω = angular velocity, a = radius of drum, and r_1 and r_2 are the radii of the inner and outer hemispheres.

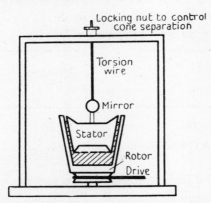

FIG. 16.5. Goodeve thixotrometer (diagrammatic).

Clearly no such simple relation obtains for non-Newtonian fluids owing to the variations in the rate of shear which such an apparatus must entail. Its principal advantages would appear to be the ease with which it can be filled, and the wide variations in shear-rate which its construction so readily permits.

(iii) *Goodeve type* (Fig. 16.5). The Goodeve 'thixotrometer', as it has been termed, is essentially a combination of the principles and desirable features of the Couette and Ungar machines, as the diagram shows. Thus the separation of the truncated cones, and hence the shear-rate, is readily variable by the micrometer head attached to the inner cone, and with a usual gear system attached to the primary drive a very wide range of shear-rates is available. On the other hand, the small angle of the (truncated) cones means that the average shear-rate does not change very considerably between the top and bottom, as it does in Ungar's apparatus. The torque on the hanging cone can be measured optically either by using a suitably thick torsion wire or, if the suspension is a rigid rod, by means of an arm and spring.

The 'viscosity' can be calculated from the measured torque, the top and bottom radii of the cones and the separation between the cones, but it is better to calibrate the apparatus with liquids of suitable and known viscosity.

(c) Capillary plastometers

These are merely capillary tube viscometers suitably modified for the systems to be studied, i.e. for much higher viscosities and pressures, and their construction and use present no new features. Their chief limitation arises from the shear-rate varying across the tube, from a maximum at the wall to zero at the centre (as discussed on p. 350), or at some intermediate point if the material shows a yield value. Instead of a parabolic curve for flow velocities a different form will be found,

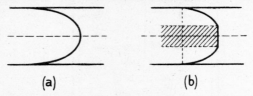

Fig. 16.6. Variation of flow-velocity across a circular tube for (a) a Newtonian fluid; (b) a plastic material with a yield-point.

as shown in Fig. 16.6 for a material with a yield value, and thus even the mean shear-rate will be uncertain. The central part of the material in the tube moves as a unit, since the stress is insufficient to overcome the yield value; hence the name 'plug-flow' for extreme cases of this behaviour. Mathematical analyses for certain simple cases (capillary with length great compared with its internal diameter) have been attempted by many workers, but even if solutions are obtained they clearly cannot be of universal validity for all systems.

(d) Penetrometers

Used particularly for the rapid assay of the (apparent) viscosity of asphalts, this instrument merely measures the penetration of a needle into a disc of the material. Under standardized conditions of time, loading, shape, and size of disc, temperature, etc., comparative results can rapidly be obtained.

(e) Farinograph

The farinograph measures the work required to stir the material, and as its name indicates, has been chiefly used to measure the consistency of doughs. Comparative results are obtained by standardizing the conditions.

Flow-rate v. applied stress curves

The two cases of true fluid and plastic solid behaviour are shown in Fig. 16.7. The former behaviour is shown, at any rate approximately, by certain bitumens; the latter, which has been much studied by

Bingham and is frequently called after him, by clays, paints, and many other pastes.

The abscissa OB is termed the 'yield-value' or better, the 'shearing strength', the slope of the linear portion the 'mobility' (analogous to

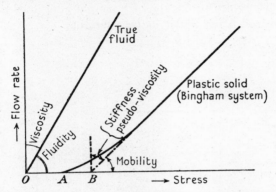

Fig. 16.7. Variation of flow-rate with applied stress for a true fluid and for a plastic solid (Bingham system).

fluidity in a true fluid), its reciprocal being then termed the 'stiffness' or 'pseudo-viscosity', being analogous to the viscosity in a true fluid.

The usual, but certainly over-simplified explanation of the Bingham curve is that part of the energy used is required to start the flow, just as in the case of static friction. After this friction has been overcome the material is then supposed to flow in the usual laminar manner.

II. *Elastometers*

For the reason given above it is only necessary to determine either E or n in the case of most colloidal systems. K is rarely measured, since it is unlikely to differ appreciably from the value for the pure dispersion medium. The methods described below are in general capable of measuring the yield strength as well as the elastic modulus of the material.

(a) *Extension or compression methods.* Systems of strength sufficient to support their own weight, and which do not flow appreciably in the time of an experiment, can be examined in extension or in compression. The material is usually fashioned into a short cylinder and then stressed by a weight (W) or by other convenient means (Fig. 16.8). The deformation (Δl) is measured, preferably with a cathetometer, and for small deformations the applied stress S is equal to $Wg/\pi r^2$, since the cross-sectional area can be regarded as constant. Eqn. (16.3) then gives E directly.

474   PLASTIC FLOW AND ELASTICITY

Gels such as gelatine are usually attached to blocks of wood and studied in extension, but for softer gels where attachment is difficult it is better to use the compression method.

One very simple modification, useful for comparative measurements of jellies, is to measure by means of a simple micrometer the sagging of a standard-sized block under its own weight.

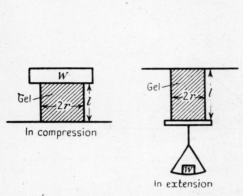
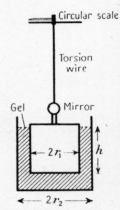

Fig. 16.8. Determination of elastic modulus of the more rigid gels.

Fig. 16.9. Couette apparatus for shear modulus of gels.

(b) *Couette method for shear modulus.* Fig. 16.9 depicts a very simple but quite accurate Couette apparatus for the shear modulus of materials which do not relax (flow) appreciably in times of the order of one second. For weak gels and many other soft rheological bodies the technique is extremely convenient.

The upper torsion head is rotated a measured amount (δ) and the deflexion (θ) of the hanging cylinder found optically by means of its attached mirror, and lamp and scale. The method is essentially that used for studying elasticity in monolayers (p. 495), and the same equation holds, namely,

$$n = \frac{K'}{4\pi h}\left(\frac{1}{r_1^2} - \frac{1}{r_2^2}\right)\frac{\delta - \theta}{\theta}, \qquad (16.13)$$

where h is the depth of cylinder immersed in the sample, r_1 and r_2 are the radii of cylinder and cylindrical container, and ($\delta - \theta$) the twist on the wire (i.e. angle turned through by torsion head minus that turned through by cylinder). K' is a constant for a particular torsion wire (the torque per radian angular displacement), and is measured by timing the oscillation of the cylinder in air, since the period is given by $2\pi\sqrt{(I/K')}$, where I is the moment of inertia of the cylinder.

This technique can also be used to study viscosities under extremely small rates of shear (of the order 0·001 sec.$^{-1}$). The upper head is rotated as before, then clamped, and the movement of the cylinder timed by means of the lamp and scale. The rate of shear is not constant but decreases throughout the experiment, but this is not entirely disadvantageous since by taking tangents at various points on the

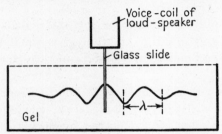

Fig. 16.10. Illustrating the principle of the photo-elastic method for the shear modulus of gels.

$(\delta-\theta)$-time curve some idea of the behaviour over a range of shear-rates can be found from a single experiment.

(c) *Photo-elastic method.* If the system relaxes too quickly for the previous technique, or if the resistance to a rapidly varying, rather than to a steady stress is required, other techniques have to be used. Of these the photo-elastic method due to Ferry (1941) would appear to be the best, although limited to transparent systems, that is, chiefly to gels and polymer systems.

Transverse waves are set up in the gel by the vibrations of a fine glass plate attached to the voice-coil of a loud-speaker unit (see Fig. 16.10) and driven by a beat-frequency oscillator and amplifier. Since isotropic gels become birefringent when strained the travelling strain waves produced by the vibrating plate appear as stationary waves when illuminated stroboscopically with the gel container between crossed nicols (or polaroids). The wavelength (λ) of these stationary waves is then measured by means of a travelling microscope.

The velocity of the travelling wave is

$$\nu\lambda = \sqrt{\frac{n}{\rho}}, \qquad (16.14)$$

where ν is the frequency (of the oscillator) and ρ the density of the gel (see, for example, Newman and Searle, 1932). The shear modulus (n) can thus be found at different frequencies, over the range of *ca.* 50–*ca.* 4,000 cycles/sec.

(d) *Other methods.* Of the remaining published methods for the detection or estimation of elasticity three require special mention. The first is particularly useful for visco-elastic materials such as polymers; the other two for detecting elasticity in sols and weak gels.

In the method due to Alexandrov and Lazurkin (1940) a disc of material is subjected by mechanical means to a stress which varies sinusoidally with time (i.e. $S = S_0 \cos \omega t$). A visco-elastic material responds with a strain which, in the steady state, also varies sinusoidally (i.e. $\sigma = \sigma_0 \cos(\omega t - \alpha)$) and the ratio S_0/σ_0 gives the dynamic modulus n. n will in general be a function of the frequency which can be varied over a wide range.

Of the remaining two methods, that due to Freundlich and Seifriz (1922), employs an extremely small nickel particle (a few μ in diameter) suspended in the system and deflected by means of a small electromagnet, the deflexion being observed in the microscope. On stopping the current the particle returns to its original position if the system exhibits elasticity, provided, of course, the elastic limit has not been exceeded. In the other, due to Bingham and Robertson (1929), the sol is contained in a capillary and made to oscillate by applying forces at opposite ends alternately. This sets up oscillations of the same frequency, but out of phase with the applied oscillation. The phase difference, which gives a measure of the elastic modulus, is obtained by observing the movement of specks of dust in the sol.

Classification of rheological properties of real materials

The behaviour of rheological materials has been classified by Scott Blair, and later by Reiner, in terms of 'essential rheological properties' as set out in Table 16.1 below. All *actual* systems can be regarded as possessing all these properties, but in very varying degree, whereas the hypothetical *ideal* materials are defined as possessing only certain of them.

TABLE 16.1

I. Viscosity (fluidity)	V. Plasticity
II. Elasticity	VI. Structural viscosity
III. Firmoviscosity	VII. Strain hardening
(plasto-elasticity)	VIII. Strength
IV. Elasticoviscosity	IX. Thixotropy
(visco-elasticity)	X. Dilatancy

Of these I and VI were considered in Chapter XIII, and of the others V, IX, and X call for special discussion here in view of their not infrequent occurrence in colloidal systems.

Plasticity

The precise definition of 'plasticity', which really means 'mouldability', has proved rather difficult. Wilson's description is 'that property which enables a material to be deformed continuously and permanently, without rupture, during the application of a force that exceeds the yield-value of the material' (Wilson, 1927). Scott Blair (1938) remarks 'Although plasticity cannot yet be defined in absolute units, yet like honesty, it is "although undefinable, associated with certain qualities"!' Plasticity implies flow at the stresses used in moulding the material, together with a yield-value such that shapes can be retained against the force of gravity. A high shear modulus is also necessary if the material is not to exhibit recoil or 'spring'.

Bosworth (1946) has recently suggested a quantitative definition of plasticity. He points out that for a plastic body

$$S = \phi(\sigma, d\sigma/dt),$$

or, as expressed by the fractional differential equation,

$$S = \phi\left(\frac{d^\alpha \sigma}{dt^\alpha}\right), \tag{16.15}$$

where ϕ denotes 'some function of' and the exponent α lies between 0 and 1, and depends upon the magnitude of S. This merely expresses the composite behaviour of plastic bodies, since for a solid body $S = \phi(\sigma)$, and for a fluid body $S = \phi(d\sigma/dt)$.

Wilson's qualitative definition implies that for a plastic body α changes fairly rapidly from 0 to 1 as the stress passes the yield-value, so that *plasticity* might be defined quantitatively as 'the reciprocal of the relative stress change required to change α from 0 to 1'. Since both these values might be approached asymptotically, a better definition might be 'the reciprocal of twice the relative stress-change required to change the exponent α from $\frac{1}{4}$ to $\frac{3}{4}$'. On this definition 'plasticity' is seen to be a dimensionless quantity, and for a Bingham system where the flow follows the law

$$\frac{d\sigma}{dt} = 0 \quad \text{for} \quad S \leqslant S_c,$$

where S_c is the critical shearing stress or yield-value,

$$\frac{d\sigma}{dt} = \frac{1}{\beta}(S - S_c) \quad \text{for} \quad S > S_c,$$

where β is a constant, its value will be infinite. (It will be indeterminate for brittle solids and for fluids.)

If the system breaks down fairly sharply from an elastic solid to a St. Venant plastic (p. 481) the plasticity has now a large positive value, being the larger the more rapid the change-over from elastic to plastic behaviour (see Fig. 16.7).

To measure plasticity in terms of the above definition the material is divided into a number of samples and to each sample a different stress is applied and the strain measured as a function of time (t). Curves of $\log \sigma$ against $\log t$ are then drawn, which will be straight lines parallel to the $\log t$ axis up to the yield point, but at higher stresses $(d \ln \sigma)/(d \ln t)$ will show positive slopes. The expression $(d \ln \sigma)/(d \ln t)$ will not always be independent of t (the time after applying the stress), but each $\log \sigma$ v. $\log t$ curve will be characterized at constant S by the fixed initial value of the slope, i.e.

$$\{(d \ln \sigma)/(d \ln t)\}_{t=0}.$$

This quantity is then plotted against the stress S, and the plasticity found from the relation

$$\text{plasticity} = \frac{S_{\frac{3}{4}}+S_{\frac{1}{4}}}{S_{\frac{3}{4}}-S_{\frac{1}{4}}}, \qquad (16.16)$$

where $S_{\frac{1}{4}}$ and $S_{\frac{3}{4}}$ are the values of S required to give slopes of $\frac{1}{4}$ and $\frac{3}{4}$ respectively.

Previous to this, measurements of plasticity have necessarily been on an empirical basis, though widely employed for comparative purposes. A great variety of methods have been used, such as the extent to which the size of cylinders could be reduced by rolling, squashing a sphere until cracks appear, squashing cylinders under constant stress, and so on (see Scott Blair, 1944).

Thixotropy

Gel systems which can be liquefied merely by shaking and revert to their original 'solid' state on standing were termed 'thixotropic' by Freundlich. Dilute sols (1–5 per cent. as a rule) of ferric oxide, alumina, and many clays show this behaviour when their stability has been somewhat lowered by the suitable addition of salts (p. 588).

Subsequently much discussion has centred around the use of this term, particularly in relation to the phenomenon known as 'false-body', a self-explanatory term used particularly in paint technology. Some regard 'false-body' as a kind of imperfect thixotropy, the undisturbed state not being a truly rigid gel, nor the disturbed condition that of a truly fluid sol. Scott Blair (1944) reviews the question and concludes that there are, in extreme cases, two quite distinct phenomena which can with advantage be distinguished from one another. In view of the uncertain state of this question we shall not pursue it further here.

Thixotropy differs from structural viscosity in that time enters explicitly as a determinant, and as would be imagined this makes its quantitative characterization extremely difficult.

Goodeve (1938), using the 'thixotrometer' described above, finds in certain thixotropic systems a linear relationship between the apparent viscosity and $1/G$, where G is the rate of shear. He therefore writes

$$S = \eta_0 G + \theta, \qquad (16.17)$$

calling θ, which is a measure of the anomaly, the 'coefficient of thixotropy' and η_0 the 'residual viscosity'.

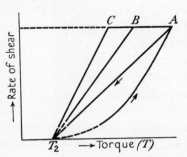

FIG. 16.11. Flow curve of thixotropic plastic material in a rotational viscometer. T denotes torque on hanging cylinder.

Assuming that, when shearing occurs, links in the loose structure are stretched, broken, and reformed, and that during the stretching process an impulse is passed from a fast moving layer to a slower moving one, Goodeve developed his 'impulse theory' of thixotropy. The chief distinction between thixotropic and Newtonian behaviour is attributed to the possibility of shear influencing the average life of a link.

The flow-curve of a thixotropic plastic material examined by means of a rotational viscometer shows the general behaviour set out in Fig. 16.11. Curve A is obtained by first increasing the rate of shear up to point A and then immediately decreasing it to zero. Curves B, C are obtained after different time intervals at the same constant top rate of shear. If this top rate of shear is maintained for a sufficiently long time an equilibrium curve CT_2 is obtained, where the structure will not break down further.

Weltmann (1943), who studied by this method a number of pigment suspensions in various oils, found that the product of the rate of breakdown of the structure and the time of agitation at any fixed shear-rate was a constant. This he has termed 'the time coefficient of thixotropic breakdown'. It is measured in absolute units, having the dimensions of viscosity (dynes.sec./cm.^{2}).

Dilatancy

Dilatancy, or 'inverse plasticity' as it has sometimes been termed, is the phenomenon of increasing resistance to deformation with increasing rate of shear. For example, a thick dispersion of starch in cold water can be stirred slowly without much effort, but if the rod is rapidly withdrawn the whole container tends to follow.

It arises from the volume changes produced by shear in liquid dispersions containing close-packed solid (or semi-solid) particles, and thus is shown particularly by stable, or deflocculated, dispersions (p. 612). Uniform spheres when close-packed would occupy *ca.* 74 per cent. of the total volume, but if they move into open packing, as is necessary if flow is to occur, this would fall to *ca.* 34 per cent. In actual systems, of course, neither uniformity of particle size nor complete sphericity is encountered.

The well-known phenomenon of wet sand becoming 'dry' when walked upon arises from these volume changes, since the particles, in their tendency to flow under the applied stress, move out of their close-packed state into a more open arrangement which absorbs all the water present.

Interpretation of rheological data

Having found experimentally certain relations between S, σ, and t for any given material, it is natural to attempt to analyse the results further. The ultimate goal will be an interpretation in terms of the shapes and sizes of the particles, and of the forces between them and between particle and medium. This desirable aim being at the moment far off, two other approaches have been developed:

(a) Discussion in terms of ideal materials and mechanical models, termed by Scott Blair the 'analytical' treatment.

(b) The 'integralist' treatment in terms of empirical mathematical equations (power laws).

An outline of these is given below: for a fuller account see Scott Blair (1944), Reiner (1943, 1945), Alfrey and Doty (1945), Alfrey (1948).

(a) *The analytical approach*

The classification of the ideal materials, whose properties can be simulated by means of mechanical models, can be done in two stages, (i) qualitatively, by a 'structural formula', (ii) quantitatively, by means of a 'rheological equation'.

(i) *Qualitative classification.* This starts from the three simplest rheological bodies or elements which are as follows:

(1) The Hooke solid, denoted by H, and represented by a perfectly elastic spring (see Fig. 16.12). Its only rheological property is its elasticity (property II), measured by the modulus of elasticity, E.

(2) The Newtonian liquid, denoted by N, and represented by a viscous element or *dash-pot* (see Fig. 16.12). Its sole rheological property is its viscosity (property I), measured by the coefficient of viscosity, η.

(3) The St. Venant plastic (deriving its name from St. Venant, who first described it mathematically), denoted by St. V. and only showing

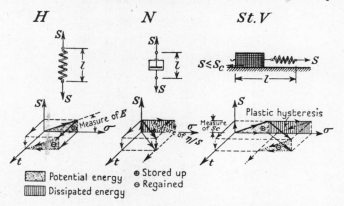

Fig. 16.12. Models of the three fundamental rheological bodies; the Hooke solid (H), the Newtonian liquid (N), and the St. Venant plastic (St. V). The dotted lines represent the behaviour on removal of the load. Elastic strain ($\sigma = \Delta l/l$) is assumed to follow instantaneously the applied stress (S). Linear stress-strain relations are depicted, but the models can be applied in a general manner to non-linear relations also.

property V. This is a solid which has a yield-point below which its deformation is elastic, but above which it flows plastically at constant stress. It is represented by a weight resting on a table with 'solid' friction between both (Fig. 16.12). Its *plastic* resistance is measured by the critical yield stress, or the 'friction constant', S_c.

Fig. 16.12 shows the behaviour of H, N, and St. V elements in terms of the three coordinates, stress, strain, and time (after Reiner, 1945). Potential energy and dissipated energy are indicated and the different behaviour of the three elements in loading and unloading can readily be followed.

By combination of these three structural elements various other model bodies can be constructed, of which the most important are the Kelvin solid, the Maxwell liquid, and the Bingham body.

(4) The Kelvin solid (Fig. 16.13) combines the H and N elements in parallel, and is denoted by K or H/N, the vertical dash (/) indicating parallel coupling. It shows rheological property III, i.e. firmoviscosity or plasto-elasticity.

This is the model for a solid which is elastic but which also exhibits time lags in the appearance and disappearance of elastic strains (elastic fore- and after-effects), as well as elastic hysteresis and hence viscous damping of oscillations.

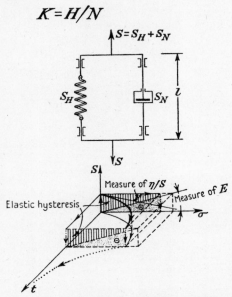

FIG. 16.13. Model of the Kelvin solid (K).

(5) The Maxwell liquid (Fig. 16.14) is denoted by M or H—N, the horizontal dash (—) indicating series coupling. It exhibits rheological property IV (elasticoviscosity or visco-elasticity).

This model serves for a liquid which shows 'instantaneous' elasticity, and also for a 'solid', the elastic stresses of which relax and which exhibits 'creep'.

(6) The Bingham body B (Fig. 16.15) has the structural formula N—St. V, the elements being in series and the dash-pot connected to the 'weight'. (The reverse arrangement—V. St. would indicate series coupling to the 'spring' end of the St. V element.)

The flow-rate (or $d\sigma/dt$) v. stress curve for this 'ideal' Bingham body is also shown in Fig. 16.15.

By suitable combinations of the three basic elements the qualitative classification of any material can be written down from the information provided by the tests to which the material has been subjected.

In the case of 0·5 per cent. gelatine sol, for example, Schwedoff found that to maintain a constant deflexion (using the coaxial cylinder apparatus, described above), the torque had gradually to be reduced as in a Maxwell liquid, but a small residual deformation remained permanently. Hence

the structural formula = M—St. V = H—N—St. V = H—B.

In a similar way Reiner (1945) has considered the rheological studies

$M = H-N$

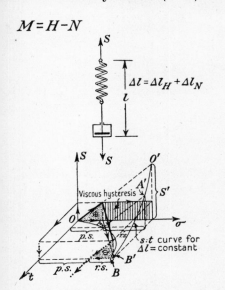

$B = N-St.V$

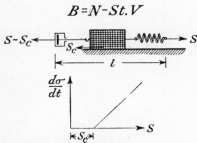

FIG. 16.14. Model of the Maxwell liquid (M).

FIG. 16.15. Model for the Bingham body.

OAB = elastic recovery for extension Δl.
$O'A'B'$ = relaxation curve for stress S'.
p.s. = permanent strain.
r.s. = recovered strain.

of flour-dough (FD) carried out under various conditions by Schofield and Scott Blair, and shows that to a fourth approximation

FD = H—N—St. V—K.

Many other rheological bodies can be imagined and they can all be arranged in a hierarchy, starting with the three *groups* of 'solids', 'plastics', and 'liquids', each one divided into the *families* of Newtonian, Maxwellian, etc., liquids, St. Venant, Bingham, etc., plastics, and Hooke, Kelvin, etc., solids, respectively (Reiner, 1945).

(ii) *Quantitative classification by means of 'rheological equations'.*
A 'rheological equation' is the relation connecting shearing stress with shear and their derivatives with respect to time: it is assumed that all rheological behaviour is a function of the four variables S, dS/dt, σ, and $d\sigma/dt$.

If the shear stress and strain are denoted by S and σ as before, the fundamental relations are:

$$\text{Newtonian (viscous) liquid} \quad S = \eta \frac{d\sigma}{dt}, \quad (16.18)$$

$$\text{Hooke (elastic) solid} \quad S = n\sigma, \quad (16.19)$$

$$\text{St. Venant (plastic) solid} \quad S = S_c. \quad (16.20)$$

The ideal materials built up of two elements are then:

$$\text{Maxwell liquid} \quad \frac{d\sigma}{dt} = \frac{S}{\eta} + \frac{1}{n_l}\frac{dS}{dt}, \quad (16.21)$$

$$\text{Kelvin solid} \quad S = n\sigma + \eta_s \frac{d\sigma}{dt}, \quad (16.22)$$

$$\text{Bingham body} \quad S - S_c = \eta_{\text{pl}} \frac{d\sigma}{dt}, \quad (16.23)$$

where n_l denotes the elastic modulus of the 'liquid', η_s the 'solid' viscosity, and η_{pl} the plastic viscosity, called by Bingham the 'stiffness'.

Eqns. (16.21) and (16.22), being of the form $dy/dx + My = N$, can be integrated, giving

$$S = e^{-n/\eta t}\left(S_0 + n \int_0^t \frac{d\sigma}{dt} e^{n/\eta t}\, dt\right) \quad (16.24)$$

and

$$\sigma = e^{-n/\eta t}\left(\sigma_0 + \frac{1}{\eta} \int_0^t S e^{n/\eta t}\, dt\right) \quad (16.25)$$

for the Maxwell liquid and Kelvin solid respectively.

Reiner (1945) has shown how, if the structural formula is known, the rheological equation can be derived, and exemplifies this point by consideration of the flour-dough experiments of Schofield and Scott Blair mentioned above. (For further discussion see Scott Blair, 1944; Reiner, 1943, 1945; and Alfrey and Doty, 1945.)

(b) *The integralist approach*

The first use of power laws to relate the S, σ, and t, which forms the basis of the integralist treatment, was by Nutting in 1921. Subsequently it was used as a convenient way of plotting rheological data, and more recently some theoretical treatments have been attempted.

Nutting's equation is
$$\sigma = \psi^{-1} t^k S^\beta \quad (16.26)$$

(where ψ, k, and β are constants), or by taking logarithms and differentiating,

$$\frac{d\sigma}{\sigma} = k\frac{dt}{t} + \beta\frac{dS}{S}. \quad (16.27)$$

From the viewpoint of classical physics eqn. (16.26) has the great disadvantage of being dimensionally incorrect. Scott Blair (1944) discusses this point at some length and concludes that it does not necessarily impair the usefulness of power laws for characterizing rheological behaviour.

Nutting found k to be always less than unity but approaching this value for soft materials, showing that none of his materials accelerated under constant stress. The constant β was evaluated by taking arbitrary times, being the slope of the $\ln \sigma$ v. $\ln S$ curves when t is constant, and was found to be either greater or less than unity. A high value indicated a rubbery texture and a low value properties like those of quicksand.

Instead of Maxwell's assumption that

$$dS/dt \propto S, \quad \text{or that} \quad (d \ln S)/dt = 1/\tau,$$

i.e. a constant independent of stress, the power law relation leads, at constant strain (σ), to

$$\frac{d \ln S}{dt} = -\frac{k}{\beta t} = -\frac{k}{\beta} (\text{constant} \times S^{\beta/k}), \qquad (16.28)$$

i.e. $(d \ln S)/dt$ is a function of the stress. Thus eqn. (16.26) really postulates for the condition of constant strain a 'relaxation time' proportional to a power of the stress or proportional to the time itself.

Maxwell's suggestion that 'relaxation time' may be a function of stress has been followed up by several workers. For example, de Bruyne (1941) considers that the molecular process involved in deformation under stress can be regarded as involving passage over an energy barrier (U), and that the elastic strain energy arising from the external shear stress S can contribute towards the necessary activation. He obtains the relation

$$S = nGAe^{(S_0-S)^2 V/2nRT}, \qquad (16.29)$$

where the energy barrier (or activation energy) $U = (1/2n)(S_0-S)^2 V$, S_0 is the ultimate cohesive stress, V is the volume of a gram mole of material, G the velocity gradient (or shear-rate), and A a constant. The general significance can be seen by writing eqn. (16.29) in the form

$$x = ye^{(1-x)^2/p} \qquad (16.30)$$

(i.e. $x = S/S_0$ and $y = \text{constant} \times G$), and plotting x against y for arbitrary values of p. As Fig. 16.16 shows, this equation is capable of representing the main types of viscous and plastic flow.

Recently Scott Blair and his co-workers have developed the use and interpretation of these power laws in connexion with their studies of 'firmness' and other rheological phenomena. The subject is still of

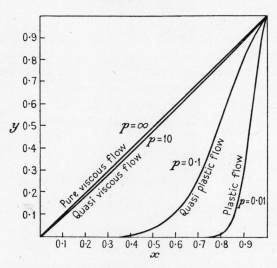

FIG. 16.16. Curves showing change from pure viscous or Newtonian flow to plastic or Bingham flow as the temperature is reduced. In viscous flow there is no yield-point, but in plastic flow there is an apparent yield-point.

considerable complexity, but Scott Blair in his recent book (1944) provides an excellent critical survey.

GENERAL REFERENCES

First and Second Reports on Viscosity and Plasticity. Prepared by a committee for the Academy of Sciences at Amsterdam, 1935 and 1938.
SCOTT BLAIR, *Introduction to Industrial Rheology*, 1938, Churchill, London.
—— *A Survey of General and Applied Rheology*, 1944, Pitman, London.
REINER, *Ten Lectures in Theoretical Rheology*, 1943, Rubin Mass, Jerusalem.
ALFREY, *Mechanical Properties of High Polymers*, 1948, Interscience, New York.
'Bingham Memorial Symposium', *J. Colloid Sci.* 1947, **2**, 1.

OTHER REFERENCES

ALEXANDROV and LAZURKIN, 1940, *Acta Physicochim, U.R.S.S.* **12**, 647.
ALFREY and DOTY, 1945, *J. applied Physics*, **16**, 700.
BINGHAM and ROBERTSON, 1929, *Kolloidzschr.* **47**, 1.
BOSWORTH, 1946, *Nature*, **157**, 447.
DE BRUYNE, 1941, *Proc. Phys. Soc.*, London, **53**, 251.
FERRY, 1941, *Rev. Sci. Instrum.*, **12**, 79.
FREUNDLICH and SEIFRIZ, 1922, *Z. phys. Chem.* **104**, 233.
GOODEVE, 1938, *Rep. Progress Physics*, **5**, 20.

MAXWELL, 1867, *Philos. Trans*, **157**, 52.
NEWMAN and SEARLE, 1932, *The Properties of Matter*, 2nd edn., Ernest Benn.
NIEUWENBURG, 1938, *Second Report on Viscosity* referred to above, p. 241.
REINER, 1945, *Rheology Bulletin*, **16**, 53.
WELTMANN, 1943, *J. applied Physics*, **14**, 343.
WILSON, 1927, *Ceramics, Clay Technology*, McGraw-Hill, New York.

Section B
STUDY OF THE INTERFACE
General introduction

THE importance of a detailed understanding of the interface in colloidal systems has been mentioned earlier. Of the possible types of interface which may theoretically exist, three are of paramount importance, namely, the gas/liquid, the liquid/liquid, and the liquid/solid. (The gas/solid interface is better dealt with as an aspect of surface chemistry rather than of colloids in general.) The results from interfacial studies can be applied directly to certain colloidal systems, e.g. foams, emulsions, suspensions, and pastes, and in addition provide a useful indirect approach to some aspects of the soaps, proteins, and gels. These applications will be discussed in Part III.

There are certain fundamental problems which clearly arise at all the interfaces here considered; such as the packing of the molecules at the interface and its effects upon the energy and electrical conditions at the phase boundary, and the mechanical properties of the interfacial film. For adsorbed films there is the further question of the rates of adsorption and desorption when the interfacial area is altered. Here we will consider the principal experimental techniques available for the study of the interface, from the point of view of their usefulness for the study of colloidal systems rather than as a detailed account of the physics and chemistry of surfaces. As far as techniques and interpretation of results go, the gas/liquid and liquid/liquid interfaces have much in common, and a more convenient subdivision is into insoluble and adsorbed monolayers, leaving the solid/liquid interface to be considered separately.

XVII

INSOLUBLE MONOLAYERS AT THE AIR/WATER AND OIL/WATER INTERFACES

As the title indicates, we shall restrict ourselves to these special cases of the gas/liquid and liquid/liquid interfaces, since for obvious reasons these are encountered most frequently in colloidal systems.

AIR/WATER MONOLAYERS

Insoluble monolayers, such as are given by oleic acid and many other long-chain compounds, have one great advantage over adsorbed films, namely, that the number of molecules per unit area (generally termed the *molecular density* in surface film work) can be determined directly without recourse to Gibbs's equation or to an assumed equation of state. It is therefore a simple matter to relate the molecular surface density to the reduction in surface energy as measured by the lowering of surface tension. In addition, the mechanical properties of the monolayer, such as viscosity and elasticity, and the change in phase boundary potential produced by the monolayer, are measurable quantities of great importance.

It may be pointed out that practically all these observable quantities are *differences between the clean water and film-covered surfaces*; for example, the 'surface pressure' (Π) is the difference in surface tension, the 'surface potential' (ΔV) is the change in phase boundary potential, and the 'surface viscosity' (η_s) the change in viscosity of the surface layer produced by the film.

Surface pressure and surface area measurements

With compounds giving insoluble (or slowly soluble) monolayers the area per molecule (A) can be measured directly by spreading on a known area of surface a small volume of a dilute solution of the film-forming compound. The micrometer syringe used for this purpose is as shown in Fig. 18.4 (p. 517) except for the addition of a fine needle tip; it reads to 0·0001 c.c. The solvents for spreading purposes are preferably petroleum ether (boiling-point 60°–80°) or benzene, although for compounds soluble only in aqueous media, such as proteins, 60 per cent. propyl alcohol, 0·5 M. sodium acetate is very suitable (Ställberg and Teorell, 1939). The chief criteria which the spreading solvent must satisfy are that it shall spread the film-forming substance completely, and shall then be rapidly lost, either by evaporation or by solution.

One precaution of great importance in this work is to ensure quantitative spreading. This can as a rule be adequately checked by examining the effect of diluting the spreading solution, or by trying other solvents. In cases of doubt it is advantageous to examine the film under dark-field

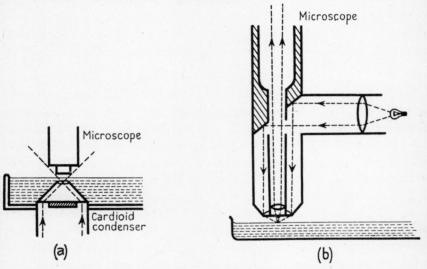

FIG. 17.1. Dark-field examination of surface films. (a) Cardioid condenser in trough. (b) 'Ultropak'.

illumination, when unspread or collapsed material usually becomes visible. (It is, of course, impossible to detect the monolayer by this means.) Two common methods are shown in Fig. 17.1. One uses a cardioid condenser sealed into the bottom of the trough by paraffin wax, the other (the Ultropak) examines the surface by reflected light. Of these the latter is much to be preferred, since it reduces the risk of contaminating the surface.

Dark-field examination is also very useful for examining condensed films for signs of collapse, the onset of which is usually shown quite readily by the appearance of a myriad of bright points, the so-called 'point structure'.

The surface pressure (Π), defined as the reduction of surface tension produced by the monolayer, is commonly measured directly by means of a Langmuir–Adam film balance, although for some purposes the Wilhelmy plate is to be preferred. The ring and other detachment methods have also been used with the more attenuated films.

The type of film balance developed in the Department of Colloid Science, Cambridge, is shown in Fig. 17.2. The dish is preferably of pyrex or hard glass, the float of mica, and the movable barriers of glass, in order to reduce impurities, particularly grease and heavy metal ions, to a minimum. The float is attached by a light metal frame to the centre of the torsion wire, so that any force upon it due to the monolayer can be compensated by a suitable rotation of the torsion head. (The torsional constant of the wire is determined in the usual way by placing

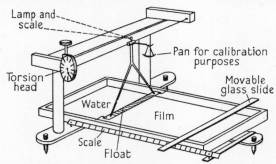

FIG. 17.2. Single wire surface balance based on the Langmuir-Adam principle.

weights in the pan shown in the figure.) Before use, the float, barriers, and sides of the trough are coated with a thin layer of pure paraffin-wax, using for this purpose a solution of wax in petroleum ether. The film is prevented from spreading past the float either by thin strips of platinum foil or by vaselined silk threads attached to the sides by paraffin-wax. Having cleaned both front and back surfaces by sweeping, the monolayer is deposited as above and then compressed by means of the movable glass slide, the torsion head being rotated so as to keep the float in a fixed position, as indicated by the optical arm. From the reading on the torsion head and the torsional constant the force exerted by the monolayer per unit length of the float (i.e. Π, in dynes/cm.) can be obtained directly as a function of A (usually expressed in A^2 per molecule, and calculated directly from the amount of substance spread and the actual film area). (For a very simple direct-reading film balance see Alexander, 1947.)

In the method due originally to Wilhelmy, of which a modern type is shown in Fig. 17.3, a glass plate (e.g. a microscope slide or cover glass) is partially immersed in the aqueous phase by suitable adjustment of the torsion balance. The position of the optical arm is first noted for a clean-water surface, and the monolayer then spread as usual. The monolayer will reduce the surface tension and the plate will accordingly move slightly upwards until a new position of equilibrium is attained, unless the torsion is suitably changed to keep it in its original position.

The change in force due to the monolayer equals $l\Pi$, where l is the perimeter of the slide, so that by compressing the monolayer as in the film balance the Π–A relationship can be found. As will be clear from Fig. 17.3 this method readily lends itself to automatic recording of Π–A curves (e.g. Dervichian, 1936; Andersson, Ställberg-Stenhagen, and

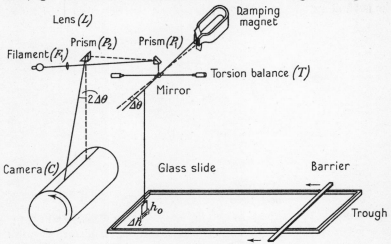

Fig. 17.3. Recording surface balance based on the principle of the Wilhelmy plate.

Stenhagen, 1944), thus permitting uniform rates of compression or expansion. For studying condensed films, or slow-time effects in monolayers, automatic recording has certain advantages over the usual manual method.

Surface potential

The 'surface potential' (ΔV) of an insoluble monolayer is defined as the difference in phase boundary potential between clean-water and film-covered surfaces, i.e. $\Delta V = V_{\text{water}} - V_{\text{film}}$. Two principal methods are in use for measuring these phase boundary potentials—the ionization method, using a static air electrode (Schulman and Rideal, 1931) and the vibrating plate method (Yamins and Zisman, 1933).

The first of these is depicted in Fig. 17.4. The 'air-electrode' consists of a well-insulated metal wire with its tip about 1–2 mm. above the water surface, the air-gap being rendered conducting either by polonium deposited on the wire or by ca. 1 mg. of mesothorium bromide in a narrow glass tube attached to the wire. A reversible electrode (e.g. Ag|AgCl) dips into the solution and is connected to a standard potentiometer box as shown. The potential with a clean-water surface is first measured by connecting the 'air electrode' to the grid of the valve by the switch and then changing the applied potential until the current in the anode circuit, as measured by the galvanometer, is at its previous value, i.e. when the grid was earthed. The potential V_{water} is then equal and opposite to the value registered by the potentiometer. The monolayer is then spread and the process repeated, the difference giving the

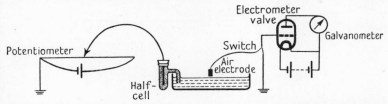

Fig. 17.4. Diagrammatic set-up for surface potentials by the radioactive air-electrode method.

surface potential (ΔV). The monolayer is compressed and the ΔV–A curve built up point by point. This technique is superior to the vibrating plate method described below in two respects; the setting of the tip

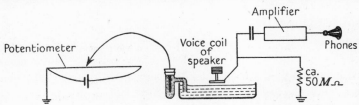

Fig. 17.5. Diagrammatic set-up for surface potentials by the vibrating electrode method.

with respect to the surface is not as critical, and owing to the small area of the tip the monolayer can be explored for homogeneity. In both methods screening of valve and leads from electrical disturbances is of great importance.

The vibrating plate method is shown diagrammatically in Fig. 17.5 (for details of suitable circuits see Yamins and Zisman, 1933; also Rosenfeld and Hoskins, 1945). The vibrating plate consists of a thin disc, preferably of gold or of gold-plated brass, actuated by attachment to the voice-coil of a loud-speaker at a frequency of about 200 cycles/sec., and placed as close to the water surface as possible. The applied potential is varied until a minimum of sound in the telephone indicates that the water and metal surfaces are at the same potential. The monolayer is then spread and the determination repeated, giving ΔV and the ΔV–A curve as before.

It is usual to write ΔV in the form

$$\Delta V = k'n \quad \text{or as} \quad \Delta V = 4\pi n \mu \tag{17.1}$$

(formally analogous to the Helmholtz equation), where k' or μ is a constant characterizing the polar group and to a first approximation independent of n, the molecular density (in molecules/cm.^{2}). μ is termed

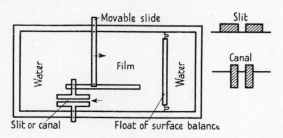

Fig. 17.6. Canal or slit viscometer.

the 'vertical component of the apparent surface moment' and is usually given in milli-Debyes (i.e. in 10^{-21} e.s.u.).

Mechanical properties

The 'surface viscosity' (η_s) is defined as the change in viscosity of the surface layer brought about by the insoluble monolayer, and is measured in one of two principal ways depending upon the system. For gaseous, expanded, and relatively attenuated films the two-dimensional capillary-flow method is most suitable, whereas for relatively close-packed films in the condensed state and for proteins and polymers, where we have high surface viscosities, the damping of a suitable oscillating system can be used.

The 'surface canal' or 'surface slit' type of viscometer is depicted in Fig. 17.6, and by analogy with three-dimensional Poiseuille flow the quantity of film passing through in unit time (Q) should be given by the relation

$$Q = \frac{\Pi d^3}{12\eta_s l}, \tag{17.2}$$

where d and l are the width and length of the canal, η_s is the surface viscosity, and Π the surface pressure of the film. This equation is obeyed experimentally only for very narrow channels, Q then increasing less rapidly than expected as d increases.

In the second method the oscillating system consists of a circular glass disc, or a glass plate such as a microscope slide, attached in each case through a heavy metal bob to a torsion wire (see Fig. 17.7). The oscillations are timed and the logarithmic decrement λ (i.e. the logarithm of the ratio of successive amplitudes) found, first with a clean water surface and then after spreading the monolayer. Using the slide as shown in Fig. 17.7 (a), this modification being preferred to the disc as

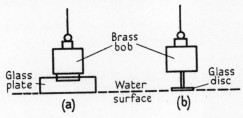

FIG. 17.7. Oscillating systems for surface viscosity. (a) Glass slide just in contact with surface. (b) Glass disc just in contact with surface.

it eliminates slip at the edges, the surface viscosity is calculated from the expression

$$\eta_s = 2 \cdot 3(\lambda_{\text{film}} - \lambda_{\text{water}}) \times \frac{4I}{\tau l^2}, \qquad (17.3)$$

where λ = logarithmic decrement (to base 10), τ = period of oscillation, l = length of slide, I = moment of inertia of system (found from oscillation experiments in air).

In addition the above technique indicates whether the flow of the film is Newtonian or anomalous, since for the former case τ is a constant independent of the amplitude, whereas if the film shows any elastic properties τ decreases as the amplitude decreases.

The shear rigidity modulus of a monolayer, as we might expect by analogy with three-dimensional systems, is only measurable if the film is in the condensed state, and it is usual to employ the same circular disc apparatus as described above for viscosity determinations. This is essentially the same method as used to determine the rigidities of rheological materials such as gels (p. 474), except that h in the equation below now refers to the thickness of the monolayer. The usual expression for the shear modulus n (in dynes/cm.²) is

$$n = \frac{K'}{4\pi h} \left(\frac{1}{r_1^2} - \frac{1}{r_2^2} \right) \frac{\delta - \theta}{\theta}, \qquad (17.4)$$

where h = depth of cylinder immersed in gel, r_1 = radius of cylinder, r_2 = radius cylindrical container, $\delta - \theta$ = torsion on wire (i.e. angle (δ) turned through by torsion head minus that turned through by cylinder), θ = angle turned through by cylinder, K' = a constant for the torsion wire (torque per radian angular displacement), and is obtained from oscillation experiments (period $= 2\pi\sqrt{(I/K')}$, where $I =$ moment of inertia of cylinder).

In addition to the shear modulus the yield-point can in principle also be determined, but in many cases the determination is complicated by plastic flow occurring.

Another quantitative method for the shear modulus is to exert a torque on the monolayer by means of a disc rotating just below the surface, and to observe in a microscope the deflexion of dust particles in the film (Mouquin and Rideal, 1927). Plastic flow is here indicated by a particle not returning to its original position on stopping the rotating disc.

There are also a number of ways in which the presence of elasticity in a monolayer can be detected qualitatively—for example, by dusting on talcum powder (unscented!), blowing gently, and observing any elastic recoil, by the same method with a piece of paper on the film, or by observing the type of expansion pattern given by a drop of an oxidized mineral oil when dropped on the film.

Finally we may mention the compressibility of the monolayer, defined as $(1/A)dA/d\Pi$. This is readily obtained from the force-area curve, and is a quantity related to the structure of the molecule as well as to the molecular density of the monolayer.

Multilayers (built-up films)

Under certain conditions, the chief of which is that the film is in the condensed state, monolayers can, by a simple dipping process, be transferred from the water surface to a slide which may be of metal, glass, mica, nitrocellulose, etc., or even of fine wire-mesh. The slide is usually made hydrophobic before use by polishing with a little ferric stearate, and the monolayer is forced on to the slide by means of a drop of a 'piston-oil' such as oleic acid, triolein, or castor oil, separated from the monolayer by a vaselined silk thread (Fig. 17.8). (A 'piston-oil' is a liquid giving a constant surface pressure, e.g. oleic acid gives about 33 dynes/cm.) This procedure may be repeated indefinitely, and films of more than a thousand monolayers have been made.

When some 25–30 monolayers have been deposited the resultant multilayer is usually thick enough to exhibit an interference colour, from which its thickness, and hence that of the original monolayer, can be obtained. The ability of this technique to provide very thin films of accurately known thickness has been put to good purpose for the study of boundary lubrication, for preparing thin absorbing screens for studying nuclear processes, and for preparing 'non-reflecting' glass. Another point of considerable interest is the use of the high degree of orientation which obtains in multilayers to induce crystallization in

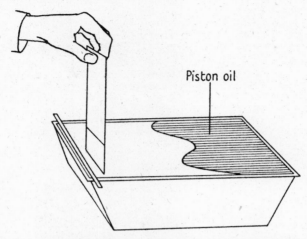

FIG. 17.8. Transfer of monolayers to solid surfaces (built-up films or multilayers).

normally intractable substances, such as the unsaturated fatty acids and amorphous polymers.

(For more detailed reviews see Schulman, 1939; Alexander, 1944.)

Principal results and applications of air/water monolayers

Having considered the experimental study of insoluble monolayers at the air/water interface it is convenient to outline the principal findings as well as their more important applications to other branches of science. This is necessary as although conclusions from such systems can be applied directly to certain colloidal systems, yet in many others the argument is more of an inferential nature.

(I) *Phases and phase changes*

The principal monolayer phases, with their properties as shown by the techniques outlined above, are given in Table 17.1 and Figs. 17.9–17.12 inclusive. Some marked analogies with three-dimensional matter are clearly evident. Only the main points of interest can, however, be included here.

(a) *Gaseous and vapour films.* An ideal two-dimensional gas would, by analogy, obey the relation $\Pi A = kT$. Like the gas law $pV = RT$

TABLE 17.1

Phase	Chief characteristics
I. Gaseous and vapour	If ideal $\Pi A = kT$, i.e. Π–A curve is rectangular hyperbola. More usually the relation $\Pi(A - A_0) = xkT$ is obeyed over most of the curve.
II. Liquid expanded (Adam). Liquid L_1 (Harkins)	Liquid of high compressibility, obeying relation $(\Pi - \Pi_0)(A - A_0) = C$.
III. Intermediate or transition	Liquid of extremely high compressibility. Marked hysteresis is usually observed.
IV. Condensed:	
(a) L_2. (Close-packed heads (Adam); liquid L_2 (Harkins).)	Newtonian liquid of low compressibility. Π–A curve usually linear.
(b) L_2'.	..
(c) LS.	..
(d) S. (Close-packed chains (Adam); solid (Harkins, Dervichian).)	Compressibility very low. May show anomalous viscosity, plasticity, or rigidity. Π–A curve linear. Packing approaches that in crystalline solid.
(e) CS. (Close-packed solid, Stenhagen.)	A the same as in crystalline solid.

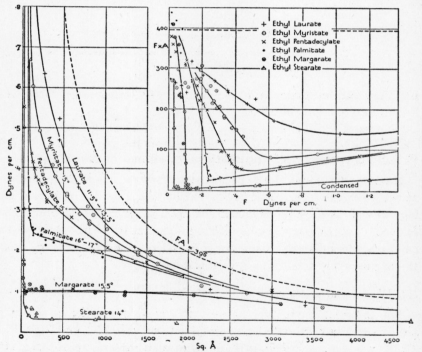

FIG. 17.9(a). Some typical force-area curves for long-chain esters. Vapour and gaseous films. Ideal gaseous behaviour indicated by dotted line.
(Note: Surface pressure (force) is here denoted by F instead of Π.)

this is an ideal to which actual systems may approximate but never attain. Some of the most nearly perfect gaseous monolayers are given by the long-chain esters, the behaviour of a series of which is shown in Fig. 17.9 (a). It is seen that at very low surface pressures and large areas (ca. 10,000 A^2 per molecule) the ideal limit $\Pi A = kT \simeq 400$ (at room

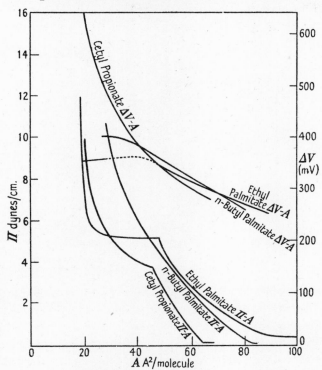

FIG. 17.9 (b). Expanded films, showing transition to condensed state in two cases. (Also ΔV-A curves.)

temperature) is approached. At higher pressures the experimental points fit the equation of state, $\Pi(A-A_0) = xkT$, suggesting that the monolayer resembles an imperfect gas, since a similar equation holds for imperfect gases at high pressures. (A_0 is the *co-area* of the molecule and x a constant.) This is brought out very clearly by the ΠA v. Π curves of Fig. 17.9 (a) (cf. Amagat's curves for compressed gases).

Experimentally this region has not been easy to investigate on account of the very low pressures (ca. 0·01 dyne/cm. and less) to be measured and of the difficulties in avoiding minute amounts of contamination.

(b) *Liquid-expanded films.* The 'liquid-expanded' phase, as it has been termed by Adam, is shown in Figs. 17.9 (b) and 17.10. Unlike the other phases considered here it has no three-dimensional counterpart, as will be obvious when its structure is considered.

Langmuir pointed out that the Π–A curve was quite accurately a rectangular hyperbola, obeying the relation $(\Pi - \Pi_0)(A - A_0) = kT$. From the obvious resemblance which this has to van der Waals's

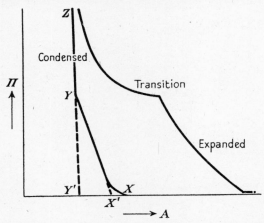

FIG. 17.10. General force-area diagram for condensed and expanded films.

equation, and from a study of duplex films (see p. 523), he formulated the theory for the structure of expanded films which is now generally accepted. The monolayer in the expanded state is regarded as a duplex oil film of extreme thinness (15–20 A), the upper surface being hydrocarbon in contact with air, the lower one consisting of the polar headgroups in two-dimensional kinetic agitation. The expanding tendency of the heads produces a spreading pressure Π_h, and in addition to this we have to consider the attractive forces between the long chains which give rise to a (negative) spreading pressure Π_0. Clearly the *observed* spreading pressure $\Pi = \Pi_h + \Pi_0$, so that if the heads obey the equation of state $\Pi_h(A - A_0) = kT$, which seems reasonable, then we obtain $(\Pi - \Pi_0)(A - A_0) = kT$ as is found experimentally. A_0 is accordingly to be identified as the 'co-area' of the molecule, and for simple straight-chain compounds has values of around 15 A^2.

Further confirmation for Langmuir's suggested structure has been obtained from the study of oil/water monolayers, as will be pointed out below.

(c) *Condensed films.* The molecular packing in the condensed phases (see Figs. 17.10 and 17.11) closely approaches that observed in the crystalline state; for example, stearic acid gives a coherent condensed monolayer from *ca.* 23 A^2 to *ca.* 20 A^2, whilst its cross-sectional area from X-ray diffraction measurements varies between 18·5 and 20 A^2 depending on the temperature. The Π–A curve is commonly, but not invariably, made up of two approximately linear portions, the precise interpretation of which is still a matter of discussion.

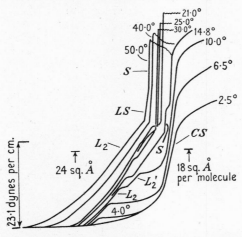

FIG. 17.11. Force-area curves for *n*-docosanoic (behenic) acid on $N/100$ HCl at various temperatures. (Ställberg-Stenhagen and Stenhagen, 1945.)

In the most compressed region (YZ in Fig. 17.10), it seems most likely that the long paraffin chains are vertically orientated and close-packed, although not quite as tightly as in the three-dimensional crystal. This view is supported by the close agreement between the compressibility of condensed monolayers and of long-chain hydrocarbons in bulk, by surface potential measurements upon ethyl stearate monolayers which can only be interpreted satisfactorily by a configuration involving vertical chains, and by consideration of the X-ray structure of the higher paraffins near their melting-points (Alexander, 1942 *b*).

The second linear portion (XY in Fig. 17.10) has a much higher compressibility, and in this case the picture is more complex, since the limiting area at $X(A_x)$ appears to be determined by a variety of factors, of which the size and nature of the polar group and the packing of the paraffin chain appear to be the most important. The experimental values of A_x lie between *ca.* 22 and 30 A^2 for simple paraffin-chain derivatives. In a number of cases it seems that A_x is determined by close-packing of the head-groups; for example, the p-alkyl phenols, anilines, and anisoles all give *ca.* 24 A^2, a value consistent with the cross-sectional area of the phenyl nucleus on edge, as it would be in the monolayer. In other cases the value of A_x appears to arise from a more directed type of association between the polar head-groups; this is shown particularly by the long-chain amides and acetamides where cross-hydrogen bonding seems to occur. The importance of the hydrocarbon chain packing is evidenced most clearly by the cis- and trans-unsaturated compounds (e.g. erucic and brassidic acids); in these compounds, as a

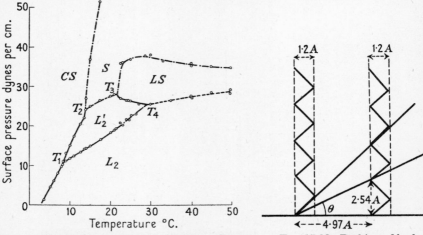

FIG. 17.12. Surface pressure–temperature diagram for n-docosanoic acid. (Unbroken heavy lines indicate first-order transitions and dotted heavy lines second-order transitions.)

FIG. 17.13. Packing of hydrocarbon chain zigzags.

model clearly shows, the chains in the trans-isomer can pack together much the more readily, and this is reflected in the monolayer behaviour. In two examples only (long-chain methyl ethers and methyl ketones) there is some indication that at the point X the chains are tilted at ca. $26\frac{1}{2}°$ to the vertical, this being the smallest angle from the vertical at which the zigzags in the hydrocarbon chain can again interlock (see Fig. 17.13).

The two condensed regions also exhibit different mechanical properties, although in this respect it is now the XY region which appears to show the uniform behaviour. In the XY part all films so far examined appear to behave as two-dimensional Newtonian liquids, obeying the equation
$$\log \eta_s = \log \eta_0 + k\Pi,$$
where k and η_0 are constants (see Fig. 17.14, due to Boyd and Harkins, 1939). An equation of this type can be deduced from Eyring's theory of the liquid state.

Condensed monolayers in the YZ region appear to show less uniformity in their mechanical properties; to qualitative tests they range from relatively fluid (e.g. methyl ketones and esters), to extremely rigid (e.g. amides, fatty acids on neutral substrates containing heavy metal ions). All the *fluid* films examined quantitatively (alcohols, fatty acids, and

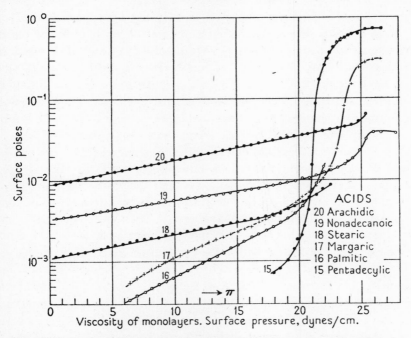

FIG. 17.14. Surface viscosity of various long-chain fatty acids as a function of surface pressure.

esters) show anomalous viscosity in this region, and the surface viscosities may be extremely high. (A monolayer of octadecyl alcohol, for example, can give a surface viscosity of up to 1 surface poise; as the thickness is only *ca.* 20 A this corresponds to a bulk viscosity of *ca.* 5×10^6 poises, as compared with values of *ca.* 10^6 for 'solid' waxes and *ca.* 10^{10} for pitch). Some of these films may be classed as 'plastic solids', others appear to show the behaviour of a normal solid body with a definite shear modulus and yield-point.

(*d*) *Thermodynamics of surface phases.* If a crystal is placed on a clean-water surface spreading occurs until an equilibrium is reached at a definite surface pressure. Such equilibrium spreading pressures are of fundamental importance, for not only do they give the latent heat of spreading, but also the conditions under which compressed monolayers are thermodynamically stable.

The latent heat of spreading (λ) is found from the two-dimensional Clapeyron equation (see Chapter IV)

$$\lambda = T\left(\frac{\partial \Pi}{\partial T}\right)(A_F - A_S), \qquad (17.5)$$

where A_F and A_S are the respective surface areas occupied by film and by solid crystals, the latter being in general negligible. For accurate determination of energy changes very precise Π–A and Π–T measurements are essential. Values of λ so found are considerable, for example, myristic acid at 15° C. gives 6·5 Cals./gm. mole, a value over half its latent heat of fusion. A very marked increase in entropy is also observed on going from the crystal to the monolayer, arising in part, no doubt, from the greater flexibility of the chains when in the latter state.

The above considerations can readily be extended to a more general study of the thermodynamics of monolayers, such as the energies, entropies, and heats of spreading and expansion, as well as to the question of the order of phase changes. (For a more detailed discussion, see Adam, 1941; Alexander, 1944; Harkins, 1944.)

(*e*) *Kinetics of spreading.* Brief mention may be made of the study of the *rate* at which liquid drops or crystals spread to form a monolayer. The normal method of measurement has been by photographic registration using a ciné camera, the edge of the advancing film being indicated by an inert powder such as talc or lycopodium. An alternative method with the advantage of eliminating the powder altogether, utilizes a series of air electrodes, as used for ordinary surface potential measurements, placed radially and attached to an oscillograph with photographic registration (Alexander and Teorell, unpublished). Both techniques show that the velocity of spreading from oil drops can be very rapid indeed (10–20 cm./sec.) and is of a different order of magnitude from the rate of dissolution from the surface which is frequently extremely slow, even if the substance is quite soluble.

(II) *Applications to other fields*

The applications of the techniques, and results, of insoluble monolayers to specific colloidal systems such as the foams and emulsions are given in Part III, but it is convenient to outline here certain aspects of more general interest.

(*a*) *Structure of complex organic molecules.* Determination of the detailed molecular architecture often becomes a matter of difficulty with complicated molecules such as the proteins, sterols, and carbohydrates. As the monolayer technique has occasionally been of assistance in structure determinations its advantages and limitations may be mentioned. Its principal limitation arises from the necessity of being able to obtain a monolayer of sufficient stability for measurement (i.e. one which neither dissolves nor goes over into a crystal too rapidly); this entails the presence of one or more polar groups in the molecule as well as a hydrocarbon portion of sufficient magnitude to ensure insolubility. (The methods for ensuring quantitative spreading were given earlier.) Whilst it is impossible to generalize, the technique can in suitable cases indicate not only the general shape of the molecule, but also the relative positions of the polar groups if more than one is present. For example, a molecule built up of a complex ring system and having two polar groups in close proximity would tend to give a condensed film of close-packed rings edge-on to the surface, thus allowing a direct comparison between the monolayer area and that calculated from models of likely structures. Should the polar groups be widely separated, however, then the molecule would tend to lie flat on the surface, giving the less closely packed structure of a gaseous or expanded film. Information as to the nature of the polar groups can be obtained from surface potential measurements, since as already pointed out above, the ΔV–n curve is a characteristic property of the polar portion of the molecule.

One great advantage is the minute amount (a few mg.) of substance required for a complete monolayer study, as well as for optical and X-ray measurements of the multilayers if the substance can be induced to form a condensed film.

The work of Adam and his co-workers on sterols appears to be the first application of monolayer methods to structure determination; since then many more complex organic molecules have been investigated (for review see Alexander, 1944). Two recent approaches to the question of the structure of proteins and protein monolayers may be mentioned, one from the study of long-chain compounds containing that fundamental protein linkage, the peptide group, —CO.NH— (Alexander, 1942a), the other from comparison of protein monolayers with those given by synthetic linear polymers (Crisp, 1946). This question of protein monolayers is clearly of great relevance in connexion with stabilization of foams and emulsions by proteins (see p. 635 and p. 655).

(b) *Diffusion through interfaces.* The extent to which monolayers can influence diffusion across a phase boundary is of great biological importance; in addition it has considerable practical interest in connexion with attempts to reduce water evaporation from reservoirs, etc., in arid climates. So far this latter aspect has been the only one thoroughly investigated.

It has been found that only monolayers in the condensed state retard water evaporation appreciably, and with octadecyl and other long-chain alcohols giving condensed monolayers it is possible to get as much as 95 per cent. reduction by using high film pressures. Reductions approaching this have also been obtained with thin oil films (see duplex films, p. 523), but neither method has yet proved feasible for the solution of the practical problem.

(c) *Reactions at interfaces.* For interfacial reactions, such as the hydrolysis of fats and other insoluble esters by aqueous alkali, the study of the bulk reaction is complicated by changing area of interface due to the soap liberated and the varying surface activity of the partial hydrolysis products. By measuring the changes in A or ΔV (at constant surface pressure) of a monolayer of ester spread upon an alkaline substrate the true velocity constant for the hydrolysis can readily be obtained free from these complications. In this way the dependence upon surface pressure, molecular orientation, concentration of alkali, temperature, and so on, has been worked out.

Most interfacial reactions are very sensitive to the molecular orientation, a point of considerable chemical as well as biological interest. The above alkaline hydrolysis of a long-chain ester is a case in point. For example, ethyl palmitate is found to hydrolyse some eight times more slowly in the condensed state ($A = 20$ Å^2) than it does in the expanded ($A > 72$ Å^2), and this is ascribed to the short ethyl chain being compressed beneath the polar group as the film is condensed and so forming a true 'steric' barrier to the OH' ions. With a long-chain acetate, on the other hand, it is clear from Fig. 17.15 that condensation could not give such a protection, and experimentally it is found that the reaction is not retarded when the film is condensed. The validity of these configurations is substantiated by a calculation of their 'surface moments' (μ), using the Thomson-Eucken method of vectorial summation, which are seen from Fig. 17.15 to agree well with those found experimentally (Alexander and Schulman, 1937).

Other reactions sensitive to molecular orientation are the oxidation of unsaturated compounds by dilute permanganate, the lactonization of hydroxy fatty acids on acid substrates, and the photochemical hydrolysis of stearic anilide. (For a recent review of reactions in monolayers, see Alexander, 1944.)

(d) *Biological problems.* The application of surface techniques to biological systems is still in its infancy, but mention may be made here of the general method of approach; a few definite examples appear at various points in Part III.

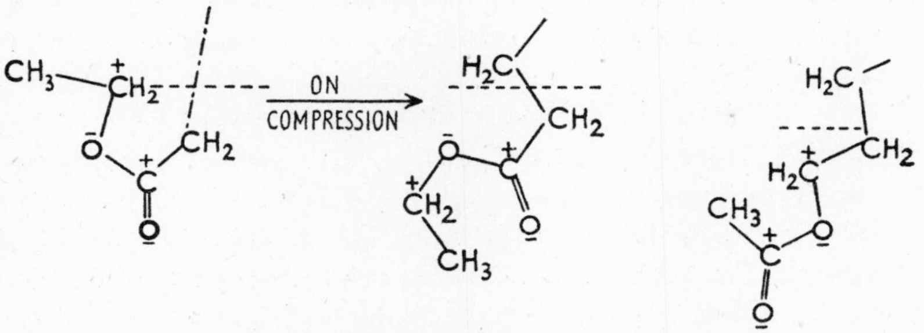

μ_{obs} = 535 milli-Debyes	μ_{obs} = 193 milli-Debyes	μ_{obs} = 502 milli-Debyes
μ_{calc} = 525 ,,	μ_{calc} = 198 ,,	μ_{calc} = 525 ,,
$A > 72$ A^2	$A = 20$ A^2	$A = 23$ A^2
$K = 0.04$ min.$^{-1}$	$K = 0.005$ min.$^{-1}$	$K = 0.15$ min.$^{-1}$
I	II	III

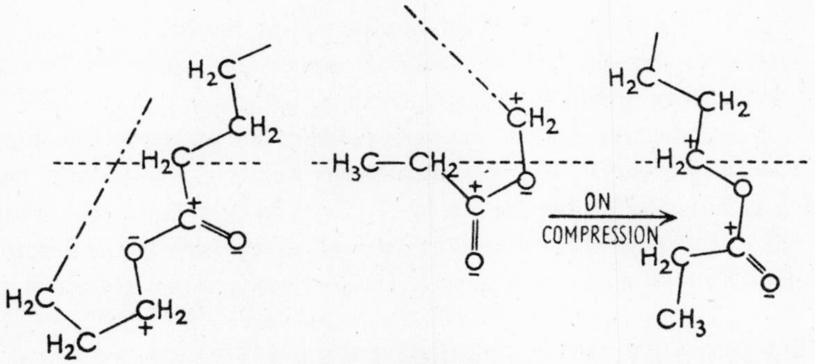

$\mu_{obs} \simeq 0$ milli-Debyes	μ_{obs} = 506 milli-Debyes	μ_{obs} = 334 milli-Debyes
μ_{calc} = 7 ,,	μ_{calc} = 525 ,,	μ_{calc} = 329 ,,
$A = 41$ A^2	$A = 65$ A^2	$A = 21$ A^2
$K \simeq 0.18$ min.$^{-1}$	$K = 0.084$ min.$^{-1}$	$K = 0.021$ min.$^{-1}$
IV	V	VI

Fig. 17.15. Configurations of the ester group and the effect upon the 'apparent surface moment' (μ) and velocity constant (K) of the hydrolysis by NaOH.

One general problem is the interaction between dissolved organic compounds known to have certain biological action (e.g. anaesthetics) and monolayers of different cell constituents such as proteins, phospholipids, waxes, carbohydrates, etc. In addition the effect upon even simpler monolayers can be investigated, such as long-chain amines, acids, amides, alcohols, esters, etc. With some systems the converse method of approach can be used—the substance is spread as a monolayer and cell constituents such as proteins injected beneath it. In both cases any interaction between monolayer and dissolved substance is readily detected by changes in surface pressure, surface potential, and mechanical properties.

In some cases the use of an oil/water interface has advantages over an air/water; in others the techniques of adsorbed films, measuring interfacial tensions and ζ potentials, may provide a better approach.

Oil/Water Monolayers

The requisite conditions for a compound to give an insoluble monolayer at an oil/water interface are clearly much more stringent than at the air/water, since the monolayer must now be insoluble (or only very slowly soluble) in both phases. The types of compounds which have so far been examined as oil/water monolayers are relatively few: phospholipoids (e.g. lecithin, lysolecithin, kephalin), proteins, long-chain detergents (e.g. sodium cetyl sulphate), long-chain amides, and a number of polar synthetic polymers such as nylon.

The study of oil/water monolayers has presented more difficulty than those previously discussed in this chapter, but nevertheless this interface possesses certain inherent advantages—the presence of the oil phase facilitates spreading (e.g. proteins, polymers), and with some types of substances which cannot readily be studied as insoluble air/water monolayers (e.g. paraffin-chain salts) the oil/water monolayer is sufficiently stable for its Π–A curve to be measured. Further, such oil/water monolayers are clearly of more direct application to emulsions as well as to biological systems than are the air/water type.

Surface pressure and surface area measurements

The Π–A curve for an insoluble oil/water monolayer can in principle be determined by the methods given above for air/water systems, for example, the Langmuir-Adam film balance, detachment methods, Wilhelmy plate, etc. Of these experiment has shown the ring method to be far superior to the others, since it eliminates the major difficulties obtained with the film balance, namely, leaks and unstable contact angles at the float (Alexander and Teorell, 1939).

The experimental set-up is shown in Fig. 17.16. The water surface is first cleaned by sucking off around the periphery with a glass capillary attached to the water-pump. (Addition of a few drops of benzene at the centre of the trough assists in sweeping any contamination to the edges.) The purified oil phase (e.g. benzene) is then added to a depth of about 1 cm., and the interfacial tension (γ_0) determined from the maximum pull on a platinum ring. The tip of a micrometer syringe filled with a solution of the film-forming compound is then lowered into the interface by means of a racking device, and a suitable small volume expelled. The interfacial tension (γ) is determined as before, giving the surface pressure Π by difference ($\Pi = \gamma_0 - \gamma$). The area per molecule (A)

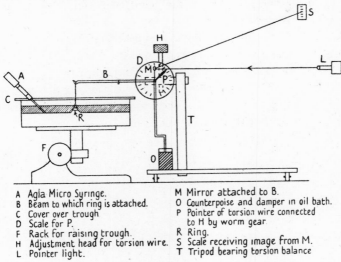

A Agla Micro Syringe.
B Beam to which ring is attached.
C Cover over trough
D Scale for P.
F Rack for raising trough.
H Adjustment head for torsion wire.
L Pointer light.

M Mirror attached to B.
O Counterpoise and damper in oil bath.
P Pointer of torsion wire connected to H by worm gear.
R Ring.
S Scale receiving image from M.
T Tripod bearing torsion balance

FIG. 17.16. Ring balance method for oil/water monolayers.

follows at once from the area of the dish and the volume of spreading solution expelled. The interfacial molecular density is then increased, not by reducing the area as at the air/water interface, but by injecting further volumes of spreading solution. In this way the Π–A curve can be constructed with an accuracy not inferior to that of the usual film balance with air/water monolayers.

The solvent used for spreading purposes is of some importance; it should have a density intermediate between the oil and water, and can with advantage be capillary active. 60 per cent. propyl or isopropyl alcohol has been found suitable for many compounds, and after addition of 0·5 M. sodium acetate will dissolve proteins as well as lipoids.

One further practical point may be mentioned—the use of a stop to limit movement of the torsion arm, so that the ring is never allowed to break through the interface and thus to rupture the monolayer.

Interfacial potentials

An oil/water monolayer might be expected to produce a change in the phase boundary potential analogous to the 'surface potential' discussed above for air/water films. Complications, both theoretical and practical, are to be found, however, as will be appreciated from the discussion of the potential changes produced by *adsorbed* films at the oil/water interface (Chapter XVIII). Nevertheless a little progress has been made in the study of 'interfacial potentials' as they may, from analogy, be termed.

The most promising results so far have been obtained using a thin film (1–2 mm.) of benzene, and a very large polonium electrode (1 cm.^{2}) placed just above the benzene phase, measuring the potentials with a Lindemann or valve electrometer as usual. Preliminary results for lecithin at the benzene/water interface indicate that for a given molecular surface density the ΔV values (and hence the surface moments μ) are close to those given by this compound at the air/water interface. Further work in this field is clearly desirable.

Mechanical properties

The mechanical properties of oil/water films can be measured by methods similar to those given above for air/water systems, although little quantitative work has been done. For viscosity measurement the oscillation method appears to be the best; and it is convenient to replace the disc or plate by a needle suspended in the interface (see Alexander, 1941).

Principal results and applications of oil/water monolayers

There is little to record in this section, the small amount of work so far done upon oil/water monolayers being in marked contrast to their air/water analogues.

The Π–A curves are in general of the vapour-expanded type, a not unexpected result since the oil medium would considerably reduce the van der Waals attractive forces between the paraffin chains. At high values of Π it would seem that the oil is largely eliminated from the interface since the molecules may reach the same packing as at the air/water interface (see Fig. 17.17 (a) which shows this with proteins). In the case of simple compounds the Π–A curves obey the same equation of state

TABLE 17.2

(from Alexander and Teorell, 1939)

	Benzene/water interface			Air/water interface		
	$-\Pi_0$ (dynes/cm.)	A_0 (Å^2)	C	$-\Pi_0$ (dynes/cm.)	A_0 (Å^2)	C
Lysolecithin	0·05	40·4	575	12·8	42·8	686
Lecithin	0·36	58	737	8·7	51·2	728
Sodium cetyl sulphate	0·15	13·9	729	..	..	..

as at the air/water interface, i.e. $(\Pi-\Pi_0)(A-A_0) = C$ (a constant). The results in Table 17.2 show that Π_0 is practically eliminated on going to the oil/water interface, whereas A_0 is unaffected, a behaviour to be expected from Langmuir's theory of expanded films (p. 500).

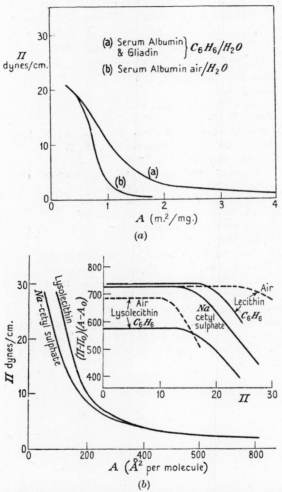

FIG. 17.17. Force-area curves at benzene/water interface. (a) Gliadin and serum albumin. (b) Lecithin, lysolecithin, and sodium cetyl sulphate. Inset: $(\Pi-\Pi_0)(A-A_0)$ v. Π.

With these compounds the constant C is always much greater than kT ($\simeq 400$). This seems to be a characteristic of ionized compounds, and theoretical reasons for it being equal to $\frac{3}{2}kT$ (i.e. $\simeq 600$) have been advanced (Cassie and Palmer, 1941).

Mention may be made of the use of oil/water monolayers in the study of the interaction between kephalin and salts in solution (Alexander, Teorell, and Åborg, 1939), and of synthetic polymers (Crisp, 1946).

GENERAL REFERENCES

ADAM, 1941, *The Physics and Chemistry of Surfaces*, 3rd edn., Oxford.

OTHER REFERENCES

ALEXANDER, 1941, *Trans. Faraday Soc.* **37**, 117.
—— 1942 a, *Proc. Roy. Soc.* A **179**, 470.
—— 1942 b, ibid., p. 486.
—— 1944, *Ann. Rep. Chem. Soc.* **41**, 5.
—— 1947, *Nature*, **159**, 304.
ALEXANDER and SCHULMAN, 1937, *Proc. Roy. Soc.* A **161**, 115.
—— and TEORELL, 1939, *Trans. Faraday Soc.* **35**, 727.
—— —— and Åborg, 1939, ibid., p. 1200.
ANDERSSON, STÄLLBERG-STENHAGEN, and STENHAGEN, 1944, article in *The Svedberg*, Almqvist and Wiksell, Uppsala.
BOYD and HARKINS, 1939, *J. Amer. chem. Soc.* **61**, 1188.
CASSIE and PALMER, 1941, *Trans. Faraday Soc.* **35**, 156.
CRISP, 1946, *J. coll. Science*, **1**, 49 and 161.
DERVICHIAN, 1936, Thesis, Paris.
HARKINS, 1944, article in J. Alexander's *Colloid Chemistry*, vol. v, Reinhold.
MOUQUIN and RIDEAL, 1927, *Proc. Roy. Soc.* A **114**, 690.
ROSENFELD and HOSKINS, 1945, *Rev. Sci. Instrum.* **16**, 343.
SCHULMAN, 1939, *Ann. Rep. Chem. Soc.* **36**, 94.
—— and RIDEAL, 1931, *Proc. Roy. Soc.* A **130**, 259.
STÄLLBERG and TEORELL, 1939, *Trans. Faraday Soc.* **35**, 1413.
STÄLLBERG-STENHAGEN and STENHAGEN, 1945, *Nature*, **156**, 239.
YAMINS and ZISMAN, 1933, *J. chem. Phys.* **1**, 656.

XVIII
ADSORBED FILMS AT THE GAS/LIQUID AND LIQUID/LIQUID INTERFACES

As in the previous chapter we shall restrict ourselves chiefly to the two interfaces which are most relevant here, namely, the air/water and the oil/water. It is convenient in this case to consider both types of interface simultaneously, as they have so much in common, both in the experimental techniques used and in the results obtained. As with insoluble monolayers we have to consider the surface energy and phase boundary potential changes brought about by adsorption, as well as the mechanical properties of the adsorbed film. The relation between bulk concentration and the boundary tension is fundamental in the study of adsorbed films, since from it the molecular surface density can frequently be calculated by means of Gibbs's equation (see below).

It may be pointed out that with both adsorbed and insoluble monolayers we are primarily concerned with the free surface energy (γ) rather than the total surface energy (ϵ_s). These are, however, connected by the relation $\epsilon_s = \gamma - T(d\gamma/dT)$ (see also Chapter IV).

Surface and interfacial tension methods

As the experimental methods for determining boundary tensions for air/water and oil/water systems are so well known it seems adequate here to attempt only an outline from the practical point of view. (For an excellent detailed account of the standard methods, see Adam, 1941.)

Methods are usually subdivided into 'static' and 'dynamic' according to whether the interface is in equilibrium with the bulk or not, and in general it is the former which are more relevant in the study of colloids since it is only to equilibrium measurements that Gibbs's equation is applicable. The principal 'static' methods are as follows: the capillary rise, the sessile drop, the pendent drop, the maximum bubble pressure, the ring and related detachment methods, and the drop-weight or drop-volume methods.

It should be emphasized that, except with the first three, these methods can only be classed as 'static' if the final detachment is made sufficiently slowly to maintain equilibrium between surface and bulk. That this factor may be of great importance is clear from recent work upon the 'surface-ageing' phenomenon (p. 529), which shows that with many paraffin-chain compounds the attainment of true equilibrium may take hours or even days. With simple un-ionized compounds such as the aliphatic acids or alcohols up to a chain-length of six carbon atoms, the times required are small (a few seconds or so), but even with some not particularly large molecules, such as hydrocinnamic acid

$$C_6H_5.CH_2.CH_2COOH,$$

several hours may be necessary. The soaps and similar paraffin-chain salts may require several hours if the concentration is below the micelle point (Chapter XXIV); when micelles are present equilibrium appears to be attained very rapidly. These points are discussed in more detail later.

In theory all methods are as equally applicable to an oil/water interface as they are to the more usual air/water, but from a practical point of view certain troubles may arise, particularly if the method involves a known (usually zero) contact angle. Where such limitations arise some reference will be made to them.

(a) *Static methods*

Capillary rise method. The capillary rise method is simple in theory and, provided that zero contact angle can be ensured and that accurately cylindrical, uniform capillaries are available (these are now made commercially†), is often convenient in practice. Measurement involves no disturbance of the surface so that slow time-effects can readily be followed.

For accurate work allowance must be made for the meniscus effect, but for most purposes (using a narrow capillary and zero contact angle) the equation $\gamma = rh\rho g/2$ suffices (h is the height of liquid in the capillary of radius r, ρ the density difference between the two phases, and g the acceleration due to gravity). If the contact angle (θ) is not zero (see Fig. 18.1), then the equation becomes

$$\gamma = \frac{rh\rho g}{2\cos\theta}. \tag{18.1}$$

In practice the method is rarely used unless θ is zero, owing to variability, and difficulty of measurement, of contact angles.

Zero contact angle at an air/water interface is usually obtained without difficulty with well-cleaned glass capillaries and most aqueous solutions, although with soap solutions, particularly if the long-chain ion carries a positive charge (cationic detergents), the surface tends to become hydrophobic and this may vitiate the method. The method is seldom used for interfacial tension work owing to the difficulty of maintaining a zero contact angle at an oil/water interface.

† 'Veridia' tubing, made by Messrs. Chance Bros., Ltd

Sessile drop method. Measurement of the dimensions of sessile drops or bubbles gives the boundary tension without the necessity of knowing the contact angle, so that the method can be used for both types of interface as well as with detergent solutions. The difficulty of ensuring

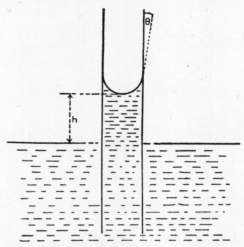

FIG. 18.1. Rise of liquid in a capillary tube.

a clean surface has been one of its limitations. The relevant dimensions are the diameter of the drop and the height between the equatorial diameter and the apex; these are measured using a travelling microscope.

Pendent drop method. The pendent drop method is useful for both surface and interfacial tensions. In it the dimensions of the drop are obtained photographically or by projecting an image upon graph-paper or a ground-glass screen. The boundary tension can be obtained in several ways, of which two have been recommended for rapid but accurate work (Hauser, 1942). In the first method (see Fig. 18.2) the point of inflexion is located and its distance (x) from the axis and the radius of curvature (r) at this point measured. The surface tension is then given by the expression

$$\gamma = \frac{Vr\rho g}{\pi x^2}, \qquad (18.2)$$

where V is the volume of the drop and ρ the density difference between the two phases. The second method is less direct, and uses the equation

$$\gamma = \frac{\rho d_e^2 g}{H}, \qquad (18.3)$$

where d_e is the equatorial diameter of the drop and H a function which can be calculated. Using photographic recording this technique is useful for rapid measurements, as well as for following slow changes.

Maximum bubble pressure method. Sugden's two-tube modification of the maximum bubble pressure method (see Fig. 18.3) eliminates the

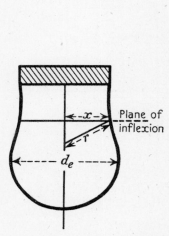

Fig. 18.2. Pendent drop.

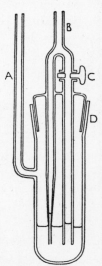

Fig. 18.3. Sugden's two-tube modification of the maximum bubble pressure method.

corrections which are otherwise necessary, and is convenient for rapid and accurate surface tension determinations. Within an accuracy of about 0·2 per cent. the surface tension is given by the simple relation

$$\gamma = k(h_N - h_W), \qquad (18.4)$$

where h_N and h_W are the differences in the manometer levels when bubbles are blown from the narrow and wide capillaries respectively, and the constant k is found by calibrating with pure water. This method has recently been extended by Hutchinson (1943) to liquid/liquid interfaces.

Ring and other detachment methods. Determination of the force (F) required to detach a ring, a thin vertical plate, a horizontal flat wire or other frame from an interface, gives the interfacial tension in absolute units only if suitable correcting factors are applied. With a ring, which is usually most convenient, the surface tension is obtained from the equation

$$\gamma = \frac{\beta F}{4\pi r}, \qquad (18.5)$$

where r is the radius of the ring and β a correcting factor which depends upon the dimensions of the ring and the volume of liquid raised by it (see Harkins and Jordan, 1930). A graph of γ against F is prepared and unknown values read off graphically.

A simple torsion balance with a platinum or platinum/iridium ring, similar to that previously described for oil/water monolayers (Fig. 17.16) is very suitable and is capable of an accuracy of ca. 0·2 per cent. With an oil/water interface zero contact angle is readily obtained owing to the movement of the ring relative to the interface. By means of a stop for the torsion arm it is a simple matter to measure the maximum pull on the ring without detaching the ring from the interface, so that slow changes can be followed. Complete wetting of the ring presents no difficulties except with cationic detergents, which may tend to render it hydrophobic.

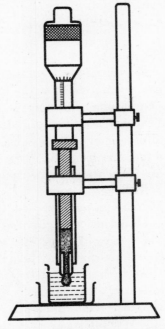

FIG. 18.4. Micrometer syringe method for surface or interfacial tensions.

Drop-weight and drop-volume methods. Measurement of the weight or volume of a drop slowly detached from a suitable tip is probably the most widely used method for both surface and interfacial tensions, but this also needs correction factors to give quantitative results. The equation for quantitative work is

$$\gamma = \frac{mg}{r}\beta, \qquad (18.6)$$

where m is the apparent mass of the drop, r the radius of the dropping tip, and β the correction factor which also depends upon the dimensions of drop and tip (see Harkins and Brown, 1919). For small quantities of solution (about 0·2 c.c.) the Agla micrometer syringe is rapid and usually sufficiently accurate, both for surface and for interfacial tensions (see Fig. 18.4). A stainless steel tip of about 1 mm. diameter, tapered as shown to ensure detachment from the same periphery, has given excellent results in this laboratory. For low interfacial tensions some difficulty may be encountered owing to the aqueous phase spreading up the sides of the tip. This can be prevented by rubbing the outside of the tip with a little ferric stearate.

Wilhelmy plate. The last static method which might be included here, namely, the Wilhelmy plate, has been discussed already in connexion with its use for insoluble monolayers. It is rarely used for surface or interfacial tension work.

(b) *Dynamic methods*

The surface of a solution at the moment of its creation will have the same composition as the bulk phase, but this will seldom correspond to equilibrium conditions, so that solute will move into or out of the interface until the system is once more in equilibrium. With a dilute solution of a capillary-active substance the surface tension will therefore show a fall, and from the variation with time the rate of movement into the interface of the solute molecules can be found.

Although for most colloidal systems equilibrium tension values are of more importance, yet these dynamic values do have quite definite applications, as in the study of the preparation of foams and emulsions, the wetting of fibres, etc., where the rapidity of movement of molecules to the interface may be a most important factor.

So far the study of dynamic tensions has been confined almost entirely to an air/water interface owing to experimental difficulties with other systems. Some preliminary results with the oscillating jet method at oil/water interfaces have, however, recently been reported by Addison (1945).

Jet methods. The two principal dynamic methods, which enable the tension of a surface to be measured after an extremely short time (ca. 0·01 sec.) from its moment of creation, are those of the oscillating and impinging jets. The former, an old method recently developed by Addison (1943, 1945), measures the wavelength of the oscillations imposed on a stream of liquid by emergence through an elliptical nozzle, whereas the impinging jet method, due to Bond (1935), measures the diameter of the coherent sheet of liquid formed by two jets meeting head-on. As might be expected, the theory in both methods is rather complicated, and for details the original papers should be consulted.

Air electrode method. Another dynamic method may be mentioned, namely, a modification of that used to study rapid surface spreading of insoluble oils (Alexander and Teorell, unpublished). In this a number of radio-active metal probes (see the measurement of 'surface potential' in the previous chapter) are placed at suitable intervals close to the surface of a jet issuing from a circular nozzle, and the potential difference between each probe and the reservoir measured by means of a suitable electrometer or oscillograph. From static calibration measurements these potential differences can be translated directly into adsorptions, as well as into dynamic surface tensions. Preliminary results are very promising, and indicate that the dynamic surface tension can be measured for any period down to *ca.* 0·001 sec.

The dynamic methods above only give the rate of *adsorption*, but the question of rates of *desorption* from surfaces is also of importance. For example, in the breaking of a foam or emulsion the surface area is continually diminishing and hence the interface is always supersaturated with respect to the bulk phase. Removal of the large molecules normally employed to stabilize such systems appears to be a relatively hindered process, and various methods have been used to study the rate of desorption. Of these we might mention the contracting of a pendent drop, the use of a rapidly spreading oil such as oleic acid to compress the adsorbed film, and the Wilhelmy plate method using a trough with a movable barrier (see Fig. 17.3). Of these the last is probably the most suitable for quantitative measurements.

Surface area determination from Gibbs's adsorption isotherm

Having measured, by any suitable method, the boundary tension as a function of bulk concentration, we have then to apply the Gibbs equation in order to calculate the surface excess and hence the area per adsorbed molecule (A). For a two-component system of solute i and solvent s this equation is usually expressed in the form

$$-d\gamma = \Gamma_i \, d\mu_i + \Gamma_s \, d\mu_s, \qquad (18.7)$$

where γ denotes boundary tension, Γ the 'surface excess', and μ the chemical potential (see Chapter IV).

Placing the geometrical surface so that the surface excess of solvent Γ_s is zero (the physical significance of this is discussed below), then

$$-d\gamma = \Gamma_i^{(s)} \, d\mu_i \qquad (18.8)$$
$$= \Gamma_i^{(s)} RT \, d \ln N_i f_i, \qquad (18.9)$$

since $\mu_i = \mu_i^0 + RT \ln N_i f_i$.

In colloidal systems we are frequently concerned with relatively dilute solutions of capillary-active materials, in which case eqn. (18.9) reduces to

$$\Gamma_i^{(s)} = -\frac{1}{RT} \frac{d\gamma}{d \ln c_i f_i}, \qquad (18.10)$$

where c_i and f_i are the concentration and activity coefficients of the solute.

Hence by plotting γ against the logarithm of the bulk activity we can determine the *surface excess* $\Gamma_i^{(s)}$. (If R is in ergs/gm. mole/degree (i.e. $8 \cdot 3 \times 10^7$) then $\Gamma_i^{(s)}$ will be in gm. moles/cm.^{2}). For dilute solutions of capillary-active substances the surface concentration is very much greater than the bulk so that we can identify $\Gamma_i^{(s)}$ as the amount of solute per cm.^{2} in the adsorbed monolayer. (The question whether adsorbed films are invariably monomolecular is discussed later.) Since $\Gamma_i^{(s)} N A \times 10^{-16} = 1$, the area per molecule (A, in A^2/molecule) of the adsorbed monolayer, and hence the Π–A curve, can be found. From

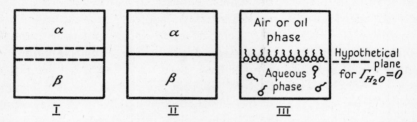

FIG. 18.5. Mathematical and physical interpretation of the 'surface excess' Γ in the Gibbs adsorption isotherm. I. Actual system with finite transition layer. II. Idealized system in which phases touch without transition layer. III. Molecular picture of interface in the case of a very capillary-active solute.

$\Gamma_i^{(s)}$ and an assumed thickness of the monolayer (say 5 A), the true surface concentration in gm. moles/cm.^{3} can readily be calculated.

For solutions of un-ionized capillary-active substances it is often sufficiently accurate to replace activities by concentrations, so that eqn. (18.10) becomes

$$\Gamma_i^{(s)} = -\frac{1}{2 \cdot 303 RT} \frac{d\gamma}{d \log_{10} c_i}. \tag{18.11}$$

With ionized molecules, on the other hand, it is clear from Debye-Hückel theory that the range of concentrations for which eqn. (18.11) is even approximately valid will be very much lower. This is one of the complications which arise in applying the Gibbs equation to solutions of colloidal electrolytes (see Chapter XXIV).

The physical meaning of the term 'surface excess', for these capillary-active substances in which we are interested, may be clarified by considering Fig. 18.5, which depicts the two systems I and II considered in the thermodynamic proof as well as the molecular structure of the interface (in III). (The solute is considered as soluble only in the aqueous (β) phase.) Gibbs compares the actual system (I) with the hypothetical one (II) in which the phases touch without a transitional layer, and defines the 'surface excess of component i', i.e. Γ_i, by the relation

Γ_i = amount of i in I − amount of i in II.

Consider an extreme case of capillary activity, in which the adsorbed monolayer is in a closely packed state with no water molecules present in it, and in which the bulk concentration is negligible in comparison with the true surface concentration. Then the hypothetical plane for $\Gamma_{H_2O} = 0$ clearly corresponds very closely with that drawn tangentially just at the bottom of the polar groups (see Fig. 18.5 III). Hence $\Gamma_i^{(H_2O)}$, the 'surface excess of solute i when the hypothetical plane is placed such that $\Gamma_{H_2O} = 0$', corresponds closely with the amount adsorbed in the monolayer.

In the case of concentrated solutions it is clear that the simple picture above will no longer hold. A number of suggestions for the interpretation of the 'surface excess' in such systems have been put forward, but space precludes their discussion here (see, for example, Adam, 1941).

Phase boundary potentials

In the case of films adsorbed from solutions we have to consider the zeta potential (ζ) as well as the 'surface or interfacial' potential (ΔV). Most of the work upon zeta potentials has been carried out with oil droplets rather than with air bubbles, owing to the inherent experimental difficulties with the latter. The experimental methods have been given earlier (Chapter XII) and some applications of zeta potential measurements are given in Chapters XX and XXIV dealing with suspensions and colloidal electrolytes.

The 'surface' or 'interfacial' potential (ΔV) is defined as usual by the relation $\Delta V = V_{\text{water}} - V_{\text{film}}$, i.e. the difference in phase boundary potentials between clean and film-covered interfaces. It is convenient to discuss the two cases (i.e. air/water and oil/water) separately as different techniques are involved.

Although few data are available, it is known that the surface potential of an adsorbed film can be readily measured by the methods given above for insoluble monolayers. It is also clear that values of the 'surface moments' (μ) obtained from the relation $\Delta V = 4\pi n\mu$ (where $n = 1/A$) agree closely with those from insoluble monolayers of compounds with the same polar group. (The values of n or A are obtained as we have just shown from surface tension measurements and use of the Gibbs equation.)

Turning now to the analogous quantity for an oil/water interface it should be possible to measure this by the same technique as described in the previous chapter for oil/water insoluble monolayers, but this does not so far appear to have been tried. The usual method has been to interpose an oil layer between two aqueous phases, one of which contains the capillary-active substance under examination. A simple type of apparatus is shown in Fig. 18.6. With oils of not too high resistance

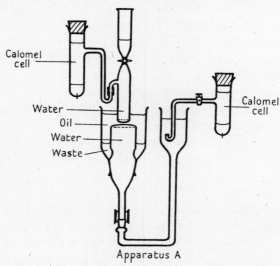

FIG. 18.6. 'Hanging-drop' apparatus for the study of oil/water potentials. (Dean.)

(e.g. octyl alcohol, amyl acetate) the method works reasonably well, but it breaks down with hydrocarbon liquids of higher resistance, such as benzene. As above, the ΔV–n relation can be obtained from surface tension measurements on the same system.

Mechanical properties of adsorbed films

As would be expected, the mechanical properties (viscosity and elasticity) of an adsorbed film are more difficult to measure than for insoluble monolayers, owing to the relatively attenuated state of the film. It seems unlikely, from the study of equilibria in condensed insoluble monolayers, that adsorbed films in true equilibrium can ever exist in the condensed state, so that signs of a solid or even plastic solid behaviour are not to be expected; we have only therefore to consider the measurement of surface viscosity. It may be pointed out that there is definite evidence for elasticity in adsorbed protein films (e.g. the apparent γ as measured by the bubble method depends on the size of the bubble), but such systems are not in thermodynamic equilibrium (see also Chapter XVII). The question of elasticity in adsorbed soap films is discussed in Chapter XXII.

Duplex films

Before considering the results of the study of adsorbed films mention should be made of 'duplex' films, the name given to those thin oil films which are formed initially when a non-spreading oil (e.g. medicinal paraffin) containing a 'spreader' such as oleic acid is placed on a water surface of limited area. The vividly coloured films formed from droplets of motor-oil on puddles and wet surfaces is a common example, the spreader in this case being either polar compounds formed by oxidation during use, or 'dopes' deliberately added to the oil. Interference colours are often observed with duplex films since the thickness is frequently only a few μ. The application by Langmuir of these thin films (they are still very many *molecules* thick!) to the study of adsorbed films at the oil/water interface and hence to the structure of liquid-expanded monolayers has been mentioned earlier, but their principal interest lies in other directions. One point, of great practical as well as theoretical interest, concerns the thermodynamic stability of these duplex films, since they are used in combating mosquito larvae in malarial regions, and have shown promise in the problem of hindering water evaporation (see also p. 505).

With simple long-chain compounds as spreaders, e.g. fatty acids, alcohols, esters, the original duplex film breaks up after a time into a number of lenses surrounded by a monolayer of the spreader. However, duplex films which appear stable have been obtained with some polymerized spreaders (e.g. oxidized oleic acid), and with certain dyestuffs (e.g. malachite green), and it is significant here that the interfacial films are rigid, suggesting that these systems are not in true thermodynamic equilibrium.

The *initial* spreading coefficient (Π_s) is defined as the decrease in surface energy when the film spreads over 1 cm.^{2} of water surface, i.e. $\Pi_s = \gamma_w - (\gamma_o + \gamma_{ow})$. As γ_w is high (*ca.* 72 ergs/cm.^{2}), and exceeds the sum of γ_o and γ_{ow} which usually have values of the order of 20 and 30 dynes respectively, initial spreading occurs. The *final* spreading coefficient (Π'_s) is given by $\Pi'_s = \gamma'_w - (\gamma_o + \gamma_{ow})$, where γ'_w is the surface tension of a film-covered surface and is therefore much lower than γ_w. (This assumes that γ_o and γ_{ow} are the same in both equations which is very often the case.) The condition that a duplex film shall remain stable, uniformly covering the surface area available to it, and not break up into lenses and monolayer, is that Π'_s shall be positive. Experiment, however, shows it to be negative in general since the duplex film does in point of fact go over into lenses and monolayer in most cases.

According to Antonoff's rule Π'_s should always be zero, i.e.
$$\gamma'_w = \gamma_o + \gamma_{ow},$$
and if this were the case then the upper and lower surfaces of the air/water monolayer would be identical energetically with macroscopic oil and oil/water interfaces respectively (see Fig. 18.7). With liquid-expanded monolayers this is not seriously in error, but nevertheless Π'_s

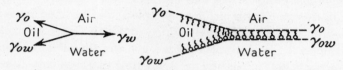

Fig. 18.7. Diagram to illustrate the similarity in surface energy of the 'duplex film and the monolayer of 'spreader'.

does seem to be negative in general. The negative values of Π'_s have been ascribed to the adhesion between spreader molecules being less at the interface than at the water surface, presumably owing to the action of the oil molecules in the former case.

Thermodynamics of adsorption

The standard free energy of adsorption (ΔG^0) can be found from the equilibrium between surface and solution. Provided that both regions are sufficiently dilute for concentration to replace activity then

$$\Delta G^0 = -RT \ln \frac{C_{\text{surface}}}{C_{\text{soln}}}, \quad \text{and} \quad C_{\text{surface}} = \Gamma/\tau,$$

where τ is the thickness of this adsorbed film. Taking a reasonable value for this (usually 5 A), and knowing Γ from the Gibbs equation, then ΔG^0 can be calculated. ΔG^0 will be negative for positive adsorption, since C_{surface} is then $> C_{\text{soln}}$, showing that adsorption is accompanied by a decrease in free energy of the system.

The heat of adsorption (ΔH^0) can be found by measuring the free energy of adsorption at various temperatures, since

$$\Delta H^0 = \Delta G^0 - T \frac{\partial (\Delta G^0)}{\partial T}.$$

Finally from ΔG^0 and ΔH^0 the entropy of adsorption (ΔS^0) is obtained, since $\Delta H^0 = \Delta G^0 + T \Delta S^0$ (see also Chapter IV).

Principal results and applications of adsorbed films

(a) *Experimental tests of Gibbs's adsorption isotherm*

As shown above the $\Pi-A$ relations for adsorbed films have to be obtained indirectly from the observed γ-concentration measurements by means of the Gibbs equation. This equation is derived by rigid proof from thermodynamics and hence is subject to the same basic premisses as all thermodynamic formulae. It has therefore the great advantage of being independent of any particular molecular hypothesis of the surface or bulk phases but, on the other hand, can refer only to conditions in which surface and bulk are in *true* equilibrium. Accordingly, great importance has been attached to its experimental confirmation, as evidenced by the number of investigations, beginning with the work of Donnan and his collaborators over thirty years ago, with this as object.

The technique used in the early work was to blow a stream of fine bubbles up a long column filled with a solution of capillary-active substance (e.g. nonylic acid). The bubbles passed out into a collecting vessel in which they were allowed to collapse, the resulting solution being analysed. From the amount of solute carried over by a counted number of drops of known size the surface excess of solute Γ_i could be found and compared with that calculated from surface tension measurements using Gibbs's equation. The observed and calculated values were found to be in moderate agreement with nonylic acid but differed widely with saponin. Later workers often used oil bubbles instead of air, and various other substances, particularly the dyestuffs, on account of their ease of estimation. Varied experimental results were obtained, some in good agreement with calculated values, others in complete disagreement.

Such varied behaviour gave rise to considerable doubts concerning the validity of the original Gibbs equation and various modifications and correcting terms were put forward. However, with our present knowledge of surface behaviour we now know that in the apparently anomalous systems there are good grounds for suspicion. Thus saponin and many dyestuffs give rigid adsorbed films which are certainly never in reversible equilibrium and are thus automatically ruled out; in other cases slow attainment of surface equilibrium may mean that the bubble and surface tension results refer to surfaces of different age and hence to different adsorptions (see below).

The problem has been attacked recently in a very direct and ingenious fashion by McBain and his co-workers. They used a long rectangular trough full of solution and skimmed off the surface layer by a carriage carrying a microtome blade shot along at high speed, on the lines of a non-stop train taking in water. The difference in concentration between the 'skim' and the bulk was then measured by an interferometer. Some idea of the agreement between observed and calculated values will be seen from those obtained with hydrocinnamic and lauryl sulphonic acids and quoted below in Table 18.1. In addition to other cases showing positive adsorption there was also fair agreement in the case of negative adsorption using a simple salt. Other less direct methods, such as the rapid compression of a surface into the path of an interferometer, give further support to the validity of Gibbs's equation.

A final confirmatory point is that the force-area curves of insoluble monolayers as determined directly by the film-balance merge without a break into those of adsorbed films determined from surface tension measurements and the Gibbs equation. In a few cases (e.g. lauric acid) the same substance can be examined by both methods and the results agree closely.

It seems reasonable to conclude therefore that the Gibbs adsorption isotherm, in the form given above, is correct, and where deviations from it appear to occur (e.g. colloidal electrolytes) these should be ascribed to other factors.

(b) *The molecular structure of the surfaces of solutions*

The reader has probably noticed that in our discussion of adsorbed films the words 'monolayer' and 'film' have been used indiscriminately. Langmuir suggested some thirty years ago that the transition layer between the gaseous and the liquid phases of a solution was monomolecular in thickness, but this view has been adversely criticized, particularly for those systems in which the Gibbs equation gave results in complete disagreement with experiment.

We have seen that, broadly speaking, there are three ways in which the surface adsorption (or excess) can be obtained:

(1) by assuming that the transition layer is monomolecular and obeys an equation of state of the form $\Pi(A-A_0) = xkT$, where A_0 and x can be assigned likely values from experiments upon insoluble monolayers (see also section (c) below). At any given Π it is clear that the area per molecule (A), and hence the surface adsorption, can be found;

(2) by measuring surface tensions and applying Gibbs's equation;

(3) by direct experimental measurement as with the microtome method.

If the transition layer were not monomolecular we should expect that (1) would give values in disagreement with the others; in point of fact they agree closely with the two solutes most accurately studied, namely, hydrocinnamic and lauryl sulphonic acids (see Table 18.1).

TABLE 18.1. *Calculated and Observed Adsorption for Two Solutes*

	Hydrocinnamic acid	Lauryl sulphonic acid
Method (1), from monolayer and assumed equation of state	8.3×10^{-8} gm./cm.^{2}	5.7×10^{-10} gm. mole/cm.^{2}
Method (2), from surface tension and Gibbs's equation	8.8×10^{-8} ,,	..†
Method (3), experimentally from microtome method	7.7×10^{-8} ,,	5.2–5.7×10^{-10} ,,

† The application of the Gibbs equation to solutions of colloidal electrolytes such as lauryl sulphonic acid is discussed in Chapter XXIV.

The conclusion that with capillary-active substances the adsorbed film is monomolecular seems therefore verified, *provided, of course, that the surface and bulk phases are in true reversible equilibrium.* This last proviso is required since, as pointed out previously, solutions of saponin, proteins, and some dyestuffs can give adsorbed films many molecules in thickness, but it is quite certain that in these systems the surface and bulk phases are not in reversible equilibrium.

It was shown by Langmuir from a consideration of Traube's rule that in the *dilute* gaseous films the long-chain molecules lie flat on the surface. (Traube's rule states that in an homologous series the concentration required to produce an equal lowering of surface tension decreases threefold for every CH_2 group added to the chain.) Thus if ΔG_n^0 and ΔG_{n-1}^0 are the standard free energies of adsorption for members with n and $n-1$ carbon atoms, then

$$\Delta G_n^0 - \Delta G_{n-1}^0 = -RT \ln \frac{(\Gamma/\tau C)_n}{(\Gamma/\tau C)_{n-1}} = -RT \ln 3 = -640 \text{ cals./gm. mole}$$

at room temperature. Thus the free energy of adsorption changes by a constant amount for each CH_2 added to the chain, which must mean that all CH_2 groups are similarly situated with respect to the surface. This suggests that the chains are parallel to the surface (cf. Ward, 1946).

(c) *Equation of state and phases of adsorbed monolayers*

The force-area (Π–A) relation for the adsorbed monolayer can be found as described earlier by the use of Gibbs's equation and identifying the surface excess with an adsorbed monolayer. The results are frequently plotted in the form ΠA against Π as with insoluble monolayers.

A number of typical results are shown in Fig. 18.8 and they show very clearly the close parallel between adsorbed and insoluble monolayers. At low pressures the ideal two-dimensional gas equation $\Pi A = kT$ is approached; at higher pressures an equation of the Amagat type $\Pi(A-A_0) = xkT$ is followed. Thus, as with some insoluble monolayers, these adsorbed films behave as imperfect two-dimensional gases. Values of A_0 and x are set out in Table 18.2 for a series of fatty acids of varying chain-length.

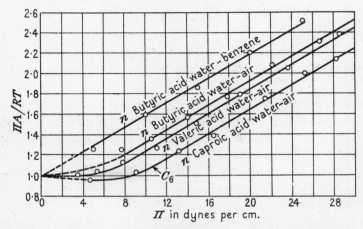

FIG. 18.8. ΠA–Π curves for soluble fatty acids at the air/water and benzene/water interfaces.

It will be seen that A_0, termed the 'co-area' from analogy with the 'co-volume' term of van der Waals' equation, is independent of chain-length, whereas x decreases as the chain-length increases, so that the deviation of x from unity is some measure of the attractive forces between the long chains.

TABLE 18.2

Compound	Interface	A_0 (Å^2)	x
n-butyric acid	air/water	24·3	0·73
n-valeric acid	,,	24·3	0·63
n-caproic acid	,,	24·3	0·43
iso-butyric acid	,,	25·1	0·78
iso-valeric acid	,,	25·1	0·68
iso-caproic acid	,,	25·1	0·48
iso-amyl alcohol	,,	22·4	0·59
n-butyric acid	benzene/water	24–25	1·00–0·95

It is interesting to see from Fig. 18.8 that at an oil/water interface the equation of state becomes $\Pi(A-A_0) = kT$. This might be expected since the oil phase eliminates the attractions between the hydrocarbon chains which, as pointed out above, are responsible for the deviations of x from unity (see also Table 18.2).

We conclude, therefore, that all adsorbed monolayers in true equilibrium are in the imperfect gaseous state, and although the area per molecule may be as little as 30 A^2 when the surface becomes saturated, yet no evidence for transition into a condensed state has yet been forthcoming.

(d) The phenomenon of 'surface ageing'

Mention has been made at various points of the unexpectedly slow rate at which the surfaces of many solutions come into equilibrium with the bulk phase—the 'surface ageing' phenomenon as it is usually termed. The ratio of the observed rate to that calculated from simple diffusion theory may be as low as 10^{-6} with soaps and detergents at concentrations below the micelle point (see Chapter XXIV); above this point equilibrium is rapidly attained. Dyestuffs show the phenomenon very well, but it is not restricted to ionized compounds since the paraffin-chain alcohols and acids behave similarly. For a given polar group the discrepancy appears to increase with the chain-length.

Although several theories have been put forward to explain the phenomenon, none of them appears to be really satisfactory. The first, due to Doss (1938, 1939) and later developed by several others, suggested that diffusion of long-chain ions to the interface was hindered by the electrostatic barrier arising from the long-chain ions already adsorbed. The calculated diffusion rate has therefore to be multiplied by a factor $e^{-Q\zeta/kT}$, where Q is the charge on the long-chain ion, and ζ the electrostatic barrier due to the double layer. (Hartley and Roe (1940) have pointed out that ζ is the ordinary zeta potential and hence is capable of direct measurement.) It has been found from monolayer studies (Alexander and Rideal, 1937) of the hydrolysis by OH' ions of long-chain esters that the soap ions produced by the reaction cause a progressive diminution in reaction velocity, so that in this case an electrostatic barrier undoubtedly exists. It is obvious, however, that it cannot be the sole, or even the major, factor since surface ageing is shown by so many un-ionized compounds.

McBain has ascribed the phenomenon to the formation of a 'surface pellicle' arising from association of the polar groups, but this cannot be correct since, as shown above, it seems quite definite that the adsorbed film is monomolecular in thickness.

Some light has recently been thrown on the problem by examining interfacial as well as surface films, and a few typical curves are shown in Fig. 18.9 (Alexander, 1941). When simple surface-active substances such as ethyl acetate are added the time factor was reduced, and by covering the surface with an insoluble liquid-expanded monolayer such

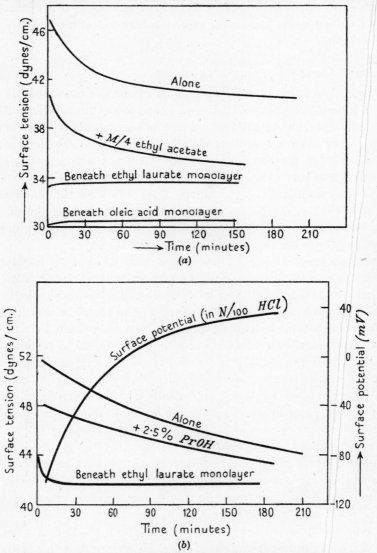

Fig. 18.9. The effect of capillary-active substances and insoluble monolayers upon the surface-ageing phenomenon. (a) Sodium dodecyl sulphate (2×10^{-3} M.) in $N/100$ HCl. (b) Hydrocinnamic acid ($2 \cdot 67 \times 10^{-2}$ M.) in $N/100$ HCl.

as oleic acid or ethyl laurate it could be eliminated entirely. With the detergent solution referred to in Fig. 18.9 the rate of adsorption appeared to show no anomaly at a benzene/water interface.

These results are clearly inexplicable upon either of the above theories alone and they suggest that the hindrance is associated with the hydrophobic portion of the molecule. It is perhaps not unexpected that a very long chain molecule should experience difficulty in penetrating the relatively close-packed surface layer. (For hydrocinnamic acid and sodium dodecyl sulphate under the conditions given, the final areas per molecule are about 30 A^2.) It must be remembered that the paraffin chains assume a coiled-up shape when forced to dissolve in water, and they must pass into a much more extended configuration in the monolayer.

Mention may be made briefly of another type of surface-ageing phenomenon, in this case from an oil phase to an oil/water interface. Many oil-soluble amphipathic compounds (i.e. those containing polar and non-polar portions) show the effect, an excellent example being, say, 1 per cent. palmitic acid in benzene or hexane. Such a solution in contact with water takes several hours to attain equilibrium, instead of the few seconds, at most, expected from diffusion theory. In such cases the rate appears to be determined by the concentration of monomeric long-chain molecules, since both rate and monomer concentration increase in the order hexane < benzene < nitro-benzene (Alexander and Rideal, 1945). That the polar groups in the fatty acid dimer should, on collision with the interface, not come sufficiently near the water molecules to bring about dissociation, is not so surprising in view of the well-known structure of the dimer.

(e) *Biological applications*

There seems little doubt that one of the principal factors associated with the biological activity of many amphipathic substances is the tendency of such compounds to be adsorbed at an interface. The degree to which there is an exact correlation between biological activity and the adsorption depends, of course, upon the particular system studied and no hard and fast rules can be given. Interfacial activity is most pronounced in the soaps and colloidal electrolytes, and it is in such systems that the correlation between biological and interfacial activities is most marked. A few examples will be given in later chapters.

Electrocapillary phenomena

Observations of the effect of electrification upon the surface tension of mercury go back almost one and a half centuries, but it was Lippmann in 1875 who first carried out any detailed investigations of the phenomenon, later studied by many others, particularly Gouy, Frumkin, Butler, and their co-workers.

A simple form of apparatus for observing these changes is shown in Fig. 18.10. The mercury is attached to the negative end of a potentiometer, and meets the aqueous solution in the fine capillary tube B. By means of the reservoir the height of the mercury in A is varied so

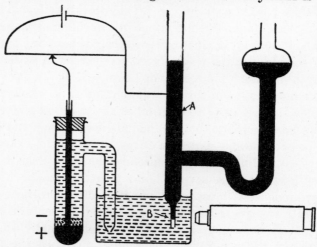

Fig. 18.10. Simple apparatus for the study of electrocapillary phenomena.

as to keep the meniscus in the capillary tube stationary, as indicated by the microscope. Changes in interfacial tension are thus reflected in, and measured by, the height of mercury in A. The aqueous solution connects through a non-polarizable electrode (e.g. a calomel half-cell or a large pool of mercury on the bottom of the vessel) to the positive end of the potentiometer. The capillary electrode is almost completely polarizable, so that any changes due to the applied potential occur solely at this interface.

The relation between the applied (negative) potential and the interfacial tension is termed the 'electrocapillary curve': some typical results are shown in Fig. 18.11. It will be seen that as the mercury is made more negative so the surface tension rises, reaching a maximum at around -1 volt for many aqueous solutions, and then falls off almost symmetrically. (Some curves approximate to a parabolic form.)

It can easily be seen that the presence of a charge at an interface tends to lower the boundary tension. Considering for simplicity a simple double layer as depicted in Fig. 18.12 (a), the electrostatic forces are clearly repulsive and thus tend to expand the interface, i.e. they oppose the ordinary surface tension forces. (For a calculation of their magnitude see Langmuir, 1932.)

The above results show that the mercury surface in absence of an applied e.m.f. must be positively charged. In the case of dilute sulphuric acid as electrolyte the interface may be as depicted in Fig. 18.12 (b).

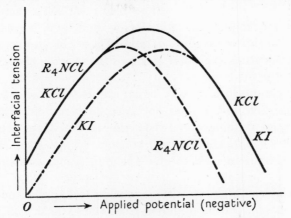

Fig. 18.11. Some typical electrocapillary curves.

Increasing the negative potential on A will decrease and finally neutralize the charge on the mercury surface, thus liberating SO_4'' ions into the solution and so increasing the interfacial tension. Beyond this

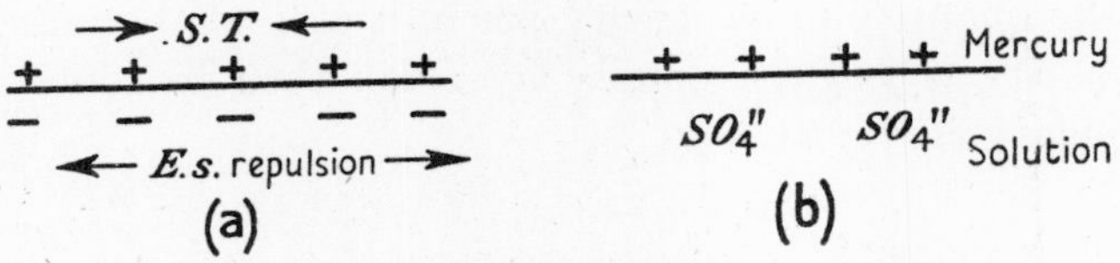

```
      → S.T. ←
  +  +  +  +  +              +   +   +   + Mercury
  −  −  −  −  −              SO₄"    SO₄"
  ← E.s. repulsion →                        Solution
       (a)                        (b)
```

Fig. 18.12. Ionic adsorption at interfaces.

point the mercury surface will become increasingly negatively charged, leading to attraction of cations (chiefly H_3O^+ in the above example) and hence to a fall in tension.

Adsorption of ions at the mercury surface is found to follow the lyotropic series (p. 16), for example, $Cl' < Br' < CNS' < I'$ is the order of increasing adsorption in the case of anions. Ions such as CNS' and I' are known to be much more capillary-active than Cl'; at equivalent concentrations they accordingly lower the surface tension more, and to get adsorption of cations (e.g. K^+ in Fig. 18.11) we have to go to more negative potentials.

With capillary-active cations as in the organo-ammonium salts $R_4\overset{+}{N}Cl'$ the opposite behaviour to that above is obtained. Before the maximum the electrocapillary curve resembles that for KCl, but the maximum is moved to lower values of E and beyond it the curve lies below that for KCl, due to capillary-activity of the R_4N^+ ions.

Addition of neutral organic substances (e.g. alcohols, ketones) to a salt such as KCl or Na_2SO_4, whose electrocapillary curve is nearly parabolic, results in the maximum being depressed, considerably flattened, and also shifted to some extent. These changes are believed to arise from adsorption, the polar group of the organic molecule being in contact with the mercury.

At the maximum in the curve equal numbers of cations and anions are there adsorbed, but it does not necessarily follow that there will be no potential difference between the mercury and the solution. In general this will not be the case, so that, despite some early claims, the method cannot be used to give absolute values of single electrode potentials.

Certain relations between surface tension (γ), applied potential (E), and concentration, can be obtained from the extension of the general thermodynamics of adsorption to systems containing charged ions. Lippmann's well-known equation (see p. 91) is

$$(\partial\gamma/\partial E)_{T,P,i,j} = -\sigma, \qquad (18.12)$$

where E is the potential applied to the mercury in A (see Fig. 18.10), and σ is the quantity of electricity per unit area of the interface.

The capacity (C) of the double layer can then be found from the electrocapillary curve, since

$$C = \partial\sigma/\partial E = -\partial^2\gamma/\partial E^2. \qquad (18.13)$$

Thus if C were constant and independent of E the curves would be parabolic in shape. As pointed out above this is approximately the case with some salts but in general C varies considerably with E. The numerical values of C are usually of the order 20–50 μF per cm.^{2}, which supports Stern's theory that part of the double layer is fixed by adsorption.

(For detailed references see Adam, 1941; Butler, 1940.)

GENERAL REFERENCES
ADAM, 1941, *Physics and Chemistry of Surfaces*, 3rd edn., Oxford.

OTHER REFERENCES
ADDISON, 1943, *J. Chem. Soc.*, p. 535.
—— 1945, *Phil. Mag.* **36**, 73.
ALEXANDER, 1941, *Trans. Faraday Soc.* **37**, 15.
—— and RIDEAL, 1937, *Proc. Roy. Soc.* A **163**, 70.
—— —— 1945, *Nature*, **155**, 18.

Bond, 1935, *Proc. Phys. Soc.* London, **47**, 549.
Butler, 1940, *Electrocapillarity*, Methuen.
Doss, 1939, *Kolloidzschr.* **86**, 205 (and earlier papers).
Harkins and Brown, 1919, *J. Amer. chem. Soc.* **41**, 499.
—— and Jordan, 1930, ibid. **52**, 1751.
Hartley and Roe, 1940, *Trans. Faraday Soc.* **36**, 101.
Hauser, 1942, article in *Advances in Colloid Science*, vol. i, Interscience.
Hutchinson, 1943, *Trans. Faraday Soc.* **39**, 229.
Langmuir, 1932, *J. Amer. chem. Soc.* **54**, 2798.
Ward, 1946, *Trans. Faraday Soc.* **42**, 399.

XIX
THE SOLID/LIQUID INTERFACE

ALTHOUGH the industrial importance of the solid/liquid interface probably exceeds that of either the gas/liquid or the liquid/liquid, yet relatively little accurate work has been done upon it, largely owing to the inherent difficulties. As with the other interfaces we are interested in how adsorption at the interface affects such properties as stability (with dilute suspensions), phase boundary potentials, and properties such as flow and sedimentation volume (with the more concentrated systems). Unlike the mobile interfaces, however, we cannot measure the interfacial energy directly, which precludes the estimation of adsorption by boundary tension measurements in conjunction with Gibbs's equation. Adsorption has therefore to be measured in other ways, as will be detailed below. Another problem is the question of the true surface area, since it is well known, to take an extreme case, that the surface area of a porous body such as charcoal may be many times its apparent geometrical area as measured by a microscope for example. Also with crystalline solids the interfacial behaviour will vary from one type of crystal face to another. One final practical problem may be mentioned, namely, the difficulty of obtaining and maintaining really clean solid surfaces.

Determination of surface area of solids

There are many methods available, of which we shall consider in any detail only those of wider application to colloidal systems. It should be emphasized at the outset that the results from different methods may not be in agreement, not necessarily due to experimental error, but owing to the difficulty of defining 'surface area' with solids containing fine pores or cracks when these approach molecular dimensions.

(a) Direct measurement

Given sufficient magnification it would be possible to measure the surface area of any solid surface or powder directly. In the case of a powder the size-frequency distribution curve would be required, and if the shape of each particle is known (or assumed), then the specific surface (i.e. surface area per unit volume), which is the quantity usually required, follows immediately. The method is a tedious one but is often carried out industrially, since the shape of the size-frequency curve is often found to be very important in ensuring the desired properties of

THE SOLID/LIQUID INTERFACE 537

the powder. The question of the correct shape factor clearly presents a major difficulty, particularly with porous absorbents.

For relatively coarse particles, down to about $0·1\ \mu$ diameter, an ordinary microscope suffices, but this clearly is of little use for the majority of colloidally dispersed solids. The advent of the electron microscope has increased the range of the method by a factor of about twenty or more in suitable cases, i.e. down to particle sizes of $ca.$ 50 A (see Chapter XV). The electron microscope is particularly useful for inorganic colloids of high scattering power such as clays, titania, etc., and has been applied as well to some organic colloids (e.g. soap fibrils, carbon-black) and to biological systems (e.g. bacteria, viruses). A few typical pictures are shown on the plates between pages 456 and 457.

(b) Sedimentation methods

For relatively coarse systems (i.e. down to about $0·1\ \mu$) a number of methods for particle size distribution were worked out early in the study of colloids, as mentioned in the historical account. Several of these are based upon the rate of sedimentation under gravity of a suspension in a suitable medium. For a powder consisting of uniform spherical particles the radius is obtained at once from the observed velocity of sedimentation by applying Stokes's law (see also Chapter IX). Two difficulties arise in applying this method to practical systems—the particles are never of uniform size and very rarely spherical. The first means that the edge of the sedimenting suspension is never sharp, the second that the calculated radius is the 'equivalent spherical radius' and this may give erroneous surface area results if the particles are very asymmetric.

The size-frequency distribution in these coarse suspensions can be measured by weighing the sediment as a function of time as shown below, and the method is still widely used for industrial purposes. Various modifications have been used, one of the best known of which is shown in Fig. 19.1 (b). For particles specifically lighter than the medium the collecting plate is, of course, placed near to the top of the liquid, and not near the bottom as shown in the figure. The other method for such particles is the sedimentation tube (see Fig. 19.1 (c)), in which sedimentation brings about changes in the level in the capillary tube which contains pure dispersion medium. The capillary is sometimes inclined to increase the sensitivity. For emulsions this modification is very useful.

A typical sedimentation curve is shown in Fig. 19.1 (a). To derive from it a particle size-distribution curve it is necessary to assume that the particles move independently and are spherical, so that Stokes's law can be applied. After a time t_1 (see Fig. 19.1 (a)) the fraction of material deposited on the plate will be given by OC, made up of two

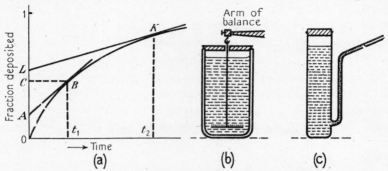

Fig. 19.1. (a) Sedimentation curve. (b) Sedimentation balance. (c) Sedimentation tube.

parts, the first arising from the *complete* deposition of all particles with radius greater than a certain value r_1, the second from the partial deposition of smaller particles. r_1 and t_1 are clearly related by the Stokes expression, since t_1 is the time for particles of radius r_1 to fall through the total height (h) of suspending medium. Hence

$$r_1^2 = \frac{9\eta h}{2(\rho_1 - \rho_2)g \cdot 60 t_1},$$

where η is the viscosity of the liquid, ρ_1 and ρ_2 the densities of solid and liquid respectively, g the acceleration due to gravity, and t the time in minutes. The rate of deposition of particles finer than $r_1 = \partial M/\partial t$ at the point t_1 and clearly will be constant up to that point. Thus the fractional amount of such particles deposited in time t_1

$$= (\partial M/\partial t)_{t_1} \times t_1 = AC.$$

Hence the fraction with radius $> r_1 = OC - AC = OA$. Similarly at a later time t_2 (point K on Fig. 19.1 (a)) the fraction with radius $> r_2 = OL$, so that the fractional amount with radius between r_1 and $r_2 = OL - OA = AL$. In this manner the size-frequency distribution can be worked out.

Some idea of the times involved in this method can be obtained by putting reasonable values into the Stokes equation as given above. Thus taking $h = 10$ cm., $\rho_1 - \rho_2 = 5$, $\eta = 0.01$ poise (water), and $g = 981$, we obtain $t = 153/r^2$ if r is in μ. Thus for values of r of 5, 2, 1, and 0.1 (μ), we find t equal to 6.1, 38.5, 153, and 15,300 minutes respectively, showing very clearly that the method becomes impossibly slow for colloidal particles below about 0·1 μ.

(c) X-ray methods

Two quite distinct methods for particle size determination by means of X-rays have been developed, but they have many theoretical as well as practical limitations (see also Chapter XV).

It is found experimentally that the diffraction rings in X-ray powder diagrams become broadened and less sharp when the crystallite size is reduced below a certain size, usually about the upper limit of the colloidal range (see plates between pp. 456–7). The effect arises from the decrease in the number of reflecting planes, and has its optical equivalent in the relationship between the number of lines in a grating and the sharpness of the spectral lines produced. Assuming the crystallites to be spherical (never strictly the case for crystalline powders), their diameter (t) can be found from the equation $t = 4\lambda/(3b \cos \theta)$, where λ is the X-ray wavelength, θ the Bragg angle, and b the breadth of the line at half-intensity maximum (Stokes and Wilson, 1942). Further assumptions involved in the above expression are that the crystallites are free from internal strain and possess a perfect three-dimensional periodic structure, both of which are unlikely in colloidal particles. Also the above equation gives the size of *single crystallites*, and this may not be the same as the size of the particles. The errors inherent in this method generally tend to reduce the value of t, so that only a lower limit for the particle size can be obtained.

The second method, which has been developed by Guinier (1939–43) in particular, is based on the low-angle scattering of X-rays by individual *particles*, which, unlike those above, need not be crystalline at all. A beam of light traversing a medium containing small opaque particles gives rise to a halo of diffracted light surrounding the incident beam, as, for example, the halo around the moon produced by small droplets of fog or the well-known lecture demonstration with a suspension of lycopodium powder. The intensity of the diffracted light has its maximum in the direction of the incident beam, decreases with increasing diffracting angle, and reaches zero at an angle θ given by the equation $\theta = \lambda/t$, where λ is the wavelength and t the average diameter of the particles. The phenomenon is observable if the diameter is of the order of 10 to 100 times the wavelength of light, and since X-rays have wavelengths of a few A, they show the effect with particles of colloidal dimensions. For details of the experimental technique the original papers should be consulted.

The above treatment, which assumed the particles to be widely separated, clearly no longer holds for loose granular structures such as carbon-black, coal, and various catalysts. As the particles come closer together the intensity of the low-angle scattering diminishes and, as soon as they are so close that their individual distances no longer deviate appreciably from the mean distance, the diffuse spot around the centre gives way to a ring-like halo. From the radius of the halo the particle distance can be estimated, although the theoretical treatment is more complex than for the dilute suspensions considered above. Fig. 19.2 shows how the diffuse central spot increases when a nickel or carbon catalyst is activated, giving a direct proof of the increase in free surface (after Guinier, 1943).

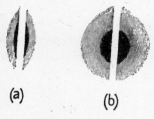

FIG. 19.2. Effect of activation of carbon upon the diffuse central spot. (a) Before activation. (b) After activation.

The errors inherent in this second method mostly tend to increase the value of the particle size, so that the method gives an upper limit.

(d) *Rates of solution or evaporation*

The former alternative is in general preferred, for obvious reasons. It is based upon the comparison of rates of removal from a surface of known area (e.g. a smooth plate or rod) and the sample of unknown surface area (usually a powder). Assuming the true and apparent areas to be identical for the smooth surface, the surface area per gram of the unknown is obtained at once. In addition to the above assumption there is the practical point of ensuring that the observed rate is determined by the actual rate of solution and not by diffusion away from the surface. This difficulty can be overcome by increasing the stirring until the rate of solution becomes independent of stirring rate. The choice of suitable solvent is also important.

Palmer and Clark (1935) used this method very successfully to obtain the surface area of vitreous silica powder, using hydrofluoric acid as solvent and following the rate by the change in electrical conductivity. For the standardization fused quartz rods were employed.

(e) *Permeability method*

When dealing with powders, sands, textile fibres, and other non-consolidated porous media, the specific surface can often be estimated from measurements of the permeability to liquids or gases. This method appears to be limited to particle sizes exceeding about $0.1\ \mu$, but is particularly useful for coarse particles ($> ca.$ 10 μ) where adsorption methods are less accurate. With consolidated media such as porous carbon or natural sandstone, or if the system contains bridges, agglomerates, or wide channels, then the amount of surface exposed to flow becomes uncertain and hence the calculated specific surface may be considerably in error.

As might be expected, a rigorous treatment for fluid flow through porous media is lacking, although approximate solutions can be obtained with idealized systems. The most fundamental law is the empirical one obtained by d'Arcy from the study of water flow through sand, namely,

$$V = kA\frac{\Delta p}{l}t, \qquad (19.1)$$

where V is the volume flowing through the bed of cross-sectional area A and length l in time t, Δp the pressure drop across the bed, and k a constant. If the viscosity is varied, and provided flow is not turbulent, then eqn. (19.1) becomes

$$V = k_p\frac{A}{\eta}\frac{\Delta p}{l}t, \qquad (19.2)$$

where k_p may be termed the permeability constant. These equations show an obvious similarity with the Poiseuille equation for the flow of a liquid of viscosity η through a circular capillary of radius r and length l, for which

$$V = \frac{\pi r^4 \Delta p\, t}{8l\eta} = \frac{r^2 A}{8\eta}\frac{\Delta p}{l}t. \qquad (19.3)$$

The equation most commonly used in the study of liquid flow through porous media is

$$V = \frac{At}{\eta S^2}\frac{\chi \Delta p}{k_0 l}\frac{\epsilon^3}{(1-\epsilon)^2} \qquad (19.4)$$

or an identical expression in which k replaces k_0/χ. Here S is the specific surface of the particles (i.e. total surface/total volume of particles), ϵ is the porosity of the medium (i.e. volume of void/gross space occupied by medium), k_0 a shape factor, and χ an orientation factor. (It is clear that with non-spherical particles the permeability will depend on their shape and upon their packing, since the latter will rarely be random.) In most cases the separation of the shape and orientation factors has not been possible so that k_0/χ is replaced by the factor k. Hence if k can be found, or predicted, then it is clear that the specific surface S can be found.

With sands, powders, etc., of known specific surface, k was found to be about 5; with beds of glass spheres about 4·5. When the porosity is less than about 0·87 and nothing is known concerning the shape or orientation of the particle surfaces, a value of $k = 5$ is used. If, however, the orientation factor χ can be calculated, as in the case of spheres or fibres within the tested range of porosities, then this should be done and used with $k_0 = 3$.

A simple apparatus for liquid-flow measurements is shown in Fig. 19.3. The powder is packed above the gauze disc, and the pressure drop across the bed measured by the manometer. The rate of flow is controlled by the tap shown.

With very fine media it is convenient to replace the fluid by air or other suitable gas, since this reduces the error arising from stagnant rings and layers.

An excellent survey of the permeability method has been given recently by Sullivan and Hertel (1942), which should be consulted for further details.

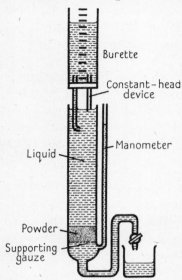

FIG. 19.3. Simple apparatus for permeability measurements using liquids.

(f) *From heats of wetting*

This method, due to Bartell and his co-workers, is based upon an equation which relates the heat of wetting $(-\Delta H_{SL})$ of a powder by a liquid of surface tension γ_L to the surface area (A_s). The equation, derived directly from the Gibbs-Helmholtz equation (p. 60), is

$$A_s = -\Delta H_{SL} \Big/ K\!\left(\gamma_L - T\frac{d\gamma_L}{dT}\right). \tag{19.5}$$

K is a constant given by the equation $\gamma_S - \gamma_{SL} = K\gamma_L$, where γ_S is the surface tension of the solid against air, and γ_{SL} the solid/liquid interfacial tension (see also Chapter XXI, p. 617).

Hence in order to find A_s we must determine $-\Delta H_{SL}$, K, γ_L, and $d\gamma_L/dT$, but of these the last two present no difficulty. The heat of wetting is done calorimetrically; K is found by an indirect method since neither γ_S nor γ_{SL} is easily measurable.

The term $\gamma_S - \gamma_{SL}$, termed the 'adhesion tension', can be found if the liquid forms a finite contact angle (θ) with the solid (see Fig. 19.4 and p. 552). Otherwise it can be found by measuring three displacement pressures with the fine powder: first the pressure required to displace air by a liquid (I) known to have zero contact angle against the solid, then air by another liquid (II) with a finite contact angle against the

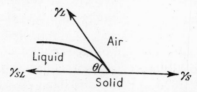

FIG. 19.4. Equilibrium at the air/liquid/solid interface.

solid, and finally that for displacement of liquid II by liquid I. These three values give the adhesion tensions (see p. 552).

(g) *Heat conductivity method*

A method for surface area measurement based upon the heat conductivity of a gas in the pores of a solid seems to give quite accurate results (Kistler, 1942), provided the heat conductivity of the solid structure is not too large. This limits its use to very light, highly porous substances such as the aerogels.

(h) *Adsorption methods*

Methods based upon adsorption, either from solution or from the gaseous phase, include almost all those in widespread use for surface area determinations. With the exception of the electrolytic method they all depend upon two assumptions: firstly, that the point on the adsorption isotherm which corresponds to the saturation of the surface with a monolayer of adsorbent can be found; secondly, that the area per molecule (A) in this adsorbed monolayer is known. For the second assumption it is usual to assume close-packing, but to assign a precise value to A is not always easy, particularly with complex molecules such as dyestuffs, as will be discussed below. There is also the usual complication with crystalline powders, namely, the differing adsorptive powers of the various crystal faces.

The general method when utilizing adsorption from solution is to shake up the powder with a solution of a substance which is readily estimated and which is adsorbed by the powder. When equilibrium is reached the powder is allowed to settle, or is centrifuged off, and the supernatant liquid analysed. This is carried out with increasing concentrations of adsorbate, until the uptake per gram becomes constant. The results are usually plotted with uptake/gram as ordinate and adsorbate concentration as abscissa, and the resultant curve is frequently of the Langmuirian type (see, for example, Fig. 21.16(a), p. 617). The uptake at the saturation point is assumed to correspond to a close-packed monolayer and by assuming a reasonable value for the cross-sectional area of the adsorbed molecule the surface area/gram follows at once. Some of the more commonly used adsorbates are considered below.

(i) *Dyestuffs*. Dyestuffs are an obvious choice since they are readily estimated colorimetrically, are usually strongly adsorbed, and sufficiently soluble. Paneth and his co-workers carried out an extensive comparison of surface areas from dye adsorption and from radio-active indicators for lead sulphate and other inorganic compounds to which the latter method is applicable (see below). In all cases dye adsorption gave the lower value, sometimes less than one-half that of the other. In this comparison the adsorbed molecules of dyestuffs were assumed to be cube-shaped, and the area per molecule calculated from $A = (M/\rho N)^{\frac{2}{3}}$, where M and ρ are the molecular weight and density of the solid, and N Avogadro's number. This is certainly rarely correct, but in the absence of precise information about the orientation of the dyestuff molecules some such assumption is essential. On general grounds it would seem desirable to use dyestuffs of relatively simple structure, not only because they can penetrate smaller surface cracks, but also because in solution they show less tendency to form aggregates (see Chapter XXIV), which might in some cases be adsorbed as such. Smaller dyestuffs would also exert less peptizing or dispersing power on the solid, which in some systems can increase the surface area. With aqueous solutions the maximum dyestuff adsorption is usually very sensitive to pH and to the presence of salts.

Despite the above limitations dyestuff adsorption is in wide use for comparative measurements.

(ii) *Organic compounds readily estimated*. Any organic compound which is strongly adsorbed and readily estimated (either by chemical or physical methods) can be used for surface area determinations. Compounds suitable for chemical estimation are phenols, fatty acids, and acetone; as examples of the physical methods we might mention surface and interfacial tension (e.g. soaps and other capillary-active substances), polarimeter (e.g. sugars), refractive index, absorption spectra, etc. A recent modification has been the use of long-chain polar compounds (e.g. cetyl alcohol, stearic acid) in organic solvents such as benzene, the concentration being found by spreading a known volume on a Langmuir trough (Hutchinson, 1947).

(iii) *Radio-active indicators.* The use of radio-active indicators is clearly very restricted, although modern advances in the preparation of radio-active isotopes will extend its scope somewhat. The classical application was to lead sulphate powders, using thorium B which is isotopic with lead. At equilibrium

$$\frac{[\text{Th B on surface}]}{[\text{Th B in solution}]} = \frac{[\text{Pb atoms on surface}]}{[\text{Pb atoms in solution}]},$$

where the square brackets denote concentrations. The amount of thorium B adsorbed is obtained from the loss in radio-activity of the solution after shaking up with the powder. (Radio-activity is measured with a gold-leaf electroscope or more usually by a Geiger counter.) This gives both terms on the left-hand side of the equation, and since the concentration of lead atoms in solution is known from chemical analysis, the number of lead atoms on the surface follows. From this the surface area of the powder is obtained by assigning a reasonable area to a lead sulphate molecule.

(iv) *Electrolytic method* (for conductors only). The potential across a polarized cathode or anode varies with the current density, giving rise to the phenomenon of hydrogen or oxygen over-potentials (polarization). At current densities sufficiently low for no bubbles of gas to be produced it was shown by Bowden and Rideal (1928) that for a cathode in dilute acid (i.e. $H^+ \to H$ atom) the relationship between the amount of electricity passed and the potential change (ΔV) is a linear one, i.e. $\Delta V = k\Gamma + \text{constant}$. Here k is a constant and Γ the amount of hydrogen deposited per cm.^{2}, which will be proportional to the number of coulombs passed. (It should be pointed out that the degree to which the surface is covered is very small, a fraction of 1 per cent.)

It was found that the constant k did not depend upon the nature of the underlying metal, but only on its surface area, since all liquid metals investigated (e.g. mercury, gallium, and low melting alloys) required the passage of approximately 6×10^{-7} coulombs per cm.^{2} of surface to change the electrode potential by 100 mV. In such cases it is reasonable to assume that the true and apparent surface areas are identical, but if this is not so the previous equation becomes $\Delta V = k\Gamma/x + \text{constant}$, where x is the ratio of the true to the apparent (geometrical) surface area. Knowing k from above, the value of x can be found. As an example of the method a smooth platinum surface gave for x a value of about 2; upon increasing the surface area as in platinum black x rose to 1,800. The increase in effective surface when metals are activated by alternate oxidation and reduction (as in the preparation of catalysts), as well as the decrease upon ageing, can be followed by this method, also the changes upon solidification, polishing, etc. In addition to metals the method seems to work for carbon in the form of rods, but clearly cannot be applied to the majority of adsorbing systems.

(v) '*Retardation volume*' *method*. The principle and use of this method, first developed by Tiselius (1942) from the technique of chromatographic

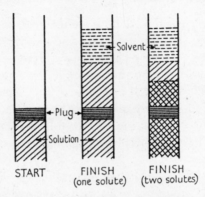

Fig. 19.5. Illustrating the principle of the 'retardation volume' technique of Tiselius.

analysis, will be clear from Fig. 19.5. Powdered solid is packed into a tube to form a plug, usually about 1 cm. thick, and a solution of suitable adsorbate allowed to rise slowly from below through the plug. If the adsorbate is strongly taken up by the solid the liquid first exuding consists of pure *solvent*; once the solid surface is saturated *solution* comes through. The boundary between solvent and solution can readily be followed by the schlieren method (p. 220), and from the volume of solvent so measured (the 'retardation volume') and the weight of powder, the amount of adsorption can readily be calculated. Recently the experimental arrangements have been modified to give automatic recording of the refractive index as a function of the volume of solution passed through the cell (Claesson, 1944). The method can in some cases be extended to two or more solutes, and has been used in the analysis of various mixtures such as amino acids and peptides. (For recent surveys see Tiselius, 1942, 1944; Claesson, 1946.)

(vi) *Methods based upon adsorption of gases or vapours*. The development of a suitable technique for surface area determinations from adsorption isotherms of gases and vapours has been due chiefly to Brunauer and Emmett (see Brunauer, 1944). Physical (sometimes termed van der Waals) adsorption is always measured, owing to difficulties encountered with chemi-sorption. The adsorption isotherm is obtained in the usual manner by taking a known weight of adsorbent and measuring the volume of gas adsorbed (converted to S.T.P.) as a function of pressure. In all cases adsorption is measured at a temperature close to the boiling-point of the gas, such as $-183°$ C. for the more

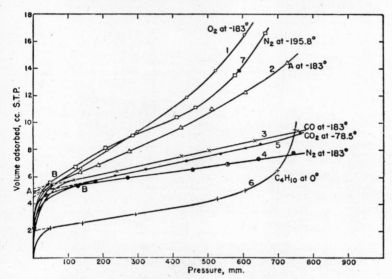

FIG. 19.6. The adsorption of various vapours upon an iron catalyst.

permanent gases nitrogen, argon, and carbon monoxide which are frequently used. A series of typical isotherms, shown in Fig. 19.6, exhibit a characteristic sigmoid shape with a well-marked linear region (from Brunauer, 1944).

To obtain a surface area from the adsorption isotherm the usual two problems arise—what point is to be identified as the close-packed monolayer, and what is the area per adsorbed molecule at that point?

For the former various suggestions have been advanced, such as the extrapolation of the linear portion to zero pressure (point A), or the beginning of the linear portion (point B). Although the question is not definitely settled, it is usual to take point B as the best approximation to the close-packed monolayer.

The cross-sectional area of the molecule may be calculated by assuming that the adsorbed film has a structure similar either to that of the solidified gas, or to that of the liquefied gas. The latter choice is more usual, since it is generally conceded that the adsorbed state has more similarity to the liquid than to the solid state. With an assumed packing (e.g. hexagonal close-packing) the area per molecule can readily be found from the molecular weight and density of the liquid.

Some idea of the agreement between this and other methods is given by Table 19.1, quoted by Brunauer (1944), from the data of Emmett and de Witt.

TABLE 19.1

Material	Surface area (in $m.^2/gm.$) from:			
	Adsorption of nitrogen	Adsorption of butane	Adsorption of salicylic acid	Radio-active indicator
Graphite	30·73	..	3·96	..
TiO_2	9·88	6·58	5·55	..
$ZrSiO_4$	2·76	..	1·33	..
$BaSO_4$	4·30	2·68	1·73	2·2

From these and other results it would seem that the gas adsorption method compares favourably for accuracy with other methods. It has the added advantages of simplicity and wide applicability, its principal limitation being that it cannot be used for coarse powders with diameters greater than about 20 μ, as the adsorption is then so small. The method can be used to find the average particle size only if the particles are smooth and possess no cracks or internal surfaces.

Adsorption isotherms

By measuring the adsorption at a series of concentrations we obtain the adsorption isotherm, examples of which are shown in Figs. 19.6 and 21.16 (a) (p. 617). Many of them are approximately Langmuirian in shape, i.e. they obey an equation of the form $x/m = k_1 k_2 c/(1+k_2 c)$, where x/m is the uptake per gram, c the concentration of the adsorbate in solution, and k_1 and k_2 are constants. From the saturation value and an assumed area for the close-packed molecules on the solid, the specific surface can be calculated as shown above.

In the case of solid/liquid systems with a finite contact angle, it is possible to calculate the force-area curve of the adsorbed film (Fowkes and Harkins, 1940). The contact angle is measured with the solid first in contact with the solvent and then with the solution, giving values θ and θ' respectively. From Newman's triangle we obtain for the two cases (see Fig. 19.4)

$$\gamma_S = \gamma_{SL} + \gamma_L \cos\theta \quad \text{(solvent only),}$$
$$\gamma'_S = \gamma'_{SL} + \gamma'_L \cos\theta' \quad \text{(solution).}$$

THE SOLID/LIQUID INTERFACE

If the solute is non-volatile then we can assume $\gamma_S = \gamma'_S$, and hence the surface pressure Π of the adsorbed film, since by definition

$$\Pi = \gamma_{SL} - \gamma'_{SL} = \gamma'_L \cos\theta' - \gamma_L \cos\theta.$$

Applying Gibbs's equation in the form $\Gamma = \dfrac{1}{RT}\dfrac{d\Pi}{d\ln c}$, we can calculate Γ, hence A, and hence the Π–A relationship. The systems studied (e.g. butyric acid at a paraffin wax/water interface) all gave the same equation of state $\Pi(A-A_0) = xkT$, i.e. similar to that of a two-dimensional imperfect gas.

Thermodynamics of adsorption

As in the previous chapter, the principal thermodynamic quantities required in the study of adsorption are the changes in free energy (ΔG^0), heat content (ΔH^0), and entropy (ΔS^0). The equilibrium constant for the distribution of solute between solution and solid is given by

$$K = \frac{a_{\text{solid}}}{a_{\text{soln}}},$$

where a_{soln} = activity of solute in solution, a_{solid} = activity of the adsorbed solute. The standard free energy of adsorption is then given by

$$\Delta G^0 = -RT \ln K = -RT \ln \frac{a_{\text{solid}}}{a_{\text{soln}}}. \tag{19.6}$$

For dilute solutions and small adsorption, where activity and concentration (c) become identical, we have

$$\Delta G^0 = -RT \ln \frac{c_{\text{solid}}}{c_{\text{soln}}} = -RT \ln \frac{\Gamma}{\tau c_{\text{soln}}}, \tag{19.7}$$

where Γ = amount of adsorbed solute per unit area surface (i.e. total uptake divided by total surface area as found from saturation adsorptions), τ = thickness of adsorbed film. For positive adsorption

$$c_{\text{solid}} > c_{\text{soln}},$$

and ΔG^0 is negative, so that adsorption is accompanied by a decrease in free energy.

In theory it is possible to determine the heat and entropy of adsorption (ΔH^0 and ΔS^0 respectively) from the temperature coefficient of ΔG^0, since $\Delta H^0 = \Delta G^0 + T\,\Delta S^0 = \Delta G^0 - T\dfrac{\partial(\Delta G^0)}{\partial T}$. From the point of view of accuracy, however, it is usually preferable to measure ΔH^0 directly, by calorimetric methods, as outlined below. The function ΔH^0

is of considerable importance, since it can be calculated theoretically in some systems.

Heats of adsorption and wetting by calorimetric methods

Three difficulties are fundamental; the preparation of a fine powder free from contamination and moisture, carefully purified liquids, and

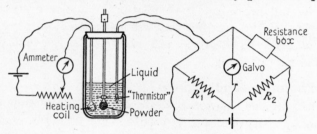

FIG. 19.7. Apparatus for the measurement of heats of wetting.

the accurate measurement of very small temperature changes (0·01–0·1° C. as a rule). For most of the accurate work described in the literature a multiple-junction thermocouple has been used; here we will describe in outline a simpler arrangement of equal accuracy, developed by Dr. E. Hutchinson of the Department of Colloid Science, Cambridge.

The apparatus and its arrangement will be clear from Fig. 19.7. The temperature changes are recorded by means of a resistance thermometer composed of a mixture of oxides, principally of uranium oxide, sealed by means of two thin platinum wires into a protective glass shield. (They are now available commercially, under the trade name 'Thermistor', from Standard Telephone Co., Ltd.) Its electrical resistance decreases exponentially with increased temperature, $R = R_0 e^{-b/T}$, and by using a Wheatstone bridge with 10:1 ratio, a temperature change of 0·001° C. can readily be measured.

The previously dried powder is contained in a thin glass bulb which is broken on the bottom of the Dewar flask and the resultant change in temperature recorded as a resistance change in the 'Thermistor'. This can be converted into calories by passing a known current for a known time through the heating coil and measuring the resultant change in resistance of the 'Thermistor', a technique obviating the need for measuring thermal capacities of powder, liquid, etc. A typical record is shown in Fig. 19.8.

For the heats of wetting to be of value it is of course necessary to measure the surface area with the same sample of powder.

Phase boundary potential changes

Measurement of the electrical changes at solid/liquid interfaces brought about by adsorbed films are confined almost entirely to ζ potentials, using the methods given earlier (Chapter XII). Some indication of the results obtained will be given in Part III.

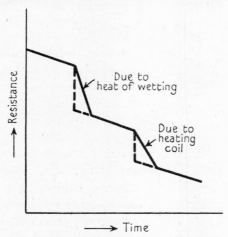

FIG. 19.8. Typical record obtained by the apparatus shown in Fig. 19.7.

Work of adhesion between solids and liquids

The work of adhesion (W_{AB}) between two liquids A and B is defined as the increase in free surface energy upon separating 1 cm.², i.e.

$$W_{AB} = \gamma_A + \gamma_B - \gamma_{AB}. \qquad (19.8)$$

Some indication of its importance has been given in earlier chapters. When one liquid is replaced by a solid, the work of adhesion W_{SL} becomes
$$W_{SL} = \gamma_S + \gamma_L - \gamma_{SL}, \qquad (19.9)$$
but in this case neither γ_S nor γ_{SL} can be determined directly. If the liquid has a finite contact angle (θ) against the solid (see Fig. 19.4), then from Newman's triangle,

$$\gamma_S = \gamma_{SL} + \gamma_L \cos\theta, \quad \text{and hence} \quad W_{SL} = \gamma_L(1 + \cos\theta). \quad (19.10)$$

Thus the work of adhesion can be measured, knowing γ_L and θ, and the methods for contact angles are accordingly given below. (The quantity $\gamma_L \cos\theta$, termed the *adhesion tension*, is often encountered, being the difference between the solid/air and solid/liquid tensions.)

In the case of solids we can also speak of the work of adhesion (W_{SS}) of the solid to itself. By analogy it would be given by the equation

$$W_{SS} = 2\gamma_S - \gamma_{SS}, \qquad (19.11)$$

but its magnitude can only be estimated indirectly as, for example, by friction measurements (see Chapter XXI). (γ_{SS} would represent the solid/solid interfacial energy.)

Measurement of contact angles

The study and use of contact angles has always been complicated by the prevalence of hysteresis effects and by the fact that an advancing interface gives a different value from a receding one. These troubles undoubtedly arise in part from grease and other contaminations, and some workers have claimed that with great care these anomalies can be avoided.

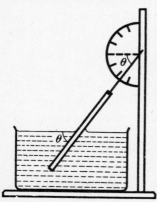

FIG. 19.9. Simple apparatus for the measurement of contact angles.

A variety of methods is available for measuring contact angles and a few may be mentioned here. Measurement is much facilitated if the solid is in the form of a plate or rod; with powders the method of displacement pressures, mentioned above, can be used. With a drop on a smooth plate the contact angle can often be measured by projecting an image on to a screen, or by measuring the size of a drop of known volume (Bikerman, 1941). Adam and Shute developed a method in which a plate is immersed in a liquid (conveniently a filled trough so that the surface can be swept) and tilted until the water surface remains undistorted right up to the line of contact (see Fig. 19.9). If the material can be made into a capillary tube (of radius r), the contact angle is readily obtained (see p. 514) from the rise (h) of the liquid in it, by the formula, $h = (2\gamma \cos \theta)/g\rho r$.

The method based upon displacement pressures is due to Bartell. The pressure (P) required to prevent a liquid of surface tension γ from entering a capillary of radius r is given by $P\pi r^2 = 2\pi r\gamma \cos \theta$, i.e. $P = (2\gamma \cos \theta)/r$, where θ is the contact angle. For powders the radius of the equivalent cylindrical capillary is unknown, but it can be eliminated by using a liquid known to give zero contact angle against the solid. In this case $P' = 2\gamma'/r$, so that the contact angle θ, or the equivalent radius r, can be deduced.

We shall have occasion to consider contact angles again in connexion with the flotation process and the stabilization of emulsions by solid powders.

Effect of adsorbed films upon macroscopic properties of solid/liquid systems

It is of considerable interest, theoretical as well as practical, to see how adsorption at the solid/liquid interface affects the macroscopic properties of suspensions, pastes, and slurries. The usual properties examined are the stability of the suspension as measured by the rate of settling or by the 'sedimentation volume', the velocity of filtration, the viscosity, or with systems such as pastes the flow under stress, and the tendency of particles to stick to solid surfaces, the so-called *adhesion numbers* of von Buzágh (see Chapter XXI for more detailed discussion). Unfortunately these properties are at best only semi-quantitative.

The *sedimentation volume*, the volume occupied by a definite weight of powder allowed to settle under standard conditions, has been widely studied. As would be expected, it depends upon the density, size, and shape of the particles, but can be varied very considerably by adsorption at the solid/liquid interface (Chapter XXI).

The state of the system depends very largely upon the relative magnitudes of the solid/solid and solid/liquid adhesions (W_{SS} and W_{SL} respectively); the solid phase being flocculated if $W_{SS} > W_{SL}$ and deflocculated (or peptized) if the converse is true. Peptization can be brought about in two ways, either by reducing W_{SS} (e.g. by the use of a boundary lubricant), or by increasing W_{SL} (e.g. by a wetting agent). In the former case the lubricant clearly must be adsorbed if it is to modify W_{SS}, and thus it must modify W_{SL} as well. Adsorbed films may increase or decrease W_{SL} depending on the system, and unless the decrease in W_{SL} is more than counter-balanced by the decrease in W_{SS} then the system will still remain flocculated.

Examples where adsorption leads to a reduction in W_{SS} and a marked improvement in fluidity and pouring properties will be detailed in Chapter XXI.

GENERAL REFERENCES

ADAM, 1941, *The Physics and Chemistry of Surfaces*, 3rd edn., Oxford.
BRUNAUER, 1944, *The Adsorption of Gases and Vapours*, vol. i, Oxford.
DEITZ, 1944, *Bibliography of Solid Adsorbents*, U.S. Cane-Sugar Refiners and the National Bureau of Standards.

OTHER REFERENCES

BIKERMAN, 1941, *Industr. Engng. Chem.* (Anal.), **13**, 443.
BOWDEN and RIDEAL, 1928, *Proc. Roy. Soc.* A **120**, 59, 80.
CLAESSON, 1944, article in *The Svedberg*, Almqvist and Wiksell, Uppsala.
—— 1946, *Ark. Kemi Min. Geol.* **23** A, 1.

FOWKES and HARKINS, 1940, *J. Amer. chem. Soc.* **62**, 3377.
GUINIER, 1943, *J. Chim. Phys.* **40**, 133, and earlier papers.
HUTCHINSON, 1947, *Trans. Faraday Soc.* **43**, 435.
KISTLER, 1942, *J. phys. Chem.* **46**, 19.
PALMER and CLARK, 1935, *Proc. Roy. Soc.* A **149**, 360.
STOKES and WILSON, 1942, *Proc. Camb. Phil. Soc.* **38**, 313.
SULLIVAN and HERTEL, 1942, article in *Advances in Colloid Science*, vol. i, Interscience.
TISELIUS, 1942, article in *Advances in Colloid Science*, vol. i, Interscience.
—— 1944, article in *The Svedberg*, Almqvist and Wiksell, Uppsala.

PART III
THE PRINCIPAL COLLOIDAL SYSTEMS
Introduction

HAVING given an account of the more important practical and theoretical techniques used in the study of colloids, some indication of the more important results can be attempted. Of the remaining nine chapters the first eight deal with colloidal systems in the generally accepted sense of the term: the last, on membranes, has been included since membranes are frequently composed of colloidal materials, and in addition are closely linked with the gels and fibres.

In discussing the various systems the following general scheme will be followed, subject to obvious additions or other modifications in some cases:

(a) method of preparation or formation;

(b) structure, both microscopic and interfacial;

(c) stability, its measurement and its origin;

(d) an outline of the more important properties.

XX
DILUTE SUSPENSIONS (SOLS)
Introduction

DILUTE lyophobic systems, termed *sols* or *suspensions*, are known in the three possible dispersion media (gas, liquid, and solid), but of these only the first two can be afforded any detailed consideration here. The continuous media considered will be practically restricted to air in the first case and to aqueous solutions in the second, and the dispersed media to liquids and solids. The term *sol* has sometimes been used to include both lyophilic and lyophobic suspensions, but it seems less confusing if it is restricted to the latter, since the extreme type of lyophilic systems (e.g. proteins in water, rubber in benzene) are merely molecular *solutions* and differ from ordinary solutions only in the size of the solute molecule.

The dispersed particles in sols are almost invariably composed of a large number of atoms or molecules, and are usually of sufficient size and refractivity to be visible in the ordinary or ultra-microscopes (*ca.* 100 A–*ca.* 1μ). This contrasts with the lyophilic systems where, in the extreme type, the particles are molecularly dispersed and invisible in the ultramicroscope.

Suspensions in liquid media (chiefly aqueous)

Most of our knowledge of lyophobic sols comes from those in aqueous dispersion media and, apart from a concluding paragraph, the following account refers only to such systems.

Stability arises almost entirely from the ζ potential, since when this is lowered to small values coagulation is very rapid. The origin of the charge with the non-ionogenic particles considered here was discussed in detail in Chapter III, and the correlation of stability with ζ potential in Chapter V.

The principal *aquasols*, as suspensions in aqueous media may be termed, are those of the metals (particularly the noble metals), insoluble paraffin-chain compounds (oils, waxes, etc.), sulphur, carborundum, nitro-cellulose, and so on. The amount of dispersed phase is always quite small, usually *ca.* 0·1 per cent. by volume, and rarely exceeds 1 per cent. for suspensions without added stabilizer, in striking contrast to the lyophilic type.

(a) Preparation

The two general ways of preparing suspensions are by breaking down particles of macroscopic dimensions, or by building up from molecular or atomic units, termed respectively dispersion and aggregation. In both cases it is essential not to go too far but to stop the process when the particles reach colloidal dimensions.

(1) *Dispersion processes*

(i) *Mechanical methods*. In principle it should be possible to break down any macroscopic solid into colloidal particles by sufficient mechanical grinding. In practice a definite limit is reached, depending on the material and the mechanical method used, owing to the tendency of the small particles to aggregate again by virtue of their high surface energy. This can be overcome to some extent by adding a protective colloid (*ca.* 5% often enables dispersions of over 10% to be made).

Dispersion is usually obtained commercially by subjecting the mixture of coarse particles and dispersion medium to intense shearing forces in a *colloid mill*, the principle of which is shown by Fig. 23.2 (p. 648). The space between the rotor and stator can be adjusted to give different average particle sizes, although the particle size is never uniform. Water ducts or cooling jackets are necessary to remove the intense heat liberated by friction.

For laboratory purposes grinding the material dry or wet in an agate mortar usually produces sufficient particles of colloidal size. Fine quartz suspensions, for example, can be prepared in this way.

(ii) *Electrical methods*. The production of colloidal metals by means of an electric arc was mentioned in Chapter I. With the d.c. arc as used originally by Bredig the method consists of both dispersion and aggregation, the former from the disintegration and vaporization at the electrodes and the latter by vapour condensing and aggregating to particles of colloidal size. Svedberg (1928) found that the method gave finer dispersions if the current density were as high as possible, and he later extended its application very considerably by the use of a high-frequency a.c. arc (frequency *ca.* 10^6 cycles/sec., from an induction coil and condenser). This so reduced the decomposition of the medium that the use of non-aqueous media became feasible, and sols of alkali metals in ether, benzene, etc., can be prepared by this means.

With the more noble metals in aqueous media, particularly gold, it is usual to add a trace of electrolyte, preferably NaOH, to stabilize the sol formed. Using highly purified water, gold gives sols of only transient stability, whereas platinum and silver oxidize sufficiently to form stabilizing electrolyte. With the more positive metals the sols contain appreciable amounts of oxide.

We may include under electrical methods a rather special case, namely the dispersion of liquids by electrostatic charging. A fine jet of water breaks up into droplets when charged to a high potential by means of a Wimshurst machine, for example—the same break-up occurring whether the water streams into air or into an oil. This is a convenient method of obtaining fine dispersions of aqueous solutions in oils, but for the converse type of dispersions (i.e. oil-in-water) the method does not appear to work, probably owing to rapid leakage of the charge when water is the outer phase. A drop is normally prevented from disintegrating by its surface energy, which leads to a pressure (directed inwards) of $2\gamma/r$ (p. 87). The presence of a charge leads to an outwardly directed pressure of $Q^2/8\pi Dr^4$ or $2\pi\sigma^2/D$ (p. 88), and if this exceeds that due to surface forces the drop will tend to disintegrate.

(iii) *Peptization*. 'Peptization' was the term given by Graham to the spontaneous dispersion of a precipitate on adding a small amount of a third substance, the peptizing agent. Examples are the peptization of clays by alkalis (due to OH' ion), AgI by KI or $AgNO_3$ (due to I' and Ag^+ respectively), fatty acid soaps by glycerol, phenols, fatty acids, etc., gelatine by iodides and thiocyanates, and many finely divided precipitates by soaps or hydrophilic colloids in general. In some cases (e.g. Prussian blue, silver halides) peptization occurs spontaneously when any electrolytes present have been washed out.

Peptization arises from adsorption, of one ion preferentially in the case of salts, of the molecule in the case of undissociated substances, and of the colloid, or colloidal ion, in the case of hydrophilic colloids. This reduces the attractive forces between the particles, and Brownian agitation, or slight mechanical movement, leads to the formation of a stable suspension.

(iv) *Other methods*. Two techniques, limited to liquids or low melting solids, employ respectively supersonic waves or a steam jet to effect dispersion. The former is outlined in connexion with the preparation of emulsions (p. 647); the latter consists in floating a small quantity of the substance on the surface of distilled water (*ca.* 1 gm. to 1 litre) and blowing steam through a fine jet situated at the surface. Dispersion is thought to be effected by ultrasonic waves produced by the collapse of steam bubbles on striking the cold water. This method is useful for electrophoretic work since it avoids the use of alcohol or mechanical methods which might contaminate the surface of the droplets.

(2) *Aggregation processes*

The formation of particles of colloidal dimensions by the aggregation

DILUTE SUSPENSIONS (SOLS) 559

of single atoms or molecules clearly requires as its first step the formation of a supersaturated solution. The supersaturation is then relieved by the formation of nuclei, or by condensation upon nuclei already present, followed by growth to larger particles which, however, must not proceed beyond the colloidal range. The three processes—supersaturation, nuclei formation, and growth of nuclei—will be considered in turn.

(i) *Formation of supersaturated solutions.* This can readily be brought about by a variety of methods such as chemical reaction, addition of a precipitant, cooling, etc.

Chemical reactions are very numerous, and as details will be found in many text-books only a few typical examples will be included here.

Hydrolysis, e.g. $FeCl_3 + H_2O$ (boiling) → ferric oxide (hydrated).
Double decomposition, e.g. $BaCl_2 + Na_2SO_4$ → $BaSO_4$.
Oxidation, e.g. $H_2S + SO_2$ → S.
Reduction, e.g. $AuCl_3 + P$ (yellow) → Au.

Addition of a precipitant is commonly used for preparing dilute suspensions of gums, waxes, polymers, etc. The disperse phase is dissolved in a low-boiling, water-miscible solvent such as alcohol or ether, and poured into hot water, the organic solvent being then removed by boiling.

Sudden chilling of dilute solutions can, if this leads to the usual diminution in solubility, throw out the solute in colloidal form. Other related examples are the formation of mist by cooling moist air, and the formation of sulphur sols by passing sulphur vapour into cold water.

(ii) *Formation of nuclei.* Before considering the process of nucleation from supersaturated solutions it is desirable to consider that from supercooled melts of pure substances. As mentioned in Chapter I this has been studied particularly by Tammann (1903 et seq.), using substances such as glycerol. Tammann measured the velocity of nucleus formation (defined as the number of nuclei formed in unit time in unit volume of the liquid) by direct microscopical counting in a thin layer, and he observed a maximum when this was plotted against temperature, as shown in Fig. 1.8 (p. 14).

The usual explanation for this maximum is based upon molecular kinetic considerations. Only molecules of low kinetic energy will be able to stay together sufficiently long to form a stable nucleus, any transient aggregate of high energy molecules breaking up rapidly (just as the recombination of atoms does not occur unless a third body is present to remove the energy liberated). As the temperature is lowered below the melting-point so the number of low energy molecules increases, and hence the rate of nucleus formation. The rate of nucleation might thus be expected to increase indefinitely as the temperature is reduced, but another factor is involved, namely the translational and rotational movement of the molecules which is necessary for the formation of a nucleus. The rate of molecular movement will be approximately proportional to the reciprocal of the viscosity, and due to the viscosity increasing exponentially with falling temperature ($\eta = \eta_0 e^{A/T}$ as a rule), it will fall off rapidly as the temperature is lowered. When this factor exceeds the previous one the rate of nucleation falls again, giving a maximum in the overall rate-temperature curve.

An explanation of Ostwald's so-called 'Law of Stages', according to which the formation, on cooling, of the unstable form is preferred (e.g. phosphorus vapour gives yellow phosphorus on condensation and not the stable red form), can be put forward on the lines of the above view of nucleus formation. The free energy of the unstable form is greater than that of the stable, and hence molecules of greater kinetic energy can take part in the formation of nuclei of the unstable form than in that of the stable. The probability of liberating the unstable form is therefore correspondingly increased, and hence such forms are commonly the first to be formed after supercooling or supersaturating.

Turning now to supersaturated *solutions* as distinct from pure substances, the same kinetic conceptions can be applied, the only essential difference being that the size of the nucleus required for crystallization is now dependent not only on the temperature but also on the degree of supersaturation. For, as pointed out on various occasions (e.g. p. 86), a small particle (or nucleus) has a higher solubility than a larger particle —it will therefore only be stable at all above a certain degree of supersaturation. In general we shall be discussing nucleus formation under isothermal conditions, since with solutions it is more convenient to produce supersaturation by chemical methods or by physical means other than lowering the temperature. The rate of nucleation can be followed by direct counting in the ultramicroscope, just as with supercooled melts. Stauff (1940) describes an arrangement of differentially connected photo-electric cells, by means of which the appearance of each new nucleus causes a discontinuity in the photo-electric current, and he has used this arrangement for studying supersaturated solutions of potassium chlorate and of a number of soaps.

The amount of material thrown out of solution is clearly proportional to $(C-C_s)$, where C is the total concentration produced momentarily by chemical (or physical) reaction, and C_s the equilibrium solubility. The initial rate of nucleus formation (dn/dt) would then be expected to be proportional to $(C-C_s)$, and also inversely proportional to the saturation solubility (C_s), i.e.

$$\frac{dn}{dt} \propto \frac{(C-C_s)}{C_s} = k\frac{(C-C_s)}{C_s}, \qquad (20.1)$$

where k is a constant for a particular system. $(C-C_s)/C_s$ is termed the 'specific supersaturation'.

According to von Weimarn, for the separation of the precipitated material in sol or gel form we need rapid formation of nuclei, i.e. dn/dt must be large (see Chapter I). This can be achieved in two ways, by making C large or by making C_s small.

C can be made large by using very concentrated solutions. However, owing to the large amount of disperse phase then separating, the nuclei are extremely close and may link up, forming a gelatinous precipitate or gel. This is shown very well by mixing concentrated solutions of barium thiocyanate and manganese sulphate, when clear cellular *gels* of barium sulphate are produced. For example, if 5 M. solutions are mixed, then

$$C = 5\text{ M.}/2 \quad \text{and} \quad C_s \simeq 4 \times 10^{-5}\text{ M.},$$

i.e. $\quad \dfrac{C-C_s}{C_s} \simeq 60{,}000.$

To make C_s small the solubility must be reduced and in the case of barium sulphate this can be done by precipitating in alcohol/water mixtures. This will also reduce C, and hence the amount of material separating, so that the nuclei will be at larger distances apart, and a sol, rather than a gel, will be obtained. For example, 0·01 M. solutions, in 50 per cent. alcohol, of barium thiocyanate and cobalt sulphate give on mixing a fine and fairly stable *sol* of barium sulphate.

Under the more usual conditions of precipitation, as for example, in qualitative analysis where colloid formation is undesirable, the barium sulphate comes down as a white precipitate, the concentration being usually *ca.* M./10 and hence the specific supersaturation *ca.* 2,000.

Both with supercooled melts and with supersaturated solutions the rate of nucleation is found to be very sensitive to the presence of added or adventitious impurities. Such effects probably arise from adsorption and will be considered in more detail in the following section upon crystal growth.

(iii) *Growth of nuclei.* As with nucleus formation it is preferable to consider supercooled melts before discussing solutions. Tammann and his pupils also studied the linear velocity of crystallization, by direct measurement in tubes, as a function of temperature. The curve obtained resembled that for nucleus formation in showing a maximum, as shown in Fig. 20.1. The maximum was generally rather flat, owing to the latent heat of crystallization having to be conducted away through the walls.

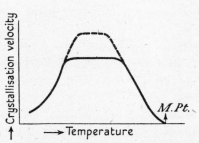

Fig. 20.1. Temperature dependence of the linear velocity of crystallization in a tube.

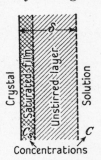

Fig. 20.2. The boundary between crystal and solution.

This was confirmed by showing that the flat portion could be reduced (as shown by the dotted curve in Fig. 20.1) by using narrower tubes.

The suggested explanation for the maximum is precisely as for nucleus formation, due to the exponential increase in viscosity so reducing molecular movement that at some stage it outweighs the inherent tendency to crystallize.

Crystallization from solutions. A good deal of work has been done upon the rate of solution of crystals and the rate of growth of crystals from supersaturated solutions. (For early references see Rideal, 1930; Freundlich, 1926.)

Consider first the solution of a solid. This will involve two processes which may be regarded as independent; the actual process of interaction of solid and solvent, and the diffusion of solvated solute away from the surface. If solvation is rapid compared with the rate of diffusion, then this latter process will determine the overall rate of solution. Diffusion can be accelerated by stirring, but even with the most vigorous agitation there is believed to be a thin film, of the order of 10^{-3} cm., unaffected. This film is termed the 'unstirred layer'. If diffusion through the unstirred layer of thickness δ (see Fig. 20.2) determines the overall rate of solution, then the Fick diffusion equation (p. 233) should be followed, the rate of solution per unit area of interface (in, say, gm./cm.^{2}) being given by

$$\frac{dm}{dt} = \frac{D}{\delta}(C_s - C), \tag{20.2}$$

where D = diffusion coefficient of solute, C_s = saturation concentration of solution, C = actual concentration of solution. It is frequently possible, by increasing the volume of the liquid phase, to make C constant and negligible, when eqn. (20.2) reduces to

$$\frac{dm}{dt} = \frac{D}{\delta} C_s, \qquad (20.2\,\text{a})$$

i.e. a constant for a given system at a given temperature.

Expression (20.2) fits the experimental results in many cases. For example the rate of solution of rods of magnesia rotating in acids such as hydrochloric, acetic, and benzoic, runs parallel with the diffusion coefficient of the magnesium salt formed, and the temperature coefficient is small (1·25 for 10° C.) as found for simple diffusion processes. The values of δ calculated from eqn. (20.2) vary somewhat with the reaction, but are of the order 0·02–0·06 mm.

The growth of a crystal might be regarded as essentially the converse of the solution process, i.e. diffusion of substance from a supersaturated solution through an unstirred layer of thickness δ to the crystal surface, followed by desolvation and incorporation into the crystal lattice. If the latter process is rapid compared with diffusion, then the rate of growth per cm.^{2} should be given by the same expression (eqn. (20.2)) as for the rate of solution, C now being the concentration of the supersaturated solution.

The early work of Marc showed that it was possible, by using rapid stirring to promote diffusion, to obtain conditions where the rate of deposition became independent of the stirring rate, as it should be on the above picture. The velocity constant, however, usually differed from that obtained on dissolution, the divergence diminishing at higher temperatures. Such a discrepancy might perhaps arise from the diffusion coefficient varying with the conditions.

Influence of foreign substances upon crystallization. Dissolution is as a rule not seriously affected by the presence of foreign substances, at any rate under conditions of rapid stirring where diffusion is the limiting factor. Crystallization, however, may be completely inhibited, or the crystal habit changed, by small amounts of impurities, and the general question is one of enormous interest, academic, industrial, and biological.

Adsorption of hydrophilic colloids (e.g. gelatine) on to growing crystals may be used to ensure stability of hydrophobic suspensions, the phenomenon being termed 'protection' (see p. 574). It is also of great importance in the laboratory and in industry, as when charcoal is used to purify solutions before crystallization. This, by removing surface active impurities, allows crystallization to proceed under conditions otherwise impossible. Modern sugar refining in particular is dependent upon suitable charcoals for facilitating crystallization from concentrated syrups. The possibility of controlling the crystal habit is frequently of industrial importance, since many operations, particularly filtration, may be very dependent upon the crystal shape; for example, long needles or plates tend to give felt-like masses which are difficult to wash and dry.

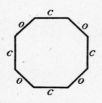

FIG. 20.3. Crystal with cubic (c) and octahedral (o) faces.

In living systems the transport of slightly soluble substances in supersaturated solutions, e.g. in bile and urine, is made possible by the hydrophilic colloidal material present. The formation of concretions in the body may be due to the breakdown of the normal protective mechanism.

Marc and his co-workers also examined the inhibiting effect of dyes upon crystallization velocity and found that only those dyes which colour the crystals had any marked effect. For a given system the inhibition increased with dye concentration, as would be expected from adsorption considerations. These supersaturated solutions containing inhibiting dyestuffs, etc., are believed to be in unstable equilibrium, otherwise the rate of dissolution would be expected to be diminished.

The ability of certain foreign substances to alter the crystal habit is also ascribed to adsorption. A well-known example is common salt crystallizing as octahedra instead of cubes when the solution contains urea. Dyestuffs have been widely examined in this connexion since adsorption can be detected and measured so readily.

Adsorption will usually take place to different extents upon the different crystal faces, depending upon the field of force and upon the nature of the adsorbate, and this may lead to a face previously of high surface energy becoming stable with respect to the other possible types: this may then radically change the habit of the growing crystal. Sodium chloride, for example, normally crystallizes as cubes owing to the high energy of the octahedral faces (consisting of Na^+ or Cl' ions exclusively) relative to the cube faces (composed of alternate Na^+ and Cl' ions), which leads to deposition of material on the latter exclusively. (That this leads to the elimination of octahedral planes can be seen from Fig. 20.3.) If, however, the relative energies of the surfaces can be reversed by adding an adsorbate taken up preferentially by the octahedral faces (as with urea and certain dyes), then octahedra will be formed.

(For a recent survey of crystal growth see Wells, 1946.)

(b) *Properties*

Three properties of suspensions in aqueous media call for special mention—the electrical mobility, the viscosity, and the optical properties.

(1) *Mobility*. Electrophoresis affords one of the most accessible and accurate means of investigating dilute suspensions in aqueous media, and for this reason has been very widely studied. An account of the experimental technique and basic theory, which enables the ζ potential and charge density to be calculated, has been given in Chapter XII.

The electrophoretic mobility observed with suspensions is found to vary widely depending upon the chemical and physical nature of the disperse phase, the pH and salt content of the medium, and the presence of capillary active ions or other material.

Non-ionogenic materials (e.g. hydrocarbons and their un-ionized derivatives, SiO_2, SiC, S, etc.) show in general a marked similarity in electrophoretic behaviour. In pure water the charge appears to be invariably negative, for reasons discussed on p. 45, and the ζ potential is usually of the order 50 mV. Several series of paraffin-chain compounds have been studied by Lewis and his co-workers, as these offer the best conditions for separating the chemical and physical factors involved. At neutral pH the mobility appears to be almost independent of the physical state (solid or liquid) as well as the nature of the polar group (provided, of course, this is non-ionogenic). For example at pH 7, and a temperature of 25° C., in the presence of 0·01 N. Na^+ ion, octadecane (solid), paraffin wax (solid), dodecane (liquid), cetyl chloride, bromide and iodide (all liquids), all gave mobilities close to 1·0 μ/sec./V/cm. Carruthers (1938), Williams (1940), and Dickinson (1941) have concluded that chain-length has no effect on mobility unless a change of state occurs. However, they state that the polar group has a definite although small effect, the differences being most marked at high OH' concentrations such as pH 12. The polar group is believed to affect the mobility by its effect upon the adsorption of the charge-determining OH' ions: the more polar the terminal atom, if negative, as it is as a rule, e.g. $—CH_2^+—Cl^-$, the lower the adsorption and hence also the mobility. Other factors, such as hydration, must enter as well.

The effect of pH upon mobility is typified by the examples given in

Fig. 20.4, showing three paraffin-chain halides in the presence of a constant cation concentration (0·01 N. Na⁺ ion). (Constancy of cation concentration is essential since the droplets are negatively charged.) The increasing negative mobility with increasing pH suggests that in

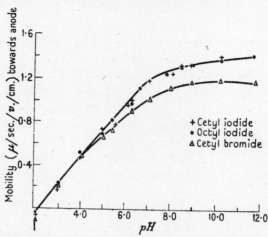

Fig. 20.4. The effect of pH on the mobility of suspensions of cetyl iodide, octyl iodide, and cetyl bromide, in the presence of 0·01 N Na⁺.

these dilute salt solutions as well as in pure water, OH' ion adsorption is solely responsible for the charge. This suggestion is strengthened by the fact that the charge density (calculated from the mobility) usually increases with pH (i.e. with OH' ion concentration) according to a Langmuir adsorption isotherm. This can be seen from Fig. 20.4, since for the systems used there the charge density is proportional to the mobility. The reason for the peculiar position occupied by the OH' ion has been discussed earlier (p. 46). It is rather striking that the 'isoelectric' point of these non-ionogenic suspensions is nearly always between pH 2·5 and pH 3·5.

Addition of salts to such suspensions leads ultimately to the ζ potential becoming very small at concentrations above *ca.* $N/10$; at lower concentrations ζ may increase, decrease, and even change its sign with increasing salt concentration. Non-ionogenic suspensions show a very uniform behaviour, and the effect of a series of salts of different valence upon the ζ potential of a paraffin oil suspension is shown in Fig. 20.5. The original negative charge is seen to be reversed by aluminium and thorium salts, but made more negative by the $[\mathrm{Fe(CN)}_6]^{4'}$ ion. The effects of these polyvalent ions have, in the past, usually been ascribed to adsorption. Such a conclusion seems quite logical at first sight but on further consideration seems much less probable. Thus aluminium and thorium salts will (unless the pH is very acid) contain suspended particles of hydroxide which will adsorb upon and modify the surface of the suspended particles. In the case of polyvalent ions such as

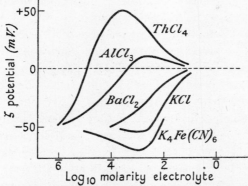

FIG. 20.5. Effect of various salts upon the ζ potential of a suspension of liquid hydrocarbon.

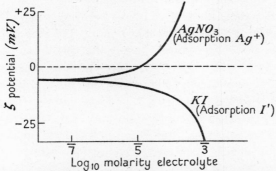

FIG. 20.6. Effect of $AgNO_3$ and KI upon the ζ potential of an AgI suspension.

ferrocyanide, hydrolysis is liable to occur, and this will increase the OH' concentration and hence the negative charge on the particles.

With suspensions of ionogenic substances, on the other hand, such as silver iodide, barium sulphate, and similar ionized solids, the electrokinetic behaviour is much more specific. Silver iodide, for example, is negatively charged in dilute potassium iodide solutions and positively charged in dilute silver nitrate, due to I' ion and Ag^+ ion adsorption respectively (see Fig. 20.6). In these systems also, just as with nonionogenic materials, ζ tends to zero as the salt concentration is sufficiently increased, and polyvalent ions, such as aluminium or thorium, can bring about a charge reversal.

If the materials classed as 'capillary-active', such as proteins, soaps, etc., are added to suspensions, the electrophoretic behaviour is primarily determined by the added substance, regardless of the nature of the dispersed material. This gives a convenient way of studying both proteins and soaps, usually by adsorbing upon suspensions of inert oils. The latter is discussed below, the former on p. 331.

Fig. 20.7 shows the mobility of suspensions of a variety of hydrocarbon derivatives in the presence of two soaps, one anionic and the other of the cationic type (see Chapter XXIV). The original negative charge is seen to be progressively increased by increasing concentrations of the former, but reversed by the cationic soap at extremely low concentrations. For a given chemical type the liquid surfaces always adsorb more strongly than the solid ones at the lower soap concentrations (just as in the case of OH' ion) but at very high soap concentrations the two practically coincide. Compounds containing —OH and —NH$_2$ groups adsorb soaps very strongly, probably due to hydrogen bonding (McMullen and Alexander, unpublished). With liquid surfaces the initial amounts of adsorption are seen to influence the mobility much more than the surface tension, the converse being the case for high soap concentrations in or near the micellar region. The reason for this is not difficult to see. The charge density is initially very small and adsorption of a few soap ions, which will not reduce the tension appreciably, can therefore affect the charge and hence the mobility quite considerably. Adsorption of a few soap ions will, however, lower the tension considerably when the adsorbed molecules approach close packing (i.e. in concentrated soap solutions), but here the fractional increase in charge will be small and hence the mobility change also slight.

(2) *Viscosity*. The viscosity of dilute suspensions is chiefly of interest in connexion with the attempts to calculate it theoretically (see Chapter XIII). For very dilute uncharged spherical suspensions the simple Einstein relation $\eta = \eta_0(1+2 \cdot 5\phi)$, or its first extension

$$\eta = \eta_0(1+2 \cdot 5\phi + 14 \cdot 1\phi^2), \quad \text{(eqn. 13.20)}$$

seem to fit reasonably; with higher concentrations deviations occur, and these are accentuated if the particles are charged, especially at low salt concentration (e.g. von Buzágh, 1943).

The extension of the simple Einstein equation to take account of the charge is due to Smoluchowski (see p. 377).

With most dispersions of solids the particles are not spherical, which introduces complications, particularly at the higher concentrations. (See Chapter XIII, and also Mardles, 1940, 1941.)

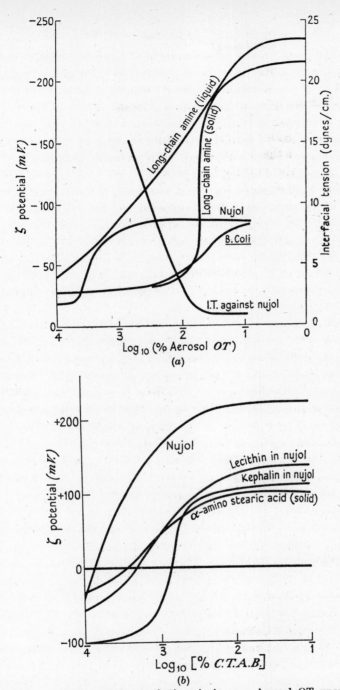

FIG. 20.7. (a) Effect of the synthetic anionic soap Aerosol OT upon the ζ potential of various suspensions. Also interfacial tension curve for comparison in the case of nujol. (b) Effect of the synthetic cationic soap C.T.A.B. upon the ζ potential of various suspensions. All measurements at 25° C. in M/20 acetate buffer, pH 6·0.

(3) *Optical properties.* Various aspects of the interaction of colloidal suspensions with light have been mentioned earlier, particularly in Chapter XV, where it was shown how the optical data can be analysed to give useful information concerning size and shape of the particles. Information additional to that from absorption, scattering, and polarization studies can be gained from the changes in optical behaviour when the particles are subjected to a velocity gradient or to an electrical or magnetic field (Chapter XIV).

Streaming birefringence has been particularly useful for indicating the shape of the particles in dilute suspensions. By this means ruby-red gold sols were shown to contain approximately spherical particles, whereas with vanadium pentoxide sols and certain dyestuffs (e.g. benzopurpurine 4B) the particles were rod-shaped. In other systems, particularly aged ferric oxide, bentonite, and Crum's alumina sol (made by boiling aluminium acetate solution), the particles are disc- or plate-like in shape.

(c) *Effects of salts, non-electrolytes, and other colloids*

The principal question which arises here is how these various additives affect the properties, chiefly the stability, of suspensions. Stability is strictly measured by the rate of growth of interface with time, which involves a size-frequency analysis of the particles in the suspension at suitable intervals, by means of the ultramicroscope. As this is very laborious, indirect methods have frequently been used instead, particularly measurements of viscosity or turbidity. The discussions of turbidity and of viscosity given on pp. 426 and 358 respectively should make it clear that such measurements can at best be only semiquantitative.

(1) *Salts.* It is found that as the ζ potential of a suspension is reduced by addition of electrolytes (see Fig. 20.5) so also is the stability, although the relationship is not a linear one. A much-cited case, due to Powis (1915), is the effect of thorium chloride upon the stability (as measured by turbidity which, as pointed out above, has serious limitations) and ζ potential of oil drops (see Fig. 20.8). The suspension is seen to be unstable when ζ is less than about 40 mV (positive or negative). Powis concluded that all suspensions possessed a certain 'critical potential', generally between 20 and 40 mV, but later work showed this to be an over-simplification. (For more recent attempts to correlate stability with the ζ potential see Chapter V.)

The concentration of electrolyte required to bring about coagulation depends, according to the Schulze-Hardy rule (see pp. 15 and 115), upon the valence of the ion of opposite sign to that of the colloid. That factors other than the valence of the coagulating ion enter the picture is clear, however, when a large number of salts and different

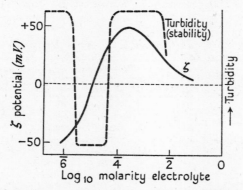

FIG. 20.8. Variation of ζ potential and turbidity (as a measure of stability) of oil droplets with $ThCl_4$ concentration.

suspensions are compared. Tables 20.1 and 20.2 below, drawn from data collected by Freundlich (1926), give some idea of the variations found.

TABLE 20.1. *Flocculation Values of some Negative Suspensions*

Electrolyte (in terms of 1 gm. mole cation)	Flocculation value (millimoles/l.)		
	As_2S_3 sol	Au sol	Pt sol
NaCl	51	24	2·5
KCl	49·5	..	2·2
½ K_2SO_4	65·5	23	..
HCl	31	5·5	..
$C_6H_5NH_2,HCl$	2·5	..	..
$CaCl_2$	0·65	0·41	..
$BaCl_2$	0·69	0·35	0·058
$Al(NO_3)_3$	0·095	..	..
½ $Al_2(SO_4)_3$	0·096	0·009	0·013
$Ce(NO_3)_3$	0·080	0·003	..

Any expression of the relative coagulating powers of ions of different valence, e.g. uni-: di-: tri- $\simeq$ 1: 100 : 1000, as frequently quoted, is seen to be at best semi-quantitative. This is not surprising since factors additional to the *charge* on the coagulating ion must clearly be involved, for example the specific nature of this ion, adsorption of the other ion, hydration of ions and of the colloidal particle, pH, etc.

It is frequently, although not invariably, found that with negative suspensions and a series of salts with a common anion the coagulating power increases with atomic weight, e.g.

$Li^+ < Na^+ < K^+ < Rb^+ < Cs^+$, and $Mg^{++} < Ca^{++} < Sr^{++} < Ba^{++}$.

This is known to be the order of decreasing hydration (p. 46), suggesting that precipitating action is favoured by low hydration of the coagulating ion.

TABLE 20.2. *Flocculation values of some Positive Suspensions*

Electrolyte (in terms of 1 gm. mole cation)	Flocculation value (millimoles/l.)		
	Fe_2O_3 sol	Al_2O_3 sol†	Al_2O_3 sol†
NaCl	9·25	77	43·5
KCl	9·0	80	46
KBr	12·5	150	··
½ $BaCl_2$	9·65	··	··
½ $Ba(OH)_2$	0·42	··	··
K_2SO_4	0·21	0·28	0·30
K_2CrO_4	··	0·60	0·95
$K_3[Fe(CN)_6]$	··	0·10	0·08
$K_4[Fe(CN)_6]$	··	0·08	0·05

† These refer to measurements by different authors.

With halide ions and positive sols the order $Cl' > Br' > I'$ has been found in some cases such as ferric and aluminium oxide sols, and the converse order in others (e.g. positive gold sols). Hydration would not, at first sight, appear to be an important factor, since these anions are thought to be little hydrated (p. 46), and from the known tendency of I' ions to be adsorbed at O/W interfaces and to act as peptizers (ascribed to their greater polarizability) the order $Cl' < Br' < I'$ might have been anticipated for all hydrophobic suspensions. Voet (1938) points out that these metal oxide sols are intermediate between the hydrophilic and hydrophobic extremes, and therefore more hydrated than the gold sols, and suggests that the order $Cl' > Br' > I'$ is to be ascribed to the greater dehydrating effect of the smaller ions. (Hydration increases as the ionic size decreases.)

The position is seen to be in need of clarification, which may come with the further study of the action of salts upon carefully purified sols.

The coagulating powers of organic ions, even simple ones such as aniline hydrochloride, are much greater than expected from simple valence considerations, and in the extreme case of soaps adsorption may take place against an electrical repulsion and so stabilize (or peptize) rather than precipitate (see Fig. 20.7). The reason lies in the tendency of the hydrocarbon portion to be expelled from the aqueous phase, as will be considered in more detail in Chapter XXIV.

(2) *Non-electrolytes.* Addition of non-electrolytes such as alcohol, ether, phenol, etc., leads in general to sensitization of the suspension towards electrolytes, the efficacy increasing with the chain-length of the compound: this suggests that the effect arises in some way from adsorption on to the particle surface.

Adsorption will give rise to a layer of reduced dielectric constant which by increasing the repulsive forces between the adsorbed ions will lead to some desorption and hence to a reduction in the charge on the particle. This will in turn lower the ζ potential and hence the stability (Chapter V). Some change in the opposite direction will arise from the decreased dielectric constant of the bulk medium (since $\zeta \simeq (Q/Dr)\{1/(1+\kappa r)\}$), but in general, with positive adsorption on to the particle surface, this will provide insufficient compensation and the overall stability will decrease.

One further effect of the reduced dielectric constant of the bulk medium is that the ionic atmospheres will be drawn closer to the particles and this, as discussed in Chapter V, will also tend to reduce the stability of the suspension.

(3) *Other colloids.* Mixing two similarly charged hydrophobic suspensions produces no effect (unless chemical reactions can occur); if they are oppositely charged mutual precipitation occurs unless one is in considerable excess. For example a negative As_2S_3 sol mixed in various ratios with a positive Fe_2O_3 sol shows mutual coagulation over a certain range. Precipitation arises primarily from electrical forces and is thus similar to that produced by electrolytes.

Several examples of mutual precipitation of similarly charged hydrosols are known, but in all such cases likely chemical reactions can be suggested to account for the apparent discrepancy. One such example is arsenious sulphide and Odén sulphur sols (both negative), where reaction occurs between H_2S present in the former and $H_2S_5O_6$ (and other polythionic acids) in the latter, e.g.

$$5H_2S + H_2S_5O_6 \rightarrow 10S + 6H_2O.$$

When a hydrophilic colloid is added to a hydrophobic suspension widely differing phenomena can occur—protection, sensitization, or even in some cases coagulation. The explanation here is also based chiefly upon adsorption concepts.

Most hydrophilic colloids show a marked tendency to be adsorbed at an air/water or oil/water interface, as indicated by the lowering of the boundary tension, so there are good grounds for expecting a similar adsorption at the solid/liquid or liquid/liquid interfaces present in suspensions. The charge carried by the suspensions would be expected to modify the extent of adsorption, particularly in the initial stages, but not to be the decisive factor as in the case of simple ions.

In general terms it can be stated that if the hydrophilic and hydrophobic colloids carry opposite charges then addition of *small* amounts of the former will tend to sensitize, or even in some cases to precipitate, the suspension. *Large* amounts generally stabilize the suspension which is then found to carry the charge of the hydrophilic colloid—this action is termed 'protection' (e.g. proteins on quartz or oil particles, p. 308).

When the charges on the two types are the same then only protection is observed.

Sensitization and protection can be shown very simply by using gelatine, or other proteins, as the hydrophilic colloid, since these can be given either a negative or positive charge merely by adjusting the pH to above or below the iso-electric point (about 4·7 in the case of most gelatines). Sols positively charged, such as alumina and ferric oxide, are thus sensitized or coagulated when small amounts of gelatine are added at neutral pH, but protected at low pH. The converse behaviour is observed with negative sols such as arsenious sulphide and most metal sols. The amounts required for maximum sensitization are generally extremely small, e.g. 5×10^{-4} per cent. gelatine for a 0·01 per cent. gold sol (Overbeek, 1938).

It seems most probable, as already discussed in Chapter V, that the sensitization or coagulation arises from the ζ potential of the suspension being reduced by the oppositely charged hydrophilic colloid. The amount of adsorption necessary to reduce ζ to very small values would be very much less than that required to cover the surface completely, since the free charges occupy such a small fraction ($\ll 1$ per cent.) of the surface. No marked compensating *increase* in stability due to the adsorbed hydrophilic colloid would therefore be expected, and the stability accordingly decreases with the fall in ζ potential. (For protection it seems that a complete monolayer is required (see below).)

As might be expected, hydrophilic colloids vary very considerably in their protective action, and various methods for comparing them have been suggested, of which Zsigmondy's 'gold-number' is the best known. An outline of this method and of the results obtained has been given earlier (p. 15). Whilst the 'gold number' does give some measure of the protective power the values vary somewhat with the actual conditions used. Also it is most improbable that they should apply equally

DILUTE SUSPENSIONS (SOLS)

to *all* types of suspensions in view of what is now known about adsorption upon solid and liquid surfaces.

When fully protected the particles are completely covered with the hydrophilic colloid, which then determines the stability, electrokinetic, and other properties of the suspension. (For a discussion of electrokinetic effects see Chapter XII.) The thickness of the adsorbed film can be calculated from the number and size of the particles, assuming complete adsorption to occur at the lowest concentration giving protection. Values of 5–10 A which have thus been calculated for gelatine and proteins on gold sols are in good agreement with the dimensions of a monolayer (probably surface denatured in the case of proteins).

(d) Kinetics of coagulation

Smoluchowski, who first worked upon the kinetics of coagulation, differentiated between the two extreme cases of 'rapid' and 'slow' coagulation, supposed respectively to occur in the region where ζ is almost zero, and in the region of the 'critical' potential which is usually about 20–40 mV.

In the 'rapid' zone, since ζ is approximately zero, there will be no repulsive forces between the particles, and every collision, resulting from kinetic (Brownian) agitation, should lead to adhesion. Smoluchowski (1917) worked out the probability of the various types of collisions which lead to aggregation, e.g. single particles to give doublets, singlets and doublets to give triplets, between doublets to give quadruplets, and so on. His final equation was

$$n_t = n_0/(1+t/\theta), \qquad (20.3)$$

where n_t = number of particles/c.c. at time t, n_0 = number originally present (i.e. when $t = 0$), θ = specific coagulation time, i.e. time when the number of particles is just halved, or

$$\theta = 1/kn_0 = 1/4\pi Drn_0,$$

where k is the velocity constant, D the diffusion coefficient, and r the radius of the sphere of action (twice the radius of the particle). For comparison with experiment eqn. (20.3) is inverted, giving

$$1/n_t - 1/n_0 = t/\theta n_0, \qquad (20.4)$$

i.e. the plot of $1/n_t$ against t should be linear on the above theory.

Eqns. (20.3) and (20.4) have been tested by a number of workers with various rapidly coagulating suspensions in both liquid and gaseous media and appear to fit the experimental data within the experimental error (e.g. Hatschek, 1921; Kruyt and van Arkel, 1920; Whytlaw-Gray and Patterson, 1932). Fig. 20.10 shows how well eqn. (20.4) is obeyed in the coagulation of certain smokes.

For 'slow' coagulation, where ζ is small but not zero, Smoluchowski introduced a factor p, the probability that a collision is followed by adhesion (i.e. the fraction of fruitful collisions). Eqn. (20.3) then becomes

$$n_t = n_0/(1+pt/\theta), \qquad (20.5)$$

or

$$1/n_t - 1/n_0 = pt/n_0\theta, \qquad (20.6)$$

i.e. the plot of $1/n_t$ against t should be linear and only differ from that of rapid coagulation in having a smaller slope (as $p < 1$). In this case, however, there is no measure of agreement between theory and experiment, the coagulation being frequently autocatalytic in character.

Suspensions in organic media

Relatively little work has been done upon suspensions in organic media, although these find a number of technical applications, such as in the electrodeposition of alkaline earth carbonates from suspensions in alcohol or ketones (for coating valve filaments), colloidal graphite in lubricating oils, etc.

When insoluble powders such as quartz or alumina, of particle size a few microns, are shaken with organic liquids no stable suspension is formed if the liquid is non-polar (e.g. CCl_4, C_6H_6). With polar liquids such as alcohols and ketones stable suspensions are obtained in some cases, depending upon the material and the liquid. The comparative stability of these suspensions has been ascribed to an electrical double layer, and the principal peptizing ions in the case of oxide particles appear to be either OH' or H^+ (Verwey, 1941), probably arising from traces of water present.

Where lyophobic suspensions of some stability have been prepared in non-polar organic media the origin of the stability has been a question of some discussion. Some ascribe it to electrical forces, the requisite charge arising spontaneously by friction between particle and medium (see p. 47), but adsorption of traces of impurities, particularly ions, seems a more plausible explanation.

Suspensions in gaseous media (aerosols)

Fogs, mists, and smokes are well-known aerosols, the disperse phase being liquid in the first two cases and solid in the third. Less obvious but equally important are the air-borne suspensions of living bacteria, viruses, and moulds, a problem of shelter-life during the recent war successfully tackled by the development of disinfectant sprays. Smokes find considerable uses in war-time for camouflage purposes and for poison-gas dispersal (e.g. 'lewisite'). There is no need to enlarge upon the undesirable aspect of most fogs and smokes which still present a major industrial problem. However, aerosols do find some practical uses in peace-time, in particular for insecticidal sprays and for frost-prevention in orchards, etc., by means of 'smudge-pots'. The action in the last case is believed to be due to the smoke particles from the burning oil reflecting back the ground radiation.

(a) *Preparation*

Just as with suspensions in liquids it is possible to use either dispersion or aggregation methods.

(1) *Dispersion methods*

In the case of liquids disintegration can be obtained by relatively simple mechanical means, as, for example, the usual scent-spray, in which a fine jet of liquid is shattered by a rapid air-stream. Modern high-pressure atomizers yield particles almost entirely in the colloidal range, a very desirable feature if a slowly sedimenting mist is required. An extreme case of mechanical disintegration is the use of explosives, as in gas-shells, a method clearly applicable to solids as well as to liquids.

Another dispersion method, applicable only to liquids, has been mentioned earlier in this chapter, namely the use of high potentials, as from a Wimshurst machine, to cause the break-up of a stream emerging from a fine nozzle. As pointed out, the effect arises from the effective reduction in surface tension brought about by the charge.

(2) *Aggregation methods*

Aggregation from the vapour phase to produce particles of colloidal size has as its first requisite the production of a state of supersaturation in a vapour, which is then relieved by condensation upon nuclei as discussed below. The more common ways in which this supersaturation has been obtained can be grouped as follows:

1. Chemical reaction, e.g. $NH_3(gas) + HCl(gas) \rightarrow NH_4Cl(solid)$.
2. Low-temperature volatilization, e.g. by passing an air-stream over organic compounds, such as stearic acid, heated to temperatures of a few hundred degrees C.
3. High-temperature volatilization, e.g. burning

$$Mg + O_2 \rightarrow MgO(solid).$$

4. Arc discharge, for metals only, giving chiefly oxides.

(i) *The formation of nuclei.* In the case of aerosols various particles are thought to act as condensation nuclei, depending upon the conditions. Air normally contains sufficient dust particles (which may be of organic or inorganic nature) to act as nuclei, but in laboratory studies of aerosols this complicating factor may be removed by careful filtration or other suitable treatment. Gaseous ions can act as condensation nuclei, and this forms the basis of the cloud-chamber technique for detecting the path of radio-active disintegration products. Ions are normally present in air, although in extremely low concentration unless augmented artificially as by an electrical discharge, X-rays, ultraviolet light, etc. If the supersaturation rises to sufficiently high values, then molecules may act as condensation nuclei. In the case of stearic acid, for example, taking its radius to be 6 A, a supersaturation of 8–10 should suffice (Whytlaw-Gray and Patterson, 1932).

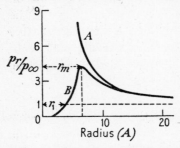

FIG. 20.9. Effect of curvature (curve A) and constant charge (curve B) upon the vapour pressure of small drops.

(ii) *Condensation upon nuclei.* Charged as well as uncharged particles have to be considered in the case of aerosols. For spherical particles of radius r, density ρ, surface tension γ, and charge Q it has been shown (p. 89) that

$$\frac{RT}{M}\ln\frac{p_r}{p_\infty} = \frac{1}{\rho}\left(\frac{2\gamma}{r} - \frac{Q^2}{8\pi D r^4}\right), \qquad (20.7)$$

where p_r = vapour pressure of drop of radius r, p_∞ = vapour pressure of plane surface (i.e. $r = \infty$), M = molecular weight of material composing the drop, D = dielectric constant of continuous phase (unity in this case).

Considering first uncharged particles ($Q = 0$), then condensation can only occur if the vapour pressure exceeds p_r. Since the ratio p_r/p_∞ increases as r decreases, as shown in Fig. 20.9 curve A for the case of water droplets, it is clear that very substantial degrees of supersaturation are necessary before small uncharged nuclei can grow (e.g. $p_r/p_\infty \simeq 3\cdot 4$ for water droplets with $r = 10$ A).

The presence of a charge reduces the surface tension effect; curve B in Fig. 20.9 shows how a constant charge modifies the equilibrium vapour pressure of water droplets of different sizes. For a given particle size condensation can then take place at lower supersaturations, and if p_r/p_∞ exceeds the value at the maximum in the curve then condensation can proceed upon droplets of any size, since the equilibrium vapour pressure diminishes as r increases or decreases.

DILUTE SUSPENSIONS (SOLS)

According to the above equation the sign of the charge should be immaterial, but it is found experimentally in the case of water vapour that condensation occurs rather more easily on negative than on positive ions. This has been ascribed to the natural double layer of the drop, which in the case of water tends to make the surface negative (Thomson, 1928). Treating the double layer as a parallel plate condenser of thickness d, where $d \ll r$, then intensity of field at surface, $I = Q/Dr^2 + 4\pi\sigma/D$, where $\sigma =$ surface density of charge. Mechanical force per unit area due to charge effects (and which opposes the surface tension) $= DI^2/8\pi$. Hence if σ is negative (as in water), I is increased if Q is negative, and decreased if Q is positive, so that condensation is assisted by the double layer in the first case (negative ions) and opposed in the second (positive ions).

If the nucleus possesses a very high affinity for water vapour, as in the case of certain inorganic compounds like SO_3 and P_2O_5, then this may reduce p_r sufficiently to enable condensation to proceed even from unsaturated air. (The stability of city fogs is generally put down to the presence of sulphuric acid and other hygroscopic nuclei arising largely from the combustion of raw coal.)

To obtain and control the degree of supersaturation in the experimental study of mists, it is usual to employ adiabatic expansion, as in Wilson's well-known cloud-chamber. For an adiabatic change

$$pv^\gamma = \text{constant},$$

and hence
$$\frac{T_1}{T_2} = \left(\frac{v_2}{v_1}\right)^{\gamma-1}, \qquad (20.8)$$

where v_1 and v_2 are the volumes at temperatures T_1 and T_2 and γ the ratio of the specific heats. As v_1, v_2, and T_1 are known, the final temperature T_2 follows, and if the air was originally saturated with water vapour, the supersaturation (i.e. p_{T_1}/p_{T_2}) also, since p_{T_1} and p_{T_2} are known from tables.

With pure air and water vapour no condensation occurs with $v_2/v_1 < ca.$ 1·25 (i.e. supersaturation $S > ca.$ 4·2). From $v_2/v_1 = 1·25$ to 1·28 (i.e. S between 4·2 and 5) condensation occurs on negative ions, and from $v_2/v_1 = 1·31$ to 1·34 (i.e. S between 5·8 and 6·8) on positive ions as well, for reasons given above. (For detecting α and β particles the range of S between 4·2 and 6·8 is therefore used.) With expansions greater than $ca.$ 1·38 ($S > 8$) condensation occurs even when extraneous nuclei and ions are believed absent, suggesting that the nucleus here is a small spontaneous aggregate of water molecules.

The character of the condensate depends very much upon the supersaturation and the number of nuclei present. In agreement with von Weimarn's principle as discussed above the dispersion is very coarse if low concentrations of nuclei are present, as in the case where the natural ionization of the air provides the nuclei and S is less than 6·8, and very fine when S is high (8–12) since large numbers of nuclei are formed spontaneously in this region.

(b) *Measurement of size, charge, and concentration of aerosols*

Determination of the number of suspended particles per unit volume, which may be of the order 10^5–10^6/c.c., is best carried out in one of two ways, for which the basic principles are given below. The first is a modification of the usual ultramicroscope method for counting colloidal suspensions, the aerosol being drawn into a special cell consisting of two parallel glass plates and the particles counted there (Whytlaw-Gray and Patterson, 1932). In the second, the condensation method, the particles are used as condensation nuclei for supersaturated water vapour, giving droplets large enough to be photographed. A continuous action apparatus for this purpose, devised by Green (1927), allows 50 counts to be completed in 100 seconds, so that the process of aggregation can be followed quite readily. This second technique is, of course, limited to suspensions of solids and non-volatile liquids.

The particles in aerosols are frequently charged, and by applying an electrostatic field between two electrodes in the observation cell of the ultramicroscope the percentage of those which are positively, negatively, and un-charged can be obtained. The electrical character of a suspension depends very much on its method of preparation, those formed by violent chemical reaction or by the electric arc being much more electrified than those prepared by less drastic means. In some cases both positively and negatively charged particles appear together, at any rate in the early stages after formation, whereas in others all the particles are similarly charged. The charge in a number of cases is believed to arise from the capture of free ions present in the atmosphere.

If the size of a particle is known (see below), then the magnitude of the charge can be found by opposing gravitational attraction with a variable and measured electrostatic field, as in the well-known Millikan method of determining the electronic charge. The particle is observed in the ultramicroscope and if held stationary

$$mg = QX, \qquad (20.9)$$

where m and Q are respectively the mass and charge of the particle,

X is the field strength (applied voltage/distance between electrodes), and g the acceleration due to gravity. It is not possible to balance very small particles at all accurately, owing to Brownian agitation, and in such cases two measurements are necessary, the rate of fall under gravity and of rise in a suitable electric field. For this case

$$\frac{QX-mg}{mg} = \frac{v_e}{v_g}, \tag{20.10}$$

where v_g and v_e are the observed velocities in the first and second cases respectively.

The most general method for determining the size of aerosol particles is from their velocity of fall (v_g) under gravity, assuming a suitable resistance law. If the particles are spherical and if the simple Stokes expression can be used then

$$mg = \tfrac{4}{3}\pi r^3(\rho_p - \rho_a)g = 6\pi\eta r v_g, \tag{20.11}$$

where ρ_p and ρ_a are the respective densities of particle and medium (air), and r is the radius of the particle (see p. 211). For particles smaller than ca. $1\,\mu$ the mean free path of the air molecules becomes comparable with the particle dimensions, and the Cunningham extension of Stokes's law is necessary, giving

$$\tfrac{4}{3}\pi r^3(\rho_p - \rho_a)g = 6\pi\eta r v_g(1+Al/r)^{-1}, \tag{20.12}$$

where l is the mean free path of the gas molecules and A a constant. (The magnitude of the term Al is ca. 9×10^{-6} cm. for air under ordinary conditions.) Difficulties arise in the case of solid dispersions owing to two factors—the departure from spherical shape which, as the electron microscope has confirmed, may be very considerable, and departure from bulk density. If, as is not uncommon, the sedimenting unit consists of an aggregate of loosely packed asymmetric particles, then the density clearly cannot be taken equal to that in bulk.

In the case of smokes (where the particles are solid) the particle size and shape can usually be obtained most directly from the electron microscope (see p. 458).

(c) Coagulation of aerosols

All aerosols appear to coagulate spontaneously although the rates differ, depending upon such factors as the size, shape, charge, concentration, and chemical nature of the dispersed material, and the viscosity of the gaseous medium. The particles are brought into contact by Brownian agitation and the principal factor which opposes coalescence is the charge (assuming its sign to be the same for all particles).

DILUTE SUSPENSIONS (SOLS)

Whytlaw-Gray and his collaborators have shown that for many systems the coagulation is given by an equation similar to eqn. (20.4) above, namely

$$\frac{1}{n_t} - \frac{1}{n_0} = kt, \qquad (20.13)$$

where n_t is the number of particles per c.c. at time t, n_0 the number at

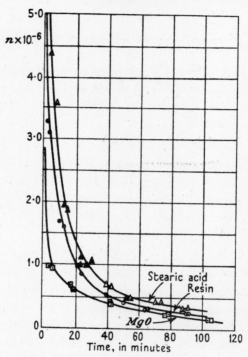

Fig. 20.10 (a). Coagulation of various smokes (MgO, resin, stearic acid), as measured by the decrease in the number (n) of particles/c.c. with time (t).

zero time, and k a constant. From Fig. 20.10 (a) and Fig. 20.10 (b) it is clear that $1/n$ is linear with t, as it should be from the above equation. ($1/n$ has been termed the 'particulate volume', since it is the average space inhabited by a particle.) Thus the Smoluchowski theory of coagulation appears to be applicable to the coagulation of aerosols as well as to suspensions in liquids (see also Whytlaw-Gray and Patterson, 1932).

From the practical point of view coagulation can be accelerated by agitation, by supersonic waves, and by electrical methods, of which the electrostatic precipitator, such as that due to Cottrell, is the most important. The principle of the electrostatic precipitator is shown diagrammatically in Fig. 20.11. Wires X are charged to a very high voltage (12,000–15,000 volts d.c.) causing considerable ionization of the air in this region; this part is therefore termed the ionization chamber.

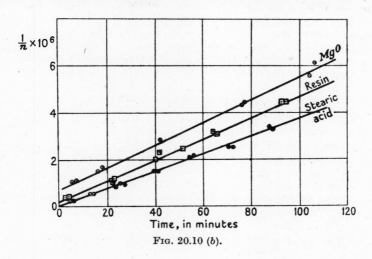

Fig. 20.10 (b).

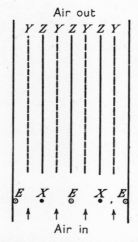

Fig. 20.11. Illustrating the principle of an electrostatic precipitator.

The aerosol then passes through the precipitation chamber, in which the alternate plates Y and Z (Fig. 20.11) are respectively charged (2,000–5,000 volts d.c.) and earthed. Owing to the considerably increased charge arising from the uptake of ions in the ionization chamber, the particles are rapidly attracted to the grounded plates, where they are discharged and deposited.

DILUTE SUSPENSIONS (SOLS)

GENERAL REFERENCES

CLAYTON, 1943, *Emulsions and their Technical Treatment*, Churchill. Includes some aspects of dilute suspensions.

FREUNDLICH, 1926, *Colloid and Capillary Chemistry*, Methuen. An excellent account of the early work on suspensions.

Symposia and discussions:
  Hydrophobic Colloids, 1938, Centen, Amsterdam.
  'The electrical double layer', 1940, *Trans. Faraday Soc.* **36**, 1 and 711.
  'The oil/water interface', 1941, ibid. **37**, 117.

OTHER REFERENCES

BUZÁGH, VON, 1943, *Kolloidzschr.* **103**, 119.
CARRUTHERS, 1938, *Trans. Faraday Soc.* **34**, 300.
DICKINSON, 1941, ibid. **37**, 140.
FREUNDLICH, 1926, *Colloid and Capillary Chemistry*, Methuen.
GREEN, 1927, *Phil. Mag.* **4**, 1046.
HATSCHEK, 1921, *Trans. Faraday Soc.* **17**, 499.
KRUYT and VAN ARKEL, 1920, *Rec. trav. Chim. Pays-Bas*, **39**, 656.
MARDLES, 1940, *Nature*, **145**, 970.
—— 1941, ibid. **148**, 345.
OVERBEEK, 1938, *Hydrophobic Colloids*, p. 121, Centen, Amsterdam.
POWIS, 1915, *Z. phys. Chem.* **89**, 186.
RIDEAL, 1930, *Surface Chemistry*, 2nd edn., Cambridge.
SMOLUCHOWSKI, 1917, *Z. phys. Chem.* **92**, 129.
STAUFF, 1940, *Z. phys. Chem.* **187 A**, 107, 119.
SVEDBERG, 1928, *Colloid Chemistry*, 2nd edn., *Chem. Catalog Co.*, New York.
TAMMANN, 1903, *Kristallisieren und Schmelzen*, Leipzig.
THOMSON, 1928, *Conduction of Electricity through Gases*, Cambridge.
VERWEY, 1941, *Rec. trav. Chim. Pays-Bas*, **60**, 618.
VOET, 1938, *Hydrophobic Colloids*, p. 109, Centen, Amsterdam.
WELLS, 1946, *Ann. Rep. Chem. Soc.* **43**, 62.
WHYTLAW-GRAY and PATTERSON, 1932, *Smoke*, Arnold.
WILLIAMS, 1940, *Trans. Faraday Soc.* **36**, 1042.

XXI
GELS AND PASTES

General introduction

As with so many colloidal systems no clear-cut line of demarcation exists between gels and pastes, nor between sols and colloidal solutions on the one hand, and gels and pastes on the other. Definitions are, however, useful, despite the impossibility of covering all the types of behaviour actually encountered. For purposes of comparison it is convenient to include here also the dilute colloidal systems of two components.

Definitions

(i) *Sol or suspension*

A dilute ($<$ 1 per cent. by volume as a rule) dispersion of lyophobic nature. Thermodynamically unstable. Two-phase system with the dispersion medium as continuous phase.

(ii) *Colloidal solution*

A dilute dispersion of lyophilic nature. Not essentially different, except in size of solute molecules, from ordinary molecular solutions. Disperse particles usually, but not invariably, molecularly dispersed. Thermodynamically stable. Single-phase system.

(iii) *Gel*

A two-component system of a *semi-solid* nature, *rich in liquid*. The *gelling component* is usually of the comparatively lyophilic type, and present in concentrations $<$ *ca.* 10 per cent. Probably bi-phasic, each phase being continuous. Thermodynamically unstable(?).

(iv) *Paste*

A concentrated ($>$ *ca.* 10 per cent. by volume) dispersion of fine solid particles in a liquid continuum. Show very definite elastic or plastic behaviour. Two-phase systems. Thermodynamically unstable(?).

The gels and pastes have one factor in common which is not usually shared by the sols and colloidal solutions, namely their frequent dependence upon the time factor, and hence upon the past history of the specimen. As we shall see this is often of marked importance and makes the characterization of these materials exceptionally difficult.

Gels

Certain gels, or jellies as they are commonly called, are of everyday occurrence. Table jellies, for example, are usually made from gelatine

(2–3 per cent.), jams are gelled by pectins which are complex carbohydrate derivatives, and junkets by milk proteins modified by the enzymatic action of rennet. Agar *slopes* or *plates*, used so extensively in bacteriological work, consist of a suitable nutrient medium gelled by the addition of 1–2 per cent. of agar, a polymer of carbohydrate nature obtained from certain sea-weeds. *Soft soaps* (i.e. potassium salts of the higher fatty acids) are frequently encountered in jelly form, and other fatty acid salts such as aluminium, calcium, and zinc oleates, stearates, etc., are the thickening agents in many greases, which may also be regarded as gels.

Less widely known are the *thixotropic* gels, in which liquefaction to the *sol* or *colloidal solution* can be produced by mechanical agitation, as by shaking (see below). Quicksand is thixotropic, as are certain clays such as bentonite, and some mineral deposits of volcanic origin, when mixed in suitable proportions with water. Thixotropic gels of ferric oxide, which have a reddish-brown colour and become more transparent in the sol state, possibly form the contents of the reliquaries used in certain religious ceremonies such as 'the liquefaction of the blood'. (The occurrence of these reliquaries in such places as Naples suggests that the materials used are probably of volcanic origin.)

Terminology and classification of gels

The terms *sol*, *gel*, and *jelly*, have been used to denote rather different systems by different workers. Here we regard the terms jelly and gel as synonymous, but prefer the latter as it is in more common scientific use. The definitions of sol and gel have been given above, but as pointed out, no hard and fast rules can be given owing to the range of systems actually encountered.

We shall refer to the change from the gel to the fluid state (or vice versa) as the *sol-gel transformation*, in conformity with accepted practice, although the sol may in some cases be a true colloidal solution rather than a sol as defined above.

Gels may be classed as *thermal* (or *non-isothermal*) and *thixotropic*, depending upon whether the sol-gel transformation can be brought about by temperature change or by mechanical agitation.

Another method of subdivision which has been frequently used is into *true* and *false* gels, depending upon whether the system remains stable or tends to exude the liquid medium. This classification will not be followed here since its validity is very doubtful (see below).

One further point of terminology warrants some consideration; that is the range of liquid content in gels. According to the definition as given above, emphasis is placed upon the outstanding property of combining a solid, or semi-solid, behaviour with a high liquid content. Unfortunately the term gel has also been applied to such systems as plasticized polymers, wood swollen by water vapour, concentrated dispersions of soap crystallites, and so on, where the liquid content is frequently much less than the gelling component. We feel this extension to be undesirable, and although it is clear that no precise figure can be given, it is useful to regard a gel as possessing not more than about 10 per cent. of gelling agent. According to this definition the systems just quoted are to be excluded.

Formation of gels

Gels are formed from colloidal suspensions or solutions (the fluid sol state) by reducing the solubility of the colloidal material to an extent sufficient to enable the particles to link together in some manner, forming a weakly solid mass entraining the dispersion medium. Reduction in solubility is produced in three principal ways: by cooling or, less frequently, warming the sol, by addition of a precipitating liquid, and, finally, by addition of salts. Closely related is the use of chemical reactions to form the gelling agent *in situ*. These will be briefly considered in turn with a few illustrative examples.

(a) By variation of temperature

In the case of simple solutions there is a definite amount of solute which can be held in stable solution at any given temperature (the saturation solubility), this amount usually decreasing as the temperature is lowered. When a saturated solution is cooled, solute separates out to maintain equilibrium, although supersaturation may be encountered if the crystalline phase or suitable crystallization nuclei are absent (see p. 559).

With solutions of many colloidal materials, particularly linear polymers, the position is rather different. Many of these polymers appear to show either a negligible or an infinite solubility in a given medium, a very striking phenomenon discussed in more detail in Chapter XXVII. The change-over from one extreme to the other can often be effected by alteration in the temperature or composition of the medium, and when this leads to a reduction in solubility the phenomenon of gelation is frequently observed.

Gelatine, for example, disperses readily in hot water, giving what is probably a molecular solution, and on cooling to room temperature this sets to a clear uniform gel if the concentration exceeds *ca.* 1 per cent. Very similar, although displaced on the concentration scale, is the behaviour of agar, polyvinyl alcohol, and the water-soluble soaps such as sodium oleate. In this last type the 'polymer' probably consists of soap molecules aggregated to form very asymmetric micelles of a plate-like or thread-like nature (see Chapter XXIV). Dibenzoyl cystine is one of a number of organic compounds with rather remarkable gelling powers, as little as 0·2 per cent. giving a very rigid gel.

Non-aqueous gels are often prepared by cooling sols of suitable strength. The gels given by a few per cent. of aluminium fatty acid salts, such as the stearate, oleate, or naphthenate, in hydrocarbon media are well known and frequently very stable.

Examples of the 'inverse' type are also known, i.e. where gelation occurs as the temperature is *raised*, in this case also due to the reduction in solubility. Certain cellulose nitrates in a number of solvents such as alcohol, and certain methyl celluloses in water, show this behaviour.

(b) *By addition of precipitating liquids*

The principle here is to add just sufficient precipitant to the sol to produce the gel state without going farther and producing complete precipitation. With polymers this frequently presents little difficulty, particularly if by rapid mixing high local concentrations of precipitant can be avoided. Solutions of ethyl cellulose, cellulose aceto-stearate, polystyrene, etc., in benzene can thus be gelled by rapid mixing with suitable amounts of a non-solvent such as petrol ether. 'Solid alcohol' is obtained by rapidly mixing saturated aqueous calcium acetate with about ten times its volume of alcohol.

In some cases the precipitant can be introduced into the sol through the vapour phase, or through a membrane impermeable to the colloid, both processes giving the requisite absence of high local concentrations of precipitant.

(c) *By addition of salts*

Addition of salts to colloidal solutions or sols can produce a variety of effects, including gel formation, depending upon the particular system. With dispersions in organic media, the effect, if any, generally arises from chemical reaction, and so comes within the purview of the following section. Aqueous dispersions can, however, be affected by salts even in the absence of chemical reactions. With the extreme type of hydrophobic suspensions quite a low concentration of an indifferent salt brings about coagulation, and gel formation is seldom or never observed, largely on account of the low concentration of disperse phase (of the order 0·1 per cent.). However, as we pass to more hydrophilic suspensions, e.g. aluminium and ferric hydroxides, vanadium pentoxide, bentonite, etc., so the amount of dispersed material can be increased, and gels can frequently be obtained by suitable additions of salts, usually about one-half of the amount needed for complete precipitation. Thus

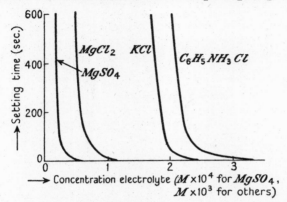

Fig. 21.1. Effect of various salts upon the setting time of an alumina sol exhibiting thixotropy.

with the positively charged hydroxide sols, divalent ions such as SO_4'' are much more potent than the univalent such as Cl'. The gels so formed are frequently thixotropic in behaviour, and this forms the standard method by which such gels are produced. Some typical data for a 2 per cent. alumina sol prepared by Crum's method (boiling the acetate with water) is given in Fig. 21.1.

The extreme type of hydrophilic suspensions, e.g. gelatine, proteins, are only affected by high salt concentrations, probably due to a salting-out effect as discussed earlier (p. 125). As a rule the material separating is in the form of solid aggregates, but under certain conditions it may separate as liquid aggregates, termed 'coacervates' by Bungenberg de Jong and his co-workers, who have made a study of this phenomenon. Coacervation between two oppositely charged hydrophilic sols is known as 'complex coacervation'.

Mention has already been made (p. 16) of the Hofmeister or lyotropic series of ions: such a series is also found to hold for the setting and liquefaction of gels. Thus in the case of anions we have the order SO_4'', acetate, tartrate, Cl', NO_3', I', CNS'; the strongly hydrated ions at the beginning favouring gelation, whilst the weakly hydrated ones at the end tend to produce liquefaction (peptization).

Gelation of sols of gelatine and other ionogenic water-soluble polymers is very dependent upon the pH as well as the nature and concentration of the salts present. Fig. 21.2 shows how the resistance to shear (taken as an indication of gelation) and the turbidity (an indication of the tendency to precipitate) vary with pH for 0·6 per cent. gelatine solutions (Kraemer, 1926). At the iso-electric point (pH *ca.* 4·9), the tendency

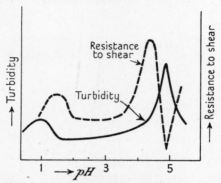

FIG. 21.2. Variation of the turbidity and shear-resistance of 0·6 per cent. gelatine with pH.

to precipitate is seen to be high (as indicated by the high turbidity) and the gelation tendency is low, but both increase on either side of the iso-electric point, i.e. as the net charge increases.

(*d*) *Chemical reactions leading to gel formation*

Some mention of this topic has been made earlier (p. 561) in connexion with von Weimarn's theory of the formation of colloids. The rapid throwing out of large amounts of material by chemical reaction was shown to lead to gel formation, e.g. barium sulphate gels from saturated solutions of barium thiocyanate and manganese sulphate. Substances which tend to separate as amorphous precipitates frequently give gels when first formed, the hydroxides of aluminium, iron, chromium, and silicon being good examples. Of these, the formation of 'silica gel' by various reactions, such as between sodium silicate and acids, or between silicon tetrachloride or various organic silicon compounds (e.g. $Si(OEt)_4$) and water, has been the most studied.

Changes occurring during gel formation

A good deal of information about the structure of gels has come from the study of the physical changes which occur in the sol–gel transformation, of which the following have been the most studied. (See Heymann, 1936, for a more detailed account of the earlier work.)

(a) Thermal

The cooling curves of solutions of low molecular weight substances usually show pronounced breaks or arrest points when solute begins to separate. With sols the onset of gelation produces a small but definite arrest, showing that the sol–gel change is accompanied by liberation of

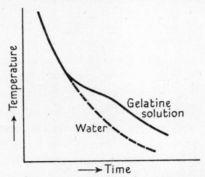

Fig. 21.3. Cooling curves for water and for gelatine solutions under identical conditions.

heat, i.e. by a diminution in heat content. Fig. 21.3 compares the cooling curves of water and strong gelatine solutions under identical conditions, the gelation and separation of the two curves both occurring between 30° and 35° C. A similar behaviour has been reported with soap solutions (e.g. Lawrence, 1935).

With thixotropic gels any thermal effects in the sol–gel transformation appear to be negligible, a not surprising result in view of the small amount of mechanical energy required to bring about liquefaction.

(b) Optical

Ultramicroscopic examination has been much used to indicate whether aggregation occurs during transition from sol to gel, the chief handicap being the small difference in refractive indices between the disperse phase and the dispersion medium in the lyophilic systems generally encountered in this work. This means that the resolving power is unlikely to approach the fine structure which is believed to exist in most gel systems.

Sols such as those of gelatine, agar, or silicic acid normally show few particles in the ultramicroscope, but during the transition to gels many more appear, these being in Brownian movement until stopped by gel formation. No continuous 'net' structure is seen and the scattering particles apparently remain separated. Threads and felt-like structures have, however, been reported in some concentrated soap gels, but the thickness of these threads must be considerable by molecular standards, at least 10–50 molecular layers.

The presence of linked fibrillar structures in soap/water systems is supported by some electron microscope pictures, although these suffer from the inherent drawback of referring to the system in the absence of dispersion medium. The photograph of sodium laurate curd, shown on Fig. 15.19, between pp. 456–7, shows these fibres extremely clearly.

The intensity of the scattered light usually increases markedly as gelation proceeds. This might arise either from an increase in particle size or from a reduced hydration increasing the refractive index difference between particles and medium. The changes in hydration are believed to be insufficient (see also below), and aggregation is therefore held to be primarily responsible, in agreement with the ultramicroscope observations.

With an ionogenic colloid such as gelatine the Tyndall effect is most marked at the iso-electric point. This again points to aggregation in agreement with the fact that the solubility is also at a minimum there.

Thixotropic alumina and ferric oxide systems are interesting in that gelation brings about an increase in turbidity; with ferric oxide some change in their reddish-brown colour occurs at the same time (Freundlich et alia, 1932).

Other optical quantities measured have been the depolarization of the scattered light, and the birefringence produced by flow, both of which can give some information about the size and shape of the particles present (see Chapters XIV and XV).

An increase in aggregation during gelation, leading to more ordered arrangements and ultimately to crystallization, should be shown up by the X-ray diffraction pattern. In the case of sols, such as alumina and iron oxide, which already contain crystalline particles the method is necessarily restricted, but with gelatine, starch, and other systems, where the sols are amorphous to X-rays, gelation leads to the development of distinct X-ray diffraction lines as would be expected (e.g. Katz et alia, 1932).

(c) *Electrical conductivity*

Changes in electrical conductivity during the formation of a gel appear at first sight to be remarkably small in relation to the enormous increase in the viscosity. Thus with iso-electric gelatine the conductivity decreases by only a few per cent. (Greenberg and Mackey, 1931), and with soaps the transition from sol to clear jelly is accompanied by no measurable change (Laing and McBain, 1919, 1920), although a marked

GELS AND PASTES

fall is found when the further change to the hard, opaque 'curd' state occurs. The reasons for such a behaviour will be considered when the structure of gels is examined later.

If the sol contains electrolytes the conductivity changes during gelation are usually negligible, the slight increase observed in some cases being ascribed to a secondary effect, the decreased ionic adsorption on the colloid arising from its aggregation.

Another rather striking result is that the electrical mobility of solid colloidal particles such as quartz, suspended in a gelatine sol, is unaffected by gelation, at any rate in the weaker gels (Freundlich and Abramson, 1928). This shows that the gel structure must break down and reform extremely readily, a type of micro-thixotropy.

(d) Dielectric constant

The dielectric constant (D) of colloidal solutions varies with the frequency or wavelength (λ) of the field in which it is measured, for reasons discussed in Chapter XIV. In region I of Fig. 21.4 only the solvent (water in this case) contributes to the polarization, in II the wavelength is now sufficiently long ($>$ a few metres as a rule) for the colloid also to contribute.

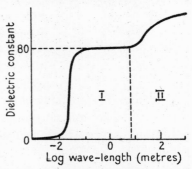

Fig. 21.4. Variation of dielectric constant of colloidal solutions with wavelength.

Considering region II it would be expected that D should fall on gelation owing to the diminished mobility of the colloidal particles; this has indeed been observed with V_2O_5 and gelatine sols. With region I, on the other hand, D will only fall when the system is gelled if an appreciable fraction of the water molecules lose their mobility by orientation around the immobilized gel particles. Kistler (1931) found experimentally that no such change occurred with bentonite, V_2O_5 or gelatine sols, the dielectric constant at a wavelength of 30 cm. remaining constant at about 80 during the sol–gel transformation. This again is not unexpected since thick orientated layers of solvent are now thought to be most improbable (see p. 125), and in any case the hydration does not appear to change much either (see also Fricke and Parker, 1940).

(e) Volume

The solution of a lyophilic colloid frequently produces a marked volume contraction. The formation of gel from sol also produces a volume change but of a much smaller order, requiring sensitive techniques, such as the use of a thermostatted dilatometer (see Fig. 21.5), for its measurement. Measurements at constant temperature are quite feasible even with thermal transformations owing to the time-lag which is so characteristic of sol–gel changes. Table 21.1, taken from Heymann (1936), shows that all types of volume change (positive, zero, or negative) can occur on gelation. These volume changes have frequently been attributed to changes in hydration, but this is by no means certain, and the question is taken up again later when the structure of gels is considered.

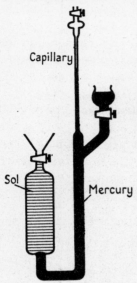

Fig. 21.5. Sensitive dilatometer for determining volume changes in the sol–gel transformation.

The volume changes are clearly very small and much less than the contractions observed when lyophilic colloids swell or dissolve in water, which for methyl cellulose and gelatine are of the order 6–8 c.c. per 100 gm. solid.

A rather different behaviour from the above is shown by silicic acid solutions. The volume change on gelation (an expansion in this case) is initially very marked, then slows down, but is still altering even after six months (see Table 21.2, taken from Heymann, 1936). It is evident that the processes responsible for the volume change and gel formation continue even in the stiff gel state.

TABLE 21.1

Sol	Volume change at 23° C.	
	in %	in c.c./100 gm. solid
Methyl cellulose 1·6%	+0·0020	+0·13
Gelatine 4·2% pH 8·8	−0·0028	−0·068
,, 6·0% pH 5·5	−0·0034	−0·057
,, 6·0% pH 3·8	−0·0033	−0·056
Ferric oxide 9·4%	0·0000	0·00

(*f*) *Mechanical properties* (*viscosity and elasticity*)

These naturally spring to mind as the most obvious changes in the sol–gel transformation. Coinciding with the marked increases in viscosity come the development of elasticity or plasticity, properties characteristic of solid bodies. (The methods used to detect and measure elasticity were given in Chapter XVI.)

TABLE 21.2

Time in hours	Volume increase		State
	in %	in c.c. per 100 gm. SiO_2	
1	0·012	0·17	liquid
3	0·028	0·41	,,
8	0·044	0·63	loose gel
24	0·057	0·83	clear jelly
92	0·073	1·06	,, ,,
240	0·084	1·22	slightly turbid
816	0·100	1·45	syneresis
1,536	0·109	1·58	,,
2,500	0·115	1·67	,,

Viscosity provides a simple way of detecting the onset of gelation, but the interpretation of such measurements is not easy, since non-Newtonian flow or 'anomalous viscosity' is invariably present during the sol–gel transformation if not before (see Chapters XIII and XVI for an account of the experimental methods and principal points of behaviour of colloidal systems). Nevertheless sols such as those of gelatine, agar, and nitrocellulose show an increase in elasticity during gelation concomitant with the increase in anomalous viscosity, suggesting that the two are intimately related.

The development of elastic properties during the change from sol to gel does not in general occur sharply, and elasticity has been detected in the sol state close to the transition point, particularly with gels of agar, gelatine, and other polymers. The question whether elasticity is detectable or not undoubtedly depends very largely upon the rate of relaxation of the 'gel structure' in relation to the time for the necessary measurements. By the earlier methods it was only possible to detect elasticity if no appreciable relaxation (flow) occurred in times of the order of a few seconds; more recently this minimum time has been reduced by a factor of a thousand or more. True solutions of high polymers have thus been shown to possess elastic properties, although the relaxation times (p. 383) are much smaller than for the common gels, for example about 10^{-4}–10^{-5} sec. for a 15 per cent. solution of polystyrene (molecular weight $ca.$ 100,000) in xylene at room temperature (Ferry, 1942).

The sharpness of the sol–gel change, which will be reflected in the development of elasticity, will also be affected by the degree of homogeneity of the gelling agent, particularly as regards chain-length in the case of polymers. In general, the shortest-chain members tend to have the greatest solubility and the wider the spectrum of chain-lengths the more diffuse the transition from the sol to the gel state. (The question of swelling and solubility of long-chain polymers is discussed in Chapter XXVII.)

The properties of gels

The above discussion of the sol–gel transformation has necessarily included a certain amount about the properties of gel systems, but there are many further points which warrant consideration.

(a) Elasticity and rigidity

The study of the elasticity under varying conditions is clearly a fundamental method for elucidating the structure of gels. (For experimental methods see Chapter XVI.) Most of the available data refers to gelatine and agar, which makes any detailed analysis impossible since the samples used were from various sources and of differing purity; nevertheless some general points emerge.

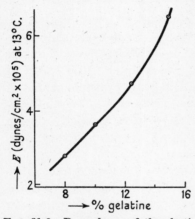

FIG. 21.6. Dependence of the elastic modulus (E) of gelatine gels upon the concentration of gelatine.

Different gelling agents differ widely in elasticity, agar gels, for example, being much more rigid than those given by gelatine at the same weight concentration.

For a given system the elastic modulus increases rapidly with concentration of gelling agent (approximately as the square of the concentration in the case of gelatine), as shown in Fig. 21.6. Addition of foreign substances may either decrease or increase the elasticity: with gelatine, for example, chlorides, iodides, and nitrates decrease the modulus, whereas sulphates and cane sugar raise it. Soaps, bile salts (e.g. sodium taurocholate), and lecithin also decrease the elasticity, whereas cholesterol increases it, observations which may have some bearing upon the problem of muscle physiology. It seems probable that compounds which lower the elastic modulus do so by peptizing some of the bonding points between the long molecules; this seems particularly likely with the soaps and iodides. Compounds with the opposite effect are ones which would be expected to increase the inter-molecular cohesion; for example SO_4'' can link chains through salt bridges such as

$$\rangle\overset{+}{N}H_3\ SO_4''\ \overset{+}{H_3}H\langle.$$

Hysteresis phenomena or permanent deformation are commonly observed, particularly at the higher deformations and longer times.

The question whether a stressed gel becomes anisotropic in its mechanical properties can be examined by utilizing the behaviour of contained gas bubbles (produced, for example, by an acetate buffer and sodium bicarbonate). If the tensile strength should be a minimum in one direction the gel will tend to split along the plane perpendicular to this, and a bubble present will become lenticular with its equatorial plane coincident with this plane. With gelatine gels Hatschek (1932) thus showed that with reversibly elongated and irreversibly compressed cylinders the tensile strength is a minimum in the direction of the axis, but with irreversibly elongated and reversibly compressed cylinders it was a minimum in directions perpendicular to the axis. The direction of minimum tensile strength during reversible deformation is thus at right angles to that after irreversible deformation, showing that some rearrangement of structure must have occurred during the requisite time interval. According to Poole (1933) the above results are to be expected if a gel possesses a fibrillar structure, which seems extremely probable in the case of gelatine.

(b) Optical properties

Gels as normally prepared are usually, although not invariably, optically isotropic, but become doubly refracting when suitably stressed as by elongation, compression, or shear. This gives rise to the so-called 'photo-elastic' effect, which can be observed in various ways, but usually under crossed Nicols.

Of the possible factors which might contribute to the observed double refraction two are identical with those considered in the case of sols (see p. 393), namely 'rod double refraction' arising from isotropic units arranged in an anisotropic manner, and 'intrinsic double refraction' due to the similar orientation of anisotropic particles. The third possibility, in which the gels show their link to isotropic solid bodies such as glass, arises from the density changes produced by tension and compression.

Gels are widely used for the photo-elastic study of stress patterns in engineering problems. Gelatine is particularly suitable on account of its extreme sensitivity from the point of view of double refraction. The usual arrangement employs crossed polaroids, giving a dark background for the unstressed gel, and from the number of fringes which

FIG. 21.7. Stress pattern from a uniform load on a small region of a large gelatine plate (13 per cent. gelatine gel). Uniform pressure obtained by a water column acting through a rubber membrane.

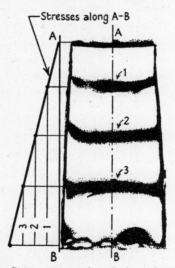

FIG. 21.8. Stress pattern of a gelatine gel (13 per cent. gelatine) subject to its own weight, producing a linear distribution of vertical compressive stresses as shown.

cross any given point on applying the load, the stress at that point can be calculated. Fig. 21.7 shows the stress pattern for a uniform load acting over a small region (indicated by the arrows) of a large plate of 13 per cent. gelatine gel, and Fig. 21.8 that for a vertical block of the same gel under the stresses produced by its own weight (from Frocht, 1941).

Attempts have been made to use the photo-elasticity of dilute gels as a means of following their relaxation under stress, but caution is necessary since some workers have reported that in the case of gelatine gels the mechanical relaxation may be almost complete without any marked diminution in the optical anisotropy.

Systems showing spontaneous birefringence have been found in the case of the aluminium soaps in organic media such as benzene (Gray and Alexander, unpublished). The intensity decreases with distance from the walls of the container, suggesting that it arises from the spontaneous orientation of the fibrillar soap micelles at the bounding surfaces. This phenomenon is probably of importance in connexion with the lubricating powers of greases.

(c) *Swelling*

If a piece of dry gelatine or agar is placed in water it first swells considerably and then either disperses to a solution, or remains as a

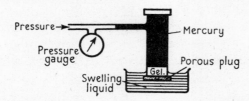

FIG. 21.9. Apparatus for the determination of the swelling pressures of gels.

coherent gel, depending upon the temperature, pH, and salts present. A similar behaviour is found with many natural and synthetic polymers in organic solvents, as will be discussed in more detail in Chapter XXVII. In other cases, however, of which silica gel has been the most studied, liquids may be absorbed (absorption occurs if the liquid wets the solid) without any marked swelling. This taking-up of a liquid by colloidal materials is often referred to as *imbibition*, and as we have seen, it may or may not be accompanied by swelling. As pointed out earlier, there is a net decrease in volume when swelling occurs.

If the colloid is confined in some way then very considerable swelling pressures may be set up; witness the use of wooden wedges to split rocks as mentioned in Chapter I. Experimentally, swelling pressures are measured by some such device as that shown in Fig. 21.9, due to Posnjak. A disc of the colloidal material is placed at the bottom of the vertical tube to which has been cemented a porous earthenware cylinder. The swelling liquid fills the outer beaker as shown. The tube and the capillary side-arm are filled with mercury and pressure is applied from a gas cylinder to maintain equilibrium as swelling occurs. The attached pressure-gauge then records the swelling pressure.

In the case of swelling pressures we are dealing with definite equilibria, since the same point can be approached from both sides, and hence thermodynamic relations can be developed. The swelling pressure is clearly only an osmotic pressure measured in a rather unusual manner, since the gel is acting as its own membrane; it is the pressure required to bring the solvent in the gel into equilibrium with free liquid. The usual osmotic treatment then holds (see p. 73), and the swelling pressure (Π) is given by:

$$-\Pi \overline{V}_1 = \Delta G^0 = RT \ln p/p_0, \qquad (21.1)$$

i.e.
$$\Pi = \frac{RT}{\overline{V}_1} \ln p_0/p, \qquad (21.2)$$

where p_0 and p are respectively the vapour pressures of pure liquid and of gel, $\overline{V}_1$ is the partial molar volume of the liquid in the gel, and ΔG^0 the free energy of dilution of the gel, i.e. the increase in free energy when one mole of liquid is added to a large bulk of the gel.

Measurements of Π and its temperature coefficient give not only the free energy but also the heat of dilution (ΔH^0), since from p. 60,

$$\Delta H^0 = \frac{\partial (\Delta G^0/T)}{\partial (1/T)}. \qquad (21.3)$$

(In some cases ΔH^0 can be found directly by calorimetric methods.) From these, the entropy of dilution (ΔS^0) follows from the usual relation

$$\Delta H^0 = \Delta G^0 + T \Delta S^0.$$

If the swelling pressure is too high for convenient measurement it can be found by measuring the vapour pressures of gel and pure liquid and inserting these values into equation (21.2) above.

The *degree* of swelling of gelatine depends upon the pH, being a minimum at the iso-electric point, and upon the concentration and nature of the salts present. For a given salt concentration the well-known lyotropic series is observed, e.g.

$$SO_4'' < H_2O < Cl' < Br' < I' < CNS'$$

in order of increasing swelling power for a series of sodium salts. The addition of sodium sulphate tends to cause a shrinkage rather than a swelling, probably due to its power to link chains through salt bridges, e.g.

$$>\overset{+}{NH_3}\ SO_4''\ \overset{+}{H_3N}<.$$

The dependence of the swelling upon the pH is, according to the theory developed by Procter and Wilson, the result of a Donnan distribution between the intermicellar and the outer fluid. Thus on adding acid (e.g. HCl) to iso-electric gelatine, protons are taken up due to the reaction

$$R\!\!<\!\!\begin{array}{c}COO'\\NH_3^+\end{array} + H^+ \rightleftarrows R\!\!<\!\!\begin{array}{c}COOH\\NH_3^+\end{array} \quad (\text{or } P^+)$$

and the equilibrium state can be represented as:

Gel	Solution
$P^+[H^+]_g[Cl']_g$	$[H^+]_s = [Cl']_s$

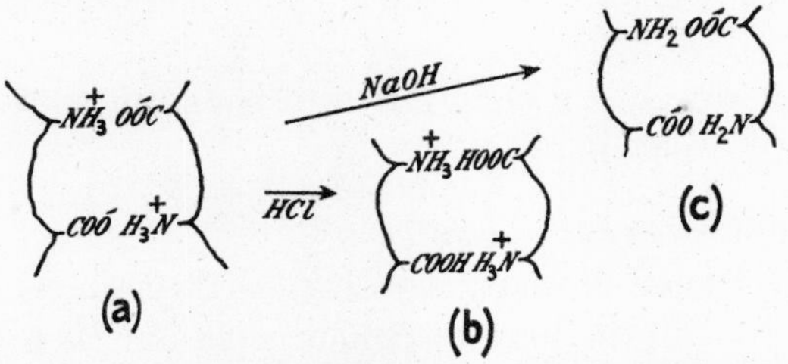

FIG. 21.10. Association of protein chains: (a) through strong salt linkages, (b) and (c) through less strong hydrogen bonds.

As has been shown in Chapters IV and VII this leads to an excess of diffusible ions inside the gel and hence to an excess osmotic (swelling) pressure.

The simple Procter-Wilson theory does not account for the swelling of iso-electric gelatine in pure water nor for the effect of simple salts. The former might be due to entropy changes although usually ascribed to soluble components present in gelatines, such as peptones and proteoses liberated by gradual hydrolysis, or to the traces of salts which are usually present. The effect of neutral salts might conceivably be due to their permitting increased binding of acid so that the concentration P^+ is increased (see below).

From the molecular viewpoint the effect of pH and salts on the swelling of gelatine is probably as follows. Iso-electric gelatine is in the zwitterion form, and it seems reasonable that oppositely charged groups on neighbouring chains will be associated, as in Fig. 21.10(a), thus preventing the chains from separating to any extent. Addition of acid or alkali will lead to the structures shown in Fig. 21.10(b) and (c); addition of neutral salts will lead to cations congregating around the —COO′ groups and anions about the —$\overset{+}{N}H_3$ groups. In all cases the forces linking the chains will diminish and thus permit an increased swelling. (In the absence of attractive forces between the chains complete solution will occur owing to the increase in entropy which results (see p. 91).) At acid pH neutral salts facilitate the uptake of protons, since the increased concentration of anions (e.g. Cl' with HCl) in the gel which proton uptake requires is thereby diminished; this means that form (b) and hence increased swelling will be favoured. Similarly the loss of protons which occurs on the alkaline side of the iso-electric point will be favoured by neutral salts, leading again to increased swelling. With ions such as I' and CNS', known to be readily adsorbed at interfaces, the increased swelling probably arises from their being adsorbed on to the chains or micelles which thereby tend to repel one another.

The above picture has clearly much in common with the uptake of acid and dyes by wool (p. 701).

(d) *Diffusion and reactions in gels*

The rate of diffusion of many crystalloids through dilute gels differs but little from that in the pure dispersion medium, as first pointed out by Graham. The use of agar gels for salt bridges to eliminate liquid-junction potentials is well known, as is the classical experiment of Lodge, published in 1886, for the direct measurement of the electrical mobility of the hydrogen ion. A dilute slightly alkaline solution containing phenolphthalein was gelled in a tube, one end of which dipped into dilute acid, and on passing a current the red colour of the indicator was progressively bleached, giving the velocity of $\overset{+}{H_3O}$ directly. Lodge thus found a value of 0·0025 cm./sec./V/cm., as compared with 0·0032 from conductivity and transport number.

With dilute gels and a series of halide salts (e.g. KCl, $BaCl_2$, $AlCl_3$) diffusion is progressively reduced with increasing valency of the cation, and for a given cation the sulphates diffuse more slowly than the chlorides.

As the concentration of gelling agent is increased the rate of diffusion falls, even with simple molecules. As the molecular size of the diffusing molecule is increased the resistance offered by the gel usually increases very suddenly at some point. This fairly sharp change is believed to occur when the size of the diffusing molecule becomes comparable with the *pore-size* of the gel, that is with the average size of the interstitial channels. Measurements of diffusion velocity of a series of small molecules of different size therefore offer a means of estimating the pore-size (or alternatively the molecular size if the pore-size is known). Actually the process is frequently a good deal more complicated than

GELS AND PASTES

this simple sieve picture suggests, particularly with ionized compounds in aqueous media where electrical effects are present, as will be further considered in Chapter XXVIII. Friedman and Kraemer (1930), for example, used urea, glycerol, and sucrose, and obtained the following values for the pore-size (in A) of gelatine gels:

	Sucrose	Glycerol	Urea
5 per cent. gel	55	57	47
10 per cent. gel	14	17	15
15 per cent. gel	5	10	8

The reason why ordinary gels completely stop the larger colloidal particles and macromolecules is thus clear.

One of the most spectacular phenomena in connexion with gels is the formation of precipitates in a rhythmic pattern. First investigated by Liesegang in 1896 and subsequently by very many others, it has still no completely satisfactory explanation. 'Liesegang Rings' can be formed by many reactions and under various conditions, a common technique being to prepare a gelatine gel of dilute potassium chromate (ca. 0·05 per cent.) and to cover it with a strong solution, or a few crystals, of silver nitrate. The usual appearance of the silver chromate precipitated under these conditions is shown by Fig. 21.11. The use of a gel does not appear to be essential but its rigid framework assists by stabilizing the periodic structure produced by the reaction.

FIG. 21.11. Liesegang rings (diagrammatic).

The time (t_n) required for the nth band to be formed, and the displacements x_n and x_{n-1} of the nth and ($n-1$)th bands from the origin, are generally related quite accurately by the expressions:

$$x_n = k\sqrt{t_n} \qquad (21.4)$$

and
$$x_n = k'x_{n-1}, \qquad (21.5)$$

where k and k' are constants.

Of the various theories advanced to explain the phenomenon the most generally accepted seems to be some modification of that first suggested by Wilhelm Ostwald shortly after Liesegang's original publication. According to this theory, diffusion of the inoculant ($AgNO_3$ in the above case) first produces a zone where the precipitate is in supersaturated solution. When the supersaturation exceeds a certain value nuclei form spontaneously and proceed to grow, not only at the expense of the supersaturation in their immediate neighbourhood but also by material diffusing back from the supersaturated zone in front. This process relieves the supersaturation and the inoculant has therefore to traverse a certain distance in front of the precipitate before the supersaturation again rises to the requisite value for nucleation, when the above process is repeated.

Considerable supersaturation in systems forming Liesegang rings has been demonstrated by various workers, although objections have been raised against Ostwald's idea of a definite supersaturation being necessary for spontaneous nucleation. With our present knowledge of the rates of diffusion and crystallization in the presence of gelatine or other gelling agent, the phenomenon can be approached in a more detailed way, and the relations obtained appear to be in reasonable accord with the experimental results given by eqns. (21.4) and (21.5) above. (See, for example, van Hook, 1944.)

Banded structures are of not infrequent natural occurrence, such as in tree rings, in gall-stones and other concretions, and in certain geological formations (e.g. agate). These all arise from disturbances in the generating system; however only certain of the geological formations obey eqn. (21.5) and can therefore be regarded as arising in the same way as Liesegang's structures.

(e) *The stability of gels. Syneresis*

The spontaneous concentration of a gel with liberation of the dispersion medium is termed 'syneresis'. The phenomenon is of frequent occurrence, being shown particularly by such gels as those of dibenzoyl cystine, 'solid alcohol' (p. 588), and dilute silica gels, as well as by some polymer systems. The extent to which syneresis occurs depends upon the system, but in general it decreases with increasing concentration of gelling agent.

The occurrence of syneresis shows that the original gel was thermodynamically unstable, but the absence of syneresis does not prove the converse to be true, owing to the marked time-lags so common in gel systems. Strictly speaking a gel could only be regarded as truly stable if the same state was obtained from both sides, e.g. by mixing the solid and liquid phases in the case of gelatine, agar, etc., or by dialysing away some of the salt in the case of an alumina sol precipitated by excess salt (p. 589).

It is a common observation that gelatine or agar may be immersed in water at room temperatures without showing any tendency to form the rigid gels which, as prepared by the ordinary process of cooling their hot solutions, show a complete absence of syneresis even over long periods of time. This suggests (although owing to possible hysteresis with these polymer molecules it does not definitely prove) that the usual gelatine and agar gels (of, say, 5–10 per cent. strength) are potentially unstable.

The question whether all gels as defined here (i.e. with less than ca. 10 per cent. gelling agent present) are thermodynamically unstable is probably best left unanswered at the moment. There would seem to be no fundamental objections, but nevertheless it seems likely that very few such gels have as yet been prepared.

(f) *'Free' and 'bound' water in gels*

The question of solvation in gels, in particular hydration in biological systems, has caused much discussion in the past. 'Bound' water was usually considered to be that fraction which did not exhibit the properties characteristic of ordinary water, the principal methods proposed for its determination being as follows:

(i) Water that has lost its solvent properties.
(ii) Water that cannot be frozen.
(iii) Water incapable of hydrating cobaltous chloride dihydrate to the hexahydrate. (This is made easily visible by the colour change, blue to purple.)
(iv) The amount of water which the solid material must possess in order that no contraction of the total volume occurs when it is brought into contact with liquid water.
(v) In certain cases (e.g. gelatine) from the side-chain X-ray spacing, since this varies with the bound water.

These methods appear to lead to rather different values in the case of gelatine (e.g. Heymann, 1941), which is not surprising, since the definitions of 'hydration' are inherently different in the different methods, and in addition there were probably variations in the gelatines used by the various workers. Thus Heymann, who developed method (iv), obtained a value of 0·6–0·7 (gm. water/gm. dry gelatine) at 23·3° C., Moran 0·51–0·58 (from the amount of water not freezing at $-18°$ C.), and Hatschek from method (iii) at 15–30°, values between 0·30 and 0·34 for a number of different gelatines.

These results are seen to be in reasonable accord with the hydration of corpuscular proteins, for which values of 0·3–0·4 are commonly found (see p. 141), and they show quite definitely that no abnormal degrees of hydration are present in gels, in contrast with some of the early theories. Similar conclusions are obtained with other colloidal systems such as the soaps, for example (see p. 126), and it would seem that the thick solvation layers favoured by some of the early workers are no more likely to be present around particles of colloidal size than around simple crystalloidal molecules. (For earlier references see, for example, Weidinger and Pelser, 1940, Heymann, 1941.)

(g) *Vapour pressures over rigid gels*

With 'rigid' gels such as silica gel, where swelling and shrinking occur only to a limited extent, the vapour pressure curve on dehydration ($ABCD$ of Fig. 21.12) differs over part of its course from that on rehydration ($DC''B'A$ in Fig. 21.12), a phenomenon first studied by van Bemmelen.

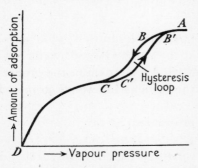

Fig. 21.12. Adsorption—vapour—pressure curve for rigid gels, showing hysteresis loop.

According to Zsigmondy the explanation is as follows: over the part AB water is being removed from between the large SiO_2 aggregates, over BC from the capillaries in these aggregates, and over CD from the polar groups on the surfaces (i.e. adsorbed water). In agreement with this the volume shrinkage on drying practically ceases at point B and opalescence first begins there. From the observed vapour pressure (p_r) over the BC region, as compared with the normal vapour pressure (p_0) for bulk liquid ($p_r < p_0$), the radius (r) of the capillaries can be calculated from the Kelvin equation $(RT/M)\ln p_r/p_0 = -2\gamma/\rho r$ (see p. 90). Values of the order 25–30 A were thus obtained by Andersson for silica gel.

The hysteresis region has generally been ascribed to the difference between the emptying and the filling of a fine capillary, due to the less complete wetting in the latter case giving rise to 'friction in the contact angle'. An alternative suggestion can be put forward if the gel consists of a network of rigid capillaries, as idealized by the structures shown in Fig. 21.13 (a) and (b) (cf. McBain, 1935). The process of dehydration and rehydration of structure (a) will be considered, as the reasoning will apply equally to structure (b). The former is seen to consist of a series of cones and inverted cones, both full with liquid when the solid has been completely wetted, and behaving independently of one another if the walls are perfectly rigid. On dehydration by pumping off the vapour the inverted cones will begin to empty first (since $r_1 \gg r_2$, and hence $p_{r_1} \gg p_{r_2}$ from the Kelvin equation), and the equilibrium vapour

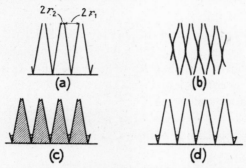

FIG. 21.13. Pore structure of rigid gels (idealized).

pressure (p_r) will decrease as the liquid is removed from the inverted cone, since r is steadily decreasing. Not until p_r has fallen to p_{r_2}, as indicated by Fig. 21.13 (c), will the cones commence to empty, but once started they will empty rapidly since r increases as dehydration proceeds. On rehydration of the completely dry gel the inverted cones will commence to fill up at lower vapour pressures than saturation (from the Kelvin equation), and at a given vapour pressure (say p_{r_2}) the state of affairs will be as represented in Fig. 21.13 (d). Thus (c) and (d) will have identical vapour pressures, but with the former having a higher liquid content, which is seen to be in accord with experiment. When allowance is made for the difference between the assumed model and an actual rigid gel, the two curves shown in Fig. 21.12 could readily be explained.

If the capillaries are not completely rigid as assumed above, it is possible that collapse may set in as dehydration proceeds (Barkas, 1947), and this may also lead to hysteresis. This effect arises from the liquid in a capillary having a reduced vapour pressure and thus being under tension, which would be transmitted to the walls of the capillary. The capillary thus tends to reduce its cross-section, which in turn will lower the vapour pressure and thus increase the tension further. Collapse will then occur if the tension of the water increases faster than the resistance of the tube to further compression.

Capillary condensation has also been shown to occur with ferric oxide gel. From the vapour pressure-temperature curves, using the break in the $\log p \sim 1/T$ linear relationship to indicate the phase change, Batchelor and Foster (1944) have recently shown that dioxan adsorbed on ferric oxide gel of pore radius $ca.$ 100 A melts sharply about 6° C. below the normal melting-point. With silica gel of pore radius $ca.$ 11 A adsorbed water remained liquid even at $-65°$ C.

The structure of gels

Having surveyed the more important properties of gels and of the changes which occur during the sol–gel transformation, we conclude with a discussion of the theories of gel structure in the light of the above facts. It is convenient to make a subdivision as follows:

(a) Rigid gels, e.g. silica gel.
(b) Elastic gels, e.g. gelatine, agar, soaps, polymers in organic media.
(c) Thixotropic gels, e.g. alumina, bentonite.

(a) *Rigid gels (silica gel)*

When mono-silicic acid is liberated, as by the action of acids upon sodium silicate solutions, it immediately begins to polymerize, e.g.

$$\text{HO—Si(OH)}_2\text{—OH} + \text{HO—Si(OH)}_2\text{—OH} \rightarrow \text{HO—Si(OH)}_2\text{—O—Si(OH)}_2\text{—OH} + \text{H}_2\text{O}.$$

It is clear that the process can be repeated in one, two, and three dimensions, leading to high molecular weight silicic acids, such as shown in Fig. 21.14, and ultimately to colloidal particles. The final state would be a three-dimensional network of Si—O links as in silica, but steric factors will prevent this from being attained except at very high temperatures.

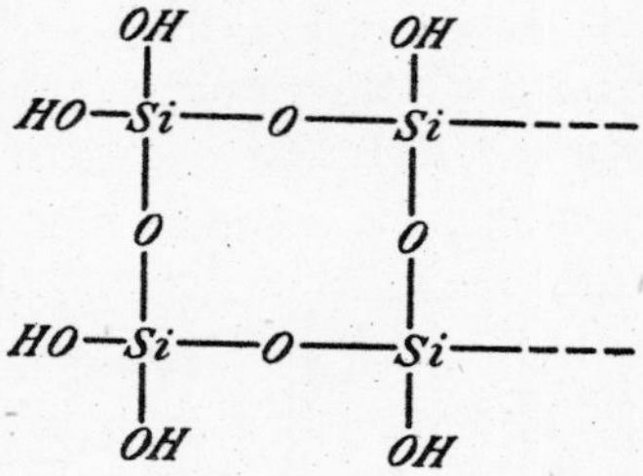

Fig. 21.14.

The linking together of the high molecular weight silicic acid molecules through primary valence bonds explains not only the great mechanical strength of silica gels, but also the absence of thixotropy, the irreversibility of the gelling process, and the small effect of drying on the gel volume. The hysteresis in the vapour pressure curves probably arises (as pointed out above) from the presence of rigid capillaries which such a structure would be expected to possess.

(b) *Elastic gels*

Both with the elastic and thixotropic types of gel it seems unlikely, owing to the ease and reversibility with which the gel structure can be produced or destroyed, that primary valence bonds are involved in linking the particles together to produce the requisite solidity. Of the theories which have been advanced, two call for special comment. The first assumes that electrical forces are responsible for fixing the particles of gelling agent in rather definite positions of minimum energy (see p. 115), thus imparting rigidity to the whole system; the second ascribes the rigidity to local attachments of the particles through weak, secondary valence forces. The former is clearly most unlikely for gels in organic media (e.g. aluminium soaps in benzene) and even for thixotropic gels in aqueous media it seems improbable (see below).

The most general, and frequently termed the 'random-mesh', theory of gel structure assumes a brush-heap arrangement of solid or semi-solid particles, these being sufficiently bonded at the points of contact to endow the whole structure with rigidity. The attractive forces at the points of contact may be of the van der Waals type, arising from the interaction of permanent or induced dipoles, and from dispersion forces, or of the coulombic type if the two points carry opposite charges. A few examples of the more important types are given below (see also Fig. 21.10).

$$-COO'\ \overset{+}{H_3N}-\ ;\ -COO'\ Ca^{++}\ 'OOC-\ ;$$
$$-SO_4'\ Ca^{++}\ SO_4'-\ ;\ -\overset{+}{NH_3}\ SO_4'\ \overset{+}{H_3N}-\ ;$$
} coulomb forces.

$$-COOH\ \overset{+}{H_3N}-\ ,\ -COO'\ H_2N-\ ,$$
$$>C=O\ H-N<\ ,$$
} hydrogen bonding.

$$\overset{-O}{\underset{+C}{}}\overset{C+}{\underset{O-}{}}\quad \overset{-Cl}{\underset{+CH_2}{}}\overset{CH_2+}{\underset{Cl-}{}}$$
} dipole forces.

In good accord with the 'random-mesh' theory is the fact that all gelation processes involve a reduction in the solubility of the solute—i.e. the attraction of the solute particles is increased relative to that between solute and solvent. If this proceeds too far then the solvent is squeezed out and syneresis results.

The linking together at a few points of a structure already present also fits in very well with the changes observed during the sol–gel transformation. The relatively small changes in volume, in conductivity and diffusion of salt solutions, and in dielectric properties, as well as the marked increase in viscosity, are clearly to be expected, also the fact that sols near the gelling point show elastic properties when subjected to a rapid stress.

In some cases, as previously pointed out, an actual framework has been observed in the ultramicroscope, but its absence in other gelled systems does not, of course, invalidate the above theory, since small or molecular aggregates would not be visible, particularly with gelling agents such as the lyophilic colloids.

Further support for the random-mesh theory is given in the following section, which also considers the principal objections which have been raised against it.

(c) *Thixotropic gels*

As mentioned above and in Chapter V, various attempts have been made to explain thixotropy by means of the potential curves for charged particles. Such theories cannot apply to gels in non-polar organic media and even for aqueous systems difficulties arise. The validity of these theories now seems doubtful (see Chapter V), and in addition, as Lawrence (1940) points out, they do not explain the re-setting after a time-lag, nor the reason why thixotropy is associated with non-spherical particles.

Rejection of such electrical theories leaves the random-mesh theory to be considered. Two principal objections may be raised against it—the small volume of disperse phase (general only a few per cent., and sometimes much less) required to impart rigidity; the second the frequent absence of visible particle/particle contacts in the ultramicroscope.

The first loses its significance when the shape of the particles present in thixotropic gels is recalled; those of alumina and bentonite, for example, are thin plates, and vanadium pentoxide particles are long rods. With such anisometric particles a simple calculation (p. 35) shows that a continuous framework can be built up even with a very small amount of dispersed material. The second objection has been discussed above in connexion with elastic gels, and the same conclusion would apply here also, although to a diminished extent, since these inorganic and more hydrophobic particles should be somewhat more easily visible.

The observed properties of thixotropic gels are not inconsistent with the random mesh theory. Thus the time-lag on setting is ascribed to rotatory Brownian agitation being necessary to bring the particles into contact, the original shaking (shearing) having caused some degree of common orientation, owing to the anisometric shape of the particles. (This orientation by shaking is the origin of the flow birefringence in the sol state, which disappears when the gel reforms.) Increase of temperature increases the Brownian agitation and thus decreases the setting time (provided, of course, no change of solubility occurs). Fig. 21.15 shows this to be the case with bentonite suspensions (Broughton and Squires, 1936).

The phenomenon of 'rheopexy', an increased rate of setting as a result of gentle motion such as rotating a test tube to and fro between the hands, has been explained by Lawrence (1940) as follows. Such a movement causes mild turbulence rather than a steady shear, and can therefore be regarded as a disorientating force which assists the disorientating action of Brownian motion.

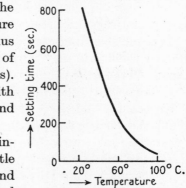

FIG. 21.15. Effect of temperature on the time of set of a 2·1 per cent. bentonite suspension.

One final, and apparently conclusive, point in favour of the random-mesh theory may be stressed. Direct confirmation of particle/particle contact has been shown in the case of mineral oils gelled by colloidal graphite. In the sol state these systems are non-conducting, but they become conducting when gelation occurs, indicating without doubt the formation of a linked graphite framework on gelation.

PASTES

According to the definition given earlier pastes are concentrated dispersions of fine solid particles in a liquid continuum, but as usual with colloidal systems there are border-line cases linking them particularly with gels on the one hand and with sols (dilute suspensions) on the other. Owing to the surface energy of the solid/liquid interface, pastes would be expected to be thermodynamically unstable systems. This may indeed be so as growth to larger particles, involving, as it does, transfer through the liquid phase, may frequently occur too slowly for detection. From the practical point of view pastes are of extreme importance, covering such diverse materials as putty, paints, clays, dough, drilling muds, tooth-paste, and so on, all of which possess peculiar and characteristic mechanical properties—as denoted by such terms as *plasticity*, *ductility*, *mouldability*, etc. These flow and deformation properties, which are so important in the practical usage of pastes, fall within the domain of the science of 'rheology' (see also Chapter XVI). Doughs and pastes are sometimes referred to as 'magmoids', and their peculiar combination of fluid and solid properties the 'parachoidal' condition (from the Greek word for thick).

It is possible from general considerations to enumerate some of the factors which will play a part in determining the flow behaviour of a paste. Of these the volume concentration of dispersed phase is obviously very important; other likely factors are as follows:

(a) the size and shape of the particles,
(b) the particle size distribution,
(c) the surface area of the particles (this is determined by (a) and (b)),
(d) the adhesions between the solid particle and the liquid, and between the solid particles, which are denoted by W_{SL} and W_{SS} respectively.

The experimental methods for studying (a), (b), and (c) have been discussed earlier (Chap. XIX, pp. 536–48), and a brief mention of adhesion between solids and liquids was given on p. 551.

Adhesion and its measurement

According to the definition given on p. 551, the adhesion of a liquid to a solid is given by
$$W_{SL} = \gamma_S + \gamma_L - \gamma_{SL} \tag{21.6}$$
$$= \Pi_{SL} + \gamma_L, \tag{21.7}$$
where the 'spreading pressure' $\Pi_{SL} = \gamma_S - \gamma_{SL}$. Similarly, for the adhesion of the solid to itself,
$$W_{SS} = 2\gamma_S - \gamma_{SS}, \tag{21.8}$$
where γ_S and γ_L are the surface tensions against air of solid and liquid respectively, and γ_{SL} and γ_{SS} the solid/liquid and solid/solid interfacial tensions. (In these systems it would strictly be better to talk of 'energies' rather than of 'tensions'.) If the liquid forms a finite contact angle (θ) with the solid, then, as shown on p. 551,
$$W_{SL} = \gamma_L(1 + \cos\theta), \tag{21.9}$$
so that W_{SL} can be determined. Even if θ is zero it is still possible to determine W_{SL} if another liquid can be found which gives a finite contact angle with the same solid, both against air and against the first liquid (see p. 542).

The adhesion (W_{SS}) between the solid particles cannot be measured directly, but it can be estimated in various ways, such as from the tendency of the particles to adhere to solid surfaces, from the sedimentation volume, and from the study of friction, as shown below. As w

GELS AND PASTES

shall see, the properties of pastes are very largely influenced by the relative magnitudes of W_{SS} and W_{SL}, the system being termed 'flocculated' or unstable if $W_{SS} > W_{SL}$ and 'deflocculated', 'peptized', or stable if the converse obtains.

The methods for obtaining information about W_{SS} and W_{SL} can now be outlined.

(a) From the sedimentation volume

A paste diluted with the dispersion medium and then allowed to settle, sooner or later shows a boundary between the almost clear dispersion medium and the subnatant suspension, the boundary moving towards the bottom of the vessel and eventually coming to a stop. Both the rate of sedimentation and the final volume of sediment—the 'sedimentation volume' as the latter is termed—are found to vary greatly with different types of suspension, even after allowing for differences in size, shape, and density, as well as of viscosity of the medium in the case of sedimentation velocity. Unstable, or flocculated, suspensions tend to give rapid settling and large sedimentation volumes; with stable (deflocculated) suspensions the converse holds.

These phenomena, the rate of sedimentation and the sedimentation volume, are actually very intimately related. If the particles tend to adhere on contact (i.e. $W_{SS} \gg W_{SL}$) the original uniform dispersion will soon contain, as a result of Brownian agitation, large, loosely aggregated units or flocs which not only sediment more rapidly than the original isolated particles, but preserving their loose structure in the sediment are also responsible for the large sedimentation volume. If, on the other hand, the particles have no tendency to adhere on contact, each will settle more or less independently of the others, and hence comparatively slowly if of colloidal size. A dense sediment, i.e. a very economical packing of the particles, requires the particles to be kept apart as long as possible. The chief factor opposing adhesion in aqueous and probably also in certain alcoholic media is the electrical double layer as discussed in detail in Chapter V. With completely non-ionizing media such as hydrocarbons, particle/particle adhesion can be reduced by adsorbed monolayers of long-chain compounds, as will be shown below.

The sedimentation volume in ionizing media is therefore affected by the nature and concentration of salts present, since these influence the ζ potential. For example, dispersions of silica powder in water, normally negatively charged, show the increase in sedimentation volume expected from the lowered ζ potential when salts, particularly those of divalent or trivalent metals, are added (p. 316).

Examples of flocculated systems are given by dispersions of very polar compounds in organic media, such as silica, titania, calcium carbonate, sodium chloride, etc., in benzene; carbon-black in water, with its polar dispersion medium and non-polar disperse phase, is another flocculated system. The above organic dispersions can readily be deflocculated by addition of long-chain polar compounds such as oleic acid or heavy metal soaps, and the carbon-black dispersion by addition of a soap or a wetting agent.

The data for titania given in Tables 21.3, 21.4, and 21.5 below, taken from Ryan, Harkins, and Gans (1932), illustrate these effects and also the order of magnitude found.

TABLE 21.3. *Sedimentation Volume (c.c.) of 8 gm. Titania in Benzene Solutions of Organic Liquids and Metallic Soaps, and in the Pure Organic Liquids*

Additive	Concentration of organic liquid (equiv./1,000 gm. benzene)				
	0	0·001	0·01	0·1	∞
Oleic acid	10·5	7·7	5·0	5·0	6·8
Methanol	10·5	7·5	8·0	7·6	7·1
Benzaldehyde	10·5	8·5	7·0	6·6	5·1
Ethyl acetate	10·5	7·9	7·2	6·7	7·2
Lead linoleate	10·5	6·4	6·3	6·9	..
Cobalt linoleate	10·5	8·2	6·7	6·7	..
Aluminium stearate	10·5	7·2	6·6	7·8	..
Titanium stearate	10·5	6·4	6·8	7·0	..

TABLE 21.4. *Effect of Water on the Sedimentation Volume (c.c.) of 8 gm. of Titania in Pure Organic Liquids*

Organic liquid	Concentration of water (gm./100 gm. dry titania)				
	0	0·63	1·9	6·3	12·5
Methanol	7·1	7·0	6·7	6·9	7·0
Oleic Acid	6·8	6·9	7·4	11·9	..
Ethyl acetate	7·2	8·3	13·4	16·2	16·7
Benzene	10·5	20·1	21·3	22·3	..
Carbon tetrachloride	11·2	20·0	22·6	22·7	..

The chief conclusions regarding polar compounds, as given by the above results for titania, may be summarized as follows:

(a) Non-polar liquids give flocculated systems of large volume; polar liquids depending upon their polarity give varying degrees of dispersion and smaller sedimentation volumes.

TABLE 21.5. *Effect of Water on the Sedimentation Volume (c.c.) of 8 gm. Titania in Benzene solutions of Oleic Acid*

Concentration of water (gm./100 gm. TiO$_2$)	Concentration oleic acid (equiv./1,000 gm. benzene)				
	0	*0·001*	*0·01*	*0·1*	∞
0	10·5	7·7	5·0	5·0	6·8
0·63	20·2	18·5	16·0	12·0	6·9
1·9	21·3	22·0	21·2	20·6	7·4

(b) Addition of polar organic liquids or metal soaps to dispersions in benzene decreases the sedimentation volume very considerably.

(c) Addition of water to non-polar liquids increases the sedimentation volume, but has little effect with very polar liquids such as methanol.

The difference between water and polar organic molecules such as oleic acid is not difficult to see. Adsorption of a monolayer will, in the latter case, leave an outer surface of non-polar hydrocarbon, whereas with an adsorbed film of water the surface will still be rather polar.

Closely related to the sedimentation volume is the maximum amount of liquid which a given quantity of powder can absorb before 'free' liquid appears. An open packing of the powder means a large *oil absorption* (a term used in paint technology for this measurement), a dense packing a small one. Thus with polar solids in pure benzene the mixture may still appear quite 'dry' with 70 per cent. of liquid (by volume), but on addition of a little oleic acid the mixture liquefies due to the reduced adhesion between the particles permitting a closer packing.

For close-packed spheres of uniform size the volumes occupied by the disperse phase and dispersion medium would be ca. 74 per cent. and ca. 26 per cent. respectively, these values being independent of the sphere radius. In pastes the particles frequently deviate considerably from the spherical, which tends to make packing more difficult and therefore to increase the 'oil absorption'. On the other hand uniformity of size is never attained, commercial materials invariably having a very wide particle-size distribution curve. Since small particles can pack into the interstices between the larger ones this tends to decrease the 'oil absorption'.

(b) *From the 'adhesion number'*

This method of estimating the particle/particle adhesion is due to von Buzágh. The suspension is contained in a closed cell with a small glass or quartz plate upon which the particles are allowed to settle.

After counting the adhering particles the cell is inverted, and the number remaining on the plate counted again. The percentage sticking is termed the *adhesion number*, and this gives a measure of the adhesion between the solid particle and the plate; this is found to run parallel to the mutual adhesion of the particles, and so gives an estimate of W_{SS}. A variation of this method is, after allowing the particles to settle as before, to tilt the plate gradually and observe the angle at which sliding begins. This measures, effectively, the coefficient of friction between the particle and the surface, which will decrease as W_{SS} decreases.

Palmer and Rideal (1939) have recently used this method to see how the adhesion of solid particles to surfaces is affected by the concentration of detergent present. Using particles of carborundum or calcium carbonate and a quartz surface they found the adhesion to be a minimum at around the critical concentration for micelle formation (Chapter XXIV).

(c) *Direct measurement of friction*

In some cases it has been possible to measure the coefficient of friction between two particles of sufficient size, and to correlate these results with the sedimentation volume and flow properties of pastes of the powdered material (Hutchinson, 1947). Thus for example, with finely powdered polar salts such as sodium nitrate in benzene W_{SS} is very high, as indicated by the large values of the sedimentation volume and coefficient of friction, but is reduced by addition of long-chain polar compounds as mentioned previously. The effect arises from adsorption, the long-chain compound being orientated with its polar group in contact with the crystal surface and its long-chain extending into the organic liquid, thus reducing the attractive forces between the solid particles. The close parallelism between the extent of adsorption (as measured by the fraction of the surface covered), and the reduction in friction is shown by Fig. 21.16 (a) and (b), where data for four long-chain compounds are included. It will be noticed that the initial amounts of adsorption appear to exert an inordinate effect on the friction, but this is to be expected, since the long-chains will begin to be effective in separating the surfaces long before they reach their final close-packed state (where $\theta = 1$ in Fig. 21.16 (a)).

As adsorption of these long-chain compounds increases and the friction falls, so the pastes become more 'pourable' and the 'oil absorption' decreases, giving further confirmation of the reduction in W_{SS}.

Another, but less direct method, of estimating the friction is to measure the power needed to rotate a paddle at a fixed speed in the paste. Using controlled conditions the effect of various additives can thus be compared. (Cf. the farinograph, p. 472.)

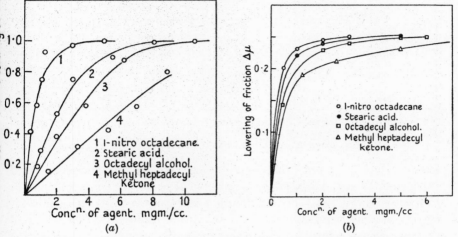

Fig. 21.16. (a) The adsorption of long-chain compounds, in benzene solution, by sodium nitrate. (b) The effect of long-chain compounds, in nitrobenzene solution, upon the friction between sodium nitrate crystals.

(d) *From heats of wetting*

Immersion of a fine powder in a liquid usually liberates heat, the amount depending a great deal upon the particular system, and for a given system being proportional to the surface area of the powder. (For experimental methods see p. 550.)

The heat of wetting ($-\Delta H_{SL}$) has sometimes been taken as a measure of the affinity between the solid and the liquid, but this can strictly be so only near the absolute zero or if $\partial \Pi_{SL}/\partial T$ is zero, since the affinity is a free energy change and is related to ΔH_{SL} by the Gibbs-Helmholtz equation (p. 60)

$$-\Delta H_{SL} = A_S\left(\Pi_{SL} - T\frac{\partial \Pi_{SL}}{\partial T}\right), \qquad (21.10)$$

where A_S is the surface area of the powder, and $\Pi_{SL} = \gamma_S - \gamma_{SL}$, as defined on p. 612. (For further discussion see p. 542.)

Most of the early results for the heats of wetting are of little value owing to the uncertainty about the purity of the solids and liquids used. For reproducible results the powder has to be dried and outgassed *in vacuo* and the liquids carefully purified and dried. This means that the term γ_S refers to the surface tension of the solid *in vacuo* rather than in air.

Table 21.6 below shows some of the results obtained with dry titania and a series of liquids by Harkins and Dahlstrom (1930).

TABLE 21.6. *Heat of Wetting*

Liquid	Cals./gm. powder	Ergs/cm.²†	Total energy‡ of adhesion (erg/cm.²)
Water	1·15	325	442
Butyric acid	0·883	250	305
Ethyl acetate	0·800	225	286
Butyl alcohol	0·769	215	264
Carbon tetrachloride	0·526	150	213
Benzene	0·390	110	179
Nitrobenzene	1·450	..	..

† Calculated from the values in the previous column and an ultramicroscopic estimation of the specific surface.

‡ Total energy of adhesion = heat of wetting + total surface energy of liquid.

The effect of polar compounds is seen to be very marked, and experiments with very dilute solutions of water or butyric acid in benzene showed that the maximum heat was evolved with approximately sufficient polar compound to form a monolayer on the solid surface. These results indicate that a monolayer suffices to satisfy the greater part of the residual affinity of the solid surface, as would be expected from the general conclusions of surface chemistry. Comparing the various organic compounds it is clear that the heat of wetting increases with the polarity of the anchoring dipole, i.e.

$$-OH < -COOEt < -COOH < -NO_2.$$

Rheological properties of pastes

As previously mentioned, it is the flow or rheological properties of pastes which so frequently determine their suitability for practical purposes, as with putty, paints, doughs in bread-making, clay 'slip' in pottery, and so on. Until quite recent times the examination of such materials was done solely by sight, touch, and their general handling qualities, the baker, for example, assessing the properties of his doughs in terms of such subjective characteristics as 'body' and 'spring'.

Much of the data on the flow behaviour of pastes is of a similar qualitative nature, but these measurements have nevertheless yielded very considerable information, and will be briefly considered.

(a) Qualitative measurements

Flocculated systems where $W_{SS} \gg W_{SL}$ (e.g. salts and polar compounds in benzene) may appear quite solid if the vessel is tilted gently (i.e. applying very small stresses), but flow readily with rather more vigorous treatment. This behaviour would be expected from the very open but adherent structure present in these systems, as indicated by

GELS AND PASTES

their large sedimentation volume and thixotropic nature with very small setting time. Electrodeposits obtained from unstable suspensions by electrophoresis have the same properties as the sediments, and are readily removed from the coated electrode.

With less unstable systems the sediment generally has a very small or negligible 'yield value', and it closely resembles the coacervates, the liquid precipitates formed when two oppositely charged hydrophilic colloids are mixed (see also p. 589). These liquid sediments are explained by Verwey (1941) on the basis of irregular distribution of the double layer charge over the particle surface. It is well known that different crystal faces may have very different structure and properties, and it seems reasonable also that crystal imperfections and edges would differ from the crystal faces in their colloid chemical behaviour. For nearly stable suspensions the particles can probably only adhere at a few points and therefore only in a restricted number of relative positions, whereas the other parts of the particles are subject to mutual repulsion due to the double layer present there. These double layer repulsions reduce the number of contacts and hence the number of adhering particles, and enable the particles to carry out restricted movements around a few contact points. The fluid properties, together with the higher density and very low 'yield value' are thus explained by Verwey.

Turning now to *stable* (deflocculated) suspensions where $W_{SS} \ll W_{SL}$, these give slow settling sediments, but once formed these adhere strongly to the bottom of the vessel (or to an electrode, if deposited by electrophoresis). The high density in both cases shows the particles to be packed together in an economical fashion, which can be attained only if adhesion between the particles is prevented for as long as possible, either by a double layer of sufficiently repulsive properties (with aqueous or ionizing media), or by surface films such as given by oleic acid with polar substances in hydrocarbon media. These thin liquid layers between the particles allow a restricted flow in the sediment, and hence the formation of a dense layer with consequent high mechanical strength to *rapid* stress. The dilatant properties which these systems frequently show also arise from the dense packing. (A further discussion of dilatancy is given in Chapter XVI.)

A practical example of how the mechanical properties of a paste may be determined by its colloid chemical stability (i.e. by the relative magnitudes of W_{SS} and W_{SL}) is given by putty, which consists of a concentrated dispersion of finely ground whiting (natural calcium carbonate) in linseed oil. When made from raw linseed oil and whiting the material has the desirable properties of being easily worked, with long 'string'. With pure precipitated calcium carbonate and acid-free linseed oil a very crumbly product with poor cohesion results; replacing the acid-free oil by raw linseed oil gives some improvement, but still not much 'string'. If, now, a little hydrated alumina is added, an excellent putty with long 'string' is obtained. The desirable plastic properties of ordinary putty thus appear to arise from the traces of free fatty acid in the linseed oil, and of hydrated aluminium (or iron) oxides in the natural chalk. This is probably due to reaction between these two constituents, leading to the formation of surface active aluminium (or iron) soaps, with the consequent reduction in W_{SS} and increase in W_{SL} for the system.

Further correlation between flow and adhesion has been provided by Bartell and Hershberger (1931), who studied the flow properties of zinc oxide pastes in a variety of organic liquids of different adhesion tensions. In agreement with the statement above, poor wetting was found to be associated with a high 'yield value' and low 'mobility', good wetting with a low yield value and high mobility.

(b) Quantitative measurements

A quantitative study of the rheological properties of a paste generally involves separating the total deformation into its non-recoverable part (viscous flow), and that which is recoverable (elastic deformation). The relevant experimental methods have been detailed in Chapter XVI. As mentioned before, the time factor is generally important, and very different behaviour can be obtained, depending upon the time scale of the experiment relative to certain inherent properties of the material (see Chapter XVI).

Pastes, particularly those with the higher concentrations of disperse phase, usually show some sort of 'yield value', which is clearly essential if the material is to exhibit plastic properties, since it has to retain its shape against the forces of gravity. In this respect they can be distinguished from substances such as glass, resins, and asphalts which, in the temperature region where a yield value to slow stresses is just not observable, still have viscosities about 10^5–10^7 that of usual pastes such as clay and paints. This is shown by the data of Table 21.7 below, also shown graphically in Fig. 21.17, giving the 'viscosity coefficient' and the lower yield value for a number of rheological materials (after Houwink, 1938). (The inverted commas are used because, in the case of pastes, it is not a true viscosity which is measured (see p. 473).)

The separation of the two types of materials is clearly shown, particularly when the different units for the ordinate scales have been emphasized.

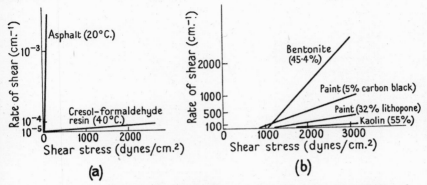

FIG. 21.17. The rate of shear as a function of shear stress for: (a) asphalt and a cresol-formaldehyde resin; (b) a number of pastes. (Note difference in ordinate scales in the two cases.)

TABLE 21.7

Substance	Temp.	'Viscosity' (poise)	Lower yield value (dyne/cm.²)	Plastometer used
Asphalt	20° C.	$2\cdot 4 \times 10^5$	Zero or very small	Rotation
Cresol-formaldehyde resin	40° C.	3×10^7	Zero or very small	Compression
Bentonite paste (45·4%)	Room	0·3	11×10^2	Capillary
Kaolin paste (55%)	Room	25·5	15×10^2	Rotation
Lithopone (32 vol%) in linseed oil	Room	5·4	12×10^2	Capillary
Paint pastes (various)	Room	~4	$\sim 1 \times 10^2$	Capillary
Carbon-black in oil: 5% concn.	Room	84	92	Rotation
Carbon-black in oil: 25% concn.	Room	1,840	2,380	Rotation

With increasing concentration of disperse phase the 'viscosity' and the yield value both increase, as shown by the last two systems in Table 21.7, and in more detail by the data of Figs. 21.18 and 21.19. The results shown in Fig. 21.18 were obtained by Lewis, Squires, and Thompson (1936) using concentrations of ground-glass in water, and measuring their flow behaviour in a Stormer viscometer. The increasing resistance to flow as the concentration approaches the value of *ca.* 40 per cent. found for loose packing of the material in air, is clearly shown. The data for carbon-black in oil varnish shown in Fig. 21.19 and due to Wolarowitsch and Borinewitsch (1936), who used a viscometer of the Couette type, are seen to be very similar in general form.

Some further data on the flow and deformation properties of pastes has been given earlier (Chapter XVI), as well as an indication of the

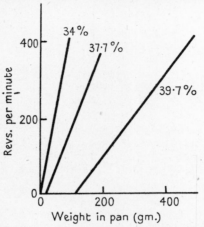

Fig. 21.18. Flow-rate—stress curves for various concentrations of ground-glass in water, using a Stormer viscometer. (Flow-rate proportional to revs. per minute, stress proportional to weight in pan.)

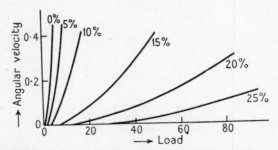

Fig. 21.19. Suspensions of carbon-black in oil varnish. Effect of concentration of disperse phase on flow (Couette apparatus).

methods used to analyse the observed behaviour. The slow progress made so far is due undoubtedly to the complexity of most of the systems used. (For further references see the general bibliography at the end of Chapter XVI.)

GELS AND PASTES

GENERAL REFERENCES (see also Chapter XVI)

FREUNDLICH, 1935, *Thixotropy*, Hermann, Paris.
HEYMANN, 1936, *The Sol–Gel Transformation*, Hermann, Paris.

OTHER REFERENCES

BARKAS, 1946, Discussion on 'Swelling and Shrinking', *Trans. Faraday Soc.* **42 B**, 137.
BARTELL and HERSHBERGER, 1931, *J. Rheology*, **2**, 177.
BATCHELOR and FOSTER, 1944, *Trans. Faraday Soc.* **40**, 300.
BROUGHTON and SQUIRES, 1936, *J. Phys. Chem.* **40**, 1041.
FERRY, 1942, *J. Amer. Chem. Soc.* **64**, 1323.
FREUNDLICH and ABRAMSON, 1928, *Colloid Symp. Monogr.* **6**, 115.
—— et al., 1932, *Z. phys. Chem.* A **160**, 469.
FRICKE and PARKER, 1940, *J. Phys. Chem.* **44**, 716.
FRIEDMAN and KRAEMER, 1930, *J. Amer. Chem. Soc.* **52**, 1295.
FROCHT, 1941, *Photo-elasticity*, vol. i, Wiley, New York.
GREENBERG and MACKEY, 1931, *J. Gen. Physiol.* **15**, 161.
HARKINS and DAHLSTROM, 1930, *Industr. Engng. Chem.* **22**, 897.
HATSCHEK, 1932, *J. Phys. Chem.* **36**, 2994.
HEYMANN, 1936, *The Sol–Gel Transformation*, Hermann, Paris.
—— 1941, *J. Phys. Chem.* **45**, 1143.
HOOK, VAN, 1944, Article in vol. v of Alexander's *Colloid Chemistry*, Reinhold.
HOUWINK, 1938, *Second Report on Viscosity and Plasticity*, Academy of Sciences, Amsterdam.
HUTCHINSON, 1947, *Trans. Faraday Soc.* **43**, 435.
KATZ et al., 1932, *Rec. trav. Chim. Pays-Bas*, **51**, 513, 835.
KISTLER, 1931, *J. Phys. Chem.* **35**, 815.
KRAEMER, 1926, *Colloid Symp. Monogr.* **4**, 102.
LAING and MCBAIN, 1919, *J. Chem. Soc.* **115**, 1279.
—— —— 1920, ibid. **117**, 1506.
LAWRENCE, 1935, *Proc. Roy. Soc.* A **148**, 59.
—— 1940, *Ann. Rep. Chem. Soc.* **37**, 99.
LEWIS, SQUIRES, and THOMPSON, 1936, *Trans. Amer. Inst. Min. (Metall.) Engrs.* **118**, 77.
MCBAIN, 1935, *J. Amer. Chem. Soc.* **57**, 699.
PALMER and RIDEAL, 1939, *J. Chem. Soc.*, p. 573.
POOLE, 1933, *Trans. Faraday Soc.* **29**, 1305.
RYAN, HARKINS, and GANS, 1932, *Industr. Engng. Chem.* **24**, 1288.
VERWEY, 1941, *Rec. trav. chim. Pays-Bas*, **60**, 618.
WEIDINGER and PELSER, 1940, ibid., **59**, 64.
WOLAROWITSCH and BORINEWITSCH, 1936, *Kolloidzschr.* **77**, 93.

XXII
FOAMS

Definitions and units

THE definition of a 'foam' as a gaseous dispersion (usually air) in a liquid continuum was given in Chapter I. 'Solid' foams, i.e. gas/solid systems, will not be discussed here, and only brief mention made of foams in other than aqueous media. Foams are usually classed as 'dynamic' or 'static', depending upon whether measurements are made whilst bubbling is proceeding or not.

The early estimations of foaming power and foam stability used arbitrary conditions and relative units to compare the various solutions, but more recently foaming capacity and stability have been established in definite units approximately independent of the apparatus used.

For dynamic foams Bikerman (1938) defines the unit of foaminess (denoted by Σ) as the average length of time that unit volume of *gas* remains in the foam. Foams were made by blowing a measured volume (V) of air in time t through a solution overlying a sintered glass disc, and it was found that vt/V, where v is the foam volume, was almost independent of the time of streaming, air pressure, and porosity and size of the disc. The value of vt/V tended to a limit, denoted by Σ, when the amount of superimposed liquid was increased. vt/V has a definite physical meaning; it is the average lifetime of a bubble in the foam, as shown below.

If the linear velocity of air in the foam is u and the foam height is h, then the lifetime of a bubble in the foam is h/u. $V = Aut$, where A is the cross-sectional area of the tube and $h = v/A$. Hence $h/u = vt/V$.

For static foams Ross (1943) suggests as unit of foaminess (L_e) the average length of time that unit volume of *liquid* can remain suspended aloft as foam. $L_e = (1/V_0) \int_{V_0}^{0} t \, dV$, where V_0 and V are the volumes of liquid in the foam at times $t = 0$ and t respectively. For foams which obey the logarithmic relation for drainage, i.e. $V = V_0 e^{-kt}$,

$$L_e = \frac{t}{2 \cdot 3 \log_{10} V_0/V} = \frac{1}{k}.$$

Foaming agents

It is generally agreed now that pure liquids do not give stable foams (e.g. Foulk and Barkley, 1943), and for any reasonable stability a third component, the foaming agent, is essential.

A very great variety of substances has been found to promote foam stability to a greater or lesser extent; the more important members can be grouped as follows:

(a) Soaps, both natural and synthetic.
(b) Proteins, including gelatine and other degraded proteins.
(c) Solid powders.
(d) Other miscellaneous compounds, e.g. polymers, saponins, certain dyestuffs.

In class (a) the fatty acid soaps such as sodium oleate or sodium stearate peptized by glycerol, are well known for their bubble-blowing capacities, and many of the synthetic type such as sodium cetyl sulphate can be made equally effective.

Proteins find a wide use as foaming agents, owing to their edible nature, in such foodstuffs as whipped cream and meringue, where the foaming agent must be edible, or at any rate non-toxic. 'Fire-foam', which is so efficacious against oil fires, consists of bubbles of carbon dioxide (liberated from sodium bicarbonate and aluminium sulphate) stabilized by dried blood, glue, or other cheap protein-containing materials.

Finely powdered solids, particularly if they have a waxy or hydrophobic surface (e.g. aluminium stearate or similar heavy metal soaps which are insoluble in water), remain at the air/water surface on agitation, and stabilize the foam so formed. Many other materials can be made to behave similarly if their surfaces are made hydrophobic by suitable treatment. This forms the basis of the 'froth-flotation' method of separating minerals, which will be discussed in some detail at the end of this chapter. Certain dyestuffs probably act as stabilizers by adsorption as fine powders at the interface.

A number of other materials require mention as foaming agents. The saponins are widely used, and as obtained commercially are complex mixtures containing triterpenoid glucosides. Beer froth is believed to be stabilized by the colloidal constituents present, which include proteins and carbohydrates. Several polymers have considerable foaming power for aqueous solutions, such as methyl cellulose (water soluble type) and more particularly polyvinyl alcohol.

Foam formation is of considerable practical importance in the churning of cream to make butter, a point which will be discussed later, and in lubricating oils for internal combustion engines, where its elimination is a great problem.

Formation and destruction of foams
Formation

Mere hand-shaking of a suitable mixture frequently suffices, but for quantitative work it is usual to bubble the gas through the liquid by means of some type of porous septum, usually of sintered glass (see Fig. 22.1). On the commercial scale beaters specially designed to entrain air are employed.

Destruction

Even under favourable conditions foams collapse slowly, although with certain 'solid' foams (e.g. egg-white foams as in meringue) the rate may be negligible. Collapse can be accelerated by suitable treatments, such as by fluctuating the pressure, by alternate heating and cooling, or by freezing, etc. These are rarely feasible on any large scale, and for commercial purposes foam-breaking is usually obtained by spraying with small amounts of organic compounds such as ether or, more usually, the intermediate chain alcohols (e.g. amyl, octyl) and fatty acids (e.g. capric). Certain crude mixtures of the intermediate fatty acids (e.g. from coconut oil) are also used. Foams in lubricating oils can be broken by addition of certain organo-silicon polymers.

Froth-breaking is usually ascribed to the added compound, owing to its greater surface activity, displacing the foaming agent, but lacking the other requisite of sufficient cohesion in the adsorbed film, it cannot yield a stable foam (see also p. 629).

Measurement of foaming power and foam stability

With the exception of the indirect method using single bubbles the methods for measuring foaming power and foam stability can be subdivided into dynamic and static, depending upon whether measurements are, or are not, made whilst bubbling is proceeding.

(a) Static methods

The usual and simplest technique is to shake a definite volume of liquid in a partially filled tube for a standard time (e.g. one minute) at a definite temperature. Alternatively the foam can be prepared by the method described below under (*b*), but this requires rather more apparatus. The foaming power of the liquid is then measured, either from the initial height of the foam, or from the volume of *liquid* held initially by the foam (this being obtained from the volume of free liquid). The stability of the resultant foam can then be measured either from its duration or, better, from its rate of collapse.

For some foams the rate of drainage is logarithmic; the foaminess is then given very simply, as pointed out above, by the expression $t/2 \cdot 3 \log_{10}(V_0/V)$, where V_0 and V are the volumes of liquid in the foam at times zero and t (see Fig. 22.4).

From the published investigations it would seem that static methods are less reproducible than the dynamic ones which will now be described.

(b) *Dynamic methods*

Most dynamic methods employ some modification of the apparatus shown in Fig. 22.1, the porous septum usually being a sintered glass disc. Bubbling conditions are standardized, and two measurements can be obtained:

(1) The rate at which foam is formed when bubbling commences. This can be taken as a direct measure of the foaming power of the liquid.

(2) The rate of decay when bubbling is stopped. This is identical with the static method above, and gives a measure of the stability of the foam under static conditions.

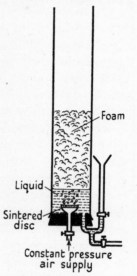

FIG. 22.1. Dynamic method for the study of foam formation.

(c) *Observations on single bubbles*

This technique, due originally to Hardy and later developed by several Russian workers, is of great interest, since it provides a link between the behaviour of adsorbed and insoluble monolayers on the one hand and that of bulk foams on the other. Although it is rarely possible to use compounds which give insoluble monolayers for bulk foam measurements (proteins are an exceptional case), yet so much is known about the structure and behaviour of insoluble monolayers that such studies are most valuable.

A single bubble is liberated just beneath the surface of the solution or insoluble monolayer under test, and its life measured, the mean life being obtained from a number of repetitions. The dependence of bubble stability upon surface tension and age of surface can thus be followed in the case of adsorbed films, and in addition its dependence upon the viscous and elastic properties of the film if insoluble monolayers (spread as usual on a Langmuir-Adam film balance—see Chapter XVII) are used.

The measurements of Rehbinder and Trapeznikov (1938), and of

Trapeznikov (1940), which appear to be the most recent, show that the stabilizing action of these insoluble monolayers is at a maximum for surface concentrations somewhat before the surface tension lowering, and hence the molecular density in the surface layer, have reached their maximum values. For example, the optima for oleic acid, palmitic acid, and cetyl alcohol were found to be at 47 A^2, 23 A^2, and 21 A^2 per molecule respectively, whereas the closest packing attainable in the monolayer is about 30 A^2, 20 A^2, and 20 A^2 respectively. These results suggest that stabilization arises not only by a mere enhancement of the molecular density in the surface film, but also by other factors, the chief being a certain degree of mobility of the stabilizing molecules, which might be advantageous for absorbing sudden strains.

Some objections have been raised against the single bubble technique for assessing foaming power, such as poor reproducibility and excessive bubble life for strongly foaming solutions, but the most important is that it clearly takes no account of the effects arising from the lamellae of liquid separating adjacent bubbles in the bulk form. In a modification suggested by Foulk and Miller (1931), which approximates more closely to the physical conditions of a foam, two bubbles are blown in juxtaposition from two opposed hollow hemispherical cups, and their stability determined.

Factors in the formation and stability of foams

From our detailed knowledge of adsorbed and insoluble monolayers it is possible to predict the relevant factors in connexion with the formation and decay of foams, as well as some idea of their relative importance. The principal factors are listed below.

(a) *The magnitude of the surface tension* (γ). Ease of formation and stability should increase as γ is lowered, since both the energy required to increase the interfacial area and the energy loss on coalescence, will be proportional to γ. (Energy barriers, as in (c), modify the picture.)

(b) *The variation in surface tension with molecular density* (n) *in the stabilizing film.* In order to withstand mechanical shocks, which set up local extensions and contractions in the bubbles, the film should be able to respond to a local extension by a rise in tension, and to a local contraction by a fall. This is the case if γ varies rapidly with n. It has been stated by many authors that the requisite is a rapid variation of γ with the bulk concentration, rather than the surface concentration (n as stated above. This, however, cannot be the relevant factor as a rule since surface studies have shown that the type of molecules necessary for stable foams enter and leave the surface film at extremely low rates, so that for sudden changes in area the film cannot maintain equilibrium with the bulk solution (see below).

(c) *The diffusion of foaming agent into and out of the surface.* The former is clearly the more relevant for dynamic foams, and the latter for static foams. As mentioned above, both ingress and egress may be remarkably hindered processes, arising in part from the very asymmetric molecular structure possessed by most foaming agents. The rate of ingress of soaps and similar molecules has been studied in great detail in connexion with the 'surface ageing' phenomenon (see p. 529).

(d) *The mechanical properties and cohesion of the stabilizing film.* High viscosity would reduce the rate of collapse by hindering the movement of molecules from the film (it would also diminish the *ease* of foam formation). If the film should possess mechanical strength (i.e. behaves as an elastic or plastic solid) then, provided the elasticity is sufficient to accommodate mechanical shocks, the foam might remain stable indefinitely. However, we know from surface studies (see p. 503) that 'solid' *monolayers* are always unstable with respect to the three-dimensional solid, so that foams of indefinite stability are unlikely to be realized with *monomolecular* stabilizing layers.

The viscous and elastic behaviour (and with monolayers the Π–A curve) give some indication of the molecular cohesion in the stabilizing film (see p. 497). With increasing chain-length the viscosity and cohesion increase, due to the increased van der Waals attraction between the chains. The van der Waals forces fall off very rapidly with distance (see p. 122), so that the cohesion between molecules in expanded and gaseous films is much less than in those of the condensed type. Electrostatic forces between the polar head groups also modify the cohesion, but the effect is only marked if the head group is ionized. With univalent soaps, e.g. $RCOO'Na^+$, the electrostatic forces in the monolayer are repulsive, leading to a reduced cohesion, but the position is reversed if polyvalent ions such as Ca^{++} or Al^{+++} are added, when the increased cohesion is sufficient to render the monolayer rigid (and unstable). The marked cohesion, as shown by their elastic properties, of monolayers of protein and other polymers, arises from entanglement of the long-chains, augmented by van der Waals attractive forces and by hydrogen bonding between $>CO$ and $>N$—H groups in the former case (see also Chapter XVII).

(e) *Ageing effects in the surface film.* It is clearly important for stability that the stabilizing film should not deteriorate with time.

For monolayers in true equilibrium with bulk solution ageing phenomena, other than those arising from chemical reactions, are unlikely, but polymolecular films, as given by the proteins, for example, appear to be unstable systems and tend to aggregate to even thicker layers.

(f) *The magnitude of the bulk viscosity.* A foam, once formed, will be stabilized if the bulk viscosity is high, since this will reduce the rate of drainage. The relevant viscosity is that in the lamellae of liquid between the bubbles, and as such lamellae may approach molecular dimensions when rupture occurs, the viscosity may not necessarily be identical with the bulk value as given by the usual methods.

(g) *The ζ potential of the surface film.* A high ζ potential will tend to oppose coalescence of the bubbles by virtue of the interaction of the two double layers as discussed in Chapter V.

The relative importance of the above factors depends very largely upon the particular foaming agent used, as will be seen when the principal types are discussed, although factors (f) and (g) are almost invariably much less important than the rest.

One conclusion can be drawn from the above analysis: the *foaming power*, as measured by the rate of foam production under standard conditions, will not in general be identical with the *foam stability* as measured by the rate of decay, since different factors are involved. This is borne out by experiment.

Discussion of the principal types of foaming agents

We will now consider in some detail how far the factors given above can explain the foaming powers of the principal types of foaming agents. It is unfortunate that the great mass of published data on foams refer to agents which, while of great practical interest, are such complicated mixtures that any detailed analysis is impossible. The other great drawback arises from the lack of reversibility of many of the adsorbed films (e.g. proteins, saponin), so that the adsorption cannot be calculated from the Gibbs equation (p. 519).

(a) *Soaps*

Following the definition given earlier, this term will include the modern synthetic detergents and wetting agents (for examples see Chapter XXIV), as well as the alkali metal salts of the fatty acids. A great deal of relevant information for foams in general has come from a study of the duplex films present in soap bubbles, so these will be briefly considered before foams in the wider sense.

Soap films in single bubbles. The fatty acid soaps are the oldest and still amongst the best foaming agents, an excellent composition for blowing stable bubbles being as follows (Lawrence, 1929):

>28·2 gm. oleic acid,
>100 ml. N.NaOH,
>300 ml. glycerine,
>1200 ml. distilled water.

The addition of three drops of concentrated ammonia is recommended. (As will be pointed out on p. 694 the function of the glycerol is to peptize the soap and *not* to increase the viscosity as usually suggested.) In other compositions the caustic soda is replaced by ammonia or tri-ethanolamine.

It is a commonplace that the thinning of a soap bubble is accompanied by the development of interference colours, the final films being 'black' with thicknesses a small multiple of the ultimate one, which has a thickness of about 50 A. (These films are 'black' owing to their extreme thinness and the phase change on reflection.) As the length of a fully extended sodium stearate molecule is about 25 A, it seems that the final film consists of a bimolecular leaflet with the polar groups almost in contact and an outer hydrophobic surface, as depicted in Fig. 22.2. (The thickness of the water film is not known, but it clearly cannot be very considerable.) Fig. 22.2 also shows the penultimate black film, and it is easy to visualize from this the transition to the final stage. These structures are clearly extreme examples of the laminar soap micelles discussed in Chapter XXIV, in which the water films have become very small. Considering their extreme thinness the stability of these black films is most surprising.

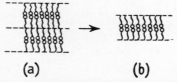

FIG. 22.2. Soap films: (a) Penultimate stage (secondary black film ca 100 A thick). (b) Final stage (primary black film ca 50 A thick).

The surface tension of a soap solution can be found by measuring manometrically the excess pressure (Δp) inside a soap bubble of radius r, since $\Delta p = 2(2\gamma/r) = 4\gamma/r$ (see p. 87). (The factor 2 arises from the fact that the soap film has two air/solution interfaces.) It is because of this excess pressure that large bubbles (where Δp is very small) are desirable when studying soap films, since diffusion of air through the film, in order to relieve the pressure, would eventually lead to the bubble collapsing.

Foams from soap solutions. Solutions of the alkali metal salts of the fatty acids such as the oleates, or the stearates after suitable peptization by glycerol, foam exceedingly well unless the dilution is too high. The surface tensions are very low, generally around 25 dynes/cm. and this is undoubtedly the major factor to which soaps owe their marked foaming powers. The molecular cohesion in the adsorbed film, which depends very much upon the structure and packing of the stabilizing molecules, is also important, however, as shown by Table 22.1, taken from Pankhurst (1941). (These results were obtained by the dynamic method given above and with the apparatus shown in Fig. 22.1.)

TABLE 22.1

Solution	0·001M			0·01M		
	θ_{50}	Stability	γ	θ_{50}	Stability	γ
Sodium oleate . . .	108	A	24·6	78	A	23·5
Sodium oleate+equivalent KMnO$_4$	183	B	43·9	120	A	25·1
Sodium ricinoleate . .	$\theta_6 = 25$	Z	44·5	130	B	32·3
Sodium sulpho-ricinoleate .	$\theta_8 = 50$	O	42·6	$\theta_{40} = 160$	Z	32·0

θ_{50} = time (in sec.) for foam to rise 50 cm. in tube, and is inversely proportional to the foaming power.
γ = surface tension in dynes/cm.
Stability units: A = no collapse in 15 minutes.
B = very slight collapse.
Z = collapse in 2 minutes.
O = immediate collapse.

It is seen that introduction of polar groups into the middle of the hydrocarbon chain, as with the ricinoleates or by oxidizing oleic acid to the dihydroxy compound with KMnO$_4$, reduces the foaming power enormously without increasing the surface tension very greatly. From monolayer studies it can be stated that such changes tend to increase the surface area occupied by each adsorbed molecule, leading to a loss of cohesion, which would in turn be reflected as a reduced foaming power.

As pointed out earlier, very little precise information about the mechanical properties of adsorbed soap monolayers is available, making it difficult for the contribution from this factor to be accurately assessed. From general considerations such films would be expected to be in the liquid state, possibly showing anomalous viscosity, but probably no plastic or elastic solid behaviour. (Where elasticity or plasticity has been reported it may have arisen from traces of heavy metal cations (e.g. Ca^{++}, Cu^{++}, Al^{+++}), or to hydrolysis liberating free fatty acid.) From insoluble monolayers it can be concluded that the surface viscosity and the molecular cohesion will increase with increasing chain-length, leading to increased stability in the case of foams.

FOAMS

TABLE 22.2

Sample	Foaming agent		Approx. concn. of active agent(%)	Foaming power (θ) and surface tension (γ)					
				0.01%			0.1%		
				θ	γ	Stability	θ	γ	Stability
A	Sulphated primary fatty alcohol (C_{16})		20	†θ_4 = 30	57.5	O	†θ_7 = 60	45.5	O
B	,, ,, ,, ,,		40	θ_8 = 90	54.8	Z	θ_{50} = 265	36.0	B
C	Sulphated secondary fatty alcohol (C_{12})		60	θ_{50} = 80	41.6	A	θ_{50} = 82	28.4	A
D	Alkyl naphthalene sulphonate		20	θ_{27} = 88	54.8	B	θ_{50} = 112	32.3	A
E	,, ,, ,,		25	θ_{25} = 80	65.9	B	θ_{50} = 85	31.5	A
F	Hydroxymethylene naphthalene sulphonate		55	No foam	70.4	O	No foam	68.4	O
G	Sulphonated ether		..	θ_{50} = 95	50.4	A	θ_{50} = 73	27.4	A
H	Oxycholesterol in sulphonated lorol.		..	θ_{50} = 110	45.8	A	θ_{50} = 81	27.9	A
I	Monosulphonated succinic ester (Na salt)	$2 \times C_4$	91.5	No foam	63.7	O	θ_9 = 180	46.9	A
J	,, ,, ,,	$2 \times C_5$	98.0	,, ,,	55.3	O	θ_{20} = 180	36.8	A
K	,, ,, ,,	$2 \times C_6$	98.8	,, ,,	47.1	O	θ_6 = 15	36.0	A
L	,, ,, ,,	$2 \times C_8$	86.7	θ_{50} = 233	28.8	Z	θ_{50} = 72	24.3	A
M	C-cetyl betaine		25	θ_{50} = 96	52.2	A	θ_{50} = 80	30.7	A
N	Quaternary ammonium salt		70	θ_{50} = 57	39.2	A	θ_{50} = 60	32.9	A
O	,, ,, ,,		90	θ_{50} = 88	34.2	A	θ_{50} = 78	32.4	A
P	,, ,, ,,		..	θ_{50} = 117	55.1	A	θ_{50} = 70	36.7	A

† θ_n where n is less than 50, indicates that it was impossible to obtain a column of foam higher than n cm.

Considering now the synthetic soaps, where the complications inherent in the fatty acid soaps (e.g. hydrolysis) are generally absent, most of the published data, such as that given in Table 22.2 on p. 633 (from Pankhurst, 1941), refer to commercial materials.

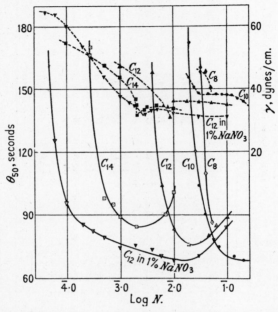

FIG. 22.3. Variation of θ_{50} and γ with concentration for various n-alkyl sodium sulphates at 38° C. θ_{50} full lines; γ broken lines.

Despite this limitation, such results show that whilst there is a general parallelism between the lowering of surface tension, the rapidity of foam formation (θ_{50}) and the stability of the resultant foam, yet wide variations exist, showing that other factors enter as well.

Fig. 22.3 shows the variation of θ_{50} (the reciprocal of which measures the rate of foam formation) with concentration and surface tension (γ) in the case of n-alkyl sodium sulphates with 8, 10, 12, and 14 carbon atoms (Pankhurst, 1941). With the C_{14} salt there is a close correlation between θ_{50} and γ, since θ_{50} is a minimum (i.e. foaming at a maximum) at the critical concentration for micelles (see p. 670), where the surface tension is also at a minimum (probably due to traces of impurities, as discussed on p. 681). The results for the C_{12} compound are not easy to explain, in that maximum foaming occurs at about twice the critical micelle concentration. Addition of 1 per cent. $NaNO_3$ permits stable foam formation at very much lower soap concentrations, due to the lowering of the critical concentration and to the increased rate of adsorption of the soap ions (p. 529).

As previously mentioned, the rate of drainage of liquid from many foams, particularly when using synthetic soaps, follows a logarithmic

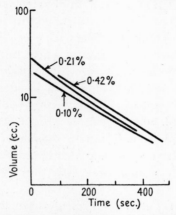

Fig. 22.4. The decay of dodecyl sulphonic acid foams with time. (Ross.)

course over the greater part of the decay, as shown in Fig. 22.4. Using a single liquid film as an approximation to a volume of froth, Ross (1943) has derived an equation for drainage arising from gravitational forces only, and which appears to be in agreement with experiment.

(b) *Proteins*

The very marked ability of proteins to stabilize foams arises, as we shall see, from a rather different balance of properties than in the above case. With the soaps the reduction in surface tension appears to be the dominating factor, although modified by others, in particular the molecular cohesion. This picture is reversed in the proteins—the surface tension factor becoming subordinate to the mechanical properties of the stabilizing film. Before attempting an analysis of the formation and stability of protein foams it is desirable to consider, first the properties of insoluble monolayers of proteins, and then protein films formed by adsorption from solutions.

Insoluble monolayers of proteins. Although proteins frequently possess a high solubility in water, being hydrophilic colloids needing relatively high salt concentrations to precipitate them, yet this property is rather easily lost by a variety of chemical and physical treatments, as will be discussed in Chapter XXVI. This 'denaturation', as it is termed, is even brought about by adsorption at an air/water interface, and as

Ramsden showed at the end of the last century, dissolved proteins can be quantitatively precipitated by bubbling air through their solutions and thus creating a large interface.

With the development of the techniques for studying insoluble monolayers (see Chapter XVII) a great deal of attention was devoted to the monolayers given by proteins. It was hoped by such studies to throw light, not only on the structure of the native protein molecule, but also on the process of surface denaturation, which may be the first step in the formation of protein membranes. Unfortunately, practically all the data published before 1939 is more or less in error, owing to the difficulties in the spreading process. By replacing the usual aqueous spreading medium by one of 60 per cent. isopropyl alcohol, 0·5 M. sodium acetate, Ställberg and Teorell (1939) overcame this difficulty, and reproducible quantitative spreading as a homogeneous monolayer now presents no difficulty, at any rate with the lower molecular weight proteins.

Fig. 17.17 (a) (p. 511) shows a typical curve for a protein, relating the lowering of surface tension (Π) to the area occupied in the surface (in m.²/mg.). The surface tension is not appreciably reduced until about 1 m.²/mg., and above about 20 dynes/cm. in most cases the monolayer collapses to a multilayer, which may be visible to the eye. On removing the pressure the film may re-expand to a greater or lesser extent, depending upon the conditions, e.g. time and extent of compression, nature of the protein, pH, and salt content of the substrate.

At the surface, owing presumably to the asymmetric force-field, the original compact native protein molecule is unfolded into its constituent polypeptide chain(s), —CO—NH—CHR—CO—NH—CHR[1]— etc., oriented so as to accommodate as many as possible of the polar side-chains (e.g. —CH_2OH, —CH_2COOH, —$CH_2.C_6H_4OH$, etc.) in the water, and as many as possible of the non-polar side-chains (e.g.—CH_3, —$CH_2.CH(CH_3)_2$, —$CH_2.C_6H_5$, etc.) out of it. It is not surprising that the protein molecule, once so spread, should find the return to its native state so difficult, even if the conditions were thermodynamically favourable. Surface denaturation seems in general to be associated with the loss of specific biological properties, such as enzymatic activity, and the ability to produce the antibody-antigen reaction.

The mechanical properties of protein monolayers, in particular their viscosity, elasticity, and compressibility as measured by the methods given in Chapter XVII, have also been investigated in some detail.

FOAMS

Qualitative differences can be seen by dusting talc on to the surface and gently blowing, or from the appearance of the 'expansion pattern' formed when a drop of oxidized motor oil is dropped on the monolayer (Langmuir, 1938). Quantitative measurements of viscosity and elasticity have usually been obtained by the oscillating disc method (p. 494),

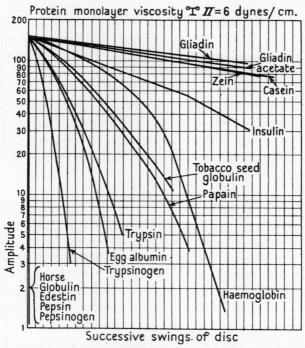

FIG. 22.5. The surface viscosities of various proteins by the oscillation method.
Surface pressure = 6 dynes/cm., pH = 5·8, T = 22° C.

and Fig. 22.5 shows the behaviour of a variety of proteins under a pressure of 6 dynes/cm. (from Langmuir, 1938). If the film shows viscosity only (i.e. no elastic properties) then log (amplitude) against the number of swings should be linear (as with the upper curves of Fig. 22.5); when this is not the case, as with egg albumin and haemoglobin, for example, then elastic forces are present, and the film is behaving more as a plastic solid than as a viscous liquid. With very coherent protein films the shear rigidity modulus can be found by the same apparatus, and its dependence upon molecular density, pH, time, etc., readily determined. The relevance of such information when the properties of bulk foams are considered needs no emphasis.

Adsorbed films of proteins. The word 'film' rather than monolayer is used advisedly here, since the surface films which form spontaneously on protein solutions are in general polymolecular in thickness, and frequently end up as visible skins or coagula.

In most cases it would seem that no true equilibrium is ever established between the dissolved and the adsorbed protein, the surface denatured monolayer folding up into a polylayer or coagulum, and being replaced from the reservoir of dissolved protein until this has all been used up. This complication makes the surface tension of a protein solution, and hence deductions from such measurements, rather meaningless. According to Bull and Neurath (1938) surface coagulation is divided into two steps—the production of surface denatured but still soluble protein, and the coagulation of this soluble surface-denatured protein. They shook egg-albumin solutions in glass bottles and found that if, after shaking, the pH was brought to the iso-electric point, precipitation occurred. This was regarded as evidence for surface-denatured protein being soluble. In view of the known insolubility of protein monolayers when fully spread, this question warrants further examination.

Wu and Ling (1927) studied the effect of surface-active materials and found that alcohol, ether, and saponin greatly decreased the rate of surface coagulation both with egg albumin and methaemoglobin. Such a behaviour can readily be ascribed to the surface-active material displacing the protein more or less completely from the surface. (This is why saponin can inhibit the churning of cream, which is discussed below.)

The mechanical properties of adsorbed protein films are, as would be expected, essentially similar to those given by protein monolayers, the mechanical strength and rigidity usually being greater on account of their greater thickness.

Foams from protein solutions. The above discussion of protein monolayers and adsorbed protein films has probably made the fundamentals of foam stabilization by proteins self-evident. As the surface tension can rarely be lowered by more than 20 dynes/cm., as compared with 40–45 in the case of soaps, the major factor is clearly the mechanical protection given by the elastic surface skins, which are usually polymolecular in thickness. It is probably always the case that these surface skins are thermodynamically unstable and tend slowly to coagulate further, but owing to the inherent time-lag arising from the high viscosity and cohesion in these thin layers, the *bulk* foams often appear remarkably stable. The reason for the rather stiff and permanent foams given, for example, by gelatine and sugar (*marshmallow*) or by egg-white (*meringue*) is thus clear.

In the case of gelatine, and probably with other proteins, the foaming power is a maximum at the iso-electric point, where the solubility is least and the molecular cohesion the greatest.

One interesting use of the denaturation of proteins in foams is in the churning of cream to make butter, which involves inverting the original fat-in-water emulsion to the water-in-fat type. Churning incorporates air bubbles, upon which the proteins present adsorb and denature. This continues until the fat particles are stripped of their protecting protein layers; the particles can then link up, and the cream emulsion inverts to butter.

(c) *Solid powders*

Certain finely divided materials (e.g. soot, heavy metal soaps), being sufficiently hydrophobic to remain at the interface, can be quite efficient foam stabilizers. Foaming or 'priming' in steam boilers arises chiefly from this cause, which also forms the basis of the 'froth flotation' process, described in more detail below. The bubbles in the foam are rather inelastic, as the interface consists of a closely-packed array of fine solid particles.

(d) *Other foam stabilizing materials*

There are a few other types of foaming agents which must be briefly mentioned, e.g. saponins, polymers (other than the proteins), and dyestuffs.

Saponins form very stable foams in extremely low concentrations (i.e. 0·005 per cent.), although their surface tension reducing properties are not considerable. The surface films formed are rather solid, and it is to their mechanical strength that the foams owe their stability, just as in the case of the proteins. They resemble proteins, further, in the apparent irreversibility of the adsorption.

Of the water-soluble polymers useful as foam stabilizers, two are commonly available—methyl cellulose (partially methylated) and polyvinyl alcohol (Chap. XXVII). Of these the latter is to be preferred for quantitative studies owing to its more definite chemical character, and as its monolayer behaviour is known, it should provide a useful system for fundamental work on foams, free from many of the complications inherent in the proteins and methyl celluloses. With these polymers, foam stability arises from essentially the same factors as in the case of proteins, although modified, of course, by the absence of ionized groups and by the different type of polar groups present. Hydrogen bonding between adjacent molecules in the film, and in the case of polyvinyl alcohol intra-molecular hydrogen bonding as well, would certainly occur.

Certain dyestuffs stabilize foams to some extent, and since the reduction in surface tension is usually quite small, other factors must be operative. Of these the adsorption of colloidally dispersed dye particles, or the spontaneous building up of polymolecular films, as shown to occur in some instances (p. 523), seem the most likely.

Flotation

The conditions under which particles are stable at an air/water interface underlie the possibility of separating minerals by the froth flotation process, which is now used industrially on a vast scale. Similarly the stabilization of emulsions by powders requires stable adsorption at the dineric surface and the two types of stabilization can therefore be treated on the same basis.

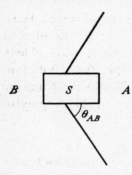

Fig. 22.6. Solid particle (S) in equilibrium at an interface between two phases A and B.

Under conditions where gravitational forces can be neglected (i.e. very small particles), the condition for stability at the interface (see Fig. 22.6) is that the contact angle θ_{AB} shall be $> 0°$ but $< 180°$. At equilibrium

$$\gamma_{BS} = \gamma_{AS} + \gamma_{AB} \cos \theta_{AB}.$$

If $\gamma_{BS} > \gamma_{AS} + \gamma_{AB}$ the particle will tend to go wholly into phase A; if $\gamma_{AS} > \gamma_{BS} + \gamma_{AB}$ then it tends to enter phase B. (If one phase is air, complete movement into it is clearly impossible owing to gravitational forces.)

If phase B be air, then θ_{AB} is the ordinary contact angle. The value of the contact angle usually depends upon whether the liquid surface is advancing or receding, and to ensure stability of the particle it is therefore desirable that θ_{AB} should not be too near zero (since θ receding $< \theta$ advancing as a rule).

Minerals differ greatly in their wettability by water, but in general the differences between the desirable and undesirable constituents of an ore have to be augmented by the use of suitable chemicals termed 'collectors'. These are usually slightly soluble organic compounds possessing a polar group by which they became anchored to the surface of the desired mineral constituent, and a non-polar group which gives the mineral an outer surface of hydrocarbon. This increases the contact angle and so stabilizes the desirable particles in the surface. For the sulphide ores (e.g. galena, zinc blende) the xanthates (I) and *aerofloats* (II) are the most important,

$$\underset{\text{(I)}}{\underset{R-O}{\overset{S}{>}}C-SM,} \qquad \underset{\text{(II)}}{\underset{S}{\overset{R-O}{>}}P\underset{SM}{\overset{O-R}{<}}}$$

where R is a hydrocarbon chain (generally ethyl) and M a metal, usually sodium.

A large air/water interface is presented to the finely ground and

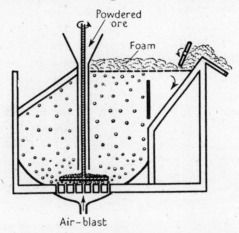

Fig. 22.7. Diagram to illustrate the principle of a flotation cell.

suitably treated ore by blowing air through the liquid in which it is suspended; the desirable constituents are thus adsorbed and carried away by the froth while the unwanted material ('gangue') remains at the bottom. A continuous process is used industrially, and some idea of the commercial plant can be seen from Fig. 22.7, which illustrates the principles involved.

In addition to the collector, it is usually necessary to add a small amount of frothing agent, commonly crude cresol, to enable a voluminous froth of fine bubbles to be formed. Originally the same material, generally a locally available one such as pine oil, was used as both collector and frothing agent.

(For a more detailed account of the surface chemistry involved in flotation see Adam, 1941; Wark, 1938, 1946.)

Foams in non-aqueous media

Stable foams can be obtained with many organic liquids, and may present industrial problems in connexion with distillation, lubrication, etc. A recent paper by King (1944) gives some data on foam formation in a variety of common organic liquids by compounds which are wetting agents in water. He finds that foaming properties usually increase with the surface tension of the solvent and with the lowering in surface tension, but there are many exceptions to any simple generalization, just as with aqueous foams. The viscosity (η) and vapour pressure are not important in foam initiation, but a high η tends to stabilize any foam that is formed.

Fractionation by adsorption on foam

The possibility of separating compounds by taking advantage of their differences in capillary activity has recently been examined by Schütz (1946), the method being clearly analogous to chromatographic adsorption on solids. The most surface active substances are carried over by the initial foam, which thus corresponds to the top of the usual chromatographic column; then comes the next surface active component, and so on, and provided the components differ sufficiently in surface activity, sharp separation should be quite feasible. For biological mixtures the method may prove quite a valuable addition to ordinary chromatography.

GENERAL REFERENCES

For an account of foams and flotation from the angle of surface chemistry see Adam, 1941, *The Physics and Chemistry of Surfaces*, 3rd edn., Oxford.

For certain technical aspects see Clayton, 1943, *Emulsions and their Technical Treatment*, 4th edn., Churchill.

OTHER REFERENCES

BIKERMAN, 1938, *Trans. Faraday Soc.* **34**, 634.
BULL and NEURATH, 1937, *J. Biol. Chem.* **118**, 163.
—— —— 1938, *Chem. Rev.* **23**, 391.
FOULK and BARKLEY, 1943, *Industr. Engng. Chem.* **35**, 1013.
—— and MILLER, 1931, ibid. **23**, 1283.
KING, 1944, *J. Phys. Chem.* **48**, 141.
LANGMUIR, 1938, *Cold Spr. Harb. Symposia Quant. Biol.* **6**, 171.
LAWRENCE, 1929, *Soap Films*, Bell.
PANKHURST, 1941, *Trans. Faraday Soc.* **37**, 496.
REHBINDER and TRAPEZNIKOV, 1938, *Acta Physicochim. U.R.S.S.* **9**, 257.
ROSS, 1943, *J. Phys. Chem.* **47**, 266.
SCHÜTZ, 1946, *Trans. Faraday Soc.* **42**, 437.
STÄLLBERG and TEORELL, 1939, ibid. **35**, 1413.
TRAPEZNIKOV, 1940, *Acta Physicochim. U.R.S.S.* **13**, 265.
WARK, 1938, *Principles of Flotation*, Melbourne.
—— 1946, *25th Meeting Austr. and N. Zealand Assoc. Adv. Sci.*
WU and LING, 1927, *Chin. J. Physiol.* **1**, 407.

XXIII
EMULSIONS
General introduction

IF a mixture of two immiscible, or partially miscible, liquids is shaken, a coarse dispersion of one in the other results, but in order to attain the degree of stability which we associate with the term 'emulsion', a third component is essential. This requisite component, usually present in amounts of 1–5 per cent., is termed the emulsifying agent. In the case of water and insoluble oils a stable dispersion of the latter can be obtained (see p. 559) without additives if the concentration of the oil is very small (*ca.* 0·1 per cent.), but such systems are to be regarded as oil hydrosols rather than as emulsions, and as such are treated in Chapter XX.

In principle the term 'emulsion' refers to any stable dispersion of one liquid in another, but the only two liquids which need concern us here are water and organic compounds ('oils') insoluble or sparingly soluble in water, since these have, for practical and theoretical reasons, been by far the most extensively studied. Biological emulsions, as well as almost all those of domestic and industrial importance, contain an oil and some aqueous solution as the two immiscible phases.

Emulsions can, both theoretically and practically, exist in two types, oil-in-water and water-in-oil, depending upon whether the aqueous or the oil phase is the continuous one. The former is referred to as O/W and the latter as W/O, the disperse phase preceding the dispersion medium. (In this respect they differ from the foams, which might formally be regarded as air-in-water emulsions and which show many points of resemblance to the O/W type of emulsions.) Thus with emulsions there are two fundamental problems: the origin of the stability brought about by the emulsifying agent, and the reason why some agents promote O/W emulsions and others the inverse type. Only the former problem arises in the case of the foams.

A simple geometrical calculation from close-packed spheres shows that the maximum amount of one phase dispersible in the other would be 74·02 per cent. of the total available volume if the droplets were uniform, undistorted spheres. For uniform spheres this figure is independent of their radius. In point of fact emulsions considerably more concentrated can readily be obtained, both of the O/W and W/O types. This is possible partly on account of the non-uniformity of droplet size but chiefly because the emulsifying agent enables the droplets to be distorted without coalescing: thus under the microscope the droplets in concentrated emulsions are seen to possess a polyhedral rather than a spherical form. In the very concentrated emulsions (80–90 per cent. or so) it is not uncommon to see several droplets one inside the other. From our experience it is difficult to make true O/W emulsions (i.e. which still have a *continuous* aqueous phase) with an oil content of much over 90 per cent. The statement that 99 per cent. by volume of oil can be emulsified in a concentrated soap solution, due originally to Pickering and repeated even in modern text-books, is certainly not correct. Lawrence showed some years ago (Lawrence, 1937) that such systems are not emulsions at all, but consist of a continuous oil phase thickened by a mass of hydrated soap crystallites.

Emulsifying agents

Emulsifying agents can be classified in two ways—either according to their general chemical or physical nature (e.g. fatty acid salts, proteins, solid powders, etc.), or according to the type of emulsion which they promote. It is more convenient in this section to use the latter criterion, but the former has some advantages and will be followed when the principal types of emulsifying agents are discussed below.

(a) Agents for oil-in-water emulsions

The principal O/W promoting agents are the hydrophilic colloids, such as the proteins, gums, carbohydrates, natural and synthetic soaps, clays, and certain hydrated oxides.

A great variety of proteins is used commercially, such as egg albumin (white of egg), haemoglobin, casein, etc.; and proteins are also responsible in whole or part for the stability of such naturally occurring emulsions as milk, rubber latex, etc. Gelatine, and the proteins generally, find extensive application in food emulsions, although in the case of gelatine its emulsifying powers depend considerably upon its source, the method of extraction, and its ash-content.

Gum acacia, gum arabic, and gum tragacanth are hydrophilic colloids of natural origin which still find widespread use as emulsifying agents, particularly in pharmaceutical preparations. Of the carbohydrates, mention may be made in this connexion of starch, dextrin, and flour.

The natural and synthetic soaps which are soluble in water form the most extensive and probably the most widely used O/W emulsifying agents. The natural soaps are chiefly the alkali metal salts of the fatty acids, but include also the bile salts (e.g. sodium cholate and glycocholate), and the phospholipoids such as lecithin. As for the synthetic soaps their number is now legion! A representative selection will be found in Chapter XXIV (p. 667). All contain a hydrocarbon tail (with straight or branched chains, or rings) and a very polar group such as $-SO_4'$, $-\overset{+}{N}(CH_3)_3$, etc. This balance of hydrophobic and hydrophilic properties endows them with marked surface activity, and they are usually much less sensitive to pH, calcium, or heavy metal ions than are the fatty acid soaps.

Of the solid emulsifiers which give the O/W type of emulsion the clays, particularly bentonite, and the hydrous oxides (particularly of aluminium) are the most commonly used.

(b) *Agents for water-in-oil emulsions*

The principal W/O promoting agents are all members of the hydrophobic class of colloids (using 'hydrophobic' in a somewhat different sense from that frequently understood), as, for example, the heavy metal salts of the fatty acids, long-chain alcohols, long-chain esters, oxidized oils, soot or lamp-black, colloidal graphite, etc.

Of the heavy metal salts of the fatty acids, those in common use are the oleates and stearates of magnesium, aluminium and calcium; those of zinc and silver are also used to some extent. (The reasons for the different types of emulsion promoted by the heavy metal and alkali metal soaps is discussed later.)

The long-chain alcohols and esters can be grouped together on the basis of a similarity in the balance between their non-polar and polar portions; the polar groups being un-ionized these compounds are much less hydrophilic than are the paraffin-chain soaps with their ionized polar groups. Common examples are cetyl ($C_{16}H_{33}OH$) and octadecyl ($C_{18}H_{37}OH$) alcohols (the usual commercial material being a mixture of these compounds), and the mixed hydroxy sterols obtained from lanolin, which consists essentially of the long-chain esters of these alcohols. Lanolin itself is widely used in pharmacy as a W/O emulsifier. Closely related are the mixed alcohol/ester type of compounds such as the mono- and di-esters of glycerol (e.g. glycerol mono-oleate) and the glycol mono- and di-esters (e.g. glycol oleate and stearate). As made and used commercially these glycerol and glycol esters are all mixtures of the mono- and di-esters.

Oxidized vegetable oils find an extensive use in the margarine industry, where the consideration of edibility limits the choice of emulsifiers.

Soya-bean oil is the usual starting material. It is oxidized at 250° C. until gelatinization occurs, and after cooling, mixed with twice its volume of the original oil. The product is termed 'Palsgaard Emulsion Oil' or 'Schou Oil', and *ca.* 1 per cent. gives stable W/O emulsions.

The principal *solid* emulsifiers promoting the W/O type which were listed above, such as lamp-black, soot, and graphite, are merely varieties of carbon particles in a suitably fine state of subdivision.

Preparation of emulsions

Before outlining the practical methods by which emulsification is achieved in the laboratory and on the industrial scale, some consideration may be given to the mechanism of the dispersion process.

Agitation of two liquids leads in general to mutual penetration in the form of lamellae and threads which then break up into drops. (It is shown in elementary text-books of physics that a cylinder of liquid is unstable when its length exceeds π times its diameter.) The extent to which the respective liquids disrupt one in the other depends to a large extent upon the volume ratio, the viscosity, and the density of the two phases. Of other factors, low interfacial tension clearly facilitates break-up.

The drops formed by agitation will rapidly coalesce unless stabilized by adsorption of the emulsifying agent, and in its absence equilibrium between break-up and coalescence is soon reached.

Several authors have studied both practically and mathematically the deformation of drops when subject to the forces arising from non-turbulent viscous flow, but this is of little relevance for practical emulsifying conditions where considerable turbulence is invariably present.

(a) Laboratory methods

A simple apparatus due to Hatschek is shown in Fig. 23.1. The oil to be emulsified is dripped slowly on to the side of the long thistle funnel, and the exit tube of the cylinder is connected to a water-pump. Under the action of the air bubbles the film of oil thus formed is broken up at the bottom of the tube where it comes into contact with the solution of the emulsifying agent.

Sumner has developed a condensation method in which the vapour of the disperse phase is injected below the surface of the dispersion medium containing the emulsifying agent. For concentrated O/W emulsions the method is clearly limited to oils of boiling-points below that of water.

EMULSIONS

Certain systems, particularly those of soap solutions and low viscosity oils, can be emulsified merely by hand-shaking. The rather unexpected finding of Briggs was that *intermittent* shaking was considerably more effective than if the shaking were continuous. He found, for example, that a benzene/ sodium oleate mixture which took 4·2 min. and 750 shakes for emulsification by continuous shaking, required only 5 shakes and less than 1 min. if a rest of 30 sec. was allowed after 2 shakes. Later workers have confirmed Briggs's findings and several theories have been advanced to explain the phenomenon. The entrainment of air bubbles, which would be much diminished by the intermittent shaking, would seem to us to be the most probable cause. Considerable entrainment of air bubbles will not only compete for the emulsifying agent but will, owing to their buffering action, considerably reduce the smashing forces between the water and oil globules.

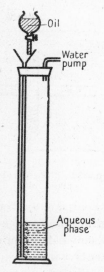

Turning now to mechanical devices for laboratory use, these can be of diverse types, ranging from simple stirrers and small hand-powered 'cream-making' machines to power-driven small-scale models of the colloid mills and atomizers used industrially. It is usual to prepare a coarse dispersion by simple mechanical agitation and then to subject this to intensive shearing forces by means of the various devices outlined in the next section.

FIG. 23.1. Hatschek's transpiration apparatus for emulsification.

Emulsification by ultrasonic waves (frequency usually 100,000–500,000 per second) may be mentioned here, as suitable ultrasonic generators are now available commercially, although at a price which excludes them from general laboratory use. The ultrasonic waves are produced by a piezo-electric quartz plate placed between two electrodes connected with a high-frequency oscillator. The quartz plate and electrodes are immersed in an oil-bath, which also contains the vessel with the mixture to be emulsified.

(b) *Industrial methods*

Only the merest outline will be attempted here to illustrate the principles involved. (For more detailed descriptions and references to the patent literature see Clayton, 1943, chap. xi.)

The simplest devices consist merely of two co-axial propellors mounted on a common shaft but reversed in pitch, so that the lower one creates a powerful upward pull from the bottom of the vessel, and the upper one an equally powerful downward flow. The two streams impinge and are then thrown violently by the centrifugal forces against the walls of the container.

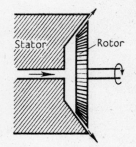

Fig. 23.2. Principle of the colloid mill.

Various arrangements utilizing centrifugal force to bring about intense shearing action have been patented. The liquid may, for example, be impelled through fine slots which produce emulsification, and may then be subject to further shear when it strikes the walls of the machine.

In 'colloid mills', which figure prominently in the literature, the coarse emulsion (or other colloid) is subjected to a combination of centrifugal and rotational shear during its passage through the narrow gap between a rotor and stator, as shown diagrammatically in Fig. 23.2. The actual mechanical arrangements vary considerably, for example the working surfaces may be roughened or corrugated, and the emulsion may be subjected to various types of shear during its passage. Provision is made for varying the size of the gap, and the rate of rotation, which are the factors chiefly determining the degree of dispersion.

Industrial emulsions (particularly milk) are usually subjected to 'homogenization' in order to decrease the particle size and so increase the stability. Emulsions of almost uniform particle size, usually around $\frac{1}{2}$–1 μ, can thus be obtained. Homogenization is usually produced by forcing the coarse emulsion through a valve which is spring-loaded so that only small clearances are possible. The pressures involved are usually considerable, usually several thousand pounds per square inch, and can be varied by adjustment of the spring. Some types of household 'cream-making' machines work on this principle, although the dispersions obtained are relatively coarse.

Determination of emulsion type

For studies of emulsifying agents, and in particular of the phase-reversal problem, it is essential to have a simple and accurate means of distinguishing between O/W and W/O emulsions. Various methods are known, but the three given below usually suffice for most purposes.

(a) *Conductivity method*

Since most oils have a much higher resistance than aqueous solutions, the W/O type of emulsions are relatively non-conducting as compared with the O/W. The simple apparatus shown in Fig. 23.3, in which the lamp lights up for O/W emulsions only, is quite suitable. Alternatively

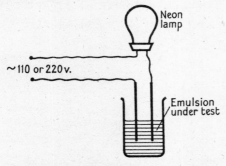

FIG. 23.3. Conductivity method for emulsion type.

the resistance can be measured with an 'Avometer' or similar commercial instrument.

(b) *The indicator method*

If an oil-soluble dye (e.g. Sudan III, fuchsin) is dusted on the surface of an emulsion it will spread, and colour the surface, only if oil is the continuous phase (W/O type). Conversely water-soluble dyes (e.g. methyl orange) will colour only the O/W type of emulsions.

(c) *The dilution method*

The principle here is the rapid miscibility of the continuous phase only with a drop of the same nature; thus O/W emulsions can only be diluted readily with water, and the W/O type only with oil. Visual observation of the effect of bringing the two drops together on a microscope slide usually suffices.

Measurement of emulsion stability

The term 'emulsion stability' has frequently been used with reference to two essentially different phenomena—the (upward) sedimentation of the emulsion, usually termed 'creaming', and the rate of agglomeration of drops to form larger ones. Creaming is due fundamentally to the density difference between the oil and aqueous phases, and is not necessarily accompanied by aggregation, although this will clearly be facilitated by the increased proximity of the drops. It will be discussed in more detail on p. 664.

Owing to its interfacial energy an emulsion should be unstable with respect to the bulk phases, the decrease in energy on coalescence thus providing a measure of its true, thermodynamic, stability. Like all thermodynamic quantities, however, this gives no information as to the *rate* at which the system will move towards equilibrium, which is the property of most practical interest. From the practical point of

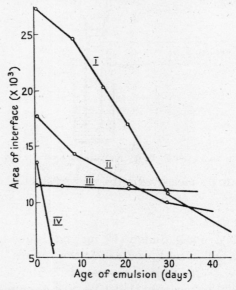

FIG. 23.4. The effect of ageing upon the interfacial area of emulsions stabilized by sodium stearate (I), by gelatine (II), by inert alumina (III), and by sodium stearate + 3 per cent. calcium chloride (IV).

view, therefore, recourse must be had to some sort of kinetic measurements to obtain the decrease of the interfacial area with time.

Various indirect methods such as turbidity have been used to study the coalescence of drops, but they are necessarily only approximate, unless calibrated against standard systems. Measurement of the volume of free oil is clearly limited as it applies only to the final stages. The only really satisfactory method is to determine the statistical distribution of particle sizes by size-frequency analysis, which gives the total surface area, and to repeat the process at suitable intervals.

A size-frequency analysis may be carried out by counting the number of particles in a small volume by means of a microscope and haemocytometer, and measuring the drop sizes by a graticule eyepiece, or by projecting the image on to a suitable screen or photographic plate. In order to diminish Brownian movement or sedimentation the samples may be diluted with glycerol or gelatine solution, provided, of course, this produces no deleterious effects. Fig. 23.4 shows the change in interfacial area with time, as obtained by the above methods, for a number of emulsions (from King, 1941).

The chief disadvantages of the above technique are the excessive time and labour involved. Another method, which has the added advantage of lending itself to automatic recording, is to determine the size-frequency distribution from sedimentation measurements, using the principle of the sedimentation tube shown on p. 538 (Kraemer, 1924).

Factors in the formation and stability of emulsions

The formation and stability of emulsions should be determined by just the same factors as in the case of foams, since the only major difference is the much greater density of the oil as compared with air. Experiment confirms that either a low interfacial tension, or the formation of elastic 'skins' around the droplets, are the principal means by which emulsions are stabilized, just as with foams (p. 628). Similarly, electrical forces are not of major importance, except in a few special systems.

Theories of emulsion type (O/W or W/O)

As previously pointed out, no analogous problem arises in the case of foams.

The early work on emulsions had shown that the alkali metal soaps tended to give the O/W type, whereas the heavy metal soaps such as of calcium, zinc, etc., showed the opposite tendency. Monolayer studies showed that the former tended to occupy much greater areas than the latter, and led Harkins to put forward the 'oriented wedge' theory of emulsion type. This theory is still quoted as correct by most text-books, although in its original form it had soon to be abandoned. It required that where the metal atom had a smaller cross-sectional area than the paraffin chain the emulsion would tend to be W/O (to achieve the greatest interfacial density of stabilizer); the converse condition would tend to give O/W. Thus monovalent soaps should tend to give O/W and di- and tri-valent soaps W/O emulsions. In point of fact several monovalent soaps (e.g. Tl, Ag) give the W/O type.

The most reasonable theory of emulsion type, which, as will be shown here, can be extended to cover molecularly dispersed stabilizers, is that originally suggested for solid powders. A solid powder, if it is to act as an efficient stabilizer, must tend to be taken up at the interface, i.e. this must be its position of minimum free energy, rather than in either of the bulk phases. As previously shown (p. 640) this requires its contact-angle (θ) to be $> 0°$ but $< 180°$ (see Fig. 23.5). If the solid is preferentially wetted by one phase, then more particles can be accommodated if the interface is curved and convex to that particular phase, i.e. preferential wetting by water ($\theta < 90°$) should tend to make the interface convex to the aqueous phase and thus to give O/W emulsions, and preferential wetting by oil ($\theta > 90°$) the inverse type.

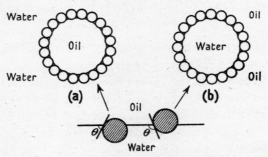

FIG. 23.5. Inert spherical particles at the O/W interface. (a) Preferential wetting by water, leading to O/W emulsions. (b) Preferential wetting by oil, leading to W/O emulsions.

Fig. 23.5 shows this principle very clearly in the ideal case of spherical solid particles. As we shall see later, this theory seems on the whole to be in good accord with experiment.

Those emulsifiers which, under the conditions of use, cannot be classed as solid powders, such as the alkali metal soaps for the O/W type and long-chain esters for the W/O, can be included in the above scheme since the former are preferentially wetted by water (being hydrophilic colloids), and the latter by oils (being soluble in oil but insoluble in water). (It is, of course, not possible to talk of a contact angle in the sense of Fig. 23.5 for molecularly dispersed particles at interfaces.)

The reader has probably noticed the qualifying phrase 'tend to give' throughout the above discussion. This has been used deliberately, since although it is primarily the nature of the emulsifying agent which determines the emulsion type, yet other factors are by no means negligible, the principal ones being the phase-volume ratio, the method of mixing, and the nature of the oil. A large amount of oil relative to the water will clearly favour the W/O type, as will addition of water to the bulk oil phase. The chemical and physical nature of the oil, particularly its viscosity, might also be expected to exert some influence.

Discussion of the principal types of emulsifying agents

Earlier we listed the principal emulsifying agents on the basis of the type of emulsion promoted. Here they will be grouped somewhat differently—as soaps, proteins, solid powders, etc.—a division which has certain advantages.

(a) *Soaps*

In the O/W emulsions given by the water soluble soaps the interfacial film is monomolecular in thickness, and just as in the foams the important factors are firstly the magnitude of the interfacial tension, and secondly the physical nature of the adsorbed film. Little is known about the latter, but from monolayer studies highly compressed gaseous films, which may show anomalous viscosity but certainly no elastic properties, would be anticipated. The variation of emulsification with interfacial tension, being readily measured, has been extensively studied, particularly with the fatty-acid soaps. The latter are not very suitable, however, as their behaviour is complicated by acid soap formation, as well as by sensitivity to CO_2 and to heavy metal cations.

Solutions of the unbranched paraffin-chain salts, such as sodium dodecyl sulphate or cetyl trimethyl ammonium bromide, give interfacial tensions of about 3–6 dynes/cm. against non-polar oils (e.g. CCl_4, nujol) which they emulsify moderately well on shaking. With many of the branched-chain compounds the tension is much lower (often less than 0·1 dyne/cm.) and the systems emulsify spontaneously. For example the straight-chain compound

$$C_{16}H_{33}-O-\underset{CH_3}{\underset{|}{\bigcirc}}-SO_3'K^+$$

cannot reduce the tension against cyclohexane much below 1·3 dynes/cm., whereas with the branched-chain member

$$C_8H_{17}-O-\bigcirc-SO_3'K^+$$
$$\underset{C_8H_{17}}{\overset{O}{\diagup}}$$

in similar concentrations the surface tension readily fell to *ca.* 0·1 dynes/cm., below which spontaneous emulsification made measurement impossible (Hartley, 1941). The reason for the difference is that the branched-chain members (owing to steric effects) need very much higher concentrations before they can aggregate to form micelles, which, as we shall see (p. 680), set a very effective limit to the surface activity. This general parallelism between interfacial activity and ease of emulsification is also followed in those mixed systems showing 'complex' formation discussed below (p. 656).

If electrical effects, arising from the repulsion between the similarly charged drops, contributed appreciably to emulsion stability, they should be most in evidence with the soap-stabilized systems, since these possess the highest known potentials due to the high density of the paraffin-chain ions in the stabilizing monolayer (p. 569). Addition of salts lowers the ζ potential, but unless this produces changes in the soap such as chemical precipitation or physical 'salting out', no great change in emulsion stability is observed, showing the relative unimportance of electrical effects.

Turning now to W/O emulsions stabilized by di- and tri-valent fatty acid salts, some interesting work by Pink (1941) has shown that for efficient emulsification the soap must be almost insoluble in both phases, must remain at the interface, and must possess some lateral adhesion between the solid particles. Benzene solutions of the metallic oleates or stearates were shaken with water, and the precipitating and emulsifying effects noted (see Table 23.1).

TABLE 23.1

Zn salts	Unaffected	No emulsifying power
Ca, Sr, and Ni salts	Partly precipitated	Fair ,, ,,
Mg salts	Wholly precipitated at interface	Good ,, ,,

Stabilization is undoubtedly due to solid particles rather than to a monolayer, since from monolayer studies we know that these heavy metal soaps can only give very unstable monolayers which would rapidly collapse to microscopic particles. Thus these soap-stabilized emulsions can, it seems, be brought into line with emulsions stabilized by solids in general.

O/W emulsions stabilized by such soaps as sodium oleate are readily inverted by addition of a few drops of saturated magnesium chloride, due to chemical reaction forming the relatively un-ionized magnesium oleate which promotes W/O emulsions. Sodium cetyl sulphate behaves similarly on addition of a soluble barium salt.

With solutions of fatty acids containing both alkali metal and divalent cations (e.g. Na^+ and Ca^{++}), the type of emulsion promoted depends upon the ratio of the ionic concentrations, as was first pointed out by Clowes. This 'ion antagonism', as it is termed, may be of some biological significance.

Further reference to the action of soaps will be found below, when the bearing of monolayer studies on emulsions is discussed.

(b) Proteins

Proteins, unless deaminated or otherwise modified, promote only the O/W type of emulsions. Before discussing their emulsifying action, some mention must be made of the insoluble monolayers given by proteins, and of the films adsorbed from protein solutions at the O/W interface, just as was required for A/W systems in the case of foams. The presence of the oil phase tends to expand all monolayers, as shown for protein films in Fig. 17.17 (a) (p. 511), but the ultimate molecular packing seems to be the same whether air or oil is the outer phase, showing that the oil can be squeezed out entirely from the interface. The mechanical properties of O/W protein monolayers have been little studied, but for a given molecular density are qualitatively the same as at the A/W interface, showing marked elastic properties at areas less than about 1 m.^{2}/mg. (Chapter XVII).

Protein films form spontaneously at an O/W interface in contact with a protein solution, the only difference from the analogous A/W system being the lower boundary tension (e.g. benzene/water is $ca.$ 35 dynes/cm. as compared with around 72 for water). This should have the effect of reducing the rate at which the initially formed monolayer folds up into the multimolecular skins or coagula which are the usual ultimate state, just as in the case of foams. Experiments by Danielli (1938) on the surface denaturation of ovalbumin at interfaces with various tensions confirm this expectation.

The stability of emulsions, as with foams, stabilized by proteins, arises from the mechanical protection given by the multimolecular elastic skins (often several μ thick), rather than by the reduction in interfacial tension which rarely exceeds 20 dynes/cm.

(c) Solid powders

The suggested theory for stabilization of emulsions by solid powders has been given above—it requires the particle to be stable at the interface and thus to have a contact angle between 0° and 180°. Packing considerations indicated that preferential wetting by one phase should lead to that particular phase being the continuous one. This conclusion accords well with experiment—for example, lamp-black, graphite, and the heavy metal soaps are preferentially wetted by oil and give W/O emulsions, whereas bentonite, hydrated alumina, and other hydrophilic colloids, preferentially wetted by water, give the O/W type. The hydrophilic colloids include the water-soluble soaps, proteins, gums, agar, and the saponins.

One interesting question arises in connexion with these solid-stabilized emulsions—namely, are they *thermodynamically* stable or not ? (cf. with soap-stabilized O/W emulsions on p. 653). The above treatment of the problem would indicate that they are thermodynamically stable, provided that any changes occurring in the system do not move the contact angle outside the range of 0° to 180°. This may be the reason why this type, although giving much coarser emulsions than given initially by the soaps, for example, yet frequently possess a much greater stability, as exemplified by the comparison between solid alumina particles and sodium stearate in Fig. 23.4.

(d) Electrolytes

Certain oils, such as amyl alcohol, can be emulsified in water without the aid of the usually accepted surface-active materials, but using instead certain electrolytes, particularly the thiocyanates and iodides (see King, 1941). As the salt concentration is increased the stability first increases, passes through a maximum, and then decreases. For a given cation the efficacy of the anions recalls the well-known lyotropic series (e.g. $CNS' > I' > Br' > Cl' > SO_4''$). Stabilization in such cases is probably due to the increase in the ζ potential brought about by the preferential adsorption of the anion, the stability finally falling since the ζ potential is always effectively eliminated by excess of salt.

These systems do not possess stability at all comparable with that given by the usual emulsifying agents, but nevertheless are of great theoretical interest.

(e) Other stabilizers

The remaining stabilizers which need some mention are the gums, carbohydrates, and saponins, which all fall into the class of hydrophilic colloids and promote O/W emulsions. Many of these materials as normally used are rather indefinite mixtures. It would seem that two principal factors are involved—an increase in viscosity of the bulk phase and adsorption at the interface, frequently accompanied by some sort of surface skin formation. With the gums (e.g. gum tragacanth, gum arabic) and carbohydrates (e.g. dextrin, agar) the bulk viscosity is probably the major factor, whereas the saponins give rigid interfacial films (probably multimolecular in thickness as with the proteins) which afford mechanical protection.

The application of monolayer studies to emulsion systems

Some interesting conclusions regarding the formation, stability, and inversion of emulsions have been obtained from the study of mixed monolayers (Schulman and Rideal, 1937; Schulman and Cockbain,

1940). When a paraffin-chain salt is injected beneath a monolayer of cholesterol or of cetyl alcohol, the surface tension rapidly falls if the film is held at constant area, or the area increases rapidly if the film is held at constant surface pressure, showing that soap molecules have

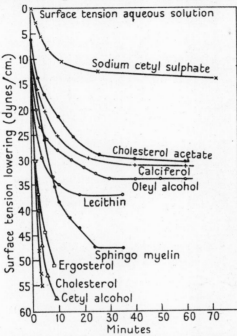

FIG. 23.6. The penetration of various monolayers at constant area by dilute sodium cetyl sulphate, as indicated by the reduction of surface tension.

penetrated the monolayer. The effect depends upon the chain-length and polar group of the soap, and for a series of compounds $C_{12}H_{25}X$ the lowering in surface tension (at constant area) was in the following order:

$$-NH_3^+ > SO_4' > SO_3' > COO' > \overset{+}{N}(CH_3)_3 > \text{bile acid anions.}$$

The stability of emulsions (all O/W type), containing cholesterol or cetyl alcohol in the oil phase and the soap in the aqueous phase, was in precisely the same order as found for the reduction of surface tension, showing that a similar phenomenon existed at the O/W interface. The mixed cholesterol-soap monolayers were much more mobile than those with cetyl alcohol, and this was paralleled by the marked difference in viscosity of the respective emulsions.

Marked differences in penetration were found if the monolayer compound was varied, keeping the soap and its concentration constant. Fig. 23.6 shows the results for a series of monolayers (held at constant area) with sodium cetyl sulphate as the soap. The fall in surface tension, which again ran parallel to the emulsion stability, was diminished if the —OH group were esterified, or if the chain were 'kinked', as in the *cis*

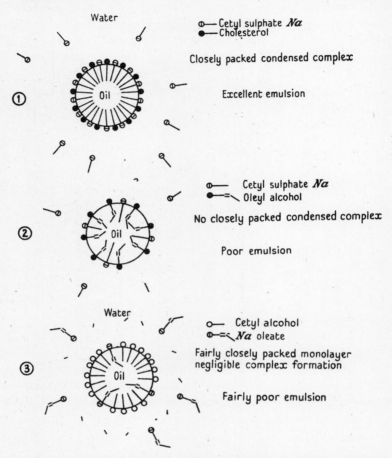

Fig. 23.7. Effect of 'kinked' chains upon the packing of complexes at interfaces and upon emulsion stability.

(but not the *trans*) unsaturated alcohols. This leads one to suggest that the 'complex' (as it has been termed) between soap and long-chain alcohol arises from a type of hydrogen bond linkage between these molecules when adlineated at an interface. The effect of 'kinked' chains upon interfacial packing and emulsion stability is depicted in Fig. 23.7.

The mechanism of phase inversion, and the factors determining the emulsion type, were also examined by Schulman and Cockbain (1940). Their results led them to suggest that for stable O/W emulsions the oil droplets must be electrically charged and the interfacial film condensed and fluid, whereas for the invert type the interfacial film must be rigid and uncharged. This change from fluid to rigid films can be brought about by addition of suitable polyvalent ions (e.g. by Ca^{++} with sodium soaps, or by SO_4'' with long-chain amine hydrochloride), which are known to invert O/W to W/O emulsions. The major objection to the above interpretation of phase inversion lies in the fact that these rigid monolayers are very unstable and under the influence of shaking would collapse to solid particles with a negligible surface vapour pressure. It is for this reason that the explanation previously suggested (p. 654) is preferred (see also Alexander, 1947).

Physical properties of emulsions

Of the physical properties of emulsions only two will be discussed here—optical (e.g. colour and light-scattering power) and viscosity.

(a) Optical properties

Most emulsions of the O/W type are white or very pale yellow in colour, and are opaque at the oil concentrations usually encountered in emulsion systems. The colour observed depends upon the opacity of the emulsion, which in turn depends upon the refractive indices and optical dispersive powers of the two phases, as well as the concentration and globule size of the internal phase.

Transparent colourless emulsions can be obtained by using two liquids having the same refractive index and optical dispersive power, e.g. glycerol and a solution of calcium oleate in carbon tetrachloride. If the liquids have the same refractive index, ensuring transparency, but different optical dispersive powers, then the droplets act as a series of lenses and the emulsion shows structural colours—hence the term 'chromatic emulsion'. An example is given by carbon disulphide/glycerol, stabilized by collodion.

(b) Viscosity

The basic theory and the methods by which viscosity is determined have been given in Chapter XIII.

For very dilute emulsions, where particle/particle interaction is negligible, the Einstein equation (see p. 358) should hold if the particles were uncharged and behaved as rigid spheres. Charged particles, as in aqueous systems unless excess salt is present (to make the ζ potential negligible), should under similar conditions obey the modified Einstein equation due to Smoluchowski (p. 377).

In the usual range of concentrations the viscosity of emulsions seems to depend very much less on the viscosity of the disperse phase than on phase-volume ratio. As pointed out earlier in this chapter, it is also affected by the viscosity of the stabilizing films around the droplets. Above a fairly definite concentration of disperse phase the viscosity becomes non-Newtonian, decreasing as the rate of shear increases, and a yield point may develop.

Certain technical and biological aspects of emulsions

No account of emulsions, however elementary, would be complete without some reference to their biological and technical uses. The following account has been largely drawn from an excellent survey given by Bennister and King (1939).

There are various reasons why oils are emulsified, but they can be grouped as follows: (a) dilution of oil, (b) increase in area of interface, (c) modification of physical properties. Table 23.2 gives further subdivisions together with examples of practical uses.

TABLE 23.2

A. *Dilution*

Purpose	*Examples*
Economy of materials or solvents (1)	(a) Collodion emulsions
	(b) Polishes
	(c) Dry-cleaning emulsions
	(d) Oil-soluble insecticides
	(e) Lubricants
Spreading of oil over a large surface (2)	(a) Road sprays (dust-laying)
	(b) Textile lubrication
	(c) Dry cleaning (see also 1 (c))
	(d) Cutting and boring oils
	(e) General lubrication (usually W/O)
	(f) Dust-laying in mines
	(g) Preservation of wood and fresh fruit (production of continuous thin film)
Dilution of effect (3)	(a) Horticultural sprays
	(b) Vitamin-containing drinks (e.g. halibut or cod-liver oil in water)

B. *Increase in area of interface*

Acceleration of chemical reaction (1)	(a) Oxidation of drying oils
	(b) Oil polymerization, e.g. emulsified dienes in synthetic rubber manufacture, also vulcanization of rubber by the action of sulphur
	(c) Emulsification of fat in the lye in soap manufacture
	(d) Hydrogenation of emulsified oils by bacterial action

Purpose	Example
Increase of physical efficiency (2)	(a) Absorption (e.g. injection of oil emulsions by Walsh and Frazer for the treatment of toxaemic conditions)
	(b) Polishes and solvents
	(c) Medical emulsions
	(d) Lubrication of leather by fat liquoring, i.e. adsorption of oil from dilute emulsions
Electrokinetic properties (3)	(a) Leather lubrication—rapid adsorption of oil from oil/water emulsion by NH_3+ groups of the tanned leather, the oil droplets being negatively charged
	(b) Electrodeposition of rubber
	(c) Electrical methods of separation of field emulsions

C. *Modification of properties*

Removal of objectionable properties (1)	(a) Taste or smell in pharmaceutical emulsions (see also A (3) (b) and B (2) (c))
	(b) Oily feel—cosmetics
	(c) Inflammability (carbon disulphide and petrol emulsions)
	(d) Reduction of 'bloom' in chocolate
Miscibility and protection (2)	(a) Addition of oil to cement
	(b) Flavouring and perfuming—cosmetics, etc.
	(c) Oil-bound distempers and paints
	(d) Incorporation of bactericidal oils in toothpaste, etc.
Fluidity (3)	(a) Emulsification of liquefiable solids (e.g. waxes, resins, and bitumen)
	(b) Emulsification of viscous oils
	(c) Rubber latex
Plasticity (4)	(a) Emulsion shock absorbers
Blended taste and other properties (miscellaneous) (5)	(a) Mayonnaise, sauces, etc.
	(b) Modification of gluten by addition of oil and fat emulsions to bread
	(c) Cream liqueurs and confectionery generally
	(d) Artificial milk and synthetic creams

The variety of applications outlined in Table 23.2 shows that the emulsifying agent is frequently subject to severe limitations, such as edibility, low cost, stability to heat, cold, or bacteria, and so on. It is because no single substance can satisfy all these requirements that such a variety are in common use.

A brief survey of biological and industrial emulsions follows.

1. *Biology and medicine*

Nature uses emulsions to facilitate transport of water-insoluble and, frequently, high-viscosity oils, as in milk, plant latices, and fat absorption from the intestine. The mechanism of fat absorption has received various explanations, of which the hydrolysis to fatty acid and glycerol, followed by diffusion through the intestinal wall and reformation of fat on the other side, was in vogue until recently. The work of Frazer, Schulman, and Stewart (1944) shows, however, that with very fine emulsions (diameter $< 0.5\mu$) even inert paraffins can be absorbed. The requisite fine emulsions are produced spontaneously in the intestine by the simultaneous presence of bile salts, oleic acid, and monoglycerides, which forms a very powerful emulsifying system.

Frazer and his co-workers (see Frazer, 1941) have shown that the presence of fine O/W emulsions can markedly influence the *in vivo* response to many biologically active materials. For example, insulin no longer produces the usual fall in blood-sugar level, and a lethal dose of diphtheria or tetanus toxin is rendered innocuous if added with the emulsion. These effects arise from the adsorption of the biologically active material on the extensive O/W interface.

2. *Pharmacy*

Most pharmaceutical emulsions are O/W, stabilized by gums, proteins, or other hydrophilic colloids; agents of an edible nature as considerations of toxicity generally arise. Their usual object is to present oily substances (e.g. cod-liver oil, liquid paraffin) in a more palatable form.

Many creams and ointments for external use are of the W/O type; for example 'hydrous lanoline', where the lanoline is both stabilizer and oil phase.

3. *Cosmetics*

Cosmetic emulsions are closely related to those just described, the permissible types of emulsifiers being considerably greater. O/W emulsions are used in 'vanishing creams' to assist penetration into the skin and to conceal the oily nature of the preparation. 'Cold cream' and hair creams are usually W/O emulsions, stabilized by cholesterol, lanoline, long-chain alcohols, or long-chain monoglycerides or monoglycol esters.

4. *Food emulsions*

Milk, cream, and butter are 'natural' emulsions in the sense that no added stabilizers are required. They have been briefly mentioned earlier.

Synthetic food emulsions comprise margarine, salad cream, mayonnaise, and fat emulsions for baking. Margarine is a W/O emulsion, usually stabilized by some polymerized vegetable oil (see p. 646). Mayonnaise and salad cream are dispersions of edible vegetable oils (e.g. olive oil) in aqueous solutions of edible acids (e.g. vinegar, lemon juice), stabilized by egg-yolk. The former contains about 60–80 per cent. of oil, the latter somewhat less.

5. *Leather and textiles*

The application of emulsions of oils and fats to leather, termed 'fat-liquoring', has two main objects—to render the leather soft and pliable and to make it water-proof. A variety of emulsifying agents is used, the oldest being egg or egg-yolk, nowadays usually replaced by negatively charged emulsifiers such as sulphonated oils or fatty acid soaps. The absorption of oil from the emulsion is believed to be due to the attraction of the positive groups ($-NH_3^+$) on the leather surface for the negatively charged emulsion droplets.

In the treatment of textiles, emulsions have been found essential for a number of purposes. Prior to spinning, the fibres are lubricated with an oil or fat to increase pliability and reduce breakages, a process only feasible by the use of dilute O/W emulsions.

The waterproofing of textile fabrics is carried out by applying a warm emulsion of a hydrophobic material (e.g. wax, rubber, or synthetic polymer in solution) stabilized by fatty acid soaps and a colloid such as casein or starch. This is then broken on the surface by means of dilute acid (e.g. formic or acetic), which precipitates the waterproofing agent in the pores of the fabric and on the colloid particles, rendering the whole fabric waterproof.

6. *Paper*

In order to prevent penetration of liquids into the final product the operation known as *sizing* is carried out, either during the beating process or after the paper has been formed.

The beater-sizing process consists in the precipitation of water-repellent material on the fibres before the actual formation of the paper. Emulsions of rosin, stabilized by rosin soap and a protective colloid such as glue or casein, are commonly used, being precipitated on to the fibres by addition of alum. The rosin may be partly or wholly replaced by certain natural waxes (e.g. paraffin, carnauba, montan, etc.), or by synthetic waxes or resins, and the rosin soap by many other emulsifying agents.

Surface-sizing of the pre-formed paper has the economic advantage of merely coating the outer layers with hydrophobic material. The usual method of application is to pass the semi-dried paper through a bath of emulsion, then through squeeze rollers and finally over drying rollers which leave the wax as a continuous surface film. In addition to the types of emulsions given above, those containing dispersed polymers, such as nitrocellulose, rubber, synthetic polymers, etc., have been used for surface-sizing and are likely to become of increasing importance.

Creaming in emulsions

The usual tendency of the oil globules in an emulsion to rise to the surface under the action of gravity (since most oils are lighter than water), is termed 'creaming'. As previously pointed out, creaming is quite distinct from the breaking of an emulsion, although it will facilitate breaking owing to the increase in particle contacts in the creamed layer. Creaming can clearly be prevented by raising the oil density up to that of water (by addition of carbon tetrachloride, for example); where this is not permissible it can be reduced by increasing the viscosity of the continuous phase (e.g. by adding gums, agar, or glycerol) or by using more concentrated emulsions. Homogenization, by forming very small particles (ca. $0 \cdot 1 - 0 \cdot 5 \mu$) which will be in active Brownian movement, will also hinder creaming unless offset by concurrent changes in the stabilizing films.

For some purposes, e.g. concentration of rubber latex, the addition of substances to facilitate creaming (without breaking the emulsion) is desirable. Normal Hevea latex consists of small rubber particles ($0 \cdot 1 \mu$–several μ in diameter) stabilized by adsorbed proteins and suspended in a serum containing electrolytes, proteins, and other organic substances. The usual rubber content is ca. 30 per cent., and this may be doubled by adding suitable creaming agents, chiefly hydrophilic colloids such as gums (arabic, tragacanth, etc.), alginic acid, polyvinyl alcohol, polyacrylic acid, methyl cellulose, and certain naturally occurring carbohydrates (van Gils and Kraay, 1942).

The first visible step in creaming appears to be agglomeration of numbers of particles to give clusters and this, by considerably reducing the Brownian agitation, allows sedimentation to proceed more rapidly. Some molecules of the creaming agent are believed to be adsorbed on to the particle surface, probably by secondary valencies, and owing to their structure are further linked to other similar molecules both in solution and adsorbed on other particles; hence the formation of clusters. (The tendency of these hydrophilic colloids, which are very asymmetric in shape, to cross-link by weak secondary forces is shown by their high viscosity and gelling powers in dilute solutions.)

If the suspended material has a higher density than the medium, addition of a 'creaming agent' will lead to formation of a sediment which, however, differs from the usual flocculated precipitate (as given by electrolytes) in being redispersed on shaking. One important technical sedimentation process is the clarification of coal-washing effluent by means of starch and other hydrophilic colloids.

Breaking of emulsions

The breaking of emulsions, or 'de-emulsification', arises in certain technical processes, the principal emulsions concerned being crude petroleum oil emulsions (W/O type), water-gas tar emulsions (W/O), suspensions of oil in condenser water (O/W), and wool-scouring wastes (O/W). A great variety of methods is used commercially; for the present purposes they can be grouped as mechanical, chemical, or electrical.

Mechanical methods include gravity or centrifugal separation, also distillation and filtration. Centrifugal separation is widely used, the system being warmed so as to increase the mobility as well as the density difference between the oil and aqueous phases. In the separation of crude petroleum emulsions by filtration, various materials, such as sand, clay, and chalk are used, made hydrophobic by initially wetting with oil. For condenser-oil suspensions active carbon is very efficacious, the bulk of the oil having previously been removed by centrifuging.

Chemical methods for breaking crude petroleum emulsions are based on the principle of antagonistic action, i.e. addition of O/W promoting agents tends to break W/O emulsions and vice versa (the work of Clowes in this connexion was mentioned on p. 654). Fatty acid soaps or sulphonated oils may be added to the emulsion, or soaps may be formed *in situ* by reacting acids present in the oil with sodium carbonate.

Dilute oil or solid suspensions in water are usually negatively charged and can therefore be readily precipitated by high valent cations, alum being the usual material used commercially. Emulsions stabilized by rigid skins (e.g. crude W/O oil emulsions) can sometimes be broken by certain surface active compounds soluble in both phases, e.g. butyric acid. Such materials probably act by dissolving or displacing the rigid interfacial skin by a mobile adsorbed monolayer which permits the drops to coalesce.

The industrial development of electrical methods of precipitating colloids is usually associated with the name of Cottrell. Aerosols (mists and smokes), as well as emulsions of both O/W and W/O types, can be coagulated by applying suitable electrical fields (p. 583). High voltages (*ca.* 10,000 volts or more) and small electrode separations are used to produce intense fields and so accelerate the rate of breaking. With O/W emulsions and dilute suspensions electrophoresis is the principal factor involved, but, with the inverse type, deformation of the water globules seems to be of more importance.

GENERAL REFERENCES

CLAYTON, 1943, *Emulsions and their Technical Treatment*, Churchill.

OTHER REFERENCES

ALEXANDER, 1947, *J. Chem. Soc.*, Tilden Lecture, 'The application of surface chemistry to problems in colloids', p. 1422.
BENNISTER and KING, 1939, *Chemistry and Industry*, **58**, 220, and earlier papers.
DANIELLI, 1938, *Cold Spring Harbor Symp. Quant. Biol.* **6**, 190.
FRAZER, 1941, *Trans. Faraday Soc.* **37**, 125.
—— SCHULMAN, and STEWART, 1944, *J. Physiol.* **103**, 306.
HARTLEY, 1941, *Trans. Faraday Soc.* **37**, 130.
KING, 1941, ibid. **37**, 168.
KRAEMER, 1924, *J. Amer. Chem. Soc.* **46**, 2709.
LAWRENCE, 1937, *Trans. Faraday Soc.* **33**, 815.
PINK, 1941, ibid. **37**, 180.
SCHULMAN and COCKBAIN, 1940, ibid. **36**, 651, 661.
—— and RIDEAL, 1937, *Proc. Roy. Soc.* B **122**, 29, 46.
VAN GILS and KRAAY, 1942, Article in *Advances in Colloid Science*, vol. i, Interscience.

XXIV
COLLOIDAL ELECTROLYTES

Some brief mention was made in Chapter II of the evolution of the term 'colloidal electrolyte', the name given to ionized compounds of relatively simple chemical constitution whose solutions over certain concentration ranges show definite colloidal characteristics.

The early studies of Krafft and of McBain, as well as of many of the later workers, dealt with aqueous solutions of the alkali metal salts of the fatty acids, but subsequently a very great number of compounds have been prepared synthetically as wetting agents, detergents, and emulsifying agents, which closely resemble the fatty acid soaps in their physical properties. A few of the more common examples are shown in Table 24.1 below. Their 'type' is also indicated, since a subdivision into anionic, cationic, and neutral types can be effected on the basis of the charge carried by the colloidal ion. This will become clearer when the physical properties are considered.

TABLE 24.1

Compound	Formula	Type
Sodium palmitate	$C_{15}H_{31}COO'\ Na^+$	Anionic
Ammonium oleate	$C_{17}H_{33}COO'\ \overset{+}{N}H_4$	,,
Sodium dodecyl sulphate	$C_{12}H_{25}.OSO_3'\ Na^+$	,,
Sodium dodecyl sulphonate	$C_{12}H_{25}.SO_3'\ Na^+$	,,
Hexadecyl sulphonic acid	$C_{16}H_{33}.SO_3'\ H^+$	,,
Heptadecyl amine hydrochloride	$C_{17}H_{35}.\overset{+}{N}H_3\ Cl'$	Cationic
Hexadecyl trimethyl ammonium bromide	$C_{16}H_{33}.\overset{+}{N}(CH_3)_3\ Br'$	,,
Dodecyl pyridinium chloride	$\langle\ \rangle\overset{+}{N}\begin{smallmatrix}Cl'\\C_{12}H_{25}\end{smallmatrix}$	,,
Long-chain esters of sugars or of condensed glycerides e.g.: $C_{12}H_{25}COO.CH_2.CH(OH).CH_2.O.CH_2.CH(OH).CH_2OH$		Neutral
Polyethylene oxides e.g.: $C_{12}H_{25}.O.CH_2.CH_2.(O.CH_2.CH_2)_2O.CH_2.CH_2OH$		,,
Sodium cholate (see below)		Anionic
Lecithin (see below)		Depends on pH
Lysolecithin (see below)		,, ,,
Kephalin (see below)		,, ,,
Orange II (see p. 697)		Anionic
Methylene blue (see p. 697)		Cationic

TABLE 24.1 (continued)

[Structure of Sodium cholate]

Sodium cholate.

$C_{17}H_{35}$—COOCH$_2$
$C_{17}H_{33}$—COOCH
  |
  CH$_2$—O—P(=O)(O$^-$)—O—(CH$_2$)$_2$—$\overset{+}{N}$(CH$_3$)$_3$

A typical lecithin.

$C_{17}H_{35}$—COOCH$_2$
HO—CH
  |
  CH$_2$—O—P(=O)(O$^-$)—O—(CH$_2$)$_2$—$\overset{+}{N}$(CH$_3$)$_3$

A typical lysolecithin.

$C_{17}H_{35}$—COOCH$_2$
$C_{17}H_{33}$—COOCH
  |
  CH$_2$—O—P(=O)(OH)—O—(CH$_2$)$_2$—NH$_2$

A typical kephalin.

(For a more detailed list see, for example, Dean, 1937; Clayton, 1943.)

The table also includes other closely-related compounds such as a typical bile salt (sodium cholate), three phospholipoids (lecithin, lysolecithin, and kephalin), and two typical dyestuffs (one acidic and one basic), which, as we shall see, are also to be classed as colloidal electrolytes.

An examination of Table 24.1 shows that the compounds listed there, despite their diverse chemical types, have one feature in common—they all have a non-polar hydrocarbon portion and a small polar group, which is usually, but not invariably, ionized. These constituent parts differ to the extreme in their affinity for water (as indicated, for example, by simple solubility considerations), the former possessing practically none, the latter a very high affinity. It is this two-sided or 'amphipathic' nature, as Hartley has termed it, of such compounds which is responsible for their colloidal properties in aqueous solution. (Only aqueous solutions are in question here unless otherwise stated.)

Hartley has proposed the term 'paraffin-chain salts' to cover the various chemical types shown in Table 24.1, but we feel this to be too long for convenient usage and therefore will use instead the term 'soap', using this generically, and referring to the alkali metal salts of the fatty acids as the 'fatty acid soaps'. On this definition the class of 'colloidal electrolytes' therefore comprises the 'soaps' and the 'dyestuffs', and it is convenient to consider these separately, since the former type show a much more uniform behaviour.

SOAPS

Considering the series of the fatty acid soaps, which might be regarded as commencing with the acetate, it is found that their solutions only exhibit colloidal properties when the chain-length exceeds about eight carbon atoms (caprylate). The postulate of the formation of colloidal aggregates was put forward to explain the low osmotic activity, as measured by freezing-point, vapour pressure (dew-point), or other methods, of the higher fatty acid soaps. Thus in the case of 1 M. potassium stearate solution at 90° C., the dew-point measurements of McBain and Salmon (1920) showed the total concentration of osmotically active particles to be 0·42 M., i.e. considerably less than if the soap were completely undissociated in solution. Conductivity measurements ruled out incomplete dissociation as a major factor since such solutions have a quite considerable conductivity, much greater than could be explicable from the amount of alkali liberated by hydrolysis as suggested by the early theories.

The combination of low osmotic activity with high electrical conductivity led McBain to suggest that the paraffin-chain ions aggregated together, forming an 'ionic micelle'. Such an aggregate would have a considerable conductivity owing to its high charge, but would behave osmotically as a single particle. The micelles were thought to form at concentrations above the minimum which the fatty acid soaps show in the conductivity concentration curve, their high conductivity explaining the rise in equivalent conductivity. McBain later suggested that two types of micelle were present in equilibrium with one another, the 'ionic micelle' mentioned above, and 'neutral colloid' composed of undissociated molecules and therefore making no contribution to the conductivity.

However, subsequent work, in particular the experimental and theoretical investigations of Hartley, have shown that all the properties of the *dilute* solutions of soaps (i.e. < 1 per cent. as a rule) are adequately explained on the basis of a single type of micelle, which is very similar to the ionic type postulated by McBain. It seems quite definite, also, that micelle formation sets in at concentrations very much lower than at the minimum in the conductivity curve.

Nevertheless in very concentrated soap solutions (i.e. > *ca.* 10 per cent.) there is some evidence, particularly from viscosity and birefringence, that another type of aggregate, the 'laminar micelle', is present. Since the two types are present in such different concentration ranges it is convenient to consider the dilute and concentrated solutions separately. The probable structure of the two types of micelle will first be given, and the behaviour and properties of the solutions then interpreted in terms of the relevant type. This is, of course, the opposite way to the historical evolution of the subject, but it enables a more connected account to be given without the necessity for a great deal of repetition.

The structure of the micelles

(a) *The spherical micelle*

As pointed out, this is the only type present in dilute soap solutions (< *ca.* 1 per cent.), and a representation of its structure is shown in Fig. 24.1. As will be shown later, it has a radius of rather less than the length of the fully extended molecule, e.g. *ca.* 20 A for the C_{16} chain salts, and in this case would contain about 50 molecules and have a molecular weight of about 14,000. (The number of constituent molecules is found by dividing the size of the micelle by the volume of a single molecule.) The hydrocarbon chains have been drawn irregularly to indicate that they are in a *liquid* state of aggregation, as will be shown particularly from the solvent properties of the micelle. The majority of the gegenions are (statistically speaking) attached to the micelle and move with it.

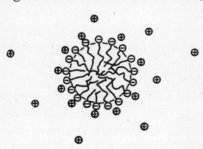

Fig. 24.1. Structure of the spherical micelle.

We will first consider briefly the physical factors which bring about the formation of aggregates, and then see how the aggregation has been analysed mathematically from the law of mass action.

Consider a single molecule of soap in dilute solution. The fully extended chain, as shown in Fig. 24.2(a), is most unlikely since, as Langmuir first pointed out, the lack of affinity between hydrocarbon and

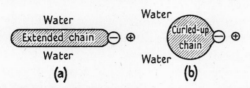

Fig. 24.2. Single soap molecule in solution. (a) Fully extended configuration; (b) Curled-up configuration.

water (as evidenced *inter alia* by the high interfacial tension (γ) between them) leads to the chain curling up so as to reduce the hydrocarbon/water interface to a minimum. On the other hand the chain will not curl up completely into a close-packed ball, since the thermal agitation will tend towards a random statistical, and therefore a more extended, arrangement. The intermediate state taken up, pictured in Fig. 24.2(b), can for most purposes, however, be taken as a close-packed ball.

Considering now the hydrocarbon/water interfacial energy, before and after aggregation, of the molecules constituting a micelle, it is clear from Figs. 24.1 and 24.2 that this interfacial energy is almost entirely eliminated on micelle formation. Taking γ as 50 ergs/cm.², and considering a C_{16} chain soap, where the micelle, as discussed above, contains ca. 50 molecules, this loss of interfacial energy would be about 10×10^{-11} ergs if the chains were originally fully extended (as in Fig. 24.2(a)), and about 6×10^{-11} ergs if they were fully curled up. Opposing the aggregation will be the electrostatic repulsions between the like charges on the soap ions, and their kinetic energy, which clearly will be very much reduced upon aggregation. The total translational energy of all 50 molecules is only ca. 0.3×10^{-11} ergs and the electrical work required is ca. 1.7×10^{-11} ergs (Hartley, 1936). This calculation, although only approximate, indicates that the decrease in potential energy suffices to stabilize the aggregate.

From the above analysis it would be expected that aggregation would be assisted by

(a) Increasing concentration, since this leads to an increase in the collision frequency and therefore also in the probability of obtaining a nucleus for further aggregation.

(b) Decreasing the temperature and hence the thermal agitation, since thermal agitation will clearly tend to break up the micelles.

(c) Increasing the hydrophobic part of the molecule. In the case of the paraffin-chain salts the loss in interfacial energy on aggregation is approximately proportional to the fourth power of the chain-length.

(d) Addition of simple salts (e.g. KCl), since these by their screening action will tend to lower the repulsive forces between the charges on the paraffin-chain ions.

Experiment shows that micelle formation in soap solutions is in conformity with the above predictions, as will be shown below.

Application of the law of mass action to micelle formation. Jones and Bury (1927) and later Murray and Hartley (1935) have considered the aggregation process from considerations of the simple law of mass action. Thus if the micelle contains m simple molecules, and C denotes the stoichiometric concentration of soap, x the fraction aggregated, and K the equilibrium constant for the single molecule/micelle equilibrium, then we have

$$m \text{ single molecules} \rightleftarrows \text{micelle}$$
(concentrations) $\quad [C(1-x)] \quad\quad [Cx/m].$

The equilibrium constant

$$K = \frac{[\text{micelles}]}{[\text{single molecules}]^m}$$
$$= \frac{Cx/m}{[C(1-x)]^m}. \tag{24.1}$$

The effect of various values of m can be found by giving K any arbitrary value (say 1) and seeing how x varies with C. It is found that when m is large (20 or more) the aggregation to micelles is almost as discontinuous as the separation of a solid phase ($m = \infty$), as will be seen from Fig. 24.3 (a) (from Murray and Hartley, 1935) and even more strikingly from Fig. 24.3 (b) (from Hartley, 1936). The above calculations can only be approximate, since the systems have been assumed to be ideal (i.e. activity coefficients unity), which is certainly very far from the truth. If allowance is made for adherence of some of the gegenions to the micelle, which certainly occurs, then the transition is sharpened, as shown in Fig. 24.3 (b) for the case $m = 20, n = 10$ (n being the number of attached gegenions).

COLLOIDAL ELECTROLYTES 673

Fig. 24.4 shows the effect of aggregation upon the concentrations of the simple long-chain ion, free gegenion, and micelle, taking the formula of the last to be $(Na_{10}X_{20})^{10-}$ (i.e. $m = 20$, $n = 10$ as above). After aggregation the concentration of simple long-chain ion is seen to remain

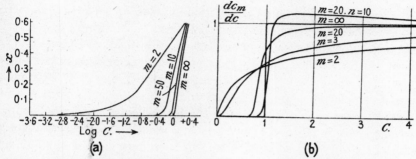

FIG. 24.3. (a) Fraction of solute in micellar form (x) plotted against the logarithm of the total concentration (C) for $m = 2$, 10, 50 and ∞. (b) Plot of $\dfrac{dCm}{dC}$ against C for $m = 2$, 3, 20 and ∞, and for $m = 20$, $n = 10$.

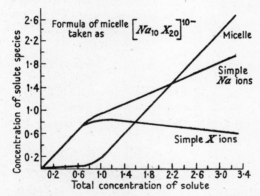

FIG. 24.4. Illustrating how aggregation affects the dependence of the concentration of various solute species upon the total solute concentration.

practically constant, all further added solute merely going to form more micelles (from Murray, 1935). As we shall see later these predictions from the law of mass action, particularly in view of the approximations involved, are in rather good agreement with experiment.

A little attention may be given to the physical picture involved in calculating the equilibrium constant from the law of mass action as above. Formally we may write

rate of association reaction = k_1^* [single molecules]^m,

and rate of dissociation reaction = k_2^* [micelles],

where k_1^* and k_2^* are constants. Since at equilibrium the two rates are equal, then

$$\frac{[\text{micelles}]}{[\text{single molecules}]^m} = \frac{k_1^*}{k_2^*} = K, \quad \text{the equilibrium constant, as above.}$$

As written, it would appear that a 20-molecular collision is involved in the formation of a micelle, and that dissociation occurs by a micelle occasionally disintegrating instantaneously into its constituent molecules. Both these are unlikely from molecular considerations. Thus we know from chemical kinetics that reactions involving only three molecules (termolecular) are very rare and higher molecularities are unknown. What probably occurs in the formation of a micelle is that a few (say two or three) molecules of low kinetic energy stick together sufficiently long to form a nucleus; this then incorporates any colliding soap molecule until the full micellar size is reached (i.e. $(m-2)$ or $(m-3)$ molecules add on to the original nucleus of 2 or 3 respectively).

Writing down the series of equilibria,

3 single molecules $\rightleftarrows$ trimer (regarded as the nucleus),

where
$$K_1 = \frac{[\text{trimer}]}{[\text{single}]^3},$$

trimer + single $\rightleftarrows$ tetramer,

where
$$K_2 = \frac{[\text{tetramer}]}{[\text{single}][\text{trimer}]} = \frac{[\text{tetramer}]}{K_1[\text{single}]^4}$$

and so on, until finally

$(m-1)$er + single $\rightleftarrows$ micelle (m molecules).

Considering now the rates of formation of the various species we have

rate of formation of tetramer $= k_1[\text{trimer}][\text{single}]$
$= k_1 K_1 [\text{single}]^4$,

and so on, until finally

rate of formation of micelle $= k_1 k_2 ... K_1 K_2 ... [\text{single}]^m$
$= k_1^*[\text{single molecules}]^m \quad$ as before.

Thus the same kinetic expression can be obtained by a series of processes each perfectly feasible from a molecular viewpoint. Similarly, if the break-up of a micelle is regarded as a series of steps, the converse of the above, rather than as a sudden 'explosion', the correct kinetic expression can likewise be obtained.

(b) The laminar micelle

From the sort of considerations discussed in connexion with the spherical type of micelle it is not difficult to see why a laminar type might be formed in very concentrated soap solutions ($\not< ca.$ 6 per cent. and usually considerably more).

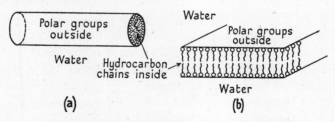

FIG. 24.5. Structure of rod-like and plate-like micelles.

Consider the spherical micelles of radius 20 A in a dilute soap solution, at a concentration of, say, $x = 0 \cdot 1$ per cent., where x is the percentage volume occupied by the soap assumed wholly in the micellar form. The average distance apart (d) of the micelles will be of the order 400 A (assuming a simple cubic arrangement), so that the repulsive forces between them, owing to their ionic atmospheres, will not be very considerable. (For a discussion of the atmosphere effects in colloidal electrolytes see Chapter V.) As x increases the micellar separation decreases, until at $x = 10$ per cent. d is now $ca.$ 86 A, and the electrostatic repulsive forces will have been enormously increased (see also Chapter V). These repulsions will therefore tend to favour other possible arrangements in which the separation of the soap ions is increased, provided, of course, that these do not bring about any overcompensating increase in the energy of the other parts of the system. Now, as we have seen, it is largely the hydrocarbon chain aggregation which causes the micelle formation, so it is clear that no considerable change in the hydrocarbon groupings is energetically feasible.

However, a little thought shows at once that the formation of long rods, or of flat plates, such as pictured in Fig. 24.5, will not materially alter the aggregation of the chains, *but will produce the requisite increased separation of the soap ions.* Table 24.2 below shows how the micellar separation (d) is increased on going from the spherical to the rod-like or plate-like micelles, for value of x of 10 per cent. and 30 per cent. A simple cubic arrangement is assumed for the spheres, a cubic one for the rods (assumed infinitely long and parallel), and a parallel disposition for the plates (also assumed infinitely large).

TABLE 24.2

	Spheres	Rods	Plates
$x = 10$ per cent	86 A	112 A	400 A
$x = 30$ per cent	60 A	65 A	134 A

Although such calculations are only approximate, they do show how the electrical repulsions can be relieved, in particular by the formation of bimolecular leaflets adlineating to form a laminated aggregate (the laminar micelle). The physical process of aggregation of spherical micelles to form first rods and then plates is not difficult to visualize, the requisite temporary adjustments being assisted by the fluid nature of the aggregated hydrocarbon chains. Further aggregation will also be helped by the greater frequency of collisions in these concentrated solutions.

Other considerations discussed above in connexion with the spherical micelle, such as the law of mass action, will apply without radical changes to the laminar type.

The physical properties of dilute soap solutions

We will now outline the more important physical properties of dilute soap solutions (i.e. $< ca.$ 1 per cent. by weight), which, as already mentioned, can be quite adequately explained on the basis of the small, Hartley type of micelle. The substances usually considered will be the paraffin-chain sulphates, sulphonates, quaternary ammonium, and pyridinium halides, etc., rather than the fatty acid soaps, owing to complications with the latter arising from hydrolysis, sensitivity to carbon dioxide, and to traces of heavy metals.

(a) Colligative properties

Measurement of any of the colligative properties (vapour pressure, osmotic pressure, freezing-point, boiling-point) enables the osmotic coefficient (g) of the solution to be calculated. Of these the boiling-point method is the least useful owing to frothing troubles.

For some of his early investigations McBain utilized the dew-point method of measuring vapour pressures, and Fig. 24.6 shows the dew-point lowering at 90° C. by the potassium salts of the fatty acids ranging from acetic to stearic (after McBain and Salmon, 1920). The marked difference in behaviour between the caprate (C_{10}) and the higher members arises from the appearance of colloidal characteristics with the longer chain-lengths.

More recently McBain and his co-workers have used the freezing-point method with a number of synthetic sulphates and sulphonates of varying chain-length, one of their typical series (after McBain and Bolduan, 1943) being given in Fig. 24.7. Plotting g against the square root of the concentration (i.e. following the Debye-Hückel treatment), the curves tend at low concentrations to that given by KCl, a typical uni-univalent

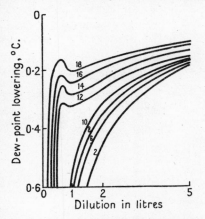

FIG. 24.6. Dew-point lowering as a function of dilution for a series of potassium fatty acid salts, at 90° C.

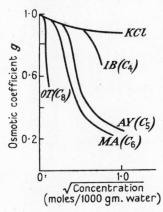

FIG. 24.7. Variation of osmotic coefficient (g) with concentration for a series of synthetic soaps of increasing chain-length. KCl included for comparison.

strong electrolyte, but show a marked deviation above a concentration which decreases as the chain-length is increased. The point of departure corresponds to the onset of micellar aggregation, and agrees approximately with that found by other methods. This method is not to be recommended for determining the critical concentration for micelles, particularly with the longer-chain compounds, as the freezing-point depressions are then so small.

(b) *Conductivity*

The equivalent conductivities (Λ) of solutions of the various types of paraffin-chain salts all have the same general form (see Fig. 24.8), although some cases, particularly the fatty acid soaps, differ a little in showing a weak minimum in rather concentrated solutions. Plotting Λ against $\sqrt{C}$ two points call for comment—the sudden deviation at low concentrations from the theoretical limiting slope of Debye-Hückel-Onsager theory, and the almost constant value of Λ at the highest concentrations. The sudden fall takes place at lower concentrations with increasing chain-length and undoubtedly corresponds to the onset of micelle formation.

Hartley (1936) has analysed the conductivity phenomena in detail, and he points out that micellar aggregation has three main consequences:

(a) the viscous drag is reduced;
(b) the ionic atmospheres of oppositely charged gegenions exert a much increased 'braking' effect;

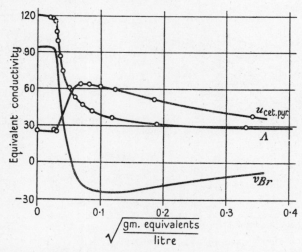

FIG. 24.8. Equivalent conductivities of the component ions, and of the whole salt, for cetyl pyridinium bromide at 35° C.

(c) some of the gegenions will adhere to the micelle, owing to its high charge, and thus be forced to travel in the direction opposite to that of the free gegenions.

Under ordinary conditions (b) and (c), which tend to reduce Λ, outweigh (a) which causes an increase, so that Λ falls on aggregation.

The fall in viscous drag follows from the application of Stokes's law to the m long-chain ions constituting the micelle, before and after aggregating. Assuming, for simplicity, that both single ions and micelles are spherical, then the ratio of viscous drag after and before aggregation $= m^{2/3}$. Thus if $m = 50$ this $\simeq 14$, i.e. the equivalent conductivity after aggregation should be increased by this factor, *assuming, of course, that the micelle carries the charge to be expected from its constituent ions* (50 in this case).

Some indication of the factors involved in (b) and (c) has already been given in Chapter V when discussing the application of Debye-Hückel theory to colloidal electrolytes. The reduction of mobility in (b) arises from the opposing electrophoretic contribution of the atmosphere, a term which increases directly with the valence of the micelle. Factor

(c) arises from the high charge and large size of the micelle which bring about the adherence of a considerable fraction of the gegenions. These adhering gegenions not only lower the charge and therefore the mobility of the micelle, but cease to play their normal part in the conduction

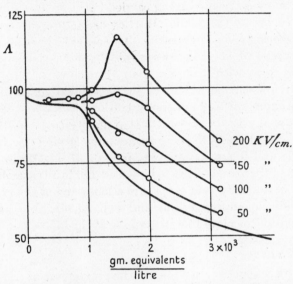

FIG. 24.9. Equivalent conductivity of cetyl pyridinium chloride at different field-strengths (25° C.).

process and carry their charges in the direction opposite to that of the free gegenions.

The essential validity of the above considerations has been shown by Hartley and his collaborators from measurements of the conductivity at high field strengths, and from transport number determinations. The latter are particularly illuminating since they enable the contribution to the conductivity of the individual ions to be ascertained.

At very high field strengths the conductivity of simple electrolytes is increased (Wien effect) and may approach the value at infinite dilution, since the ions now move so rapidly that the atmospheres cannot form sufficiently quickly. With colloidal electrolytes the reduction in atmosphere effects sets free some of the bound gegenions, and the conductivity *rises* on aggregation due to the Stokes law effect now preponderating (see the curve for 200 kV/cm. in Fig. 24.9).

The mobilities of the individual ions, as obtained from conductivity and transport number measurements at normal field strengths, are shown in Fig. 24.8 for the compound cetyl (hexadecyl) pyridinium bromide. When micelles start to form, and the total equivalent conductivity falls, that for the paraffin-chain ion rises very steeply, due to the Stokes law effect discussed above, whereas that for the gegenion (Br') falls catastrophically (due to factors (b) and (c)) and soon becomes *negative*, i.e. more Br' is now being carried in the reverse direction than travels unattached to the anode. Hartley has estimated the fraction of gegenions adhering when aggregation first sets in to be of the order 0·74, so that a micelle of 50 molecules would have a *net* charge of about 13 units, and carry with it no less than 37 gegenions.

The rather unusual behaviour of the fatty acid soaps in showing a broad minimum at rather high concentrations (e.g. about 0·1 M. in the case of sodium myristate at 60° C.) is probably to be ascribed to the possibilities of hydrolysis in these compounds, giving rise to acid soap in the micelle and liberation of some free alkali (NaOH in the case of sodium salts). Other possibilities to be considered in all cases, but possibly accentuated in the fatty acid soaps, are a loosening of the gegenions (as the atmospheres begin to overlap), and a change in the structure of the micelle (to rod-like or plate-like structures) at the higher concentrations.

(c) *Surface behaviour*

As might be expected, in view of their commercial importance as wetting agents, detergents, and emulsifying agents, the surface activity of the synthetic and natural soaps has been considerably investigated, for example by Powney and his co-workers (Powney and Addison, 1937; Powney and Wood, 1941). The question of detergent action will be discussed later (p. 686); their uses as foaming agents in Chapter XXII, as emulsifiers in Chapter XXIII, and their effects upon dilute and concentrated suspensions in Chapters XX and XXI respectively, have already been considered. This leaves the question of surface and interfacial activity, methods for the determination of which have been given in Chapter XVIII.

With increasing soap concentration the boundary tension at an air/water or oil/water interface first falls rapidly until at a fairly definite value an almost sharp transition occurs, subsequent additions causing little further lowering (see Fig. 24.10). The transition is due to the formation of micelles in the bulk solution and occurs at a lower concentration if the temperature is lowered, if salts are added, or if the chain-length is increased, for reasons already discussed (p. 672). The measurement of surface or interfacial tension (preferably the latter, as 'surface ageing' effects are less troublesome) thus provides what is probably the simplest and most accurate means of determining the 'critical concentration' for micelles. Table 24.3 shows some typical

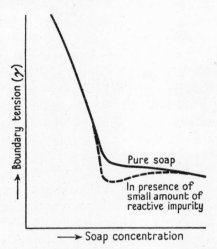

FIG. 24.10. Variation of surface or interfacial tension of a soap solution with concentration.

results (Powney and Addison, 1937), together with a comparison with values from conductivity data.

TABLE 24.3. *Critical Concentrations (Molar) of Sodium Alkyl Sulphate Solutions at 60° C.*

Salt	From surface tension	From interfacial tension	From conductivity data
$C_{12}H_{25}SO_4Na$	8×10^{-3}	7.7×10^{-3}	9×10^{-3}; 7.6×10^{-3}
$C_{14}H_{29}SO_4Na$	2×10^{-3}	2.1×10^{-3}	2.6×10^{-3}
$C_{16}H_{33}SO_4Na$	0.43×10^{-3}	0.47×10^{-3}	1×10^{-3}; 0.76×10^{-3}
$C_{18}H_{37}SO_4Na$	0.17×10^{-3}	0.22×10^{-3}	0.39×10^{-3}

It was frequently found, particularly in the early work, that the boundary tension-concentration curve showed a minimum, as indicated by the dotted curve in Fig. 24.10. The minimum arises from the presence of traces of adventitious impurities, particularly from heavy metal cations (e.g. Cu^{++}, Fe^{+++}, Al^{+++}) in the distilled water when studying fatty acid soaps and other anionic detergents (Robinson, 1937; Harkins and Myers, 1937; Mitchell *et al.*, 1937; Reichenberg, 1947). An example where the 'impurity' is deliberately added in fixed amount is shown in Fig. 24.19 (a). The reason for the tension being minimal at the micellar point is not difficult to visualize. The 'impurity' must clearly interact in some way with the adsorbed soap molecules, which has been found experimentally to reduce the boundary tension below that for the pure soap solution (see Cassie and Palmer, 1941, for the quantitative formulation of the effects of cations on anionic soap monolayers), but, once micelles start to form, the micellar surfaces, which clearly do not differ greatly from a macroscopic plane surface, compete for the fixed amount of 'impurity' present. With sufficient excess of soap, therefore, the tension will tend to the value for the pure soap solution. (This is clearly shown in Fig. 24.19 (a), where the 'impurity' is 0·025 per cent. hexyl resorcinol deliberately added in fixed concentration throughout.)

As mentioned above, the addition of indifferent salts (in amounts insufficient to cause salting out of the soap) lowers the critical concentration for micelles. It also brings about a general increase in capillary activity at all types of interfaces and an increase in detergent power. These phenomena have been quantitatively explained by showing that salts change the potential energy of the ionized soap molecules in a monolayer, and thus alter the adsorption of capillary active molecules from solution (Cassie and Palmer, 1941; Aicken and Palmer, 1944). Their approach is essentially that given in Chapter V, p. 101; involving the integration of the Poisson-Boltzmann equation for a plane uniformly charged surface, and considering the effect of added salts upon the electrical potential ψ. This treatment confirms that it is only the ion of the added salt opposite in charge to the capillary active ion which affects ψ and hence the surface activity. This question is clearly intimately connected with the Schulze-Hardy rule, dealing with the effect of salts in coagulating hydrophobic suspensions (p. 571).

The critical concentration is reduced, with increasing chain-length, to about one-quarter by addition of two —CH_2 groups (see Table 24.3), and also by decreasing temperature, the usual temperature coefficients being between 1 and 2 per cent. per degree. It is also very dependent upon the structure of the hydrophobic part of the molecule, branched chain compounds usually having a higher critical concentration than their straight chain analogues (e.g. Hartley, 1941), owing to the greater difficulty of packing the molecules together, as required in micelle formation, in such cases.

As mentioned in Chapter XVIII, one question of great importance in connexion with surface activity of soap solutions is the molecular density (or its reciprocal, the area/molecule) in the adsorbed film, which is usually monomolecular in thickness. Two methods are available for obtaining the molecular density (number molecules/cm.^{2})—the use of a suitable equation of state, or of Gibbs's adsorption isotherm. In the former method it is preferable to consider oil/water adsorbed films, for which the very simple equation of state $\Pi(A-A_0) = C$, holds (see Chapter XVIII). As pointed out there, the values of the adsorption calculated from this equation agree closely with the best experimental values, and from the value of the molecular area at the critical concentration it is also possible to make an estimate of the size of the micelle (Alexander, 1942).

The application of the Gibbs equation to colloidal electrolytes is more difficult than with un-ionized compounds since the activity coefficient deviates from unity at very much lower concentrations (p. 520), and so the use of the approximate form $\Gamma = (-1/RT)\{d\gamma/(d \ln C)\}$ is more limited. When aggregation occurs the osmotic coefficient falls very rapidly, as can be seen from Fig. 24.7, owing to adsorption of gegenions (see Chapter V), and the use of the approximate equation is even less valid. Alternatively we can regard the question from the viewpoint of the single molecules, since only these can be present in the surface layer. The exact form of the Gibbs equation is then $\Gamma = (-1/RT)\{d\gamma/(d \ln c_s f_s)\}$, where $c_s f_s$ is the activity of the single molecules. If, therefore, $c_s f_s$ (or very approximately the concentration of single molecules) is independent of the stoichiometric concentration (C), the adsorption Γ and the boundary tension γ should be constant and independent of C. This is approximately what is found experimentally (see Fig. 24.10), and the constancy of single molecule concentration when micelles are present is in accord with the predictions from the approximate law of mass action (see Fig. 24.4 and p. 672).

The great surface activity of fatty acid soaps (e.g. potassium oleate) is undoubtedly due in part to the formation of free fatty acid by hydrolysis of the adsorbed molecules, the liberated acid then coordinating with the soap ions, probably by a hydrogen bond (see p. 568). This is supported by the observation that addition of potassium carbonate (i.e. raising the pH and therefore diminishing the hydrolysis) considerably reduces the surface activity.

(d) Solubility (the Krafft phenomenon)

Soaps exhibit an unusual difference from other low molecular weight compounds (cf. however, polymers, p. 786) in that their solubility increases enormously over a small temperature range (hence the term Krafft 'point'). Some typical curves are given in Fig. 24.11 (*a*).

It has been explained (Murray and Hartley, 1935) upon the general principles of micellar aggregation as outlined above. As the temperature is raised the concentration of single molecules in solution rises

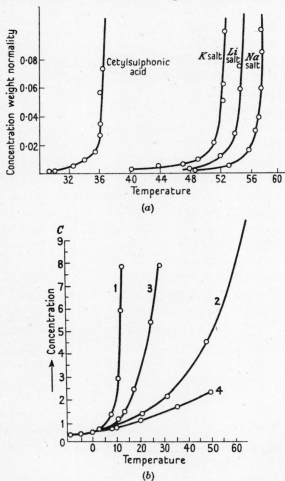

FIG. 24.11. Variation of solubility of soaps with temperature: (a) observed; (b) calculated (see text).

steadily owing to the usual increase in solubility, until it exceeds the critical concentration, when micelles start to form. From this point onwards the transfer is effectively from the solid phase to the micelle, since the concentration of single molecules only increases quite slowly once micelles are present, as pointed out earlier. Now a single molecule in the micelle is at first sight more akin to one in the solid soap phase than is a single molecule in solution, both from structural and even more from energetic considerations (cf. Fig. 24.12 with Fig. 24.1 and Fig. 24.2). However the free energy of solution to micelle and to single molecules must be identical (these being in thermodynamic equilibrium), whereas the increase in entropy to micelle would certainly be much less than to single molecules.

The energy required to transfer a molecule from the solid phase to a micelle would therefore be expected to be much less than for transfer to a molecularly dispersed solution. The 'heat of solution' should thus decrease once micelles form and this will modify the temperature coefficient of the activity. (From the thermodynamic relation $(d \ln a_i)/dT = L_i/RT^2$, where a_i is the activity of solute i and L_i its molar heat of solution) (see p. 58).

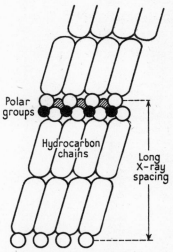

FIG. 24.12. Arrangement of ions in a soap crystal (schematic).

Fig. 24.11 (b) shows some solubility curves calculated by Murray and Hartley as follows. The saturation solubility (C) of an ideal solute is given by $\ln C = \alpha + \beta/T$, or approximately, for small temperature ranges, by
$$\ln C = A + BT, \qquad (24.2)$$
where A and B are constants.

In the case of an aggregating substance it is assumed that it is the concentration of simple molecules, $C(1-x)$, which satisfies this relation i.e.
$$\ln C(1-x) = A + BT, \qquad (24.3)$$
where x is the fraction aggregated and which can be found in terms of C from eqn. (24.1) if K and m are known.

Assuming any arbitrary value of K the *general form* of the solubility-temperature curve can thus be found for various values of m. Taking a quite normal value (0·01) for B, curve 1 of Fig. 24.11 (b) is thus obtained for $m = 20$ (i.e. a twenty molecule micelle), curves 2 and 3 after making allowance first for ionization and then for adsorption of gegenions. These curves (particularly 3) are seen to bear a very close resemblance to the experimental ones, and to differ radically from the non-aggregating case (curve 4) having the same A and B values.

(e) *Solubilization in dilute soap solutions*

The ability of soap solutions to dissolve organic compounds insoluble, or only slightly soluble, in water is another of their rather striking properties. The present discussion will be restricted to dilute soap solutions (i.e. $< ca.$ 1 per cent. concentration as a rule), although the phenomenon is also shown by concentrated solutions as discussed below. Little additional uptake is found until the micellar concentration is reached, indicating that the compounds are taken up in some way by

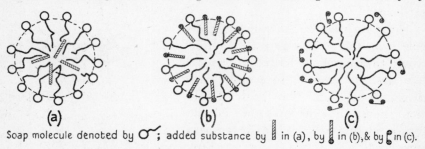

Soap molecule denoted by ⟅; added substance by ▌ in (a), by ▌ in (b), & by ▐ in (c).

FIG. 24.13. Structure of micelles in the presence of: (a) Compounds soluble only in oil (e.g. hydrocarbons). (b) Compounds soluble to some extent in both oil and water (e.g. phenols). (c) Compounds only water-soluble (e.g. glycerol).

the micelles. It is generally believed that with non-polar hydrocarbons uptake occurs by solution into the (liquid) interior of the micelle, and with partially miscible compounds such as phenol, aniline, and octyl alcohol, by adsorption on to the micelle surface with the hydrocarbon part inside and the polar group in the aqueous phase. (Compounds with a dative hydrogen atom, such as the above, are probably linked to the soap ion by a hydrogen bond.) In addition there is the question of the effect of water-soluble polar substances such as glycerol and sugars which are insoluble in oils; these are thought to be taken up on the exterior of the micelles and thus act as peptizing agents. The probable structures in the three cases are shown in Fig. 24.13.

Solubilization in soap solutions is now widely used to enable polymerization of relatively insoluble monomers such as styrene to be carried out in aqueous solution or suspension. Mention may also be made here of the interaction between soaps and proteins, gelatine, etc., which not only is of general interest but may have biological significance (e.g. Pankhurst and Smith, 1945; Bull, 1946).

(For an account of the work before 1940 see Lawrence, 1940.)

(f) *Detergent action*

Only a very brief mention of some of the fundamental factors in detergency is possible here; the technical side is so vast that even a summary is useless.

COLLOIDAL ELECTROLYTES

The removal of dirt (detergence) is essentially a dynamic process involving three separate stages: (a) access of the detergent to the dirty surface; (b) the loosening or peptizing of the dirt; (c) the removal of the dirt into the bulk of the solution.

In general the 'dirt' particles are covered with a greasy film and therefore have a hydrophobic nature, but the outer surface becomes hydrophilic after adsorption of the detergent molecules, since these will be orientated with their chains in the hydrophobic surface and their polar groups in the water. The detergent, owing to its strong adsorption on all surfaces, will also tend to displace the grease from the surface. The dirt, thus loosened, is detached upon mechanical agitation and is carried away into the solution as an emulsion or a suspension, or solubilized in the interior of the micelles.

(For a discussion see *Wetting and Detergency*, 1937.)

(g) Diffusion

It is found experimentally that colloidal electrolytes in pure aqueous solution diffuse at rates which are comparable with those given by ordinary electrolytes, but after addition of excess of simple salts the diffusion coefficient drops to values of the same order as given by uncharged colloidal particles. The unexpectedly high rate in the first case arises from the powerful electrical potential set up by the more rapidly diffusing gegenions which thus drag the colloidal ions behind them (see also p. 254). Hence to calculate the size of the micelle from diffusion measurements (see Chapter X), these must be carried out in the presence of a 'swamping' excess of simple electrolyte. Using the porous diaphragm method Hartley and Runnicles (1938), assuming spherical particles and Stokes's law to be applicable, thus obtained a value of 26 A for the micellar radius of cetyl pyridinium salts.

(For a discussion of diffusion in the absence of simple salts see Hartley, 1939.)

(h) Density

Potassium n-octoate, one of the shortest chain soaps, shows a quite marked increase in its partial molal volume over the concentration region around $0{\cdot}5N$ (Davies and Bury, 1927), where the steep fall in osmotic coefficient (obtained earlier by McBain and his co-workers from freezing-points) shows that aggregation is occurring.

A tentative explanation for this phenomenon, based on surface energy considerations, can be advanced. As pointed out on p. 671, the high hydrocarbon/water surface energy leads to the paraffin chains of molecularly dispersed molecules in soap solutions being curled up so as to reduce the surface energy to a minimum, and this must subject the chains to a considerable compressive force. On aggregation to a micelle the hydrocarbon/water surface energy is almost entirely eliminated, the chains now being surrounded by similar chains, so that this pressure disappears, the chains expand, and the partial molal volume increases (see also p. 692).

An extremely crude calculation illustrates the order of magnitude of the quantities involved. Considering first the single molecule with C_{16} chain, assumed to be curled up as a sphere of radius ca. 7 A, the excess pressure due to the surface energy $= 2\gamma/r$ (see p. 87)

$$\simeq \frac{2 \times 50}{7 \times 10^{-8}} \simeq 14 \times 10^8 \text{ dynes/cm.}^2$$

For the micelle $r \simeq 20$ A and $\gamma \simeq 5$ dynes/cm., i.e.

$$2\gamma/r \simeq 0.5 \times 10^8 \text{ dynes/cm.}^2$$

(i) Hydrolysis

It is only with salts of weak acids or weak bases that appreciable hydrolysis is to be expected, which in the case of soaps chiefly means the alkali metal salts of the fatty acids and the hydrochlorides of long-chain primary amines. Powney and Jordan (1938) made a thorough study of the former type by pH measurements with the glass electrode, and found a sharp increase in hydrolysis at about the point where other measurements (conductivity, surface tension, etc.) show that micelles are commencing to form. The reason is undoubtedly that hydrogen bond formation between undissociated fatty acid and soap ions at the micelle surface (just as at a macroscopic interface, p. 657) so displaces the equilibrium that further hydrolysis is permitted. With fatty acid salts, the formation of stable acid soaps such as $NaSt,HSt$ (where St is the stearate radical) is, of course, well known.

(j) Other properties

Very many other properties of dilute soap solutions have been measured, e.g. optical, dielectric, viscosity, ultrafiltration, ultracentrifugal sedimentation, etc. Optically these dilute solutions are isotropic and differ from the very concentrated solutions in that they give no X-ray diffraction pattern.

Ultrafiltration and sedimentation velocity measurements are subject to an inherent complication owing to the ease with which micelles are known to dissociate and re-form. Thus in the former case single molecules may pass through the filter and re-combine to micelles on the other side, showing that the method can give little definite information about the size and types of particles present in solutions of colloidal electrolytes. With sedimentation in the ultracentrifuge the

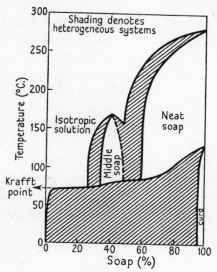

FIG. 24.14. Principal points of the phase diagram for the system sodium palmitate/water.

position is complicated by the relatively little effect which even high centrifugal fields can exert on the single soap molecules, the micelles thus sedimenting through an approximately constant concentration of single molecules.

Concentrated systems

The phase diagram soap/water at various temperatures has been thoroughly examined only in the case of certain fatty acid soaps, largely by McBain and his co-workers. Fig. 24.14 shows the principal phases given by sodium palmitate.

The hot solution of a fatty acid soap, which is optically clear, shows two changes on cooling. First a point is reached where the solubility is largely reduced and the system becomes slightly opaque and very viscous (the 'soft' soap stage); then occurs an abrupt transition to the very opaque hard white 'curd'. Potassium soaps do not reach the curd stage at ordinary temperatures (unless salted out with sufficient KCl) and thus remain as soft soaps. X-ray diffraction indicates the existence of four polymorphic forms in the case of the solid sodium soaps

(Ferguson, 1944), and that soft soaps consist of a random arrangement of ultramicroscopical crystallites (McBain and Ross, 1944). The curd state is believed to consist of a felt of micro-crystalline hydrated fibres, as suggested by electron micrographs (Fig. 15.19 between pp. 456–7).

The physical properties of concentrated soap solutions

The range of concentrations usually considered here will be about 10–50 per cent. unless otherwise stated. Examination of the physical properties of such concentrated solutions shows many striking differences from the dilute solutions previously discussed—for example the viscosity (η) is usually anomalous (i.e. the apparent η varies with the rate of shear) and the original isotropic solution becomes doubly refracting upon shear (streaming birefringence); also certain systems (e.g. ammonium oleate) even show definite elastic properties. (Such behaviour is indicative of a change from the spherical micelle to a more asymmetric type such as a rod or plate.) These systems exhibit solubilization of hydrocarbons, etc., as with the dilute solutions, and concentrated solutions of alkali metal soaps have been recently used to solubilize vinyl compounds for polymerization purposes (e.g. Harkins et al., 1945; Mark et al., 1945).

The principal technique used to elucidate the structure of these concentrated soap solutions has been X-ray diffraction, for, unlike the dilute soap solutions previously considered, these concentrated systems show a definite X-ray diffraction pattern, similar to that given by powders (p. 453). This observation was first made by Hess and Gundermann (1937) using sodium oleate, and subsequently repeated and extended by many other investigators. The fatty acid soaps have generally been used, although a very similar behaviour has been reported with phospho-lipoids such as lecithin and kephalin, and with some synthetic soaps such as dodecyl amine hydrochloride and sodium tetradecyl sulphate. (For early references see McBain, *Advances in Colloid Science*, vol. i, 1942; Hughes, Sawyer, and Vinograd, 1945.)

The long-spacing (d) is found to depend upon several factors—chiefly the concentration and chain-length of the soap, and the nature and concentration of any salts present, as well as the pH in the case of fatty acid soaps. The solution of water-insoluble materials such as benzene increases d, but does not affect the small spacing of *ca.* 4·4 A which is also present.

For a given soap the long-spacing (d) decreases linearly with soap concentration, as shown for various compounds in Fig. 24.15 (from

Harkins, Mattoon, and Corrin, 1946a), finally tending to the value found in the solid soap. This figure also shows how d varies with the chain-length of the soap molecule. Addition of indifferent salts such as potassium chloride has been reported as giving a slight increase in d

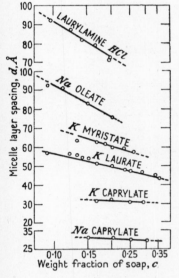

FIG. 24.15. Dependence of the long X-ray spacing (d) of various soap solutions upon the soap concentration.

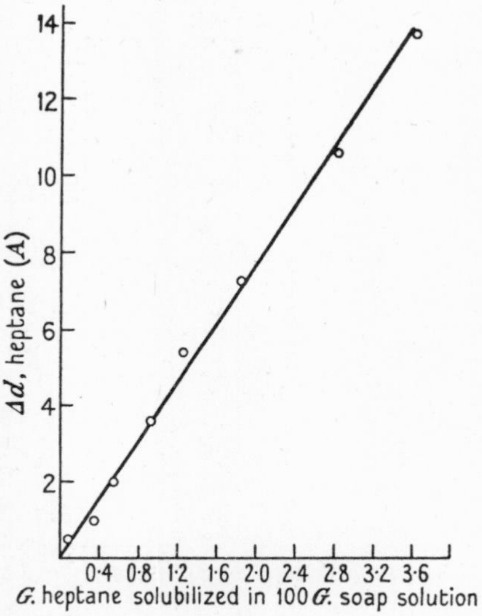

FIG. 24.16. Relationship between Δd, the increase in the long X-ray spacing due to solubilization, and the weight of n-heptane per unit weight of soap solution (25 per cent. potassium laurate).

with the fatty acid soaps, but a decrease in the case of kephalin and nerve lipoids (Bear, Palmer, and Schmitt, 1941). The latter behaviour would be expected to be the more common, since addition of salts would reduce the electrostatic forces responsible for keeping the layers apart.

One of the most widely examined aspects has been the effect of solubilized materials such as hydrocarbons, which, as first shown by Kiessig (1941), increase d but not the small spacing of ca. 4·4 A. For a fixed soap concentration the increase in long-spacing (Δd) depends upon the amount and nature of the additive, the relationship in the former case usually being a linear one as shown in Fig. 24.16 for the case of n-hexane in 25 per cent. potassium laurate (from Harkins, Mattoon, and Corrin, 1946b). Comparing different compounds on a molecular basis, Δd increases with size as indicated by the molecular volume. Fig. 24.17 shows this for various aliphatic and aromatic hydrocarbons solubilized in 22·1 per cent. potassium laurate solution (from Hughes, Sawyer, and Vinograd, 1945). Density measurements have shown that the solubilized substance, particularly the initially absorbed amounts, has a higher density than the bulk material, as

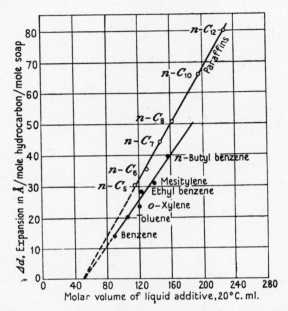

Fig. 24.17. Variation of Δd, the increase in the long X-ray spacing due to solubilization, with the molecular volume of the solubilized material.

indicated by Table 24.4, taken from Harkins, Mattoon, and Corrin (1946 b).

TABLE 24.4. *Density at 25° C. of 25 per cent. Potassium Laurate Solutions containing n-Heptane, and Apparent Density of Heptane in these Solutions*

Gm. heptane solubilized in 100 gm. of 25% KC_{12}	Density of final soln. (gm./c.c.)	Apparent density of heptane (gm./c.c.)
0·08449	1·018832	0·6944
0·36108	1·017569	0·68843
1·8632	1·010160	0·68402
3·5652	1·002008	0·67985

Density of bulk heptane = 0·67982.

These X-ray observations have been generally interpreted on the basis of a laminar type of micelle such as that depicted in Fig. 24.18 (*a*) (first suggested by Hess and Gundermann, 1937), the long-spacing (*d*) being identified with the repeat distance indicated by d in that figure. Since the thickness of each leaflet is unlikely to exceed twice the length of a fully extended soap molecule (i.e. $2 \times ca.$ 16 A for potassium laurate) it is clear that quite thick water layers are intercalated between the

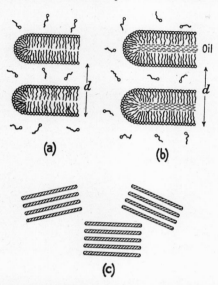

Fig. 24.18. Suggested structure of laminar micelles in aqueous solution: (a) alone; (b) containing solubilized oil; (c) laminated 'bundles' (two-dimensional).

sheets of polar groups. In the case of 25 per cent. nerve lipoid containing kephalin the water layers must be as much as 90 A thick, since two lipoid molecules could only contribute about 60 A towards the measured d of $ca.$ 150 A. The force keeping the sheets apart and approximately parallel must be electrical in origin, arising from the repulsion of the two interpenetrating and similarly charged diffuse double layers (see Chapter V).

Oil solubilized into these soap solutions is believed to be taken up between the ends of the soap molecules, as shown in Fig. 24.18(b). It seems unlikely that much penetration between the chains occurs, since the short spacing of $ca.$ 4·4 A, which is undoubtedly the usual side-chain spacing as found in many liquid hydrocarbons, is unaffected by these dissolved oils. The increase in density of the oil on solubilization can probably be explained on the same lines as suggested earlier for the hydrocarbon chains of the soap molecules themselves.

So far the size of these micelles, as regards area of each molecular sheet, and the number of sheets, has not been mentioned. It is clear from simple calculations based on the soap concentration and the d values, as well as from the observed fluidity, that the micelles are not 'infinite' in either dimension, and a suggested arrangement for the solution is given in Fig. 24.18 (c). (See, however, Hartley (1948).)

The viscosity of these concentrated soap solutions seems to fit in quite well with the above structures. For example, with sodium oleate solutions the viscosity-concentration curve turns up very rapidly around 10 per cent. soap, and from the streaming double refraction Snellman (1944) concluded that the transition to large micelles occurs over the concentration range 6–10 per cent. As previously pointed out, it is just over this range that the transition would be expected, and where the long X-ray spacing first becomes detectable.

Three-component systems

Mention has already been made of the possible ways in which a third component may interact with dilute soap solutions; this will now be considered in more concentrated systems.

A solution of sodium stearate of, say, 3 per cent. concentration, on cooling to room temperature sets to a pasty mass, due to the separation of micro-crystals of soap. Addition of suitable amounts of many organic compounds partially or completely miscible with water, e.g. glycerol, sugars, alcohols, phenols, amines, etc., results in the solution remaining clear—these compounds are therefore termed *peptizers* as they have clearly effected a dispersal of the soap crystals into molecular or micellar units. (This is the function of glycerol in the mixtures recommended for blowing bubbles, rather than to increase the viscosity which is actually diminished.) A similar behaviour is observed with other types of soaps, although most of the work on three-component systems has been done with fatty acid soaps.

With sodium stearate solutions (M./5–M./20) Lawrence (1937) found that saturation occurred at 6 mols. of n-amyl alcohol per mol. of soap, at about 4 for hexyl and heptyl alcohols, and at about 1 for the long-chain oleyl and cetyl homologues. These are incorporated into the micelle (whether spherical, rod, or plate-like) in the manner discussed above. These long-chain alcohol complexes show an obvious link with the acid soaps NaR.HR, which are stable in the solid state.

With partially miscible systems such as phenol/water or cresol/water, addition of soap lowers the critical solution (consolute) temperature, a fact known as long ago as 1886. The viscosity and conductivity of fatty acid soap solutions in the presence of phenols have been studied in particular by Angelescu and his co-workers (1941 and earlier papers). They find that at a fixed soap concentration (e.g. M./5 sodium stearate) the viscosity shows a maximum at a certain phenol concentration, the maximum becoming less marked with rise of temperature.

Four-component systems

Mention may be made of certain four-component systems which are optically clear and where oil is the continuous phase. They are frequently known as 'soluble-oil', a typical formula being:

Oil	150	parts by volume
Water	100	,, ,,
Soap	50	,, ,,
Alcohol	40	,, ,,

Hoar and Schulman (1943) have recently suggested that the water is dispersed as minute droplets of *ca.* 150 A diameter stabilized by a soap/alcohol complex. This they term the 'oleopathic-hydro-micelle', since it encloses water and tolerates an oily dispersion medium.

'Reversed' soap bubbles

Ordinary soap bubbles are discussed in connexion with foams (Chapter XXII), but mention may be made here of *reversed* soap bubbles, i.e. a spherical shell of air surrounded on both sides by liquid. These have recently been prepared by allowing drops of dense salt solutions containing surface active materials to fall into water (Rose, 1946).

The biological effects of soaps

The naturally occurring soaps include chiefly the sodium salts of the fatty acids, the bile salts, and some phospho-lipoids such as lecithin and lysolecithin. In view of the very great tendency of soaps to be adsorbed at interfaces, as well as to solubilize or interact with all sorts of organic molecules, their profound biological importance is not so surprising. The subject is an extensive one, and mention can only be made of one aspect which does, however, illustrate some of the factors frequently involved.

Fig. 24.19 shows how the biological activity of phenolic compounds can be modified by the presence of a varying concentration of soap ((*a*) from Alexander and Trim, 1946; (*b*) Alexander and Tomlinson, unpublished). The biological activity was measured in one case by the rate of penetration of hexyl resorcinol from 0·025 per cent. solution into *Ascaris lumbricoides* (the pig round worm), in the other by the time for 0·5 per cent. phenol to kill a suspension of *B. coli*. It is seen that low concentrations of soap increase the biological activity of the phenol, whereas very high concentrations reduce it enormously. (The

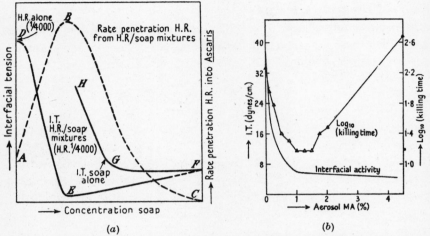

Fig. 24.19. Effect of soap concentration upon: (a) the rate of penetration of hexyl resorcinol into *Ascaris*; (b) the time required for 0·5 per cent. phenol to kill a suspension of *B. Coli*.

antiseptic value of *carbolic* soap, where the soap is in great excess over the phenol, is thus seen to be due solely to the soap, the phenol merely providing a desirable smell!) Measurement of the surface or interfacial tension of the mixtures (see Fig. 24.19 and Fig. 24.10) shows that maximum biological activity occurs at the point where micelles start to form, i.e. when the surface activity reaches its maximum. Previous to this point the biological and surface activities increase together, suggesting an intimate relation between the rate of penetration into, and the state of affairs at, the biological surface (either from increased adsorption or modified surface structure). Once micelles start to form they compete with the biological surface for the fixed amount of 'drug' (just as on p. 681), and the biological activity accordingly falls.

Soaps in non-aqueous media

Soaps, in particular some of the heavy metal salts of the fatty acids, can be dissolved or dispersed in certain organic media. Despite the technical importance of some of these systems—the lubricating greases, for example, consist of mineral oils thickened by soaps of sodium, calcium, aluminium, lead, etc.—our knowledge of the fundamentals is

COLLOIDAL ELECTROLYTES

extremely scanty. One of the difficulties has been the preparation of pure heavy metal soaps, since the presence of traces of water or free fatty acid exerts such a profound influence. This is particularly so with one of the most important commercially, namely the aluminium soaps.

Lawrence (1938, 1939) has prepared a number of pure soaps and examined the optical and mechanical properties of their dispersions in *nujol* (a high boiling hydrocarbon similar to medicinal paraffin). Clear mobile solutions can be obtained on heating to *ca.* 200°, in which the soap molecules are probably aggregated with their polar groups together (i.e. the opposite configuration to that in water). On cooling the isotropic liquid sets to a clear gel just below the dispersion temperature, and on further cooling opacity develops and the mechanical properties change. The changes depend upon the cation, and are also very much affected by alcohols, phenols, amines, fatty acids, and so on, which appear to act as peptizers in these non-aqueous systems also. Water also has a considerable effect owing to the fatty acid liberated by hydrolysis.

DYESTUFFS

The dyes or dyestuffs, which together with the soaps constitute the group of colloidal electrolytes, show a much more varied colloidal behaviour. The reason is obvious upon inspecting the structure of a few typical members, such as those given below, and comparing them with a few typical soaps as shown in Table 24.1, p. 667.

Orange II

Methylene Blue

Benzopurpurine 4B

The dye molecules clearly differ very much more amongst themselves as regards both the general shape and the relative disposition of the hydrophobic and hydrophilic parts of the molecule. The comparative simplicity of the paraffin-chain salts is lacking, and it is not surprising that aggregation is very much more specific.

Dyestuffs are usually classed as acid or basic (corresponding to anionic or cationic soaps) according to whether the dye ion carries a negative or positive charge.

Measurement of aggregation of dyes

The question of aggregation is of considerable importance in connexion with the dyeing process, so the methods of investigating it will be briefly mentioned.

The colligative properties of solutions can, of course, be used, subject to the same general limitations as in the case of soaps. Freezing-point and osmotic pressure methods have been employed to some extent, the chief difficulty arising from the difficulties associated with traces of contaminating electrolytes.

It was shown above (p. 677) how aggregation was reflected in the mobilities of the colloidal and simple ions, the individual ionic mobilities being found from conductivity and transport number measurements. Assuming the mobility to be proportional to ze/r (ze = charge and r = radius of particle), as indicated below, then some idea of the aggregation can be obtained (Robinson, 1935). (For a given field strength, the velocity of the particle is $\propto ze$ and $\propto 1/r$, assuming spherical particles and Stokes's law, i.e. mobility $\propto ze/r$.)

Diffusion is generally thought to give the best idea of aggregation in dyestuff solutions, although it is essential to carry out the determination in the presence of sufficient electrolyte to wash out anomalous electrical effects (see p. 254), and this added salt is not without influence on the aggregation. The usual assumptions of spherical particles and the validity of Stokes's law are also necessary.

Aggregation produces changes in the magnitude and wavelength of the absorption bands, and this has recently been discussed as a method for determining the degree of association (e.g. Sheppard, 1942; Vickerstaff and Lemin, 1946). Although an indirect one, this method should prove useful for relative measurements.

Ultrafiltration has also been used, but as pointed out earlier this method may be of very doubtful value owing to the ease with which aggregates can dissociate and reform.

Other physical properties which could provide some indication of the aggregating concentration and the size of the micelles are the boundary tension and the solubilization of insoluble oils. So far neither appears to have been used for this purpose.

The aggregation of dyes

Despite the experimental difficulties of measuring aggregation at all precisely the general picture seems to be fairly clear.

Dyes of relatively simple structure, such as orange II or methylene blue, are not much associated, as seen from the data given in Table 24.5 below (Valko, 1946).

TABLE 24.5

Dye	Dye concn. (%)	Temp. (°C.)	Salt concn. (M.)	Degree of association
Orange II	0·005–0·01	25	0·02	1·2
	0·002–0·1	25	0·05–0·2	1·7
	0·05	25	0·2	2·6
	0·02	60	0·05	0·82
	0·02	90	0·05	0·87
Methylene blue	0·005	25	0·01–0·02	1·8
	0·02	25	0·05	2·2

Very much higher values are obtained with some dyes, benzopurpurine 4B being an example where the degree of aggregation rises to several hundreds. In such cases the assumption that the particles are spherical is certainly invalid, benzopurpurine forming long needles which are responsible for the intense birefringence and gel-forming tendency of its solutions.

If the dye aggregates at all, this always increases with salt concentration, since the salt reduces the electrostatic repulsive forces which tend to oppose aggregation (just as with the soaps). Raising the temperature generally decreases the aggregation, as would be expected.

Colloid aspects of the dyeing process

The materials to be dyed are usually of a fibrous nature, consisting of macro-molecules of vegetable, animal, or purely synthetic origin. Vegetable fibres have cellulose as the fundamental constituent and include linen, cotton, and various modified celluloses or cellulose derivatives such as viscose, cellulose acetate, and cellulose ethers. Wool and silk are the principal fibres of animal origin. Both are polycondensates of amino acids, with keratin and fibroin respectively as the chief constituent, and they differ from vegetable fibres in possessing appreciable numbers of free acidic (chiefly carboxyl) and free basic (chiefly amino) groups. Of the purely synthetic fibres, nylon, obtained by condensing diamines with dicarboxylic acids, is the best known.

Fibres are rarely homogeneous on a molecular scale, but contain ordered regions or *crystallites* where the long-chains are parallel, and *amorphous* regions where no such regularity exists (see also Chapter XXVII). The size of these crystallites, as estimated, for example, by X-ray diffraction, does not change when the fibre absorbs water or dye, showing that these processes are confined to the amorphous regions. The question of pore-size in these amorphous regions is clearly very pertinent to the dyeing process, and permeability measurements for water-swollen cellulose and wool indicate values of the order of 40–50 A.

In practical dyeing the rate-determining step is diffusion from the surface into the interior of the fibre. The rate in the fibre, which is much less than for free diffusion in water, increases with the temperature as would be expected, whereas with increasing salt concentration the rate initially increases, passes through a maximum, and then decreases. Diffusion is complicated by adsorption of dye molecules on to the walls of the internal channels, by mechanical 'sieve' action if the dye molecule or aggregate is commensurate in size with the internal pores, and by electrical effects arising from the charges on the fibre surface and dye ion. If the dye is strongly aggregated good commercial dyeing is not possible, but the size of the dye particle has frequently been overestimated, since under the usual conditions of salt concentration and temperature, not only many acid wool dyes, but also some direct cotton dyes, form approximately molecular dispersions.

Theory of dyeing

Equilibrium measurements of dye adsorption are necessary to formulate any theory of dyeing, although in commercial practice such equilibrium is never attained. It is convenient to consider the chief types of fibres separately, as the factors involved differ considerably.

(a) Cellulose fibres

With direct azo dyes (e.g. congo red) and cotton or regenerated cellulose film (*cellophane*) the principal findings from equilibrium studies are as follows:

(i) The uptake of dye increases with the salt concentration (cf. with kinetic measurements above).

(ii) The uptake decreases with increasing temperature.

(iii) The sorption isotherm is often approximately Langmuirian, tending to a saturation value determined by the salt concentration and temperature.

COLLOIDAL ELECTROLYTES

(iv) Saturation values for regenerated cellulose are greater than for cotton under the same conditions (due to the presence of more amorphous material in the former).

(v) For the dye to be a substantive one its molecule should possess a flat or planar shape.

The affinity of the dye for the fibre in these cases is ascribed to van der Waals attractive forces arising from the interaction of permanent dipoles and induced dipoles, and from the dispersion (London) forces. Of the permanent dipole interactions, special emphasis should be laid on the hydrogen bond, since the hydroxyl groups in the cellulose may coordinate with the ionized group(s) of the dye molecule, e.g.

$$\begin{array}{cc} \text{cellulose} & \text{dye} \\ | & | \\ \text{O---H} \cdots \text{SO}_3' & \text{Na}^+, \end{array}$$

a behaviour known to occur in simpler systems (see p. 568). The presence of conjugated double bonds, essential, of course, for the colour of the dye, must facilitate adsorption, since the associated electron cloud is known to be unusually polarizable. The fact that a planar shape is essential for substantivity may be explained either on packing considerations, a flat molecule allowing closer adlineation to the fibre surface, or by its facilitating the formation of a resonance system.

Owing to the necessity of maintaining electrical neutrality in the fibre the uptake of dye anions must be accompanied by an equivalent amount of cations (and vice versa for a basic dye). This increases the concentration of cations in the fibre above that in solution and thus tends to increase the free energy of the systems. By addition of simple salts this effect can be reduced, thus increasing the uptake of dye. (The maximum in the *kinetic* measurements probably arises from the higher salt concentrations causing formation of aggregates too large to penetrate into the fine pores of the fibre.)

(b) *Protein fibres*

The equilibrium uptake of dyes by wool and silk decreases with increasing temperature, but otherwise differs fundamentally from the behaviour of cellulose fibres. With the wool and acid dyes it is found that:

(i) The amount of dye anion taken up decreases from pH 2·5 to pH 5 (the iso-electric point), although a definite sorption still occurs at pH > 5.

(ii) The maximum uptake is *ca.* 0·7–1 milli-equiv./gm. wool, i.e.

slightly more than the maximum amount of acid (e.g. HCl) which can be taken up.

(iii) The dye anions are preferentially adsorbed from a mixture of simple and dye anions.

The affinity in these cases is clearly modified by the existence of Coulomb forces between the charged groups on the fibre (e.g. $-\overset{+}{N}H_3$ if pH < 5; $-COO'$ if pH > 5), and the dye anions. The van der Waals attractive forces between dye anion and fibre will thus be assisted by the Coulomb forces at pH < 5, but opposed for pH's > 5.

The effect of salts upon the dye uptake of fibrous proteins such as wool has recently been placed on a quantitative basis by Gilbert and Rideal (1944) from a consideration of the electrical forces involved.

(c) *Cellulose acetate*

The dyeing of cellulose acetate differs from the above cases in that the dye is taken up homogeneously throughout the whole mass of the fibre, rather than on the walls of the internal sub-microscopic channels. The dyes used have therefore to be soluble in cellulose acetate, and this frequently runs parallel with solubility in organic solvents.

Some biological applications of dyes

Staining methods with dyes find considerable use in the identification of biological materials, e.g. bacterial types. These depend upon the different affinities of the constituents of the envelope (e.g. protein, cellulose, lipoid) for various dyes under certain conditions, and the principles involved are closely related to those in ordinary dyeing. The basic type of dyestuffs (e.g. gentian violet) find considerable use as antiseptics, resembling, in their bactericidal powers, the cationic soaps.

GENERAL REFERENCES

Symposium, *Colloidal Electrolytes*, 1935, Trans. Faraday Soc.
Symposium, *Wetting and Detergency*, 1937, Harvey (London).
Symposium, *Surface active agents*, 1946, N.Y. Acad. Sci.
HARTLEY, *Aqueous Solutions of Paraffin-chain Salts*, 1936, Hermann (Paris).
    1948, *Ann. Rep. Chem. Soc.* **45**, 33.
LAWRENCE, Article on 'Colloids', 1940, ibid.
MCBAIN, Article in *Advances in Colloid Science*, vol. i, 1942, Interscience.
VALKO, Article in Alexander's *Colloid Chemistry*, vol. vi, 1946, Reinhold.

OTHER REFERENCES

AICKIN and PALMER, 1944, *Trans. Faraday Soc.* **40**, 116.
ALEXANDER, 1942, ibid. **38**, 253.
—— and TRIM, 1946, *Proc. Roy. Soc.* B **133**, 220.
ANGELESCU and MANOLESCU, 1941, *Kolloidzschr.* **94**, 319.

BEAR, PALMER, and SCHMITT, 1941, *J. Cell. Comp. Physiol.* **17**, 355.
BULL, 1946, *J. Amer. Chem. Soc.* **68**, 747.
CASSIE and PALMER, 1941, *Trans. Faraday Soc.* **37**, 156.
CLAYTON, 1943, *Emulsions and their Technical Treatment*, 4th edn., p. 115, Churchill.
DAVIES and BURY, 1930, *J. Chem. Soc.*, p. 2263.
DEAN, 1937, Article in *Wetting and Detergency*, Harvey (London).
FERGUSON, 1944, *Oil and Soap*, **21**, 6.
GILBERT and RIDEAL, 1944, *Proc. Roy. Soc.* A **182**, 335; **183**, 167.
HARKINS et al., 1945, *J. Chem. Phys.* **13**, 381, 534.
HARKINS, MATTOON, and CORRIN, 1946 a, *J. Amer. Chem. Soc.* **68**, 220.
—— —— —— 1946 b, *J. Colloid Sci.* **1**, 105.
—— and MYERS, 1937, *Nature*, **139**, 367.
HARTLEY, 1936, *Aqueous Solutions of Paraffin-chain Salts*, Hermann (Paris).
—— 1939, *Trans. Faraday Soc.* **35**, 1109.
—— 1941, ibid. **37**, 130.
—— and RUNNICLES, 1938, *Proc. Roy. Soc.* A **168**, 420.
HESS and GUNDERMANN, 1937, *Ber. dtsch. chem. Ges.* **70**, 1800.
HOAR and SCHULMAN, 1943, *Nature*, **152**, 102.
HUGHES, SAWYER, and VINOGRAD, 1945, *J. Chem. Phys.* **13**, 131.
JONES and BURY, 1927, *Phil. Mag.* **4**, 841.
KIESSIG, 1941, *Kolloidzschr.* **96**, 252.
LAWRENCE, 1929, *Soap Films*, Bell.
—— 1937, *Trans. Faraday Soc.* **33**, 325.
—— 1938, ibid. **34**, 660.
—— 1939, ibid. **35**, 702.
—— 1940, *Ann. Rep. Chem. Soc.* **37**, 102.
MCBAIN and BOLDUAN, 1943, *J. Phys. Chem.* **47**, 94.
—— and ROSS, 1944, *Oil and Soap*, **21**, 97.
—— and SALMON, 1920, *J. Amer. Chem. Soc.* **42**, 426.
MARK et al., 1945, *India Rubber World*, **112**, 436.
MITCHELL, RIDEAL, and SCHULMAN, 1937, *Nature*, **139**, 625.
MORTON, 1935, *Trans. Faraday Soc.* **31**, 262.
MURRAY, 1935, ibid. **31**, 206.
—— and HARTLEY, 1935, ibid. **31**, 183.
PANKHURST and SMITH, 1945, ibid. **41**, 630.
POWNEY and ADDISON, 1937, ibid. **33**, 1243, 1253.
—— and JORDAN, 1938, ibid. **34**, 363.
—— and WOOD, 1941, ibid. **37**, 152.
REICHENBERG, 1947, ibid. **43**, 467.
ROBINSON, 1935, ibid. **31**, 245.
—— 1937, *Nature*, **139**, 626.
ROSE, 1946, ibid. **157**, 299.
SHEPPARD, 1942, *Rev. Mod. Physics*, **14**, 303.
SNELLMAN, 1944, *Arkiv Kemi Min. Geol.* B **19**, No. 5, 1.
VALKO, 1935, *Trans. Faraday Soc.* **31**, 230.
—— 1946. See general references above.
VICKERSTAFF and LEMIN, 1946, *Nature*, **157**, 373.
WARD, 1940, *Proc. Roy. Soc.* A **176**, 412.

XXV
CLAYS AND ZEOLITES
General considerations

THE inorganic constituents of soils, clays, and sands consist almost entirely of silicates and silica, so that the study of silicates is fundamental not only to soil science, with its close links to agronomy, engineering, and geology, but also to ceramics, water softening by base exchange methods, and many other everyday uses. Here we will be concerned only with the colloidal aspects of these materials, which limits us to silicates and silica with particle size less than $ca.\ 2\mu$, and to the natural and synthetic zeolites (permutites) with their large internal surfaces.

From the practical point of view the term 'clay' denotes rather different substances to the agriculturist and to the ceramist, the two principal workers so long concerned with them. To the former 'clay' is the finely divided ($< ca.\ 2\ \mu$) mineral constituents of the soil, to the latter the small laminated particles having the requisite plastic and firing properties required in the manufacture of pottery. Zeolites and permutites, the latter now a household word owing to their widespread use as water softeners, show some important differences from the clays, chiefly as regards speed and capacity for ion-exchange, although they have a marked similarity in many other respects.

Before considering clays and zeolites in more detail a brief account of their relationship to other silicates and to silica is desirable. (For more detailed accounts see, e.g., Bragg, 1937; Clark, 1940; Wells, 1945.)

The principal types of aggregation in silicates

Chemical analysis had shown the complexity of silicate minerals, but gave no clue to their structure nor to obvious similarities and differences between the members. X-ray analysis provided the solution as regards their molecular structure, and more recently the electron microscope has given some information about the size and shape of colloidal, submicroscopic particles.

The fundamental concept in silicate chemistry, due to Bragg, is of close-packed oxygen atoms (or ions), these being the largest atoms present. Pauling's rules of coordination in complex ionic crystals proved particularly helpful in the interpretation of the silicates, which are conveniently classified according to the way in which the SiO_4 tetrahedra are linked together (by sharing corners, edges, and faces), to build up larger complexes. If some of the Si is replaced by Al then the framework acquires a negative charge, since for each Si atom replaced by a 4-coordinated Al atom a univalent cation (or its electrochemical

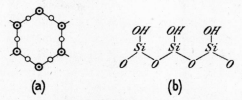

FIG. 25.1. The 6-ring type of silicon-oxygen sheet (idealized): (a) in plan; (b) in elevation. In (a) oxygen atoms lying above the Si atoms (small black circles) are drawn more heavily.

equivalent of polyvalent ion) must be taken up in some way so as to maintain electrical neutrality.

In some silicates discrete SiO_4 groups (i.e. with no common oxygen atoms) are present, but more frequently the SiO_4 tetrahedra are linked up as larger groups, Si_xO_y, of the following types:

(a) The pyrosilicate ion $[Si_2O_7]^{6-}$, where one O atom is shared by two SiO_4 groups.

(b) Closed rings or chains $[(SiO_3)_n]^{2n-}$ formed by two O atoms of every SiO_4 being shared with other such groups, e.g.

and larger rings, and

which tends to $[(SiO_3)_n]^{2n-}$ with increasing chain-length.

(c) Sheets of composition $[(Si_2O_5)_n]^{2n-}$, formed by three O atoms of each SiO_4 being shared with other tetrahedra. Two kinds of such sheet are formed in silicates of which the one shown in Fig. 25.1 is of most interest to us here as it is characteristic of the micas and the clay minerals; the other, with alternate 4 and 8 rings, is found in the mineral apophyllite.

(d) Three-dimensional frameworks formed by the sharing of all corners of each SiO_4 tetrahedron, which if continued indefinitely leads to the completely neutral silica framework SiO_2. If Al replaces some of the Si the requisite cations are taken up inside the lattice, and the *framework*, if infinite in extent, will always have the composition $(Si,Al)O_2$. The zeolites and felspars are examples of complex silicates which can be regarded as possessing such infinite three-dimensional structures.

If the linking of the SiO_4 tetrahedra does not proceed to completion (as in silica gel formed at not too elevated temperatures, owing to steric effects), then some O atoms must remain which are not linked to two Si atoms; in silica gel these will normally be present as —OH groups. The well-known drying properties of silica gel are probably due in part to these free polar groups, as well as to condensation in the submicroscopic pores and channels.

There are further possible complexes in which the environment of all the Si atoms, in terms of shared and unshared O atoms, is not identical, but these will not be considered here. (See above references.)

Clays

Principal types

The principal clay minerals, together with their chemical composition and type of crystal lattice, are given in Table 25.1 below (from Hauser, 1945).

Table 25.1

Mineral	Chemical composition†	Type of crystal lattice
Kaolinite	$Al_4(Si_4O_{10})(OH)_8$	⎫
Dickite	$Al_4(Si_4O_{10})(OH)_8$	⎪
Nacrite	$Al_4(Si_4O_{10})(OH)_8$	⎬ Kaolinite
Halloysite	$Al_4(Si_4O_6)(OH)_{16}$	⎪
Metahalloysite	$Al_4(Si_4O_{10})(OH)_8$	⎭
Attapulgite	$Mg_5[Al](Si_8O_{20})(OH_2)_4(OH)_2 . 4H_2O$	
Pyrophyllite	$Al_4(Si_8O_{20})(OH)_4$	⎫
Talc	$Mg_6(Si_8O_{20})(OH)_4$	⎪
Montmorillonite	$Al_4[Mg](Si_8O_{20})(OH)_4 . xH_2O$	⎬ Montmorillonite
Nontronite	$Fe_4[Mg](Si_8O_{20})(OH)_4 . xH_2O$	⎪
Beidellite	$Al_4[Mg](Si_6[Al]O_{20})(OH)_4 . xH_2O$	⎪
Saponite	$Mg_6(Si_8[Mg]O_{20})(OH)_4 . xH_2O$	⎭
Illite	$K_y . Al_4[Fe_4 . Mg_4 . Mg_6](Si_{8-y} . Al_y)O_{20}$‡	Illite
Muscovite	$K_2 . Al_4(Al_2Si_6O_{20})(OH)_4$	Mica

† Symbols in [ ] indicate that they may substitute for the symbol written to the left of the bracket.   ‡ According to Grim, y varies from 1 to 1·5.

Structure and shape of clay minerals

The silicon-oxygen sheets illustrated in Fig. 25.1 occur in the clays and micas as composite layers of one or two such sheets combined with layers of hydroxyl groups attached through Mg or Al atoms. These composite layers form the building units which pack together to form the crystals, and their formation can be envisaged as follows (after Wells, 1945). Fig. 25.2 (a) shows part of a sheet of linked SiO_4 tetra-

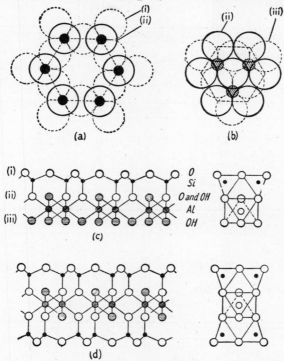

FIG. 25.2. The formation of composite Si–Al–O or Si–Mg–O layers.

hedra with three O atoms of each group shared and the vertices of all the tetrahedra pointing upwards, and Fig. 25.2 (b) the arrangement of the oxygen atoms in a layer of the $Mg(OH)_2$ or $Al(OH)_3$ structures. (Hydrogen atoms are omitted as they cannot be located by X-ray analysis.) The separation of the upper oxygen atoms in (a) (indicated by (ii)), does not differ greatly from that in (b) (also indicated by (ii)). By turning (b) over and placing it on top of (a) a composite layer can thus be formed having these oxygen atoms in common, as shown in Fig. 25.2 (c). The remaining OH groups fit in at the centres of the hexagonal rings in (a). This process can be repeated on the other side of the (b) layer giving the more complex layer shown in Fig. 25.2 (d). The compositions of these layers are

(c) $Mg_3(OH)_4Si_2O_5$ or $Al_2(OH)_4Si_2O_5$,

(d) $Mg_3(OH)_2Si_4O_{10}$ or $Al_2(OH)_2Si_4O_{10}$,

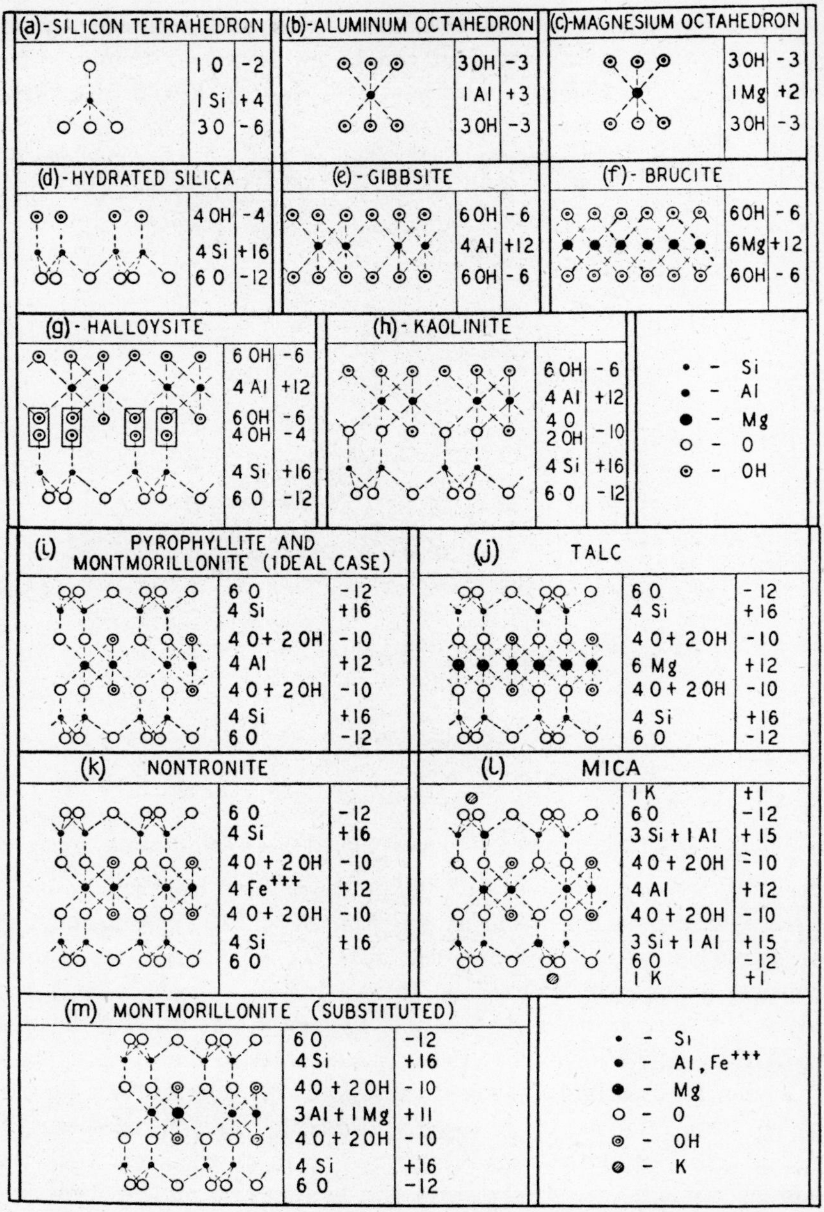

FIG. 25.3. Structural data of clay minerals (after Hauser and Grim). All atoms have been projected into one plane. The three sections in each sub-figure ((a)–(m)) show respectively the unit cell of the crystal lattice, the number and type of atom or group in every lattice plane, and the amount of available or needed electrons in every sheet. If the (+) and (−) in the third column are equal the lattice framework is electrically neutral, if not, it carries a charge.

according to whether the octahedral holes are occupied by Mg or Al.

Another way of regarding the process is to place the OH surface of the hydrated silica sheet (Fig. 25.3 (d)) in contact with one face of the Al(OH)$_3$ or Mg(OH)$_2$ layer structures (Fig. 25.3 (e) and (f) respectively),

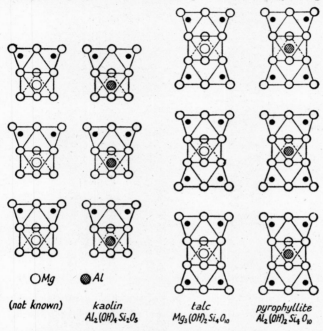

Fig. 25.4. The structures of some silicate minerals with layer structures (diagrammatic).

and then allow condensation to take place between the OH groups which are in contact. This gives the structure shown in Fig. 25.2 (c). By sandwiching the metal hydroxide sheet between two sheets of hydrated silica and condensing on both faces we similarly obtain structure (d) of Fig. 25.2.

Of the four compositions shown above, three occur naturally, Al$_2$(OH)$_4$Si$_2$O$_5$ being the basis of the kaolin minerals, Mg$_3$(OH)$_2$Si$_4$O$_{10}$ that of talc, and Al$_2$(OH)$_2$Si$_4$O$_{10}$ of pyrophyllite (see Fig. 25.4, due to Wells, 1945). Kaolinite, dickite, and nacrite are closely related, all having the same composition and the same kind of layer although differing in detail. All these layer structures are seen to be electrically neutral, so that in crystals built from them the forces between adjacent layers will be of the van der Waals type and therefore relatively weak, thus providing a simple explanation of the ready cleavage of the crystals. This is shown, for example, by talc (french chalk) used as a polishing material, due to its lubricating qualities.

Some of the Si atoms in the complex layers above can be replaced by Al, and, as previously pointed out, for each atom so replaced a

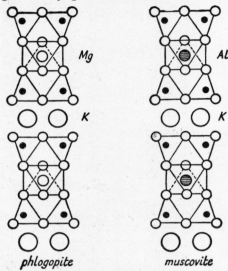

FIG. 25.5. The structures of the micas phlogopite and muscovite (diagrammatic).

univalent cation (or its equivalent polyvalent ion) must be incorporated so as to preserve electrical neutrality. Such replacement has not been found in the kaolin type layers, but it does occur with the type (d) layers of Fig. 25.2, giving rise to the micas, with typical formulae:

$$KMg_3(OH)_2Si_3AlO_{10} \quad \text{Phlogopite}$$
$$KAl_2(OH)_2Si_3AlO_{10} \quad \text{Muscovite}$$
$$CaAl_2(OH)_2Si_2Al_2O_{10} \quad \text{Margarite}$$

In the first two, one-quarter of the Si atoms have been replaced by Al, the resultant negatively charged layers being held together by cations as shown in Fig. 25.3 (l) and in Fig. 25.5 (Wells, 1945), with the K^+ ions occupying large holes between twelve oxygen atoms. The K—O electrostatic bond strength is thus only 1/12, so that the bonds between the layers are comparatively easily broken, giving rise to the well-known ready cleavage of the micas. When the layers are held together by divalent ions, as by Ca^{++} in margarite, the doubled electrostatic bond strength increases the hardness and reduces the ease of cleavage—hence their designation as 'brittle micas'.

Further examples of structures and building units present in clay minerals will be found in Fig. 25.3, which is based upon an illustration due to Hauser (1941).

Of the montmorillonites, two of which are shown in Fig. 25.3 (*i*) and (*m*), special mention may be made of the natural material bentonite, a clay with particularly interesting colloidal properties. An ideal

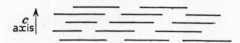

FIG. 25.6. Aggregation of plate-like particles in clay minerals as seen in elevation. The *c*-axis is perpendicular to the plates which lie normal to the plane of the paper.

montmorillonite consists of three layers: a gibbsite sheet enclosed between two silica sheets with their vertices pointing toward each other (see Fig. 25.3 (*i*)). As in kaolinites, the layers are attached by primary valence bonds, due to the sharing of O atoms in the vertex positions, and those hydroxyl groups in the gibbsite layer which could not find opposing hydroxyls from the top and bottom silica sheets remain located in the hexagons formed by the O atoms in the vertex positions of the silica sheets. The composite sheet is clearly neutral except for unsaturated edge and corner positions, so that the forces between composite sheets will be small.

Pyrophyllite, of composition $Al_4(Si_8O_{20})(OH)_4$, has the ideal montmorillonite structure, although certain difficulties arise in interpreting its properties. In general, however, *natural* montmorillonites do not conform to the ideal composition: in addition to a variable Al:Si ratio, magnesium and iron are frequently present, and not as adventitious impurities. With nontronite, for example, which from X-ray diffraction has the same structure as montmorillonite, the Al has been more or less replaced by Fe^{III}, but owing to their valencies being identical the electronic balance is undisturbed. This is not the case in beidellite, which differs from pure montmorillonite in partial replacement of Si by Al in the top and bottom layers, leading to a negatively charged framework and adsorption of cations from the environment. If Mg substitutes for Al in the gibbsite layer, as in saponite, the centre of the lattice acquires a negative charge and cation adsorption on to the surface occurs.

As the degree of hydration of the montmorillonite minerals is increased, the *c* spacing (d_{001}) increases (see Fig. 25.6), due to water molecules penetrating between the sheets and prising them apart.

According to Clark and his co-workers 1, 2, 3, or 4 layers of water molecules may be introduced in swelling to form discrete hydrates, the periodicity value along the c-axis thus increasing from 9·6 to 21·4 A, i.e. *ca.* 3 A for each added water layer. Beyond this point the layers are comparatively free. Cation substitution is believed to take place between, rather than within, the composite sheets, since the replacement of H^+, Na^+, or Ca^{++} by large cations (e.g. substituted ammonium ions, methylene blue) leads to great increases in the 001 or c spacing (see Fig. 25.6).

The structures as given above were derived very largely from X-ray studies. More recently the electron-microscope has provided data on the size and shape of colloidal clay particles, and shown that kaolinite and bentonite, for example, do actually consist of thin plates, in complete accord with the plate-like structures indicated by the X-ray data. Photographs of kaolinite are shown in Fig. 15.20 (facing p. 457).

The properties of clay dispersions

We can now consider the properties of clay dispersions, which naturally refer almost exclusively to aqueous media, and their interpretation in terms of the structure and shape of the ultimate colloidal particles.

As would be expected, owing to the polar nature of clay minerals, electrical forces play an important part. The particles are normally charged, either by preferential adsorption of ions (particularly OH') in the case of clays with no ionizable groups, or by ionic dissociation, as, for example, in the bentonites.

Considering first kaolinite, which occurs as thin hexagonal plates, a fracture along the cleavage plane does not rupture primary valence bonds, as pointed out above, whereas one perpendicular to this plane (i.e. parallel to the c-axis, see Fig. 25.6) must break bonds between Si and O or between Al and O (see Fig. 25.3 (i)). Ionic adsorption (chiefly of OH' if dispersed in pure water) will then occur at these points of unsaturation, as well as upon the oxygen atoms in the basal silica sheet. (Owing to the thin plate-like structure the latter factor would be expected to predominate.) The particle thus acquires a negative charge, and an equivalent amount of cations will be found in the neighbourhood of the particle giving rise to the usual electrical double layer. As the OH' ion concentration is increased, by addition of NaOH, for example, the negative charge on the particle, and hence the repulsive forces between them, is initially increased, and the dispersion accordingly stabilized. (Continued addition of NaOH leads eventually to precipitation, as discussed in more detail below.) Stable dispersions are said to be in the 'deflocculated', 'dispersed', or 'peptized' state.

Minerals of the montmorillonite group usually possess a substituted lattice (Fig. 25.3 (m)) rather than the ideal structure. This, as shown above, gives rise to negative charges at certain points and hence to adsorption of cations, usually Na^+. These cations are adsorbed chiefly upon the surfaces of the silica sheets, and when hydration occurs on immersion of the dry material in water these cations prise the particles apart, forming double layers of appreciable thickness.

(a) *Ion-exchange*

It has long been known that soils possess the power of removing from solution small amounts of dissolved bases such as ammonia or lime, and that when a salt such as potassium sulphate percolates through a soil some of it is removed, and an equivalent quantity of calcium sulphate appears in the solution. The phenomenon was originally termed 'base-exchange' but is now, more correctly, termed 'ion- or cation-exchange', since only ions are involved. Ion-exchange is usually expressed in milli-equivalents of cations per 100 gm. of clay, and some idea of the order of magnitude for different types of clay can be seen from Table 25.2.

TABLE 25.2

Mineral	Ion-exchange capacity (milli-equiv. per 100 gm. of clay)
Montmorillonite	60–100
Attapulgite	25–30
Illite	20–40
Kaolinite	3–15

The phenomenon is of great importance, not only in agriculture, where it enables soils to retain nutrients, such as ammonia, which would otherwise be readily washed away, and to stabilize their pH by the buffer action involved, but also in ceramics, since the physical behaviour of clay 'slips' or 'slurries' is found to be very sensitive to alteration in the 'exchangeable' cations. (For an account of its importance in agriculture see, for example, Russell, 1937.) Ion-exchange also forms the basis of the 'permutite' water-softening process, as discussed in more detail below under the zeolites.

Table 25.2 shows that the ion-exchange capacity of a clay depends very much on its structural configuration, that of the kaolinite group being particularly low on account of its essentially neutral structure. For a series of cations the exchange capacity usually follows the Hofmeister or lyotropic series (p. 16), e.g.

$$Li^+ < Na^+ < K^+ < NH_4^+ < Rb^+ < Cs^+,$$
or
$$Mg^{++} < Ca^{++} < Sr^{++} < Ba^{++};$$

this being the order of decreasing size in the hydrated state. A similar

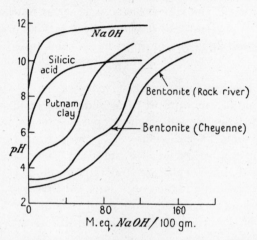

Fig. 25.7. Titration curves of electrodialysed colloidal clays, with electrodialysed silicic acid for comparison.

behaviour is found with many other types of cation exchanger, as illustrated by the data, due to Nachod and Wood (1945), shown in Fig. 25.12. Organic cations can, if sufficiently hydrophilic (e.g. organic ammonium complexes), also be introduced into clays by the ion-exchange process, and by this means water-resistant clay films have been produced.

Interchange occurs not only with salts but with dilute acids, giving a hydrogen or acid clay which can be regarded as a multivalent colloidal acid, since the anion is of colloidal dimensions. Hydrogen clays are also obtained by dialysis or preferably by electrodialysis, since under these conditions the original cations are eventually replaced by hydrogen ions.

The similarity of the hydrogen clays to a weak acid is shown in their ability to bring about the inversion of cane-sugar, in their pH-concentration curves, and in their titration curves against alkali. Addition of salts such as sodium chloride increases the acidity of the dispersion medium, but decreases that of the clay (by exchange of H_3O^+ for Na^+). Fig. 25.7, based upon the work of Bradfield, shows a number of typical titration curves for electrodialysed colloidal clays and also for silicic acid against N/10 NaOH, using the hydrogen or glass electrode. The clay acids are seen to buffer in the region of pH 5–6·5. 'Apparent pK' values, obtained from the pH for half-neutralization and assuming the clay to behave as a simple monobasic acid, are of the order 4–6, so the clay acids are about as strong as carbonic acid (first dissociation), but much stronger than silicic acid (pK *ca.* 10–11).

The titration curves with different bases (e.g. NaOH and $Ba(OH)_2$), differ in position but become coincident on addition of N/2 solution of the metal chloride. This behaviour is the reverse of that shown by the simple weak acids, where the titration curves with different bases diverge increasingly with addition of neutral salts. It can probably be explained by the excess of neutral salt eliminating electrical effects arising from the charges carried by the clay particles and their adherent double layers, which will affect the titration curve just as salts influence the surface activity of soaps (p. 682). (See also p. 215).

At pH's greater than *ca.* 9 the clay particles become chemically unstable, tending to form aluminates and silicates, thus making it difficult to decide whether or not there is a definite limit to the exchangeable hydrogen.

(b) *Dispersion, swelling, and flocculation*

The ease of dispersion would be expected to depend upon the strength of the bond holding the individual (composite) sheets together, just as in the case of cleavage. With kaolinite and an ideal montmorillonite, for example, the attraction arises from the van der Waals forces between the —O⋯HO— and —O⋯O— sheet systems respectively, and of these the former (being a hydrogen bond) is much the stronger. This is a possible explanation for the greater ease of dispersion of montmorillonites as compared with members of the kaolinite group. Bentonites frequently show spontaneous swelling and dispersion when dusted on to a water surface, the swelling taking place along the c-axis of the mineral (see Fig. 25.6), as shown by X-ray analysis.

The stability of colloidal suspensions depends on their ability to form diffuse electrical double layers, so that swelling and dispersion of clays will be assisted by increasing the ζ potential of the particles. Since in water the particles tend to become negatively charged, this entails the preferential absorption of anions, usually OH', as with sodium hydroxide and sodium carbonate, which are good dispersing agents. A number of other compounds, such as sodium silicate, phosphates, and tannates, as well as certain organic compounds (usually as sodium salts) are also very effective for this purpose (e.g. Ford, Loomis, and Fidiam, 1940), due no doubt in part to the alkalinity of their solutions, but probably also to adsorption of anions and in some cases to removal of polyvalent cations.

The extent to which the particle is spontaneously charged, and hence

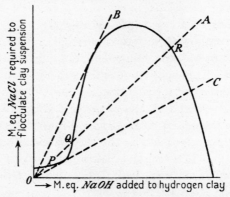

FIG. 25.8. The variation of stability of a hydrogen-clay with added caustic soda.

the swelling in pure water, will also be intimately related to its capacity for ion-exchange, since this gives a measure of the (negative) charge carried by the framework. Thus we find that the bentonites with a high exchange capacity show much greater swelling than the kaolinite minerals, with the illites occupying an intermediate position.

When a clay is shaken with dilute ammonia, sodium carbonate solution (or very dilute caustic soda), a very stable deflocculated suspension results, as shown by the very slow rate of settling, but the system rapidly flocculates if the pH is brought to neutrality by addition of dilute acid, or if sufficient neutral salt is added. The requisite concentration of neutral salt decreases with increasing valency of the cation in approximate accord with the Schulze-Hardy rule (p. 15), since the colloidal particle carries a negative charge. The ultramicroscopic study of the flocculation of dilute clay suspensions by Tuorila showed that the Smoluchowski equation of rapid coagulation (p. 575) was followed when more than a critical amount of flocculant had been added; with less than this critical amount coagulation still occurred, but not at the rate predicted by Smoluchowski's modification for slow coagulation (p. 576). The rate of coagulation was found to be closely connected with the ζ potential as determined by electrophoresis of the single clay particles.

One factor which has to be considered in the flocculation of clays is the influence of exchangeable cations in the clay. Fig. 25.8 shows how the coagulating concentration of sodium chloride varies with the amount of exchangeable Na in the clay, prepared by adding increasing amounts of NaOH to a hydrogen clay. Addition to a hydrogen clay of a given mixture of NaCl and NaOH (i.e. moving along OA) leads to flocculation at P, deflocculation at Q, and finally flocculation again at R. This 'anomalous flocculation', discovered by Oakley (1926), is due to the sigmoid shape of the curve, and can only occur with mixtures with compositions between OB and OC. As a general rule the cation of the added salt will not, as in the above case, be the same as that combined with the clay, and this will complicate matters by ion exchange. The complication is not very important in the case of neutral salts and a hydrogen clay, since little of the H^+ exchanges unless an alkali or a salt of a weak acid is present.

Suspensions of clays containing a divalent cation are less stable and have a lower ζ potential in the absence of electrolytes than those containing a univalent alkali cation or hydrogen ion. To coagulate clays with exchangeable Na^+ or K^+ ions requires high concentrations of the corresponding chlorides. The stability and ζ potential increase with the degree of hydration of the cation, i.e. as its atomic weight decreases. Addition of electrolytes was shown by Mattson to depress the ζ potential of a hydrogen clay much more rapidly than with a calcium clay, soon leading to a reversal of their stabilities.

As would be expected, the question whether the clay constituents of a soil are in the deflocculated or flocculated state is of great importance, particularly with 'alkali' soils formed by accumulation of sodium salts. Provided the salt content of the soil exceeds a certain critical amount the clay remains flocculated and no deterioration in the condition occurs, but should it fall below this threshold value deflocculation takes place, and the soil, although wet and sticky, allows water to percolate only extremely slowly. Soluble calcium salts in sufficient quantity prevent this deflocculation; hence the addition of lime, chalk, or calcium sulphate to heavy clay soils. (For further discussion of the importance of flocculation and deflocculation phenomena in soils see, for example, Russell, 1937.)

One characteristic of the coagulation of clay suspensions in alkaline media is the formation and rapid settling of large, loose aggregates, or flocs, the coagulation being more rapid than at neutral pH and much faster than calculated from the Smoluchowski theory of rapid coagulation, i.e. from Brownian movement alone. The phenomenon was shown to be due to the formation of a colloidal suspension of metal hydroxides (e.g. $Ca(OH)_2$), which, owing to their positive charges, were attracted to the oppositely charged clay particles by electrostatic forces. A loose structure with the clay particles linked through those of the metal hydroxide is thus rapidly formed, and as its size precludes Brownian motion it quickly settles out. The marked efficiency of aluminium sulphate as a coagulating agent for aqueous suspensions (e.g. in sewage purification) which are usually negatively charged, is probably due to the same cause, i.e. formation of a fine dispersion of hydrated, and positively charged, aluminium oxide particles.

Mention may be made here of the interaction of clays with proteins, such as albumins and gelatine, which can also be studied by the changes in X-ray diffraction pattern. In the system montmorillonite/gelatine, for example, the absorption is more complete at low pH, indicating that adsorption of proteins as cations is partly responsible for their combination with this clay (see Ensminger and Gieseking, 1941).

(c) Mechanical properties (viscosity, plasticity, thixotropy, dilatancy)

The rheological properties of concentrated clay dispersions are of particular importance in the ceramics industry, and some aspects have already been mentioned in Chapters XVI and XXI. Clay suspensions do not obey Poiseuille's law except at great dilution, so that the determination and representation of the flow properties of the more concentrated pastes ('slips' or 'slurries') is a matter of some difficulty. A typical graph connecting flow-rate and applied stress was given on p. 473, the flow at high pressures being expressed by two constants, one measured by the intercept on the stress (or pressure) axis (the 'yield value' or 'shearing strength'), the other analogous to viscosity, which can be calculated from the slope (the 'stiffness' or 'pseudo-viscosity').

Whether the clay is in the flocculated or deflocculated state has a very marked effect on the properties even of a concentrated paste; in the former state both the intercept and the pseudo-viscosity are higher. Deflocculants such as sodium silicate are added by the ceramists to their casting slip to keep the water content to the minimum consistent with adequate pour properties, since all the water present has ultimately to be removed. The plastic properties therefore depend a very great deal upon the electrical double layer present, which, as we have seen, depends very much upon the structure and composition of the particular clay, as well as on the nature and concentration of added salts. In the case of kaolinite, plastic properties only become evident with particles having an average equivalent spherical diameter (from sedimentation velocity and Stokes's law) of less than $ca.$ 4 μ, and they become more pronounced the smaller the particle size.

Clays containing exchangeable Na^+ or K^+ ions, flocculated by the corresponding chlorides, or clays suspended in chloride solutions of above about N concentration, give thixotropic gels. The strength of the rigid structure of the sediment depends to some extent on the exchangeable ions present, and on the nature and concentration of the added salt, although the precise relationships appear to be complex.

Thixotropic gels of clays, particularly of bentonite, have been mentioned earlier when discussing gel structures. Dialysed bentonite does not exhibit thixotropy, but does so after addition of alkali hydroxides or electrolytes. Ultramicroscopic studies by Hauser have shown that, on addition of electrolyte to a dialysed bentonite sol, first the translatory and then the rotatory Brownian movement ceases. Hauser also fractionated dispersions of crude bentonite by centrifuging, and showed that thixotropy became more pronounced the smaller the particle size. With particles of average equivalent spherical diameter of 150 A thixotropic gels were obtained at concentrations as low as 0·05 per cent., as compared with the 5–10 per cent. necessary with the usual commercial bentonites. Bentonite dispersions show a marked change in light transmission when the sol sets to a gel. These fine particle-size bentonites also show rheopexy (p. 611), the rheopectic setting time being about one-hundredth of the normal setting time when undisturbed. Addition of small amounts of iron oxide or other positive colloids to a hydrogen bentonite (which is negatively charged) also gives thixotropic gels with marked rheopectic properties. It has been suggested (Freundlich, 1935) that the formation of perfect petrifications of delicate animals such as jelly-fish in the slate of Solnhofen in Bavaria was made possible by the thixotropic nature of this material, which consists of $ca.$ 95 per cent. $CaCO_3$ with 1–2 per cent. of clay. It is probable that the animals were deposited on a shallow beach and covered by a fine dust, blown from neighbouring hills, which became a thixotropic paste on contact with sea-water. This would then solidify and so preserve the form of the encased body. In accord with this theory finely powdered Solnhofen slate, when mixed with sea-water, gives thixotropic pastes in a broad range of concentration between 40 and 60 per cent. of powder. These pastes have plastic properties like putty, and can readily be moulded, retaining their moulded shapes on drying.

Sand is normally dilatant (p. 479), but quicksand is thixotropic, due to the presence of colloidal clays of high ion-exchange capacity.

(d) *Streaming birefringence*

Of the optical properties of clay suspensions, particular mention may be made of the double refraction of dilute sols of bentonite when in flow. Bentonite is particularly suitable owing to the small size of the particles relative to other clays, and their thin, plate-like shape (small particle size is desirable to reduce the rate of sedimentation). Hauser (1945) recommends a 1 per cent. suspension of fractionated bentonite, having particles of equivalent spherical diameter between 150 and 500 A, since sedimentation is effectively absent, the sol is quite clear to the eye and shows pronounced birefringence at very low rates of flow and at any temperature up to the boiling-point of water. For the visual study of flow patterns in engineering problems these sols have proved very useful (see Fig. 25.9, due to Hauser and Dewey, 1942).

Double refraction can also be brought about by orientating the particles by means of electrical fields, and it has been suggested that inhomogeneous electric fields in liquids can thus be studied and measured by optical means.

Applied colloid chemistry of clays

Certain aspects of clay technology, particularly in relation to agriculture and to the ceramics industry, have been mentioned above, and the use of clays (particularly bentonites) as emulsifying agents in Chapter XXIII. The study of clays in relation to soil science is of great importance, not only in agriculture, but in civil engineering, particularly in the construction of roads, reservoirs, and canals. For example, the seepage of water from lakes has been overcome by changing the pervious natural soil into an impervious sodium soil (see p. 717) by flooding with sea-water, and in constructing canals a layer of bentonite is sometimes added to the sides and bottom—this swells on contact with water and so reduces loss by seepage. Clays upon which certain surface active materials such as soaps have been precipitated are, owing to their diminished hydrophilic nature, much less affected by water, a fact used in road engineering for soil-stabilization.

During the drilling of oil-wells, a so-called 'drilling mud', usually a bentonite suspension, is used. The object of the mud is to keep the rock fragments produced in suspension when the drill is at rest, so that sedimentation and caking are avoided. The thixotropic nature of bentonite suspensions makes them very suitable for this purpose, and

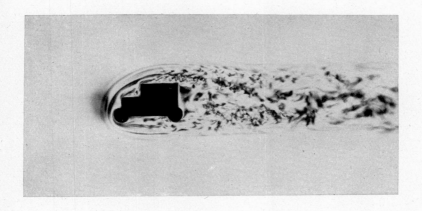

Fig. 25.9. Flow patterns around models of old-fashioned and modern streamlined cars, as indicated by means of a bentonite sol in circularly polarized light.

their low viscosity when sheared enables the liquid containing the rock fragments to be pumped readily out of the bore hole. In addition these muds, particularly if weighted by heavy powders such as barium sulphate, seal off formations containing liquids or gases which might otherwise cause interference with the well when in operation.

An interesting recent development in clay technology has been the production of flexible films by careful spreading and drying of thixotropic gels of montmorillonites, preferably those of high ion-exchange capacity. Microscopic examination shows that the clay particles aline during evaporation and form interweaving threadlike aggregates, giving rise to a micro-fabric structure. By binding the particles together with suitable ions, introduced by a process of ion-exchange, these films have been made resistant to moisture as well as to organic solvents. The structure is now comparable to that of muscovite and might therefore be termed synthetic mica (Hauser, 1945).

Thixotropic bentonite suspensions find many applications in the field of adhesives, since these pastes can easily be applied by brushing, but set immediately after application, thus avoiding flow and uneven distribution. Polymers such as rubber are used in the form of a latex to which the requisite amount of bentonite is added.

Zeolites

The remarkable occlusive properties of the zeolites, coupled with their frequently known molecular structure, make them of great interest in connexion with the general problem of sorption by porous solids.

As previously mentioned, they can be regarded as infinite three-dimensional frameworks as in the various forms of silica, except that some of the Si atoms (often about one-half) are replaced by Al, the *framework* always having the composition $(Si, Al)O_2$. This replacement of Si by Al confers a negative charge on the framework, and equivalent cations are taken up in the interstices (as distinct from the surfaces in the case of clays) to maintain electrical neutrality. In addition to the naturally occurring crystalline minerals there are amorphous gel zeolites and synthetic materials, the permutites, which are widely used in water softening. Permutite (sodium permutite) has the approximate composition $Na_2Al_2Si_2O_8 \cdot 6H_2O$, and is manufactured in various ways, such as by reacting sodium silicate and sodium aluminate, or by fusing together a mixture of sodium carbonate, china clay or alumina, and silica (sand, quartz, etc.) followed by leaching the vitreous mass with water.

Structure of principal types

A complete study of the various kinds of zeolites is lacking, but three main types of zeolite framework are now recognized. These are as follows (for further details see Bragg, 1937; Wells, 1945):

(1) Robust three-dimensional networks, typified by analcite, $Na(AlSi_2O_6).H_2O$, and chabazite, $(Ca, Na_2)Al_2Si_4O_{12}.6H_2O$.

(2) Laminar structures with rather weak bonding between the sheets, such as heulandite, $Ca_2(Al_4Si_{14}O_{36}).12H_2O$.

(3) Chain structures cross-linked by comparatively few bonds, giving the crystals a fibrous rather than a massive nature. Natrolite, $Na_2(Al_2Si_3O_{10}).2H_2O$, is an example of this group.

The water of crystallization which the zeolites possess can be driven off by heating. The extent of lattice shrinkage so caused depends upon the type, being small with type (1), but much larger, often amounting to irreversible collapse, with the laminar and fibrous zeolites. The water molecules are situated in the same interstitial holes and channels as the cations, and in the hydrated crystal they probably have definite sets of neighbours, linking cations to O atoms of the (Si, Al)O_2 framework as shown below (after Wells, 1945).

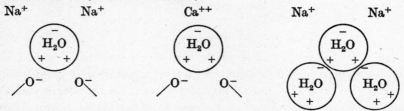

(The negative charge on the O atoms of the framework arises from the substitution of Al for Si.)

Water molecules and cations can diffuse from one lattice site to another, but whereas the former can leave the lattice readily, the latter, owing to the greater energies involved, can only do so by a replacement with an electrochemical equivalent of other cations. The behaviour of water (and other non-ionized compounds which can replace it) thus recalls the behaviour of hydrogen in solution in palladium. The gas-zeolite system is best considered as an interstitial solid solution with the occluded molecule interacting with the adsorbent by van der Waals forces (dispersion, polarization, and ion-dipole forces).

Sorptive properties

In this connexion the non-collapsing zeolites of type (1) are the most important, for after dehydration they not only regain their crystal water quite readily, but may occlude other gases and vapours in its stead. The interstitial volume available to water may be quite considerable, e.g. *ca* 0·2 c.c./gm. in the case of chabazite.

Absorption of non-polar gases (e.g. He, A, N_2, CH_4) is shown by

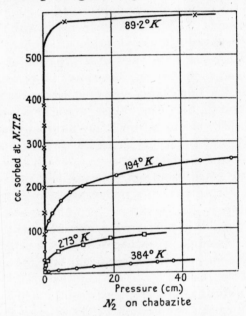

Fig. 25.10. Sorption isotherms of N_2 in chabazite.

chabazite, gmelinite, mordenite, and 'active' analcite, but not all robust network structures have this property, which depends upon the relative dimensions of absorbate and zeolitic channel. Some typical isotherms, for the nitrogen-chabazite system, are given in Fig. 25.10 (from Barrer, 1938). These show the normal decrease in sorption with increasing temperature, and obey an equation of the Langmuir type (p. 22), viz.

$$p = \text{constant} \times \theta/1-\theta,$$

where p is the equilibrium pressure and θ the fraction of all the available interstices occupied by solute molecules. The continuous nature of the usual sorption-desorption isotherm indicates, as shown on p. 10, that binary interstitial solid solution, and not compound formation, is occurring. (See also Campbell and Johnson (1948).)

Certain zeolites, e.g. chabazite, will absorb chain molecules such as the *n*-paraffins, even *n*-heptane, where the molecule must take up a stretched configuration in the interstitial channels. The consequent exclusion of many configurations possible in the gas phase leads to an additional negative entropy of occlusion.

Occlusive powers of zeolites depend very much upon the heating and degassing treatment to which they have been subjected, since these determine the extent of dehydration and of lattice changes. For example, absorption of CO_2 by chabazite increases with the degree of dehydration until *ca.* 95 per cent. dehydration, after which absorption falls off rapidly, probably due to the onset of lattice collapse. Ion-exchange has also been found to modify the absorptive properties of zeolites.

Heats of occlusion have been measured either directly or by the Clapeyron equation (p. 57). The values, even for permanent gases of small molecular size, such as nitrogen, are unusually large, due to the interstitial position of the solute molecule. With equal diameters of solute and interstice the interaction energy, which arises from van der Waals forces only (dispersion and polarization), may be some eight times that for adsorption on to a plane surface of the same material. For the *n*-paraffins in chabazite, the sorption heat is approximately linear with the number of CH_2 groups, the increment per CH_2 being about 3,000 cals. As might be anticipated, the value obtained depends not only upon the amount of gas occluded, but also upon the initial water content of the zeolite.

The rates of diffusion into zeolites ('zeolitic solution') have been measured as well as the equilibrium values discussed above, and for the usual case of simple solution Fick's laws (p. 233) are obeyed. The rate depends very much upon the particular solute–zeolite system, as well as upon the degree of dehydration, grain-size, temperature, and degree of saturation. For zeolites which have a laminar structure (class (2)), diffusion anisotropy occurs; in the case of heulandite, for example, the rate across the (001) plane is much less than across the (201) plane (each normal to the unit lamellae of the layer lattice). In the case of homogeneous zeolites the diffusion coefficient (D) is related to the fraction of sites occupied (θ), by the equation $D = D_0(1-\theta)$, where D_0 is the value of D as $\theta \to 0$.

The temperature dependence frequently observed, with activation energies of several kilocalories (e.g. *ca.* 7,000 calories for $n\text{-}C_5H_{12}$ into chabazite), has been interpreted on the basis of an activated diffusion process, the energy being needed to move the solute molecule from one position of minimum energy in the zeolitic lattice to the next similar position. As would be expected, the activation energy depends upon the nature of the solute, as well as the degree of dehydration.

The molecular sieve properties of the three types of zeolites are summarized in Table 25.3 (from Barrer, 1944).

TABLE 25.3. *Molecules Occluded or Excluded by the Three Classes of Molecular Sieve*

	Typical molecules rapidly occluded at room temp. or below	Typical molecules moderately rapidly or slowly occluded at room temp. or above in the thermal stability range	Typical molecules occluded at negligible rates, or totally excluded within the thermal stability range
Section (i): Class 1 minerals	He Ne A H_2 N_2 CO CO_2 COS, CS_2 H_2O HCl, HBr NO NH_3 $CH_3.OH$ $CH_3.NH_2$ $CH_3.CN$ HCN Cl_2 CH_3Cl, CH_3Br CH_3F CH_2Cl_2, CH_2F_2 CH_4, C_2H_6 C_2H_2 CH_2O H_2S $CH_3.SH$	C_3H_8 and simple higher n-paraffins $C_2H_5.OH$ $C_2H_5.NH_2$ C_2H_5F C_2H_5Cl C_2H_5Br I_2, HI CH_2Br_2 CH_3I $C_2H_5.CN$ $C_2H_5.SH$ $H.CO_2Me$, $H.CO_2Et$ $COMe_2$ $CH_3.CO_2Me$ $NHMe_2$, $NHEt_2$	Branched-chain hydrocarbons. Cyclo-paraffins. Aromatic hydrocarbons. Derivatives of all these hydrocarbons. Heterocyclic molecules (e.g. thiophen, pyridine, pyrrole). $CHCl_3$, CCl_4, $CHCl.CCl_2$, $CH_3.CHCl_2$, $CHCl_2.CCl_3$, C_2Cl_6, and analogous bromo- and iodo-compounds. Secondary straight-chain alcohols, thiols, nitriles, and halides. Primary amines with NH_2 attached to a secondary C atom. Tertiary amines. Branched-chain ethers, thio-ethers, and secondary amines.
Section (ii): Class 2 minerals	He Ne A H_2 O_2 N_2 CO NH_3 H_2O	CH_4, C_2H_6 $CH_3.OH$ $CH_3.NH_2$ $CH_3.CN$ CH_3Cl, CH_3F HCN CS_2 Cl_2	All classes of molecules in cols. 3 and 4 above.
Section (iii): Class 3 minerals	He Ne H_2 O_2 N_2 H_2O	A HCl NH_3	All molecules referred to in col. 4 section (ii). Also: CH_4, C_2H_6, $CH_3.OH$, $CH_3.SH$, $CH_3.CN$, $CH_3.NH_2$, CH_3Cl, CH_3F.

Class 1 is represented by well-outgassed chabazite, gmelinite, active analcite, and a synthetic crystalline zeolitic mineral; class 2 by well-outgassed natural mordenite; and class 3 by well-outgassed calcium and barium mordenites produced hydrothermally at high temperatures. At room temperature class 1 excludes iso-paraffins but occludes n-paraffins with a facility which decreases with chain-length; class 2 excludes paraffins above ethane and larger molecules; class 3 occludes nitrogen, oxygen, and smaller molecules but excludes methane and ethane.

Barrer has shown how these differences in occlusive and exclusive properties permit separation of molecular mixtures. Thus by the use of chabazite the simple n-paraffins may be separated from all iso-paraffins and aromatic hydrocarbons, and the C_1 and C_2 hydrocarbons from propane and higher hydrocarbons by using mordenite. Separation of non-polar from polar organic compounds, or of polar mixtures, is also possible in those cases which possess suitable differences in molecular size and shape, upon which the method primarily depends.

These inorganic lattices with such definite uniform structure and characteristic sieve properties stand in rather striking contrast with the man-made membranes of varying porosity, with which we shall be concerned in the last chapter. The synthesis of structures with pore-sizes as uniform as those found in simple zeolitic minerals, but of larger pore-size (say in the range 10 A–100 A), would represent a valuable addition to ultra-filtration, osmometry, and membrane studies generally.

Ion-exchange

The phenomenon of base or ion-exchange by zeolites cannot properly be included under their adsorptive properties, since the process involved is purely one of exchange. The *total* number of ions in the lattice may vary somewhat, as when one Ca^{++} ion is replaced by two Na^+ ions, but, as previously pointed out, the essential factor is the necessity of maintaining electrical neutrality.

Water softening by the 'permutite' process depends upon the fact that when present in equal concentrations, calcium ions are taken up by the permutite in preference to sodium. Although clays also show ion-exchange, the materials used practically are always zeolites, of synthetic or natural origin, owing to their much greater exchange capacity (400–500 m.eq./100 gm. as compared with up to 100 for the most active bentonites (see Table 25.2)). Equilibrium in the clays is reached almost instantaneously, whereas with permutites many hours may be required, indicating an external surface phenomenon in the first case and capillary permeation in the zeolites, where the exchangeable ions are believed to occupy positions in the crystal lattice itself. Support for this interpretation comes from the fact that a large cation, such as methylene blue, can quantitatively displace all the exchangeable base from a clay, but is not taken up appreciably by a zeolite.

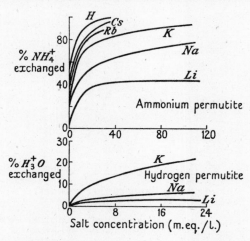

FIG. 25.11. Variation of the degree of ion-exchange with salt concentration for an ammonium-permutite and for a hydrogen permutite.

The relation between the amount of salt added and that taken up by the zeolite (i.e. the amount of cation exchange) is frequently adequately covered by the Freundlich adsorption isotherm (p. 22), although this empirical equation does not predict the saturation which in point of fact must exist. Other isotherms, such as that of Langmuir (p. 22), have also been employed with varying success. Some typical data due to Jenny (1932) are shown in Fig. 25.11, representing the percentage displacement of NH_4^+ and H_3O^+ from ammonium and hydrogen permutites with the equilibrium concentration of various salts.

The replacing power of different cations follows the order Li^+, Na^+, K^+, Rb^+, Cs^+, H_3^+O, the last being the strongest replacer (see Fig. 25.11 and Fig. 25.12). Thus a given concentration of a certain cation will displace more Li^+ from a lithium permutite than Na^+ from an equivalent amount of sodium permutite and so on. These results have been interpreted on the ionic hydration theory, since in the alkali metal series the Li^+ ion is the most hydrated (and therefore the largest) and the Cs^+ is the least hydrated. This explanation was strengthened by the fact that in media such as 80 per cent. alcohol, where the ions would be less hydrated, the relative difference in replacing power became less. In confirmation of the 'sieve theory' it is found that, with the substituted ammonium ions, $\overset{+}{\text{N}}\text{H}_3(\text{CH}_3)$ is much less absorbed than $\overset{+}{\text{N}}\text{H}_4$, and the more fully substituted ions such as $\overset{+}{\text{N}}\text{H}(\text{CH}_3)_3$ and $\overset{+}{\text{N}}(\text{CH}_3)_4$ are effectively excluded. For the alkaline earths, Mg^{++} is the

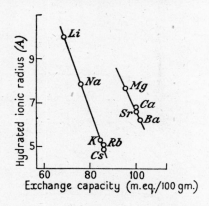

FIG. 25.12. Variation of exchange capacity of a sulphonated coal type cation exchanger with the hydrated size of the cation.

weakest replacer, followed by Ca^{++}, Sr^{++}, and Ba^{++}. Here, however, a given concentration of a definite cation displaces more Ba^{++} than Ca^{++} and more Ca^{++} than Mg^{++}, showing that further factors in addition to size are involved in this replacement reaction. According to Wiegner and Jenny, the differences arise from the ions being held to the permutite by hydroxyl bindings, and since $\text{Mg}(\text{OH})_2$ is less soluble than $\text{Ca}(\text{OH})_2$ the Mg^{++} is assumed to be more tightly bound to the permutite surface than Ca^{++}.

Recently Nachod and Wood (1944, 1945) have measured the kinetics of ion-exchange reactions in a very simple manner merely by shaking small samples of the solid with the salt solution and analysing the solution at suitable intervals. With both natural and synthetic exchangers, the ion-exchange reaction was found to be bimolecular and second order. Their results for the dependence of the equilibrium exchange upon the hydrated cation radius, which were mentioned earlier, are shown in Fig. 25.12.

GENERAL REFERENCES

Clays

References to the more important earlier work will be found in Schofield, *Ann. Rep. Chem. Soc.* 1931, **28**, 351.

Some aspects, chiefly of the later work, are given by Hauser, in *Chem. Rev.*, 1945, **37**, 287; and by Hauser and le Beau, article in J. Alexander's *Colloid Chemistry*, vol. vi, 1946, p. 191, Reinhold.

Zeolites

BARRER, 1944, *Ann. Rep. Chem. Soc.* **41**, 31.

OTHER REFERENCES

BARRER, 1938, *Proc. Roy. Soc.* A **167**, 392, 406.
BRAGG, 1937, *The Atomic Structure of Minerals*, Oxford.
CAMPBELL and JOHNSON, 1948, (in the press).
CLARK, 1940, *Applied X-rays*, McGraw-Hill, 3rd edn.
ENSMINGER and GIESEKING, 1941, *Soil Sci.* **51**, 125.
FORD, LOOMIS, and FIDIAM, 1940, *J. Phys. Chem.* **44**, 1.
FREUNDLICH, 1935, *Thixotropy*, Hermann, Paris.
HAUSER, 1941, *J. Amer. Ceram. Soc.* **24**, 179.
—— and DEWEY, 1942, *J. Phys. Chem.* **46**, 212.
JENNY, 1932, ibid. **36**, 2217.
NACHOD and WOOD, 1944, *J. Amer. Chem. Soc.* **66**, 1380.
—— —— 1945, ibid. **67**, 629.
OAKLEY, 1926, *Nature*, **118**, 661.
RUSSELL, 1937, *Soil Conditions and Plant Growth*, 7th edn., Longmans.
WELLS, 1945, *Structural Inorganic Chemistry*, Oxford.

XXVI
PROTEINS

As yet, the principal colloidal systems have been classified and discussed largely according to their physical type, e.g. suspensions, gels and pastes, foams, emulsions. Within these categories, certain chemical groups of material form colloidal systems of such great importance that it is necessary to consider them separately, and in this and the following chapter we shall therefore discuss the proteins and the polymers respectively.

As already pointed out in Part II, the proteins form ideal test materials as a result of their reproducible character and the frequently monodisperse and corpuscular nature of their molecules. Added to these properties is their obvious biological importance. The decision to devote a chapter to the consideration of the proteins seems therefore justifiable. It is intended in this chapter to consider broadly the results of the application of the newer experimental techniques, with a view to illustrating these techniques as well as to describing the physico-chemical properties of the proteins, about which many detailed works have been concerned (Lloyd and Shore, 1938; Schmidt, 1938; Cohn and Edsall, 1943). Some aspects, e.g. the properties of protein films at air/water and oil/water interfaces, have already been discussed (pp. 635, 655) and will not be further considered: this chapter deals chiefly, therefore, with the physico-chemical properties in solution of proteins which are in a state as near as possible to that in the living system from which they were taken (i.e. 'native' proteins). However, in considering the internal structure of the native protein, it will be necessary to consider also the X-ray studies of solid crystalline and fibrous proteins.

Definition and general features

As first used, the term protein referred generally to the complex nitrogen-containing organic substances occurring in animal and plant tissues. Subsequently, however, it acquired its more precise modern meaning as that group of naturally occurring substances which on hydrolysis give rise to a large proportion of α-amino acids or closely related substances (e.g. imino acids like proline). The protein molecule may be regarded, therefore, as built up from the condensation together of a large number of different α-amino acids, all of which may be written as

$$\begin{array}{c} R \\ H \end{array} C \begin{array}{c} NH_3^+ \\ COO^- \end{array}$$

R, representing the 'side chain', may vary widely in different amino acids, e.g. it may be a non-polar paraffin chain, or it may contain another amino group (as in basic amino acids) or carboxyl group (as in acidic amino acids) as well as sulphur or halogen atoms.

The widely different physical properties of the proteins are accounted for by their different amino-acid compositions and the large number of ways in which these acids may be arranged. In spite of such diversity in chemical composition and structure, features common to the whole protein kingdom exist. Thus:

1. All naturally occurring α-amino acids possess the same spatial arrangement of the groups R—, H—, NH_3^+, and COO^- about the central carbon atom; an arrangement which is opposite to that about the fifth carbon atom of d-glucose, and therefore known as the l form.

2. The principal bond linking the amino acids is the peptide bond, in which the amino group of one amino acid condenses with the carboxyl group of an adjacent amino-acid molecule, liberating one molecule of water:

$NH_3^+.CHR'COO^- + {}^+H_3N.CHR''.COO^- \rightarrow NH_3^+.CHR'CO.NH.CHR''COO^-.$

Repetition of this condensation gives a skeleton structure which, when stretched out, is of the type

$$\begin{array}{c} \text{[peptide chain structure with alternating } R^I, R^{II}, R^{III}, R^{IV}, R^V \text{ side chains]} \end{array}$$

$\longleftarrow$ Fibre axis $\longrightarrow$

in which, owing to the occurrence only of l-amino acids, alternate side chains (or H atoms on the same carbon) are all either above or below the plane of the paper.

Evidence for the predominance of the peptide link in linking the unit amino acids comes from several sources. The small number of free amino and carboxyl groups in proteins, the strong protein biuret reaction, and the equivalence of the acid and basic groups liberated on hydrolysis, indicate the condensation together of amino and carboxyl groups into a bond of the type —NH—CO—, which is known with certainty to occur in simpler naturally occurring compounds. The fact that synthetic polypeptides can be hydrolysed by the very specific naturally occurring enzymes responsible for the breakdown of proteins in the digestive tract further suggests the extensive occurrence of the peptide link in protein molecules.

Classification

Proteins which contain amino acids only (as shown by hydrolysis) are known as 'simple proteins'. It seems likely in such cases, as we have indicated, that the peptide bond is almost entirely responsible for linking up the amino acids forming the polypeptide chain, though the actual configuration of the chain may be determined by other bonds (see p. 747).

Many proteins occur, however, in which, in addition to a large percentage of amino acids, the products of hydrolysis contain a non-protein or prosthetic group or its breakdown products. Such proteins are known as 'conjugated proteins', and the strictly protein portion of the molecule may be associated with a variety of prosthetic groups, e.g. metal containing pigments (as in the respiratory proteins), carbohydrate groups (as in glucoproteins), or nucleic acid (as in nucleoproteins).

A third group is known as the 'derived proteins', since they are usually obtained from the other groups by partial hydrolysis, and are to be regarded as the first stages in the breakdown of proteins. Gelatin derived from the tissue protein, collagen, is a common example.

Each of the above groups is usually subdivided, simple proteins in terms of solubility, and conjugated proteins according to the prosthetic group. Since the solubility of the simple proteins, with which we are largely concerned here, is determined by their amino-acid composition, the classification is also partly chemical. The elements of this classification, based upon the scheme of Lloyd and Shore, are given in Table 26.1. (For details see Lloyd and Shore, 1938; Schmidt, 1938; or biochemical text-books.)

Isolation of proteins

Before attempting to examine the properties of a protein in solution, it is usually necessary to separate it from other extraneous matter alongside which it occurs in the living system. For this purpose it is usual to precipitate the protein by some means and to repeat the precipitation until a specimen of the required purity is obtained. Precipitation is usually accomplished by:

(*a*) addition of soluble salts, e.g. ammonium or sodium sulphate;

(b) removal of salts, e.g. by dialysis against pure water;
(c) change of pH by addition of alkali or acid.

TABLE 26.1. *Classification of Simple Proteins*

Class	Origin	Solubility	Characteristic property
Albumins	Animal and vegetable	Soluble in water and dilute salts, acids, alkalis.	Insoluble in strong salt solution.
Globulins	,,	Soluble in dilute salts, acids, and alkalis.	Insoluble in water or strong salt solutions.
Gliadins (or prolamines)	Cereal seeds	Soluble in dilute acids or alkalis.	Soluble in 70–90% alcohol.
Glutelins	,,	Soluble in dilute alkali.	Forms dough with water.
Protamines	Reproductive cells	Soluble in water and dilute acid—precipitated by ammonia.	Strongly basic.
Histones	,,	Soluble in water, salt, acid, or alkaline solutions.	Basic.
Keratins	Animal	Insoluble in aqueous solutions.	Resistant to enzyme action.
Connective-tissue proteins	,,	,,	,,

Since we are mainly interested in the properties of the native protein, no drastic procedure (e.g. the use of reducing or oxidizing agents) which would involve an irreversible change of the proteins is permissible, and it is essential that the procedures suggested above should be shown not to involve alteration of the protein components. In some systems it is by no means certain that this is true of recommended procedures. Lloyd and Shore (1938, p. 144) have discussed clearly the possible changes which may occur on precipitation, e.g. aggregation or disaggregation of proteins and protein complexes, transformation of a soluble protein into a different insoluble one, precipitation of more than one protein owing to overlapping of precipitation zones or adsorption. Sorensen's (1930) view, that proteins exist in solution as dissociable complexes, containing possibly several components, that the equilibrium state in such a system is dependent upon circumstances (e.g. composition of liquid), and that the components precipitated are not necessarily to be identified with those which occur in solution, is clearly another expression of these views. It is important, therefore, to consider briefly the evidence available on these questions.

If the proteins in solution exist as labile aggregates of smaller stable units A, B, C, etc., and if alteration of the liquid medium affects the equilibrium state, then it might be expected that isolated proteins should have a chemical composition, and possess physical properties varying with the exact method of isolation. On the other hand, constancy of chemical composition and physical properties would indicate the existence of a well-defined chemical individual, and that Sorensen's views are not universally valid.

The isolation in recent years of several proteins (e.g. lactoglobulin, chymotrypsinogen) of almost constant chemical composition and solubility, as well as of uniform molecular size and charge (see Kekwick and McFarlane, 1943) would seem to demonstrate that these individual proteins, at least, cannot be regarded as labile aggregates, but are distinct chemical entities. The universal validity of the equilibrium theory would thus seem to be disproved.

In many other cases, however, it is proved experimentally that a change in the solvent or in the conditions does result in definite changes in the protein species in solution (e.g. see Johnson, 1946). Thus, mere dilution of aqueous solutions of carboxy-haemoglobin appears to cause dissociation of the molecule and, on the other hand, association has been observed on adding ammonium sulphate to a solution of the globulins of the ground-nut when one protein species disappears to give another of probably twice the molecular weight. Dissociation is observed in the latter system on dilution. In the case of thyroglobulin, Lundgren (1936) has shown that increased temperature and high dielectric constant as well as low protein and salt concentration favour dissociation of the molecule, which under appropriate conditions may be reformed. Svedberg (1937) has also reported the effects of changes in protein and salt concentration, and of traces of added amino acids upon the state of aggregation of proteins. It appears, therefore, that reversible protein dissociation-association systems are of fairly general occurrence,† and that changes in the state of aggregation in a system may occur as a result of the usually-considered mild treatments involved in isolating proteins. In general, proteins of higher molecular weight seem to be the more prone to dissociation into sub-units, and some structural feature, yet unknown, may be involved here (see p. 736).

Clearly the experimental procedures of protein isolation cannot be regarded as without effect on the nature of the protein molecule, and in seeking information about the native protein, they require careful checking. Especially is it desirable, where feasible, to compare the

† Recently, Ogston and Johnston (1946) have suggested an alternative explanation of certain association-dissociation phenomena, but it is unlikely that this is applicable to the bulk of such observations.

properties of isolated proteins with those within their natural environment.

THE PHYSICO-CHEMICAL PROPERTIES OF PROTEINS IN SOLUTION

Table 11.1, due to Pedersen (1940), contains the results of sedimentation, diffusion, electrophoretic and other relevant measurements, as well as the calculated molecular weights and frictional ratios for several of the better-known proteins. Partial specific volumes are seen to vary between the relatively narrow limits of 0·70–0·75, a result, as we have seen, of the amino-acid composition of the proteins. Values outside these limits do, however, occur, especially in the case of the conjugated proteins. A good example is the 'X' protein of serum thought to contain an appreciable proportion of lipoid, which, being of higher partial specific volume than pure protein, gives rise to a value for the protein complex of 0·96 (Pedersen, 1945).

Molecular weights

The molecular weights M_s and M_e, obtained from sedimentation-diffusion and sedimentation equilibrium measurements respectively, show a substantial measure of agreement, a fact giving encouraging confirmation of the basic theory of both procedures. It should be noted also that these values compare favourably with those obtained by osmotic, X-ray, and light-scattering methods, a selection of which, suitably rounded, is contained in Table 26.2.

TABLE 26.2. *A Comparison of the Differently Derived Molecular Weights of Certain Proteins*

Protein	Method				
	Osmotic pressure	Sedimentation equilibrium	Sedimentation-diffusion	X-rays	Light-scattering
Pepsin	36,000	39,000	35,500	..	..
β-Lactoglobulin	..	38,000	41,500	(40,000)	..
Egg albumin	45,000	40,500	44,000	..	38,000
Haemoglobin (horse)	67,000	68,000	68,000	66,700	..
Serum albumin (horse)	72,000	68,000	70,000	(82,800)	74,000
Serum globulin (horse)	175,000	150,000	167,000	..	..
Excelsin	..	..	295,000	306,000	280,000
Amandin	206,000	330,000	330,000	..	Standard

N.B. Values in brackets not corrected for residual water.

Although discrepancies undoubtedly occur in Table 26.2 and further work is required, in general, where the different methods have been carried out under similar conditions of pH, salt concentration, temperature, etc., agreement as good as that of Table 26.2 is observed. A conspicuous example of discrepancies arising from the different conditions of measurement is the case of the haemocyanins (the respiratory protein of many lower animals), which are notoriously prone to dissociate to different extents according to the conditions.

A surprising result, emerging first from the ultracentrifugal work of Svedberg and his collaborators, was the molecular homogeneity of many protein systems. Thus, such systems contain, if carefully prepared, one or, at the most, a limited number of different molecular species, in striking contrast to the polydispersity of many other polymers whether of natural (e.g. carbohydrates, rubber, etc.) or synthetic (e.g. polystyrene, vinyl polymers, etc.) origin. This conclusion was obtained chiefly from the fact that the boundary spreading observed in the ultracentrifugal examination of protein solutions could be correlated with separately measured diffusion coefficients. The presence of molecular species with properties very similar to, but not entirely identical with, those of the mean is not entirely excluded, however, on this evidence, and indeed it has been suggested (Bresler and Talmud, 1944) that a monodisperse protein really contains a distribution of molecular weight with a very sharp maximum. However, the high degree of order existing in protein crystals and the immunological properties of proteins (see p. 752) strongly oppose this view and support that of a monodisperse protein containing molecules identical in every way.

The Svedberg multiple law and the Bergmann-Niemann hypothesis

After several ultracentrifugal determinations of molecular weight, Svedberg noted that the latter appeared to fall around certain well-defined values, which were approximate multiples of 17,600. Under the heading M_{calc}, Svedberg's values for the different multiples are given in Table 11.1, and from a glance at the table it would appear that the measured molecular weights do indeed group themselves about these values. Such a conclusion, if substantiated, must arise from some feature basic to protein structure, and considerable interest in it has therefore been aroused. However, it is clear from a closer inspection that several experimental values deviate from the multiples of 17,600, by an amount considerably greater than experimental error. The device (Fig. 26.1) by Bull (1941) of plotting a 'logarithmic spectrum' of molecular weights shows clearly the frequency of such deviations, and, as this author states, 'all numbers, if they are large enough, are approximate multiples of 17,600'.

A statistical examination of molecular weight data (Johnston and others, 1945) also provided no support for the hypothesis. The 'multiple law' can only be regarded, therefore, as a rough working rule for the lower molecular weight proteins.

A modified form of the law was, however, involved in the Bergmann-Niemann (1936, 1938) hypothesis, which suggests that the total number

FIG. 26.1. A logarithmic 'Spectrum' of the molecular weights of proteins. The line of numbers beginning with 1 and ending with 384 is the supposed number of the 17,600 units in the proteins, i.e., the molecular weight classes. The lower line of numbers is an arithmetic scale of logarithms, while the top line indicates the position of the logarithm of the molecular weights taken from *The Ultracentrifuge* by T. Svedberg and I. O. Pedersen, Oxford, 1940.

of amino-acid residues in protein molecules (rather than the protein molecular weight) is given by one of a series of simply related numbers, expressible in the form $2^m \times 3^n$, where m and n are integers (not equal to zero). Further, the number of residues of any one type is given by a number, $2^{m'} \times 3^{n'}$, m' and n' being integers and possibly zero. Table 26.3 illustrates the hypothesis in the case of silk fibroin (Bergmann and Niemann, 1938), and it has been applied to several other well-known proteins with apparent success (e.g. see Chibnall's (1942) analytical results for edestin).

TABLE 26.3. *The Amino-acid Content of Silk Fibroin*

(From Bergmann and Niemann, 1938.)

Amino acid	Weight, %	Molecular weight	Gm. molecule per 100 gm. protein		Relative nos. of gm. molecules	Fraction of total residues (frequency)
			Exptl.	Calcd.		
	(1)	(2)	(3)	(4)	(5)	(6)
Glycine	48.3	75	0.584_0	0.584_0	1296	2
Alanine	26.4	89	0.296_6	0.292_0	648	4
Tyrosine	13.2	181	0.072_9	0.073_0	162	16
Arginine	0.95	174	0.005_5	0.005_4	12	216
Lysine	0.25	146	0.001_7	0.001_8	4	648
Histidine	0.07	155	0.0004_5	0.0004_5	1	2592

Minimum total number of residues per molecule = $2592 = 2^5 \times 3^4$.

In view of variations from protein to protein in the mean molecular weight of the constituent amino acids, protein molecular weights for a given value of $2^m.3^n$ are expected to vary considerably, in agreement with observation. The hypothesis therefore overcomes a major difficulty associated with the multiple law.

Bergmann and Niemann postulated that the numerical relationships arose from the regular arrangement of the amino acids. Thus if glycine (G) and alanine (A) respectively represent a half and a quarter of the total residues then they considered that the arrangements of these residues must be of the type

$$-G-X-G-X-G-X-G-X-$$
and
$$-A-X_3-A-X_3-A-X_3-A-,$$

where X represents any residues other than glycine and alanine respectively.

Such structural regularity does not necessarily follow from the Bergmann-Niemann numerical formulae. though numerical consequences of some type must arise from a regular structure. Ogston (1943, 1945) has, however, pointed out that the numbers 2 and 3 hold no unique position in a repeating structure of the type we have considered, though they may arise from certain three-dimensional structures.

However, Neuberger (1939) has stated that, as a result of the large error in amino-acid determinations, there is a very high probability that a random distribution of amino acids should appear to fit the frequency formula. Further, in the light of more accurate analytical data (e.g. Chibnall, 1942) several well characterized proteins (e.g. β-lactoglobulin, haemoglobin, insulin) do not fit the formula. Chibnall pointed out that one reason might be that the molecule of such proteins contains several different polypeptide chains, and though each chain might comply with the formula, the whole molecule would not be expected to do so. It would seem, therefore, that the Bergmann-Niemann hypothesis, like its predecessor the multiple law, is to be regarded with considerable doubt.

The shape of protein molecules

Owing to uncertainty in the contribution of hydration to the frictional ratio (see p. 214) and to the complexity of the problem of determining molecular shape, our information on this subject is much less reliable than that on molecular weights.

Assuming the hydration to be 0·30 gm. of water per gram of dry protein, and treating the molecules as rigid ellipsoids of revolution, the values of the axial ratio, shown in Table 11.2, were obtained as the most likely. Accepting these values, we see that though no protein can be regarded as truly spherical, many are only moderately asymmetrical ($a/b < 5$), and few have pronounced asymmetry ($a/b > 15$–20). It follows also that, in the main, all dimensions of the molecules are comparable amongst themselves, but are considerably greater than those of solvent molecules. One of the conditions required for the validity of Stokes law (p. 210) is thus obeyed. This result is to be contrasted with the case of some extended linear polymer molecules discussed in the next chapter. It should, however, be realized that the most asymmetric proteins, e.g. the keratins, in which the polypeptide chain is nearly fully extended, are usually insoluble and are therefore not represented in the table.

In general, a choice between the flattened and elongated forms of the ellipsoid of rotation is not possible, though in isolated cases one form may appear more likely. Thus, from measurements of the rotational diffusion coefficient, Foster and Edsall (see p. 402) deduced that the zein molecule is well represented only by an elongated ellipsoid of axial ratio *ca.* 20. Dervichian (1943), on the other hand, favours the oblate ellipsoidal form, and has calculated that, on this view, the thickness of the simple proteins is constant at about 15 A. Further confirmation is, however, required before this can be generally accepted.

Owing largely, no doubt, to the uncertainties involved in axial ratio determinations, the values obtained by the different methods are not in quantitative agreement. Oncley (1941) has given a very illuminating diagrammatic comparison of the different methods, which is reproduced in Fig. 26.2. In this figure it is clear that different combinations of hydration and axial ratio are in accordance with experimental results, the shaded areas corresponding with the uncertainty arising from the appropriate experimental error. Clearly, however, different methods do not cover the same areas, and the most likely combinations of hydration and asymmetry are given by those regions of the diagrams on which the greatest number of methods is represented. Thus, in the case of serum albumin, the most likely values of the hydration and axial ratio would appear to be *ca.* 0·2 gm./gm. and 4 respectively. However, an unambiguous set of values is clearly not always obtained from the diagrams. The agreement demonstrated by the different methods is, however, encouraging.

A valuable general conclusion arises from the fair agreement between methods involving rotational diffusion and other methods. Before such agreement was known it was conceivable that rotation in a velocity gradient might involve distortion of the protein molecule, and that the application of an alternating electrical field might cause the local rotation of small polar portions of the whole molecule about flexible

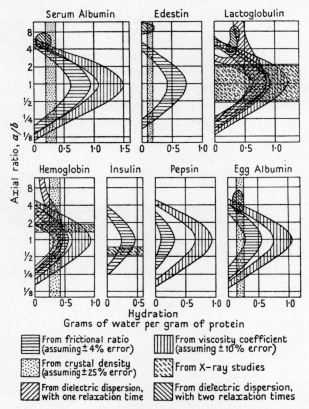

Fig. 26.2. Asymmetry and hydration of certain protein molecules. (Oncley (1941).)

links rather than of the entire molecule. As indicated in Chapter XIV, it would appear that the molecules of the corpuscular proteins are rigid structures, whose overall dimensions are not appreciably affected by velocity gradients or electrical fields. Some such effects might, however, occur with the more elongated molecules and in interpreting new results this should be remembered.

One further point arising from the by no means negligible asymmetry of the proteins should be added. The dynamic properties (sedimentation constant, diffusion coefficient, etc.) are expected to be, to some extent, concentration dependent, and, especially in the case of the more asymmetric molecules, it is necessary to extrapolate to zero concentration when obtaining data for use in precise molecular weight and shape determinations (cf. p. 289).

Applications of the electrophoretic method

The ultracentrifuge, we have seen, provides a test of the homogeneity of protein preparations with respect to sedimentation constant, itself dependent chiefly upon molecular weight and dimensions. An additional test of homogeneity is provided by electrophoresis, in which the criterion is electrophoretic mobility, involving, chiefly, electrical charge and molecular dimensions. The two tests are thus by no means identical, and should in fact be considered complementary, for it is quite conceivable that proteins of the same molecular weight should have different electrical and chemical properties (especially if there is any truth in the multiple law) and vice versa. A test of monodispersity should therefore involve, at least, electrophoretic and ultracentrifugal examination.

Fig. 26.3. Ultracentrifugal analysis of normal human serum. Phosphate buffer at pH 8 and $I = 0.1$. $\Delta n = 0.00300$. Interval between exposures, 10 mins. Field strength, 270,000 g. (1) Cell index, (2) globulin, (3) albumin, (4) meniscus. (Kekwick (1939).)

Blood serum provides a very clear demonstration of the different nature of the two methods. In addition to the electrophoretically different components shown in Fig. 12.15 (b) (albumin, α-, β-, and γ-globulins) further subdivision has been reported. Thus the existence of two α-globulins, α_1 and α_2 has been suggested, and both β-globulin and albumin have shown the presence of two components (see review by Kekwick and McFarlane, 1943). However, the subdivision indicated in Fig. 12.15 (b) is a very convenient one and is still often used. On the other hand, sedimentation velocity examination (Fig. 26.3) under similar conditions shows the presence of only two major components.

From these diagrams it is clear that at least one of the components (probably the slower) separating out in the ultracentrifuge must contain electrophoretically different fractions. An obvious requirement, therefore, is the careful isolation by the preparative technique (p. 326) of the electrophoretically different components and their examination in the ultracentrifuge (the reverse procedure of isolating ultracentrifugally different fractions is not usually practicable). Although a large amount of work has been thus directed (e.g. Cohn, 1941, and previous papers;

Pedersen, 1945) it is not yet possible to give the complete correspondence of ultracentrifuge and electrophoretic diagrams.

As might be anticipated, the existence, within an apparently homogeneous electrophoretic peak, of molecules of different sedimentation constants is also possible. Thus Svedberg (1937 b) reports that the products of dissociation of the haemocyanin from Helix Pomatia do not separate in electrophoresis, though with molecular weights 6,740,000, 3,370,000 and 842,000 their sedimentation constants are quite different. A more recent case has been observed in the ground-nut globulins. Under certain conditions (Johnson, 1946) two well-defined species sediment in the ultracentrifuge at quite different rates. Under identical conditions, however, these species separate in electro-phoresis only with considerable difficulty, if at all.

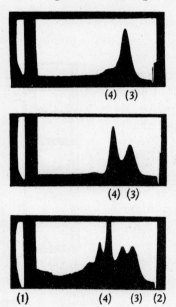

FIG. 26.4(a). Sedimentation diagrams of normal and pathological sera in phosphate buffer at pH 8 and $I = 0.1$. $\Delta n = 0.00300$.

Sedimentation diagrams. Field strength: 270,000 g. Photographs taken 45 mins. after reaching full speed. (1) Cell index, (2) meniscus, (3) albumin, (4) globulin. *Top*: Normal human serum. *Centre*: Myelomatosis serum 3. *Bottom*: Myelomatosis serum 5.
(Kekwick (1940).)

As shown in Chapter IX, the area of any one peak on a schlieren diagram is a measure of the increase in refractive index caused by that component and, for known refractive index increments, of its concentration. Electrophoretic and sedimentation diagrams therefore provide a convenient quantitative measure of the composition of a protein system. In the case of the serum and plasma proteins, for subjects in normal health, the diagrams show only minor variations, but with various pathological conditions of the subject, characteristic changes have been observed, in which the area relationships are chiefly affected, though mobilities may also be involved (see Figs. 26.4 (a) and (b)). Usually such changes take the form of an increase in one (or more) of the globulin fractions relative to the albumin, the particular fraction concerned and the extent of its change varying with the pathological condition (see review by Abramson *et al.* (1942), p. 186). So much has been done on these lines in recent years that the examination of sera now forms an additional method of diagnosis.

Closely related to the question of pathological sera is the investigation of immune sera and the location of antibody activity. Although antigens (i.e. substances whose introduction into the animal organism stimulate

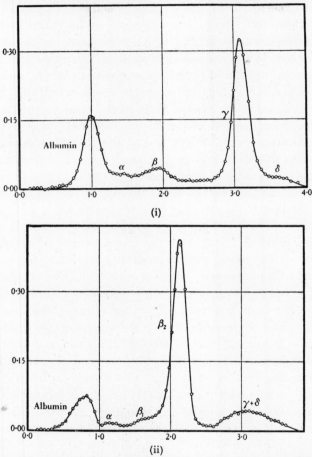

FIG. 26.4 (b). Electrophoresis diagrams of normal and pathological sera in phosphate buffer at pH 8 and $I = 0.1$. $\Delta n = 0.00300$.

Electrophoresis diagrams. Anode limbs. Potential gradient: 5 V./cm. Photographs taken 120 mins. after starting current. Abscissa: distance in U-tube in cm. Ordinate: scale line displacement in mm. (i) Myelomatosis serum 3; (ii) Myelomatosis serum 5.

(Kekwick (1940).)

the production of protective antibodies, e.g. diphtheria toxin which stimulates the formation of diphtheria anti-toxin) are not necessarily protein in nature, it appears that the antibodies produced in solution are universally so. Tiselius and co-workers (1937, 1939), especially, have examined immune sera (containing antibodies) from several animals, and have shown that the antibody frequently migrates electrophoretically with the γ-globulin fraction (e.g. in pig, rabbit, and monkey). In other cases (e.g. horse) an electrophoretically new component may appear, whose removal was shown to occur by reaction with antigen. Ultracentrifugal examination of these antibodies has also shown, in some cases, the presence of new rapidly sedimenting species. During the recent war, alongside the extensive work on the preparations required for blood transfusions, such antibody investigations have been considerably extended and several components of serum, antibody and otherwise, have been prepared as stable powders suitable for clinical use (e.g. see Cohn et al. 1944). It is likely that, in the near future, such serum fractionation, followed at different stages by electrophoretic and other examination, will be much further developed, and there is every reason to expect that the products obtained will be of considerable importance in medicine.

Much attention has been given to the serum protein system since it has been studied so thoroughly electrophoretically. Electrophoretic techniques have, in addition, been applied to a multitude of protein systems, e.g. bacterial, virus, and plant, etc., to give information on the composition of the whole system as well as on the properties of the components. Comparison of electrophoretic behaviour alone and in mixtures has begun to provide information about interactions between different proteins, and between proteins and other substances, which will undoubtedly throw light upon the role of proteins in the living organism.

One of the important uses of electrophoresis is as a control in the isolation and purification of proteins by salt precipitation and crystallization methods. In the isolation of the serum proteins it has been shown by Cohn and co-workers (1940) that the precipitation ranges of the different fractions overlap considerably, and that monodisperse preparations can, in general, only be obtained if detailed knowledge of the stages in purification, such as is provided by electrophoresis, is available. Protein preparations, previously considered monodisperse after numerous re-crystallizations, have thus been shown more complex. For instance, crystalline egg albumin was shown to possess two components (Longsworth and MacInnes, 1940), and serum albumin, several times re-crystallized, contains appreciable quantities of globulin-like material (Johnson and Shooter, 1946). Serious doubts as to the reliability of crystallinity as a criterion of monodispersity therefore exist, a conclusion strongly supported by Pirie (1940).

Accurate determination of pH-mobility curves for well-defined protein preparations (e.g. egg albumin, serum albumin, β-lactoglobulin) have now been made by both the ultramicroscope and moving boundary methods. Using the mobility determinations for the proteins in the dissolved state, and making assumptions (see p. 332) regarding the size and shape of the molecules, absolute values of the net charge as a function of pH have been calculated and shown to be in not unreasonable agreement with values from titration curves and membrane potential measurements. Further support for the relatively asymmetric nature of the egg and serum albumin molecules has (as already stated on p. 334) been provided by this work.

As to the location of the electrical charge on protein molecules, some indication is given by the dipole moment obtained from dielectric dispersion methods (see p. 422), and associated with a given value of the net molecular charge (i.e. at a given pH and ionic strength). Given approximate dimensions for the molecule it is thus possible to say that for certain molecules (serum albumin, haemoglobin, and edestin) the charged groups are fairly symmetrically distributed over the whole volume of the molecule.

The Structure of the Protein Molecule

As yet we have been concerned with the external properties of size and shape of protein molecules, but it is necessary now to consider their internal structure with a view to explaining these and other characteristic properties.

Certain important features have already emerged. Thus, even disregarding the multiple law for protein molecular weights and the Bergmann-Niemann hypothesis, the general occurrence of reversible dissociation reactions, especially of the larger protein molecules (e.g. haemocyanins, plant globulins), would seem to indicate the existence of sub-units within the protein molecule. The mode of linkage of the units within the parent molecule is not definitely known, but the ease of the dissociation in many cases indicates a weak type of bond. Of these, hydrogen bonds, particularly $>C=O\ldots H-N<$, and salt linkages such as $-COO^-\ {}^+H_3N-$ are probably involved. Pedersen has suggested that certain types of non-protein cementing substances (e.g. carbohydrates and nucleic acids) may be responsible, but other less specific forms of linkage seem more probable in general (see p. 747). In other cases dissociation may not occur so readily nor be reversible; in this category must be placed those dissociation reactions occurring as a result of rather large pH changes, irradiation (Svedberg and Brohult, 1939), and high concentrations of urea. The effect of urea, which has been particularly studied, can be seen from Table 26.4, selected from the work of Burk (1938, and earlier papers).

TABLE 26.4

Protein	Molecular weight (normal)	Molecular weight in iso-electric urea solution (6·66M)
Egg Albumin	45,000	36,000
Haemoglobin (horse)	67,000	34,300
Serum Albumin	72,000	73,800
Serum Globulin	175,000	173,000
Amandin	206,000	30,300
Edestin	310,000	49,500

Whilst molecular weights obtained in urea solutions probably do not have the quantitative significance of the most recent measurements in aqueous solution, it is clear that there are two classes of protein, in one of which members are considerably affected in molecular weight by the presence of urea (haemoglobin, amandin, edestin), whilst in the other they are little affected, if at all (egg albumin, serum albumin, and serum globulin).

It would appear, in general, that dissociation reactions, however caused, are more probable the higher the protein molecular weight; the absence of dissociation with the smaller molecules is readily explained if such molecules constitute the basic units. However, this must be tested experimentally. In the absence of precise information concerning the extent and nature of dissociation reactions these ideas regarding substructure can only be treated as likely though not completely proven.

Leaving the approach to protein structure from considerations of molecular size and shape in solution, we now consider it from the very different viewpoint of X-ray methods. Under suitable conditions, protein crystals yield sharp X-ray diagrams of considerable complexity (Chapter XV), and, in the case of certain of the better-known proteins (see Table 15.4), these have yielded the unit-cell dimensions. Although interesting conclusions have also been reached regarding molecular weight and shape as well as the existence of the 'Svedberg molecule' in the crystal, the complexity of even the simplest protein molecule has yet defeated any attempt at a complete structure determination involving the exact location of every atom within the molecule (as accomplished, for example, in the phthalocyanines, p. 450). Further, it seems unlikely that a full unambiguous structural determination is possible by the present methods. It would rather seem that the limits of the present Fourier methods are being approached, and that far-reaching developments in technique and interpretation will be essential for complete solutions.

However, the progress made in structure determinations of fibrous proteins has an important bearing on corpuscular proteins, since, as pointed out above, there is a strong case for supposing that the predominant link in both types is the same (the keto-imino or polypeptide link) (see p. 731). Further, Astbury, Dickinson, and Bailey (1935) have shown that X-ray photographs of stretched denatured films and threads, formed from typical corpuscular proteins (edestin, egg-white), are analogous to those of stretched animal hairs (i.e. β-keratin), and they consider that protein denaturation (see p. 753) 'involves the liberation or generation of peptide chains which aggregate on coagulation into parallel bundles like those found in the structure of β-keratin, and similar fibres'. In the case of excelsin crystals, an intermediate form was detected which gave, on one and the same photograph, indications of both normal crystalline and fibrous behaviour. The increased asymmetry arising on denaturing by urea (Neurath and Saum, 1939) provides evidence in favour of this view of denaturation. It thus appears that in corpuscular proteins the polypeptide chain occurs as such, but in order to explain their corpuscular and highly organized nature and the predominant monodispersity of the molecules, it is necessary that the chain should be folded in some very unique and stable manner. As to the bonds involved and the nature of this folding, nothing is known with certainty, though speculation has by no means been lacking!

Several links probably contribute in stabilizing the folding, amongst which, one of the strongest is probably the covalent disulphide link of cystine ($[SCH_2.CH(NH_2).COOH]_2$), which may be imagined to be formed by the oxidation of adjacent sulphydryl groups:

748   PROTEINS

Such links almost certainly occur in hair and wool, both of which contain large amounts of cystine.

Electrostatic links ('salt links'), which may form between basic and acidic side chains as a result of the attraction of NH_3^+ and COO^- or any other pairs of oppositely charged groups, would also have considerable stability.

A much weaker bond which probably occurs to some extent in all proteins is the hydrogen bond joining two electro-negative atoms in a bond with an energy varying between 5,000 and 8,000 cals. The commonest form of the bond is usually considered to be that occurring between —NH— and —CO— groups in neighbouring polypeptide chains, as below:

$$\begin{array}{c}
\vdots\quad\vdots\quad\vdots\\
O\quad H\quad O\\
\diagdown C \diagup \diagdown N \diagup \diagdown C \diagup \\
\diagup N \diagdown \diagup CHR \diagdown C \diagup \diagdown N \diagdown \diagup CHR \diagdown \diagup N \diagdown \\
H\quad O\quad H\\
\vdots\quad\vdots\quad\vdots\\
O\quad H\quad O\\
\diagdown N \diagup \diagdown C \diagup \diagdown N \diagup \diagdown C \diagup \\
\diagdown CHR \diagdown \diagup N \diagdown \diagup C \diagdown \diagup CHR \diagdown \diagup N \diagdown \diagup C \\
O\quad H\\
\vdots\quad\vdots
\end{array}$$

Although individually weak, an assembly of such bonds may, it is thought, give considerable stability to a molecule (see Pauling and Niemann, 1939).

Other bonds may occur, but definite experimental evidence is lacking.

Particular prominence was given to the 'cyclol' theory of protein structure developed by Wrinch (1937), in which the amino acids were supposed to form a system of closed hexagons in three dimensions, the basic unit being of the type:

$$\begin{array}{c}
-CO\quad\quad N-\\
\diagdown\quad\diagup\\
N-\!\!-C(OH)\\
\diagup\quad\quad\diagdown\\
CHR\quad\quad CHR\\
\diagdown\quad\diagup\\
CO-NH.
\end{array}$$

This unit was originally postulated by Astbury for the α-keratin structure. The cyclol theory has, however, been severely criticized, especially by Pauling and Niemann (1939), on chemical, energetic, and steric grounds. Other workers have confirmed that the cyclol theory does not allow the packing of the side chains, and Huggins (1940) has suggested alternative forms of folding for the polypeptide chain which avoid this difficulty. However, a direct experimental test of any such theory on the basis of X-ray measurements is, as yet, quite impossible.

The suggestions by Astbury (1941, 1944) for the folding of the polypeptide chain in corpuscular proteins, based upon his conclusions regarding the folding of the chain in α-keratin, deserve close attention for reasons already mentioned. We have seen that the repeat distance in a α-keratin, corresponding to three or more amino-acid residues, is 5·1 A, and to explain it Astbury suggests the following folded form (Fig. 26.5 (a)) of the chain, in which alternate side chains point either all up or all down, a result of the universally l-configuration of the amino acids and the nature of the folding. Fig. 26.5 (b) gives this folded chain diagrammatically, emphasizing the triad close-packed grouping of the side chains which is a prominent feature. This form of folding is also satisfactory, in that, in agreement with experiment, it makes the α (i.e. unstretched) form about half as long as the extended β form, the densities of the two forms are practically the same, and there is no steric interference between the side chains, a common source of trouble in earlier theories. For the first time, the side chains receive detailed consideration in the proposed structure, and Astbury considers that their close-packing on both sides of the folded chain is probably also important in corpuscular proteins. It is clear, however, that such close-packing is dependent on the nature of the side chains, stable close-packing being impossible if the side chains in a triad mutually repel one another. Thus they must be all polar or all non-polar. Alternate amino acids in the polypeptide chain must therefore have all polar or all non-polar side chains, a severe restriction on the amino-acid composition and arrangement. Astbury has, however, provided evidence that the amino-acid composition of several proteins does not violate this restriction.

In corpuscular proteins the polypeptide chains are supposed folded into parallel laminae by means of loops of the type:

$$\begin{pmatrix} \cdots \diagup \text{CO} \diagdown \diagup \text{CHR} \diagdown \diagup \text{NH} \diagdown \diagup \text{CO} \diagdown \diagup \text{CHR} \diagdown \cdots \\ \text{NH} \diagup \text{CO} \diagup \text{CHR} \diagup \text{NH} \diagup \\ \cdots \diagdown \text{NH} \diagup \diagdown \text{CHR} \diagup \diagdown \text{CO} \diagup \diagdown \text{NH} \diagup \diagdown \text{CHR} \diagup \cdots \\ \text{CO} \diagdown \text{CHR} \diagdown \text{NH} \diagdown \text{CO} \diagdown \\ \cdots \diagup \phantom{\text{CO}} \diagdown \text{NH} \diagup \diagdown \phantom{\text{CHR}} \diagdown \text{CO} \diagup \diagdown \text{CHR} \diagup \end{pmatrix}$$

again with all polar side chains on one side of the laminae and non-polar on the other. Adjacent laminae, packed together with polar side to

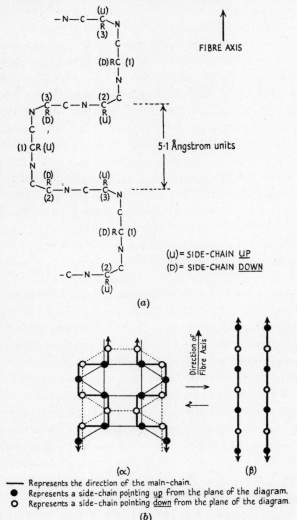

FIG. 26.5. (a) Proposed basis of intramolecular fold in α-keratin and α-myosin. (b) Close-packing of side-chains in the α-fold of keratin and myosin.

polar side and non-polar to non-polar, form stable aggregates owing to interactions of various types, such as those already discussed (p. 747). In aqueous solution, low interfacial energy will result if polar groups only interact with the solvent and if the non-polar side chains are hidden between laminae.

Such a picture, first suggested by Astbury (1936) for egg albumin, has several advantages. It explains the approximately corpuscular nature of corpuscular proteins, the similarity in density of all forms of protein, the easy transition from corpuscular to fibrous forms, and the magnitude of the area of a protein like egg albumin on spreading as a monolayer. Further, it gives a ready explanation of the reversible dissociation of proteins (as the separation of the polar faces of adjacent laminae) and of molecular weight groups. The suggestions of Perutz (see p. 456) for the haemoglobin molecule, based upon a detailed Patterson and Fourier analysis of X-ray measurements, are in striking agreement with Astbury's view, for which there is undoubtedly much support.

Viruses

Before concluding this short review of X-ray structure determinations of the proteins some mention of the work involving plant viruses, mainly tobacco mosaic virus, must be made (Bernal and Fankuchen, 1941). Solutions of tobacco mosaic virus of medium concentration (e.g. 5 to 10 per cent.) spontaneously separate into two layers on standing, the top layer, an isotropic liquid containing 1·6 to 4 per cent. of the virus, and the bottom layer, a spontaneously birefringent sol or gel according to the virus concentration (4 to 34 per cent.). Orientation of the long, rod-like virus particles in the top dilute layer can be accomplished in a velocity gradient (e.g. see p. 400), but the parallel orientation of the particles in the bottom layer occurs spontaneously and is responsible for the birefringence observed. Bottom layer gels can be made to swell or contract according to salt content and pH.

X-ray methods have shown that in such gels two types of spacing, intermolecular and intramolecular, occur. The former varies from 152 A for air-dried gels to as much as 600 A in dilute birefringent sols, the orientation of the particles in a plane perpendicular to their length being highly regular and probably maintained by their ionic atmospheres. It would appear from X-ray data that the particles are approximately 150 A in diameter and not less than 1,500 A in length.

Inside the particles there is also a high degree of order, as indicated by the intramolecular spacings, which, unlike intermolecular spacing, are not affected by concentration changes. Some of the reflections observed do not obey Bragg's law and are to be interpreted on the theory of gratings of limited size. It would appear that cubic sub-units of 11 A side occur in a hexagonal (or pseudo-hexagonal) lattice with dimensions $a = 87$ A, $c = 68$ A. This structure, which is unaffected by drying, shows certain resemblances to both fibrous and crystalline proteins, but is probably simpler than the latter.

Not all virus particles are, however, elongated, those of tomato bushy-stunt virus, which are spherical and of diameter *ca.* 276 A, being a notable exception. The virus crystal contains the particles in body-centred cubic close packing.

It is necessary now to consider certain of the general properties of the proteins in the light of the suggestions put forward for protein structure.

Important General Properties of Proteins

One very common and almost universal feature of the reactions involving native proteins is their specificity. Thus a considerable number of proteins crystallize in a characteristic form which, with other physical properties, differs even amongst similar proteins from members of closely related animal species. For example, the haemoglobins of different species differ considerably in crystallizability, solubility, affinity for oxygen and carbon monoxide, and absorption spectrum.

In the enzymic properties of certain proteins a marked specificity is also shown. Thus, especially amongst the hydrolysing enzymes, a given enzyme will usually attack only one type of bond, which must also occur in a definite environment. Thus pepsin, for instance, hydrolyses the first of the peptides shown below at the position indicated, but a similar bond in a slightly different environment in the second peptide is not affected:

$$\text{carbobenzoxy—1-glutamyl—1-tyrosine}$$

$$\text{1-glutamyl—1-tyrosine.}$$

Many similar cases are known (for details see Bergmann and Fruton, 1941).

In their immunological properties also an even more striking specificity is shown. Thus the antibody formed by the organism in response to the introduction of a given antigen is specific to this antigen. It seems likely that antibodies usually consist of one or other of the serum globulins modified in some way by the influence of the antigen so that further antigen can specifically combine with it (cf. Ehrlich 'lock and key' picture). The protective power of an antibody arises from its capacity to react with the specific antigen to form an antigen-antibody complex which, though it may or may not precipitate, effectively protects the organism from the effects of free antigen. Antibody formed against a given protein will usually give such a complex only with the original antigen, although proteins from closely different species may cross-react to some extent. Thus anti-serum against hen egg albumin will react with duck egg albumin, demonstrating the close similarity of the two proteins, but it can be shown quantitatively that the two antigens do react differently.

Such general specificity in reactions and properties is undoubtedly a result of protein structure, and it strongly supports the view that a given protein molecule is unique in structure as well as in composition. Thus the amino-acid residues must occur in a perfectly definite sequence in the polypeptide chain which is folded and held in a manner which does not vary from one molecule to another. As we have seen, the X-ray examination of single crystals supports this view, and it would appear that the view of Bresler and Talmud (see p. 736) that a monodisperse protein really contains a distribution of molecular weights with a sharp maximum, is unlikely.

Since the specificity of a protein arises, in our view, from its unique composition and structure, it is to be expected that any departure from this 'native' state will tend to destroy the specific properties. Thus, agencies causing the breaking of the stabilizing side-chain linkages (hydrogen bonds and others) and a freeing of the polypeptide chain, would give rise to alternative configurations of the molecule, and remove the basic cause of specificity. It is very likely that the much-invoked phenomenon of 'denaturation' involves such a molecular change. The common agents of denaturation, acids, alkalis, alcohol, urea, salicylate, heat, ultra-violet light, adsorption at interfaces, etc., are all such as would be expected to modify hydrogen bonds or other weak stabilizing links. The multitude of changes covered by the word 'denaturation' are those to be expected from a freeing of the main polypeptide chain, possibly followed later by its random folding. Thus loss of crystallizability, of enzymatic and immunological properties, as well as modification of physical properties so that species differences disappear, all occur on denaturing a protein. The considerable decrease in solubility associated with denaturation is also readily explained by the freeing of the polypeptide chain leading to the exposure of the non-polar groups.

Change in the reactivity of the various side-chain groups on denaturation also finds a ready explanation in the partial unfolding of the polypeptide chain. Most proteins contain a finite proportion of basic and acidic amino acids, which can be estimated either by analysing the completely hydrolysed protein, or by titration methods applied to the intact molecule. The good agreement between the two methods demonstrates that basic and acidic groups are normally available in the native molecule. Not so, however, the disulphide, sulphydryl, and phenolic groups; of these, a small fraction only (if any) is available in native proteins, whilst on denaturation (which, as we have seen, ultimately involves the complete liberation of the polypeptide chain) the remainder become detectable (see Mirsky and Pauling, 1936). Accepting the view of a native protein molecule as a pile of laminae with two exposed faces containing polar side chains, it is clear that within the pile many more polar groups must tend to be hidden especially from reagents of considerable molecular size. The availability of acidic and basic groups is probably to be attributed to the small size and mobility of hydroxyl and hydrogen ions.

Quantitative measurements of the temperature coefficient of denaturation by heat have allowed a calculation of the entropy change, and it was shown, for example, that denatured trypsin has an entropy greater by 100 E.U. than the native form (Mirsky and Pauling, 1936). This is interpreted, in agreement with the picture of protein structure developed above, as being due to the much larger number of configurations available in the denatured state, which results from the destruction of the stabilizing side-chain links (about 30 hydrogen bonds, according to Mirsky and Pauling).

Summarizing, it seems that the layer structure of proteins as suggested by Astbury, or some modification of this, is capable of explaining the chief properties of proteins and in particular their physico-chemical properties, their specificity, and their tendency under various influences to undergo denaturation. Much, however, remains to be done, and the complete structure determination of even the simplest protein molecule still seems a very long way off.

REFERENCES

ABRAMSON, MOYER, and GORIN, 1942, *Electrophoresis of Proteins*, Reinhold.
ASTBURY, 1936, *Nature*, London, **137**, 803.
—— 1941, *Chem. and Industr.* **60**, 491.
—— 1944, in *Colloid Chemistry*, vol. v, ed. by J. Alexander, Reinhold, p. 529.
—— DICKINSON, and BAILEY, 1935, *Bio-chem. J.* **29**, 2351.
BERGMANN, 1938, *Chem. Rev.* **22**, 423.
—— and NIEMANN, 1937, *J. biol. Chem.* **118**, 301.
—— —— 1938, ibid. **122**, 577.
—— and FRUTON, 1941, in *Advances in Enzymology*, vol. i, Interscience, p. 63.
BERNAL and FANKUCHEN, 1941, *J. gen. Physiol.* **25**, 111, 147.

BRESLER and TALMUD, 1944, *C.R. Acad. Sci. U.R.S.S.* **43**, 310, 349.
BULL, 1941, in *Advances in Enzymology*, vol. i, Interscience, p. 1.
BURK, 1938, *J. biol. Chem.* **124**, 49.
CHIBNALL, 1942, *Proc. Roy. Soc.* B **131**, 136.
COHN and co-workers, 1940, *J. Amer. Chem. Soc.* **62**, 3386.
—— 1941, *Chem. Rev.* **28**, 395.
—— and EDSALL, 1943, *Proteins, Amino Acids, and Peptides*, Reinhold.
—— et alia, 1944, *J. clin. Invest.* **23**, 417, and the following papers.
DERVICHIAN, 1943, *J. chem. Phys.* **11**, 236.
HUGGINS, 1940, ibid. **8**, 598.
JOHNSON, 1946, *Trans. Faraday Soc.* **42**, 28.
—— and SHOOTER, 1946, unpublished results.
JOHNSTON and others, 1945, *Trans. Faraday Soc.* **41**, 588.
KEKWICK, 1939, *Biochem. J.* **33**, 1122.
—— 1940, ibid. **34**, 1248.
—— and MCFARLANE, 1943, *Ann. Rev. Biochem.* **12**, 93.
LLOYD and SHORE, 1938, *Chemistry of the Proteins*, 2nd edn., Churchill.
LONGSWORTH and MACINNES, 1940, *J. Amer. Chem. Soc.* **62**, 705.
LONGSWORTH, 1942, *Chem. Rev.* **30**, 323.
LUNDGREN, 1936, *Nature*, London, **138**, 122.
MIRSKY and PAULING, 1936, *Proc. nat. Acad. Sci. Wash.* **22**, 439.
NEUBERGER, 1939, *Proc. Roy. Soc.* A **170**, 64.
NEURATH and SAUM, 1938, *J. biol. Chem.* **128**, 347.
OGSTON, 1943, *Trans. Faraday Soc.* **39**, 151.
—— 1945, ibid. **41**, 670.
—— and JOHNSTON, 1946, ibid. **42**, 789.
ONCLEY, 1941, *Ann. N.Y. Acad. Sci.* **41**, 121.
—— 1943, in *Proteins, Amino Acids, and Peptides*, by Cohn and Edsall, Reinhold.
PAULING and NIEMANN, 1939, *J. Amer. Chem. Soc.* **61**, 1860.
PEDERSEN, 1940, in *The Ultracentrifuge*, by Svedberg and Pedersen, Oxford University Press.
—— 1939, *Proc. Roy. Soc.* A **170**, 59.
—— 1945, *Ultracentrifugal Studies on Serum and Serum Fractions*, Uppsala.
PIRIE, 1940, *Biol. Rev.* **15**, 377.
SCHMIDT, 1938, *The Chemistry of the Amino Acids and Proteins*, Baillière, Tindall, & Cox.
SORENSEN, 1930, *C.R. Lab. Carlsberg*, **18**, No. 5.
SVEDBERG, 1937 *a*, *Nature*, London, **139**, 1051.
—— 1937 *b*, *Chem. Rev.* **20**, 81.
—— and BROHULT, 1939, *Nature*, London, **143**, 938.
TISELIUS, 1937, *Bio-chem. J.* **31**, 1464.
—— and KABAT, 1939, *J. exp. Med.* **69**, 119.
WRINCH, 1937, *Proc. Roy. Soc.* A **161**, 505.

XXVII
POLYMERS

Introduction

The last twenty years have seen a rapid increase in the use of polymeric materials both of natural and synthetic origin. Parallel developments in research have gone far towards elucidating polymer structure and properties as well as the reaction mechanisms involved in synthesis. In this chapter dealing with the more important aspects of polymer structure and properties, it is necessary, in view of the extent of the polymer field, to concentrate on the various types of organic polymers (excluding proteins). Amongst inorganic polymers, the silicates are discussed in Chapter XXV; for other polymers, and for more detailed aspects of organic polymers, the account by Meyer (1942) is recommended.

High polymers may be defined as compounds in which each molecule contains a large number of similar (but not necessarily identical) units, linked by primary valencies. The large molecular weight and dimensions resulting from this constitution are responsible for the colloidal properties shown. The units of structure may be single atoms or, more commonly, well-defined groups of atoms, and since these are strongly linked in the polymer molecule, the concept of polymer molecular weight is quite unambiguous. Simple polymers containing only identical units of structure are well known (e.g. rubber, cellulose), but in recent years co-polymers (e.g. buna rubber) containing usually two or more related units have, by reason of their controllable properties, become very important industrially.

For a simple molecule to give rise to a polymer it is essential that it should be at least bi-functional, i.e. be capable of reacting with at least two similar molecules which in turn may react further, i.e.

$$nM \rightarrow \text{---M---M---M---M---M---}.$$

An α-amino acid, the repeating unit in a polypeptide chain, which contains a reactive amino and carboxyl group is clearly bifunctional, and it can undergo polycondensation to give a polypeptide (see p. 731). The polymer resulting from a bifunctional 'monomer' is known as a linear chain polymer, since all the units are united in a single chain of valencies. Monomers which are capable of further reaction (i.e. tri-, tetra-, and polyfunctional) can clearly build up more complex branched structures in two or three dimensions. Such structures are in fact known and will be mentioned here.

Two reactions leading to the production of a high polymer may be distinguished: (1) Polymerization, in which no change in chemical composition occurs, e.g.

$$(CH_2=CH_2)_n \rightarrow -CH_2-CH_2-CH_2-CH_2-;$$

(2) Condensation, in which two monomer molecules react with the elimination of water or other simple substance as in the formation of a polypeptide (p. 731). Being chiefly concerned with the properties rather than the origin of a polymer, we shall not be further concerned with such possible differences in synthesis.

The physical properties of materials, in general, usually depend upon the size and shape of the constituent molecules, but in the case of polymers, certain properties (e.g. density, refractive index, dielectric constant) become essentially independent of molecular weight when the end groups form a negligible fraction of the total. Other properties (e.g. osmotic pressure, solubility, and many properties in solution), however, remain closely dependent upon molecular weight, and an attempt to explain this dependence will be made in this chapter.

Principal Types of Polymers

(a) Saturated hydrocarbons

Polythene

Probably the simplest of all polymers is the saturated straight chain hydrocarbon $-CH_2-CH_2-CH_2-CH_2-$, known commonly as polythene. Formed by the polymerization of ethylene at about 200° C. and 1,000 atmospheres pressure, the polymer, of molecular weight usually *ca.* 5,000–50,000, exists at room temperature as a tough opaque white solid, insoluble in organic solvents. There is a rapid increase in solubility with rise in temperature at about 70° C. As a result of its non-polar constitution, polythene possesses almost unrivalled dielectric properties, and the successful developments of many branches of radar during World War II were in no small measure due to its use. It melts at *ca.* 115° C., a temperature lower than that obtained by extrapolation from the melting-points of the lower, more crystalline paraffins, the discrepancy probably being attributable to the more amorphous nature of the high polymer. The melted polymer varies from a viscous liquid for the lower molecular material to a rubber-like solid for the higher molecular weight types.

Having such a simple composition polythene has been the subject of an unusually detailed X-ray examination. Bunn (1939; see also review, 1946) examined cold-drawn threads and rolled sheets, and found that, in the main, the polymer consisted of a planar zigzag chain of CH_2 groups as in the lower crystalline paraffins, with a repeat distance of 2·53 A:

$$\diagdown CH_2 \diagup ^{CH_2} \diagdown CH_2 \diagup ^{CH_2} \diagdown CH_2 \diagup ^{CH_2} \diagdown .$$

←— 2·53 A →

Electron density diagrams obtained from a Fourier synthesis indicated that the electron clouds of the CH_2 groups were not spherical (Fig. 27.1), a result which cannot wholly be attributed to crystal distortions and thermal vibrations which occur preferentially perpendicular to the chain length. Real distortion of the electron clouds may possibly, according to Bunn (1946), occur. Such an effect, if real, must be a serious complication in interpreting X-ray results for polymers of more complex constitution.

FIG. 27.1. Section through the polyethylene molecule showing contours of equal electron density. (Bunn (1939).)

Infra-red absorption measurements have indicated the presence, to a small extent, of methyl and carbonyl groups (p. 440), and further work is required to clarify both the origin of these groups and their effect on the polythene structure.

Many of the derivatives of ethylene form straight-chain polymers of interest, and, as might be anticipated, the different substituents have important effects on the properties of the final polymer. Such polymers will now be considered.

Poly-isobutene (and poly-iso-olefines)

Isobutene, $CH_2\!\!=\!\!C(CH_3)_2$, and other iso-olefines of the type

$$CH_2\!\!=\!\!CXY$$

(monomers of the type $CHX\!\!=\!\!CHY$ do not usually polymerize) polymerize in the presence of boron fluoride or aluminium chloride at low temperature (-60 to $-80°$ C.) to form highly extensible rubber-like products (Oppanol and Vistanex) of the structural formula:

$$\diagup ^{CH_2} \diagdown _{C} \diagup ^{CH_2} \diagdown _{C} \diagup ^{CH_2} \diagdown _{C} \diagup ^{CH_2} \diagdown .$$
$$\quad CH_3 \; CH_3 \; CH_3 \; CH_3 \; CH_3 \; CH_3$$

On stretching, poly-isobutene gives one of the sharpest and most detailed of all fibre patterns with an identity period of 18·5 A, a distance smaller than eight times the repeat distance (2·5 A) for an unsubstituted paraffin chain. It has, therefore, been suggested that instead of a planar paraffin chain, each $C(CH_3)_2$ group is rotated 45° with respect to the preceding one, leading to sufficient shortening of the chain to accommodate eight structural units in the observed repeating distance. The details of the structure do not, however, appear to be definitely settled.

Polyvinyl polymers

The polyvinyl polymers are formed by the polymerization of the vinyl derivatives, $CH_2{=}CHX$, where X may be any of the groups Cl, Br, I; —$OCOCH_3$, —COOH, CN, OCH_3; —C_6H_5, or other aromatic residue, and even a grouping involving further double bonds.

Vinyl chloride, $CH_2{=}CHCl$, polymerizes rapidly in the light to give a product of the structural formula:

$$-CH_2-CHCl-CH_2-CHCl-CH_2-CHCl-$$

in which the structural units (—CH_2—CHCl—) are probably joined 'head to tail'. At room temperature *polyvinyl chloride* is a hard solid, with limited solubility in non-polar liquids, but in the plasticized condition (i.e. swollen slightly by some involatile organic solvent, e.g. tri-cresyl phosphate) it has moderate flexibility. Partially orientated fibres can be obtained by stretching, but the X-ray diffraction spots are diffuse, indicating very imperfect crystallinity. The repeat distance is *ca.* 5 A, suggesting a planar zigzag chain with alternate chlorine atoms in corresponding positions.

Polyvinyl acetate, $[-CH_2-CH(O.CO.CH_3)-]_n$, is formed by polymerization of the monomer under the influence of light or heat, either alone or in solution, the molecular weight and general character of the final product depending upon the conditions of the polymerization. The very high molecular weight material is not easily soluble, but lower fractions swell and may dissolve in benzene and chloroform. At room temperature a solid, almost transparent and glassy, the polymer softens at *ca.* 80° C. to give a very extensible rubber-like solid. Only a slight sharpening of the very diffuse rings characteristic of the unstretched material occurs on stretching the softened polymer, suggesting an ill-defined repeat distance of 7 A. Although this indicates some parallel alinement of the molecular chains, it is shown by a study of the elastic force-temperature curve (see p. 98) that no crystallization has occurred. In this respect polyvinyl acetate is to be contrasted with *polyvinyl alcohol*, $[-CH_2-CHOH-]_n$, the solid formed by hydrolysis of the acetate polymer, which, after softening by warming, may be crystallized by stretching. The presence of —OH groups would appear to be responsible for its insolubility in benzene, and its solubility in water, the latter being unusual in decreasing with increasing temperature.

X-ray diagrams of the stretched polyvinyl alcohol give a repeat distance of 2·57 A, not significantly different from that of simple paraffins: it therefore follows that the carbon chain is a plane zigzag and that OH groups occupy corresponding positions on alternate carbon atoms. Details of the structure are not yet available.

Polystyrene, $-[CH_2-CH(C_6H_5)-]_n$, formed by the polymerization under the influence of light and heat of styrene (vinyl benzene) has become important commercially as a result of its excellent dielectric properties. It is a transparent, glass-like solid, swollen and sometimes dissolved by non-polar solvents; at about 100° C. it softens to an extensible rubber-like solid, which on stretching gives (like polyvinyl acetate) only feeble signs of orientation. It has been suggested that this absence of crystallizability may be due to a mixture of 'head-to-tail' and 'head-to-head' polymerization leading to an irregular structure of the type

$$-CHR-CH_2-CHR-CH_2-CH_2-CHR-,$$

but the question has not been settled. A repeat distance of 10 A is indicated for the polymer. Solutions of polystyrene under the influence of velocity gradients become feebly birefringent and Signer and co-workers (see p. 405) provided evidence that distortion of the linear molecules can occur at suitably high gradients.

Acrylic acid, $CH_2=CH.COOH$, *methacrylic acid*, $CH_2=C{\overset{CH_3}{\underset{COOH}{\diagdown}}}$

and their esters polymerize readily to give high molecular weight products, some of which are of considerable commercial importance. By varying the type of ester chosen as monomer and the conditions of polymerization, colourless transparent products with a wide range of properties can be obtained. *Polymethyl methacrylate*,

$$[-CH_2-C(CH_3)(CO.OCH_3)-]_n,$$

a hard glass-like solid, has found widespread use as the non-splintering window material 'perspex'. X-ray diffraction indicates no appreciable orientation on stretching, but Coumoulos (1943), from electron diffraction measurements (see p. 462) on multilayers of polyvinyl acetate, polyacrylates, and methacrylates suggests that the main chain is a planar zigzag with the side chains approximately normal to it and alternately right and left.

Many other vinyl polymers are known, e.g. polyvinyl ethers from monomers of the type $CH_2{=}CH(OR)$, polymeric vinylamines from $CH_2{=}CHN(R_1 R_2)$, and polymers of the type of polyindene from indene and other aromatic substituted indenes. Details may be obtained from Meyer (1942) or Ellis (1935).

Amongst the vinyl compounds there are many known examples of co-polymerization. Thus vinyl chloride and vinyl acetate co-polymerize to give an important product of structural formula:

$$-CH_2-CHCl-CH_2-CH(OCOCH_3)-CH_2-CHCl-.$$

Similar co-polymers are formed between vinyl chloride and the acrylic esters. Not all pairs of the different monomers co-polymerize, however, as is shown by the pair styrene and vinyl acetate, which polymerizes to give only polystyrene. Further examples of co-polymerization will be mentioned later. In general, since co-polymer molecules are of irregular structure, crystallization does not occur, and they therefore soften at temperatures lower than polymers from the individual monomers. Other properties of the mixed polymer result from the types and degree of occurrence of the different groups. Careful choice of the types and quantities of monomers used thus gives a useful means of controlling the physical properties of the final product.

Silicone Polymers

Formed by the hydrolysis and condensation of the methyl-chlorosilanes, saturated silicone polymers (or methylpolysiloxanes) varying from the linear to the three-dimensional type are obtained according to the chloride content of the original material. Perhaps the best known is the linear polymer, formed by the hydrolysis of dimethyldichlorosilane $[(CH_3)_2SiCl_2]$, with the structural formula

$$CH_3-\underset{\underset{CH_3}{|}}{\overset{\overset{CH_3}{|}}{Si}}-O-\left[\underset{\underset{CH_3}{|}}{\overset{\overset{CH_3}{|}}{Si}}-O\right]_x-\underset{\underset{CH_3}{|}}{\overset{\overset{CH_3}{|}}{Si}}-$$

each Si atom being linked to the next by an oxygen atom and to two methyl groups. Clearly cyclization of such chains may occur, and the occurrence of two- or three-dimensional structures is to be expected if the starting material contains more chlorine.

Chemically stable and inert substances varying from liquids to elastic solids are known, all of which have an unusually low temperature coefficient of viscosity; their value as lubricants and sealing compounds is therefore outstanding. Osmotic pressure measurements have indicated molecular weights varying from *ca.* 300,000 for plastic material to 3×10^6 for the more elastic varieties (Scott, 1946). Thin films of silicones confer strong hydrophobic properties, and considerable further development and utilization of these polymers is to be anticipated.

(b) Unsaturated hydrocarbons

Rubber and gutta-percha, two well-known members of this group of polymers, are different non-interconvertible forms of polyisoprene, whose formation from single isoprene molecules may be written

$$n[CH_2{=}C(CH_3){-}CH{=}CH_2] \rightarrow -CH_2-C(CH_3){=}CH-CH_2-CH_2-C(CH_3){=}CH-CH_2-.$$

It is generally believed that rubber is the cis-form, i.e.

$$-CH_2\underset{CH_3}{\overset{}{\diagdown}}C{=}C\underset{H}{\overset{}{\diagup}}CH_2-CH_2\diagup\underset{}{\overset{CH_3}{C{=}C}}\diagdown\underset{}{\overset{H}{}}CH_2-CH_2\underset{CH_3}{\overset{}{\diagdown}}C{=}C\underset{H}{\overset{}{\diagup}}CH_2-$$

$\longleftarrow$ repeat distance $\longrightarrow$

and gutta-percha, of which modifications (α and β) occur, the trans-form, of the polymer, i.e.

$$CH_2\diagup\underset{CH_3}{\overset{H}{C{=}C}}\diagdown CH_2 \diagup CH_2 \diagdown\underset{CH_3}{\overset{H}{C{=}C}}\diagdown CH_2 \diagup$$

$\leftarrow$ repeat distance $\rightarrow$

Despite their chemical similarity, the two polymers have quite different physical properties.

Rubber

Rubber occurs naturally as an aqueous dispersion (latex) containing proteins, fats, and soaps (see p. 664), which is obtained commercially from the plant Hevea Brasiliensis. After separation and purification from other components (e.g. see Ward, 1945), the crude rubber is usually milled between hot rolls in order to reduce the chain length and to cause enough structural breakdown to allow of ease in working it.

Rubber is not usually completely soluble in hydrocarbon solvents, the soluble portion (*sol-rubber*) and insoluble portion (*gel-rubber*) probably differing only in the presence of a relatively small number of cross-linkages (see below) leading to a weak three-dimensional network in the latter. Molecular weights of crêpe rubber, as given by osmotic methods, are of the order of 200,000 to 400,000, degraded products giving smaller values.

The chief property of rubber, upon which its great importance rests, is its great reversible extensibility. Undoubtedly this arises from its peculiar molecular structure; the information obtained from X-ray studies is therefore of special importance. X-ray examination of the unstretched material shows it to be a typically amorphous liquid-like body, but, on stretching, the X-ray pattern changes strikingly, well-defined 'reflections', characteristic of considerable crystallinity, appearing, from which the repeat distance of 8·1 A is obtained. For planar isoprene units in the cis configuration the repeat distance for two isoprene units is 9·13 A (Bunn, 1946), so that the rubber molecules are evidently non-planar in the crystals. On this basis Bunn has suggested structures which are compatible with the positions and intensities of the observed X-ray reflections; the structure is, however, by no means settled, and space forbids a treatment of the different possibilities here.

Estimates of the crystallite size in stretched rubber have been made from the sharpness of the diffraction spots. Hengstenberg and Mark (1928) thus obtained: length along axis $>$ 600 A; width and breadth, 500 and 150 A respectively. This order of magnitude has been confirmed by further work (e.g. Gehman and Field, 1944). It is of interest that the crystallite size appears to be roughly independent of the degree of stretching, of temperature variations, and of molecular weight; the number of crystallites, however, increases with the amount of stretching, with a decrease in temperature, and with increasing molecular weight, and is dependent upon the extent of cross linking or vulcanization.

We have already seen that 'gel' rubber owes its comparative insolubility to the presence of a small number of cross-linkages, which, being readily broken (e.g. by addition of butyl alcohol), are probably hydrogen bonds or dipolar interactions arising, for example, from small numbers of $C=O$, $C-OH$, and $COOH$ groups introduced through oxidation (which cannot usually be completely avoided). However, the presence of a small proportion of more permanent cross-links in rubber is essential for most practical purposes, for without them, on stretching, the molecular chains tend to slide over one another, giving irreversible extension. In the process of vulcanization the main chemical reaction is believed to be the formation of sulphur bridges between neighbouring molecular chains according to a scheme of the type

$$\begin{array}{c}\diagdown\!\!\diagup{\rm CH}_3\quad {\rm H}_3{\rm C}\diagdown\!\!\diagup\\ {\rm C}\qquad\quad {\rm C}\\ \|\quad +\quad \|\quad +{\rm S}_2{\rm Cl}_2\rightarrow\\ {\rm C}\qquad\quad {\rm C}\\ \diagup\,\diagdown{\rm H}\quad {\rm H}\diagup\,\diagdown\end{array}\quad\begin{array}{c}\diagdown\!\!\diagup{\rm CH}_3\quad {\rm H}_3{\rm C}\diagdown\!\!\diagup\\ {\rm C}\!-\!{\rm Cl}\qquad {\rm Cl}\!-\!{\rm C}\\ |\qquad\qquad\quad |\quad +{\rm S.}\\ {\rm C}\!\!-\!\!-\!\!{\rm S}\!\!-\!\!-\!\!{\rm C}\\ \diagup\,\diagdown{\rm H}\quad {\rm H}\diagup\,\diagdown\end{array}$$

It is likely that reactions, other than that depicted, occur simultaneously (see Farmer, 1946; Sheppard and Sutherland, 1945); other methods of vulcanization are also employed. The resulting changes in physical properties are readily explicable in terms of the introduction of cross-linkages.

On this basis vulcanization would be expected to lead to decreased plastic flow, solubility, and swelling, and, in the early stages, to an increased crystallinity, as is observed experimentally. It is of interest that if unvulcanized rubber is induced to crystallize (by stretching, and lowering the temperature) it is similar in certain properties to vulcanized material, since the crystallization introduces inter-chain linkages similar to those of vulcanized rubber. Foreign particles, particularly carbon, also strengthen the structure in some way and add to the effect of cross-linkages. Ordinary rubber tyres, for example, contain *ca.* 50 per cent. of rubber and up to 50 per cent. of carbon black.

By rolling vulcanized rubber between hot rollers the effects of vulcanization may be removed, since the three-dimensional structure is broken down into an aggregate of small branched fragments of the original chains, which can show plastic flow.

Gutta-percha

As we have indicated, gutta-percha is a naturally occurring trans-isomer of polyisoprene. At room temperature it is solid, and only slightly extensible reversibly. It is completely soluble in benzene, chloroform, and other organic solvents, indicating its freedom from cross-linkages (cf. rubber). Osmotic pressure measurements in solution have indicated a molecular weight *ca.* 30,000, a value considerably smaller than that for rubber.

On heating above 60–70° C. the solid becomes amorphous and rubber-like though possessing plastic properties as a result of its lack of cross-linkages and comparatively short chain-length. Rapid cooling of the rubbery material leads to crystallization of the β form of gutta-percha, with an X-ray repeat distance of 4·7 A, whilst slow cooling yields the α-form with its longer repeat distance of 8·8 A. From their conditions of formation and their melting-points (for the α-form 65° C. and for the β-form 56° C.) it would appear that the α-form is the stable modification at room temperature (this form also occurs naturally); the β-form is, however, permanently metastable at room temperature.

X-ray examination of cold drawn fibres and rolled sheet of β-gutta-percha has shown that the unit cell is orthorhombic and contains 4 molecules. There is some doubt about the arrangement of the molecular chains in the lattice, alternative arrangements having been proposed to give the established repeat distance of 4·7 A. Less is known of α-gutta-percha, for which the repeat distance of 8·8 A would suggest two non-planar isoprene units. As might be expected from its chemical constitution, gutta-percha can be vulcanized like rubber, with similar effects.

Interesting derivatives of rubber and gutta-percha, possessing modified properties, are known. Thus catalytic hydrogenation of rubber or gutta-percha at low temperatures gives the saturated polymer, *hydrorubber*, which though reversibly extensible (like rubber) does not crystallize on stretching. Addition compounds of rubber and the halogens also occur, which are solid at room temperature, softening only on heating. *Rubber hydrochloride*, the addition compound of rubber and hydrochloric acid, melts at 115° C. and, unlike other such compounds, crystallizes on stretching. An X-ray examination and a structure have been reported by Bunn and Garner (1942).

Synthetic unsaturated polymers

Many attempts to synthesize rubber, using, as monomer, isoprene and similar di-olefines, have been made, with only partial success. For, although high molecular weight polymers resembling rubber in constitution and properties have been made, no polymer containing only polyisoprene in the cis-configuration has yet been made. We can mention only the more important synthetic rubbers.

Butadiene polymers

Butadiene, $CH_2=CH-CH=CH_2$, forms the starting-point of several high molecular weight commercial polymers and co-polymers. In the presence of metallic sodium, butadiene polymerizes to give a product with solubility and swelling properties similar to those of rubber, which is moderately extensible though without crystallizing. On vulcanizing with sulphur it very much resembles lightly vulcanized rubber, and it is thought that the unvulcanized product contains branched chains and therefore a more three-dimensional structure than natural rubber.

Emulsion polymerization of butadiene is said to yield an unbranched polymer (Meyer, 1942, p. 182) which on stretching gives a fibre diagram.

More important are the co-polymers of butadiene, especially with styrene (buna—S, or GR—S) and acrylic nitrile (perbunan), both of which are less soluble and harder than rubber. No crystallization occurs on stretching. In certain respects (e.g. resistance, to wear and to hydrocarbons) such co-polymers are superior to rubber, and, though more expensive, can compete economically.

Polychloroprene (neoprene)

Chloroprene, $CH_2\!=\!C.Cl\!-\!CH\!=\!CH_2$ (cf. isoprene), may be polymerized in emulsion to give tough, rubber-like, high molecular weight polychloroprene or neoprene. On stretching, an X-ray fibre diagram resembling that of β-gutta-percha, with a repeat distance of 4·8 A, is obtained, indicating the trans-configuration of the polymer. In molecular structure, polychloroprene may be regarded as β-gutta-percha with the methyl group replaced by chlorine, though in properties (e.g. solubility and elasticity) it closely resembles rubber. It is, however, harder than rubber at room temperature, softening only at 70–80° C., and is more resistant to oils. Possessing unsaturated double bonds, cross-linkages ('vulcanization') can be performed, not with sulphur, but usually with metallic oxides according to the equation

$$\!>\!C\!-\!Cl\;\;Mgo\;\;Cl\!-\!C\!<\;\;\rightarrow\;\;\!>\!C\!-\!O\!-\!C\!<\;\;+Mg\,Cl_2,$$

anticipated changes in properties occurring.

(c) Carbohydrates

Of the many naturally occurring carbohydrate polymers we can only deal here with cellulose (and its derivatives), starch, and glycogen—more important members of the group (for others see Meyer, 1942).

Cellulose and its derivatives

Cellulose occurs widely as the chief structural material of the plant and vegetable world. In the plant it is not usually found in the pure state but commonly occurs alongside or combined with other plant constituents, e.g. lignin, hemicellulose, pectic matter, waxy materials, and mineral constituents (see Marsh and Wood, 1942). In this section we are concerned with a somewhat idealized material freed from combination with other substances. Cotton, containing approximately 90 per cent. of pure cellulose, is the chief source of such material.

The occurrence of cellobiose as a major decomposition product of cellulose led Haworth and co-workers (see e.g. Haworth, 1939) to

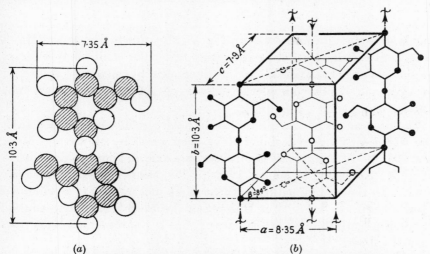

FIG. 27.2. (a) Three-dimensional model of a cellulose unit. (b) Diagrammatic representation of the unit cell of native cellulose (Meyer and Misch).

suggest that cellulose was a chain of glucose residues, joined as in cellobiose by the 1:4 linkage:

This formula, which is in agreement with the chemical properties of cellulose, has been strongly supported by X-ray measurements, and is now generally accepted. Modifications of the glucose residues, e.g. by oxidation of $-CH_2OH$ to $-COOH$ and esterification of OH groups, may occur without leading to serious modification of the main cellulose chain.

A considerable volume of work has been devoted to the X-ray examination of native cellulose fibres, which, being crystalline to a considerable extent, provide sharp X-ray diagrams. The repeat distance of 10·3 Å along the fibre axis corresponds closely with the length of two alined glucose (or one cellobiose) units (see Fig. 27.2 (a)), and it follows that in the crystal the cellulose chain must be nearly straight (Meyer, 1942, p. 241, Fig. 79). The monoclinic unit cell, which contains four glucose residues, possesses the following dimensions:

$$a = 8\cdot 35 \text{ A}, \qquad \beta = 84°,$$
$$b = 10\cdot 3 \text{ A},$$
$$c = 7\cdot 9 \text{ A},$$

and is shown diagrammatically in Fig. 27.2 (b), due to Meyer and Misch (1937). It is believed that neighbouring molecular chains may interact in the plane perpendicular to the fibre axis by means of hydroxyl or hydrogen bonds, but precise details of this interaction are not known.

The dimensions of the crystallites in native cellulose, obtained from the width of X-ray reflections, have been given by Meyer and Mark (1929) as 60–70 A perpendicular to the fibre axis and not less than 600 A along it. According to Meyer (1942, p. 254) these values are in agreement with those obtained from surface area measurements as obtained from adsorption data. For this reason he claims that the cellulose crystallite, with its high lattice energy, has a greater significance than that of rubber.

The suggested structure for cellulose fibres clearly anticipates pronounced anisotropy of properties, as confirmed experimentally by observations of thermal expansion, elasticity, optical rotatory power, and birefringence. The model is also in agreement with the known chemical properties of cellulose. Thus from the structure of the cellulose chains and their packing in the crystallites it is clear that three hydroxyl groups are available in each glucose residue (except the terminal ones which contain four such groups); the existence of up to tri-esterified cellulose esters (e.g. cellulose tri-nitrate and tri-acetate) is therefore explained. The well-known end-group method, used by Haworth in estimating the chain-length of cellulose (see p. 133), is based upon these views of the availability of hydroxyl groups.

It is unlikely that any solvent exists which dissolves native cellulose without some rupture of the main chain linkages. Molecular weight and size as determined in solution do not therefore refer to native material. The chief solvent employed in the examination of cellulose in solution has been cuprammonium (i.e. a solution of copper oxide in ammonia) and one of the most recent and reliable estimates of molecular weight in this solvent has been made by Gralen (1944), who gives for Georgia cotton and unbleached American linters values around 2×10^6

(see Table 27.1). All processes used in the commercial treatment of cellulose result in further degradation and lowered molecular weight in cuprammonium. It will be noted that the highest molecular weight values for cellulose derivatives are of the same order as those for native cellulose, though modification of the cellulose structure is usually attended by some degradation (see p. 770).

Of the cellulose derivatives, we can mention only the more important esters and ethers.

Cellulose nitrates, prepared usually by nitration of cellulose, in the form of paper or cotton, with nitric acid-sulphuric acid mixtures, are of considerable importance, especially in the explosives and lacquer industries. The nitration is an equilibrium process, its extent depending upon the nitric acid content of the acid mixture. A maximum of three nitrate groups per glucose residue may be introduced according to the equation:

$$-COH + HNO_3 + H_2SO_4 \rightleftharpoons -CONO_2 + H_2SO_4 + H_2O.$$

The sulphuric acid appears to act both as an accelerator of the reaction, removing water as it is formed, and as a swelling agent, making accessible the inner parts of the cellulose fibres.

For material with up to 8 per cent. nitrogen content, good X-ray diagrams of hydrated cellulose (a modified cellulose) are obtained, indicating the surface character of the nitration; between 8 and 13 per cent. of nitrogen the X-ray reflections are diffuse except those corresponding to the normal cellulose repeat distance along the fibre axis which has not yet changed. Above 13 per cent. nitrogen the X-ray reflections become sharper in a new diagram corresponding to cellulose trinitrate (i.e. 14·14 per cent. nitrogen) which has the following lattice dimensions:

$$a = 13\cdot0 \text{ A}, \quad \beta = 90°,$$
$$b = 25\cdot6 \text{ A} \quad (5 \text{ glucose units}),$$
$$c = 8\cdot9 \text{ A}.$$

Comparison with the lattice dimensions of cellulose shows that the introduction of nitrate groups has caused a separation of the molecular chains along the a-axis (see Fig. 27.2 (b)) in addition to giving a new repeat distance along the fibre axis corresponding to five glucose residues, a point which has not yet been adequately explained. The diffuseness of the X-ray spots at between 8 and 13 per cent. nitrogen has been taken to indicate the randomness with which nitrate groups are introduced at this stage.

The introduction of nitrate groups into cellulose, as might be expected, modifies profoundly its solubility and interactions with organic and other solvents. Esters, ketones, and ethers interact strongly with nitrate groups, and many workers have considered the existence of molecular compounds, e.g. between nitrocellulose and acetone (see Meyer, 1942, p. 310). The evidence advanced in support of this view is by no means conclusive, however (Campbell and Johnson, 1945). In general, the solubility in strongly interacting solvents increases with increasing nitrogen content; in the case of alcohol and ether, however, the solubility passes through a maximum with increasing nitrogen content, and for cellulose nitrates of between 2 and $2\frac{1}{2}$ nitrate groups per glucose residue only suitable mixtures of the two solvents dissolve the polymer.

Cellulose nitrates in solution have been considerably investigated. All samples are polydisperse, the mean chain-length depending on both the original cellulose and its treatment during nitration; most commercial materials are degraded considerably during the latter process. Gralen (1944) attempted to reduce degradation by the use of nitric-phosphoric acid nitration mixtures, obtaining, by the sedimentation-diffusion procedure, the molecular weight values of Table 27.1.

For comparison the molecular weight of the original celluloses in cuprammonium, derived similarly, are included.

TABLE 27.1. *Molecular Weights of Cellulose and Derived Cellulose Nitrates*

(Gralen, 1944.)

Original cellulose	Cellulose in cuprammonium, $M. 10^{-3}$			Derived cellulose nitrate	
	Cellulose-copper complex	Cellulose	D.P.†	$M. 10^{-3}$	D.P.†
Unbleached American linters	2,100	1,500	9,300	780	2,700
Bleached American linters .	690	490	3,000	400	1,360
Chlorite bleached linters .	1,700	1,200	7,300	680	2,300
Sulphate cellulose . .	560	400	2,500	420	1,450
Sulphite cellulose . .	700	500	3,100	430	1,470

† D.P. = Degree of Polymerization.

Clearly, even with special precautions during nitration, degradation is considerable.

Cellulose nitrate is extensively used with suitable plasticizing materials (e.g. with camphor to give celluloid) in the manufacture of moulded articles, as well as in the explosives and lacquer industries.

Cellulose acetate, another important cellulose ester, is prepared by acetylation by acetic acid or anhydride in the presence of sulphuric acid. The higher chain-length material (comparable with the higher chain-length cellulose nitrates) is not readily soluble: on reducing the chain-length and acetyl content by hydrolysis, more soluble materials are obtained, the usual acetone-soluble material having $2-2\frac{1}{2}$ acetyl groups per glucose residue.

As with nitration, the course of acetylation has been followed by X-rays, and similar observations have been reported. Up to about $1\frac{1}{2}$ acetyl groups per glucose residue, the unaltered cellulose diagram indicates surface reaction; with increasing acetyl content only the meridian spots remain sharp, corresponding to unaltered periodicity along the fibre axis but penetration of the cellulose micelles; towards 62 per cent. acetyl, the new lattice of cellulose triacetate is indicated by a new sharp diagram.

Cellulose acetate is often used in the manufacture of films, lacquers, artificial (acetate) silk, and in general as a substitute for cellulose nitrate where the inflammability of the latter is a disadvantage. When plasticized (e.g. with dibutyl phthalate, tricresyl phosphate) it may also be used in injection moulding.

Other cellulose esters and mixed esters are also known in which the introduction of the different ester groupings causes modified properties. Details of their preparation and properties are given in text-books on cellulose chemistry (e.g. Marsh and Wood, 1942, or Ott, 1943).

Cellulose ethers

Cellulose ethers possess the advantage over the esters of being resistant to hydrolysis by both acids and alkalis. The most important members are the methyl and ethyl celluloses, obtained by treating cellulose, in the presence of caustic soda, with the relevant alkyl chloride or sulphate. Fibres of completely methylated cellulose give good X-ray diagrams in which a repeat distance of 10·3 A, identical with that of native cellulose, occurs. The completely methylated compound is soluble in organic solvents but insoluble in water: it is therefore used as a water-resistant plastic. Lower methylated ethers are soluble in cold (but insoluble in hot) water, and they are therefore used in the preparation of pastes for the printing of fabrics.

Starch and glycogen

Glucose is commonly stored naturally in the form of hydrolysable polysaccharides, such as starch in plants and glycogen in animals.

772 POLYMERS

Unlike cellulose, neither are structural materials, their chief role being as energy reserves.

Starch occurs in the plant cell in the form of small grains ($2\,\mu$ to $170\,\mu$) possessing a concentric layer-like structure, and containing as the chief constituent a mixture of carbohydrates to which the name starch applies. The water soluble portion, amylose, and the water insoluble fraction, amylopectin, account for most of the carbohydrate; other forms of subdivision are, however, employed (Meyer, 1942, p. 388). Table 27.2, taken largely from Meyer (1942), enumerates the chief properties of the two components.

TABLE 27.2. *The Main Fractions of Starch*

Property	Unbranched portion (amylose)	Branched portion (amylopectin)
Fraction of starch	10–20%	80–90%
Molecular weight	10,000–60,000	50,000–1,000,000
End groups	1 per molecule	1 per 20–30 glucose units
Phosphorus content	None	Some forms contain free, but others only combined phosphorus
Action of β-amylase	100% hydrolysed	60% hydrolysed
Reaction with iodine	Blue colour	Violet to red-violet colour

From enzymatic breakdown (yielding maltose) and methylation studies it appears that the glucose units of amylose and amylopectin are linked by 1:4 linkages (occurring also in maltose), a conclusion which is confirmed by a study of optical rotations, and of the kinetics of hydrolysis of starch and maltose. The main structure is therefore

in accordance with the non-reducing properties of undegraded starch. Amylose is considered to consist of a single unbranched chain of glucose residues of this type. However, from the viscosity and other properties of starch and its derivatives, it is considered that amylopectin contains a branched structure with 24–30 glucose units per branch. Osmotic molecular weights are in agreement with those determined chemically on the basis of this structure. From enzyme studies and from the fact that experiment has not confirmed the presence of reducing groups which should occur at one end of a chain of glucose residues (Haworth,

1939), it appears that the bonds involved in branching are of the 1:6 variety: the absence of reducing properties in amylopectin is thereby explained.

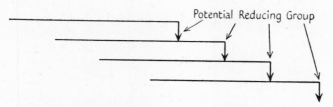

1:6 link suggested for branching of amylopectin.

According to Haworth, the amylopectin molecule may contain up to 80 or 90 chains so linked, each chain containing 24–30 glucose residues (Fig. 27.3).

FIG. 27.3. The structure of amylopectin (diagrammatic).

As in the case of cellulose, esters and ethers of starch can be prepared, but, especially in the preparation of esters, hydrolysis occurs more readily.

Glycogen is, chemically, closely related to starch, occurring partly as an insoluble material (though not in granules), and partly in a soluble form, both probably in combination with albuminous protein. Like amylopectin (see Table 27.2) it gives a red to violet colour with iodine, typical of a branched chain of glucose residues. From methylation and enzymatic work, the basic linear chain of glucose residues joined by the 1:4 link, already used for starch, is postulated Further, from viscosity, the high osmotic molecular weight $(1-2 \times 10^6)$ and small number of glucose units per end group (12–18), it is evident that branching must occur, probably with branching links similar to those of starch. From the comparative shortness of the branches and the large number required to explain the high molecular weight of glycogen, the molecule cannot be asymmetrical to the same extent as cellulose or even starch. Its association with proteins in the animal is not yet understood.

Other reserve carbohydrates (e.g. inulin, konjak-mannan) are known, but space forbids their consideration here.

(d) Linear poly-esters and poly-amides

Long chain poly-esters are formed by the condensation together of dicarboxylic acids ($R_c(COOH)_2$) and di-alcohols ($R_A(OH)_2$), or of ω-hydroxy acids alone, according to schemes (a) and (b) respectively:

(a) $HO.R_A.OH + HOOC.R_c.COOH \rightarrow —O.R_A.O.OC.R_c.CO—.$

(b) $HO.(CH_2)_n.COOH + HO.(CH_2)_n.COOH$
$\rightarrow —O.(CH_2)_n.CO.O.(CH_2)_n.CO—.$

Linear poly-amides are formed by the condensation of di-carboxylic acids and di-amines, $R_M.(NH_2)_2$, or of ω-amino-carboxylic acids according to schemes (c) and (d) respectively:

(c) $H_2N.R_M.NH_2 + HOOC.R_c.COOH$
$\rightarrow —HN.R_M.NH.OC.R_c.CO—.$

(d) $H_2N.(CH_2)_n.COOH + H_2N.(CH_2)_n.COOH$
$\rightarrow —HN.(CH_2)_n.CO.NH.(CH_2)_n.CO—.$

Such poly-esters and poly-amides, if of sufficiently high chain-length, can be drawn into fibres whose examination by X-rays (e.g. Fuller, 1938) has provided useful information on their molecular structure. One such polymer formed from adipic acid, ($HOOC.(CH_2)_4.COOH$), and hexamethylene diamine, ($H_2N.(CH_2)_6.NH_2$), known as Nylon, has quickly achieved fame as a result of its elasticity, high tensile strength, resistance to wear and to chemical attack.

(e) Branched or three-dimensional polymers

If the monomer possesses more than two active groups (i.e. is polyfunctional), instead of the linear type of polymer so far considered, branched polymers in two or three dimensions may be formed. The condensation products of phenol and formaldehyde, and of related compounds, known commercially as bakelite, are of such a type, and the reaction, usually catalysed by acid or alkali, is probably to a large extent represented by:

[reaction scheme showing phenol + formaldehyde condensation to cross-linked phenol-formaldehyde network]

The properties of the products can be varied by variation of the reaction conditions, the lower condensation products being soluble and somewhat elastic whilst the higher ones are hard, very resistant to heat, and almost completely insoluble.

Formaldehyde may also be condensed with urea and other dibasic amides, to give a linear chain of the type:

$$-CH_2-NH-CO-NH-CH_2-NH-CO-NH-CH_2-.$$

With excess formaldehyde, three-dimensional nets may be formed through the reaction of the —NH hydrogen atoms in the chain:

$$\begin{array}{c}-CH_2-N-CO-NH-CH_2-N-\\ ||\\ CH_2CH_2\\ ||\\ -N-CO-NH-CH_2-N-\end{array}$$

Polymerization is effected by heat, and as with the bakelites, moulding powders, consisting of partly polymerized powders mixed with suitable filling and colouring materials, are further polymerized at high temperature and pressure in the mould to give the finished moulded article.

Other three-dimensional polymers (e.g. vulcanized rubber) are considered with the linear polymers from which they were derived (pp. 763, 766) and will not be discussed here.

The Properties of Polymers

The behaviour of polymers and plastic materials generally in their thermal stability range may be described in terms of four states (see Treloar, 1943):

(1) The amorphous glass-like state.

(2) The crystalline state.
(3) The rubber-like state.
(4) The plastic or viscous liquid state.

Not all of these states are of equal prominence in any given polymer; for instance, since crystallization does not readily occur in the methacrylate polymers (p. 760), the properties associated with the crystalline state do not therefore appear. Similarly, the shorter chain polymers (e.g. unvulcanized gutta-percha) do not show good rubber-like properties since the latter require the presence of long-chain molecules. However, a general treatment of polymers must cover the four states.

Of these states, three ((1), (2), and (4)) have quite close analogies amongst non-polymeric systems, but the rubber-like state is peculiar to high polymers, and for this reason it will have chief consideration here. At sufficiently low temperature in all polymers, the energy of thermal agitation is insufficient to cause appreciable movement of the molecules against the forces of cohesion, the polymer then possessing properties analogous to those of the solid state, amorphous or crystalline. In the amorphous state the molecular chains are in a state of disorder, and the hard amorphous glass-like state characteristic of polymethyl methacrylate and polystyrene at room temperature, or of other non-crystallizing polymers at suitably low temperature, is characteristic. Rubber, and other polymers which can be made to crystallize (e.g. by stretching and cooling), may also be obtained in the amorphous condition by cooling the unstretched material (to about $-70°$ C. for unvulcanized rubber). The amorphous state is characterized by its loss of elastic or plastic properties, its isotropic nature, and its absence of order as determined by X-rays.

If, on the other hand, the molecular chains are regularly arranged before cooling, as in highly stretched rubber, then the polymer takes up the crystalline state. That a change of state has occurred is clear, since if the tension on the rubber specimen is removed, the latter no longer reverts to its original length, but remains at its stretched length, having lost its original rubber-like elasticity. The sharp X-ray diagram and the marked anisotropy in the mechanical and optical properties associated with the crystalline state result from the adlineation of the molecular chains. The crystalline state of high polymers does, however, differ in certain ways from that associated with low molecular weight materials. Thus, crystallinity in rubber is not associated with any regularity of external form, and the ordered crystalline regions do not necessarily extend throughout the specimen. A single molecular chain may indeed pass alternately through regions of order and disorder. The size of the crystalline regions or crystallites may, as we have seen, be estimated by X-ray methods (p. 453), and values for various types of polymer have already been given. Fig. 27.4, due to Treloar, shows diagrammatically the nature of the crystalline polymer state, and compares with the amorphous state for the same material.

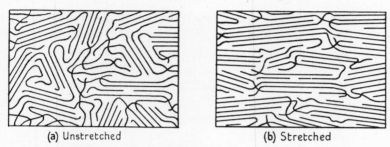

(a) Unstretched   (b) Stretched

FIG. 27.4. Molecular structure of crystalline rubber (diagrammatic). The parallel bundles represent crystallites.

A significant phenomenon associated with the crystallinity of flexible chain molecules is the linear contraction on warming, instead of the usual expansion in all three dimensions common to low molecular weight crystals. This is believed to arise from the partially regained flexibility of the chains on being freed by a rise in temperature from the forces stabilizing them in the crystalline order. In general, stretching and decrease in temperature lead to an increase in the number (though not in the individual size) of the crystalline regions.

Another feature peculiar to crystallization in high polymers is its slowness, which arises from the high viscous resistance offered to the motion of molecules or parts of molecules in the partly crystalline solid. Changes in the degree of crystallinity may continue for periods of several years, as shown in the case of rubber (Meyer, 1942, p. 166). Clearly such systems are not necessarily in true thermodynamic equilibrium (see p. 92).

For a polymer to occur in the crystalline state it would appear necessary that the chain molecules should possess a regular repeating structure; the failure of co-polymers to crystallize is thus readily understandable since the necessary regularity could only be expected if the polymerizing units were very similar chemically and geometrically, a condition which rarely holds. The failure to crystallize of polymers of apparently regular repeating structure is not so easily explained, but although a definite explanation is lacking, a likely reason is some form of packing difficulty; for example, the occurrence of right-hand and left-hand groups suggested by Staudinger (1932), or the steric interference resulting from a large side group like C_6H_5— in styrene. Structural irregularities arising out of the polymerization (e.g. a mixture of 'head-to-head' and 'head-to-tail' polymerization) are not yet, however, ruled out in many cases.

The cohesive forces in the amorphous and crystalline states may be of various types; from weak van der Waals forces (as in polythenes) and hydrogen bonding, to strong dipolar interactions and even covalent bonding. Clearly, the stronger the forces the greater the thermal energy required to overcome them, and the higher the melting-point. A crystalline polymer, with its higher internal regularity and closer packing, thus possesses a higher melting-point than an otherwise similar amorphous material, and since crystalline polymers always contain an appreciable portion of amorphous material, melting does not occur sharply, but over a considerable temperature range, which varies in position and extent with the degree of crystallinity of the material. Thus, although raw rubber usually melts at $0°$ C.$\pm 10°$, those specimens which are highly crystalline as a result of storage below $0°$ C. may melt at a considerably higher temperature (even up to $43°$ C.).

Poly-methyl methacrylate, although non-crystalline, contains strong dipoles and its melting-point ($> 80°$ C.) is considerably higher than that of rubber. Cellulose and its derivatives, by reason of strong inter-chain forces, do not melt before decomposition. On introducing plasticizing materials into such polymers, the molecules are separated and their interactions reduced, so that a lowering in melting-point is observed.

In spite of these qualitative agreements between theory and observation, it must be acknowledged that certain observations give difficulty. For instance, the high melting-point of polythene (*ca.* $115°$ C.) as compared with that of rubber (*ca.* $0°$ C.) is not easily explained (see Bunn, 1946, p. 136).

The dependence of melting-point on chain-length for a homologous series of polymers is of some importance, in spite of its lack of sharpness in so many cases. According to Meyer and van der Wyk (1937), the melting-point, (T_f), on the absolute scale, of straight-chain crystalline paraffins as a function of chain-length is given by the expression:

$$\frac{1}{T_f} = a + \frac{b}{Z}, \tag{27.1}$$

where a and b are constants with the values $2 \cdot 395 \times 10^{-3}$ and $17 \cdot 1 \times 10^{-3}$ respectively, and Z is the number of carbon atoms in the chain. Clearly, with increasing values of Z, T_f approaches a limiting value of 145°. (N.B.—This value is considerably higher than the observed melting-point of polythene, a fact probably to be attributed to the partially amorphous character of the latter.) Similarly, in the case of other more complex chain molecules, when the proportion of end groups becomes small, the melting-point (as well as other physical properties, e.g. density, refractive index, see p. 757) becomes independent of chain length.

Although formally similar to the melting of low molecular weight materials, the melting of high polymers gives rise not to a mobile liquid, but to a rubber-like material. On further heating (provided chemical decomposition does not occur) the material loses its typically rubber-like properties, becoming more plastic and ultimately indistinguishable from a normal viscous liquid. The rubber-like state may therefore be considered as an intermediate between solid and liquid, not only with regard to its occurrence on the temperature scale, but also since it partakes of certain properties of both solids and liquids. Thus, though rubber-like materials possess a definite external form like typical solids, their ready deformability, incompressibility, and coefficient of thermal expansion are much more typical of liquids.

The rubber-like state

Before considering the molecular structure associated with the rubber-like state, it is necessary to mention its chief properties. The most obvious are its capacity to undergo large reversible extensions (up to 800–900 per cent.) and the low initial Young's modulus (*ca.* 10^6 dynes per sq. cm.) as compared with that (*ca.* 10^{10} dynes per sq. cm.) for most solids. At great extensions, however, the modulus increases to values which are typical of normal solids and fibrous materials.

Less obvious, but none the less important, are the thermodynamic properties of rubber. As already pointed out (p. 98), the tension at constant length is almost proportional to the absolute temperature (Meyer and Ferri, 1935), and it follows, where complications arising from crystallization and plastic flow are removed, that the change in internal energy with extension is small compared with the entropy change. It is helpful here to note the analogy with a perfect gas, for which the internal energy is independent of the volume, and the entropy reduced by a decrease in volume. The pressure of a perfect gas, arising from the thermal motion of its molecules, is analogous to the tension (strictly, $-Z$) of a mass of stretched rubber, arising from the thermal motion of the repeating units in the molecular chain and the tendency to achieve more compact forms. The proportionality between tension and absolute temperature is not therefore so surprising.

This behaviour of rubber-like substances is to be contrasted with that

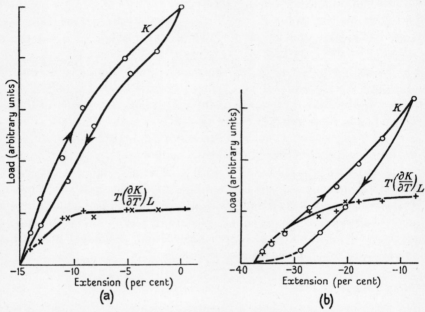

FIG. 27.5. Showing the total load (K) and the entropy contribution ($+ - + -$ stretching; $\times - \times -$ contracting) for supercontracted wool fibres in water. The internal energy load is the difference of these,

$$(\delta E/\delta L)_T = K - T(\delta K/\delta T)_L.$$

The temperature range for which $(\delta K/\delta T)_L$ was observed was $20°-40°$ C.

of wool (Woods, 1946) in which the increase in internal energy on extension is considerably more important than the decrease in entropy (Fig. 27.5 (a)). An exception is, however, observed for the early stages of the extension of largely supercontracted wool fibres (Fig. 27.5 (b)). This is readily understood in view of the decreased orientation of the supercontracted form which, to some extent, is comparable with unstretched rubber.

In seeking an explanation of these properties of rubber-like materials, their invariable association with long-chain polymeric compounds must be noted. Thus, rubber-like properties are shown by saturated (e.g. poly-isobutylene) as well as unsaturated long-chain hydrocarbons, by the polymerized forms of sulphur and selenium, by polyphosphonitrilic chloride, by aluminium soaps in organic solvents (p. 696), and by many other compounds. Evidently the rubber-like state is independent of the precise composition and structure of the main chain or of the occurrence of side groups, but it requires long-chain molecular character. A satisfactory theory must therefore explain the rubber-like nature of a wide array of high polymer compounds.

A theoretical explanation of the rubber-like state can, however, most simply be attempted in terms of the simplest of polymers, a saturated hydrocarbon chain, though the detailed behaviour of certain polymers may require for its explanation the consideration of more complicated molecules. We are, here, concerned only with the general properties of rubbers, and shall therefore proceed to consider the properties of simple unbranched paraffin chains.

The statistical length of flexible chain-like molecules

Owing to free rotation about each of the main chain bonds of a saturated hydrocarbon chain, the number of possible and energetically equivalent configurations of the chain is very large. A single long molecule will therefore usually exist, not as an extended chain, but in an irregular kinked or coiled-up form, which changes continuously owing to its thermal energy and for which the distance between the ends of the chain is only a small fraction of the outstretched length. The kinked chain can be stretched and straightened only by the application of some force opposing the tendency to contraction arising from the lateral vibrations of the atoms in the chain.

Guth and Mark (1934) and Kuhn (1934) have treated the long paraffin chain statistically, obtaining for the probability $p(r)$ that the ends of a chain will be separated by a length (in any direction) within the range r to $r+dr$,

$$p(r)\,dr = \frac{\beta^3}{\pi^{\frac{3}{2}}} 4\pi r^2 e^{-\beta^2 r^2}\,dr, \qquad (27.2)$$

where β is defined by

$$\frac{1}{\beta^2} = \frac{2}{3} l_c^2 n \frac{1+\cos\theta}{1-\cos\theta}, \qquad (27.3)$$

l_c being the C—C bond length, n the number of links in the chain, and $180°-\theta$ the valence angle. The distribution function, which is valid only for large values of n, is illustrated in Fig. 27.6, the maximum of the curve occurring at $r = 1/\beta$.

Thus, for a system of isolated molecules, the statistically most probable length is $r = 1/\beta$, and it is readily shown that the ratio of the outstretched length to this most probable length for a paraffin chain is approximately $0.7 \times n^{\frac{1}{2}}$, a large number for long-chain molecules.

Clearly, a single paraffin chain molecule, with its capacity for undergoing long range, reversible extension without increase of energy, has

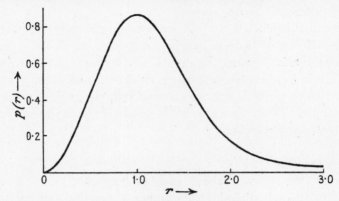

FIG. 27.6. Theoretical distribution of molecular lengths according to equation 27.2.

properties analogous to those of rubbers, and it is now necessary to consider the behaviour of an assembly of such molecules.

If the individual molecules are to retain their rubber-like properties in the assembly, the forces between them must be weak so as not to interfere with the freedom of movement of each molecular chain, and some interlocking of the molecules at a few points along their length (junction points) is required if plastic or irreversible deformation is to be avoided. The rubber-like solid is therefore to be imagined as a rather complex irregular network of long-chain molecules, held by entanglements or chemical bonds at certain points, but in which large portions of the chains are free to take up a multitude of spatially different, but energetically similar configurations. In treating such an assembly it has become customary to speak of that part of a molecular chain between adjacent junction points as the molecule, and in the following treatment this special use of the word will be observed.

Kuhn (1936), in treating such a network system, introduced the following simplifying assumptions:

(1) All molecules (i.e. portions of the molecular chain between junction points) have the same chain length.

(2) The distances separating the two ends of the chains are given by the distribution function.

(3) On extension, the components of length of each molecule change

in the same ratio as the length components of the macroscopic sample.

(4) At the highest extensions considered, the chains are not nearly fully stretched.

(5) There is no volume change on deformation.

He then calculated the entropy change on extending a sample of rubber, and, assuming $(\partial E/\partial L) = 0$, utilized eqn. (27.2) to obtain a relation between tension and extension. Kuhn's treatment was subsequently improved by several workers (see review by Guth, James, and Mark, 1946), amongst whom we mention Wall (1942), who considered a long-chain network of the type described above, and made assumptions similar to those of Kuhn. On the basis of eqns. (27.2) to (27.3) describing the behaviour of single long-chain molecules, Wall obtained the relative probabilities (P) of the stretched and unstretched states of the macroscopic sample, and making use of Boltzmann's equation ($S = k \ln P$) obtained the stress-strain relation

$$Z = nkT\left(\alpha - \frac{1}{\alpha^2}\right) = \frac{\rho RT}{M}\left(\alpha - \frac{1}{\alpha^2}\right), \qquad (27.4)$$

Z being the tension per unit cross-sectional area of the original specimen, n the number of molecules per c.c., $\alpha = l/l_0$, the ratio of the stretched to the unstretched lengths of the specimen, ρ the density of the material, and M the molecular weight of the chain between junction points. Applying to both extension and compression, this equation is clearly different from Hooke's law, for which the stress Z is proportional to the strain α, which may, however, be shown, by similar methods, to hold for shear deformations. Eqn. (27.4) is somewhat surprising in that it predicts that the elastic behaviour of a rubber-like solid is determined by only one molecular constant, M, the molecular weight between junction points. Within the limits of the assumptions, therefore, eqn. (27.4) should be applicable to any form of long-chain molecule, irrespective of its structure or chemical composition. In spite of the assumptions involved, it corresponds well with the behaviour of lightly vulcanized rubber at moderate extensions and compressions, as Fig. 27.7 (a), due to Treloar (1944), clearly shows. For other materials (e.g. latex rubber) the agreement at similar extensions was not so good and it is probable that in these cases the structure does not approximate so closely to the ideal form required. At great extensions (> 400–500 per cent. for lightly vulcanized rubber) all rubber-like materials deviate considerably from eqn. (27.4), as is shown well by Fig. 27.7 (b), again due to Treloar, in which the experimental data of Sheppard and Clapson (1932) are compared with the theoretical

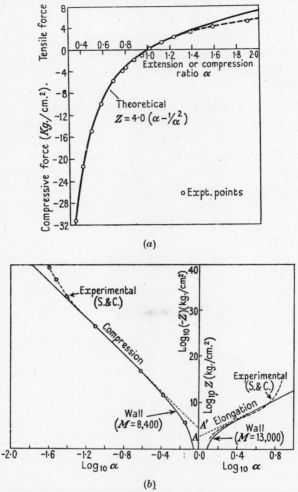

FIG. 27.7. (a) Tensile and compressive force as a function of α. 8% S. rubber, 20° C. (b) Comparison of theoretical stress-strain relationship with Sheppard and Clapson's experimental data for vulcanized rubber in elongation and compression.

curve derived from Wall's equation (27.4). Other such comparisons have indicated limitations of the network model, and a reconsideration of the assumptions involved is clearly necessary.

Assumption (1) is clearly invalid since short chains as well as long ones undoubtedly occur in any rubber-like network. Further, during extension, the short chains approach the fully extended form whilst the long ones are still relatively unstretched, so that assumption (4) no longer applies also. This would be expected to introduce deviations from eqn. (27.4) of the type illustrated in Fig. 27.7 (b). Neglect of steric hindrance, of van der Waals' forces, and of crystallization phenomena, the latter of considerable importance at the higher extensions, may also be involved, and it is likely that future theories will consider these factors.

Cross-linkages and their effects

We have seen that an ideal rubber-like solid requires a small number of junction points to prevent the occurrence of permanent deformations resulting from the flow of the molecular chains across one another. Such junction points may be merely mechanical entanglements of the kinked molecular chains, crystalline nuclei influenced possibly by particles of amorphous foreign materials (e.g. carbon or zinc oxide), or in other cases may be strong chemical bonds or cross-links formed between active points on individual chains (as in vulcanized rubber). Since the number of such cross-links determines to a large extent the mechanical properties of linear polymers, their effects will be mentioned now.

When only a few random cross-links occur, large portions of all the molecular chains are unaffected, and it is justifiable to apply formula (27.2) for the distribution of the molecular lengths of an isolated molecule. As the number of cross-linkages grows, however, the lengths of unaffected chain diminish and it becomes invalid to apply the same distribution formula, for the intra-molecular Brownian motion is seriously impeded. The extensibility of the material decreases correspondingly and ultimately approaches that of ordinary solids, and since, at such a stage, extension involves distortion of bond lengths and angles, i.e. energy effects in place of the predominant mechanism involving entropy changes for the rubber-like state, the elastic moduli are now typical of ordinary solids. Bakelite, the well-known three-dimensional condensation polymer of phenol and formaldehyde, clearly illustrates the properties associated with almost unlimited cross-linking.

The crystallizability of linear polymers is also radically affected by cross-linking. Thus weakly vulcanized rubber on stretching readily shows its crystallinity by its X-ray pattern, but with further vulcanization the crystalline pattern appears only at higher extensions and is then less complete. Finally, the 'bakelite' type of polymer shows no crystallinity on stretching.

Another interesting result of cross-linking is the change (usually decrease) in solubility of linear polymers which occurs. Thus, the addition to fluid solutions of cellulose nitrate of suitable cross-linking agents may under certain circumstances cause the gelling of the solution, followed by syneresis. The addition of hexamethylene di-isocyanate, in the presence of certain bases, to a solution of cellulose nitrate in nitrobenzene (Campbell, 1945), causes a pronounced increase in viscosity followed by gel formation, the cross-links in this case probably being covalent bonds formed between residual cellulosic hydroxyls and the isocyanate groups:

Cellulose—OH + OCN.(CH$_2$)$_6$.N.CO + HO—Cellulose

→ Cellulose—O.CO.NH.(CH$_2$)$_6$.NH.CO.O—Cellulose.

The addition of basic oxides (e.g. CaO) to acetone solutions of cellulose nitrate gives rise to the formation of very high molecular weight aggregates in solution, as shown by the ultracentrifuge, followed by the gelling of the solution (Johnson, 1948). Similar gelling phenomena have also been reported by Signer and von Tavel (1943) for methyl cellulose.

From these varied effects it would certainly appear that the formation of cross-linkages offers a useful means of modifying the physical properties of the basic polymeric materials.

The Interactions of High Polymers with Low Molecular Weight Liquids

The solubility of a low molecular weight solute in a given solvent is a well-defined quantity which may be varied continuously with variations in the solvent or the temperature. The solubility behaviour of most high polymers is, however, strikingly different. Thus a given polymer is usually either completely soluble in a solvent or almost completely insoluble, although in the latter case it may be swollen to considerable extents. If completely soluble, a well-defined critical solution temperature (for which there is no counterpart in low molecular weight chemistry) often exists, below which the polymer is quite insoluble but above which it is completely soluble. A slight lowering of temperature may therefore cause complete precipitation of the polymer. An explanation of this peculiar behaviour is clearly required.

As shown on p. 92, a polymer dissolves in a solvent only if this involves a decrease in free energy. If, on the other hand, pure solvent is in equilibrium with a mixture of solvent and polymer (swollen polymer), then it is clear that, using the nomenclature of Chapter IV, $\Delta \bar{G}_1 = 0$. (Similarly if pure polymer is in equilibrium with polymer-solvent mixture, then $\Delta \bar{G}_2 = 0$, though, as we have seen, in practice if a polymer is incompletely miscible with a liquid, the liquid phase is usually practically pure solvent.) Clearly, therefore, if $\Delta \bar{G}_1$ (and $\Delta \bar{G}_2$) is known as a function of the composition and temperature of the system, by equating to zero, the polymer-solvent composition in equilibrium with free solvent (or polymer) at any temperature can be determined, and the effect of temperature on the nature of the equilibrium system can readily be assessed. An outline of the calculation of the free energies of dilution and solution, which resolves itself into the calculation of the corresponding heat content and entropy changes (ΔH and ΔS), is therefore given here. (For details see Gee, 1942, 1946; Flory, 1941, 1942, 1943; Huggins, 1942, 1943.)

It is assumed, in the calculation of the entropy and heat of mixing, that the liquid and polymer are randomly distributed throughout the system. For low heats of mixing this assumption is reasonable, but for high values, non-random mixing may occur since certain configurations will be preferred energetically. This difficulty has been considered by Orr (1944) and Guggenheim (1945), but it appears that its effects are small compared with other uncertainties in the theory.

The polymer molecule is regarded as being made up of a large number of spherical sub-molecules, flexibly joined and each identical with or bearing a simple volume relation to a single spherical solvent molecule (see also Chapter VII). In calculating the entropy of mixing, the solvent-polymer system is considered as a regular lattice, at each point of which a solvent molecule or a polymer segment of equal volume may occur. For a mixture containing n_1 solvent molecules and n_2 polymer molecules, each containing x segments, the number of lattice points required is (n_1+xn_2). After introducing n_1 polymer molecules, the fraction of the lattice sites occupied is

$$i = \frac{xn_i}{(n_1+xn_2)}$$

and the number available for the terminal segment of the next polymer molecule is $(1-i)(n_1+xn_2)$. If Z is the coordination number of the lattice, the probable number of sites available for adjacent segments of the same $(i+1)$th molecule is less than Z, since there is a chance that some of these sites are already filled by a segment of the n_i molecules already present. Flory gives the probable number of sites for the second segment of this molecule as $Z(1-i)$ and for the third segment by similar reasoning $(Z-1)(1-i)$, since one of the coordinating sites of the second segment is definitely filled by the first segment. Neglecting the possible filling of other sites which might be caused by the curling back of a long chain-like molecule, Flory gives for the number of sites available for each subsequent segment of the molecule the same value $(Z-1)(1-i)$. The number of configurations which the chain can assume without moving its terminal segment is thus

$$Z(1-i) \times [(Z-1)(1-i)]^{x-2}, \qquad (27.5)$$

and the total number of its configurations (allowing for the possible positions of the first segment, and putting $Z = Z-1$) is

$$\nu_{i+1} \cong \tfrac{1}{2}(1-i)^x(n_1+xn_2)(Z-1)^{x-1}, \qquad (27.6)$$

the factor $\tfrac{1}{2}$ being introduced to eliminate overcounting due to the indistinguishability of the ends of the polymer.

The total number of possible configurations for the system of n_1 solvent and n_2 polymer molecules is then given by

$$W = \frac{1}{n_2!} \prod_{n_i=1}^{n_2} \nu_{n_2}, \qquad (27.7)$$

the factor $1/n_2!$ being introduced to eliminate identical configurations arising from the indistinguishability of the polymer molecules. Introducing (27.6) into (27.7), simplifying, and introducing into the Boltzmann equation ($S = k \ln P$), we obtain for the entropy of mixing (ΔS_M^x) n_1 solvent and n_2 perfectly arranged polymer molecules

$$\Delta S_M^x = -k\left[n_1 \ln\left(\frac{n_1}{n_1+xn_2}\right) + n_2 \ln\left(\frac{n_2}{n_1+xn_2}\right)\right] + \\ + k(x-1)n_2[\ln(Z-1)-1] - kn_2 \ln 2. \qquad (27.8)$$

To obtain the real entropy of mixing ΔS_M, it is necessary to subtract the configurational entropy of a non-oriented polymer (obtained by putting $n_1 = 0$ in the equation above) when we have

$$\Delta S_M = -k\left[n_1 \ln\left(\frac{n_1}{n_1+xn_2}\right) + n_2 \ln\left(\frac{xn_2}{n_1+xn_2}\right)\right]. \qquad (27.9)$$

Flory's final expression for the entropy of mixing is

$$\Delta S_M = -k[n_1 \ln \phi_1 + n_2 \ln \phi_2], \qquad (27.10)$$

where ϕ_1 and ϕ_2, the volume fractions of solvent and polymer respectively, replace $n_1/(n_1+xn_2)$ and $xn_2/(n_1+xn_2)$ respectively. It should be recalled that the ideal entropy of mixing for low molecular weight materials is similar, mole fractions merely taking the place of the volume fractions of eqn. (27.10).

Differentiating with respect to n_1 we obtain for the partial molar entropy of dilution

$$\Delta \bar{S}_1 = R\left[\ln \frac{1}{\phi_1} - \phi_2\left(1 - \frac{1}{x}\right)\right] \qquad (27.11)$$

and, with respect to n_2, the partial molar entropy of solution

$$\Delta \bar{S}_2 = R\left[\ln\frac{1}{\phi_2} + \phi_1(x-1)\right]. \qquad (27.12)$$

For polymer segments equal in volume to β solvent molecules Flory obtains

$$\Delta \bar{S}_1 = \frac{R}{\beta}\left[\ln\frac{1}{\phi_1} - \phi_2\left(1 - \frac{1}{x}\right)\right]. \qquad (27.13)$$

It should be noted that the coordination number Z does not occur in these entropy expressions ((27.9)–(27.13)). In more rigorous determinations, however, it occurs prominently (Huggins, 1942; Miller, 1943), and calculations of $\Delta \bar{S}$ values as a function of the composition of the system therefore require the assumption of a suitable value of Z.

The heat of mixing of a liquid and a polymer is not different in principle from that for two simple liquids, which may be expressed in the form of a power series:

$$\Delta \bar{H}_1 = k_2' \phi_2^2 + k_3' \phi_2^3 + ..., \qquad (27.14)$$

the first term containing the square of the volume fraction ϕ_2. In mixing two liquid components (1 and 2), the energy required to separate molecules of the pure components as well as that evolved when two different molecules come together affects the heat changes observed. On this basis the heat of mixing may be expressed in terms of the cohesive energy density E (Hildebrand, 1936), which, in the case of a simple liquid, is merely the latent heat of evaporation at constant volume per c.c., i.e.

$$E = \frac{L}{V} - RT, \qquad (27.15)$$

where L is the molar latent heat of evaporation and V the molar volume. For rubber a value of the cohesive energy density may be estimated from those of the related low molecular weight hydrocarbons. It may then be shown for the polymer-solvent system, if volume changes on mixing are small, that

$$\Delta \bar{H}_1 = \left(\sqrt{\frac{E_1}{V_1}} - \sqrt{\frac{E_2}{V_2}}\right)^2 V_1 \phi_2^2 = \alpha_1 \phi_2^2, \qquad (27.16)$$

and

$$\Delta \bar{H}_2 = \left(\sqrt{\frac{E_1}{V_1}} - \sqrt{\frac{E_2}{V_2}}\right)^2 V_2 \phi_1^2 = \alpha_2 \phi_1^2, \qquad (27.17)$$

where $\Delta \bar{H}_1$ and $\Delta \bar{H}_2$ are the increases in heat content on adding one mole respectively of components 1 and 2 to the mixture.

Introducing eqns. (27.11) and (27.16) into eqn. (4.24), the definition of $\Delta \bar{G}_1$, we obtain

$$\Delta \bar{G}_1 = \frac{RT}{\beta}\left[\ln(1-\phi_2)+\phi_2\left(1-\frac{1}{x}\right)+\frac{\alpha_1}{RT}\phi_2^2\right], \qquad (27.18)$$

slightly different forms being obtained by the use of the alternative

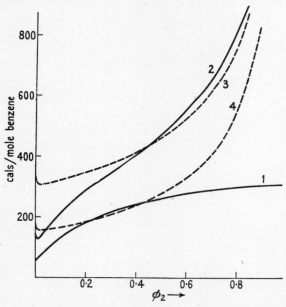

Fig. 27.8. Thermodynamic data for natural rubber plus benzene. (Gee (1947).)

Curve 1. $\Delta \bar{H}_1/\phi_2^2$ } Experimental.
 " 2. $T\Delta \bar{S}_1/\phi_2^2$
 " 3. $T\Delta \bar{S}_1/\phi_2^2$. Theoretical ($Z = \infty$).
 " 4. $T\Delta \bar{S}_1/\phi_2^2$. Theoretical ($Z = 4$).

expressions of Huggins and Miller for the entropy of dilution. Eqn. (27.18) or similar ones are basic to the thermodynamic treatment of polymer-liquid systems. Its simplification at low concentrations (p. 154) to give the osmotic pressure equation for polymer solutions has already been indicated, and its partial experimental confirmation at low polymer concentrations has thus been described. It is, however, desirable to test eqn. (27.18) over a wider range of composition than is possible by osmotic pressure measurements.

The most complete data refer to the rubber-benzene system, studied extensively by Gee and Treloar (1942), for which the free energy of dilution has been obtained (by methods indicated in Chapter IV) for volume fractions of rubber (ϕ_2) varying from 0·005 to 0·999. Fig. 27.8, due to Gee (1947), shows the variation of heat and entropy of dilution with composition, curves 1 and 2 giving experimental results and 3 and 4 being calculated from eqn. (27.13), and a modified equation involving a finite value ($Z = 4$) of the coordination number. The considerable

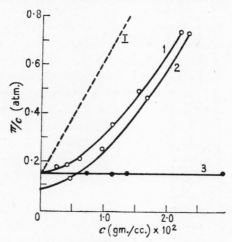

FIG. 27.9. Π/c v. c curves.

Curves 1 and 2. Natural rubber in benzene.
Curve 3. Natural rubber in benzene plus 15 per cent. methanol. } Experimental.
Curve I. Theoretical curve for natural rubber in benzene. ($Z = 5$; $\alpha_1/RT = 0.16$).
(*From* data by Gee (1946).)

variation in $\Delta \bar{H}_1/\phi_2^2$ with composition is clearly contrary to eqn. (27.16) and may arise from neglect of intramolecular contacts of the segments of a polymer molecule (see below). Measured values of $T\Delta \bar{S}_1/\phi_2^2$ are in good agreement with eqn. (27.13) at the higher volume fractions of rubber, but fall to considerably lower values at lower volume fractions (< 0.4), whilst curve 4 is in good agreement with the experimental values only at very low volume fractions. The divergence of theory and observation is also demonstrated by the osmotic data of Fig. 27.9, also due to Gee, in which curve I, calculated for rubber and benzene by an equation involving the coordination number, clearly deviates from experimental results. The reduced slope of the Π/c v. c curve for benzene plus 15 per cent. methanol illustrates well the advantages of choosing such a mixed solvent in molecular weight determinations (see p. 155).

In this outline of the statistical and thermodynamical treatment of high polymers, a field which is rapidly advancing from day to day, it has been necessary to omit the detailed consideration of certain difficulties. Below, however, we mention some of the more important and certain directions in which improvements may be expected.

Aqueous systems for which $\Delta \bar{H}_1$ and $\Delta \bar{S}_1$ are usually negative do not as a rule comply with the theory. The reason is probably to be sought in the polar character of such systems, water molecules being to a considerable extent (see p. 140) bound to soluble polymers by electrostatic forces or hydrogen bonds, giving an ordered system and possibly a negative entropy of mixing. The volume changes also occurring, which have been neglected in the theory, probably also contribute. Other polar systems (e.g. cellulose nitrate in acetone) are also expected to conform less than the typical rubber-benzene non-polar system.

The assumption of random mixing has already been mentioned as strictly invalid where the heat of mixing is not zero, but for moderate heats its effects are probably small relative to others. In assessing the number of configurations of the polymer molecules, Flory's treatment assumes that the probability of a lattice site near a segment of a molecule being filled is the same as the overall probability of a site being filled (i.e. $(1-i)$). However, in view of the undoubted occurrence, in some systems, of polymer-polymer contacts within the molecule, the probability of a site being occupied near a polymer molecule must be greater than the overall probability. Clearly the difference must vary with the particular system, the polymer flexibility and heat of mixing being important. On the basis of molecular volumes, derived from viscosity measurements, Flory (1945) has attempted to correct for the effect, which being more important at the lower concentrations, is probably responsible for discrepancies which have been observed there (e.g. see Fig. 27.8).

One further complication in the entropy calculation arises from the presence of cross-links within the molecule, which must reduce considerably the number of possible configurations and, therefore, the entropy of mixing. Flory and Rehner (1943) thus obtained for the reduction in entropy

$$\delta(\Delta \bar{S}_1) = \frac{R \rho_2 V_1}{M_c} \phi_2^{\frac{1}{3}}, \qquad (27.19)$$

M_c being the molecular weight between junction points. The corresponding increase in the free energy change is obtained by introducing into (4.24). As $\phi_2 \to 0$, $\Delta \bar{G}_1$ tends to become positive, indicating the limited swelling of cross-linked polymers.

Uncertainty in the treatment of heat of reaction also occurs. Thus, although experimental values are subject to a large error, it is doubtful if eqn. (27.16) does satisfactorily fit the facts. On the other hand, it would seem reasonable to expect that the type of configuration of polymer molecules (as well as the volume fraction), which may vary with concentration and the type of solvent, must affect the heat effects observed; on these grounds an equation as simple as (27.16) appears unlikely. The uncertainty in the calculation of heats of mixing is relatively greater than in those of entropy, and for this reason the most satisfactory comparisons with experiment are those involving only small heat effects.

The solubility and swelling of high polymers

The simplified treatment of solubility presented here, like the above statistical treatment, assumes a completely random and disordered arrangement in the solvent, polymer, and mixture, and therefore neglects the possible effects of crystallite size (p. 776) on polymer solubility.

One of the first quantitative treatments of solubility was made by Brönsted (1938), who suggested the empirically useful expression

$$\ln(c'/c'') = -\frac{\lambda M}{RT}, \tag{27.20}$$

c' and c'' being the polymer concentration in the two phases, λ being a constant which varies from positive through zero to negative values according to the particular solvent and polymer. For large values of M, c'/c'' may take on extreme values, according to the value of λ, in agreement with observation (for $\lambda > 0$, $c'/c'' \to 0$, and for $\lambda < 0$, $c'/c'' \to \infty$). The equation is, however, based upon the assumption of a difference in potential energy between the two phases, whilst, as we have seen, entropy rather than energy differences largely control solubility behaviour. On more reasonable grounds, however, Gee (1942) derived a similar equation differing only in the more complex temperature dependence of solubility:

$$\ln\left(\frac{\phi_2'}{\phi_2''}\right) = M(a-b/RT), \tag{27.21}$$

where $\quad a = (1-\phi_1'')\dfrac{1}{M_1}, \quad \text{and} \quad b = \alpha_1\dfrac{\{1-(\phi_1'')^2\}}{M_1},$

the phase ' denoting the dilute solution. The success of the Brönsted equation is therefore more understandable.

It is convenient here to develop further the statistical treatment of high polymers already given. Consider therefore a polymer-liquid system in which the two phases (of composition ϕ_2 and ϕ_2') are in equilibrium. It follows (see p. 94) that

$$\Delta \bar{G}_1(\phi_2') = \Delta \bar{G}_1(\phi_2'') \quad \text{and} \quad \Delta \bar{G}_2(\phi_2') = \Delta \bar{G}_2(\phi_2''). \tag{27.22}$$

From eqn. (27.18), if $\alpha_1 = 0$ (i.e. $\Delta \bar{H}_1 = 0$), $\Delta \bar{G}_1$ decreases continuously with increasing polymer concentration ϕ_2, and two-phase equilibrium is impossible, as has already been indicated (see p. 92) from experimental curves for rubber and benzene.

If, however, $\Delta \bar{H}_1 \gg 0$, the curve of $\Delta \bar{G}_1$ against ϕ_2 will possess a

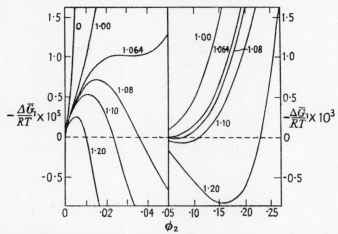

FIG. 27.10. Partial molal free energies as functions of the volume fraction of polymer for various values of K in the critical region for partial miscibility.

maximum and a minimum. Differentiating eqn. (27.18) with respect to ϕ_2, putting $K = 2\alpha_1/RT$, and equating to zero, we obtain

$$Kx\phi_2^2 - \phi_2(Kx - x + 1) + 1 = 0, \qquad (27.23)$$

a quadratic equation whose roots become equal as the condition of incipient precipitation. Applying this condition, we obtain

$$K_{\text{critical}} = \frac{(1+\sqrt{x})^2}{x}, \qquad (27.24)$$

and the solution of the equation becomes

$$\phi_2(\text{critical}) = \frac{1}{1+\sqrt{x}}. \qquad (27.25)$$

(The alternative value of K_{critical}, $\frac{(1-\sqrt{x})^2}{x}$, gives negative values of ϕ_2 and is therefore discarded.) By substituting $K = 2\alpha_1/RT$, eqn. (27.24) may be written in the alternative form

$$T_{\text{critical}} = \frac{2\alpha_1 x}{R(1+\sqrt{x})^2}. \qquad (27.26)$$

For large values of x, ϕ_2(critical), the concentration of polymer at which the two phases become identical is small, K_{critical} being nearly unity and approaching unity as $x \to \infty$. Fig. 27.10 gives curves, calculated by Flory (1942), for the Gibbs free energy of dilution ($\Delta \bar{G}_1$) for a polymer with $x = 1000$, and values of K around unity. Above the critical curve, for which $K = 1\cdot 0644$ (by eqn. (27.24)), $\Delta \bar{G}_1$ decreases continuously with concentration of polymer, whilst below it ($K > 1\cdot 0644$), the maximum in $\Delta \bar{G}_1$ indicates the separation of two phases, since two

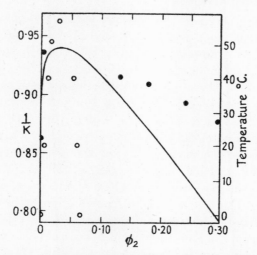

FIG. 27.11. Phase composition curve calculated for a solution of a polymer composed of 1,000 segments.
●: n-propyl laurate-polystyrene.
○: chloroform-cellulose acetate. (Flory (1942).)

different compositions (ϕ_2' and ϕ_2'') exist which satisfy eqn. (27.22). For given values of K the volume fractions ϕ_2' and ϕ_2'' may be calculated by utilizing the equilibrium condition (27.22), as has been done by Flory (1942) in deriving the results of Fig. 27.11.

Since $K = 2\alpha_1/RT$, variations in K arise from changes in the absolute temperature T, which is therefore given as an alternative ordinate. Clearly, below the critical temperature ($\approx 50°$ C.), the dilute polymer phase is almost pure solvent though the other phase contains appreciable quantities of both components. For a decrease of a few degrees, the polymer which is completely soluble at temperatures above 50° C. becomes almost insoluble (though considerably swollen) in the same solvent, for though the polymer at temperatures below the critical is almost insoluble in the solvent, the solvent is readily soluble in the polymer. The experimental data included by Flory in Fig. 27.11 demonstrates these points clearly.

The statistical and thermodynamical treatment of high polymers thus explains satisfactorily the general features of the behaviour of non-polar polymer-solvent systems. Discrepancies exist, and the simple theory outlined here undoubtedly requires further modifications (see p. 792, also Gee, 1946), but further treatment here is impossible.

Fractionation of polymers

In treating high polymers theoretically we have so far assumed them to be completely monodisperse (i.e. in molecular size and composition). Most polymers are, however, polydisperse at least with respect to chain-length and in some cases (e.g. rubber) with respect to chemical composition as well. We are concerned here only with the former type of polydispersity, chemical homogeneity being assumed.

From eqns. (27.25) and (27.26) it is evident that such polydispersity must cause a blurring of the critical phenomena, reducing the apparent temperature coefficient of solubility of the polymer and its 'all or nothing' solubility. Thus, under certain conditions, the polymer of lower chain-length may be soluble, whilst the longer chain-lengths are quite insoluble. This is the basis of fractionation procedures. Clearly polymers of different chain-lengths possess different critical constants, and, on the basis of the Flory treatment of critical phenomena, it should be possible to work out the conditions of solubility and precipitation for each individual chain-length. However, to understand the difficulties involved it is necessary to consider the more practical aspects of fractionation.

Theoretically the processes leading to fractionation are to be regarded as changes in the value of the constant K, which, from its definition $(K = 2\alpha_1/RT)$, can be accomplished merely by a change in temperature or in solvent composition, mixed solvents being commonly employed. In practice, fractionation by the latter process has mainly been employed, suitable mixtures of solvent and non-solvent (or precipitant) being used. The most satisfactory method is that of fractional precipitation from a *dilute* solution (< 1 per cent.) in a good solvent, usually by careful addition of the non-solvent, local regions of high precipitant concentration being avoided by vigorous stirring. After adding sufficient non-solvent to precipitate a fraction, the turbidity thus produced may usually be removed by slight warming, after which slow cooling and subsequent thermostatting for several hours cause precipitation of a fraction which is less contaminated with soluble lower chain-length material than in straightforward rapid precipitation. The fractions of cellulose nitrate for which η_{sp}/c. v. c is plotted in Fig. 11.12 were thus obtained from 1 per cent. acetone solution by the addition of water.

Fractional solution may also be employed, a series of solvents of increasing solvent power being used to extract the unfractionated polymer. However, it would seem that the slowness of reaching equilibrium, which is inseparable from the use of a solid or condensed gel structure, severely limits the success of the method. The method of using a single good solvent for extraction, relying on the different rates at which polymers of differing chain-length diffuse into the solvent, must inevitably achieve a very incomplete separation also.

Temperature variation of a dilute polymer solution which is brought to the point of incipient precipitation by the addition of non-solvent provides the possibility of a good fractionation. Careful thermostatting for several hours at a series of decreasing temperatures allows the recovery of several well-defined fractions.

It must be understood, however, that no single-stage fractionation can produce a really homogeneous fraction, and that re-fractionation from very dilute solution ($\ll 1$ per cent.) is recommended wherever possible. The work of Flory (1943) on polyisobutylenes was performed thus, and probably represents one of the best fractionations yet accomplished. The whole fractionation procedure, especially if attended by quantitative measurements, is extremely tedious, and very little complete data is yet available.

The use of mixtures of solvent and non-solvent is widespread in fractionations and from this arises one of the difficulties in the theoretical treatment. In the two phases separating out in fractionation the two liquids are usually differently distributed; the mixture cannot therefore be treated as a single liquid with intermediate properties. Further, as the conditions (solvent composition or temperature) are changed during fractionation, the molecular configuration of the polymer, and therefore its entropy of solution, must vary. The theory outlined, however, considers only one such value appropriate to the completely flexible molecule. Chemical inhomogeneities (e.g. the presence of protein and nitrogenous matter in rubber) also undoubtedly add complication. The difficulties of treating fractionation in detail theoretically are therefore very great, but some progress has been made.

Schulz (1937) extended Brönsted's eqn. (27.20) to mixed liquids, assuming the constant λ to be a linear function of the solvent composition, and obtained the equation

$$\gamma = A + B/M, \tag{27.27}$$

where γ is the proportion of non-solvent which dissolves a polymer of molecular weight M, A and B being constants. This equation has proved useful experimentally in several polymer fractionations, but like the Brönsted equation it was originally based upon erroneous assumptions.

For each molecular species, i, in a mixed solvent, Schulz modified the Brönsted equation thus:

$$\ln\left(\frac{c_i'}{\alpha c_i''}\right) = -\lambda \frac{M_i}{RT}, \qquad (27.28)$$

α being a constant independent of M_i but depending on the polymer-solvent system. By assuming a range of M_i values and appropriate values of the constants, Schulz showed the need in fractionations of using only dilute polymer solutions. A more correct approach has been used by Gee (1942 a), who assumed an equation similar to (27.10) for the entropy of mixing and deduced for the distribution of two molecular species, i and j, the following relation:

$$\frac{1}{M_i}\ln\left(\frac{\phi_i'}{\phi_i''}\right) = \frac{1}{M_i}\ln\left(\frac{\phi_j'}{\phi_j''}\right). \qquad (27.29)$$

Using this relation, Gee confirmed with Schulz the need for dilute polymer solutions in fractionating, and the inadequacy of single-stage fractionations in providing homogeneous fractions.

The present theoretical treatment of fractionation is clearly unsatisfactory, and further advances, together with more quantitative experimental data, are to be expected.

Polymers in Solution

In this section we shall summarize the properties and behaviour in solution of those polymers whose molecules are predominantly asymmetric; various aspects have already been discussed in some detail individually. The greater part of the natural and synthetic polymers fall into this class. The behaviour of non-asymmetrical polymers in general is comparable in many features with that of the corpuscular proteins and will not therefore be discussed here.

Most polymers are normally polydisperse over wide limits, and although the mean molecular weight and range are of interest for particular samples of material, these quantities do not have the same general significance as in the case of the monodisperse corpuscular proteins. Measurements of such molecular weights by several methods, particularly with carefully fractionated materials, is, however, of importance in testing the basic assumptions of the methods for molecules of complex form. In addition to polydispersity we have the further complication of non-specific molecular configurations, fluctuations about a mean configuration occurring continuously as well as variations in this mean with the experimental conditions. The treatment, in terms of a simple geometrical model, of such complex molecular forms is thus open to grave doubts, and requires the most careful testing.

The difficulties arising from marked molecular asymmetry have already been stressed in earlier chapters. Reliable molecular weight data are not therefore very plentiful, much of that published, especially from viscosity methods, being only of a semi-quantitative nature. Further, as a result of the polydispersity of most polymers, it is not usually possible to compare differently derived values unless deliberately obtained by the same worker under carefully controlled conditions.

The work of Gralen (1944) on cellulose and its derivatives, which has already been mentioned, is amongst the most careful yet published. Table 27.3 summarizes the results of sedimentation velocity and diffusion measurements on cellulose nitrates prepared with the minimum of degradation from different celluloses (see also Table 27.1).

TABLE 27.3. *Molecular Weight and Shape of Cellulose Nitrates in Acetone*

(From Gralen, 1944.)

Nitrate of:	$s_0 \times 10^{13}$	$D_0 \times 10^7$	$M \times 10^{-3}$	f/f_0	a/b	$2b$ in A units	$2a$ in A units
Unbleached American linters	19·0	1·00	780	12·2	870	11·3	9,800
Bleached American linters	14·0	1·44	400	10·6	670	9·8	6,600
Chlorite-bleached linters	18·5	1·11	680	11·5	780	11·2	8,800
Sulphate cellulose	16·2	1·56	420	9·6	550	10·7	5,900
Sulphite cellulose	16·4	1·56	430	9·6	550	10·8	5,900
Holocellulose from spruce	21·6	2·57	340	6·3	250	13·1	3,200

Some of the nitrates referred to in Table 27.3 were also examined by the sedimentation equilibrium method, and molecular weights were calculated by the corrected equilibrium eqn. (8.17); for the nitrates from unbleached American linters and sulphite cellulose in amyl acetate the molecular weights 618,000 and 454,000 were respectively obtained, as against 780,000 and 430,000 from sedimentation and diffusion. Although the agreement thus quoted is by no means perfect, the corrected equation is shown to be a great advance on the ideal equation (8.6). A considerable amount of published data derived from the ideal equation is of very doubtful quantitative value.

The axial ratios (a/b) of column 5, derived from the frictional ratios of the previous column, were calculated upon the basis of unsolvated elongated ellipsoids of rotation; from the molecular volumes (derived from molecular weights), the dimensions (a and b) of the ellipsoids were also obtained. In view of the assumptions and approximations involved, a close analysis of these dimensions is not justifiable but, if these values can be accepted, the lengths ($2a$) are somewhat shorter than those of the completely stretched-out molecules. This effect may arise spuriously from neglect of solvation, for the molecular weights of column (3) refer to the unsolvated molecule, whilst frictional ratios refer to the solvated entity. The molecular volumes to be compared with axial ratios are the solvated values, which, for cellulose nitrate in acetone, are likely to be considerably in excess of the unsolvated ones. Increase in both calculated dimensions would thus result from consideration of this increased molecular volume. In the absence of quantitative data on solvation, it is not possible to decide between the alternatives of a solvated stretched chain and a slightly folded unsolvated one. Although no check by other methods of the molecular weights and dimensions of the cellulose nitrates is available, Gralen did confirm that the Staudinger law does not apply to cellulose in cuprammonium, a fact that has now been demonstrated for several systems (e.g. see p. 369).

Wissler (1940) also found that the molecular lengths of cellulose nitrates, as determined by viscosity and streaming birefringence methods, were smaller than the maximum outstretched lengths, a conclusion which finds some confirmation not only in the work of Gralen (1944) but also in that of Mosimann (1943). A similar conclusion was also obtained (p. 180) from a light-scattering investigation of cellulose acetate fractions, the discrepancy being greater with the higher molecular weight fractions ($M > 80,000$).

The conclusions of Signer and Gross (1933), later confirmed by Signer and Sadron (1936), on the flexibility of the high molecular weight polystyrene molecule in solution deserve mention here (see p. 405). The discontinuity in the curve of birefringence against velocity gradient (G) at a definite value of $G\eta$ is reasonably interpreted as an indication of the straightening of originally curled molecular chains. The range of relaxation times required to explain the dielectric properties of solid polystyrene and other polymers is also evidence in support of the motion of segments of a polymer chain (rather than of the whole molecule) and therefore of chain flexibility.

The lowered intrinsic viscosity of certain linear polymers (including polystyrene) in 'poor' solvents also indicates considerable molecular flexibility, increasingly stretched configurations being obtained (on an average) in solvents which interact strongly with the polymer. Light-scattering confirmed such variations in configuration, though the actual changes in molecular length so derived were of considerably smaller magnitude than to be expected from parallel viscosity measurements— a further clear demonstration of the need for examining a system not by one but by several methods, under parallel conditions. Factors other than the nature of the solvent-polymer interaction are also involved, of course, and clearly such variation in molecular configuration may be prevented by the nature of the polymeric chain, e.g. as in cellulose and its derivatives where the —O— bonds linking adjacent glucose residues are somewhat inflexible, or in certain polymers by cross-links of various kinds.

Although quantitative agreement is largely absent, the generally asymmetrical nature of most polymer molecules in solution is undoubted and certain general conclusions for their future experimental investigation have become clear. The interactions between dissolved polymer molecules are important at concentrations (e.g. $\approx$ 1 per cent.) normally considered dilute for high molecular weight corpuscular molecules or low molecular weight solutes. The study of sedimentation velocity thus shows particularly clearly how the behaviour of asymmetrical molecules at the higher concentrations (> 0.5 per cent.) is typical no longer of the single molecules but of a complex aggregate. Only at lower concentrations (< 0.5 per cent.) do molecules of different chain-length begin to show characteristic differences in behaviour which are more accentuated the lower the concentration. The complications arising from intermolecular interaction undoubtedly differ according to the property studied, but are never absent. Attempts to calculate their effects have been made with some success, as in the equilibrium methods, but a satisfactory treatment even in these cases is not yet available. In other cases, especially amongst the dynamic methods, the problem is more complex, and a solution more distant. Immediate progress depends, therefore, on the elimination of interaction effects through the use of low concentration measurements and reliable extrapolation procedures, which, in their turn, are dependent upon a substantial refinement of experimental techniques.

REFERENCES

BRÖNSTED, 1938, *C.R. Lab. Carlsberg*, Ser. chim., **22**, 99.
BUNN, 1939, *Trans. Faraday Soc.* **35**, 482.
―― 1946, in *Advances in Colloid Science*, vol. ii, Interscience, p. 95.
―― and GARNER, 1942, *J. chem. Soc.*, p. 654.
CAMPBELL, 1945, Thesis, Cambridge University.
―― and JOHNSON, 1945, *Ministry of Supply Report*.
CAROTHERS and HILL, 1932, *J. Amer. Chem. Soc.* **54**, 1579.
COUMOULOS, 1943, *Proc. Roy. Soc.* A **182**, 166.
ELLIS, 1935, *The Chemistry of Synthetic Resins*, Reinhold.
FARMER, 1946, in *Advances in Colloid Science*, vol. ii, Interscience, p. 299.
FLORY, 1941, *J. chem. Phys.* **9**, 660.
―― 1942, ibid. **10**, 51.
―― 1943, *J. Amer. chem. Soc.* **65**, 375.
―― 1945, *J. chem. Phys.* **13**, 453.
―― and REHNER, 1943, ibid. **11**, 521.
FULLER, 1938, *Industr. Engng. Chem.* **30**, 472.
GEE, 1942, *Rep. Progr. Chem.* **39**, 7.
―― 1942 a, *Trans. Faraday Soc.* **38**, 276.
―― 1946, in *Advances in Colloid Science*, vol. ii, Interscience, p. 145.
―― 1947, *J. chem. Soc.*, p. 280.
―― and TRELOAR, 1942, *Trans. Faraday Soc.* **38**, 147, and later papers.
GEHMAN and FIELD, 1944, *J. Applied Phys.* **15**, 371.
GRALEN, 1944, Thesis on *Sedimentation and Diffusion Measurements on Cellulose and Cellulose Derivatives*, Uppsala.
GUGGENHEIM, 1945, *Trans. Faraday Soc.* **41**, 107.
GUTH, JAMES, and MARK, in *Advances in Colloid Science*, vol. ii, Interscience, p. 253.
―― and MARK, 1934, *Mh. Chem.* **65**, 93.
HAWORTH, 1939, *Chem. and Industr.* **58**, 917.
HENGSTENBERG and MARK, 1928, *Z. Kristallogr.* **69**, 271.
HILDEBRAND, 1936, *Solubility*, Reinhold.
HUGGINS, 1942, *J. phys. Chem.* **46**, 1; *Ann. N.Y. Acad. Sci.* **43**, 1.
―― 1943, *Industr. Engng. Chem.* **35**, 216.
JOHNSON, 1948, *J. Polym. Sci.*, in the press.
KUHN, 1934, *Kolloidzschr.* **68**, 2.
―― 1936, ibid. **76**, 258.
MARSH and WOOD, 1942, *Introduction to the Chemistry of Cellulose*, Chapman & Hall, 2nd edn.
MEYER, 1942, *Natural and Synthetic High Polymers*, Interscience.
―― and MARK, 1929, *Z. phys. Chem.* B **2**, 115.
―― and MISCH, 1937, *Helv. chim. Acta*, **20**, 232.
―― and VAN DER WYK, 1937, ibid., **20**, 1313.
MOSIMANN, 1943, ibid., **26**, 61.
ORR, 1944, *Trans. Faraday Soc.* **40**, 306.
OTT, 1943, *Cellulose and Cellulose Derivatives*, Interscience.
SCHULZ, 1937, *Z. phys. Chem.* A **179**, 321.
SHEPPARD and CLAPSON, 1932, *Industr. Engng. Chem.* **24**, 782.
―― and SUTHERLAND, 1945, *Trans. Faraday Soc.* **41**, 261.

STAUDINGER, 1932, *Die Hochmolecularen Organischen Verbindungen*, Springer, Berlin, p. 114.
TRELOAR, 1943, *Rep. Progr. Phys.* **9**, 113.
—— 1944, *Trans. Faraday Soc.* **40**, 59.
WALL, 1942, *J. chem. Phys.* **10**, 485.
WARD, 1945, *Colloids, Their Properties and Applications*, Blackie, p. 82.
WISSLER, 1940, Thesis, Bern.
WOODS, 1946, *Nature*, London, **157**, 229.

XXVIII
MEMBRANES

So far membranes have been mentioned solely in connexion with their use in the study of colloidal phenomena, such as dialysis and ultrafiltration, osmotic pressure, membrane equilibria, etc. They also have to be included in their own right, for not only does one dimension (the thickness) normally fall within the colloidal range ($ca.$ 50 A–$ca.$ 10 μ), but the constituent materials are frequently natural or synthetic polymers. In addition membranes are closely related to the gels (Chapter XXI), and their permeability is quite frequently determined by what happens at the boundary surfaces, for the further analysis of which an approach through surface chemistry is essential.

A little thought shows at once that with membranes it is the question of 'permeability' which far outweighs all others. This is clearly so with the artificial membranes used in the various colloidal phenomena mentioned above, and in biological systems the permeability behaviour of certain membranes may be of decisive importance in maintaining the biological function and integrity of the living cell. It is with this question always in mind that the structure and general behaviour of both natural and artificial membranes is considered below. The separation of membranes into natural and artificial is not essential but it is convenient, since with the former more is known about their composition, and the production of uniform specimens suitable for experimentation is easier.

ARTIFICIAL MEMBRANES

The principal laboratory uses of this type of membrane in connexion with colloidal systems have already been mentioned: in addition, the study of their structure and permeability is one obvious means of approach towards an analysis of the behaviour of biological membranes. As regards their widespread commercial and industrial applications it should suffice merely to recall to mind a few household words such as motor tyres, balloons, waterproof fabrics, etc.

(a) Preparation of principal types

Artificial membranes may be subdivided further according to whether they are composed of inorganic or organic materials. The former include some of the earliest studied membranes such as copper ferrocyanide (deposited on some porous support), also the zeolites, whose 'sieve' properties were discussed in Chapter XXV, and sintered glass membranes. Compared with those made from organic materials, these inorganic membranes have been little studied and are much less akin to biological membranes; for these reasons they will not be considered in any detail here.

Considering, then, organic membranes only, these can be subdivided again into (i) those consisting of polymeric molecules, whether of natural or synthetic origin, and (ii) thin oil films.

(i) *Membranes from natural and synthetic polymers*

The natural polymers and their derivatives have, up to now, been by far the most commonly used. Cellulose, which forms the basis of the principal group, is sometimes employed in the form of parchment paper or as 'cellophane' (regenerated cellulose in the form of sheet), but more often as the nitro, acetyl, or ethyl derivative, of which nitrocellulose undoubtedly takes pride of place. Of other natural and derived polymers, rubber, gelatine (frequently after hardening by chemicals such as formaldehyde), and alginic acid and its derivatives also require mention.

The synthetic polymers offer some advantages over the natural polymers and their derivatives, in that their molecular groupings (repeat units) can be more readily and more systematically varied. Such variations will influence certain molecular properties such as the flexibility of the long chains (which in turn affect the ease of packing to form micelles or crystallites and hence the crystalline/amorphous ratio (p. 777)), as well as the solubility of substances in the membrane. Synthetic polymers with a wide range of 'polarity' are now available, ranging from the completely non-polar materials such as polythene and polystyrene, through the somewhat polar methacrylates and acrylates, to the more polar nylons and the organo-silicon polymers (Chapter XXVII). Again, in each of these classes systematic changes can be brought about by suitable substitutions, as in the series methyl, ethyl, n-propyl, etc., methacrylates.

Two component membranes, in which the preponderating component is a polymer, have also been used for certain purposes. For example, plasticizers or swelling compounds may be added to the original polymer solution, and these, if not removed during the formation of the membrane, will modify the properties of the resulting film. Dyestuffs form another group of substances commonly added to modify permeability properties.

The preparation of membranes from polymeric materials would not, at first sight, appear to be difficult, since films are readily obtained by removing the solvent from a solution of the polymer. An *ideal* membrane should, however, not only be uniform in thickness but also in regard to 'pore-size', i.e. the spaces between the single molecules, micelles, or crystallites which constitute the membrane. Unless the membrane is very thin ($<$ *ca.* 0·1 mm.) the former in general presents little difficulty, the whole problem resolving into the ways of obtaining uniform pore-size. The production of uniformly permeable membranes has been advanced considerably in connexion with ultrafiltration, due largely to the work of Zsigmondy, Bechhold, and Elford.

Elford (1937) and Grabar (1937) have examined in some detail the preparation of strong reproducible nitrocellulose membranes having graded porosities ('gradocol' membranes). The change from the initial sol state (solution) to the requisite gel state (see below) may be brought about in various ways. When using a volatile solvent (e.g. ether/alcohol) a thin layer is spread in a dish and evaporation allowed to occur under controlled conditions of temperature and humidity until a gel film forms, which is then equilibrated against water. With non-volatile solvents such as acetic acid, a thin film is held in a suitable support (e.g. filter paper) and subjected to the precipitating action of water. Of these methods the use of volatile solvents seems to give superior uniformity in structure.

For the preparation of good 'gradocol' membranes it is essential to obtain first of all regular gels, and to preserve this regularity during the 'hardening' of the gel by water. To obtain a gel the original nitrocellulose solution must contain 'gelifying liquids' (i.e. liquids which cause swelling but not complete solution), such as ether, ethyl or propyl alcohols. (If only solvent liquids such as ethyl acetate or acetone are used a very compact and relatively impermeable *film*, rather than a *gel*, will result.) By varying such factors as solvent/non-solvent ratio, concentration of nitrocellulose, conditions of evaporation, etc., the porosity of the resultant membrane can be controlled between wide limits (from *ca.* 1 μ down to ordinary molecular dimensions).

(ii) *Thin oil films*

Oil films constitute a different class of artificial membranes, and as models for biological membranes their permeability behaviour has been studied by Overton, Osterhout, and others. The wide variety of 'oils' used are usually classed as 'neutral' (e.g. benzene, long-chain paraffins such as 'nujol', higher alcohols such as amyl or octyl, esters such as octyl acetate, etc.), 'acid' (e.g. phenols, oleic acid), or 'basic' (e.g. aniline, long-chain amines). For some purposes mixtures of oils may be preferable, or the oil may be 'thickened' by solution of a suitable polymer.

Clearly the difficulties of the previous section concerning the preparation of films of uniform thickness and porosity do not arise with these oil layers. Thin films may be prepared and studied by the 'hanging drop' technique as shown in Fig. 18.6 (p. 522), although with thicknesses below about 1 μ they become as a rule rather too fragile for easy study. This is rather a limitation since biological membranes (see below) are frequently believed to be much thinner (of the order 0·01 μ).

(b) Structure

The question of structure only arises in the case of rigid membranes. Originally these were represented diagrammatically as continuous bodies traversed by pores, but with the development of spatial models of polymeric substances they can better be considered as a network of primary valence chains with (in general) lyophobic and lyophilic side-chains in lateral positions. For example, gelatine contains polypeptide chains
$$-CO.NH.\overset{\overset{\text{R}}{|}}{CH}-CO.NH-\overset{\overset{}{|}}{\underset{\underset{\text{R}^1}{|}}{CH}}-,$$ where the side-chains R, R^1, etc., may be neutral (e.g. —CH$_3$, —C$_6$H$_5$OH), acidic (e.g. —CH$_2$.COOH), or basic (e.g. —CH$_2$.NH$_2$). As we shall see, the presence of these ionizable groupings plays an important part in controlling the permeability of membranes to ions.

The primary valence chains forming the network of the membrane will frequently be aggregated in the form of bundles (micelles or crystallites), as depicted in two-dimensions in Fig. 28.1, the same molecular chain possibly forming part of two (or more) such bundles. The constituent primary chains will be held together by secondary forces, and in the case of the very porous membranes (e.g. *ca.* 1 μ pore-size) it is clear that each bundle or micelle must be several molecules in length. With the membranes used for ultra-filtration the length of the channels may be about 100 to 10,000 times the average pore diameter, and since the latter is of ultramicroscopical dimensions, surface effects (e.g. adsorption and denaturation) are frequently important.

'Gradocol' membranes are graded in terms of their 'average pore diameter', which is found from the rate of flow of water through the membrane. In addition it is important to know the degree of uniformity, a convenient criterion for which is the ratio of the maximum pore-size (from the critical air pressure) to the average pore-size (from the rate

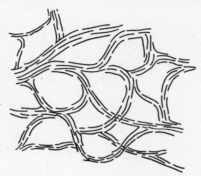

Fig. 28.1. Structure of porous membrane (diagrammatic).

of flow of water) (see Elford, 1937). For gradocol membranes this ratio is about 2.

(c) Permeability

As stressed earlier, the problem of diffusion through membranes, particularly in its bearing upon biological systems, is of paramount importance. Depending upon its type and upon the prevailing conditions a membrane may be more or less permeable to a great variety of substances, e.g. gases and vapours, either in the gaseous state or in solution, simple organic molecules, salts, protein molecules, viruses, and so on. It is convenient as before to consider the rigid (chiefly polymeric) and non-rigid (oil) types of films separately.

The process of diffusion across any membrane is usually regarded as composed of three consecutive events which can be treated separately: the molecule first jumps from the solution (or gas phase) into the membrane, it then diffuses through the membrane, and finally it jumps from the membrane into the phase on the far side.

Quantitatively the 'permeability' or 'permeability coefficient (P) of a membrane to a substance X' may be defined by the equation

$$Q = P\frac{A}{d}t\Delta c, \qquad (28.1)$$

where Q is the amount of substance X diffusing in time t across an area A of membrane of thickness d, under a concentration difference Δc (see also Chapter X).

For biological membranes, where d is seldom known at all accurately, eqn. (28.1) is often replaced by
$$Q = P' A t \Delta c, \tag{28.2}$$
where $P' = P/d$.

(i) *Permeability of rigid membranes*

Traube, in his early studies of osmosis, made much use of precipitation membranes (e.g. copper ferrocyanide on a porous support), and ascribed their semi-permeability to their having pores sufficiently large to permit the passage of solvent (water) but too small for the solute molecules (e.g. sugar). This selectivity is termed 'sieve' action, and from the work of Collander (1924) it does appear to suffice for certain membranes such as copper ferrocyanide.

Later work showed, however, that permeation through membranes frequently involved other factors as well, of which, as we shall see, solubility in the membrane and the charge carried by the membrane frequently predominate. The substances used in permeability studies can be grouped as follows: gases (and vapours), non-electrolytes, electrolytes, colloidal molecules, and organisms. Their principal points of interest are outlined below.

Gases (and vapours). For diffusion of gases or vapours eqn. (28.1) becomes
$$Q = P \frac{A}{d} t \Delta p, \tag{28.3}$$
where Δp is the pressure drop across the membrane.

If the permeation process involves solution in the membrane and if Henry's law is obeyed then
$$c = Sp, \tag{28.4}$$
where c is the solubility, p the pressure, and S the solubility coefficient.

From Fick's first law (p. 233)
$$Q = DAt\left(\frac{\partial c}{\partial x}\right)$$
and for a state of steady flow
$$\frac{\partial c}{\partial x} = \frac{\Delta c}{d}.$$

From eqn. (28.4) $\Delta c = S \Delta p$, and hence from eqns. (28.1) and (28.3)
$$P = DS. \tag{28.5}$$

Thus, on this basis, the permeability coefficient is the product of two factors, the solubility coefficient (S) and the diffusion coefficient (D).

If the permeation process involves diffusion through capillaries then Graham's law should hold, i.e.

$$P \propto \frac{1}{\sqrt{M}}, \quad \text{or} \quad PM^{\frac{1}{2}} = \text{constant}, \tag{28.6}$$

where M is the molecular weight of the diffusing substance.

For experimental study the membrane, of measured area and thickness, is supported between wire gauze or perforated metal sheets, as shown in Fig. 28.2. The pressure drop across the membrane is measured by suitable manometers (depending on the pressures involved). If one side is evacuated then a single pressure measurement and the rate of diminution in volume of the non-evacuated side (at constant p) suffice for the determination of P. (For details see, for example, Barrer, 1941; Doty, Aiken, and Mark, 1944.)

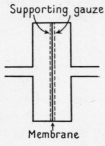

FIG. 28.2. Set-up for permeability studies (diagrammatic).

The permeability of dry and moistened collodion membranes to a number of gases, such as H_2, N_2, O_2, CO_2, HCl, and NH_3, has been measured by Northrop (1929). For the dry membrane and the gases H_2, N_2, O_2, and CO_2 he found the ratio of permeabilities to be $7 \cdot 5 : 0 \cdot 45 : 1 \cdot 1 : 7 \cdot 0$, and that saturating the membrane with water had practically no effect on the rates. These results showed conclusively that solution in the membrane, involving a distribution coefficient of the permeating molecule, was involved, for if diffusion occurred only through pores then by eqn. (28.6) the rates for $H_2 : N_2 : O_2 : CO_2$ should be in the ratio

$$7 \cdot 5 : 1 \cdot 7 : 1 \cdot 6 : 1 \cdot 4.$$

Also any pore diffusion would be enormously reduced by saturating the membrane with water, since this effectively blocks the pores.

According to eqn. (28.3) the permeability coefficient P should be independent of thickness. This is so in many cases, as shown for example by the data for water vapour and polythene (cast) shown in Table 28.1 (from Doty, Aiken, and Mark, 1944). With systems in which the gas has a high affinity for the membrane, e.g. water–cellulose, P in general varies not only with membrane thickness, but also with the pressure difference.

P usually increases exponentially with temperature ($P = P_0 e^{-E/RT}$) as found for the permeation through various polymer films of the permanent gases (see Barrer, 1941), and of water vapour, some data for

TABLE 28.1

Thickness (μ)	$P \times 10^8$ (Q in c.c., A in cm.^{2}, d in mm., t in sec., Δp in cm. Hg)
58	2·38
105	2·46
163	2·40

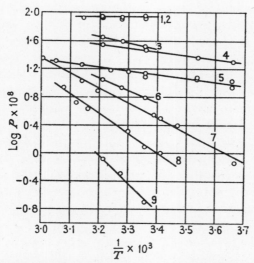

FIG. 28.3. Variation with temperature of the permeability coefficient (P) for water vapour and various polymer films.

the latter system being shown in Fig. 28.3 (from Doty, Aiken, and Mark, 1946). The 'activation energies' (E), for permeation of water vapour through the various polymer films of Fig. 28.3, are given in Table 28.2.

TABLE 28.2

Film No.	Material	E (cal./mole)	$Log\, P_0$
1, 2	Polystyrene	0	−6·08
4	Polyvinyl chloride-acetate	2,350	−4·80
5	Polyvinyl chloride	2,350	−5·10
6	Polyethylene (calendered)	8,000	−1·34
7	,, (cast)	10,200	+0·13
8	Rubber hydrochloride	12,800	+1·63
9	Polyvinylidene chloride	17,500	+4·20

In addition to P, D and S also vary exponentially with temperature, i.e. $D = D_0 e^{-E_D/RT}$ and $S = S_0 e^{-\Delta H/RT}$. Here E_D is the energy of activation for diffusion, and since S is an equilibrium quantity (as contrasted with P and D which are essentially rate constants), ΔH may be identified as the heat of solution. P_0 and D_0 correspond to frequency factors in the Arrhenius equation of chemical kinetics.

The transfer of gases across these polymer membranes is believed to be an activated diffusion process, just as in the zeolites (Chapter XXV). Gas molecules dissolve in the surface layers and move about by jumping into holes which form and disappear in their immediate neighbourhood owing to the thermal agitation of the segments of the polymer molecules. The concentration gradient in the film then causes a drift of the dissolved gas towards the side of lower concentration. From an application of the 'transition state' approach to reaction kinetics Barrer (1941) concluded that the activated state for diffusion involves many degrees of freedom, indicative of the formation of a 'hole' for the diffusing molecule to move into, a conclusion supported by some recent results and calculations of Doty (1946).

The effect of plasticizers, at any rate for the permeation of water vapour through linear polymers, was shown by Doty (1946) to be due primarily to their lowering the heat of solution, and the changes in the entropy of solution support the very reasonable hypothesis that the water molecules have more freedom in the plasticized than in the unplasticized polymer.

Non-electrolytes (*in solution*). The apparatus shown in Fig. 28.2, after the addition of a suitable stirring mechanism, is clearly applicable to the study of permeating molecules in solution, and by suitable analytical methods the permeability coefficient may be determined.

As with gases and vapours, permeation may occur either through the pores, or by solution in the membrane. In the former case transfer is clearly an ordinary diffusion process, and if the Einstein equation of diffusion, $D = (RT/N)(1/6\pi\eta r)$ is applicable, then for a series of spherical molecules

$$P\left(\frac{M}{\rho}\right)^{\frac{1}{3}} = \text{a constant,} \qquad (28.7)$$

(M is the molecular weight and ρ the density). The temperature coefficient will be mainly that of the viscosity of the medium (i.e. the same for all molecules and usually small). Einstein's equation is certainly valid for spherical colloidal molecules but deviations occur with asymmetric colloids and with small molecules such as urea, glycerol, etc. Disagreement with eqn. (28.7) in the case of a series of molecules, particularly if coupled with high temperature coefficients, indicates permeation by solution in the membrane.

The process of diffusion through a thin organic layer (e.g. polymer or oil film) lying between two aqueous solutions (as is commonly the case) will be somewhat as follows. Under conditions of rapid stirring on both sides of the membrane the whole resistance to diffusion can be regarded as located at the interfaces and within the bulk of the membrane. Let us consider two molecules of comparable size but which differ to the extreme in their affinities for the aqueous and organic phases, for example a short-chain hydrocarbon on the one hand and a polar molecule such as urea or glycerol on the other. The weak attachment of the former to the water molecules enables it to leave the aqueous phase very readily, whereas with polar molecules such as glycerol relatively strong links (hydrogen bonds in this case) have to be broken. Also it seems very probable that the more polar types of organic molecules, such as the alcohols, fatty acids, esters, etc., will diffuse through the membrane as aggregated (frequently dimeric) molecules, whereas in water they are molecularly dispersed. Such changes, the former involving a high energy barrier, the second a low probability (or large 'steric factor'), will tend to reduce the rate of permeation of polar molecules relative to those of the non-polar type.

Electrolytes. Compacted polymer films, as prepared, for example, by evaporation from a good solvent (cf. p. 806), would be expected to be practically impermeable to salts, since even water with its relatively much less polar character permeates only extremely slowly. On the other hand the type of porous membranes used in osmotic pressure and ultra-filtration work, which allow water to pass very readily, would be expected to be equally permeable to dissolved salts. The behaviour of compact films accords with expectation, but with the porous type, on the other hand, many examples are known which do not allow the two ions of a salt to permeate equally readily. Parchment paper and collodion, for example, allow cations to pass more readily than anions; a gelatine membrane is respectively anion or cation permeable depending upon whether the pH is below or above its iso-electric point. Dyeing a membrane with acid dyes renders it cation permeable; with basic dyes it becomes permeable to anions. Such observations clearly rule out any explanation based on 'sieve' action, and point to the electrical charge on the membrane as the decisive factor (Wilbrandt, 1935; Teorell, 1935; Meyer and Sievers, 1936).

The charge is believed to arise primarily from ionogenic groups fixed (by primary bonds as a rule) in the membrane material, for example —COO' in cellulose, —SO$_4'$ and —COO' in nitrocellulose. Such cases as these, with fixed anions and mobile cations, will be cation permeable, since the cations may be displaced if a supply of others is maintained from one side, whereas anions cannot enter owing to the electrostatic repulsive action of the negatively charged framework. (Similarly, a positively charged framework will be anion permeable.) The concentration of fixed ions (in gm. equiv./l. imbibed liquid) is a characteristic quantity termed the 'selectivity constant', A. (For the preparation of 'permselective' collodion membranes, which combine extreme ionic selectivity with high permeability, see Gregor and Sollner, 1946.)

If the membrane separates two identical salt solutions, then on passing a current across it the ratio of the transport due to cations and anions respectively (i.e. n_K/n_A) will depend upon the relative numbers of cations and anions traversing the membrane (n_K and n_A have been termed 'traversal numbers' to distinguish them from 'transport numbers' for the more usual aqueous media, although they may be determined by the same methods). n_K and n_A will be proportional to the number and mobilities in the membrane (U_K and U_A respectively) of the mobile ions, hence

$$\frac{n_K}{n_A} = \frac{U_K(y+A)}{U_A y}, \qquad (28.8)$$

where y = concentration of salt (assumed uni-univalent) in the membrane and A the selectivity constant defined as above.

Considering first a membrane with wide pores so that 'sieve' effects can be neglected, it can be assumed that the distribution of salt between the interior of the membrane and the bulk liquid will be governed by a Donnan equilibrium (Chapter VII), which gives (neglecting activity coefficients)

$$y(y+A) = C^2,$$

where C = concentration of salt in the surrounding liquid.

(The state of affairs in the case of NaCl and a negatively charged membrane is shown in Fig. 28.4.) Hence

$$\frac{n_K}{n_A} = \frac{U_K}{U_A}\frac{\sqrt{(4C^2+A^2)}+A}{\sqrt{(4C^2+A^2)}-A} = \frac{U_K}{U_A}R. \qquad (28.9)$$

The 'ionic selectivity', which may be expressed by the ratio n_K/n_A, is thus given by eqn. (28.9). If $A = 0$, then $R = 1$, and the membrane will not influence ionic diffusion, but if $A \neq 0$, then it will do so to an extent depending upon the ratio C/A, as shown by the following figures:

C/A	.	10	1	0·1	0·01	0·001
R	.	1·1	2·6	101	10^4	10^6

The selectivity thus increases as the ratio of the added salt to the fixed ions decreases, thus providing an explanation for the long-known observation that ionic selectivity is much greater at lower than at higher salt concentrations.

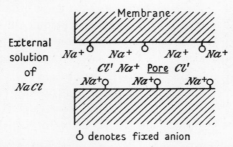

FIG. 28.4. Structure of pore in a membrane containing fixed anions in the presence of NaCl solution.

Ionic selectivity is not, however, the sole factor governing the permeability of membranes to ions; 'sieve' and 'solubility' effects may also enter. The former will become increasingly important with decreasing pore-size, and for a given pore-size with increasing salt concentration, since the 'ionic' effect discussed above decreases with the salt concentration.

Even if charge effects were absent the presence of attractive forces (e.g. van der Waals) between the interior surfaces of the membrane and the mobile ions would tend to increase the concentration of salt in the membrane above that in the bulk solution. 'Solubility' of salts in the membrane thus becomes of particular importance with organic ions. Denoting the partition coefficient by S, the membrane phase by M and the aqueous phase by W, we have $S_K = C_K^M/C_K^W$ and $S_A = C_A^M/C_A^W$ for cation and anion respectively.

Applying the Donnan condition, we have
$$C_K^M C_A^M = C_K^W C_A^W,$$
or
$$y(y+A) = S_K S_A C^2, \qquad (28.10)$$
where C, y, and A have the same meanings as before.

Eqn. (28.10) can readily be transposed into the same form as eqn. (28.9) with which it is identical, except that A has now become

$A/\sqrt{(S_K S_A)}$, thus showing that an increased solubility in the membrane (i.e. $S_K S_A > 1$) leads to a decrease in selectivity. With very hydrophilic membranes, with consequent high water content, this solubility effect is small.

Allowing for both sieve and solubility effects eqn. (28.9) thus becomes

$$\frac{n_K}{n_A} = \frac{U_K}{U_A} \frac{\sqrt{(4C^2 S_K S_A + A^2)} + A}{\sqrt{(4C^2 S_K S_A + A^2)} - A}, \qquad (28.11)$$

where U_K/U_A is the ratio of migration rates in the membrane under the influence of the sieve effect, and S_K and S_A the partition coefficients of the ions with respect to membrane and water.

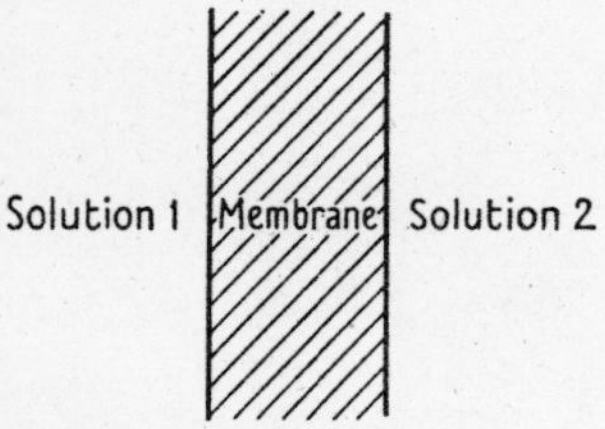

These two essential properties of the membrane—its sieve action with respect to different ions (as expressed by the ratio U_K/U_A) and its selectivity constant (A), can be found by measuring the traversal numbers at different concentrations, preferably by the potentiometric method outlined below.

FIG. 28.5. Membrane separating two salt solutions.

The total membrane potential (E) for the simple case of two salt solutions of different concentrations separated by a membrane containing immobile anions is given by

$$E = E_1 + E_2 + E_D. \qquad (28.12)$$

Here E_1 and E_2 are Donnan potentials arising at the two membrane/solution boundaries and E_D the diffusion potential in the interior of the membrane (see Fig. 28.5).

Assuming solubility effects to be absent,

$$E_1 = \frac{RT}{F} \ln \frac{C_1}{y_1 + A} \quad \text{and} \quad E_2 = -\frac{RT}{F} \ln \frac{C_2}{y_2 + A}.$$

If the concentration across the membrane falls linearly, we can calculate E_D from Henderson's formula, i.e.

$$E_D = \frac{RT}{F} \frac{({}_I V_K - {}_I V_A) - ({}_{II} V_K - {}_{II} V_A)}{({}_I V_K + {}_I V_A) - ({}_{II} V_K + {}_{II} V_A)} \ln \frac{{}_{II} V_K + {}_{II} V_A}{{}_I V_K + {}_I V_A},$$

where ${}_I V_K$ = sum of mobility of cations in solution I (i.e. concentration × mobility).

For the above system

$${}_I V_K = U_K(y_1 + A), \qquad {}_I V_A = U_A y_1.$$
$${}_{II} V_K = U_K(y_2 + A), \qquad {}_{II} V_A = U_A y_2.$$

Substituting for E_1, E_2, and E_D we have

$$E = \frac{RT}{F}\left[u\ln\frac{x_2+Au}{x_1+Au} + \tfrac{1}{2}\ln\frac{(x_1+A)(x_2-A)}{(x_1-A)(x_2+A)}\right], \qquad (28.13)$$

where $\qquad u = \dfrac{U_K - U_A}{U_K + U_A} \quad$ and $\quad x = \sqrt{(4C^2 + A^2)}.$

This treatment of membrane potentials is of great interest in connexion with the interpretation of the potentials observed in biological systems. (For more general cases see Meyer and Sievers, 1936.)

When $A = 0$, i.e. the membrane is non-selective, eqn. (28.13) reduces to

$$E_{(A=0)} = \frac{RT}{F}\cdot\frac{U_K-U_A}{U_K+U_A}\ln\frac{C_2}{C_1}, \qquad (28.14)$$

the ordinary Nernst equation for diffusion potentials. In this case a single determination of E gives U_K/U_A and therefore n_K/n_A directly.

In the more general case ($A \neq 0$) at least two determinations of E are necessary to find the two unknowns A and U_K/U_A from eqn. (28.13). Direct numerical calculation is not easy, but Meyer and Sievers (1936) give a graphical method entailing measurement of the potential when the concentrations are varied but the ratio C_1/C_2 is kept constant. (See also Meyer and Bernfeld, 1945, and Sollner and Carr, 1944.)

Colloidal molecules and organisms. The preparation and structure of the highly porous 'gradocol' membranes used for ultra-filters were outlined earlier in this chapter.

It might be imagined that ultra-filters would function solely by virtue of a 'sieve' action, and indeed this has to be assumed when estimating the size of colloidal particles or small organisms by the ultra-filtration method. However, owing to the large internal surfaces in such membranes, surface effects (e.g. adsorption, which may be followed by denaturation in the case of proteins) are frequently dominating. To overcome this Elford (1937) recommends as medium Hartley's broth, an extract of digested animal tissue containing a somewhat capillary active constituent which diminishes adsorption troubles. Since most ultra-filtration work is carried out in the presence of salts (e.g. physiological saline) complications due to charge effects on the membrane are unlikely to be of great importance.

The 'filtration end-point' for a particular suspension is determined by filtering the system through membranes of progressively finer pores until one which just retains all the disperse phase is reached. Unfortunately no simple relation appears to exist between the mean pore-size

(d) of the limiting membrane and the particle diameter (σ). Elford has expressed the relation in the form $\sigma = \beta d$, where the factor β varies with d as shown below.

Limiting pore diameter (d)	Factor β
100–1000 A	0·33–0·50
1000–5000 A	0·50–0·75
5000–10,000 A	0·75–1·0

(For a detailed discussion of this relation see Elford, 1937.)

Subsequently the same author has provided a more detailed survey of filtration end-point data for systems of known particle size. Table 28.3 shows the relevant data for a number of systems (from Elford, 1937).

TABLE 28.3. *Filtration End-point Data for Systems of Known Particle Size*

System	'End-point' L.P.D. (A)	Particle size (A)	Ratio particle size/L.P.D.	Remarks
B. Prodigiosus	7,500	5,000–10,000	1·0 (av.)	Microscopical measurement.
Vaccinia virus	2,500	1,500–1,800	0·66 (av.)	Size by ultra-violet photography.
Infectious Ectromelia	2,000	1,300–1,400	0·67 (av.)	Size by ultra-violet light photography.
Haemocyanin (Helix)	550	240	0·44	Size by ultra-centrifugal analysis.
Edestin	180	80	0·44	Size by ultra-centrifugal analysis.
Serum pseudo-globulin (horse)	110–120	69	0·60	Size by ultra-centrifugal analysis.
Serum albumin (horse)	90–100	54	0·57	
Oxyhaemoglobin	100	50	0·50	Size by ultra-centrifugal analysis.
Egg albumin	60	43·4	0·72	Size by ultra-centrifugal analysis.

Each system when filtered was in Hartley's broth medium at pH 7·6–7·8.

It is clear that ultra-filtration is unlikely to approach the more precise methods such as sedimentation and diffusion as a means for size and shape of colloidal particles. Nevertheless, for biological organisms, and for exploratory work generally, it has undoubted advantages owing to its simplicity. Of the other uses of ultra-filtration dependent upon sieve-action mention may be made of the concentration and purification of viruses, toxins, enzymes, etc., and certain bacteriological uses such as sterilization of fluids, resolution of mixed cultures, etc.

(ii) *Permeability of thin oil films*

As far as permeability to non-electrolytes is concerned, oil membranes do not differ fundamentally from the compact type of polymer films. Rates of permeation will in general be greater owing to lower viscosity and higher solubility.

With electrolytes, on the other hand, liquid films cannot exhibit the general 'cation' or 'anion' permeability possible with rigid membranes since they lack the requisite 'fixed ions'. In the membrane the number of 'mobile' cations will be the same as the number of 'mobile' anions, and the ratio of transport numbers of the ions will equal the ratio of migration rates in the oil film. 'Sieve' effects are also absent from liquid membranes, so that the migration of ions through oil films is less specific than for membranes with a structure. Nevertheless the transport numbers in oils may differ considerably from those in water, probably due to solvation effects.

Accordingly, differences in the permeability of oil films to ions are believed to arise from differences in (*a*) mobilities, (*b*) partition coefficients, in the membrane.

The potentials set up when an oil layer separates two aqueous salt solutions of different concentration or different composition has been the subject of much study and discussion, since it bears directly upon the problem of bio-electric potentials. If the oil layer is acidic in character (e.g. contains oleic acid or cresol) then it behaves as a preferentially cation-permeable membrane—if basic as preferentially anion-permeable. For example, the system

$$M./10 NaCl | Oil | M./10 MX$$

gives a potential of 100 mV if the oil is cresol and MX is sodium oleate; if MX is aniline hydrochloride the potential is 50 mV and reversed in sign. According to Beutner the reason is that acidic substances tend to increase the solubility of cations in the membrane; basic substances will have the same tendency with anions.

For the simple system $NaCl(C_1) | Oil | NaCl(C_2)$ the steady state can, from the considerations of diffusion in oil layers advanced above, be considered as follows. In the oil layers immediately adjacent to the aqueous phases the salt concentration will be $C_1\sqrt{(S_A S_K)}$ and $C_2\sqrt{(S_A S_K)}$ respectively, where S_A and S_K as before are the partition coefficients for anion and cation. As in the case of the rigid membrane the observed potential (E) can be regarded as the sum of three potentials, one at each O/W interface and a diffusion potential (E_D) between the two concentrations of salt in the membrane, i.e.

$$E = E_1 + E_D + E_2$$
$$= \frac{RT}{F}\ln\frac{C_1\sqrt{(S_A S_K)}}{C_1} + \frac{RT}{F}\frac{U_K - U_A}{U_K + U_A}\ln\frac{C_2\sqrt{(S_A S_K)}}{C_1\sqrt{(S_A S_K)}} + \frac{RT}{F}\ln\frac{C_2}{C_2\sqrt{(S_A S_K)}}$$
$$= \frac{RT}{F}\frac{U_K - U_A}{U_K + U_A}\ln\frac{C_2}{C_1}. \tag{28.15}$$

As before, U_K and U_A are the mobilities of cation and anion in the membrane. Eqn. (28.15) is, as it should be, identical with eqn. (28.14), which was obtained from eqn. (28.13) for the special case $A = 0$.

If the membrane is only permeable to one type of ion, then U_A or U_K is zero. For the former case (only cation permeable) eqn. (28.15) reduces to
$$E = \frac{RT}{F}\ln\frac{C_2}{C_1}. \tag{28.16}$$

Natural Membranes

The membranes in living matter are as a rule quite thin, frequently only 100 A or so in thickness, and yet upon their powers of discrimination the ability of the biological system to maintain its normal function frequently depends. Systems which have been much investigated include erythrocytes, marine and other eggs, yeast, bacteria, plant cells, muscle cells, chitin membranes, frog skin, kidney, etc. Of these, space precludes discussion of more than a few representative examples. (For a further account see, for example, Davson and Danielli, 1943.)

Transport across biological membranes is usually divided into 'passive' and 'active', according to whether movement occurs from a region of higher to one of lower concentration (strictly activity), or conversely. The latter behaviour, which does not, of course, arise in artificial systems, is ascribed to metabolic processes in the cell supplying the requisite energy, although little is known concerning the mechanism.

(a) Structure

In connexion with the structure of natural membranes, knowledge of their chemical composition would clearly be invaluable. The extreme thinness of most natural membranes, however, usually renders direct analysis too difficult and an inferential approach is necessary. A number of lipoidal materials have been shown to be of wide occurrence in natural membranes, in particular cholesterol and phospholipoids such as lecithin, kephalin, and sphingomyelin. Some form of insoluble protein is usually present to impart the requisite degree of mechanical strength to the membrane framework.

The early work of Overton towards the close of the last century showed that fatty substances penetrated cell membranes much more readily than substances insoluble in fats, suggesting that the membrane consists largely of lipoidal material. Gorter and Grendel (1925), from

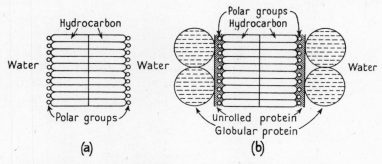

FIG. 28.6. Structure of natural membrane: (a) according to Gorter and Grendel; (b) according to Danielli.

the total lipoid of erythrocytes of several species (found by extracting with solvents and then measuring as a monolayer with the film balance), concluded that the membrane consisted essentially of a bimolecular leaflet, as depicted in Fig. 28.6 (a).

An alternative suggestion ascribed selectivity to 'sieve' action, the requisite mesh structure, with pores of molecular dimensions, arising from the polypeptide chains of proteins.

These two basic ideas have been drawn upon for other postulated arrangements, for example a mosaic made up of lipoidal and protein areas (i.e. in parallel), or one type superimposed on the other (i.e. in series). From the molecular and surface energy point of view the most plausible arrangement is that shown in Fig. 28.6 (b), suggested by Danielli (1938) from considerations of surface films. The denatured protein layer beneath, or more probably intermingled with, the polar groups of the lipoid is probably of great importance in conferring suitable mechanical rigidity to the whole membrane; its influence on diffusion is more problematical (see also Hass, 1939).

The thickness of the lipoid layer appears to be of the order of 50–100 A. This is indicated by the work of Gorter and Grendel mentioned above, as well as by electrical impedance and optical studies.

The electrical impedance of cells to high-frequency alternating currents can provide some information about the capacity and resistance of their membranes. It has thus been shown that the membrane has a very low conductivity, in contrast with the fairly high conductivity of the cell contents. A thin layer of lipoid would possess such a low conductivity, whereas a similar thickness of protein would show a much higher value. Cell membranes appear to have 'static' capacities of $ca.$ $1\,\mu F$ per cm.^{2}, giving a thickness of 30–150 A depending upon the value taken for the dielectric constant. (The value of 3, which is commonly taken, is reasonable for large fatty molecules.) The electrical resistance of most normal cells (in the resting, unexcited, state) lies between 10^3 and 10^6 ohms/cm.^{2} of membrane. If the specific resistance were 10^{10} ohms (the value for olive-oil in contact with salt solution) a membrane of resistance 10^4 ohms/cm.^{2} would have a thickness of $10^4/10^{10}$ cm., i.e. 100 A, in agreement with the values from capacity measurements. (For impedance studies on synthetic membranes see, for example, Dean, Curtis, and Cole, 1940; Goldman, 1943.)

Of the various optical methods for estimating the membrane thickness mention may be made of an ingenious utilization of multilayers (p. 496), used by Waugh and Schmitt (1940) for red cells. The intensity of light reflected from a dried red cell 'ghost' (i.e. after loss of cell contents) is matched against that from a step-film of barium stearate of known and graded thickness. Values from 120 to 200 A (depending upon the pH of the haemolysing solution) were thus found for the thickness of the red-cell membrane. After extraction with organic solvents (to remove lipoids) the values were approximately halved.

Plant cells differ from animal cells in usually possessing an outer, rigid, cellulose wall, permeable to both salts and water (see Fig. 28.7). The plasma membrane, which is relatively salt impermeable, separates this cellulosic wall from the central vacuole of sap contents (essentially a salt solution). When the cell is placed in a hypertonic salt solution osmosis results in the plasma membrane leaving the cell wall; this phenomenon is termed 'plasmolysis' (see Fig. 28.7).

(b) Permeability

Again it is only possible to consider a few of the comparatively simple systems, such as erythrocytes and certain plant cells, and to survey some of the better-known aspects of their permeability behaviour.

The erythrocyte has been a favourite subject for permeability work, not only on account of its intrinsic interest in relation to respiration, but because of its ready availability, and since swelling beyond a certain point, causing rupture (haemolysis), is readily detected by the liberation of the coloured pigment (e.g. haemoglobin).

Certain plant cells, those of *Valonia* and *Nitella* being particularly well known, are sufficiently large for permeability to be measured by chemical means on a single cell. As a rule, however, as with yeasts and bacteria, a suspension containing a large number of cells is essential.

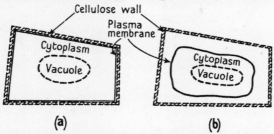

FIG. 28.7. Structure of plant cell: (a) in hypotonic saline; (b) in hypertonic saline (plasmolysed).

The experimental techniques used to follow permeability depend very much upon the particular system being studied; an indication of the more common ones is given below.

Gases

Owing to the rapid diffusion of gases through cell membranes (an exception is the swim bladder of fishes which is practically impermeable to O_2 and N_2), high-speed techniques are necessary.

Hartridge and Roughton (1927) utilized the changes in absorption spectrum of haemoglobin (Hb) upon combination with O_2 and CO to follow the diffusion of these gases through the red-cell membrane. A suspension of oxygen-free red cells is rapidly mixed with isotonic saline saturated with oxygen and then forced at high speed through an observation tube. Changes in absorption spectra along the tube can be converted into percentages of HbO_2 and from the known rate of flow, the total gas-combining power of the Hb, and the initial amount of oxygen, the rate of combination of O_2 with Hb can be found. Using first intact cells and then haemolysed cells the effect of the membrane upon diffusion can readily be found, despite the rapidity of the changes (small fractions of a second). From Fig. 28.8 it is seen that the membrane cuts down diffusion by about a factor of ten, and that O_2 penetrates more rapidly than CO.

Non-electrolytes (in solution)

Penetration of non-electrolytes into erythrocytes is usually determined from the rate of haemolysis, which is followed by an optical method based on the changes in scattering power of the cells on bursting. As haemolysis proceeds the suspension becomes less opaque and the depth of suspension through which a glowing filament can just be seen increases. Frequently the times required for a series of substances to reach a given degree of haemolysis (e.g. 75 per cent.) are taken as a measure (inverse, of course) of the relative rates of penetration.

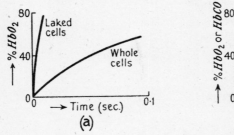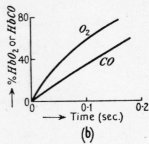

FIG. 28.8. (a) Rate of combination with red cells, before and after laking; (b) rate of combination of O_2 and CO with red cells.

Of non-electrolytes water is clearly of vital importance. Its penetration into erythrocytes can be found from the above technique using either water or hypotonic solutions of non-penetrating substances such as sucrose or NaCl. Compared with other cells erythrocytes appear to be very permeable to water. Addition of low concentrations of salts decreases the permeability, Ca^{++} and Mg^{++} being much more effective than Na^+.

Non-electrolytes other than water were found by the early workers to differ to a wide degree in their ease of penetrating erythrocytes. The early work of Hedin, for example, showed that with polyhydric alcohols the rate of penetration decreased with the number of OH groups in the molecule, and that as a rule the most lipoid soluble substances penetrated the most readily. Later work has confirmed these general conclusions but has also shown that different species of erythrocyte differ in permeability. Molecular volume appears to be a factor but in addition more specific factors seem to be necessary to account for variations between isomeric molecules and between those of similar size and lipoid solubility.

With plant cells the plasmolysis method was widely used by the early workers. If the external osmotic pressure is raised above that of the cell contents, water is withdrawn (see Fig. 28.7(b)) and a new equilibrium results provided the added substance is non-penetrating. If it does penetrate, however, then the plasma membrane will return to its former position, and the rate of de-plasmolysis will give an indication of the rate of penetration.

The early work of Overton and the later and more extensive studies of Collander and his school have shown that the permeability of plant cells increases in a general way with the oil-water partition coefficient of the penetrating substance. Table 28.4 drawn from Collander's work, shows some typical results for a series of molecules which differ widely in their permeability behaviour.

TABLE 28.4. *Olive-oil/Water Partition Coefficient and Permeability to Leaf Cells of* Chara Ceratophylla

Substance	Permeability	Partition coefficient ($\times 10^2$)
Methyl alcohol	0.99	0.78
Propylene glycol	0.087	0.57
Ethylene glycol	0.043	0.049
Glycerol	0.00074	0.007
Erythritol	0.000046	0.003

Electrolytes

The cell contents of erythrocytes usually differ from their surrounding plasma particularly as regards cation composition. Human cells, for example, contain *ca.* 110 millimoles of K^+ per 1,000 gm. but practically no Na^+, and most other species are very similar in this respect except a few such as the cat, where the position is almost reversed. At the same time erythrocytes appear to be freely permeable to anions. In order to account for the above cation concentration differences it is therefore necessary to assume either a membrane impermeable to cations or that some metabolic process is involved. Recent work with radio-active isotopes suggests that red cells are normally somewhat permeable to cations, and other investigations indicate that some metabolism is necessary for the continued uptake of K^+ by the cells of stored blood. Whilst these lend support to the second of the above alternatives, no definite decision is yet possible.

On the assumption of cation impermeability the equilibrium across the red-cell membrane can be pictured as shown in Fig. 28.9. Owing to the presence of the non-diffusible and negatively-charged haemoglobin (Hb^-) the system can be treated by the Gibbs-Donnan equilibrium (Chapter VII), giving

$$\frac{[Cl']_i}{[Cl']_0} = \frac{[HCO_3']_i}{[HCO_3']_0} = \frac{[OH']_i}{[OH']_0} = x.$$

The ratio x is about 0.8 under normal conditions.

Many plant cells also appear to be able to concentrate a particular cation, for example the concentration of K$^+$ in *Valonia* is some forty times that of its normal sea-water environment. Osterhout (1943) has recently described a model system which simulates in many respects the potassium effect found in these plant cells.

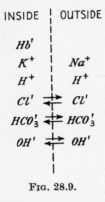

Fig. 28.9.

Ion antagonism, particularly between Na$^+$ and Ca^{++}, has been found in some cases. Osterhout, for example, showed that *Spirogyra* cells plasmolysed in a mixture of NaCl and CaCl$_2$ in molal ratio 20:1, whereas pure solutions of NaCl or CaCl$_2$ were unable to do so. Such effects are probably due to changes in the permeability of the membrane, brought about by adsorption of cations upon fixed anions in the surface.

Weak acids and bases, such as carbonic acid, ammonium hydroxide, lactic acid, adrenaline, etc., frequently penetrate cells much more rapidly than an equivalent concentration of a strong acid or strong base. This suggests that they do so in the un-ionized rather than in the ionized form, since at physiological pH both forms are present.

The differences in the ease with which various dyes penetrate cells was advanced by Overton in support of his lipoid theory of the cell membrane, since those staining fats and phospholipoids were just those which penetrated readily. With certain plant cells this generalization appears to break down, but it seems probable that in such cases the outer wall of cellulose provides a barrier to penetration. Dyes existing in both ionized and un-ionized forms resemble the weak acids and bases in entering most readily in the un-ionized state.

(c) Bio-electric potentials

Arising from the unequal distribution of various ions a potential is usually observed across biological membranes. By examining the effect on the bio-electric potential of variations in the ionic composition on one or both sides of the membrane, some interesting information about permeability has been obtained.

With nerve- or muscle-fibres, for example, a considerable potential, known as the injury or resting potential, is found between two electrodes, one placed on the undamaged outer surface, the other connected with the interior by damaging the tissue beneath that electrode. On varying the KCl concentration, keeping the osmotic pressure constant to avoid osmosis, the injury potential (E) was found to be a linear function of $\log[\text{K}^+]$, or $E = \text{constant} \times \log\dfrac{[\text{K}^+]_{\text{outside}}}{[\text{K}^+]_{\text{inside}}}$. This expression is very similar to eqn. (28.14) when $U_A = 0$, suggesting that if the Teorell-Meyer theory is applicable, then $A = 0$ (i.e. no fixed ions in the membrane) and that the membrane is permeable only to K^+. The constant in the above expression is always smaller than RT/F, a discrepancy which has been ascribed to some short-circuiting by the intercellular tissue spaces.

When the nerve is stimulated and conducts an impulse it becomes more permeable to K^+, since K^+ is found to leak out in amount approximately proportional to the number of impulses.

The experimental study of a series of inorganic and organic salts shows that the membrane in nerve and muscle (like that of red cells) is permeable to cations, particularly to K^+, and to a much smaller extent to anions. Organic ions with relatively low oil/water partition coefficients (e.g. $\overset{+}{\text{N}}(\text{CH}_3)_4$, $\text{CH}_3\text{COO}'$) resemble inorganic ions, but with increasing hydrocarbon chain-length the activity (measured by the diminution in resting potential) increases rapidly, approximately parallel to the oil/water partition coefficient. With the more capillary active ions the effect becomes less reversible, suggesting that the reduction in potential arises from a loosening of the membrane structure.

The nature of the membrane in nerve and muscle still seems very undecided. According to some workers it is porous and acts as a molecular sieve, the pores being of such a size as to permit passage of the small ion K^+ but not larger ions such as Na^+, Ca^{++}, etc.; according to others it is a homogeneous non-aqueous layer.

A number of large plant cells have also been shown to resemble nerve and muscle in being much more permeable to K^+ than to other ions, but here again no very detailed conclusions about membrane structure seem to be possible.

(d) Theories of cell permeability and specificity

As previously mentioned, one of the chief reasons for studying the permeability of natural membranes is to ascertain their structure and the way in which they function in the living organism.

The early investigations of Overton towards the end of the last century, some mention of which has been given above, led him to the conclusion that the plasma membrane was largely lipoidal in character, and that selective permeability was determined primarily by lipoid solubility. Subsequently Traube pointed out that permeability ran parallel to surface activity, so that the dominating factor might be the concentration built up by adsorption on the cell surface. In most cases surface activity and partition coefficient increase together, so that it was difficult to decide between the two theories. Traube's theory certainly breaks down in some systems and according to Davson and Danielli (1943) it has now fallen into general discredit.

From the brief survey of permeability given above it would seem that Overton's theory is satisfactory for a large number of natural membranes, although not of universal validity. In a number of cases selectivity appears to be due to 'sieve' action rather than to lipoid solubility; in some others the membrane appears to be largely lipoidal, but to contain a few pores, and thus shows an intermediate type of behaviour. 'Mosaic' membranes, consisting of areas of different properties (e.g. lipoid and protein), have also been postulated to account for selectivity, but definite proof of their existence in biological systems appears to be lacking.

There is no doubt that great advances have been made towards understanding the structure and function of biological membranes; nevertheless only in rare cases can the picture be said to be other than qualitative. One of the fundamental reasons is undoubtedly the lack of a sufficiently comprehensive study of synthetic membranes with structures more closely approximating to those of natural membranes.

GENERAL REFERENCES

A general discussion, 'The properties and functions of membranes, natural and artificial', *Trans. Faraday Soc.* 1937, **33**, 911–1151.
Symposium on natural membranes. *Cold Spr. Harb. Symp. Quant. Biol.* 1940, **8**.
BARRER, 1941, *Diffusion In and Through Solids*, Cambridge.
DAVSON and DANIELLI, 1943, *The Permeability of Natural Membranes*, Cambridge.

OTHER REFERENCES

COLLANDER, 1924, *Kolloidchem. Beih.* **19**, 72.
DANIELLI, 1938, *Cold Spr. Harb. Symp. Quant. Biol.* **6**, 190.
DEAN, CURTIS, and COLE, 1940, *Science*, **91**, 50.
DOTY, 1946, *J. Chem. Phys.* **14**, 244.
—— AIKEN, and MARK, 1944, *Industr. Engng. Chem.* (Anal.), **16**, 686.
—— —— —— 1946, *Industr. Engng. Chem.* **38**, 788.
ELFORD, 1937, *Trans. Faraday Soc.* **33**, 1094.
GOLDMAN, 1943, *J. Gen. Physiol.* **27**, 37.
GORTER and GRENDEL, 1925, *J. exp. Med.* **41**, 439.
GRABAR, 1937, *Trans. Faraday Soc.* **33**, 1104.

GREGOR and SOLLNER, 1946, *J. Phys. Chem.* **50**, 53.
HARTRIDGE and ROUGHTON, 1927, *J. Physiol.* **62**, 232.
HASS, 1939, *Arch. Path.* **28**, 177.
MEYER and BERNFELD, 1945, *Helv. Chim. Acta*, **28**, 962, 972, 980.
—— and SIEVERS, 1936, ibid. **19**, 649, 987.
NORTHROP, 1929, *J. Gen. Physiol.* **12**, 435.
OSTERHOUT, 1943, ibid. **27**, 91.
SOLLNER and CARR, 1944, ibid. **28**, 1.
TEORELL, 1935, *Proc. Soc. Exp. Biol. N.Y.* **33**, 282.
WAUGH and SCHMITT, 1940, *Cold Spr. Harb. Symp. Quant. Biol.* **8**, 233.
WILBRANDT, 1935, *J. Gen. Physiol.* **18**, 933.

APPENDIX I

The Relation between Displacement and Diffusion Coefficient

CONSIDER a dilute suspension and its diffusion along the x-axis of the cylinder of unit cross-sectional area shown below.

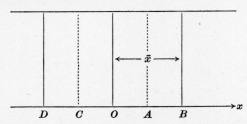

All particles are assumed to undergo the same displacement $\bar{x}$ in a short interval of time t, so that the changes in distribution of the particles are also small ($\bar{x}$ is strictly the root-mean-square displacement). It is further assumed that as many move to the right $(+x)$ as to the left $(-x)$.

In the time t one-half of the particles lying between the planes O and D will pass across the plane O (since only half have positive displacement), i.e. number diffusing from left to right across $O = \frac{1}{2}n_1\bar{x}$, where $n_1 =$ mean concentration in the volume OD, or the concentration in the mean plane C. Similarly, the number diffusing from right to left across $O = \frac{1}{2}n_2\bar{x}$, where $n_2 =$ concentration in the mean plane A. Hence the *net* transfer across $O = \frac{1}{2}\bar{x}(n_1-n_2)$. By Fick's first law (p. 233)

$$\text{net transfer/unit area} = -D\frac{dn}{dx}t$$

$$= D\left(\frac{n_1-n_2}{\bar{x}}\right)t,$$

since

$$-\frac{dn}{dx} = \frac{n_1-n_2}{\bar{x}}$$

if $\bar{x}$ is small. Hence

$$\tfrac{1}{2}\bar{x}(n_1-n_2) = D\frac{(n_1-n_2)}{\bar{x}}t,$$

i.e.

$$\bar{x}^2 = 2Dt.$$

APPENDIX II

The Preparation of Thimble-type Cellulose Nitrate Membranes

A SOLUTION of the cellulose nitrate, HL 120/170, of the composition already indicated (p. 166), stored in a well-stoppered vessel, is recommended for the preparation of the membranes. The cylindrical glass moulds whose hemispherical rounded ends are perforated by a small pin-hole, blown without distorting the ends, are chosen to be of a size matching the rubber pressure tubing support. The mould is rotated about a horizontal axis at about 15 r.p.m. and, whilst rotating, the cellulose nitrate solution, free from all air bubbles, is poured over in a continuous thick stream from a test-tube, beginning at the rounded end and working steadily along until a suitable length of mould is covered. The solution dripping from the mould is collected in another test-tube placed below, and after allowing time for the air bubbles to disappear it may be used for subsequent coats of the membrane.

After a coat is applied, it is allowed to dry on the rotating mould in front of a small electric heater (e.g. a 'bowl fire') so placed that the heating is gentle and uniform. Two or three minutes should be required for a coat to become dry enough for the next to be applied without the first being removed. Three coats are normally required in making membranes suitable for polymer work, suggested times of drying in front of the heater being 2, 3, and 5 minutes respectively.

After the last coat has been applied and exposed to the heater, the membrane on its mould is allowed to stand in the air at room temperature, rounded end downwards. Short times of standing (e.g. $\frac{1}{2}$ hour) tend to give increased permeabilities and much longer times (e.g. 24 hours) much reduced values. The mould is then filled with, and immersed in, water, and after $\frac{1}{2}$ hour, the membrane, which is usually trimmed on the mould at its open end, is forced from it by water pressure. It is then mounted (see p. 160) upon closely fitting and supported rubber pressure tubing.

Membranes may readily be tested for freedom from pin-holes and for permeability by inserting a burette (without the tap) into the mount, by filling with water, and by applying a pressure from a mercury manometer. For general protein work and for membranes with dimensions similar to those indicated in Fig. 7.5 (a), the penetration of water under a pressure of 450 mm. of mercury should not be appreciably greater than 0·03 c.c. per minute. For polymers of very high molecular weight, higher permeabilities are permissible, and lower ones are required for very low molecular weight polymers.

As already mentioned, permeabilities are increased by short times of standing in the atmosphere after application of the last coat to the membrane. Increased permeabilities are also obtained by using higher initial concentrations of ethylene glycol (which dissolves out of the membrane in water) and by the use of more dilute and less viscous cellulose nitrate solutions. If the cellulose nitrate used is very different from HL 120/170 it may be necessary, not only to make use of solutions of different concentration, but also to use more or fewer coats than suggested above. If, however, the directions above are carefully followed at normal room temperature, membranes of a generally suitable permeability and mechanical strength are obtained.

APPENDIX III

A Method for the Partial Denitration of Cellulose Nitrate Membranes

IN preparing cellulose nitrate membranes which are permeable to organic solvents it is usually necessary to denitrate partially the high nitrogen content, soluble cellulose nitrate from which the membranes are originally caste. One method, in which the denitration is performed by means of ammonium sulphide, has already been mentioned (see p. 174) but there is another which has been successfully applied in the authors' laboratory to the denitration of the thimble-type membranes suitable for the Dobry type of osmometer. This makes use of the fact that the nitration of cellulose by mixtures of nitric and other acids is usually an equilibrium process in which the nitrogen content of the cellulose nitrate produced depends largely on the final concentration of nitric acid. Thus if a membrane of high initial nitrogen content ($>$ 11 per cent.) is immersed in a mixture of relatively low nitric acid concentration, the nitrogen content of the membrane may be considerably reduced. The following mixture (by weight) is recommended for denitration to a stage suitable for permeability to acetone and other organic solvents:

37.17 per cent. of HNO_3 (specific gravity 1.5),
34.41 per cent. of H_2SO_4,
28.42 per cent. of H_2O.

These proportions of acid refer to the anhydrous materials and not to ordinary laboratory reagents.

The membrane mounted upon its glass support (see p. 170) is immersed in the acid mixture at $0°$ C. (to minimize degradation) for 2 hours, enough of the mixture being used to prevent any appreciable change in the nitric acid concentration during reaction. The acid mixture is then run off and the membrane is immersed in ice-cold water before washing for several hours in running water. After storing in absolute alcohol for 12 hours, it is stabilized by three boils of 3 hours each in absolute alcohol.

The membrane and support are then mounted in the osmometer tube containing acetone, the liquid being changed periodically until zero osmotic pressure indicates the removal of all soluble matter. Membranes so prepared contain *ca.* 6 per cent. of nitrogen and though insoluble in acetone are considerably swollen by, and permeable, to it.

APPENDIX IV

Dimensions of Standard U-tube Viscometers (see Fig. 13.3)

All linear dimensions are given in centimetres.
All volumes are given in millilitres.

Viscometer No.	0	1	2	3	4
Range (cs)	0·5–2·0	1·5–6·0	5–40	30–250	200–1,500
Capillary (de)—Length ± 5 per cent.	12	12	10	10	10
Int. diam. ± 5 per cent.	0·038	0·060	0·115	0·23	0·38
Tube (aB)—Length ± 5 per cent.	6·0	6·0	7·0	7·0	7·0
Int. diam. ± 10 per cent.	0·5	0·5	0·40	0·7	0·7
Bulb (BC)—Ext. diam. ± 10 per cent.	2·2	2·2	2·1	2·8	3·4
Capacity ± 5 per cent.	6·5	6·5	5·5	16·0	26·0
Bulb (Cd)—Capacity ± 10 per cent.	0·4	0·4	0·4	1·2	1·4
Bent tube (ef)—Min. int. diam.	0·5	0·5	0·5	0·7	0·8
Tube (Gh)—Int. diam. ± 10 per cent.	0·5	0·5	0·5	0·7	0·8
Bulb (fG)—Ext. diam. ± 10 per cent.	2·2	2·2	2·1	2·8	3·4
Min. capacity	8·0	8·0	7·0	20	30
Dimension (X) — ± 10 per cent.	8·0	8·0	5·5	5·5	7·0
Distance between vertical axes — ± 10 per cent.	1·7	1·7	1·5	2·0	2·5

Extracted from B.S. 188: 1937—'Determination of Viscosity of Liquids in Absolute (C.G.S.) Units', by permission of the British Standards Institution.

AUTHOR INDEX

Major sections involving an author are indicated by page numbers in bold face type

Åborg, 512.
Abramson, 29, **297**, **299–300**, **306**, **310**, 314, 315, 316, 317, 331, 338, 593, 742.
Adair, G. S., 81, 126, 141, 142, 145, 148, **160–9**, **183–6**, 204, 220, 337.
Adair, M. E., 126, 141, 142, 145, 148, 169, 183, 185, 186, 337.
Adam, 24, 499, 504, 505, 513, 520, 534, 552.
Addison, 518, 680, 681.
Affens, 437, 438.
Aicken, 682.
Aiken, 810, 811.
Aitken, 247.
Alexander, 47, 497, 501, **504–6**, **508**, **510**, **512**, 518, 529, 531, 568, 599, 659, 683, 695.
Alexandrov, 476.
Alfrey, 175, 369, 375, 480, 484.
Anderson, 460, 461.
Andersson, 492, 606.
Angelescu, 695.
Anson, 239.
Ascherson, 2, 3.
Astbury, 28, 456, 457, **747–9**, 751.

Bailey, 747.
Balarev, 37.
Balkin, 339.
Barker, 22.
Barkley, 624.
Barnes, 428, 431.
Barr, 350.
Barrer, 723, 725, 810, 812.
Bartell, 542, 552, 620.
Bartovics, 175, 369, 375.
Bastow, 126.
Batchelor, 608.
Beams, 267.
Bear, 691.
Bechhold, 8, 805.
Beckmann, 253.
Bennister, 660.
Bergmann, 737, 738, 739, 752.
Berkley, 75.
Bernal, 751.
Bernfeld, 817.
Berzelius, 3.
Beutner, 819.
Bingham, 476.
Bjerrum, 106.
Blaschke, 372.
Blekkingh, 37.
Boeder, 390, 393.
Bolduan, 677.
Bolland, 135.
Boltzmann, 238.
Borinewitsch, 621.
Bourdillon, 165.
Bowden, 126, 545.
Bradbury, 313.

Bradfield, 714.
Bragg, 443, 446, 704, 723.
Bredig, 6, 9, 47.
Bresler, 736, 753.
Briggs, 647.
Brohult, 745.
Brönsted, 793.
Brosteaux, 434, 435, 436.
Brown, 2, 517.
Bull, 146, 147, 165, 169, **299**, 315, 316, **342–3**, 378, 638, 686, 736.
Bunn, 440, 442, 444, 448, 458, 757, 758, 763, 765.
Bunzle, 364.
Burgers, 364, 369.
Burk, 159, 169, 746.
Burton, 75, 318.
Bury, 31, 672, 687.
Butler, 50, 137, 531, 534.
Bywater, 176.

Campbell, 171, 242, 283, 364, 372, 723, 770, 786.
Cannon, 335, 337.
Carothers, 134.
Carr, 817.
Carruthers, 565.
Carter, S. R., 174.
Carter, W. C., 369.
Casperson, 428.
Cassie, 109, 511, 682.
Chapman, 117.
Chibnall, 737, 738.
Claesson, 243, 546.
Clapson, 784.
Clark, G. L., 440, 540, 704, 712.
Clark, R. E. D., 540.
Clowes, 654.
Cockbain, 656, 658.
Cohen, 37.
Cohn, **131–2**, **138–9**, 169, 273, 423, **730**, **741**, **744**.
Cole, 822.
Collander, 809, 825.
Cooper, 147.
Corrin, 691, 692.
Cottrell, 582.
Couette, 470, 471, 474.
Coumoulos, 462, 760.
Crisp, 505, 512.
Crowfoot, 146, 456.
Cruickshank, 174.
Curtis, 822.

da Vinci, 6.
Dahlstrom, 617.
Danielli, 655, 820, 828.
Davies, 687.
Davson, 820, 828.

AUTHOR INDEX

de Bruyne, 485.
de Rosset, 404.
de Witt, 548.
Debye, 7, 21, 29, 72, **101–2, 104–9,** 178, 179, 383, 413, 419, 427.
Derjaguin, 116.
Dervichian, 492, 739..
Devaux, 23.
Dewey, 720.
Dickinson, S., 747.
Dickinson, W., 318, 565.
Dobry, 170, 175.
Donnan, 22, 25, 525.
Dorn, 12.
Doss, 529.
Doty, 176, 178, 180, **430, 432–4, 436–8,** 480, 484, 810, 811, 812.
Douglas, 339, 340.
Dundon, 36.

Edsall, 138, 139, 169, 273, 354, **394–5, 397–400, 402–3,** 407, 423, 730, 739.
Einstein, 3, 17, 18, 27, 178, 257, 358, 659.
Eirich, 359, 364.
Eisenschitz, 364.
Elford, 805, 808, 817, 818.
Ellis, 761.
Emmett, 546, 548.
Ensminger, 718.
Ewing, 464.
Eyring, 502.

Fahraeus, 204.
Fairbrother, 339.
Fankuchen, 454, 455, 751.
Faraday, 6, 7, 15, 115, 204, 426.
Farmer, 764.
Ferguson, 690.
Ferri, 98, 779.
Ferry, 414, 475.
Fick, 233, 234, 235, 258.
Fidiam, 716.
Field, 763.
Flory, 153, 174, 175, 369, 370, 787, 788, 792, 795, 796, 797.
Ford, 716.
Fordyce, 369.
Foster, 402, 608, 739.
Foulk, 624, 628.
Fourier, 751.
Fowkes, 548.
Fowler, 152.
Frazer, 662.
Freundlich, 22, 28, 299, **428–30, 434,** 476, 478, 562, 571, 593, 719.
Fricke, 593.
Friedman, 603.
Frisman, 410, 411.
Frocht, 598.
Frumkin, 25, 531.
Fruton, 752.
Fuoss, **170–1, 173–4,** 372.

Gabor, 460.
Garner, 765.

Garnett, 428.
Gans, 430, 432, 434, 614.
Gee, 94, 175, **787, 790, 791, 793, 796, 798.**
Gehman, 763.
Gibb, 219.
Gibbs, 10, 22, 23, 25, 519, 520, 524, 525, 526.
Gieseking, 718.
Gilbert, 702.
Gils, 664.
Glasstone, 69, 140.
Goldman, 822.
Goodeve, 471, 479.
Gordon, 241.
Gorin, 297, 299, 300, 305, 307, 316, 317, 331, 332, 333, 337, 338.
Gorter, 821.
Gortner, 16, 316, 341, 342, 343.
Gouy, 12, 101, 117, 119, 531.
Grabar, 805.
Graham, 1, 2, 3, 4, 602.
Gralen, 201, 202, 205, 250, 253, 283, 768, 770, 799, 800.
Gray, 599.
Green, 580.
Greenberg, D. M., 133, 157, 592.
Gregor, 814.
Grendel, 821.
Grieff, 107.
Grim, 708.
Gronwall, 107.
Gross, D., 255.
Gross, H., 205, 249, **397, 404–5, 407,** 410, 800.
Guggenheim, 50, 73, 83, 99, 152, 787.
Guinier, 539, 540.
Gundermann, 690, 692.
Guth, 360, 372, 781, 783.

Ham, 45, 339, 340.
Hamaker, 115, 122.
Hardy, 3, 15, 22, 23, 115, 318, 377, 571, 627.
Harkins, 22, 25, 502, 504, 517, 548, 614, 617, 651, 681, 690, 691, 692.
Harris, 47.
Hartley, E. G. T., 75.
Hartley, G. S., 31, **104, 110–13,** 239, 240, **241–2,** 529, 653, 669, **670, 672,** 676, **677, 679–80, 682, 684,** 685, 687, 694.
Hartridge, 823.
Hartshorne, 398.
Hass, 821.
Hatschek, 575, 597, 605, 646.
Hauk, 98.
Hauser, 515, 706, 708, 711, 719, 720.
Haworth, 133, 134, 766, 767, 768, 773.
Hedin, 824.
Helmholtz, 11, 12.
Hendry, 131, 132.
Hengstenberg, 763.
Henriot, 267.
Henry, 114, 302, 303, 304, 314, 315, 332, 333.
Hershberger, 620.

Hertel, 542.
Herzog, 261.
Hess, 690, 692.
Hewitt, 169.
Heymann, 590, 594, 605, 606.
Hibbert, 369.
Hildebrand, 151, 789.
Hoar, 695.
Hofmeister, 16.
Hollingshead, 241.
Horsfall, 323.
Hoskins, 493.
Houwink, 620.
Hoyrup, 141, 142.
Hückel, 21, 29, 72, 101, **102, 104–9**, 315.
Huggins, 95, 153, 154, **362, 363, 366, 368,** 372, 749, 787, 789, 790.
Hughes, 690, 692.
Huguenard, 267.
Hulett, 36.
Hutchinson, 516, 544, 550, 616.

Illig, 261.
Ingelman, 461.

James, 241, 783.
Jeffery, 362, 390.
Jenny, 728.
Jilk, 171.
Johnson, 39, 171, 176, 242, 281, 283, 290, 326, 364, 372, 723, 734, 742, 744, 770, 786.
Jones, 31, 672.
Jordan, D. O., 313, 688.
Jordan, H. F., 517.
Jullander, 177, 292.

Katz, 592.
Kaufman, 433, 434.
Kausche, 461.
Kekwick, 205, 325, 734, 741, 742, 743.
Kibrick, 337.
Kiessig, 691.
King, 642, 656, 660.
Kistler, 543, 593.
Kopp, 139.
Kraay, 664.
Kraemer, 138, 205, 261, 292, 361, 370, 603, 651.
Krafft, 31, 667.
Krasny-Ergen, 377.
Krishnan, 432, 434.
Kruyt, 575.
Kudar, 261.
Kuhn, 362, 363, 410, 781, 782, 783.

La Mer, 107, 428, 431.
Labrouste, 23.
Laing, 592.
Lamm, 230, 231, 242, 243, 248, 254.
Langmuir, 22, 23, 24, 500, 523, 526, 527, 532, 637.
Lansing, 205.
Lauffer, 287, 461.
Lawrence, 591, 610, 611, 644, 686, 694, 697.

Lazurkin, 476.
Lea, Carey, 6, 8, 9.
Lemin, 698.
Levine, 116, 120, 121, 124.
Lewis, G. N., 62, 71, 138.
Lewis, W. C. M., 318, 565.
Lewis, W. I., 621.
Liesegang, 603, 604.
Linder, 318.
Ling, 638.
Lippmann, 12, 531.
Lipsett, 39.
Lisse, 313.
Lloyd, 730, 732, 733.
Lodge, 317, 602.
Loeb, 182, 183.
Loomis, 716.
Longsworth, 223, 228, 317, 318, 335, 336, 744.
Lorenz, 9.
Lundgren, 734.

Maas, 39.
Machemer, 134.
MacInnes, 317, 318, 744.
Mack, 36.
Mackey, 592.
Magat, 369.
Marc, 563, 564.
Marcelin, 23.
Mardles, 568.
Margaretha, 364.
Mark, 175, 347, 359, 364, 369, 370, 373, 375, 690, 763, 768, 781, 783, 810, 811.
Marrack, 169.
Marsh, 458, 766, 771.
Martin, 176, 373.
Masson, 174.
Mattoon, 691, 692.
Mattson, 29, 310.
Maxwell, 467, 468.
McBain, 23, 31, 148, 525, 529, 592, 606, 665, 669, 670, 676, 677, 689, 690.
McFarlane, 205, 734, 741.
McMeekin, 142, 143, 144, 146.
McMullen, 47, 568.
Mead, 170, 171, 173, 174, 372.
Mehl, 366, 400.
Meissner, 39.
Melsens, 3.
Melville, 174.
Menzies, 174.
Meyer, 98, 157, 175, 369, **756, 761, 766,** **768, 770, 772, 777–9**, 813, 817.
Mie, 7, **428–9, 431, 437**.
Miller, A. R., 789.
Miller, J. N., 628.
Millikan, 18.
Mirsky, 754.
Misch, 768.
Mitchell, 681.
Montonna, 171.
Moran, 147, 605.
Mosimann, 205, 283, 800.
Mouquin, 496.

AUTHOR INDEX

Moyer, 297, 299, 300, 305, 316, 317, 331, 332, 333, 337, 338.
Mueller, 300.
Müller, 107, 112.
Muralt, 399, 400.
Murray, 672, 673.
Muthmann, 6, 7, 8.
Myers, 681.

Nachod, 714, 728.
Nernst, 12, 15.
Neuberger, 738.
Neumann, 98.
Neurath, 146, 243, 248, 638.
Newman, 211, 350, 467, 475.
Nichols, 198, 218.
Niemann, 736, 737, 738, 748, 749.
Nieuwenburg, 469.
Northrop, 239.
Noyes, 15.

Oakley, 165, 717.
Oncley, 260, 262, 366, 367, **413, 414, 416, 418,** 422, 739, 740.
Onsager, 104, 109.
Orr, 787.
Osborn, 373.
Osterhout, 806, 826.
Ostwald, Wilhelm, 603, 604.
Ostwald, Wolfgang, 13.
Ott, 771.
Overbeek, 116, 127, 574.
Overton, 806, 821, 825, 827, 828.
Owen, 447.

Palmer, A. H., 337.
Palmer, K. J., 691.
Palmer, R. C., 109, 511, 616, 682.
Palmer, W. G., 540.
Pankhurst, 632, 633, 634, 686.
Parker, 593.
Partington, 50.
Patterson, A. L., 452, 453, 751.
Patterson, H. S., 576, 578, 580, 582.
Pauli, 6, 7, 31, 47.
Pauling, 748, 749, 754.
Pedersen, **193–4, 196–7,** 202, 203, 219, 221, 231, 236, 254, 256, 281, 292, 293, 735, 747.
Pelser, 606.
Pennycuik, 31, 47.
Perrin, F., 213, 260, 261, 333, 384, 385, 386, 420, 421.
Perrin, J., 3, 13, 17, 18, 190, 380, 384.
Perutz, 456, 749.
Peterlin, 393, 394, 395.
Pfeiffer, 373.
Philpot, 223.
Pickels, 267.
Picton, 318.
Pink, 654.
Pirie, 744.
Pockles, 23.
Polson, 249, 252.
Poole, 597.

Posnjak, 599.
Powis, 115, 570.
Powney, 680, 681, 688.
Prentiss, 131, 132.
Price, 318.
Procter, 601.
Putzeys, 434, 435, 436.

Quincke, 12.
Quist, 343.

Ramsden, 3, 636.
Randall, 71, 138.
Rayleigh, 6, 7, 23, **427–8, 430, 432**.
Record, 174.
Rehbinder, 627.
Rehner, 792.
Reichenburg, 681.
Reiner, 476, 480, 481, 483, 484.
Reuss, 2, 11.
Reychler, 30.
Rideal, 25, 492, 496, 529, 531, 545, 562, 616, 656, 702.
Riley, 146.
Robertson, J. K., 398.
Robertson, J. M., 445, 449, 450, 451, 452.
Robertson, J. W., 476.
Robinson, C., 681, 698.
Robinson, J. R., 355, 376, 391, 392, 400, 401, 402.
Robinson, M. E., 169, 220.
Roche, 169.
Roe, 529.
Rose, 695.
Rosenberg, 253.
Rosenfeld, 493.
Ross, 624, 635, 690.
Roughton, 823.
Runnicles, 239, 240, 241, 242, 687.
Russell, 713.
Ryan, 614.

Sadron, 397, 399, 405, 406, 409.
Salmon, 669, 676.
Sandved, 107.
Sawyer, 690, 692.
Scatchard, 104.
Schmidt, 730, 732.
Schmitt, 691, 822.
Schofield, 483.
Schulman, 492, 497, 656, 658, 662, 695.
Schulz, 797, 798.
Schulze, 3, 15, 115, 571.
Schütz, 642.
Schwedoff, 482.
Scott, D. W., 762.
Scott, R. L., 369.
Scott-Blair, 464, 469, 476, 477, 478, 480, 483, 484, 485, 486.
Searle, 211, 350, 467, 475.
Seifriz, 476.
Selmi, 3, 15, 115.
Shack, 414.
Sheppard, J. R., 784.
Sheppard, N., 764.

AUTHOR INDEX

Sheppard, S. E., 698
Shooter, 326, 744.
Shore, 730, 732, 733.
Shute, 552.
Siedentopf, 16.
Siegbahn, 461.
Sievers, 813, 817.
Signer, 203, 205, **397**, **404–7**, **410**, 760, 786, 800.
Simha, 360, 363, 364, 366.
Singer, 176.
Slater, 46.
Smith, M. E., 313.
Smith, R. C. M., 686.
Smoluchowski, 8, 15, 17, 302, 303, 314, 315, 332, 377, 568, 575, 576, 582, 659.
Snellman, 397, 692.
Sollner, 817.
Sookne, 47.
Sorenson, **141–2**, **144**, **147–8**, 160, 733.
Spurlin, 165, 176.
Squires, 611, 621.
Ställberg, 489, 636.
Ställberg-Stenhagen, 492, 501.
Stamm, 205.
Stanley, 461.
Staudinger, 28, 368, 369, 778.
Stauff, 560.
Stein, 178, 180, 430, 432, 434, 436.
Stenhagen, 492, 501.
Stern, 117.
Stewart, 662.
Stokes, A. R., 539.
Stokes, G. G., 210, 213, 384.
Stormer, 470.
Strutt, 427.
Stuart, A., 398.
Stuart, H. A., 393, 394, 395.
Sugden, 516.
Sullivan, 542.
Sutherland, 440, 764.
Svedberg, 3, 9, 17, 18, 19, 26, 177, **191**, **193–4**, **196–8**, **202–4**, **218–19**, **221**, 231, 236, 254, 256, 259, 267, 281, 292, 293, 318, 734, 736, 742, 745.
Svensson, 227, 228, 317.

Talmud, 736, 753.
Tamman, 14, 559, 561.
Taylor, G. I., 354.
Taylor, G. L., 169.
Tennent, 176.
Teorell, 489, 504, 508, 510, 512, 518, 636, 813.
Thompson, H. W., 440.
Thompson, W. I., 621.
Thomson, J. J., 11, 579, 584.
Tiselius, 29, 245, 255, **318–19**, **321–2**, 331, 546, 743.
Tomlinson, 695.
Töpler, 29.
Torkington, 440.
Trapeznikov, 627, 628.
Traube, 22, 139, 527, 828.

Treloar, 175, 777, 783, 784, 790.
Trim, 695.
Tyndall, 6, 28, 426.

Ungar, 470, 471.

Valko, 699.
van Arkel, 575.
van Bemmelen, 9, 606.
van Hook, 604.
van der Wyk, 369, 778.
Verwey, 116, 117, 120, 121, 122, 124, 619.
Vickerstaff, 698.
Vinograd, 690, 692.
Voet, 572.
von Buzágh, 568, 615.
von Tavel, 786.
von Weimarn, 14, 15, 561.

Wagner, 176.
Ward, A. F. H., 527.
Ward, A. G., 762.
Warner, 142, 143, 144, 146.
Washburn, 343.
Watson, 402.
Waugh, 822.
Weaver, 203.
Weber, 393.
Weidinger, 606.
Weissberger, 165.
Weltmann, 479.
Wertheim, 175.
Whitney, 15.
Whytlaw-Gray, 575, 578, 580, 582.
Wiegner, 728.
Wiener, 17, 393.
Wilbrandt, 813.
Wilhelmy, 491, 518.
Williams, J. W., 402.
Williams, G. C., 565.
Williams, R. C., 460.
Wilson, A. J. C., 539.
Wilson, C. T. R., 32, 39.
Wilson, H., 477.
Wilson, J. A., 601.
Wissler, 404, 407, 408, 409.
Wolarowitsch, 621.
Wood, F. C., 766, 771.
Wood, L. J., 680.
Wood, W., 714, 728.
Woods, 780.
Wrinch, 748.
Wu, 638.
Wyckoff, 460.
Wyman, 416, 419, 422.

Yamins, 492, 493.

Zawidzki, 152.
Zimm, 178, **430**, **432**, **434**, **437–8**.
Zisman, 492, 493.
Zsigmondy, 15, 16, 606, 805.
Zvetkov, 410, 411.

SUBJECT INDEX

Main sections are indicated by page numbers in bold face type

Acetamides, 502.
Activity, 68, 74.
— coefficients, **69**, 71, 196.
Adhesion, **612**.
—, measurement of, 613.
—, —, from adhesion number, 615.
—, —, from friction, 616.
—, —, from heats of wetting, 617.
—, —, from sedimentation volume, 613.
— number, 553, 615.
— tension, 542, 551, **612**.
—, work of, 551, 612.
Adsorbed films, 29, 462, **513**, 637.
Adsorption, 3, 10, 45, 543.
— at air/water and oil/water interfaces, 513.
— at gas/solid interface, 546.
— at solid/liquid interface, 543.
—, entropy of, 524, 549.
—, free energy of, 73.
—, heat of, 524, 549, 550.
— isotherms, 548.
— —, Freundlich, 22.
— —, Gibbs, 10, 22, 84, **519**, **524**, 683.
— —, Langmuir, 22, 548, 723.
— of dyestuffs, 544.
— of gases, 23, 24, 546, 723.
— of ions, 45.
—, thermodynamics of, 524, 549.
Aerogels, 5.
Aerosols, 5, 32, **576**.
—, coagulation of, 581.
Affinity, standard, 73.
Agar, 30, 586, 596, 602, 608.
Ageing of surfaces, 529.
Aggregation (coagulation), 15, **570**, 575, 582, 716.
— methods for preparation of colloidal solutions, **558**.
Air electrode, 492.
Air/water interface, 23, **488**, 513.
Air/water monolayers, 489.
Albumin, egg, 139, 141, 169, 198, 205, 288, 293, 367, 735, 740, 744, 751.
— serum, 42, 145, 169, 205, 256, 288, 293, 299, 324, 331, 333, 367, 735, 740.
Alcogel, 5.
Alcosol, 5.
Alkyl phenols, 501.
Aluminium hydroxide, 5, 9.
Amides, 501.
Amino acids and protein structure, 139, 330, **730**.
Amylopectin, 772.
Amylose, 772.
Analcite, 722.
Analytical methods, 130.
— —, combining weight, 132.
— —, end group, 133.
— —, trace element or group, 131.

Angle of isocline, *see* Extinction angle.
Anilines, 501.
Anisoles, 501.
Anomalous flow, **347**, 376, 380.
— viscosity, **347**, 376, 380, 401.
Antibodies, 27, 743.
Antigens, 743.
Apophyllite, 705, 706.
d'Arcy equation, 541.
Area, cross-sectional, of groups, 24.
—, surface, 24, 34–40, 489, **497**, 508, 519, 524. *See also under* Surface area.
—, —, of solids, 536.
Arsenious sulphide sols, 15.
Asymmetric particles, effect on viscosity, 361.
— —, motion of, **212**, 259, 283.
Asymmetry, *see* Axial ratios.
Attapulgite, 706.
Averages, different types of, **42**, 370.
Avogadro's number, 17, 18.
Axial ratios, **212**, 261, 293, 363, 384, 400, 420, **738**, **799**.

Bacterial cellulose membranes, 174.
Bakelite, 774.
Balance, film, 24, **490**.
Barium sulphate, 37.
Barrier, electrostatic, 529.
—, glass, 23.
Base-exchange, 713, 726.
Beidellite, 706.
Bentonite, 30, 711, 720.
Bergmann-Niemann hypothesis, 736.
Bingham flow, 466.
— body, 482.
Bio-colloids, 32.
Bio-electric potentials, 826.
Biological problems and surface techniques, 506.
Birefringence of flow, 27, 28, **387**, **393**, 570, 599, 720 (streaming birefringence). *See also* Double refraction of flow.
Bitumen, 30, 464.
Boltzmann equation, 101, 109.
Bonds, hydrogen, 47, 748.
—, —, in proteins, 747.
Bound water in gels, 605.
Boundary, method of moving, 29, **317**.
Branched or three-dimensional polymers, 774.
Brassidic acid, 502.
Brownian motion, 2, 12, Appendix I.
— —, rotational (or rotary), 17, **380**.
— —, —, determination of Avogadro number from, 18.
— —, translational, 17, 256. *See also* Appendix I.
Brucite, 708.

SUBJECT INDEX

Built-up films, 496.
Bulk modulus, 466.
Buna rubber, 766.
Bushy-stunt virus, 205, **752**.
Butadiene polymer, 765.

Calcium sulphate, 36.
Capacity factors, 63.
Capillarity, 21, 32, 488.
—, electro-, 12, 22, 90, **531**.
Capillary condensation, 607.
— plastometer, 472.
— rise, 514.
— viscometer, 350.
Carbohydrates, **766**.
Carbon black, 29, 764.
Castor-oil, 23, 496.
Catalysts, metal, 35.
Cataphoresis, 2, 11, **295**. See also Electrophoresis.
Cell permeability, 827.
Cellulose, **766**.
—, chain-length of, 133.
— derivatives, 769.
— —, acetate, 180, 771.
— —, hydrated, 769.
— —, nitrate, 201, 283, 769.
— —, sedimentation of in solution, 283.
Centrifugal force, 26, 27, 53, 54, **266, 288**.
Chabazite, 722.
Charcoal, 22, 35.
Charge density, **38, 39**, 48, **306**, 315.
—, distribution, for two interacting double layers, 116.
—, distribution of, 48, 423.
—, effect of, on solubility, 37.
— effects, 215.
— — in diffusion, 254.
— — in osmosis, 157.
— — in sedimentation equilibrium, 202.
— — — velocity, 276.
— of colloidal particles, 26, 31, 48, **332**, 566.
—, origin of, **43**.
Cheese, 1, 30, 464.
Chemical potential, 35, 37, 38, 53, 59, 66, 67, **73, 86**, 92, 232.
— reactions in gels, 603.
— reactivity and particle size, 34, 35, 39.
Chemi-sorption, 547.
Clapeyron-Clausius equation, 58, 71.
Clay pastes, 26.
Clays, 1, 26, 29, 30, 464, **704, 706**.
—, dispersion of, 715.
—, dispersions of, 712.
—, flocculation of, 715.
—, ion exchange in, 713.
—, mechanical properties of, 718.
—, streaming birefringence of, 720.
—, swelling of, 715.
Cloud, 1.
— chamber, 32.
Coacervation, 589, 619.
Coagulation, 3, 6, 13, 15, 16, **115**, 124, 315, **570**, 582, 716. See also Flocculation.

Coagulation, kinetics of, 15, 575.
— of smoke, 581.
Co-area, 500.
Coefficient of diffusion, 17, **233**, 381.
— of spreading, 523.
— of viscosity, 468.
Collagen, 457, 732.
Colligative methods, 136.
Colloid mill, 557.
Colloidal electrolytes, 30, 110, **667**. See also Soaps.
— ions, mobility of, 111.
— particles, charge of, 26, 43, 48, **332**, 566.
— —, shape of, 26.
— —, size of, 26.
— prussian blue, 3, 5, 558.
— silver chloride, 3.
— solutions, 585.
— state, 1, 5, 34.
— sulphur, 3, 7, 40, 47, 559, 565, 573.
Colloids, 1, 4, 5, 34. See also Gels, Sols, and Suspensions.
—, hydrophilic, 13, 44, **124**.
—, hydrophobic, 13, 47, **114**. See also Sols.
—, inorganic, 3.
—, lyophilic, 13, **124**, 585.
—, lyophobic, 13, 47, **115**, 315, **556**, 585. See also Lyophobic colloids.
—, stability of, 15, 29, **114**, 315, **570**, 604, 626, 649, 715.
Colour, 7, 34, 40, **426**.
Compression plastometer, 469.
Concentration, determination by optical methods, **216**.
—, surface, 84.
Concentric cylinder apparatus, **354, 387**, 396, 399.
— — —, double refraction in, **387**, 391, **393**.
— — —, orientation in, **387**, 390, **395**.
Condensation, 757
—, upon gaseous ions, 576.
Condensed films, 500.
Conductivity, electrical, 31.
—, —, dispersion of, 415, 421.
—, —, of dyes, 698.
—, —, of gels, 602.
—, —, of soaps, 677.
—, of heat, 543.
Conductors and non-conductors, colour of sols of, 7, **426**.
Conjugated proteins, 732.
Conservation of energy, law of, 50.
Contact angle, 552.
Copolymers, 761, 766.
Correction of diffusion constants, 263.
— of sedimentation constants, 290.
Cotton, 1, **767**, 799.
Cottrell precipitator, 582.
Couette viscometer, 28, **354**, 387, 470, 474. See also Concentric cylinder apparatus.
Counter-ions, see Gegenions.
Cream, 464.
Crystallites, 453, 539.
Crystallization, 14, 15, 562.

Crystallization in polymers, 776–9.
— nuclei, 9, 14, 55,9, 578.
—, rate of, 14, 15.
Crystalloids, 4.
Cyclol theory, 748.

Dalton's Law of Partial Pressures, 62.
Debye-Hückel theory, 101.
Decantation, electro-, 7.
Deformation, 465.
—, ideal, 465.
—, non-ideal, 465.
—, Hookean, 465.
—, non-Hookean, 465.
Dehydration, 10.
Denaturation of proteins, 3, 636, 753.
Density of charge, *see* Charge density.
Depolarization ratios, 432.
Detergents, 30, **667**.
— and detergent action, 686.
Dextrin, 16.
Dialysis, 5, 7, 13.
—, electro-, 7.
Diamond, 40.
Dickite, 706, 709.
Dielectric constants, **412**, 416.
— — and dipole moments, 422.
— —, dispersion of, 413.
— —, increments, 416.
— — of gels, 602.
Diffraction of electrons, 27, 28, 29, 461.
— of X-rays, 21, 27, 28, **440**, 501, 539.
— — and crystallite size, 453, 539.
— — by crystalline proteins, 454.
— — by fibrous materials, 456.
— — by soap solutions, 458, 690.
— — in colloidal systems, 454.
— —, use of intensity measurements in, **445**.
Diffuse double layer, 12, **116, 296,** 529, 534.
Diffusion, 4, 17, 26, 27, 42.
— and membrane permeability, 808.
— and osmotic pressure, 256.
—, coefficient of, 17, **233**, 381.
— in gels, 602.
— of soaps and dyes, 687, 698.
— potentials, 215.
—, rotational, 48, **380**, 413.
—, translational, **233**.
Dilatancy, 30, 479.
Dimerization, 40.
Dipole angle, 420.
— interaction, 24, 46.
— moment, 423.
— — and dielectric increments, 422.
Disperse phase, 1.
Dispersion medium, 1.
— methods for preparation of colloidal solutions, 557.
— of dielectric constant, 413. *See also* Dielectric constants.
— — — and relaxation times, 413.
— of conductivity, 421.
Dissociation of proteins, 255, 281, 745.
Distribution of molecular size, 27, 41.

Distribution of molecular size, for proteins, 736; for polymers, 781, **796.**
Donnan equilibria, 25, **76, 157,** 182, 215, 815.
— — and osmotic pressure, 157.
Double layer, 12, **296.** *See also* Diffuse double layer.
— —, capacity of, 534.
— —, interaction of, 116.
Double refraction of flow, 387, **393,** 720. *See also* Birefringence of flow.
— — —, and molecular orientation, 387.
— — — and viscosity, 376, 401.
— — —, effect of molecular deformation on, 410.
Dough, 1, 30, 464.
Drop weight method, 517.
— volume method, 517.
Ductility, 30.
Duhem-Margules equation, 64, 69.
Duplex films, 500, 523.
Dyeing, 699.
— of cellulose acetate, 702.
— of cellulose fibres, 700.
— of protein fibres, 701.
Dyestuffs, 30, 697.
—, adsorption of, 544.
—, aggregation of, 698.
Dynamic methods, 208.

Egg white, 3.
Einstein viscosity equation, 358.
Elastic gels, 608.
Elastic modulus, 466.
Elasticity, 30, 466.
— of gels, 596.
Elastomers, 473.
Electrical charge on protein molecules, 44, **332,** 741.
Electrical double layer, 12, **116, 296,** 534. *See also* Double layer.
Electrocapillarity, 12, 22, 90, **531.**
Electrochemical potential, 53, 76, 298.
Electro-decantation, 7.
Electro-dialysis, 7.
Electrokinetic phenomena, 11, 29, **295.**
— potential, 12, 15, 29, 38, 48, 100, 111, 114, 124, **296,** 521, 529, 565, 570. *See also* Zeta potential.
— — and colloidal stability, 15, 29, **114,** 315, 556, **570,** 573.
— — and electrical double layer, 12, 100, 296.
— — and electro-viscous effects, 377.
— — and thermodynamic (Nernst) potential, 298.
— —, comparison from different methods, 299, 339.
Electrolytes, colloidal, 30, 110, **667.** *See* Colloidal electrolytes.
Electron diffraction, 27, 28, 29, 461.
— microscope, 29, **458,** 537.
Electro-osmosis, 2, 11, 295, **338.**
— in closed tubes, 310.
—, stationary levels in, 310.

Electro-osmotic mobility, 302, 312.
— — comparison with electrophoretic mobility, 315.
— — determination of, 338.
Electrophoresis, 2, 11, 29, 111, 295, **300**.
—, micro-method, 308, 313.
—, —, results of, 314.
—, moving boundary method, 29, **317**, 326.
—, — — —, anomalies in, 328.
—, — — —, results of, **330**, 741.
—, — — —, results of, in monodisperse systems, 323.
—, — — —, results of, in pauci- and polydisperse systems, 324.
—, theoretical, **300**.
—, —, Henry's equation, 303.
—, —, Hückel's equation, 301.
—, —, Smoluchowski's equation, 302.
Electrophoretic mobility, definition of, 300.
— —, determination of, 309, 324.
— —, influence of solute concentration, 326.
— — of asymmetric particles, 306.
— — of colloidal ions, 111, 315, **330**, 565.
— velocity, 12.
Electrostatic barrier, 529.
Electro-viscous effects, 377.
Ellipsoidal particles, motion of, **212, 259**, 283.
Emulsifying agents, 644, **652**.
— —, electrolytes, 656.
— —, for O/W, 644.
— —, for W/O, 645.
— —, proteins, 655.
— —, soaps, 653.
— —, solid powders, 655.
Emulsions, 25, 31, **643**, 651.
—, agents for, 644, **652**.
—, application of, 661.
—, breaking of, 665.
—, creaming of, 649, 664.
—, determination of type, 648.
—, preparation of, 646.
—, properties and uses of, 659, 660.
—, stability, 649.
—, viscosity of, 659.
Endosmosis, 2, 11, 295, **338**. See Electro-osmosis.
Energy, 50.
—, free, 35, 44, 55, 72. See Free energy.
—, —, of adsorption, 73, 549.
—, —, of Gibbs, 55.
—, —, of Helmholtz, 55.
—, —, standard, 73.
—, surface, **34**, 54, 84, 86.
—, —, free, 35, 44.
Entropy, 51.
— change, 51, 125.
— — in polymer systems, 92, 788.
— of adsorption, 524, 549.
— surface, 84.
Enucic acid, 502.
Enzymes, 27, 35, 752.
Equation of state, 499 et seq., 527.

Equation of state, rheological, 483.
Equilibria, in osmosis, 167.
—, membrane. See Donnan equilibrium.
—, thermodynamic criteria of, 56, 65.
Equilibrium, Donnan membrane, 25, **76, 157**, 182, 815.
—, heterogeneous, 72.
—, membrane, see Donnan membrane.
—, sedimentation, 18, 42, 43, 83, **188**, 735.
—, —, complicating factors in, 199.
—, —, in general, 188.
—, —, in the ultracentrifuge, 191.
—, —, — of monodisperse systems, 197.
—, —, — of polydisperse systems, 198.
—, —, —, results of, 204.
—, —, —, theoretical, 194.
— systems, investigation of, by diffusion, 255.
— —, — of, by sedimentation velocity, 279.
—, thermodynamic criteria of, 56, 65.
Erythrocytes, 821–8.
Ethyl stearate, monolayers of, 501.
Evaporation, rate of, 540.
Excess, surface, **84**, 519.
Exchange, of ions, 713, 726.
Extinction angle, 391–410.
— —, and angle of isocline, 399.
— —, and molecular orientation, 391, **395**.

Farinograph, 472.
Fats, 3.
Fatty acid soaps, 31, 44, 625, 630, 653, **669**.
Felspar, 706.
Fibres, 1, 456, 747, 750, 767, 780.
Fick's Laws of Diffusion, **233, 234**, 258, 381, 562.
— — — and Membrane Permeability, 809.
Film balance, Langmuir-Adam, 24, **490**, 508.
— —, Wilhelmy, **490**, 508, 518.
Films, see also Monolayers.
—, adsorbed, 29, **513**, 637.
—, built up, 496.
—, condensed, 500.
—, duplex, 500, 523.
—, gaseous, 497, 527.
—, insoluble, 23, **489**.
—, liquid expanded, 499.
—, oil, 806, 819.
—, soap, 31, 630, 653.
—, vapour, 497.
Filtration, ultra-, 8, 806, 817.
First Law of Thermodynamics, 50.
Flocculation, 15, **570**, 582, 716. See also Coagulation.
Flocculation values, 571, 572.
Flotation, 22, 640.
Flow, 464.
—, anomalous, **347**, 376, 380.
—, Bingham, 466.
—, Newtonian, 346, 353, 466.
—, non-Bingham, 466.
—, non-Newtonian, 347, 356. See also Anomalous flow.

SUBJECT INDEX

Flow, plastic, 464, 466, 779.
—, plasto-elastic, 466.
—, plasto-inelastic, 466.
—, viscous, 466.
Fluidity, 346.
Foaming agents, 624, **630**.
Foaming power, measurement of, 626.
— —, —, by dynamic methods, 627.
— —, —, by single bubble method, 627.
— —, —, by static method, 626.
Foams, 22, 25, 31, **624**.
—, destruction of, 626.
—, formation of, 626.
—, in non-aqueous media, 641.
—, stability of, 628.
Fogs, 1, 32, 576.
Forces, electrical, 101, 115, 140, 296, 748.
—, van der Waals, 115, 140.
Fractionation of polymers, 796.
Free energy, 35, 44, 72.
— —, Gibbs, 55.
— —, —, and interaction of colloidal particles, 120.
— —, —, in polymer systems, 92.
— —, Helmholtz, 55.
— —, —, in polymer systems, 97.
— —, of adsorption, 73, 549.
— —, standard, 73.
Freundlich adsorption isotherm, 22.
Friction, coefficient of, 616.
—, —, and properties of pastes, 616.
Frictional coefficient, 17, 27. *See also* Frictional constants.
— constants, **210**.
— —, of rotation, 212.
— —, of translation, 211, 212.
— ratios, 213, **260**, 292.
— — and molecular properties, 213, **259**, 293.
— —, effect of asymmetry on, **260**, 385.
— —, effect of solvation on, 260, 293.

Gas adsorption, 23, 24, 546, 723.
— ions, condensation upon, 578.
Gaseous films, 497, 527.
Gases, imperfect, 62.
—, perfect, 60, 62.
Gegenions, 44, 48, 101, 110, 678.
Gelatine, 4, 6, 16, 30, 35, 474, 587.
Gelation, 115, **586**.
— changes during, 590.
Gels, 5, 26, 30, 35, 464, 474, **585**.
—, aero-, 5.
—, alco-, 5.
—, dielectric constant of, 593.
—, diffusion in, 602.
—, elastic, 608.
—, elasticity of, 594, 596.
—, electrical conductivity of, 592.
—, formation of, 587.
—, hydro-, 5.
—, hysteresis in, 597, 606.
—, optical properties of, 597.
—, reactions in, 602.
—, rigid, 608.

Gels, silica, 22, 591, 608.
—, structure of, 608.
—, swelling of, 26, **599**.
—, thixotropic, 30, 586, **610**.
—, viscosity of, 594.
—, volume of, 593.
—, X-ray diffraction of, 592.
Gestalt, 464.
Gibbs adsorption isotherm, 10, 22, **84**, **519**, 524, 683.
Gibbs Free Energy, 55.
— — —, and interaction of colloidal particles, 120.
— — —, in polymer systems, 92.
Gibbs-Helmholtz equation, 60, 542, 617.
Gibbsite, 708, 711.
Globulins, 733.
—, ground-nut, 281, 734.
—, serum, 42, 145, 169, 205, 256, Table 11.1, 293, 299, 324, 735, 741.
Glycogen, 773.
Gold number, 15, 574.
— sols, 2, 7, 15, 31, 40, 47, 259, **428**, 460, 571.
Goodeve Plastometer, 471.
Graphite, 30, 645, 655.
Gravitational sedimentation, 1, 18, 26, 188, 266, **537**, 581.
Gum arabic, 16, 644.
Gutta percha, 765.

Haemocyanins, 132, 403, 745.
Haemoglobin, 42, 131, 255, 281, 456.
—, dissociation of, 255, 281.
Haemolysis, 823.
Halloysite, 706, 708.
Heat capacity, at constant pressure, 51.
— —, at constant volume, 55.
— conductivity, 543.
— content, 54.
— of adsorption, 524, 549, 550.
— of solution, 39, 685.
— of wetting, 542, 550, 617.
Helmholtz Double Layer, 12.
— free energy, 55.
Heterogeneity of polymers, 42, 198, 250, 275, 409, 796.
Heterogeneous equilibrium, 72.
Hexadecyl sulphonic acid, 30, 667.
High polymers, *see* Polymers.
Hofmeister series, 16, 125, 589, 714.
Homogeneity of polymers, 197, 248, 271, 323, 409.
Honey, 464.
Hooke solid, 480.
Hooke's Law, 467.
Hydrated cellulose, 769.
Hydrates, 10, 140.
Hydration, 44, 46, 125, 126, **140**. *See also* Solvation.
—, effect on frictional ratio, 260, 293.
—, effect on viscosity, 366.
— in gels, 605.
— of colloidal particles, 126.

Hydration of proteins, **140**, 262, 293, 738.
— of simple ions, 46, 140, 571.
Hydrocarbons, saturated, 757.
—, unsaturated, 762. *See also* Paraffins.
Hydrogel, 5.
Hydrogen bonds, 47, 658.
— — in proteins, 745, 748.
Hydrophilic colloids, 13, 16, **124**.
Hydrophobic colloids, 13, 15, 47, **114**. *See also* Sols.
Hydrosol, 5.
Hysteresis, 597, 606, 608.

Illite, 706.
Immune sera, 743.
Imperfect gases, 62.
Increment of refractive index, 220.
Infra-red absorption, 439, 758.
Inks, 1.
Insoluble films, 23.
— monolayers, 23, **489**, 508, 635.
— —, mechanical properties of, **494**, 510, 636.
Intensity factors, 63.
Intensity measurements in X-ray methods, 445.
Interfaces, air/water, 23, **488**, **513**.
—, diffusion through, 505.
—, gas/liquid, *see* Air/water.
—, liquid/liquid, 14, 22, **488**, **508**, **513**, 643.
—, mercury/water, 22, 531.
—, oil/water, 22, 48, 488, **508**, **513**.
—, orientation at, *see* Orientation.
—, solid/liquid, 14, 22, **536**.
—, thermodynamics of, 25, **84**, 503, 519.
Interfacial potentials, **295**, 509, 521, 551.
— techniques, and molecular structure, 504.
— —, and biological problems, 506, 531.
— tension, 12, 508, 513–19.
— reactions, 21, 506.
Intrinsic viscosity, 358.
— —, and molecular shape, 361.
— —, and molecular weight, 367.
— —, effect of temperature and solvent on, 374.
Inverse plasticity, *see* Dilatancy.
Ion atmosphere, 10, 297.
— —, thickness of, 109, 297.
— antagonism, 826.
— exchange, 713, 726.
Ionic activities, 71.
— strength, 72, 102, 108.
— —, and thickness of double layer, 298.
Ionization, 44, 330.
Iso-electric point, 45, 158, 334, 377, 566.
Iso-ionic point, 334.
Isothermal change, 60.
— —, non-, 61.
Isotherm, adsorption, 548.
—, —, Freundlich, 22.
—, —, Gibbs, 10, 22, **84**, **519**, 524, 683.
—, —, Langmuir, 22, 548, 723.
Isotopes, radioactive, 545, 825.

Jellies, *see* Gels.

Kaolin, 709.
Kaolinite, 709.
Kelvin solid, 481.
Kephalin, 508, 512.
Keratin, 456, 733, **747**, 749.
Kinetic theory, 2, 17, Appendix I.
Kinetics of coagulation, 575, 582.
— of spreading, 504.
Krafft Point, 31, 683.

Lactoglobulin, **142**, 293, 367, 735, 740, Table 11.1.
—, hydration of, **142**.
Langmuir-Adam Film Balance, 24, **490**, 508.
Langmuir adsorption isotherm, 22, 548, 723.
Law of conservation of energy, 50.
— of partial pressures, 62.
—, Poiseuille's, 350, 353, 494.
—, Raoult's, 68, 69, **151**.
—, Stokes's, 17, 111, **210**, 537.
Laws, of thermodynamics, 50.
Lecithin, 508, 510.
Liesegang rings, 603.
Light, polarized, 28, **387**, 393, 395, **430**, 597.
—, —, and light scattering, **430**.
—, —, and streaming birefringence, **387**, 393, 395 et seq.
—, scattering of, 28, **178**, **426**, 434.
—, —, and molecular properties, 178, **434**.
—, —, and osmotic pressure, 179.
—, —, in colloidal solutions, 178, 426, 434.
Linear polymers, *see* Polymers.
Lipoids, 458, 507, 508, 820, 826, 828.
Liquid diffusion, 4, **233**. *See under* Diffusion.
— expanded films, 499.
—, Maxwell, 482.
Liquid/liquid interface, 14, 22, 488, **508**, **513**, 643.
Liquids, Newtonian, 481.
Lyophilic colloids, 13, 124, 585.
— —, properties of, 13.
— —, stability of, **124**.
Lyophobic colloids, 13, 16, **115**, 315.
— —, preparation of, 3, 7, **556**.
— —, properties of, 13, **565**.
— —, stability of, **115**, 315, 570.
— —, viscosity of, 568.
Lyotropic series, 16, 125, 589, 714.
Lysolecithin, 508, 510.
Lubrication, 23.

Margarite, 710.
Mass, molecular and particle, 40, 41.
Maximum bubble pressure method, 516.
Maxwell liquid, 482.
Mechanical properties of insoluble monolayers, **494**, 510, 636.
Medium, disperse, 1, 30.
—, dispersion, 1.

SUBJECT INDEX

Melting-point and chain-length, 778.
Membrane equilibria, *see also* Donnan equilibria.
— —, 25, **73**, 76, 157, **180**, 815.
— —, criteria for, 167.
— —, diffusible ions and, 158.
— —, rate of attainment of, 166.
— potentials, 25, 79, **180**, 816.
— —, ion pressure differences and, 184.
— —, measurement of, 182.
— —, protein charge and, 185, 337.
— —, protein concentration and, 183.
Membranes, 10, 21, 25, 73, 157, 160, 165, **804**, 820.
—, and Bioelectric potentials, 826.
—, artificial, 804.
—, bacterial cellulose, 174.
—, diffusion through, 809.
—, for osmotic pressure determinations, 160, 165, 171, 174, Appendixes II and III.
—, natural, 820.
—, permeability of, 808, 822.
—, porous, 11.
—, semi-permeable, 66. *See also* 'for osmotic pressure determinations'.
—, structure of, 807, 820.
Metal sols, 2, 6, 7, 9, 16, 31, 40, 47, **556**. *See also* Gold, platinum, and silver sols.
Methyl cellulose, 771.
Micelles, 31, 45, **669**.
—, formation and law of mass action, 672.
—, structure of, 670.
—, —, spherical, 670.
—, —, laminar, 675.
Micro-electrophoresis, 308, 313. *See also* under Electrophoresis.
—, results of, 314.
Micrometer syringe, 489, 509.
Microscope, electron, 29, **458**, 537.
—, ultra, **16**, 18, 26, 28, 29, 308.
—, —, examination of clay suspensions by, 716.
—, —, — of gels by, 591.
Microtome, 23, 525.
Milk, 1, 644.
Mill, colloid, 557.
Mists, 1, 32, **576**, 665.
Mobility, 111, 565. *See also* Electrophoretic mobility.
Modulus, bulk, 466.
—, elastic, 466.
—, shear, 466, 470.
—, Young's, 466.
Molecular constants for proteins, Table 11.1, 293, 367, 735, 738.
— kinetic theory, 2, 17, Appendix I.
— orientation, *see* Orientation.
— shape, 260, 292, 334, 361, 367, 384, 419, 434, 458, **738**, 799.
— —, effect of solvent on, 374, 801.
— —, effect of velocity gradient, 410, 800.
— sieve, 725.
— weight, 26, 40, 41, Table 11.1, 735, 798.
— —, averages, 41.
— —, from analytical methods, 130, 133.

Molecular weight, from diffusion and sedimentation velocity, 289.
— —, from osmotic pressure, 156, 169.
— —, from sedimentation equilibrium, 205.
— —, from viscosity, 367.
Monodispersity, criteria of, 197, 248, 271, 323, 409.
Monolayers, 23, 24, **489**.
—, air/water, 489.
—, insoluble, 23, 24, **489**, 508, 635. *See also* Insoluble monolayers.
—, oil/water, 508.
—, protein, 505, 508, **635**.
—, viscometer for, 494, 510.
Montmorillonites, 706, 708, **711**.
Moving boundary, 29, **317**, 326. *See also* under Electrophoresis.
Multilayers, 496.
Multiple law, of Svedberg, 736.
Muscovite, 706, 710.
Mutual action of sols, 15, 573.
Myristic acid, 504.

Nacrite, 706, 709.
Neoprene, 766.
Nephelometry, 426.
Nernst potentials and Zeta potentials, 298.
Newtonian flow, **346**, 353, 380, 466.
—, liquids, 481.
Nitro-cellulose, *see* Cellulose nitrate.
Non-Bingham flow, 466.
Non-conductors, 7.
Non-Newtonian flow, **347**, 356, 380, 466.
Nuclei, 9, 14, **559**, 578.
—, formation of, 14, **559**, 577.
Number, average molecular weight, **42**, 292, 370.
—, Avogadro's, 17, 18, 190.
Numerical aperture, 459.
Nylon, 508, 774.

Octadecyl alcohol, 503.
Oil, castor, 23, 496.
— droplets, 3.
— films, 806, 819.
—, olive, 3, 23.
Oil/water interface, 22, 48, 488, **508, 513**.
— monolayers, 508.
Oleic acid, 496.
Optical methods, 216.
— —, light absorption, 217.
— —, refractive index, 219.
— —, — —, scale, 230.
— —, — —, schlieren, **220**, 546.
— properties of colloids, 6, 7, 13, 16, 28, 40, 178, **426**, 570, 591, 597, 659.
— — —, depolarization ratios, 432.
— — —, light scattering, 28, **178, 426,** 434.
— — —, polarization of scattered light, 430.
— — —, turbidity, 178.
— — —, Tyndall effect, 6, 13, 28, 426.
Ore flotation, 640.

SUBJECT INDEX

Orientation, at interfaces, 22, 24, 497, **500**, 506, 527, 636, 658.
—, in electrical fields, 412.
—, — —, limitations of, 424.
—, in velocity gradients, 387.
—, — —, limitations of, 412.
Osmotic balance, 177.
— coefficient, 69.
— pressure, theoretical, **73**, 150.
— —, —, and calculation of molecular weights, 156.
— —, —, and diffusion, 256.
— —, —, and Donnan equilibria, 157.
— —, —, and light scattering, 178.
— —, —, and membrane equilibria, 79, 157.
— —, —, — potentials, 81.
— —, —, and thermodynamics, 73.
— —, —, calculation for linear polymers, 153.
— —, —, equations of, 150–7.
— —, experimental, **160, 169**.
— —, —, and membrane potentials, 184.
— —, —, criteria of equilibria for, 167.
— —, —, dependence on concentration for linear polymers, 156, 175; for proteins, 156.
— —, —, dynamic method of measurement, 167.
— —, —, extrapolation procedures, 155, 175.
— —, —, in aqueous systems, 160.
— —, —, in aqueous systems, osmometers for, **160**.
— —, —, in non-aqueous systems, 169.
— —, —, in non-aqueous systems, osmometers for, **170**.
— —, —, limitations of method, 176.
— —, —, membranes for, 165, 174, Appendix II.
— —, —, results, 167, 175.
Ostwald's Law of Stages, 560.

Paints, 1, 26, 29, 30, 464.
Paraffins, 757. *See also* Hydrocarbons.
—, melting-point and chain-length of, 778.
Partial molar free energy, 59, 63. *See also* Free energy.
— — quantities, 62.
— — volume, 64, 74, 137.
— specific volume, **137**, 194, 201, 287, Table 11.1, 291.
Particle and molecular weight, 40.
Particle size, 10, 13, 18, 26, 28.
— —, and effect on properties, 10, 34.
Particles, anisometric, 35.
Pastes, 464, 585, **611**.
—, clay, *see* Clay pastes.
—, Rheological properties of, 618.
Pathological sera, 742.
Paucidispersity, 198, 273, 324.
Pendent drop method, 515.
Penetration, 657.
Penetrometers, 472.
Pepsin, Table 11.1, 293, 367, 735, 740.

Peptization, 3, 5, 558, 589.
Perfect gases, 60, 62.
Permeability, 166, 540, **808**, 822.
—, of cells, 822, 827.
Permutite, 721. *See also* Zeolites.
Phase boundary potential, **295**, 509, 521, 551.
—, disperse, 1.
Phases, surface, 84, **497**, 527.
—, —, and stability of emulsions, 651.
—, —, and stability of foams, 630.
—, —, and thermodynamics of, **84**, 503, 519.
Phenol-formaldehyde polymers, 774.
Phlogopite, 710.
Photography, 9.
Phospholipoids, 508, 820. *See also* Lipoids.
Piston oil, 496.
Plasma proteins, 741.
Plasmolysis, 822.
Plastic, St. Venant, 478, **481**.
— flow, 464, 466, 779.
Plasticity, 30, 476, **477**.
Plasticizer, 30, 778, 805, 812.
Plasto-elastic flow, 466.
— -inelastic flow, 466.
Plastometers, 468.
—, capillary, 472.
—, Couette, 470, 474.
—, compression, 469.
—, Goodeve, 471.
—, recovery, 469.
—, rotation, 469.
—, Stormer, 470.
—, Ungar, 471.
Plate, Wilhelmy, 490, 518.
Platinum sols, 31, 47, 557.
Point, Krafft, 31, 683.
Poiseuille's Law, 350, 353, 494.
Poisson's ratio, 466.
Polarizibility, 46, 572.
Polarization of light, by colloidal solutions, 427, **430**, 751.
— —, effect of molecular properties on, 433.
Polarized light, 28, 387, 393, 395, 430, 597. *See also under* Light.
Poly-amides, 774.
Poly-butadiene, 765.
Poly-chloroprene, 766.
Poly-esters, 774.
Poly-isobutene, 758.
Poly-isoprene, 762.
Poly-methyl chlorosilanes, 761.
Poly-styrene, 760.
—, flow birefringence of, 404, 800.
—, viscosity of solutions of, 375.
Poly-thene, 757.
Poly-vinyl polymers, 759.
Poly ω-hydroxydecanoic acid, 135, 774.
Polydispersity, *see* Heterogeneity of polymers.
Polymers, 17, 21, 26, 28, 41. *See also* Proteins.
—, definition of, 756.

SUBJECT INDEX

Polymers, different types of, 757.
—, in dilute solution, 798.
—, — —, flexibility of molecule, 800.
—, — —, molecular weight and shape, 799.
—, interaction with organic liquids, 786.
—, linear, 28, 40, 42, **756.**
—, mechanical properties of, 97.
—, properties of, 775.
—, solubility of, 94, 786, **793.**
—, thermodynamics of, 91, 97.
Potential, chemical, 35, 37, 38, 53, 59, 66, 67, **73, 86,** 92, 232.
—, distribution of for two interacting double layers, 116.
—, electrochemical, 53, 76, 298.
—, electrokinetic, see Electrokinetic potential.
—, interfacial, **295,** 509, 521, 551.
—, membrane, see Membrane potentials.
—, phase boundary, 489. See Phase boundary potential.
—, sedimentation, 11, 295, 338, 343.
—, streaming, 11, 295, 338, 341.
—, surface, 25, 489, **492,** 521.
—, zeta, see Electrokinetic and Zeta (ζ) potentials.
— curves, 122, 298.
Preparation of colloids, 3–6, 7, **556,** 576, **587,** 626, 646.
— —, by aggregation, 558, 577.
— —, by dispersion, 557, 577.
Pressure, osmotic, 25, 42, 73. See also Osmotic pressure.
—, —, of polymer solutions, 153, **169.**
—, —, of protein solutions, 156, 167.
—, partial, Dalton's Law of, 62.
—, surface, **489,** 508.
—, swelling, 1, 30, 94, **599,** 793.
—, vapour, 34, 35, 39, 86.
Protection of sols, 6, 15, 573.
Protein reactions, specificity of, 752.
Proteins, 3, 16, 17, 21, 25, 26, 28, 41, 42, 44, 48, 124, **730.**
—, adsorbed films of, 637.
—, as emulsifying agents, 655.
—, as foaming agents, 635.
—, composition and structure of, 330, **730, 745.**
—, denaturation of, 3, 636, 753.
—, dissociation of, 255, 281, 745.
—, electrical charge of, 44, **332,** 741.
—, hydration of, **140,** 262, 293, 738.
—, isolation and purification of, 732.
—, investigation by electrophoresis, 317, **330,** 741.
—, — by light scattering, 435, 735.
—, — by membrane potentials, 180.
—, — by osmotic pressure, 157, 168, 735.
—, — by rotational diffusion, 399.
—, — by sedimentation equilibrium, 198, 204, 735.
—, — — velocity, 292, 735.
—, — by translational diffusion, 255, 735.
—, — by viscosity, 366.

Proteins, investigation by X-rays, 454, 735, 746.
—, molecular constants for, Table 11.1, 293, 366, 735, 738.
—, monolayers of, 505, 508, **635.**
—, serum, see Serum proteins.
Prussian blue, 3, 5, 558.
Purple of Cassius, 2, 6.
Putty, 464.
Pyrophyllite, 706, 708, 711.

Radioactive isotopes, 545, 825.
Raoult's Law, 68, 69, **151.**
Rayleigh equation, 427.
Reactivity, chemical and surface area, 34, 35, 39.
Recovery plastometer, 469.
Refractive index increment, 220.
— — methods, 219.
— — —, scale, 230.
— — —, schlieren, **220,** 546.
Relaxation time, 111, **383, 413,** 415, 420, 423, **467,** 485, 595.
— —, and rotational diffusion, 383, 413.
— —, and molecular properties, 419.
Retardation volume, 546.
Rheological equation, 483.
Rheology, 30, **464,** 618.
Rheometer, 468.
Rheopexy, 611.
Rigid gels, 608.
Rigidity, 466, 596.
Ring method, for measuring interfacial tension, 508, 516.
Rotational diffusion, 17, 48, **380.**
— —, and Brownian motion, 18, 382.
— —, and concentric cylinder apparatus, 395.
— —, and frictional coefficients, 382.
— —, and molecular dimensions, 384.
— —, and polydispersity, 409.
— —, and relaxation times, 383, 413.
— —, and viscosity, 376, 401.
— —, coefficient of, 381.
— —, —, and relaxation times, 383.
— —, for polymer solutions, 404.
— —, for protein solutions, 399.
Rotation Plastometer, 469.
Rubber, 30, 41, 176, 251, 253, **762,** 776, 779, 785, 790, 796.
— –benzene system, 94, 790.
—, Buna, 766.
Ruby glass, colour of, 6.

'Salting-out', 13, 16, 125, 732.
Sand, 30.
Saponite, 706.
Scale method, 230.
Schlieren methods, **220,** 546.
Schultze-Hardy rule, 3, 15, 109, **115,** 123, **571,** 716.
Second Law of thermodynamics, 50, 51.
Sedimentation, gravitational, 1, 18, 26, 188, 266, **537,** 581.
— balance, 537.
— constants, 272, Table 11.1, 799.

SUBJECT INDEX

Sedimentation constants, and effect of solvation, 290.
— — and molecular properties, 287.
— — and molecular weight calculations, 289.
— —, complicating factors in, 276.
— —, extrapolation of, 289.
— equilibrium, see Equilibrium, sedimentation.
— potentials, 11, 295, 338, 343.
— velocity, 18, 26, 42, **266.**
— —, and ultracentrifuge, 267.
— —, anomalies in, 278.
— —, complicating factors in, 276.
— —, effect of concentration on, 276.
— —, effect of diffusion on, 272.
— —, in equilibrium systems, 279.
— —, in interacting systems, 283.
— —, in non-interacting systems, 271.
— —, in polydisperse systems, 274, 285.
— volume, 553, **613.**
Semi-permeable membranes, 66. See also under Membranes.
Sensitization, 125, 573.
Series, Hofmeister or Lyotropic, see under Hofmeister and Lyotropic.
Serum proteins, 3, 41, 145, 169, 205, 256, Table 11.1, 293, 299, 324, 331, 333, 367, 735, 740. See also under Albumins and Globulins.
Sessile drop method, 515.
Shape of colloid particles, 26–30. See also Molecular shape.
— — —, and gelling, 35, 609.
— — —, investigation of, 214, **260,** 292, 334, 361, 384, 419, 434, 458, **738, 799.**
Shear modulus, 466, 470.
Shearing strength, 473.
Silica gel, 22, 590, 599, 604, 606, 608.
Silicates, 548, **704.**
Silicic acid, 594, 608.
Silicone polymers, 761.
Silk, 28, 737.
Silver sols, 2, 6, 7, 9, 557.
Size distribution, 27, 30, **537.**
Smoke, 1, 6, 31, 32, **576.**
—, coagulation of, 581.
—, formation of, 577.
—, investigation of, 580.
Soap bubbles, 22, 31, **630.**
— —, reversed, 695.
— films, 31, **630,** 653.
— solutions, concentrated, 689.
— — —, —, phase diagrams for, 689.
— — —, —, solubilization in, 691.
— — —, —, viscosity of, 694.
— — —, —, X-ray investigation of, 458, **690.**
— — —, dilute, colligative properties of, 676.
— — —, —, critical concentrations, 681.
— — —, —, density of, 687.
— — —, —, detergent action of, 686.
— — —, —, diffusion in, 687.
— — —, —, electrical conductivity of, 677.
— — —, —, hydrolysis in, 688.
— — —, —, solubilization in, 685.

Soaps, 22, 25, 30, 44, 458, 667, **669.**
—, as emulsifying agents, 653.
—, as foaming agents, 630.
—, biological effects of, 695.
—, fatty acid, 31, 44, 625, 630, 653, **669.**
—, four component systems, 695.
—, heats of solution of, 685.
—, heavy metal, 645, 654, 696.
—, in non-aqueous media, 696.
—, mobility of suspensions in, 568.
—, solubility of, 683.
—, surface behaviour of, 108, **680.**
—, three component systems, 694.
Sodium cetyl-sulphate, 508, 510.
Soil, 1, 30, **704.**
Sol–gel transformation, 30, 586, **590.**
Solid, Kelvin, 481.
Solid/liquid interface, 14, 22, **536.**
Sols, 5, **556,** 584.
—, aero-, 5.
—, alco-, 5.
—, aqua-, 556.
—, arsenious sulphide, 15, 571–3.
—, electrophoretic mobility of, 315, 565.
—, gold, see Gold Sols.
—, hydro-, 5.
—, in gaseous media, 576–83.
—, in organic media, 576.
—, metal, 2, 6, 7, 9, 16, 31, 40, 47, **556.**
—, methods of preparation, 3–7, **556,** 576.
—, — —, aggregation processes, 558.
—, — —, dispersion processes, 557.
—, — —, dispersion processes, electrical, 9, 557.
—, — —, dispersion processes, mechanical, 557.
—, — —, dispersion processes, peptization, 558.
—, optical properties, 6, 7, 13, 16, 28, 40, **426,** 570.
—, platinum, 31, 47, 557.
—, precipitation by salts, 3, 6, 13, 15, **115,** 124, **570.**
—, protection of, 6, 15, 573.
—, sensitization, 125, 573.
—, silver, 2, 6, 7, 9, 557.
—, size-frequency distribution, 537.
—, stability of, 13, 15, 29, **114,** 315, 566, 570–5, 715.
—, streaming birefringence of, 28, 399, 570.
—, sulphur, 3, 7, 40, 47, 559, 565, 573.
—, vanadium pentoxide, 28, 399, 570.
—, viscosity of, 13, 568.
Solubility, 70.
—, effect of particle charge on, 37, 89.
—, — — size on, 35, 86.
—, of colloidal electrolytes, 683.
—, of polymers, 94, 786, **793.**
Solution, rate of, 15, 540, 562.
Solvation, 125, **140.**
—, effect on frictional ratios, **260,** 293.
—, — on viscosity, 360, 366. See also under Hydration.

SUBJECT INDEX

Specific surface, 35, **536**.
— viscosity, **358** et seq.
Spreading, 22, 503, 523.
— coefficient, 523.
Stability, of colloids, 13, 15, 29, **114**, 315, **570**, 715.
—, of emulsions, 649.
—, of foams, 626.
—, of gels, 604.
Standard affinity, 73.
— free energy, 73.
Starch, 45, **791**.
State, colloidal, 1, 5, 34.
—, equation of, 499 et seq., 527.
Staudinger equation, 28, **368**.
Stearate, ethyl, 501.
Stearic acid, 39, 500.
Stokes's Law, 17, 111, **210, 213**, 537.
Stormer Plastometer, 470.
Strain, 465.
Streaming birefringence, 27, 28, **387, 393**, 570, 599, 720 (Double refraction of flow).
— potential, 11, 295, 338, 341.
Stress, 464.
Structure factors, 446–53.
— —, determination of, 447.
— —, use in electron density calculations, **445**.
St. Venant plastic, 481.
Sulphur sols, 3, 7, 40, 47, 559, 565, 573.
Supercooling, 14, 559.
Supersaturation, 15, 559 et seq.
Surface ageing, 513, **529**.
— area, 24, **34**, 40, 489, **497**, 508, 519.
— —, and physical properties, 34.
— —, of solids, 536.
— —, —, methods for determination of, 536–48.
— concentration, 84.
— energy, **34**, 54, 84, 86.
— entropy, 84.
— excess, 84, 519.
— phases, 84, **497**, 527. See also under Phases.
— potentials, 25, 489, **492**, 521.
— pressure, **489**, 508.
—, specific, 35, **536**.
— tension, 10, 13, 21–5, 35.
— —, determination of, 513.
— —, by dynamic methods, 518.
— —, by static methods, 514.
— viscosity, 25, 489, **494**, 510, 522, 636.
Surfaces, structure of, 526.
Suspensions, 2, 13, 47, **556**, 585. See also Sols.
Swelling, 1, 30, 94, **599, 793**.
— pressure, 1, 599.
Syneresis, 5, 604.
Synthetic rubber, 765.
Syringe, micrometer, 489, 509.

Talc, 637, 706–10.
Talcum powder, 496.
Tension, adhesion, 542, 551, **612**.
—, interfacial, 12, 508, 513–19.

Tension, surface, 10, 13, 21–5, 35. See also under Surface tension.
Thermistor, 550.
Thermodynamics, 50.
—, and colloidal systems, 73.
—, Laws of, 50, 51.
—, of adsorption, 524, 549.
—, of interfaces, 25, **84**, 503, 519.
Thin oil films, properties of, 819.
Thixotropic gels, 30, 586, **610**.
Thixotropy, 30, 115, **478**.
Time of relaxation, see Relaxation time.
Titration curves for proteins, 334.
Tobacco mosaic virus, 361, 376, 399–402, 438, 461, 751.
Translational diffusion, 4, 17, 26, 27, 42, **233**.
— — and boundary anomalies, 254.
— — and charge effects, 254.
— — and chemical potential, 233.
— — and frictional constants, 258.
— — and osmotic pressure, 256.
— —, experimental methods of, 239.
— — in equilibrium systems, 255.
— — in interacting systems, 238, 250.
— — in non-interacting systems, 248.
— —, laws of, 233, 234.
— — coefficient, 233.
— — — and molecular properties, 255, **259**.
— — —, effect of solvent and temperature on, 263.
Traube's Rule, 22, 527.
Turbidity, 178.
Tyndall effect, 6, 13, 28, 426.

Ultracentrifuge, boundary anomalies in, 278.
—, cell for, 191, 270.
—, experimental, 267.
—, for sedimentation equilibrium, 191.
—, — velocity, 267.
—, investigation of equilibrium systems in, 279.
—, — of asymmetric macromolecules in, 283.
—, — of molecular weight and shape in, 287.
—, optical systems for, 192, **216**.
—, sedimentation of interacting systems in, 283.
—, — of non-interacting systems in, 195, 271.
Ultrafiltration, 8, 806, 817.
Ultramicroscope, **16**, 18, 26, 28, 29, 308.
—, examination of clay suspensions by, 716.
—, examination of gels by, 591.
Ungar Plastometer, 470.
Urea, action on proteins, 745.
Urea-formaldehyde polymers, 775.

Vanadium pentoxide sols, 28, 399, 570.
Van der Waals forces, 115.
Vapour films, 497.
— pressure of colloidal particles, 34, 39, **86**.

Velocity gradient, 355.
— —, and streaming birefringence, **387**, 393.
— —, orientation in, 28, 376, **387, 395**.
— of sedimentation, 18, 26, 42, 195, **271**, 288.
Viruses, 27, 28, 456, **751**, 818.
—, Bushy Stunt, 205, 752.
—, tobacco mosaic, 361, 376, 399–402, 438, 461, 751.
Viscometers, 349.
—, capillary, 350.
—, concentric cylinder (Couette), 28, 354, 387.
—, falling sphere, 211.
—, surface, 494, 510.
Viscosity, 13, 17, 27, 42, **345**.
— and concentration, 371.
— and electro-viscous effects, 377.
— and molecular orientation, 364.
— and molecular weight, 367.
— — shape, **362**.
—, anomalous, **347**, 376, 380, 401.
—, changes on gel formation, 594.
—, coefficient of, 345, 467.
—, effect of solvation on, 360, 366.
—, effect of solvent on, 366, 374.
—, effect of temperature on, 375.
—, Einstein equation, 358.
—, functions of, 357.
—, intrinsic, 358, 367.
—, Newtonian, and non-Newtonian, **347**, 356, 465–86. *See also* Anomalous viscosity.
—, of colloidal solutions, 358.
—, of sols, 13, 568.
—, of solutions of asymmetric particles, 361.
—, — of spherical particles, 358.
—, structural, 347.
—, surface, 25, 489, **494**, 510, 522, 636.
Viscous flow, 466.
Volume, partial specific, **137**, 194, 201, 287, Table 11.1, 291.

Volume, sedimentation, 553, **613**.
Vulcanization, 764, 785.

Water, adsorption of, 44, 46, 125, **140**.
—, —, by colloidal particles, 125.
—, —, by ions, 46, **140**, 571.
—, —, by proteins, **140**, 262, 293, 738.
Weight, molecular and particle, 26, 40, Table 11.1, 735, 798. *See also* Molecular weight.
Weight-average molecular weight, 43, 196, 370.
Wetting, heat of, 542, 550, 617.
— agents, 30, **667**. *See also* Detergents.
Wilhelmy plate, method of, 490, 518.
Wool, 1, 28, 456, 780.
Work of adhesion, 551, 612.

X-rays, diffraction of, 21, 27, 28, **440**, 501, 539. *See also* Diffraction of X-rays.
—, —, estimation of crystallite size, 453.
—, —, use of intensities in, 445.

Yield value, 473.
Young's modulus, 466.

Z-average molecular weights, 43, 196.
Zein, 402, 739.
Zeolites, 704, 721.
—, ion exchange in, 726.
—, sorptive properties of, 722.
—, structure of, 722.
Zeta (ζ) potential, 12, 15, 29, 38, 48, 100, 111, 114, 124, **296**, 521, 529, 565, 570. *See also* Electrokinetic potential.
— — —, and calculation of charge, 304.
— — —, and electrophoretic mobility, 300.
— — —, and electro-osmotic mobility, 302, 338.
— — —, and streaming potentials, 341.
— — —, and thermodynamic potentials, 298.
— — —, determination of, 308.

Tim Haddad
Randy Markon

(signatures)

Harry B. Kircher, Ph.D.

CHAIRMAN, DEPARTMENT OF EARTH SCIENCE,
  GEOGRAPHY AND PLANNING
SOUTHERN ILLINOIS UNIVERSITY
  AT EDWARDSVILLE

OUR NATURAL

Donald L. Wallace, M.S.

CERTIFIED PROFESSIONAL SOIL SCIENTIST
SOUTHWEST ILLINOIS METROPOLITAN AND
  REGIONAL PLANNING COMMISSION

Our natural resources

The symbol on the left and on the cover was drawn by Pam Decoteau as a logo for the Environmental Studies Program at Southern Illinois University at Edwardsville and is used with permission of the University. It represents the four major spheres of the earth's ecosystem: atmosphere, hydrosphere, lithosphere and biosphere.

FIFTH EDITION

RESOURCES

Consultant

DOROTHY GORE, Ph.D.

*Professor
Department of Geology,
Hardin-Simmons University
Abilene, Texas*

The INTERSTATE
PRINTERS & PUBLISHERS, *Inc.*

DANVILLE, ILLINOIS

OUR NATURAL RESOURCES
Fifth Edition

Copyright © 1982 by The Interstate Printers & Publishers, Inc. All rights reserved. Prior editions: 1954, 1964, 1970, 1976. Printed in the United States of America.

Library of Congress Catalog Card No. 80-83584

ISBN 0-8134-2166-7

To the youth of America

May this book help you realize a richer heritage

PREFACE

This text has now undergone four revisions since the first edition, written by P. E. McNall in 1954. In the 2½ decades since then, the people of the United States have been forced into a new awareness of the influence of natural resources on their lives. At the publication of the first edition, we were in a consumer economy. Growing consumption of goods and services was seen as the formula for success. With increasing industrialization and prosperity, we began to lose sight that our entire economy is dependent upon the natural resource base. But, then, we were awakened by various shortages and ecological disasters. It became more apparent that, while we have the power to increase or decrease our population and to make technical and societal adjustments, we cannot restore some of our most important exhaustible resources, nor can we prevent the degeneration of all energy to a state of entropy. Thus, we renewed our efforts to increase our usage of inexhaustible resources and of replaceable exhaustible resources and to use energy more efficiently.

This fifth edition addresses itself not only to analyzing the basic resource position of the United States but also to suggesting the best alternative for resource use in the future. Although the text has been extensively revised, especially the chapters on energy and minerals, the basic principles continue to be emphasized. Some of the illustrations have been changed. A selected list of general sources for environmental information, for use in expanding upon and updating the text, has been added.

Designed to accompany the fifth edition of OUR NATURAL RESOURCES is a student laboratory manual entitled INVESTIGATIONS IN CONSERVATION OF NATURAL RESOURCES (also available from The Interstate Printers & Publishers, Inc.). In addition to exercises which relate to topics in *OUR NATURAL RESOURCES*, the laboratory manual contains much additional information about conservation, including numerous maps, diagrams, charts, etc.

This edition, like its predecessors, is dedicated to the youth of our nation.

H.B.K.

ACKNOWLEDGMENTS

The authors gratefully acknowledge the help received during the preparation of the new edition of this text. Dr. Kircher, the principal author, especially appreciates the encouragement and help given him by Professors Melvin Kazeck, Robert Koepke, Panos Kokoropoulos and Halsey Miller, Southern Illinois University at Edwardsville, and by Robert Ring, former Director of Conservation Education, State of Illinois; the inspiration of his former teachers, Professors Raymond Murphy and Henry Warman, Clark University, Erich Zimmermann (now deceased), the University of North Carolina, and Lewis Thomas (now deceased), Washington University; and the assistance of his former colleague, Norman Bowsher, Economist, Federal Reserve Bank of St. Louis, and of Robert Craft, Public Information Officer, USDA Soil Conservation Service.

The authors are grateful for the many additions made by Dr. Dorothy Gore, a geologist, Hardin-Simmons University, Abilene, Texas, who was a consultant for the chapters on minerals, and for the many valuable suggestions given by R. A. Lytle, the principal author of *INVESTIGATIONS* . . . ; by students in the M.S. in Environmental Studies Program at Southern Illinois University at Edwardsville, especially Daniel Clampitt, Thomas Davis, Diana Friesz, David Gordon and Carol Sutherland; and by Richard Nicol, a biology graduate student.

The authors also are indebted to various government agencies, private businesses, nonprofit private research organizations and associations and the news media. Various sources have been credited in the text.

CONTENTS

	Page
Preface	vii
Acknowledgments	ix

Chapter

1 THE CHARACTER OF NATURAL RESOURCES — 1

Natural resources manifest energy, 3
    Forms of energy, 3
    Animate and inanimate energy, 3
    Inexhaustible resources, 4
    Exhaustible resources, 4
Distribution of natural resources, 7
Summary, 9

2 THE EVALUATION OF NATURAL RESOURCES — 11

Supply factors and natural resources, 12
    It may cost to make resources available, 12
Demand factors and natural resources, 14
    Elasticity of demand, 15
The market system, externalities and resources, 16
Wants and abilities and resources, 16
    Human wants and social objectives, 17
    Technological and societal arts, 18
Our human resources and nature's system, 18

Chapter	Page

      The science of ecology, 19
      Environmental perception, 21
    Freedom and natural resources, 22

3 The Increasing Use of Natural Resources   25

The population explosion, 25
Early use of animate energy, 26
Relatively few resources demanded, 27
Change to inanimate energy use, 28
    Stepped-up demand for resources, 28
Energy (potential power) most necessary, 28
    Macro unit of energy, 29
    Amounts of energy used in the United States, 29
    Sources of energy for power, 31
    New sources of energy for power, 32

4 Wind and Water Power   35

Wind first used for sailing, 35
    The compass expanded navigation, 36
    Steam replaced sails, 36
    Different types of sailing vessels, 36
Windmills became important, 37
    Windmills on the prairies, 38
    Pumping water a slow process, 39
    Reasons for the disappearance of windmills, 39
New proposals for using wind power, 40
Winds responsible for rainfall, 40
Water power from flowing streams, 41
Costs of conventional sources of water power, 42
    Silting back of dams, 46
    Variable water supply limited in use, 47
Benefits of conventional water power, 47
    Distribution of water power, 48
New sources of water power, 48
    Tidal power from the sea, 49
    Heat interchanges in the sea, 49
    Use of the Gulf Stream current, 50

Chapter	Page
5 Coal	53

    Location of coal reserves, 55
    How coal is formed, 55
    Kinds of coal, reserves and uses, 56
        Characteristics of different types of coal, 57
    Coal determined the sites for industry, 59
    Reducing costs and increasing availability of coal, 60
        Gasification and liquefaction, 63
    Other uses of coal, 64
    Why coal development has lagged, 64
        Strip mining, 65
    Coal and the energy crisis, 65
        Air pollution problems, 66

6 Oil and Gas   69

    Early petroleum use, 69
    Oil a nuisance, 70
    Many useful products from petroleum, 70
    Source of oil and gas, 71
    Where is oil found?, 71
    Reserves of oil and gas, 72
        Alaskan and outer continental shelf reserves, 74
        Relief from foreign supplies?, 75
        Some new gas fields, 75
    Better recovery developed, 77
    Prospecting and mining development, 78
    Mineral reserves and their estimates, 79
    Oil shales and tar sands, 82
    What is the future?, 83

7 Nuclear Energy   85

    Energy from nuclear fission, 85
    The question of safety, 86
    Breeder reactors, 87
    Adequate fuel supplies in the United States, 88
    Nuclear fusion, 89
    Industrial significance of nuclear energy, 90

Chapter	Page

8 Other Sources of Energy and Our National Energy Dilemma — 93

Solar energy, 93
Geothermal energy, 96
Hydrogen, 97
Solid waste energy, 98
The national energy dilemma, 99

9 Nonfuel Minerals — 105

Nonfuel minerals—the muscles of the Iron Age, 105
Nonfuel minerals—a source of chemicals, 105
Nonfuel minerals—a source of building materials and other products, 106
Nonfuel minerals not as critical a resource as fuels, 106
Principal metals, 107
    Iron, 108
    Aluminum, 115
    Copper, 120
    Lead, 124
    Sulfur, 126
    Zinc, 129
Scarce metals, 131
    Manganese, 133
    Chromium, 133
    Tungsten, 133
    Columbium, 134
    Cobalt, 134
Plant life needs minerals, 134
    The puzzle of how plants grow, 134
    Real need of plants for minerals, 135
Nonfuel minerals—enough for everyone?, 139

10 Water Supply for Industrial and Private Uses — 143

Why are we short of water?, 143
    The hydrologic cycle and unequal distribution, 144
    Abuse of the water supply, 144
    Misuse and increasing use of the water supply, 146

Chapter	Page

Cleaning up our water supply, 147
    Progress in water cleanup, 148
Increasing the availability of our water supply, 149
    Atmospheric moisture, 149
    Surface water supplies, 151
    Vadose water and ground water supplies, 162
Salt water conversion, 164
    The use of saline waters, 164
    Removing salts by distillation, 164
    Other processes, 164
    Desalinization plants, 165
    Cost of desalted water, 166
Reclaiming waste water, 167
Plenty of water, if . . . , 167

11 THE NATURE OF FORESTS AND HISTORY OF THEIR GROWTH   169

Woods and wildlife, 169
Different woods meet different needs, 171
Woodlands a handicap to early farming, 171
Forest soils not as productive as prairie soils, 172
Our virgin forests almost gone, 173
Saw timber grows slowly, 175
Nature produces forests slowly, 175
How much timber have we?, 177

12 THE MANAGEMENT AND USE OF FORESTS   181

Managed woods grow faster, 181
Management practices, 181
    Planting selected varieties of trees, 181
    Breeding supertrees, 185
    Controlling insects and diseases, 186
    Selective harvesting trees, 191
    Clearcutting forests plots, 192
Importance of small landowners, 193
Recreational uses of forests and woods, 195
    Multiple use of national forests, 195
    Fantastic increase in forest visits, 196
Wilderness, 198
Better use of forest resources, 200

Chapter	Page

    The keys to our forest resources, 201
        Don'ts for forest visitors, 202

13   THE ORIGIN AND DEPLETION OF SOIL   205

    Formation of soil, 205
    Soil formation—a long process, 208
        How man upsets nature's balance, 209
        Why some soils take much longer to form than others, 209
        How limestone rock makes soil, 210
    What is erosion?, 211
    Kinds of erosion, 211
        Water erosion, 212
        Wind erosion, 213
    Soil depletion, 216
    Building up, tearing down, 216

14   THE EXTENT OF OUR SOIL RESOURCES   219

    Soil resources once considered inexhaustible, 219
    Cheap land persisted, 220
    The extent of our soil resources, 220
        Much of our land is forest soil, 222
        Marshes and swamps make more soil, 223
        Nearly one-third is grazing land, 224
        About one-fourth is cropland, 227
    Use of soils in urban areas, 229
    Good soil is limited, 230

15   SOIL CONSERVATION ON FARMS AND RANCHES   233

    Current loss of soil still large, 233
    Good conservation, 233
        Special soil conserving practices, 234
    Land-use capability classes, 235
        Class I, 235
        Class II, 235
        Class III, 236
        Class IV, 236
        Class V, 236

Chapter

    Class VI, 236
    Class VII, 237
    Class VIII, 237

16 THE USE OF VEGETATIVE COVER IN SOIL CONSERVATION — 247

Beneficial effects of vegetative cover, 247
Annual crops only partially effective cover, 249
Hay crops offer better soil protection, 249
Special considerations in using vegetative cover, 249
Pasture and range land improvement, 252
    Grazing lands owned by the federal government, 252
    Production increased from pasture lands, 253
Cover crops, 257
    Limitations to the use of cover crops, 259
    Crops used as cover crops, 259

17 STRIP CROPPING, CONTOURING AND TERRACING IN SOIL CONSERVATION — 263

Effect of water speed, 263
Crops differ in slowing down water run-off, 265
Strip cropping, 266
    Shorter slopes, 267
    Value of strip cropping, 267
    Width of strips, 268
    Kinds of strip cropping, 268
Contouring, 270
Terracing, 271
    What are terraces?, 272
    Why should terraces be used?, 273
    Types of terraces, 274
    Terraces for nonfarm uses, 278
    Laying out and building terraces, 279
    Outlets for terrace channels, 279

18 CROP ROTATIONS AND OTHER PRACTICES IN SOIL CONSERVATION — 283

Importance of crop rotation, 283
    Definite rotation attempted, 284

Chapter	Page

Length of rotations, 284
Rotations help control erosion, 285
Rotations add organic matter, 285
Rotations and strip cropping, 286
Beneficial use of fertilizers, 286
    Organic or inorganic fertilizer?, 287
Use of herbicides and insecticides, 288
The problem of gullies, 289
    How gullies are made, 289
    Control of gullies, 292
    Use of flumes, 293
Stream bank erosion, 294
    Vegetative growth prevents intermittent stream bank cutting, 294
    Many stream banks require the use of jetties, 296
    Types of jetties, 297
Rip-rap, 297
Protecting adjacent land, 298
Channelization of streams and rivers, 298
Many beneficial ways to manage our soil resources, 299

19 THE CITY AND NATURAL RESOURCES — 301

City growth, 301
    The city seen from space, 302
    City influence beyond boundaries, 303
    Cities and man, 304
    Need for cities, 305
The green city, 305
    Providing cluster zoning and parks and preserving natural areas, 306
The CBD and the central city, 308
Air quality and the city, 310
Planning city growth, 311
    Planned Unit Development, 311
    What will happen to farm land?, 314
Soil, sewage and the city, 314
    Suitability of soils, 314
    Adaptability for sewage, 315
New towns, 316
The place of small cities and towns, 317
Cities are natural—we have made them unnatural, 318

Chapter		Page
20	**Why Wildlife Is Important and How to Save It**	321

   Justification of wildlife, 322
   Not all wildlife beneficial, 325
   Providing habitat for wildlife, 325
      Wildlife need more than waste acres, 330
   The struggle for survival, 331
   Wildlife management, 334
      Hunting restrictions, 335
      Game refuges provide valuable help, 338
      These measures only partially effective, 339
      All landowners must provide habitat, 341
   Development of wildlife research and education
      program, 342
   We're not out of the woods, 342

| 21 | **Providing Wildlife Habitat** | 345 |

   Wildlife habitat on farms and ranches, 345
   Ponds and lakes for wildlife, 347
      Managing ponds and lakes, 347
      Stocking with desirable fish, 348
      Avoiding weeds and siltation in lakes and ponds, 349
      Harvesting the fish, 349
      Providing for fish, waterfowl and fur bearers, 350
   Wetlands for wildlife, 351
      Much of our wetland should never be drained, 352
   Stream banks can afford wildlife shelter, 353
   Wildlife habitat in the city, 353
   Wildlife's future depends on you and me, 354

| 22 | **Natural Resources from the Seas** | 357 |

   Sea water—a remarkable resource, 357
      Many minerals in sea water, 358
   Plant and animal life of the sea, 358
   Fish—a desirable food, 359
   The early colonists used fish, 359
      Best fish required by foreign markets, 360
      Fish traded for molasses, and rum for slaves, 360

Chapter	Page

  Fish supply small share of our nation's food, 362
  Many fish species disappearing, 362
  Relative importance of fresh water fish, 363
  Fish farming in the sea or on "land," 363
  Fish catch may increase, 365
  Cycle of life in the sea, 366
    Importance of plankton production, 367
  Our nation's underwater boundary extended, 368
  More regard for nature's cycles needed, 368

23   The Ecological Approach to Natural Resources   371

  The science of ecology, 371
  Air pollution and resources, 373
  Insecticides—for better or worse?, 376
  Solid waste disposal, 377
  Recreation and open space, 378
  Philosophy of ecology, 380
  Conservation versus ecology, 381
  Environmental problems worldwide, 382

24   Conservation—Everybody's Opportunity   385

  Beauty as a goal, 386
  NEPA and EIS, 387
  Provisions for open space, 388
  The business cleanup, 388
  The agricultural turnabout, 388
  Land-use planning, 390
    Slow progress in land-use planning, 390
  Ecoregions—an ecological system, 391
  Other problems still to be solved, 393
  History of philosophy of conservation, 394
    Saving for future use, 394
    Prevention of waste, 396
    Wise use, 396
  The new conservation philosophy, 397
  Two views of our natural resource base, 399

Chapter	Page

Don't kill the goose, 400
What can I do?, 400
The battle for natural resources, 401

25 THE HUMAN BODY AND NATURAL RESOURCES 403

The human body, 404
Air and health, 404
Water and health, 407
Energy from the sun and the human body, 408
Pesticides, metals and health, 409
Noise and humans, 410
Radiation and the body, 410
The body, other resources and the future, 411

Recommended General Sources for Environmental Information 415

Index 419

The character of natural resources

ATLAS, THE GREEK DEITY of gigantic size and enormous strength, is depicted as bearing the world on his shoulders. That may be possible for gods. But, for mankind, it is the world, this one and only earth of ours, which bears us on its "shoulders." Our natural resources are the basis of all life on earth, and they provide the materials essential to cultural development.

When Terry Bradshaw completes a 40-yard pass, or Chris Evert Lloyd wins the fifth set at Wimbledon, or Lee Trevino socks a 250-yard drive, or Jane Fonda wins an Oscar, or Eric Heiden skates his way to five gold medals, or Barbra Streisand has a million-selling album or you and I pass a test—be it our ability to coordinate or to cogitate—none of us are thinking "Thank our natural resources!" But, we should. We all should!

A man plowing corn or cutting down pine trees knows he is working with natural things. But, how many of us realize that performance in any line—sports, diplomacy, scientific exploration, art, writing, etc.—is dependent upon a natural resource base? Everyone needs food for energy, air to breathe, water to drink. And aren't the muscles, nerves, veins and brain, along with other parts of the body, natural resources?

But, how about all the equipment needed? And, how about the great stadiums, meeting halls and libraries required? Aren't these all man-made? Yes, they are, we admit. But, they are made of fine hardwoods, or limestone and sand, or iron hardened into special steels or other rare metals and earth substances. They are formed from natural resources.

Under traditional classification, resources were put into three categories: natural, human and cultural (economists use the terms *land, labor* and *capital*). Earth materials and life forms, except man, which are "free gifts of nature" were classified natural. These included the air above and the water beneath, soils, natural (wild) vegetation, every creature except man and all the rocks and minerals.

Man was classified as a human resource. To state it in doggerel:

> Bushman or Bulgar, Kirghiz or Kurd,
> No matter how different sounding the word,
> Is really the same kind of species of course,
> Called *homo sapiens*, a human resource.

All the works of man were classified traditionally as cultural resources. Here was included everything made by man, from abacus to zwieback.

The preceding division of resources into three groupings is both convenient and useful for many purposes. But, it has one serious flaw which we should recognize at once. Such a division tends to make us forget the interrelatedness of all resources. We become labor specialists and boast that little would be produced without man's brawn. We develop a love affair with cultural hardware, such as electronic computers. "Our computers will solve all problems," we say. And, sometimes we become natural resource specialists. We say, "Don't change nature!" But, as good natural scientists, we should know that Nature herself is constantly changing.

Most of the world's phenomena are natural, but they are not necessarily resources. To be resources, they must be available for use by man. A mild, sunny climate, productive soil, valuable minerals and the "cool, cool" water sung of on our western plains are all natural resources. Likewise, a harsh, cloudy climate, boggy lowlands, craggy mountains and warm sulfur springs are natural resources *if they are available for use by man*. Water which is trapped far beneath the earth's surface and inaccessible to man, however, is not ordinarily considered a resource. Nor is the common plantain weed. It is important to realize, however, that such "neutral" or "nuisance" materials or forms of life may well be essential to our earth life system and thus, indirectly, to our own lives, and some of them may even become valuable. Oil was once more of a curiosity than a fuel.

It becomes clear that we mortal men and our works are not superior to nature. We are a part of it. As we travel down the superhighway in our compact cars, eyes glued to the ribbon of concrete and ears attuned to the radio, we feel secure and isolated in a man-made world. But, this is only an illusion. We are riding through, on and by reason of natural resources. They are the air we breathe, the light we see and the ground over which we move. They are the sources of energy in our cars' motors and in our bodies. Natural resources constitute the basis of all life.

☐ NATURAL RESOURCES MANIFEST ENERGY

All resources are interrelated. This becomes clear when we think of resources as manifestations of energy. We are accustomed to thinking of matter and energy as different things. They are really just different expressions of the same thing, energy. These expressions differ in form, animation and exhaustibility.

Forms of energy

In form, energy may be manifested as solid, liquid or gaseous, or as a wave of energy—heat, light or electricity. These forms are often convertible or substitutable, one for the other. This transferability is very significant in evaluating our resource base. For example, energy from coal can be used directly to furnish heat energy or converted as steam-electric power; or natural gas or oil may be used instead of coal. Our analysis of resources available for heat energy should take into account these and other forms of energy capable of doing this work.

Animate and inanimate energy

A second important distinction of natural resources is their manifestation as animate (living) or inanimate (nonliving) energy. Coal expresses inanimate energy. Animal and plant life are animate energy. The advantage of substituting the one form of energy for another to carry out physical work can be readily understood. A horse may pull a plow. Or, a tractor may pull it. The first is the use of animate energy, animal muscle power. The second is the use of inanimate energy, gasoline, diesel oil or propane gas operating through solid energy forms of the tractor. In both cases, natural resources are being employed. The tractor, inanimate energy, will be far more productive than the horse.

Conversion to use of inanimate energy on a large scale, which began over two hundred years ago, wrought such a change in our productive methods and capacity and even in our way of living that it has been called the Industrial Revolution. Machines permitted increased specialization and the division of labor, the basis of the factory system. Much of our population moved from farm to city, yet the farmers produced more than ever before. Manufacturing output fairly "leaped forward." The increased productivity and freedom from many of the limitations of time and space were a great blessing to mankind. But, these benefits were accompanied by an accelerating and severe drain on and abuse of our natural resources. To solve this dilemma has not been easy.

Inexhaustible resources

Another classification of energy is according to its apparent exhaustibility. Is its supply diminishing with use? Or, is it being constantly replenished?

Three natural resources which are classified as inexhaustible are solar energy, radiant energy renewed by the sun; the atmosphere renewed constantly by sun interaction with plants, animals and bacteria; and water replenished constantly as water is carried from the oceans as clouds and vapor overland where it is "dumped," renewing inland water supplies in a movement known as the *hydrologic cycle* (see Figure 1-1). Unfortunately, man has so contaminated the air and water in recent years that these resources may have been impaired even though they have proved inexhaustible over billions of years. Plant and animal life under natural conditions are also constantly replenishing themselves and thus may be considered inexhaustible. However, since such conditions are becoming the exception rather than the rule, we class them exhaustible *replaceable* resources.

Exhaustible resources

The exhaustible resources are those which are available in more or less limited quantities or which must be processed for use from

FIGURE 1-1. Water power is an inexhaustible resource as long as rain falls and there is surface run-off. *(Courtesy, Wisconsin Conservation Department)*

somewhat limited sources. Aluminum, for example, is one of the exhaustible natural resources. It may be found in clay soils all over the world, which makes the total amount of aluminum in its natural form practically inexhaustible; yet, the aluminum we use comes mostly from a few mines where there is a high concentration of bauxite ore. It is practical to recover aluminum from bauxite, but it is not economical or feasible to recover aluminum from the tremendous deposits of common clay. Nor is aluminum, like the sun's energy, being replenished as it is used.

There are two classes of exhaustible resources: (1) irreplaceable resources and (2) replaceable or renewable resources.

6 ☐ OUR NATURAL RESOURCES

IRREPLACEABLE RESOURCES

This class includes those resources which cannot be replaced or renewed when once used up. Our mineral resources, such as iron, copper, nickel, sulfur, phosphate, potash and salt, are in this class. Nor can coal, oil and natural gas be renewed when present supplies are exhausted.

In some instances even our soils must be classified as irreplaceable because once they are eroded (washed away) they cannot be renewed or replaced during our lifetime or even during the lifetime of our nation. The only way to build up a soil again when once it has been lost is through the slow action of weather and time.

It has been estimated that two thousand years are required for nature to make one inch of soil from the original rock of the earth. Whether the time be one-half or several times two thousand years, it is so long that for all practical purposes a soil once eroded cannot be replaced.

REPLACEABLE RESOURCES

The second general class of exhaustible natural resources consists of those which can be renewed after being wasted or used up, providing some seed stock still exists. Our forests, our grasslands and some kinds of wildlife are examples of this class of resources. Even though these plant and animal resources are completely destroyed in certain areas, they can generally, to some extent, be renewed. For example, a few decades ago, the moose population was almost wiped out. It has been replenished in parts of its former range in numbers greater than 12,000, with another 120,000 in Alaska. In 1925, the number of pronghorn antelope ranged between 13,000 and 26,000. According to the Wildlife Management Institute, the population is now up to 500,000. White-tailed deer in 1900 were estimated at ½ million, while in 1976, they were estimated at nearly 12½ million, thus establishing the deer as the most abundant large wild animal. In 1900, elk were estimated at 41,000; presently there are close to a million in 16 states. These facts are impressive, considering the rate at which our wildlife habitat is disappearing.

Even some elements of our soil are replaceable. The depletion or using up of plant nutrients from a soil may result in its abandonment as a source of production for plants. As long as the soil has not eroded, the elements used up may be successfully replaced in some cases, making the soil to become productive again.

☐ DISTRIBUTION OF NATURAL RESOURCES

Another important characteristic of natural resources, especially minerals, is that they are unevenly distributed. The United States east of the one-hundredth meridian has generally adequate rainfall for abundant vegetation, but west of this line much of the region is arid. Productive brown forest soils prevail in the northeastern United States, but infertile red desert soils prevail in the Southwest. Mountains and associated plateaus and valleys occupy the western third of our country, and lowlands, the central section. Major bituminous coal deposits are found in the Appalachian and central states, while lignite fields are characteristic of the western states.

The unequal distribution of resources throughout the world often severely limits their availability to the United States. In the beginning of the energy crisis in 1973, many Americans who were faced with empty gasoline pumps at filling stations for the first time in their lives refused to believe that our country had a limited supply. They thought it was all a scheme of the oil companies. The fact is, the Arab nations sat on the world's greatest proved oil reserves and, for political reasons, chose not to release them.

Because of erratic distribution, the United States imports many other basic minerals and raw materials essential for our standard of living. In late 1974, John Keil, then the Assistant Secretary of the Interior, forecasted that we were heading into a shortage of minerals and materials that would "make our present fuel crisis look like a Sunday School picnic!" A major reason for this, besides heavy use and waste, is that we have become so heavily dependent upon imports. No nation today has everything it needs.

Some nations have demonstrated great ability to overcome deficiencies of natural resources through trading manufactured goods for raw materials. But, history teaches that they may lose their ability to maintain this complementary relationship. England, one hundred years ago, was known as the "workshop of the world." Drawing upon local reserves of coal and iron, she obtained from abroad the raw materials, such as cotton, that she needed and became an immensely wealthy nation. Today, her coal and iron reserves are less accessible, and the factories of other nations are outproducing hers. She has had one financial crisis after another and is a debtor nation whose imports exceed exports.

The relationship between the location of natural resources and the location of industry may be quite indirect and often is, even when both are within the same country. The iron and steel industry in the United States is a good example of this. After an early start in Massachusetts in colonial days, based on bog iron deposits and wood for fuel, the center of production shifted westward to the Scranton-Wilkes-Barre area, based on local iron deposits and anthracite for fuel. The next center was the Pittsburgh-Youngstown district. Here coking coal proved superior to the anthracite, and local iron deposits were adequate at first. Besides, newly developing markets of the Midwest were more easily reached from Pittsburgh than from farther east. As Pittsburgh grew, a very large body of iron ore was discovered west of Lake Superior in the Mesabi range. Why not center the steel industry there? It would not have been economical. While a steel plant was eventually located at Duluth, Minnesota, not far from the iron mines, it never became a serious competitor of Pittsburgh. Shipment of the ore to Pittsburgh was found to be more efficient, in view of its good market position and nearness to fuel. And, since it takes the equivalent of 1½ tons of coal to process 1 ton of iron, it was less expensive to move iron to Pittsburgh than coal to the Mesabi. In other instances, the location of a steel complex at Sparrow's Point, Maryland, was related to markets and convenient access by ship to ores from South America; steel manufacture at Birmingham, Alabama, developed on the basis of local iron and coal deposits and markets; manufacture of steel and steel products in the Chicago area, now the U.S. production leader, began at Gary, Indiana, because of the large market there for steel and steel products.

Among the most generally used and widespread resources are our forests. Their distribution is shown in Figure 1-2. Note that even this "common" resource is unavailable locally in some regions, such as the Great Plains.

The Midwest is known as the "breadbasket" of the nation, reflecting its outstanding natural resources of productive soil, humid climate and abundant water supply which have made possible an enormous output of crops and livestock products. Other natural resources have helped it become a leading manufacturing region as well. But, even this diverse region must import raw materials and fuels from other parts of the United States and from foreign nations in order to meet its needs.

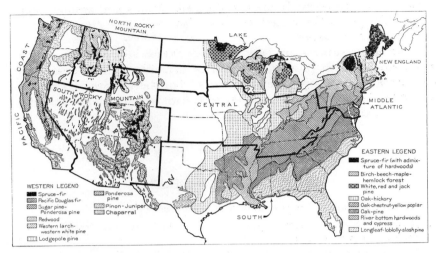

FIGURE 1-2. Forest areas of the United States. Note that most of the area east of the Mississippi River is classed as forest areas, while practically no forest land is shown in the plains and prairie area east of the Rocky Mountains. *(Courtesy, U.S. Forest Service)*

□ SUMMARY

Natural resources are useful manifestations of energy that occur in nature. Man and his works, although separately classified as human and cultural resources, are interlinked with and dependent upon natural resources.

Natural resources are distinguished from one another in form, animation and exhaustibility. In a different form, one resource may sometimes be converted to or substituted for another, affording flexibility in resource use and conservation. The substitution of inanimate energy for animate energy on a large scale has brought about a radical change in our way of life. While the resultant outpouring of goods and the release from drudgery has benefited us, the accelerated drain of our natural resources and the pollution of our environment threaten our very lives.

Inexhaustible resources do not diminish with use. The sun's energy, the atmosphere and water are principal examples. But their effectiveness has been reduced by man's pollution of the environment.

Exhaustible resources are either irreplaceable or replaceable.

10 □ OUR NATURAL RESOURCES

Minerals, including fuels, are irreplaceable. Plants and animals may be replaceable. Soil is largely irreplaceable, but some elements in it are replaceable.

Natural resources are very unevenly distributed. Nations or any given areas which possess a wealth of available natural resources have an advantage over the "have nots." For the welfare of a region depends to a great extent upon the ease with which it can obtain and develop natural resources.

□ □ □

QUESTIONS AND PROBLEMS

1. When does energy become a natural resource?
2. What advantage might be gained by substituting one resource for another? Give an example.
3. Why is our performance at school and work dependent upon natural resources?
4. Is there any need to be concerned about the future of inexhaustible resources? They can't be used up, can they? Explain your answer.
5. Give an example of an exhaustible, but nevertheless renewable, resource.
6. What is the most basic resource of all?
7. Is all matter a resource? Why or why not?
8. What effect does the uneven distribution of natural resources have on the wealth of the United States?
9. How did the Industrial Revolution affect the relationship between man and natural resources?
10. Name three important irreplaceable exhaustible resources. Are any of them important to your community?
11. Name three natural resources classified as inexhaustible.
12. What is the difference between natural, human and cultural resources? What do they have in common?
13. What areas of the United States have been developed because of the use of the following resources:
    a. Coal?
    b. Iron?
    c. Coal and iron?
    d. Soil?
14. Which one of the resources found in your area is most important to your community? Why?

2 □ □ □

The evaluation of natural resources

THE WORTH OF A RESOURCE to society or to an individual is often quite different from its *economic* value. How much is Independence Hall in Philadelphia, the beauty of the free-flowing Klamath River in Oregon or even a single rose worth to us? How much is a dog worth? Author Kircher's Dachshund, Schatzie, who like thousands of other lovable pets has grown more and more droop-eared and overweight, is priceless to her masters, but worthless to anyone else.

The economic worth of a resource is dependent upon its quality, accessibility and scarcity. Consider the value of a piece of land, for example. Farm land with excellent soil, thus of high quality, is ordinarily relatively valuable, even though it may be located remote from farm markets and be relatively inaccessible, because such quality land is scarce. By contrast, a very poor quality land—with highly acidic soil and poor drainage—might be very valuable if in the center of a city because it is readily accessible and because building sites for that particular city are scarce. In the third case, a piece of land with poor quality soil may be rather inaccessibly located on the outskirts of a rapidly growing city, but it may sell for a relatively high price since land for urban expansion has become scarce.

One measure of the economic worth of a good is its market value—that price for which a good will sell. If market forces work efficiently, they are reliable indicators of the value of commercial goods at a particular time. But, they do not work for pets like Schatzie. Nor, is the market place a good judge of value for many other things, such as those which have great historic, scenic or biologic value, but little cash value. And, unfortunately, even for commercial goods, the market is often imperfect in its appraisal of value. Despite these difficulties, we must try to determine the "true" worth of resources if we are to use them wisely.

□ SUPPLY FACTORS AND NATURAL RESOURCES

The potential of a supply of a natural resource to meet demand is determined by its availability. Availability is dependent upon the way the resource is manifested in nature and its distribution, as discussed in Chapter 1, and the utilities given it by man.

It may cost to make resources available

Because an inexhaustible resource, such as air, is freely available, it generally has no economic value. No one markets the air we breathe. However, in view of growing pollution, "pure" air has become scarce and thus has acquired economic value. It is, for example, a major asset of resorts in the mountains (see Figure 2-1). In certain circumstances, such as in underground mining operations, submarines or air-conditioned buildings, it costs to supply air to people. But, this is not because air itself is scarce; rather, it is the cost of getting the air to the restricted area. Resources that abound in excess of human needs or desires have value only when making these resources accessible to individuals in isolated or out-of-the-way places, or at unusual times, is costly.

We pay high prices for drinking water in desert places, not because the cup of water is any different in taste or quality from millions of cups of water found in other places. The reason we pay so much for it is because it costs to transport the water to a desert spot, and we would rather pay the cost charged for the drink than make the long trip to the free supply of water.

An illustration that all of us are familiar with is the pumping of water from wells and piping it to our houses. We are willing to pay whatever costs are necessary to bring the water to us in our homes where we wish to use it. We sometimes dig wells hundreds of feet deep just to get a continuous supply of pure, clear water, and we are willing to pay this cost. The value to us in this situation is what economists call *place utility*. If we go to the stream or spring to get

THE EVALUATION OF NATURAL RESOURCES   □   13

FIGURE 2-1. A snow-covered ridge of the high Sierras some 150 miles southeast of San Francisco. The magnificent scenery, the "pure" mountain air, the snow, the streams and lakes, the wildlife and the forests are all important resources here. People in adjacent valleys depend upon snow melt for their water supply throughout the dry summer. *(Photo by Susan Kircher)*

our drink of water, we pay no money for it. It costs us only the time and energy used in going to the spring. It is only when the water is moved from its source of supply to us in our homes or at our work that it has a value to us which makes us willing to pay money for it.

Economists use different names to describe or to qualify the type of use of any article or service that satisfies our wants. This process is spoken of as the creation of utility.

We have already mentioned *place utility*. It costs to move an article or a service from one place to another, and this cost is usually added to the price we as consumers pay.

The common method of creating utility is to change the form of a resource. This is called *form utility*. Changing the form from a product that is of no use to us to something that we as consumers or users want and are willing to pay for is the basis of our industrial development.

Cotton, wool and flax are made into bedding, draperies, rugs, clothing, etc. Iron, copper, aluminum and other metals are converted into engines, automobiles, structural steel and a thousand and one other articles used in our everyday living. These are illustrations of *form utility*.

A farmer harvests grain after it ripens in the summer or early fall. When that process is completed, economists speak of the farmer as having created *elementary utility*.

The grain is then moved to some mill where it is worth a little more because of the shipping cost. *Place utility* enters here. The miller produces flour, bran and germ meal from the grain, and *form utility* is created. More costs are added, and more *place utility* is created, when the flour is shipped to some organization where it is stored until a consumer wants to buy it for his own use. Holding or storing the product until it is bought by a consumer creates *time utility*.

All these functions and services are for our convenience and add to the cost. The creation of these various utilities results in our so-called higher standard of living, as well as in adding to the cost of living. When evaluating a resource, we need to take into account its scarcity and accessibility.

☐ DEMAND FACTORS AND NATURAL RESOURCES

Demand for a resource can be defined as "the amount that is wanted at a given price." To create effective demand, we must have the ability to pay. Simple Simon of nursery fame did not create an effective demand for pie, for when the pieman asked Simon for his penny, the latter replied, "Indeed, I have not any!" Simple Simon went hungry. And, so do over half the people in the world today—not because they don't want to eat—they can't afford to!

Total effective demand for a good at any given time is figured by multiplying the number of effective demand units by the amount each consumes. Of great significance to our natural resource base is that we can reduce the claims on our natural resources from a *demand* standpoint either by reducing the total number of demand units or by reducing the amount each requires or, of course, by reducing both. Thus, if we have a sufficiently smaller population, we can maintain or even increase the amount each of us uses with no greater drain on our natural resources. But, should our population rise sharply within the next few decades, as world population is doing at present rates, we would need to limit greatly our use of resources to maintain the same total demand. New discoveries, greater efficiencies and avoidance of waste might help alleviate the situation, but they would not change the basic relationship.

Elasticity of demand

Another important factor in the nature of the demand for a resource is the flexibility of its response to price changes, which economists call its *elasticity*. For example, the demand for items made primarily from metal, such as automobiles and home appliances, is generally *elastic*. If buying lags, it can generally be revived by lowering prices. But, a rise in prices causes buying to fall off. Since goods such as automobiles and appliances generally continue to operate for a long time, we can postpone buying new ones without having to "do without."

The demand for certain resources, especially staple foods, is *inelastic*. "Enough potatoes is enough potatoes" may be incorrect English, but it is correct economics. If we are wealthy, we eat all the potatoes we want and will buy no more just because the price drops. Even if we are poor, assuming that potatoes are the mainstay of our diet, we will buy no more if the price drops because we are already eating all we can. Vice versa, ordinarily higher prices will not cause the rich or the poor to purchase fewer potatoes. If we are wealthy, cost makes no difference. If we are poor, we need to buy potatoes regardless of how much the price rises since they are still probably the cheapest food we can get.

These flexibilities (elasticities) of demand have an important effect upon the use of our natural resource base and our ability to conserve it. In times of prosperity, we drive around in high-powered, gadget-equipped sportscars which have low passenger-carrying capacity and high fuel consumption. The manufacturer steps up the assembly of his supersports in order to sell while the market is "hot." Prices rise, but sales continue to climb, for cost seemingly is no object. Rising prices at this point are ineffective to control the waste of our metal and fuel resources in what has been called "conspicuous consumption." But, few of us are aware of this. Should a severe economic recession set in, however, the cars would be repossessed by the finance companies and could not be sold to anyone at any price. The waste of resources because of changes in taste would become obvious to everyone.[1]

Inflexible (inelastic) demand, when not taken into account in resource development, can also lead to waste. To help support the price of agricultural commodities, our government has paid farmers to plow under crops and to destroy livestock. The Brazilian government at times has destroyed more than $1 billion worth of coffee to prevent it from glutting the American market, where most of it is sold.

☐ THE MARKET SYSTEM: EXTERNALITIES AND RESOURCES

While the market system is an extremely valuable help in the allocation of resources, we can see from the previous discussion that the distribution may not be the best one for man or the resource base. In addition, there may be various side effects to the environment, called "externalities" by economists, which are not taken into account by the market place. These may be good or bad effects. A forest is selectively harvested in order to sell its timber. By opening up the forest in this process, we increase the food supply for wildlife—an external, good effect. However, one part of the forest, through carelessness, is overcut. Although the timber cut from it brings a good profit, the soil is exposed by the overcutting and severe erosion results—a bad effect. In using resources, we need to provide incentives or penalties to minimize the "bads" and maximize the "goods."

☐ WANTS AND ABILITIES AND RESOURCES

We have seen how the mechanics of supply and demand—the number of buyers, response to price, and so on—have an important

---

[1]Most sportscars are available in relatively low-powered models which may provide maximum efficiency and safety for small families. Heavy sedans may provide safety and comfort, but they seldom effect economy of resource use.

effect upon our resource base. But, we have not considered, except by a few random examples, what gives us our ability to convert natural resources into useful products (supply) nor why we desire to do so (demand). One of the leading authorities on resource evaluation, the late Professor Erich Zimmermann, classified the motivations for our desires under the headings of *wants* and *social objectives*.[2] Our abilities, he stated, depend upon the state of our *technological* and *societal arts*. Let us consider how these factors affect the appraisal of natural resources.

Human wants and social objectives

Zimmermann explained that our wants may be divided into two groups: *creature* and *culture* wants. Creature wants are the basic needs of food, shelter and clothing. Culture wants are the desires for things in addition to those that meet our basic needs. Books, television sets and air-conditioned houses are examples. The early North American Indians used relatively few natural resources. They satisfied their creature wants and little else. Modern society demands more and more culture wants. Accordingly, the demand on our resource base has multiplied many times. The impact of this expanding demand upon our natural resources is so significant that the next two chapters are devoted largely to explaining it.

Maintaining a healthful, peaceful and productive community is a social objective. To achieve this objective, we must frequently modify our culture wants, and especially our individual culture wants. For one family or a small group of families in a village, the quaint outhouse privies, with or without a halfmoon on their doors, were adequate and generally of little danger to health. But, for one million families living in a city, they would be disaster! Likewise, the exhaust fumes from a hundred automobiles are insignificant. From a hundred million, they are deadly! The conflict between what we would like to have as individuals or small, isolated groups of people and what we can have in our close-knit, industrial world has to be resolved in favor of wise group objectives if we are to have a life-

---

[2]Erich Zimmermann, *World Resources and Industries*, rev. ed. (New York: Harper & Brothers, 1951). The first edition of this text appeared in 1933. A paperback version (Henry L. Hunker, ed.) was published in 1964. Thus, Zimmermann's concepts have been taught for more than a generation! The new edition is called *Introduction to World Resources*.

supporting environment. For example, we will need to reduce our use of the large family automobile and increase the use of mass transit facilities, such as buses or passenger trains, if we are to prevent air pollution and to save space. That this is becoming recognized is illustrated by the millions of dollars which have been granted by the federal government for research in rapid transit systems and which, by federal law, require automobile manufacturers to reduce the fuel consumption of their products drastically.

Technological and societal arts

While wants and social objectives set our course in resource development, technological and societal abilities determine how well we are able to use our natural resources to reach our objectives. The *technological arts* consist of both our ability to operate machinery and the machinery itself. The *societal arts* are our abilities to organize for effective action; in short, they are the arts of living together harmoniously.

The space program, from Apollo to Voyager, illustrates successful performance of both these arts. Technologically, we had to master the art of building and operating large rockets, guidance systems, space capsules, lunar landing ships and other complicated space hardware. From a societal standpoint, we had to organize a working force of thousands of technicians and administrators, collect and disburse billions of dollars and have a political system capable of doing all this.

Technically, we communicate with ease, whether by radio, telephone or jet plane. But, socially, we have great difficulty in understanding one another. Unfortunately, our ability to destroy each other exceeds our ability to live together peacefully; many times our natural resources go up in smoke and flame instead of in houses and schools.

☐ OUR HUMAN RESOURCES AND NATURE'S SYSTEM

As has been noted in both this and the preceding chapter, when we develop a resource, we need to be concerned about the effect of this development on the rest of the world, especially the world of nature. All things are interrelated. As an example, a new highway was planned to cut across the Great Smoky Mountains National Park. It would have relieved traffic congestion. But, at the same time, it would have cut across the headwaters of streams, polluting them; it would have destroyed a wildlife habitat; and it would have threatened the forest through which it passed. To try to assure full consideration of the environment by federal government agencies, Congress passed the Environmental Protection Act in 1970. This act requires federal agencies to prepare Environmental Impact Statements to justify all proposals for legislation and other major federal actions that significantly affect the quality of the environment. (A further discussion of this act and its enforcement is included in Chapter 24.) Here we will consider some of the relationships between man and nature, according to the science of ecology.

The science of ecology

Ecology is the study of how living organisms and their nonliving environment function together. One approach to an understanding of this interrelationship is to begin with the sun. From it comes most of the energy on earth. But, this energy is largely unavailable to animals directly. It must be transmitted to them by green vegetation through a process known as *photosynthesis*. In this process, the sun's energy is transferred through a substance in the vegetation called *chlorophyll* (from the Greek, *chloros,* "green," and *phyllos*, "leaf") in the presence of water to become free oxygen and food sugar. Animals can obtain their energy by eating plants or other animals (which have eaten plants at some stage). As plants and animals decay, with the help of bacteria and fungi, they release chemicals in the earth which, in turn, help feed plants. This cycle is illustrated by the diagram in Figure 2-2. While this explanation is oversimplified, it reliably illustrates the general principle of a cycle in nature.

Such a circular movement as just described may be viewed as one complex cycle or as a series of interlocked smaller cycles. This circulation enables the earth's basic substances—carbon, oxygen, nitrogen and water—and others to move between the earth's major stratums: air (the *atmosphere*), water (the *hydrosphere*), soil and rocks (the *lithosphere*) and living organisms (the *biosphere*).

Consequently, the earth is one great interlinked system. Ecologists sometimes refer to this entire earth system as an *ecosystem*.

20 ☐ OUR NATURAL RESOURCES

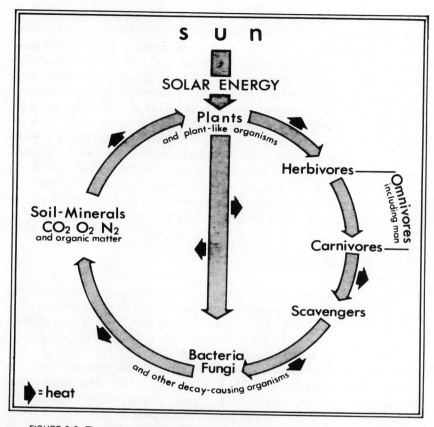

FIGURE 2-2. The energy cycle. All life on earth depends upon the sun's energy. The diagram represents the flow of energy in both living and nonliving matter. Solar energy is transferred through the chlorophyll of plants to serve as food for animals, which may be preyed upon by other animals. Animal wastes and plant and animal remains succumb to decay organisms. They may be converted to fossil fuels or become concentrated as other minerals. Or, they may aid in the formation of soils. Heat is given off as the cycle proceeds.

However, just as there are smaller cycles of life within larger ones, so there are smaller ecosystems within one large earth system. Thus, a forest, pond or field may be an ecosystem. It must contain sufficient living organisms and nonliving environmental components of certain characteristics for it to operate as a system. This is a scientific problem for ecologists to determine. But assuming it has enough variety to qualify as an ecosystem, the ecosystem itself goes through various stages, or cycles, of development. The forest is young and full of grasses, bushes and a variety of saplings. By maturity, the forest floor is relatively clear, and a few species of trees, now grown tall, dominate. Finally, the old trees die out, and the cycle repeats itself.

As long as man met all his needs by hunting, fishing and gathering, he was part of a particular natural ecosystem. The early North American Indians were an example. But, we modern men, with our many inventions, have made this interplay with nature often indirect. Is this an ecological relationship? The scientific relationship of cause and effect is difficult to prove. Nevertheless, however remote, the relationship is there. Even such a highly man-altered system as the city can be considered an ecosystem; but it may function very poorly. There is one important distinction to bear in mind. While natural ecosystems such as forest, field and pond can survive without us, we cannot survive without them! The problem of resource development by us, therefore, is to work in harmony with these natural ecosystems so that they are life-sustaining as long as possible.

Environmental perception

What does a mountain mean to you? Is it a great place to hike, or a poor place to farm? A storehouse for minerals, or a wall from the world? Or, do you think it is all of these?

One of the problems in resource development and the environment lies not in the limits of the ecosystem, nor in questions of our greed or short-sightedness, nor in the imperfections of our economic system. The problem is in comprehension. No two people think of the environment in the same way or have the same levels of knowledge. Some individuals' views may be colored by superstitions.

The study of why we think and act the way we do toward the environment, as well as toward each other, is carried out by behavioral scientists such as psychologists and geographers who differ as to the reasons for various behaviors and attitudes toward our resources and the impact of the environment on us. We can leave the arguments to them. But, it is fit for all of us as we develop resources to realize, with humility, the limits of our own environmental perceptions and, with tolerance, those of our neighbors. Meanwhile, we should work toward more understanding for all.

☐ FREEDOM AND NATURAL RESOURCES

One of the most cherished social goals in our nation is to assure freedom of the individual. Our business system is called the "free enterprise" system. Largely freed of government interference, it helped develop the country much faster than might otherwise have been possible. We rejoiced in an unequaled increase in the output of goods. But, we came to realize that many natural resources had been wasted and spoiled in the process. Continuation of abuses would mean the end of our economic system and our personal freedoms as well. We would be prisoners of want and despair. Our experience showed us that our liberties could be maintained only if individuals and organizations were willing to assume responsibility for and cooperate in a national resource policy set according to the total requirements of the environment.

Thus, we have begun to put these lessons of history into action. We have adopted an economic system in which business, labor and government are asked to cooperate. Economic history teaches that productivity is stimulated by the incentives of a competitive market and the private profit system. But, we have also found that it will not always operate to our advantage without the establishment and enforcement of standards for business and labor and control of non-market factors (externalities). We, as consumers, have accepted some controls over our rate of resource use—for example, the Pure Food and Drug Act and the Environmental Protection Act. The situation can be likened to that of the sport of baseball. The game would be chaos, as it often was in the old sandlot days, without well-planned and agreed upon rules and regulations, observed in a cooperative manner by the players with the help of the umpires and a well-marked baseball diamond.

Markets are valuable, but imperfect, allocators of resources. And, many of the most important benefits or costs to many from resource use are external to the market. Therefore, we need to consider more than economic incentives in developing our natural resources. We must consider goals such as health, happiness and peace. We should consider beauty and nature's system—the ecosystem. We should take into account the limits of man's environmental perception. And, we need to consider the impact of resource development on individual freedom. Such considerations, along with sound economic goals supported by laws which are enforced and practices which are carried out, will help to conserve our environment.

□ □ □

QUESTIONS AND PROBLEMS

1. Give an example of how the economic worth of a resource is different from its value in other ways.
2. Is water a free resource, or does it cost to make it available? Explain your answer.
3. Why didn't Simple Simon get any pie? What does this example have to do with the demand for resources?
4. Is the demand for staple food products generally elastic or inelastic? Why might inelastic demand result in a waste of resources?
5. Distinguish between creature and culture wants. What effect do these have upon our resources?
6. How does the space program illustrate the successful carrying out of both technological and societal arts?
7. Are we generally successful in getting our technological and societal abilities to work together to achieve our goals? Why or why not?
8. What are some worthwhile conservation goals for your neighborhood, town, city or rural area? What societal abilities are needed to reach them?
9. How can maintaining our individual freedoms and achieving our economic goals be likened to the playing of a game of baseball?
10. What is meant by "environmental perception?" How do you think yours differs from the perceptions of your parents?
11. Think of a good example to illustrate how living organisms and their nonliving environment are dependent upon one another. Why is this important to you?
12. What act did Congress pass in 1970 to protect the environment? What does the law require government agencies to do to help protect the environment?

3 □ □ □

The increasing use of natural resources

WE, AS A NATION, enjoy one of the highest standards of living the world has ever known. Not only do we have more food and better shelter than most other countries, but also we have more things to lighten our work and add to our enjoyment.

Many foreign students who come to the United States and live here for a few years no longer want to return home. They want to share in our exceptional national wealth and personal liberties. Despite its problems, the United States is still the great land of opportunity. The immense expansion of our national wealth, however, has placed an increasing strain upon our natural resource base. If this nation is to remain a land of opportunity, we must learn to use resources more wisely and to abuse them less.

□ THE POPULATION EXPLOSION

One reason that we use more resources than ever before is the accelerating rate of population growth, called the population explosion. On the average, in the past five years, over 11,000 people have been born every hour in the world, while many less than that die. World population increased by 700 million people during the 19 centuries from the time of Christ to 1850. Since then, in less than two centuries, it has jumped by some 3.3 billion people to about 3.7 billion in 1975 and over 4.2 billion in 1980. Thus, population is increasing, and at an *increasing* rate. If such population growth rates were to continue, world population would double shortly after the turn of the century and triple within another 40 years!

While the rate of population increase in the United States has been only about half that of the world as a whole, there were about 15 million more people in the United States in 1980 than there were in 1970. Furthermore, in a world where from one-half to two-thirds of the population presently have an inadequate diet, the impact of overcrowding in other lands will be felt in the United States.

Population growth has a much greater impact on resources than is represented by the increase in numbers of people. In our country especially and in other advanced nations as well, the tremendous per capita consumption of goods has been multiplying the impact exponentially. *Fortune* magazine commented in a special issue in February, 1970:

> If the population declined and technology continued to breed without any improvement in the arrangements for its prudent use, a small fraction of the present U.S. population could complete the destruction of the physical environment while justling one another for room.[1]

☐ EARLY USE OF ANIMATE ENERGY

The best use of our natural resources could not be made as long as most of our efforts had to be used in obtaining food, clothing and shelter. It was only after we learned to harness the running stream, unlock the power in our coal and release the energy in our oil and gas resources and use this knowledge to create new things for our enjoyment that our standard of living began to expand beyond the three elemental necessities of life—food, clothing and shelter.

Man as a nomad depended for his living mostly upon grass and the animals which ate grass. When he forsook his wanderings and settled down to a more sedentary life, he immediately began making greater use of natural resources. Wooden tools were among the earliest inventions of our ancestors. As they learned to use various natural resources, our forefathers enjoyed life more and increased their standard of living.

---

[1] "The Environment, National Mission of the Seventies," *Fortune*, February, 1970, p. 101.

FIGURE 3-1. The first source of energy used by the colonists was wood from surrounding forests, which supplied not only building material but also the fuel so necessary to keep them warm. The second source was flowing streams, which turned stones for grinding grain and spindles for making cloth. *(Courtesy, National Park Service)*

☐ RELATIVELY FEW RESOURCES DEMANDED

Even as late as colonial times, our ancestors had only the power of horses or oxen to produce the things they wanted. The tools used then were crude and ineffective when compared with present standards. The colonists had tapped none of the natural resources except wood, water and the soil to help them in their struggle for the better things of life (see Figure 3-1).

It required the total effort and planning of the whole family to supply the bare necessities—food, clothing and shelter. The typical family in those times had two horses, or the equivalent of 20 manpower, to create all the material things available for its use and enjoyment.

If we turn back the pages of history from colonial times to the time of the height of Athenian culture (500 B.C.), which was built almost exclusively on slave labor, we find it also required around 20 men (slaves) to serve one free family. This means that the progress made during the intervening centuries to colonial times amounted to substituting two horses per family for the 20 slaves. Although this was truly a remarkable development for mankind, it meant that the colonial family still had available for its use about the same amount of power as the free Athenian family.

☐ CHANGE TO INANIMATE ENERGY USE

Our present standard of living is built upon more than 40 horsepower per family, or the equivalent in physical work of 400 men. Another comparison that shows the progress resulting from the enormously expanded use of our natural resources is that at the present not more than 3½ percent of this country's population is needed to produce in abundance the food and clothing for some 225 million people; whereas five decades ago, around 30 percent of the population was so employed, and during colonial times, 90 percent of the 4 million people were occupied in producing those necessities for living in much less abundance than is now enjoyed.

Stepped-up demand for resources

This immense amount of inanimate energy which is used to make life easy and pleasant comes from nature. We can continue to have all these things, and more too, as long as our reserves of natural resources are not used up. It is the use of resources such as iron, copper, coal, oil, gas and water power that has freed us from the dreary drudgery of the past. But, what assurance have we that these various resources will continue to be available for our use and enjoyment into the indefinite future? Will we always have enough iron and coal to produce these labor-saving wonders of our machine age? Will our oil and gas reserves continue to meet our needs in the future? Will other resources continue to be plentiful as long as we need them?

☐ ENERGY (POTENTIAL POWER) MOST NECESSARY

It is *power*—horsepower, wind power, water power, power from coal, oil and gas—that turns our other natural resources into usable tools and items which make for our comfort. The immense amount of energy used in this country contributes to making life pleasant, easy and secure. Without this power (energy), our other resources would be practically useless.

Since the iron, copper and other metals that we need for making the machines, tools and other appliances of our present machine age would be worthless for these purposes were it not for the sources of power, let us look at our supplies of energy to determine if there will be enough available to meet not only our present needs but also our use of energy in the future.

Macro unit of energy

It is not possible to compare directly unlike quantities of energy, or sources of power, such as tons of coal, barrels of oil, cubic feet of gas and feet of water head. Some unit of energy common to all sources of power must be used, and for this purpose the British thermal (heat) unit—B.T.U.—is used. The British thermal unit is a technical term meaning the amount of heat required to raise the temperature of a pound of water 1°F. at or near its point of maximum density, 39.1°F. This amount of heat is 252 calories.

The amount of energy used in this country runs into exceedingly large numbers of British thermal units. Because of this, let us call one trillion B.T.U. a "macro unit" (M.U.) of energy.

Amounts of energy used in the United States

Since the turn of the century, while our population has increased about 3 times, the total use of energy has increased about 10 times. Thus, use of energy in the United States has increased more than three times as fast as population. While more efficiency of use must and will be achieved, the continued existence of our modern, dynamic production system is dependent upon the ready availability of a vast supply of energy for power. Furthermore, there will be more consumers. Even if low fertility rates are maintained and immigration is controlled, we will have about 20 million more citizens by the year 2000.

In 1900, this nation was using approximately 8,000 M.U. (8

FIGURE 3-2. The tremendous increase in our use of energy and our ability to make energy available is graphically illustrated by this powerful steam-electric generating plant compared with the water wheel shown in Figure 3-1. It would probably require over a million small mills like that one to have the capacity for work of the Baldwin Plant of the Illinois Power Company. Its summer capacity is rated at 1,800,000 kilowatts. The plant also evidences industry's increasing concern for the environment. The lake in the foreground, of 2,000 acres, was man-made by the company to recycle water for cooling the power station's condensers. As an extra bonus, the warmed lake provides year-round fishing. The metal hopper-like structures between buildings and stack are electrostatic precipitators which remove 99.6 percent of the solid particulate matter from the flue gases. *(Courtesy, Illinois Power Company, Decatur, Illinois)*

quadrillion B.T.U.) of energy for all purposes. By 1930, the amount had increased to 22,000 M.U. The depression in the early 1930's resulted in a drop in energy use to 16,000 M.U. By 1940, however, business had recovered so that 25,000 M.U. were produced, while during World War II this total became more than 30,000 M.U. Since then, there has been a gradual expansion in energy use until, in 1975, more than 75,800 M.U. were used. Some reduction in the rate of energy use is anticipated in the 1980's, reflecting the impact of energy shortages and conservation.

Sources of energy for power

Coal, along with some water, was the source of practically all the energy that was used for power in this country in 1900 (see Table 3-1). By 1920, oil and gas were supplying slightly more than one-sixth of our energy, while the use of coal dropped nearly 13 percent. The energy obtained from water power remained at practically the same percentage.

By 1940, coal was supplying only about half the power, while the proportion from oil and gas had increased approximately three times. This trend has continued. In 1980, the proportion obtained from oil and gas is estimated to have dropped from its high of 75 percent in 1977. Power derived from coal is on the rise again after having declined to supplying only one-quarter of our energy.

TABLE 3-1. Sources of energy used in the United States, in percentages

Year	Coal	Petroleum (Oil)	Gas	Total from Minerals	Water	Nuclear	Total Energy
1900	89	5	3	97	3	–	100
1910	85	8	4	97	3	–	100
1920	80	12	4	96	4	–	100
1930	63	24	10	97	3	–	100
1940	53	31	12	96	4	–	100
1950	42	33	20	95	5	–	100
1960	25	46	28	99	1	–	100
1970	23	43	33	99	1	[1]	100
1977	18	46	29	93	3	4	100
1980	25^2	42^2	25^2	92	3^2	5^2	100

[1]Less than 1 percent.
[2]Estimates by authors. About 12 percent of electrical energy was nuclear-powered.

(Sources: U.S. Bureau of Mines, 1960, 1970, 1980 [estimated] converted by authors from report by Joint Committee on Atomic Energy [see reference, end of Chapter 8]. Figures rounded to nearest tenth. *U.S. Statistical Abstract of the United States, 1977.*)

Percentages alone, however, do not always tell just what happened in the use of energy (see Figure 3-2). For example, from 1900 to 1930, when the proportion of energy supplied by coal was reduced from 89 to 63 percent of the total power used, the amount of coal used practically doubled. But, during this time, the energy derived from oil and gas increased from 623 M.U. to 7,356 M.U., an increase of nearly 1,200 percent.

In 1979, Sam H. Schurr, *et al.*, prepared one of the most comprehensive studies of our future energy needs undertaken in recent years for Resources for the Future, a private research organization. "Except for coal and uranium (as extended by breeders [reactors])," the study observes, "domestic mineral resources seem small when compared with cumulative energy consumption for the rest of this century. However, with coal as a source of liquid and gaseous fuels—and both coal and uranium as a basis for generating electricity—the long-term domestic outlook for mineral energy is good. Costs (in the broadest sense) are the major constraints."[2]

New sources of energy for power

The consumption of power discussed in the preceding paragraphs involves those energy sources that have furnished most of our power in the past and are expected to continue to do so during the next few decades. But, shifts to other sources may occur sooner than have been projected. There is a new awareness of pollution problems from the use of extraction of a number of the conventional sources of supply. And, we have a renewed concern about our dependence on foreign markets for oil. Therefore, the pressure is on to develop potentially cleaner or domestically available power sources that have been barely tapped. These now include geothermal energy (heat from underground); various new water sources, such as tides and the waves of the ocean; the heat of the Gulf Stream; the heat interchanges caused by the upwelling of water in the seas; oil shales; and energy from combustion of wastes. Also, we may reintroduce some of the power sources which were important at one time but which have become largely neglected today, such as wind power.

Chapters 4 through 9 consider the present status and future prospects for our fuel and nonfuel mineral reserves, including new sources.

---

[2] Sam H. Schurr, *et al.*, *Energy in America's Future: The Choices Before Us*, Resources for the Future, The Johns Hopkins University Press, Baltimore, 1979, p. 25.

QUESTIONS AND PROBLEMS

1. Illustrate how the use of natural resources has resulted in the creation or invention of new things which add to our material well-being.
2. What natural resources did primitive man use?
3. What are the elemental and most important needs of man?
4. Approximately how much power did a family during colonial times have to supply it with the necessities and luxuries of that period?
5. How did this compare with the amount of power each family used during the time that Athenian culture was at its height?
6. How much power per family is now available in this country, and how does this affect the living standards of our people?
7. In what ways would our minerals and forests be useful to us if there were no power from coal, gas, oil and water with which to convert the raw resources to finished products?
8. How much did the energy used in this country increase while the population was increasing 2½ times?
9. Which of our natural resources could you do without most easily in your locality? Why?
10. In what ways are the natural resources found in your area useful to your community?
11. What is meant by the "population explosion"? How does it affect the United States?
12. What are our major sources of energy for power today? Which ones are being used the most?

Wind and water power

EVERYONE KNOWS that water from running streams and from lakes is the source of a considerable amount of power, but few of us appreciate the importance that wind power had to the ancient peoples of the world, as well as to sections of this country in its earlier development. In recent years, because of the need to conserve energy, to be less dependent upon foreign supplies and to have cleaner power, there has been renewed interest in wind and water power.

☐ WIND FIRST USED FOR SAILING

Wind was one of the early used resources upon which no monetary values have been placed, mostly because of the vagaries of this universal yet most undependable of Nature's gifts to man. Its early use was most commonly in connection with water traffic. Sailing vessels were practically the only form of water transportation for hundreds of years before and following the old galleys of Roman days when men heaved and sweated and died over the tiers of oars in the dark holds of slow-moving, clumsy boats.

The Phoenicians of about three thousand years ago were the first people to build a great maritime trade through coastline commerce, and sailing vessels were their most important means of water transportation. Cedars of Lebanon were moved by sails from the Syrian coast to Egypt. Sails replaced oars early in the history of the Phoenicians and added to their sphere of influence as well as to the speed and volume of their business.

Greece followed Phoenicia in developing an empire through the

36 □ OUR NATURAL RESOURCES

use of sails, and it in turn was followed by Rome. Although not dependent upon sails for her control, Rome, too, disintegrated, and pieces of the empire were picked up by other nations which continued to use sails in sending and receiving the products of the soil from the then known world.

The compass expanded navigation

It was not until after the invention of the compass, about A.D. 1400, and its application to ocean navigation that sailing vessels could head for distant ports out of the sight of land. During the next four hundred years, sailing vessels moved into every nook and corner of the known water world. They even explored and opened up new worlds beyond the dreams of earlier navigators, and it was only through the use of sails that the commerce of the world was carried on.

Steam replaced sails

Steam-driven vessels appeared around 1820 and continued to increase in numbers until they finally all but eliminated sailing vessels from the trade lanes of the world. In the meantime, all ocean traffic increased in volume. Sailing vessels carrying the U.S. flag dotted the seven seas. The sailing ship reached its greatest perfection as a seagoing vessel about 1860. The clipper ship represented the sailing vessel at its best, but with variable winds, no winds at all or winds which had to be driven against much of the time, even this vessel was no match for a ship powered by steam in dependability of delivery, in speed and in size of cargo carried. Between 1860 and 1880, sailing vessels made the most use of this natural resource (wind) and carried their greatest amounts of world shipping. But, steam gradually took over, and, by 1900, steam tonnage outstripped sailing vessel tonnage. Since that time, sailing vessels have disappeared from the trade lanes of the world, and this once valuable resource (wind) no longer is important as a source of oceangoing power.

Different types of sailing vessels

During this period of sail domination, American ingenuity was responsible for the development of different types of vessels that won world recognition for their usefulness in different situations. In 1745, Andrew Robinson of Gloucester, Massachusetts, built a square-sterned vessel with two masts, a sloop sail and a bowsprit with jib. As this vessel was launched, it sped over the water so fast after coming from the ways that a bystander remarked, "See how she scoons." (*Scoon* is a word used to express the skipping of a flat stone over the surface of the water.) Mr. Robinson heard the statement and, not having named his newly rigged outfit, immediately called the vessel a "schooner." Schooners were used as fishing boats, merchant ships and even privateers.

Another type of ship, that for combination freight and passenger traffic, came into use before the Civil War. This was called the "packet." Coastwise traffic was first developed, but later various packet lines were established between U.S. and European ports. These were fast sailing ships, frequently making the trip from the United States to Europe in two weeks.

A third type of vessel, the clipper ship, was developed after the packet and was intended primarily for moving freight. Fast freight vessels were needed in trading halfway around the world in order to prevent heavy business losses and even sometimes ruin, which resulted from a decline in the market for goods while they were in slow transit.

The first of these vessels was made in Great Britain. U.S. ship builders improved on the construction so as to make for greater speed. A ship called the *Rainbow*, built in this country in 1843, was claimed to be the fastest sailing ship in the world. This ship made the round trip to Canton, China, from New York City around Cape Horn, in 6 months and 14 days. Three weeks of this time were spent in loading and unloading the cargo. Very few sailing ships ever equaled this record.

The total tonnage of ocean-borne freight and passenger traffic now carried by sailing vessels is practically negligible, and the world lost a touch of color and glamor when these full-sailed, picturesque white birds disappeared from our seas.

☐ WINDMILLS BECAME IMPORTANT

A second use of wind power was to turn the wheels of windmills for grinding grain and pumping water. Windmills were used in

European countries twelve hundred years ago, but despite the improvements made in them since that time, they still cannot compete efficiently with oil, gas and electricity in serving our people.

When Mark Twain visited the Azores in 1867, he was impressed by the use made of windmills on those islands. He wrote that these mills would grind 10 bushels of corn a day if the wind did not change direction. Whenever that happened, it was necessary to hitch some donkeys to the upper half of each mill and turn it around until the sails were again in position to catch the wind. This illustrates one of the obvious shortcomings of the early windmills. The Dutch overcame this handicap by building mills with rotating tops. Even so, these mills were difficult to control in variable winds, and, although the grinding was much more rapid than if done by hand, it was exceedingly slow when compared with present-day rates of grinding by other means.

FIGURE 4-1. The most spectacular type of wind is the tornado or twister, or cyclone as it is frequently called. Storms of this type generally occur in the warmer seasons of the year. More than 250 individual tornadoes have been recorded by the U.S. Weather Bureau in one season. Tornadoes unleash immense destructive power. Steady, reliable winds are needed for efficient wind power.

Windmills on the prairies

Windmills probably reached the peak of their usefulness and their numbers in the settling of the prairie states east of the Rockies.

In fact, but for the use of windmills, it is doubtful that farming and the hundreds of small towns on the prairies would have developed so early or so rapidly. Water was needed for livestock as well as for people. Surface water was available only in the creeks and rivers, and, although most of the early towns were built on or near streams, the vast multitude of farms had no surface water available for their use. Farmers had to depend on either hand-dug or bored wells and hand-operated pumps for water, both for their houses and for their livestock. It was in this region that windmills became most useful.

Pumping water a slow process

The rate of pumping water by wind power was not great, but with some wind blowing almost every day, it was possible to store enough water in stock tanks and reservoirs to meet all farm needs.

A 10-foot diameter wheel in a 16-mile wind would raise in one hour:

19.0 gallons	25 feet
9.5 gallons	50 feet
6.6 gallons	75 feet
4.7 gallons	100 feet

Reasons for the disappearance of windmills

Two events probably accounted for the rapid disappearance of windmills from our towns and farms—the development of gasoline and electric engines and the demand for greater volumes of water. It occasionally happened that for many successive days no wind would blow. Storage tanks would become empty, and in order to supply livestock with the necessary water, someone in the family would have to spend hours pumping water by hand for a herd of thirsty animals. This inconvenience always seemed to occur when there were many other kinds of farm work to be done. The use of gas engines and, later, the wide availability of the more convenient electricity were welcome substitutes for wind.

A second reason for doing away with this intermittent wind power was the desire of both city and farm dwellers for continuous water supplies. When homes had running water, they no longer had to depend upon the vagaries of the wind to keep up their water supplies. Coal, gas and even electricity replaced the windmill.

☐ NEW PROPOSALS FOR USING WIND POWER

Wind is no longer a valuable source of power for any large segment of our society. Enthusiasts of wind power, however, have not given up trying to find a cheap way to use this abundant resource to supplement power needs. Continuing research is going on at the National Aeronautics and Space Administration (NASA), at private agencies, at some universities and by numerous individuals.

Present research is along three lines: (1) small wind turbines with small electrical outputs, (2) intermediate-scale systems of 100- to 500-kilowatt capacities and (3) larger-scale systems designed to produce more than 500 kilowatts. Some intermediate-scale systems are in operation now, and multiunit systems supplying up to 10 megawatts are set for demonstration in the 1980's.

Wind industry sources estimate that by the year 2000 between one-half million to one million small windmills will be generating 1 percent of U.S. electrical needs and that windmills overall will account for 3 to 5 percent of all electrical power generated. If federal and state financial incentives for individuals and businesses and large research contracts continue, these expectations may be realized.

☐ WINDS RESPONSIBLE FOR RAINFALL

Of course, winds will continue to be necessary in carrying the moisture-laden air from over seas and oceans to the soils of the world. If winds should fail to blow even for one year, the vast soil areas of the country would produce nothing; and if no winds blew for a series of years, the productive lands would become deserts. Our soils would become useless, and all life would disappear except along the shores of the oceans. Winds, when considered from this point of view, will continue to be an essential resource. They are not considered an economic resource, however, because, even though necessary for our existence, they continue to move uncontrolled and undirected across continents. Our knowledge of wind currents has increased greatly in recent years, but we still cannot predict with accuracy just when, where and how much winds will blow, especially in the middle latitudes. Nor have we mastered the technology of controlling winds.

□ WATER POWER FROM FLOWING STREAMS

Let us look at water power from flowing streams, the earliest important resource of power in this country. Water power is thought of as a continuing resource—one not depleted with use. It was one of the most commonly used sources of power during our early history. Water wheels were placed in streams for grinding grain and sawing lumber. These operations represented the principal industrial purposes for which power was used at that time.

FIGURE 4-2. Lake Hokah in Minnesota was constructed for power nearly a hundred years ago, and it was used as a resort by the citizens of nearby communities. Railroad shops and a flour mill were built here to take advantage of the power. (See also Figures 4-3 and 4-4.) *(Courtesy, Soil Conservation Service)*

Later on, as inventions added to our power requirements, the water power of streams was used to turn spindles for the textile industry and to generate electricity. Its use did not continue to keep pace with the need for power, however, so that by 1900 only a small proportion, less than 4 percent, of the power used in this country was derived from water. Water will continue to supply a relatively small portion of the power used in the United States, even though great strides have been made since 1900 in storing our flood waters for use in industry and agriculture, as well as in reducing flood water losses. This has resulted in water power production's keeping pace with the overall increase in power used, and if all the proposed watershed developments materialize into flood water control areas, much more power will be derived from this source.

☐ COSTS OF CONVENTIONAL SOURCES OF WATER POWER

One of the misconceptions about water power is that, since it is derived by harnessing free-flowing water in nature, it is relatively cheap. But, this "ain't necessarily so," to use the phrase made famous by composer George Gershwin many years ago. There are many costs involved in making water power useful to industry. Some of these, such as costs of installation, are more or less common to all sources of power, but others pertain only to the use of water power.

Let us consider these costs under the following headings: (1) installation costs, (2) operating costs, (3) costs because of irregularity of water supply, (4) costs of loss of benefits under natural stream conditions and (5) non-economic environmental costs. The latter two types of costs are apt to be borne by others than the development agency.

1. *Installation costs.* The costs of constructing dams and of installing water wheels, generators and other equipment necessary to convert water into usable power are substantial for most streams. The size of streams cannot be increased, and, since most streams are small, only a small amount of power can be produced from each installation. Installation costs per unit of power generated usually are disproportionately large for the smaller sources of water power, and in many instances, these costs are relatively so great that it is not profitable to harness these small streams. Larger streams, especially those which are fairly fast-flowing, ordinarily can be harnessed profitably. When once the installation has been made, keeping it in repair is not an expensive task.

2. *Operating costs.* The costs of carrying electrical energy are

fairly high because copper wire, used because it is a good conductor, is costly. Even where aluminum-coated steel wire is used, the cost remains high. Also, losses of power into the air, along the transmission lines, are considerable, especially where this energy is carried long distances.

Most water power installations made in earlier times used the power directly through water wheels and gears to run spindles, grinding stones, etc., so that the power of small streams could be used effectively. With the coming of big industry, large-scale production and lower relative costs, the installation of more of these small units became unprofitable. Those already installed were used either until the water storage space back of dams filled with silt or until the machinery needed to be replaced. Many of these early installations were then abandoned, either because replacement costs were so great that power could be supplied by coal or some other source more cheaply than through the local water supply or because the specific product being made could be shipped in at less cost than it could be produced locally. An illustration of this change is the har-

FIGURE 4-3. Clearing of timber from the steep hillslopes, plowing up and down hills and absence of vegetation on the surface to slow the water run-off filled Lake Hokah with silt. All that was left of the little lake when this picture was taken was a swamp with a small stream running through the middle. (See also Figures 4-2 and 4-4.) *(Courtesy, Soil Conservation Service)*

nessing of small streams all over Kansas and other sections of the Wheat Belt. Seventy-five years ago, the power produced from these small streams was used to make wheat flour and grind feed. The area served by each of the mills was limited to a few townships or at most a few counties. If an owner wished to expand his operations, he could do so only by installing a supplemental power unit wherein the additional energy was supplied by coal or oil shipped in over the railways which stretched across the prairies. Such a mill continued to operate as long as the milling or power machinery did not have to be replaced or the mill pond back of the dam did not fill with sediment.

Competition became so severe, however, because of the cheapness with which flour could be made in large mills that these local mills could not meet all necessary repairs and improvement costs, pay for labor and still make any money. It is for these reasons that whenever any catastrophe overtook the local mill it was not replaced; instead, it went out of existence.

3. *Costs because of irregularity of water supply.* In addition to obtaining power from the drop of water from the distance between

FIGURE 4-4. Fifteen years after the photo in Figure 4-3 was taken, the swamp had filled in enough to make possible the raising of some garden produce on a part of the old lake bed. These fields continue to be subject to flooding so that it is not always possible to use this small area for crop production. (Compare with Figures 4-2 and 4-3.) *(Courtesy, Soil Conservation Service)*

FIGURE 4-5. The Grand Coulee Dam is one of the largest concrete structures in the world. Potential capacity of the dam is almost 10 million kilowatts. It is also one of the most powerful hydroelectric installations in the world. Irrigation and recreation benefits are provided as well by this and other dams in the Columbia River Basin. *(Courtesy, Bureau of Reclamation)*

spillway and generator levels, storing water in back of dams is done in order to obtain a more nearly constant supply of power. If rains fell once or twice a week and about the same quantity of water fell within every four- to six-week period throughout the year, there would always be a fairly constant supply of water in the smaller streams. This does not happen, however, because not only do storms vary in frequency, but also they vary greatly in intensity. Rainfall varies also with the different seasons, so that during one part of the year an area may have many times the precipitation of another season.

The larger rivers have more nearly constant water supplies because they drain much larger areas. They, too, are subject to seasonal variation, however, so that immense dams are built to impound flood waters of one season in order to feed generators during seasons of little rainfall.

4. *Costs of loss of benefits under natural stream conditions.* When a river flood plain is flooded by dams, the use of the river bottomland for agriculture, as a base on which to locate highways and railroads, and for many other uses, will be effectively precluded for tens, perhaps hundreds, of year—or even forevermore. The dam builders point out that the protection of the flood plain below the dam helps offset such losses. But, such protection may be more theoretical than real. If rain continues after water has filled the reservoir to capacity, the impounded water must be released. No more can be stored. Thus, severe flooding may occur below the dam. Calculation of the cost of any dam should include a charge-off against "lost" resources.

5. *Non-economic environmental costs.* The non-economic environmental costs of damming a stream may be the most critical to society. But, they are generally difficult to assess because they have no clear market value. Who can express in dollars and cents the value of the loss of the beauty of the canyons now under the waters impounded by the Glen Canyon Dam on the Colorado River? Or, who can weigh the cost when the wild, rushing torrent of water in a stream is brought to a quiet halt? How much are the delicate wildflowers and exotic wildlife worth that may be destroyed by damming a stream? And, how do we evaluate the loss of free-flowing water for recreation—canoeing, kayaking or rafting? Economists have devised ways to estimate such "external" costs. Such figures appear likely to gain increasing acceptance in the 1980's.

Silting back of dams

One important problem in connection with practically all dams is the filling with silt of the reservoirs back of them as fine soil particles are dropped from slowed-down stream waters. This silting takes place most rapidly in streams which carry the most sediment.

Most of the streams that drain cultivated land are muddy, silt-carrying streams. Practically all the streams of the prairie states, as well as of Indiana and Ohio and many of the southern and southeastern states, are muddy. Even when these streams originate in mountainous or forested areas and start as clear water, they become dirt filled as soon as they pass through cultivated land. Dams built on these streams after they reach cultivated land may be expected to fill with silt and to lose their water-holding capacities within a few decades.

Whenever a reservoir is filled with silt, it is no longer a continuing source of power. The amount of water that can be stored becomes so limited that only a fraction of the seasonal water run-off can be stored so as to be available later on in the year. No practical method has yet been devised for flushing the deposited silt out of the reservoirs. Silting can be reduced and the life of the reservoir lengthened, however, by controlling erosion on the banks of the stream and adjoining watershed, by building dams on small tributary streams to help precipitate silt before it reaches the main stream and to aid in controlling flood waters and by providing a siltation basin at the upstream end of the reservoir.

Those dams which are built on clear water streams may not fill with silt for hundreds of years. Snow-fed streams and those which drain forest and woodland areas carry the least amount of sediment. Also, many streams that drain mountain country are clear water streams.

Variable water supply limited in use

In considering the availability of water for power, remember that not all water power is continuously available to turn wheels or to generate electrical energy. For example, if all stream resources in the United States were harnessed for power, there would be available 45 million horsepower for 90 percent of the time, while approximately 60 million horsepower would be available 50 percent of the time.

This means that although 60 million horsepower of energy can ultimately be developed from stream power when all available sources are harnessed, this amount will be available only half of the time. If the power were needed 90 percent of the time during the year, only three-fourths of this amount could be used. Or, if the power were needed 100 percent of the time, somewhat less than 75 percent of the 60 million horsepower could be used.

☐ BENEFITS OF CONVENTIONAL WATER POWER

The benefits of using free-flowing streams as a source of power for manufacturing in colonial days are clear. Wind was the only alternative source of inanimate energy power until steam engines became available. Today, many other sources of energy are available and, also, are more efficient. Where multiple use of dams for flood control, navigation and power is effectively done, as by the Tennessee Valley Authority (see Chapter 9), the benefits of power are a sort of bonus received when carrying out the other desirable functions. Water power thus figured as a share of other costs is relatively cheap.

In one respect, hydroelectric power remains outstanding. It is clean power. No air pollution, super-heated water or waste-disposal problems, such as are typical of other power plants, occur.

Distribution of water power

The availability of water power from free-flowing streams is greatly limited by its uneven distribution in the country, as illustrated by the following:

	Percentage
New England States	2.4
Middle Atlantic States	10.1
East North Central States	1.9
South Atlantic States	6.5
East South Central States	5.9
West South Central States	2.1
Mountain States	25.5
West North Central States	4.2
Pacific Coast States	41.4
Total	100.0

Thus, water power has only limited prospects as a natural resource. At the present time, only a little over 3 percent of the power used in this country is from water sources. Even if all prospective sources from free-flowing streams were to be developed, the total amount of electrical power produced would probably only about double, assuming present technology.

□ NEW SOURCES OF WATER POWER

Sources of water power other than free-flowing streams have

been little researched and even less used in the United States. Yet, they offer astonishing possibilities to supplement our energy needs. New sources of water power that have been proposed include tidal power from the sea, heat interchanges in the sea and use of the Gulf Stream current.

Tidal power from the sea

Use of tidal power as an energy source is especially appealing because the rising and falling of the tides are as regular in their occurrences as the movement of the moon around the earth and the rotation of the earth itself. Here is a never-failing source of energy. However, the problems are to: (1) find areas where the rising and falling of the tides are enough to generate a worthwhile amount of power; (2) store or provide for alternate power in times of change in the tides or low tides; and (3) devise a profitable system for harnessing the surge of water.

The first proposal to use the power of the tides in this country was for Passamaquody Bay, located off the Bay of Fundy on the Maine coast. This bay has an effective tide of 18 feet. Schemes to harness its immense power were developed in the 1930's through encouragement of President Franklin Delano Roosevelt, who had vacationed in that picturesque area as a young man. But a tidal plant still had not been realized as of 1980 in the United States, even though France has had a successfully functioning tidal plant on its northern seacoast for several years.

The two technological developments that made this venture seem more possible in recent years than a quarter of a century ago are refinements in turbines to be used in generating the power and economies in extra high voltage transmission of the power.

The project is massive. One estimate was that it would require 7½ miles of dams and 15 years to complete. It would trap some 70 billion cubic feet of sea water and generate a million kilowatts of power. But, if successful, the benefits would also be great. For it would bring cheaper power to a traditionally power-short region—New England.

Heat interchanges in the sea

One possibility for water power that has yet to be demonstrated commercially, but has such vast potential it is worth noting, is to obtain energy by exploiting the great heat differences that exist between surface waters and those deep down in the seas. Off the Florida coast, for example, the ocean surface may be 75° to 90°F., while several thousand feet down, it is near freezing. A fluid with a low boiling point would be heated near the surface to drive a turbine, then condensed in the cold water below.

Use of the Gulf Stream current

The Gulf Stream is a warm current of water which flows eastward out of the Gulf of Mexico and then northward along our eastern coast at a rate of about 4 miles per hour, gradually diminishing in velocity. Some scientists have suggested that a huge underwater paddle be installed to use the force of this current to drive a generator for electrical power. Such a project is still visionary. But, we need to dream more about the possibilities of our ocean resources. The oceans' potential as a power source has been largely overlooked.

□ □ □

QUESTIONS AND PROBLEMS

1. What ancient nations used a common natural resource to develop their trade and increase their influence and power?
2. What invention about A.D. 1400 greatly extended the use of this resource?
3. At about what time did sailing vessels reach their peak of usefulness, and why were they replaced by steam-powered ships as the carriers of the oceangoing freight of the world?
4. What types of sailing vessels were developed in this country, and for what was each noted?
5. Name a second common use of wind as a power and name the countries which became noted because of the use of this power.
6. What caused windmills to disappear from the farms and the cities of our prairies?
7. In what way will the winds of the continent continue to be useful to the nation?
8. In what ways is water power better than wind power?

9. What are some of the costs that must be met in harnessing the water power of our thousands of streams throughout the land?
10. What are non-economic environmental costs that may result from the damming of a stream?
11. What are some of the benefits of water power? In what respect is water power outstanding?
12. What are three possibilities of obtaining more power by harnessing the energy of ocean waters?
13. Will water power ever be able to meet all our energy needs? Why or why not?
14. How much water power is developed in the county in which you live, and for what purposes is it used?
15. Does present research indicate that wind power may again be used extensively in the United States? Explain your answer.

5 □ □ □

Coal

COAL, WHICH IN 1980 was supplying over half of the nation's total thermal generation (all power generation except hydroelectric), is the most widely distributed of our current major energy resources. Yet, very little of it was used in this country one hundred years ago. China used a little coal as long ago as 100 B.C., and there is evidence to show that the Romans used it in Britain in 400 B.C. It was not used commonly as fuel, however, until about seven hundred years ago, or about A.D. 1200. The English people themselves used coal as a household fuel as early as A.D. 850.

The early colonists had little need for coal because they had plenty of wood. Not only did these pioneers use firewood as a source of fuel, but clearing the land of timber served the double purpose of supplying fuel and making available more land upon which to produce food and fibre crops. Fathers Marquette and Joliet reported the discovery of coal in the "Illinois Country" around 1673, while anthracite coal was found in northeastern Pennsylvania in 1790. At that time these discoveries were of little importance.

The growth of the coal-using industry in this country took place during the past hundred years. In 1824, only about 81,000 tons of coal were mined in the United States, while, in 1947, the total output was 676 million tons—about 8,300 times as much. Coal production then dropped off to about 420 million tons in 1960. Since then, demand has again risen; output was about 600 to 650 million tons per year in the early 1970's and may double by 1985.

A grand total of more than 40 billion tons of coal has been mined in the United States since the beginning of coal-mining operations about two hundred years ago. While this amount represents less than 2 percent of the estimated coal reserves in this country,

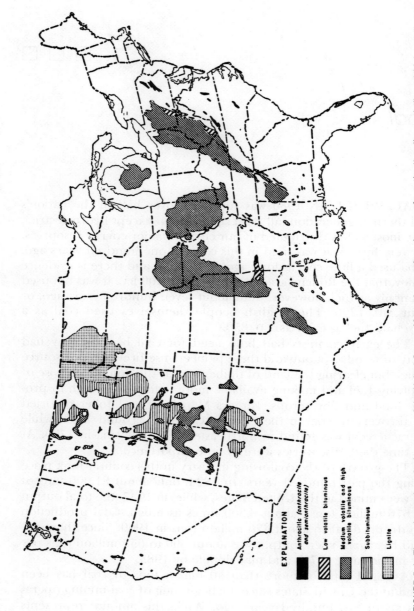

FIGURE 5-1. The major coal-producing areas of the United States. Most of our soft coal comes from the "medium volatile and high volatile bituminous" areas. (Courtesy, U.S. Geological Survey)

some of the reserves are not easily accessible. As is explained in the following paragraphs, the potential for economically feasible reserves depends upon many variable factors.

☐ LOCATION OF COAL RESERVES

The wide distribution of coal makes it available to every corner of the country. Over 30 states have mines, and coal is available to practically any community with no more than a 500-mile haul. There are three general coal regions which mine enough to be recognized as coal-producing areas: the Appalachian field, the Interior field and the Rocky Mountain field.

The largest and most extensive of the three is the Appalachian field. This field is found in nine states, stretching from Pennsylvania in the North to Alabama in the South. This field has the highest grade coal to be found in North America and is the largest soft coal field in the world.

The Interior field is next in importance, though it is much less extensive than the Appalachian field. There are several separate producing areas in this general location, the most important of which is the area of Illinois, Indiana and western Kentucky. Another field touches the five states of Iowa, Nebraska, Kansas, Oklahoma and Arkansas, while smaller fields are found in Texas and Michigan.

The third field, the Rocky Mountain field, extends brokenly from Montana through Wyoming, Colorado and Utah to New Mexico. It has vast reserves of lignite, sub-bituminous and generally low-grade bituminous coals with a B.T.U. potential estimated to be nearly as much as, or possibly even more than, that of the fields east of the Mississippi River. Production has increased greatly in this field because of the lower sulfur content of coal than in the Applachian and Interior fields. During the 1980's, more coal is expected to be produced in the West. But, impending changes in laws controlling pollution, in technology and in use of alternative fuels make the future difficult to predict.

A little coal is found in the three Pacific Coast states, but mining operations are not extensive.

☐ HOW COAL IS FORMED

Coal is the fossilized remains of trees and other vegetation which formed immense, dense forests millions of years ago. The clay or other material over which coal is always found was once the soil which supported the vegetation that later became coal.

Many coal beds are separated by layers of clay, shale, sandstone or limestone. These layers indicate that not all the coal-forming vegetation grew in one age. Each bed of coal between the different layers of clay, limestone, etc., represents a period of thousands of years wherein new trees and other vegetation grew and thrived. The layers of non-carbonaceous material between the coal beds were deposited on land or in shallow seas as the land slowly subsided and rose after each coal bed was deposited. The deposition of each layer took thousands of years. It is estimated that it required at least one thousand years to form a layer of coal one foot thick.

☐ KINDS OF COAL, RESERVES AND USES

Coal is made of hydrocarbons (compounds of hydrogen and carbon) and moisture. When coal is burned, that which burns with the clearest flame and leaves the least ash has the highest carbon content. The moisture, mineral content and sulfur compounds produce gaseous sulfur compounds, while the mineral compounds produce the ash.

At any given site, assuming the coal is equally accessible, coal's economic value and use is dependent upon the ratio of hydrogen to carbon, other mineral content and moisture.

Although there is enough coal in this country to last several hundred years, not all of it is of sufficient quality or in thick enough beds for commercial recovery. Also, some coal beds have been moved by earthquakes or offset by slippage of rocks along fractures so that the layers are not continuous. Other beds lie so deep in the earth that the costs of recovery are too high at the present time to be economically feasible. The U.S. Geological Survey considers the following thicknesses of coal beds as minimum for most commercial production or recovery:

Bituminous coal	30 inches
Sub-bituminous coal	24 inches
Lignite	36 inches
Anthracite	36 inches

Underground production equipment has recently been designed to work either bituminous or anthracite coal seams of less thickness than those just listed, and operators using small-scale, hand-loading equipment in underground production sometimes mine from beds no thicker than 24 inches.

The coal should not be more than 6,000 feet below the earth's surface, and no coal with an ash content of more than 30 percent is considered as a part of the commercial supply. With these qualifications, it is estimated that there are still unmined over 3,000 billion tons of coal in this country, with about one-half of it at depths to 3,000 feet in mapped and explored areas. Actual usable reserves under our present capabilities are less than 10 percent of this amount, however, according to estimates of the U.S. Bureau of Mines. Almost one-third of these usable reserves are close enough to the surface to be recoverable by surface mining methods.

The quality of coal mined varies greatly, and each type of coal has its best uses as well as certain drawbacks. Some coals burn more completely than others (that is, with less ash), while others give off more volatile gases from which by-products can be made, and still others have considerable waste in the form of moisture. This means that coals must be carefully selected for different purposes if they are to be most effectively used.

Characteristics of different types of coal

The different types of coal and the uses for which they are especially suitable are as follows:

1. *Peat*. The first step in the development of coal from plants is the formation of peat. This always takes place under water. Not much is known about just how peat becomes coal, but it is believed that the various types of coal are formed by different amounts of heat and pressure. Large peat bogs occur in southern Europe, as well as in the more northern regions. Some peat is found in the northern part of the United States, although much more is found in Canada. It is used for fuel in countries such as Ireland, Scotland and England, as well as in the Netherlands, Denmark and in some parts of northern East and West Germany. Charcoal is frequently made from peat, and this product is used for working and tempering the finer kinds of cutlery.

2. *Lignite*. A great deal of the coal on the European continent, as well as much of Canada's supply, is lignite. The word *lignite* is

taken from the Latin word *lignum* meaning "wood." Thus, lignite is "half-made" coal and usually is quite woody. Most lignite is brown in color, although some is so black it is called jet. This black lignite is sometimes used for making buttons, toys, jewelry, etc., because it takes a high polish.

3. *Bituminous.* Bituminous (soft) coal is next in hardness after lignite. One of its chief varieties is called coking coal, which swells up into a mass when heated. Non-coking coal, which does not melt into a mass when heated, is the kind used in home furnaces. Most of the coal used in this country is bituminous coal.

4. *Anthracite.* Anthracite coal is the hardest and most brittle of coals. It is jet black and practically free from pitch or bitumen, and, when ignited, it burns slowly with a clear blue flame. It generates a great amount of heat and burns with little or no smoke. It is so difficult to ignite that one of its early uses was not as fuel but as road building material. Later, anthracite was used for heating homes in the picturesque base-burners. A pail of coal would burn for 10 to 12 hours in one of these stoves, and it gave out enough heat to warm the normal-sized living room. The important use of anthracite now is for making gas, which is used primarily for heating and cooking.

When burned, both bituminous and anthracite coal produce dangerous fumes that should not be allowed to escape into rooms or workshops.

The different types of coal with their ranges in fixed carbon, volatile matter and moisture are shown in Table 5-1.

TABLE 5-1. Ranges in fixed carbon, volatile matter and moisture of the different types of coal in percentage by weight

Type of Coal	Range in		
	Fixed Carbon	Volatile Matter	Moisture
	............................(%)............................		
Lignite	30–55	30–10	43–30
Bituminous	47–65	41–32	72– 3
Low-volatile bituminous	75–83	22–12	5– 3
Anthracite	85–96	2– 1	6– 2

As the fixed carbon content becomes greater, the volatile matter becomes less. The final product is graphite, which is practically pure carbon. Anthracite coal, which was burned in the old base-burners, burns with very little color and leaves very little ash, since about 90 percent of it is carbon. On the other hand, bituminuous coal varies greatly in the way it burns and in the amount of ash it leaves after the carbon is burned out of the coals.

FIGURE 5-2. A new section is lifted into an auger mining machine. Like a giant carpenter's bit, these sections push a cutting head 200 feet into a hillside coal seam, drawing the coal out as they revolve. As the coal reaches the outside, it is loaded into trucks by the conveyor at the left. This is a highly efficient production method. Auger mining in 1974 produced coal at the rate of 45 tons per man-day, compared with 36 tons for strip mines and 11.4 tons for deep underground mines. *(Courtesy, National Coal Association)*

☐ COAL DETERMINED THE SITES FOR INDUSTRY

The importance of coal to the development of factories is shown in every country where there is much manufacturing. In England, the important industries are carried on in the North Country and the Midlands where coal is found. Areas with no coal have few factories. The Ruhr development on the European mainland is the result of the coal and iron deposits located there and nearby. In the

United States, the first great concentration of industries was in the coal-producing regions of Pennsylvania. Birmingham, Alabama, has many factories because of its coal.

While ready access to an energy source is still important today for high energy-using manufacturers, the low cost (to the shipper) with which coal is moved by barge or, after conversion, sent over power lines as electricity and the availability of other sources of power have greatly lessened the significance of coal fields as location sites for industry.

☐ REDUCING COSTS AND INCREASING AVAILABILITY OF COAL

The ability of coal to compete with other energy sources has been enhanced by reduced costs. And, in new forms, it may soon be more readily available than ever before.

FIGURE 5-3. Big Muskie, the world's largest dragline, cost $20 million to build and has a reach the length of a football field. It weighs 27 million pounds and lumbers back and forth on mammoth "feet" at a central Ohio strip mine. Its 220-cubic yard bucket can move 325 tons of overburden at a single pass. Note how the men and even cars and trucks are dwarfed by comparison. *(Courtesy, Bucyrus Erie Company)*

One proposal to obtain cheaper power from coal is to burn the coal underground at the source and use the heat for power. The process remains to be perfected.

Another way of reducing mining costs is by the use of more efficient machinery. In some mines a three-man team with one of the newer push-button machines can turn out coal at the rate of 250 tons an hour and work seams no thicker than 3½ feet.

Allowing for changes in the value of the dollar, it costs less to produce coal than it did 25 years ago, even though wages have risen sharply. Strip mining has especially low costs, with giant shovels lifting over 200 tons of overburden (the cover of soil or stone) at a bite (see Figure 5-3)! Smaller shovels then load the coal into trucks (see Figure 5-4). Also, strip mining is safer than shaft mining and thus less costly in terms of human lives.

FIGURE 5-4. After topsoil and overburden are removed, a loading shovel scoops coal from a seam averaging 55 feet thick at the Cordero Mine south of Gillette, Wyoming. This surface mine is operated by Sunoco Energy Development Co. (SUNEDCO), a unit of Sun Oil Company, Inc., and produces 3 million tons of low sulfur coal a year. *(Courtesy, Sun Oil Company, Inc.)*

62 □ OUR NATURAL RESOURCES

One of the newest methods of underground mining is long-wall mining (see Figures 5-5 and 5-6). This method permits subsidence at the surface of the ground, if it is to occur, to take place under control and in a relatively short time, thus reducing the problem.

Environmental costs are critical to society. These costs include the cost of land restoration. One reason coal development has lagged in recent years has been due to the abuses of mining in the past.

FIGURE 5-5. Miner activates movable hydraulic steel roof props as long-wall mining system operates in a central Pennsylvania coal mine. Cable in background pulls a shearing machine back and forth along a coal face several hundred feet long; then the loosened coal drops into a conveyor and is carried away. Alternate props are lowered, moved forward and raised again, pushing the conveyor forward. Behind them, the mine roof is allowed to collapse. *(Courtesy, Joy Manufacturing Co.)*.

FIGURE 5-6. A giant shearer cuts a 30-inch slice off a coal face 7 feet high and 600 feet long. This machine set a world record of more than 12,000 tons of coal in three shifts during a 24-hour period. Heavy steel shields protect the operator, moving forward as the machine advances deeper into the coal. Behind them, the mine roof is allowed to collapse. *(Courtesy, Consolidation Coal Co.)*

Underground mining has resulted in serious land subsidence problems in some areas. Both surface and underground mining have often been the cause of acid water drainage and of large areas of land made unproductive because they were covered with mine wastes.

Gasification and liquefaction

The development of processes to produce coal-derived fuels is now underway. Gasification processes would produce either a high heat-content, pipeline-quality substitute for natural gas or a low heat synthetic fuel for utility boilers. The liquefaction process (also called *hydrogenation*) would wring crude oil out of coal, which in turn could be converted into gasoline or lubricating oils.

☐ OTHER USES OF COAL

When bituminous coal is heated without access to air, it does not burn but gives off gases. The remaining product is no longer called coal; it is coke, which is mostly carbon. Coke is used for fuel, but its most important use is for changing (reducing) iron oxide to iron, essential to the production of iron and steel. Since not all coals will form satisfactory coke, those that will should not be diverted to other uses. In past years, the gases were used for lighting dwellings and businesses but this use has been largely discontinued.

Another by-product in the manufacture of gas from coal is coal-tar. This is useful in many ways because of the products that can be made from it. It is used as a paint and as a preservative of timber in its crude form. It is also used in making tar paper. Practically all aniline dyes are made from coal-tar, and the shades of color obtained from these dyes compete with the rainbow. Ammonia, carbolic acid, benzine, alum, picric acid, used in the manufacture of explosives, and saccharine, which is three hundred times sweeter than sugar, all are by-products of coal. Few would expect to find that some delicate perfumes are made from coal, yet that is the case.

☐ WHY COAL DEVELOPMENT HAS LAGGED

In view of the many advantages of coal as a source of power, why have we reduced our use of it, relative to total use of other fuels? Why have we turned from a resource with reserves within our national boundaries that, by even the most conservative estimates, will last many decades and is readily available at economical costs? By contrast, as the following chapter relates, oil reserves have become more costly to obtain domestically, and we are dependent upon foreign supplies. Nuclear energy, through fission—the current process—has serious security and environmental risks. Other promising sources of clean power, such as nuclear fusion and geothermal energy, are still in the developmental stages.

There are two major reasons why coal has been put on the back burner in development and research. First, coal doesn't blow up cities! Our major effort, nuclear fission development, was spurred by its military potential. Second, the coal industry has been a notoriously poor husbandman of our other earth resources. The exploitation of no other mineral has resulted in so much ravaged land or polluted air.

Strip mining

Today, however, there is no need to repeat the disastrous abuses of the past in coal mining and burning. Strip-mined land can be reclaimed to productive use as experiments here and abroad have demonstrated, and the costs of reclamation are bearable.

The Surface Mining Control and Reclamation Act of 1977 was intended to correct abuses of the environment by strip mining operations. An interim program was put into effect, pending agreement on regulations that would implement the law. Following court injunctions obtained by a number of coal companies, implementation has been further forestalled into mid-1981 or beyond. The Conservation Foundation comments in its *Conservation Foundation Letter:* ". . . [T]he whole strip-mine regulatory program is beset by antipathy and resistance . . . it is a revealing example of the way conflicts between federal and state governments are triggered by federal environmental protection laws and efforts to implement them."[1]

Strip mining is prohibited in certain areas, such as prime agricultural land and land with steep slopes. Mining legislation needs to be coordinated with other land-use policy legislation.

□ COAL AND THE ENERGY CRISIS

There appears to be no way to eliminate the energy crisis in the United States immediately, or even in the near future, by greatly increasing the production of fuels, and there is little likelihood of obtaining substantially larger supplies from abroad. Any short-run relief will have to come from less waste of energy and cuts in usage.

There is one way, however, to help assure that the energy crisis

---

[1]Conservation Foundation, *Conservation Foundation Letter*, November, 1980, p. 1.

does not worsen. On the consumption side, the United States must continue to progress in conserving energy. On the production side, it should keep full steam behind programs of coal research and development. Such programs would anchor our energy resource base to a reliable source of proven abundance within our own national boundaries. They would lessen our dependence on oil and natural gas and would help hold down prices by affording more competition between fuels. We would then be in a better bargaining position to obtain the oil we need from abroad. Without the threat of an inbred nuclear disaster, we would gain time to develop other promising sources of power, aiming for that day when we can tap directly, on a massive scale, the radiant energy of the sun—the ultimate answer to the need for clean, unlimited power.

Air pollution problems

The elimination of air pollution from the combustion of coal has proven difficult. Present processes have not been successful enough in removing sulfur oxide gases and particulate matter, or in recovering and using heat, gases and particulate matter, to help offset filtering costs. While more research is needed, it appears that if available devises were in general use, air pollution from coal burning could be reduced and cost increases could be held down sufficiently to be tolerable during the transition period to better controls or other fuels.

Another problem of controlling air pollution has been how to measure it and set up equitable regulations. The Environmental Protection Agency, acknowledging that its system of measuring air pollution from coal-burning plants is flawed, has been considering rewriting the rules that govern smokestack emissions.

The problem is that emissions of sulfur dioxide combine with moisture in the atmosphere to form a "rain" of weak sulfuric acid. Furthermore, sulfur content in the coal can vary from mine to mine, day to day and even lump to lump.

The new method under consideration is called "expected exceedences." It would take into account the varying sulfur quality of the coal being burned. The result would be mathematical predictions—and legal allowances—for days of both high and low emissions.

Meanwhile, advanced stack gas-cleaning processes are being developed by the Environmental Protection Agency. And, direct com-

FIGURE 5-7. A stripping shovel operator's view of the pit, Sinclair Mine, Peabody Coal Company, Kentucky. The dipper of the big shovel holds 125 cubic yards of earth and rock. The loading shovel in the background scoops coal into 135-ton trucks. *(Courtesy, National Coal Association)*

bustion processes using fluidized bed boilers are being jointly developed by the Environmental Protection Agency and the Department of the Interior.

☐ ☐ ☐

QUESTIONS AND PROBLEMS

1. What portion of the total energy used in the United States comes from coal?
2. How many tons of coal were used in this country in the early 1970's in comparison with the amount in 1824? How much more is expected to be used by 1985?
3. Would you say that coal is widely available? How do reserves west of the Mississippi River compare to those east of it?
4. How is coal formed?
5. Why is coal always found over a layer of clay or shale?

6. Name the three kinds of coal commonly found in this country and tell how they differ.
7. Which coal burns with the least flame and gas and leaves the least ash?
8. Name some industrial centers which owe their existence to the presence of coal. Is coal as important to manufacturing locations today as formerly? Explain your answer.
9. Name some of the things that are made from by-products of coal.
10. What is one system of coal use that is being tried out in an effort to reduce the cost of converting coal to energy?
11. What percent of the total estimated coal reserves are actually usable under our present capabilities? Why is this percentage so small?
12. What kind of coal, if any, is found in your state? Name several uses to which this coal is put.
13. How will the processes of gasification and liquefaction extend the use of coal?
14. How is strip mining cheaper, and how is it apt to be more expensive than shaft mining? Are there any disadvantages to shaft mining?
15. Why has coal development lagged, since it is such an abundant and a useful fuel?
16. How does long-wall mining deal with the problem of land subsidence?
17. Why does the method of providing for "expected exceedences" appear to be a fairer system than the present one for establishing rules for permissible levels of sulfur emissions?

6 ▢ ▢ ▢

Oil and gas

NO ENERGY SOURCES have been more a matter of public concern in our nation in recent years than oil and gas. For, these are the fuels we use to provide three-fourths of our energy needs. Yet, in the early 1970's, we suddenly discovered that we were in short supply. For the first time since the introduction of the automobile, millions of Americans found themselves waiting at the gas pumps. More alarming, a sudden increase in the costs of these fuels and our dependence upon foreign supplies to fulfill our needs threatened to wreck our economy. It was only about a century ago, however, that oil and gas were little-known substances. The story of their development is a dramatic example of the functional concept of resources, described earlier in this text. In response to our wants, we used our technological and societal abilities to develop oil and gas from the uncommon to the common—the major fuels of our energy base.

▢ EARLY PETROLEUM USE

One of the earliest known oil products was asphalt. Asphalt is a black or brownish solid substance which, when heated, becomes somewhat runny and sticky. The smell suggests that it is derived from petroleum. It is formed from the evaporation and oxidation of petroleum on or near the surface of the earth. There are many sources of this substance over the world; the most noted source in this hemisphere is the island of Trinidad, off the coast of Venezuela, which has a lake of asphalt, or pitch, at La Brea.

Asphalt was used seven thousand years ago to embalm bodies in ancient Egypt, where it also was used to waterproof various river craft. The baby Moses was set adrift amidst the rushes on the Nile in a basket made waterproof through the use of this same product.

Oil, or petroleum,[1] from Sicily was burned in the Temple of Jupiter at Rome about twenty-five hundred years ago. We know that natural gas was used about that time because the "eternal fires" of Baku, Persia, were natural gas burning in what became one of the major oil fields in the world.

☐ OIL A NUISANCE

The development of our oil and gas industry is the story of unappreciated resources. Early in the history of this country, the finding of oil and gas was distinctly a nuisance. Drilling was done to obtain water and brine for salt making, and wherever oil or gas was encountered, the location was no longer useful.

It was not until 1859 that a salt-well driller was hired to put down a well for oil. Crude petroleum at that time was selling for $18 a barrel, but since it was used primarily for medicinal purposes, only a small quantity was needed. Even so, the Drake well at Titusville, Pennsylvania, was put down in the hopes of finding some of this scarce, precious petroleum. This famous strike proved so productive that the market was soon flooded, and the price of crude petroleum dropped from $18 to 10 cents a barrel—a price so low that the containers used to take it away were worth more than the product itself.

From that small start about one hundred years ago, oil and gas have become the fastest growing of the energy-producing industries. Oil and gas rank above all other sources of energy used in the United States.

☐ MANY USEFUL PRODUCTS FROM PETROLEUM

Besides its importance as a fuel, petroleum has become one of the most useful of our natural resources because of the vast number of products that can be made from it. It is broken down by a process called distillation, whereby various parts, or fractions, of the petroleum are driven off and collected. In addition to gasoline, kerosene

---

[1] The words *oil* and *petroleum* are used interchangeably in this chapter.

and lubricants, the list of other useful products derived from petroleum is long and varied. We know that paraffin and petroleum jelly come from this source, but it is surprising to learn that in the list, which is too long to enumerate, are enamels, waterproof cloth, electric insulating material, plastics, rayon, nylon, radiator antifreeze, a food moistener and mold growth retarder for baking products and even a blackstrap molasses for cattle feeding.

□ SOURCE OF OIL AND GAS

Nature's method of making oil is not definitely known. The more common belief among geologists is that oil came from the deposits of both animal and vegetable life. The principal evidence for this belief is that petroleum contains certain complex chemical compounds called *porphyrins* which are derived from hemin and chlorophyll—the materials giving the red color to blood in animals and the green color to plants. Most petroleum hydrocarbons have the power to rotate the plane of polarization of polarized light, a property known to occur in sugars and cholesterol, which are produced by living organisms as part of their biochemical processes.

□ WHERE IS OIL FOUND?

The organic matter from which petroleum and natural gas was derived was deposited as small particles along with clay mud in layers on the bottom of shallow seas. These muds were slowly buried by later deposited sediments, sands, shell fragments and other muds. The weight of the overlying sediment compressed the organic-containing mud (the "source beds" of oil and gas), the temperature increased somewhat to a maximum of about 392°F. (probably less) with depth of burial, and the plant-animal tissues contained in the source bed were chemically or biochemically altered to hydrocarbons in a liquid and/or a gaseous state.

Bacterial activity is believed to be a major factor in producing petroleum compounds from plant and animal tissues. These were squeezed out of the source bed by further compression, much as water is squeezed from a saturated sponge. The petroleum then entered an adjacent porous rock and migrated through it until it reached the land surface where it seeped out or until it reached some type of trap which stored it underground.

72 □ OUR NATURAL RESOURCES

Oil and gas must be trapped in order for them to be sufficiently concentrated for drilling. Traps consist of a number of different types of geological structures. All, however, feature a porous reservoir rock (in which the oil and gas can be stored) underneath a non-porous or impermeable "cap rock" through which the fluids cannot migrate. The cap rock is shaped like an upside-down bowl and may be several miles in width and several hundreds of feet in depth. Oil and gas source beds and reservoir rocks formed into suitable traps are found in rocks ranging in age from 600 million years (Cambrian Period) to only a few million years (Tertiary Period). The younger rocks have produced much more oil than the older ones.

□ RESERVES OF OIL AND GAS

The known recoverable reserves of oil and gas are sparse compared to those of coal. Domestic coal reserves have a prospective life of hundreds of years, as has been noted, even with expanded demand. By contrast, our *proved* reserves of oil (verified by testing, recoverable by existing methods and economically feasible to develop [see section on "Mineral Reserves and Their Estimates,"]) of some 27 billion barrels in 1980[2] were sufficient to last less than seven years at current rates of use and our continued dependence on imports. If imports were cut off, they would last only about half that time.

Of course, our oil reserves will last much longer than suggested by the above figures because we are making new discoveries and thus adding to our reserves. But, even estimated *potential* reserves are not great. *Potential* reserves can be estimated by taking into account all the rock layers thought to be oil-bearing, estimating the amount of oil they likely hold and then multiplying this by the rate of recovery from such rock experienced in past years (around 30 percent). Other methods allow for the declining yields experienced with additional exploration and drilling activity. Such estimates of undiscovered, recoverable reserves in this country vary widely, as shown by Table 6-1. The estimates vary from 50 to 127 million barrels by the U.S. Geological Survey to 30 to 100 million by the Shell Oil Company. If we assume that national demand for oil could be

---

[2]"IOCC (International Oil Compact Companies): Excise Tax to Cut Reserves 490 Billion Barrels," *Oil and Gas Journal*, August 4, 1980, p. 30. (Figure is for January 1, 1980.)

OIL AND GAS □ 73

FIGURE 6-1. This $4 million gas conservation plant was built in an oil field to take gas which would otherwise be wasted and make it into useful products. In the past, it was considered more economical to let such isolated fields of gas escape into the air than to construct such plants. *(Courtesy, Standard Oil of New Jersey)*

TABLE 6-1. Estimates of U.S. undiscovered recoverable oil reserves

Estimator	Year of Estimate	Billion Barrels
American Association of Petroleum Geologists	1972	120
U.S. Geological Survey	1975	50–127
National Academy of Science	1975	105–120
Shell Oil Company	1979	30–100

(Sources: See Robert Gillette, "Oil and Gas Resources: Academy Calls USGS Math 'Misleading,'" *Science*, AAAS, February 28, 1975, pp. 723–727; Betty Miller, et al., *Geological Estimates of Undiscovered Recoverable Oil and Gas Reserves in the United States*, Geological Circular 725, U.S. Geological Survey, 1975 (estimates are as of January 1, 1975); "U.S. Petroleum Industry Will Face Monumental Task in Next Decade," *Oil and Gas Journal*, November 12, 1979, pp. 170–184. For an additional discussion of reserves, see Sam H. Schurr, et al., *Energy in America's Future: The Choices Before Us*, The Johns Hopkins University Press, Baltimore, 1979, pp. 226–229.)

held at a 7-billion-barrel annual level, we would have only some 27 years' supply at the most optimistic estimate and but 13 years' supply at the lowest, if we rely entirely upon our own reserves. (Note: Proved and inferred reserves of 62 billion barrels [see Figure 6-3] are added to undiscovered recoverable reserves for these calculations.)

Alaskan and outer continental shelf reserves

The question is often asked as to how important Alaskan and offshore (called *outer continental shelf*—OCS) reserves are as part of the total potential reserves. This resource supplies limited but important amounts of oil and a larger share of our natural gas. OCS oil production under federal jurisdiction in 1977 was about 830,000 barrels per day, or roughly 5 percent of domestic oil consumption. Gas production in 1977 was about 3.7 million cubic feet, nearly 20 percent of U.S. total supplies. Alaska and the OCS are thought to be the nation's last frontiers for major new oil and gas finds. Both government and industry estimate that about 60 percent of our undiscovered oil and natural gas resources may be located there.

Department of Energy preliminary estimates of oil and gas production for the period from 1985 to 1995 suggest, however, that in the near future the OCS contribution to our energy needs will be rather small. It is projected that oil output may almost double, but natural gas production will decline.[3]

While OCS reserves are some of the most accessible, environmental objections need to be overcome. The Council on Environmental Quality has studied the problem of developing these offshore reserves for several years and has recommended improvements in these areas: (1) in equipment design and operating practices to protect the people involved; (2) in technology to meet harsher conditions in the Atlantic and Alaskan OCS than in many others; (3) in technology and practices to minimize the impact in undeveloped areas. Environmentalists have been disappointed because the council has not flatly ruled out exploration in high risk areas. The oil companies feel that adverse economic impacts have been overestimated.

---

[3]Council on Environmental Quality, *Environmental Quality, The Tenth Annual Report of the Council on Environmental Quality*, U.S. Government Printing Office, Washington, D.C., 1979, p. 333.

And, the governors of some of the coastal states have opposed development in part because of disagreement over state versus federal jurisdiction of the areas.

Perhaps societal problems such as these enumerated can be resolved. Over 17,000 wells have already been drilled in waters off U.S. coasts (see Figure 6-2). While only a few disastrous oil spills have resulted, the damage from even a single break is potentially great. And, experience indicates that some oil spills are unavoidable. We cannot afford to have our beaches fouled and marine fishery resources destroyed by oil. Hopefully, this can be avoided. We have learned how to mitigate or eliminate impacts by proper siting, environmental controls, careful construction and operation and studies and monitoring to identify areas that need to be avoided. The question is, "Will we do it?"

Relief from foreign supplies?

Even foreign supplies of oil, were they to become readily available, afford no long-term prospect for meeting our fuel requirements. The world's undiscovered recoverable oil amounted to about 2,000 billion barrels in 1980, according to various estimates reported in *The Global 2000 Report to the President*.[4] *Proved* world reserves were about one-third *potential* reserves. With world production running 25 to 30 billion barrels a year, the world supply of *proved* reserves would last perhaps 25 years, and *potential* would extend that life another 50 to 100 years. Since world demand is expected to become much greater, the reserves will probably last even fewer years. World gas reserves likewise afford no long-term answer to our fuel needs.

Despite promising potentials for oil discoveries, the fact that our proved reserves lead our annual consumption requirements by only a few years indicates the more critical nature of oil reserves than coal. Some potential reserves may never "prove out," while developing others may be unwise.

Some new gas fields

The quantities of known recoverable gas supplies in the United

---

[4]Council on Environmental Quality and the U.S. Department of State, *The Global 2000 Report to the President: Entering the Twenty-First Century*, Vol. II, U.S. Government Printing Office, Washington, D.C., 1980, pp. 186, 187.

FIGURE 6-2. The large drilling platform shown here is representative of those used to work offshore oil deposits. This one is located in one of the new Gulf of Mexico fields, known as Mississippi Canyon Block 194 (Cognac), where production began in 1979. By year-end, it was averaging 10,000 barrels of oil daily. The field, in which Shell is operator and has a 29 percent interest, will cost more than $800 million to bring to full production. The cost includes leases, exploratory wells, building and installing a three-section drilling and production platform in an industry-record water depth of 1,025 feet, 28 miles of pipeline and 62 development wells. Cognac is expected to produce 50,000 barrels of oil daily by 1983 and 150 million cubic feet of gas daily by 1992. It is estimated that the field eventually will produce a total of at least 100 million barrels of oil and 500 billion cubic feet of gas. *(Courtesy, Shell Oil Company)*

States, like those of oil, depend on the discovery of new fields and changes in the process of withdrawal from the underground reserves. As new locations are found and improved recovery methods are applied, more underground reserves are brought to the surface.

The *proved* reserves of natural gas in the United States at the beginning of 1980, according to the *Oil and Gas Journal*, were more than double those 30 years ago. But, the increase in proved reserves has been more than offset by an increase in the demand for natural gas. The U.S. Geological Survey has estimated *potential* reserves to be about seven times as great as *proved*. But, because of the physical nature of the source and cost-price relationships in recent years, such estimates are apt to be unreliable. A 1973 report of the Joint Committee on Atomic Energy stated: "[Domestic gas] . . . is probably the hardest [of energy resources] to estimate. Most engineers will agree that the nation will be fortunate if long range projections of gas availability are 50% either way of actual production."[5]

□ BETTER RECOVERY DEVELOPED

Early methods of recovering both oil and gas from their underground reservoirs were most wasteful. In many instances as little as one-fourth of the oil in an underground pool was brought to the surface, and this rate of recovery was considered good. A recovery of 50 percent of the oil in a field is now average.

In many instances a much larger percentage of the underground reserves could be recovered if a field could be controlled and operated as one producing unit rather than as many independent units. When neighboring wells are owned and operated by different producers, it is ordinarily to the advantage of each producer to take out all the oil he can from this commonly owned underground pool. If he does not, a neighboring producer will profit by removing it through his wells.

In order to conserve these underground supplies better, most of the oil-producing states, as well as the federal government, have developed plans for allotting to individual producers certain proportions of the permitted production of a given reservoir or "field." The rate of depletion also is regulated because the total supply of oil that

---

[5]*Understanding the National Energy Dilemma*, Joint Committee on Atomic Energy, Washington, D.C., 1973, p. 17.

may be taken is frequently increased greatly by slowing down the current rate of removal.

Various devices and procedures have been developed to recover a larger proportion of the oil and gas reserves, but in no instance is it economically feasible or even physically possible to recover the complete supply of fuel trapped in the underground reservoirs.

☐ PROSPECTING AND MINING DEVELOPMENT

Since certain types of deposits were formed along with and are regularly associated with certain types of rocks, broad-scale geological mapping is usually the first tool used in estimating the possible and probable amount of ore or fuel deposits. Then, for detailed exploration and prospecting, close examination of geological structures and of the mineralogical, chemical and physical (such as electrical, magnetic, and seismic) properties of the rock bodies is frequently done. Some examination is done on the surface by collecting samples for analysis and by carrying measuring instruments to the field. Some is accomplished by taking measurements from airborne instruments, by using air photos or satellite images and other data, which have proved quite useful in collecting mineral prospect locations.

When several favorable indicators are found in an area, then subsurface testing must be done by means of drilling. Analysis of the chips or cylinders of rock cut out by the drill and the fluid, such as oil or gas, found in the rock answers the question "Is there anything valuable here?" If there is, further drilling is done to outline the size and shape of the deposit.

If it is an oil or a gas deposit, a plan for rate of withdrawal and location of withdrawal points is made which will result in the most production under available technology. Too rapid a withdrawal rate or failure to maintain fluid pressure in the oil and gas reservoir underground results in some otherwise producible oil or gas being left unobtainable in the ground. Under present technological and economic conditions, half or more of the oil in every oil field reservoir cannot be produced. The rapidly rising price of oil may alter this in the 1980's.

If an ore deposit has been located, the underground and surface workings necessary to produce it must be constructed. The mine plan will depend on the size, shape and ore grade of the deposit. First, a shaft must be sunk. This is costly and time-consuming because only a few workers and a small amount of equipment can get into the small area at the bottom of the shaft and because all the broken rock must be lifted to the surface by some type of temporary hoisting equipment. When the shaft has reached its planned depth, then horizontal and other vertical openings required in the mine can be advanced more rapidly as there is now a large permanent hoist available which fills a large proportion of the shaft. Ore processing buildings, waste disposal sites, offices, laboratories, dressing room and showers for the miners and transport facilities must all be built at the surface.

Prospecting through test drilling often requires two to five years. Mine construction frequently requires an additional four to five years for a large mine. There will be anywhere from hundreds of thousands to many millions of dollars invested before one pound of ore is shipped!

□ MINERAL RESERVES AND THEIR ESTIMATES

A mineral reserve is a valuable mineral material which has not been produced and is still in the ground. It is not available for use until it is mined.

While most geological processes at some time or other concentrate mineral materials useful to man, it is often difficult to determine their location and amount. Most mineral deposits are in bedrock, but some are in loose, weathered materials found on the surface of the ground. Almost all are covered with soil and vegetation. Therefore, there are built-in uncertainties in all mineral reserve estimates.

The U.S. Geological Survey and the U.S. Bureau of Mines have adopted this ranking of reserve categories:

proved — These amounts are *measured* from drilling data.
probable — These are *indicated* to be there by a favorable combination of geological factors adjacent to known deposits.
possible — These are *inferred* to be present from the presence of some evidence and favorable geological factors.

Perhaps we should add one more category, namely, "*undiscovered.*"

There is greater and greater unreliability in tonnage estimates

80 ◻ OUR NATURAL RESOURCES

from *proved* through *probable, possible* and *undiscovered* reserves. U.S. reserves of oil and gas in these categories determined by geologists for the U.S. Geological Survey are shown in Figure 6-3.

The diagram shows 34.25 billion barrels of *proved* reserves, an additional 4.636 billion barrels of *probable* reserves, still another 23.1 billion barrels of *possible* reserves and, finally, 50 to 127 billion barrels of *undiscovered* reserves. Some confusion in interpretation results because whereas the U.S. Geological Survey separates *measured* and *indicated* as *proved* and *probable,* the oil industry commonly calls the total of the two *"proved"* (shown as *"demonstrated"* on the chart). All of these are further classified as *"economic"* which means they can be recovered at a reasonable cost. Thus, in Table 6-1, they are called "undiscovered recoverable" reserves. (The figure for the U.S. Geological Survey 1975 estimate is the one shown in Table 6-1.)

Figure 6-3 also indicates a large amount of sub-economic reserves—an additional 166 to 251 billion barrels altogether. Such reserves are not ordinarily counted in any projections as being capable of meeting demands in the future, for it is assumed it would be prohibitively expensive to do so. But, there is a probability that they exist.

A similar scheme of reserves is shown for natural gas.

The cumulative figure provided below each chart shows how much oil and gas production there had been in the United States from the first well in 1859 to January 1, 1975. It is from comparing these kinds of figures that one can make the observation that the United States has more petroleum reserves underground now than all the oil and gas this nation has ever consumed. This is true—if *undiscovered recoverable* reserves are included. As the analysis of Table 6-1 shows, however, even if the undiscovered recoverable reserves indeed prove out, we can continue present consumption rates for only a few more decades. And, this is also true for the world as a whole.

But, one may say, "The chart was prepared in 1975. I have just read [December, 1980] of new finds in the Overthrust Belt of the West that might double our proved reserves and in the Ob Valley of the Soviet Union that might equal those of the Middle East—over one-half of worldwide reserves. Doesn't this indicate that the entire earlier figure is too pessimistic?"[6]

---

[6]"Rocky Mountain High," *Time,* December 15, 1980, p. 31; and Thomas Watterson, "Bearish Soviet 'News' Routs Investors," *The Christian Science Monitor,* December 8, 1980, p. 11.

CRUDE OIL RESOURCES OF THE UNITED STATES
(BILLION BARRELS)

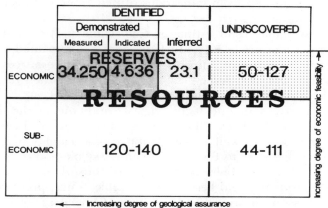

Total U.S. Cumulative Oil Production 106 Billion Barrels

NATURAL GAS RESOURCES OF THE UNITED STATES
(TRILLION CUBIC FEET)

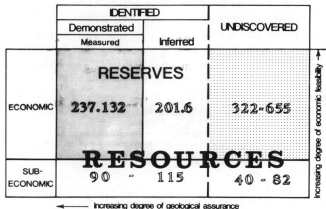

Total U.S. Cumulative Gas Production 481 Trillion Cu Ft

FIGURE 6-3. Geological estimates are of January 1, 1975, but they are representative of the situation in 1980. *(Source: Betty Miller, et al.,* **Geological Estimates of Undiscovered Recoverable Oil and Gas Resources in the United States**, *Geological Circular 725, U.S. Geological Survey, 1975)*

In fact, the 1975 estimate did allow for, by including the undiscovered recoverable reserves, just such additions to our supplies. But, it is a long step from indications of new reserves to actual production.

The available 1980 estimates (see Table 6-1) indicated that proved reserves were less than in 1975, and they remained conservative in their estimates of undiscovered recoverable resources.[7]

All the published reserve figures, whether for fuels or nonfuel minerals, and the concepts of *recoverable*, *submarginal*, etc., are based entirely on economic and technological conditions at the time the studies were made. They tell the degree of certainty of the existence of the commodity as well as the economic practicability of its recovery. If someone were to find a cheap and easy method of extracting aluminum from clay, billions of billions of tons of aluminum which were submarginal would become recoverable. Thus, progress in science (such as an improved *understanding* of how ores form and the geological factors governing their deposition), technology (such as development of drilling methods capable of going deeper than at present) and exploration (such as a method of sensing the presence of oil thousands of feet underground without drilling), as well as economic considerations, would change reserve information.

Let us consider an example of how this principle works. Suppose the price of a metal is $2.00 per pound. A mine producing it is able to drill, blast, move to the surface and extract the ore mineral from 20 tons of rock at a cost of $1.95 per pound. If ore containing less metal per ton is utilized, the cost would exceed $2.00 per pound because more rock would have to be handled, and this costs more. There is much low-grade ore down below which doesn't have enough metal in it to pay for the cost of mining and processing. If the price of the metal rises to $3.00 or more, the low-grade ore will become profitable to mine.

In general, then, the total amount of metal available increases as the economically minable ore grade decreases.

☐ OIL SHALES AND TAR SANDS

If oil shales in our own western states and tar sands nearby in

---

[7]See Sam H. Schurr, *et al.*, *Energy in America's Future: The Choices Before Us*, Resources for the Future, The Johns Hopkins University Press, Baltimore, 1979, p. 229, for other comments on estimating reserves.

Canada could be made to produce all the oil theoretically possible from their reserves, we would have double, triple or even more reserves than we now have. Colorado oil shales alone have some 600 billion barrels of oil, according to U.S. Geological Survey estimates. The tar sands, which are located adjacent to Lake Athabasca in Canada, are said to have a potential of some 300 billion barrels of oil.

Oil shale development poses serious environmental problems. Oil shale may be surface mined, shaft mined or processed underground. Surface mining would despoil the landscape. Shaft mining would leave large piles of spent shale on the surface. Processing underground might disturb and contaminate the ground water supply. Another obstacle to development is the large water requirement for mining. A 1973 study by the Committee on Science and Astronautics, House of Representatives, estimated that a 3- to 5-million-barrel-per-day industry using current technology might require all the surface water in the oil shale regions. Nevertheless, in early 1974, large tracts of public lands were leased from the Department of the Interior at a cost of almost half a billion dollars. At this writing no one can reliably forecast what the outcome will be. In subsequent development, many problems of a financial and technical nature have arisen.

The Canadian tar sands have also experienced a flurry of renewed development. Although known about for many years, the tars are so locked in the sands that extraction had seemed unlikely. Then, in the early 1960's a pilot operation was begun at a cost of around $200 million. Development costs proved to be more than anticipated, and while some oil was produced, the operation ran up losses in the millions before finally making a profit in 1974. Since then, renewed financing of $1 billion, including Canadian government funds, has been undertaken. Whether or not the venture will become a major producer remains to be seen.

□ WHAT IS THE FUTURE?

Briefly, the known reserves of oil and gas in this country are definitely limited. The development of offshore oil wells has become increasingly costly, and the pollution threat may limit their potential. On balance, in the short run, we should be conserving our use of oil. Otherwise, we will become hopelessly dependent upon foreign

84 □ OUR NATURAL RESOURCES

supplies, for in the next 10 to 15 years, Americans will continue to depend heavily on oil and natural gas for their fuel needs. In the long run, we will need to turn to other sources of energy to meet our needs.

□ □ □

QUESTIONS AND PROBLEMS

1. How does the development of oil and gas illustrate the functional concept of resources (that is, that resources are a function of our wants and abilities)?
2. Discuss the various problems of prospecting for minerals.
3. How many millions of years have coal and oil been in the process of forming? What does this suggest about our ability to replace them?
4. What is meant by *proved* petroleum reserves? What is the difference between the U.S. Geological Survey definition and the oil industry definition?
5. In general, how do reserves of oil and gas compare to those of coal?
6. Do you think we can rely on *undiscovered recoverable* reserves for our supplies of oil and gas? Explain your answer.
7. Can we expect to discover new oil fields as rapidly as we deplete the older ones, or should we expect the demand for both oil and gas to outrun the supply? Explain your answer.
8. List some of the products that are now being made from oil.
9. If oil and/or gas is found in your state, find out when either or both were first developed and what has happened since then.
10. How large are the proved reserves of these two sources of power in our nation compared with annual use?
11. How long will these reserves last if they continue to be used at the present rate?
12. What should be our long-run national policy regarding dependence upon oil and gas? Explain why we should or should not base our strategy on the amount of *undiscovered recoverable* resources?

7

Nuclear energy

THE MOST STRIKING DEVELOPMENT in sources of power in recent years is the release of the energy in the atom. We are apt to think, therefore, that the knowledge of atoms is very recent. The fact is, the ancient Greeks included atomic theory as one of their philosophical studies. Systematic knowledge about atoms was developed in the 17th century. In the early 1900's, theory became clarified. Then, in the 1940's, the atom was split, and the resulting energy was put to work. The first large-scale use of fission power was in the atomic bombs used by the United States in 1945 to knock Japan out of World War II. Since then, increasing attention has been given to the peaceful uses of nuclear energy. In 1980, the amount of electrical power generated by nuclear power plants was about 11 percent of the total electrical supplies in the United States, down from 13 percent in 1979.

□ ENERGY FROM NUCLEAR FISSION

Atomic energy in the United States today is produced by the process known as *nuclear fission*. In this process, the nuclei of heavy atoms are split under bombardment by neutrons. When a sufficient amount of fissionable material is brought together, a chain reaction occurs, splitting the atoms and releasing tremendous heat. About 20,000 times as much heat and other energy is released from uranium fuels, for example, as from an equivalent amount of coal.

The fission reactors are attractive to electric power companies. The most commonly used are light water reactors (LWR's), so called since they use ordinary water. Such reactors have been in commercial use since 1957 and have been found to be workable. They produce electricity at competitive costs. And, most important, LWR's are free from the environmental problems of coal-fired electric power plants, which the utilities had begun to despair of overcoming. The principal disadvantages are the potential of a reactor explosion and/or contamination from the atomic wastes left by the plants, which have led to widespread public objection to them. Also, soaring construction costs and technical problems have slowed their growth.

☐ THE QUESTION OF SAFETY

In 1974, the Ford Foundation brought out a preliminary report of a three-year study undertaken by responsible authorities in a number of fields. Regarding nuclear power, the report noted that "its use poses serious environmental issues, including reactor safety, radioactive waste management and nuclear theft. . . . The wisdom of a commitment to nuclear power ultimately rests on the capability of our technology and institutions to manage, perhaps indefinitely, a very hazardous enterprise."[1] The report also noted that reasonable men differ substantially in their judgement about the safety of nuclear power.

Also in 1974, 47 American scientists and engineers completed a two-year study for the Atomic Energy Commission in which they assessed the possibilities of accident risks in atomic reactors. The report concluded that "the odds against an American dying from a nuclear-power accident are 300 million to one." Subsequently, a list of 34 eminent scientists signed and released for publication a policy statement calling for nuclear development because ". . . the benefits of a clean, inexpensive, inexhaustible domestic fuel far outweigh the risks."

Since these early reports, an alarming nuclear accident at Three Mile Island, a large power installation near Harrisburg, Pennsylvania, on March 28, 1979, shook public confidence and appears to have tipped the balance in favor of the opponents of nuclear power in the United States. A chain of events, involving both mechanical and human failure, led to the release of a considerable amount of

---

[1] *Exploring Energy Choices*, The Ford Foundation, Washington, D.C., 1974, pp. 49, 51.

radioactivity and the evacuation of preschool children and pregnant women within 5 miles of the plant.[2]

Nevertheless, the 67 nuclear facilities producing electricity in this nation are expected to continue operating, and 90 more under construction are to be ready for production by 1990.

Three Mile Island has brought about a new effort on the part of both business and government to develop and carry out better safety. A special commission, chaired by John Kemeny, was set up to analyze the mishap and to make recommendations. The commission found that both reactor improvements and better operation training were needed. Utility executives formed two new safety institutions: one, the Nuclear Safety Analysis Center, which monitors performance of safety, and the other, the Institute of Nuclear Power, which inspects existing reactors and evaluates them.

Nuclear power may be down, but it is not out. Some European countries are expanding their facilities. *Time* magazine reported in its March 24, 1980, issue: "After conducting a four-year $4.1 million review of U.S. energy requirements, the National Academy of Sciences and the Department of Energy predict that in the next two decades the U.S. will still need to rely on nuclear power."[3]

Some risks for a nuclear-powered society remain to be solved. No one knows how safe reactors would be in case of deliberate acts of sabotage, or how vulnerable a United States largely dependent upon atomic power would be to potential destruction by a hostile power. Furthermore, safe atomic waste disposal is still a technological problem. Only in 1980 did the federal government set forth a coordinated program for waste disposal. It remains to be seen how quickly and effectively it will go forward.

☐ BREEDER REACTORS

Fast breeder reactors, which are more efficient to operate than LWR's, are undergoing experiment. Heat output of these nuclear reactors per unit compared with that of coal would be as high as 1,500,000 to 1. There are two types of fast breeder reactors: the

---

[2] For a full account of the event, see the Council on Environmental Quality, *Environmental Quality, The Tenth Annual Report of the Council on Environmental Quality*, U.S. Government Printing Office, Washington, D.C., December, 1979, pp. 361–363.

[3] "We're Fighting for Our Lives," *Time*, March 24, 1980, p. 62.

liquid metal-cooled fast breeder reactor (LMFBR) and the gas-cooled fast breeder reactor (GCFBR). Called "breeders" because they produce more fuel than they consume, these reactors are much more efficient than LWR's and can use thorium more readily than LWR's and U^{238} without enrichment. They can use 50 to 70 percent of the uranium mined, in contrast to only 2 percent by the LWR's. Thus, they could greatly expand the life of ore reserves if they were to replace the latter.

The Atomic Energy Commission anticipates the possibility of commercial-sized plants by the mid-1980's. But, the safety problems of LMFBR's may prove to be more severe than those of the LWR's.[4]

ADEQUATE FUEL SUPPLIES IN THE UNITED STATES

The major mineral sources for nuclear power are uranium^{238} and thorium^{232}, from which fissionable fuels U^{235} and plutonium^{239} can be produced (see Figure 7-1). In our present processes, U^{238} is preferred over thorium^{232} as a mineral source, and is used almost exclusively for the production of fissionable fuels.

In the early stages of development, the United States relied heavily on imports of uranium from South Africa and Canada. Then, during the 1950's and 1960's, western U.S. deposits were developed to the extent that the U.S. Bureau of Mines was able to report: ". . . domestic production is adequate to supply present and near-future domestic requirements."

The United States in 1981 was still the world's leading miner of uranium. Reserves have been predicted adequate because it has been assumed that if shortages were to develop, lower-grade ores would be available to meet needs. Professor Eric S. Cheney, however, casts doubt on this assumption. He warns that "If we are to avoid the consequences of being dependent upon foreign ores, some very large domestic reserves must be discovered almost immediately."[5] He suggests, fortunately, that adequate reserves may be discovered (in this country) in heretofore untested rocks of the pre-Paleozoic

---

[4]Allen L. Hammond, William D. Metz and Thomas H. Maugh II, *Energy and the Future*, American Association for the Advancement of Science, Washington, D.C., 1973, p. 36.

[5]Eric S. Cheney, "The Hunt for Giant Uranium Deposits," *American Scientist*, January–February, 1981, p. 37.

FIGURE 7-1. This is the site of the first mill in the United States built for the refining of uranium ore. The community of Uravan is on the Dolores River in the mountain region of Colorado and is 90 miles from a railroad. Uranium for the first atomic bomb came from this mill. *(Courtesy, International Harvester Co.)*

Age. Present sources in the United States are Mesozoic sandstones or Cenozoic volcanic rocks. Should breeder reactors, which require small amounts of fuel, become the principal source of nuclear power, almost unlimited supplies of high-cost uranium would become economical.

☐ NUCLEAR FUSION

Dwarfing *nuclear fission* in possibilities of almost unlimited energy is *nuclear fusion*. Fusion is a *combining together*. The atoms are fused together rather than split apart. The problem is that the process is so difficult to control that it is questionable whether commercial adaptation will ever become economically feasible. Fusion requires extreme pressure and temperatures as high as 100 million degrees! Such heat was achieved in the hydrogen bomb by first setting off a fission explosion. While control of the process is a problem, fuel is not. Fusion reactors would be fueled by deuterium, an isotope of hydrogen, available in almost unlimited supply in sea water.

The U.S. government has been pumping about half a billion dollars a year into the development of several different fusion reactor programs and hopes to have a "demonstration" reactor in operation by the year 2015.

□ INDUSTRIAL SIGNIFICANCE OF NUCLEAR ENERGY

The move toward more peaceful use of nuclear energy has been striking. Progress has been made in the development of underseas and space nuclear power, medical use of radioisotopes and research in nuclear physics. But, the industrial development of nuclear energy, while showing great promise, has become ensnarled by many problems—technological, economic and social.

In the 1960's, the Atomic Energy Commission forecasted that by the year 2000 all new power plants under construction would be atomic powered and that nuclear energy would account for one-half of all electricity generated. But, this forecast does not seem likely to be realized. It takes 10 to 12 years to get a nuclear plant in operation from the time it is first planned. And, no new plant orders were signed in early 1980, due to the Three Mile Island accident and its aftermath. Although the U.S. government is spending vast sums to develop breeder reactors, fossil-fueled plants will continue to supply the bulk of our electrical power for some time to come.

□ □ □

QUESTIONS AND PROBLEMS

1. How long ago did man speculate about atomic particles? When did commercial development of nuclear energy begin?
2. Look up the description of the atom in a reference book in your library. Explain what is meant by *nuclear fission*.
3. What are the advantages and disadvantages of LWR's? How many are planned for your state? Where will they be located?
4. What are *fast breeder reactors*? What are their advantages and disadvantages compared with LWR's?

5. Are nuclear energy fuel supplies a problem? Why or why not?
6. What is the difference between nuclear fission and nuclear fusion? Why don't we convert all atomic plants to nuclear fusion now?
7. What are some of the nonmilitary uses of atomic energy? (Consult the library to add to your list.)
8. Will atomic energy plants replace those fueled by coal, oil or gas? Explain your answer.
9. Write an essay comparing the environmental hazards of nuclear plants to the pollution problems of coal-fired steam plants.
10. Find out from your local power company how your electrical power is produced. How does this compare with how it is produced in your state? With the nation as a whole?

8

Other sources of energy and our national energy dilemma

SO FAR, WE HAVE CONSIDERED wind and water power, the fossil fuels—coal, oil and gas—and nuclear energy. While our full potential from these resources is far from realized, they are nevertheless well known. They are conventional sources of power. Four unconventional sources of power are now considered: solar energy, geothermal energy, hydrogen and energy from solid wastes. Also, we shall take an overall look at our national energy situation.

□ SOLAR ENERGY

The sun, as well as the tides of the oceans, remains as a source of largely unutilized energy. Yet, the energy from either or both of these sources of power may sometime in the future be more fully harnessed as a servant of man. If ways are found to convert this energy to usable power, there would be enough power from solar radiation alone to meet all our needs.

Let us consider the power that could be developed in one of our areas that has little rainfall and much sunshine. Suppose that only $1/10$ of the area of the state of New Mexico were covered by heat collectors and that only 15 percent of the energy thus produced were converted to mechanical work. Allowing for cloudy weather and for nighttime, this state could supply from solar radiation alone over 10 trillion horsepower per year of mechanical power. Imagine

94 ☐ OUR NATURAL RESOURCES

one source of power from one state producing all the energy used in the United States for heat, light and power combined! It can easily be seen that if and when it becomes feasible to harness the energy of the sun, we will not want for power.

The large expense associated with a venture of this type is the installation of heat collectors. After this is done, operating expenses will be light. Solar energy has the handicap of any intermittent source of power because no power can be generated at night or on cloudy days. Power which is used as it is transformed from the sun's heat will involve small expense, but if it becomes necessary to store the power in order to permit continuous use, the expense of storage may be very great.

One of the reasons we have neglected research on solar power is because we have thought of it in too grandiose a fashion. Relatively simple solar heat exchanger devices are being used to heat water in thousands of Japanese homes today. In India, reflectors are used by thousands of families for cooking. We could all employ solar heat to

FIGURE 8-1. This four-bedroom house illustrates that practical solar heating is here. The steep roof, which faces south, is made of corrugated aluminum covered by glass. Water passing over the metal becomes heated by the sun and is stored to heat the house. About 75 percent of the heat is supplied by this solar heat unit, with a small oil furnace supplementing it when needed. The system, which has been tested for 15 years, is now being marketed as the *Solaris System*. Operating costs, figured over the life of the house, are said to be cheaper than conventional heating. The house also has a "natural" cooling system. *(Courtesy, Dr. Harry E. Thomason, Thomason Solar Homes, Inc., 6802 Walker Road, Washington, D.C. 20027)*

more advantage in many cases, simply by the judicious planting of trees, by the use of awnings, and by the use of properly oriented glass areas and reflectors. Think of the savings if 10 million homeowners were to do this!

A number of U.S. companies are now beginning to manufacture solar heating components for homes. While the costs of heating a home with such units are still much higher than for conventional units, the costs should come down rapidly if builders and home buyers create a large market for solar heating. In a major study by the TRW Systems group, it was concluded that solar heating will become a billion-dollar-a-year industry. Another study by General

FIGURE 8-2. Solar energy can serve commercial buildings too. Scientists are monitoring this model solar energy building at PPG Industries, Harmanville, Pennsylvania, research center near Pittsburgh as a prelude to construction of "solar skyscrapers." The 1½-story structure, which has 27 solar collectors in its roof and walls, is a project of Aluminum Company of America, Oliver Tyrone Corp., Phelps Dodge Brass Co., Standard Oil of Ohio, Sun Oil Company, Inc., and PPG Industries to demonstrate that solar energy systems can help furnish heating, cooling and hot water for office and commercial buildings. The project's next phase will probably be construction of an actual sun-fueled office building. *(Courtesy, PPG Industries, One Gateway Center, Pittsburgh, Pennsylvania 15222)*

Electric estimated that 40 million buildings will be heated by solar energy by the year 2000.

Martin Wolf, research associate professor of electrical and mechanical engineering and an associate of the National Center for Energy Management and Power at the University of Pennsylvania, reported on solar energy utilization by physical methods in the April 19, 1974, volume of *Science* (AAAS). His projections of energy use to the year 2020 indicate that by then solar energy will rank about equal with nuclear and will exceed natural gas and coal.

Because of the renewed interest and excitement over solar energy, it seems possible development is accelerating. Figures 8-1 and 8-2 illustrate that solar-heated buildings are no wild dream, but rather a present reality.

In space heating by solar energy, we must avoid creating an overwhelming demand for supplemental supplies during very cold or prolonged cloudy periods, especially if they are supplied by electricity. In the latter case, a sudden demand by many consumers could strain available generating capacity so that many people would be literally "left out in the cold."

□ GEOTHERMAL ENERGY

Geothermal energy is power from heat from rocks or molten magmas in the interior of the earth. Characteristically, it is transferred to the surface by means of heated water.

The amount of geothermal energy near the earth's surface is almost inconceivable. It has been estimated that the heat in the top 10 miles of the earth's crust is equivalent to 2,000 times the amount of heat potential from the earth's total reserves of coal. But, much of this geothermal energy is impossible to reach either physically or economically. In the United States, likely sources are believed to occur in the 11 westernmost states. Some geothermal resources might be accessible beneath other states in the East as well.

The most promising U.S. area of proved geothermal resources today is the Salton Trough of the Imperial Valley in California. Unfortunately, however, the water from it issues as a corrosive brine. Another problem in some areas is keeping the liquid hot till it reaches the surface.

In some areas, energy can be developed by sending fresh water to the hot rocks below the earth's surface, whence it returns as steam or very hot—but still nonmineralized—water. Such a dry steam field is in successful operation north of San Francisco, running a power plant with a capacity of about 400 megawatts.

One report of the Department of the Interior forecasted that geothermal power has the potential to provide 20 percent of the nation's electrical power by 1985. While it seems unlikely that this potential will be realized, there is little question but that we could be using more of this power.

□ HYDROGEN

With thousands of compounds composed of hydrogen and carbon (hydrocarbons) and some three-fourths of the earth's surface covered by water (H_2O), it would appear that pure hydrogen should be a major energy resource today. A great advantage of using hydrogen as a fuel is that the product of its combustion is water, a non-polluting material.

However, the problem is that hydrogen, being not freely available in nature like natural gas, for example, must be separated from its compounds and that this separation, under present processes, uses more energy than the hydrogen will subsequently produce. Thus, there is a net loss from conversion to hydrogen.

If, however, unused or excess power from such inexhaustible resources as the sun, wind and tides can be used to convert hydrogen and the hydrogen stored, then the conversion loss is not significant and the usefulness of these sources would be greatly extended. Hydrogen could also store up power from electric power plants during their otherwise idle periods.

A major problem of hydrogen conversion is that of technology. Most hydrogen being used today is produced from the hydrocarbon of natural gas, itself in short supply. Other methods proposed require great heat and pressure. Electrolytic conversion requires some 200 times normal atmospheric pressure and heat up to 350°F. to 400°F. Direct thermal decomposition of water would require 4500°F.

Not only is hydrogen difficult to produce in large volume, but also it is difficult to handle and store. It will "eat through" metals and is highly explosive under certain conditions, though it can be controlled.

In balance, much additional research is needed before hydrogen potentials can be realized.

☐ SOLID WASTE ENERGY

Getting energy from solid wastes is one of the most desirable ways of expanding our energy base. On the one hand, we are developing a "new" resource. On the other hand, we are helping solve a serious problem—solid waste disposal! An EPA spokesman has estimated that there is enough energy in solid wastes from homes, businesses and industries in the United States "to light every home and office building in the country." It is estimated that the energy value per pound of mixed solid waste is 4,500 B.T.U.'s. Refuse derived fuel (waste which has undergone further treatment) has a value of 7,500 B.T.U.'s per pound. This compares to 10,500 B.T.U.'s per pound for bituminous coal, on the average.

Several areas have already begun to use wastes for energy. In Connecticut, for example, there is a recycling plant operating experimently at Bridgeport. Two others, one at Hartford and one at New Haven, were still in the planning stage in 1980, and the project may not be completed within the decade.

In Nashville, Tennessee, the National Thermal Plant, a garbage-to-steam plant, has been in operation since 1974, generating heat and air-conditioning for downtown buildings. According to the plant's general manager, the solid waste system is reasonably competitive with conventional fuels. Another steam plant using wastes may be built in the 1980's by the Nashville Electric Service.

There are other promising potentials for energy from solid wastes. In Guymon, Oklahoma, wastes from 100,000 cattle from feedlots have been developed into an energy source. Some 40 percent of Hawaii's electricity comes from burning sugar cane wastes to drive steam generators. And, ambitious plans are being developed by the federal government to increase gasohol (a mixture of alcohol and gasoline) production and to obtain methane gas from biologic sources.

Other plants have been established, but the general use of solid waste appears to be developing rather slowly. A waste-burning program by Union Electric Company in St. Louis, for example, was discontinued because of citizen opposition to the proposed location of the waste treatment plant.

☐ THE NATIONAL ENERGY DILEMMA

The resource and use situation for energy in the United States is exceedingly complex. But, our dilemma is clear. We have been using energy at a faster rate than we can supply it from our own resources; meanwhile, foreign sources of supply have been becoming more and more unreliable. We must increase our national supply of energy resources and the efficiency of our use. Otherwise, our nation shall move from one energy crisis to another.

To help us visualize the overall energy situation, we refer to Figures 8-3, 8-4, 8-5 and 8-6 from the *Natural Energy Plan II, Appendix B: U.S. Energy Projections* (U.S. Department of Energy, 1979, pp. 22, 23, 26 and 47).[1]

The estimate of U.S. primary and end-use consumption to the year 2000 is shown in Figure 8-3, based upon assumption of medium oil prices. While end-use consumption (the total figure minus conversion losses) will rise over 30 percent from 1980 to the year 2000, this will be at a much lower rate of increase than during the previous two decades when it increased about 50 percent. The reason that the rate of use has been recently dropping (note the dip in the line on the graph, Figure 8-3, prior to 1980) and is expected to continue at a lower rate is because of more efficiencies of use, such as in achieving greater miles per gallon in automobiles and less waste of heat through better design and insulation.

That we must continue to press this conservation effort is illustrated by Figure 8-4, showing projections for the U.S. oil supply. Imports probably will still supply about 45 percent of our needs.

---

[1]The U.S. Department of Energy cautions that the user of these charts and supporting data "should not be beguiled by the complexity and detail of computer models, nor assume that they are 'black boxes' from which the truth will somehow emerge, or which will conceal bias." The computers perform the calculations, but it is the insights into the behavior of energy markets and the forces that will shape them in the future which derive the results. The Department points out, "These insights and perceptions must rely heavily on judgement, since basic truths and detailed data are lacking. The world's and Nation's energy future is far from deterministic."

In short, the Department is saying that prediction is an art as well as a science, fraught with uncertainties. The forecast attempts "to deal with uncertainties by identifying the most important ones and then using a range of estimates to bracket the range of responsible opinions."

For explanations of the various assumptions and complexities, the reader is referred to the original report, U.S. Department of Energy, *National Energy Plan II*, U.S. Government Printing Office, Washington, D.C., May, 1979, and its three appendices. The projections used here are based upon assumption that a "medium" world oil price will prevail—rather than the highest or the lowest expectations.

100 □ OUR NATURAL RESOURCES

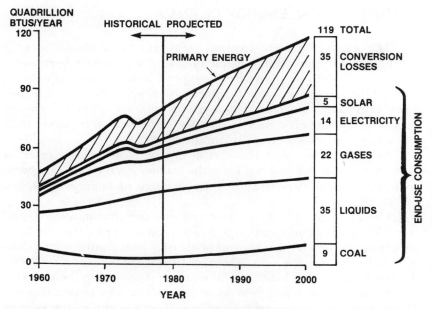

FIGURE 8-3. U.S. Primary and end-use consumption. *(Courtesy, U.S. Department of Energy,* **Natural Energy Plan II, Appendix B: U.S. Energy Projections,** *1979)*

Conventional oil and natural gas liquids (NGL) will supply only about 27 percent. Relatively new sources, Alaska (10 percent), shale oil (5 percent), secondary recovery—identified as enhanced oil recovery (EOR) (5 percent), and synfuels—synthetic fuels from coal, identified as coal liquids (about 3 percent). The remaining 5 percent is accounted for by refinery loss, stock changes and "unaccounted for" crude.

The place that nuclear energy and renewables used for electricity (principally hydroelectricity, biomass, wind power and geothermal energy) is expected to play can be determined by studying Figures 8-3 and 8-5. Figure 8-3 shows sources for total end-use consumption, and Figure 8-5 shows a breakdown of that part generated by electricity.

Referring again to Figure 8-3, we see that electricity use is expected to be about 16 percent of total natural end-use consumption by the year 2000. Figure 8-5 shows that these renewables would be less than 2 of these percentage points and nuclear power about 5. The balance, 9 percent, will be accounted for by coal, oil and gas.

OTHER SOURCES OF ENERGY AND OUR ENERGY DILEMMA □ 101

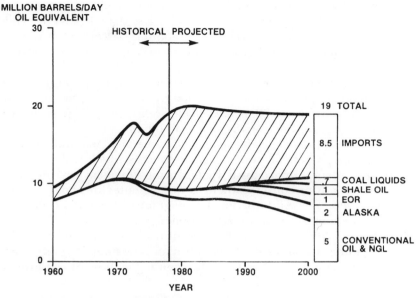

FIGURE 8-4. U.S. oil supply. *(Courtesy, U.S. Department of Energy,* **Natural Energy Plan II, Appendix B: U.S. Energy Projections,** *1979)*

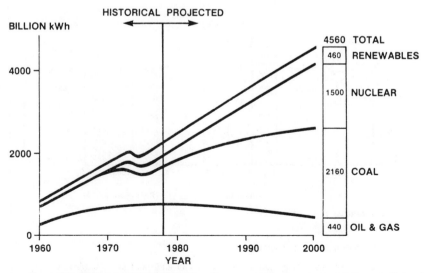

Figure 8-5. U.S. electricity generation. *(Courtesy, U.S. Department of Energy,* **Natural Energy Plan II, Appendix B: U.S. Energy Projections,** *1979)*

102 □ OUR NATURAL RESOURCES

Direct use of energy from the renewables production (too small to be shown in the graph) will account for less than 1 percent, while solar power is expected to account for about 6 percent, according to these Department of Energy estimates assuming a medium world oil price.

Adding direct and indirect (through electrical power) use, total end-use consumption by the year 2000 of renewable energy would then amount to about 23 percent. The remaining 77 percent will come from our conventional fuels—coal, oil and gas.

To sum up, there are not expected to be any "magical" technological developments or new natural resources which will relieve us from dependence upon fossil fuels during the next two decades. Whether or not to conserve them is not a matter of choice. It is a matter of necessity. And, the need is heightened because of our continued dependence upon foreign supplies.

This view of expectations for the future would not be complete without some accounting of the ultimate consumers. Figure 8-6 shows that the industrial sector will account for most of the growth

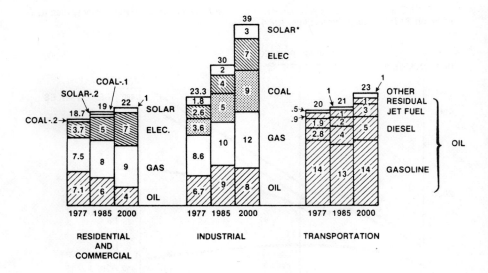

*INCLUDES 1.8 QUADS OF BIOMASS

FIGURE 8-6. End-use energy consumption by sector and fuel type, quadrillion B.T.U.'s per year (medium world oil price case). *(Courtesy, U.S. Department of Energy,* **Natural Energy Plan II, Appendix B: U.S. Energy Projections,** *1979)*

in end-use energy consumption by the year 2000. By comparison, residential and commercial use and transportation use are expected to increase only marginally.

Note that electricity and direct coal use, which will increase by 75 to 80 percent, will supply most of the anticipated new requirements of industry. More natural gas, electricity and solar power will meet the slight increase in residential and commercial demand, offset by a decline in use of fuel oil. Strikingly, transportation will continue to be almost entirely dependent upon petroleum-based fuels.

The Joint Committee on Atomic Energy's statement of the U.S. energy dilemma is as significant now in its implications for the 1980's as when it was issued in 1973.

> The United States with about 6% of the world's population is now consuming over 35% of the planet's total energy and mineral production. The average American uses as much energy in just a few days as half of the world's people on an individual basis consume in one year. This Nation has literally been developed without any significant restrictions due to lack of energy or mineral resources. However, we now see ever-increasing indications of the fact that the United States cannot long maintain the growth rate of recent years in our energy consumption without major changes in our energy supply patterns.

High use of energy is identified with a high standard of living. However, wasteful use is both unethical and costly. Through conservation we would use less energy and still be better off. In the highly acclaimed Harvard Business School Report, *Energy Future*, Robert Stobaugh and Daniel Yergin observe:

> The United States might use 30 or 40 percent less energy than it does, with virtually no penalty for the way Americans live—save that billions of dollars will be spared, save that the environment will be less strained, the air less polluted, the dollar under less pressure, save that the growing and alarming dependence on OPEC oil will be reduced, and Western society will be less likely to suffer internal and international tension. These are benefits Americans should be only too happy to accept.[2]

---

[2]Robert Stobaugh and Daniel Yergin, eds., *Energy Future, Report of the Energy Project at the Harvard Business School* (New York: Ballantine Books, 1979), p. 229.

QUESTIONS AND PROBLEMS

1. It seems as though solar energy should be very cheap. What factors make it expensive to develop on a large scale?
2. What are some relatively simple applications of solar energy?
3. How could you use solar energy more efficiently in your home? At school?
4. What is geothermal energy? Where is it most apt to be applied in the United States?
5. Since hydrogen is so common as a part of water (H_2O), why isn't it being used by all of us for fuel today?
6. What sources of power might be available that would make it economical to produce hydrogen?
7. How is solid waste being disposed of in your community? How could it be used to help meet our energy needs?
8. What is our national energy dilemma? Try to state it in terms of supply and demand for resources.
9. What kind of developments could help resolve our energy dilemma?
10. Suggest five ways by which you personally could help reduce the intensity of the national energy problem.

9 □ □ □

Nonfuel minerals[1]

□ NONFUEL MINERALS—THE MUSCLES OF THE IRON AGE

OUR ENERGY SUPPLIES would do us little good if we did not have the other geological resources of rocks and minerals to house and contain them and to convert their gaseous and liquid energy into mechanical energy. The reference to our historic period as the "Iron Age" speaks of the basic importance of metals to our industrial world. They might be called "the muscles of the Iron Age." Without iron, copper, lead, zinc and various combining metals such as manganese, chromium and tungsten, we would not be able to use effectively the mineral fuels. We would not be able "to build the engines to drive the machinery to make the parts to build the factories to make the goods that 'Jack' wants"—to twist an old nursery rhyme a little. In other words, it takes metal to harness our energy resources, to drive machine tools, to create other tools to build the complicated machines to produce the airplanes and automobiles, calculators and computers and telephones and televisions of our modern world.

□ NONFUEL MINERALS—A SOURCE OF CHEMICALS

Besides metals, nonfuel minerals are also an important source of

---

[1]The 1977 data for nonfuel minerals used rather extensively in this chapter are from the standard official reference source, *Minerals Yearbook*, U.S. Department of the Interior, Bureau of Mines. The 1977 yearbook was published and distributed in 1980. Other reports, as noted, have been used to supplement and update this material because of current developments. The major one is *Minerals and Materials, A Monthly Survey*, by the same source.

the chemicals which keep us in business. Of course, all chemicals are not listed here. You see, over 700,000 have been identified!

Chemicals can be classed as *organic* or *inorganic*. The organic are derived from plants and animals whose availability depends upon environmental factors described in other chapters. The inorganic chemicals, which are obtained from nonliving sources, can be grouped according to source as follows: those obtained from (1) air and water, (2) the mineral fuels and (3) nonfuel minerals.

One of the most widely used nonmetallic elements is sulfur. When we strike a match, write on paper, paint or use a plastic container, we may be using sulfur. Sulfur is also used, of course, to produce sulfuric acid which is required in so many industrial processes that the volume produced is used as an indicator of the level of activity of all business.

☐ NONFUEL MINERALS—A SOURCE OF BUILDING MATERIALS AND OTHER PRODUCTS

Another, and no less important, group of nonfuel minerals is the nonmetallic mineral building materials such as stone, clay, sand and gravel. These are the materials for surfacing our highways with concrete, constructing buildings, making pipes for sewage and drainage and building dams. Such minerals are also the raw materials for the china we eat off of and the glasses we drink from, among other uses.

Mineral building materials are at last a mineral resource for which we can say, "Don't worry, we have plenty." That is to say, for the country as a whole, our rocks, such as limestone, sandstone, and the gravels and sands, are, from a practical standpoint, unlimited. But, for various localities, this is not necessarily so. Even sand and gravel may have to be transported many miles to them. And, because of quality considerations, as well as availability, we may get our marble from Vermont and our granite from Georgia. Even suitable clays for making bricks are hard to get in some places.

☐ NONFUEL MINERALS NOT AS CRITICAL A RESOURCE AS FUELS

We noted in Chapter 6 that *proved* reserves of oil and gas were those which have been verified by testing, are recoverable by existing methods and are economically feasible to develop. The same considerations hold for nonfuel minerals, but there are some important differences in supply and demand characteristics between fuels and nonfuels. It is generally easier to locate and determine the extent of deposits of nonfuel minerals; nonfuel minerals do not deteriorate once extraction has begun; they generally last much longer in use; in contrast to fuels, they are often readily reused; and they are easier to store. Both, however, face competition from substitutes.

To evaluate in detail the effects of all the above factors is far beyond the scope of this book. But it all boils down to this: Nonfuel minerals are not as critical a resource as fuels. There is not the frantic concern to develop nonfuel minerals that there is for fuels. Mining companies are more apt to "prove out" new resources only as needed. For world reserves, the estimated "life" (obtained by dividing the volume of use during a year into the volume of proved reserves) should generally be regarded only as a signal which must be carefully interpreted. It is useful, for example, in indicating how much more abundant the developed resources of one mineral are in comparison to those of another. It does give a good idea of how one nation stands relative to another in developed reserves.

The character of the supply and demand for nonfuel minerals means that the index of self-sufficiency for our country (net imports as a percent of consumption) shows how heavily we are relying upon supplies from abroad. But, this does not necessarily mean that we could not develop supplies from our own resources if we wanted to. When the statistics are checked against the geological facts, we *are* definitely lacking in certain minerals.

With these reservations and considerations in mind, let us now take a look at the status of a number of our most important minerals.

☐ PRINCIPAL METALS

The relative amounts of five principal metals produced in this country, as well as their share of our total use, are given in Table 9-1. The table makes obvious the overall dominance of iron in our present-day machine age. It does not show the importance of other metals, such as those used in making steel stronger, harder, more elastic, rustproof or capable of a high permanent polish.

TABLE 9-1. Production and self-sufficiency of five important minerals, 1949 and 1979[1]

	1949		1979	
	Production	Self-sufficiency[1]	Production	Self-sufficiency[1]
	Million Tons	Percent	Million Tons	Percent
Iron ore	94.2	95	87.0	72
Bauxite	1.3	43	2.2	07
Copper	1.1	78	1.4	87
Lead	0.8	81	0.8	92
Zinc	0.6	76	0.4	38

[1]Self-sufficiency is the ratio of production (from both mines and scrap) to total use. Total use includes exports. Column 3 gives production figures for 1979, except for bauxite, which is for 1977.

(Sources: U.S. Bureau of Mines, *Minerals Yearbook*, Vol. I and II for 1949, and Vol. I for 1977, U.S. Government Printing Office, Washington, D.C., 1952 and 1980; and *Minerals and Materials, A Monthly Survey*, U.S. Government Printing Office, Washington, D.C., August, 1980)

Table 9-1 also shows that our nation is dependent upon imports to meet our mineral needs despite larger production in some cases. Columns 2 and 4, showing self-sufficiency, indicate the percentage of our needs we have met with our own production (this includes both mine output and recovery from scrap metal). The deficit was made up by imports. For these important minerals, except copper and lead, we met less of our total needs with our own production in 1979 than we did in 1949. The general trend shown is more significant than the specific figures. Annual changes in production reflect market and labor conditions as well as the supply potential. Furthermore, the total needs include metals imported but used in exports, which account for a considerable share of the use of minerals such as copper.

Iron

Iron, a metal which is one of the most abundant and most generally used all over the world, was probably first used by the Egyptians about 5,500 years ago. It was not known in Europe until 3,500 years later, or about 1000 B.C. This iron probably was obtained from meteorites, and only very small quantities were used.

The first evidence of obtaining iron from smelting, that is melting the iron out from the ore with heat, dates back to about 700 B.C. in Ethiopia. Its early use there was limited to items such as hinges, bolts, keys, chains and nails. The first ore was refined in the United States on the banks of the Saugus River near Lynn, Massachusetts, in 1646 (see Figure 9-1). The ore was obtained from a nearby bog, and charcoal was used as fuel to reduce the ore to metal.

Later processes used coke instead of charcoal as a source of heat (see Figure 9-2). After the ore has been brought to melting temperature, it is necessary for the mixture to remain open so that the melted iron will trickle through the mass and collect at the bottom of

FIGURE 9-1. The first iron ore in this country was produced in commercial quantities near Lynn, Massachusetts, about 1646. The ore was dug from a swamp along the Saugus River, and the charcoal was made from neighboring forests. The giant waterwheel shown above supplied the power to run the huge leather-and-wood bellows for the forced air draft. This Saugus smelter has been restored to commemorate the beginning of the U.S. iron and steel industry. *(Courtesy, U.S. Steel Company)*

FIGURE 9-2. "Cornwall furnace" in Pennsylvania is the oldest blast furnace in the United States. During the Revolutionary War, cannons for the Continental Army were cast here. It was in operation for one hundred years and is now preserved by the Pennsylvania State Historical Society. *(Courtesy, Bethlehem Steel Co.)*

the furnace. Many fuels will supply the necessary heat to melt the iron, but coke is still the only fuel found that will form an open structure in the lower part of the mixture so that the metal melted from the ore can move down through the coke and accumulate on the hearth below.

Making coke from coal is an involved process, and when carried on so as to save the various by-products, it results in the making of other valuable products in addition to coke. One ton of coal will make:

    1,425.0 pounds of coke
       19.0 pounds of sulfate of ammonia
        2.0 gallons of crude, light oil
        7.0 gallons of tar
       10.5 cubic feet of gas

It was not until after 1857, after the invention of the Bessemer process, that steel, a refined commercial form of iron, was produced cheaply enough to make its widespread use practical. The Bessemer process of making steel from cast iron consists of burning out the carbon and other impurities by use of a stream of air which is forced through the molten metal. All the carbon ordinarily is removed, and a definite amount is again added, along with other ingredients, to make the desired kind and quality of steel.

How Steel Is Produced

There are several general steps which are necessary to produce steel. When a body of ore is discovered and "blocked out" so that the approximate quantity of ore is known, development steps are taken. These consist of sinking shafts, driving tunnels or stripping overburden.

The next step is the actual mining of the ore. This work is done either by hand or by machinery and involves loosening the ore from the general rock formation and moving it to some mill, where the iron ore is separated from much of the rock with which it is intermixed.

This separation process is known as "concentrating," "milling," "ore dressing" or "benefication." The resulting product is called a "concentrate," because all that actually is done here is to concentrate the ore by removing a part of the waste. Concentrating the ore usually is done near the mine, reducing the weight and bulk which must be hauled to smelters.

Smelters and other plants which are used to remove various impurities and unwanted elements with which the metal or metals are chemically combined from the concentrate are usually located near the source of heat rather than near the mine. One reason for this arrangement is obvious when we realize that it requires 2 tons of coal to make 1 ton of steel.

The iron which has been smelted, or refined, is cast into bars known as "pigs," "ingots" or "bars." These are then ready to be made into various manufactured articles by remelting and adding the correct kinds and amounts of other metals before rolling, drawing or otherwise processing into forms such as castings, forgings, sheets, rods, tubes, wire, etc.

Increased Use of Steel

The growth of the steel industry in the United States was very slight before 1900, when approximately 15 million tons of pig iron were made. From that date on, there was a steady growth. By 1980, annual iron production had increased to an annual rate of over 80 million tons, and the steel-producing capacity of the nation had increased to around 150 million tons. Actual annual production of steel, however, averaged 136 million tons from 1970 to 1979. It would be unusual to produce at full capacity. The per capita consumption of steel increased from 46 pounds in 1880 to over 1,200 pounds in 1979. This is more than enough to supply each of our families with a car each year.

Do we have enough iron to keep up this high rate of use? Practically 85 percent of the iron ore used in the United States before 1950 came from the Lake Superior area (see Figure 9-3). Mining operations were concentrated there because of the immense supply, its availability and the high quality of the ore. This tremendous store of high-grade ore has nearly run out, but the area has been successful in developing plants to use lower-grade ores.

Lower-Grade Ores Abundant

Because the reserves of the lower-grade ore of iron are practically inexhaustible, production in the United States has turned largely to the lower-grade magnetic taconite rock. According to geologists, the supply of this ore is so great that all iron mining for generations could come from this source without exhausting the supply. The ore mineral is *magnetite*, an iron oxide formed from pre-existing iron minerals in the iron formation when an intrusion of hot melted rock, the Duluth Gabbro, entered the area near the west end of Lake Superior. This magnetic taconite is so tough that the development of a new drilling technique has been necessary in order to mine it successfully. This technique involves using a blast of fuel oil and oxygen which is burned to heat the rock and then quenched with a spray of water which shatters it as it is quickly cooled. Because the magnetic grains are so small, they must undergo four steps in processing before they can be used in the furnace. First, the rock is ground to fine powder. Second, the magnetite is extracted by passing the powder over large electromagnets. Third, the magnetic powder is formed into pellets about ½ inch in diameter. Fourth, the pellets are baked hard in a furnace. Then, the taconite can be handled for shipping and blast furnace charging.

The expense of concentrating the iron from lower-grade ore is,

FIGURE 9-3. Before 1950, the Mesabi range was the largest iron ore operation in the world. This and other mines in Minnesota have shipped almost 2½ billion tons of iron ore to the steel mills of the nation. The Lake Superior district is still producing a substantial proportion of the nation's yearly supply of iron ore. This area has mined its lower-grade taconite ore to maintain its position in the increasing market for iron.

of course, passed along to the consumer in higher prices of refrigerators, automobiles, tools, nails and any other products containing iron.

FINDING NEW ORE DEPOSITS

The alternative to using lower-grade ores is the development of new fields of high-grade ores or the importation of more ore, specifically from Canada or South America, or perhaps from Liberia which has exported some iron ore to us.

North America's Major Iron Ore Reserves.—Lake Superior and the Canadian Labrador-Quebec iron ores are the most important in North America. The rich direct shipping ores of the Mesabi fed the mills of the United States for decades, but today they are mostly mined out. However, the Labrador Trough has the same high-grade ore as that formerly in abundance in the Mesabi range. This area and other Canadian mines provided approximately one-half of the

U.S. iron imports from 1975 to 1980. Some geologists believe that Labrador may contain as much as a billion tons of high-grade ore, from which Canada could meet the iron needs of the United States for years to come.

Some sizeable iron ore reserves are found in New York, Alabama, Michigan, Missouri, Utah and Texas, while smaller reserves are found in 10 or 12 other states.

South America's Major Iron Ore Reserves.—A second source of high-grade ores for our smelters is South America. Suppliers include Brazil, Peru and Venezuela. U.S. capital has helped develop a large field of high-grade ore in Venezuela and has helped other South American countries to explore and develop their ore resources.

Although reserves of iron ore of lower grade are ample to meet all U.S. demands for steel for an indefinite length of time and although high-grade Canadian and South American ores will be available for the United States to use for generations to come, the importation of these ores is based upon a continuation of friendly relations and mutual good will between these countries and the United States. These supplies, along with the immense reserves of coal within the United States, assure us of the base materials—coal and iron ore—which are necessary to continue to produce all the machines and mechanical developments of our present age.

Occurrence in Nature

The Lake Superior and Canadian Labrador-Quebec iron ores occur in sedimentary rock approximately 2,000 million years old, formed during the Huronian Age (late Pre-Cambrian). These rocks contain about 30 percent iron, too little to be at present utilized as ore. However, they have been locally enriched to about 55 percent iron, and these zones have been extensively mined.

The source of iron in the Huronian iron formations was the rock of a series of large mountain ranges which eroded down to a vast lowland. These rocks were granites and metamorphosed sediments in the form of gneisses, schists and slates. Most of these rocks contained a small percent of iron in the form of iron silicate minerals such as biotite, amphibole and pyroxene. At that time, only primitive plants, such as algae, mosses and lichens, existed on land, but their activity, plus chemical reactions of the iron silicates with oxygen, carbon dioxide and water in the atmosphere, produced solutions and colloidal particles containing iron, silica, calcium and other chemicals found in the rocks.

How did the iron become concentrated? Most of the silicon and iron were *colloids* (particles only a few microns in diameter), which carry positive electrical charges. These colloids did not stick together because of their light weight, like electrical charges and a protective coating of organic material. Thus, they were carried to a marine basin where the salt water neutralized the electrical charges and allowed the colloids to form into masses and to settle to the bottom. This produced a sediment (iron hydroxide, iron silicate, silica, calcium and magnesium carbonates) which became a cherty iron formation known as taconite, which contains about 30 percent iron.

After many more thousands of years, the concentrations of iron ore varied greatly. The iron formations occurred as layers up to a few hundred feet thick, interlayered with sand and mud. These layers were folded and faulted to various degrees, depending on their location relative to the source of deforming forces. The westernmost area was less deformed than those to the east in Michigan, Quebec and Labrador; thus, it became North America's major iron range, known as the Mesabi. Called a range because the iron formation, which is more resistant to erosion than associated rocks, outcrops as a low ridge, the Mesabi is about 100 miles long and 1 to 3 miles wide.

The Mesabi had some of the richest ores in the world. It became enriched when the ore was exposed at the surface and subject for over millions of years to movement of oxygenated water from the land surface downward through it. (A few geologists believe the oxidizing agent was hot water derived from large igneous intrusions moving upward to the surface.) This process accomplished oxidation of the iron hydroxides and iron silicates and removal by solution of the silica and carbonate. In many instances, this left millions of tons of nearly pure iron oxide at or near the surface, an ore requiring no processing whatsoever. It could be dug from open pits or underground, loaded onto railroad cars and shipped directly to blast furnaces for reduction to pig iron.

Aluminum

It has already been mentioned that aluminum is one of the commonest metals. It is second only to iron in its abundance over the

116 □ OUR NATURAL RESOURCES

earth and in its usefulness. Aluminum comprises one-seventh of the earth's crust and occurs in many forms and combinations. Some of the better known forms are bauxite, corundum, turquoise, kaolin, feldspar, mica, ruby and sapphire. It also occurs in all our clay soils.

The main source of aluminum is bauxite ore, although many other ores carry a considerable quantity. In most minerals in which aluminum is found, it occurs with silicon, and both are very tightly bound to oxygen. It is possible to extract aluminum from these minerals, but not yet practical from an economical standpoint, since very great amounts of energy must be used to free the aluminum from the strong chemical bonds with which it is held. Bauxite, though a scarcer mineral than many aluminum silicates, is used as the main aluminum ore because the aluminum can be extracted from hydrous aluminum oxides of bauxite at much less cost. Bauxite is a product of chemical reaction between aluminum silicate minerals and oxygen, water and carbon dioxide in the atmosphere. This reaction will take place on the earth's surface only in tropical and subtropical climates where there is an abundant supply of water during at least one season of the year. Thus, world bauxite deposits are confined to regions which, during some period of time, have been in these climatic zones.

OCCURRENCE IN NATURE

Aluminum ore occurs in pockets as well as in layers, or sheets. If the deposits of ore are large enough, it pays to mine them. Because bauxite is formed as a product of weathering, its deposits are confined to the thin zone of contact between rock and the atmosphere. They are generally confined to a sheet-like layer at the earth's surface which may be a few to a hundred or so feet thick. In some cases, oxygenated and carbonated water may penetrate fractures in rocks containing aluminum silicates. This produces pod- or wedge-shaped bauxite masses just under the earth's surface. Because of the mode of formation, residual bauxite is always strip mined from shallow pits dug into the surface deposits. Underground mines may be required if, as in Dalmatia, Yugoslavia, the residual deposits were buried, then tilted steeply during the Alpine mountain-building deformation. Occasionally, the bauxite from the weathered residue is eroded, transported and deposited elsewhere as an ordinary layer of sediment. It may be buried later by other sediment layers. If bauxite content is

FIGURE 9-4. This is one of several open pit mines of the Aluminum Company of America at Bauxite, Arkansas, where about 90 percent of our domestic bauxite is produced. The 30-cubic-yard dragline, mounted on crawlers, removes the overburden. The ore is then drilled, blasted and loaded onto 54-ton trucks for transport to the processing plant. *(Photo by Dorothy Gore)*

high enough, it may be mined at the surface or underground, but such instances are not very common.

The principal bauxite-producing area in the United States is located in Arkansas, where there is a town named Bauxite because of these deposits. Approximately 85 percent of the ore mined in the United States is from this location. Bauxite was first discovered in this country in Rome, Georgia, in 1883, while the Arkansas deposits were not discovered until about 1900. Smaller deposits are found in Alabama, Tennessee and Mississippi.

All the U.S. deposits are in the South. This is no accident of nature, because all bauxite deposits were formed and laid down

118 □ OUR NATURAL RESOURCES

under warm climatic conditions. Also required are nearly continuous rainfall and relatively good drainage over long periods of time. Some deposits are found in the cooler parts of the world because of changes in climate which occurred after the aluminum ore was formed. Bauxite is quarried or mined in about the same way as shallow iron or copper ore.

Aluminum is extracted from bauxite by dissolving it in a hot melted salt and passing an electrical current through it. The positively charged aluminum ions in the solution are attracted to the

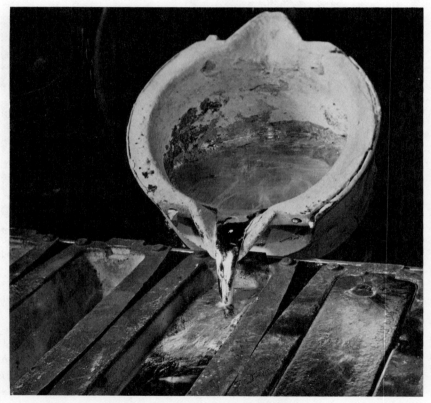

FIGURE 9-5. After the ore has been processed in electrolytic cells, the melted aluminum is drawn off into large ladles and poured into molds. When cool, the metal is ready to be shipped to various factories for production of useful articles. In 1979, about 25 percent of the aluminum produced was the result of the recycling program initiated by the industry. This conserved existing aluminum ores and saved 95 percent of the electricity that otherwise would have been needed to extract aluminum from bauxite. *(Courtesy, Aluminum Company of America)*

negatively charged electrode. The temperature is sufficient enough to melt the aluminum. It collects at the bottom of the electrolytic cell and is tapped off into huge ladles and poured into molds from which it is rolled into sheets, drawn into wire or converted into one of many chemical compounds of aluminum, which have proved useful in industrial or consumer goods (see Figure 9-5).

USES OF ALUMINUM

The uses of aluminum are many because of its lightness, freedom from rust and the ease with which it can be combined with other metals. We are familiar with its use in kitchen utensils, automobile parts, airplane construction and even building materials. A building made entirely of aluminum stands on the campus of Purdue University in West Lafayette, Indiana. It was constructed so that the advantages and disadvantages of a structure made of aluminum could be studied.

About 25 percent of all the aluminum used in the United States in 1977 was for building construction. Siding, roofing, flashing, windows, screening, doors, curtain walls, insulation and builders' hardware are all made of aluminum. Another 20 percent was used in motor vehicles—automobiles, buses, trucks, motor homes, etc. Containers and packaging and electrical manufacturing used about 21 percent. Another 10 percent was used in making durable consumer goods such as cooking and serving utensils and various small household appliances.

Alumina (Al_2O_3) is used in making grinding and polishing compounds, bricks for lining high temperature furnaces, bearings for revolving shafts and a cement which is resistant to chemicals as well as to heat.

SUPPLY AND DISTRIBUTION OF BAUXITE

The known world reserves of bauxite are estimated at 5.7 billion short tons, and the total resource potential is estimated at some 3.8 trillion tons.[2] The reserves are distributed widely, with 27 countries

---

[2]Council on Environmental Quality and the U.S. Department of State, *The Global 2000 Report to the President: Entering the Twenty-First Century*, Vol. I, U.S. Government Printing Office, Washington, D.C., 1980, p. 31. (Figures converted from metric tons to short tons by authors.)

producing bauxite in 1977. Australia has the largest proven supply and is also the largest producer by far, accounting for about one-third of the world total. Other major producers in recent years have been, in order of importance, Jamaica, Guinea, Surinam, the Soviet Union, Hungary, Greece, Guyana and Yugoslavia.

We import more than four-fifths of our yearly supply of bauxite and will probably continue to do so. Our reserves are only 40 million tons, and we are already using about one-third of this amount annually. To conserve reserves, we have been limiting domestic production to less than 2 million tons and are importing the balance of our needs.

Copper

Copper is a beautiful metal which was probably the second metal used by man, gold being the first. Archeological studies lead us to believe that its discovery was about 6000 B.C. Ancient Egypt used a great deal of the metal, much of which came from the island of Cyprus. This island gave copper its name—*cyprium* or *cuprum*. Copper mining in Cyprus began about 2500 B.C.

Copper was produced in the American colonies in 1709 from ore deposits in Connecticut. Deposits were later found in Vermont and New Mexico. However, it was not until the discovery of the ore in northern Michigan in the early 1840's that the U.S. production exceeded a few hundred tons a year. Still later, extensive ore deposits were found in Arizona, Utah and Montana. It was these later discoveries that gave the great impetus to copper production.

Most Versatile Metal

Copper is among the most versatile of metals. It is distinguished from all others by its peculiar red color. It has many of the characteristics of gold in that it can be drawn into a fine wire (ductile), is easily hammered or pressed into sheets (malleable) and yet is very tenacious. When 11 parts of copper are mixed with 2 parts of zinc, the combination may be hammered into foil comparable in appearance to goldleaf. In this form it is called "Dutch metal."

Copper becomes very soft and malleable when heated to red heat and immediately immersed in water. This is opposite to the behavior of steel under similar treatment. On the other hand, if it is heated to a higher temperature than red heat, it becomes brittle, another trait which shows its versatility.

Copper is the only metal which occurs in relatively pure or native form abundantly and in large masses. It is also found chemically combined with carbonate and sulfur in many ores and gives to these ores their beautiful shades of green and blue.

Mining Development of Copper and Other Metals

Minerals are sometimes widely dispersed in rocks, or they may be concentrated. Some understanding of their occurrence helps us appreciate why they may be costly and difficult to mine.

Copper, lead, zinc, manganese, chromium, tungsten, columbium, cobalt and many other metals, including gold and silver, occur in similar types of deposits. In fact, most rocks contain slight traces of these metals in them. When a rock mass is pushed deep within the earth's crust (at least several thousands of feet up to perhaps 30,000 or 40,000 feet) and sufficient heat is encountered to melt or recrystallize the rock, the metals are freed to migrate. None of these metals (except manganese) are of the same size as iron, aluminum, calcium, sodium and potassium atoms, which are much more abundant and go into making up the rock minerals. Thus, the metals first noted do not fit into the crystal structure of any common mineral. Furthermore, there is not a sufficient amount of them to form minerals of their own. So, they collect in residual hot water which is freed from the melted or recrystallized rock mass. This hydrothermal (hot water) solution moves laterally or upward from the new rock mass, in whatever direction the resistance is least. It moves through small interconnected pores which occur in some rocks, through fractures or faults, or along contacts between different rock masses.

The ore minerals are precipitated when the hydrothermal solutions either react chemically with rocks through which they seep (this is especially true of limestone) or cool sufficiently to saturate the solution with some particular ore mineral. In either case, the ore minerals may fill pre-existing openings in the rocks (filling) or take the place of some rock mineral which is dissolved by the hydrothermal solution (replacement).

Deposits may occur in many different shapes and grades of ore, each requiring a different mining method. Massive or scattered replacement deposits may be irregular or blanket shaped and range in size from a few hundred to many thousands of feet in dimension.

Deposits made in fractures and faults are commonly narrow veins in near vertical position. These are the lode or vein deposits frequently mentioned in popular fiction. One mining area may contain several sets of intersecting veins which are sometimes broken and offset by rock movements which occurred after the ore was deposited.

The development of any mineral prospect is an expensive and a financially risky business. Sufficient test hole drilling must be done to outline the volume and shape of the ore body. Extensive physical and chemical tests must be made in order to plan the mining, processing and waste disposal methods. The mine shafts, hoists and underground equipment must be installed. From initial ore discovery to first ore shipment frequently takes as long as 10 years and an investment of hundreds of millions of dollars. Mines of any kind—metal or coal, underground or surface, new or reactivated—cannot be quickly developed. Long-range planning, stable markets and consistent government policies (regulations, taxes, tariffs, etc.) are all required for profitable mining ventures to exist.

Many Uses for Copper

The uses to which copper is put are as varied as its colors in its many different mineral ore compounds. Ornaments, tools and weapons of copper and of bronze, an alloy of copper and tin, have been found in the ruins of prehistoric dwellings. The metal probably was used by man before iron. Early North American Indians used it as evidence of wealth—Samuel de Champlain in 1610 was given a sheet of copper a foot long by the Indians who reportedly had found the metal on the bank of a river near one of the Great Lakes. The Indians gathered the copper in lumps, and after melting it, they spread it into sheets and smoothed it with stones.

A mixture of copper and zinc, called brass, is one of the most commonly used copper alloys at the present time.

Copper coins were common throughout the world for many years. Kitchen utensils have long been made of copper, and roofing, screens and other building materials are made of this metal. Copper piping is used where some non-corrosive metal is necessary.

The ability of copper to conduct electricity makes it an essential metal in most electrical installations. Fully one-half of all copper used is in connection with electricity and electrical appliances. Electric power production, transmission and use would be most seriously handicapped, if not impossible, without copper. There would be no

FIGURE 9-6. Copper smelter of Anaconda, Montana, where ore is concentrated and refined. The water used by plants at Anaconda is brought in 80 miles from northwest of the area, while the ore is moved by rail from Butte, which is 50 miles to the northwest. *(Courtesy, Anaconda Copper Co.)*

extensive wire or radio communication, and the life of power machinery and equipment would be greatly reduced if these were made without copper.

Nothing has yet been found to take the place of copper in making shell cases, as well as other necessary military weapons. In medicine it is used as an astringent (a shrinking agent) and occasionally as an emetic to cause vomiting. It is also used as an antiseptic. Care must be exercised in the use of copper because it is poisonous in all soluble salt forms.

OCCURRENCE IN NATURE

Practically all the important deposits of copper were concentrated by hot water from the earth. Much of this was closely related to volcanic activity. Copper was deposited in veins or crevices of rock, in pockets or occasionally in sheets. It combines quite readily with other chemical elements, which results in a variety of copper-bearing ores. Native (pure) copper is found in the Lake Superior region where there are pockets of various sizes, while veins of the ore are found in the Butte, Montana, area (see Figure 9-6). The deposits of Arizona, Nevada and Utah contain copper minerals scattered through the mass of large igneous rock intrusions. Some pyritic, (copper-iron) deposits are located in Tennessee as well as in northern California.

Reserves Not Large

World copper reserves are estimated to be about 500 million short tons, with the resource potential about four times this amount. Known reserves are located in these principal areas: the Andes in South America, central Africa, the Ural and Kazakstan regions of the Soviet Union, central Canada and the western United States. A number of new mine projects have been opened in Papua (New Guinea), western Canada, Indonesia, South Africa and Zaire during the past decade.

Until 1940, the United States produced all the copper it used and had some for export. This situation rapidly changed during World War II when we imported approximately two-thirds as much as we produced. Since then, with the return of more nearly normal living conditions, our imports have been from one-fifth to one-third of our production, and as much as 45 percent of our needs have been recovered from recycling.

Annual mine production was 1.5 million short tons in 1979, with domestic consumption of 3.2 million tons. Even though copper is one of the metals with a high reusable rate, and our production and reuse will undoubtedly increase in the years ahead, we will probably continue sizeable imports to meet our future needs, reflecting the advantages of complementary trade.

Lead

Lead is one of the few metals mentioned in the earliest known writings of civilized man. King Midas, the miser of history, is credited with its discovery one thousand years before Christ was born. There is evidence, however, to indicate that lead was used as money by the Chinese a thousand years before that, and some authorities claim that the Egyptians used lead as long ago as 7000 B.C.

Even in those early times, lead was used for many purposes. In addition to employing it for money, the Chinese are credited with using it to mix with and debase (cheapen) the more valuable forms of metallic money. It was used by the Egyptians for glazing pottery and as one of the ingredients of a solder. Ancient Babylonians held iron clamps in sockets with lead, and the Hanging Gardens, one of the Seven Wonders of the Ancient World, was floored with sheet lead. Other people used it as a roofing material for cathedrals and as a covering of basement pillars to prevent moisture damage. White lead was used as an ointment by the Egyptians, while red lead was the source of a cosmetic.

DISTRIBUTION IN THE UNITED STATES

Lead was mined and used in Virginia in 1621, and its discovery in the Upper Mississippi Valley was reported as early as 1690. The French attempted to establish lead-mining operations in the valley region, but it was not until the latter half of the century that production was on a permanent basis. This area was the important source of lead for this country until the latter part of the 19th century.

Lead ores were worked by the French in what is now the Southeast Missouri Lead District during the early 1780's, but production was negligible. In 1867, discoveries at new depths led to the development of one of the most productive lead-mining areas in the world.

At the present time, approximately 98 percent of our lead comes from four states. Missouri continues to be the largest producer, accounting for 84 percent of the total mine output in 1977. Idaho, Colorado and Utah produce most of the remainder.

STILL MANY USES OF LEAD

Around 60 percent of the lead that is used yearly in this country is reused. Present uses include paint pigments, chemicals, storage batteries, cable covering, type metal and building construction. Approximately 10 percent of the lead is used primarily because of its weight; 30 percent because of its softness, malleability and corrosion resistance; 25 percent because of its alloying properties (tendency to melt and blend with other metals); and 35 percent because of its chemical character in various compounds. Although it can be rolled to a foil so thin that it requires 2,000 sheets to make an inch in thickness, it does not have tensile strength, so it cannot be drawn into a wire. It is both the softest and the heaviest of our common metals.

SUPPLIES IN THIS COUNTRY LIMITED

The United States has depended upon some imports of lead to meet its yearly needs since 1931. Imports were relatively unimportant during the early part of the 1930's, when the imports of lead never exceeded 10,000 tons, or 2 percent of our total yearly supply. Beginning in 1940, however, this country sharply increased its supplies from abroad. During the 1970's, we were importing about 17 percent of our annual lead used, but imports dropped sharply in 1980 to about 10 percent of our needs.

Two important facts about lead are its indestructibility and its reusability. It will not corrode or rust away, and in most cases at the end of the useful life of the product containing it, it can be recovered and used again. This helps account for the fact that, although we use three or four times our yearly production, we do not depend upon foreign lead to the extent that our production figures suggest.

The use of lead in the United States varied from 1.2 to 1.5 million tons a year during the 1970's, while mine production was from one-third to nearly one-half this amount. Imports were about one-half as great as mine production. The difference was made up from our reused lead.

Thus, although we will continue to look to foreign countries for a considerable part of our lead, our supply does not present a critical problem in the near future because of our large mine resources and quantities of reusable lead. But, with reserves of only 166 short tons, as of 1976, life expectancy at the rate of use during that year would only be 37 years, according to *The Global 2000 Report to the President*.[3]

Sulfur

Sulfur is another mineral found in antiquity. It was used to bleach linen, and it was used in religious rites about 2000 B.C. Egyptians used it as a paint component four hundred years later. About 1000 B.C. it was used as a fumigant. Five hundred years later, it appeared as a component of the explosives of that time.

Commercial production of sulfur began in Italy early in the 15th century, but it was three hundred years later, with the development of a process for making sulfuric acid, that it acquired indus-

---

[3]*Ibid.*, p. 29.

trial importance. Sicily was the major, and practically the only, source of sulfur until near the end of the 19th century.

WHERE SULFUR IS FOUND

Sulfur is found in nature as free sulfur and combined or mixed with many other elements. Until the beginning of the present century, practically all sulfur was obtained by separating it from gypsum and other minerals it was mixed with in sedimentary deposits such as those in Sicily.

Since early in this century, much of the sulfur has been obtained from the rock caps of salt domes. The largest known of these deposits are in the United States and Mexico. There are also notable deposits in Iraq, Poland and the Soviet Union.

These formations are dome-shaped bulges of salt, probably forced up into denser sedimentary rock while the salt was in plastic form. They vary in size and percent of recoverable sulfur. Only a few of the salt domes discovered in this country to date contain enough sulfur to warrant development. Of more than 230 of these deposits found in Texas and Louisiana, only 21 have been commercially productive. The process of mining, known as the Frasch process, uses superheated water (see Figure 9-7). All the native (not naturally combined chemically with any other element and not recovered secondarily, as from coal, oil or smelting of ore) sulfur produced in this country is from Frasch mines in Texas and Louisiana.

Frasch process sulfur was 52 percent of the domestic production of sulfur in all forms in 1979. Recovered sulfur, a by-product from natural gas and petroleum and coking operations, sulfide ore smelting and utility plants, accounted for 33 percent of the total domestic production in that year. The balance of our needs was obtained from imports, except a small amount (about 2 percent) produced from pyrite, hydrogen sulfide and sulfur dioxide.

USES OF SULFUR

Both free sulfur and sulfuric acid are required in various industries. Much more sulfuric acid than free sulfur is used, and approximately 80 percent of all sulfur is used to make acid. The U.S. Bureau of Mines lists nearly a hundred uses of sulfur in industry, but it notes that the list is far from complete. Figure 9-8 shows the wide range of application of the two forms of sulfur.

128 □ OUR NATURAL RESOURCES

FIGURE 9-7. This is part of a half-million-ton mountain of sulfur which was pumped from underground beds after being melted by superheated water at about 260°F. Sulfur is one of the most important of our mineral resources. It is used in making items such as matches, gunpowder, fireworks, medicines, fertilizers, paints, insecticides, paper, rayon, vulcanizing rubber and sulfuric acid. When the price of sulfur justifies recovery from smaller, deeper and marsh land deposits, known supplies in this country will probably meet our needs for the next generation. (*Courtesy, Texas Gulf Sulphur Company*)

WHAT OF THE FUTURE FOR SULFUR?

Sulfur and sulfur-bearing materials in the United States and the rest of the world are ample to supply "any demand," the U.S. Bureau of Mines assures us. Native deposits in the Gulf Coast area alone may exceed 200 million tons. Recoverable reserves from fossil fuels, sea water and other secondary sources are in the billions of

NONFUEL MINERALS □ 129

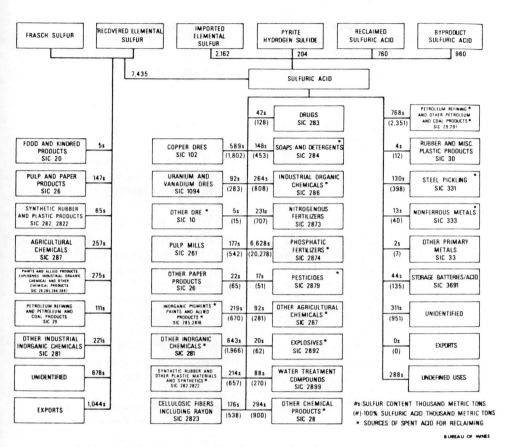

FIGURE 9-8. Sulfur–sulfuric acid supply and end-use relationship in 1977. Frasch and recovered sulfur from domestic sources provided most needs. About 2.5 million tons of the sulfur went directly into food, paper products, chemicals and other uses (column 1). Over 7 million tons were first shipped as sulfuric acid which was then consumed as shown in the other columns. A relatively small amount was exported. *(Courtesy, U.S. Bureau of Mines)*

tons. An orderly transition to these alternate sources of supply appears possible.

Zinc

This metal was little known until around 200 B.C. when both the Romans and the Greeks produced brass by melting together copper and zinc. The metal itself was apparently forgotten or unknown from that time until the 16th century when the Portuguese brought zinc back from China. The Chinese were evidently familiar with this metal before Greek and Roman times. Zinc was first smelted in England about 1740, and the first continental European smelter was erected in 1807.

Production in the United States

The commercial production of zinc in the United States began in 1858 when smelters were built in Pennsylvania and Illinois. The industry grew rapidly from the beginning, and by 1890, the yearly average production of zinc was 64,000 tons.

From 1900 to the time of the depression in the early 1930's, production had increased to where the yearly average was 725,000 tons. One-half of this production was from the tristate district of Missouri, Kansas and Oklahoma. At that time our production met all our needs and furnished some zinc for export.

During World War II, our production of zinc reached a peak of more than 942,000 tons. This was the largest production in our history. Since then, mine production has dropped to less than 500,000 tons annually with it being only 267,000 tons in 1979. Secondary production has exceeded mine production during recent years, including the first eight months of 1980. We also have been importing over one-half of our requirements.

The source of supply also has changed. Nineteen states are now in commercial production. The tristate district that formerly produced one-half of our zinc is now accounting for a negligible amount, although other deposits have been developed in Missouri. Tennessee now leads the nation in zinc output. Other leading states are Missouri, New York, Colorado, New Jersey, Idaho, Pennsylvania and New Jersey.

Uses of Zinc

Currently, most zinc is used in galvanizing steel products to prevent rusting. Rust damage to unprotected steel products amounts to billions of dollars each year. Yet, in 1977, over 436,000 tons of zinc, or about 40 percent of the total consumption of slab zinc, were used for galvanization.

The second most important use is in zinc alloys. In this category of uses, 37 percent of our total consumption of 1,102,000 tons, was used. Automobile parts such as radiator grills, fuel pumps, carburetors, hydraulic brakes, instrument panel parts, door and window accessories, heaters and air-conditioners are commonly zinc die castings, as are business machines, electrical appliances, building hardware and tools. The aircraft industry also uses die-cast alloys. The amount of zinc in brass ranges from 3 to 45 percent, with the average brass containing about 30 percent. Zinc is also used in wet batteries, weather stripping, radios, washers, oil burners, vacuum cleaners, slide projectors and kitchen appliances.

Domestic Reserves Not Large

Our present rate of consumption, including ores and secondary recovery, is between 1.5 and 2 million tons per annum. The present mine production of less than 500,000 tons a year is expected to continue. Even with secondary recovery, we will still have to look abroad for a major part of our zinc. The chief sources of imports are expected to continue to be Canada, Mexico and Peru and other South American countries. It is quite possible if prices increase enough over costs and new smelters become available that new deposits will be developed to add to our present known reserves. The Council on International Economic Policy has stated that "world reserves are ample to provide zinc in the quantity likely to be demanded at current price levels for the rest of the century."

□ SCARCE METALS

There are many other metals which, though used in relatively small quantities, are as essential as coal and iron for making the various types and kinds of metal alloys, as well as the many machines so necessary in this age of mechanical and industrial development. A modern automobile, for example, requires about 20 different metals for its construction and operation. An electric motor requires the use of many of these same materials, and the more nearly automatic these machines become, the greater the number of metals required in their construction.

We need metals such as nickel, tungsten, chromium, cobalt and manganese for making the various kinds of metal mixtures (alloys). Some alloys are harder than steel, others are more elastic or tougher

132 □ OUR NATURAL RESOURCES

while still others are brilliantly colored, rustproof or capable of a high permanent polish.

Table 9-2 lists 35 minerals and metals according to the percent we imported in 1979 in order to meet our needs. Table 9-2 also shows that we depended upon other countries for over one-half of our supplies of 19 of these metals and minerals and that we even imported almost all of some of them.

TABLE 9-2. Net import reliance of selected metals and minerals used in the United States, 1979 (net import reliance[1] as a percent of consumption[2])

More than 50 Percent Reliance[3]		Less than 50 Percent Reliance	
Metals and Minerals	Percent	Metals and Minerals	Percent
Columbium	100	Gold	56
Mica (sheet)	100	Titanium (ilmenite)	46
Strontium	100	Silver	45
Titanium (rutile)	100	Antimony	43
Manganese	98	Barium	40
Tantalum	96	Selenium	40
Bauxite and alumina	93	Gypsum	33
Chromium	90	Iron ore	28
Cobalt	90	Vanadium	25
Platinum group metals	89	Copper	13
Asbestos	85	Iron and steel products	11
Tin	81	Sulfur	11
Nickel	77	Cement	10
Cadmium	66	Salt	9
Potassium	66	Aluminum	8
Mercury	62	Lead	8
Zinc	62	Pumice and volcanic cinder	4
Tungsten	59		

[1]Net import reliance = imports − exports.
[2]Apparent consumption = U.S. primary and secondary production + net import reliance.
[3]Substantial quantities are imported for fluorspar, graphite, rhenium and zircon. Data withheld to avoid disclosing company proprietary data.

(Source: U.S. Bureau of Mines, *Minerals and Materials, A Monthly Survey*, U.S. Government Printing Office, Washington, D.C., August, 1980)

Since the United States is dependent upon imports for many minerals and metals that are necessary for making machinery, equipment and utensils, if we expand the use of these, we must of necessity depend upon imports for the increased supplies more than we do now. This means that we are limited in the use and enjoyment of many products by imports of strategic minerals and metals to the United States from foreign countries.

Manganese

More manganese is used in making various kinds of steel than any other metal except iron. Yet, in 1979, only about 2 percent of the amount used in this country was supplied from within our boundaries. Other uses of this metal are for electric dry batteries, driers for varnish, Japanese lacquer and printing ink, glass, porcelain enamel, some building brick, welding rods, pottery and tile. We require over 2 million tons of manganese ore yearly to meet our normal needs, of which only a small amount is supplied from sources within the United States.

Chromium

Chromium is familiar to all of us because of its use on the bumpers and trim of our automobiles. It is used in these places because it carries a high polish and will not rust. It is called the "work horse" of the alloy steel industry because it makes steel harder and tougher, does not rust and is the key ingredient in practically all stainless and heat-resisting steel. It is also important to the chemical industry. Unfortunately, domestic mine production of chromite essentially ceased in 1961, and we must now import almost our entire supply of this metal.

Tungsten

Tungsten is one of our most needed metals. As an alloying metal, it helps steel retain its hardness and toughness at high temperatures. It is especially useful in alloys employed in cutting tools which must keep sharp edges, even when red hot. An important military use is in making armor plate, as well as gun barrels and armor-piercing shells. Tungsten is used in about 1,500 different military items. During recent years, we have imported or drawn from stocks around two-thirds of all tungsten needed in this country.

However, domestic production during normal periods of the past 40 years has accounted for about two-thirds of our consumption. The supply for the immediate future appears adequate, but for a longer time it is questionable whether imports can be obtained at reasonable cost.

Columbium

Columbium is used to make the steel needed for the walls of jet engines which will withstand the terrific heat of the burning gases. Columbium, first discovered in 1801, was not commercially used in significant amounts until recently. It is also employed in making stainless steel, fast cutting tools, nuclear reactors, heavy construction equipment and other implements needing to withstand heavy strain or intense heat. Very little is found in this country, and no production has been reported since government contracts were terminated in 1959. Paying quantities have been found in Canada, Norway, Portugal, Malaysia, the Congo and Nigeria. We must look to foreign countries for most of our columbium.

Cobalt

Cobalt is another rare metal found mainly in other countries. It is necessary in making electric generators, as well as in making porcelain enamels, pigments and hard facing steel. Like most of the other scarce alloying metals, it has important uses in atomic energy projects. Canada has some deposits of cobalt, as do the Congo, New Caledonia, Morocco, northern Rhodesia and the United States. Use in the United States must be met by imports, except for secondary recovery.

□ PLANT LIFE NEEDS MINERALS

The puzzle of how plants grow

Early records of man show he was interested in and puzzled by the growth of plants. How did plants grow from small seeds and become large and woody? Where did the material come from out of which the woody, more permanent parts of plants were made? It was known that water was necessary to grow plants, but the question of where the solid or firm parts of plants came from remained unanswered.

It was thought that all plant growth came from water alone through some mysterious transformation by nature. One of the earliest recorded experiments attempting to solve this puzzle was carried on by John Woodward in London about 350 years ago. He grew spearmint in pots containing identical kinds and amounts of soil. In these experiments he supplied each plant for 77 days with all the water needed for satisfactory growth. The water for each pot was taken from a different source.

The difference in growth of the various pots of plants was so much as to be startling, as is shown by the following figures:

Source of Water	Gain in Weight After 77 Days
Rain	17.5 grains
River Thames	26.0 grains
Hyde Park conduit water	139.0 grains
Hyde Park conduit water plus 1.5 oz. of garden mold	284.0 grains

There was also a very slight decrease in the weight of the soil used in the different pots.

Real need of plants for minerals

Woodward's conclusion was that, since the soil lost only a very small proportion of its weight, the increase in weight of the spearmint was not caused by any ingredient in the soil. Neither was it caused by water alone, but by some "peculiar" terrestrial matter in the different waters. His explanation was that there was more of this terrestrial matter in some waters than in others and that the rain water contained the least. No thought was given to the very small loss in soil weight following the experiment.

Although the reason for plant growth as given by Woodward was incorrect, it did open the way for other experiments that finally resulted in showing the need of plants for various plant nutrients, such as phosphates, potash, nitrogen and numerous other elements, in exceedingly small amounts, as well as for water and sunshine.

We now know that the three plant foods needed by most plants for healthy growth and seed production are phosphates, potash and nitrogen, which frequently are not readily available in the soil in sufficient amounts. These three elements are needed in the largest amounts and are found in practically all fertilizers.

Phosphates

Phosphorus is never found free in nature. It is usually combined with other elements as phosphates.

Most deposits have been laid down by water and are of organic origin. Sea water contains traces of phosphates that are absorbed by the myriads of marine organisms as well as by the larger animals of the sea. When these creatures die, they accumulate on the floor of the sea. The phosphates are concentrated through the decay of the organic parts of these organisms and animals and through the dissolving of the more soluble parts.

The bones of land animals of past ages are also sources of phosphates. Beds of the remains of sea organisms and prehistoric land animals are found in various locations throughout the South, especially in Florida, Tennessee and the Carolinas. Arkansas, Idaho, Kentucky, Montana, Utah and Wyoming have workable deposits, with Idaho having the largest known reserves.

Potash

Potassium is found almost everywhere in nature. It is estimated that 2½ percent of the earth's crust is composed of potassium. It is found, for example, in rocks, soils and salt lakes as well as in the waters of oceans, fresh water lakes and streams.

Where Guano Is Found.—Guano (droppings of birds and bats) is a source of phosphates and nitrogen as well as potash. Deposits are found in regions of limited rainfall and in caves where the soluble elements are not leached out.

Peruvian natives were probably the first people to use guano as an aid to crop production. They had been using it for generations before Baron Humboldt took some of it to Europe in 1804. That was the Europeans' introduction to guano as a fertilizer.

Guano Used in the United States.—Peruvian guano was first brought to this country for use as a fertilizer in 1843 when a shipload arrived in Baltimore. It was widely used from then until the late 1850's and was the first commercial fertilizer to gain widespread use in the United States. After the 1850's its use declined rapidly because of the increasing availability of mixed chemical fertilizers. Guano from the Mammoth Cave in Kentucky was used during the Civil War in the manufacture of explosives.

Early sources of guano were the islands off the west coast of South America and off the African coast. These deposits are becoming depleted, but other nitrates of a mineral source are found in various regions throughout the world.

How Potash Deposits Were Formed.—Most U.S. deposits of potash, as well as those of Europe, originated in past geological times. Some of the deposits are the result of sea water having been cut off from the ocean and gradually drying up through evaporation. At times of heavy ocean storms, additional sea water was washed into these salt lakes so that over the centuries deposits of potash and other salts became deeper.

In other situations inland lakes with no outlets have become so filled with salts as to be workable for potash as well as for other salts. The Great Salt Lake in Utah and the Dead Sea between Israel and Jordan are the two bodies of water with the heaviest concentrations of salts. Each has about six times the concentration of sea water.

Ages ago when the Great Salt Lake was formed, it was larger than Lake Huron. Geologists call it Lake Bonneville. As it dried up over the centuries, several smaller lakes were formed in addition to the Great Salt Lake.

Potash from Wood Ashes.—One of the earliest sources of potash commonly used in the past was from wood ashes. It is from this source that potash obtained its name. It was originally obtained by running water slowly through wood ashes and boiling the resulting liquid down in open kettles. The solid white residue was called potash because it was made in pots from ashes. Potash was first produced in this country at Jamestown, Virginia. One of the reasons given for England's establishing colonies here was to obtain potash from wood ashes to meet her growing need for this product.

Where Potash Is Found.—The main source of potash for the United States as well as for the rest of the world from 1860 until the beginning of World War II was Germany. Our first serious effort to develop our potash resources was during World War I. As soon as that war was over, however, and foreign potash again became available, we turned to the use of these supplies and lagged in the development of our own deposits.

Our present supplies are obtained mostly from mines near Carlsbad, New Mexico and the Williston Basin of North Dakota and imported from Saskatchewan, Canada. The mineral occurs in beds like coal and is mined in a somewhat similar manner. Smaller quantities are obtained from California and Utah. Other possible sources are kelp (sea weeds), recovery from industrial waste, Georgia shales and subterranean deposits in Texas and a few other states.

NITROGEN

Nitrogen is one of the most nearly universal of our natural resources. Seventy-eight percent of the air is nitrogen. It is estimated that 400 million tons of nitrogen are removed from the air each year by plants, rainfall and electrical discharges and that a like amount is added to the air through decaying organic matter and other ways. This is an abbreviated statement of the process called the nitrogen cycle. This yearly transfer represents only one-millionth of the total nitrogen in the air.

Nitrate Deposits.—The world's most extensive deposits of natural nitrates are in northern Chile. These deposits supplied most of the world's needs for nitrogen until in the 1920's when German chemists developed a cheap and efficient method of manufacturing synthetic nitrate from the air.

The Chilean nitrates accumulated along the western coast of South America ages ago as a result of the decay of the plant life of that time. They escaped being washed away when the climate changed from a moist climate that encouraged the growth of luxuriant vegetation to one of almost perpetual drought.

Nitrogen from the Air.—Nitrogen compounds were synthesized from the air in Germany before World War I for use mainly in explosives. After the war, demand for commercial fertilizers showed the need for more phosphates and nitrates. This need was partially met when the Muscle Shoals Dam was constructed in the Tennessee Valley and a part of the power was used to produce fertilizer. The first fertilizer produced in the United States for agriculture was made from phosphate rock. Before World War II, however, the prospective need for nitrates both for the military and for agriculture hastened the conversion of phosphate plants to nitrate production.

Nitrogen from Natural Gas.—Today, most nitrogen in this country is produced from natural gas. Therefore, with the fuel crisis (commented on in Chapter 6), prices in recent years have risen sharply and there has been fear of scarcity. The production of ammonia used for fertilizers accounts, however, for only a small part of total natural gas consumption (about 2 percent in recent years). The price increases and shortages appear to reflect the general increases in oil and gas prices, that plants are operating at near capacity and that there is an increasing demand by farmers for fertilizers.

The production of no other element has increased as rapidly as has that of various forms of nitrogen during the last few years. We are now using about 15 million tons of nitrogen in various forms annually, and the nitrogen in fertilizers for agriculture accounts for nearly three-fourths of this amount. Other uses include explosives, resins, fibers, plastics and animal feeds. Two decades ago we were producing about one-tenth as much nitrogen, and only about one-half was used for agriculture.

The use of fertilizers on the farm is discussed in Chapter 18.

☐ NONFUEL MINERALS—ENOUGH FOR EVERYONE?

Our nation's current resource position on nonfuel minerals, in sharp contrast to that on fuels, is a strong one. We have abundant resources of many of the most basic and widely used minerals. And, we have convenient, friendly access to many of the others. But, we are dependent upon foreign sources to meet our needs for all or a very large share of a number of nonfuel minerals critical to our industrial production and national defense.

The Global 2000 Report to the President foresees "steady increases in demand and consumption" of nonfuel minerals on a worldwide basis. But, the study adds that the "projections point to no mineral exhaustion problems but . . . further discoveries and investments will be needed to maintain reserves and production of several mineral commodities at desirable levels. In most cases, however, the potential is still large . . . , especially for low grade ores."[4]

---

[4]*Ibid.*, p. 27.

140 □ OUR NATURAL RESOURCES

While we should continue developing our domestic reserves and strengthening our international friendships, we must not neglect to consider our role as consumers. If we practice wise use instead of misuse and abuse, our nonfuel minerals resource position will remain strong.

□ □ □

QUESTIONS AND PROBLEMS

1. Try to name 20 important products we get from nonfuel minerals.
2. Why are nonfuel mineral resources said to be less critical than fuel minerals?
3. Indicate the steps necessary to produce iron from the ore in the parent rock.
4. What is steel, and how much did its use expand from 1900 to 1960?
5. Is the United States in danger of not having enough iron to meet its needs in the future? Give reasons for your answer.
6. In what part of the country are aluminum ores found? How do you explain this distribution of these ores?
7. How is aluminum ore mined?
8. What are the characteristics of aluminum that make it so useful for various purposes? Name some of the common uses of aluminum.
9. Is aluminum likely to become a scarce metal during the next 50 years? Give reasons for your answer.
10. Why is copper called one of the most versatile of metals?
11. Why do you suppose copper was used by prehistoric man long before there was any evidence of his using iron or other metals?
12. Name some of the more common uses for copper and tell why you think the metal is especially adapted to various special uses.
13. Where are the more important concentrations of copper in this country, and what one of nature's activities is responsible for most of the deposits?
14. Discuss the possibility of this country's having all the copper it needs during the next generation.
15. What are some of the reasons why lead was so useful in early times?
16. Where does most of our lead come from at the present time?
17. Name some of the ways in which sulfur is used.
18. What is the outlook for the sulfur resources of the United States?
19. What are some of the special uses for zinc?

NONFUEL MINERALS □ 141

20. What are the functions of the scarce metals that are used in relatively small amounts and yet are so important to our mechanized development?
21. Name the important uses to which the following metals are put:
    a. Manganese
    b. Chromite
    c. Tungsten
    d. Columbium
    e. Cobalt
22. Name some metals that are found in your state and tell how they are used.
23. Describe John Woodward's efforts to find the reasons for plant growth.
24. What are the sources of phosphates used in fertilizers?
25. How did potash get its name?
26. What is guano and where is it found?
27. Tell how mineral deposits of potash were formed.
28. What country first synthesized nitrogen from the air?
29. What limits the production of nitrogen? What is the principal use?
30. In 1979 we depended upon foreign sources for over half of our supply of metals. Name five of the metals. Describe the use of one of them.
31. Compare the resource position of fuels to nonfuel minerals. Which is the most critical? Explain why.

10 □ □ □

Water supply for industrial and private uses

IN THE PAST we have thought of water for the home, the farm and the various industries as an inexhaustible resource. How can we be short of water when three-fourths of the earth's surface is covered with water; when rain or snow falls every few days and brings new supplies of clear, cool water from our oceans and lakes; and when below us, under the surface, are streams and lakes of water which are continuously being replenished from these rains and snows? Yet, we find cities that are critically short of water, not only for industries but also for street cleaning, yard watering and household purposes such as laundering, cooking and cleaning. New York City has rationed its water during several droughts. Miami, Florida, has had problems in keeping salt water out of its system, and some of the West Coast cities are having to use water supplied a hundred or more miles away in order to meet the increasing demand for this invaluable resource.

□ WHY ARE WE SHORT OF WATER?

The reasons for our water shortages can be grouped under three major headings: (1) unequal distribution, (2) abuse and (3) misuse. Intensifying these problems has been the strain put upon our supply by increased use.

The hydrologic cycle and unequal distribution

The unequal distribution of water occurs because of the vagaries of the hydrologic cycle, which we first noted in Chapter 1. This process by which water is evaporated from the oceans by the sun, moved overland by the winds and precipitated over the surface of the earth is a very imperfect one.

Distribution of water is uneven both from place to place and from time to time. The map of areas vulnerable to drought (see Figure 10-1[1]) shows that while the eastern half of our nation is humid, most of the West is dry. Even in the humid regions, however, droughts can occur. And, cloudbursts are not uncommon in the desert regions.

Water is also unequally distributed after initial precipitation. For the United States, exclusive of Alaska and Hawaii, it is estimated that of the water precipitated overland some 70 percent is either evaporated or transpired (called, *evapotranspired*) back into the atmosphere. Of the remaining 30 percent, which stays on the surface or sinks into the ground, about one-fourth is diverted for human use.

Abuse of the water supply

Adding to the problems of Nature's erratic disposal of water is man's pollution and often wasteful allocation of water. Water pollution problems have become very serious in the United States. Many of our major streams have become open sewers. The Cuyahoga River near Cleveland actually caught fire on the surface because of accumulated waste. The Connecticut River was called by one nauseated native "the world's most beautifully landscaped cesspool." George Washington's Potomac River, which featured a popular "bathing" beach at Washington, D.C., in the 1920's, has become so

---

[1]While the U.S. Geological Survey has not classified drought areas for Alaska and Hawaii on the map, the general situation is well known. In Alaska, annual rainfall is heavy (up to 200 inches) in the southern and western areas and much lighter (under 25 inches) in the northern areas. In Hawaii, the reverse generally holds true. The northeast coast and the mountainous areas usually have heavy annual amounts (up to 200 inches), and the southern and western areas, light amounts (less than 20 inches). The effect of these variations in rainfall on vegetation and human occupation is quite different because of great temperature differences, reflecting the northern, more continental location of Alaska and the southern, trade wind location of Hawaii. Moisture deficiencies occur in the drier areas of both states.

WATER SUPPLY FOR INDUSTRIAL AND PRIVATE USES   □   145

FIGURE 10-1. Drought potential. Drought occurs when precipitation is less than the long-term average and when this deficiency is great enough to hurt mankind. In humid regions, a drought of a few weeks is quickly reflected in soil moisture deficiencies and other water resources. In arid regions, the inhabitants protect themselves from short droughts by depending upon surpluses of ground or surface water, and a drought becomes critical when it is sufficiently prolonged to reduce these supplies. Prolonged droughts occur rarely in humid regions, but they reduce the normal ground or surface water supplies. In semi-arid regions, some people may be affected by every drought, whether of short or long duration. *(Courtesy, U.S. Geological Survey)*

polluted that swimming is prohibited. Instead, its foul odor offends sightseers who come to visit the seat of government of our nation. Lake Erie is so filled with industrial and sewage wastes that scientists despair that it can ever be returned to desirable purity.

Pollution of streams must be stopped at its source. Industry has been the chief offender over the years, accounting for twice as much organic waste in streams as the sewage of all cities combined. The organic chemicals that wash from farm lands are another serious source of pollution of streams.

We are all aware of some of the properties of water—its color, odor, transparency, taste, hardness, saltiness, foaming qualities and temperature. Farmers, engineers and public health officials are concerned with the dissolved oxygen content, acidity, dissolved salts, plant nutrients and toxic substances. They are especially concerned—as are not we all—with the amounts of suspended matter and with disease-producing bacteria, the worst offenders of all.

FIGURE 10-2. To help compensate for the uneven distribution of water in California, water from Mt. Shasta, and other sources in the northern part of the state, is carried by aqueducts to the southern part of the great valley—a distance of over 400 miles. *(Courtesy, Bureau of Reclamation)*

Misuse and increasing use of the water supply

Misuse of water can mean that we are wasting it and/or that we are not using it efficiently. Or, in the case of ground water, we may be drawing it down faster than it is being recharged.

Waste of water appears to be a way of life in the United States. We use twice as much water as actually needed to wash the car, take a shower, clean the porch, etc. Industry also has not paid sufficient attention to water economy. Savings could be made especially by recycling—treating the water that has been used in a process to remove impurities and using it over again. With better national and state clean water laws and better enforcement, we can anticipate much reclamation of waste water in the future.

Misuses of other resources also often critically affect the water supply. Our resources are interdependent. To destroy one resource frequently is detrimental to, or wasteful of, some other resource, or even the wise use of one resource may result in detriment to some other resource. We drain our marshes in order to produce more food, and in so doing we ruin the homes and feeding places of many of our wild animals and birds. We break up our prairies in order to feed our growing population, and we lose the buffalo, the antelope and the prairie chicken. We cut our forests, thus making our clear, cool, bowered streams to become stark-naked, dirty, mud-filled water courses.

Another reason why water shortages will become more critical in the future, unless corrective steps are taken now, is the expanding use of water for all purposes. Much more water for domestic use is needed by our growing population than in other countries of the world. A study of 10 of the largest cities of the United States showed them to be using nearly three times as much water per person as was being used by 10 cities of equal size in Europe. All new industrial techniques, such as those associated with the development of new explosives, synthetic fuel production and chemical industries, necessitate the use of more water. Even the bringing of new land into crop production requires more water, because much of this undeveloped land is in those parts of the country which need irrigation to produce crops.

□ CLEANING UP OUR WATER SUPPLY

A Task Group on Coordinated Water Resources Research of the Federal Council for Science and Technology, made up of leaders in government, business and science, has suggested how we should go about cleaning up our water supply. To improve the quality of our water supply, we must first determine the quality requirements for various uses. Water used by humans must be free of disease organisms. Water used for recreation must look clean as well. Water used by industry must be relatively free of damaging chemicals and abrasive particles. Water for agriculture, as well as that serving as a habitat for fish and other creatures, must not contain damaging toxic substances.

After we have determined the various requirements, we must find out what effect the various substances we now add to our water supply have on it. These additives include detergents, pesticides, chemical fertilizers and various industrial wastes.

After successfully researching the problems of requirements and the causes of pollution, we can then establish water quality standards and a plan to achieve them. Finally, we must enforce the standards and management practices.

Since the Task Group report, the question of water pollution and our requirements has received renewed public interest and action. The Federal Water Pollution Control Administration (now part of the U.S. Environmental Protection Agency), established in 1965, requires the state governments to submit water quality standards to it. In the last few years, it has authorized hundreds of millions of dollars to help municipalities build sewage treatment plants.

Progress in water cleanup

Congress, reacting to the serious water quality problems, has enacted legislation and funding to clean up our waters. The Water Pollution Control Act of 1972 was amended in 1977. Its goal is to make all waters of this country fishable and swimmable by 1985. It provides funds for the identification of all sources of pollution and for the development of implementable plans to control these sources.

The act also was intended to help local communities solve their water quality problems. Areawide water quality plans as well as state water quality plans were established. The areawide plans cover relatively small areas within a state, usually consisting of several counties centered about a common problem such as acid water run-off from mines, urban storm water run-off or erosion of agricultural lands. The state plans usually address all water quality problems of the state outside the areawide planning areas. As this effort proceeds, all plans within a state must be compatible. The entire water quality program is often referred to as the 208 Program because Section 208 of the act spells out the actions that are to be taken by the states.

The establishment and enforcement of a uniform set of standards is an absolute essential to control of industrial and municipal pollution. All industries need to be treated alike under the law if they are to be able to compete with one another. Unevenly enforced laws favor one producer at the expense of another. If one industry cleans up and another does not, the one cleaning up is apt to lose its competitive ability to stay in business. Its costs will rise, while its negligent competitor will keep costs down and be able to undersell it. If, on the other hand, a uniform code of water management is enforced, the costs will be fairly borne by all firms. The higher costs will eventually be passed on to us as consumers. The cost of pollution control should be recognized as a necessary cost of production.

CASE OF THE ILLINOIS RIVER

How pollution of a stream develops is illustrated by the case of the Illinois River. This is the largest river within the state of Illinois, flowing over 300 miles from near Chicago to enter the Mississippi near St. Louis, Missouri. The Illinois was noted by the early American Indians for its fishing. In the 1600's, Marquette and Joliet, the famous French explorers, remarked on its beauty. By the 1800's, it had become the most important commercial fishing stream in the United States. But, this distinction came to an end when, beginning in 1901, all the sewage of the Chicago area was dumped into the Illinois through the newly constructed 37-mile sanitary district canal. The water in the river became so contaminated that, even with the diversion of 10,000 cubic feet of water per second from Lake Michigan to the river, all fish were destroyed for 150 miles downstream. After 1930, Chicago installed more efficient sewage treatment facilities and, more recently, banned pollution. Fish life has recovered somewhat, especially in the southern part of the river basin. Nevertheless, pollution of the river bottom mud, thus destroying the clams, snails and other organisms and aquatic vegetation upon which fish feed, affected the quality of the fish and may have made the northern basin sterile for generations.

□ INCREASING THE AVAILABILITY OF OUR WATER SUPPLY

The hydrologic cycle provides us with four sources of water supply: (1) atmospheric moisture, (2) surface water, (3) vadose water (that held in the topmost layer of earth) and (4) ground water. Let us consider briefly the nature of these and the possibility of increasing their availability.

Atmospheric moisture

Atmospheric moisture is water carried in the atmosphere in various forms of water vapor or in condensed form. As noted earlier, some 70 percent of the water is returned to the atmosphere. Moisture-laden air is indeed an important potential source of supply. Changing the weather to induce precipitation is one of the most intriguing ways of obtaining this supply.

Modifying the Weather

If the needed water supply in a particular place could be obtained from the atmosphere, this would certainly solve many water shortage problems. The idea reminds us of the rain dances of the Indians. Modern methods, however, are based on scientific reasoning. A common method has been to release fine particles of chemicals into a moisture-laden cloud from an airplane. The particles form a base around which water droplets form, thus starting rainfall.

Modern rainmakers have been quite successful in inducing rainfall in this manner. But, regardless of how successful the triggering of rain from clouds may be, there is one critical factor which suggests that artifically induced rainfall can never be a panacea for water shortages. The limiting factor is that there must be moisture in the air before moisture can be obtained from it. Our severe drought conditions are generally characterized by relatively dry air masses. With little water in the air, there is little for the ground.

Another problem of changing the weather is the danger of loss of control. Would light showers become "gully washers"? Would the rain fall where unintended? These questions are particularly thorny ones. They might give rise to thousands of lawsuits against rainmakers by irrate landowners. And, even with perfect control, *who* is going to decide *who* shall get the rain?

Reducing Evapotranspiration

Since so much of the water in the hydrologic cycle is evaporated and transpired, we could greatly increase our available supply by preventing the return of moisture to the atmosphere. Removal of undesirable phreatophytes (plants which send their roots deep in the earth to reach water) is one method. Other methods include covering or lining with cement canals which transport water for irrigation purposes; spreading a thin film of chemicals on reservoirs; irrigating less wastefully, and with the distribution pipes placed underground; and fallowing (plowing and tilling but leaving unseeded) fields in semi-arid areas.

Surface water supplies

The use of surface water is the most obvious way to help meet our water needs. This source is large, enough to provide water to meet most needs in the foreseeable future—if it is accessible and if excessive water pollution is stopped. Water consumption figures are deceptively large because, of course, it is the same water used again and again in its progress to the ocean. The water of the Wisconsin River, for instance, is harnessed by 13 dams along its rather short course and therefore is counted 13 times in computing the amount used for power. Even after passing over the last dam, the water can still be used for industrial or other purposes.

Practically all the water that flows again into the ocean does so through our rivers, with only a small portion returning through ground water. By damming streams, the availability of their water supply is greatly increased. Many factors should be considered in addition to water supply, however, before a reservoir is created, as was noted in Chapter 4 which dealt with water power. The most successful large-scale surface water control project in the United States, and probably in all the world, is the Tennessee Valley Authority.

THE TENNESSEE VALLEY AUTHORITY (TVA)

In 1933, Congress passed the Tennessee Valley Authority Act "to improve the navigability and provide for the flood control of the Tennessee River; to provide for the reforestation and the proper use of marginal lands in the Tennessee Valley; to provide for the agricultural and industrial development of said valley; and for other purposes." The main river channel is over 600 miles long and with its tributaries the TVA system extends into seven states (see Figure 10-3).

Multiple-Purpose River Basin Development.—As a result of this act of Congress, steps were taken to place dams in the watershed of the Tennessee Valley so as to control water run-off and thus reduce the damage done by flood waters as they race down these valleys. When the Ohio River is in flood stage, the flow of the Tennessee can actually be stopped, thus alleviating additional flood damage. The whole watershed was included in the project; otherwise, complete control of the waters of the river could not have been achieved.

152 □ OUR NATURAL RESOURCES

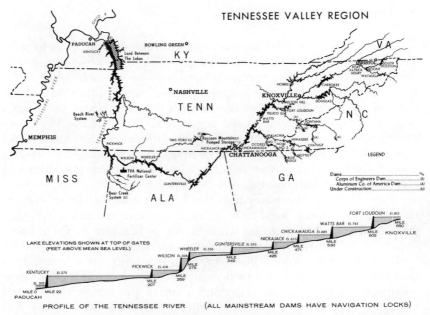

FIGURE 10-3. Map of the Tennessee Valley showing the location of dams in the system and the drop in elevation between the dams. Potential flood waters are stored behind the dams during wet seasons and released during dry seasons. This provides for flood control and maintains a sufficiently deep channel for navigation. The dams are also sites of hydroelectric plants, and the reservoirs furnish recreation. Flood damages of over a billion dollars have been averted by the TVA system since its first project went into operation in 1936. *(Courtesy, Tennessee Valley Authority)*

Not only were flood waters tamed by the work done in this valley, but land was conserved as soil from the uplands and hillsides was saved and rebuilt; forests were replanted, wildlife was restored and more water was made available for various industrial and home uses. Besides, the general level of living was improved for the residents of the watershed. An offset to soil saving in this type of program is the loss of soils in the fertile bottomlands which now are lake bottom.

Improved navigation was another asset created by the TVA. The formerly sandbar-choked, meandering stream now became navigable from one end to the other during most of the year for commercial barge tows and large and small pleasure craft, as reservoirs became filled and locks were opened for by-passing the dams.

Another major development was the installation of hydroelectric plants at the dams. Within a few years after the project began, cheap electrical power became available to part of a region which President Franklin Delano Roosevelt had described as "Our Number One Economic Problem."

The unified multiple-purpose development of an entire large river basin carried out by the TVA was a new visionary concept that had never been tried. The basic development activities—flood control, navigation and power production—were not new in themselves, but this was the first time these related parts of resource development had been brought together in one comprehensive effort for a major region.

Cumulative flood damage of over a billion dollars has been averted since 1936. Flood control benefits have been almost five times the total outlay for the flood control system since its inception.

Tennessee River commercial freight tonnage has set new records from year to year, and savings to shippers have exceeded the federal costs of maintaining and operating the waterway.

In this region, which was once as depressed as surrounding Appalachia, private investment in waterfront manufacturing plants and terminals is in the billions of dollars. Electrical power use is common on practically all farms of the area as well as for city dwellers and industrial plants.

Forestry programs continue to expand in the TVA region. Over one million acres have been reforested. The TVA research and field workers have become leaders in developing uses for forest products, including the use of saw mill waste for pulp and paper making. The forest products industry has millions of dollars in plant investments in the area. Wildlife habitats and demonstration areas are located throughout the forests.

Not the least of the many benefits derived from the TVA program is the development of recreational facilities. From an area which once had very few recreational developments of any kind, the Tennessee Valley has become one of the best developed playgrounds in the United States. About half a billion dollars has been spent in beach developments, camps and resorts, fishing docks and other private and public works. Over 90 public parks, 400 public access areas and 100 group camps and club sites have been created, and over 345 boat docks have been built. There are some 46,000 moored boats and 13,000 private residences on lakefront property. The TVA region has received over 60 million visitors for recreation annually in recent years. Many other reservoirs similar to those of the TVA have been built in the United States (see Figure 10-4).

154 □ OUR NATURAL RESOURCES

FIGURE 10-4. Some private water developments rival those of the government. While from appearances this might well be one of the TVA reservoirs, it is instead the Lake of the Ozarks, created in 1931 by a private power company dam across the Osage River in Missouri. The lake has a length of 129 miles with 1,375 miles of shoreline. For four decades it has been supplying hydroelectric power to St. Louis and other communities. It has also become a popular resort area. *(Photo by H. B. Kircher)*

The TVA system is not without its problems. Stream bank erosion adds to the turbidity of the water and silting up of the reservoirs. Coal-burning electric generating plants create air pollution and draw upon land-destroying strip mines, although the TVA now requires reclamation projects of its suppliers. Natives claim that damming of tributaries has destroyed priceless natural beauty. But, economically, the TVA has proved an old adage wrong by turning a sow's ear into a silk purse. And, in 1979, the TVA staged its first annual environmental conference in coordination with the Environmental Protection Agency. Both economically and environmentally, the TVA is striving to do a good job.

CALIFORNIA STATE WATER PROJECT

Rivaling the TVA in size, although different in purpose and construction, is the California State Water Project, which was chosen

WATER SUPPLY FOR INDUSTRIAL AND PRIVATE USES   155

as the outstanding civil engineering project of 1972 by the American Society of Civil Engineers. Unquestionably, it is the largest and most complex water-moving project in history. This project is a 685-mile aqueduct system consisting of dams and flood control reservoirs and hydroelectric power stations like the TVA, but unlike it, having huge concrete-lined canals and pumping plants for irrigation and domestic water supplies (see Figures 10-5, 10-6 and 10-7). The major purpose of this system is to trap excess run-off in the Feather River watershed in northern California for multiple uses in southern California, the San Joaquin Valley and the San Francisco Bay area (see Figure 10-5). The project also provides flood control on the

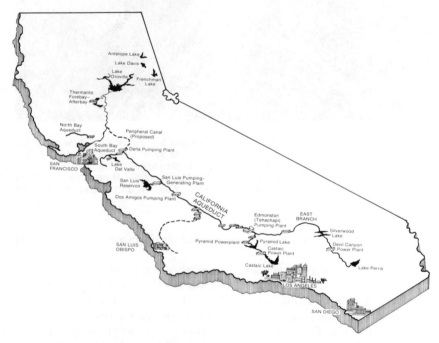

FIGURE 10-5. This map depicts the location and features of the California State Water Project, a major part of the vast and complex system of rivers, canals and aqueducts which serve California. This project extends some 685 miles, and its initial facilities, completed in 1973, include 18 reservoirs, 15 pumping plants, 5 power plants and 540 miles of aqueducts, pipelines and tunnels. This map does not show the Federal Central Valley Project or the several regional water transfer systems which, together with the state project, form a network which taps water from the mountain ranges on both sides of the Sacramento and San Joaquin valleys, their streams and the Colorado River. *(Courtesy, California State Department of Water Resources)*

FIGURE 10-6. This view is downstream as the California Aqueduct of the State Water Project moves water along its 600-plus–mile journey from north to south. Adjacent orchards and field crops are irrigated by the Delta Mendota Canal, a part of the Federal Central Valley Project. The two projects share facilities and storage at nearby San Luis Reservoir. At the upper right of the picture are the foothills of the Diablo Range. *(Courtesy, California State Department of Water Resources)*

Feather River, recreation facilities and conservation of fish and wildlife.

While engineers have been working on various aspects of the California State Water Project since the 1950's (construction began in 1957), it was not until October 7, 1971, that water was pumped to southern California by the new project. Initial facilities were completed in 1973, but the whole project is not expected to be completed until after the turn of the century. A program of additional reservoirs, a canal to convey project water around the Sacramento–San Joaquin rivers delta, water conservation and recycling and additional ground water storage is now pending before the California legislature.

Ecologists have expressed concern over the project. The project

FIGURE 10-7. This view looking downstream on the California Aqueduct of the State Water Project is about 200 miles south of the view in Figure 10-6. The Wheeler Ridge and Wind Gap pumping plants, which keep the water flowing, and an outcrop of the Tehachapi Mountains are in the foreground. The latter also stretch across the middle background. Beyond is the Antelope Valley, and far to the south, the San Gabriel Mountains. *(Courtesy, California State Department of Water Resources)*

transports water through channels of the Feather and Sacramento rivers to the delta and then diverts some of it to the south. Some ecologists believe the project threatens to pollute the delta and San Francisco Bay by diminishing the rivers' seaward flow. Others decry the increase in population that is almost sure to follow the increased availability of water.

Greenwood and Edwards declare: "The movement of water to

FIGURE 10-8. Here water is being turned into an almond grove. It has been carried 300 miles to supply the hot, thirsty soil. *(Courtesy, Bureau of Reclamation)*

people rather than people to water has been an ecological disaster."[2] However, it should be recognized that all over the world, since the beginning of the Industrial Revolution, the movement of water to people has facilitated industry, quality water supply, building of cities and disposal of wastes. And, the project engineers of the California State Water Project point out that the environment is being considered in the project. Under state law and by order of the California Water Resources Control Board, the project must maintain good water quality in the delta. According to the project engineers, the proposed canal around the delta will be the best way to maintain delta water quality and to restore delta fish and wildlife to pre-project levels. On the basis of prior water projects and experience with this one, the engineers have tried to provide for land subsidence, ground water pollution, earthquakes, floods and other

---

[2] Ned H. Greenwood and J. M. Edwards, *Human Environments and Natural Systems* (Scituate, Massachusetts: Duxbury Press, 1973), p. 173.

hazards. And, they are proud that this and earlier water projects have provided the water needed by the millions who live in southern California (Figure 10-8) and have enabled the state to become the most productive of all the 50 states in agricultural output.

OTHER RIVER BASIN DEVELOPMENTS

While the TVA remains unique in the scope and unified control of planning and management of an interstate river basin's resources and the California State Water Project is unexcelled in size and complexity, unified efforts are being made today in many other river basins. Projects in the arid West particularly involve development of hydroelectric power and irrigation facilities, while in the humid East, the emphasis is on hydroelectric power, flood control and navigation.

The Pacific Northwest has one of the greatest potentials for hydroelectric development. Its Columbia River, for example, nearly twice as long as the Tennessee River, and its main channel and tributaries have over five times the hydroelectric capacity of the Tennessee Valley, according to current projections. Grand Coulee Dam, on the Columbia River, alone has a potential hydroelectric capacity of 9.7 million kilowatts—nearly twice that of the Tennessee Valley (see Figure 4-5). The Columbia Basin has over 30 dams. Like the river systems of California, it has extensive irrigation works.

Other river development projects benefit our populace from coast to coast. The Colorado River has been dammed in numerous places, providing flood control, irrigation and hydroelectric power, while, fortunately, still preserving the scenic wonders of the Grand Canyon. The Missouri River basin has some 100 dams, 8 of them large, to provide for flood control, navigation and power production and to aid in development of the watershed. The Mississippi-Missouri River System, including the Ohio River and its tributaries, carries over one-third of our inland waterway commerce. It is exceeded in this regard only by the Great Lakes System, with which it is connected. Dams up north and levees on the lower Mississippi protect millions of acres of farm land in Mississippi and Arkansas from flooding, although whether the benefits have been sufficient to justify the costs is questioned by some, and the management problems are great as noted in the section dealing with flood plains.

A remarkable water development in the northeastern and lake states is the St. Lawrence Seaway, which connects lake ports such as

Chicago and Milwaukee with the Atlantic trade routes. It also provides electrical power. In New England, water power utilization directly powered the early industrial development, as we have noted. Today, even hydroelectric power there runs a poor second to other sources of power, but control of the rivers to prevent floods is vital to industry.

The North American Water and Power Alliance Plan.—Still in the proposal stage is the most incredible water diversion project of all time, known as the North American Water and Power Alliance. It would hook up streams and lakes of Canada, the United States and Mexico at a cost of over $100 billion. In view of the environmental upset this would cause and the questionable economic benefits, compared with alternative projects, it appears that this proposal will never be accepted.

WILD RIVERS

A few rivers in our country still remain in a relatively natural condition. Their courses have been little affected by man. These streams are known as *wild rivers*.

What is a wild river worth? What is it worth to taste the cool spray; wade in the rippling rapids; hear and see the white-green-blue waters of the roaring falls; hike along the bank heavy in grasses or hemlock, pine and birch; contemplate the still pools or quiet, deep, clear waters of a wild river—a river left in its unspoiled state?

The great tows moving down the Mississippi or the Tennessee, the mammoth dams and power plants of the Columbia or the Colorado, the exotic ocean freighters plying the St. Lawrence are evidence of vital industry and trade and express the beauty of science and technology. The wealth they bring is welcome.

But, if we destroy every wild river, we will destroy the value of those we have tamed. If we do, we will destroy one more place where many of us can go to renew our own spirits and, perhaps, help lift up that of others. Of what avail outward pomp and circumstance if inwardly we lead lives of bitterness, frustration and jealousy? In a message of the same vein universal to all peoples in all times, the gospel according to St. Mark (Mark IX: 36) declares: "For what shall it profit a man, if he shall gain the whole world and lose his soul?"

The Allagash, Buffalo, Current, St. Croix and Rogue are a few of some 75 rivers now protected by or being considered for inclusion in our Wild Rivers System, established in 1968. We must preserve these few free-flowing streams. Out nation's economy can afford them. We cannot afford to be without them.

RIVER FLOOD PLAINS

No discussion of our inland surface water resource would be complete without calling attention to the plight of flood plains. In the United States, as in the rest of the world, man has settled on these river lowlands because of the natural conveniences—flat land and flowing water. Here, as everywhere, we have paid the price—catastrophic floods with loss of property and life.

Gilbert F. White, one of America's leading authorities on flood plains, points out that, as their name implies, they will be flooded. We must provide a place for the water to go. Ian McHarg, the noted ecologist and land-use planner, says in his film *Multiply and Subdue the Earth*, "... the 50-year flood plain by definition should be not occupied by any residential development, the 100-year flood plain might, the 5-year flood plain absolutely never." Dr. Lincoln Brower, Professor of Biology, Amherst College, at a flood plain conference in St. Louis on February 4, 1974, pointed out that "Flood damage is bad, but the overflowing is good." In other words, used for agricultural and ecological purposes, river bottomlands benefit from floods, but, where unwisely occupied by man, they become casualty areas.

The advantages and problems of damming streams have been pointed out in earlier paragraphs of this chapter, and especially in Chapter 4. How about levees (embankments raised to prevent the river from overflowing)? In limited areas they are helpful. They can protect part of a given urban area from flooding. They are useful for protection against moderate flooding of farm land or wild life areas. But, Brower and others have called the attempt to levee an entire river system, as is being done on the lower Mississippi, a disaster. Not only may the levees be breached, but also water may seep underneath—and emerge on the flood plain in sand boils. Major flooding can be caused behind the levees by water rushing down from the adjacent uplands through tributaries and/or ditches, only to be blocked by the water of the main stream, now cresting at a high level between the levees. This situation has occurred many times on the state of Illinois side of the Mississippi River at St. Louis, although it has been "protected" by a river system for half a century.

Levees are useful. But, adequate spillways and ponding areas must be provided, and critical areas of the flood plain must be freed of permanent structures. A river serves us best when it is allowed to do what comes naturally—flood!

Vadose water and ground water supplies

Water under the surface of the earth is classified into two major types: vadose water—that held in the topmost layer of earth, which is the source of most soil moisture—and ground water—water found considerably deeper beneath the surface in the pores and fissures of the rocks, including clays, sands and gravels.

VADOSE WATER IS ESSENTIAL

Vadose water is essential to most plant life. Using soil moisture efficiently or being able to help drain saturated soils or to add to those deficient in water is essential to getting the maximum production from soils used for agriculture. Refraining from disturbing soil moisture conditions may be equally important to conserving wildlife habitats or natural vegetation. Such management practices are discussed in the chapters on soils, forests and wildlife.

GROUND WATER SUPPLIES ARE VAST

Potential resources of ground water are enormous. The volume of underground fresh water has been estimated to be at least 10 times the average annual precipitation of 30 inches. Unfortunately, it is not always available in the place where it is needed the most. And, ground water supplies can be easily polluted or overdrawn.

Ground water is replenished by water seeping into water bearing rocks or underground channels from some source above the ground. Or, it may be recharged by stream and river water which seeps into underlying permeable materials. The water pumped from wells for a community may have moved hundreds of miles through sand, gravel and porous rock before it reached that point from which it is pumped, perhaps several hundred feet below the surface. It is estimated, for example, that the water now being pumped out of 800-foot wells for the city of Madison, Wisconsin, fell as rain and snow in the northern part of the state several hundred years ago. Because of the slow movement of these underground waters, it can easily be imagined that their replenishing may be too slow to maintain the supply if water is pumped too rapidly.

Ground Water Use Is Increasing.—According to studies made during the past few years, the water supplies of two-thirds of our urban communities are from ground water. Only one-fifth of the total amount of water used by the nation is ground water, however. The other four-fifths is supplied from our rivers and lakes, except for a very small quantity obtained from desalted sea water.

One-fifth of all water from ground water may seem like a relatively small amount. Yet, we are drawing from these ground supplies twice as much water as we were 10 years ago, and the effects of this increased use are showing up all over the country.

Water Table Varies.—The water table, that is, the upper level of ground water, is gradually getting lower in some regions, and, consequently several problems have developed. In some sections of Ohio, the water table has dropped 100 feet in the past 50 years. In the Central Valley of California, it has been dropping around 10 feet per year, and, if no other source of water had been made available to this area, approximately 400,000 acres of highly productive land would have returned to a semi-arid condition before now.

In some coastal cities, the dropping of the water table has resulted in so reducing the water pressure that instead of fresh water flowing from the land to the sea the reverse is taking place and the salty ocean water is flowing through the underground channels toward the land. Can you imagine the distress of finding salt water coming from all the faucets in the house?

In some places, the ground water supply has been recharged by recycling water after it has been used for cooling purposes, as in Louisville, Kentucky, for example. In others, as in Memphis, Tennessee, the supply of clear, cool ground water has never been wanting, and the city gets its supply from this source rather than the surface waters of the Mississippi River, on which it is located. In fact, the coastal plains facing the Atlantic Ocean and the Gulf of Mexico, including a large area extending northward up the Mississippi River Valley to the tip of Illinois, and the flood plains of our eastern rivers, all have excellent ground water potentials.

□ SALT WATER CONVERSION

The use of saline waters

When saline water is mentioned, we quickly think of sea water. The water of the various oceans represents inexhaustible supplies if and when the salinity (saltiness) can be economically removed. There are other waters in many parts of the country that cannot be used for human consumption even though they are not nearly as salty as sea water. These brackish waters contain from $1/30$ to $1/2$ as much salt, or dissolved solids, as does sea water. Sea water contains some 35,000 parts per million, or $3\frac{1}{2}$ percent, of dissolved salts, while the various brackish underground waters contain from 1,000 to 15,000 parts per million. The dissolved salts in these waters consist of many minerals besides calcium, such as magnesium, sodium, bicarbonates and sulfides. The greater the number of salts and the heavier the concentration in the water, the more costly it is to remove enough to make the water usable. Drinking water should not contain more of these salts than 500 parts per million, or one-half of one percent.

Removing salts by distillation

Sea water distillation in small quantities has been used on ships for years as a means of maintaining a supply of fresh water during the time at sea. Until fairly recently, the method employed was not economical enough to use when larger quantities were needed for whole communities. Since then, much progress has been made, with Congress appropriating millions of dollars to subsidize experiments in the desalinization of water and the building of plants.

The techniques of distillation have been so improved during the past few years and the costs so reduced that they have been widely adopted by many communities requiring larger supplies of fresh water. The two major items governing the cost of water supplied by this process are installation and fuel costs.

Other processes

Some of the other processes that are being experimented with are (1) electro-dialysis, (2) reverse osmosis and (3) freezing. The first two processes achieve separation of excessive salt from the water by diffusion of the sea water through a semi-permeable membrane. The third process achieves separation by freezing, since ice carries a relatively low percentage of dissolved salts.

Dr. John Hult, a physicist, has a really "cool" proposal to increase the water supply of southern California. He suggests that icebergs be transported from Antarctica to be melted offshore in California and the water piped inland. Hult believes that this is feasible and that the costs of delivered water would be less than half as much as the cost of that delivered over the California State Water Project.

Desalinization plants

According to an inventory published in 1977 by the Office of Water Research and Technology, there are about 800 water desalting plants in the world today. Plants are operating in the United States in California, New Mexico, North Carolina, South Dakota,

FIGURE 10-9. This vertical tube salination unit on St. Thomas in the Virgin Islands is one of four currently operated by the V.I. Water and Power Authority. With two on St. Croix and two on St. Thomas, the units have a combined capacity of approximately 5 million gallons of water per day. *(Courtesy, V.I. Water and Power Authority)*

166 □ OUR NATURAL RESOURCES

Texas and Florida and in the Caribbean area (see Figure 10-9). An early plant built at San Diego, California, was moved to our naval base at Guantánamo Bay, Cuba, when Fidel Castro cut off the water supply, and it has successfully met all the needs of the base for years. Another plant, built at Key West, Florida, in 1967, supplies 2.6 million gallons of water per day. This amount could meet the household needs of 75,000 to 100,000 people. A proposed new plant at Yuma, Arizona, will help to improve the quality of the Colorado River water, supplying water to many more people than the Key West plant (see Figure 10-10).

FIGURE 10-10. An artist's conception of the 96-million-gallon-per-day desalting plant being installed on the Colorado River near Yuma, Arizona, to process highly mineralized water after irrigation use. After its installation is completed, the desalinized water will be added to the river to improve its quality before it flows into Mexico, thus honoring an agreement between the United States and Mexico. The plant will be the largest membrane plant in the world. It is scheduled to begin operation in 1985. *(Courtesy, U.S. Department of the Interior, Water and Power Resources Service)*

Cost of desalted water

The cost of fresh water from usual or conventional sources has been increasing during the past few years for several reasons. One of the reasons for these higher costs is the greater distance, either horizontal or vertical, that communities must go to get enough fresh water for their needs. Also, other items of cost, such as fuel, labor and equipment, are continuously increasing. Furthermore, because the pollution load introduced into streams from all sources is much greater than formerly, maintaining the higher water quality standards which are now required costs more. In the past few decades, the cost of reclaiming saline water has declined sharply. However, the common experience of all of us with inflated prices suggests costs may soon be rising again just as sharply as they declined a few years ago.

Although we have the ability to purify sea water, we cannot afford to neglect effective water management of our fresh water supplies, for use of saline water is a costly alternative which may never be feasible at many inland locations.

☐ RECLAIMING WASTE WATER

Most water is not "consumed" with use, but is eventually returned to the water supply. But by then, as we have noted, it is often badly polluted. A major way to increase our water supply is to recycle the water. By this is meant that after water had been used in a process, it would be treated for undesirable impurities and used again. With better national and state clean water laws and better enforcement, we can anticipate much reclamation of waste water in the future.

☐ PLENTY OF WATER, IF . . .

The question of the adequacy of our water supply is what can be called an "iffy" question. There appears to be plenty of water for everybody . . . *if* we increase its availability by the various methods related and *if* we clean up our waters and stop inordinate polluting of them. In addition, we must stop wasting the water we use and start using the water we waste. Finally, we will meet our requirements far more easily and cheaply *if* we move people to water instead of water to people.

☐ ☐ ☐

QUESTIONS AND PROBLEMS

1. Give some of the reasons why water for use by cities, in homes and for industry, is becoming a scarce resource in various places throughout the United States.

2. What are the principal sources of water supply, and which ones must be depended upon at the present time to supply water for the aforementioned uses?
3. What are the two principal reasons we are short of water, not counting our expanding use of it?
4. Describe the hydrologic cycle. Why is it called an "imperfect" system?
5. Cite several examples of water pollution given in the text. Cite any similar situations in or near your community.
6. Why are uniform water control standards essential from the standpoint of the costs of businesses?
7. Illustrate how good water conservation affects other natural resources.
8. Whose responsibility is it to see that water is properly conserved as it makes its way from the small piece of land where it falls as rain or snow through ravine, lake and stream to its place again as a drop of water in the ocean?
9. Under what conditions can ocean water be used for industrial or home use?
10. Where does your city, or the one nearest you, get its water supply? What are the advantages and/or disadvantages?
11. There is so much moisture in the air. Why isn't *changing the weather* an ideal solution to our water shortage problems?
12. Should levees ever be used on flood plains? Explain your answer.
13. What is meant by describing the TVA as a "unified, multiple-purpose development?" How has it been successful? How does it differ in character and purpose from the California State Water Project?
14. What is vadose water, and why is it essential?
15. Since ground water supplies are so vast, why can't they meet all our supply problems?
16. Why is the adequacy of our water supply called an "ify" question?
17. Isn't keeping a river in a wild state a waste of our surface water resources? Explain your answer.

11

The nature of forests and history of their growth

PROBABLY NO OTHER RESOURCE has been as useful to man over the centuries as have trees. Our native Indians always established their camp sites near wood and water. Not only was wood the source of fuel to keep them warm during the cold weather, but it also supplied them with birch bark to build canoes, not to mention shade to keep them comfortable during the warm summer months. Hickory, elm and other woods were used to make bows and arrows for hunting or for war purposes, as well as for furnishing poles, walls and roofs for shelters. The Indians' form of religion showed that these primitive people also had an appreciation of the beauty and inspiration that unspoiled forests can give, just as we ourselves are now coming to realize how important these non-material values are to us.

☐ WOODS AND WILDLIFE

Some woods are composed of dense forest trees whose tops form a canopy over the whole area and completely shade the ground. Few animals roam these woods because there is no low growth of shrubs or grass upon which herb- or grass-eating animals can live. Neither are there many insects or small seeds to support small bird life. The vast forests of pine, hemlock and spruce are of

FIGURE 11-1. This forest of virgin lodgepole pine in Wyoming will support very little wildlife. There is practically no low undergrowth for their food supply and protection. *(Courtesy, U.S. Forest Service)*

this type, called evergreen, coniferous, cone-bearing or softwood forests.[1]

Other woods are not as dense and usually consist of a mixture of different kinds of trees. Forests of mixed hardwoods or of hardwoods and cone-bearing trees together are ordinarily open enough; that is, the trees are far enough apart to permit shrubs and some kinds of grass to flourish. It is in these woods that most of our animal and bird life is found. Grass and shrubbery are available for the animals to eat and hide in and for insects to thrive. These woods

---

[1]This term *softwood* is used by foresters to designate cone-bearing trees, while the term *hardwood* refers to the broad-leaved trees which shed their leaves each fall.

attract birds. Wherever food is available, animals and birds will be there to live on it.

☐ DIFFERENT WOODS MEET DIFFERENT NEEDS

We are told that when the white man first settled in this country, more than three hundred years ago, about 43 percent of the land area was in woods. As more people came and settled, they found the forests both a help and a handicap in making homes and producing food. Practically all the buildings put up by the colonists and later settlers were made of timber cut from the land upon which the individual families settled. Fuel was also supplied from the same source, and household furniture, as well as most parts of farm implements and machinery, was made from the different kinds of wood that grew on the land. But still, the job of clearing the land of trees and preparing it for crops was really uphill work.

The variety of trees was such that some kind of wood was found for every purpose. Trees such as the pines, basswoods and hemlocks supplied soft, easily worked woods for fashioning kitchen and other household utensils and cheap furniture, while the oaks supplied wood for heavy use, sturdy construction and dignified appearance. Fine-grained woods, such as cherry, hard maple and walnut, where carefully finished, gave beautiful luster and color to furniture. Many of these pieces are still in use after several generations of wear. Farm use required tough, hard wood, such as elm, ash and hickory.

All these woods were found growing in greater or less profusion on practically all the land east of the Mississippi River. Only the land comprising central Illinois was free from a covering of virgin forest growth.

☐ WOODLANDS A HANDICAP TO EARLY FARMING

The early settlers did not regard these benefits of the forests as an unmixed blessing, however. They had to clear the land of trees before crops could be planted. It required months of work for a family to cut the trees with axe and saw and later to clear the stumps from a single acre of land before it could be used for crop production. And, because of the immense amount of work required to clear

172 ☐ OUR NATURAL RESOURCES

FIGURE 11-2. As our plane dips over the region known as the Wheaton Moraine Country of Illinois, preparing to land at Chicago, we see that farms have taken over most of the land, with only a small section (foreground) left or grown back into forest. At one time, even the prairie state of Illinois was over 40 percent forested. *(Photo by H. B. Kircher)*

such small areas, these people considered the forests as enemies to progress in settling this country.

☐ FOREST SOILS NOT AS PRODUCTIVE AS PRAIRIE SOILS

Another discouraging factor about forest areas is that, in general, most of their soils are not as productive of farm crops as are most prairie soils; so that even after an acre of land is cleared, the production from this care is not large, and after it has been farmed many years, crop yields will decline even to the point that the farms will be abandoned. These soils were not as productive as the soils in the homelands from which the colonists came. George Washington once said that the early colonists found it was better to clear and cultivate new land, of which there was an abundance, than to try to maintain crop production on land already cleared and cropped for a few years. This was true despite the great cost in time required to clear land during colonial times.

There are three obvious reasons why the productivity of the cropland was not maintained at that time. The first is that New England soils generally were rather infertile soils, formed under woodland. The organic matter, as well as plant nutrients, was quickly used up or washed away when once planted to the crops usually grown at that time.

The second and more fundamental reason for the loss in fertility of these nearly virgin soils is that little was known of the value of crop rotations in keeping up soil productivity and less yet was known of the requirements of plants for the various plant food elements, or nutrients. The need of plants for lime, phosphorus, potash and nitrogen is so generally recognized at the present time that one can scarcely believe there was a time when farmers did not know that crops would not grow without the presence of those nutrients in usable form in the soil.

A third reason why the productivity of those soils was not maintained is associated with the experiences of the immigrants in their homelands before coming to this country. The rains here are of a thunderstorm type, in which from one to several inches of water may fall within a few hours. Land cleared of forest growth and depleted of leaf mold cannot absorb these heavy rainfalls, and soils are carried away with the water run-off. The rainfalls of the northern European countries usually are gentle, steady rains most of which are absorbed in the soils instead of being flushed off over the surface. It was only recently that farmers in this country became aware of the seriousness of these soil losses and began to take measures to control erosion.

It was only a little more than a century ago that chemists discovered the dependence of plants upon certain plant food elements. And, it is less than a century that artificial plant nutrients in the form of fertilizers and lime have been used to supplement the application to soils of farmyard manures, leaf mold and river bottom sediment as sources of food for plants.

☐ OUR VIRGIN FORESTS ALMOST GONE

The early attitude that forests were obstructions to progress continued almost to the present generation. As a result of excessive

FIGURE 11-3. This shows a part of our remaining virgin forests. These mature white pines are growing near Pierce, Idaho. The next generation will see none of this type of timber unless the trees now here are preserved. *(Courtesy, U.S. Forest Service)*

cutting and burning, this country now has left not more than a small percent of its once magnificent stands of virgin forests.

Even though these virgin forests are nearly gone (a great esthetic loss), it is not the economic tragedy usually visualized by the uninformed. Trees are growing continuously so that what was a sapling 150 years or even 100 years ago may now be a mature tree. Yet, this is scarcely half the age of our country. Also, contrary to some commonly held views, wood from these later grown trees may be as good if not better than wood from the virgin forests for practically all the uses devised by man.

□ SAW TIMBER GROWS SLOWLY

Why is there such a loss in standing timber when most trees grow to lumber size in 75 to 150 years? The problem is one of slow growth relative to use and losses. Some species of softwoods, such as white pines, planted 100 years ago are now large enough to cut for lumber. It is also true that trees such as walnut, maple, elm and gum will grow to maturity within 120 years. A glance at Table 11-1 shows tree growths over the country, with different lengths of time in different localities needed to produce what lumbermen call "saw timber," which is quality timber of usable size for lumber and woodworking. Whereas it requires only 30 to 40 years to grow saw log pines in the South, it takes 150 to 180 years to grow western pines of like size in the Rocky Mountain states.

TABLE 11-1. Years required to grow saw log timber

Area	Type of Timber	Years to Grow
Southern states	Pine	30– 40
Great Lakes and	Pine	60–100
Northeastern states	Hardwoods	100–120
Western states	Fir and hemlock	100
Rocky Mountain states	Western pine	150–180

□ NATURE PRODUCES FORESTS SLOWLY

The natural reproduction of a forest is a slow process. Trees do not grow quickly. The history of any forest shows that it requires

FIGURE 11-4. One of the great enemies of both forests and wildlife is fire. The continued burning of undergrowth destroys all young trees, burns out the leaf mold and leaves no place for wildlife to find food or shelter. *(Courtesy, Soil Conservation Service)*

hundreds of years to develop a forest soil after the virgin cover has once been destroyed either by cultivation or by fire. The fine leaf mold which carpets all hardwood forests contains large amounts of organic matter which, when dry, burn readily. Even when plowed and cultivated for ordinary farm crop production, this organic matter and the organic layer of soils are quickly lost through oxidation and erosion. Oxidation in this sense is a slow process of burning. It is so slow that it may take several years for the oxygen of the air to combine with the organic matter of the soil and thus destroy it, whereas a fire will produce the same result in a day or two. The loss by erosion is the most serious loss of all, not only because the plant nutrients are lost but also because the soil itself is eroded away.

Nature must rebuild this soil before it will again produce the kind of timber that was there as a virgin or original crop at the time white men first saw it. The process used by nature in rebuilding forest soils requires a cycle (sequence) of different plants. When a pine forest is logged or cut over or burned, the first growth to start again is not pine. Usually some quick-growing shrubbery, berry vines or annual plants will immediately spring up to cover the ground.

The quick growth of these berry-producing plants is one reason why Indians used to burn small forest areas. Among these low-growing shrubs, and aided by the protection they afford, some quick-growing trees will start. These trees usually are hardwoods. They include elm, ash, basswood, birch, poplar and soft maples, which grow quickly and soon outstrip in growth the shrubs, vines and grasses.

As they grow, they produce more shade. This shade reduces the growth of shrubs and vines which must have direct sunlight on their leaves in order to remain vigorous and strong.

As this smaller growth becomes less dense, it offers the opportunity for the development of those trees which start and make their early growth best under some protection from direct sunlight. Many of the conifers are among these trees.

The taller growing conifers continue to thrive in the protection afforded by the broad leaves until they ultimately grow up through and outstrip their early protectors. The conifers then cut off the direct sunlight from the hardwood trees, which in turn die off as did the shrubs and grasses a few years earlier.

Thus, nature's way of rebuilding a forest is slow. The stand also may be very irregular, being much too thick in some parts of the area for the production of sturdy, heavy-trunked trees, while in other spots, being so sparse that the trees throw out spreading branches and grow laterally almost as much as they do upward. This type of tree is desirable for shade or parks but not for timber production, which requires tall branchless trunks if timber free of knots is to be developed.

☐ HOW MUCH TIMBER HAVE WE?

Since the first settlement in this country, more than 2,700 billion board feet of timber have been removed from our forests. One-third of the timber removed was used by man. Two-thirds was destroyed by fires, diseases and insects. Over half of this drain has occurred since 1906. We started out with approximately a billion acres of forest land. It contained over 8,000 billion board feet of potential saw timber.

According to U.S. Forest Service statistics, with Alaska and Hawaii added, we have about three fourths as much forest land left, with just over one-half as much saw timber. Furthermore, in balance, the quality is poorer since much of the Alaskan forests are of poor quality naturally and many of the other forest lands have been poorly managed.

The great reduction in our forest resources does not mean that we are critically short of supplies. It does suggest that without wise management the forests will be seriously depleted in a few decades. The demand for forest products is expected to double by the year 2000. The U.S. Forest Service concluded as a result of its study that these future demands could be met with more intensive forest management and utilization. The next chapter discusses ways in which these goals might be met.

The growth of all timber has been exceeding the drain from all causes. This speaks well for the work done by our foresters in combatting fires, diseases and insects and in using reforestation methods and tree species that make for faster growth than when "Nature took its course."

☐ ☐ ☐

QUESTIONS AND PROBLEMS

1. Define the terms *hardwood* and *softwood* as used by lumbermen.
2. What types of forest growth offer food and shelter to various forms of wildlife?
3. Why did the early settlers in the United States look upon forest land both as a blessing and as a handicap to settling the country?
4. Name some of the trees that were found in the areas settled by the early colonists and tell how they were used in the home or on the farm.
5. Why did the colonists find the soils of the Atlantic Coast disappointing when compared with productions from the cultivated lands of Europe from which they came? Give reasons why the production of these soils was not maintained.
6. How much of our virgin forests was still left in this country in 1945? How much standing timber do we have now in comparison with the original forests, even though there have been 100 to 150 years in which to grow new forests after the virgin forests were destroyed?
7. Are we growing forests faster than they are being used up, or is the reverse true?
8. What part of the country will grow saw timber most quickly, and in what area will it develop most slowly?
9. What is nature's way of regrowing a forest?

10. Give the names of 10 trees that grow in your neighborhood.
11. Tell how each of five of these trees is used.
12. What uses can you think of for forest products?

12

The management and use of forests

☐ MANAGED WOODS GROW FASTER

MOST OF OUR PRESENT WOODLAND GROWTH does not require the slow process of nature, explained in the previous chapter, to develop salable timber, even though we cannot approximate the size, dignity and majesty of our "forests primeval." We have learned that we can plant some kinds of trees to replace those cut or lost through fire or wasteful handling. Today's forests respond to a system of management which results in greater and more uniform growth than when the trees were permitted to develop "according to Nature's plan."

☐ MANAGEMENT PRACTICES

Common management practices to increase the productivity of forests include planting selected varieties of trees, breeding supertrees, controlling insects and diseases, providing fire control, selective harvesting trees and clearcutting forest plots.

Planting selected varieties of trees

The kinds of hardwoods which are most valuable for certain purposes are listed in Table 12-1. This table shows that some trees may be used for several purposes; and the rate of growth, as well as the use for which the wood is grown, should determine the variety of tree to plant. There is little doubt that lumber and veneer will continue to be two of the most important uses of wood, while some of the other uses shown in the table may be less important outlets in the future than they have been in the past.

TABLE 12-1. Uses for hardwoods

Lumber	Veneer	Poles and Posts	Excelsior	Windbreaks
Ash	Basswood	Catalpa	Aspen	Green ash
Basswood	Beech	Coffee tree	Basswood	Box elder
Beech	Yellow birch	Red elm	Cottonwood	Cottonwood
Birch	Black cherry	Eucalyptus	Willow	Eucalyptus
Black cherry	Sugar maple	Black locust	Yellow poplar	Hackberry
White elm	Oaks	Honey locust		Silver maple
Hickory	Red gum	Russian mul-		Russian mulberry
Sugar maple	Sycamore	berry		Osage orange
Oaks	Black walnut	Oaks		Russian olive
Red gum	Yellow poplar	Osage orange		White willow
Black walnut		White willow		Yellow willow

Barrels	Tool Handle Stock	Railroad Ties and Mine Timbers	Hardwood Distillation
Ash	Ash	Black locust	Beech
Basswood	Beech	Honey locust	Black birch
Beech	Birch	Red oak	Yellow birch
White elm	Hickory	White oak	Sugar maple
Sugar maple	Sugar maple		
White oak	White oak		
Red gum			

(Source: U.S. Forest Service)

Since the time required to grow salable timber is always a factor in its planting and care, it may pay to consider this in any planting which is to be made. The fastest-growing trees ordinarily have fewer uses than those which are slower growing. When species of trees are planted in climates most suitable to their growth, we find the eucalyptus to be among the fastest-growing trees, while the oak, birch, beech and hard maples are among the slowest (see Table 12-2).

TABLE 12-2. Average height growth of hardwood trees from seed

Kind	10 Years	20 Years	30 Years	50 Years
		(feet)		
Ash, green	26	41	52	70
white	19–25[1]	34–50[1]	45– 67[1]	62– 83[1]
Aspen	18–21	27–40	28– 55	50– 75
Basswood	—	16-32	23– 44	37– 63
Beech	—	8–19	13– 28	22– 42
Birch, paper	13	30	44	62
yellow	—	30	18– 39	15– 54
Box elder[2]	20	—	—	—
Catalpa, hardy[2]	19	27	33	—
Chestnut	7	17	33	64
Cottonwood[2]	56	97	115	136
Elm, white	—	21	28	40
Eucalyptus[2]	24–80	70–90	85–160	—
Gum, red	35	66	88	108
Hickory, shagbark	3– 7	8–18	15– 32	32– 51
Locust, black	15–20	28–45	—	44– 65
honey[2]	18	27	35	—
Maple, silver[2]	22	44	60	80
sugar	—	18	29	40– 48
Oak, burr[2]	—	—	40	60
red	13	32	46	72
white	11–12	22–25	32– 38	53– 63
Osage orange[2]	—	15–25	37	—
Poplar, yellow	20–27	36–50	50– 64	78– 83
Walnut, black	18	30	40	60
black[2]	50	73	89	109
Willow, white[2]	15–24	35–50	—	—

[1]Where a range in height is given, the lower figure in each case is in poor growth situations.
[2]Plantation grown or grown under more favorable than average conditions. Others forest grown.
(Source: U.S. Forest Service)

FIGURE 12-1. Practically the only virgin forests now left in this country are in the western part of the United States. It is only through the vigilant efforts of conservation-minded individuals and organizations that these forests have not been sacrified. Selective cutting, such as is being practiced in this forest, is beneficial to both the remaining stand and the users of wood, especially if the selection and felling of the timber are properly supervised. *(Courtesy, National Park Service)*

Most so-called timber species, when planted close enough for forest planting, will grow quite tall before showing much growth in diameter. They will then increase in diameter at a variable rate, again depending on the variety of trees grown (see Table 12-3).

TABLE 12-3. Average diameter growth of hardwood trees

Kinds	Average Number of Years to Grow One Inch in Diameter
Eucalyptus[1]	½–3
Cottonwood,[1] white willow,[1] honey locust,[1] black locust,[1] black willow	2–4
Silver maple,[1] white elm,[1] Russian mulberry,[1] hardy catalpa,[1] red gum, yellow poplar, chestnut	3–6
White ash,[1] green ash,[1] box elder,[1] black walnut,[1] butternut,[1] burr oak,[1] osage orange,[1] red oak, black oak, aspen, basswood	4–7
Hard maple,[1] hickory, white oak, chestnut oak, paper birch, yellow birch, beech	5–10

[1]The growths of these trees were measured from plantation growth on farm lands, while the species not marked were grown in natural forests.
(Source: U.S. Forest Service)

To grow a 15-inch diameter black willow, eucalyptus or cottonwood will require from 40 to 50 years. If the species grown is black walnut, one of the oaks or a hard maple, it may require about 100 to 125 years to grow a tree of the same diameter.

Plantations of conifers, as well as those of hardwoods, are common throughout the country. Wherever they are planted, they make satisfactory growth as windbreaks and for ornamental and commercial use.

Breeding supertrees

In addition to selecting the proper varieties of trees, timber producers achieve maximum timber production by the breeding and planting of quality stock. To achieve this, timber producers select

OUR NATURAL RESOURCES

FIGURE 12-2. Modern machinery makes forestry much more efficient today. This "Super-Skidder" moves and loads logs as if they were matches. The operator is well protected and has numerous power-assists to give him greater maneuverability and ease of operation. *(Courtesy, International Harvester Co.)*

seeds of the fastest-growing, tallest, straightest and healthiest trees and reproduce them in nurseries. The superseedlings are then set out to produce supertrees. Millions have been planted. One corporation alone, the International Paper Company, has planted "two supertrees for every man, woman and child in the country."

Controlling insects and diseases

Enemies of forests are many and are increasing. Some diseases and insects affect certain varieties of trees without damaging others,

FIGURE 12-3. This is a forest of West Virginia where a continuous stand of good trees will persist as long as trees are cut when they reach maturity and the young growth is protected during the selective cutting. It requires good forest management to accomplish the results shown here. *(Courtesy, U.S. Forest Service)*

and when once a disease or insect pest becomes well established, it is practically impossible to eliminate it completely or even partially. The trees attacked must develop an immunity to or a toleration for the disease or pest, or they will be completely destroyed.

INSECT, FUNGAL AND BACTERIAL ENEMIES

Such has been the history of the chestnut tree of the eastern United States. The chestnut blight has caused the complete destruction of our commercial chestnut from Canada to the Gulf of Mexico (see Figure 12-4). Not only has this removed a commercially valuable and beautiful tree from our landscape, but it also has deprived this generation of children of the thrill and delight of gathering and roasting the tasty nuts in the fall of the year and has removed from the woods a native food for the wildlife.

FIGURE 12-4. Wildlife requires both food and protection in order to survive. This forest of native hardwoods in North Carolina, with its open spaces filled with undergrowth of shrubbery and grasses, supported some wildlife. However, since the early 1900's, when this picture was made, the chestnut trees in this woods and all over the United States have been killed by the chestnut blight. *(Courtesy, U.S. Forest Service)*

Another tragic loss has been the death of most of our magnificent ornamental and shade elms. The loss of these trees is the result of the combined destructive work of the European elm bark beetle, which makes holes in the bark, and the Dutch elm fungus, which thrives in the protection offered by these holes.

The white pine blister rust entered this country over 50 years ago. This disease does not spread from pine to pine but must live a part of its life cycle on an alternate host, such as wild gooseberries or currants. The only known way to control this disease is to destroy its alternate host. It appears now that the white pine may be saved for future generations to enjoy, but only by the destruction of our wild currants and gooseberries.

The story of the gypsy moth, which has been a pest in our forests for over 60 years, is the story of a well-intentioned man who introduced this moth from Europe in the hope of developing a hardy, silk-producing insect. The result is the widespread prevalence of an insect which eats the leaves of both hardwood and evergreen trees.

There are dozens of other insect and fungus enemies in the United States, and there are hundreds of these pests waiting for a chance to get a foothold in this country and further destroy our forests, upset the balance of nature and add to the costs of control.

FIGURE 12-5. Two or three men with a tree planting machine will plant 12 to 15 acres of trees in a day if the land is fairly level and free of stumps and rocks. This machine greatly reduces the labor required to plant trees and thereby reduces the cost. *(Courtesy, Soil Conservation Service)*

190 ☐ OUR NATURAL RESOURCES

PROVIDING FIRE CONTROL

The methods of fighting forest fires have radically changed. Forest managers no longer send in workers to fight every fire. First, they carefully consider which ones to fight and which ones to let burn. And, with caution, they may even light a few, for they have come to realize that fires are nature's way of controlling distribution and succession in forests. A study of a 300-year period of part of the Boundary Waters Canoe Area, which is a protected wilderness, shows, for example, that, on the average, the forest has been completely burned over by Nature every 100 years.

Fires are essential to the growth of some species of trees and provide better habitat for some wildlife. Jack pine cones, for example, do not release their seeds until exposed to intense heat. Fires may destroy harmful insects and turn ground cover into beneficial

FIGURE 12-6. The sign "Tree Farm" on this planting of pines indicates that the owner is managing and protecting the stand in accordance with standards of the American Forest Institute. The American Tree Farm System began in the Douglas fir region in 1941. Today, there are over 100 million acres of Tree Farms on locations from coast to coast. Most of them are on small, privately owned forests or farm woodlands. The Soil Conservation Service promotes this program with advice on plantings and management and certification of those complying. *(Courtesy, Soil Conservation Service)*

ash. Most helpful to wildlife is that fires open up the woods, providing new leaves, young shoots and other vegetation for them to browse on, and birds find more nesting and feeding areas. Dense forest is not attractive to birds or beasts.

Fire Prevention and Control.—Many forest fires are, of course, devastating. The prevention and control of fires continues to be a cornerstone of good forest management. The record of control by the U.S. Forest Service over the past few decades has been a good one (see Table 12-4).

TABLE 12-4. Area burned over by forest fires in the United States since 1930 (10-year intervals and 1978)

Year	Number of Acres
	(million)
1930	52.3
1940	25.8
1950	15.5
1960	4.5
1970	3.3
1978	3.9

(Sources: U.S. Bureau of the Census, *Historical Statistics of the United States, Colonial Times to 1957*, Washington, D.C., 1960, Table L98–105, p. 318; *Statistical Abstract of the United States, 1979*, 100th edition, 1980, Table 1270, p. 728)

Controlled burning of forests should be practiced only with expert advice and preferably under actual supervision of a forester. Many fires, deliberately started to thin out brush and kill insects, have done the job thoroughly—and killed the trees, too.

Sad and destructive are fires started by carelessness, such as the flip of a cigarette out the car window. The resultant fire, spreading with little warning, may trap game and even man. All of us can help prevent these woodland disasters by being careful with fire.

Selective harvesting trees

Selective harvesting means that trees are thinned out each year in accordance with a plan so that the remaining trees grow faster, taller and healthier. The wood that has been removed is marketed.

FIGURE 12-7. The above shows a poorly managed forest of white spruce. Though these trees are but 6 inches in diameter, they are overmature and many are dead. The areas thus becomes a fire hazard and an eyesore. Proper management would have made this a desirable young forest. *(Courtesy, National Park Service)*

If this program is carried out year after year, a steady source of income is provided, and the forest is said to be under *sustained yield management*.

Clearcutting forest plots

Clearcutting is the practice of cutting off a sizeable block of trees during one harvest period. The wisdom of this practice, which leaves the land temporarily almost denuded of vegetation, is highly debated among foresters. On the one hand, clearcutting is essential for successful commercial reforestation of some species of trees, can provide good forage for wildlife and is more profitable to the timber companies. On the other hand, when steep slopes of thin soil are stripped of timber, severe gullying or sheet erosion results. If large areas are saturated with unwise clearcutting, heavy rains can send their waters unchecked down the slopes to turn streams into roaring, destructive torrents. Furthermore, whether clearcutting is profitable to the timber companies over the long run is questionable.

THE MANAGEMENT AND USE OF FORESTS □ 193

FIGURE 12-8. When cattle have the free run of woodland, they trample and eat down the growth of young trees. They obtain very little feed from the wooded area. For best woodlot management, do not let livestock graze the area. *(Courtesy, Soil Conservation Service)*

□ IMPORTANCE OF SMALL LANDOWNERS

Since some 70 percent of commercial forest land is held by private owners, the destiny of our present timber supply is in private hands. This does not mean, however, that its destiny is in the hands of large corporations, important as they are. For records of about a quarter of a century (1952–1975) show that more than one-half of the saw timber in the United States has been grown on land owned by farmers and other small landowners, while less than one-fifth has been grown on land owned by large timber companies.

Small private timberland owners appear to be doing a better job of forest management than large companies, as the following comment from the Council on Environmental Quality indicates:

An increasingly important national problem in the future may be overcutting of privately held forest land, particularly land held by large companies. A study released in January 1979 by the Library of Congress concluded that the Forest Service, farm, and other small rural landowners appear to be cutting their softwood timber at a rate closer to a long-term sustained yield level than timber companies.

This conclusion is based on Forest Service data on national changes in softwood timber growth, mortality, and harvest over a 25-year period from 1952 to 1977. The report compares the amount of growth for each type of landowner to the amount of timber cut. In 1952, for farm and other private ownerships, which included 58 percent of the total U.S. forest land acreage, the softwood timber cut exceeded growth by 3 percent. Yet, in 1977, despite a 2.5 percent reduction in the amount of land in the farm and other private ownerships category, the relationship was reversed: growth exceeded cut by 65 percent.

In contrast, the timber industry land has shown a softwood growth deficit over the past 25 years. In 1952, the softwood cut exceeded growth by 32 percent. In 1977, even though the timber industry's land holdings had increased by 14 percent since 1952, the softwood timber cut still exceeded growth by 21 percent.[1]

The reason that small landowners are cooperating more is probably because they have learned that forest conservation is profitable. A number of studies carried out over the past several decades have shown that with good management small woodlands can be profitable. In 1954, the Federal Reserve Bank of St. Louis published a study by Clifton Luttrell of a small 208-acre farm in Tippah County, Mississippi, with 92 acres in timber. The study showed that over a 30-year period, cumulative gains from enlightened forestry management were four times that of unplanned management. Luttrell concluded that like increases could be realized in the 30 million acres of timber in Mississippi and Arkansas if given similar improved care. Some 14 years later, the Bureau of Business Research, Indiana University, published an article, "Conservation for the Southern Indiana Uplands," by Robert Menke. Menke pointed out that 92 percent of the woodlands in Indiana were in small, privately owned

---

[1]Council on Environmental Quality, *Environmental Quality, The Tenth Annual Report of the Council on Environmental Quality*, U.S. Government Printing Office, Washington, D.C., 1979, p. 420. The Library of Congress Report cited is Congressional Research Service, *Overview of U.S. Forest Situation*, report from Robert E. Wolf, Assistant Chief Environment and Natural Resources Policy Division to Senator Herman E. Talmadge, January 19, 1979.

tracts and then, echoing the conclusion of Luttrell, wrote: ". . . with little effort the volume of wood in Indiana could be tripled."

Today, potentials for increasing productivity from small woodlands are even more attractive, for although lumber prices have risen, the costs of management involve few cash outlays and thus are affordable. In forest care, farmers can generally use the same equipment they use for farming, with the exception of a chain saw. They can do most of the work, such as thinning, harvesting and cutting fire lanes, in the off-season for crops. And, they can draw upon nearby state or national foresters or experts employed by timber companies for free advice in most cases. If laws were changed to tax woodlands only when they are harvested, and perhaps if some incentive payments were made by the federal government to get the program started, these lands might enable us to meet our forest production goals.

☐ RECREATIONAL USES OF FORESTS AND WOODS

Our forests, as well as smaller wooded areas, are becoming more popular each year for recreational uses. Apparently, the more the living habits of people take them away from forests, woods and streams, the greater becomes their desire to spend time in the woods or near streams or lakes where they can fish, examine trees and shrubs, study wildlife or just rest and let the world go by. We as a people agree with Byron that:

> There is a pleasure in the pathless woods,
> There is a rapture on the lonely shore,
> There is a society, where none intrudes,
> By the deep sea, and music in its roar.

Multiple use of national forests

For many years our national forests were managed primarily with timber production in mind. But, as our traditional recreation areas, the national parks, became more and more crowded, we realized that the national forests, which were often nearby, could provide relief. They had space and scenery which would attract and accommodate people and help take pressure off the parks. In June,

1960, the Forest Service Multiple Use Act was passed by Congress, providing that:

> ... the national forests are established and shall be administered for outdoor recreation, range, timber, watershed, and wildlife and fish purposes.

Fantastic increase in forest visits

There is only one word to describe the increase in visitors to forested areas in recent years—*fantastic!* Annual visits to national forests have increased from about 27 million in 1950, shortly after World War II, to 205 million in 1978. Overnight stays increased at about the same rate (see Table 12-6). Visits to these forests for recreational purposes are increasing at a rate of 10 percent a year, *The Recreation Imperative*, a benchmark report published in September, 1974, by Senator Henry M. Jackson's Interior and Insular Affairs Committee, observed. The Jackson committee predicted that recreation visits will quadruple by the year 2000.

Another forecast, made in various activity categories, and covering range land and waters associated with forests and range lands, is more conservative, but still predicts increases of over three times by the year 2030 (see Table 12-5).

TABLE 12-5. Projected increase in demand for forest and range products

Product	Base Year	Projected Increase in Demand[1]	
		2000	2030
Dispersed camping	1977	133	205
Hiking	1977	117	159
Downhill skiing	1977	178	334
Waterfowl hunting	1977	133	169
Fresh water fishing	1977	139	190
Range grazing	1976	135	141
Timber	1976	164	207
Water consumption	1975	127	159

[1]Medium level: base year equals 100.

(Source: U.S. Forest Service, *An Assessment of the Forest and Range Land Situation in the United States*, Washington, D.C., 1980, p. ix)

TABLE 12-6. U.S. recreational statistics: Visits to recreational areas, 1950–1978 (in thousands)

Recreational Areas	1950		1960		1970		1978	
	Visits	Overnight Stays	Visits	Overnight Stays	Visits	Overnight Stays	Visits	Overnight Stays
State parks	111,291	6,079	259,001	20,569	482,536	50,572	856,000[5]	NA[1]
National parks[2]	33,253	4,501	79,229	9,365	172,005	16,160	283,090	NA[1]
National forests	27,368	5,760[3]	92,592	15,454	172,559	38,477	218,494	NA[1]
Total United States	171,912[4]	16,340	430,822	45,388	827,100[4]	105,209	1,357,584[5]	NA[1]

[1] NA—Not available.
[2] Visits for calendar year 1950 not adjusted for comparability with counting system as modified in 1960. Prior to 1965 excludes visits to the White House.
[3] No category for overnight stays in national forest data for 1950. Includes campground use and hotel and resort visits.
[4] Adjusted by authors.
[5] Estimated by authors by projecting state figures and recent annual rates.

(Sources: National Recreation and Park Association, *State Park Statistics*, 1970; U.S. National Park Service, annual reports on recreational use; Outdoor Recreation Review Commission, *Outdoor Recreation for America*, 1962; U.S. Forest Service, *Recreation Use of the National Forests*, annual; unpublished data from the U.S. Department of the Interior, as cited in the U.S. Bureau of the Census, *Statistical Abstract of the United States, 1973*, Tables 323, 326 and 328; Council on Environmental Quality, *Environmental Quality, The Tenth Annual Report of the Council on Environmental Quality*, Washington, D.C., 1979, p. 700; and U.S. Forest Service, *An Assessment of the Forest and Range Land Situation in the United States*, Washington, D.C., 1980, p. 113)

The U.S. Forest Service points out that these demands are growing faster than the capacity to meet them. "Thus," it concludes, "the Nation is faced with a growing imbalance between supply and the quantity of forest, range, and water products that people would like to consume."[2]

☐ WILDERNESS

Vast tracts of mostly forested land in this nation that contribute little in commercial productivity but contribute immeasurable values in other ways are *wilderness areas*. A wilderness is "an unsettled, uncultivated natural region . . . a piece of land that is set aside to grow wild," the *American Heritage Dictionary of the American Language* tells us. To many Americans today, *wilderness* has come to mean an area set aside by, or proposed for inclusion in, The Wilderness Act. This act, signed into law on December 19, 1964, gives permanent protection to millions of acres of nature sanctuaries formerly subject to the whims of various administrators. Proposals for additions are to be reviewed by citizens before acceptance into The Wilderness System. This is the first law in our nation's history which gives the general public a direct input into federal land-use decision-making. Also, all forms of commercial activities, except for mining by special permit and for a number of years, are prohibited in the wilderness areas.

On December 2, 1980, President Jimmy Carter signed the Alaska National Interest Lands Conservation Act of 1980 into law, thus increasing the wilderness area protection in the United States to 56.6 million acres (see Table 12-7). Conservation System units (parks, refuges and monuments) add another 100 million acres set aside.

Additional acreage is under study for wilderness designation. Late in 1980, the Bureau of Land Management completed a review of 174 million acres in the contiguous United States to be considered for possible inclusion. Of these, 24 million acres would be designated as study areas. The Bureau of Land Management has until 1991 to complete the process. According to The Wilderness Society, although conservationists objected to the exclusion of some areas from wilderness designation, they felt that, except in Utah, the Bureau of Land Management had made a "good faith" effort. In Utah, one-

---

[2]U.S. Forest Service, *An Assessment of the Forest and Range Land Situation in the United States*, U.S. Government Printing Office, Washington, D.C., 1980, p. x.

TABLE 12-7. Lands set aside by Alaska National Interest Lands Conservation Act of 1980 (thousands of acres)

National Park System		National Wilderness Preservation System	
Parks	24,599	Parks and preserves	32,355
Preserves	18,986	Refuges	18,860
Total	43,585	Forests (including forest monuments)	5,362
National Wildlife Refuge System	53,720	Total for the Wilderness System	56,577
National forest monuments	3,206		
Total for Conservation System units (except Wilderness)	100,511		
National Forest System additions	3,350		
BLM national conservation and recreation areas	2,220		

(Sources: Alaska Coalition, *Alaska Status Report*, December 5, 1980, p. 2. [This was the final report of this organization.] For future reference and interpretation of wilderness areas, see periodic reports by The Wilderness Society in various issues of its periodical, *The Living Wilderness*. Environmental impact statements on all areas are also available from federal government sources.)

third less land was set aside for wilderness than conservationists identified as qualifying for such status.[3]

The setting aside of wilderness has had a mixed reception. Conservationists are jubilant. The Wilderness Society has been credited with being primarily responsible for the successful passage of the legislation. Conservationists and ecologists point out the inspirational worth of "getting back to nature," as related for wild rivers in Chapter 10. But, more important, they note, are the opportunities such areas afford to draw upon the unique communities of diverse forms of plant and animal life found there. Such life forms constitute a storehouse upon which we may draw for possible breeding or medicinal uses. They are a base against which we can measure biotic changes. They help us better understand the intricacies of the web of life.

Opponents of wilderness preservation include people living near the areas to be set aside, land developers and timber and mining companies. Local residents fear loss of jobs or loss of potential job opportunities. Businessmen view wilderness preservation as a

---

[3]"24 Million Acres Selected for BLM Wilderness Study," *The Living Wilderness*, Vol. 44, No. 151, December, 1980, pp. 38–39.

"lock up" of resources. One nationally known wildlife expert was against an eastern wilderness proposal because better habitat for wildlife could be provided there by managing the land. Besides, it was not a "pure" (never used commercially by man) wilderness. Opposition to eastern wilderness proposals is especially strong on the latter count—that they are not "pure."

Each wilderness area is unique and should be considered for protection under the act on the basis of its own merits. Each of us, either on his own competence or through being represented by an organization, can participate in the citizen's review process to help determine if protection is justified.

Wilderness is a trust established for all of us. How much respect we give it can be shown by how we act when in the woods. It is only in the wilderness areas that the last vestige of virgin woods, beautiful ferns and rare native flowers persists in abundance. It is apparently impossible for the average vacationist, intent upon enjoying the landscape about him, to appreciate the rare beauties of nature without picking them or digging them up. He thus destroys the beauty he came to enjoy.

Most of us do not possess the information necessary to propagate or even keep alive these spring beauties, lady slippers, trilliums, gentians, arbutus and other rapidly disappearing gems of our woods. We, in our eager, selfish desire to have these beautiful plants with us, do not seem able to carry this beauty as a picture in our memories. We must pick them, and as we pick, we see them wilt and wither in our hands. Thus, both the flowers and the picture are gone.

☐ BETTER USE OF FOREST RESOURCES

As in the case of all of our other resources, we can increase greatly the adequacy of our forest resources by less waste and more efficient use. Progress is being made. More wood chips and other forms of waste products are being used today than formerly. Once-discarded material is being made into particle board. Houses are being redesigned to use lumber more efficiently. Waste paper and newspapers are being recycled. And, plastics and other renewable resources are being substituted for wood in many products. But, we have a long way to go before we anywhere near reach the potentials for savings.

FIGURE 12-9. The woods on the right were grazed by cattle up to the time this picture was taken. The woods on the left had been protected from cattle for 10 years. Young trees were making headway in the ungrazed woods, whereas there were no young trees to replace old or mature trees in the pastured woods. *(Courtesy, Soil Conservation Service)*

☐ THE KEYS TO OUR FOREST RESOURCES

The key to a more abundant timber resource in our nation is good management. The ones who could turn this key to the greatest advantage are the small landowners who control over half of the forest land, which has by far the best potential for higher yields of wood products.

If all our forests were well managed and if economy of use, including substitution, were practiced, our nation might eventually become a net exporter of timber.

No less important than its value as timber is our forest resources' recreational and ecological value in parks and wilderness areas. The largest amount of such land use is in the public domain. We are the trustees. The key to maintaining these values is for us to live up to our individual responsibilities as concerned citizens.

202 □ OUR NATURAL RESOURCES

Don'ts for forest visitors

Whether people are picnicking, camping or traveling through the woodlands and forests as tourists, it is well for them to have a few fundamental rules in mind for their own safety and for the preservation of woods, shrubs and plants:

1. *Don't be careless with fires.* Drench every camp fire with water before leaving it. Be sure to extinguish cigarette or cigar butts before dropping them anywhere in the woods or along roadsides.
2. *Don't leave papers, tin cans and other signs of camping.* Leave every camp site looking as you would like to find it if you were to return to the same camp.
3. *Don't hack, mark, deface or place names on trees or stones.* Remember there are 20 million other people who may follow your example.
4. *Don't stray from marked trails when exploring unless you are "woods wise."* Being lost is frightening, and it can be fatal.
5. *Don't pick any flowers.* Most woods flowers wilt as soon as picked, and the picking of many of our rare flowers weakens and ultimately kills the plants.
6. *Don't dig specimens to plant in your garden.* Many of the rare wild flowers will grow only in special soils, and even when a ball of earth is taken with a plant, the plant soon loses its woodland quality when "potted" in other soil.
7. *Don't disturb anything in the woods.* Even the removal of old logs which cross trails may destroy the home of some woods animal or plant.
8. *Don't depend on someone else to do what you should do.*

□ □ □

QUESTIONS AND PROBLEMS

1. Which makes the faster growth—managed forests or forests which are permitted to restock and grow naturally?
2. Approximately how much of our original billion acres of forest land is left today? How much of it is commercial? (See Chapter 11.)
3. How many management practices to increase productivity of forests can you list? Describe any one of these.

THE MANAGEMENT AND USE OF FORESTS □ 203

4. Name some of the trees that grow in your part of the country and list some of their various uses.
5. Name three enemies of trees, other than fire, which have destroyed completely the species attacked or which are continually being fought to keep the trees from destruction.
6. How can forest fires possibly do any good? Why should they be controlled?
7. What is the difference between clearcutting and selective harvesting?
8. What are the advantages and disadvantages of clearcutting?
9. Why are small landowners so important to our forest resources? What arguments would you use to try to persuade them to improve management?
10. What is meant by "multiple use" of forests? How can multiple use in a national forest help a national park?
11. What is a wilderness area? How does The Wilderness Act protect it? How can you help provide for and take care of wilderness areas?
12. What are some of the ways by which we can extend our forest resources by better use?
13. Name some substitutes for wood that are being used in your community.
14. List the eight "don'ts for forest visitors." Which two would you place at the top of the list? Why?

13 □ □ □

The origin and depletion of soil

IN OUR LARGELY INDUSTRIAL SOCIETY, with less than 5 percent of our labor force directly engaged in agricultural production, the study of our soil resources is apt to be neglected, except by farmers and a few specialists. Yet, two-thirds of all businesses are said to be agri-businesses—businesses whose economic activities are dependent upon agricultural products in one way or another. And, no resource is more essential to our well-being, and to life itself, than soil. All of us need to appreciate and to understand the nature of this resource so that we can support and carry out the practices that will help preserve it. If those of us who are nonfarming citizens, for example, fail to demand that the federal government support conservation farming programs, not only will our soil be lost, but our nation will too! The first six of the following chapters describe soil processes and soil conservation in agriculture. The seventh chapter of this group, Chapter 19, which deals with the city, indicates the significance of soil in urban situations.

□ FORMATION OF SOIL

Soil, most simply defined, is that part of the earth's mantle which supports plant growth. Overlying the earth's crust, which is of rock, is a covering of loose materials called *regolith* (derived from the Greek; *rega*, meaning "covering" or "blanket," and *lith*, meaning "stone"). From this loose material soil is formed, eventually consist-

FIGURE 13-1. The immense sand dunes in the southwestern part of the United States represent one of the many long steps taken by nature in reducing rock to soil. These dunes will not become soil for many thousands of years. *(Courtesy, National Park Service)*

ing of a combination of air, water, minerals and decayed plant materials (humus).

Soil formation is a complex process. The top part of the rock mantle of the earth is subject to the action of rain, heat, cold, sunshine and the mild acids of plant roots so that its composition and structure are changed until it is quite different from the lower parts (see Figure 13-1). The roots of the plants which grow on the surface penetrate the upper part of this mantle in all directions. When the roots die, they become part of the mantle, and the surface growth falls to form a covering over the mantle.

The upper part of the mantle is honeycombed with small burrows caused by angleworms, beetles, moles, gophers and a host of other insects and animals as they search for food or seek protection from enemies. Even smaller forms of life, such as bacteria, molds and fungi, live in and on the mantle. They help in the decay of plant roots as well as in the process of making smaller particles out of the larger pieces of the mantle. As a result of all these processes, the surface layer of the soil is formed.

The mantle, or combination of topsoil and subsoil, was formed by the disintegration of the underlying rock or was transported to its present location in one of several ways.

The sedentary or residual deposits, or soils, which were formed over the rock from which they were derived, have two general sources. The most usual source of these soils is the disintegration of underlying rock. The rock below was broken up into fine particles or disintegrated through the combined action of water, heat, cold, etc., to form a covering. Much of this covering was eroded, while other parts were dissolved in water and leached (drained off) through this covering. The material that was left forms the present residual deposits, or residual soil.

The second source of residual soil is the result of decaying plant growth in the flatter, poorly drained places. Peat, muck and swamp soils result from this process and form a small part of our general soil covering.

FIGURE 13-2. The soil in this picture is alluvial, or water formed. The soil was deposited from quiet water, from a lake or a pond. It is called "lacustrine." The strip of grass between the onions reduces wind erosion. *(Courtesy, Soil Conservation Service)*

FIGURE 13-3. Colluvial deposits at the base of the Great Organ formation in the Capital Reef Monument, New Mexico. Deposits of this type are more common in areas of low rainfall. *(Courtesy, National Park Service)*

The transported soils, or soils which have been moved by wind or water, were formed by one or more of the following processes: (1) transportation by water—alluvial deposits (see Figure 13-2); (2) transportation by winds—aeolian, or loessial, soils; (3) transportation by glaciers—glacial soils; (4) accumulation of deposits at the base of cliffs—colluvial deposits (see Figure 13-3).

☐ SOIL FORMATION—A LONG PROCESS

It takes millions of years to form the rock mantle, for it requires the disintegration of many feet of the original, or parent, rock to form one foot of mantle. The formation of the top part of the mantle, or soil, which supports plant growth, requires a greater thickness of the parent rock and a much longer time to form than any other part of the mantle of equal thickness.

When nature is left to make and rework its soil without the interference of man, the rock mantle continues slowly to increase in thickness through the disintegration and breaking up of the rock upon which it rests. This accumulation at the bottom of the mantle tends to increase at about the same rate at which soil is lost from the top of the mantle. Thus, the thickness of the mantle under natural conditions tends to remain about the same, century after century. This balance of losses in soil from the top of the mantle with increases at the bottom of the mantle is spoken of as a condition of equilibrium in nature. This equilibrium, or balance, in nature is possible only because of the continued presence of plant growth and accumulation on the surface of the mantle.

How man upsets nature's balance

When man begins to use the soil for agricultural production, he destroys the balance which natural erosion has set up because he substitutes a sparce-growing plant for the natural denser growth and he takes the growth off the soil instead of letting it drop and accumulate over the top of the soil. He uses up the organic matter of the soil by his plowing and seeding operations and makes the soil less porous. His crops take out the plant nutrients which have accumulated in the soil; thus less plant growth is produced, making a thinner covering for the soil. The final result is a washing away of the soil by rains which previously were absorbed or a blowing away through dust storms in the drier parts of the country.

Why some soils take much longer to form than others

No one knows how many thousands of years it takes for nature to pulverize the original rock of the earth in order to form the mantle, or covering of loose material, from which the soil finally is derived. We do know that scratches left on the face of our harder quartzite rocks by the glaciers are still plainly visible after weathering for 25,000 to 50,000 years. More than that, we know there are polished surfaces on rock which have not lost this polish in 100,000 years. Gutzon Borglum, the sculptor who created the Mount Rushmore National Memorial in the Black Hills of South Dakota, once said that erosion will reduce the surface of this sculpture by not more than one inch in 100,000 years.

The rates of disintegration of some of the softer rocks of the earth, such as limestone, are much faster than that of quartzite. The rates of breaking up these rocks are much faster, too, in climates with alternate freezing and thawing temperatures than in one in which the temperature is constantly above freezing or is almost constantly below freezing throughout the year. Another factor is moisture. A moist climate is thought to lead to faster disintegration of rock than a dry climate because both chemical and mechanical weathering take place under moist conditions. In a dry climate the disintegration of rock is largely a mechanical process—the result of the action of wind.

How limestone rock makes soil

Some geologists estimate that the face of limestone rock disappears at the rate of about one foot in a thousand years. Most of the limestone, or calcium carbonate, is slowly soluble in water and washes away with the water run-off. The remaining impurities form the residual mantle. Let us assume that the limestone rock has 5 percent impurities. If none of the impurities are carried away, it will require 20,000 years to build up one foot of mantle. But, it is more likely that most of the impurities will wash away. If we assume three-fourths of the impurities are carried away, it will require around 80,000 years to make one foot of rock mantle from limestone rock. This may seem like a long time, but it requires a much longer time to make an inch of soil from this mantle than it does to make several inches of the mantle itself.

Whether these estimates are high or low, once the precious mantle of soil is gone from limestone rock, this valuable material is gone forever as far as our generation or even the life of our nation is concerned. We think of limestone soils as among our most productive soils. Soil making from the original rock formation of the earth is such a slow process that we can safely say that no significant amounts of soil have been created since the beginning of human history.

□ WHAT IS EROSION?

The formation of soil is one of the steps taken by Nature in reducing mountains and leveling hills. It is a "way station used by the rock of the earth as it journeys to the sea." The vehicle used to make this transition is erosion. Soil is built by erosion. It is also destroyed by erosion. Soil is built through the ages by geological erosion. It is destroyed in years by man-made, or agricultural, erosion. The difference between geological and man-made erosion is essentially one of degree, not of kind.

The same forces of nature cause both kinds of erosion. Wind and water, heat and cold are the agencies which throughout the ages have made soil building possible. They are also the agencies which can destroy the land in a few short hours. Erosion began when the first rains fell. It cannot be stopped, although it can be retarded and held in check. The discussion of the formation of the rock mantle in the early part of this chapter is in reality a discussion of the results of erosion. Erosion carried away much of the original rock, leaving only a fraction of the total quantities of disintegrated rock. This residual fraction created the mantle; the eroded and leached parts were carried away to the sea where they were deposited and later became rock. Thus, wherever erosion occurs, deposition also occurs. Whenever soil is washed away, it must be deposited in some other place. In ages to come, this deposition may again be raised above the water from which it was deposited and so later become soil.

□ KINDS OF EROSION

There are two types of erosion which are destroying our soils and adding to our waste land—wind erosion and water erosion. Wind erosion takes place where there is little rainfall or where the soil does not support a good vegetative growth. The most disastrous effects of wind erosion are produced when land is plowed and left bare in areas of limited rainfall. Water erosion, on the other hand, takes place wherever rain falls. It, too, has resulted in most serious damage where the lands have been cultivated. The term *soil erosion* or *erosion* as used here refers mostly to man-made erosion and not to the small amount of normal or geological erosion which occurs with good vegetative cover and proper soil management.

212 ▢ OUR NATURAL RESOURCES

Water erosion

Wherever there is rainfall, there is the possibility of water erosion, and wherever water runs off over the surface of the soil, there is sure to be erosion. The amount of soil washing, or soil losses, depends upon the total amount as well as the intensity of individual rains, the length and steepness of the slope, the type of soil, the kind and density of the crops covering the soil and the method of cultivating the soil (see Figure 13-4).

The surface run-off, roughly estimated to be about 30 percent of the total precipitation (rainfall and snow), results in either sheet or gully erosion. Where the water moves off over the surface as a sheet, it takes off a complete layer of soil and is true sheet erosion. Very little of the water actually runs off any field in smooth sheets. It usually collects in numerous little rills which may become larger and

FIGURE 13-4. The lower, darker 8 feet of soil were laid down by water (alluvial soil) thousands of years ago. Grass growth has added organic matter to the top 2 feet of this layer, causing that area to be darker in color than the soil lower down. The top 4 or 5 feet of lighter, stratified soil, also alluvial, were washed down from nearby hills after being broken up and put to crops. The cut of 11 to 12 feet, which shows the soil profile, is the result of recent erosion. *(Courtesy, Soil Conservation Service)*

less numerous as the bottom of the slope is neared. As long as the rills can be plowed across and the field smoothed again for the next crop, this rilling is designated as sheet erosion. The concentration of the water run-off into larger channels which cannot be filled in by plowing across the fields is the start of gully erosion. These gullies or channels become larger with each rainfall and ultimately may so cut up a field that it has to be abandoned.

Gully erosion is conspicuous and is more startling in appearance than sheet erosion. It was the appearance of gullies in fields that wakened the public to the seriousness of our erosion problem. Great have been the losses to agriculture by this type of erosion.

Sheet erosion, on the other hand, is more insidious and more dangerous than gully erosion. It is not so readily recognized, and much of the topsoil may be washed away before the farmer is conscious of the damage done. Because it is possible to plow across fields which are subject to sheet erosion and because the evidence of the damage is thus hidden by each plowing, the process may continue until practically all the topsoil is gone. Where sheet erosion does not lead later to the formation of gullies, it frequently is difficult to alert the farmer to the type of problem he must face if he is to continue to produce crops. We all have seen the smooth, rounded, light-colored knolls or hilltops in a field which is covered everywhere else with a darker soil. When this field was first broken and planted to crops, the soil on these higher points was the same color as the other parts of the field. The change in the soil color on these knolls is the result of sheet erosion. The darker-colored topsoil which once covered these higher parts of the field is now on the lower parts of the field or has washed off the farm completely.

Wind erosion

Wind erosion takes place only where the soil is dry and little or no vegetative cover is maintained on the field. A soil which is very loose and fine requires more vegetative cover to keep it from blowing away than one which is coarser. Some land that is too closely pastured, or that has been burned over, is subject to wind erosion. Wind erosion is especially noticeable in the Great Plains area of the United States because of the semi-droughty character of the country and the velocities of the prevailing winds. High winds and dry soil may come at any season of the year. If they come after the ground is

FIGURE 13-5. This approaching dust storm, or black blizzard, sometime during the 1930's, shows how our loessial soils were made. Many feet of wind-formed soils are deposited throughout the semi-arid parts of the country. It required hundreds of years of dust storms, such as the one shown above to deposit the loessial soils of the prairies of the Midwest. All wind-formed soils must first be made by some other forces of nature before they can be picked up by the wind and transported elsewhere. *(Courtesy, Soil Conservation Service)*

plowed and before a good growth of crop is produced, the whole depth of the plowed soil may be blown away.

DRY FARMING WITH DUST MULCH ENCOURAGES WIND EROSION

A generation ago farmers in the semi-arid areas of the Midwest were encouraged to make a dust mulch of the soil surface in order to keep down weed growth and to conserve moisture. Although in this condition no weeds grew to use up the moisture of the soil, the flour-like dust mulch also prevented the water from being absorbed by the soil. When large areas adopted this practice, the situation was set for wind erosion and the historic dust storms that occurred.

The strong winds of the prairie broke up the slight crust of the surface, and "black blizzards" followed (see Figure 13-5). It is estimated that during one day's dust storm in the "dust bowl" of the Oklahoma, Texas, Colorado and Kansas wheat lands in 1936, as much as 300 million tons of fertile topsoil were blown away. Dust from this area was deposited over the eastern half of the United States, and dust clouds were observed 300 miles out over the Atlantic. Not only were these storms destructive of soil, but they were also a most terrifying experience to live through. It was almost impossible to breathe, the air was so full of dust. Children could not find their way home from school, and all traffic was completely stopped because no light could penetrate the blackness of a mid-day storm.

Drifts as large as winter snow drifts buried fences, smothered buildings and sifted through windows and doors in unbelievable quantities. Some attics were so filled with dust that the plaster of the ceilings below gave way. The winds took the soil from where it was most useful and left it where it could do the most harm. It is not surprising that many people, now destitute, gathered up a few possessions and left this desolation (see Figure 13-6).

FIGURE 13-6. Wind erosion following soil depletion caused the loss of this farm, its neighbors and the community. Only a forsaken church bell and a few abandoned buildings remain to tell the story of lost effort and shattered hope. *(Courtesy, Soil Conservation Service)*

□ SOIL DEPLETION

Another aspect of soil erosion which adds to the problems of crop production is the depletion of the soil of its organic matter and plant nutrients. Dr. H. H. Bennett is the authority for the statement that more than 90 million tons of the five principal elements of plant food are contained in the soils that are lost each year through erosion. Forty-three million tons of these consist of phosphorus, potash and nitrogen—the three plant nutrients used in commercial fertilizers. If this quantity of fertilizer were supplied to our croplands, it would be equivalent to 225 pounds on every acre of cropland throughout our country. Calcium and magnesium, though they are necessary for the best production of most crops, are not as scarce or as valuable as the three aforementioned elements.

This loss of plant nutrients is in addition to the loss through the depletion by crops, which is not replaced, and through leaching by the water that drains through the soil. It is estimated that the amount of plant food lost through leaching is equal to that lost through erosion and that the amount lost through erosion is more than 20 times the amount removed by farm crops. Thus, we see that losses through leaching and erosion are many times the losses through crop production.

□ BUILDING UP, TEARING DOWN

Each time a piece of rock falls from a cliff or is forced from its parent rock by the action of water or frost, the first step is being taken by nature in the ultimate formation of new soil. On the other hand, every muddy stream or every cloud of dust carried away by some vagrant wind is evidence of the erosion of the soil that took Nature ages to create. Thus, the great cycle goes on through eons of time.

□ □ □

QUESTIONS AND PROBLEMS

1. What is the rock mantle, or regolith, of the earth?
2. How is this mantle formed?

3. What happens to the mantle so that the upper part becomes soil?
4. What is meant by equilibrium in nature?
5. What happens to the soil when man destroys the balance, or equilibrium, of nature?
6. How long did Gutzon Borglum estimate it will take for the natural forces of wind, rain, heat and cold to cause the loss of one inch of the surface of the Mount Rushmore National Memorial?
7. About how many years does it take to reduce the face of limestone rock by one foot?
8. If we assume that limestone rock has 5 percent impurities and that none of these impurities are carried away by the water which dissolves the limestone, about how many years will it take to build a foot of mantle?
9. Does it take more or less time to make soil from the mantle than it does to make mantle from the limestone rock?
10. What is the difference between natural erosion and man-made erosion?
11. What are the two kinds of water erosion, and how do they differ?
12. What conditions make for wind erosion?
13. What is a dust mulch, and is it desirable to maintain such a mulch? Give reasons for your answer.
14. How does soil depletion differ from soil erosion? Give the proportions of the loss of plant nutrients by (a) erosion, (b) leaching and (c) soil depletion.
15. Was the soil near your home made from limestone, sandstone or granite?
16. Was the soil in your community deposited there by wind or water? If by water, was it deposited by the slowed-down action of running streams, or was it formed by still lake waters?

14

*The extent of
our soil resources*

THERE IS NO MORE IMPORTANT NATURAL RESOURCE than our land, that is, the solid portion of the earth's surface. Soil is defined as that part of the land surface of the earth which supports plant growth. Not only is the soil the source of all our vast forest resources, but from it also comes practically all our food and clothing. Thus, our present living standard is the result of the productivity of our soil.

□ SOIL RESOURCES ONCE CONSIDERED INEXHAUSTIBLE

In colonial times there seemed to be an inexhaustible quantity of land, and the only cost to the colonists of obtaining cropland was that necessary to clear the land of its tree growth. The cost of clearing a piece of land was all that made it valuable. The first president of our nation, George Washington, in writing to Arthur Young of England concerning the use of land for the production of crops in the colonies, made the statement that it was easier to clear new land for farming than it was to attempt to maintain the productivity of land which had been used several years for crop production.

Even as late as 1909, Dr. Milton Whitney, Chief of the Bureau of Soils of the U.S. Department of Agriculture, wrote that "the soil is the one indestructible, immutable asset that cannot be exhausted, that cannot be used up." And, this mistaken view of soil resources as virtually unlimited unfortunately persisted for decades.

220 □ OUR NATURAL RESOURCES

□ CHEAP LAND PERSISTED

Until about 1862, when the Homestead Act was passed, any man could get land for farming from the various states or from the federal government by moving onto it and taking possession. After that time public land could be homesteaded by paying a small fee, $1.25 per acre, and by living on the piece for a very few years.

Since those early days, farm land has risen in price until farmers now pay many times as much per acre as a whole 160-acre farm cost during the latter part of the 19th century. This means that soil is now recognized as a valuable but limited natural resource. Unfortunately, however, industrial, commercial and residential uses still continue to take over prime farm land.

□ THE EXTENT OF OUR SOIL RESOURCES

The total land area of the United States, exclusive of its outlying possessions, is approximately 2,271,600,000 acres. The cropland is divided into the uses shown in Table 14-1, the latest such comparison available.

The 413 million acres classed as cropland is an average of 1.93

TABLE 14-1. Cropland use in 1977, compared to the *1967 Conservation Needs Inventory*

Cropland Type	1967	1977	%Change
	(millions of acres)		
Row crops	160.4	203.2	+27
Close-grown crops	100.6	104.9	+ 4
Rotation hay and pasture	50.4	30.2	−40
Hayland	27.2	33.1	+22
Orchards, vineyards, etc.	5.1	5.5	+ 8
Other cropland	61.9[1]	6.8	−89
Summer fallow	32.0	29.3	− 8
Total	437.6	413.0	− 5.6

[1]Other cropland in 1967 included conservation acres and idle cropland (set-aside), as well as "other" cropland.

(Source: USDA, Soil Conservation Service, *1977 National Resources Inventory*, U.S. Government Printing Office, Washington, D.C., December, 1977)

acres per person; whereas, in 1900, the average was 4.50 acres. Our cropland use has been decreasing in the past 10 years (see Table 14-1), while our population has continued to increase. Population is expanding at the present time at the rate of 2 to 3 million per year. If this rate of increase continues, there will be some 30 million, or 15 percent, more people to feed and clothe within a decade.

Our food supplies will probably be large enough to feed this number of people, but even now we are faced with the problem of supporting our larger population and helping meet world food shortages. How can this be done? The three possibilities usually mentioned are to (1) import more food and export less, (2) consume less food or eat more cereals and less livestock products and (3) produce more per acre and increase the amount of land in cultivation.

The problem with trying to import more food is that most of the other countries of the world have less agricultural productivity per person than we have. With more than half the people of the world already on an inadequate diet, other countries are looking to us to export, not import, vital food supplies. Of course, we will have to limit exports, if we need the foodstuffs at home.

Consumption of less food is a desirable goal for most Americans. We are the first great nation in the world whose population as a whole has suffered from eating too much rather than too little! While expert opinion may be divided on just how much we can cut meat consumption without impairing our health and vitality, it seems reasonable that some reduction could be made. Some of the world's strongest animals, such as gorillas, are vegetarians; and at least one of its most witty men, George Bernard Shaw, was a vegetarian. Brawn and brains are thus not exclusively the product of red meat ingestion!

The third way to increase our food supplies, through producing higher yields or putting more land in cultivation, deals with the resource base. The balance of our discussion of agriculture is largely concerned with examining these possibilities.

If all our swamp land could be drained and all our plowable pasture plowed up, we would add about 10 percent to our total cropland. To drain all the swamps and wetlands would be folly. Migratory fowl and some animals depend on these lands for habitat, and such lands help store and purify our water supply. We also know that it is poor conservation to plow up all our plowable pasture because much of this is on the rougher, more sloping parts of farms. We may think it possible to obtain more cropland from some of the grazing land, but this land is practically all in the semi-arid part of the country where crop production is low because of the limited and variable rainfall. Besides, the area is subject to violent thunderstorms and strong winds which greatly increase erosion hazards when the land does not have permanent vegetation cover. If properly managed, grazing land will produce as much feed without destructive erosion and losses as it would if plowed up for crops. So, we cannot expect to increase food production much, if any, in this way. The best way to increase our food supplies is to make better use of our present cropland.

It now appears that by using the 438 million or more acres that can be cropped and by more intensive use of all our cropland, we can so increase our production as to feed our growing population for several decades to come. This seems possible if we use new, improved varieties of seed, proper and timely application of fertilizers and efficient weed and pest control and if we make more efficient use of our water resources.

Over 58 million acres of our good cropland are now being irrigated as a supplement to rainfall and other forms of precipitation. Another 155 million acres are having their excess soil moisture removed, and an additional 50 million acres of flood land receive flood protection. There are also another 115 million acres of flood land that ultimately may be protected from flooding. But, as is noted in the following paragraphs, much of our soil is being lost to soil erosion, and much of our cropland is being diverted to nonagricultural uses.

Much of our land is forest soil

About one-third of our national area, including Alaska, is forest land. This figure includes woodland on farms, forest areas set aside for water management and commercial timberland. Forest land ordinarily is less subject to erosion than are lands used for other purposes. Land which carries a good forest cover will absorb most of the rainfall with little or no direct run-off (see Figure 14-1).

The sponge-like nature of these soils, resulting from many years' accumulation of decaying leaves, twigs and bark, makes an ideal absorbent for water. The excess water requires days to drain off by slow percolation through these soils, rather than rushing off over the surface within a few hours. In unpastured woodlands and

 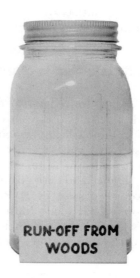

FIGURE 14-1. Water run-off from land carries with it the finest parts of the soil and much of the plant nutrients. This gives the water the dark color shown in the jar labeled "run-off from corn-field." Run-off from protected woodlands does not carry soil particles and plant nutrients and is practically clear, as is shown in the right-hand jar. *(Courtesy, Soil Conservation Service)*

forest areas which still have fairly dense stands of trees, good soil condition is thus actually being built up instead of being lost through erosion.

Marshes and swamps make more soil

There is about one-fourth as much land in marshes, swamps and other low, flat areas as in forest land. Little or no soil losses take place in these areas. The land in these marshes and swamps actually may be adding to its small soil content as soil-laden water drains into it from neighboring hills. The depositing of soil in these low lying areas in this manner is sometimes called sedimentation, or silting. Also, the deposits of organic matter left by the rank plant growth in these low spots finally fill the areas, and swamps first become peat, and later muck soil. This is an exceedingly slow process, however, requiring hundreds of years to fill in a marsh or a swamp. Even

224 □ OUR NATURAL RESOURCES

then, there must be some way to control the water level, either by drainage outlets or by dams if such an area is to be used for crop production.

Nearly one-third is grazing land

Nearly one-third of our acreage is public and private grazing land. These lands are in regions of low rainfall, so that without irrigation they will not support ordinary crop production. It is for this reason that they are not classed as cropland. About one-fourth as much land now in farms is classed as pasture land.

These lands are all subject to severe erosion if overgrazing takes place (see Figure 14-2). Practically all these grazing lands are subject to occasional heavy rainfalls and flash floods. Since there is little grass growth on these heavily grazed areas, the flood waters rush off in torrential streams which cut channels deeply into the surface soil.

GRASSES STOP GROWING

The damage to these lands is not limited to erosion, even though that is bad enough. The loss of moisture is just as serious.

FIGURE 14-2. One of the big failings of farmers and ranchmen is to overgraze pasture land. The first effect of this condition is poor cattle. Later, when rains come, the soil moves off with the water. *(Courtesy, Soil Conservation Service)*

FIGURE 14-3. Gravel, sand and dirt from overgrazed pasture land washed down into this broad valley, covering the rich soil and reducing its value for crop production. Much of the pasture land was also destroyed. *(Courtesy, Soil Conservation Service)*

Before the lands were overgrazed and all grass cover was removed, these broad, level or slightly rolling prairie areas with their abundant covering of green or dry grasses absorbed most of the meager rainfall, even though much of it fell during thunderstorms, which are always of short duration but frequently of severe intensity.

Since most of the rainfall was absorbed into the soil where it fell, a good growth of grass followed. This growth was not cut or heavily grazed by the animals that roamed the region, so that it remained to catch the next rainfall. The result was a continuous covering of grass which easily survived moderate wildlife grazing.

Wherever one of these broad basins or stretches of grass is overgrazed and gullies are cut through, then rainfall immediately rushes off so that even the areas between the gullies, or ravines, absorb very little water. The result is no grass growth on any of the land (see Figure 14-3). Useless, sparse weeds or inedible shrubs replace the grass. Thus, a whole grazing area is lost.

PASTURES LOSE SOIL

Lands classed as pasture lands ordinarily are found on farms rather than on ranches. These areas receive from 25 to 50 inches of rainfall yearly, whereas the ranch country, where most of the grazing lands are located, receives less than 20 inches.

Much of the pasture land on our farms is eroding or will erode if not handled correctly. The same situation that causes grazing lands to erode is causing soil losses on these lands. Summer rains frequently come as thunderstorms of high intensity similar to the storms of the ranch country. Since these pasture lands receive more rainfall than the public and private grazing lands, greater care must be taken to keep these soils from washing.

These lands usually represent the steeper slopes of farms, or they may have stony outcrops or other hindrances to cultivation which place them in the pasture land class. Some fields that are now classed as pasture land were originally good farm lands. They have lost their topsoil or have been so cut with gullies that they can no longer be used effectively for crop production.

Another reason for large soil losses on these lands is the practice of pasturing them so early and so heavily that little or no plant growth remains on the surface. In most instances, farmers mistakenly think they must do this in order to "make the lands pay." The result is that most of the rainfall runs off in rivulets which come together farther down the slope to cut gullies across the fields. As Table 14-2 shows, the Soil Conservation Service (SCS) estimates that the condition of the nation's nonfederal range lands has improved markedly since a similar range survey was done by the Soil Conservation Service in 1963. The 60 percent that still remains in fair or poor condition remains a challenge, however, in light of increasing demands for red meat and pressures to produce more livestock on grass and less on grain.

Improving all our range lands to good or excellent condition

TABLE 14-2. Trends in range condition class on nonfederal range lands

Condition	1963	1977	Change
		(%)	
Excellent	5	12	+140
Good	15	28	+ 87
Fair	40	42	+ 5
Poor	40	18	− 55

(Source: USDA, Soil Conservation Service, *1977 National Resources Inventory*, U.S. Government Printing Office, Washington, D.C., December, 1977)

not only would improve environmental quality and reduce soil erosion and water pollution but also would provide grazing enough to produce an additional 600 to 800 million pounds of meat annually.

About one-fourth is cropland

The most valuable of all our agricultural resources is our cropland. It must supply us with practically all of our food and most of our clothing, as well as a considerable quantity of material used in industry. Dr. H. H. Bennett, the "father" of the Soil Conservation Service, stated that civilizations of the past have risen and fallen with food supplies. An example of this process is the country of Somalia, where under pressure of drought and war, 1.5 million refugees have fled from a desert area.[1]

FIGURE 14-4. These small gullies were made on a field in Iowa after the crop was taken off. The slope of the field is 8 percent, and fields with that slope are commonly used for crop production. The loss of surface soil here is estimated at 100 tons per acre. *(Courtesy, Soil Conservation Service)*

---

[1]Robert Paul Jordan, "Somalia's Hour of Need," *National Geographic Magazine*, Vol. 159, No. 6, June, 1981, p. 757.

228 □ OUR NATURAL RESOURCES

How well are we taking care of our farm cropland? Is it losing its surface soil as rapidly as our grazing and pasture lands? The story here is brief and tragic (see Figures 14-4 and 14-5). In the past two hundred years, which is a short time in the life of a nation, this country has lost approximately one-third of its topsoil, according to SCS estimates.

The 1977 *National Resources Inventory (NRI)* of recent estimates of erosion losses is shown in Table 14-3.

The survey indicates an average annual soil loss of 7.8 tons per acre. Other state and regional studies indicate, however, that this is a very conservative erosion estimate.

There is also evidence of low estimates in states west of the Rocky Mountains. For example, in the Palouse region of eastern Washington, the *NRI* estimate on Class IVe cropland was a 3-ton loss per acre, while a new River Basin study estimated a 24-ton loss. On Class VIe cropland, the *NRI* still estimated a 3-ton loss per acre, while the field study showed a 55-ton loss. (For a description of vari-

FIGURE 14-5. A 2-inch rain that fell in two hours caused the damage shown above to a Michigan field which was being summer fallowed for wheat. Many tons of valuable topsoil were forever lost from this field during that two-hour period. *(Courtesy, Soil Conservation Service)*

TABLE 14-3. Estimated average annual sheet and rill erosion on cultivated cropland, by selected capabilities

Capability Class	Row Crops		Close-Grown Crops		Total Cultivated Cropland	
	Nonirr.	Irr.	Nonirr.	Irr.	Nonirr.	Irr.
	...........(average tons of soil loss per acre per year)...........					
I	4.1	2.2	2.2	1.3	3.6	2.0
IIe	7.8	3.4	2.9	1.7	5.5	2.9
IIIe	15.1	4.7	3.6	1.6	7.8	3.3
IVe	20.6	4.6	5.9	1.2	10.7	3.1
VIe	37.3	8.4	10.0	1.5	19.4	6.2
Average All classes	7.8	2.9	3.1	1.4	5.5	2.3

(Source: USDA, Soil Conservation Service, *1977 National Resources Inventory*, U.S. Government Printing Office, Washington, D.C., December, 1977)

ous classes and subclasses, see the next chapter. The *e* indicates that erosion is the main limitation.)

Average annual wind erosion loss on cropland in the 10 Great Plains states is 5.3 tons per acre. These annual rates for each state range from a low of 1.3 tons per acre in Nebraska to a high of 14.9 tons per acre in Texas. Marginal cropland is the major problem. For example, Texas has 2 million acres of Class IVe cropland, losing 57.5 tons per acre per year. The erosion rate for Class VIe is almost double at 101.6 tons per acre per year.

□ USE OF SOILS IN URBAN AREAS

Our soil resources are also being used for many nonagricultural uses. These land uses include housing, schools, industry, recreation and transportation. Seventy-five percent of the land converted to urban uses has been from former cropland, including high value irrigated cropland. In 1967, 61 million acres were used in this way; by 1977, these had increased to 90 million. If this trend continues, by 1985, urban and built-up land use will occupy twice the acreage it did in 1967.

When land is changed from agricultural use to urban use, the problems of erosion are always present. When the surface is covered with roofs, parking lots, roads and other like construction, the water does not enter the soil. Water running off these surfaces causes local flooding and contributes to inundation in natural flood-prone areas. When houses, roads, etc., are being constructed, the soil is disturbed, and the vegetative cover is often removed. The soil in this condition is easily eroded when rains occur. This erosion contributes greatly to the sediment (silt) carried in our rivers and streams. Such soil losses can be as large as 200 tons of soil per acre per year.

□ GOOD SOIL IS LIMITED

In this chapter we have called attention to the vast soil resources of this country, but only about one-fifth is cropland. Also, losses of valuable soil have been enormous and will continue to be large for some years to come, even though many steps are now being taken to reduce them. We must think of land as a most valuable heritage for our children and our children's children and no longer permit it to be wasted or misused according to the whims or vagaries of individual owners.

□ □ □

QUESTIONS AND PROBLEMS

1. Do you agree with Dr. Whitney, Chief of the Bureau of Soils, who wrote in 1909 that "the soil is the one indestructible, immutable asset that cannot be exhausted, that cannot be used up"? Explain your answer.
2. We have approximately 2,271,600,000 acres of land in the United States. How much of this is cropland?
3. How many acres of cropland per person do we have?
4. How much will we have to add to our present cropland if, by 1990, we keep the same number of crop acres per person we now have?
5. What are the problems we must solve before we can increase our production by adding our swamp land or our grazing land to our cropland?
6. Would it be desirable to plow up our forests and use that land for cropland? Give reasons for your answer.
7. Tell how our marshes and swamp land ultimately will make for more cropland, even though it may be a few hundred years hence.
8. What do you consider to be the best use for the grazing land we now have in this country? Explain your answer.

9. What happens to the land when these grazing lands are overgrazed?
10. How many acres of pasture land do we have on our farms? What causes soil losses on these lands?
11. Since our cropland is our most valuable land resource, one would logically expect it to be conserved more than our other resources. Tell what has happened to this resource during the past two hundred years.
12. Can we continue to use this cropland to produce food for us to eat and feed for our livestock and still conserve the soil? Explain your answer.
13. What is needed to make the soil in your community produce more?
14. What percent of the land converted to urban land use has been from former cropland?

15 □ □ □

Soil conservation on farm and ranches

□ CURRENT LOSS OF SOIL STILL LARGE

YOU ALREADY KNOW that we must consider our soil as a resource which is not permanent and which is renewable only over a very long time. Dr. H. H. Bennett pointed out a generation ago that we were losing our soil at the rate of ½ million acres a year. We still are. Every year we are losing more than 3.5 billion tons of soil, the equivalent of 6 inches of soil on 5.5 square miles of land surface. At present prices, this would be several hundred billion dollars' worth of fertility.

The public is becoming aware of the seriousness of this loss and is giving its support to programs to educate and train people about the need for conserving this resource. Hopefully, these programs not only will reduce the yearly loss of soil but ultimately will build back into the soil some of the productivity which it has already lost.

Conservation specialists, as well as country agricultural agents, are employed in every part of the country to bring this problem to the attention of farmers and to show them ways of controlling erosion and of again building up their soils.

□ GOOD CONSERVATION

Any measure which helps to keep the land productive is useful in conserving the soil. Soil conservation includes any practice, structure or device which will keep the land in place and make it produce more. This is a restatement of the definition often cited that "good conservation means the use of a resource in such a manner as best to serve the needs of mankind."

The three general problems which must be met in conserving our soils are how best to (1) keep the topsoil from washing or blowing away, (2) keep the soils from becoming depleted of their plant nutrients and lime and (3) build up soils so as to improve their water-holding capacities and their plant nutrients. When these things are accomplished, better crop production will be inevitable.

Special soil conserving practices

Let us look at soil conservation as a series of practices which, when put into effect, will slow down or stop erosion or will build up the soil. Different soils will require the use of different practices. Also, one soil which is on a gentle slope may require practices which are quite different from those needed for the same soil if it happens to lie on a steeper slope. And, soils on long slopes may require somewhat different practices from identical soils on very short slopes.

Some practices can be adopted by individual farmers regardless of what their neighbors do. Strip cropping, terracing, grassed waterways, pasture renovation and conservation rotations are examples of these practices. Other problems in soil conservation can be solved only, or best, through project action. This might be two landowners working together or many landowners in a large watershed working in concert.

Preventing surface run-off from a watershed (blocking large gullies as they cut their ways across farms and highways) and arresting the filling up of stream beds with silt may all require the action of a number of neighboring farmers. Some such actions by neighbors may require some legal organization or approval in order to keep within the law, while in other instances the neighbors may act jointly without legal support or sanction.

In any case, it will be necessary to obtain technical advice from those qualified to give the desired information. Such specialists are employed by the Soil Conservation Service of the U.S. Department of Agriculture and are located in nearly every county.

Most of the work done by these specialists is by request, through local soil and water conservation districts, of individual farmers. A study of each farm is made showing the lay of the land, the size and slope of each field and the erosion which has already taken place. When this study is completed and reported to the farmer, he receives a map of his farm showing how each piece of land should be handled, what practice or practices should be adopted to keep the soil in place and what crops should be grown.

☐ LAND-USE CAPABILITY CLASSES

In order to simplify the work of the specialist, the farm land of the nation has been grouped into eight separate classes called "capability classes." The purpose of this is to group into a single class those soils which can be treated in the same way (with the least loss of soil) when used for agricultural purposes. Subclasses are also designated to indicate major limitations within each class.

It is encouraging to know that from a review of studies compiled by the Soil Conservation Service and other government agencies up to 1980, the National Association of Soil Conservation Districts concluded that the majority of the nation's lands are used in keeping with their inherent capabilities. A study of the classifications which follow will show why crop yields were found to drop sharply as less suitable land was brought into production.

Class I

Soils in Class I have few limitations that restrict their use. They can be safely used for the production of the commonly grown field crops with conventional farming methods. No special soil conservation practices need be followed on these soils except for maintenance of tilth and fertility.

Class II

Soils in Class II have some limitations that reduce the choice of plants or require moderate conservation practices. The moderate conservation practices may be one or more of the following: contour tillage (performing various tillage operations as nearly as possible on the level), protective cover crops, conservation tillage, tile drainage and conservation crop rotations which may include grasses and legumes.

Class III

Soils in Class III have severe limitations that reduce the choice of plants or require special conservation practices, or both. The special conservation practices may be one or more of the following: contour strip cropping, terraces, drainage of excess water, conservation tillage and/or conservation crop rotations which may include grasses and legumes.

Class IV

Soils in Class IV have very severe limitations that restrict the choice of plants, require very careful management or both. Soils in this class are usually marginal cropland and probably best used as hayland. Small grains can be grown if followed by several years of grasses and legumes.

Class V

Soils in Class V have little or no erosion hazard but have other limitations, impractical to remove, that limit their use largely to pasture, range, woodland or wildlife food and cover. Soils in this class have water at or on the surface most of the year and are not practical or economical to drain.

Class VI

Soils in Class VI have severe limitations that make them generally unsuited for cultivation and limit their use largely to pasture or range, woodland or wildlife food and cover. Soils in this class may be steep, droughty or shallow to rock, or they may contain moderate amounts of salts which are toxic to plants or they may be combinations of these factors.

Class VII

Soils in Class VII have very severe limitations that make them unsuited to cultivation and that restrict their use largely to grazing, woodland or wildlife. Soils in this class may be very steep, rocky or extremely droughty, or they may contain an excess amount of soils which are toxic to plants or they may be combinations of these factors.

Class VIII

Soils and landforms in Class VIII have limitations that preclude their use for commercial plant production and restrict their use to recreation, wildlife, water supply or esthetic purposes. Soils or landforms in this class may be extremely rocky, steep or sterile, or they may be combinations of these factors.

CLASSES RANKED BY CULTIVABILITY

The foregoing classification divides land according to both productive capability and land-use practices. The classification ranging from cultivation to nonagricultural use is as follows:

Those lands suitable for cultivation with:

    I. No special erosion control or other conservation practices.
    II. Simple conservation practices.
    III. Intensive erosion control practices.

Those lands suitable for limited or occasional cultivation with:

    IV. Limited agricultural use and intensive erosion control practices.

Those lands not suitable for cultivation but which can be used for permanent vegetation with:

    V. No special erosion control practices.
    VI. Moderate restrictions in use.
    VII. Severe restrictions in use.

Those lands not suitable for cultivation, grazing or forestry.

> VIII. These lands may be extremely rough, sandy, wet or arid and not suitable for cultivation, grazing or forestry. Some forms of wildlife may be found on all this land.

LAND-USE CAPABILITY SUBCLASSES

In order for users of soil survey information to understand the major limitations of the capability classes, these classes are further defined by the addition of a small letter—*e, w, s* or *c*—to the class numeral. (An example is IVe.) The letter *e* shows that the main limitation is risk of erosion unless close-growing plant cover is maintained. The *w* shows that water in or on the soil interferes with plant growth or cultivation (in some soils the wetness can be partly corrected by artificial drainage). The *s* shows that the soil is limited mainly because it is shallow, droughty or stony. And, the *c* shows that the chief limitation is a climate that is too cold or too dry.

Figures 15-1 through 15-12 represent each of the eight classes of land.

We see from Table 15-1 that the amount of land which can safely be used for crop production without the use of one or more soil erosion control practices is definitely limited. Even these lands will benefit by the use of fertilizers and the rotation of crops.

Chapter 16 deals with some of the most widespread and commonly used practices for the control of erosion.

TABLE 15-1. Distribution of the classes of farm land in the United States (including Alaska and Hawaii), 1977

Class	Million Acres	Approximate Percentage
I	39	2.7
II	286	19.9
III	287	20.0
IV	188	13.1
V	30	2.1
VI	267	18.6
VII & VIII	338	23.6
Total	1,435	100.0

(Source: USDA, Soil Conservation Service)

SOIL CONSERVATION ON FARMS AND RANCHES □ 239

FIGURE 15-1. Looking past the storm cellar and trees on the farmstead, we see that Class I land appears to extend to the horizon in this central Illinois farm scene. Hedges which formerly stood between fields have been removed. Many of the fields are tiled to assist drainage. Class I land must be level to nearly level and well enough drained, either naturally or artificially, to permit the growing of ordinary crops. It must not show more than the slightest erosion, regardless of treatment. *(Photo by H. B. Kircher)*

FIGURE 15-2. Class II land is suitable for continuous cultivation with the use of simple practices to control erosion, conserve water, drain wet soils, supply irrigation water, remove stones or other obstacles to cultivation or add to soil productivity. Erosion control or moisture conservation practices commonly used are contour tillage, strip cropping, cover crops, rough tillage, basin listing and surface mulch. Occasionally terracing is recommended. This land ordinarily has a slight slope. *(Courtesy, Soil Conservation Service)*

240 □ OUR NATURAL RESOURCES

FIGURE 15-3. Class III land usually needs a combination of practices for safe and continuous cultivation. Frequently this land requires more careful application or more intensive use of the practices recommended for Class II land: longer crop rotations, narrower strips for strip cropping, terrace outlets with terraces, buffer strips, diversion ditches and cover crops. Needed drainage or irrigation systems may be more difficult to install, or the land may be droughty or the slope of the land may be slightly greater than for Class II land. *(Courtesy, Soil Conservation Service)*

FIGURE 15-4. Another example of Class III land.

SOIL CONSERVATION ON FARMS AND RANCHES □ 241

FIGURE 15-5. Class IV land usually is steeper than Class III land. Or, it may be more eroded or susceptible to erosion, less fertile, more droughty, more open or more difficult to drain or to irrigate. Nearly level land of low productivity that is so wet in late spring that crop yields are reduced may be regarded as Class IV land. *(Courtesy, Soil Conservation Service)*

FIGURE 15-6. An example of Class IV land that is too wet for best production. This wet land has been pastured over the years, and the hummocky condition is mostly the result of tramping by the grazing livestock. *(Courtesy, Soil Conservation Service)*

242 □ OUR NATURAL RESOURCES

FIGURE 15-7. Class V land is fairly level and not subject to erosion. It is suitable only for grazing land or for woodland because of wetness, dryness or stoniness. The only restrictions on its use are not to overgraze or to deforest the area completely so as to reduce grass or tree growth. This land is frequently recommended for wildlife. *(Courtesy, Soil Conservation Service)*

FIGURE 15-8. Another example of Class V land.

SOIL CONSERVATION ON FARMS AND RANCHES □ 243

FIGURE 15-9. Class VI land includes steep, stony, sandy or shallow soils. It is usually steeper or more subject to water or wind erosion than is Class IV land. It is best suited for pasture, woodland or wildlife. Pasture areas which are free from stones may be renovated to increase grass growth. *(Courtesy, Soil Conservation Service)*

FIGURE 15-10. Class VII land is usually very steep, rough, stony, eroded or sandy. It usually produces some wood or forage. Where the land is not so steep or rough, it may still be classed as VII land if there is scant natural vegetation and if the soil is easily subject to wind erosion. It is best suited for woodland or wildlife. *(Courtesy, Soil Conservation Service)*

FIGURE 15-11. Another example of Class VII land.

FIGURE 15-12. Class VIII land may be rough, stony or barren, or it may be a swamp or marsh land that cannot be used in any way for the production of forest products or grass. Some forms of wildlife may find food and shelter here. *(Courtesy, Soil Conservation Service)*

QUESTIONS AND PROBLEMS

1. How many acres of farm land are being lost each year through soil erosion?
2. What three conditions must be met if we are to continue to use our croplands for food production?
3. Name some of the soil conserving practices that individual farmers can put into effect. What are some practices that must be used through groups of farmers acting together or by local units of government in order to reduce soil erosion?
4. Why has the Soil Conservation Service grouped the soils of farms into different soil classes? Give the characteristics of each soil class and tell what the land in each class is best suited for. Where it is desirable to use one or more soil conserving practices for a soil class, name the practice or practices which are usually recommended.
5. What percent of our farm land can safely be used with no supporting soil conserving practices?
6. What classes of land are most common on the farm lands near your home? What soil conserving practices are being used for these classes?

16 □ □ □

The use of vegetative cover in soil conservation

PROBABLY THE SIMPLEST WAY of controlling erosion is to maintain cover with some plant growth. This includes the use of native or adapted grasses, close-growing biennial or perennial crops, shrubs or a mixture of grasses, shrubs and trees. Where this vegetative cover is to be used in connection with farming operations, the crops must be close-growing and the crop residues should be left as cover for most of the year; otherwise, they will not protect the soil from eroding.

□ BENEFICIAL EFFECTS OF VEGETATIVE COVER

A healthy vegetative cover reduces soil losses three ways:

1. Heavy plant growth intercepts falling rain and reduces the impact on the surface of the soil. If one were to stand under a maple or pine tree during a rain, he would find that rain would not come through the foliage to wet him during the early part of the rain. If the rain continued, he would be wet by small droplets of water as they dripped or flowed from leaves and twigs. The heavy force of the rain would not reach him. This is similar to the effect upon soil of any covering, whether it be growing or dead grass, weeds or shrubs.

When rain drops strike bare soil, some of the finer particles of

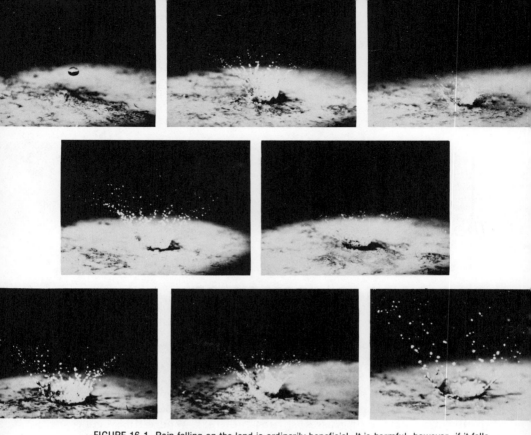

FIGURE 16-1. Rain falling on the land is ordinarily beneficial. It is harmful, however, if it falls on bare soil as this drop does. Every direct splash on bare soil dissolves in the rain drop some of the finer parts of the soil and carries them away with the water run-off. If there had been some grass growth or other surface cover, the force of the falling rain drop would have been broken and the finer parts of the soil would not have been detached in the splash of rain. *(Courtesy, Soil Conservation Service)*

the soil are loosened and suspended in drops (see Figure 16-1.) This is the first step in soil erosion.

2. The speed of water run-off from the surface is reduced. The fallen leaves, stems and litter on the surface slow down the rate at which the water runs off. This reduced speed carries less of the soil with it. It also allows more time for the soil to absorb a greater portion of the rainfall.

3. The growth and the decay of plant roots make the soil more porous. Thus, it absorbs more of the rainfall in a shorter time. The decaying roots also add some organic matter to the soil. Organic matter readily absorbs water.

Farmyard manure, when spread over the soil, helps to break the flow of water as well as any plant residue does and later, when turned under, opens up the soil and adds a little organic matter.

☐ ANNUAL CROPS ONLY PARTIALLY EFFECTIVE COVER

All growing crops provide some vegetative cover to the soil. Any of this cover is more or less effective in conserving soil. The more completely the soil is covered, the better it is protected from eroding. One of the problems in connection with most cultivated crops is the length of time the soil is left unprotected by any crop growth. Crops such as corn, soybeans, potatoes, cotton, spring wheat, oats, barley and even our truck crops are planted in the late spring and harvested from three to six months later.

In preparing land for most cultivated crops, the farmer prepares the soil for planting by plowing, followed by other tillage practices, leaving the soil bare prior to planting. It remains essentially unprotected for two to four weeks after planting because the young plants offer very little ground cover. From the time the soil is plowed and until the planted crop provides good ground cover, the soil is subject to severe erosion. After the crop has matured and has been harvested, the crop residues protect the soil if they are not plowed under. However, in many parts of the country, the remaining crop residues are plowed under. This leaves the soil bare from six to nine months of the year, so that there is little surface growth or litter to break the force of the rainfall or to decrease the water run-off during that time.

☐ HAY CROPS OFFER BETTER SOIL PROTECTION

Biennial or perennial hay crops give much better protection to the soil from erosion (see Figure 16-2). This growth offers the much-needed surface protection during times of heavy rainfall. And, it offers protection even though the plants are not growing. Seed bed preparation and planting takes place with these crops less frequently than with crops planted each year. The hazards of washing rains coming while the soil is open and exposed thus are reduced.

☐ SPECIAL CONSIDERATIONS IN USING VEGETATIVE COVER

Although the use of vegetative cover is one of the easiest

FIGURE 16-2. Plant cover can stop erosion on hillsides. This is a two-year-old stand of crown vetch seeded on cut slopes of a highway in Illinois. Seed and mulch were applied by blower. Fertilizer was applied later. The lower picture, a close-up taken at an Indiana nursery, shows the solid cover of this plant. *(Courtesy, Soil Conservation Service)*

methods of controlling erosion, it may be very ineffective unless it fits the specific conditions that make it suitable. There are many kinds and combinations of crops and other vegetation that may be employed to help control erosion. One type of cover may succeed where another will fail. This means that each crop or type of vegetation will do best in reducing erosion if used in the right situation. Also, some methods of erosion control by plant cover apply to several sections of the country, while others are useful in one part of the country but are of no value in other parts. The best erosion control procedure in every area and on practically every farm is most often based on the use of a combination of practices. Ordinarily, no one practice alone will control erosion while the field is used for crop production. The kind of vegetative cover to employ and the best way to use this cover can be determined only after considering the following conditions:

1. The part of the country in which the crop is to be used. The length of growing season, the average as well as extreme tempera-

FIGURE 16-3. The absence of vegetation started this gully which made the pasture lot an ideal place for weed growth. *(Courtesy, Soil Conservation Service)*

tures, the prevailing winds and the season and severity of the heavy rains all must be considered.

2. The kind of soil—whether it is sandy or clayey, well or poorly drained, etc.

3. The physical condition of the soil. Is it a fertile soil, or is it depleted of plant nutrients, organic matter and lime?

4. The steepness and length of the slope of the land.

5. The likelihood of the farm's using the crops grown. The farmer should use these crops to advantage. They should fit into his system of farming.

6. Adaptability of the crop to local conditions.

The use of vegetative cover as a part of farming includes the following practices: grazing land improvement, cover crops, strip cropping, conservation crop rotations, contour tillage, gully erosion and stream bank erosion control, terraces, diversions and windbreaks. We will consider some of these practices in more detail in Chapters 17 and 18.

☐ PASTURE AND RANGE LAND IMPROVEMENT

Any improvement in the grass growth on our pasture lands will serve two purposes. It will increase the ground cover, thus reducing the surface run-off, and it will increase the amount of water absorbed by the soil. When these two things are accomplished, there will be better grazing for livestock. Of course, when the heavier growth of grass is obtained, owners should use caution not to increase the livestock numbers enough to overgraze the lands. Overgrazing will destroy the vegetative cover, leaving the surface subject to erosion.

Grazing lands owned by the federal government

Extensive areas of grazing lands are owned by the federal government. The policy of the government is to lease various areas to individual cattle or sheep owners. Restrictions on grazing these areas are intended to prevent overgrazing. As a result, not only is the value of the land maintained for grazing purposes, but also the soil is not subject to erosion, which in many instances causes losses that cannot be repaired.

FIGURE 16-4. Grazing on this federally owned land has not been controlled, and excessive grazing has killed off the desirable forage grasses. Unpalatable sagebrush has taken over. Unless the management of this land is changed, severe erosion will occur. *(Courtesy, Bureau of Land Management)*

The U.S. government has the power to restrict grazing in these areas, and only it can exercise this power (see Figure 16-4). The government leases grazing rights to individuals through local organized grazing associations. Part of the fee charged is returned to the local governments to support schools, roads and other governmental concerns. The remainder of the fee helps maintain and improve the quality of the grazing lands. The federal government limits the number of sheep or cattle for each range. It also supervises the grazing to prevent damage to the land from early grazing or overgrazing in local plots.

Production increased from pasture lands

There are several ways in which the growth on pasture and grazing land can be increased. They are delayed, rotational and controlled grazing; application of fertilizers; and the use of different grasses or combinations of grasses which will make for a thriftier and more continuous cover.

254 □ OUR NATURAL RESOURCES

FIGURE 16-5. Work Unit Conservationist, John Conroy, and farm operator, Gene Miller, and his daughter, Marilyn, are standing on the edge of a beautifully maintained grassed waterway as they discuss the corn crop on this southern Illinois farm. This is the principal waterway for drainage on this farm. The corn is planted on the contour. *(Courtesy, Soil Conservation Service)*

DELAYED GRAZING

Most of our grazing land, as well as our farm pastures, is overgrazed. Livestock frequently are turned out so early in the spring that the grass cannot maintain sufficient root reserves to provide a vigorous growth. When livestock are turned out this early, they usually are kept on pasture continually from this time on, and the grass is grazed too closely. The results are that the total growth for the season is not as much as it should be and the individual grass plants become less able to make quick recovery growth when rains come.

Often, in addition to early grazing, too many cattle are placed on an area, and it soon becomes overgrazed. If the grazing lands of the Great Plains area are to be protected from overgrazing, it will be necessary to limit the number of cattle or sheep that are placed on any range. Where the land is under private ownership, the owner will benefit by placing only the number of livestock on the range that it will support throughout the growing season. The grass also should have a good spring growth before any grazing is permitted.

ROTATION OF PASTURE LANDS

Experiments have shown that a greater amount of good grazing can be obtained if a pasture or range is divided into several fields and one field pastured at a time. Each field is numbered and pastured in turn. Every field is pastured once in four to six weeks. The length of time the herd or flock is left on a field is determined by the amount of grass growth and the number of animals pastured. The animals should be moved from a field before it is completely eaten off, and each field should have a good growth of grass before the animals are turned into it. Of course, there should be water and salt available in each field if this arrangement is to work out satisfactorily.

While it is possible for a farmer to divide a pasture into three or four fields with the use of electric fences, it is not so easy to divide a hundred thousand acres of grazing land; but where rotation is possible, livestock will come off the grass in better condition, and in case of summer drought, it will not be necessary to sell cattle or sheep to reduce starvation losses. Even on the open range, herds or flocks can be moved from place to place. Grazing pressures can be reduced or transferred from area to area by providing widely spaced watering facilities and predetermined spacing of salt blocks.

Another long-term benefit of rotation of pasture lands is a better permanent stand of grass with fewer weeds and good erosion protection. As discussed earlier, the grazing lands of the nation showed little or no erosion damage until they were grazed too heavily and practically all grass growth was eaten off. The remedy for this situation on those areas which are not too badly eroded is again to keep some grass growth on the land all year. This is especially true during the growing season when heavy rains occur. If a good growth of grass is left at the end of the growing season, there will be some cover of dead grass to protect the soil throughout the remainder of the year.

MANAGEMENT OF FARM PASTURE LANDS

Pasture lands on farms are frequently as badly overgrazed and poorly managed as range lands. They are pastured too early in the

FIGURE 16-6. Well-managed grazing land of the Southwest. Grazing on this land has been regulated according to the grass growth and the season. Water run-off has been checked through simple earthen structures. The result is a fine growth of desirable grasses with practically no sagebrush. *(Courtesy, Bureau of Land Management)*

spring and are overstocked later in the year when grasses are short and the weather is hot and dry.

It is much easier to improve most farm pastures than range lands. Most of our pasture land is in areas having enough rainfall to mature commonly grown farm crops, while most of our range land has an erratic rainfall distribution with annual rates of less than 20 inches. This means that rainfall is usually not the important factor limiting the establishment and growth of grass in pastures. Also, pasture lands are much smaller than range lands.

A pasture field on a farm may be 10 to 40 acres and at most will not exceed a very few hundred acres. A section (640 acres) of pasture land is an exceptionally large field. Range land, on the other hand, ordinarily consists of thousands of acres. To manage range land may mean to look after 100,000 acres or more of land under adverse conditions.

APPLICATION OF FERTILIZERS

Permanent pasture lands over much of the country have been

pastured ever since the farms were homesteaded. In most instances, thousands of pounds of livestock growth have been taken from these lands and nothing in the form of fertilizers (plant nutrients) has been returned. The result is that most pastures produce less feed than they originally produced.

The carrying capacity of pastures may be greatly increased if the land is properly fertilized and limed. Frequently it is necessary to re-establish a pasture in order to work calcium, nitrogen, phosphorus and potash again into the soil to replace what has been drawn out by the growing grass over the years. Where pastures have been limed, fertilized and reseeded to adaptable grasses, the carrying capacity may be doubled and in some instances trebled, when measured by the number of livestock that can be pastured satisfactorily.

USE OF DIFFERENT GRASSES OR COMBINATIONS

We usually think that our native grasses are the ones best adapted, not only to survive under local climatic conditions but also to produce the largest amount of feed while adequately protecting the soil. We find, however, that many new varieties of grasses have been introduced and many of the local strains have been improved to the point where they not only greatly outyield native grasses but also survive climatic conditions just as well.

We also find that combinations of legumes and non-legume grasses many times outyield the pure seedings, as well as protect the soil against erosion (see Figure 16-7). Here again, the best combination of grasses for an area should be obtained from the local agricultural authority. New combinations of grasses are constantly being brought out that are proving more satisfactory than previously recommended mixtures.

□ COVER CROPS

Soil erosion impacts may be reduced by growing some crops on the land when the regular crops have been removed. When these crops are grown primarily to keep the soil from being exposed to wind and rain during the off-crop season, they are called cover crops. It is sometimes difficult to distinguish between green-manure crops and cover crops because all green-manure crops serve the

258 □ OUR NATURAL RESOURCES

FIGURE 16-7. A close-up of an alfalfa-brome-ladino growth in a meadow. Where the soil is covered with a growth like this, no rain will damage it. *(Courtesy, Soil Conservation Service)*

purpose of cover crops during the time they are growing and before they are turned under. The real reason for growing green-manure crops is to conserve fertility and to improve the soil condition. Cover crops, on the other hand, are grown primarily to hold the soil against erosion. They also will improve the condition of the soil, as well as hold the nitrogen which might otherwise be leached from the soil.

Many crops may be used for cover crops, and it often happens that regularly grown cereal crops are used for this purpose. The usual crops grown only to protect soils, however, are not those grown in the regular cropping system.

Fewer cover crops are grown in the northern part of the United States than in the southern part because the ground is frozen and there is less erosion in the winter. In the Great Plains, protection against wind erosion is most effective when the crops are seeded on the contour and across the prevailing winds, especially if there is any significant slope.

Limitations to the use of cover crops

Probably the time of year when most soils need protection against erosion is during the winter months. In some parts of the country, however, where regular crop production is limited by the yearly rainfall, it may be more desirable to conserve all moisture for the regular crops and depend upon other practices for reducing erosion instead of planting cover crops. Cover crops should not be attempted when the late fall and winter months are so cold or snowy that little growth is made by the crops.

IMPROVED TILLAGE PRACTICES AS A SUBSTITUTE

In many parts of the country, improved tillage practices are used to reduce soil losses. After grain crops are harvested, the fields are chisel plowed (see Figure 16-8). This implement consists of tines (prongs) which can penetrate about 14 inches into the soil. The tines are spread about 15 inches apart. When pulled by a tractor, the chisel plow shatters or fractures the soil. This improves water intake, yet leaves the crop residue on the surface. This tillage practice, known as conservation tillage, helps reduce erosion by wind and water.

Crops used as cover crops

Winter rye is the principal cover crop used in the upper Midwest, while the New England area may use hairy vetch along with winter rye. In the southern Appalachian area, crimson clover, vetch, bur clover, Austrian winter peas and red clover, either alone or in various mixtures, are commonly grown to provide winter protection.

In the Atlantic and Gulf coastal plains, Austrian winter peas, hairy vetch, crimson clover, blue lupine and bur clover are the legumes used, while in the upper Great Plains, red clover, alfalfa, biennial sweet clover and legume-grass mixtures constitute the principal winter covers. Rye grass has been used as winter cover in parts of Oklahoma and Texas. In the southern part of the Great Plains, winter grains are the most effective winter cover crops.

The Pacific coastal areas use wild oats, bur clover, alfalfa, vetches, sour clover and mustard for soil protection. It is possible to add

FIGURE 16-8. *Upper:* A deep-working chisel plow with wide and deep trash tunnels is well suited to primary tillage operations. The trash (pieces of corn stalk, etc.) left on the surface acts as a wind breaker to help prevent wind erosion of the soil. *Lower:* This machine, called a conservation planter, can spread insecticides and fertilizer and sow the seeds in the field. The discs cut a furrow into which seeds are drilled, leaving the surface relatively undisturbed. *(Courtesy, Deere and Company)*

many legumes, grasses and cereals to the crops already listed as cover crops. However, the choice of a cover crop depends upon the cost of the crop and the use to which it may be put, as well as the local climatic and soil conditions. The best information regarding the local use of cover crops can be obtained from an agriculture teacher, a soil conservation specialist or a county agricultural agent.

It must be remembered that any growth on the soil affords better protection from erosion than no growth. Ordinarily, the finer the growth and the longer it grows during the year, the better it protects. Any surface cover or crop residue, whether dead or growing, helps to slow down surface run-off and water absorption by the soil.

□ □ □

QUESTIONS AND PROBLEMS

1. In what ways does the growth of grasses or shrubs help reduce soil erosion?
2. What are the disadvantages of using crops such as small grains, cotton, corn or potatoes as soil conserving crops?
3. Why are perennial hay crops better for keeping soil from washing away than annual crops such as those mentioned in question 2?
4. What are some of the points to keep in mind in using vegetative cover to control erosion?
5. Name two soil conservation benefits to be derived from improved pastures.
6. What are the ways that the growth on pasture lands can be increased?
7. In what ways are government grazing lands managed to maintain good cover and forage?
8. Why is it easier to improve pasture lands on farms than to improve large grazing areas?
9. Is it more important to use fertilizers for improving farm pastures or to sow other grasses than those which grow naturally? Explain your answer.
10. What are cover crops, and how can they be used to reduce erosion? What are the limitations to the use of cover crops? What crops are commonly used in your part of the country for cover crops?
11. What kinds of vegetative cover are being used in your part of the state to keep the soil from washing away?

17

Strip cropping, contouring and terracing in soil conservation

THE DAMAGE POTENTIAL of water rushing down a slope is so great that much of the soil conservation work in the United States, and throughout the world, is done to prevent or at least to hold such damage to a minimum. Such erosion can be reduced by shortening the length of the slope, reducing the velocity (speed) of the water as it flows downhill. Or, it can be combinations of the two. Common methods of accomplishing this are strip cropping, contouring and terracing.

□ EFFECT OF WATER SPEED

We know that water increases in speed and in its capacity to move and carry away soil as it flows down a slope. A steep slope will lose more soil than a gentle slope of equal length, even though the amount of water run-off is but very little greater (see Table 17-1). This is because the speed of the water is greater than it is on the gentler slope. Practically every long slope loses soil because more water flows off and the velocity of the water run-off increases with the length of the slope.

The effect of increased speed of water run-off and its ability to wash and carry away soil can be very damaging to our soil resource. When water flows at the rate of over 2 feet per second, it can loosen and carry away some topsoil from fields containing no green cover or litter.

Assuming that the speed of water flowing down a slope increases from 2 feet per second at about 50 feet from the top of the slope to 4 feet per second at 100 feet farther down, what effect will this increase in speed have upon the ability of the run-off to do damage to the slope? (Four feet per second is only about 2.7 miles per hour, or the speed of a horse walking slowly. We can walk at that rate and call it a leisurely pace.)

TABLE 17-1. Soil losses from one soil type but with different slopes, five-year average[1]

Slope	Water Run-off per Acre	Soil Loss per Acre
(%)	(inches)	(tons)
3	14	5.1
8	14	10.8
13	16	22.6
18	20	28.6

[1]The above figures were obtained from the Upper Mississippi Valley Soil Conservation Experiment Station at LaCrosse, Wisconsin. The plots were 72.6 feet long. The amount of water run-off was not greatly influenced by slope until the 18 percent slope was reached. The velocity with which the water ran from the plots was greatly increased with steeper slope, however, and this increased speed of water flow accounts for the greater soil losses.

The first point to consider is that, as the speed of the water flow is doubled, its ability to cut, or to tear away, soil is increased four times. In other words, we should expect to see four times as much damage done to the soil when the speed of the run-off is doubled. The capacity of the water for carrying away soil is even greater than its cutting ability would suggest. It can carry around 30 times as much soil as it could with half the speed. It is because of these relationships that the velocity of water run-off should be as slow as possible.

The volume or total quantity of water run-off from a given area of land also affects the amount of erosion. If the quantity of water run-off is doubled without increasing its velocity, the rate of erosion as well as the amount of soil carried in the run-off will also be doubled. This means that the volume of water run-off may be greatly increased without doing much damage if the speed is kept low. An increase in the velocity of water run-off is much more damaging to the soil than an increase in the quantity of run-off.

☐ CROPS DIFFER IN SLOWING DOWN WATER RUN-OFF

Some crops are very effective in slowing down water run-off, while others reduce the rate very little. Corn, soybeans, potatoes, tobacco, cotton and many of the truck crops, which are planted in rows far enough apart to permit cultivating or working between them, do very little to retard the flow of water (see Table 17-2). Small grains are considerably more effective in slowing down the rate of water run-off during the three to six months the crops are on the land. During the remainder of the year, they offer little resistance to water erosion. Even so, farmers should expect only about one-seventh as much loss of soil from fields which are continuously in grain as from ones on which corn is grown continuously.

TABLE 17-2. Soil losses from different cropping practices, seven-year average[1]

Crops Grown	Soil Lost Yearly	
	Tons per Acre	Inches of Soil
Fallow	162	1.12
Corn, every year	112	0.77
Grain, every year	16	0.10
Bluegrass, every year	0	0.00
Corn, 3-year rotation	53	0.37
Grain, 3-year rotation	30	0.21
Hay, 3-year rotation	1	0.007
Average, 3-year rotation	28	0.30

[1]The above figures were obtained from the Upper Mississippi Valley Soil Conservation Experiment Station at LaCrosse, Wisconsin. They show that for this soil it requires less than one year to lose one inch of surface soil under continuous fallow land, while it takes 1⅓ years to lose one inch of surface soil from a field which grows corn continuously. It can be seen that conditions which cause a large loss of soil under continuous cropping of corn will cause only one-seventh as much loss under continuous cropping of small grain. No doubt other soils, when growing similar crops, will vary from these figures, but the general relationship of losses should hold true for most soils.

The biennial or perennial hay or grass crops, on the other hand, show little or no loss of soil over a series of years. They furnish some surface cover to retard water run-off during the whole year, and

FIGURE 17-1. An airplane view of strip cropped fields in western Wisconsin. Each alternate strip should be in grass in order to retard soil losses from the strip above. *(Courtesy, Soil Conservation Service)*

because of this, they are exceedingly useful in any program which depends upon vegetative cover to control erosion.

☐ STRIP CROPPING

Strip cropping is a system of growing common farm crops in which the crops are planted in strips across the slope of a field. Effective strip cropping usually requires that every other strip on the slope be a hay or grass strip (see Figure 17-1). The water run-off from the corn or grain strip tends to collect from the smaller rivulets into larger streams as it flows down the slope. The hay or grass strip will slow down and spread the water as it comes from the grain or corn strip above, and, as the water slows down, it drops a part of the silt it has picked up from the strip above. This is the reason why it is beneficial if every other strip on the slope is in hay or grass.

Shorter slopes

Strip cropping maintains a vigorous grass and legume cover on at least one-half of the field. The strips, to be effective, must always be worked and planted at right angles to the slope or on the level across the slope.

Value of strip cropping

About 22.6 million acres of farm land have been strip cropped, and this fact in itself shows the importance attached to this practice by farmers. Strip cropping reduces soil losses by slowing down water run-off. It also results in a reduction in run-off and in an increase in the amount of water absorbed into the soil. These two benefits are enough to justify the practice, because without them crop yields

FIGURE 17-2. These strips of rye are planted to protect against wind erosion on muck soil. Strips are planted at right angles to the direction of the prevailing winds. *(Courtesy, Soil Conservation Service)*

would decrease. Not only may production be maintained but also, in some instances, it may actually be increased as a result of strip cropping a field.

Width of strips

The width of strips is influenced by many factors, and no one width can be considered best for all conditions. The length of the slope, steepness of the slope, soil type, degree of erosion, length of rotation, amount and intensity of individual rain storms and type of cultivation all help to determine the necessary width of strips. Other factors being equal, the wider the strips, the greater the soil losses.

Theoretically, the strips should be quite narrow so that the flow of water across one strip of a grain or corn crop can gain very little speed and can pick up very little silt before it is slowed down by the hay or grass strip. It is not practical, however, to make strips too narrow. They should be wide enough to be easily worked with ordinary tillage and harvesting machinery. This is the reason that the smallest practical width is frequently set at 40 or 50 feet. On the other hand, strips for peanut growing are frequently as narrow as 12 feet.

The greatest width should be set by the largest amount of soil the field can lose and still maintain satisfactory yields. Field practice throughout much of the Midwest suggests 6 rods or 100 feet as a reasonable maximum width. Some conditions are found where it is safe to make strips 150 or 160 feet wide. This is usually on nearly level land or on soil types which absorb water quite readily.

In much of the Great Plains area where dry farming is practical, the strips may be as much as 300 or 320 feet wide. These widths are suggested for the control of wind erosion. In areas where both water and wind erosion are prevalent, the greatest width of strip should not exceed the safe limit for either type of control when used alone.

In any case, it is assumed that when definite maximum widths are set, it is a compromise between some soil loss and the more efficient use of machinery and labor.

Kinds of strip cropping

There are several kinds of strip cropping. The differences are based upon the purpose for which the strip is laid out, as well as the character of the strip itself. The four general types of strip cropping are (1) field strip cropping, (2) wind strip cropping, (3) contour strip cropping and (4) buffer strips.

1. *Field strip cropping* refers to strips of crops which are parallel across the field. They need not necessarily remain on the exact level, or contour, because if contours were followed across a field with uneven slopes, the two sides of the contours could not remain parallel. The most frequent use of field strip cropping is on long, smooth, gentle slopes where water run-off is not gathered into a few streams as it moves down the slopes. This method is used throughout the Great Plains area.

2. *Wind strip cropping* is used wherever the land is fairly level and the soil is subject to wind erosion. These strips are usually parallel across a field. They are laid out at right angles to the direction of the prevailing winds and do not necessarily follow contours (see Figure 17-2).

3. *Contour strip cropping* is the most commonly used form of strip cropping because of its importance in controlling water erosion. These strips are laid out at right angles to the natural slope of the land, or on the contour (see Figure 17-3). When the slope is more than 5 or 6 percent but not greater than 12 percent, strip cropping is frequently used with terracing, which will be discussed later. Although these strips are used mostly for the control of water erosion, they may be used effectively on sloping land where wind erosion also may be serious.

4. *Buffer strips* is a term applied to strips of a more nearly permanent nature. They may be wide or narrow, long or short, and they may vary in width from one end to the other. A small part of one field may be badly eroded, or it may be a very steep, short slope. In either instance, the piece should be plowed only occasionally, if at all. In this situation buffer strips of grass, legumes or shrubs may be put in and left for several years. Buffers are also used in small irregular areas, odd corners and small triangular pieces which are inconvenient to work with ordinary field implements or which have special erosion problems.

Buffer strips may be used to break up long slopes or to give protection from erosion where a field is planted to a single crop, such as grain or some intertilled crop. In these circumstances, the buffer strips may effectively replace the grass or hay strips. When used in this way, the buffer strips should be as wide as the grass strips for which they are substituted.

270 □ OUR NATURAL RESOURCES

FIGURE 17-3. The fields on this 270-acre Illinois dairy farm, located in Stephenson County, are contour strip cropped to better accommodate slopes of from 3 to 14 percent. Crop rotation will help further and will accommodate the characteristics of the eight silt loam soils found on the farm. *(Courtesy, Soil Conservation Service)*

When used on the contour, the buffer strips facilitate contour field operations. They are sometimes used to prevent wind erosion. They may well be called field strips when used in these ways, unless they are not grazed or harvested and are there for more than one year.

Strip cropping is a valuable method of soil conservation where conditions are adapted to its use, and the great spread of its use in recent years is witness to the wide acceptance of this practice.

□ CONTOURING

Contour tillage is the practice of performing all field operations on the contour, or level. Plowing, planting and other tillage opera-

FIGURE 17-4. Contour tillage has some of the advantages of strip cropping in that all tillage operations are performed on the level. Water from small showers and light rains is more readily held in place to be absorbed by the soil. Most erosion comes from the less frequent, heavier rains, as contour tillage does little to retard losses in these situations. *(Courtesy, Soil Conservation Service)*

tions are at right angles to the slope of the land (see Figure 17-4). This is not strip cropping because the whole field may be used for one crop, such as cotton, corn or grain (see Figure 17-5). This method of controlling erosion is recommended where soil erosion is not severe enough to need the use of other practices.

☐ TERRACING

Terraces have been used for centuries by different people to help control erosion and to make it easier to cultivate sloping land. It is believed that terracing for rice fields in the Philippines was begun two thousand years ago and that the Incas in Peru terraced their steep hillsides more than four thousand years ago. Stone terraces built in France and in Phoenicia of Biblical times are still effective in controlling erosion on steep slopes after hundreds and even thousands of years of use.

272 □ OUR NATURAL RESOURCES

FIGURE 17-5. A field of soybeans planted on the contour. Rains will not cause serious erosion on this field. *(Courtesy, Soil Conservation Service)*

What are terraces?

When properly installed and maintained, terracing is a positive conservation practice which will reduce soil erosion. The practice of terracing in this country grew out of the practice of contour farming and was started along our eastern seaboard at about the time of the Revolutionary War. A terrace was originally defined as a raised, level stretch of earth which is kept in place on a hillside by a wall or a bank of turf. Terraces were made so that the steep hillsides could be cultivated. Terracing is now done, not only to hold the land in place on a slope but also to cause rainfall to run off a field slowly or to hold it on the field until it is absorbed into the soil. The main feature of most terraces built today is a channel or a ditch to carry or hold water.

The original idea that a terrace was for the purpose of protecting some leveled out piece of land from washing away does not completely describe all terraces now. Many kinds of terraces are recognized today. A channel with a ridge on the lower side, a channel with no ridge and a lister furrow may be classed as terraces. We must remember that the important function of a terrace is to reduce the length of slope and to reduce the volume and speed of any water run-off by intercepting it and carrying it across the slope of land, rather than letting it run down the slope as in the case of strip cropping, contour tillage or grass cover. Or, it may be for the purpose of holding rainfall rather than carrying it slowly off the field.

Why should terraces be used?

Terraces reduce soil losses. Studies covering a 14-year period at the Upper Mississippi Valley Soil Conservation Experiment Station at LaCrosse, Wisconsin, show that there was only one-sixth as much loss of soil from the experimental terraced areas as from the unterraced. There were 11 heavy rainstorms during this 14-year period, which accounted for 60 percent of the soil loss on the unterraced fields. Terraces controlled most of the erosion, even during these critical storms. In these experiments, both the terraced and the unterraced areas were treated the same.

Terraces increase crop yields (see Figure 17-6). If soil is saved and there is less water run-off from terraced areas, it is reasonable to expect some increase in crop yields over the years. Experiments show this to be true. Where longer rotations were used in tests and fertilizers were applied to both the terraced and the unterraced fields, crop yields were 7 percent greater on the terraced than on the unterraced fields.

Terraces allow a more flexible cropping system. The entire field can be used for one or many crops, depending on the yearly desires of the operator. Strip cropping, in contrast, must have alternate strips of some grass or hay crop to be effective. Most present-day terraces are so constructed that farm machinery can be used on them. Even machinery such as drills, combines and corn pickers ordinarily are able to pass over the terraces without dragging or cutting into the ridges. The ridges are not so high or narrow or the channel sides so steep that it is hard to drive over them. Water channels ordinarily should be from 15 to 25 feet across from the top of the ridge to the farther edge of the channel. The depth of a channel usually does not exceed 10 inches, although some channels with ridges are 12 inches deep.

FIGURE 17-6. There are 18 miles of terracing on this farm, and the luxuriant growth of brome–clover–crested wheat mixture speaks for the effectiveness of this type of conservation. *(Courtesy, Soil Conservation Service)*

Types of terraces

Most terraces can be classed under one of four headings: (1) bench terraces, (2) ridge terraces, (3) diversion terraces or (4) lister terraces. However, variations found in terrace construction are so great that a classification cannot always be complete in every detail.

BENCH TERRACES

Bench terraces are constructed much like the terraces of a thousand years ago and most nearly exemplify the original meaning of the word. They are used on very steep land or hillsides in order to reduce the sharp slope to a series of narrow, level or nearly level strips. Each terrace has a retaining wall of stones or grass-covered earth to keep the soil from eroding away, and because the benches are constructed on the level, water run-off is greatly retarded. The purposes served by the bench terraces, then, are to reduce water run-off, retard erosion and make crop production possible.

STRIP CROPPING, CONTOURING AND TERRACING □ 275

FIGURE 17-7. Even gently sloping fields require conservation practices if soil erosion is to be reduced. A district conservationist and a landowner are looking over parallel tile outlet terraces on this eastern Illinois farm. The terraces will greatly reduce soil erosion. *(Courtesy, Soil Conservation Service)*

Bench terraces are generally used in densely populated areas where land is scarce and where most of the land is hilly or mountainous. There are only a few places in this country where bench terracing is in place. In the southern and southeastern parts of the United States, the steep hillsides have been bench terraced for generations. It is not unusual to see corn or cotton being grown on these narrow strips. In the productive citrus and avocado areas of southern California, bench terraces are also installed.

These terraces usually cost more than other forms of terraces. Since they are used mostly on steep slopes, they must of necessity be narrow and very close together.

LEVEL RIDGE TERRACES

A ridge terrace, as the name indicates, has a ridge of earth on the lower side of the channel. The earth from the channel goes to make a part of the ridge, and the rest of the necessary earth is taken from below the ridge.

Ridge terraces are constructed primarily to hold the water that falls rather than to let it drain off. This means these terraces are used more in the Great Plains where water is the important factor limiting production. They are built to spread the water over large surfaces so as to facilitate absorption by the soil.

Ridge-Channel Terraces

One modification of the ridge terrace, sometimes called the channel terrace, is constructed similarly to the regular ridge terrace except that no earth is pulled up from below to form the ridge. It is all taken from the channel. Also, the channel of the channel terrace ordinarily is not made as deep as that of the ridge terrace.

Both ridge and channel terraces are built on gentle slopes. The channel terrace is best adapted to soils that are relatively impervious to water and in areas with a reasonably good distribution of rainfall throughout the growing season and enough rainfall to justify draining off the surplus through surface drainage. These conditions are found throughout the Southeast, the Middle Atlantic states and the Tennessee and Ohio valleys, as well as parts of the Mississippi Valley where there is a fairly good distribution of rainfall throughout the growing season.

Both the ridge and channel terraces are left open at the ends so that excess rainfall can escape before it overtops the rim and cuts it away.

A modification of the channel terrace is built with no ridge on the lower side. In very heavy soils where the land is nearly level, it is desirable to remove excess water in order to improve the drainage. Such a heavy soil frequently has small shallow pot holes which fill with water every spring and after each rain. Because of the imperviousness of the soil, the water is very slow to soak in or to evaporate, thus delaying any field work. The result is a delay of timely seeding and cultivation. Machinery becomes mired in these wet spots too, and crop production is reduced.

The channel with no ridge permits the use of the earth from this channel to fill the pot holes. This serves the double purpose of doing away with the wet spots and improving yields. In very heavy, nearly level soils, the ridgeless type of channel is superior to the ridge-channel type of terrace in several ways: (1) The water is carried in a channel below the ordinary surface of the ground, which means that overtopping from an extremely heavy rain does not cause severe damage. (2) Channel terraces are but little obstruction to the passage of machinery. (3) They are easier to align in a true parallel pattern. (4) Earth removed by the excavation of the channel can be used to fill objectionable pot holes and draws in the field, thereby eliminating ponds and soft spots.

DIVERSION TERRACES

Another modification of the channel terrace is the diversion terrace. Diversion terraces are built on the steep slopes (usually more than 10 percent slope—10-foot drop in 100 feet) where the erosion problem is more serious than on the gentler slopes. The channels are deeper than those of the usual terraces, and they may have

FIGURE 17-8. Wallace Cole cultivates soybeans on his farm of parallel terraces in Moultrie County, Illinois. The terraces help reduce soil erosion. *(Courtesy, Soil Conservation Service)*

steeper sides. They are constructed to carry away more water and to do it more quickly than the channels of standard terraces. For this reason, the grade or slope of a channel may be increased to as much as 2 feet per 100 feet along the channel. In order to protect the channels from cutting and washing away, farmers should keep them permanently in sod.

Diversion terraces cannot be easily crossed by farm machinery and should never be plowed up with the intention of using the land in the rotation. All tillage operations should be on the contour between the terraces, which may be placed two to four times as far apart as standard terraces. Diversion terraces are built primarily to carry water off the steeper slopes with the least possible amount of erosion.

LISTER TERRACES

Lister terracing is essentially contour tillage. The implement used to make this type of channel is ordinarily the lister, which throws the earth both ways, thus creating a furrow which is not filled in by the next furrow slice as in ordinary plowing operations. In some parts of the country, this implement is called a "middle-buster" or "middle-breaker."

Lister terracing is a temporary land conditioning operation for the purpose of holding the water where it falls until it is absorbed by the soil. Frequently, the lister has an attachment that throws up a small dam or dike every few feet across the furrow (channel). This attachment is especially useful where it is not convenient to work the soil on the absolute contour.

When it is time to plant the crop, the field is smoothed out if grain is to be planted. If the field is to be planted to corn, the lister may be used to split the temporary ridges, thus creating new ridges in the old channels.

Terraces for nonfarm uses

Erosion can be equally serious in nonfarm areas. In urban areas soil losses can be very high. High soil losses can affect houses, streets, bridges and many other urban structures because such losses produce sediment as well as remove soil from lawns, house and bridge foundations and other critical areas.

To control soil losses and to reduce erosion, terraces are often applied to intercept water and carry it slowly to a non-erosive watercourse, or channel. The design and function is similar to that of farm terraces; those terraces in an urban setting usually have permanent grass cover. As nonfarm areas generally have buildings, streets, water lines and other like developments throughout, the designs of terraces in these areas require unusual considerations. Therefore, expert assistance should be obtained from soil specialists.

Laying out and building terraces

All terraces should be laid out by competent persons. Usually soil specialists or conservationists are available in every county in the United States for this work. They also will designate the type of terraces to construct, the height of the ridges, the width and depth of the channels, the distance between the channels and the type of equipment best adapted to the work. When once done properly, this work should not have to be done over again, so it is very important that it be done correctly the first time.

Outlets for terrace channels

Wherever there is to be water run-off from a field through terraces or other constructed channels across the slope of a field, the water must be discharged into an outlet of some sort. Outlets for terraces may be either open grassed or closed tile. The open grassed outlets may be located in natural waterways or in constructed grassed waterways.

Any open water outlet must be sodded or grassed with a growth of fine-leaved grass (see Figures 17-9 and 17-10). It should be large enough to carry the run-off from the heavier rains. Wherever the outlet is to be crossed by machinery, the sides should slope gently. In any situation the slope of the outlet should be such that the flowing water will not cut away the sides. One common standard for size of outlet in regions with not more than 18 or 20 inches of rainfall during the growing season is 2 square feet of area in the cross section of the channel for every acre of land that drains into the channel. If very heavy storms occasionally occur, the carrying capacity of the

280 □ OUR NATURAL RESOURCES

FIGURE 17-9. This picture was taken from inside a large city-type bus as it was driven down a grass waterway through the center of a corn field on a rainy day. This ride, part of a conservation tour sponsored by the Soil Conservation Service, dramatically proves how effective such waterways can be in preventing erosion. *(Photo by H. B. Kircher)*

FIGURE 17-10. A waterway that is covered with a heavy growth of grass will carry fast-moving surface run-off with no washing. You will notice that the water is clear. It is free of silt, yet it is flowing rapidly. *(Courtesy, Soil Conservation Service)*

outlet should be larger; otherwise, the water run-off will overflow the channel and cause greater soil losses.

In the northern part of the United States, where the winters are long and cold, one of the problems in connection with water outlets is keeping the channels from freezing closed. Wherever an outlet can be made on a southern or southwestern exposure, this problem will be greatly reduced. In no case should the outlet be used as a road. Neither should the channel be open for grazing by cattle, because the longer the grass blades, the better the protection afforded the channel.

The closed tile outlets are a special type of outlet consisting of an open inlet, for the water in the terrace to enter, connected to a closed circuit (usually farm tile). This form of terrace outlet should only be installed when properly designed.

Not only does the use of terraces reduce soil losses and increase crop yields, but also it makes for a more flexible cropping system. The many types of terraces are useful under differing soil conditions. Proper outlets for terraces are important, so that no soil erosion occurs as the water moves from the field.

☐ ☐ ☐

QUESTIONS AND PROBLEMS

1. If the steepness, or grade, of a slope is increased from 3 percent (3-foot drop in 100 feet of horizontal distance) to 8 or 10 percent, how much more water run-off would you expect to find?
2. What causes the greater soil losses if the amount of water run-off is not increased?
3. If the steepness of the slope is increased to 18 percent or more, what is likely to happen to water run-off?
4. Which will cause the larger soil loss: (a) A definite volume of water run-off at the rate of 2 miles per hour, or (b) One-half this volume of run-off at the rate of 4 miles per hour? Explain your answer.
5. If the volume or quantity of water run-off is doubled without increasing the velocity, or speed, of the water: (a) How much more cutting or erosion will take place? (b) How much more soil will the run-off carry?
6. If the velocity of the water run-off is doubled without increasing the volume: (a) How much more cutting or erosion will take place? (b) How much more soil will the run-off carry?

7. Which crops slow down water run-off the most? Which the least?
8. What is strip cropping, and how does it reduce soil washing?
9. What factors help determine the width to make crop strips across a slope?
10. Name the kinds of strip cropping that are used in different situations and indicate some use for each. Which are used in your area?
11. What is the difference between contour tillage and strip cropping?
12. What is the definition of a terrace as used during early times? How does this differ from the definition used at the present time?
13. Name the advantages of using terraces.
14. Describe the difference between bench, ridge-channel and diversion terraces and tell where each is most useful.
15. Tell how these terraces are made. What types of terraces are used near you?

18 □ □ □

Crop rotations and other practices in soil conservation

THE AGRICULTURE of some parts of the United States, such as the Wheat Belt of the Midwest and the Cotton Belt of the South, was built upon the production of one crop. The agriculture of most of the country today, however, depends upon the production and use of several crops; therefore, the former one-crop regions are becoming more diversified. When various crops are grown on a farm, they tend to follow a definite sequence. For example, in those areas where corn, small grain and hay are produced, a field that is planted to corn one year will ordinarily be planted to some small grain, such as oats, barley, rye or wheat, the second year. Some hay crop may be planted along with the grain or immediately following its harvest. This may be any legume or nonlegume that will survive the winter and make a hay or pasture crop the third year. A common example of a four-year rotation in the Midwest is corn, soybeans, wheat and clover, in a four-year succession. In this rotation the fifth year the clover is plowed and corn planted, thus starting again on the four-year program. A crop rotation may be simply defined as a sequence of crops grown in recurring succession on the same field or area of land.

□ IMPORTANCE OF CROP ROTATION

Crop rotations are vital to agriculture. They aid in the efficient use of plant nutrients found in the soil. This is so because each crop has differing needs for plant food, and what one crop doesn't use before being lost to leaching, another can. As a result, loss of nutrients to leaching is diminished. Crop rotations also aid in the reduction of insect losses. Insects do not have as great an opportunity to increase in number if the host crop is rotated. Often, introduction of another crop may bring in insects which prey upon the insects of previous crops. Third, crop rotations reduce soil and wind erosion when used to maintain maximum ground cover.

Crop rotations are generally a useful conservation practice. If ill-conceived, however, they can reduce and eventually destroy our natural resource base; if well conceived, they can strengthen it.

Definite rotation attempted

Most farmers try to follow a definite rotation and even lay out their fields with this thought in mind. Misfortunes such as winter killing of the hay crop, drought damage at certain critical periods during the growing season and an attack of crop-destroying pests and diseases (chinch bugs, Hessian flies, corn borders, smut, rust, etc.) often break into the rotation. When misfortunes strike, the crop sequence is altered to fit the circumstances. Thus, the rotation is different for that period. Conservation-minded farmers, however, soon return to the planned sequence.

Length of rotations

Crop rotations vary in length from 2 or 3 years to 10 or 12. The usual rotations throughout the Midwest are three or four years, although rotations of five or six years are not uncommon. Longer rotations may consist of two or more years of corn, followed by a year or two of some small grain and two or more years of hay if the land is not subject to erosion, that is, if it is Class I land. Or, rotations may be lengthened by increasing the number of years the fields are in hay or grass if the soil erodes, as in Class IV land.

The need for special crops in some sections of the country may determine the length of rotations. In the eastern part of the United States, there is greater acreage of hay than of other crops. This results from the expansion of the dairy industry close to the consuming East. Farmers there find it cheaper to ship in 20 pounds of grain for every 100 pounds of milk produced and to raise more hay than to attempt to raise both. This rotation, consisting essentially of hay and pasture, may be no longer, however, than many of the rotations used on Midwest farms which ship the grain to the East. Each section of the country has its rotation or system of cropping which meets its special need, either for feed or fibre crops, or for the control of erosion. Here again, the desirable rotation for any area not only must meet the needs of the area for food and feed but also must safeguard the soil.

Rotations help control erosion

It has already been mentioned that soil losses are greatest with intertilled crops such as corn, soybeans, cotton, tobacco and others, while losses with hay crops are practically negligible. A rotation ordinarily combines into a sequence of years those crops which show small soil losses with those which show greater. The average yearly loss for the complete rotation is thus smaller than losses which occur with the production of corn or small grain alone.

Figures from the five states of Oklahoma, Ohio, Missouri, North Carolina and Wisconsin show the average yearly loss of soil in a rotation of cotton or corn, small grain and meadow or hay crops to vary from a little more than 4 tons per acre for Oklahoma to 28 tons for Wisconsin. Water run-off losses did not show as great a range as did soil losses. The lowest amount of water run-off, slightly over 9 percent of the rainfall, was in North Carolina, which had approximately a 15-ton soil loss per acre, while the highest percent of water run-off loss, over 21 percent, was from Ohio, which showed only around a 13-ton soil loss per acre. The slope of land on which these losses were measured ranged from 7 percent for Oklahoma to 16 percent for Wisconsin (see Table 18-1). The soil loss studies from which these figures were taken tend to bear out the conclusion that the steepness of the slope of the land has more to do with soil losses in a rotation than does the amount of water run-off.

Rotations add organic matter

In addition to reducing soil losses, a crop rotation which includes legumes and grasses also adds organic matter and so loosens the soil that frequently the loss of the soil from the following corn crop is less than that of the second year after plowing in preparation for small grains. As already noted, under continuous cropping, soil losses from corn are much greater than those from small grain.

It should be remembered, however, that good crop rotations alone will not reduce soil losses to a safe point, except on those fields with the slightest slope. Other practices, such as strip cropping and terracing, must be used in connection with crop rotations if they are to be effective in controlling erosion.

TABLE 18-1. Average yearly water run-off loss and soil loss with a three-year rotation

Area	Soil per Acre	Water, Portion of Total Rainfall	Slope of Fields
	(tons)	(%)	(%)
Statesville, N.C.	15.1	9.3	10
Guthrie, Okla.	4.2	10.2	7
Bethany, Mo.	9.1	16.2	8
LaCrosse, Wis.	28.0	16.9	16
Zanesville, Ohio	13.3	21.2	12

Rotations and strip cropping

Strip cropping makes possible the use of crop rotations on fields which are subject to erosion. Also, if it were not for crop rotations, strip cropping would not be as useful as it is in controlling erosion.

Since one of the first requisites for the use of strip cropping is that every other strip be in some hay crop or grass, it is easily understood why experts so generally recommend a four-year rotation of one year each for some intertilled crop and grain, followed by two years of hay or hay and pasture. If more hay and pasture are needed, or if hay is needed for longer periods of time, a six-year rotation may be desirable.

□ BENEFICIAL USE OF FERTILIZERS

Soil losses tend to be cumulative. They remove plant nutrients which are essential for crop production. This reduces the quality of crop growth. And, this reduction in quality of plant growth increases the erosion hazard.

An effective way to reduce soil erosion in combination with crop rotation and other conservation products is to add fertilizers necessary to maintain and increase crop yields. When this is done, it aids in reducing erosion by providing vigorous vegetation.

Fertilizers, which include green manures and animal wastes, should be applied to the soil in accordance with needs. Deficiencies can be determined by taking soil samples and having them tested. Local county agents of the USDA Extension Service should be contacted as to availability and location of soil-testing facilities. When nutrients are applied according to these tests and other conservation measures are employed, the productivity of the soil is increased. This increase in soil productivity not only increases crop yields but also increases the volume of organic matter produced which, in turn, helps increase or maintain the organic matter of the soil. This increase in organic matter improves the soil's workability (tilth) and makes the soil more resistant to erosion. Care should be taken not to overfertilize. Not only is this costly and wasteful, but it also could be harmful to the crops. And, if there is an excess, it may pollute water supplies.

Organic or inorganic fertilizer?

Fertilizers generally applied to crops are nitrogen, phosphorus and potassium. Law requires that the fertilizer components be indicated on the bag. They are generally stated as pounds available per hundred pounds of material. There are many forms of nitrogen, phosphorus and potassium used in the fertilizer industry. These forms are both organic and inorganic. The form applied is immaterial to the green plant as it absorbs plant food in the elemental form (ions) and thus does not differentiate between organic and inorganic forms. Enthusiasts for using "natural" fertilizers claim, however, that when organic materials are used, the plants receive more trace elements since the nutrients are absorbed more slowly.

ORGANIC FARMING AND ORGANIC GARDENING

There is nothing wrong with organic farming—using only so-called "natural" fertilizers such as manure and decayed vegetation—except for the idea that it would be practical to return to it today. We farmed that way 75 years ago. But, wastes used to fertilize the organic way are generally very low in fertility compared with chemical fertilizers. Also, plowing crops under to restore fertility is a costlier method than applying commercial fertilizers. We simply could not meet our food needs today without a large input of chemicals and fertilizers. The idea that nitrogen added to the soil as a chemical is somehow harmful to plants, whereas nitrogen derived from a legume is somehow "purer" and harmless, has no basis in fact. Dr. Sam Aldrich, agronomist and former member of the Illinois Pollution Control Board, comments: "Nitrogen from plant residues and manure is first converted from organic form to ammonium and then to nitrate. Nearly all nitrogen in fertilizer is in ammonium form and it, too, converts to nitrate." A more justifiable objection is that nitrogen and other nutrients of commercial fertilizers, being water soluble, can add to water pollution if applied unwisely.

Organic gardening can be fun, and productive enough for home needs. Furthermore, it is a convenient and useful way to dispose of many wastes. Whether or nor organic vegetables are more healthful than those raised the conventional way is debatable. However, there is no question that the debates themselves are beneficial, for they spur more agricultural research.

☐ USE OF HERBICIDES AND INSECTICIDES

In recent years, farmers have been using a variety of herbicides and insecticides to combat weeds and insect pests and to maintain conservation crop rotations. When suitable chemicals are used, weeds and insects are controlled, resulting in greater crop yields and farm production efficiency. Much has been said about the environmental pollution caused by their use. This criticism has some merit because the chemicals developed first had long half lives (long lasting—slow to disintegrate) and their residual effects plagued the environment. The insecticide DDT, which is toxic to man and absorbed through the system, is an example.

Wildlife with traces of DDT were found from Florida to Alaska. One conservation organization said DDT really stood for: "Dead Ducks Tomorrow." Fortunately, it is largely outlawed today. Also, indiscriminate use of agricultural chemicals has had a bad effect on the environment. Again, a change for the better has been made. The chemicals now being developed and used have short half lives, with little or no residual environmental effects.

Agricultural chemicals are needed for top agricultural output. They should not be used indiscriminately, however. Advice as to kind and amount of insecticide or herbicide to use can be obtained from the USDA Extension Service, which has offices in nearly all counties.

☐ THE PROBLEM OF GULLIES

It frequently becomes necessary to use a special device or type of construction in order to retard or stop erosion. This is especially true where the water run-off has already done a great deal of washing and where the cuts or gullies are large.

Various practices or devices are used to protect against further cutting away, and frequently the use of a mechanical or special aid, as well as the more commonly used soil conserving practices, is needed to do the best job of keeping the remaining soil in place.

How gullies are made

The most conspicuous damage that we see when erosion is severe is in the form of gullies. Gullies are found almost everywhere. In most cases where gullying starts, there will be little or no grass covering the surface, and practically no other plant growth will be found to slow down water run-off.

The depth and shape of gullies, as well as the rate at which they are made, depend upon any one or more of several factors. Some soils, such as wind-deposited (loess) soils and certain types of sandy soils, seem to erode fairly readily, leaving straight-walled cuts which are easily widened to form "U"-shaped gullies. Any soil with a soft subsoil which erodes readily will tend to form gullies of this shape.

If, on the other hand, the subsoil is heavy and absorbs little water, the gullies usually are not as deep as those formed in the lighter subsoils. They ordinarily are "V"-shaped, and their progress in cutting up a field is not as striking as in the case of the formation of the "U"-shaped gullies. If the soil is shallow and underlaid with rock, the depth of washing will be only a few inches through the thin soil to the rock, but the width of the wash may be very great.

FIGURE 18-1. The bench terraces on this Iowa farm enable the farmer to produce crops on this land without severe erosion. The earthen dam prevents further gullying and provides a water supply. *(Courtesy, Soil Conservation Service)*

A "U"-shaped gully is usually formed by water falling over the edge of a bank. The falling water at the edge of the eroding soil washes away the underlying subsoil. The topsoil then caves in, and the lip of the gully thus gradually moves up the slope. Many times we have watched small "Niagaras" of this type after heavy rains, as they drop to a lower level and undercut the lip of the watercourse as they fall. We probably were not conscious of the damage done to a field as the water cut deeper and deeper channels which backed slowly up the hill. If there were no drop at the edge of the field or if a good growth of grass were maintained, there would be no waterfalls and no gully cutting.

It is much easier to prevent the formation of gullies than to control them and heal the scars after they are once started. Sometimes this type of erosion may be prevented by the relocation of roadways or cattle lanes so as more nearly to follow the contour.

CROP ROTATIONS AND OTHER PRACTICES □ 291

FIGURE 18-2. Gullies cut on the side hill in a cattle lane. Wherever cattle wear a path smooth, there is every chance for erosion. (See also Figure 18-3.) *(Courtesy, Soil Conservation Service)*

FIGURE 18-3. The eroded area shown in Figure 18-2 was planted to trees which made enough growth to protect the slope from further washing. The trees are five years old. *(Courtesy, Soil Conservation Service)*

Most gullying is preceded by sheet erosion so that if the land is farmed so as to prevent this form of erosion, it is possible to keep many of the gullies from forming. When gullies are once started, they require more drastic measures, however, if their devastating progress is to be controlled.

Control of gullies

As mentioned previously, it is sometimes possible to stabilize a gully by planting the entire ravine to some fast-growing, dense vegetation. Different grasses, vines or shrubs may be used for this purpose, depending upon the soil, the climate and the topography of the area.

Where vine or grass growth is difficult to establish, it may be possible to put in a temporary dam of earth, grass, brush, wire or stones in order to slow down the water run-off until vegetative cover is well started. On the other hand, if the volume of run-off is too large for satisfactory vegetative control, it may be necessary to make a more nearly permanent dam which is large enough to hold the water that accumulates as a result of any heavy rain. The sloweddown water run-off will drop its excess load of soil back of the dam. In order to let most of the water escape from the pond after it has been slowed down and much of the suspended soil has been deposited in the bottom of the pond, a structure known as a drop inlet, which is constructed as an integral part of the dam, will carry the excess water off through a metal or clay tube in the lower part of the dam (see Figure 18-4).

In some situations, diversion ditches are built across the waterway and above the gully. These ditches do not let the water flow into the washed-out gully, but they carry it around and to one side. Diversion ditches should be permanent features of the land and should be so put in that they will not, in turn, be the source of further erosion.

It doesn't take long to fill back of a dam, and if the dam is built high enough, a small acreage of new soil will be created. Unless steps are also taken to stop the erosion from the hillsides, a dam of this sort is but a temporary device for holding back soil and relieving flood conditions. The larger the dam, of course, the longer it will serve as a pond for holding flood waters and reducing flood and silting damage farther down the ravine.

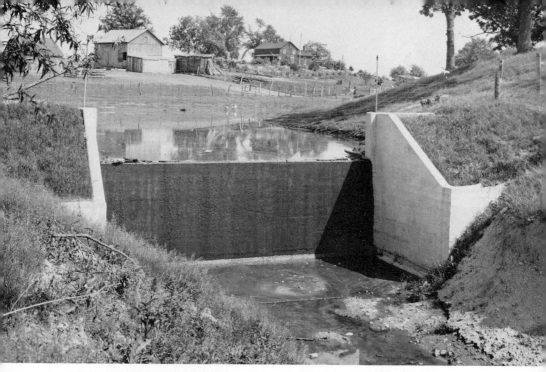

FIGURE 18-4. This 10-foot-deep gully drains a 75-acre area. The concrete dam makes a pond which will ultimately be filled with silt. The flume drops the water 9 feet and slows the rate of flow so that no washing will take place. *(Courtesy, Soil Conservation Service)*

Use of flumes

Flumes (chutes) are used to carry water run-off down steep hillsides or over the lips of gullies to lower levels. They must be built so as to resist erosion or cutting by the run-off water.

Sod-covered flumes are ordinarily used for dropping the water less than 10 feet (perpendicularly) and for draining small areas of probably not more than 25 acres. In areas where Kentucky bluegrass or Bermuda grass grows well, sod flumes have been used to drain larger watersheds into gullies that have only a few feet drop. Flumes are not built to fill gullies with silt or to stabilize either the gully channels or sides, but only to lower the water to another level. The steepness, or slope, at which a flume may be put in depends not only upon the soil type and the size of the watershed but also upon the height of the waterfall and the type and quality of sod used. The steepest slope for a sod flume is about a 1-foot drop, or vertical distance, to a 4-foot horizontal distance. This means that if the water is to be dropped 4 feet, the flume should be at least 16 feet long. It also should be wide in proportion to its depth because shallow water will not run as rapidly down a slope as will deeper water. In general, the flume should be at least 15 inches wide for every acre of the watershed, and the depth of the water at its maximum run-off should not exceed 12 inches.

Concrete- or stone-constructed flumes are used where the water run-off is too great for sod flumes to hold or where the drop to the lower level is too much. Occasionally, wooden or metal flumes are as satisfactory as concrete. All of these flumes are of a more permanent nature than the grass-sodded flumes, and the costs are much greater. Their slopes may be twice as steep as those of sodded flumes, and they will carry more water run-off if they are properly constructed.

☐ STREAM BANK EROSION

Streams never travel in straight lines. They wind and twist, always seeking out the easiest way to move to lower levels. The greatest force of the water in a stream is exerted against the bank on the outside of each bend, and any deposit of sand or gravel occurs on the inside where the current is quietest. As this process continues, the bends in the stream become more abrupt, and the current cuts more deeply into the bank. The final result is that much of the bottomland along the stream bed becomes so cut up as to be of little use for crop production. Stream bank cutting also may take place on farms where water flows only during times of heavier rainfalls or in the spring when melting snow causes water run-off.

Vegetative growth prevents intermittent stream bank cutting

Ordinarily, it is possible to control the bank cutting that takes place in a stream which flows only intermittently by the use of grasses, shrubs and trees. Before adequate growth can be established, it is frequently necessary to use some temporary obstructions along the eroded bank to check water velocities enough to redeposit the soil lost from the location. After the bank cutting has been checked and enough silt or other sediment has been deposited, it may be necessary to plant the desired vegetation if sufficient cover is not provided by natural growth.

CROP ROTATIONS AND OTHER PRACTICES ☐ 295

FIGURE 18-5. This stream started to wash away the bank, and, if it had not been stopped, it would have gradually cut across the productive land to the left, forming an ox bow. The bank as shown here has been worked down, and willow cuttings have been planted between the poles held down by posts and wire. (See also Figure 18-6.) *(Courtesy, Soil Conservation Service)*

FIGURE 18-6. The willow cuttings planted on the eroded stream bank shown in Figure 18-5 have made enough growth to protect the bank from further washing. The cuttings have been in two years. *(Courtesy, Soil Conservation Service)*

Many stream banks require the use of jetties

Jetties are extensions built into a stream from the bank to deflect the water flow. These obstructions, when built correctly in the right places on the curves, direct the stream flow away from the bank and permit some filling with sediment between the jetties. The Ohio State University conducted tests some years ago to determine the best location of the jetties for the control of stream bank cutting. According to this authority, the first jetty should be placed at the point where the flow line of the stream intersects the eroding stream bank, "A" on Figure 18-7. A second line should then be drawn parallel to the flow of the stream, but passing through the outer point (toe) of the first jetty. The point where this line intersects the stream bank, "B," should be half the distance between the first jetty and the second jetty, "C." Each succeeding jetty around the bend should be placed at the point where a line projected across the toes of the next two upper jetties intersects the bank, "D," "E," etc. These jetties

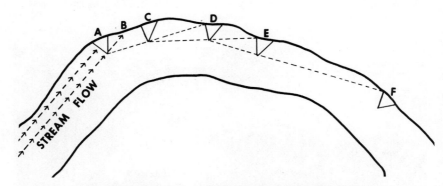

FIGURE 18-7. Jetties for a stream bank to keep it from washing away.

should point down stream, and the upper edge of each should form about a 45 degree angle with the bank. The top of each jetty should slope downward toward the stream from the bank at about the angle expected for the ultimate bank. It was also found desirable to use low jetties which cost less to build than higher ones but which are as effective in stopping the erosion.

The distance that the jetties should project into the stream depends upon the swiftness of the current and the volume of water in the stream. In no instance should they project more than one-fourth to one-third the width of the stream at flood time. When extended too far into the stream, the jetties themselves may be washed away.

Types of jetties

Jetties may be made of various materials, depending upon the type of stream which is doing the cutting. On shallow, slow-moving streams, piles of alternate layers of rock and brush have proven satisfactory. The faster a stream moves, the more securely the jetties must be anchored. Loose rocks held in place by woven wire and forming sausage-shaped jetties have proved effective in streams where the rock-brush jetties would not hold. Any one of the fastest-moving streams requires jetties which are firmly tied into the bank and anchored to the bed of the stream so as to be protected from undercutting. These jetties frequently are built as cribs which are then filled with stone, pieces of cement and the like. Jetties of this type are most expensive to build and are scarcely ever put up by individual farmers to protect their property. Ventures of this size usually are community sponsored and financed for the protection of highways.

☐ RIP-RAP

A rip-rap, or revetment, completely covers the area to be protected and differs from a wall or fence in that it is laid on the slope of the land. It is used where other means of control are practically as expensive without giving the complete protection afforded by a rip-rap. Because it is so expensive, it is used sparingly and is supplemented as far as possible with vegetative cover. Most rip-raps are of a permanent nature, however, and as such may be made of stone laid in cement or mortar. Occasionally, temporary protection is desired until the area can be covered with some tree or shrub growth. In such instances, brush matting or a covering of loose stones is used. When trees or shrubs completely cover the area, the protecting rip-rap is no longer needed.

□ PROTECTING ADJACENT LAND

Continuously flowing streams usually have a more persistent erosion problem than is found in the ravines which carry water only periodically. One important factor, both as a preventive and a control measure, is the protection of the strip of land immediately adjacent to a stream. As livestock will eat off the protective vegetation, they should have only limited access to it, and that access should be on level stretches of the stream bank if the best erosion control practices are to be used.

□ CHANNELIZATION OF STREAMS AND RIVERS

Channelization is the practice of deepening, widening and/or straightening an existing channel in a stream in order to increase its capacity to carry greater volumes of water. This is done to reduce the frequency of flood damage to lowlands adjacent to the stream, to reduce the severity of the meanders (windings) of a river and to improve the navigability of streams.

When meanders are removed, the total length of the river is reduced and the gradient (the amount of fall expressed in feet per mile) is increased. This aids navigation, but it can speed the water downstream faster to add to flooding.

Channelization is a rather controversial practice from a conservation viewpoint. In some places, after the channels had been dug, it was found that poor planning had resulted in a deterioration of the environment. Dredged materials were left in huge piles, with nothing done to prevent them from adding sediment, in great amounts, to the stream. Aquatic life was destroyed, and water quality was reduced. In other areas, the stream banks, which had been stable, became unstable and produced little or no habitat for wildlife.

Today, before any sizeable channelization project is begun by our government, the wildlife and environment of the stream are evaluated. And, the project is fully reviewed at public hearings. Environmental considerations are then provided for in the project plans. A channelization project, before approval, should be worthwhile. It should add to our resource base by minimizing flood damage, providing drainage of lands needed for agriculture or providing better transportation potential, and should not result in costly losses of irretrievable environmental assets.

MANY BENEFICIAL WAYS TO MANAGE OUR SOIL RESOURCES

Crop rotation is a useful practice that increases output and has many conservation benefits. Fertilizers, besides adding to higher crop yields, also can help save the soil if used widely. Herbicides and insecticides have been improved and greatly increase farm efficiency. But, they too must be used with care.

Gullies are a sign of neglect of the land. They are easier to prevent than to control. However, there are many ways to deal with them. Stream bank erosion is a natural process. Various devices can be used to prevent streams from washing away valuable land and property. Channelization of streams and rivers should be done only when it adds to our resource base, considering all environmental costs.

Conservation does pay. Soil losses are reduced, yields are maintained or increased and the soil is conserved to produce food and fibre in the future.

□ □ □

QUESTIONS AND PROBLEMS

1. What is crop rotation? Why is a definite rotation not followed by most farmers?
2. Under what conditions is it safe to depend upon rotations alone to control erosion?
3. What is a common rotation for your part of the state? How is it benefiting the soil?
4. How can adding fertilizers wisely help the soil? What are some of the problems in using commercial fertilizers? What are the advantages? Why would an "organic farming enthusiast" not agree with you?
5. What are some of the pros and cons of using insecticides and herbicides? How can you find out which ones are the best to use?
6. Tell how gullies are formed. Where a "U"-shaped gully is formed in one field and a "V"-shaped gully in another, what difference in the soil of the two fields is indicated?
7. No doubt you have seen a gully in the process of being cut or formed. Describe the conditions which apparently are responsible for the gully—for example, a lane along the side of a hill where cattle go to and

from a pasture, a cut to reduce the grade of a highway or a plowed field on the side of a hill.

8. What structures or devices are used to control gullies? Under what conditions is it possible to use sod or other vegetative growth to stop or control gullies?
9. What is a drop inlet? What ultimately happens to the pond in back of a dam when a drop inlet is put in?
10. What is a diversion ditch? Should it ordinarily be built as a permanent or a temporary structure? Explain your answer.
11. What is a flume, and what is the reason for using a flume—to fill a gully, to stabilize the channel or to lower water to another level?
12. How can the erosion caused by streams or rivers be controlled?
13. How does rip-rap differ from jetties? Out of what materials may rip-rap be constructed? Which types of construction are commonly used for permanence?
14. Is channelization of streams and rivers a "good" or a "bad" conservation practice? Explain your answer.

19 □ □ □

The city and natural resources

So FAR, WE HAVE CONSIDERED the nature of wind and water power, fuel and nonfuel minerals, water and forests and soil and vegetation, and we have discussed the conservation of these. None of these, we have learned, is a resource until it is of use to us. It has become evident that the most important fact about the "life" of a resource is often not so much the stock of that resource provided us by nature, as the way in which we use it and the rate of use. Nowhere is this problem more focused than in the city. The city is the caldron in which pressures on resources come to a boil. Most of us Americans live in urban places. And, large cities send their influence far beyond their boundaries. Our "way of life" in the city has a tremendous impact upon our entire resource base. Some say that the city threatens to destroy it. Others see the city as the generator of resources. Let us take a look.

□ CITY GROWTH

From 1960 to 1978, the number of supercities[1] in the United States increased from 209 to 279. The total number of residents in supercities increased from 113 million to 158 million. This was a relative increase of 10 percent—in 1960, 63 percent of the U.S.

---

[1] By *supercities* we mean the Standard Metropolitan Statistical Areas (SMSA's) as defined by the census. They are central cities which, when combined with suburban neighborhoods, exceed 50,000 population and have close interconnections.

population lived in supercities; in 1978, the number had increased to 73 percent. The 1980 census preliminary statistics show that the growth has continued, although the relative size of the central cities of some metropolitan areas has declined sharply.

The city seen from space

So great is the impression of our cities on the landscape that they were clearly visible to the astronauts of Skylab when they circled the earth some 270 miles above it in space. Dr. Robert Holz, a member of the scientific team that trained Skylab 4[2] crewmen in earth observations and who conducted debriefings afterwards, told the American Association of Geographers at its annual meeting in Milwaukee on April 21, 1975: "The development of the city significantly changes the environment and produces a new signature on the landscape."

The photographs taken by the crewmen on Skylab showed that cities are spreading across the countryside (see Figure 19-1). Analyzing the photographs, Dr. Holz commented:

> Austin, Texas, is growing rapidly. . . . The gray tone signature of the urban build-up has spread along the transportation arteries leading out of the city. . . . [at Denver, Colorado] The combination of man-made structures, snow melt from the urban heat island, snow removal and snow pollution, and increased woody vegetation . . . gives a signature that clearly delineates the city. . . . [at San Diego, California] The level areas are densely built, but the steep ravines and ridge tops are unoccupied. . . . [at Miami, Florida] Perhaps the most startling area of contrast is the continuous nature and density of urban build-up which extends about 125 kilometers from Perrine . . . northward to West Palm Beach.

Looking down from Skylab, the crewmen could see an inner city at Chicago, surrounded by several zones or subzones with "tendrils that follow transportation routes into the suburban fringe."

---

[2] Skylab 4 was launched November 19, 1973, and was in space 84 days. Earlier teams had occupied the Skylab for 28 days and 59 days, respectively. On July 12, 1979, Skylab plunged to earth with final disintegration in southwestern Australia. It was the largest, most complex piece of equipment ever sent into orbit. See Tom Riggert, "Skylab's Fiery Finish," *National Geographic Magazine*, October, 1979, pp. 581-584. Now, the space shuttle flights, begun in 1981, have introduced a second space age—the age of space exploitation—in which even larger vehicles capitalize on the exploratory work of Skylab and its predecessors.

THE CITY AND NATURAL RESOURCES ☐ 303

FIGURE 19-1. The Chicago-Evanston-Gary area, with Lake Michigan at the top of the picture, taken by Skylab 3. Note how the urban growth radiates outward from the central city across the countryside. A ring of smaller satellite cities surrounds Chicago. All are tied together like a vast web, as the land is occupied along transportation routes connecting the cities. *(Photo by Skylab 3, in flight July 28 to September 25, 1973, National Aeronautics and Space Administration)*

City influence beyond boundaries

The economic, the social and even the physical impact of cities upon our resources extend far beyond city boundaries. Some urban geographers and sociologists divide our entire country into city regions. What goes on in Troy, Illinois, and Heber Springs, Arkansas, depends very much upon developments in their city regions centered in St. Louis and Little Rock. No place is really isolated from the impact of the city.

Even agriculture receives its impulse from cities, according to Jane Jacobs, a sociologist who has become well known for her three books on cities and as an editor of *Architectural Forum*. "Cities came first—rural development later," she writes. She relates that fodder crops were developed in the city gardens of France a century before they were adopted in rural farming. Today, she notes, fattening beef on corn before slaughter began in the city stockyards of Kansas City and Chicago. Now, the fattening is taking place in rural areas—a transplant from the city. Ms. Jacobs predicts that cities will become ". . . more intricate, comprehensive, diversified, and larger than today's. . . ."[3]

Cities and man

A concern of many of us is that cities threaten not only our natural resources but also our own being. We fear that we will become unhealthy and mentally unbalanced if confined to urban living. René Dubois, a famous microbiologist, stated: "Wherever he goes, and whatever he does, man is successful only to the extent that he functions under environmental conditions . . . under which he evolved . . . man does not really 'master' the environment . . . (he creates) sheltered environments within which he controls local conditions."[4] He suggests that there are limits beyond which our lives cannot safely be altered by social and technological innovations. Since by the end of the century, most human beings will be born in urban situations, Dubois points out that "the future will therefore depend upon our ability to create urban environments having the proper biological qualities."[5] He stresses that while man has remarkable adaptive capacity, he still requires immense environmental diversity if his full potentials are to be reached.

To those of us reared on the farm or in rural communities, it seems "natural" to assume that living in the country is best for us. But, the city may well be more satisfactory to some people. Herbert Gans tells us that while, for some people, being outdoors provides satisfactions ". . . I have known other people who derive similar

---

[3]Jane Jacobs, *The Economy of Cities* (New York: Random House, Inc., 1969), pp. 17, 250.

[4]William R. Ewald, Jr., ed., *Environment for Man: The Next Fifty Years* (Bloomington: Indiana University Press, 1967), p. 15.

[5]*Ibid.*, p. 21.

benefits from walking through the streets of Manhattan. . . ."[6] He cites the case of tenement residents who, when taken to the wave- and wind-swept beaches of Cape Cod, wanted to get back as quickly as possible to their own neighborhood. He comments, "They come from a culture which does not prepare them for being alone and for becoming immersed in nature."

Need for cities

The reasons for the growth of our metropolitan areas, our supercities, vary. They may be primarily economic. Many Americans would prefer to live in the country. But, they cannot afford to do so. The city is where they as consumers and producers can meet most readily and thus at less cost to the one and more profit to the other. This is sometimes referred to as the *advantages of aglomeration*—the advantages of having ready access to a great mass of things all jumbled together. For others, the attractions of the city are what may be called the *amenities*, the nonfinancial attractions that make life more fun—major league baseball, national hockey, all kinds of films, stage plays, and operas, bowling, ice skating, tennis and golf to choose from and restaurants serving anything from the sukiyaki and curried rice of the East to the *sauerbraten* and *kartoffelkloss* of the West.

From the days of Catal Hüyük, the earliest known city which existed in the region of present-day Turkey some 8,000 years ago, cities have been with us. The question is not, What can we do to disperse city dwellers into villages, nor, how can we get along without them? The question is, How can we make cities better? Can we build them so they will enrich our resource base and us?

☐ THE GREEN CITY

First, let us consider what can be done about our present cities. One of the most obvious ways we can make them more healthful is by providing more vegetation and more usable open space. Vegetation is essential to helping purify the air. Open space is needed to permit the sunlight to reach the earth's surface and us, as well as to

---

[6]Davis W. Fischer and John E. Lewis, eds., *Land and Leisure: Concepts and Methods in Outdoor Recreation* (Chicago: Maaroufa Press, 1974), p. 19.

provide a place for the plants to grow and for us to move around in. Ways of providing more trees and other greenery are to plant them along streets, in residential areas and in parks and recreation areas and to set aside outstanding natural areas.

The addition of trees or gardens to the downtown or commercial districts of cities is being done today in many cities. Not only do the trees and flowers add beauty, but they also are a good indicator of air quality. If the plants do not survive, it may be because of air pollution.

In residential areas, trees and grass are often neglected by the city. Trees have been cut down to make way for utility lines and, at least in one town, have been removed so that buses can conveniently pull up to the curbing. In other towns, trees have been removed because they were diseased—although a good idea, the trees have not been replaced. One reason for this neglect of vegetation is that it saves the city money. We should see to it that curbside plantings and parkways are restored or introduced anew.

Providing cluster zoning and parks and preserving natural areas

In new residential areas, green space can be provided by the use of cluster zoning (see Figure 19-2). Here, houses are constructed closely together on one part of a parcel of land, leaving more space for recreational and park areas on the rest of the parcel. Under the traditional system, house lots are too small to be of much use.

The word today on parks is "bring parks to the people." Neighborhood parks make the city more liveable. They provide a place for sunlight and fresh air. Vest-pocket parks are small islands of trees and flowers that relieve the monotony of asphalt, brick and cement and provide diversity which helps give identity to neighborhoods. A ball diamond or a few tennis courts provide open space and needed recreation areas that can supplement the parks.

Some cities have unique natural areas within their metropolitan boundaries. These should be preserved. For example, giant saguaro cactuses standing just outside Tucson, Arizona, have been set aside as a national monument. These cactuses are up to 50 feet in height. Beaches, canyons and unique bogs and marshes are other examples of the kinds of natural areas which exist near some of our major cities and should be saved.

THE CITY AND NATURAL RESOURCES ☐ 307

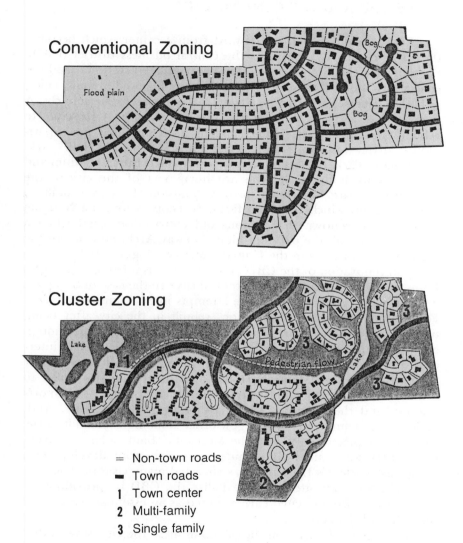

FIGURE 19-2. Echo Hill, a 250-acre development in Amherst, Massachusetts, was first planned as a typical grid-type community *(top)*. But then, builder-developer William E. Aubin switched to Planned Unit Development *(bottom)* to provide green spaces for community "breathing room." *(The above diagram appeared with an article by Charles Dole, "Proof of P.U.D. Is in the Living." TCSM, April 12, 1968. Adapted by permission from* **The Christian Science Monitor** © *1968, The Christian Science Publishing Society. All rights reserved.)*

☐ THE CBD AND THE CENTRAL CITY

Cities customarily grow outward from a center which becomes the focus of activity known as the central business district (CBD). Surrounding this core area is a densely populated section called the central city. Both of these areas have become blighted in many cities, as both businesses and private citizens have moved to the suburbs. A major urban problem is to restore this rotted core. A program to rebuild the CBD in St. Louis, Missouri, has provided new apartments, office buildings and recreation centers. The ball park, Sportsman's Park, was fashioned anew, renamed Busch Stadium and moved many miles, from the northern part of the city to the downtown. A large apartment project, Mansion House, was built on the riverfront, which the CBD faces. In cooperation, the National Park Service renovated a famous old courthouse, established a museum and built the 630-foot-high Gateway Arch, now one of the top tourist attractions in the United States (see Figure 19-3).

Can renovation of the CBD of a city, such as is being done in St. Louis, really revitalize the entire central city? In the case of St. Louis, it is too soon to tell. Some of the attempts in St. Louis have failed. For example, the Spanish Pavilion, rebuilt in the city after being moved from the New York World's Fair, failed to draw large enough crowds to support its activities and was taken over by private interests to be converted to a motel with a completed addition of a new wing in 1981. The Mansion House apartments likewise failed to draw sufficient tenants and were converted to a motel. The sports stadium and the Gateway Arch have been great successes. And, building still continues. The largest office building ever built in the city was completed in 1977 by the Mercantile Bank, a large convention center was built near it and a new riverfront development, known as "Laclede's Landing," has attracted many visitors. The program illustrates an attempt to revitalize the CBD by providing diversity, which Dubois, cited earlier, noted as an essential if man is to reach his full potential.

To upgrade housing in the St. Louis area, the city, several decades ago, undertook the largest, in terms of a single area, land-clearance project in the nation. Some 2 square miles of buildings were razed. Subsequent building has proceeded with a combination of residential and commercial buildings and at least one large manufacturing plant, a printing concern. This project, known as the Mill Creek Valley development, indeed has a variety. Slab housing, mod-

THE CITY AND NATURAL RESOURCES □ 309

FIGURE 19-3. A supersonic jet fighter, an F-15 Eagle, flying over the St. Louis waterfront and famous Gateway Arch illustrates how the changing use of natural resources has affected our cities. In the early 1800's, when water transport dominated the movement of freight, St. Louis was known as the "Gateway to the West," as symbolized by the Arch. Its superb waterway position near the confluence of the Missouri, Mississippi and Illinois rivers was unmatched for commercial access to the West. But, transportation modes changed. While the volume of river freight by barge today far exceeds the amount carried by steamboats years ago, its relative significance has declined, as trains, trucks and planes have taken over. The jet flying over the city represents not only the new status of St. Louis but also that of all the large metropolitan centers in the United States as ports of the world, which have access to resources inconceivable to all but a few visionaries a few centuries ago. *(Courtesy, McDonnell Douglas–St. Louis)*

ular housing, apartments, duplexes, single-family dwellings—all are represented in the residential section. Commercial structures likewise are varied. And, change is still going on. A major question is whether or not the various projects, which were added one by one instead of as part of a comprehensive plan, will ever really "jell" as a community.

Other attempts to clear blighted areas in the central city have met with mixed success. Notably successful developments include those by Leon Strauss, who was named St. Louis Construction Industry Man of the Year in 1980, and who has also received the St. Louis award for outstanding community service because of his redevelopment housing programs. His formula for success was a package of attractive financing arrangements and development schemes. The latter based development on ability to control whole city blocks and thus renovate an entire area rather than isolated units. He also effected innovative plans for rehab housing which considered both aesthetic and functional goals.

Most disastrous, and since having gained national infamy as a failure in public housing, is the St. Louis Pruitt-Igoe project. Hailed when built in 1955–56 as a model of housing to rehabilitate low-income families, the apartments had become quagmires of vice and violence by the 1960's. Most of the $41 million project to house 10,000 people has now been torn down. Lack of proper design was one reason for the failure. Oscar Newman, in his book, *Defensible Space*, suggests that the major physical flaw in the design of public housing can be fixed. Pruitt-Igoe was a monolithic type of high-rise apartments with no streets through the center. Newman states that projects must be open to view from the outside. This provides onlookers, which adds to safety. Pruitt-Igoe had skip-stop elevators and long hallways and inadequate playgrounds. There was no feeling of territoriality—a sense of pride and responsibility for one's own area of the project, which Newman points out is essential. He also recommends that new public housing projects should be built in middle-income areas and should be kept small (500 units) and low (under 7 stories).

Another problem with Pruitt-Igoe was a lack of social planning and guidance. Harborplace, Baltimore, Faneuil Hall Marketplace, Boston, and Columbia, Maryland, are successful developments by Master Planner James Rouse. (See Michael Demarest, reported by Robert T. Grieves and Peter Stoler/Baltimore, with other U.S. bureaus, "He Digs Downtown," *Time*, August 24, 1981, pp. 42–53.)

□ AIR QUALITY AND THE CITY

All of us have been concerned about the quality of air in the city, but how many of us have realized that the impact of the city on air masses extends far beyond its boundary? Here again, we shall refer to St. Louis as an example, for it happens to have been the center of a unique national project—a five-year project, 1971–1976, called METROMEX (Metropolitan Meteorological Experiment). The project involved 35 scientists from a number of different private and public research agencies.

The project found that two counties to the east, and thus downwind from St. Louis, get 10 to 30 percent more rainfall in the summer than comparable regions outside the city's influence. Its net effect is to increase crop yields by 2 to 5 percent. Individual rainstorms can dump as much as 300 percent more water than those formed elsewhere. While the added rainfall is found to be beneficial to crops, the thunderstorms that often accompany it—especially those with hail—are harmful.

An adverse effect of the city on air quality has been increases in ozone levels sometimes far from the central city. In 1980, Edwardsville, Illinois, led the state in EPA ozone readings—exceeding both state and federal standards. Yet, Edwardsville lies east of the major industrial areas of St. Louis and has virtually no industry of its own. In recent years, high ozone-level records have been associated with communities in Lake County, Illinois, well north of the Chicago hub. The readings, of course, depend upon various factors of the specific location and the number of air monitoring stations. Even though ozone diffuses rapidly, all parts of a community do not necessarily have uniform air quality.

While some of these effects are because of added heat from the city, as well as air pollution, we must wonder about the potential impact of our other large metropolitan areas on the surrounding countryside. The battle against air pollution in cities, where auto exhausts and factories are concentrated, is a matter of life and death (see "Air Pollution and Resources," Chapter 23).

☐ PLANNING CITY GROWTH

Planned unit development

Rehabilitating the old sections of cities is one opportunity to better use our natural resources. It is sensible and economical to continue to use for urban use those land areas now occupied by cities. Both in redeveloping large sections of our old cities and in providing for new growth, a new, unified type of development is being practiced known as Planned Unit Development (PUD).[7] This is one way of avoiding the urban sprawl of the past, in which building often proceeded without regard for environmental conditions or for total community needs.

In this type of development, a large area of land, say, several square miles or more, is developed as a unit over a span of years. The development may include commercial and industrial as well as residential sections (see Figure 19-4). The character of the physical conditions of the land—drainage, existing vegetation, soils, etc.—is

FIGURE 19-4. The swimmer in the foreground is standing beside a community swimming pool overlooking a lake in the Cottonwood Village section of the Cottonwood Planned Unit Development. Modular housing is in the background. This is one of the lower-cost communities of the development. *(Courtesy, Merrill Ottwein, The Cottonwood Companies, Edwardsville, Illinois)*

taken into account in the planning. Thus, a more harmonious adjustment to the resource base is possible than under the traditional piecemeal method in which residential subdivisions, shopping centers and industrial establishments were sometimes located so that they compounded problems of adjustment to the landscape rather than solving them (see Figure 19-5).

---

[7]For a technical definition, see Robert W. Burchell, *Planned Unit Development: New Communities, American Style*, Center for Urban Policy Research, Rutgers University, New Brunswick, New Jersey, 1972.

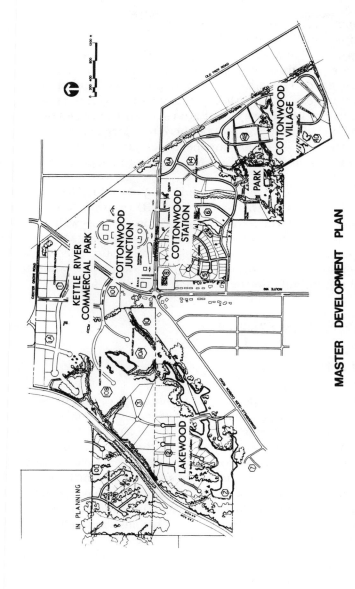

MASTER DEVELOPMENT PLAN

FIGURE 19-5. This master development plan of the 813-acre Cottonwood Planned Unit Development, located near Edwardsville, Illinois, illustrates how unit planning can provide variety and natural areas. Cottonwood Village features modular homes and several different styles of conventionally built single-family homes and townhouses. Cottonwood Station features homes in a medium-price range, clustered to provide more usable open space. Cottonwood Junction is a shopping center with an enclosed sports complex. Kettle River is subdivided for commercial ventures. Lakewood and its annex, Ginger Creek (not identified in drawing), have luxury single-family homes and townhouses. Many acres of heavily wooded parkland with nature trails and numerous lakes provide the entire community with a natural setting. (*Courtesy, Merrill Ottwein, The Cottonwood Companies, Edwardsville, Illinois*)

What will happen to farm land?

Several million acres of farm land on the edge of cities are taken out of agriculture and put into buildings and streets and other urban uses each year (see Chapter 14). Will PUD's continue to take over the farms? One way of preventing this is being tried out in Suffolk County, Long Island. The citizens of the county have decided to buy the right of farmers to sell their acreage to developers. *Time* magazine points out how it works: "If a farm is worth say $6,000 per acre to a developer but only $1,500 per acre to a farmer, the county will pay the difference—$4,500—for the 'development rights.' In return, farmers who join the program must agree to keep the land in farms forever."[8]

Another method to save farm land, proposed for the Green Spring Valley of Maryland, was having the PUD try to include in its boundaries those farms which were to be kept in crops as well as the area that was thought environmentally suited for housing. Whether used for crops or housing, the land would then be purchased at the same price per acre by the developer. And, the developer would be legally bound not to build on the farm acreage.

□ SOIL, SEWAGE AND THE CITY

Suitability of soils

Too often urban development has ignored the requirements of soil. Probably the most important investment most of us make in our lifetime is the house in which we live. The soil on which we build may be very undesirable for this use and may thus create many problems. Soil characteristics that should be considered are the seasonal water table, flooding potential, slope, shrink-swell potential, potential frost action and depth to bedrock. These factors have an influence on the ability of the soil to support a house without its having cracked walls, foundations and driveways. They also indicate potential for wet basements as well as flooded houses. Obviously, we should investigate soil conditions prior to building; otherwise, our "dream home" can become a financial disaster.

---

[8]"Planning City Growth," *Time*, April 21, 1975, p. 48.

Adaptability for sewage

Another matter that must be taken into account besides the suitability of soils for supporting buildings is their adaptability for sewage. Otherwise, our home as a place to live may become very disappointing.

Centralized sewage systems are the best. But, many areas are beyond public sewers and must dispose of household sewage on site through septic tanks into a buried filter field, if development is permitted. The success or failure of these systems depends on the soil. Soil characteristics important for this use are (1) permeability (the ease with which air and water move down through the soil), (2) height of the water table, (3) flooding potential, (4) slope of the land and (5) depth to bedrock.

Soil may be not permeable enough or too permeable to accommodate septic systems satisfactorily. If a soil is not permeable water and the effluent will not be absorbed by the soil. When this occurs, the disposal area becomes fully saturated, and the area becomes unsightly and unhealthy. If the soil is too permeable, another hazard may exist. The effluent will flow rapidly downward and possibly contaminate ground water.

The level of the water table has a direct bearing on the efficiency of waste disposal. Soils vary considerably in seasonal wetness and seasonal water tables. If a soil has a very high water table, additional water in the form of sewage effluent cannot be added to it without overflow. Generally, soils having seasonal water tables below 6 feet have few or no problems, while soils having seasonal water tables less than 4 feet have serious problems. Soils have inherent properties that can be mapped so as to indicate seasonal water tables, enabling us to avoid problems.

Flooding is an obvious hazard for any form of sewage disposal. When soils become flooded, sewage-disposal fields associated with septic tanks become inoperative, creating a serious health hazard as well as making household plumbing useless.

Slope of the land is also critical to efficient sewage disposal in septic tank systems because it affects the layout of the disposal area. The steeper the slope, the greater the chance of the sewage effluent migrating to the surface by lateral movement. This is offensive to the eyes and nose and could be fatal.

If bedrock is at the surface, it would be difficult to install a septic tank disposal field as well as to have the bedrock absorb the discharge. If the bedrock is fractured, the discharge might move downward as well as laterally to contaminate ground water supplies. It is important to have at least 4 feet of soil to filter the effluent.

Septic systems generally are not a satisfactory way of disposing of sewage. In many areas, especially urban, they should be prohibited.

☐ NEW TOWNS

When PUD's are developed to such an extent that they actually are small cities, they are referred to as "new towns." Such complete communities have been built in the United Kingdon for over three decades. And, others have developed on the continent of Europe. Over a decade ago, we in this country began building what we called "new towns," but only recently have some of the projects really qualified for this designation.

A new town may be defined as a new community planned to be largely self-contained, providing homes, services and jobs for its citizens, in a physically, socially and economically balanced environment. New towns differ from PUD's in that they are much more ambitious in extent—often proposing to accommodate 100,000 or more people eventually—and generally require planning for a much longer period of time and massive government financing.

Both new towns and PUD's are adaptable to meet the requirements for the rehabilitation of central cities. They could be located smack in the center of the city, for example, after land clearance. Thames-Meade, a British new town, is located in a former factory area on the Thames and within the London Metropolitan Area, for example (see Figure 19-6).

New towns are worth considering. In 1972, a three-year private study commissioned by the late Laurence Rockefeller concluded that our government ought to adopt a national policy calling for the creation of new towns. The study suggested free-standing new towns as well as those to be located within cities or on the edge of cities, or located to expand growth of small towns. There is no question that the chance to plan for diversity, including the amenities and reducing the adverse environmental effects, is appealing. However, experience with new town development in Britain has been mixed. Those towns built nearest metropolitan centers have been found most successful. But, in a free society, while government can encour-

FIGURE 19-6. Thames-Meade is a new town located within the London metropolitan area. Note the high-rises in the background and the two- and three-story dwellings on the right. An office building is in the center background. To the left is a lagoon for run-off water which also is used for recreation. Some 40,000 people live here. *(Photo by H. B. Kircher)*

age people to occupy new towns and establish business there, it cannot make them do so. Ways need to be found to keep alive opportunities for private initiative and preferences in planned communities in order to ensure their success.

☐ THE PLACE OF SMALL CITIES AND TOWNS

While our discussion here has been of supercities because they are the major problem areas today, there is no intention to deny the importance of smaller cities and towns in accommodating resource development. In a study of such communities in the Eighth Federal Reserve District, a region comprised of all of Arkansas and parts of six other states, author Kircher showed how these cities met the manufacturing trade, recreation and other needs of the region. He concluded that they were a vital part of the region's economy. As these cities grow or reconstruct, application of unit development principles will help them relate better to the natural resource base.

If the tide of population movement from country to city could be reversed, some of the pressures would be taken off the city. In fact, some of the social maladjustments of the city have occurred just because country job seekers became "lost" in the metropolis.

The National Council on Development for the 1980's, agreed in the first phase of its work that physical development should include the following two points: (1) communities should be more compact, both in developing and existing populated areas, and (2) a balance should be developed between employment and residential uses in existing urban centers as well as in new satellite communities. The Council recommended that by various land-use controls, ". . . 'urban villages' [be encouraged] within towns and cities where diverse housing and related services lay in close proximity to a commercial and industrial core."[9]

☐ CITIES ARE NATURAL—WE HAVE MADE THEM UNNATURAL

In this chapter, we have been able to touch upon only a few of the ways to deal with cities so that they will be more of an element of strength in use and development of our natural resource base.

Cities are here to stay. Barring a nuclear war which might wipe us all out, cities are bound to grow, and their influence on our resources is bound to increase.

We can improve our present cities. We must. Otherwise, chaos threatens to overwhelm us. City and regional planning are absolutely essential to the wise use of our urban resources and the protection of our rural resources. Well-designed and implemented plans will protect not only the city dwellers but the farmers, ranchers, lumbermen and other rural dwellers as well.

Planning does not necessarily mean *government* planning. It is a trick, but one that must be performed if we are to develop resources to the best advantage, to preserve wise local discretion in planning, while at the same time assuring that reasonable regional and national land-use needs are not being shortcut. And, we need to preserve the ingenuity and initiative of private businessmen, while still protecting our irreplaceable resources. PUD's are generally private, not government, ventures. And, even government projects are generally

---

[9]"Policy Trends, Harmonizing Design with Policy," *AGORA*, Newsletter of the Landscape Architecture Foundation, Vol. 1, No. 1, Autumn, 1980, p. 3.

planned with the help of private consulting firms and are built by private builders.

City planning should proceed on the basis of sound ecological assessment of the soil, water, air and other resources, Modern schemes of interrelated design and long-range development, such as those of PUD's and new towns, are promising ways of helping check the chaotic, wasteful metropolitan growth of the past.

Despite the importance of elements of design and the necessity of building in harmony with nature, it must be noted that some of the most pressing problems of our cities are social. Our resources, it will be recalled, depend just as much upon our societal as upon our technological abilities. In the city, our societal abilities especially appear to be most wanting: "... many cities large and small are now prowled by cold-eyed youths who mug and kill without emotion or remorse."[10]

We have discovered cities over 8,000 years old. Who knows how many were washed away in the muds of the Yangtze Kiang, blown away by desert winds or drowned in the rising seas? Like all other living things, man has his communities. The largest of them are cities. If we can preserve their health and vigor, we shall thrive. If we let them disintegrate into chaos, our entire resource base will be threatened.

□ □ □

QUESTIONS AND PROBLEMS

1. How do cities influence our resources in the country? Give two examples.
2. Where would you rather live—in the country or in a city? Why?
3. Why do people want to live in cities?
4. How can we make our cities "greener"?
5. How does cluster zoning make better use of our land resource?
6. How is St. Louis rejuvenating its central business district (CBD)?
7. What are some problems of trying to rebuild the central city?
8. What was the Pruitt-Igoe project? What did its failure teach us about better planning for cities?

---

[10] "The Menace of Any Shadow," *Time*, December 22, 1980, p. 32.

320 □ OUR NATURAL RESOURCES

9. How can a Planned Unit Development (PUD) result in better use of our land resource than we had without it?
10. What is a method of planning or of buying land that can help prevent agricultural land from being taken over by nonfarm development?
11. Why is it important to know about the soil in cities?
12. Can new towns solve the problem of rehabilitating cities? Explain your answer.
13. Are small towns and cities still needed for resource development? Explain your answer.
14. Do you agree or disagree with the conclusion that "cities are natural"? Explain your answer.

20 ◻ ◻ ◻

*Why wildlife
is important and
how to save it*

WHEN THE TERM *WILDLIFE* is mentioned, some of us immediately think of deer, bears or water birds which can be hunted during the "open seasons"; to others, it brings to mind song birds, rare plants of the woods, the virgin forests or lovely flowers that can be enjoyed only as long as they are left in their native habitat; while, to a third group, it means the fish of our streams, lakes and seas. The term *wildlife*, indeed, includes all nondomesticated plants and animals: mammals, birds, reptiles, amphibians and fishes—all *vertebrates*—and *invertebrates*, such as shellfish, crabs, starfish, insects, spiders, worms, corals, sponges and numerous one-celled animals. Wildlife is so rich and varied a resource that we can only skim the surface here. We shall concentrate on the wild animals and the way in which non-specialists can help conserve them.

"Life today, despite the process of elimination [which has occurred throughout geologic time] is not poorer but richer and more varied than in any previous epoch," Marion Newbigin, a zoologist, botanist and geographer, has written in *Plant and Animal Geography*.[1] How much longer our rich and varied wildlife will remain so appears to be in question, however, unless we do a better job of taking care of it. Within the last century, 9 mammals, 31 birds and 6 fishes have been exterminated in the United States. Since colonial days, 62

---

[1] Marion Newbigin, *Plant and Animal Geography* (London: Methuen, 1950), p. 24.

animals have become extinct. Today, over 100 animals are threatened with extinction. And, 1 out of every 10 plants in the world is said to be endangered. It is sad that these animals are gone, for the record shows that we can do wonders in restoring wildlife. Deer are said to be more plentiful in North America now than when Europeans first settled here. Antelope, elk and turkeys are examples of wild game that have been brought back from dangerously low levels to number in the hundreds of thousands.

The United States is a world leader in programs to rescue endangered species by controlled breeding, *U.S. News & World Report* observed in its April 17, 1978, issue. Since 1966, when Congress passed the Endangered Species Preservation Act, the federal Office of Endangered Species has set up recovery teams to bring back over 50 species of wildlife from near extinction.

☐ JUSTIFICATION OF WILDLIFE

Why save wildlife? To some of us, this seems an absurd question. However, for those of us who do not see its absurdity, the question must be asked so that we will understand why wildlife is important.

Wildlife can be justified for reasons we have chosen to call "the four E's." These are *E*conomic, *E*sthetic, *E*thical and *E*cological.

1. *Economic value of wildlife.* Wildlife has a tremendous dollar-and-cents value. There is income from sales of fish and furs and savings from eating wild game. But, this is only a small part of their worth. Billions of dollars are spent each year for hunting, fishing and other forms of recreation related to wildlife. More billions are saved by the control of the environment by wildlife.

According to a 1975 survey by the U.S. Fish and Wildlife Service, residents of one state alone, Illinois, which is noted for its agriculture and manufacturing, spent nearly $1 billion in 1975 for hunting and fishing—over $400 million of it within the state. And, according to a study by wildlife biologist Allen Farris, in 1971, $22 million was expended by resident and non-resident hunters in the great farming state of Iowa just for pheasant hunting alone, exclusive of license fees.

It was estimated over a decade ago that in the more humid parts of the eastern United States, the value of wildlife averaged about 14 cents per acre per year for meat and 23 cents per acre for the destruction of insects and other pests which hindered or retarded crop production. In the more arid parts of the country, the meat value was given as 4 cents per acre and the pest control value as 13 cents. Since then, values in dollars and cents have at least doubled. If we assume an average total food and insect-destroying value of 50 cents per acre for the whole United States, this places a yearly value of over $1 billion on our wildlife. Even this is a modest estimate of the money value of our insect-eating birds alone, for if they should fail for just one summer to consume the myriads of insects and insect eggs, we would find our country without feed for our livestock and food for our people.

2. *Esthetic value of wildlife.* No one can measure the value of beauty, but we all can recognize it. Women spend billions of dollars each year on cosmetics to "make themselves more beautiful." We pay fortunes to architects to design beautiful buildings. For many years, American-made automobiles were sold on the basis of their beautiful lines. President Lyndon Johnson proclaimed "beauty" a national goal in the 1960's. What is more beautiful than wildlife? We value birds for their beautiful plumage, song and flight. We value antelope and deer for the beauty of their graceful movements. Much of the joy of fishing is in the beauty of the leaping trout. How beautiful are flickering fireflies on a quiet summer night—even the deep "hrumm, hrumm" of Mr. Bullfrog is beautiful. How drab life would be without the beauty of wildlife!

3. *Ethical reasons to save wildlife.* We have a "moral" responsibility to pass along to our children and grandchildren our wildlife heritage. Furthermore, we should not be wasteful. The Indians' use of buffalo for food, shelter and clothing was a wise use of a resource. But, our reducing the numbers of buffalo in wholesale slaughter from over 50 million to less than 1,000 in a few years was unethical.

4. *Ecological value of wildlife.* Last considered of "the four E's," but the most important to our material well-being, is the ecological worth of wildlife. As explained in Chapter 23, in which some of the principles of ecology are discussed, all life forms on earth are interdependent in what is referred to as the "web of life." When we wipe out various species of wildlife, we are destroying some of the richness of our own life. Carried far enough, it means total destruction for us too.

In *food chain* relationships, animals eat plants and man eats animals. If the balance of nature becomes upset, man will be also. When ranges were overgrazed by cattle, the soil was exposed and eroded away, and sagebrush took over from the grasses or sandy, rocky wastes developed. Then, the land could support neither cattle nor wild animals, and certainly not man. Insects and even reptiles can be as important to our well-being as large animals. We stamp them out without thought.

The interrelationship of plants, animals and land to man is illustrated by problems of development of the Florida Keys—the small islands stretching southward from the Florida Peninsula. Land developers wanted to fill mangrove swamps with fill from the shallow sea nearby, thus providing sites for hotels. Dr. Arthur Weiner, a marine biology professor at Florida Keys Community College, told *The Christian Science Monitor* on March 2, 1973, that this would upset the ecosystem. "Mangroves are vital to preserving the marine ecology and the fish population. As the leaves of the mangroves decay, they turn into food for microscopic animals, which in turn provide food for worms and larger animals which feed the fish. Young fish hide from their predators [enemies] in 'nurseries' of the tangled roots of the trees which also hold sediment during torrential rainstorms, slowing erosion and helping to preserve the clear water for which the Keys are famous. . . . Killing the swamps by dredging and filling will kill the tourist trade."

Fortunately, our coastal areas are being increasingly protected against harmful development. The Coastal Zone Management Act of 1972 (CZMA) provided that 35 coastal states and territories were eligible to receive funding for planning coastal zone management. Such management has as one of its principal objectives the protection of unique and significant natural resources. Between 1974 and September, 1979, $70 million was distributed to these states to help them develop suitable programs. Thirty-one of the 35 eligible units either adopted new statutes and regulations protecting these wetlands or improved the implementation of existing laws during the period.[2]

Another reason for preserving wildlife, one not included in the four "E's," is their inspirational value. Whose spirit has not been uplifted by the cheerful chirping of chickadees on a dreary winter's day or by the sight of the wavering "V" of Canada Geese high in the air

---

[2]Council on Environmental Quality, "Protecting the Coastal Zone," *Environmental Quality, The Tenth Annual Report of the Council on Environmental Quality*, U.S. Government Printing Office, Washington, D.C., 1979, pp. 498–512.

as they wing southward? We characterize our friends by referring to them as "courageous as a lion," "sly as a fox" or "wise as an owl."

□ NOT ALL WILDLIFE BENEFICIAL

A few years ago, author Kircher and some friends were camping near Fishing Bridge in Yellowstone National Park. The cook was assigned the job of sleeping in the back of the stake truck, which had the food supplies. Late one night, a black bear decided to investigate some interesting scent he had caught, so he clambered up the tailgate of the truck. The cook, suddenly awakened, was almost startled out of his wits to find himself suddenly face to face with a black bear. Fortunately, he just happened to have a baseball bat handy. He proceeded to give Mr. Bear a "friendly" tap on the nose, with the fervent prayer that he had convinced him to dine elsewhere. At that point, the cook was convinced that at some times and some places wildlife are not especially beneficial. It took man many years to domesticate certain animals successfully. A wild animal cornered, desperately hungry or sick (as with rabies) can be dangerous. A word to the wise is sufficient!

Some wildlife which have gotten out of balance with nature appear to be mostly destructive. Large flocks of blackbirds—redwinged blackbirds, common grackles, cowbirds, rusty blackbirds and starlings—are noisy, have an odor and can be disease carriers. They are a problem to farmers. One theory for their destructiveness is that they have been forced out of their habitats to become a nuisance in cultivated areas. Another is that the flocks have grown out of size because the birds' natural predators—hawks and crows—have been so greatly reduced in numbers. No conclusive theory or solution to the problem has been found. The situation emphasizes the point that in our world, so heavily settled by man, more wildlife management than ever before is needed.

□ PROVIDING HABITAT FOR WILDLIFE

The *American Dictionary of the English Language* defines *habitat* as "the area or type of environment in which an organism or biological population lives or occurs." What kind of habitat is it that enables wild animals to survive? It is an environment which includes shelter,

FIGURE 20-1. The beaver, the wild turkey and the Canada goose are examples of the kinds of wildlife that have increased in numbers in response to management practices in recent years. *(Courtesy, U.S. Fish and Wildlife Service)*

food, water and range—a place to move around in. The greater the diversity of habitat, the greater will be the diversity of the wildlife there.

Vegetative cover provides shelter for wild animals, so a first step in building up an area for animal habitat is to restore cover, if needed. This step also reduces erosion and helps hold the rain on the soil until it can be absorbed.

Generally, vegetation cover benefits both wildlife and soil. Even weed growth will afford temporary shade and some food for birds and small animals. In an area with sufficient rainfall, where vegetative cover has been lost, nature itself will start the process of restoration, although it may be relatively unsuccessful if problems such as acid mine drainage, wastes or steep erosion slopes are present. Under favorable conditions, grass and shrubs will start fairly soon from seeds carried to the area by wind or birds. Seedlings from trees such as cottonwoods and maples may also drift into the area and start growing at about the same time as the shrubs.

In an area with little rainfall, the problem of establishing vegetation cover is more difficult. Furrows plowed on the contour help hold the rainfall on hillsides, and the use of baffle strips in freshly cut ravines slows down water run-off. Check dams accomplish the same result, or even sodding some strips across the watercourse may help in reducing the speed of water run-off and in restoring some plant growth.

Another consideration in wildlife preservation is to provide suitable places for animals to build homes and to live securely from their enemies. This again means shrubs and trees for some animals, good grass for others and, for still others, a soil in which they can make holes or runways.

Habitat must also provide food. We should realize, however, that some types of plants are much more beneficial to wildlife than others. Some plants are valuable for domestic livestock and erosion control but are of little value to wildlife for either food or cover. Planting beneficial species makes more sense than planting just anything to hold the soil in place. Landowners should ask the USDA Extension Service advisor in their areas for recommendations of specific plants.

Farmers can increase wildlife habitat by leaving a few rows of unplowed crops next to fence rows or wooded areas. Food plots which, after plowing, provide stubble which will stand up above the snow are most valuable for wildlife. Good intentions don't feed wildlife if the food is buried.

FIGURE 20-2. Every form of life lives off some other form. Without predators, other enemies and disease, any one species could well populate the earth. It is only because of this dependence of one form of life on another that a balance in nature is kept. Without predators such as mountain lions and lynxes, our deer population would require hunting seasons on does to help keep the deer population in check; and without foxes, hawks and owls, rabbits would be a worse plague than the locusts of Biblical times. *(Courtesy, U.S. Fish and Wildlife Service)*

An adequate range for certain wild animals is beyond the capacity of all but a few extremely wealthy persons to provide. Lord Estes, for whom the town of Estes Park, Colorado, was named, once controlled an area of many square miles in the very heart of the Colorado Rockies. While a number of large farms and ranches are still held in the United States today, most of us must leave the provision of range for wildlife such as antelope, buffalo, deer and elk to the federal government. Because these animals move freely from public to private lands, unaware that they may be "tresspassing," we must make sure that any small landholdings we may have or may use enhance habitat for wildlife.

The situation in Iowa, as described in a letter of February 27, 1980, from Ron George, Wildlife Research Biologist with the Iowa Conservation Commission, provides a good example of the habitat problem.

> Many species of upland wildlife thrive in habitat provided by diversified Iowa farmlands. Ring-necked pheasants, bobwhite quail, Hungarian partridge, cottontail rabbits, white-tailed deer, waterfowl and many types of songbirds prosper in farmlands that provide their basic needs for survival. In Iowa, two of the most important limiting factors for wildlife are the lack of undisturbed nesting and winter cover. Connie Mohlis, an Iowa State University Wildlife student, reported that comparison of 1939 and 1972 aerial photographs for 27 counties in north-central Iowa (Iowa's traditional pheasant range) revealed a 76% decline in good quality pheasant nesting cover and a 33% decline in winter cover. Winter flush counts conducted on the Winnebago Study Area in north-central Iowa by Thomas Scott and Thomas Baskett in 1941 revealed 180–200 pheasants per square mile, and an intensive search by Iowa Conservation Commission personnel failed to locate any pheasants on the entire 1,520 acre area in 1976.

Small animals such as foxes, oppossums and raccoons may travel only a few acres or a few miles in search of food, while others such as beavers and muskrats may occupy a particular pond. In these cases, small landowners may be able to meet all their needs within their own landholding.

Third, there must be a supply of water available. Animals in the desert or on the semi-arid plains have special systems which permit them to adapt to a very low water supply, or they obtain it from somewhat unusual sources, such as cactuses. Or, nature has equipped them to travel long distances in search of water. In the

FIGURE 20-3. This land was completely overbrowsed by deer. No food is left within reach of even the largest animal. *(Courtesy, Wisconsin Conservation Department)*

former case, the water-gathering plants upon which they depend must be protected or provided. In the latter case, an adequate range, perhaps hundreds of square miles, needs to be provided. In the humid eastern part of the country or in the mountains and on the seacoasts of the West, the water supply needs to be kept uncontaminated by poisonous wastes and chemicals. Small landowners may be able to increase the availability of water by damming creeks or even drilling wells to create man-made lakes.

More information on the specific requirements for providing habitat is given in the next chapter.

Wildlife need more than waste acres

The large number and great variety of our wildlife require many kinds of food and shelter. Because of these varied needs, desirable numbers of the different species cannot be maintained by limiting them to the use of severely eroded land alone. It is for this reason that wildlife conservation must include many localities and environments for the maintenance of a satisfactory wildlife population. There are many small pieces of land in all agricultural areas that are not useful for crop production but which are more or less ideally adapted as homes for wildlife. These small acreages serve most wildlife purposes best if they are fenced so as to keep domestic animals from rooting up, grazing or trampling the vegetation. Where this is done, the owners in many areas will be rewarded within a few years by the return of animal wildlife as well as the woods or prairie flowers.

☐ THE STRUGGLE FOR SURVIVAL

The importance of predators (enemies) to the survival of any one kind of wildlife should not be overlooked. One provision of na-

FIGURE 20-4. During some winters the number of deer lost by starvation has been enormous. These trees show they have been so heavily browsed during the summer that little growth is left for winter deer browse. *(Courtesy, Wisconsin Conservation Department)*

332 ☐ OUR NATURAL RESOURCES

FIGURE 20-5. The cottontail, the coyote and the raccoon are illustrations of wildlife that have adapted to the ways of man. They survive in excellent farming conditions. *(Courtesy, U.S. Fish and Wildlife Service)*

FIGURE 20-6. If you look closely, you will see a day-old baby elk which hid itself in this manner twice within an hour. Hiding places were selected by this little fellow where protective mottling of sunshine and shadow was most effective. Its mother was not near. Open range is required for natural perpetuation of elk. *(Courtesy, National Park Service)*

ture is that each form of life becomes food for some other form. The simplest, smallest forms of aquatic (water) plants are consumed by exceedingly small, or microscopic, animals. These, in turn, are devoured by somewhat larger animals which later are eaten by still others. When the wildlife population maintains a fairly definite proportion year after year, it is said to be in natural, or biological, balance.

This struggle for survival serves two purposes. First, the stronger, more alert of each group survive, making a hardier race, since the weaker or less cautious individuals are the ones that become food for others.

The second purpose served by this struggle for existence is to prevent the dominance of any one species. The increase in insect life during one summer would swarm the earth were it not for the enemies of insects—parasites, weather and disease. Disease is an enemy that is always a threat to overabundance of animal life.

Rabbits could take over the country except for predators such as foxes, coyotes, hawks and owls. Nature's balance among the various kinds of life is maintained mostly because of the continuous pressure of wildlife upon its food supply—other living things.

THE KAIBOB DEER INCIDENT

One of the great lessons in game management was learned on the Kaibob Plateau, "high country" just north of the Colorado River in northwestern Arizona. Here a mesa, relatively cut off from the rest of the countryside, was the home of some 3,000 Rocky Mountain mule deer. In 1906, to preserve this herd from the fate which was rapidly befalling other wild game in the United States, President Theodore Roosevelt set aside one million acres of this area as a national game preserve. About 6,000 predators of the deer—coyotes, mountain lions, bobcats and wolves—were destroyed. The management program appeared to be a great success. Within about 10 years, the deer herd doubled. Two years later, it doubled again. By 1923, the herd had reached 100,000 head, according to some estimates. But, there was no glory in the numbers. By then, it was obvious that something had gone wrong. Deer carcasses were lying around by the thousands. Thousands of animals were gaunt from malnutrition and disease. The trouble was easy to find. There was little to eat. The trees and bushes had been chewed up. In what the National Wildlife Federation has called "The Terrible Lesson of the Kaibob," two sterling principles of wildlife management were learned: (1) that predators along with plant eaters must be protected on game refuges and (2) that when man upsets the "balance of nature," he must actively manage the habitat. Failure to carry out either of these principles means that game refuges become death traps.

☐ WILDLIFE MANAGEMENT

Among the approaches that have been made to the problem of conserving our wildlife, two of the commonly recognized methods date back to colonial times. They are hunting restrictions and wildlife refuges. Today, we recognize that these alone cannot do the job. All landowners must provide habitat where possible.

Hunting restrictions

The idea that all game belong to the sovereign or the state originated in Europe, where only the rulers or owners of vast estates were privileged to hunt. Before the Magna Carta, which in 1215 granted definite privileges and liberties to the English day-workers, the killing of a stag cost the offender his hands, his eyes or even his life. Or, he could be imprisoned for "a year and a day." Even after this date, severe penalties continued to be enforced against the laborers for the possession of any kind of wildlife, though they no longer lost life or limb because of this.

These drastic restrictions to hunting were not brought to this country, though the idea that all wildlife belongs to the state was accepted. The earliest game regulations were for the purpose of legalizing the hunting by colonists that was taking place after they once learned how to live "off the land."

FIGURE 20-7. Canada geese over Crab Orchard Refuge in Illinois. Refuges help in perpetuating wildlife but should not be looked upon as a solution to abundance. *(Courtesy, U.S. Fish and Wildlife Service)*

Early Colonists Not Hunters

It is interesting that the early colonists were faced with a lack of food amounting to starvation conditions, even though they were surrounded by an abundance of wildlife. Some say that they were too "civilized" to make use of the plentiful wildlife of the woods, the seashore and the streams. The chances are that neither English nor Dutch colonists knew anything about hunting or fishing, because these privileges belonged only to the rulers in their parent countries. Few, if any, had ever caught a fish or killed a wild animal. They knew nothing about "living off the land." It was only with the help of the Indians and through the severe school of experience that they finally learned to adapt themselves to the use of food supplied by the abundant wildlife around them.

Hunting Permits Required

Hunting privileges were granted as early as 1629 by the Dutch West Indies Company, and in the Massachusetts Bay Colonial Ordinance of 1647, provisions were included for the right of hunting. In 1677, Connecticut passed laws regulating seasons for hunting and prohibiting the export of game, hides or skins. As early as 1700, all the colonies, except Georgia, had established closed seasons on deer. In Virginia, for example, the killing of deer out of season provoked a fine of 500 pounds of tobacco. Deer were not to be hunted by firelight, and running deer with dogs was prohibited in New York as early as 1788. The early colonists also regulated the hunting of game such as bobwhites, wild turkeys, woodcocks, ruffed grouse, heath hens (now extinct) and muskrats.

Closed Seasons

There are so many different kinds of wildlife protected today that we can scarcely realize the vast numbers of all kinds of birds and animals that were once killed just for the fun of killing. Yet, it was only about a hundred years ago that closed seasons, or complete protection of certain birds and animals, became common. In 1818, Massachusetts declared a closed season on snipes; in 1821, New Hampshire protected beavers, minks and otters. Maine protected her moose in 1830, while the first prohibition of spring duck shooting was enacted by the Rhode Island assembly in 1846. In 1850, both

WHY WILDLIFE IS IMPORTANT, AND HOW TO SAVE IT □ 337

FIGURE 20-8. These white cedar woods are good winter deer browse. As long as green growth is found in quantity within easy reach of deer, there will not be starving deer in the area. Is it more humane to let deer starve, protect predators which in turn will keep down the numbers or permit hunters to reduce the deer population? *(Courtesy, Wisconsin Conservation Department)*

Connecticut and New Jersey provided general protection to all small song birds and to the "small owl" for the first time. The first hawk to receive protection was the osprey, or fish hawk, in New York in 1886. These regulations are evidence that, although early in our colonial history we were conscious of the importance of conserving the wildlife of the country, most of the laws protecting it were enacted only quite recently.

Obviously, game laws are not effective unless they are observed. If hunters disregard these regulations, they threaten the very survival of various species of wildlife and of the sport of hunting itself. A good hunter will do even more than the law requires by following the old saying: "Limit your kill; don't kill your limit!"

HUNTERS PAY FOR WILDLIFE PROTECTION TODAY

Paradoxical as it may seem, it is the hunters and fishermen today who support game protection programs. They are the ones who buy hunting and fishing licenses and pay for special permits and stamps which largely fund state wildlife programs. And, thanks to the Pittman-Robertson Federal Aid in Wildlife Restoration Act passed in 1937, 11 cents of every dollar they spend on arms and ammunition is siphoned off by federal tax and given to the states for use in wildlife restoration. Over one-half billion dollars has been distributed to the states from this source. And, at least another quarter billion dollars for wildlife has been realized from duck stamp sales to hunters and fishing equipment purchases by fishermen (Dingell-Johnson Act, 1950).

Hunters organizations have also provided valuable direct help and management aids for wildlife. For example, the Migratory Waterfowl Hunters Association has set out thousands of nesting boxes for wood ducks, helping bring them back from critically low numbers. An excellent booklet on wildlife management principles and practices is published by the National Rifle Association and sold for a very low price (1979).

Game refuges provide valuable help

Another approach to the problem of conserving our wildlife is through game refuges. This is one of the systems used by some of our Indian tribes in their efforts to have plenty of game. Theodore Roosevelt told of the Creek Indians putting aside large tracts of land where persimmons, haws, chestnuts, muscadines and fox grapes were plentiful and where bears could feed and fatten unmolested. They called these tracts "Beloved Bear Grounds." Then at certain times the tribe would move in and kill large numbers of bears, which were their main source of animal food and oil.

Most of the wildlife refuges established during colonial days were private preserves, such as the deer parks of Virginia, North Carolina and Maryland. These refuges accomplished little except within restricted areas, and, by 1800, most of them had ceased to exist.

The first state game refuge was established in California in 1870, and the second was established in Indiana 33 years later in 1903. This was also the date of the establishment of the first federal wildlife refuge for colony-nesting birds, that is, birds that live and nest in groups. Among the birds protected during the next five years were the gulls, terns and pelicans. The first federal refuges for migratory waterfowl were set up in 1908 in Oregon—the Lower Klamath and Malheur Lake refuges.

The story of some of these large refuges is a story of both hope and disillusionment. Large areas were drained for agricultural use at great expense and later were again partially restored as refuges when it was found that they were not profitable agricultural ventures. Other refuges were left in their original states and provided nesting places, feeding grounds and shelter for the migratory birds.

According to the Bureau of Sport Fisheries and Wildlife, in 1978 there were 392 refuges in the Wildlife Refuge System, occupying 46 million acres of land. These refuges have been established to accommodate various kinds of wildlife as shown by Figure 20-9.

These measures only partially effective

Closed seasons and bag limits are useful in controlling the numbers of the various species of game which are killed yearly, and because of the power to change both seasons and bag limits, it is possible to adjust the yearly kill of game by hunters to the change in population of the various species. This is only a partial answer to the wildlife conservation problem, however, for it has not resulted in an abundance of game.

The same is true of game refuges or sanctuaries. They are especially useful in offering some protection to a few of the vanishing species of birds by furnishing them with safe havens during their nesting seasons. A second advantage of refuges is the security they afford various forms of migratory wildlife while they are rearing their young.

The hope was that these refuges would provide surplus numbers of the various kinds of wildlife, which would move out from the refuges to populate other areas. While this does take place to a limited extent, we cannot expect that less than 1 percent of the total land area of the country can supply game for the other 99 percent.

One of the reasons why this kind of program cannot succeed is the impossibility of raising sufficient numbers of wildlife in these restricted areas. Many forms of wildlife will not thrive in restricted areas, while others do not tolerate crowding. They demand ranging, feeding and nesting space out of proportion to their size and capacity for eating. For example, the production from 2 or 3 acres of cropland will furnish a year's supply of feed for a cow, or another

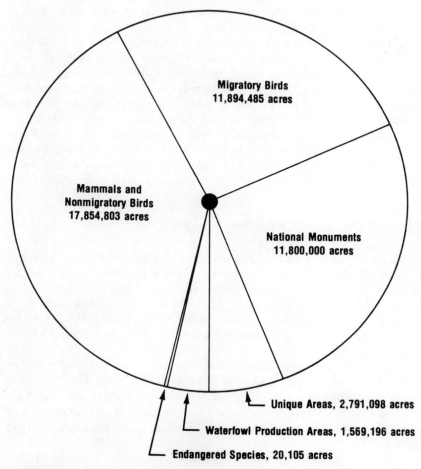

FIGURE 20-9. Acreage, by type, of national wildlife refuges, 1978. *(Courtesy, U.S. Fish and Wildlife Service)*

animal of like size, but the number of pheasants to be found on this same acreage of land will ordinarily not exceed one bird. Yet, this bird can make use of only a fraction of the feed and protection afforded by this amount of land.

Another question about the value of refuges is raised by E. H. Graham in the instance of certain species being so reduced in numbers that their existence is in real jeopardy. He states that "such species may have lost the vitality or 'biological potential' necessary to perpetuate themselves under any circumstances." For example, many authorities on the subject believe that once passenger pigeons were reduced in numbers to a certain point, they were headed for extinction because they then seemed to have lost their desire to nest.

All landowners must provide habitat

Probably the main disadvantage in the use of refuges as a solution to wildlife problems is failure to give proper consideration to the best use of the land. Wildlife is part and parcel of nature's balance. Its conservation cannot be limited to federal or even state or local government ownership of all the land that should be used for wildlife. If the problem is to be handled effectively, it must be the responsibility of every landowner, whether that owner be a public agency or a private individual. Some farm land should be used.

Although we already have 46 million acres of land in national wildlife refuges and 50 million acres in state refuges, there are still more than 30 million acres in privately owned farms and ranches that would serve their best use if devoted to wildlife. These acreages could not be combined into larger pieces even if the government were willing to buy them. Besides that, our government[3] now owns approximately two-fifths of all the land of the country (see Table 20-1). It is a question of whether the welfare of our people is best

TABLE 20-1. Our land and its ownership

Ownership	Acres (in millions)	Percent
Private	1,316	58.1
Indian land	51	2.3
Public land	897	39.6
Federal	761	33.6
State and county	116	5.1
Municipal	20	0.9
Total	2,264	100.0

(Source: U.S. Bureau of the Census, *Statistical Abstract of the United States, 1979*, 100th edition, 1980)

---

[3]"Our government" in this sense includes all federal, state, county and municipal units.

served by further concentration of land ownership in these various units of government, except in the case of wetlands.

DEVELOPMENT OF WILDLIFE RESEARCH AND EDUCATION PROGRAM

Our government has developed an exceptional system for education about wildlife in the past five decades. The nation's first fish and game management teaching program was set up at Iowa State University (then, a college) at Ames in 1931. Such programs have since become established at many other state universities. Cooperative wildlife research units have been set up at various locations in the country. Fish and wildlife specialists have been added to the staff of the USDA Extension Service in many states. Both the Extension Service and the Soil Conservation Service have specialists in plant life. Many states have upgraded the status of their wildlife field personnel from that of game wardens, largely enforcing hunting laws, to conservation officers who spend much of their time researching and teaching conservation.

A serious weakness in the program has been its neglect of nongame species of wildlife. Some progress has been made in the last few years.

WE'RE NOT OUT OF THE WOODS

Wildlife is important to us for both its economic and noneconomic value. There is little question about that. But, as far as "saving" wildlife is concerned, it is perhaps appropriate to use a well-known expression and say, "We're not out of the woods, yet." You will recall that we have pointed out several times that dense woods are *not* the best habitat for animals. We still haven't found all the answers to "saving" wildlife, nor are we doing nearly as much as we should be. The next chapter suggests one way in which we may improve the record by providing better habitats.

☐ ☐ ☐

QUESTIONS AND PROBLEMS

1. What different forms of life are included as wildlife?

2. Give some comparison that may show the economic value of our bird life.
3. Does wildlife serve a better purpose because of its economic value to us or because of the non-economic values we derive from the various forms of wildlife? Explain your answer.
4. List the requirements for the continued support of wildlife.
5. Describe the struggle for survival among wildlife. What is meant by natural, or biological balance? What are predators? Explain their purpose or use in maintaining the biological balance.
6. The statement is frequently made that certain pieces of waste, badly eroded or cut up land, are useful only for wildlife. Explain why this statement is or is not true.
7. In attempting to "manage wildlife," what are two restrictions that limited the killing of game as long ago as colonial times?
8. List the probable reasons why the early colonists of this country were faced with starvation, even though the woods and streams around them supported an abundance of wildlife.
9. Why are closed seasons, reduced bag limits and game refuges only partially effective in maintaining the wildlife of this country? How much land is there in this country that would serve its best use if devoted to wildlife?
10. If we are to have the variety and amount of wildlife which the land of this country will support, who must assume the responsibility of bringing this about?
11. Name some game birds and wild animals that are surviving in your agricultural neighborhood.
12. What game birds and animals have disappeared in the last century? What wild flowers? (Check the library for information.)
13. What is "The Terrible Lesson of the Kaibob" deer incident? What principles of wildlife management did we learn from it?
14. How do hunters help pay for conservation? How can nonhunters help?
15. What responsibility does every landowner have if we are to have an effective conservation program?

21 □ □ □

Providing wildlife habitat

AS STRESSED in the last chapter, providing habitat—a space in which wildlife can find adequate shelter and food—is an especially critical conservation problem today. So great is the pressure for development of land for houses, factories and commercial buildings that little thought is apt to be given to providing a home for wildlife. It's often thought that productive land use and wildlife habitat are incompatible—that both cannot exist in the same place. This is not necessarily so. We are not proposing to put a herd of buffalo on the plaza at Rockefeller Center, New York, nor grizzly bears in the Chicago Loop. But, song birds are at home in Central Park, and gulls thrive along the Chicago River. Conversely, wide open spaces that are inhospitable to man may be equally undesirable for wildlife. Unreclaimed strip mines with acid waters and barren soil are an example. This chapter is devoted to a consideration of habitat possibilities.

□ WILDLIFE HABITAT ON FARMS AND RANCHES

The management of wildlife areas on farms and ranches need not reduce total crop production. It need not require the diversion of land now used for crops, except for small acreages that are best suited for use by different forms of wildlife. It does require a knowledge of the best use of certain pieces of land and a program that provides for their most suitable utilization.

In many cases the only steps required are protection from fire and from grazing or use by domestic animals, while in others additional steps should include some tree or shrub planting. A third con-

346 □ OUR NATURAL RESOURCES

servation measure pertains to the control and use of water, that is, ponds and streams on the farms.

Millions of acres of potential wildlife land are located on privately owned farms. Much of this land could be improved by relatively inexpensive treatment or handling. The major categories listed by the Soil Conservation Service in which land-use practices valuable to wildlife could be improved are the better management of marshes, ponds, stream banks, field borders, mined-over areas and various odd acreages that are often otherwise unused.

A survey of all states conducted in 1978 by Charles Deknatel indicated that 44 states offer some type of program for wildlife habitat on private lands. Such programs focus on the landowner or operator as decision-maker or implementor. Most of them are voluntary and small in scale, particularly as applied to farms. "They in-

FIGURE 21-1. Idle corners, when left to grow to brush and weeds, and especially when planted to corn, sorghum or other grains, make ideal food patches and cover spots for quail, pheasants, Hungarian partridges, rabbits, etc. It costs little to encourage wildlife on good farm land. *(Courtesy, Wisconsin Conservation Department)*

clude various types of plantings, borders, shelter-belts, seeding of marginal lands, food or cover plots, and possibly water related practices," Deknatel notes.[1]

Wildlife can be increased on well-developed farm and ranch land even though the land contains no waste corners, wet spots or stony knolls. Windbreaks of trees, which help reduce wind erosion, furnish shelter for small game and song birds; strip cropping in many instances provides nesting spots for birds; fence rows, if not cleaned of all grass and shrub growth, are havens for both game and song birds. Even on ranches and in range country, if range land is so managed that a good growth of native grasses is maintained, prairie chickens, sage hens and even the elusive antelope may again appear.

One of the problems of wildlife in ranch country is the competition for forage. Elk, deer and antelope compete with cattle for grass and other forage. A buffalo requires about as much roughage as a steer, so that for every animal of this type, the feed for one steer is gone. The ratio of other animals to steers are two elk, four deer, four bighorns and four antelope. Unless the rancher is getting a handsome income from dude ranching, he may find the wildlife a great expense.

☐ PONDS AND LAKES FOR WILDLIFE

During the past few years, tens of thousands of ponds and small lakes have been built in this country. They are built ordinarily to provide water for farm livestock, to slow down or stop water run-off from farms or to encourage scenic water sports—not to promote the production of fish and other wildlife. Yet, with but little change, these water bodies could be made to produce some fish and to become the homes of ducks, muskrats, shore birds, etc.

Managing ponds and lakes

It is well to seek the advice of the local agent, soil conservation agent or state game representative when planning a pond or lake,

---

[1]Charles Deknatel, "Wildlife Habitat Development on Private Lands: A Planning Approach to Rural Land Use," *Journal of Soil and Water Conservation*, November–December, 1979, pp. 260–261.

FIGURE 21-2. This farm pond in West Virginia is made with a dirt dam. It is the source of many meals of pan fish caught by young folks during times of recreation and play. *(Courtesy, U.S. Fish and Wildlife Service)*

since there are so many factors to be considered. Requirements for the plains area of the Midwest will be used as an example here.[2]

A pond in the Midwest should have a water area of at least ½ acre, a depth of 8 to 10 feet in 25 percent of the area; a spillway; a watershed of sufficient size to maintain the water level unless water can be pumped in from a nearby stream; a well-constructed dam; no erosion and pollution in the area; and no excessive brush, trees or aquatic vegetation.

Stocking with desirable fish

The pond should be stocked with fish that can maintain themselves in balance with the food supply and other fish. In the Mid-

---

[2] The following information is summarized, in part, from Alvin C. Lopinot, *Fish Conservation Teacher's Manual* (Illinois Resource Management Series), Division of Conservation Education, State of Illinois, Springfield, 1965.

west, one of the successful combinations has been largemouth bass and bluegill. The bluegill use the microscopic plankton and insects as food. The bass eat the bluegill. If bluegill are stocked before bass, they may become so overcrowded that the bass cannot control them. They will prevent the bass from multiplying by eating bass spawn and fingerlings.

Certain species of fish are undesirable for stocking ponds. There may be exceptions, but, in general, bullhead, carp, green sunfish and crappie will spoil the pond or lake for better game fish. The bullhead and carp keep the water muddy as they feed on the bottom. The other two varieties multiply so rapidly that they generally soon overpopulate the pond. And, they eat bass eggs, young bass and the food that bass need.

Avoiding weeds and siltation in lakes and ponds

Midwest lakes generally do not need to be fertilized. Unfertilized lakes should support from 200 to 400 pounds of fish per acre of water each year. This is more fish than will generally be caught in ordinary family sport fishing. The water does require weed control, however. Most pond vegetation is of little value for fish production because it uses nutrients in the water that would otherwise produce phytoplankton as food for the fish. Besides, the resultant "scum" (algae) not only looks bad but also often smells like rotten eggs and may actually be poisonous to fish, birds and even livestock. Pond weeds may also favor mosquito breeding. Chemical control of algae and weeds is possible. USDA Extension Service advisors should be consulted.

A pond must be protected from siltation, or it will not last long. Besides, muddy water is not favorable to game fish. Worst of all, the fish can't see the bait on our hooks! The erosion controlling practices we have discussed in connection with soils should be practiced, and livestock should be kept away from the shore and off the dam.

Harvesting the fish

Finally, harvesting the fish should be adequate to maintain balanced pond conditions. When we catch small panfish in a well-established lake, we should not throw them back in. Keeping them

350 □ OUR NATURAL RESOURCES

will leave more food for the remaining fish and will help prevent overstocking. Of course, if the lake is understocked, it would be wise to throw back the medium-size fish which will provide better breeding stock. Also, in ponds stocked with bluegill, sunfish and bass, we should fish out the first two species, as compared to the number of bass, in the ratio of four to one. In any case, we should not take more than the daily limit allowable by law.

Providing for fish, waterfowl and fur bearers

If, rather than fish, the main purpose of the pond is to provide for fur-bearing animals, water and shore birds, fencing the pond is desirable. This will allow undisturbed nesting, resting and feeding grounds for wildlife. In this case, there should be some shrub and tree growth at or near the shore. It is possible, of course, to have

FIGURE 21-3. A group of students on a conservation field trip look over a wildlife conservation area on a Corn Belt farm. The area was established on part of an acreage too wet to farm in most seasons. The pond helped drain the land so part of it could be put back into production. It now provides good fishing. Multiflora rose and a mixed planting of maples, pines and other trees, shrubs and grasses provide feed and cover for wildlife. State of Illinois conservation agent Don Hastings led the group. *(Photo by H. B. Kircher)*

both fish and other wildlife in the same area. The number of each will naturally be smaller than if the whole emphasis is on the production of one or the other, because conditions for the production of both kinds of wildlife are a compromise of the best requirements for either form of life if produced alone.

☐ WETLANDS FOR WILDLIFE

Our first thought about marsh lands and inland swamps is that here are sources of virgin soil which will produce well when once the water is removed. We become convinced that their value for producing cultivated crops far outweighs their value as producers of muskrats, waterfowl and shore birds. Drainage projects follow. Water is removed, and the water table is lowered. Farms are set up, and crops are produced. But, in many instances, the costs of operating these units are high, and productions are disappointingly low.

Next, we condemn all marsh and swamp drainage projects as failures. There are moves toward restoration of the marsh areas and many drainage projects are given up. Farmers move out of the areas and wildlife again takes over. Examples of this change in attitude are found in the restoration of Lake Mattamuskeet in North Carolina, Thief Lake in Minnesota, Horicon Marsh in Wisconsin, Seney Marsh in Michigan, Lower River Flood Plain in North Dakota and Lake Malheur in Oregon. It was found after several years of disappointing experience that these areas are more valuable as wildlife habitats than for intensive agricultural use.

It happens that neither of these two extreme viewpoints is entirely correct and neither is entirely wrong. Some marshes have been drained and are being used profitably as farm land. Some marsh lands in Louisiana, which for some unexplained reason were poor places for wildlife, were drained and found to be poor cropland. Later, when they were restored as wildlife habitats, they proved to be excellent homes for fish and other wildlife. Others have been dismal failures when drained and cannot now be restored to their original use as habitats for wildlife.

Probably the outstanding example of this experience was the Lower Klamath National Wildlife Refuge in southern Oregon and northern California. Fifty years ago, this area contained over 80,000 acres of marsh land, and it was one of the greatest nurseries for waterfowl and shore birds in the western states. Literally hundreds of varieties of birds were found nesting there, and millions of birds were estimated to have made it their summer home.

Then, this wetland was destroyed to accommodate the popular belief that once freed of surface water, this area would add to the agricultural wealth of the West. A by-pass was built above the lake so that no more water flowed in, and, within four years, the lake had dried up. Peat fires started and in many instances burned to a depth of 6 feet. The result was that instead of having myriads of wildlife or thousands of dollars' worth of agricultural production, "a vast alkaline ashy desert" was formed "from which clouds of choking dust arose, often obscuring the sun." A waste area was the result.

Each marsh or swamp, whether large or small, must be studied carefully, and its most profitable long-time use determined only after all possibilities have been explored. It is for this reason that no general statement can safely be made as to the best use of all marsh land. Undoubtedly, it will be best to leave some of it to wildlife and to use some of it for agricultural production.

Much of our wetland should never be drained

The Soil Conservation Service in a comprehensive natural resource study in 1977 indicated that there are about 270 million acres of wet soils (those soils that are classed as having problems of seasonal or permanent water tables serious enough to inhibit cultivation), of which 104 million acres are cropland, 56 million are pasture and 110 million are forest and other land. Of the total, about 41.5 million acres of land are classed as permanent wetland. These are lands that have not been and are not likely to be drained for other uses. But, many of the other wet soils' acreages serve as wetlands for game habitat.

The Soil Conservation Service pointed out that wetlands support 19 species of small game, 7 species of big game, 11 species of fur animals and hundreds of nongame mammals, birds, amphibians and reptiles. Wetlands provide many of the basic food elements for the nation's commercial fisheries as well.

Unfortunately for wildlife, wetlands have been lost through drainage, dredging and filling at an estimated rate of some 318,000 acres per year since 1956.[3]

---

[3]*A Summary: Non-federal Natural Resources of the United States*, National Association of Conservation Districts, Washington, D.C., circa 1979, p. 11.

☐ STREAM BANKS CAN AFFORD WILDLIFE SHELTER

Most of our stream banks continue to wash away, and the beds of streams move from one location to another. Stream bank erosion has other bad effects in addition to its destruction of excellent cropland. It frequently undercuts trees, bridges and houses. It destroys highways and railroads. It covers good land with sand and gravel, and it loosens silt, which is carried downstream to be deposited in back of dams, thus destroying the capacities of the dams for impounding water.

Probably the easiest and cheapest way to control most stream bank erosion is through the use of grass, shrubs or tree growth along the banks. Almost any bank which is to be kept from eroding must be sloped. There are few exceptions to this rule. The slope should not be greater than a 1-foot rise for each 1½ feet of horizontal distance. After the bank has been sloped, it should be planted without delay. Shrubs and grass offer better protection for the slope than does tree growth.

Occasionally, it is found necessary to direct the force of the stream away from a bank by laying stone rip-rap from the bank several feet out into the bed of the stream. Work of this kind is quite technical and should be attempted only under the direction of a conservation specialist. It is only when stream banks furnish shelter and security, along with some food supply, that wildlife is encouraged to stay there. All of these steps, helpful to the perpetuation of wildlife, are also desirable soil conservation measures.

☐ WILDLIFE HABITAT IN THE CITY

If we are to provide for wildlife in large metropolitan areas, we need to maintain food, shelter and water. It is not generally possible, as in some rural areas, to establish a natural area and then just help it regenerate itself. Potentials for wildlife habitat vary from every man's back yard to large city parks or the type of village commons set aside in many modern residential developments.

A homeowner's own back yard can be a wildlife habitat. His provision may be as simple as filling a bird feeder and providing a bird bath. Requiring more space and time is the planting of bushes and trees for nesting and food. He can also provide bird houses, by giving careful thought to the type needed and the best location. In this planning, he can become an expert by obtaining publications such as the National Wildlife Federation's book, *Gardening with Wildlife*.

Planning for larger urban open-space areas should be done with the help of landscape gardeners and wildlife experts. Management practices similar to those for farms and ranches and lakes and ponds are often applicable. All of us can take a part as citizens by seeing to it that our public lands are managed to the best of their wildlife potential. Private abuses of habitat can be stopped by good zoning and anti-pollution laws properly enforced.

☐ WILDLIFE'S FUTURE DEPENDS ON YOU AND ME

No other resource is so dependent on the action of each of us as individuals as is our wildlife resource. Whether landowners or not, we are directly influencing wildlife by our support or lack of support of environmental programs. How each of us acts toward wild plants and animals wherever and whenever we encounter them will determine how rich and varied wildlife will remain. Private owners, of course, have a special responsibility. They must provide habitat if we are not to suffer continued losses of wildlife. But, let us remember, as citizens we are all stewards of the one-third of our nation's land in the public domain.

An excellent example of citizens' support for wildlife is the program of the Iowa Conservation Commission. The Iowa Conservation Commission has initiated a private lands incentive program for the 1980's to encourage the establishment of an agricultural crop that is both economically desirable to private landowners and beneficial to nesting wildlife. The Commission is planning to spend a total of $1 million derived from the sale of wildlife habitat stamps to cost-share the establishment and proper management of switchgrass pastures.

In the *National Wildlife* issue devoted to endangered species referred to earlier, the federation expressed the conclusion we have arrived at very well in these words:

> ... in the final analysis, it all boils down to how strongly individual people feel about endangered wildlife—and how effectively they translate that concern into action.

QUESTIONS AND PROBLEMS

1. How much is it necessary to decrease agricultural production in order to develop good shelter and food patches for wildlife on well-improved farm land? How much of the 33 million acres of potential wildlife land now in farms can be fairly readily improved so as to support some form of wildlife? List the kinds of treatment and the acreage of each.
2. On what other parts of the farm besides marsh lands, stream banks, odd acreages, etc., can wildlife be given shelter and still not reduce the production of the farm?
3. Ponds and lakes have served what purposes in the past? What kinds of wildlife can be supported in and around these water bodies?
4. Why is it desirable not to have silt-filled, muddy water drain into ponds and lakes?
5. How would you go about managing a pond or lake for better fishing?
6. What kind of information do you think you need to obtain in order to manage a water body properly?
7. How can you have both fish and waterfowl or fur-bearing animals in the same pond area?
8. Under what conditions is it desirable to confine the access of cattle or other livestock to a limited part of the shore?
9. Whom should you consult for information on management of lakes and ponds for wildlife?
10. How many of the 127 million acres of wetland should not be drained for agricultural purposes, according to the Soil Conservation Service?
11. What is the simplest way to control erosion on most stream banks and to make them useful as wildlife coverage?
12. In order to secure the best land use, which specialists should share in the planning of wildlife habitat?
13. What game birds or game animals are found in your area? What non-game species?
14. How can their numbers be increased?
15. How can wildlife habitat be created in the city?
16. What, in the final analysis, is going to really determine how well wildlife management is provided for?

22

Natural resources from the seas

WE ARE VERY FORTUNATE in the United States in having excellent access to the seas of the world. Our shores are lapped by the waves of three out of five of the world's oceans—the Atlantic, the Pacific and the Arctic. We also border on the Gulf of Mexico. Our long coast line is another advantage, for the most abundant sea life is to be found in coastal waters. Also, here we can most easily reach the bottoms, if desired, to tap mineral wealth, as in the case of offshore oil deposits. In previous chapters we have discussed how the oceans may be thought of as the basic reservoir for the hydrologic cycle, and thus the basic source of all our water supply; how ocean currents, waves and tides might be harnessed for energy; and how there is a possibility of desalting ocean water and drinking it. This leaves unanswered the question of extracting minerals from the sea water and of the ocean's contributing to our food supply.

□ SEA WATER—A REMARKABLE RESOURCE

All of us who live near or visit the sea marvel at the differing colors of the water, enjoy the fun of splashing or swimming in it and appreciate its ability to support us in boats on its surface and to maintain sea life within its depths. But, how many of us have thought of it as a remarkable combination of minerals, organic substances and gases, whence all life on earth is said to have issued?

Many minerals in sea water

Among the elements present in sea water are sodium, chlorine, magnesium, sulfur, calcium, bromine and some 40 more. The list is so long that the sea would appear to be an almost unlimited source for minerals. The trouble is the cost. Most of these minerals can be obtained more easily on land by mining beds laid down millions of years ago in ancient seas now vanished. A notable exception is salt, a combination of sodium and chloride, which is easily gotten by evaporation. Magnesium also is obtained from sea water. Both of these are also gotten from mines. Commercial mining of magnesium taken from the sea floor has been under consideration for a number of years. Sands containing potassium-bearing minerals have been used as fertilizers.

☐ PLANT AND ANIMAL LIFE OF THE SEA

The plant and animal life in the waters of the sea occur in an amazing variety and number of shapes which are distributed very unequally from place to place. This is not surprising when we recall that the seas cover 71 percent of the earth's surface. They are said to provide some three hundred times more living space than all land surfaces and fresh waters combined. Of all this life, we see only a small part. And, we use directly for food only the larger forms of this rich sea life.

Life in the seas falls into three fairly distinct major categories, according to ability to move about, as stated by Professor Robert E. Coker.[1] These are (1) the sitters and creepers, which, in general, move slowly along the bottom of the oceans (the *benthos*, Greek for "depth of the sea"); (2) the swimmers, which are able to move through the waters at will (the *nekton*, Greek for "swimmers"); and (3) the drifters, which drift with the currents or other outside forces (the *plankton*, Greek for "wanderers"). The skimmers (the *neuston*, Greek for "boat" or "ship"), a fourth major group of water life, are mostly found in lakes and ponds, for they rest on the film at the water's surface, which must be relatively smooth.

Higher plant life is completely absent in the open ocean. A

---

[1] The discussion of the basic resources of the seas is drawn from Robert E. Coker, *This Great and Wide Sea*, 3rd ed. (New York: Harper & Row, Publishers, 1957).

flowering plant of the pond-weed family, called "eelgrass," however, lives on most ocean shores where wave action is not severe. But, this is an exception. The oceanic equivalent of our abundant land vegetation is found in microscopic plants located in the upper zone, reached by the sunlight.

Two of the aquatic life groups are very familiar to all of us, for we eat them for dinner. In the first group are oysters, clams and crabs. The swimmers, the second group, include fish, shrimp and whales. The third group, the drifters, is by far the largest in volume and numbers. But, many of them are almost too small to be seen by the naked eye. One of the most important varieties of animal plankton is only 5 millimeters in length. One, however, with the impressive name of *Giant ostracod*, is a centimeter or more long. Plant plankton may resemble those that form green "blankets" on our fresh water lakes and ponds. Plant plankton are at the base of the food pyramid for other forms of aquatic life. They are the principal photosynthetic agents of the seas.

☐ FISH—A DESIRABLE FOOD

Earliest history records the importance of fish as a human food, and in times of shortage of other foods, fish have saved people from starvation. It was found following World War I that the children of some European countries showed the effects of a poorly balanced diet in their retarded physical growth. The children of Finland, on the other hand, appeared robust and healthy, even with a restricted diet.

A study of the reason for this satisfactory child development in contrast to that of other countries with no worse diet showed that the whole fish diet was the answer. The people who ate the whole of the smaller fish apparently were supplied with those trace elements in food which are necessary for early normal human development.

☐ THE EARLY COLONISTS USED FISH

The New England colonies turned early to fishing as a source of food and also in order to have something to sell to foreign countries for the items they in turn wished to buy from those countries.

There are many reasons why the New England colonists became

fishermen. The rapidly increasing demand for fish both in America and in Europe opened a market for anyone with fish to sell, and the New England colonists were much nearer the banks along the American side of the Atlantic where the fish were found than were European fishermen. Thus, it was much easier for them to reach these new fishing grounds and start fishing operations between the severe storms which swept the banks than for European fishermen who required weeks of oceangoing travel to reach the banks.

It was mentioned earlier that the colonists took up shipbuilding because of an abundance of timber and a familiarity with maritime activities. It was natural, therefore, for them to combine fishing with their other work, such as farming, lumbering and shipbuilding.

Best fish required by foreign markets

As a result of the large catches of fish, the colonists developed some unusual trade relationships. By the middle of the 18th century, the New England fishermen found it profitable to salt a catch so it would keep. It was then sorted into three classes for the three markets then available. The largest fattest fish comprised the first class, and they were sold locally, probably because they were most difficult to cure thoroughly. Fish at this time were salted and sold whole. No attempt was made to dress the fish before salting.

The second class of fish were the best quality, were a nice size for curing and carried the proper amount of salting. These fish were sold on the European markets where demand for this quality of fish was good.

The third class, which amounted to nearly one-half the whole catch, included those fish which were small, bony, too salty, broken or otherwise damaged in handling or those which were off flavor because they did not have enough salt. The market for these fish was the West Indies where the manual labor for raising sugar cane was done by Negro slaves brought in from Africa. These slaves seldom lived longer than six or seven years in this work, so the owners did not pay much for the food supplied to them.

Fish traded for molasses, and rum for slaves

A second unusual feature of this fish trade was that the West

Indies planters needed a market for their by-product known as molasses. A deal was made whereby the third class of fish was traded for the useless molasses. The New Englanders found that a gallon of rum could be made from a gallon of molasses, and the rum would sell for 10 times the cost of the molasses. So, the New England trader sold his stale or oversalted fish for molasses. He then traded the rum, which was made from the molasses, for gold or slaves in Africa.

FIGURE 22-1. These fish, caught in the Atlantic, are being taken from the pocket of a pound net and loaded onto a boat. Fishing in our coastal waters can be further increased. *(Courtesy, U.S. Fish and Wildlife Service)*

He then sold the slaves to plantation owners in this country who needed cheap labor for cotton production. Thus, the early fishing industry, through these ramifications, was a source of both wealth and human degradation, the latter being a major cause of the Civil War.

☐ FISH SUPPLY SMALL SHARE OF OUR NATION'S FOOD

Fish are a valuable addition to our diet, as has been noted. But, they provide a relatively small share of our total food supply. The total yearly catch of fish in this country amounts to some 5 billion pounds, or 26 pounds for every person in the United States. With imports, it is possible that Americans get somewhere near the world average of 10 percent of their food consumption from fish. Most of this supply comes from the oceans on both sides of this country, rather than from inland lakes and rivers.

☐ MANY FISH SPECIES DISAPPEARING

We think of the vast oceans which cover almost three-fourths of the surface of the earth as containing inexhaustible quantities of sea food. This is far from true. Many of our fish species are getting scarce.

The halibut of the Atlantic were once a most important sea food. Now only negligible quantities are caught, and they have all but disappeared from our tables.

The shad, also of the Atlantic Coast, are spoken of as among the most prized fish for the table. Until recently, they were among the three most abundant species of fish on the whole Atlantic Coast. A year's catch of nearly 50 million pounds was not uncommon. Not only were shad used fresh for immediate consumption, but also many were salted and kept for winter use. Now, the great shad runs are history. Many factors have contributed to their disappearance. Overfishing, the construction of dams which blocked them from their spawning grounds in the tributaries of all the Atlantic Coast streams and water pollution by industrial waste, together, have practically exterminated this valuable species.

The haddock, which were quite numerous in 1930, have declined so much that the current catch is less than half of what it was then.

Other fish, such as salmon, spend the greater part of their lives in the ocean depths. It is only at spawning time that they leave these unknown ocean haunts and return to fresh water streams where eggs are deposited and hatched, the young returning to the ocean.

During these migrations, the fish have been taken in such great numbers that there is fear of destroying the industry.

The salmon industry was known as the lifeblood of Alaska until the establishment of our vast military network there and the development of petroleum resources. The salmon industry alone is worth many times as much to this state as the production of gold. The permanence of the industry depends upon a continuation of the migration of these yearly supplies of salmon to the streams of the country.

Only a few thousand whales remain of the millions in the seas when Herman Melville wrote his famous novel *Moby Dick*. The decline is the result of waste and greed. It is indicative of man's ability to destroy animals, no matter how large and powerful they are. Whales are the largest mammals in the world and range the oceans from the north to the south polar regions. Studies have indicated that if international limits on whaling could be agreed upon, whalers would be able to harvest tens of millions of dollars worth of whales annually without endangering the species. But, the agreements arrived at so far have not been effective.

☐ RELATIVE IMPORTANCE OF FRESH WATER FISH

Most of the fish from our fresh water lakes and streams are caught in recreational or sports activities. The value of these activities to those of us who spend time at them cannot be measured in terms of the food value of the catch. Sport fishing is a relaxation for tired, tense Americans from coast to coast.

The amount of fresh water fish is less than 0.02 that of ocean species. About two-thirds of the salt water catch are *nekton*, the swimmers. The catch of shrimp, crabs, oysters, clams and related species amounts to about one-third the weight of the *nekton*.

☐ FISH FARMING IN THE SEA OR ON "LAND"

The raising of fish in fresh water ponds or the providing of beds for shellfish in bays or estuaries (those parts of the river where the tides mix with the river current) for the market has been increasing in importance in the United States.

FIGURE 22-2. These carp are being taken from Lake Mattamuskeet, North Carolina, by a commercial fisherman. Carp represent one of the many types of fish caught in inland waters. *(Courtesy, U.S. Fish and Wildlife Service)*

Fish farming in ponds 3 to 5 feet deep, and ranging in size from less than an acre to more than 15 acres, has been practiced in China and India for centuries. The waters of these ponds must be fertilized in order to produce the plankton so necessary for the life cycle of water life. In Japan, where quite favorable conditions prevail, the yearly production of fish farms is estimated to be about 6 percent of all seafood sold there.[2]

It should not be assumed, however, that fish farming is as sim-

---

[2] T. V. R. Pillay, "The Role of Aquaculture in Fishery Development and Management," *Journal of Fisheries Research Board of Canada*, Vol. 30, 1973, pp. 2202–2217.

ple as putting fish in the water and just letting them multiply. Owners of large fish farms must use many chemicals and equipment to keep the fish productive and healthy and, of course, fed.

Aquaculture, or salt water farming, in the United States is important even though relatively small. In 1975, aquaculture production was only about 3 percent of U.S. fish and shellfish landings, according to *The Global 2000 Report*, but this still constituted about one-quarter of our salmon production, two-fifths of our oyster production and one-half of our catfish and crawfish output.

The report forecasts an upswing in aquaculture.

> There is cause for reasoned optimism, when considering food production from aquaculture . . . global yields are increasing. . . . The 1976 FAO World Conference on Aquaculture concluded that even with existing technology a doubling of world food production from aquaculture will occur within the next decade and that a 5–10 fold increase by the year 2000 is feasible. . . .[3]

☐ FISH CATCH MAY INCREASE

New devices have been developed for locating schools of fish by the noise they make. The adoption of other special equipment and up-to-date techniques in fishing has resulted in some of the largest catches of fish known. In fact, the world's fish catch has tripled since World War II, according to the UN Food and Agricultural Organization. But, after World War II, the commercial catch in the United States declined for several years and then remained relatively static at about 6 billion pounds a year for the last decade through 1980.

On December 22, 1980, President Carter signed the American Fisheries Promotion Act. Martha Blaxall, director of the Office of Utilization and Development at the National Marine Fisheries Service, maintains that "the act will expand foreign trade and seafood products and provide more effective monitoring of foreign fish operations.[4] U.S. fleets could support a catch about three times as large as at present, the National Marine Fisheries Service reports.[5]

---

[3]Council on Environmental Quality and the U.S. Department of State, *The Global 2000 Report to the President: Entering the Twenty-First Century*, Vol. II, U.S. Government Printing Office, Washington, D.C., 1980, p. 112.

[4]D. K. Plot, "U.S. Fish Industry Lands a Whopper—New Law to Haul Up Productivity," *The Christian Science Monitor*, December 29, 1980, p. 7.

[5]*Ibid*.

□ CYCLE OF LIFE IN THE SEA

The problem of trying to increase the food resource from the sea is not so much that of catching fish as it is of knowing just where to break into the cycle of life in the sea. There is a balance which over the years is maintained between the production of plankton, which grow in greater or lesser abundance in the sea, and of other higher forms of life, which in turn are fed upon by still higher forms. Through a series of what scientists call "predations," these lower forms of plant life are transformed into successively higher forms of life.

The plankton composed of plants is the only part of the sea cycle which creates organic matter from inorganic materials in the open seas. This is done in the presence of sunlight in a manner similar to that used by any green-leaved land plant. Animals do not possess this ability, so they must feed either upon plants or upon other animals that have first fed on plants.

There is a good deal of confusion over the term *plankton*. The drifting life of the sea is called *plankton*. There are both plant and animal drifters. *Algae*, a group of plants that make their own food from inorganic substances and generally contain chlorophyll, comprise the bulk of the *plant plankton*. Not all algae are plankton. Some are fastened to the sea bottom or coasts, and, of course, some are common to us on inland water bodies or moist surfaces. Some algae are quite large. Kelp, a brown seaweed, for example, may be 100 feet in length. *Animal plankton*, which are generally minute in size, feed upon the plant plankton. Some plankton have such a mixture of plant and animal qualities that both botanists and zoologists claim them.

Each time some form of life consumes some other form, there is a wastage or loss of organic matter. The amount of the lowest form of animal life which is produced from eating plankton is much less than the amount of plankton consumed, and every time another form of animal life consumes some simpler form, there is another loss. The volume, or poundage, of fish produced may not be more than 0.001 of the volume of the original plankton which was consumed by the lowest form of animal life in this final step in the cycle of life in the sea.

If we wish to increase the total quantity of food produced from an acre of water, we should catch the smaller fish. Where this is systematically done, the total poundage of fish produced per acre of

NATURAL RESOURCES FROM THE SEAS □ 367

FIGURE 22-3. This 78-acre lake on the campus of Southern Illinois University at Edwardsville serves a multiple purpose. It provides for fishing. The school's heating and cooling plant uses its water. Up the lake from this spot there is a swimming beach. In addition, the lake serves as a scenic outlook for student housing. *(Photo by H. B. Kircher)*

water can be greatly increased. And, if some economical way could be found to make human food out of plankton, the total production of food from water could be increased. But, if we want the larger fish, we must be content with less food than if we use the smaller-sized fish. In any case, conditions must be such that a large amount of plankton is produced if a large fish crop is to be obtained.

Importance of plankton production

The preceding paragraphs stressed the importance of having plankton in water where fish are to be produced. If an acre of the ocean bottom produces a yearly supply of 50 pounds of fish, the overlying waters must produce from 25,000 to 50,000 pounds of plankton during the same time. And, since plankton are not produced farther below the surface than sunlight actually penetrates, it is readily understood why ponds need not be deeper than 5 or 6 feet to produce immense amounts of this microscopic plant life.

A second requirement for plant growth is heat. Also, where it is hoped to produce the largest quantities of fish, the proper kinds and amounts of plant food or nutrients must be present in the water.

If it were possible to convert plankton directly to our food supply, and we could harvest tremendous quantities, where would the higher forms of fish life get their supply of food? Thus, if we break the food chain relationship, we might lose our present fish resource. We might imperil the entire ecology of the seas, resulting in eventual loss of the plankton. We might end up with less food than ever before.

□ OUR NATION'S UNDERWATER BOUNDARY EXTENDED

Our nation's underwater boundary was greatly extended in 1964. At a United Nation's law-of-the-sea conference in 1958, it had been agreed that the limit of sovereignty over the sea bottom should be at whatever line is drawn by a 600-foot depth of water. Sovereignty implies not only the ownership but also the right to develop. By this agreement, which became legal in 1964, the offshore land area of our nation was extended by one-third its former total extent!

Our territorial control has been extended even further. In 1976, the Fisheries and Conservation Act set up a mechanism to control all fishery resources within 200 miles of our shores. Research to supply needed information for this purpose and advice on policies are supplied by the National Oceanic and Atmospheric Agency (NOAA), established in 1969.

□ MORE REGARD FOR NATURE'S CYCLES NEEDED

Modern technology has made possible more thorough study of the oceans' depths and marine life. New research efforts are being made in this direction. We have extended our territorial waters and are attempting to stop the pirating of them by other nations. Also, more fish farming may add to our food supplies only if it is not polluted. Thus, the sea may eventually yield more than its present share of our food supply. However, this is going to require more regard for nature's cycles of life than we have shown in the past. Otherwise, it will yield less.

QUESTIONS AND PROBLEMS

1. Why don't the minerals in sea water presently constitute our major minerals resources for industry?
2. What are the three major groups of plant and animal life in the seas? Give an example of a form of life in each.
3. Which form of plant or animal life in the sea is most basic? Why?
4. Why is fish a desirable food?
5. What are the reasons why the colonists took up fishing along the Atlantic shore?
6. How did the colonists dispose of the three classes of salted fish?
7. Is our present resource of fish increasing or decreasing? Give some cases which support your answer.
8. Do we get a major portion of our food supply from fish of the sea? How important are fresh water fish compared with those from the oceans?
9. What is meant by fish farming?
10. Explain how the cycle of life in the sea operates. If we could convert plankton to a commercial food crop, how might it threaten life in the sea?
11. How have our national underwater boundaries been extended?
12. If we should let our soil erode away so that our agricultural production fell off by a third or more, would we be able to make up the deficiency by increasing our output of food from the sea? Explain your answer.

23

The ecological approach to natural resources

ALTHOUGH VARIOUS RESOURCES have already been discussed, we have learned that they are all closely interrelated. The study of how living organisms and the nonliving environment function together as a whole is called *ecology*. Scientists call this interrelated whole an *ecosystem*. We can only develop natural resources for the maximum benefit of man by considering man's place as part of this ecosystem.

☐ THE SCIENCE OF ECOLOGY

Ecological relationships are a fundamental basis of life on earth. One principle of ecology is that living organisms—animals, including human beings, and plants—and their nonliving environment—rocks; minerals; gases, including air; and water—are inseparably related. The dependence of animals upon one another was expressed well many years ago by the famous writer, Jonathan Swift:

> Big fleas have little fleas
> Upon their backs to bite 'em
> And little fleas have lesser fleas
> And so, *ad infinitum*.

With apologies to Mr. Swift, we might add in order to include the full concept of ecology:

> Big fleas like to live on dogs
> And dogs are man's best friend;
> But, without grass and sun and sky
> They'd all die in the end.

The face of the moon gives us a good idea of what an unbalanced natural system looks like. With no atmosphere, there is searing heat or numbing cold—nothing in between. No processes have allowed photosynthesis. Without this process and other factors, there are no life forms. The moon's environment is entirely unsuitable to life.

We don't need to go to the moon, however, to find examples of unbalanced systems. While human ecology is still very imperfectly understood, it seems a reasonable assumption that the crowding of man into vast, box-like apartments, without recreation space or elbow room from his neighbors, does not provide a balanced environment. These prison-like structures destroy a sense of belonging and home, which exists even in some of the poorest, run-down small dwelling units which they have replaced. This imbalance apparently has helped account for a rise in crime and delinquency.

Another principle of ecology is that all life is dependent upon the continuous inflow of the concentrated energy of light and radiation. We need the sun! To the extent that we limit energy receipts from the sun by air pollution, or in other ways, we are limiting life.

A third principle of ecology is that most of the world obtains its food through the process of plant photosynthesis and microorganism photosynthesis. Thus, the survival of this minute plant and animal life in our environment is one of the keys to our survival.

Ecologists note a fourth principle, that all living matter has limiting conditions. Some of these limiting substances may be very small and essential, such as the traces of minerals needed in our diet. Others may be large and essential, such as rainfall and light. Development of an ecosystem is limited by the essential factor available in the *least* quantity. To put it simply, a chain is as strong as its weakest link.

The final principle of ecology that we will note here might be called the *community* principle. A community—that is, a group of animals or plants living together in the same environment—has a community character that is dependent upon the total relationships of the community. That is to say, the community is more than the sum of its individual members. When one member of a community is destroyed, it affects the character of the entire community. Under balanced conditions, each species finds its place and a diversity of species assures that no single one will dominate The various life cycles support one another. A pond which has a balanced plant and animal life, as explained in an earlier chapter, is a good example of a small community. Our whole world may be thought of as a large community.

That nature tends to strike a balance between living things does not mean that change is not going on constantly. As we have studied various resources, we have learned that all nature is changing. The land is eroding; ponds are drying up; and plants and animals reach a peak of growth and then die. A succession of various types of communities follow one another. We cannot wholly prevent change. But, we may be able to slow it down or to speed it up to work in man's favor.

Unfortunately, man has been ignorant of or has chosen to ignore the principles of ecology. As a result, we are facing the many resource problems which we have already noted and others which we shall now consider.

☐ AIR POLLUTION AND RESOURCES

A key problem of our environment is that we are destroying the availability of our remaining resources by polluting them. We had once thought of atmosphere as an unlimited, self-renewing resource. Now, we realize that it is a relatively thin film of gases subject to contamination.

Excessive air pollution could result in the flooding of most of our seacoast cities. If carbon dioxide reaches the atmosphere faster than it can be absorbed by the ocean or converted back into carbon and oxygen by plants, some scientists believe that we may experience a gradual warming trend which would melt the polar ice caps and raise the ocean levels 100 feet or more. New York and Los Angeles would be among the cities drowned out.

Another view of air pollution is that solid particle emissions may be helping form a cloud cover on earth which would cool the atmosphere and bring continental glaciation and cold climate far south of its present limits. Much of the northern United States would become uninhabitable. We would again be in the Ice Age.

Whatever the long-term effect of air pollution, the immediate effect has apparently, among other things, shortened human life. At Donora, Pennsylvania, on October 26, 1948, a heavy, soot-laden fog began to blanket the town. Before it was washed away by rain four days later, over 5,000 of the town's 14,000 residents had become ill. Twenty died! On December 5, 1952, a great black smog began to envelop London, England. During the five days before it lifted, 4,000 more deaths took place than would have normally occurred in that period of time. In the 1950's and the 1960's, both New York and Chicago experienced apparent "killer" smogs. While these are now historic accounts, their lessons should not be forgotten, for severe air quality problems still plague us. Of 41 SMSA's studied from 1975 to 1977 by the Council on Environmental Quality 5—New York, Chicago, Denver, Los Angeles and San Bernardino–Riverside–Ontario—had "unhealthful" ratings (over 100 on the Pollution Standard Index of the EPA) for 30 percent of the year. There were 11 cities that had over 100 days in this category, and 20 cities with over 50 days.

In 1970, Congress passed the Clean Air Act which was designed to control industrial sources through federal administration. But, there was not sufficient participation by state and local authorities to control area sources of pollution effectively. Nor were actual control and maintenance aspects satisfactorily addressed.

In 1977, the act was amended to become known as the 1977 Clean Air Amendments which put emphasis on state responsibilities for clean air. Specific goals were set for attainment, and more attention was given to controlling non-industrial sources. As the Environmental Quality Index of the National Wildlife Federation shows (see next chapter), the new standards have been enforced with some success. Studies of national trends, conducted by the Environmental Protection Agency and the Council on Environmental Quality, support this conclusion.[1]

More recently, acid rainfall has been causing great concern. The Council on Environmental Quality in its *Tenth Annual Report* declared:

> Acid rain is a major environmental problem on both sides of the Atlantic Ocean. Although the problem is currently confined to the Northern Hemisphere, it appears likely that the problem will also occur in the Southern Hemisphere. It is considered a primary environmental threat in the Scandinavian countries and a source of marked concern in Japan and

---

[1] Council on Environmental Quality, *Environmental Quality, The Tenth Annual Report of the Council on Environmental Quality*, U.S. Government Printing Office, Washington, D.C., 1979, pp. 17–50.

THE ECOLOGICAL APPROACH TO NATURAL RESOURCES □ 375

Canada. As required by the National Energy Plan in 1977, the President commissioned a study on potential environmental impacts of increased coal use. The study report, known as the Rall Report, identified acid rain in the United States as one of the six environmental problems requiring closer scrutiny. . . . Although acid rain may have some unquantified benefits, the adverse effects of acid rain are of great concern. In the eastern half of the United States, the average pH of rainfall is now between 4.0 and 4.5. Some rainfall has a pH as low as 3.0. This

FIGURE 23-1. The orbiting Skylab photographed from the detached Command and Service Module during the Skylab 2 mission. Four solar array wings provided power to the Apollo Telescope Mount. Thus, Skylab observed, monitored and recorded the structure and behavior of the sun. The orbital workshop provided living quarters for the crew. Monitoring from such a spacecraft or unmanned satellite can help us make better use of our natural resource base by identifying problems and opportunities as they develop. *(Photo by Skylab 2, in flight May 25 to June 22, 1973, National Aeronautics and Space Administration)*

is about equivalent to the acidity of lemon juice. In the eastern half of the United States, the acidity of rainfall appears to have increased about 50-fold during the past 25 years.[2]

Industry is often accused of being the main culprit in causing air pollution, but we are all guilty. While the fumes of manufacturing plants are certainly a major air polluter, so are those of burning dumps, back yard and municipal incinerators, jet planes and especially automobiles. Whatever the causes, it is clear that clean air can no longer be taken for granted. Nor, should dirty air be tolerated. Cigarette smoking has become a national health hazard. Now, concerns are being expressed about the effect of heavy concentrations of tobacco smoke upon nonsmokers. Federal, state and local laws are being adopted which require that public places provide non-smoking areas.

What can be done about air pollution? Three principal steps that might be taken, suggested by a committee of the American Association for the Advancement of Science, are (1) filter the effluents, (2) modify processing to eliminate pollution and (3) cooperate better with nature. The third method is one in which we can all cooperate. For example, awnings installed over south and west facing windows can be raised to permit more direct sun heat to warm the house in winter and lowered to keep rooms cooler so less air-conditioning is used in summer. Since less power will then be used from the power company, such practices will cut down on their emission fumes. We can mulch leaves instead of burning them, returning to nature the organic matter and eliminating smoke. We can keep the engines of our motor cars tuned up, thus burning fuel more efficiently and emitting less exhaust gas.

☐ INSECTICIDES—FOR BETTER OR WORSE?

Chemical insecticides have greatly reduced disease and increased crop and livestock production through their reduction of insect pests. These pests damage one-third of our crops and can carry death-dealing diseases. Allen Boraiko in a *National Geographic Magazine* article, "The Pesticide Dilemma,"[3] reports that U.S. farm-

---

[2]*Ibid.*, pp. 70–71.

[3]Allen Boraiko, "The Pesticide Dilemma," *National Geographic Magazine*, February, 1980, pp. 145–183.

ers and foresters, plagued by 2,000 detrimental insects, weeds and plant diseases, devote more than $2.25 billion a year to pest control.

However, insecticides have become a major worry because of their ability to persist in the environment. Traces of these poisons have been found all over the world. Although the insecticide DDT was banned in 1972, it has been found in 584 of 590 samples of fish taken from 45 U.S. rivers and lakes, according to a study conducted by the Bureau of Sport Fisheries and Wildlife. Insecticides such as the banned DDT and its chemical relatives are nerve poisons. Their particular threat to human life is that they do not break down readily into less lethal forms. Thus, they *persist*. They last from insect to fish, to animal, to man, and so on, however the "food chain" might operate. Opponents argue that they will become concentrated in harmful amounts in humans, causing pesticide-related illnesses and death. "Our ability to detect infinitesimal traces of pesticides in food, air, and water surpasses our understanding of how they may affect our bodies,"[4] Boraiko writes after interviews with chemists at the Harvard Biological Laboratories and with other experts in government and business.

Therefore, pesticides must be used with caution. Information on pesticide use is available from the various USDA extension offices which are located in every state in the United States.

Progress must be made in improving alternative methods of pest control. One such method is biological warfare in which harmless insects may be used, for example, to eliminate ones harmful to man.

☐ SOLID WASTE DISPOSAL

For thousands of years, man has regarded the land surface of the earth and its underground extent as practically limitless. It was inconceivable to think that there would not always be a site suitable for a city location, that highways would ever cover enough ground to affect agriculture or that we would not have plenty of waste land or underground areas in which to dispose of our garbage and trash. But, that day has already arrived! A million or more acres a year of farm land are being buried beneath our superhighways each year. While we are talking of building new cities in the middle of open plains—a most costly process, our solid wastes are cluttering the countryside.

---
[4]*Ibid.*, p. 183.

The amount of solid waste generated is increasing each year. Total U.S. municipal waste, which includes industrial and commercial wastes, was estimated at 1,400 pounds per person in 1978. It rose at a rate of 2 percent per year from 1970 to 1978. The amount of municipal waste generated per person also rose (about one percent annually) during this period to an average level of 3.85 pounds per day. Assuming no new federal policies to reduce waste generation, the Office of Solid Waste of the Environmental Protection Agency projects that solid waste will continue to increase at over one percent a year through 1985, based on trends from 1960 to 1978.[5]

Matter cannot be destroyed. It merely becomes transformed, as we noted in the first chapter of this text. Simple burning of trash is not the answer to the waste disposal problem, for this would simply be turning solid waste pollution into air pollution. But, controlled burning with proper equipment has been providing useful energy, as was noted in an earlier chapter. We need to find additional methods of turning waste products into useful resources, thus taking care of their disposal in a beneficial manner. Recycling is such a method; however, from 1960 to 1978, on the average, less than 10 percent of our consumer solid waste was recycled. No substantial improvement in the relative share is forecast by the Environmental Protection Agency through 1985.[6]

☐ RECREATION AND OPEN SPACE

The availability of our space resources cannot be properly evaluated simply by calculating density figures, that is, the number of persons or machines, or some other units, per given area. The question always to be asked is, What kind of area? Is the area productive land, snow-covered waste or rocky, sandy desert?

Certain kinds of land use are much more demanding of space than others. And, our space requirements per person are constantly changing. We can stack people layer on layer in apartments. Little land is used per person. But, if too many apartments are placed side by side, where will the children play? With more urban growth, we have more concentrations of people in small areas and thus more need to provide recreation space near these concentrations. The

---

[5]Council on Environmental Quality, *op. cit.*, pp. 256–257.
[6]*Ibid.*, p. 257.

move into the city may provide some available nearby space if farm land is abandoned on the outskirts of town, or at least becomes available for recreation because it is hilly and not very productive agriculturally. This has happened on our East Coast. But, such areas may have limited use for recreation. They may be flat, uninteresting lowlands, suitable for ball parks, but not attractive to the eye. We need to consider quality as well as quantity. Scenic spots and clear streams are apt to be destroyed by suburban sprawl as well as by industrial blight unless definite conservation plans are made and carried out to preserve them.

The crowding of humanity has spread far beyond our cities. Wallace Stegner, a novelist and historian who has written widely

FIGURE 23-2. Vacationers Tom Watson and Nancy Branch look over the crystal clear waters of a lake in the high Sierras of California. Even such apparently remote areas must be protected from excessive use and from abuse, or their natural systems will be destroyed. Predictions call for 250 million annual visits to national parks by the year 2000. *(Photo by Susan Kircher)*

about conservation, describes the plight of our national parks in this way:

> ... Old Faithful Yellowstone National Park at eruption time is a mob scene in a parking lot; Yosemite Valley on a summer weekend is like Times Square without electric signs. In California's splendid system of state parks and beaches a summer traveler can rarely get a camp site, and if he does he may well find himself squatting in a rural slum.... You cannot bring 40,000 people and 12,000 cars into the eight square miles of Yosemite Valley in a single day without consequences.[7]

Fortunately, the concern about overcrowding has been reflected in public policy. More camp grounds have been provided in national forests. Visitors have been bused into some parks to reduce the load of cars. Despite accelerating travel costs, people continue to visit the parks in near-record numbers.[8] In 1980, the National Park Service Director, Russ Dickeson, urged that growth of the national park system be slowed and that only areas with true national significance be added. He pointed out that expansion of the system has been so rapid (52 natural and recreational areas in 20 years) that maintenance is lagging behind.

□ PHILOSOPHY OF ECOLOGY

As we review the contradictory record of havoc and heaven that our resource strategy has wrought, we are forced to the realization that man, while a molder of nature, is also part of nature. He has sought out many inventions which have freed him from drudgery and helped compress time and space. We would not turn the clock back to the day when we had to haul our water by bucket from the nearest stream, to fire up the cook stove with wood and to saddle up the family horse and race for the doctor every time there was an emergency. On the other hand, we wish we could bring back the tall trees, clear streams and clear sky. We are part of nature. We are inseparably related to other living organisms, dependent upon a continuous flow of energy from the sun and food energy through photosynthesis.

---

[7]Wallace Stegner, "Whatever Happened to the Great Outdoors?," *Saturday Review of Literature*, May, 1965, p. 37.

[8]"Wildlife Digest," *National Wildlife*, October–November, 1980, p. 28-A.

We must operate within limits our systems can tolerate. Of course, a few of us can exist for a limited time in space capsules like the astronauts. And, at least one scientist has visualized that we may build great cities in like manner. Cities would be enclosed in great balls which might float on the ocean or even in space. To most of us, not only is such an idea fantastic, but it is also repugnant. Even if it were possible, which it is not, would life be worth living without the wonders of life in its natural state—without flowers and trees and birds and wild animals, the sky above and the earth beneath our feet? Ecology teaches us that even an artificial system cannot go on forever outside nature's cycles. It must at some point depend upon nature's reserves. And, in wholly unnatural surroundings, how long could man remain sane?

□ CONSERVATION VERSUS ECOLOGY

What is the difference between a conservational and an ecological viewpoint? The basic difference is in the action implied by the words. *Conservation* implies doing something—"conserving." The "doing" may be a "setting aside." For wilderness areas, for example, we noted that the action of withholding these from commercial and industrial use is a conservation action. The history of conservation and the development of modern concepts are covered in the next chapter. The point we want to make here is that while conservation involves education and research, finally something must be done—or there is no conservation.

By contrast, *ecology* simply means the study of the environment. The word itself does not imply that there will be a follow-through to action, even though ecologists are sometimes leaders in conservation.

Professional conservationists should understand, at least within the limits of their specialty, the principles of ecology. They should always be conscious that one change sets another in action—that the web of life is, indeed, fragile for some communities. It is only as strong as the weakest strand that supports it.

Ecologists, on the other hand, do not need to understand how to bring about effective conservation. They do not need to know how to correct erosion problems, build ponds, economize in the use of minerals, and so on. Nor do they need to practice conservation to be ecologists. They may choose to analyze the "balance of nature" of an endangered natural community, such as a particular pond or woods, without worrying about how it might be saved, or whether it should. The world of human action can remain entirely outside their field of study. Hopefully, however, their studies will be used by conservationists both in establishing goals and policies and in setting a plan for action.

To some people, whom we shall call "Simon-pure ecologists," man is an intruder. He is apart from nature. He upsets nature's balance. To these zealots, conservation is a bad word. Certain conservation practices may encourage tilling the soil, diverting water, harvesting wild animals and cutting forests. But, to the zealots, God made trees and little green apples, but not tractors, houses and barns. They see man as just a bad actor.

Conservationists know that conservation means the best use of our natural resources for man's well-being, taking into account the natural system. This may involve setting some resources aside. But, it is a positive concept that finds proper use a good idea. Without modern agriculture, supported by good conservation, even our country would be short of food today.

Conservation does not mean going back to the illusory "good old days." As noted, we would not want to. Firing the furnace, shoveling manure, milking cows or digging ditches by hand never was great sport. New methods and instruments must be adapted to achieve more, not less, productivity while still not abusing our resources. Under good conservation, corn yields have jumped from 50 to 150 bushels an acre on the same land after a half century of cropping, range land and stripped land have been reclaimed, smoke abatement has been achieved in some of our cities and forest land in both the East and the West has been restored.

Conservationists must stamp out the idea that nature without man is a benevolent system in which all works together in harmony and love. A slogan for Lucky Strike cigarettes many years ago, even though we use it in a different context, fits well: "Nature in the raw is seldom mild!" Man may be an irresponsible rogue at times, but he is not irredeemable. He can be redeemed by practicing conservation. *"Conservation"* is a good word.

☐ ENVIRONMENTAL PROBLEMS WORLDWIDE

The problem of maladjustment to the environment is not ours alone in the United States. It faces industrialized nations everywhere and has spread to the developing nations. Pollution and waste disposal problems face the Soviet Union and Japan, East and West Germany and the United Kingdom. Thus, the problem is not a matter of the particular kind of government and economy we have. The problem is more basic. We must change our goals. We must tailor our resource planning to the interconnectedness of all things. We must view the world as a unit. Pollution of air and water, spoilation of the landscape and needless waste of mineral resources may hit certain parts of the earth especially hard. But, like the waves from a stone dropped in a pond, they eventually reach to every shore.

□ □ □

QUESTIONS AND PROBLEMS

1. How might an ecological approach to a natural resource such as coal differ from a strictly economic approach?
2. What is the ecological relationship between grass, beef cattle and man?
3. How does the environment on the moon differ from that on earth? What does this suggest about our care of our environment?
4. Make a list of 10 ways in which we benefit directly or indirectly from the light and radiation of the sun. Give a satisfactory substitute for as many of these as you can.
5. Look up the word *photosynthesis* in a good reference book. Why is this process essential to our lives?
6. How does the life in a well-balanced farm pond illustrate the community principle of ecology?
7. Why does the population explosion have a greater impact on our resource base than just the increase in the numbers of people suggests?
8. What particular actions would you suggest to reduce air pollution in your neighborhood?
9. In what ways would you suggest solid waste might be put to use?
10. How would you suggest that recreation facilities be improved in your county?
11. What is the difference between a conservational and an ecological viewpoint? Are they in conflict?
12. Why do we need to be concerned about ecological problems outside our own country?

24 □ □ □

Conservation—
everybody's opportunity

THE LAST DECADE was one of remarkable progress in increasing environmental awareness and action. In 1970, the National Environmental Protection Act, one of the most far-reaching pieces of environmental legislation ever passed by any nation, was enacted. Also, in 1970, an unprecedented nationwide expression of environmental concern by citizens was reflected in the first Earth Day. At the end of the 10-year period, there were reasons to believe that we, as a nation, were more concerned about our natural resources.

The National Wildlife Federation in its February–March, 1980, issue of *National Wildlife* commented:

> After a decade of frustration and doubt, the U.S. may finally be turning the corner in its campaign to blot out water pollution. There were signs of real progress last year, including some new statistics that show several of the worst pollutants have not increased at all since 1975. That means many industries and communities are now moving toward, rather than away from, the Clean Water Act's goal of "fishable, swimmable" waters by 1985.
>
> It's debatable whether that objective is achievable nationwide, but there have been some remarkable successes. . . .[1]

An overall indicator of the "progress" we are making is the Environmental Quality Index (EQI), computed annually by the National Wildlife Federation. This index tries to measure the nation's

---

[1]Copyright © 1980 by the National Wildlife Federation and reprinted from the February-March, 1980, issue of *National Wildlife*, p. 33.

386 □ OUR NATURAL RESOURCES

environmental status in terms of quality of air, water, minerals, wildlife, living space, forests and soil. The chart and comments on the index, used with permission of the National Wildlife Federation, are given in Figure 24-1, which presents the 1981 index, with estimates through 1980. In the last five years, the overall rate of deterioration in environmental quality appears to have lessened.

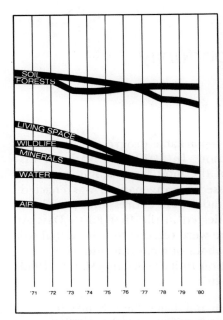

Wildlife: Worse
The land and water wildlife needs to survive are still going down the drain, but now conservationists have a new tool to help stem the flow: a hard-won law promising broader programs, fresh sources of funds.

Minerals: Same
Uncle Sam has embarked on a costly "synthetic fuels" program to produce more oil, but millions of Americans are stressing conservation and taking a hard look at alternative energy sources.

Air: Same
Air quality is not deteriorating in many of the nation's cities, but pollution in some rural areas is getting worse as a result of a rapidly increasing phenomenon: acid rainfall.

Water: Worse
Over much of the country, polluters have stopped dumping wastes directly into rivers and lakes, but now there's an appalling new threat: toxic dumps are contaminating the nation's ground water supplies.

Forests: Same
Wood demand is down for now, but there's mounting pressure to tap more riches of our national forests.

Living Space: Worse
A mutiny for the bounty on public western lands could challenge U.S. jurisdiction nationwide.

Soil: Worse
The U.S. is losing a million acres of farmland annually to development, and despite some holding action in a few states, cropland protection remains almost nonexistent throughout most of America.

FIGURE 24-1. The 1981 Environmental Quality Index (EQI) summary. The annual EQI is a subjective analysis of the state of the nation's natural resources. The judgements on resource trends represent the collective thinking of the editors and staff members of the National Wildlife Federation, based on consultation with government experts, private specialists and academic researchers. *(Courtesy, National Wildlife Federation. Copyright © 1981 by the National Wildlife Federation and reprinted from the February–March, 1981, issue of* **National Wildlife***, p. 36.)*

□ BEAUTY AS A GOAL

During the last decade, the value of establishing the preservation and attainment of beauty as a national conservation goal was further implemented by the restoration of historic buildings and by city beautification projects. President Lyndon Johnson, with the ardent support of his wife, led a crusade to preserve and enhance America's beauty in 1965. In the *Conservation Yearbook* of the Department of the Interior for that year, he is quoted as follows:

> A few years ago we were greatly concerned about the ugly American. Today we must act to prevent an Ugly America. For once the battle is lost, once our national splendor is destroyed, it can never be recaptured. And once man can no longer walk with beauty or wonder at nature, his spirit will wither and his sustenance be wasted.

□ NEPA AND EIS

President Richard Nixon recognized the environmental crisis of our nation in his State of the Union Address, January 22, 1970, saying:

> The great question of the '70's is: shall we surrender to our surroundings or shall we make peace with nature and begin to make reparation for the damage we have done to our air, to our land and to our water?

Shortly before the speech, he had signed the National Environmental Protection Act (NEPA) into law on New Year's Day. The purpose of the act was "to declare a national policy which will encourage productive and enjoyable harmony between man and his environment. . . ." The heart of the law is the requirement that all government agencies prepare Environmental Impact Statements (EIS) on any proposed actions which might significantly affect the quality of the environment.

As the decade of the 1970's began, starts had been made in cleaning up the air. Congress set up a National Center for Air Pollution Control and granted money to carry out improvements. Exhaust control devices had been installed on new automobiles. In New York, Chicago and other cities, laws on air pollution were being enacted and enforcement was begun. Young people were demonstrating in high schools and colleges for an environmental cleanup. As noted earlier, the Clean Air Act was passed in 1972 and amended in 1977 to make it more effective.

We began to try to control noise pollution. Construction of a jet airport that would have threatened wildlife of the Florida Everglades was stopped. We began to consider how we might control the sonic booms of high-speed aircraft or at least limit their flight patterns so that they would not shatter the glass and structure of houses, not to mention the nerves of their inhabitants.

☐ PROVISIONS FOR OPEN SPACE

Steps were taken during the 1970's to preserve and care for more open space for recreation, enjoyment of beauty and possible scientific values. Additional areas of redwood forests were made national parks. A number of streams were designated wild rivers, to be preserved in a natural state. Eight new river segments (a total of about 696 miles) were added to the National Wild and Scenic Rivers System in 1978. Magnificent stretches of beach on Cape Cod on the Atlantic and at Point Reyes on the Pacific and other areas were set aside as new national parks. We extended our system of national trails. And, the capstone for the decade was the conservation protection given millions of acres of land in Alaska, as noted in Chapter 12.

☐ THE BUSINESS CLEANUP

In the 1970's, business also made progress in avoiding waste and pollution. Some steel companies and electric power companies installed devices to eliminate most of the air pollutants caused by their heating operations. Water was being recycled more than ever. Some mine operators built settling basins to precipitate wastes in the sludge from mines rather than dumping it into streams. Oil companies captured and used much of the excess gas escaping from oil wells which they formerly had burned off.

Abating pollution and in other ways improving the relation of industry to the environment should be looked upon as an opportunity by businessmen. The equipment to process wastes will need to be manufactured. The ultimate costs of such improvements will be paid by the consumer, in any case.

☐ THE AGRICULTURAL TURNABOUT

In the past four decades, agriculture made the dramatic turnabout from robbing the soil to farming on the basis of land

FIGURE 24-2. *Upper:* The wheat field being harvested in the foreground is on a reclaimed surface coal mine in Randolph County, Illinois. The yield in 1980 averaged 37 bushels per acre, with some plots yielding as high as 60 bushels, as verified by the University of Missouri. Nine inches of topsoil and 39 inches of reconditioned soil known as "agricultural root medium" have been replaced. The structure barely visible on the horizon is the shaft of an underground mine, which is a slope-type mine reaching coal too deep to strip. *(Courtesy, Peabody Coal Company)* *Lower:* This watery playground was once a coal mine. The area, near Huntsville, Missouri, was strip mined and three years later revegetated and turned into a private development. This picture was taken 16 years after reclamation. *(Courtesy, National Coal Association)*

capabilities. Forestry turned the corner from "cut out and get out" policies to management on a sustained yield basis. State and federal agencies adopted rules and recommended practices from pesticide use that lessened the danger to our biotic system. Farmers and foresters supported these programs. Our western range land policy was under complete re-evaluation, with conservation goals in view.

□ LAND-USE PLANNING

By the 1980's, states had begun ro recognize the need for statewide land-use planning. We, as a nation, had found out that helter-skelter private development based solely on the profit motive and governmental planning based largely upon political expediency threaten our prime agricultural soils, our forests and wilderness areas, our free-flowing streams, our Great Lakes and coastal waters and other natural resources. They cost us money and even our lives. Houses had been washed over cliffs by torrential rains in California, engulfed by swirling flood waters in Illinois, Mississippi and Pennsylvania and smashed by waves under the fury of hurricanes in Texas, Louisiana and Florida.

It was becoming increasingly clear that many of our losses were unnecessary. They could have been prevented or, at least, modified, had good land-use plans been made and implemented.

Slow progress in land-use planning

Congress is well aware of the need for land-use planning, but progress has been slow, In 1980, after many years of debate, it still had not succeeded in enacting a land-use bill into law. A bill proposed would have provided $100 million a year to help states set up a nationally coordinated state planning program. The bill was endorsed by the National Association of Counties, the National Governors' Conference and all major environmental groups. However, opponents misrepresented the bill as undermining state and personal property rights. And, the administration said we couldn't afford it because of the unbalanced budget.

Meanwhile, at the state level, land-use legislation has continued to face opposition as the complications of such programs have become evident. On the other hand, almost all the states have considered, initiated a study of or enacted some kind of state land-use law.

Many citizens, unfortunately, still see land-use legislation more as a threat to private property than as a protection of resources. Zoning does affect land prices, of course, and some ordinances could be very inequitable in the eyes of landowners. Ways need to be found, such as those pointed out in Chapter 19, to reward those who forego selling their land at higher prices for residential, industrial and commercial use, in order that it may remain in a needed environmental use. We can hardly blame a landowner for feeling it unfair if his neighbor is able to sell his 100 acres of, say, Class III or Class IV land for $500,000, while he is told that he can sell his Class I land only for agricultural purposes which would bring a price of $200,000 or less.

☐ ECOREGIONS—AN ECOLOGICAL SYSTEM

The ecoregions classification adopted by the U.S. Forest Service (see map in Figure 24-3) illustrates how an ecological system can be developed for land-use planning.[2]

The U.S. Forest Service explains:

> There are a number of different kinds of regions depending upon objective or purpose. Just as a region based on agriculture is an agricultural region, one based on ecosystems is an ecosystem region or *ecoregion*.
>
> To date, most work based on the ecosystem concept of resource management is at a detailed level. There are at least two reasons why a regional view of the ecosystem is needed: (1) to permit detailed data to be aggregated into more generalized units for decisionmaking at higher levels; (2) to provide an integrating frame of reference needed to fully interpret the more detailed information.
>
> This map was developed to meet these needs. Maps based on classification of climatic types, vegetation associations, and soil groups have been widely used, but no comparable broad-

---

[2] A more detailed map and a booklet explaining the system are available from the U.S. Forest Service, Rocky Mountain Forest and Range Experiment Station, 240 West Prospect Street, Fort Collins, Colorado 80526. The authors are indebted to Dr. Robert G. Bailey, who is credited with the design of the map in Figure 24-3, and who cooperated in supplying other information.

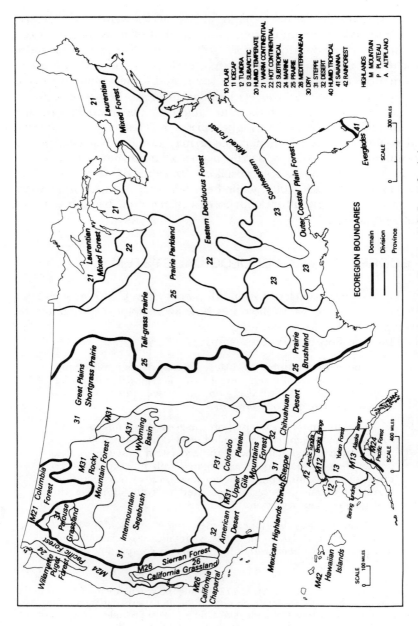

FIGURE 24-3. Ecoregions of the United States. (*Courtesy*, U.S. Forest Service)

scale synthesis of these maps has been commonly accepted. This map is an attempt to fill that gap.

Because land is a complex of surface attributes, in other words, an ecosystem, the classification should reflect spatial patterns as well as properties. How a piece of land will behave cannot be predicted fully in terms of local controls or single factors acting in isolation, but is in part determined by relationships with adjoining areas. There is thus intrinsic advantage of assessing land in terms of interacting units at various scales of grouping. The process of grouping objects on the basis of spatial relationships rather than solely on similarity of taxonomic properties is called *regionalization*.

☐ OTHER PROBLEMS STILL TO BE SOLVED

Many conservation problems remain to be solved. Water pollution from offshore oil seepage or leakage from oil tanker vessels still occurs. Conservationists and lumbermen are in disagreement as to the best policies to protect our forests. Many of our farmers still have not signed up for conservation programs. Most of the many small privately owned woodlands in the nation are still not being well-managed for productivity of timber. Municipalities continue to be faced with crime, housing and traffic problems. Sewage improvement programs, however, have moved toward completion.

The position of the administration under President Gerald Ford was dedicated to getting the nation through the energy crisis with a minimum of monetary inflation. Environmental actions, the President felt, had to be shelved for the time being, if necessary, to accomplish these other objectives. In a message prefacing the *Fifth Annual Report of the Council on Environmental Quality*, in December, 1974, he wrote:

> ... today, millions of our citizens share a new vision of the future in which natural systems can be protected.... We must also recognize that, even with a strong conservation program, we will still have to mine more coal, drill for more oil and gas, and build more power plants and refineries....

Beautifying the landscape received more publicity than action in the first half of the decade. A bill for highway beautification was stripped of its effective provisions by the time it passed Congress. Cities were having trouble passing bond issues to provide money for buying land for parks and other recreation facilities before the land was built up. Thousands of miles of abandoned strip mines remained unreclaimed.

President Jimmy Carter began his term of office with a strong conservation stand. And, in the following years, he supported wilderness preservation and preservation of public lands in Alaska. But, as the nation entered an economic recessionary period in 1980, he proposed sharp cuts in many conservation programs, including those supporting solar energy, habitat protection and acquisition of mass transit.

Emphasizing productivity in an attempt to reverse an alarming rise in inflation and unemployment, President Ronald Reagan continued cutbacks in conservation programs. However, James Watt, the Secretary of the Interior, declared early in 1981 that "we can protect the environment—and bring on development."[3]

☐ HISTORY OF PHILOSOPHY OF CONSERVATION

As we consider the many resource problems we have studied, it sometimes seems as though the battle to save our environment has just begun. But, as far back as 1748, a book on conservation farming practices by Jared Elliott, a minister, was published in America. Elliot's book and several others which soon followed it, however, had little impact at that time.

Saving for future use

The earliest concept of conservation evident in the policies of our nation was that of "saving for future use." In the 1800's, the word "conservation" was applied mostly to our forests. National concern over forest conservation was aroused by the report in 1871 of a committee appointed by the Commissioner of Agriculture, that in less than two centuries nearly one-half of our original forest land had been cleared. Up to that time, it was the common belief that our natural resources were practically inexhaustible. Our forests, in particular, were assumed to be large enough to meet all future needs

---

[3]"We Can Protect Environment—And Bring On Development," *U.S. News & World Report*, May 25, 1981, p. 49.

FIGURE 24-4. Conservation includes preservation and wise use of natural resources. This is a view overlooking the Estes Park, Colorado, area, with peaks of the Rocky Mountain National Park in the background. The Rocky Mountains provide not only excellent habitat for wildlife and recreational sites for humans but also forests, range land, water and minerals. *(Photo by Harry B. Kircher)*

for lumber and wood. Trees were as common as weeds and were destroyed with little more regard. It came as a shock to learn that we might not always have timber for every use!

The first step as a follow-up of this report was the creation of a Division of Forestry within the Department of Agriculture. This division immediately began to publicize the disappearance of our forests and to call attention to the need for definite government action to preserve this most useful resource.

In 1889, an act was passed by Congress giving the President of the United States power to set aside land for forest preserves. This power was used sparingly before Theodore Roosevelt became President, but it did much to stress the importance of the fast disappearing forests to our national welfare, and he set aside more land for national forests than had all preceding presidents.

As time passed, we became aware of the need for saving other natural resources. We began to fear that coal and iron ore reserves would not last forever. New oil fields were discovered, but the production from the older fields dwindled. Finally, we became conscious of the loss of our soil and of its possible effect upon future production of food. Thus, gradually, over the period of two generations, we enlarged the meaning of the word *conservation* from its limited use in connection with the preservation of our forests to include preservation of all those natural resources which are useful to us in our complex way of living (see Figure 24-4).

Prevention of waste

A second meaning we have come to associate with the word *conservation* is *prevention of waste*. We Americans have been prodigal with our natural resources, as the previous chapters of this text have shown. We have come to realize that, in addition to saving our forests for use in the future, we need to eliminate present waste in harvesting and use of wood, which has been estimated as amounting to two-thirds of all timber cut. We have come to realize that besides concern about the extent of our natural gas reserves, we should worry about the amount we are losing in the wells (see Figure 24-5). A few years ago the daily waste of natural gas from one oil field alone was enough to supply the needs of 10 cities the size of our national capital.

Wise use

A third concept of conservation is that it deals not only with savings and prevention of waste but with *wise use* as well. In fact, good conservation, in some cases, requires that a use rather than a savings be made. Thus, many mines when once opened must be kept operating, or they will fill up with water, or the timbers supporting the structures above will give way, and the remaining reserves will be lost or at best recovered only at greatly increased costs. When timber of our commercial forests reaches what foresters call maturity, it should be cut, or it will deteriorate and become worthless for commercial purposes.

☐ THE NEW CONSERVATION PHILOSOPHY

Conservation today not only incorporates the three principles of savings, waste prevention and wise use but also stresses the interaction of the whole environment, including man. It also emphasizes the need to be concerned with quality. Quality needs to be set above quantity as a goal in developing resources. A wealthy man who is seriously ill is the object of pity, not envy. Our national wealth of resources will do us little good if badly polluted.

FIGURE 24-5. Prevent waste. This picture symbolizes the time, mostly past, when the sky was filled with fumes from escaping gas or the wasteful burning of gas from wells to enable them to produce oil more readily. The belching smoke is caused by the burning of waste from drilling operations—oil, mixed with mud, water and the chemicals used in drilling a new well.

FIGURE 24-6. Wise use. Some land can be used for the production of cultivated crops and still not be lost through erosion, while other land can best be kept in place through the production of grass or woods. The stripped areas are growing crops such as corn, oats and hay, while the land that shows no strips is in grass. Still other land is wooded, and the accumulating leaf mold and small undergrowth in those woods, if they are not pastured, serve as sponges to prevent surface run-off and erosion. *(Courtesy, Soil Conservation Service)*

The "new" conservation stresses the human benefits to be derived from resource management. Professor Ruben Parson, a resource geographer and writer of a popular conservation text, has stated: "Unless man is served, there is no conservation." Benefits to man include "life, liberty, and the pursuit of happiness."

In conclusion, we can define conservation as *the management, including either preservation or use, of our natural resources for the well-being of man, taking the total environment into account.*

☐ TWO VIEWS OF OUR NATURAL RESOURCE BASE

Those who are concerned with our natural resource base fall into two major groups: optimists and pessimists. The optimists emphasize the limitless nature of energy and the potentials of man's intelligence. They believe that man's ingenuity can meet resource needs and that man is too intelligent to continue plundering the earth to his own ultimate destruction. The pessimists emphasize the exhaustible nature of our natural resources and the "creature" nature of man. They fear that man cannot halt the rising rate of population growth and of abuse to his environment.

It is helpful to consider both the optimistic and the pessimistic outlooks. A Pollyanna attitude leads to disregard of unpleasant facts, postponement of action and overconfidence. An alarmist view of our energy potential leads to fear, paralysis of action and a defeatist attitude. What is needed is a concerned yet confident view that inspires firm action and perseverance. The accomplishments of genius are said to be 1 percent inspiration and 99 percent perspiration. In our highly industrialized nation, adjustment to our natural environment does not come naturally. It requires inspiration and perspiration. But, the job can be enjoyable and rewarding.

"Reports on the future of energy by the Andrew W. Mellon Foundation, the Ford Foundation, the Resources for the Future Project, and Harvard Business School agree on just one point: The 'key energy resource' is conservation," Mel Maddocks writes in *The Christian Science Monitor*. "Alternatives are what make the world 'thinkable.' Alternatives, multiple choices, second and third ways break the fatalism produced by Fearful Prognostications. One begins to subscribe to patience, to look for improvements rather than The Answer—to think small, if you will."[4]

What needs to be done is to realize the land ethic of Aldo Leopold, co-founder of The Wilderness Society and father of modern game management in the United States. He wrote:

> All ethics so far evolved rest upon a single premise: that the individual is a member of a community of interdependent parts. His instincts prompt him to compete for his place in the community, but his ethics prompt him also to co-operate....

---

[4]Mel Maddocks, *The Christian Science Monitor*, November 14, 1979, p. 13.

The land ethic simply enlarges the boundaries of the community to include soils, waters, plants and animals, or collectively: the land.

We can be ethical only in relations to something we can see, feel, understand, love and otherwise have faith in. A land ethic, then, reflects the existence of an ecological conscience, and this in turn reflects a convention of individual responsibility for the health of the land.[5]

☐ DON'T KILL THE GOOSE

Since profitability should not be the sole reason for resource allocations, some environmentalists have hinted that if we are to conserve our environment, perhaps the solution is to do away with the private profit system. Such a policy would be like doing away with "the goose that lays the golden egg." While it is true that our country's productivity would not have been possible without an outstanding natural resource base, it should be remembered that the incentive that sparked that development was private profit. The abuses that went along with it are shocking. But the federal government has done much to throttle some of the uninhibited rapaciousness evident in earlier years. And, business policy and action show new environmental awareness. If adequate laws were equitably administered, then business could without loss add in environmental costs, such as maintaining water purity and air quality, as a part of the cost of their products. We should all pay for the cost of environmental quality.

It seems strange to hear some people say that the answer to the environmental problem is to have *more* government. We need *better* government, not more. And, we need honest business to go along with it. What is needed is an acceptance of moral responsibility on the part of all.

☐ WHAT CAN I DO?

But, a person asks, "What can I do? I am so small and the world so large!" First of all, we can practice conservation at home and in our jobs, be it school or work; and we can practice conservation at play. Secondly, we should be advocates of conservation in the groups

---

[5] Aldo Leopold, *Sand County Almanac*, as reported in *Toward an American Land Ethic*, a report of The Wilderness Society, Washington, D.C., 1980, p. 3.

to which we belong—church or club, fraternity or sorority, labor union or chamber of commerce, League of Women Voters or National Association for the Advancement of Colored People, and to our personal friends and acquaintances. Third, we should join and support conservation organizations such as the National Wildlife Federation, the National Audubon Society or our state society, the Sierra Club and The Wilderness Society. It is through these and like organizations that we can gain the clout to achieve conservation goals. Already such organizations have helped save San Francisco Bay from pollution by dumping, wild rivers from being tamed, wilderness areas from being scarred by mines, treasured wildlife from being wiped out and all of us from being exterminated by DDT.

Finally, we may be able to help conserve our resources by preparing ourselves and finding jobs directly in conservation work. Whether forest ranger, lawyer specializing in environmental law or regional planner, we would then have the satisfaction of knowing that our job was helping all of us to survive.

□ THE BATTLE FOR NATURAL RESOURCES

Our young people will inherit the resource base. They must be willing to fight for it. Battles are generally destructive of life and property. The battle to conserve natural resources is an exception. It is a battle to preserve life.

It is a battle to benefit all men, not just a few. It is a battle which upholds rather than ridicules youthful ideals. Unfortunately, it is youth who will "pay the piper" should the battle not be won—should man continue an arrogant disregard for the laws of nature. Youth will also reap the harvest if a better conservation is achieved.

The sciences, physical and social, should coordinate their activities to tackle the problem of the total environment and to help establish realistic national goals. All of us should support conservation programs, both in the spirit and in the letter of the law. Conservation is *everybody's opportunity!*

□ □ □

QUESTIONS AND PROBLEMS

1. Did the EQI show that our environment was improving or deteriorating? What are some of the problems?

2. How did NEPA and EIS attempt to protect the environment?
3. What are several conservation actions taken by government, business and agriculture during recent years to help protect the environment?
4. What are the advantages of statewide planning? Why should anyone object to it?
5. What was the early meaning of *conservation*, and to what resources did it mostly apply?
6. In what circumstances is it good conservation to continue to use a resource?
7. Which methods of conserving a resource would you suggest for: (a) a forest, (b) an oil or a gas well, (c) a river and (d) the soil?
8. Name four natural resources found in your area and tell how you think each could be better conserved.
9. How does our present-day concept of conservation differ from that of a century ago?
10. What can you do to help conserve our resources?

25

The human body and natural resources[1]

MANKIND IS CLASSIFIED as a human resource (see Chapter 1). But, that unique collection of organic and inorganic matter and gases which we call the body is also a natural resource. Its elements are free gifts of nature. And, however great our wealth of "other" natural resources may be—of what value is this to us humans if we ourselves are weak and sickly? So, we conclude our study of natural resources with a chapter noting the interlinkage of this resource and other natural resources.

Humans are, of course, more than physical bodies. Without the marvelous spark of energy we call life, the body is just a collection of earth materials—it is no longer a human being. And, without the inspiration which ignites our mental processes and the power to think, which enables us to store and increase knowledge, humans would be just another animal species living largely by instinct. Life may go on after death, but while we are here on earth, it is through the human body and the agents it manufactures that we are able to express ourselves, creating or destroying resources. The health of our bodies is dependent upon the extent to which we intelligently manage and live with our other resources.

---

[1]The authors are especially indebted to Diana Friesz, an environmental control specialist in private industry, and Richard Nicol, a graduate biologist, for help in researching this chapter.

☐ THE HUMAN BODY

As we noted in Chapter 2, the earth has four major systems or spheres—the atmosphere, hydrosphere, lithosphere and biosphere—which have their smaller systems. The human body is a system within the biosphere with 11 major systems of its own: the skeletal, muscular, nervous, vascular, digestive, respiratory, endocrine, lymphatic, urinary, integumentary and reproductive systems. Each system integrates with the others in an attempt to maintain a constant internal state, despite variations in the external environment—a process known as *homeostasis*.

Effects of the external environment upon homeostasis, and subsequently upon human health, can, of course, be beneficial or destructive. Good health occurs when the processes of adaptation to changing environmental factors proceed within the limits of the homeostatic mechanisms' ability to adjust (see "The Science of Ecology," Chapter 23). Many such adjustments are readily made, such as constriction of blood vessels to prevent heat loss, rapid breathing to compensate for rarified air and scarring to aid the healing process. Disease occurs when the steady state between organisms and the physiochemical environment (air, water, food), biological environment (bacteria, viruses, plants and animals, including other humans) or social environment (work, leisure, cultural habits and patterns) is upset. Homeostatic mechanisms may be taxed beyond their limits by the rapid and severe changes possible in the external environment as by pollution, insecticides, or harmful radiation.

Although death from infectious diseases has shown a precipitous decline since 1900, death and sickness from environmentally related respiratory diseases and cancer have shown an equally precipitous rise.[2]

☐ AIR AND HEALTH

Air is essential to life on earth. Most cells of the body derive the bulk of their energy from chemical reactions involving oxygen obtained from the environment by breathing. Carbon dioxide, despite its low concentration in air (less than $1/600$ that of oxygen), plays a

---

[2]John H. Dingle, "The Life of Man," *Scientific American*, September, 1973, Vol. 229, pp. 82–83.

vital role in the cellular process. It is a waste product of cellular respiration, and it also serves to regulate the rate of breathing. The presence of carbon dioxide in the form of carbonic acid functions to control the acidity-alkalinity balance of the blood. All of the major components of air, and some of the minor ones, are of biological importance in maintaining homeostatic conditions in man.[3]

Unfortunately, as we are only too aware, air pollution has become common (see Chapters 19 and 23). The principal components of air pollution are carbon monoxide, sulfur oxides, nitrogen oxides, hydrocarbons, photochemical oxidants such as ozone and particulate matter.

A common disease attributed to air pollution is emphysema. This disease attacks the membranous walls of the alveoli. Alveoli are clusters of tiny cup-shaped hollow sacs, lined with blood capillaries serving as the sites of gas exchange in the lungs. Oxygen diffuses inward and carbon dioxide outward across the delicate membranes. Pollutants cause the thin alveolar walls to lose their elasticity and tear apart. Nonfunctioning air spaces are left. The area in which gas can exchange is restricted. Oxygen deficiency results.

Bronchial asthma and chronic constrictive ventilatory disease, a condition related to emphysema, are also aggravated by air pollution. Some researchers even believe that there is a link between air pollution and the common cold. Air pollution is also a contributor to lung cancer. Carbon monoxide is another pollutant which is a killer if in sufficient concentration because it deprives the blood of oxygen. The oxygen-carrying capacity of the blood is the result of the affinity of the hemoglobin, an oxygen-bearing protein, in the red blood cells for oxygen. Carbon monoxide combines with hemoglobin even more readily than oxygen, and once combined, the new "carboxyhemoglobin" molecule is no longer able to carry oxygen. Thus, a person may die from asphyxiation even in the presence of oxygen if the carbon monoxide concentration is high enough. People trapped in heavy rush-hour automobile traffic are exposed to much higher than normal concentrations of carbon monoxide. Drivers may suffer loss in reaction time, alertness and visual acuity.[4]

---

[3] To amplify this discussion of air and water and health, see the following texts, which served as sources: Charles Robert Carroll, Dean Miller and John C. Nash, *Health, The Science of Human Adaptation* (Dubuque, Iowa: William C. Brown Company, Publishers, 1st ed., 1976, 2nd ed., 1979); Norman S. Hoffman, *A New World of Health* (New York: McGraw-Hill Book Company, 1977); and Kenneth E. Maxwell, *Environment of Life* (Encino, California: Dickenson Publishing Co., Inc., 1973).

[4] Hoffman, *op. cit.*

Sulfur oxides and nitrogen oxides may also result in air pollution. While sulfur oxides are derived from the combustion of coal which contains sulfur, nitrogen oxides are the product of the high-temperature combustion of coal, oil and gasoline. Both of these types of oxides damage human health by causing irritation of the eyes and respiratory passages, which may be permanently damaged by long exposure.

Hydrocarbons are another air pollutant we have cited. Although there is no evidence linking hydrocarbons directly to human disease, hydrocarbons are a major constituent of smog. And, smog can bear many irritants besides contributing to mental depression and accidents related to poor visibility.

Photochemical oxidants, the most well known of which are ozone and peroxyacetyl nitrates (PAN), are known to be irritants to the eyes, nose and throat. They may cause headaches, coughs, shortness of breath and thickening of the bronchiolar walls.

Any matter in the air, which is not a gas, but which is either a solid or a liquid, is referred to as particulate matter. Many substances can adhere to particulates so that often it is not the particulates themselves but what they carry with them into the lungs that is harmful. Particulate matter can damage especially the cilia which line respiratory passages. Cilia are microscopic hairlike processes. Their whiplike motion cleanses the respiratory tract of pathogenic microorganisms and dirt. If damaged, they lose their capacity to function properly, and the underlying cells of the respiratory system structures are vulnerable to disease.

Among other extremely hazardous pollutants carried in the air which may impede homeostasis are asbestos, beryllium and mercury. Asbestos has been widely used for many industrial purposes, including brake linings in cars. The inhalation of asbestos fibers has been linked to malignant diseases of the lungs such as bronchogenic cancer and mesothelioma.

The metal beryllium is used in rocket fuels, missile guidance systems, nuclear reactors and atomic weapons. Inhaled beryllium leads to progressive lung disease and death.

Mercury, commonly used in the manufacturing of paint, pulp and paper, batteries and mildew-proofing, may also pollute the air. Air-borne mercury can affect the central nervous system and cause weight loss, insomnia, tremors and psychological disturbances.

As was widely publicized upon the installation of catalytic converters in U.S. automobiles, high concentrations of lead in the air are associated with emissions from motors burning leaded gasoline. Excessive lead ingestion can seriously impair the nervous system. Lead poisoning in ghetto children who eat paint and plaster which contains lead has evidenced this problem.

☐ WATER AND HEALTH

In the chapter on water (Chapter 10), we noted that despite its apparent abundance, we often face shortages, especially water of a desirable quality. No substance is more essential to the human body. The human body contains up to 90 percent water, depending upon age, with the body generally dehydrating as it ages. If we lose more than 20 percent of the water of our bodies, we will die. Even a healthy person can live only a few days without water.

But, it is more than just the quantity of water that is vital to our bodies. It is its quality! Ordinarily, water has many dissolved salts, sugars and many other substances. These substances in the water are what makes it so valuable in the human body. Water transports enzymes into the digestive tract and provides the lubrication for circulation, digestion and excretion. Food must be dissolved before it can enter the bloodstream of animals. Even carbon dioxide and oxygen must be dissolved in the liquid portion of the blood before it can be used by the body.

Water should be free of harmful bacteria and excesses of various substances for healthy use. It is impossible for the human body to maintain homeostasis while utilizing polluted water. Illnesses such as typhoid, cholera, dysentery, gastroenteritis and infectious hepatitis can result from the action of microorganisms present in the water. The bacterial contamination of water supplies is a common disease hazard in the United States today (see Chapter 10).

Some cardiovascular conditions are related to the consumption of contaminated waters. Increased sodium intake from polluted waters puts a strain on the heart and circulatory system. Nitrates, which may enter the human system through contamination by sewage and wastes from domestic animals and by run-off from agriculture, are especially harmful to newborn babies. The digestive tract of infants contains bacteria which convert otherwise harmless nitrates into toxic substances. The result is a blood disease, methemeoglobinemia, that can lead to death due to suffocation.

ENERGY FROM THE SUN AND THE HUMAN BODY

In Chapter 2, we noted that without plant life, our bodies could not obtain the sun's energy. Energy is transmitted to humans from the sun indirectly through plants by the process of photosynthesis. The carbohydrates (carbon, hydrogen and oxygen) thus synthesized are essential to our metabolic processes. Some are quick energy sources for the body. Others are stored and broken down for later use.

The simple sugars (monosaccharides) may reach our system through candies or baked goods, or fruits and vegetables. Two sugar units per molecule (disaccharides) may be derived from sugar cane or sugar beets (sucrose), corn, potatoes, wheat and rice (maltose) and milk (lactose).

Another important product of photosynthesis is cellulose which forms the basic structure for wood and cotton, so useful to shelter and clothes for our bodies.

Another family of natural resources essential to life are proteins and amino acids, consisting primarily of carbon, oxygen, hydrogen and nitrogen. Amino acids are fundamental units of proteins. Some, but not all, amino acids can be synthesized by our bodies. Others, however, must be derived from the food supply. In volume, meat and eggs are principal protein foods in the United States, but grain and dairy products are also major contributors.

Fats and oil are also vital to human life. Traditionally, we obtained our supplies from butter and other animal products. In recent decades, many consumers have switched to vegetable oils, which were made into solids by the addition of hydrogen. Such oils include cottonseed, peanut, corn germ, soya bean, coconut and sunflower seed. Too much saturated fat can increase cholesterol, a soapy-like substance which is a universal constituent of tissue and which is believed by some to promote heart disease. Thus, the heightened interest in replacing fats with oils.

Finally, the health of our bodies is dependent upon a proper balance of enzymes, vitamins and hormones. These resources comprise less than one percent of the weight of the body. But, they are extremely critical to our health.

Enzymes are proteins produced by living organisms which function as biochemical catalysts. Essential to certain body processes, enzymes enable the body processes to perform with less energy. Hormones are substances formed by one organ and conveyed to another which is stimulated to function by their chemical activity. Vitamins are complex carbon, hydrogen, oxygen compounds which are essential in small amounts for the control of metabolic processes.

In sum, our bodies are critically dependent upon the natural resource base for food, shelter and clothing and the complex chemical compounds which determine health.

□ PESTICIDES, METALS AND HEALTH

Manufactures from natural resources may be either a boon or a bane to humans. Insecticides have benefited human health by controlling insects and increasing food production (see Chapter 18). Malaria, which is transmitted by the *Anopheles* mosquito, and typhus, carried by the human body louse, have been virtually eliminated from the United States with the help of insecticides.

Insecticides have been a hazard to human health by interfering with the functioning of the central nervous system, damaging the respiratory and digestive tracts, skin and eyes and adversely affecting mucous membranes, the rate of metabolism and the condition of visceral organs. It is believed that insecticide poisoning is responsible for up to 200 deaths per year.[5]

Among chemical compounds that have been in global food chains are high concentrations of a group of industrial herbicides, known as the polychlorinated biophenyl or PCB's. PCB's are insoluble in water, soluble in fats and oils and very resistant to either chemical or biological degradation. In short, they tend to persist. In 1979, officials were still discovering new victims of a 1973 incident in which PCB was accidentally mixed with animal feed. Thousands of domestic animals and possibly millions of humans were affected.[6]

Only small quantities of some substances can cause serious damage to the body. Asbestos fibers, for example, are suspected of causing cancer. And even low levels of cadmium, a widely used heavy metal, can be a threat to health.

---

[5]G. Tyler Miller, Jr., *Living in the Environment*, 2nd ed. (Belmont, California: Wadsworth Publishing Co., Inc., 1979), pp. E90–E91.

[6]Lawrence H. Hall, "A Plague of Poisons," *National Wildlife*, August–September, 1979, pp. 29–32. This paragraph and the following two are based on this excellent article by Mr. Hall.

410 □ OUR NATURAL RESOURCES

The toxicity of some chemical wastes was dramatically revealed when chemicals from an old canal channel, called Love Canal, erupted to the surface of the ground in a residential area in New York State. Children there developed unexplained sores, and a number of babies developed unexplained birth defects. In a subsequent investigation, some 80 dangerous substances were identified in the area which the state finally declared a disaster area. Love Canal has proved to be just one of thousands of potentially hazardous waste sites in the United States identified by the Environmental Protection Agency.

□ NOISE AND HUMANS

At a certain point, even noises can become hazardous to human health and hearing. For urban dwellers, sounds from automobiles, jets, construction activities, power mowers, motorcycles, office machines, televisions and stereos may combine to create a harmful level of noise. The Environmental Protection Agency has summarized some of the physical effects of exposure to high levels of sound in the following symptoms:[7]

1. Blood vessels in the brain dilate.
2. Blood pressure rises.
3. Blood vessels in other parts of the body constrict.
4. Pupils of the eyes dilate.
5. Blood cholesterol levels rise.
6. Various endocrine glands pour additional hormones into the bloodstream.

It is difficult to link noise directly with a specific disease or mental disorder, but there appears to be little doubt that it is a contributing factor to such disorders.

□ RADIATION AND THE BODY

Today, the advent of nuclear power as an energy source has brought a new threat of radiation exposure to humans (see Chapter

---

[7]U.S. Environmental Protection Agency, *Report to the President and Congress on Noise*, U.S. Government Printing Office, Washington, D.C., 1971.

7). As has been noted, improper radioactive waste disposal and the vulnerability of nuclear power plants to sabotage by terrorist groups are some of the potential threats to human health from radiation. And, there is the potential holocaust of nuclear war!

Assessing the risk of exposure to radioactivity involves extremely complex relationships. A number of theories and measurements have been explored. But, the various assumptions and conclusions are still controversial. That is to say, while there is no question as to the carcinogenic property of radiation, there are questions as to the ways to assess its potential threat.

Radiation may be transported by many means. Radon, the product of the decay in uranium tailings, may end up in the lungs. Water may also be an important pathway for radionuclides which move into ground or surface waters. Humans may be exposed to them directly by eating fish, or they may be exposed indirectly by drinking the water or by bathing in it. A third transport medium for radionuclides is plants which humans may eat either directly or indirectly by their eating the meat of animals that have consumed such plants.

As to health effects, most theories attribute damage to breakage or mutalism of the DNA within the cell nucleus. If sufficient energy is deposited in an organ, excessive cell death may occur. Even low levels of radiation may lead to leukemia, cancer or shortened life span.[8]

☐ THE BODY, OTHER RESOURCES AND THE FUTURE

Will mankind sink or swim? Will man be able to overcome the resistances to more abundant resources, including a healthier body, or will he be overcome by them? The world's futurists, those scholars and others who make it their business to predict the future, appear moderately optimistic. After reviewing many of the reports of scholarly, government and business organizations, including the 50,000-member World Future Society, *U.S. News & World Report* concluded:

> Despite the downbeat nature of many current forecasts, many scholars come to a loose consensus that humanity's his-

---

[8]Douglas J. Crawford and Richard W. Leggett, "Assessing the Risk of Exposure to Radioactivity," *American Scientist*, September–October, 1980, pp. 524–536.

toric ability to cope with perils will emerge and the world will somehow survive.[9]

As for the human body specifically, it should be recalled that life spans have been extended dramatically within recent years, from an average of 40 years to an average of 70 years-plus within the experience of all the older people alive at present.[10]

Prognosticators, reviewed by *U.S. News & World Report* in the preceding article, believe that many of the most dramatic breakthroughs in technology in the future will be in medicine. Already on the drawing boards or in the testing stage are a diet supplement that can improve memory, a totally safe and nonfattening artificial sweetener and a drug that retards the aging process.

Some researchers believe, according to the report, that severed limbs may, within a few decades, be regenerated by using electrical impulses to spur the growth of new tissue. And, more and more defective body parts will be replaced by artificial implants.

David Lilienthal, former chairman of the TVA, first chairman of the Atomic Energy Commission and a successful businessman, who died in January, 1981, at the age of 81, was passionate about America's future, the *St. Louis Post-Dispatch* reports:

> "If we make up our mind, if we get the lead out, we'll find this is the greatest underdeveloped country in the world, and that it doesn't have to be timid and fearful of growth," he told the *Post*.
>
> "Human energy, 'drive, brainpower, creativity, imagination,' puts other energy to work for human use," he said. "Utilize this energy and you solve the problems of producing power for factories and homes, as well as the other essentials of life, including the development of a rich culture and sound government."[11]

We sympathize with Mr. Lilienthal's sentiments. But, there is a catch. It is true that if we develop our highest human energy potential, we will realize the highest potential for our other resources. But, the catch is that unless we develop our other resources wisely, we cannot reach our highest human energy potential.

---

[9]"What Next 20 Years Hold for You," *U.S. News & World Report*, December 1, 1980, pp. 51–53.

[10]Ronald Blythe, "Living to Be Old," *Harper's*, July, 1979, p. 35.

[11]"To Lilienthal, Human Energy Was Nation's Greatest Asset," *St. Louis Post-Dispatch*, January 18, 1981, Sec. 8-F, p. 1.

QUESTIONS AND PROBLEMS

1. Identify some major natural resources which aid the respiratory system. Explain how each of these resources actually affects the system.
2. Define *homeostasis* in your own words, after consulting several sources to help you better understand the process. Why must homeostasis be maintained by the body?
3. Name the six principal components of air pollution. Which appears to be the least serious to the human body in direct effects? Can any component be the most serious? Explain.
4. Cigarette smoking is said to be a contributing factor to emphysema. Describe how it may affect the alveoli of the lungs.
5. What is particulate matter? Describe how this substance may affect the lungs.
6. Name a substance which affects the nervous system. Check the *Reader's Guide to Periodical Literature* and bring in a report to class relating at least one instance of mercury poisoning in humans.
7. Why is water so important to the body?
8. How does polluted water threaten our health?
9. Insecticides can be beneficial or harmful. Give two examples, one illustrating, on the one hand, how an insecticide has benefited humans and, on the other, how it has harmed them.
10. Give an example of a chemical which persists in the human system.
11. Is it possible that noise alone can be harmful? Explain your answer.
12. Why is excessive radiation a threat to the body?
13. What facts lead us to be encouraged about the future prospects for human life, despite the problems of pollution?
14. What must we do if we are to develop our highest human energy potential?

RECOMMENDED GENERAL SOURCES FOR ENVIRONMENTAL INFORMATION

No text can cover in detail the encyclopedic array of information available today on the environment. From the hundreds of sources consulted by the authors, the following publications, selected because they present in a forthright manner a wide coverage of environmental concerns, are recommended for the establishment of a small reference section in the library for use by students. This listing is not meant to discourage those who have the means of supplementing these publications with others appropriate to their fields of interest and geographic location. A more complete list of references and aids is given in Lytle and Kircher, *Investigations in Conservation of Natural Resources* (Danville, Illinois: The Interstate Printers & Publishers, Inc., 1979), the student laboratory manual designed to accompany this text.

☐ **American Association for the Advancement of Science (AAAS),** *Science,* **1515 Massachusetts Avenue, NW, Washington, D.C. 20005**

This weekly journal reports the latest scientific experiments. It contains excellent book review and informative editorials and letters to the editor.

☐ **Conservation Foundation,** *Letter, A Monthly Report on Environmental Issues,* **Conservation Foundation, 1717 Massachusetts Avenue, NW, Washington, D.C. 20036**

Published monthly, this letter analyzes specific environmental topics in detail.

416 ☐ OUR NATURAL RESOURCES

☐ **Council on Environmental Quality,** *Annual Report,* **Council on Environmental Quality, 722 Jackson Place, NW, Washington, D.C. 20006**

> This annual report is probably the most comprehensive report available on environmental actions during the past year.

☐ **National Wildlife Federation,** *National Wildlife,* **National Wildlife Federation, 1412 16th Street, NW, Washington, D.C. 20036 (also publishes *International Wildlife* and special reports)**

> These monthly magazines carry beautifully illustrated articles on timely environmental concerns. They include other topics besides wildlife.

☐ **Resources for the Future,** *Annual Report* **and** *Resources* **(a brief journal published three times a year), Resources for the Future, 1755 Massachusetts Avenue, NW, Washington, D.C. 20036**

> This non-profit private organization studies the most pressing resource problems. Its journal and report discuss the highlights of the organization's findings and review the books subsequently published.

In addition, students should have readily available the following:

☐ ***The National Atlas of the United States of America,* U.S. Geological Survey, Department of the Interior, Washington, D.C., 1970 (for sale by the Superintendent of Documents, Washington, D.C. 20402)**

> In this magnificent atlas, students can find many of the basic geographic relationships essential to understanding environmental interrelationships.

☐ ***Statistical Abstract of the United States, 1979,* 100th ed., U.S. Bureau of the Census, Department of Commerce, Washington, D.C., 1980 (for sale by the Superintendent of Documents, Washington, D.C. 20402, or from any Department of Commerce district office)**

> This 100th edition is exceptionally comprehensive. It is an over–1,000-page national inventory. Social and socio-economic data, such as status of the aged, women and minority groups; characteristics of unemployed; number of households comprise about one-third of the statistics. Business and industry and associated statistics comprise about one-fourth of

the data. Agricultural, transportation, communications and other data and a section of recent trends and methodology and reliability complete the volume. (Of course, new editions also should be used when they become available.)

INDEX

A

Air pollution
  acid rainfall and, 374
  carbon dioxide and, 373
  cities and, 374, 376
Alaskan national interest lands ☐ 198, 199
Algae ☐ 368
Aluminum
  exhaustible nature of, 5
  occurrence of, 116–118
  source of, 115, 116
  uses of, 119
Anthracite ☐ 58, 59
Atomic energy
  (See **Nuclear energy)**

B

Bauxite
  occurrence of, 116, 117
  processing of, 118, 119
  producing areas, 117–119
Beauty and conservation ☐ 386, 387
Bennett, Dr. H. H. ☐ 216, 227
Bessemer process ☐ 111
Bituminous coal
  (See **Coal)**
British thermal unit (B.T.U.) ☐ 29, 30
Building materials ☐ 106

C

California State Water Project ☐ 154-159
Carbon ☐ 58, 59
Carter, President Jimmy ☐ 198, 394
Chemicals
  (See also **Insecticides,**
    Pesticides) ☐ 105, 106
Channelization of streams ☐ 298
Chestnut blight ☐ 187
Chromium ☐ 133
Cities
  air quality and, 310, 311
  CBD and central city, 308–310
  farm land and, 314
  green city, 305, 306
  growth of, 301, 302
  influence of, 303, 304
  new towns, 316, 317
  man and, 304, 305
  planning, 318
  sewage and, 314–316
  small cities and towns, 317, 318
  soil and, 314, 315
Clean Air Act of 1970 and 1977 Clean Air
  Amendments ☐ 374
Coal
  air pollution and, 66, 67
  availability, costs of, 60–63
  coal-tar, 64
  coke from coal, 64, 110
  energy crisis and, 65, 66

☐ 419

formation of, 55, 56
gasification and liquefaction, 63, 64
history of industry, 53, 55
industrial location and, 8, 58, 59
lagging development of, 64
long-wall mining, 62
quality of, 57
reserves, location of, 55
reserves of, 56, 57
strip mining, 61, 65
types of, 57-59

Coastal Zone Management Act of 1972 □ **324**
Cobalt □ **134**
Columbium □ **134**
Conservation
agriculture and, 388, 389
beauty and, 386, 387
business and, 388, 400
definition of, 398
ecology versus, 381, 382
energy and, 399
everybody's opportunity, 401
history of, 385–396
land-use planning and, 380, 391
of soil defined, 233, 234
personal responsibility for, 400, 401
philosophy of, 394–398
practices for soil, 234, 235
problems worldwide, 382, 383
progress in, 385, 386

Contour tillage □ **270, 271**
Copper
history of, 120
mining of, 121, 122
occurrence of, 121, 123, 124
reserves of, 124
uses of, 122, 123
versatility of, 120, 121

Cover crops □ **257–261**
Cropland □ **220–222, 227–229**
Crop rotation
adds organic matter, 285, 286
effect on erosion, 285
importance of, 283–284
length of, 284, 285
strip cropping and, 286

D

Dams
(*See* **Water power**)
Deuterium □ **90**
Dingell-Johnson Act □ **338**
Dutch metal □ **120**

E

Earth Day, first □ **385**
Ecology
conservation versus, 381, 382
philosophy of, 380, 381
science of, 19, 371–373

Ecosystem □ **19-21**
Elm beetle □ **188**
Energy
(*See* **Oil and gas**, *etc.*)
animate and inanimate, 3, 4, 6, 26–28
consumption in the United States, 29, 30
exhaustible and inexhaustible, 4–7
future needs, 99–103
geothermal, 96, 97
hydrogen as source of, 97, 98
macro unit, 29
national dilemma, 99–103
resources and, 3–7
solar, 93-96
solid waste energy, 98
sources of, 31, 32
use in colonial times, 27

Environmental Impact Statements □ **19, 387**
Environmental perception □ **21**
Environmental Protection Act □ **19**
Erosion
cropland and, 228, 229
crop rotation and, 285
gully, 213, 289–292
inventory of soil, 229, 230
kinds of, 211
sheet, 213
soil formation and, 211
stream bank, 294-297
water and, 213–215
water run-off and, 265

water speed and, 263, 265
wind, 213-215

F

Farming
    dry, 214, 215
    fish, 363–365
    organic, 287, 288
    wildlife and, 345–347
Fertilizers
    beneficial use of, 286, 287
    organic or inorganic, 287, 288
Fish
    as food, 359
    catch may increase, 365
    early use of, 359-361
    farming, 363-365
    fresh water, 363
    national territory and, 368
    share of nation's food, 362
    species disappearing, 362
Fisheries and Conservation Act □ 368
Flood plains □ 161, 162
Flumes □ 293, 294
Food chains □ 366–368, 409
Food supply adequacy □ 221, 222
Ford, President Gerald □ 393
Forests
    better use needed, 200
    clearcutting, 192
    controlling insects and diseases in, 186–189
    demand for, 178
    early uses of, 171
    farming and, 171, 172
    fire control, 190, 191
    hardwoods in, 182
    management practices, 181, 201
    multiple use of, 195, 196
    recreational uses of, 195–198
    rules for visitors to, 202
    saw timber and, 175, 177
    selective harvesting of, 191, 192
    small landowners and, 193–195
    succession, 176, 177
    supertrees in, 185, 186
    virgin, 173, 175

visitors to, 196
wildlife and, 169–171
Frasch process □ 127

G

Game refuges
    (*See* **Wildlife habitat**)
Gas
    (*See* **Oil and gas**)
Geothermal energy □ 96, 97
Grazing land
    delayed grazing on, 254, 255
    improvement of, 252, 253
    of federal government, 252, 253
    soil loss and, 224, 225
Ground water
    and water table, 163
    occurrence of, 162, 163
    potential resources, 162
    use, 163
Gypsy moth □ 189

H

Haddock decline □ 362
Halibut decline □ 362
Hardwoods
    diameter of, 185
    height growth, 183
    uses of, 182
Hay crops □ 249
Herbicides □ 288, 289
Homestead Act □ 220
Human body
    air and, 404–407
    energy potential and, 412
    homeostasis and, 404
    noise and, 410
    pesticides and, 409
    radiation and, 410, 411
    resources and, 403, 411, 412
    sun and, 408, 409
    systems of, 404
    water and, 407
Human resources and nature □ 18
Hunting
    (*See* **Wildlife**)

422 □ OUR NATURAL RESOURCES

Hydrogen □ 97, 98
Hydrologic cycle □ 144, 150

I

Illinois River pollution □ 149
Insecticides □ 288, 289, 376, 377
Iron
    early history of, 108–111
    occurrence of, 114, 115
    ore reserves, 112–114

J

Jacobs, Jane □ 304
Jetties □ 296, 297
Johnson, President Lyndon □ 386, 387

K

Kaibob deer incident □ 334

L

Land ethic □ 399, 400
Land-use capability classes □ 235–244
Land-use planning
    Congress and, 390
    ecoregions and, 391–393
    private property and, 391
    states and, 390, 391
Lead
    history of, 124, 125
    production of, 125
    supplies of, 126
    uses, 125, 126
Leopold, Aldo □ 399, 400
Levees □ 161, 162
Lignite □ 55–58
Lilienthal, David □ 412
Lower Klamath National Wildlife Refuge □ 351, 352

M

McHarg, Ian □ 161
Manganese □ 133
Marshes □ 223
Mesabi iron range □ 112–115

Metals, principal □ 107, 108
    (See also separate headings)
Metals, scarce □ 131–133
    (See also separate headings)
Minerals and fuels □ 26–32, 53–103
    (See also separate headings: Coal, Oil and gas, Nuclear energy, *etc.)*
Minerals, nonfuel □ 105–141
    (See also separate headings: Building materials, Chemicals, Iron, Metals, *etc.)*
Minerals, use by plants □ 134–136

N

National Environmental Protection Act □ 385, 387
Natural resources
    classification of, 1–7
    contrasting views, 398–400
    definition of, 2
    demand factors and, 14–16
    distribution of, 7, 8
    economic worth of, 11
    energy and, 3, 4
    exhaustible, 4, 5
    externalities and, 16
    freedom and, 22, 23
    goals and, 22
    industrial location and, 7, 8
    inexhaustible, 4, 12
    irreplaceable, 6
    of the ocean, 357 ff.
    perception and, 21
    replaceable, 6, 7
    social objectives and, 18
    societal arts and, 18
    supply factors and, 11–14
    technological arts and, 18
    wants and, 15–18
    wise use of, 396, 397
    youth and, 401
Newbigin, Marion □ 391
Newman, Oscar □ 310
New towns □ 316, 317
Nitrogen □ 138, 139
Nixon, President Richard □ 387
Noise pollution □ 387, 388
Nonfuel minerals
    building materials and, 106

chemicals and, 105
principal, 107–108
sufficiency of, 138–139
Nuclear energy
breeder reactors, 87, 88
fission, 85, 86
fusion, 89, 90
historical development, 85
industrial significance, 90
safety of, 86, 87

O

Oil and gas
Alaskan reserves of, 74
early use of, 69, 70
exploration and development of, 78, 79
foreign supplies of, 75
oil shales and, 82, 83
outer continental shelf (OCS) and, 74, 75
products from, 70, 71
recovery methods, 77, 78
reserves, estimation of, 79–82
reserves, gas, 75, 77
reserves, oil, 72–75
source of, 71
tar sands and, 82, 83
Open space
provisions for, 388
recreation and, 378–380
Organic farming, 287, 288

P

Pacific Northwest □ 159
Passenger pigeons □ 341
Pastures
improvement of, 252, 253
management of, 255, 257
rotation on, 255
soil loss and, 225–227
Peat □ 57
Pesticides □ 276, 277, 288, 289
Petroleum
(*See* Oil and gas)
Phosphates □ 136
Photosynthesis □ 19

Pittman-Robertson Federal Aid in Wildlife Restoration Act □ 338
Plankton
Cycle of life and, 366–368
varieties of, 359, 366
Planned Unit Development □ 311–313
Pollution
(*See various headings, such as* **Air pollution**)
Population explosion □ 25, 26
Potash □ 136–138

R

Range land
rotation on, 255
soil losses and, 226, 227
Reagan, President Ronald □ 394
Recreational area visits □ 197, 198
Resources
(*See* **Natural resources**)
Rip-rap □ 297
Rivers
(*See also* **Water, Water power**)
basin development of, 151 ff.
channelization of, 298
flood plains of, 161
referred to specifically
Allagash, 160
Buffalo, 160
Colorado, 159
Columbia, 159
Connecticut, 144
Current, 160
Cuyahoga, 144
Feather, 155–157
Illinois, 149
Mississippi, 149, 159
Missouri, 159
Ohio, 159
Potomac, 144, 145
Rogue, 160
Sacramento, 157
San Joaquin, 155, 156
St. Croix, 160
St. Lawrence, 159, 160
Tennessee, 151–154, 159
Wisconsin, 151
wild, 160, 161

424 □ OUR NATURAL RESOURCES

Roosevelt, President Theodore □ 334, 395, 396

S

Sailing vessels, use of □ 35–37
Salmon □ 363
Salt water
    desalinization process, 165, 166
    desalting costs, 166
    distillation of, 164
    electro-dialysis of, 164, 165
    freezing of, 164, 165
    reverse osmosis of, 164, 165
    use of, 164
Saw timber
    (*See also* **Forests**)
    amount standing, 177
    growth and drain, 175, 177, 178
Scarce metals
    imports of, 132
    need for, 131, 132
Sea water
    minerals in, 358
    plant and animal life in, 358–363
    resource, 357
Shad decline □ 362
Skylab □ 302, 303, 375
Small towns and conservation □ 317
Soil
    conservation defined, 233, 234
    depletion, 216
    erosion of, 211
        (*See* Erosion)
    exhaustibility of, 219, 233
    extent of, 220, 222
    formation of, 205–210
    formation of in forests, 172, 173, 176
    marshes, swamps and, 223, 234
    practices, 234, 235
    urban areas and, 229, 230
    vegetative cover and, 247–252
Solar energy □ 93–96
Solid waste disposal □ 377, 378
Solid waste energy □ 98
Sources of environmental information □ 415–417

Steel
    how produced, 111
    use of, 111, 112
St. Lawrence Seaway □ 159, 160
St. Louis □ 308–310
Strip cropping
    defined, 266
    kinds of, 268–270
    value of, 267, 268
    width of, 268
Sulfur
    future of, 128, 129
    history of, 126, 127
    source of, 127
    uses of, 127, 128
Swamps □ 223, 224

T

Taconite □ 112
Tar sands □ 82, 83
Technology and society □ 18
Terraces
    construction of, 279
    defined, 272
    nonfarm use of, 278, 279
    outlets for, 279, 281
    purpose of, 273
    types of, 274–279
Thorium □ 88
Tides, power from □ 49
Timber
    (*See* **Forests** *and* **Saw timber**)
Tungsten □ 133

U

Uranium □ 88, 89
Utility and resources □ 12–14

V

Vadose water □ 162

W

Washington, President George □ 219
Waste disposal □ 377, 378

Water
(See also Rivers, Sea water, Water power)
   abuse of supply, 144, 145
   adequacy of, 157
   atmospheric, 149, 150
   clean up of, 147–149
   erosion and, 187, 188
   evapotranspiration and, 150
   ground, 162, 163
   hydrologic cycle and, 144
   misuse of, 146, 147
   Pollution Control Act, 148
   saline, 164–167
   shortage of, reasons for, 143
   sources of, 129
   speed, effect of, 235–238
   surface, 151
   table, 163
   use of, 146, 147
   vadose, 162
   variability of supply, 44–46, 144
   waste water, 167
Water power
   amount of, 41, 42, 48, 167
   cost-benefits, 42–48
   distribution of, 48
   Gulf Stream and, 50
   new sources of, 48–50
   tidal, 49
Weather modification □ **150**
Whales □ **363**
White, Gilbert F. □ **161**
White pine blister rust □ **188, 189**
Wilderness □ **198–200**
Wildlife
(See also Wildlife habitat)
   classification of, 321
   closed seasons and bag limits, 339
   disappearance of, 321, 322
   education and, 342
   endangered species, 322
   food chain and, 293
   future of, 354
   game laws, 336, 337
   harvesting fish, 349, 350
   hunters pay for, 337, 338
   hunting restriction on, 335–338
   individual responsibility for, 354
   justification of, 322–325
   management of, 334–342
   needs of, 297, 298
   non-beneficial sometimes, 325
   nongame species, 309
   predators and, 331–334
   research and, 342
   success in restocking, 292
   survival of, 331–334
   whales vanishing, 328, 329
   why important, 292–294
Wildlife habitat
(See also Wildlife)
   city and, 353, 354
   farms and, 345–347
   forage for antelope, bighorns, buffalo, deer, elk versus cattle, 347
   fur bearers and ponds, 350, 351
   game refuges, 338, 341
   Kaibob Plateau, lesson of, 334
   landowners and, 341, 342
   ponds and lakes for, 347–351
   providing, 325–330
   ranches and, 347
   stocking ponds, 348
   stream banks and, 353
   wetlands for, 351, 352
   wastelands and, 330, 331
   woods and, 149–151
Wind power
   new proposals for, 40
   rainfall and, 40, 41
   sailing, use of for, 35–37
   windmills, use of for, 37–40

Z

Zimmermann, Erich □ **17**
Zinc
   history of, 129, 130
   production of, 130
   reserves, 131
   source of, 130
   uses of, 130, 131